## Physical Constants

| | | | |
|---|---|---|---|
| Gravitational constant | $G$ | $=$ | $6.673(10) \times 10^{-11}$ N m$^2$ kg$^{-2}$ |
| Speed of light (exact) | $c$ | $\equiv$ | $2.99792458 \times 10^8$ m s$^{-1}$ |
| Permeability of free space | $\mu_0$ | $\equiv$ | $4\pi \times 10^{-7}$ N A$^{-2}$ |
| Permittivity of free space | $\epsilon_0$ | $\equiv$ | $1/\mu_0 c^2$ |
| | | $=$ | $8.854187817\ldots \times 10^{-12}$ F m$^{-1}$ |
| Electric charge | $e$ | $=$ | $1.602176462(63) \times 10^{-19}$ C |
| Electron volt | 1 eV | $=$ | $1.602176462(63) \times 10^{-19}$ J |
| Planck's constant | $h$ | $=$ | $6.62606876(52) \times 10^{-34}$ J s |
| | | $=$ | $4.13566727(16) \times 10^{-15}$ eV s |
| | $\hbar$ | $\equiv$ | $h/2\pi$ |
| | | $=$ | $1.054571596(82) \times 10^{-34}$ J s |
| | | $=$ | $6.58211889(26) \times 10^{-16}$ eV s |
| Planck's constant $\times$ speed of light | $hc$ | $=$ | $1.23984186(16) \times 10^3$ eV nm |
| | | $\simeq$ | 1240 eV nm |
| Boltzmann's constant | $k$ | $=$ | $1.3806503(24) \times 10^{-23}$ J K$^{-1}$ |
| | | $=$ | $8.6173423(153) \times 10^{-5}$ eV K$^{-1}$ |
| Stefan–Boltzmann constant | $\sigma$ | $\equiv$ | $2\pi^5 k^4/(15c^2 h^3)$ |
| | | $=$ | $5.670400(40) \times 10^{-8}$ W m$^{-2}$ K$^{-4}$ |
| Radiation constant | $a$ | $=$ | $4\sigma/c$ |
| | | $=$ | $7.565767(54) \times 10^{-16}$ J m$^{-3}$ K$^{-4}$ |
| Atomic mass unit | 1 u | $=$ | $1.66053873(13) \times 10^{-27}$ kg |
| | | $=$ | $931.494013(37)$ MeV/$c^2$ |
| Electron mass | $m_e$ | $=$ | $9.10938188(72) \times 10^{-31}$ kg |
| | | $=$ | $5.485799110(12) \times 10^{-4}$ u |
| Proton mass | $m_p$ | $=$ | $1.67262158(13) \times 10^{-27}$ kg |
| | | $=$ | $1.00727646688(13)$ u |
| Neutron mass | $m_n$ | $=$ | $1.67492716(13) \times 10^{-27}$ kg |
| | | $=$ | $1.00866491578(55)$ u |
| Hydrogen mass | $m_H$ | $=$ | $1.673532499(13) \times 10^{-27}$ kg |
| | | $=$ | $1.00782503214(35)$ u |
| Avogadro's number | $N_A$ | $=$ | $6.02214199(47) \times 10^{23}$ mol$^{-1}$ |
| Gas constant | $R$ | $=$ | $8.314472(15)$ J mol$^{-1}$ K$^{-1}$ |
| Bohr radius | $a_{0,\infty}$ | $\equiv$ | $4\pi\epsilon_0 \hbar^2/m_e e^2$ |
| | | $=$ | $5.291772083(19) \times 10^{-11}$ m |
| | $a_{0,H}$ | $\equiv$ | $(m_e/\mu)a_{0,\infty}$ |
| | | $=$ | $5.294654075(20) \times 10^{-11}$ m |
| Rydberg constant | $R_\infty$ | $\equiv$ | $m_e e^4/64\pi^3 \epsilon_0^2 \hbar^3 c$ |
| | | $=$ | $1.0973731568549(83) \times 10^7$ m$^{-1}$ |
| | $R_H$ | $\equiv$ | $(\mu/m_e)R_\infty$ |
| | | $=$ | $1.09677583(13) \times 10^7$ m$^{-1}$ |

Note: Uncertainties in the last digits are indicated in parentheses. For instance, the universal gravitational constant, $G$, has an uncertainty of $\pm 0.010 \times 10^{-11}$ N m$^2$ kg$^{-2}$.

# AN INTRODUCTION TO MODERN ASTROPHYSICS

## Second Edition

Bradley W. Carroll
*Weber State University*

Dale A. Ostlie
*Weber State University*

San Francisco   Boston   New York
Cape Town   Hong Kong   London   Madrid   Mexico City
Montreal   Munich   Paris   Singapore   Sydney   Tokyo   Toronto

Editor-in-Chief: Adam R. S. Black
Senior Acquisitions Editor: Lothlórien Homet
Assistant Editor: Deb Greco
Editorial Assistant: Ashley Taylor Anderson
Executive Marketing Manager: Christy Lawrence
Managing Editor: Corinne Benson
Production Supervisor: Shannon Tozier
Manufacturing Manager: Stacey Weinberger
Project Management and Composition: Techsetters, Inc.
Illustration: Techsetters, Inc.
Cover Illustration: Kenneth Xavier Probst
Cover Design: Seventeenth Street Studios
Text Printer: R. R. Donnelley, Crawfordsville
Cover Printer: Phoenix Color

**Library of Congress Cataloging-in-Publication Data**

Carroll, Bradley W.
  An introduction to modern astrophysics / Bradley W. Carroll, Dale A. Ostlie.–2nd ed.
    p. cm.
  Includes bibliographical references and index.
  ISBN 0-8053-0402-9 (alk. paper)
 1. Astrophysics–Textbooks. I. Ostlie, Dale A. II. Title.
  QB461.C35 2007
  523.01–dc22

                                              2006015391


ISBN 0-8053-0402-9

2 3 4 5 6 7 8 9 10—DOC—09 08 07
www.aw-bc.com/astrophysics

*For Lynn,*
*and*
*Candy, Michael, and Megan*
*with love*

# Preface

Since the first edition of *An Introduction to Modern Astrophysics* and its abbreviated companion text, *An Introduction to Modern Stellar Astrophysics*, first appeared in 1996, there has been an incredible explosion in our knowledge of the heavens. It was just two months before the printing of the first editions that Michel Mayor and Didier Queloz announced the discovery of an extrasolar planet around 51 Pegasi, the first planet found orbiting a main-sequence star. In the next eleven years, the number of known extrasolar planets has grown to over 193. Not only do these discoveries shed new light on how stars and planetary systems form, but they also inform us about formation and planetary evolution in our own Solar System.

In addition, within the past decade important discoveries have been made of objects, within our Solar System but beyond Pluto, that are similar in size to that diminutive planet. In fact, one of the newly discovered Kuiper belt objects, currently referred to as 2003 UB313 (until the International Astronomical Union makes an official determination), appears to be larger than Pluto, challenging our definition of what a planet is and how many planets our Solar System is home to.

Explorations by robotic spacecraft and landers throughout our Solar System have also yielded a tremendous amount of new information about our celestial neighborhood. The armada of orbiters, along with the remarkable rovers, Spirit and Opportunity, have confirmed that liquid water has existed on the surface of Mars in the past. We have also had robotic emissaries visit Jupiter and Saturn, touch down on the surfaces of Titan and asteroids, crash into cometary nuclei, and even return cometary dust to Earth.

Missions such as Swift have enabled us to close in on the solutions to the mysterious gamma-ray bursts that were such an enigma at the time *An Introduction to Modern Astrophysics* first appeared. We now know that one class of gamma-ray bursts is associated with core-collapse supernovae and that the other class is probably associated with the merger of two neutron stars, or a neutron star and a black hole, in a binary system.

Remarkably precise observations of the center of our Milky Way Galaxy and other galaxies, since the publication of the first editions, have revealed that a great many, perhaps most, spiral and large elliptical galaxies are home to one or more supermassive black holes at their centers. It also appears likely that galactic mergers help to grow these monsters in their centers. Furthermore, it now seems almost certain that supermassive black holes are the central engines responsible for the exotic and remarkably energetic phenomena associated with radio galaxies, Seyfert galaxies, blazars, and quasars.

The past decade has also witnessed the startling discovery that the expansion of the universe is not slowing down but, rather, is actually accelerating! This remarkable observation suggests that we currently live in a dark-energy-dominated universe, in which Einstein's cosmological constant (once considered his "greatest blunder") plays an important role in our understanding of cosmology. Dark energy was not even imagined in cosmological models at the time the first editions were published.

Indeed, since the publication of the first editions, cosmology has entered into a new era of precision measurements. With the release of the remarkable data obtained by the Wilkinson Microwave Anisotropy Probe (WMAP), previously large uncertainties in the age of the universe have been reduced to less than 2% ($13.7 \pm 0.2$ Gyr). At the same time, stellar evolution theory and observations have led to the determination that the ages of the oldest globular clusters are in full agreement with the upper limit of the age of the universe.

We opened the preface to the first editions with the sentence "There has never been a more exciting time to study modern astrophysics"; this has certainly been borne out in the tremendous advances that have occurred over the past decade. It is also clear that this incredible decade of discovery is only a prelude to further advances to come. Joining the Hubble Space Telescope in its high-resolution study of the heavens have been the Chandra X-ray Observatory and the Spitzer Infrared Space Telescope. From the ground, 8-m and larger telescopes have also joined the search for new information about our remarkable universe. Tremendously ambitious sky surveys have generated a previously unimagined wealth of data that provide critically important statistical data sets; the Sloan Digital Sky Survey, the Two-Micron All Sky Survey, the 2dF redshift survey, the Hubble Deep Fields and Ultradeep Fields, and others have become indispensable tools for hosts of studies. We also anticipate the first observations from new observatories and spacecraft, including the high-altitude (5000 m) Atacama Large Millimeter Array and high-precision astrometric missions such as Gaia and SIM PlanetQuest. Of course, studies of our own Solar System also continue; just the day before this preface was written, the Mars Reconnaissance Orbiter entered orbit around the red planet.

When the first editions were written, even the World Wide Web was in its infancy. Today it is hard to imagine a world in which virtually any information you might want is only a search engine and a mouse click away. With enormous data sets available online, along with fully searchable journal and preprint archives, the ability to access critical information very rapidly has been truly revolutionary.

Needless to say, a second edition of BOB (the "Big Orange Book," as *An Introduction to Modern Astrophysics* has come to be known by many students) and its associated text is long overdue. In addition to an abbreviated version focusing on stellar astrophysics (*An Introduction to Modern Stellar Astrophysics*), a second abbreviated version (*An Introduction to Modern Galactic Astrophysics and Cosmology*) is being published. We are confident that BOB and its smaller siblings will serve the needs of a range of introductory astrophysics courses and that they will instill some of the excitement felt by the authors and hosts of astronomers and astrophysicists worldwide.

We have switched from cgs to SI units in the second edition. Although we are personally more comfortable quoting luminosities in ergs $s^{-1}$ rather than watts, our students are not. We do not want students to feel exasperated by a new system of units during their first encounter with the concepts of modern astrophysics. However, we have retained the natural units of parsecs and solar units ($M_\odot$ and $L_\odot$) because they provide a comparative context for numerical values. An appendix of unit conversions (see back endpapers) is included for those who delve into the professional literature and discover the world of angstroms, ergs, and esu.

Our goal in writing these texts was to open the entire field of modern astrophysics to you by using only the basic tools of physics. Nothing is more satisfying than appreciating the drama of the universe through an understanding of its underlying physical principles. The advantages of a mathematical approach to understanding the heavenly spectacle were obvious to Plato, as manifested in his *Epinomis*:

> Are you unaware that the true astronomer must be a person of great wisdom? Hence there will be a need for several sciences. The first and most important is that which treats of pure numbers. To those who pursue their studies in the proper way, all geometric constructions, all systems of numbers, all duly constituted melodic progressions, the single ordered scheme of all celestial revolutions should disclose themselves. And, believe me, no one will ever behold that spectacle without the studies we have described, and so be able to boast that they have won it by an easy route.

Now, 24 centuries later, the application of a little physics and mathematics still leads to deep insights.

These texts were also born of the frustration we encountered while teaching our junior-level astrophysics course. Most of the available astronomy texts seemed more descriptive than mathematical. Students who were learning about Schrödinger's equation, partition functions, and multipole expansions in other courses felt handicapped because their astrophysics text did not take advantage of their physics background. It seemed a double shame to us because a course in astrophysics offers students the unique opportunity of actually using the physics they have learned to appreciate many of astronomy's fascinating phenomena. Furthermore, as a discipline, astrophysics draws on virtually every aspect of physics. Thus astrophysics gives students the chance to review and extend their knowledge.

Anyone who has had an introductory calculus-based physics course is ready to understand nearly all the major concepts of modern astrophysics. The amount of modern physics covered in such a course varies widely, so we have included a chapter on the theory of special relativity and one on quantum physics which will provide the necessary background in these areas. Everything else in the text is self-contained and generously cross-referenced, so you will not lose sight of the chain of reasoning that leads to some of the most astounding ideas in all of science.[1]

Although we have attempted to be fairly rigorous, we have tended to favor the sort of back-of-the-envelope calculation that uses a simple model of the system being studied. The payoff-to-effort ratio is so high, yielding 80% of the understanding for 20% of the effort, that these quick calculations should be a part of every astrophysicist's toolkit. In fact, while writing this book we were constantly surprised by the number of phenomena that could be described in this way. Above all, we have tried to be honest with you; we remained determined not to simplify the material beyond recognition. Stellar interiors,

---

[1] Footnotes are used when we don't want to interrupt the main flow of a paragraph.

stellar atmospheres, general relativity, and cosmology—all are described with a depth that is more satisfying than mere hand-waving description.

Computational astrophysics is today as fundamental to the advance in our understanding of astronomy as observation and traditional theory, and so we have developed numerous computer problems, as well as several complete codes, that are integrated with the text material. You can calculate your own planetary orbits, compute observed features of binary star systems, make your own models of stars, and reproduce the gravitational interactions between galaxies. These codes favor simplicity over sophistication for pedagogical reasons; you can easily expand on the conceptually transparent codes that we have provided. Astrophysicists have traditionally led the way in large-scale computation and visualization, and we have tried to provide a gentle introduction to this blend of science and art.

Instructors can use these texts to create courses tailored to their particular needs by approaching the content as an astrophysical smorgasbord. By judiciously selecting topics, we have used BOB to teach a semester-long course in stellar astrophysics. (Of course, much was omitted from the first 18 chapters, but the text is designed to accommodate such surgery.) Interested students have then gone on to take an additional course in cosmology. On the other hand, using the entire text would nicely fill a year-long survey course (and then some) covering all of modern astrophysics. To facilitate the selection of topics, as well as identify important topics within sections, we have added subsection headings to the second editions. Instructors may choose to skim, or even omit, subsections in accordance with their own as well as their students' interests—and thereby design a course to their liking.

An extensive website at http://www.aw-bc.com/astrophysics is associated with these texts. It contains downloadable versions of the computer codes in various languages, including Fortran, C++, and, in some cases, Java. There are also links to some of the many important websites in astronomy. In addition, links are provided to public domain images found in the texts, as well as to line art that can be used for instructor presentations. Instructors may also obtain a detailed solutions manual directly from the publisher.

Throughout the process of the extensive revisions for the second editions, our editors have maintained a positive and supportive attitude that has sustained us throughout. Although we must have sorely tried their patience, Adam R. S. Black, Lothlórien Homet, Ashley Taylor Anderson, Deb Greco, Stacie Kent, Shannon Tozier, and Carol Sawyer (at Techsetters) have been truly wonderful to work with.

We have certainly been fortunate in our professional associations throughout the years. We want to express our gratitude and appreciation to Art Cox, John Cox (1926–1984), Carl Hansen, Hugh Van Horn, and Lee Anne Willson, whose profound influence on us has remained and, we hope, shines through the pages ahead.

Our good fortune has been extended to include the many expert reviewers who cast a merciless eye on our chapters and gave us invaluable advice on how to improve them. For their careful reading of the first editions, we owe a great debt to Robert Antonucci, Martin Burkhead, Peter Foukal, David Friend, Carl Hansen, H. Lawrence Helfer, Steven D. Kawaler, William Keel, J. Ward Moody, Tobias Owen, Judith Pipher, Lawrence Pinsky, Joseph Silk, J. Allyn Smith, and Rosemary Wyse. Additionally, the extensive revisions to the second editions have been carefully reviewed by Bryon D. Anderson, Markus J. Aschwanden, Andrew Blain, Donald J. Bord, Jean-Pierre Caillault, Richard Crowe, Manfred A. Cuntz, Daniel Dale, Constantine Deliyannis, Kathy DeGioia Eastwood, J. C. Evans,

Debra Fischer, Kim Griest, Triston Guillot, Fred Hamann, Jason Harlow, Peter Hauschildt, Lynne A. Hillenbrand, Philip Hughes, William H. Ingham, David Jewitt, Steven D. Kawaler, John Kielkopf, Jeremy King, John Kolena, Matthew Lister, Donald G. Luttermoser, Geoff Marcy, Norman Markworth, Pedro Marronetti, C. R. O'Dell, Frederik Paerels, Eric S. Perlman, Bradley M. Peterson, Slawomir Piatek, Lawrence Pinsky, Martin Pohl, Eric Preston, Irving K. Robbins, Andrew Robinson, Gary D. Schmidt, Steven Stahler, Richard D. Sydora, Paula Szkody, Henry Throop, Michael T. Vaughn, Dan Watson, Joel Weisberg, Gregory G. Wood, Matt A. Wood, Kausar Yasmin, Andrew Youdin, Esther Zirbel, E. J. Zita, and others. Over the past decade, we have received valuable input from users of the first-edition texts that has shaped many of the revisions and corrections to the second editions. Several generations of students have provided us with a different and extremely valuable perspective as well. Unfortunately, no matter how fine the sieve, some mistakes are sure to slip through, and some arguments and derivations may be less than perfectly clear. The responsibility for the remaining errors is entirely ours, and we invite you to submit comments and corrections to us at our e-mail address: `modastro@weber.edu`.

Unfortunately, the burden of writing has not been confined to the authors but was unavoidably shared by family and friends. We wish to thank our parents, Wayne and Marjorie Carroll, and Dean and Dorothy Ostlie, for raising us to be intellectual explorers of this fascinating universe. Finally, it is to those people who make *our* universe so wondrous that we dedicate this book: our wives, Lynn Carroll and Candy Ostlie, and Dale's terrific children, Michael and Megan. Without their love, patience, encouragement, and constant support, this project would never have been completed.

And now it is time to get up into Utah's beautiful mountains for some skiing, hiking, mountain biking, fishing, and camping and share those down-to-Earth joys with our families!

Bradley W. Carroll
Dale A. Ostlie
Weber State University
Ogden, UT
`modastro@weber.edu`

# Contents

# PART
# I

# The Tools of Astronomy

# CHAPTER

# 1

# The Celestial Sphere

## 1.1 ■ THE GREEK TRADITION

Human beings have long looked up at the sky and pondered its mysteries. Evidence of the long struggle to understand its secrets may be seen in remnants of cultures around the world: the great Stonehenge monument in England, the structures and the writings of the Maya and Aztecs, and the medicine wheels of the Native Americans. However, our modern scientific view of the universe traces its beginnings to the ancient Greek tradition of natural philosophy. Pythagoras (ca. 550 B.C.) first demonstrated the fundamental relationship between numbers and nature through his study of musical intervals and through his investigation of the geometry of the right angle. The Greeks continued their study of the universe for hundreds of years using the natural language of mathematics employed by Pythagoras. The modern discipline of astronomy depends heavily on a mathematical formulation of its physical theories, following the process begun by the ancient Greeks.

In an initial investigation of the night sky, perhaps its most obvious feature to a careful observer is the fact that it is constantly changing. Not only do the stars move steadily from east to west during the course of a night, but different stars are visible in the evening sky, depending upon the season. Of course the Moon also changes, both in its position in the sky and in its phase. More subtle and more complex are the movements of the planets, or "wandering stars."

### The Geocentric Universe

Plato (ca. 350 B.C.) suggested that to understand the motions of the heavens, one must first begin with a set of workable assumptions, or hypotheses. It seemed obvious that the stars of the night sky revolved about a fixed Earth and that the heavens ought to obey the purest possible form of motion. Plato therefore proposed that celestial bodies should move about Earth with a uniform (or constant) speed and follow a circular motion with Earth at the center of that motion. This concept of a **geocentric universe** was a natural consequence of the apparently unchanging relationship of the stars to one another in fixed constellations.

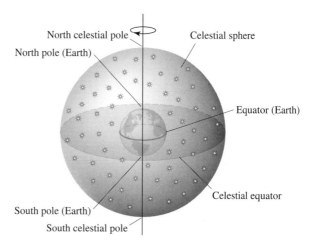

North celestial pole

North pole (Earth)

Celestial sphere

Equator (Earth)

Celestial equator

South pole (Earth)

South celestial pole

**FIGURE 1.1**    The celestial sphere. Earth is depicted in the center of the celestial sphere.

If the stars were simply attached to a **celestial sphere** that rotated about an axis passing through the North and South poles of Earth and intersecting the celestial sphere at the **north** and **south celestial poles**, respectively (Fig. 1.1), all of the stars' known motions could be described.

### Retrograde Motion

The wandering stars posed a somewhat more difficult problem. A planet such as Mars moves slowly from west to east against the fixed background stars and then mysteriously reverses direction for a period of time before resuming its previous path (Fig. 1.2). Attempting to understand this backward, or **retrograde**, **motion** became the principal problem in astronomy for nearly 2000 years! Eudoxus of Cnidus, a student of Plato's and an exceptional mathematician, suggested that each of the wandering stars occupied its own sphere and that all the spheres were connected through axes oriented at different angles and rotating at various speeds. Although this theory of a complex system of spheres initially was marginally successful at explaining retrograde motion, predictions began to deviate significantly from the observations as more data were obtained.

Hipparchus (ca. 150 B.C.), perhaps the most notable of the Greek astronomers, proposed a system of circles to explain retrograde motion. By placing a planet on a small, rotating **epicycle** that in turn moved on a larger **deferent**, he was able to reproduce the behavior of the wandering stars. Furthermore, this system was able to explain the increased brightness of the planets during their retrograde phases as resulting from changes in their distances from Earth. Hipparchus also created the first catalog of the stars, developed a magnitude system for describing the brightness of stars that is still in use today, and contributed to the development of trigonometry.

During the next two hundred years, the model of planetary motion put forth by Hipparchus also proved increasingly unsatisfactory in explaining many of the details of the observations. Claudius Ptolemy (ca. A.D. 100) introduced refinements to the epicycle/deferent

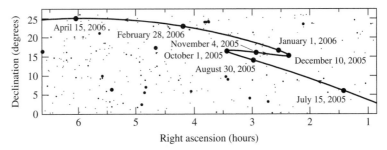

FIGURE 1.2    The retrograde motion of Mars in 2005. The general, long-term motion of the planet is eastward relative to the background stars. However, between October 1 and December 10, 2005, the planet's motion temporarily becomes westward (retrograde). (Of course the planet's short-term daily motion across the sky is always from east to west.) The coordinates of right ascension and declination are discussed on page 11 and in Fig. 1.13. Betelgeuse, the bright star in the constellation of Orion, is visible at $(\alpha, \delta) = (5^h55^m, +7°24')$, Aldebaran, in the constellation of Taurus, has coordinates $(4^h36^m, +16°31')$, and the Hyades and Pleiades star clusters (also in Taurus) are visible at $(4^h24^m, +15°45')$ and $(3^h44^m, +23°58')$, respectively.

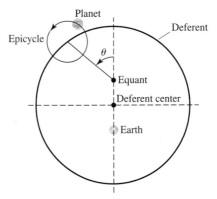

FIGURE 1.3    The Ptolemaic model of planetary motion.

system by adding **equants** (Fig. 1.3), resulting in a constant *angular* speed of the epicycle about the deferent ($d\theta/dt$ was assumed to be constant). He also moved Earth away from the deferent center and even allowed for a wobble of the deferent itself. Predictions of the Ptolemaic model did agree more closely with observations than any previously devised scheme, but the original philosophical tenets of Plato (uniform and circular motion) were significantly compromised.

Despite its shortcomings, the Ptolemaic model became almost universally accepted as the correct explanation of the motion of the wandering stars. When a disagreement between the model and observations would develop, the model was modified slightly by the addition of another circle. This process of "fixing" the existing theory led to an increasingly complex theoretical description of observable phenomena.

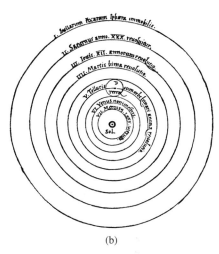

(a) (b)

**FIGURE 1.4** (a) Nicolaus Copernicus (1473–1543). (b) The Copernican model of planetary motion: Planets travel in circles with the Sun at the center of motion. (Courtesy of Yerkes Observatory.)

## 1.2 ■ THE COPERNICAN REVOLUTION

By the sixteenth century the inherent simplicity of the Ptolemaic model was gone. Polish-born astronomer Nicolaus Copernicus (1473–1543), hoping to return the science to a less cumbersome, more elegant view of the universe, suggested a **heliocentric** (Sun-centered) model of planetary motion (Fig. 1.4).[1] His bold proposal led immediately to a much less complicated description of the relationships between the planets and the stars. Fearing severe criticism from the Catholic Church, whose doctrine then declared that Earth was the center of the universe, Copernicus postponed publication of his ideas until late in life. *De Revolutionibus Orbium Coelestium* (*On the Revolution of the Celestial Sphere*) first appeared in the year of his death. Faced with a radical new view of the universe, along with Earth's location in it, even some supporters of Copernicus argued that the heliocentric model merely represented a mathematical improvement in calculating planetary positions but did not actually reflect the true geometry of the universe. In fact, a preface to that effect was added by Osiander, the priest who acted as the book's publisher.

### Bringing Order to the Planets

One immediate consequence of the Copernican model was the ability to establish the order of all of the planets from the Sun, along with their relative distances and orbital periods. The fact that Mercury and Venus are never seen more than 28° and 47°, respectively, east or west of the Sun clearly establishes that their orbits are located inside the orbit of Earth. These planets are referred to as **inferior planets**, and their maximum angular separations east or west of the Sun are known as **greatest eastern elongation** and **greatest western**

---

[1] Actually, Aristarchus proposed a heliocentric model of the universe in 280 B.C. At that time, however, there was no compelling evidence to suggest that Earth itself was in motion.

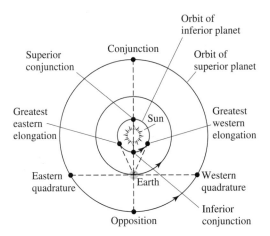

**FIGURE 1.5**    Orbital configurations of the planets.

**elongation**, respectively (see Fig. 1.5). Mars, Jupiter, and Saturn (the most distant planets known to Copernicus) can be seen as much as 180° from the Sun, an alignment known as **opposition**. This could only occur if these **superior planets** have orbits outside Earth's orbit. The Copernican model also predicts that only inferior planets can pass in front of the solar disk (**inferior conjunction**), as observed.

### Retrograde Motion Revisited

The great long-standing problem of astronomy—retrograde motion—was also easily explained through the Copernican model. Consider the case of a superior planet such as Mars. Assuming, as Copernicus did, that the farther a planet is from the Sun, the more slowly it moves in its orbit, Mars will then be overtaken by the faster-moving Earth. As a result, the apparent position of Mars will shift against the relatively fixed background stars, with the planet seemingly moving backward near opposition, where it is closest to Earth and at its brightest (see Fig. 1.6). Since the orbits of all of the planets are not in the same plane, retrograde loops will occur. The same analysis works equally well for all other planets, superior and inferior.

The relative orbital motions of Earth and the other planets mean that the time interval between successive oppositions or conjunctions can differ significantly from the amount of time necessary to make one complete orbit relative to the background stars (Fig. 1.7). The former time interval (between oppositions) is known as the **synodic period** ($S$), and the latter time interval (measured relative to the background stars) is referred to as the **sidereal period** ($P$). It is left as an exercise to show that the relationship between the two periods is given by

$$
1/S = \begin{cases} 1/P - 1/P_{\oplus} & \text{(inferior)} \\ 1/P_{\oplus} - 1/P & \text{(superior),} \end{cases}
\tag{1.1}
$$

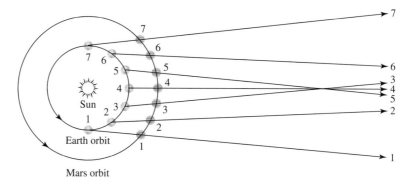

FIGURE 1.6   The retrograde motion of Mars as described by the Copernican model. Note that the lines of sight from Earth to Mars cross for positions 3, 4, and 5. This effect, combined with the slightly differing planes of the two orbits result in retrograde paths near opposition. Recall the retrograde (or westward) motion of Mars between October 1, 2005, and December 10, 2005, as illustrated in Fig. 1.2.

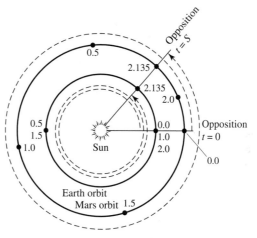

FIGURE 1.7   The relationship between the sidereal and synodic periods of Mars. The two periods do not agree due to the motion of Earth. The numbers represent the elapsed time in sidereal years since Mars was initially at opposition. Note that Earth completes more than two orbits in a synodic period of $S = 2.135$ yr, whereas Mars completes slightly more than one orbit during one synodic period from opposition to opposition.

when perfectly circular orbits and constant speeds are assumed; $P_{\oplus}$ is the sidereal period of Earth's orbit (365.256308 d).

Although the Copernican model did represent a simpler, more elegant model of planetary motion, it was not successful in predicting positions any more accurately than the Ptolemaic model. This lack of improvement was due to Copernicus's inability to relinquish the 2000-year-old concept that planetary motion required circles, the human notion of perfection. As a consequence, Copernicus was forced (as were the Greeks) to introduce the concept of epicycles to "fix" his model.

Perhaps the quintessential example of a scientific revolution was the revolution begun by Copernicus. What we think of today as the obvious solution to the problem of planetary motion—a heliocentric universe—was perceived as a very strange and even rebellious notion during a time of major upheaval, when Columbus had recently sailed to the "new world" and Martin Luther had proposed radical revisions in Christianity. Thomas Kuhn has suggested that an established scientific theory is much more than just a framework for guiding the study of natural phenomena. The present **paradigm** (or prevailing scientific theory) is actually a way of *seeing* the universe around us. We ask questions, pose new research problems, and interpret the results of experiments and observations in the context of the paradigm. Viewing the universe in any other way requires a complete shift from the current paradigm. To suggest that Earth actually orbits the Sun instead of believing that the Sun inexorably rises and sets about a fixed Earth is to argue for a change in the very structure of the universe, a structure that was believed to be correct and beyond question for nearly 2000 years. Not until the complexity of the old Ptolemaic scheme became too unwieldy could the intellectual environment reach a point where the concept of a heliocentric universe was even possible.

## 1.3 ■ POSITIONS ON THE CELESTIAL SPHERE

The Copernican revolution has shown us that the notion of a geocentric universe is incorrect. Nevertheless, with the exception of a small number of planetary probes, our observations of the heavens are still based on a reference frame centered on Earth. The daily (or **diurnal**) rotation of Earth, coupled with its annual motion around the Sun and the slow wobble of its rotation axis, together with relative motions of the stars, planets, and other objects, results in the constantly changing positions of celestial objects. To catalog the locations of objects such as the Crab supernova remnant in Taurus or the great spiral galaxy of Andromeda, coordinates must be specified. Moreover, the coordinate system should not be sensitive to the short-term manifestations of Earth's motions; otherwise the specified coordinates would constantly change.

### The Altitude–Azimuth Coordinate System

Viewing objects in the night sky requires only directions to them, not their distances. We can imagine that all objects are located on a celestial sphere, just as the ancient Greeks believed. It then becomes sufficient to specify only two coordinates. The most straightforward coordinate system one might devise is based on the observer's local horizon. The **altitude–azimuth** (or **horizon**) **coordinate system** is based on the measurement of the azimuth angle along the horizon together with the altitude angle above the horizon (Fig. 1.8). The **altitude** $h$ is defined as that angle measured from the horizon to the object along a great circle[2] that passes through that object and the point on the celestial sphere directly above the observer, a point known as the **zenith**. Equivalently, the **zenith distance** $z$ is the angle measured from the zenith to the object, so $z + h = 90°$. The **azimuth** $A$ is simply the angle

---

[2]A great circle is the curve resulting from the intersection of a sphere with a plane passing through the *center* of that sphere.

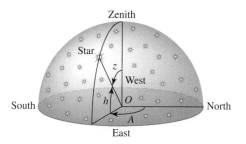

**FIGURE 1.8** The altitude–azimuth coordinate system. $h$, $z$, and $A$ are the altitude, zenith distance, and azimuth, respectively.

measured along the horizon eastward from north to the great circle used for the measure of altitude. (The **meridian** is another frequently used great circle; it is defined as passing through the observer's zenith and intersecting the horizon due north and south.)

Although simple to define, the altitude–azimuth system is difficult to use in practice. Coordinates of celestial objects in this system are specific to the local latitude and longitude of the observer and are difficult to transform to other locations on Earth. Also, since Earth is rotating, stars appear to move constantly across the sky, meaning that the coordinates of each object are constantly changing, even for the local observer. Complicating the problem still further, the stars rise approximately 4 minutes earlier on each successive night, so that even when viewed from the same location at a specified time, the coordinates change from day to day.

### Daily and Seasonal Changes in the Sky

To understand the problem of these day-to-day changes in altitude–azimuth coordinates, we must consider the orbital motion of Earth about the Sun (see Fig. 1.9). As Earth orbits the Sun, our view of the distant stars is constantly changing. Our line of sight to the Sun sweeps through the constellations during the seasons; consequently, we see the Sun apparently move through those constellations along a path referred to as the **ecliptic**.[3] During the spring the Sun appears to travel across the constellation of Virgo, in the summer it moves through Orion, during the autumn months it enters Aquarius, and in the winter the Sun is located near Scorpius. As a consequence, those constellations become obscured in the glare of daylight, and other constellations appear in our night sky. This seasonal change in the constellations is directly related to the fact that a given star rises approximately 4 minutes earlier each day. Since Earth completes one sidereal period in approximately 365.26 days, it moves slightly less than 1° around its orbit in 24 hours. Thus Earth must actually rotate nearly 361° to bring the Sun to the meridian on two successive days (Fig. 1.10). Because of the much greater distances to the stars, they do not shift their positions significantly as Earth orbits the Sun. As a result, placing a star on the meridian on successive nights requires only a 360° rotation. It takes approximately 4 minutes for Earth to rotate the extra 1°. Therefore a given star rises 4 minutes earlier each night. **Solar time** is defined as an *average* interval of

---

[3]The term *ecliptic* is derived from the observation of eclipses along that path through the heavens.

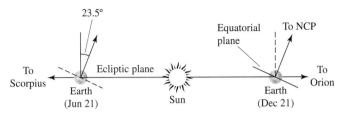

**FIGURE 1.9**   The plane of Earth's orbit seen edge-on. The tilt of Earth's rotation axis relative to the ecliptic is also shown.

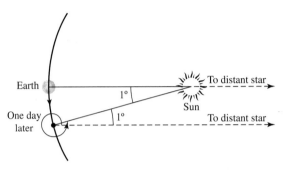

**FIGURE 1.10**   Earth must rotate nearly 361° per solar day and only 360° per sidereal day.

24 hours between meridian crossings of the Sun, and **sidereal time** is based on consecutive meridian crossings of a star.

Seasonal climatic variations are also due to the orbital motion of Earth, coupled with the approximately 23.5° tilt of its rotation axis. As a result of the tilt, the ecliptic moves north and south of the **celestial equator** (Fig. 1.11), which is defined by passing a plane through Earth at its equator and extending that plane out to the celestial sphere. The sinusoidal shape of the ecliptic occurs because the Northern Hemisphere alternately points toward and then away from the Sun during Earth's annual orbit. Twice during the year the Sun crosses the celestial equator, once moving northward along the ecliptic and later moving to the south. In the first case, the point of intersection is called the **vernal equinox** and the southern crossing occurs at the **autumnal equinox**. Spring officially begins when the center of the Sun is precisely on the vernal equinox; similarly, fall begins when the center of the Sun crosses the autumnal equinox. The most northern excursion of the Sun along the ecliptic occurs at the **summer solstice**, representing the official start of summer, and the southernmost position of the Sun is defined as the **winter solstice**.

The seasonal variations in weather are due to the position of the Sun relative to the celestial equator. During the summer months in the Northern Hemisphere, the Sun's northern declination causes it to appear higher in the sky, producing longer days and more intense sunlight. During the winter months the declination of the Sun is below the celestial equator, its path above the horizon is shorter, and its rays are less intense (see Fig. 1.12). The more direct the Sun's rays, the more energy per unit area strikes Earth's surface and the higher the resulting surface temperature.

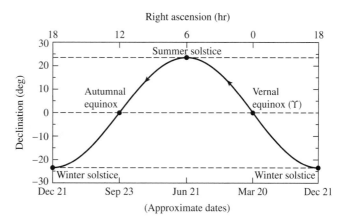

FIGURE 1.11  The ecliptic is the annual path of the Sun across the celestial sphere and is sinusoidal about the celestial equator. Summer solstice is at a declination of 23.5° and winter solstice is at a declination of −23.5°. See Fig. 1.13 for explanations of right ascension and declination.

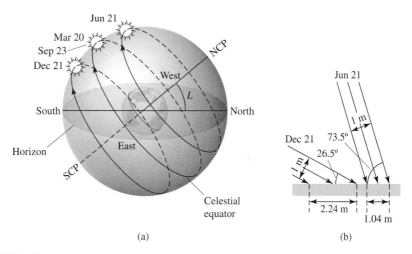

FIGURE 1.12  (a) The diurnal path of the Sun across the celestial sphere for an observer at latitude $L$ when the Sun is located at the vernal equinox (March), the summer solstice (June), the autumnal equinox (September), and the winter solstice (December). NCP and SCP designate the north and south celestial poles, respectively. The dots represent the location of the Sun at local noon on the approximate dates indicated. (b) The direction of the Sun's rays at noon at the summer solstice (approximately June 21) and at the winter solstice (approximately December 21) for an observer at 40° N latitude.

## The Equatorial Coordinate System

A coordinate system that results in nearly constant values for the positions of celestial objects, despite the complexities of diurnal and annual motions, is necessarily less straightforward than the altitude–azimuth system. The **equatorial coordinate system** (see Fig. 1.13) is based on the latitude–longitude system of Earth but does not participate in the planet's rotation. **Declination** $\delta$ is the equivalent of latitude and is measured in degrees north or

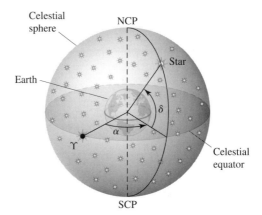

**FIGURE 1.13**    The equatorial coordinate system. $\alpha$, $\delta$, and $\Upsilon$ designate right ascension, declination, and the position of the vernal equinox, respectively.

south of the celestial equator. **Right ascension** $\alpha$ is analogous to longitude and is measured eastward along the celestial equator from the vernal equinox ($\Upsilon$) to its intersection with the object's **hour circle** (the great circle passing through the object being considered and through the north celestial pole). Right ascension is traditionally measured in hours, minutes, and seconds; 24 hours of right ascension is equivalent to $360°$, or 1 hour $= 15°$. The rationale for this unit of measure is based on the 24 hours (sidereal time) necessary for an object to make two successive crossings of the observer's local meridian. The coordinates of right ascension and declination are also indicated in Figs. 1.2 and 1.11. Since the equatorial coordinate system is based on the celestial equator and the vernal equinox, changes in the latitude and longitude of the observer do not affect the values of right ascension and declination. Values of $\alpha$ and $\delta$ are similarly unaffected by the annual motion of Earth around the Sun.

The **local sidereal time** of the observer is defined as the amount of time that has elapsed since the vernal equinox last traversed the meridian. Local sidereal time is also equivalent to the **hour angle** $H$ of the vernal equinox, where hour angle is defined as the angle between a celestial object and the observer's meridian, measured in the direction of the object's motion around the celestial sphere.

### Precession

Despite referencing the equatorial coordinate system to the celestial equator and its intersection with the ecliptic (the vernal equinox), **precession** causes the right ascension and declination of celestial objects to change, albeit very slowly. Precession is the slow wobble of Earth's rotation axis due to our planet's nonspherical shape and its gravitational interaction with the Sun and the Moon. It was Hipparchus who first observed the effects of precession. Although we will not discuss the physical cause of this phenomenon in detail, it is completely analogous to the well-known precession of a child's toy top. Earth's precession period is 25,770 years and causes the north celestial pole to make a slow circle through the heavens. Although Polaris (the North Star) is currently within $1°$ of the north celestial pole, in 13,000 years it will be nearly 47° away from that point. The same effect also causes a 50.26″ yr⁻¹ westward motion of the vernal equinox along the ecliptic.[4] An additional precession effect due to Earth–planet interactions results in an eastward motion of the vernal equinox of 0.12″ yr⁻¹.

Because precession alters the position of the vernal equinox along the ecliptic, it is necessary to refer to a specific **epoch** (or reference date) when listing the right ascension and declination of a celestial object. The current values of $\alpha$ and $\delta$ may then be calculated, based on the amount of time elapsed since the reference epoch. The epoch commonly used today for astronomical catalogs of stars, galaxies, and other celestial phenomena refers to an object's position at noon in Greenwich, England (**universal time, UT**) on January 1, 2000.[5] A catalog using this reference date is designated as J2000.0. The prefix, J, in the designation J2000.0 refers to the **Julian calendar**, which was introduced by Julius Caesar in 46 B.C.

Approximate expressions for the changes in the coordinates relative to J2000.0 are

$$\Delta\alpha = M + N \sin\alpha \tan\delta \qquad (1.2)$$

$$\Delta\delta = N \cos\alpha, \qquad (1.3)$$

where $M$ and $N$ are given by

$$M = 1°\!.2812323T + 0°\!.0003879T^2 + 0°\!.0000101T^3$$

$$N = 0°\!.5567530T - 0°\!.0001185T^2 - 0°\!.0000116T^3$$

and $T$ is defined as

$$T = (t - 2000.0)/100 \qquad (1.4)$$

where $t$ is the current date, specified in fractions of a year.

---

**Example 1.3.1.**   Altair, the brightest star in the summer constellation of Aquila, has the following J2000.0 coordinates: $\alpha = 19^h50^m47.0^s$, $\delta = +08°52'06.0''$. Using Eqs. (1.2) and (1.3), we may precess the star's coordinates to noon Greenwich mean time on July 30, 2005.

Writing the date as $t = 2005.575$, we have that $T = 0.05575$. This implies that $M = 0.071430°$ and $N = 0.031039°$. From the relations between time and the angular

*continued*

---

[4] 1 arcminute = 1′ = 1/60 degree; 1 arcsecond = 1″ = 1/60 arcminute.
[5] Universal time is also sometimes referred to as **Greenwich mean time**. Technically there are two forms of universal time; **UT1** is based on Earth's rotation rate, and **UTC** (**coordinated universal time**) is the basis of the worldwide system of civil time and is measured by atomic clocks. Because Earth's rotation rate is less regular than the time kept by atomic clocks, it is necessary to adjust UTC clocks by about one second (a *leap second*) roughly every year to year and a half. Among other effects contributing to the difference between UT1 and UTC is the slowing of Earth's rotation rate due to tidal effects.

measure of right ascension,

$$1^h = 15°$$

$$1^m = 15'$$

$$1^s = 15''$$

the corrections to the coordinates are

$$\Delta\alpha = 0.071430° + (0.031039°) \sin 297.696° \tan 8.86833°$$

$$= 0.067142° \simeq 16.11^s$$

and

$$\Delta\delta = (0.031039°) \cos 297.696°$$

$$= 0.014426° \simeq 51.93''.$$

Thus Altair's precessed coordinates are $\alpha = 19^h51^m03.1^s$ and $\delta = +08°52'57.9''$.

---

### Measurements of Time

The civic calendar commonly used in most countries today is the **Gregorian calendar**. The Gregorian calendar, introduced by Pope Gregory XIII in 1582, carefully specifies which years are to be considered leap years. Although leap years are useful for many purposes, astronomers are generally interested in the number of days (or seconds) between events, not in worrying about the complexities of leap years. Consequently, astronomers typically refer to the times when observations were made in terms of the elapsed time since some specified zero time. The time that is universally used is noon on January 1, 4713 B.C., as specified by the Julian calendar. This time is designated as JD 0.0, where JD indicates **Julian Date**.[6] The Julian date of J2000.0 is JD 2451545.0. Times other than noon universal time are specified as fractions of a day; for example, 6 PM January 1, 2000 UT would be designated JD 2451545.25. Referring to Julian date, the parameter $T$ defined by Eq. (1.4) can also be written as

$$T = (\text{JD} - 2451545.0)/36525,$$

where the constant 36,525 is taken from the **Julian year**, which is defined to be exactly 365.25 days.

Another commonly-used designation is the **Modified Julian Date (MJD)**, defined as $\text{MJD} \equiv \text{JD} - 2400000.5$, where JD refers to the Julian date. Thus a MJD day begins at midnight, universal time, rather than at noon.

---

[6]The Julian date JD 0.0 was proposed by Joseph Justus Scaliger (1540–1609) in 1583. His choice was based on the convergence of three calendar cycles; the 28 years required for the Julian calendar dates to fall on the same days of the week, the 19 years required for the phases of the Moon to *nearly* fall on the same dates of the year, and the 15-year Roman tax cycle. $28 \times 19 \times 15 = 7980$ means that the three calendars align once every 7980 years. JD 0.0 corresponds to the last time the three calendars all started their cycles together.

Because of the need to measure events very precisely in astronomy, various high-precision time measurements are used. For instance, **Heliocentric Julian Date** (HJD) is the Julian Date of an event as measured from the center of the Sun. In order to determine the heliocentric Julian date, astronomers must consider the time it would take light to travel from a celestial object to the center of the Sun rather than to Earth. **Terrestrial Time** (TT) is time measured on the surface of Earth, taking into consideration the effects of special and general relativity as Earth moves around the Sun and rotates on its own axis (for discussions of special and general relativity, refer to Chapter 4 and Section 17.1, respectively).

### Archaeoastronomy

An interesting application of the ideas discussed above is in the interdisciplinary field of **archaeoastronomy**, a merger of archaeology and astronomy. Archaeoastronomy is a field of study that relies heavily on historical adjustments that must be made to the positions of objects in the sky resulting from precession. It is the goal of archaeoastronomy to study the astronomy of past cultures, the investigation of which relies heavily on the alignments of ancient structures with celestial objects. Because of the long periods of time since construction, care must be given to the proper precession of celestial coordinates if any proposed alignments are to be meaningful. The Great Pyramid at Giza (Fig. 1.14), one of the "seven wonders of the world," is an example of such a structure. Believed to have been erected about 2600 B.C., the Great Pyramid has long been the subject of speculation. Although many of the proposals concerning this amazing monument are more than somewhat fanciful, there can be no doubt about its careful orientation with the four cardinal positions, north, south, east, and west. The greatest misalignment of any side from a true cardinal direction is no more than $5\frac{1}{2}'$. Equally astounding is the nearly perfect square formed by its base; no two sides differ in length by more than 20 cm.

Perhaps the most demanding alignments discovered so far are associated with the "air shafts" leading from the King's Chamber (the main chamber of the pyramid) to the outside. These air shafts seem too poorly designed to circulate fresh air into the tomb of Pharaoh, and

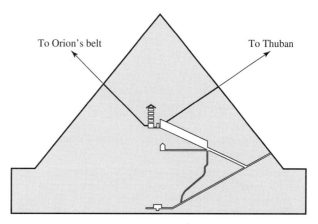

**FIGURE 1.14**   The astronomical alignments of the Great Pyramid at Giza. (Adaptation of a figure from Griffith Observatory.)

it is now thought that they served another function. The Egyptians believed that when their pharaohs died, their souls would travel to the sky to join Osiris, the god of life, death, and rebirth. Osiris was associated with the constellation we now know as Orion. Allowing for over one-sixth of a precession period since the construction of the Great Pyramid, Virginia Trimble has shown that one of the air shafts pointed directly to Orion's belt. The other air shaft pointed toward Thuban, the star that was *then* closest to the north celestial pole, the point in the sky about which all else turns.

As a modern scientific culture, we trace our study of astronomy to the ancient Greeks, but it has become apparent that many cultures carefully studied the sky and its mysterious points of light. Archaeological structures worldwide apparently exhibit astronomical alignments. Although some of these alignments may be coincidental, it is clear that many of them were by design.

### The Effects of Motions Through the Heavens

Another effect contributing to the change in equatorial coordinates is due to the intrinsic velocities of the objects themselves.[7] As we have already discussed, the Sun, the Moon, and the planets exhibit relatively rapid and complex motions through the heavens. The stars also move with respect to one another. Even though their actual speeds may be very large, the apparent relative motions of stars are generally very difficult to measure because of their enormous distances.

Consider the velocity of a star relative to an observer (Fig. 1.15). The velocity vector may be decomposed into two mutually perpendicular components, one lying along the line of sight and the other perpendicular to it. The line-of-sight component is the star's **radial velocity**, $v_r$, and will be discussed in Section 4.3; the second component is the star's

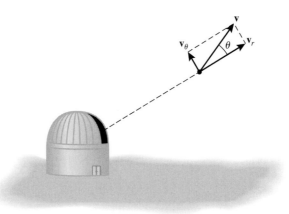

**FIGURE 1.15**   The components of velocity. $\mathbf{v}_r$ is the star's radial velocity and $\mathbf{v}_\theta$ is the star's transverse velocity.

---

[7]Parallax, an important periodic motion of the stars resulting from the motion of Earth about the Sun, will be discussed in detail in Section 3.1.

**transverse** or **tangential velocity**, $v_\theta$, along the celestial sphere. This transverse velocity appears as a *slow, angular change* in its equatorial coordinates, known as **proper motion** (usually expressed in seconds of arc per year). In a time interval $\Delta t$, the star will have moved in a direction perpendicular to the observer's line of sight a distance

$$\Delta d = v_\theta \Delta t.$$

If the distance from the observer to the star is $r$, then the angular change in its position along the celestial sphere is given by

$$\Delta\theta = \frac{\Delta d}{r} = \frac{v_\theta}{r}\Delta t.$$

Thus the star's proper motion, $\mu$, is related to its transverse velocity by

$$\mu \equiv \frac{d\theta}{dt} = \frac{v_\theta}{r}. \tag{1.5}$$

**An Application of Spherical Trigonometry**

The laws of spherical trigonometry must be employed in order to find the relationship between $\Delta\theta$ and changes in the equatorial coordinates, $\Delta\alpha$ and $\Delta\delta$, on the celestial sphere. A spherical triangle such as the one depicted in Fig. 1.16 is composed of three intersecting segments of great circles. For a spherical triangle the following relationships hold (with all sides measured in arc length, e.g., degrees):

*Law of sines*

$$\frac{\sin a}{\sin A} = \frac{\sin b}{\sin B} = \frac{\sin c}{\sin C}$$

*Law of cosines for sides*

$$\cos a = \cos b \cos c + \sin b \sin c \cos A$$

*Law of cosines for angles*

$$\cos A = -\cos B \cos C + \sin B \sin C \cos a.$$

Figure 1.17 shows the motion of a star on the celestial sphere from point $A$ to point $B$. The angular distance traveled is $\Delta\theta$. Let point $P$ be located at the north celestial pole so that the arcs $AP$, $AB$, and $BP$ form segments of great circles. The star is then said to be moving in the direction of the **position angle** $\phi$ ($\angle PAB$), measured from the north celestial pole. Now, construct a segment of a circle $NB$ such that $N$ is at the same declination as $B$ and $\angle PNB = 90°$. If the coordinates of the star at point $A$ are $(\alpha, \delta)$ and its new coordinates at point $B$ are $(\alpha + \Delta\alpha, \delta + \Delta\delta)$, then $\angle APB = \Delta\alpha$, $\overline{AP} = 90° - \delta$, and $\overline{NP} = \overline{BP} = 90° - (\delta + \Delta\delta)$. Using the law of sines,

$$\frac{\sin(\Delta\theta)}{\sin(\Delta\alpha)} = \frac{\sin[90° - (\delta + \Delta\delta)]}{\sin\phi},$$

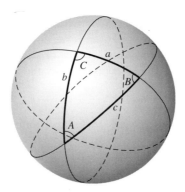

**FIGURE 1.16**    A spherical triangle. Each leg is a segment of a great circle on the surface of a sphere, and all angles are less than 180°. $a$, $b$, and $c$ are in angular units (e.g., degrees).

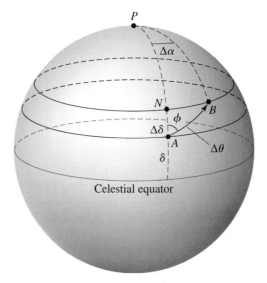

**FIGURE 1.17**    The proper motion of a star across the celestial sphere. The star is assumed to be moving from $A$ to $B$ along the position angle $\phi$.

or

$$\sin(\Delta\alpha)\cos(\delta + \Delta\delta) = \sin(\Delta\theta)\sin\phi.$$

Assuming that the changes in position are much less than one radian, we may use the small-angle approximations $\sin\epsilon \sim \epsilon$ and $\cos\epsilon \sim 1$. Employing the appropriate trigonometric identity and neglecting all terms of second order or higher, the previous equation reduces to

$$\Delta\alpha = \Delta\theta\frac{\sin\phi}{\cos\delta}. \tag{1.6}$$

The law of cosines for sides may also be used to find an expression for the change in the declination:

$$\cos[90° - (\delta + \Delta\delta)] = \cos(90° - \delta)\cos(\Delta\theta) + \sin(90° - \delta)\sin(\Delta\theta)\cos\phi.$$

Again using small-angle approximations and trigonometric identities, this expression reduces to

$$\Delta\delta = \Delta\theta\cos\phi. \tag{1.7}$$

(Note that this is the same result that would be obtained if we had used plane trigonometry. This should be expected, however, since we have assumed that the triangle being considered has an area much smaller than the total area of the sphere and should therefore appear essentially flat.) Combining Eqs. (1.6) and (1.7), we arrive at the expression for the angular distance traveled in terms of the changes in right ascension and declination:

$$\boxed{(\Delta\theta)^2 = (\Delta\alpha\cos\delta)^2 + (\Delta\delta)^2.} \tag{1.8}$$

## 1.4 ■ PHYSICS AND ASTRONOMY

The mathematical view of nature first proposed by Pythagoras and the Greeks led ultimately to the Copernican revolution. The inability of astronomers to accurately fit the observed positions of the "wandering stars" with mathematical models resulted in a dramatic change in our perception of Earth's location in the universe. However, an equally important step still remained in the development of science: the search for *physical causes* of observable phenomena. As we will see constantly throughout this book, the modern study of astronomy relies heavily on an understanding of the physical nature of the universe. The application of physics to astronomy, *astrophysics*, has proved very successful in explaining a wide range of observations, including strange and exotic objects and events, such as pulsating stars, supernovae, variable X-ray sources, black holes, quasars, gamma-ray bursts, and the Big Bang.

As a part of our investigation of the science of astronomy, it will be necessary to study the details of celestial motions, the nature of light, the structure of the atom, and the shape of space itself. Rapid advances in astronomy over the past several decades have occurred because of advances in our understanding of fundamental physics and because of improvements in the tools we use to study the heavens: telescopes and computers.

Essentially every area of physics plays an important role in some aspect of astronomy. Particle physics and astrophysics merge in the study of the Big Bang; the basic question of the origin of the zoo of elementary particles, as well as the very nature of the fundamental forces, is intimately linked to how the universe was formed. Nuclear physics provides information about the types of reactions that are possible in the interiors of stars, and atomic physics describes how individual atoms interact with one another and with light, processes that are basic to a great many astrophysical phenomena. Condensed-matter physics plays a role in the crusts of neutron stars and in the center of Jupiter. Thermodynamics is involved everywhere from the Big Bang to the interiors of stars. Even electronics plays an important role in the development of new detectors capable of giving a clearer view of the universe around us.

With the advent of modern technology and the space age, telescopes have been built to study the heavens with ever-increasing sensitivity. No longer limited to detecting visible light, telescopes are now capable of "seeing" gamma rays, X-rays, ultraviolet light, infrared radiation, and radio signals. Many of these telescopes require operation above Earth's atmosphere to carry out their missions. Other types of telescopes, very different in nature, detect elementary particles instead of light and are often placed below ground to study the heavens.

Computers have provided us with the power to carry out the enormous number of calculations necessary to build mathematical models from fundamental physical principles. The birth of high-speed computing machines has enabled astronomers to calculate the evolution of a star and compare those calculations with observations; it is also possible to study the rotation of a galaxy and its interaction with neighboring galaxies. Processes that require billions of years (significantly longer than any National Science Foundation grant) cannot possibly be observed directly but may be investigated using the modern supercomputer.

All of these tools and related disciplines are used to look at the heavens with a probing eye. The study of astronomy is a natural extension of human curiosity in its purest form. Just as a small child is always asking why this or that is the way it is, the goal of an astronomer is to attempt to understand the nature of the universe in all of its complexity, simply for the sake of understanding—the ultimate end of any intellectual adventure. In a very real sense, the true beauty of the heavens lies not only in observing the stars on a dark night but also in considering the delicate interplay between the physical processes that cause the stars to exist at all.

*The most incomprehensible thing about the universe is that it is comprehensible.* — Albert Einstein

## SUGGESTED READING

### General

Aveni, Anthony, *Skywatchers of Ancient Mexico*, The University of Texas Press, Austin, 1980.

Bronowski, J., *The Ascent of Man*, Little, Brown, Boston, 1973.

Casper, Barry M., and Noer, Richard J., *Revolutions in Physics*, W. W. Norton, New York, 1972.

Hadingham, Evan, *Early Man and the Cosmos*, Walker and Company, New York, 1984.

Krupp, E. C., *Echos of the Ancient Skies: The Astronomy of Lost Civilizations*, Harper & Row, New York, 1983.

Kuhn, Thomas S., *The Structure of Scientific Revolutions*, Third Edition, The University of Chicago Press, Chicago, 1996.

Ruggles, Clive L. N., *Astronomy in Prehistoric Britain and Ireland*, Yale University Press, New Haven, 1999.

Sagan, Carl, *Cosmos*, Random House, New York, 1980.

*SIMBAD Astronomical Database*, http://simbad.u-strasbg.fr/

*Sky and Telescope* Sky Chart,
   http://skyandtelescope.com/observing/skychart/

### Technical

Acker, Agnes, and Jaschek, Carlos, *Astronomical Methods and Calculations*, John Wiley and Sons, Chichester, 1986.

*Astronomical Almanac*, United States Government Printing Office, Washington, D.C.

Cox, Arthur N. (ed.), *Allen's Astrophysical Quantities*, Fourth Edition, Springer-Verlag, New York, 2000.

Lang, Kenneth R., *Astrophysical Formulae*, Third Edition, Springer-Verlag, New York, 1999.

Smart, W. M., and Green, Robin Michael, *Textbook on Spherical Astronomy*, Sixth Edition, Cambridge University Press, Cambridge, 1977.

## PROBLEMS

**1.1** Derive the relationship between a planet's synodic period and its sidereal period (Eq. 1.1). Consider both inferior and superior planets.

**1.2** Devise methods to determine the *relative* distances of each of the planets from the Sun given the information available to Copernicus (observable angles between the planets and the Sun, orbital configurations, and synodic periods).

**1.3** **(a)** The observed orbital synodic periods of Venus and Mars are 583.9 days and 779.9 days, respectively. Calculate their sidereal periods.

   **(b)** Which one of the superior planets has the shortest synodic period? Why?

**1.4** List the right ascension and declination of the Sun when it is located at the vernal equinox, the summer solstice, the autumnal equinox, and the winter solstice.

**1.5** **(a)** Referring to Fig. 1.12(a), calculate the altitude of the Sun along the meridian on the first day of summer for an observer at a latitude of $42°$ north.

   **(b)** What is the maximum altitude of the Sun on the first day of winter at the same latitude?

**1.6** **(a)** Circumpolar stars are stars that never set below the horizon of the local observer or stars that are never visible above the horizon. After sketching a diagram similar to Fig. 1.12(a), calculate the range of declinations for these two groups of stars for an observer at the latitude $L$.

   **(b)** At what latitude(s) on Earth will the Sun never set when it is at the summer solstice?

**(c)** Is there any latitude on Earth where the Sun will never set when it is at the vernal equinox? If so, where?

**1.7 (a)** Determine the Julian date for 16:15 UT on July 14, 2006. (*Hint:* Be sure to include any leap years in your calculation.)

**(b)** What is the corresponding modified Julian date?

**1.8** Proxima Centauri ($\alpha$ Centauri C) is the closest star to the Sun and is a part of a triple star system. It has the epoch J2000.0 coordinates $(\alpha, \delta) = (14^h 29^m 42.95^s, -62°40'46.1'')$. The brightest member of the system, Alpha Centauri ($\alpha$ Centauri A) has J2000.0 coordinates of $(\alpha, \delta) = (14^h 39^m 36.50^s, -60°50'02.3'')$.

**(a)** What is the angular separation of Proxima Centauri and Alpha Centauri?

**(b)** If the distance to Proxima Centauri is $4.0 \times 10^{16}$ m, how far is the star from Alpha Centauri?

**1.9 (a)** Using the information in Problem 1.8, precess the coordinates of Proxima Centauri to epoch J2010.0.

**(b)** The proper motion of Proxima Centauri is $3.84''$ $\text{yr}^{-1}$ with the position angle 282°. Calculate the change in $\alpha$ and $\delta$ due to proper motion between 2000.0 and 2010.0.

**(c)** Which effect makes the largest contribution to changes in the coordinates of Proxima Centauri: precession or proper motion?

**1.10** Which values of right ascension would be best for viewing by an observer at a latitude of 40° in January?

**1.11** Verify that Eq. (1.7) follows directly from the expression immediately preceding it.

# CHAPTER

# 2

# Celestial Mechanics

## 2.1 ■ ELLIPTICAL ORBITS

Although the inherent simplicity of the Copernican model was aesthetically pleasing, the idea of a heliocentric universe was not immediately accepted; it lacked the support of observations capable of unambiguously demonstrating that a geocentric model was wrong.

### Tycho Brahe: The Great Naked-Eye Observer

After the death of Copernicus, Tycho Brahe (1546–1601), the foremost naked-eye observer, carefully followed the motions of the "wandering stars" and other celestial objects. He carried out his work at the observatory, Uraniborg, on the island of Hveen (a facility provided for him by King Frederick II of Denmark). To improve the accuracy of his observations, Tycho used large measuring instruments, such as the quadrant depicted in the mural in Fig. 2.1(a). Tycho's observations were so meticulous that he was able to measure the position of an object in the heavens to an accuracy of better than $4'$, approximately one-eighth the angular diameter of a full moon. Through the accuracy of his observations he demonstrated for the first time that comets must be much farther away than the Moon, rather than being some form of atmospheric phenomenon. Tycho is also credited with observing the supernova of 1572, which clearly demonstrated that the heavens were not unchanging as Church doctrine held. (This observation prompted King Frederick to build Uraniborg.) Despite the great care with which he carried out his work, Tycho was not able to find any clear evidence of the motion of Earth through the heavens, and he therefore concluded that the Copernican model must be false (see Section 3.1).

### Kepler's Laws of Planetary Motion

At Tycho's invitation, Johannes Kepler (1571–1630), a German mathematician, joined him later in Prague [Fig. 2.1(b)]. Unlike Tycho, Kepler was a heliocentrist, and it was his desire to find a geometrical model of the universe that would be consistent with the best observations then available, namely Tycho's. After Tycho's death, Kepler inherited the mass

(a)                                                    (b)

**FIGURE 2.1**    (a) Mural of Tycho Brahe (1546–1601). (b) Johannes Kepler (1571–1630). (Courtesy of Yerkes Observatory.)

of observations accumulated over the years and began a painstaking analysis of the data. His initial, almost mystic, idea was that the universe is arranged with five perfect solids, nested to support the six known naked-eye planets (including Earth) on crystalline spheres, with the entire system centered on the Sun. After this model proved unsuccessful, he attempted to devise an accurate set of circular planetary orbits about the Sun, focusing specifically on Mars. Through his very clever use of offset circles and equants,[1] Kepler was able to obtain excellent agreement with Tycho's data for all but two of the points available. In particular, the discrepant points were each off by approximately 8′, or twice the accuracy of Tycho's data. Believing that Tycho would not have made observational errors of this magnitude, Kepler felt forced to dismiss the idea of purely circular motion.

Rejecting the last fundamental assumption of the Ptolemaic model, Kepler began to consider the possibility that planetary orbits were elliptical in shape rather than circular. Through this relatively minor mathematical (though monumental philosophical) change, he was finally able to bring all of Tycho's observations into agreement with a model for planetary motion. This paradigm shift also allowed Kepler to discover that the orbital speed of a planet is not constant but varies in a precise way depending on its location in its orbit. In 1609 Kepler published the first two of his three laws of planetary motion in the book *Astronomica Nova*, or *The New Astronomy*:

> **Kepler's First Law**    A planet orbits the Sun in an ellipse, with the Sun at one focus of the ellipse.

> **Kepler's Second Law**    A line connecting a planet to the Sun sweeps out equal areas in equal time intervals.

---

[1] Recall the geocentric use of circles and equants by Ptolemy; see Fig. 1.3.

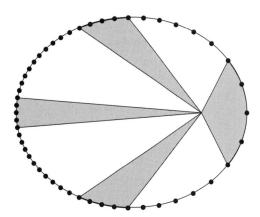

**FIGURE 2.2**   Kepler's second law states that the area swept out by a line between a planet and the focus of an ellipse is always the same for a given time interval, regardless of the planet's position in its orbit. The dots are evenly spaced in time.

Kepler's first and second laws are illustrated in Fig. 2.2, where each dot on the ellipse represents the position of the planet during evenly spaced time intervals.

Kepler's third law was published ten years later in the book *Harmonica Mundi* (*The Harmony of the World*). His final law relates the average orbital distance of a planet from the Sun to its sidereal period:

**Kepler's Third Law**   *The Harmonic Law.*

$$P^2 = a^3$$

where $P$ is the orbital period of the planet, measured in *years*, and $a$ is the average distance of the planet from the Sun, in *astronomical units*, or AU. An **astronomical unit** is, by definition, the average distance between Earth and the Sun, $1.496 \times 10^{11}$ m. The graph of Kepler's third law shown in Fig. 2.3 was prepared using data for each planet in our Solar System as given in Appendix C.

In retrospect it is easy to understand why the assumption of uniform and circular motion first proposed nearly 2000 years earlier was not determined to be wrong much sooner; in most cases, planetary motion differs little from purely circular motion. In fact, it was actually fortuitous that Kepler chose to focus on Mars, since the data for that planet were particularly good and Mars deviates from circular motion more than most of the others.

### The Geometry of Elliptical Motion

To appreciate the significance of Kepler's laws, we must first understand the nature of the **ellipse**. An ellipse (see Fig. 2.4) is defined by that set of points that satisfies the equation

$$r + r' = 2a, \tag{2.1}$$

where $a$ is a constant known as the **semimajor axis** (half the length of the long, or major axis of the ellipse), and $r$ and $r'$ represent the distances to the ellipse from the two **focal**

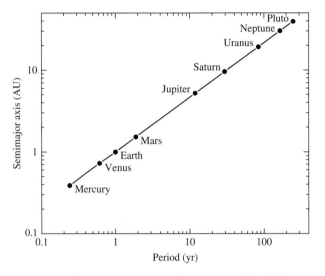

**FIGURE 2.3**    Kepler's third law for planets orbiting the Sun.

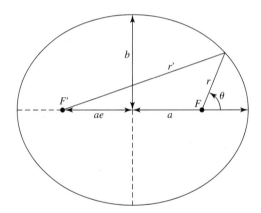

**FIGURE 2.4**    The geometry of an elliptical orbit.

**points**, $F$ and $F'$, respectively. According to Kepler's first law, a planet orbits the Sun in an ellipse, with the Sun located at one focus of the ellipse, the **principal focus**, $F$ (the other focus is empty space). Notice that if $F$ and $F'$ were located at the same point, then $r' = r$ and the previous equation would reduce to $r = r' = a$, the equation for a circle. Thus a circle is simply a special case of an ellipse. The distance $b$ is known as the **semiminor axis**. The **eccentricity**, $e$ ($0 \leq e < 1$), of the ellipse is defined as the distance between the foci divided by the major axis, $2a$, of the ellipse, implying that the distance of either focal point from the center of the ellipse may be expressed as $ae$. For a circle, $e = 0$. The point on the ellipse that is closest to the principal focus (located on the major axis) is called **perihelion**; the point on the opposite end of the major axis and farthest from the principal focus is known as **aphelion**.

A convenient relationship among $a$, $b$, and $e$ may be determined geometrically. Consider one of the two points at either end of the semiminor axis of an ellipse, where $r = r'$. In this case, $r = a$ and, by the Pythagorean theorem, $r^2 = b^2 + a^2 e^2$. Substitution leads immediately to the expression

$$b^2 = a^2 \left(1 - e^2\right).$$

(2.2)

Kepler's second law states that the orbital speed of a planet depends on its *location* in that orbit. To describe in detail the orbital behavior of a planet, it is necessary to specify where that planet is (its position vector) as well as how fast, and in what direction, the planet is moving (its velocity vector). It is often most convenient to express a planet's orbit in polar coordinates, indicating its distance $r$ from the principal focus in terms of an angle $\theta$ measured counterclockwise from the major axis of the ellipse beginning with the direction toward perihelion (see Fig. 2.4). Using the Pythagorean theorem, we have

$$r'^2 = r^2 \sin^2 \theta + (2ae + r \cos \theta)^2,$$

which reduces to

$$r'^2 = r^2 + 4ae(ae + r \cos \theta).$$

Using the definition of an ellipse, $r + r' = 2a$, we find that

$$r = \frac{a \left(1 - e^2\right)}{1 + e \cos \theta} \qquad (0 \le e < 1).$$

(2.3)

It is left as an exercise to show that the total area of an ellipse is given by

$$A = \pi a b.$$

(2.4)

---

**Example 2.1.1.** Using Eq. (2.3), it is possible to determine the variation in distance of a planet from the principal focus throughout its orbit. The semimajor axis of Mars's orbit is 1.5237 AU (or $2.2794 \times 10^{11}$ m) and the planet's orbital eccentricity is 0.0934. When $\theta = 0°$, the planet is at perihelion and is at a distance given by

$$r_p = \frac{a \left(1 - e^2\right)}{1 + e}$$

$$= a \left(1 - e\right)$$

(2.5)

$$= 1.3814 \text{ AU}.$$

*continued*

Similarly, at aphelion ($\theta = 180°$), the point where Mars is farthest from the Sun, the distance is given by

$$r_a = \frac{a\left(1 - e^2\right)}{1 - e}$$

$$= a\left(1 + e\right) \tag{2.6}$$

$$= 1.6660 \text{ AU}.$$

The variation in Mars's orbital distance from the Sun between perihelion and aphelion is approximately 19%.

---

An ellipse is actually one of a class of curves known as conic sections, found by passing a plane through a cone (see Fig. 2.5). Each type of conic section has its own characteristic range of eccentricities. As already mentioned, a circle is a conic section with $e = 0$, and an ellipse has $0 \le e < 1$. A curve having $e = 1$ is known as a **parabola** and is described by the equation

$$r = \frac{2p}{1 + \cos\theta} \qquad (e = 1), \tag{2.7}$$

where $p$ is the distance of closest approach to the parabola's *one* focus, at $\theta = 0$. Curves

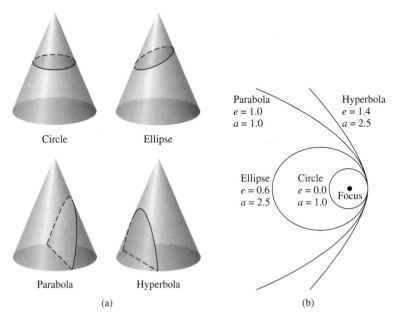

(a)                                            (b)

**FIGURE 2.5**    (a) Conic sections. (b) Related orbital paths.

having eccentricities greater than unity, $e > 1$, are **hyperbolas** and have the form

$$r = \frac{a\left(e^2 - 1\right)}{1 + e \cos\theta} \qquad (e > 1). \qquad (2.8)$$

Each type of conic section is related to a specific form of celestial motion.

## 2.2 ■ NEWTONIAN MECHANICS

At the time Kepler was developing his three laws of planetary motion, Galileo Galilei (1564–1642), perhaps the first of the true experimental physicists, was studying the motion of objects on Earth [Fig. 2.6(a)]. It was Galileo who proposed the earliest formulation of the concept of inertia. He had also developed an understanding of acceleration; in particular, he realized that objects near the surface of Earth fall with the same acceleration, independent of their weight. Whether Galileo publicly proved this fact by dropping objects of differing weights from the Leaning Tower of Pisa is a matter of some debate.

### The Observations of Galileo

Galileo is also the father of modern observational astronomy. Shortly after learning about the 1608 invention of the first crude spyglass, he thought through its design and constructed his own. Using his new telescope to carefully observe the heavens, Galileo quickly made a number of important observations in support of the heliocentric model of the universe. In particular, he discovered that the band of light known as the Milky Way, which runs from horizon to horizon, is not merely a cloud, as had previously been supposed, but

(a)  (b)

**FIGURE 2.6**  (a) Galileo Galilei (1564–1642). (b) Isaac Newton (1642–1727). (Courtesy of Yerkes Observatory.)

actually contains an enormous number of individual stars not resolvable by the naked eye. Galileo also observed that the Moon possesses craters and therefore is not a perfect sphere. Observations of the varying phases of Venus implied that the planet does not shine by its own power, but must be reflecting sunlight from constantly changing angles relative to the Sun and Earth while it orbits the Sun. He also discovered that the Sun itself is blemished, having sunspots that vary in number and location. But perhaps the most damaging observation for the geocentric model, a model still strongly supported by the Church, was the discovery of four moons in orbit about Jupiter, indicating the existence of at least one other center of motion in the universe.

Many of Galileo's first observations were published in his book *Sidereus Nuncius* (*The Starry Messenger*) in 1610. By 1616 the Church forced him to withdraw his support of the Copernican model, although he was able to continue his study of astronomy for some years. In 1632 Galileo published another work, *The Dialogue on the Two Chief World Systems*, in which a three-character play was staged. In the play Salviati was the proponent of Galileo's views, Simplicio believed in the old Aristotelian view, and Sagredo acted as the neutral third party who was invariably swayed by Salviati's arguments. In a strong reaction, Galileo was called before the Roman Inquisition and his book was heavily censored. The book was then placed on the *Index* of banned books, a collection of titles that included works of Copernicus and Kepler. Galileo was put under house arrest for the remainder of his life, serving out his term at his home in Florence.

In 1992, after a 13-year study by Vatican experts, Pope John Paul II officially announced that, because of a "tragic mutual incomprehension," the Roman Catholic Church had erred in its condemnation of Galileo some 360 years earlier. By reevaluating its position, the Church demonstrated that, at least on this issue, there is room for the philosophical views of both science and religion.

### Newton's Three Laws of Motion

Isaac Newton (1642–1727), arguably the greatest of any scientific mind in history [Fig. 2.6(b)], was born on Christmas Day in the year of Galileo's death. At age 18, Newton enrolled at Cambridge University and subsequently obtained his bachelor's degree. In the two years following the completion of his formal studies, and while living at home in Woolsthorpe, in rural England, away from the immediate dangers of the Plague, Newton engaged in what was likely the most productive period of scientific work ever carried out by one individual. During that interval, he made significant discoveries and theoretical advances in understanding motion, astronomy, optics, and mathematics. Although his work was not published immediately, the *Philosophiae Naturalis Principia Mathematica* (*Mathematical Principles of Natural Philosophy*), now simply known as the *Principia*, finally appeared in 1687 and contained much of his work on mechanics, gravitation, and the calculus. The publication of the *Principia* came about largely as a result of the urging of Edmond Halley, who paid for its printing. Another book, *Optiks*, appeared separately in 1704 and contained Newton's ideas about the nature of light and some of his early experiments in optics. Although many of his ideas concerning the particle nature of light were later shown to be in error (see Section 3.3), much of Newton's other work is still used extensively today.

Newton's great intellect is evidenced in his solution of the so-called brachistochrone problem posed by Johann Bernoulli, the Swiss mathematician, as a challenge to his colleagues. The brachistochrone problem amounts to finding the curve along which a bead could slide over a frictionless wire in the least amount of time while under the influence only of gravity. The deadline for finding a solution was set at a year and a half. The problem was presented to Newton late one afternoon; by the next morning he had found the answer by inventing a new area of mathematics known as the calculus of variations. Although the solution was published anonymously at Newton's request, Bernoulli commented, "By the claw, the lion is revealed."

Concerning the successes of his own career, Newton wrote:

> I do not know what I may appear to the world; but to myself I seem to have been only like a boy, playing on the seashore, and diverting myself, in now and then finding a smoother pebble or a prettier shell than ordinary, while the great ocean of truth lay all undiscovered before me.

Today, classical mechanics is described by Newton's three laws of motion, along with his universal *law of gravity*. Outside of the realms of atomic dimensions, velocities approaching the speed of light, or extreme gravitational forces, Newtonian physics has proved very successful in explaining the results of observations and experiments. Those regimes where Newtonian mechanics have been shown to be unsatisfactory will be discussed in later chapters.

Newton's first law of motion may be stated as follows:

**Newton's First Law** *The Law of Inertia.* An object at rest will remain at rest and an object in motion will remain in motion in a straight line at a constant speed unless acted upon by an external force.

To establish whether an object is actually moving, a reference frame must be established. In later chapters we will refer to reference frames that have the special property that the first law is valid; all such frames are known as **inertial reference frames**. Noninertial reference frames are accelerated with respect to inertial frames.

The first law may be restated in terms of the momentum of an object, $\mathbf{p} = m\mathbf{v}$, where $m$ and $\mathbf{v}$ are mass and velocity, respectively.[2] Thus Newton's first law may be expressed as *"the momentum of an object remains constant unless it experiences an external force."*[3]

The second law is actually a definition of the concept of force:

**Newton's Second Law** The *net* force (the sum of all forces) acting on an object is proportional to the object's mass and its resultant acceleration.

If an object is experiencing $n$ forces, then the net force is given by

$$\mathbf{F}_{\text{net}} = \sum_{i=1}^{n} \mathbf{F}_i = m\mathbf{a}. \tag{2.9}$$

---

[2]Hereafter, all vectors will be indicated by boldface type. Vectors are quantities described by both a magnitude and a direction. Some texts use alternate notations for vectors, expressing them either as $\vec{v}$ or $\overrightarrow{\mathbf{v}}$.

[3]The law of inertia is an extension of the original concept developed by Galileo.

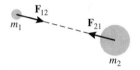

**FIGURE 2.7**    Newton's third law.

However, assuming that the mass is constant and using the definition $\mathbf{a} \equiv d\mathbf{v}/dt$, Newton's second law may also be expressed as

$$\mathbf{F}_{\text{net}} = m\frac{d\mathbf{v}}{dt} = \frac{d(m\mathbf{v})}{dt} = \frac{d\mathbf{p}}{dt}; \qquad (2.10)$$

the net force on an object is equal to the time rate of change of its momentum, $\mathbf{p}$. $\mathbf{F}_{\text{net}} = d\mathbf{p}/dt$ actually represents the most general statement of the second law, allowing for a time variation in the mass of the object such as occurs with rocket propulsion.

The third law of motion is generally expressed as follows:

**Newton's Third Law**    For every action there is an equal and opposite reaction.

In this law, *action* and *reaction* are to be interpreted as forces acting on different objects. Consider the force exerted *on* one object (object 1) *by* a second object (object 2), $\mathbf{F}_{12}$. Newton's third law states that the force exerted on object 2 by object 1, $\mathbf{F}_{21}$, must necessarily be of the same magnitude but in the opposite direction (see Fig. 2.7). Mathematically, the third law can be represented as

$$\mathbf{F}_{12} = -\mathbf{F}_{21}.$$

## Newton's Law of Universal Gravitation

Using his three laws of motion along with Kepler's third law, Newton was able to find an expression describing the force that holds planets in their orbits. Consider the special case of *circular* orbital motion of a mass $m$ about a much larger mass $M$ ($M \gg m$). Allowing for a system of units other than years and astronomical units, Kepler's third law may be written as

$$P^2 = kr^3,$$

where $r$ is the distance between the two objects and $k$ is a constant of proportionality. Writing the period of the orbit in terms of the orbit's circumference and the constant velocity of $m$ yields

$$P = \frac{2\pi r}{v},$$

and substituting into the prior equation gives

$$\frac{4\pi^2 r^2}{v^2} = kr^3.$$

Rearranging terms and multiplying both sides by $m$ lead to the expression

$$m\frac{v^2}{r} = \frac{4\pi^2 m}{kr^2}.$$

The left-hand side of the equation may be recognized as the centripetal force for circular motion, so

$$F = \frac{4\pi^2 m}{kr^2}$$

must be the gravitational force keeping $m$ in its orbit about $M$. However, Newton's third law states that the magnitude of the force exerted on $M$ by $m$ must equal the magnitude of the force exerted on $m$ by $M$. Therefore, the form of the equation ought to be symmetric with respect to exchange of $m$ and $M$, implying

$$F = \frac{4\pi^2 M}{k'r^2}.$$

Expressing this symmetry explicitly and grouping the remaining constants into a new constant, we have

$$F = \frac{4\pi^2 Mm}{k''r^2},$$

where $k = k''/M$ and $k' = k''/m$. Finally, introducing a new constant, $G \equiv 4\pi^2/k''$, we arrive at the form of the **Law of Universal Gravitation** found by Newton,

$$\boxed{F = G\frac{Mm}{r^2},} \tag{2.11}$$

where $G = 6.673 \times 10^{-11}$ N m$^2$ kg$^{-2}$ (the **Universal Gravitational Constant**).[4]

Newton's law of gravity applies to any two objects having mass. In particular, for an extended object (as opposed to a point mass), the force exerted by that object on another extended object may be found by integrating over each of their mass distributions.

---

**Example 2.2.1.** The force exerted by a *spherically symmetric* object of mass $M$ on a point mass $m$ may be found by integrating over rings centered along a line connecting the point mass to the center of the extended object (see Fig. 2.8). In this way all points on a specific ring are located at the same distance from $m$. Furthermore, because of the symmetry of the ring, the gravitational force vector associated with it is oriented along the ring's central axis. Once a general description of the force due to one ring is determined, it is possible to add up the individual contributions from all such rings throughout the entire volume of the mass $M$. The result will be the force on $m$ due to $M$.

*continued*

---

[4]At the time this text was written, the uncertainty in $G$ was $\pm 0.010 \times 10^{-11}$ N m$^2$ kg$^{-2}$.

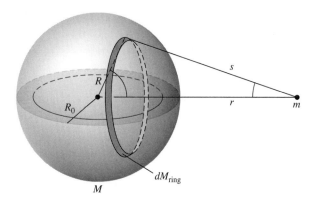

**FIGURE 2.8**   The gravitational effect of a spherically symmetric mass distribution.

Let $r$ be the distance between the centers of the two masses, $M$ and $m$. $R_0$ is the radius of the large mass, and $s$ is the distance from the point mass to a point on the ring. Because of the symmetry of the problem, only the component of the gravitational force vector along the line connecting the centers of the two objects needs to be calculated; the perpendicular components will cancel. If $dM_{\text{ring}}$ is the mass of the *ring* being considered, the force exerted by that ring on $m$ is given by

$$dF_{\text{ring}} = G\frac{m\, dM_{\text{ring}}}{s^2}\cos\phi.$$

Assuming that the mass density, $\rho(R)$, of the extended object is a function of radius only and that the volume of the ring of thickness $dR$ is $dV_{\text{ring}}$, we find that

$$dM_{\text{ring}} = \rho(R)\, dV_{\text{ring}}$$
$$= \rho(R)\, 2\pi R \sin\theta\, R\, d\theta\, dR$$
$$= 2\pi R^2 \rho(R) \sin\theta\, dR\, d\theta.$$

The cosine is given by

$$\cos\phi = \frac{r - R\cos\theta}{s},$$

where $s$ may be found by the Pythagorean theorem:

$$s = \sqrt{(r - R\cos\theta)^2 + R^2\sin^2\theta} = \sqrt{r^2 - 2rR\cos\theta + R^2}.$$

Substituting into the expression for $dF_{\text{ring}}$, summing over all *rings* located at a distance $R$ from the center of the mass $M$ (i.e., integrating over all $\theta$ from 0 to $\pi$ for constant $R$), and then summing over all resultant *shells* of radius $R$ from $R = 0$ to $R = R_0$ give the total force of gravity acting on the small mass $m$ along the system's line of symmetry:

$$F = Gm \int_0^{R_0} \int_0^{\pi} \frac{(r - R\cos\theta)\rho(R)2\pi R^2 \sin\theta}{s^3} \, d\theta \, dR$$

$$= 2\pi Gm \int_0^{R_0} \int_0^{\pi} \frac{r R^2 \rho(R) \sin\theta}{\left(r^2 + R^2 - 2rR\cos\theta\right)^{3/2}} \, d\theta \, dR$$

$$- 2\pi Gm \int_0^{R_0} \int_0^{\pi} \frac{R^3 \rho(R) \sin\theta \cos\theta}{\left(r^2 + R^2 - 2rR\cos\theta\right)^{3/2}} \, d\theta \, dR.$$

The integrations over $\theta$ may be carried out by making the change of variable, $u \equiv s^2 = r^2 + R^2 - 2rR\cos\theta$. Then $\cos\theta = (r^2 + R^2 - u)/2rR$ and $\sin\theta \, d\theta = du/2rR$. After the appropriate substitutions and integration over the new variable $u$, the equation for the force becomes

$$F = \frac{Gm}{r^2} \int_0^{R_0} 4\pi R^2 \rho(R) \, dR.$$

Notice that the integrand is just the mass of a *shell* of thickness $dR$, having a volume $dV_{\text{shell}}$, or

$$dM_{\text{shell}} = 4\pi R^2 \rho(R) \, dR = \rho(R) \, dV_{\text{shell}}.$$

Therefore, the integrand gives the force on $m$ due to a spherically symmetric mass shell of mass $dM_{\text{shell}}$ as

$$dF_{\text{shell}} = \frac{Gm \, dM_{\text{shell}}}{r^2}.$$

*The shell acts gravitationally as if its mass were located entirely at its center.* Finally, integrating over the mass shells, we have that the force exerted on $m$ by an extended, spherically symmetric mass distribution is directed along the line of symmetry between the two objects and is given by

$$F = G\frac{Mm}{r^2},$$

just the equation for the force of gravity between two point masses.

When an object is dropped near the surface of Earth, it accelerates toward the center of Earth at the rate $g = 9.80$ m s$^{-2}$, the local acceleration of gravity. Using Newton's second law and his law of gravity, an expression for the acceleration of gravity may be found. If $m$ is the mass of the falling object, $M_\oplus$ and $R_\oplus$ are the mass and radius of Earth, respectively, and $h$ is the height of the object above Earth, then the force of gravity on $m$ due to Earth is given by

$$F = G\frac{M_\oplus m}{(R_\oplus + h)^2}.$$

Assuming that $m$ is near Earth's surface, then $h \ll R_\oplus$ and

$$F \simeq G \frac{M_\oplus m}{R_\oplus^2}.$$

However, $F = ma = mg$; thus

$$\boxed{g = G \frac{M_\oplus}{R_\oplus^2}.} \tag{2.12}$$

Substituting the values $M_\oplus = 5.9736 \times 10^{24}$ kg and $R_\oplus = 6.378136 \times 10^6$ m gives a value for $g$ in agreement with the measured value.

### The Orbit of the Moon

The famous story that an apple falling on Newton's head allowed him to immediately realize that gravity holds the Moon in its orbit is probably somewhat fanciful and inaccurate. However, he did demonstrate that, along with the acceleration of the falling apple, gravity was responsible for the motion of Earth's closest neighbor.

---

**Example 2.2.2.**   Assuming for simplicity that the Moon's orbit is exactly circular, we can calculate the centripetal acceleration of the Moon rapidly. Recall that the centripetal acceleration of an object moving in a perfect circle is given by

$$a_c = \frac{v^2}{r}.$$

In this case, $r$ is the distance from the center of Earth to the center of the Moon, $r = 3.84401 \times 10^8$ m, and $v$ is the Moon's orbital velocity, given by

$$v = \frac{2\pi r}{P},$$

where $P = 27.3$ days $= 2.36 \times 10^6$ s is the sidereal orbital period of the Moon. Finding $v = 1.02$ km s$^{-1}$ gives a value for the centripetal acceleration of

$$a_c = 0.0027 \text{ m s}^{-2}.$$

The acceleration of the Moon caused by Earth's gravitational pull may also be calculated directly from

$$a_g = G \frac{M_\oplus}{r^2} = 0.0027 \text{ m s}^{-2},$$

in agreement with the value for the centripetal acceleration.

---

**Work and Energy**

In astrophysics, as in any area of physics, it is often very helpful to have some understanding of the energetics of specific physical phenomena in order to determine whether these processes are important in certain systems. Some models may be ruled out immediately if they are incapable of producing the amount of energy observed. Energy arguments also often result in simpler solutions to particular problems. For example, in the evolution of a planetary atmosphere, the possibility of a particular component of the atmosphere escaping must be considered. Such a consideration is based on a calculation of the escape speed of the gas particles.

The amount of energy (the work) necessary to raise an object of mass $m$ a height $h$ against a gravitational force is equal to the change in the *potential energy* of the system. Generally, the change in potential energy resulting from a change in position between two points is given by

$$U_f - U_i = \Delta U = -\int_{\mathbf{r}_i}^{\mathbf{r}_f} \mathbf{F} \cdot d\mathbf{r}, \qquad (2.13)$$

where $\mathbf{F}$ is the force vector, $\mathbf{r}_i$ and $\mathbf{r}_f$ are the initial and final position vectors, respectively, and $d\mathbf{r}$ is the infinitesimal change in the position vector for some general coordinate system (see Fig. 2.9). If the gravitational force on $m$ is due to a mass $M$ located at the origin, then $\mathbf{F}$ is directed inward toward $M$, $d\mathbf{r}$ is directed outward, $\mathbf{F} \cdot d\mathbf{r} = -F\,dr$, and the change in potential energy becomes

$$\Delta U = \int_{r_i}^{r_f} G\frac{Mm}{r^2}\,dr.$$

Evaluating the integral, we have

$$U_f - U_i = -GMm\left(\frac{1}{r_f} - \frac{1}{r_i}\right).$$

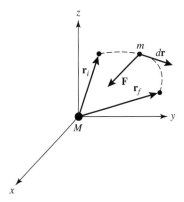

**FIGURE 2.9** Gravitational potential energy. The amount of work done depends on the direction of motion relative to the direction of the force vector.

Since only *relative changes* in potential energy are physically meaningful, a reference position where the potential energy is defined as being identically zero may be chosen. If, for a specific gravitational system, it is assumed that the potential energy goes to zero at infinity, letting $r_f$ approach infinity ($r_f \to \infty$) and dropping the subscripts for simplicity give

$$U = -G\frac{Mm}{r}. \tag{2.14}$$

Of course, the process can be reversed: The force may be found by *differentiating* the gravitational potential energy. For forces that depend only on $r$,

$$F = -\frac{\partial U}{\partial r}. \tag{2.15}$$

In a general three-dimensional description, $\mathbf{F} = -\nabla U$, where $\nabla U$ represents the *gradient* of $U$. In rectangular coordinates this becomes

$$\mathbf{F} = -\frac{\partial U}{\partial x}\hat{\mathbf{i}} - \frac{\partial U}{\partial y}\hat{\mathbf{j}} - \frac{\partial U}{\partial z}\hat{\mathbf{k}}.$$

Work must be performed on a massive object if its *speed*, $|\mathbf{v}|$, is to be changed. This can be seen by rewriting the work integral, first in terms of time, then speed:

$$
\begin{aligned}
W &\equiv -\Delta U \\
&= \int_{\mathbf{r}_i}^{\mathbf{r}_f} \mathbf{F} \cdot d\mathbf{r} \\
&= \int_{t_i}^{t_f} \frac{d\mathbf{p}}{dt} \cdot (\mathbf{v}\,dt) \\
&= \int_{t_i}^{t_f} m\frac{d\mathbf{v}}{dt} \cdot (\mathbf{v}\,dt) \\
&= \int_{t_i}^{t_f} m\left(\mathbf{v} \cdot \frac{d\mathbf{v}}{dt}\right) dt \\
&= \int_{t_i}^{t_f} m\frac{d\left(\frac{1}{2}v^2\right)}{dt}\,dt \\
&= \int_{v_i}^{v_f} m\,d\left(\frac{1}{2}v^2\right) \\
&= \frac{1}{2}mv_f^2 - \frac{1}{2}mv_i^2.
\end{aligned}
$$

We may now identify the quantity

$$K = \frac{1}{2}mv^2 \qquad (2.16)$$

as the **kinetic energy** of the object. Thus work done on the particle results in an equivalent change in the particle's kinetic energy. This statement is simply one example of the **conservation of energy**, a concept that is encountered frequently in all areas of physics.

Consider a particle of mass $m$ that has an initial velocity $\mathbf{v}$ and is at a distance $r$ from the center of a larger mass $M$, such as Earth. How fast must the mass be moving upward to completely escape the pull of gravity? To calculate the **escape speed**, energy conservation may be used directly. The total initial mechanical energy of the particle (both kinetic and potential) is given by

$$E = \frac{1}{2}mv^2 - G\frac{Mm}{r}.$$

Assume that, in the critical case, the final velocity of the mass will be zero at a position infinitely far from $M$, implying that both the kinetic and potential energy will become zero. Clearly, by conservation of energy, the *total* energy of the particle must be identically zero at all times. Thus

$$\frac{1}{2}mv^2 = G\frac{Mm}{r},$$

which may be solved immediately for the initial speed of $m$ to give

$$v_{\text{esc}} = \sqrt{2GM/r}. \qquad (2.17)$$

Notice that the mass of the escaping object does not enter into the final expression for the escape speed. Near the surface of Earth, $v_{\text{esc}} = 11.2$ km s$^{-1}$.

## 2.3 ■ KEPLER'S LAWS DERIVED

Although Kepler did finally determine that the geometry of planetary motion was in the more general form of an ellipse rather than circular motion, he was unable to explain the nature of the force that kept the planets moving in their precise patterns. Not only was Newton successful in quantifying that force, he was also able to generalize Kepler's work, deriving the empirical laws of planetary motion from the gravitational force law. The derivation of Kepler's laws represented a crucial step in the development of modern astrophysics.

### The Center-of-Mass Reference Frame

However, before proceeding onward to derive Kepler's laws, it will be useful to examine more closely the dynamics of orbital motion. An interacting two-body problem, such as

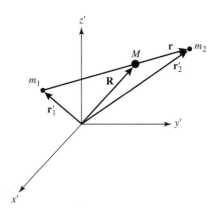

**FIGURE 2.10**    A general Cartesian coordinate system indicating the positions of $m_1$, $m_2$, and the center of mass (located at $M$).

binary orbits, or the more general many-body problem (often called the $N$-body problem), is most easily done in the reference frame of the **center of mass**.

Figure 2.10 shows two objects of masses $m_1$ and $m_2$ at positions $\mathbf{r}'_1$ and $\mathbf{r}'_2$, respectively, with the displacement vector from $\mathbf{r}'_1$ to $\mathbf{r}'_2$ given by

$$\mathbf{r} = \mathbf{r}'_2 - \mathbf{r}'_1.$$

Define a position vector $\mathbf{R}$ to be a *weighted* average of the position vectors of the individual masses,

$$\mathbf{R} \equiv \frac{m_1 \mathbf{r}'_1 + m_2 \mathbf{r}'_2}{m_1 + m_2}. \tag{2.18}$$

Of course, this definition can be immediately generalized to the case of $n$ objects,

$$\mathbf{R} \equiv \frac{\sum_{i=1}^{n} m_i \mathbf{r}'_i}{\sum_{i=1}^{n} m_i}.$$

Rewriting the equation, we have

$$\sum_{i=1}^{n} m_i \mathbf{R} = \sum_{i=1}^{n} m_i \mathbf{r}'_i.$$

Then, if we define $M$ to be the total mass of the system, $M \equiv \sum_{i=1}^{n} m_i$, the previous equation becomes

$$M\mathbf{R} = \sum_{i=1}^{n} m_i \mathbf{r}'_i.$$

Assuming that the individual masses do not change, differentiating both sides with respect to time gives

$$M\frac{d\mathbf{R}}{dt} = \sum_{i=1}^{n} m_i \frac{d\mathbf{r}_i'}{dt}$$

or

$$M\mathbf{V} = \sum_{i=1}^{n} m_i \mathbf{v}_i'.$$

The right-hand side is the sum of the linear momenta of every particle in the system, so the total linear momentum of the system may be treated as though *all* of the mass were located at $\mathbf{R}$, moving with a velocity $\mathbf{V}$. Thus $\mathbf{R}$ is the position of the center of mass of the system, and $\mathbf{V}$ is the center-of-mass velocity. Letting $\mathbf{P} \equiv M\mathbf{V}$ be the linear momentum of the center of mass and $\mathbf{p}_i' \equiv m_i \mathbf{v}_i'$ be the linear momentum of an individual particle $i$, and again differentiating both sides with respect to time, yields

$$\frac{d\mathbf{P}}{dt} = \sum_{i=1}^{n} \frac{d\mathbf{p}_i'}{dt}.$$

If we assume that *all* of the forces acting on individual particles in the system are due to other particles contained within the system, Newton's third law requires that the total force must be zero. This constraint exists because of the equal magnitudes and opposite directions of action–reaction pairs. Of course, the momentum of individual masses may change. Using center-of-mass quantities, we find that the total (or net) force on the system is

$$\mathbf{F} = \frac{d\mathbf{P}}{dt} = M\frac{d^2\mathbf{R}}{dt^2} = 0.$$

Therefore, the center of mass will not accelerate if no external forces exist. This implies that a reference frame associated with the center of mass must be an inertial reference frame and that the $N$-body problem may be simplified by choosing a coordinate system for which the center of mass is at rest at $\mathbf{R} = 0$.

If we choose a center-of-mass reference frame for a binary system, depicted in Fig. 2.11 ($\mathbf{R} = 0$), Eq. (2.18) becomes

$$\frac{m_1\mathbf{r}_1 + m_2\mathbf{r}_2}{m_1 + m_2} = 0, \tag{2.19}$$

where the primes have been dropped, indicating center-of-mass coordinates. Both $\mathbf{r}_1$ and $\mathbf{r}_2$ may now be rewritten in terms of the displacement vector, $\mathbf{r}$. Substituting $\mathbf{r}_2 = \mathbf{r}_1 + \mathbf{r}$ gives

$$\mathbf{r}_1 = -\frac{m_2}{m_1 + m_2}\mathbf{r} \tag{2.20}$$

$$\mathbf{r}_2 = \frac{m_1}{m_1 + m_2}\mathbf{r}. \tag{2.21}$$

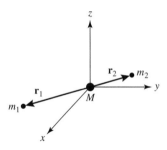

**FIGURE 2.11** The center-of-mass reference frame for a binary orbit, with the center of mass fixed at the origin of the coordinate system.

Next, define the **reduced mass** to be

$$\mu \equiv \frac{m_1 m_2}{m_1 + m_2}. \tag{2.22}$$

Then $\mathbf{r}_1$ and $\mathbf{r}_2$ become

$$\mathbf{r}_1 = -\frac{\mu}{m_1}\mathbf{r} \tag{2.23}$$

$$\mathbf{r}_2 = \frac{\mu}{m_2}\mathbf{r}. \tag{2.24}$$

The convenience of the center-of-mass reference frame becomes evident when the total energy and orbital angular momentum of the system are considered. Including the necessary kinetic energy and gravitational potential energy terms, the total energy may be expressed as

$$E = \frac{1}{2}m_1 |\mathbf{v}_1|^2 + \frac{1}{2}m_2 |\mathbf{v}_2|^2 - G\frac{m_1 m_2}{|\mathbf{r}_2 - \mathbf{r}_1|}.$$

Substituting the relations for $\mathbf{r}_1$ and $\mathbf{r}_2$, along with the expression for the total mass of the system and the definition for the reduced mass, gives

$$E = \frac{1}{2}\mu v^2 - G\frac{M\mu}{r}, \tag{2.25}$$

where $v = |\mathbf{v}|$ and $\mathbf{v} \equiv d\mathbf{r}/dt$. We have also used the notation $r = |\mathbf{r}_2 - \mathbf{r}_1|$. The total energy of the system is equal to the kinetic energy of the reduced mass, plus the potential energy of the reduced mass moving about a mass $M$, assumed to be located and fixed at the origin. The distance between $\mu$ and $M$ is equal to the separation between the objects of masses $m_1$ and $m_2$.

Similarly, the total orbital angular momentum,

$$\mathbf{L} = m_1\mathbf{r}_1 \times \mathbf{v}_1 + m_2\mathbf{r}_2 \times \mathbf{v}_2$$

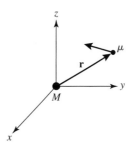

**FIGURE 2.12**   A binary orbit may be reduced to the equivalent problem of calculating the motion of the reduced mass, $\mu$, about the total mass, $M$, located at the origin.

becomes

$$\mathbf{L} = \mu \mathbf{r} \times \mathbf{v} = \mathbf{r} \times \mathbf{p}, \tag{2.26}$$

where $\mathbf{p} \equiv \mu \mathbf{v}$. The total orbital angular momentum equals the angular momentum of the reduced mass only. *In general, the two-body problem may be treated as an equivalent one-body problem with the reduced mass $\mu$ moving about a fixed mass $M$ at a distance $r$* (see Fig. 2.12).

### The Derivation of Kepler's First Law

To obtain Kepler's laws, we begin by considering the effect of gravitation on the orbital angular momentum of a planet. Using center-of-mass coordinates and evaluating the time derivative of the orbital angular momentum of the reduced mass (Eq. 2.26) give

$$\frac{d\mathbf{L}}{dt} = \frac{d\mathbf{r}}{dt} \times \mathbf{p} + \mathbf{r} \times \frac{d\mathbf{p}}{dt} = \mathbf{v} \times \mathbf{p} + \mathbf{r} \times \mathbf{F},$$

the second expression arising from the definition of velocity and Newton's second law. Notice that because $\mathbf{v}$ and $\mathbf{p}$ are in the same direction, their cross product is identically zero. Similarly, since $\mathbf{F}$ is a central force directed inward along $\mathbf{r}$, the cross product of $\mathbf{r}$ and $\mathbf{F}$ is also zero. The result is an important general statement concerning angular momentum:

$$\frac{d\mathbf{L}}{dt} = 0, \tag{2.27}$$

*the angular momentum of a system is a constant for a central force law.* Equation (2.26) further shows that the position vector $\mathbf{r}$ is always perpendicular to the constant angular momentum vector $\mathbf{L}$, meaning that the orbit of the reduced mass lies in a plane perpendicular to $\mathbf{L}$.

Using the radial unit vector $\hat{\mathbf{r}}$ (so $\mathbf{r} = r\hat{\mathbf{r}}$), we can write the angular momentum vector in an alternative form as

$$\mathbf{L} = \mu \mathbf{r} \times \mathbf{v}$$

$$= \mu r \hat{\mathbf{r}} \times \frac{d}{dt} (r\hat{\mathbf{r}})$$

$$= \mu r \hat{\mathbf{r}} \times \left( \frac{dr}{dt} \hat{\mathbf{r}} + r \frac{d}{dt} \hat{\mathbf{r}} \right)$$

$$= \mu r^2 \hat{\mathbf{r}} \times \frac{d}{dt} \hat{\mathbf{r}}.$$

(The last result comes from the fact that $\hat{\mathbf{r}} \times \hat{\mathbf{r}} = 0$.) In vector form, the acceleration of the reduced mass due to the gravitational force exerted by $M$ is

$$\mathbf{a} = -\frac{GM}{r^2} \hat{\mathbf{r}}.$$

Taking the vector cross product of the acceleration of the reduced mass with its own orbital angular momentum gives

$$\mathbf{a} \times \mathbf{L} = -\frac{GM}{r^2} \hat{\mathbf{r}} \times \left( \mu r^2 \hat{\mathbf{r}} \times \frac{d}{dt} \hat{\mathbf{r}} \right) = -GM\mu \hat{\mathbf{r}} \times \left( \hat{\mathbf{r}} \times \frac{d}{dt} \hat{\mathbf{r}} \right).$$

Applying the vector identity $\mathbf{A} \times (\mathbf{B} \times \mathbf{C}) = (\mathbf{A} \cdot \mathbf{C})\mathbf{B} - (\mathbf{A} \cdot \mathbf{B})\mathbf{C}$ results in

$$\mathbf{a} \times \mathbf{L} = -GM\mu \left[ \left( \hat{\mathbf{r}} \cdot \frac{d}{dt} \hat{\mathbf{r}} \right) \hat{\mathbf{r}} - (\hat{\mathbf{r}} \cdot \hat{\mathbf{r}}) \frac{d}{dt} \hat{\mathbf{r}} \right].$$

Because $\hat{\mathbf{r}}$ is a unit vector, $\hat{\mathbf{r}} \cdot \hat{\mathbf{r}} = 1$ and

$$\frac{d}{dt}(\hat{\mathbf{r}} \cdot \hat{\mathbf{r}}) = 2 \hat{\mathbf{r}} \cdot \frac{d}{dt} \hat{\mathbf{r}} = 0.$$

As a result,

$$\mathbf{a} \times \mathbf{L} = GM\mu \frac{d}{dt} \hat{\mathbf{r}}$$

or, by referring to Eq. (2.27),

$$\frac{d}{dt}(\mathbf{v} \times \mathbf{L}) = \frac{d}{dt}(GM\mu \hat{\mathbf{r}}).$$

Integrating with respect to time then yields

$$\mathbf{v} \times \mathbf{L} = GM\mu \hat{\mathbf{r}} + \mathbf{D}, \tag{2.28}$$

where $\mathbf{D}$ is a constant vector. Because $\mathbf{v} \times \mathbf{L}$ and $\hat{\mathbf{r}}$ both lie in the orbital plane, so must $\mathbf{D}$. Furthermore, the magnitude of the left-hand side will be greatest at perihelion when the velocity of the reduced mass is a maximum. Moreover, the magnitude of the right-hand side is greatest when $\hat{\mathbf{r}}$ and $\mathbf{D}$ point in the same direction. Therefore, $\mathbf{D}$ is directed toward perihelion. As shown below, the magnitude of $\mathbf{D}$ determines the eccentricity of the orbit.

We next take the vector dot product of Eq. (2.28) with the position vector $\mathbf{r} = r \hat{\mathbf{r}}$:

$$\mathbf{r} \cdot (\mathbf{v} \times \mathbf{L}) = GM\mu r \hat{\mathbf{r}} \cdot \hat{\mathbf{r}} + \mathbf{r} \cdot \mathbf{D}.$$

Invoking the vector identity $\mathbf{A} \cdot (\mathbf{B} \times \mathbf{C}) = (\mathbf{A} \times \mathbf{B}) \cdot \mathbf{C}$ gives

$$(\mathbf{r} \times \mathbf{v}) \cdot \mathbf{L} = GM\mu r + rD\cos\theta.$$

Finally, recalling the definition of angular momentum (Eq. 2.26), we obtain

$$\frac{L^2}{\mu} = GM\mu r \left(1 + \frac{D\cos\theta}{GM\mu}\right),$$

where $\theta$ is the angle of the reduced mass as measured from the direction to perihelion. Defining $e \equiv D/GM\mu$ and solving for $r$, we find

**Kepler's First Law (revisited)**

$$r = \frac{L^2/\mu^2}{GM(1 + e\cos\theta)}. \tag{2.29}$$

This is exactly the equation of a conic section, as may be seen by comparing Eq. (2.29) with Eqs. (2.3), (2.7), and (2.8) for an ellipse, parabola, and hyperbola, respectively. *The path of the reduced mass about the center of mass under the influence of gravity (or any other inverse-square force) is a conic section.* Elliptical orbits result from an attractive $r^{-2}$ central-force law such as gravity when the total energy of the system is less than zero (a bound system), parabolic trajectories are obtained when the energy is identically zero, and hyperbolic paths result from an unbounded system with an energy that is greater than zero.

When Eq. (2.29) is translated back to a physical reference frame on the sky, we find that Kepler's first law for bound planetary orbits may be stated as: *Both objects in a binary orbit move about the center of mass in ellipses, with the center of mass occupying one focus of each ellipse.* Newton was able to demonstrate the elliptical behavior of planetary motion and found that Kepler's first law must be generalized somewhat: The center of mass of the system, rather than the exact center of the Sun, is actually located at the focus of the ellipse. For our Solar System, such a mistake is understandable, since the largest of the planets, Jupiter, has only $1/1000$ the mass of the Sun. This places the center of mass of the Sun–Jupiter system near the surface of the Sun. Having used the naked-eye data of Tycho, Kepler can be forgiven for not realizing his error.

For the case of closed planetary orbits, comparing Eqs. (2.3) and (2.29) shows that the total orbital angular momentum of the system is

$$L = \mu\sqrt{GMa\left(1 - e^2\right)}. \tag{2.30}$$

Note that $L$ is a maximum for purely circular motion ($e = 0$) and goes to zero as the eccentricity approaches unity, as expected.

## The Derivation of Kepler's Second Law

To derive Kepler's second law, which relates the area of a section of an ellipse to a time interval, we begin by considering the infinitesimal area element in polar coordinates, as

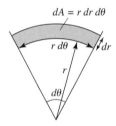

**FIGURE 2.13**   The infinitesimal area element in polar coordinates.

shown in Fig. 2.13:

$$dA = dr\,(r\,d\theta) = r\,dr\,d\theta.$$

If we integrate from the principal focus of the ellipse to a specific distance, $r$, the area swept out by an infinitesimal change in $\theta$ becomes

$$dA = \frac{1}{2}r^2\,d\theta.$$

Therefore, the time rate of change in area swept out by a line joining a point on the ellipse to the focus becomes

$$\frac{dA}{dt} = \frac{1}{2}r^2\frac{d\theta}{dt}. \tag{2.31}$$

Now the orbital velocity, $\mathbf{v}$, may be expressed in two components, one directed along $\mathbf{r}$ and the other perpendicular to $\mathbf{r}$. Letting $\hat{\mathbf{r}}$ and $\hat{\boldsymbol{\theta}}$ be the unit vectors along $\mathbf{r}$ and its normal, respectively, $\mathbf{v}$ may be written as (see Fig. 2.14)

$$\mathbf{v} = \mathbf{v}_r + \mathbf{v}_\theta = \frac{dr}{dt}\hat{\mathbf{r}} + r\frac{d\theta}{dt}\hat{\boldsymbol{\theta}}.$$

Substituting $v_\theta$ into Eq. (2.31) gives

$$\frac{dA}{dt} = \frac{1}{2}rv_\theta.$$

Since $\mathbf{r}$ and $\mathbf{v}_\theta$ are perpendicular,

$$rv_\theta = |\mathbf{r} \times \mathbf{v}| = \left|\frac{\mathbf{L}}{\mu}\right| = \frac{L}{\mu}.$$

Finally, the time derivative of the area becomes

**Kepler's Second Law (revisited)**

$$\boxed{\frac{dA}{dt} = \frac{1}{2}\frac{L}{\mu}.} \tag{2.32}$$

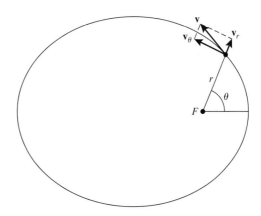

**FIGURE 2.14**   The velocity vector for elliptical motion in polar coordinates.

It has already been shown that the orbital angular momentum is a constant, so *the time rate of change of the area swept out by a line connecting a planet to the focus of an ellipse is a constant*, one-half of the orbital angular momentum per unit mass. This is just Kepler's second law.

Expressions for the speed of the reduced mass at perihelion ($\theta = 0$) and aphelion ($\theta = \pi/2$) may be easily obtained from Eq. (2.29). Since at both perihelion and aphelion, **r** and **v** are perpendicular, the magnitude of the angular momentum at these points simply becomes

$$L = \mu r v.$$

Eq. (2.29) at perihelion may thus be written as

$$r_p = \frac{(\mu r_p v_p)^2/\mu^2}{GM(1+e)},$$

whereas at aphelion

$$r_a = \frac{(\mu r_a v_a)^2/\mu^2}{GM(1-e)}.$$

Recalling from Example 2.1.1 that $r_p = a(1-e)$ at perihelion and $r_a = a(1+e)$ at aphelion, we immediately obtain

$$v_p^2 = \frac{GM(1+e)}{r_p} = \frac{GM}{a}\left(\frac{1+e}{1-e}\right) \tag{2.33}$$

at perihelion and

$$v_a^2 = \frac{GM(1-e)}{r_a} = \frac{GM}{a}\left(\frac{1-e}{1+e}\right). \tag{2.34}$$

at aphelion.

The total orbital energy may be found as well:

$$E = \frac{1}{2}\mu v_p^2 - G\frac{M\mu}{r_p}.$$

Making the appropriate substitutions, and after some rearrangement,

$$E = -G\frac{M\mu}{2a} = -G\frac{m_1 m_2}{2a}. \tag{2.35}$$

The total energy of a binary orbit depends only on the semimajor axis $a$ and is exactly one-half the time-averaged potential energy of the system,

$$E = \frac{1}{2}\langle U \rangle,$$

where $\langle U \rangle$ denotes an average over one orbital period.[5] This is one example of the *virial theorem*, a general property of gravitationally bound systems. The virial theorem will be discussed in detail in Section 2.4.

A useful expression for the velocity of the reduced mass (or the relative velocity of $m_1$ and $m_2$) may be found directly by using the conservation of energy and equating the total orbital energy to the sum of the kinetic and potential energies:

$$-G\frac{M\mu}{2a} = \frac{1}{2}\mu v^2 - G\frac{M\mu}{r}.$$

Using the identity $M = m_1 + m_2$, this simplifies to give

$$v^2 = G(m_1 + m_2)\left(\frac{2}{r} - \frac{1}{a}\right). \tag{2.36}$$

This expression could also have been obtained directly by adding the vector components of orbital velocity. Calculating $\mathbf{v}_r$, $\mathbf{v}_\theta$, and $v^2$ will be left as exercises.

### The Derivation of Kepler's Third Law

We are finally in a position to derive the last of Kepler's laws. Integrating the mathematical expression for Kepler's second law (Eq. 2.32) over one orbital period, $P$, gives the result

$$A = \frac{1}{2}\frac{L}{\mu}P.$$

Here the mass $m$ orbiting about a much larger *fixed* mass $M$ has been replaced by the more general reduced mass $\mu$ orbiting about the center of mass. Substituting the area of an ellipse,

[5]The proof that $\langle U \rangle = -GM\mu/a$ is left as an exercise. Note that the time average, $\langle 1/r \rangle$, is equal to $1/a$, but $\langle r \rangle \neq a$.

$A = \pi a b$, squaring the equation, and rearranging, we obtain the expression

$$P^2 = \frac{4\pi^2 a^2 b^2 \mu^2}{L^2}.$$

Finally, using Eq. (2.2) and the expression for the total orbital angular momentum (Eq. 2.30), the last equation simplifies to become

**Kepler's Third Law (revisited)**

$$P^2 = \frac{4\pi^2}{G(m_1 + m_2)} a^3. \tag{2.37}$$

This is the general form of Kepler's third law. Not only did Newton demonstrate the relationship between the semimajor axis of an elliptical orbit and the orbital period, he also found a term not discovered empirically by Kepler, the square of the orbital period is inversely proportional to the total mass of the system. Once again Kepler can be forgiven for not noticing the effect. Tycho's data were for our Solar System only, and because the Sun's mass $M_\odot$ is so much greater than the mass of any of the planets, $M_\odot + m_{planet} \simeq M_\odot$. Expressing $P$ in years and $a$ in astronomical units gives a value of unity for the collection of constants (including the Sun's mass).[6]

The importance to astronomy of Newton's form of Kepler's third law cannot be overstated. This law provides the most direct way of obtaining masses of celestial objects, a critical parameter in understanding a wide range of phenomena. Kepler's laws, as derived by Newton, apply equally well to planets orbiting the Sun, moons orbiting planets, stars in orbit about one another, and galaxy–galaxy orbits. Knowledge of the period of an orbit and the semimajor axis of the ellipse yields the total mass of the system. If relative distances to the center of mass are also known, the individual masses may be determined using Eq. (2.19).

---

**Example 2.3.1.**   The orbital sidereal period of Io, one of the four Galilean moons of Jupiter, is 1.77 days = $1.53 \times 10^5$ s and the semimajor axis of its orbit is $4.22 \times 10^8$ m. Assuming that the mass of Io is insignificant compared to that of Jupiter, the mass of the planet may be estimated using Kepler's third law:

$$M_{Jupiter} = \frac{4\pi^2}{G} \frac{a^3}{P^2} = 1.90 \times 10^{27} \text{ kg} = 318 \text{ M}_\oplus.$$

---

Appendix J describes a simple computer program that makes use of many of the ideas discussed in this chapter. `Orbit` will calculate, as a function of time, the location of a small mass that is orbiting about a much larger star (or it may be thought of as calculating the motion of the reduced mass about the total mass). Data generated by `Orbit` were used to produce Fig. 2.2.

---

[6]In 1621 Kepler was able to demonstrate that the four Galilean moons also obeyed his third law in the form $P^2 = ka^3$, where the constant $k$ differed from unity. He did not attribute the fact that $k \neq 1$ to mass, however.

### 2.4 ■ THE VIRIAL THEOREM

In the last section we found that the total energy of the binary orbit was just one-half of the time-averaged gravitational potential energy (Eq. 2.35), or $E = \langle U \rangle /2$. Since the total energy of the system is negative, the system is necessarily bound. For gravitationally bound systems in equilibrium, it can be shown that the total energy is always one-half of the time-averaged potential energy; this is known as the **virial theorem**.

To prove the virial theorem, begin by considering the quantity

$$Q \equiv \sum_i \mathbf{p}_i \cdot \mathbf{r}_i,$$

where $\mathbf{p}_i$ and $\mathbf{r}_i$ are the linear momentum and position vectors for particle $i$ in some inertial reference frame, and the sum is taken to be over all particles in the system. The time derivative of $Q$ is

$$\frac{dQ}{dt} = \sum_i \left( \frac{d\mathbf{p}_i}{dt} \cdot \mathbf{r}_i + \mathbf{p}_i \cdot \frac{d\mathbf{r}_i}{dt} \right). \tag{2.38}$$

Now, the left-hand side of the expression is just

$$\frac{dQ}{dt} = \frac{d}{dt} \sum_i m_i \frac{d\mathbf{r}_i}{dt} \cdot \mathbf{r}_i = \frac{d}{dt} \sum_i \frac{1}{2} \frac{d}{dt} \left( m_i r_i^2 \right) = \frac{1}{2} \frac{d^2 I}{dt^2},$$

where

$$I = \sum_i m_i r_i^2$$

is the *moment of inertia* of the collection of particles. Substituting back into Eq. (2.38),

$$\frac{1}{2} \frac{d^2 I}{dt^2} - \sum_i \mathbf{p}_i \cdot \frac{d\mathbf{r}_i}{dt} = \sum_i \frac{d\mathbf{p}_i}{dt} \cdot \mathbf{r}_i. \tag{2.39}$$

The second term on the left-hand side is just

$$-\sum_i \mathbf{p}_i \cdot \frac{d\mathbf{r}_i}{dt} = -\sum_i m_i \mathbf{v}_i \cdot \mathbf{v}_i = -2 \sum_i \frac{1}{2} m_i v_i^2 = -2K,$$

twice the negative of the total kinetic energy of the system. If we use Newton's second law, Eq. (2.39) becomes

$$\frac{1}{2} \frac{d^2 I}{dt^2} - 2K = \sum_i \mathbf{F}_i \cdot \mathbf{r}_i. \tag{2.40}$$

The right-hand side of this expression is known as the *virial of Clausius*, named after the physicist who first found this important energy relation.

If $\mathbf{F}_{ij}$ represents the force of interaction between two particles in the system (actually the force on $i$ due to $j$), then, considering all of the possible forces acting on $i$,

$$\sum_i \mathbf{F}_i \cdot \mathbf{r}_i = \sum_i \left( \sum_{\substack{j \\ j \neq i}} \mathbf{F}_{ij} \right) \cdot \mathbf{r}_i.$$

Rewriting the position vector of particle $i$ as $\mathbf{r}_i = \frac{1}{2}(\mathbf{r}_i + \mathbf{r}_j) + \frac{1}{2}(\mathbf{r}_i - \mathbf{r}_j)$, we find

$$\sum_i \mathbf{F}_i \cdot \mathbf{r}_i = \frac{1}{2} \sum_i \left( \sum_{\substack{j \\ j \neq i}} \mathbf{F}_{ij} \right) \cdot (\mathbf{r}_i + \mathbf{r}_j) + \frac{1}{2} \sum_i \left( \sum_{\substack{j \\ j \neq i}} \mathbf{F}_{ij} \right) \cdot (\mathbf{r}_i - \mathbf{r}_j).$$

From Newton's third law, $\mathbf{F}_{ij} = -\mathbf{F}_{ji}$, implying that the first term on the right-hand side is zero, by symmetry. Thus the virial of Clausius may be expressed as

$$\sum_i \mathbf{F}_i \cdot \mathbf{r}_i = \frac{1}{2} \sum_i \sum_{\substack{j \\ j \neq i}} \mathbf{F}_{ij} \cdot (\mathbf{r}_i - \mathbf{r}_j). \tag{2.41}$$

If it is assumed that the only contribution to the force is the result of the gravitational interaction between massive particles included in the system, then $\mathbf{F}_{ij}$ is

$$\mathbf{F}_{ij} = G \frac{m_i m_j}{r_{ij}^2} \hat{\mathbf{r}}_{ij},$$

where $r_{ij} = |\mathbf{r}_j - \mathbf{r}_i|$ is the separation between particles $i$ and $j$, and $\hat{\mathbf{r}}_{ij}$ is the unit vector directed from $i$ to $j$:

$$\hat{\mathbf{r}}_{ij} \equiv \frac{\mathbf{r}_j - \mathbf{r}_i}{r_{ij}}.$$

Substituting the gravitational force into Eq. (2.41) gives

$$\sum_i \mathbf{F}_i \cdot \mathbf{r}_i = -\frac{1}{2} \sum_i \sum_{\substack{j \\ j \neq i}} G \frac{m_i m_j}{r_{ij}^3} (\mathbf{r}_j - \mathbf{r}_i)^2$$

$$= -\frac{1}{2} \sum_i \sum_{\substack{j \\ j \neq i}} G \frac{m_i m_j}{r_{ij}}. \tag{2.42}$$

The quantity

$$-G \frac{m_i m_j}{r_{ij}}$$

is just the potential energy $U_{ij}$ between particles $i$ and $j$. Note, however, that

$$-G \frac{m_j m_i}{r_{ji}}$$

also represents the same potential energy term and is included in the double sum as well, so the right-hand side of Eq. (2.42) includes the potential interaction between each pair of particles twice. Considering the factor of $1/2$, Eq. (2.42) simply becomes

$$\sum_i \mathbf{F}_i \cdot \mathbf{r}_i = -\frac{1}{2}\sum_i \sum_{\substack{j \\ j \neq i}} G\frac{m_i m_j}{r_{ij}} = \frac{1}{2}\sum_i \sum_{\substack{j \\ j \neq i}} U_{ij} = U, \qquad (2.43)$$

the total potential energy of the system of particles. Finally, substituting into Eq. (2.40) and taking the average with respect to time give

$$\frac{1}{2}\left\langle \frac{d^2 I}{dt^2}\right\rangle - 2\langle K\rangle = \langle U\rangle. \qquad (2.44)$$

The average of $d^2 I/dt^2$ over some time interval $\tau$ is just

$$\left\langle \frac{d^2 I}{dt^2}\right\rangle = \frac{1}{\tau}\int_0^\tau \frac{d^2 I}{dt^2}\,dt \qquad (2.45)$$
$$= \frac{1}{\tau}\left(\left.\frac{dI}{dt}\right|_\tau - \left.\frac{dI}{dt}\right|_0\right).$$

If the system is periodic, as in the case for orbital motion, then

$$\left.\frac{dI}{dt}\right|_\tau = \left.\frac{dI}{dt}\right|_0$$

and the average over one period will be zero. Even if the system being considered is not strictly periodic, the average will still approach zero when evaluated over a sufficiently long period of time (i.e., $\tau \to \infty$), assuming of course that $dI/dt$ is bounded. This would describe, for example, a system that has reached an equilibrium or steady-state configuration. In either case, we now have $\langle d^2 I/dt^2\rangle = 0$, so

$$\boxed{-2\langle K\rangle = \langle U\rangle.} \qquad (2.46)$$

This result is one form of the virial theorem. The theorem may also be expressed in terms of the total energy of the system by using the relation $\langle E\rangle = \langle K\rangle + \langle U\rangle$. Thus

$$\boxed{\langle E\rangle = \frac{1}{2}\langle U\rangle,} \qquad (2.47)$$

just what we found for the binary orbit problem.

The virial theorem applies to a wide variety of systems, from an ideal gas to a cluster of galaxies. For instance, consider the case of a static star. In equilibrium a star must obey the virial theorem, implying that its total energy is negative, one-half of the total potential energy. Assuming that the star formed as a result of the gravitational collapse of a large cloud (a nebula), the potential energy of the system must have changed from an initial value of nearly zero to its negative static value. This implies that the star must have lost energy in the process, meaning that gravitational energy must have been radiated into space during the collapse. Applications of the virial theorem will be described in more detail in later chapters.

## SUGGESTED READING

### General

Kuhn, Thomas S., *The Structure of Scientific Revolutions*, Third Edition, University of Chicago Press, Chicago, 1996.

Westfall, Richard S., *Never at Rest: A Biography of Isaac Newton*, Cambridge University Press, Cambridge, 1980.

### Technical

Arya, Atam P., *Introduction to Classical Mechanics*, Second Edition, Prentice Hall, Upper Saddle River, NJ, 1998.

Clayton, Donald D., *Principles of Stellar Evolution and Nucleosynthesis*, University of Chicago Press, New York, 1983.

Fowles, Grant R., and Cassiday, George L., *Analytical Mechanics*, Seventh Edition, Thomson Brooks/Cole, Belmont, CA, 2005.

Marion, Jerry B., and Thornton, Stephen T., *Classical Dynamics of Particles and Systems*, Fourth Edition, Saunders College Publishing, Fort Worth, 1995.

## PROBLEMS

**2.1** Assume that a rectangular coordinate system has its origin at the center of an elliptical planetary orbit and that the coordinate system's $x$ axis lies along the major axis of the ellipse. Show that the equation for the ellipse is given by

$$\frac{x^2}{a^2} + \frac{y^2}{b^2} = 1,$$

where $a$ and $b$ are the lengths of the semimajor axis and the semiminor axis, respectively.

**2.2** Using the result of Problem 2.1, prove that the area of an ellipse is given by $A = \pi a b$.

**2.3** (a) Beginning with Eq. (2.3) and Kepler's second law, derive general expressions for $\mathbf{v}_r$ and $\mathbf{v}_\theta$ for a mass $m_1$ in an elliptical orbit about a second mass $m_2$. Your final answers should be functions of $P$, $e$, $a$, and $\theta$ only.

    (b) Using the expressions for $\mathbf{v}_r$ and $\mathbf{v}_\theta$ that you derived in part (a), verify Eq. (2.36) directly from $v^2 = v_r^2 + v_\theta^2$.

**2.4** Derive Eq. (2.25) from the sum of the kinetic and potential energy terms for the masses $m_1$ and $m_2$.

**2.5** Derive Eq. (2.26) from the total angular momentum of the masses $m_1$ and $m_2$.

**2.6** **(a)** Assuming that the Sun interacts only with Jupiter, calculate the total orbital angular momentum of the Sun–Jupiter system. The semimajor axis of Jupiter's orbit is $a = 5.2$ AU, its orbital eccentricity is $e = 0.048$, and its orbital period is $P = 11.86$ yr.

**(b)** Estimate the contribution the Sun makes to the total orbital angular momentum of the Sun–Jupiter system. For simplicity, assume that the Sun's orbital eccentricity is $e = 0$, rather than $e = 0.048$. *Hint:* First find the distance of the center of the Sun from the center of mass.

**(c)** Making the approximation that the orbit of Jupiter is a perfect circle, estimate the contribution it makes to the total orbital angular momentum of the Sun–Jupiter system. Compare your answer with the difference between the two values found in parts (a) and (b).

**(d)** Recall that the moment of inertia of a solid sphere of mass $m$ and radius $r$ is given by $I = \frac{2}{5}mr^2$ when the sphere spins on an axis passing through its center. Furthermore, its rotational angular momentum may be written as

$$L = I\omega,$$

where $\omega$ is the angular frequency measured in rad s$^{-1}$. Assuming (incorrectly) that both the Sun and Jupiter rotate as solid spheres, calculate approximate values for the rotational angular momenta of the Sun and Jupiter. Take the rotation periods of the Sun and Jupiter to be 26 days and 10 hours, respectively. The radius of the Sun is $6.96 \times 10^8$ m, and the radius of Jupiter is $6.9 \times 10^7$ m.

**(e)** What part of the Sun–Jupiter system makes the largest contribution to the total angular momentum?

**2.7** **(a)** Using data contained in Problem 2.6 and in the chapter, calculate the escape speed at the surface of Jupiter.

**(b)** Calculate the escape speed from the Solar System, starting from Earth's orbit. Assume that the Sun constitutes all of the mass of the Solar System.

**2.8** **(a)** The Hubble Space Telescope is in a nearly circular orbit, approximately 610 km (380 miles) above the surface of Earth. Estimate its orbital period.

**(b)** Communications and weather satellites are often placed in *geosynchronous* "parking" orbits above Earth. These are orbits where satellites can remain fixed above a specific point on the surface of Earth. At what altitude must these satellites be located?

**(c)** Is it possible for a satellite in a geosynchronous orbit to remain "parked" over any location on the surface of Earth? Why or why not?

**2.9** In general, an *integral average* of some continuous function $f(t)$ over an interval $\tau$ is given by

$$\langle f(t) \rangle = \frac{1}{\tau} \int_0^\tau f(t)\, dt.$$

Beginning with an expression for the integral average, prove that

$$\langle U \rangle = -G \frac{M\mu}{a},$$

a binary system's gravitational potential energy, averaged over one period, equals the value of the instantaneous potential energy of the system when the two masses are separated by the distance $a$, the semimajor axis of the orbit of the reduced mass about the center of mass. *Hint:* You may find the following definite integral useful:

$$\int_0^{2\pi} \frac{d\theta}{1 + e\cos\theta} = \frac{2\pi}{\sqrt{1 - e^2}}.$$

**2.10** Using the definition of the integral average given in Problem 2.9, prove that

$$\langle r \rangle \neq a$$

for the orbit of the reduced mass about the center of mass.

**2.11** Given that a geocentric universe is (mathematically) only a matter of the choice of a reference frame, explain why the Ptolemaic model of the universe was able to survive scrutiny for such a long period of time.

**2.12** Verify that Kepler's third law in the form of Eq. (2.37) applies to the four moons that Galileo discovered orbiting Jupiter (the Galilean moons: Io, Europa, Ganymede, and Callisto).

(a) Using the data available in Appendix C, create a graph of $\log_{10} P$ vs. $\log_{10} a$.

(b) From the graph, show that the slope of the best-fit straight line through the data is $3/2$.

(c) Calculate the mass of Jupiter from the value of the $y$-intercept.

**2.13** An alternative derivation of the total orbital angular momentum can be obtained by applying the conservation laws of angular momentum and energy.

(a) From conservation of angular momentum, show that the ratio of orbital speeds at perihelion and aphelion is given by

$$\frac{v_p}{v_a} = \frac{1 + e}{1 - e}.$$

(b) By equating the orbital mechanical energies at perihelion and aphelion, derive Eqs. (2.33) and (2.34) for the perihelion and aphelion speeds, respectively.

(c) Obtain Eq. (2.30) directly from the expression for $v_p$ (or $v_a$).

**2.14** Cometary orbits usually have very large eccentricities, often approaching (or even exceeding) unity. Halley's comet has an orbital period of 76 yr and an orbital eccentricity of $e = 0.9673$.

(a) What is the semimajor axis of Comet Halley's orbit?

(b) Use the orbital data of Comet Halley to estimate the mass of the Sun.

(c) Calculate the distance of Comet Halley from the Sun at perihelion and aphelion.

(d) Determine the orbital speed of the comet when at perihelion, at aphelion, and on the semiminor axis of its orbit.

(e) How many times larger is the kinetic energy of Halley's comet at perihelion than at aphelion?

## COMPUTER PROBLEMS

**2.15** Using Orbit (the program is described in Appendix J and available on the companion website) together with the data given in Problem 2.14, estimate the amount of time required for Halley's comet to move from perihelion to a distance of 1 AU away from the principal focus.

**2.16** The code `Orbit` (Appendix J) can be used to generate orbital positions, given the mass of the central star, the semimajor axis of the orbit, and the orbital eccentricity. Using `Orbit` to generate the data, plot the orbits for three hypothetical objects orbiting our Sun. Assume that the semimajor axis of each orbit is 1 AU and that the orbital eccentricities are:

(a) 0.0.

(b) 0.4.

(c) 0.9.

*Note:* Plot all three orbits on a common coordinate system and indicate the principal focus, located at $x = 0.0$, $y = 0.0$.

**2.17** (a) From the data given in Example 2.1.1, use `Orbit` (Appendix J) to generate an orbit for Mars. Plot at least 25 points, evenly spaced in time, on a sheet of graph paper and clearly indicate the principal focus.

(b) Using a compass, draw a perfect circle on top of the elliptical orbit for Mars, choosing the radius of the circle and its center carefully in order to make the best possible approximation of the orbit. Be sure to mark the center of the circle you chose (note that it will not correspond to the principal focus of the elliptical orbit).

(c) What can you conclude about the merit of Kepler's first attempts to use offset circles and equants to model the orbit of Mars?

**2.18** Figure 1.7 was drawn assuming perfectly circular motion and constant orbital speeds for Earth and Mars. By making very slight modifications to `Orbit` (Appendix J), a more realistic diagram can be created.

(a) Begin by assuming that Mars is initially at opposition and that Earth and Mars happen to be at their closest possible approach (aphelion and perihelion, respectively). Use your modified version of `Orbit` to calculate the positions of Earth and Mars between two successive oppositions of Mars. Graph the results.

(b) How much time (in years) elapsed between the two oppositions?

(c) Does your answer in part (b) agree precisely with the results obtained from Eq. (1.1)? Why or why not?

(d) Would you have obtained the same answer to part (b) if you had started the calculation with Earth at perihelion and Mars at aphelion? Explain your answer.

(e) From the results of your numerical experiment, explain why Mars appears brighter in the night sky during certain oppositions than during others.

# CHAPTER

# 3

# The Continuous Spectrum of Light

## 3.1 ■ STELLAR PARALLAX

Measuring the intrinsic brightness of stars is inextricably linked with determining their distances. This chapter on the light emitted by stars therefore begins with the problem of finding the distance to astronomical objects, one of the most important and most difficult tasks faced by astronomers. Kepler's laws in their original form describe the *relative* sizes of the planets' orbits in terms of astronomical units; their actual dimensions were unknown to Kepler and his contemporaries. The true scale of the Solar System was first revealed in 1761 when the distance to Venus was measured as it crossed the disk of the Sun in a rare transit during inferior conjunction. The method used was **trigonometric parallax**, the familiar surveyor's technique of triangulation. On Earth, the distance to the peak of a remote mountain can be determined by measuring that peak's angular position from two observation points separated by a known baseline distance. Simple trigonometry then supplies the distance to the peak; see Fig. 3.1. Similarly, the distances to the planets can be measured from two widely separated observation sites on Earth.

Finding the distance even to the nearest stars requires a longer baseline than Earth's diameter. As Earth orbits the Sun, two observations of the same star made 6 months apart employ a baseline equal to the diameter of Earth's orbit. These measurements reveal that a nearby star exhibits an annual back-and-forth change in its position against the stationary background of much more distant stars. (As mentioned in Section 1.3, a star may also change its position as a consequence of its own motion through space. However, this *proper motion*, seen from Earth, is not periodic and so can be distinguished from the star's periodic displacement caused by Earth's orbital motion.) As shown in Fig. 3.2, a measurement of the **parallax angle** $p$ (one-half of the maximum change in angular position) allows the calculation of the distance $d$ to the star.

$$d = \frac{1\,\text{AU}}{\tan p} \simeq \frac{1}{p}\,\text{AU},$$

57

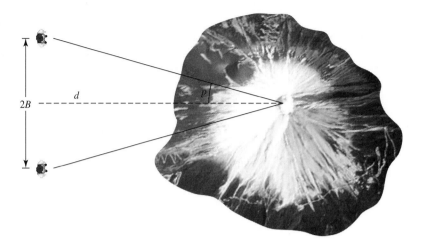

**FIGURE 3.1**   Trigonometric parallax: $d = B/\tan p$.

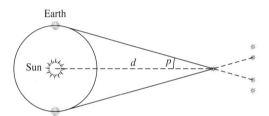

**FIGURE 3.2**   Stellar parallax: $d = 1/p''$ pc.

where the small-angle approximation $\tan p \simeq p$ has been employed for the parallax angle $p$ measured in *radians*. Using 1 radian $= 57.2957795° = 206264.806''$ to convert $p$ to $p''$ in units of *arcseconds* produces

$$d \simeq \frac{206,265}{p''} \text{ AU}.$$

Defining a new unit of distance, the **parsec** (**par**allax-**sec**ond, abbreviated pc), as 1 pc $= 2.06264806 \times 10^5$ AU $= 3.0856776 \times 10^{16}$ m leads to

$$\boxed{d = \frac{1}{p''} \text{ pc.}} \tag{3.1}$$

By definition, when the parallax angle $p = 1''$, the distance to the star is 1 pc. Thus 1 parsec is the distance from which the radius of Earth's orbit, 1 AU, subtends an angle of $1''$. Another unit of distance often encountered is the **light-year** (abbreviated ly), the distance traveled by light through a vacuum in one Julian year: 1 ly $= 9.460730472 \times 10^{15}$ m. One parsec is equivalent to 3.2615638 ly.

Even Proxima Centauri, the nearest star other than the Sun, has a parallax angle of less than $1''$. (Proxima Centauri is a member of the triple star system $\alpha$ Centauri, and has a parallax angle of $0.77''$. If Earth's orbit around the Sun were represented by a dime, then Proxima Centauri would be located 2.4 km away!) In fact, this cyclic change in a star's position is so difficult to detect that it was not until 1838 that it was first measured, by Friedrich Wilhelm Bessel (1784–1846), a German mathematician and astronomer.[1]

---

**Example 3.1.1.**   In 1838, after 4 years of observing 61 Cygni, Bessel announced his measurement of a parallax angle of $0.316''$ for that star. This corresponds to a distance of

$$d = \frac{1}{p''} \text{ pc} = \frac{1}{0.316} \text{ pc} = 3.16 \text{ pc} = 10.3 \text{ ly},$$

within 10% of the modern value 3.48 pc. 61 Cygni is one of the Sun's nearest neighbors.

---

From 1989 to 1993, the European Space Agency's (ESA's) Hipparcos Space Astrometry Mission operated high above Earth's distorting atmosphere.[2] The spacecraft was able to measure parallax angles with accuracies approaching $0.001''$ for over 118,000 stars, corresponding to a distance of $1000 \text{ pc} \equiv 1 \text{ kpc}$ (kiloparsec). Along with the high-precision Hipparcos experiment aboard the spacecraft, the lower-precision Tycho experiment produced a catalog of more than 1 million stars with parallaxes down to $0.02'' - 0.03''$. The two catalogs were published in 1997 and are available on CD-ROMs and the World Wide Web. Despite the impressive accuracy of the Hipparcos mission, the distances that were obtained are still quite small compared to the 8-kpc distance to the center of our Milky Way Galaxy, so stellar trigonometric parallax is currently useful only for surveying the local neighborhood of the Sun.

However, within the next decade, NASA plans to launch the Space Interferometry Mission (SIM PlanetQuest). This observatory will be able to determine positions, distances, and proper motions of stars with parallax angles as small as 4 *micro*arcseconds ($0.000004''$), leading to the direct determination of distances of objects up to 250 kpc away, assuming that the objects are bright enough. In addition, ESA will launch the *Gaia* mission within the next decade as well, which will catalog the brightest 1 billion stars with parallax angles as small as 10 microarcseconds. With the anticipated levels of accuracy, these missions will be able to catalog stars and other objects across the Milky Way Galaxy and even in nearby galaxies. Clearly these ambitious projects will provide an amazing wealth of new information about the three-dimensional structure of our Galaxy and the nature of its constituents.

---

[1]Tycho Brahe had searched for stellar parallax 250 years earlier, but his instruments were too imprecise to find it. Tycho concluded that Earth does not move through space, and he was thus unable to accept Copernicus's model of a heliocentric Solar System.

[2]**Astrometry** is the subdiscipline of astronomy that is concerned with the three-dimensional positions of celestial objects.

## 3.2 ■ THE MAGNITUDE SCALE

Nearly all of the information astronomers have received about the universe beyond our Solar System has come from the careful study of the light emitted by stars, galaxies, and interstellar clouds of gas and dust. Our modern understanding of the universe has been made possible by the quantitative measurement of the intensity and polarization of light in every part of the electromagnetic spectrum.

### Apparent Magnitude

The Greek astronomer Hipparchus was one of the first sky watchers to catalog the stars that he saw. In addition to compiling a list of the positions of some 850 stars, Hipparchus invented a numerical scale to describe how bright each star appeared in the sky. He assigned an **apparent magnitude** $m = 1$ to the brightest stars in the sky, and he gave the dimmest stars visible to the naked eye an apparent magnitude of $m = 6$. Note that a smaller apparent magnitude means a brighter-appearing star.

Since Hipparchus's time, astronomers have extended and refined his apparent magnitude scale. In the nineteenth century, it was thought that the human eye responded to the difference in the *logarithms* of the brightness of two luminous objects. This theory led to a scale in which a difference of 1 magnitude between two stars implies a constant *ratio* between their brightnesses. By the modern definition, a difference of 5 magnitudes corresponds exactly to a factor of 100 in brightness, so a difference of 1 magnitude corresponds exactly to a brightness ratio of $100^{1/5} \simeq 2.512$. Thus a first-magnitude star appears 2.512 times brighter than a second-magnitude star, $2.512^2 = 6.310$ times brighter than a third-magnitude star, and 100 times brighter than a sixth-magnitude star.

Using sensitive detectors, astronomers can measure the apparent magnitude of an object with an accuracy of $\pm 0.01$ magnitude, and *differences* in magnitudes with an accuracy of $\pm 0.002$ magnitude. Hipparchus's scale has been extended in both directions, from $m = -26.83$ for the Sun to approximately $m = 30$ for the faintest object detectable.[3] The total range of nearly 57 magnitudes corresponds to over $100^{57/5} = (10^2)^{11.4} \simeq 10^{23}$ for the ratio of the apparent brightness of the Sun to that of the faintest star or galaxy yet observed.

### Flux, Luminosity, and the Inverse Square Law

The "brightness" of a star is actually measured in terms of the **radiant flux** $F$ received from the star. The radiant flux is the total amount of light energy of all wavelengths that crosses a unit area oriented perpendicular to the direction of the light's travel per unit time; that is, it is the number of joules of starlight energy per second (i.e., the number of watts) received by one square meter of a detector aimed at the star. Of course, the radiant flux received from an object depends on both its intrinsic **luminosity** (energy emitted per second) and its distance from the observer. The same star, if located farther from Earth, would appear less bright in the sky.

---

[3]The magnitudes discussed in this section are actually *bolometric* magnitudes, measured over all wavelengths of light; see Section 3.6 for a discussion of magnitudes measured by detectors over a finite wavelength region.

Imagine a star of luminosity $L$ surrounded by a huge spherical shell of radius $r$. Then, assuming that no light is absorbed during its journey out to the shell, the radiant flux, $F$, measured at distance $r$ is related to the star's luminosity by

$$F = \frac{L}{4\pi r^2},$$
(3.2)

the denominator being simply the area of the sphere. Since $L$ does not depend on $r$, the radiant flux is inversely proportional to the square of the distance from the star. This is the well-known **inverse square law** for light.[4]

---

**Example 3.2.1.** The luminosity of the Sun is $L_\odot = 3.839 \times 10^{26}$ W. At a distance of 1 AU $= 1.496 \times 10^{11}$ m, Earth receives a radiant flux above its absorbing atmosphere of

$$F = \frac{L}{4\pi r^2} = 1365 \text{ W m}^{-2}.$$

This value of the solar flux is known as the **solar irradiance**, sometimes also called the **solar constant**. At a distance of 10 pc $= 2.063 \times 10^6$ AU, an observer would measure the radiant flux to be only $1/(2.063 \times 10^6)^2$ as large. That is, the radiant flux from the Sun would be $3.208 \times 10^{-10}$ W m$^{-2}$ at a distance of 10 pc.

---

### Absolute Magnitude

Using the inverse square law, astronomers can assign an **absolute magnitude**, $M$, to each star. This is defined to be the apparent magnitude a star would have *if* it were located at a distance of 10 pc. Recall that a difference of 5 magnitudes between the apparent magnitudes of two stars corresponds to the smaller-magnitude star being 100 times brighter than the larger-magnitude star. This allows us to specify their flux ratio as

$$\frac{F_2}{F_1} = 100^{(m_1 - m_2)/5}.$$
(3.3)

Taking the logarithm of both sides leads to the alternative form:

$$m_1 - m_2 = -2.5 \log_{10}\left(\frac{F_1}{F_2}\right).$$
(3.4)

### The Distance Modulus

The connection between a star's apparent and absolute magnitudes and its distance may be found by combining Eqs. (3.2) and (3.3):

$$100^{(m-M)/5} = \frac{F_{10}}{F} = \left(\frac{d}{10 \text{ pc}}\right)^2,$$

---

[4]If the star is moving with a speed near that of light, the inverse square law must be modified slightly.

where $F_{10}$ is the flux that would be received if the star were at a distance of 10 pc, and $d$ is the star's distance, measured in *parsecs*. Solving for $d$ gives

$$d = 10^{(m-M+5)/5} \text{ pc.} \tag{3.5}$$

The quantity $m - M$ is therefore a measure of the distance to a star and is called the star's **distance modulus**:

$$m - M = 5 \log_{10}(d) - 5 = 5 \log_{10}\left(\frac{d}{10 \text{ pc}}\right). \tag{3.6}$$

---

**Example 3.2.2.**   The apparent magnitude of the Sun is $m_{Sun} = -26.83$, and its distance is 1 AU $= 4.848 \times 10^{-6}$ pc. Equation (3.6) shows that the absolute magnitude of the Sun is

$$M_{Sun} = m_{Sun} - 5 \log_{10}(d) + 5 = +4.74,$$

as already given. The Sun's distance modulus is thus $m_{Sun} - M_{Sun} = -31.57$.[5]

---

For two stars at the same distance, Eq. (3.2) shows that the ratio of their radiant fluxes is equal to the ratio of their luminosities. Thus Eq. (3.3) for absolute magnitudes becomes

$$100^{(M_1 - M_2)/5} = \frac{L_2}{L_1}. \tag{3.7}$$

Letting one of these stars be the Sun reveals the direct relation between a star's absolute magnitude and its luminosity:

$$M = M_{Sun} - 2.5 \log_{10}\left(\frac{L}{L_\odot}\right), \tag{3.8}$$

where the absolute magnitude and luminosity of the Sun are $M_{Sun} = +4.74$ and $L_\odot = 3.839 \times 10^{26}$ W, respectively. It is left as an exercise for you to show that a star's apparent magnitude $m$ is related to the radiant flux $F$ received from the star by

$$m = M_{Sun} - 2.5 \log_{10}\left(\frac{F}{F_{10,\odot}}\right), \tag{3.9}$$

where $F_{10,\odot}$ is the radiant flux received from the Sun at a distance of 10 pc (see Example 3.2.1).

The inverse square law for light, Eq. (3.2), relates the intrinsic properties of a star (luminosity $L$ and absolute magnitude $M$) to the quantities measured at a distance from

---

[5]The magnitudes $m$ and $M$ for the Sun have a "Sun" subscript (instead of "$\odot$") to avoid confusion with $M_\odot$, the standard symbol for the Sun's mass.

that star (radiant flux $F$ and apparent magnitude $m$). At first glance, it may seem that astronomers must start with the measurable quantities $F$ and $m$ and then use the distance to the star (if known) to determine the star's intrinsic properties. However, if the star belongs to an important class of objects known as *pulsating variable stars*, its intrinsic luminosity $L$ and absolute magnitude $M$ can be determined *without* any knowledge of its distance. Equation (3.5) then gives the distance to the variable star. As will be discussed in Section 14.1, these stars act as beacons that illuminate the fundamental distance scale of the universe.

## 3.3 ■ THE WAVE NATURE OF LIGHT

Much of the history of physics is concerned with the evolution of our ideas about the nature of light.

### The Speed of Light

The speed of light was first measured with some accuracy in 1675, by the Danish astronomer Ole Roemer (1644–1710). Roemer observed the moons of Jupiter as they passed into the giant planet's shadow, and he was able to calculate when future eclipses of the moons should occur by using Kepler's laws. However, Roemer discovered that when Earth was moving closer to Jupiter, the eclipses occurred earlier than expected. Similarly, when Earth was moving away from Jupiter, the eclipses occurred behind schedule. Roemer realized that the discrepancy was caused by the differing amounts of time it took for light to travel the changing distance between the two planets, and he concluded that 22 minutes was required for light to cross the diameter of Earth's orbit.[6] The resulting value of $2.2 \times 10^8$ m s$^{-1}$ was close to the modern value of the speed of light. In 1983 the speed of light *in vacuo* was formally *defined* to be $c = 2.99792458 \times 10^8$ m s$^{-1}$, and the unit of length (the meter) is now derived from this value.[7]

### Young's Double-Slit Experiment

Even the fundamental nature of light has long been debated. Isaac Newton, for example, believed that light must consist of a rectilinear stream of particles, because only such a stream could account for the sharpness of shadows. Christian Huygens (1629–1695), a contemporary of Newton, advanced the idea that light must consist of waves. According to Huygens, light is described by the usual quantities appropriate for a wave. The distance between two successive wave crests is the **wavelength** $\lambda$, and the number of waves per second that pass a point in space is the **frequency** $\nu$ of the wave. Then the speed of the light wave is given by

$$c = \lambda \nu. \qquad (3.10)$$

---

[6]We now know that it takes light about 16.5 minutes to travel 2 AU.

[7]In 1905 Albert Einstein realized that the speed of light is a universal constant of nature whose value is independent of the observer (see page 88). This realization plays a central role in his Special Theory of Relativity (Chapter 4).

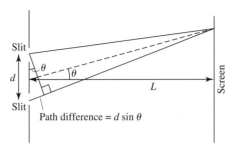

**FIGURE 3.3**    Double-slit experiment.

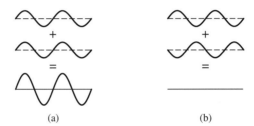

(a)                              (b)

**FIGURE 3.4**    Superposition principle for light waves. (a) Constructive interference. (b) Destructive interference.

Both the particle and wave models could explain the familiar phenomena of the reflection and refraction of light. However, the particle model prevailed, primarily on the strength of Newton's reputation, until the wave nature of light was conclusively demonstrated by Thomas Young's (1773–1829) famous double-slit experiment.

In a double-slit experiment, monochromatic light of wavelength $\lambda$ from a single source passes through two narrow, parallel slits that are separated by a distance $d$. The light then falls upon a screen a distance $L$ beyond the two slits (see Fig. 3.3). The series of light and dark *interference fringes* that Young observed on the screen could be explained only by a wave model of light. As the light waves pass through the narrow slits,[8] they spread out (diffract) radially in a succession of crests and troughs. Light obeys a *superposition principle*, so when two waves meet, they add algebraically; see Fig. 3.4. At the screen, if a wave crest from one slit meets a wave crest from the other slit, a bright fringe or maximum is produced by the resulting **constructive interference**. But if a wave crest from one slit meets a wave trough from the other slit, they cancel each other, and a dark fringe or minimum results from this **destructive interference**.

The interference pattern observed thus depends on the difference in the lengths of the paths traveled by the light waves from the two slits to the screen. As shown in Fig. 3.3, if $L \gg d$, then to a good approximation this path difference is $d \sin \theta$. The light waves will arrive at the screen *in phase* if the path difference is equal to an integral number of wavelengths. On the other hand, the light waves will arrive 180° *out of phase* if the path difference is equal to an odd integral number of half-wavelengths. So for $L \gg d$, the angular

---

[8]Actually, Young used pinholes in his original experiment.

positions of the bright and dark fringes for **double-slit interference** are given by

$$d \sin \theta = \begin{cases} n\lambda & (n = 0, 1, 2, \ldots \text{ for bright fringes}) \\ \left(n - \frac{1}{2}\right)\lambda & (n = 1, 2, 3, \ldots \text{ for dark fringes}). \end{cases} \qquad (3.11)$$

In either case, $n$ is called the **order** of the maximum or minimum. From the measured positions of the light and dark fringes on the screen, Young was able to determine the wavelength of the light. At the short-wavelength end, Young found that violet light has a wavelength of approximately 400 nm, while at the long-wavelength end, red light has a wavelength of only 700 nm.[9] The diffraction of light goes unnoticed under everyday conditions for these short wavelengths, thus explaining Newton's sharp shadows.

### Maxwell's Electromagnetic Wave Theory

The nature of these waves of light remained elusive until the early 1860s, when the Scottish mathematical physicist James Clerk Maxwell (1831–1879) succeeded in condensing everything known about electric and magnetic fields into the four equations that today bear his name. Maxwell found that his equations could be manipulated to produce wave equations for the electric and magnetic field vectors **E** and **B**. These wave equations predicted the existence of *electromagnetic waves* that travel through a vacuum with speed $v = 1/\sqrt{\epsilon_0 \mu_0}$, where $\epsilon_0$ and $\mu_0$ are fundamental constants associated with the electric and magnetic fields, respectively. Upon inserting the values of $\epsilon_0$ and $\mu_0$, Maxwell was amazed to discover that electromagnetic waves travel at the speed of light. Furthermore, these equations implied that electromagnetic waves are *transverse* waves, with the oscillatory electric and magnetic fields perpendicular to each other *and* to the direction of the wave's propagation (see Fig. 3.5); such waves could exhibit the polarization[10] known to occur for light. Maxwell wrote that "we can scarcely avoid the inference that light consists in the transverse modulations of the same medium which is the cause of electric and magnetic phenomena."

Maxwell did not live to see the experimental verification of his prediction of electromagnetic waves. Ten years after Maxwell's death, the German physicist Heinrich Hertz (1857–1894) succeeded in producing radio waves in his laboratory. Hertz determined that these electromagnetic waves do indeed travel at the speed of light, and he confirmed their reflection, refraction, and polarization properties. In 1889, Hertz wrote:

> What is light? Since the time of Young and Fresnel we know that it is wave motion. We know the velocity of the waves, we know their lengths, and we know that they are transverse; in short, our knowledge of the geometrical conditions of the motion is complete. A doubt about these things is no longer possible; a refutation of these views is inconceivable to the physicist. The wave theory of light is, from the point of view of human beings, certainty.

---

[9]Another commonly used measure of the wavelength of light is the **angstrom**; 1 Å = 0.1 nm. In these units, violet light has a wavelength of 4000 Å and red light has a wavelength of 7000 Å.

[10]The electromagnetic wave shown in Fig. 3.5 is *plane-polarized*, with its electric and magnetic fields oscillating in planes. Because **E** and **B** are always perpendicular, their respective planes of polarization are perpendicular as well.

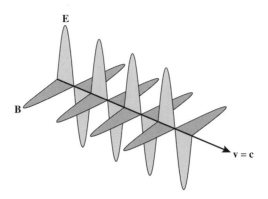

**FIGURE 3.5**   Electromagnetic wave.

**TABLE 3.1**   The Electromagnetic Spectrum.

| Region | Wavelength |
|---|---|
| Gamma ray | $\lambda \; < \; 1 \text{ nm}$ |
| X-ray | $1 \text{ nm} < \; \lambda \; < \; 10 \text{ nm}$ |
| Ultraviolet | $10 \text{ nm} < \; \lambda \; < \; 400 \text{ nm}$ |
| Visible | $400 \text{ nm} < \; \lambda \; < \; 700 \text{ nm}$ |
| Infrared | $700 \text{ nm} < \; \lambda \; < \; 1 \text{ mm}$ |
| Microwave | $1 \text{ mm} < \; \lambda \; < \; 10 \text{ cm}$ |
| Radio | $10 \text{ cm} < \; \lambda$ |

### The Electromagnetic Spectrum

Today, astronomers utilize light from every part of the **electromagnetic spectrum**. The total spectrum of light consists of electromagnetic waves of all wavelengths, ranging from very short-wavelength gamma rays to very long-wavelength radio waves. Table 3.1 shows how the electromagnetic spectrum has been arbitrarily divided into various wavelength regions.

### The Poynting Vector and Radiation Pressure

Like all waves, electromagnetic waves carry both energy and momentum in the direction of propagation. The rate at which energy is carried by a light wave is described by the **Poynting vector**,[11]

$$\mathbf{S} = \frac{1}{\mu_0}\mathbf{E} \times \mathbf{B},$$

where $\mathbf{S}$ has units of W m$^{-2}$. The Poynting vector points in the direction of the electromagnetic wave's propagation and has a magnitude equal to the amount of energy per unit time that crosses a unit area oriented perpendicular to the direction of the propagation of

[11]The Poynting vector is named after John Henry Poynting (1852–1914), the physicist who first described it.

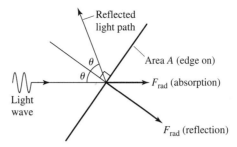

**FIGURE 3.6** Radiation pressure force. The surface area $A$ is seen edge on.

the wave. Because the magnitudes of the fields **E** and **B** vary harmonically with time, the quantity of practical interest is the *time-averaged* value of the Poynting vector over one cycle of the electromagnetic wave. In a vacuum the magnitude of the time-averaged Poynting vector, $\langle S \rangle$, is

$$\langle S \rangle = \frac{1}{2\mu_0} E_0 B_0, \tag{3.12}$$

where $E_0$ and $B_0$ are the maximum magnitudes (amplitudes) of the electric and magnetic fields, respectively. (For an electromagnetic wave in a vacuum, $E_0$ and $B_0$ are related by $E_0 = c B_0$.) The time-averaged Poynting vector thus provides a description of the radiant flux in terms of the electric and magnetic fields of the light waves. However, it should be remembered that the radiant flux discussed in Section 3.2 involves the amount of energy received *at all wavelengths* from a star, whereas $E_0$ and $B_0$ describe an electromagnetic wave of a specified wavelength.

Because an electromagnetic wave carries momentum, it can exert a force on a surface hit by the light. The resulting **radiation pressure** depends on whether the light is reflected from or absorbed by the surface. Referring to Fig. 3.6, if the light is completely absorbed, then the force due to radiation pressure is in the direction of the light's propagation and has magnitude

$$F_{\text{rad}} = \frac{\langle S \rangle A}{c} \cos\theta \qquad \text{(absorption)}, \tag{3.13}$$

where $\theta$ is the angle of incidence of the light as measured from the direction perpendicular to the surface of area $A$. Alternatively, if the light is completely reflected, then the radiation pressure force must act in a direction perpendicular to the surface; the reflected light cannot exert a force parallel to the surface. Then the magnitude of the force is

$$F_{\text{rad}} = \frac{2\langle S \rangle A}{c} \cos^2\theta \qquad \text{(reflection)}. \tag{3.14}$$

Radiation pressure has a negligible effect on physical systems under everyday conditions. However, radiation pressure may play a dominant role in determining some aspects of the behavior of extremely luminous objects such as early main-sequence stars, red supergiants, and accreting compact stars. It may also have a significant effect on the small particles of dust found throughout the interstellar medium.

## 3.4 ■ BLACKBODY RADIATION

Anyone who has looked at the constellation of Orion on a clear winter night has noticed the strikingly different colors of red Betelgeuse (in Orion's northeast shoulder) and blue-white Rigel (in the southwest leg); see Fig. 3.7. These colors betray the difference in the surface temperatures of the two stars. Betelgeuse has a surface temperature of roughly 3600 K, significantly cooler than the 13,000-K surface of Rigel.[12]

### The Connection between Color and Temperature

The connection between the color of light emitted by a hot object and its temperature was first noticed in 1792 by the English maker of fine porcelain Thomas Wedgewood. All of his ovens became red-hot at the same temperature, independent of their size, shape, and construction. Subsequent investigations by many physicists revealed that any object with a temperature above absolute zero emits light of all wavelengths with varying degrees of efficiency; an *ideal emitter* is an object that absorbs *all* of the light energy incident upon it and reradiates this energy with the characteristic spectrum shown in Fig. 3.8. Because an ideal emitter reflects no light, it is known as a **blackbody**, and the radiation it emits is called **blackbody radiation**. Stars and planets are blackbodies, at least to a rough first approximation.

Figure 3.8 shows that a blackbody of temperature $T$ emits a **continuous spectrum** with some energy at all wavelengths and that this blackbody spectrum peaks at a wavelength $\lambda_{max}$, which becomes shorter with increasing temperature. The relation between $\lambda_{max}$ and

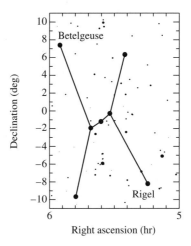

**FIGURE 3.7**    The constellation of Orion.

---

[12]Both of these stars are pulsating variables (Chapter 14), so the values quoted are *average* temperatures. Estimates of the surface temperature of Betelgeuse actually range quite widely, from about 3100 K to 3900 K. Similarly, estimates of the surface temperature of Rigel range from 8000 K to 13,000 K.

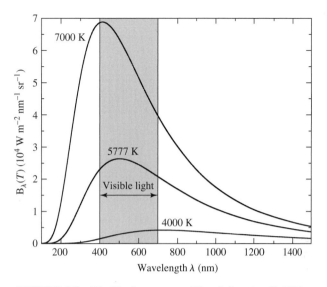

**FIGURE 3.8** Blackbody spectrum [Planck function $B_\lambda(T)$].

$T$ is known as **Wien's displacement law:**[13]

$$\lambda_{\text{max}} T = 0.002897755 \text{ m K.} \tag{3.15}$$

---

**Example 3.4.1.** Betelgeuse has a surface temperature of 3600 K. If we treat Betelgeuse as a blackbody, Wien's displacement law shows that its continuous spectrum peaks at a wavelength of

$$\lambda_{\text{max}} \simeq \frac{0.0029 \text{ m K}}{3600 \text{ K}} = 8.05 \times 10^{-7} \text{ m} = 805 \text{ nm},$$

which is in the infrared region of the electromagnetic spectrum. Rigel, with a surface temperature of 13,000 K, has a continuous spectrum that peaks at a wavelength of

$$\lambda_{\text{max}} \simeq \frac{0.0029 \text{ m K}}{13,000 \text{ K}} = 2.23 \times 10^{-7} \text{ m} = 223 \text{ nm},$$

in the ultraviolet region.

---

**The Stefan–Boltzmann Equation**

Figure 3.8 also shows that as the temperature of a blackbody increases, it emits more energy per second at *all* wavelengths. Experiments performed by the Austrian physicist

---

[13]In 1911, the German physicist Wilhelm Wien (1864–1928) received the Nobel Prize in 1911 for his theoretical contributions to understanding the blackbody spectrum.

Josef Stefan (1835–1893) in 1879 showed that the luminosity, $L$, of a blackbody of area $A$ and temperature $T$ (in kelvins) is given by

$$L = A\sigma T^4. \tag{3.16}$$

Five years later another Austrian physicist, Ludwig Boltzmann (1844–1906), derived this equation, now called the **Stefan–Boltzmann equation**, using the laws of thermodynamics and Maxwell's formula for radiation pressure. The Stefan–Boltzmann constant, $\sigma$, has the value

$$\sigma = 5.670400 \times 10^{-8} \text{ W m}^{-2} \text{ K}^{-4}.$$

For a spherical star of radius $R$ and surface area $A = 4\pi R^2$, the Stefan–Boltzmann equation takes the form

$$\boxed{L = 4\pi R^2 \sigma T_e^4.} \tag{3.17}$$

Since stars are not perfect blackbodies, we use this equation to *define* the **effective temperature** $T_e$ of a star's surface. Combining this with the inverse square law, Eq. (3.2), shows that at the surface of the star ($r = R$), the *surface flux* is

$$F_{\text{surf}} = \sigma T_e^4. \tag{3.18}$$

---

**Example 3.4.2.**   The luminosity of the Sun is $L_\odot = 3.839 \times 10^{26}$ W and its radius is $R_\odot = 6.95508 \times 10^8$ m. The effective temperature of the Sun's surface is then

$$T_\odot = \left( \frac{L_\odot}{4\pi R_\odot^2 \sigma} \right)^{\frac{1}{4}} = 5777 \text{ K}.$$

The radiant flux at the solar surface is

$$F_{\text{surf}} = \sigma T_\odot^4 = 6.316 \times 10^7 \text{ W m}^{-2}.$$

According to Wien's displacement law, the Sun's continuous spectrum peaks at a wavelength of

$$\lambda_{\text{max}} \simeq \frac{0.0029 \text{ m K}}{5777 \text{ K}} = 5.016 \times 10^{-7} \text{ m} = 501.6 \text{ nm}.$$

This wavelength falls in the *green* region (491 nm $< \lambda <$ 575 nm) of the spectrum of visible light. However, the Sun emits a continuum of wavelengths both shorter and longer than $\lambda_{\text{max}}$, and the human eye perceives the Sun's color as yellow. Because the Sun emits most of its energy at visible wavelengths (see Fig. 3.8), and because Earth's atmosphere is transparent at these wavelengths, the evolutionary process of natural selection has produced a human eye sensitive to this wavelength region of the electromagnetic spectrum.

Rounding off $\lambda_{\text{max}}$ and $T_\odot$ to the values of 500 nm and 5800 K, respectively, permits Wien's displacement law to be written in the approximate form

$$\lambda_{\text{max}} T \approx (500 \text{ nm})(5800 \text{ K}). \tag{3.19}$$

---

**The Eve of a New World View**

This section draws to a close at the end of the nineteenth century. The physicists and astronomers of the time believed that all of the principles that govern the physical world had finally been discovered. Their scientific world view, the *Newtonian paradigm*, was the culmination of the heroic, golden age of classical physics that had flourished for over three hundred years. The construction of this paradigm began with the brilliant observations of Galileo and the subtle insights of Newton. Its architecture was framed by Newton's laws, supported by the twin pillars of the conservation of energy and momentum and illuminated by Maxwell's electromagnetic waves. Its legacy was a deterministic description of a universe that ran like clockwork, with wheels turning inside of wheels, all of its gears perfectly meshed. Physics was in danger of becoming a victim of its own success. There were no challenges remaining. All of the great discoveries apparently had been made, and the only task remaining for men and women of science at the end of the nineteenth century was filling in the details.

However, as the twentieth century opened, it became increasingly apparent that a crisis was brewing. Physicists were frustrated by their inability to answer some of the simplest questions concerning light. What is the medium through which light waves travel the vast distances between the stars, and what is Earth's speed through this medium? What determines the continuous spectrum of blackbody radiation and the characteristic, discrete colors of tubes filled with hot glowing gases? Astronomers were tantalized by hints of a treasure of knowledge just beyond their grasp.

It took a physicist of the stature of Albert Einstein to topple the Newtonian paradigm and bring about two revolutions in physics. One transformed our ideas about space and time, and the other changed our basic concepts of matter and energy. The rigid clockwork universe of the golden era was found to be an illusion and was replaced by a random universe governed by the laws of probability and statistics. The following four lines aptly summarize the situation. The first two lines were written by the English poet Alexander Pope (1688–1744), a contemporary of Newton; the last two, by Sir J. C. Squire (1884–1958), were penned in 1926.

> Nature and Nature's laws lay hid in night:
> God said, *Let Newton be!* and all was light.

> It did not last: the Devil howling "Ho!
> Let Einstein be!" restored the status quo.

## 3.5 ■ THE QUANTIZATION OF ENERGY

One of the problems haunting physicists at the end of the nineteenth century was their inability to derive from fundamental physical principles the blackbody radiation curve depicted in Fig. 3.8. Lord Rayleigh[14] (1842–1919) had attempted to arrive at the expression by applying Maxwell's equations of classical electromagnetic theory together with the results

---

[14]Lord Rayleigh, as he is commonly known, was born John William Strutt but succeeded to the title of third Baron Rayleigh of Terling Place, Witham, in the county of Essex, when he was thirty years old.

from thermal physics. His strategy was to consider a cavity of temperature $T$ filled with blackbody radiation. This may be thought of as a hot oven filled with standing waves of electromagnetic radiation. If $L$ is the distance between the oven's walls, then the permitted wavelengths of the radiation are $\lambda = 2L, L, 2L/3, 2L/4, 2L/5, \ldots$, extending forever to increasingly shorter wavelengths.[15] According to classical physics, each of these wavelengths should receive an amount of energy equal to $kT$, where $k = 1.3806503 \times 10^{-23}$ J K$^{-1}$ is Boltzmann's constant, familiar from the ideal gas law $PV = NkT$. The result of Rayleigh's derivation gave

$$B_\lambda(T) \simeq \frac{2ckT}{\lambda^4}, \qquad \text{(valid only if } \lambda \text{ is long)} \tag{3.20}$$

which agrees well with the long-wavelength tail of the blackbody radiation curve. However, a severe problem with Rayleigh's result was recognized immediately; his solution for $B_\lambda(T)$ grows without limit as $\lambda \to 0$. The source of the problem is that according to classical physics, an infinite number of infinitesimally short wavelengths implied that an unlimited amount of blackbody radiation energy was contained in the oven, a theoretical result so absurd it was dubbed the "ultraviolet catastrophe." Equation (3.20) is known today as the **Rayleigh–Jeans law**.[16]

Wien was also working on developing the correct mathematical expression for the blackbody radiation curve. Guided by the Stefan–Boltzmann law (Eq. 3.16) and classical thermal physics, Wien was able to develop an empirical law that described the curve at short wavelengths but failed at longer wavelengths:

$$B_\lambda(T) \simeq a\lambda^{-5}e^{-b/\lambda T}, \qquad \text{(valid only if } \lambda \text{ is short)} \tag{3.21}$$

where $a$ and $b$ were constants chosen to provide the best fit to the experimental data.

### Planck's Function for the Blackbody Radiation Curve

By late 1900 the German physicist Max Planck (1858–1947) had discovered that a modification of Wien's expression could be made to fit the blackbody spectra shown in Fig. 3.8 while simultaneously replicating the long-wavelength success of the Rayleigh–Jeans law and avoiding the ultraviolet catastrophe:

$$B_\lambda(T) = \frac{a/\lambda^5}{e^{b/\lambda T} - 1},$$

In order to determine the constants $a$ and $b$ while circumventing the ultraviolet catastrophe, Planck employed a clever mathematical trick. He assumed that a standing electromagnetic wave of wavelength $\lambda$ and frequency $\nu = c/\lambda$ could not acquire just any arbitrary amount of energy. Instead, the wave could have only specific allowed energy values that

---

[15]This is analogous to standing waves on a string of length $L$ that is held fixed at both ends. The permitted wavelengths are the same as those of the standing electromagnetic waves.

[16]James Jeans (1877–1946), a British astronomer, found a numerical error in Rayleigh's original work; the corrected result now bears the names of both men.

were integral multiples of a minimum wave energy.[17] This minimum energy, a **quantum** of energy, is given by $h\nu$ or $hc/\lambda$, where $h$ is a constant. Thus the energy of an electromagnetic wave is $nh\nu$ or $nhc/\lambda$, where $n$ (an integer) is the number of quanta in the wave. Given this assumption of quantized wave energy with a minimum energy proportional to the frequency of the wave, the entire oven could not contain enough energy to supply even one quantum of energy for the short-wavelength, high-frequency waves. Thus the ultraviolet catastrophe would be avoided. Planck hoped that at the end of his derivation, the constant $h$ could be set to zero; certainly, an artificial constant should not remain in his final result for $B_\lambda(T)$.

Planck's stratagem worked! His formula, now known as the **Planck function**, agreed wonderfully with experiment, but *only* if the constant $h$ remained in the equation:[18]

$$B_\lambda(T) = \frac{2hc^2/\lambda^5}{e^{hc/\lambda kT} - 1}. \tag{3.22}$$

**Planck's constant** has the value $h = 6.62606876 \times 10^{-34}$ J s.

### The Planck Function and Astrophysics

Finally armed with the correct expression for the blackbody spectrum, we can apply Planck's function to astrophysical systems. In spherical coordinates, the amount of radiant energy per unit time having wavelengths between $\lambda$ and $\lambda + d\lambda$ emitted by a blackbody of temperature $T$ and surface area $dA$ into a solid angle $d\Omega \equiv \sin\theta\, d\theta\, d\phi$ is given by

$$B_\lambda(T)\, d\lambda\, dA\, \cos\theta\, d\Omega = B_\lambda(T)\, d\lambda\, dA\, \cos\theta\, \sin\theta\, d\theta\, d\phi; \tag{3.23}$$

see Fig. 3.9.[19] The units of $B_\lambda$ are therefore W m$^{-3}$ sr$^{-1}$. Unfortunately, these units can be misleading. You should note that "W m$^{-3}$" indicates power (energy per unit time) per unit area per unit wavelength interval, W m$^{-2}$ m$^{-1}$, *not* energy per unit time per unit volume. To help avoid confusion, the units of the wavelength interval $d\lambda$ are sometimes expressed in nanometers rather than meters, so the units of the Planck function become W m$^{-2}$ nm$^{-1}$ sr$^{-1}$, as in Fig. 3.8.[20]

At times it is more convenient to deal with frequency intervals $d\nu$ than with wavelength intervals $d\lambda$. In this case the Planck function has the form

$$B_\nu(T) = \frac{2h\nu^3/c^2}{e^{h\nu/kT} - 1}. \tag{3.24}$$

---

[17]Actually, Planck restricted the possible energies of hypothetical electromagnetic oscillators in the oven walls that emit the electromagnetic radiation.

[18]It is left for you to show that the Planck function reduces to the Rayleigh–Jeans law at long wavelengths (Problem 3.10) and to Wien's expression at short wavelengths (Problem 3.11).

[19]Note that $dA\,\cos\theta$ is the area $dA$ projected onto a plane perpendicular to the direction in which the radiation is traveling. The concept of a solid angle will be fully described in Section 6.1.

[20]The value of the Planck function thus depends on the units of the wavelength interval. The conversion of $d\lambda$ from meters to nanometers means that the values of $B_\lambda$ obtained by evaluating Eq. (3.22) must be divided by $10^9$.

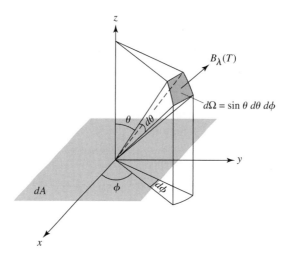

**FIGURE 3.9**   Blackbody radiation from an element of surface area $dA$.

Thus, in spherical coordinates,

$$B_\nu \, d\nu \, dA \, \cos\theta \, d\Omega = B_\nu \, d\nu \, dA \, \cos\theta \, \sin\theta \, d\theta \, d\phi$$

is the amount of energy per unit time of blackbody radiation having frequency between $\nu$ and $\nu + d\nu$ emitted by a blackbody of temperature $T$ and surface area $dA$ into a solid angle $d\Omega = \sin\theta \, d\theta \, d\phi$.

The Planck function can be used to make the connection between the observed properties of a star (radiant flux, apparent magnitude) and its intrinsic properties (radius, temperature). Consider a model star consisting of a spherical blackbody of radius $R$ and temperature $T$. Assuming that each small patch of surface area $dA$ emits blackbody radiation *isotropically* (equally in all directions) over the outward hemisphere, the energy per second having wavelengths between $\lambda$ and $\lambda + d\lambda$ emitted by the star is

$$L_\lambda \, d\lambda = \int_{\phi=0}^{2\pi} \int_{\theta=0}^{\pi/2} \int_A B_\lambda \, d\lambda \, dA \, \cos\theta \, \sin\theta \, d\theta \, d\phi. \tag{3.25}$$

The angular integration yields a factor of $\pi$, and the integral over the area of the sphere produces a factor of $4\pi R^2$. The result is

$$L_\lambda \, d\lambda = 4\pi^2 R^2 B_\lambda \, d\lambda \tag{3.26}$$

$$= \frac{8\pi^2 R^2 hc^2/\lambda^5}{e^{hc/\lambda kT} - 1} \, d\lambda. \tag{3.27}$$

$L_\lambda \, d\lambda$ is known as the **monochromatic luminosity**. Comparing the Stefan–Boltzmann equation (3.17) with the result of integrating Eq. (3.26) over all wavelengths shows that

$$\int_0^\infty B_\lambda(T) \, d\lambda = \frac{\sigma T^4}{\pi}. \tag{3.28}$$

In Problem 3.14, you will use Eq. (3.27) to express the Stefan–Boltzmann constant, $\sigma$, in terms of the fundamental constants $c$, $h$, and $k$. The monochromatic luminosity is related to the **monochromatic flux**, $F_\lambda \, d\lambda$, by the inverse square law for light, Eq. (3.2):

$$F_\lambda \, d\lambda = \frac{L_\lambda}{4\pi r^2} \, d\lambda = \frac{2\pi hc^2/\lambda^5}{e^{hc/\lambda kT} - 1} \left(\frac{R}{r}\right)^2 \, d\lambda, \qquad (3.29)$$

where $r$ is the distance to the model star. Thus $F_\lambda \, d\lambda$ is the number of joules of starlight energy with wavelengths between $\lambda$ and $\lambda + d\lambda$ that arrive per second at one square meter of a detector aimed at the model star, assuming that no light has been absorbed or scattered during its journey from the star to the detector. Of course, Earth's atmosphere absorbs some starlight, but measurements of fluxes and apparent magnitudes can be corrected to account for this absorption; see Section 9.2. The values of these quantities usually quoted for stars are in fact corrected values and would be the results of measurements above Earth's absorbing atmosphere.

### 3.6 ■ THE COLOR INDEX

The apparent and absolute magnitudes discussed in Section 3.2, measured over all wavelengths of light emitted by a star, are known as **bolometric magnitudes** and are denoted by $m_{\rm bol}$ and $M_{\rm bol}$, respectively.[21] In practice, however, detectors measure the radiant flux of a star only within a certain wavelength region defined by the sensitivity of the detector.

#### UBV Wavelength Filters

The *color* of a star may be precisely determined by using filters that transmit the star's light only within certain narrow wavelength bands. In the standard *UBV* system, a star's apparent magnitude is measured through three filters and is designated by three capital letters:

- $U$, the star's *ultraviolet* magnitude, is measured through a filter centered at 365 nm with an effective bandwidth of 68 nm.

- $B$, the star's *blue* magnitude, is measured through a filter centered at 440 nm with an effective bandwidth of 98 nm.

- $V$, the star's *visual* magnitude, is measured through a filter centered at 550 nm with an effective bandwidth of 89 nm.

#### Color Indices and the Bolometric Correction

Using Eq. (3.6), a star's absolute color magnitudes $M_U$, $M_B$, and $M_V$ may be determined if its distance $d$ is known.[22] A star's $U - B$ **color index** is the difference between its ultraviolet

---

[21]A *bolometer* is an instrument that measures the increase in temperature caused by the radiant flux it receives at all wavelengths.

[22]Note that although apparent magnitude is not denoted by a subscripted "*m*" in the *UBV* system, the absolute magnitude is denoted by a subscripted "*M*."

and blue magnitudes, and a star's $B - V$ color index is the difference between its blue and visual magnitudes:

$$U - B = M_U - M_B$$

and

$$B - V = M_B - M_V.$$

Stellar magnitudes *decrease* with increasing brightness; consequently, a star with a smaller $B - V$ color index is *bluer* than a star with a larger value of $B - V$. Because a color index is the difference between two magnitudes, Eq. (3.6) shows that it is independent of the star's distance. The difference between a star's bolometric magnitude and its visual magnitude is its **bolometric correction** $BC$:

$$\boxed{BC = m_{\mathrm{bol}} - V = M_{\mathrm{bol}} - M_V.} \qquad (3.30)$$

---

**Example 3.6.1.**   Sirius, the brightest-appearing star in the sky, has $U$, $B$, and $V$ apparent magnitudes of $U = -1.47$, $B = -1.43$, and $V = -1.44$. Thus for Sirius,

$$U - B = -1.47 - (-1.43) = -0.04$$

and

$$B - V = -1.43 - (-1.44) = 0.01.$$

Sirius is brightest at ultraviolet wavelengths, as expected for a star with an effective temperature of $T_e = 9970$ K. For this surface temperature,

$$\lambda_{\max} = \frac{0.0029 \text{ m K}}{9970 \text{ K}} = 291 \text{ nm},$$

which is in the ultraviolet portion of the electromagnetic spectrum. The bolometric correction for Sirius is $BC = -0.09$, so its apparent bolometric magnitude is

$$m_{\mathrm{bol}} = V + BC = -1.44 + (-0.09) = -1.53.$$

---

The relation between apparent magnitude and radiant flux, Eq. (3.4), can be used to derive expressions for the ultraviolet, blue, and visual magnitudes measured (above Earth's atmosphere) for a star. A *sensitivity function* $S(\lambda)$ is used to describe the fraction of the star's flux that is detected at wavelength $\lambda$. $S$ depends on the reflectivity of the telescope mirrors, the bandwidth of the $U$, $B$, and $V$ filters, and the response of the photometer. Thus, for example, a star's ultraviolet magnitude $U$ is given by

$$U = -2.5 \log_{10}\left(\int_0^\infty F_\lambda S_U \, d\lambda\right) + C_U, \qquad (3.31)$$

where $C_U$ is a constant. Similar expressions are used for a star's apparent magnitude within other wavelength bands. The constants $C$ in the equations for $U$, $B$, and $V$ differ for each of these wavelength regions and are chosen so that the star Vega ($\alpha$ Lyrae) has a magnitude of *zero* as seen through each filter.[23] This is a completely arbitrary choice and does *not* imply that Vega would appear equally bright when viewed through the $U$, $B$, and $V$ filters. However, the resulting values for the visual magnitudes of stars are about the same as those recorded by Hipparchus two thousand years ago.[24]

A *different* method is used to determine the constant $C_{\text{bol}}$ in the expression for the bolometric magnitude, measured over all wavelengths of light emitted by a star. For a *perfect* bolometer, capable of detecting 100 percent of the light arriving from a star, we set $S(\lambda) \equiv 1$:

$$m_{\text{bol}} = -2.5 \log_{10} \left( \int_0^\infty F_\lambda \, d\lambda \right) + C_{\text{bol}}. \tag{3.32}$$

The value for $C_{\text{bol}}$ originated in the wish of astronomers that the value of the bolometric correction

$$BC = m_{\text{bol}} - V$$

be negative for all stars (since a star's radiant flux over all wavelengths is greater than its flux in any specified wavelength band) while still being as close to zero as possible. After a value of $C_{\text{bol}}$ was agreed upon, it was discovered that some supergiant stars have *positive* bolometric corrections. Nevertheless, astronomers have chosen to continue using this unphysical method of measuring magnitudes.[25] It is left as an exercise for you to evaluate the constant $C_{\text{bol}}$ by using the value of $m_{\text{bol}}$ assigned to the Sun: $m_{\text{Sun}} = -26.83$.

The color indices $U - B$ and $B - V$ are immediately seen to be

$$U - B = -2.5 \log_{10} \left( \frac{\int F_\lambda S_U \, d\lambda}{\int F_\lambda S_B \, d\lambda} \right) + C_{U-B}, \tag{3.33}$$

where $C_{U-B} \equiv C_U - C_B$. A similar relation holds for $B - V$. From Eq. (3.29), note that although the apparent magnitudes depend on the radius $R$ of the model star and its distance $r$, the color indices do not, because the factor of $(R/r)^2$ cancels in Eq. (3.33). Thus the color index is a measure solely of the temperature of a model blackbody star.

---

**Example 3.6.2.** A certain hot star has a surface temperature of 42,000 K and color indices $U - B = -1.19$ and $B - V = -0.33$. The large negative value of $U - B$ indicates that this star appears brightest at ultraviolet wavelengths, as can be confirmed with Wien's displacement law, Eq. (3.19). The spectrum of a 42,000-K blackbody peaks at

$$\lambda_{\text{max}} = \frac{0.0029 \text{ m K}}{42,000 \text{ K}} = 69 \text{ nm},$$

*continued*

---

[23] Actually, the average magnitude of several stars is used for this calibration.

[24] See Chapter 1 of Böhm-Vitense (1989b) for a further discussion of the vagaries of the magnitude system used by astronomers.

[25] Some authors, such as Böhm-Vitense (1989a, 1989b), prefer to define the bolometric correction as $BC = V - m_{\text{bol}}$, so their values of $BC$ are usually positive.

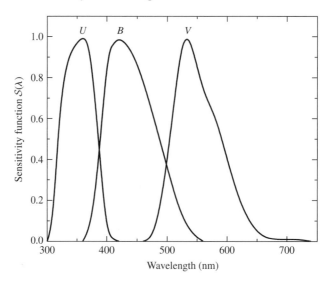

**FIGURE 3.10**    Sensitivity functions $S(\lambda)$ for $U$, $B$, and $V$ filters. (Data from Johnson, *Ap. J.*, *141*, 923, 1965.)

in the ultraviolet region of the electromagnetic spectrum. This wavelength is much shorter than the wavelengths transmitted by the $U$, $B$, and $V$ filters (see Fig. 3.10), so we will be dealing with the smoothly declining long-wavelength "tail" of the Planck function $B_\lambda(T)$.

We can use the values of the color indices to estimate the constant $C_{U-B}$ in Eq. (3.33), and $C_{B-V}$ in a similar equation for the color index $B - V$. In this estimate, we will use a step function to represent the sensitivity function: $S(\lambda) = 1$ inside the filter's bandwidth, and $S(\lambda) = 0$ otherwise. The integrals in Eq. (3.33) may then be approximated by the value of the Planck function $B_\lambda$ at the center of the filter bandwidth, multiplied by that bandwidth. Thus, for the wavelengths and bandwidths $\Delta\lambda$ listed on page 75,

$$U - B = -2.5 \log_{10} \left( \frac{B_{365}\,\Delta\lambda_U}{B_{440}\,\Delta\lambda_B} \right) + C_{U-B}$$

$$-1.19 = -0.32 + C_{U-B}$$

$$C_{U-B} = -0.87,$$

and

$$B - V = -2.5 \log_{10} \left( \frac{B_{440}\,\Delta\lambda_B}{B_{550}\,\Delta\lambda_V} \right) + C_{B-V}$$

$$-0.33 = -0.98 + C_{B-V}$$

$$C_{B-V} = 0.65.$$

It is left as an exercise for you to use these values of $C_{U-B}$ and $C_{B-V}$ to estimate the color indices for a model blackbody Sun with a surface temperature of 5777 K. Although the resulting value of $B - V = +0.57$ is in fair agreement with the measured value of $B - V = +0.650$ for the Sun, the estimate of $U - B = -0.22$ is quite different from the measured value of $U - B = +0.195$. The reason for this large discrepancy at ultraviolet wavelengths will be discussed in Example 9.2.4.

### The Color–Color Diagram

Figure 3.11 is a **color–color diagram** showing the relation between the $U - B$ and $B - V$ color indices for main-sequence stars.[26] Astronomers face the difficult task of connecting a star's position on a color–color diagram with the physical properties of the star itself. If stars actually behaved as blackbodies, the color–color diagram would be the straight dashed line shown in Fig. 3.11. However, stars are not true blackbodies. As will be discussed in detail in Chapter 9, some light is absorbed as it travels through a star's atmosphere, and the amount of light absorbed depends on both the wavelength of the light and the temperature of the star. Other factors also play a role, causing the color indices of main-sequence and supergiant stars of the same temperature to be slightly different. The color–color diagram in Fig. 3.11 shows that the agreement between actual stars and model blackbody stars is best for very hot stars.

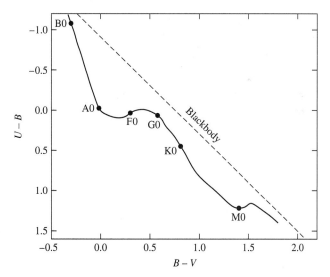

**FIGURE 3.11** Color–color diagram for main-sequence stars. The dashed line is for a blackbody. (The data are taken from Appendix G.)

[26]As will be discussed in Section 10.6, main-sequence stars are powered by the nuclear fusion of hydrogen nuclei in their centers. Approximately 80% to 90% of all stars are main-sequence stars. The letter labels in Fig. 3.11 are *spectral types*; see Section 8.1.

## SUGGESTED READING

### General

Ferris, Timothy, *Coming of Age in the Milky Way*, William Morrow, New York, 1988.

Griffin, Roger, "The Radial-Velocity Revolution," *Sky and Telescope*, September 1989.

Hearnshaw, John B., "Origins of the Stellar Magnitude Scale," *Sky and Telescope*, November 1992.

Herrmann, Dieter B., *The History of Astronomy from Hershel to Hertzsprung*, Cambridge University Press, Cambridge, 1984.

Perryman, Michael, "Hipparcos: The Stars in Three Dimensions," *Sky and Telescope*, June 1999.

Segre, Emilio, *From Falling Bodies to Radio Waves*, W. H. Freeman and Company, New York, 1984.

### Technical

Arp, Halton, "$U - B$ and $B - V$ Colors of Black Bodies," *The Astrophysical Journal, 133,* 874, 1961.

Böhm-Vitense, Erika, *Introduction to Stellar Astrophysics, Volume 1: Basic Stellar Observations and Data*, Cambridge University Press, Cambridge, 1989a.

Böhm-Vitense, Erika, *Introduction to Stellar Astrophysics, Volume 2: Stellar Atmospheres*, Cambridge University Press, Cambridge, 1989b.

Cox, Arthur N. (ed.), *Allen's Astrophysical Quantities*, Fourth Edition, Springer-Verlag, New York, 2000.

Harwit, Martin, *Astrophysical Concepts*, Third Edition, Springer-Verlag, New York, 1998.

*Hipparcos Space Astrometry Mission*, European Space Agency,
`http://astro.estec.esa.nl/Hipparcos/`.

Lang, Kenneth R., *Astrophysical Formulae*, Third Edition, Springer-Verlag, New York, 1999.

Van Helden, Albert, *Measuring the Universe*, The University of Chicago Press, Chicago, 1985.

## PROBLEMS

**3.1** In 1672, an international effort was made to measure the parallax angle of Mars at the time of opposition, when it was closest to Earth; see Fig. 1.6.

    **(a)** Consider two observers who are separated by a baseline equal to Earth's diameter. If the difference in their measurements of Mars's angular position is $33.6''$, what is the distance between Earth and Mars at the time of opposition? Express your answer both in units of m and in AU.

    **(b)** If the distance to Mars is to be measured to within 10%, how closely must the clocks used by the two observers be synchronized? *Hint:* Ignore the rotation of Earth. The average orbital velocities of Earth and Mars are 29.79 km s$^{-1}$ and 24.13 km s$^{-1}$, respectively.

**3.2** At what distance from a 100-W light bulb is the radiant flux equal to the solar irradiance?

**3.3** The parallax angle for Sirius is 0.379″.

    **(a)** Find the distance to Sirius in units of (i) parsecs; (ii) light-years; (iii) AU; (iv) m.

    **(b)** Determine the distance modulus for Sirius.

**3.4** Using the information in Example 3.6.1 and Problem 3.3, determine the absolute bolometric magnitude of Sirius and compare it with that of the Sun. What is the ratio of Sirius's luminosity to that of the Sun?

**3.5** **(a)** The Hipparcos Space Astrometry Mission was able to measure parallax angles down to nearly 0.001″. To get a sense of that level of resolution, how far from a dime would you need to be to observe it subtending an angle of 0.001″? (The diameter of a dime is approximately 1.9 cm.)

    **(b)** Assume that grass grows at the rate of 5 cm per week.

        **i.** How much does grass grow in one second?

        **ii.** How far from the grass would you need to be to see it grow at an angular rate of 0.000004″ (4 microarcseconds) per second? Four microarcseconds is the estimated angular resolution of SIM, NASA's planned astrometric mission; see page 59.

**3.6** Derive the relation

$$m = M_{\text{Sun}} - 2.5 \log_{10}\left(\frac{F}{F_{10,\odot}}\right).$$

**3.7** A $1.2 \times 10^4$ kg spacecraft is launched from Earth and is to be accelerated radially away from the Sun using a circular solar sail. The initial acceleration of the spacecraft is to be $1g$. Assuming a flat sail, determine the radius of the sail if it is

    **(a)** black, so it absorbs the Sun's light.

    **(b)** shiny, so it reflects the Sun's light.

    *Hint:* The spacecraft, like Earth, is orbiting the Sun. Should you include the Sun's gravity in your calculation?

**3.8** The average person has $1.4 \text{ m}^2$ of skin at a skin temperature of roughly 306 K (92°F). Consider the average person to be an ideal radiator standing in a room at a temperature of 293 K (68°F).

    **(a)** Calculate the energy per second radiated by the average person in the form of blackbody radiation. Express your answer in watts.

    **(b)** Determine the peak wavelength $\lambda_{\text{max}}$ of the blackbody radiation emitted by the average person. In what region of the electromagnetic spectrum is this wavelength found?

    **(c)** A blackbody also absorbs energy from its environment, in this case from the 293-K room. The equation describing the absorption is the same as the equation describing the emission of blackbody radiation, Eq. (3.16). Calculate the energy per second absorbed by the average person, expressed in watts.

    **(d)** Calculate the net energy per second lost by the average person via blackbody radiation.

**3.9** Consider a model of the star Dschubba ($\delta$ Sco), the center star in the head of the constellation Scorpius. Assume that Dschubba is a spherical blackbody with a surface temperature of 28,000 K and a radius of $5.16 \times 10^9$ m. Let this model star be located at a distance of 123 pc from Earth. Determine the following for the star:

    **(a)** Luminosity.

(b) Absolute bolometric magnitude.

(c) Apparent bolometric magnitude.

(d) Distance modulus.

(e) Radiant flux at the star's surface.

(f) Radiant flux at Earth's surface (compare this with the solar irradiance).

(g) Peak wavelength $\lambda_{max}$.

**3.10** (a) Show that the Rayleigh–Jeans law (Eq. 3.20) is an approximation of the Planck function $B_\lambda$ in the limit of $\lambda \gg hc/kT$. (The first-order expansion $e^x \approx 1 + x$ for $x \ll 1$ will be useful.) Notice that Planck's constant is not present in your answer. The Rayleigh–Jeans law is a *classical* result, so the "ultraviolet catastrophe" at short wavelengths, produced by the $\lambda^4$ in the denominator, cannot be avoided.

(b) Plot the Planck function $B_\lambda$ and the Rayleigh–Jeans law for the Sun ($T_\odot = 5777$ K) on the same graph. At roughly what wavelength is the Rayleigh–Jeans value twice as large as the Planck function?

**3.11** Show that Wien's expression for blackbody radiation (Eq. 3.21) follows directly from Planck's function at short wavelengths.

**3.12** Derive Wien's displacement law, Eq. (3.15), by setting $dB_\lambda/d\lambda = 0$. *Hint:* You will encounter an equation that must be solved numerically, not algebraically.

**3.13** (a) Use Eq. (3.24) to find an expression for the frequency $\nu_{max}$ at which the Planck function $B_\nu$ attains its maximum value. (*Warning:* $\nu_{max} \neq c/\lambda_{max}$.)

(b) What is the value of $\nu_{max}$ for the Sun?

(c) Find the wavelength of a light wave having frequency $\nu_{max}$. In what region of the electromagnetic spectrum is this wavelength found?

**3.14** (a) Integrate Eq. (3.27) over all wavelengths to obtain an expression for the total luminosity of a blackbody model star. *Hint:*

$$\int_0^\infty \frac{u^3 \, du}{e^u - 1} = \frac{\pi^4}{15}.$$

(b) Compare your result with the Stefan–Boltzmann equation (3.17), and show that the Stefan–Boltzmann constant $\sigma$ is given by

$$\sigma = \frac{2\pi^5 k^4}{15 c^2 h^3}.$$

(c) Calculate the value of $\sigma$ from this expression, and compare with the value listed in Appendix A.

**3.15** Use the data in Appendix G to answer the following questions.

(a) Calculate the absolute and apparent visual magnitudes, $M_V$ and $V$, for the Sun.

(b) Determine the magnitudes $M_B$, $B$, $M_U$, and $U$ for the Sun.

(c) Locate the Sun and Sirius on the color–color diagram in Fig. 3.11. Refer to Example 3.6.1 for the data on Sirius.

**3.16** Use the filter bandwidths for the $UBV$ system on page 75 and the effective temperature of 9600 K for Vega to determine through which filter Vega would appear brightest to a photometer [i.e., ignore the constant $C$ in Eq. (3.31)]. Assume that $S(\lambda) = 1$ inside the filter bandwidth and that $S(\lambda) = 0$ outside the filter bandwidth.

**3.17** Evaluate the constant $C_{bol}$ in Eq. (3.32) by using $m_{Sun} = -26.83$.

**3.18** Use the values of the constants $C_{U-B}$ and $C_{B-V}$ found in Example 3.6.2 to estimate the color indices $U - B$ and $B - V$ for the Sun.

**3.19** Shaula ($\lambda$ Scorpii) is a bright ($V = 1.62$) blue-white subgiant star located at the tip of the scorpion's tail. Its surface temperature is about 22,000 K.

    **(a)** Use the values of the constants $C_{U-B}$ and $C_{B-V}$ found in Example 3.6.2 to estimate the color indices $U - B$ and $B - V$ for Shaula. Compare your answers with the measured values of $U - B = -0.90$ and $B - V = -0.23$.

    **(b)** The Hipparcos Space Astrometry Mission measured the parallax angle for Shaula to be $0.00464''$. Determine the absolute visual magnitude of the star.

(Shaula is a pulsating star, belonging to the class of Beta Cephei variables; see Section 14.2. As its magnitude varies between $V = 1.59$ and $V = 1.65$ with a period of 5 hours 8 minutes, its color indices also change slightly.)

# CHAPTER

# 4

# The Theory of Special Relativity

## 4.1 ■ THE FAILURE OF THE GALILEAN TRANSFORMATIONS

A *wave* is a disturbance that travels through a medium. Water waves are disturbances traveling through water, and sound waves are disturbances traveling through air. James Clerk Maxwell predicted that light consists of "modulations of the same medium which is the cause of electric and magnetic phenomena," but what was the medium through which light waves traveled? At the time, physicists believed that light waves moved through a medium called the **luminiferous ether**. This idea of an all-pervading ether had its roots in the science of early Greece. In addition to the four earthly elements of earth, air, fire, and water, the Greeks believed that the heavens were composed of a fifth perfect element: the ether. Maxwell echoed their ancient belief when he wrote:

> There can be no doubt that the interplanetary and interstellar spaces are not empty, but are occupied by a material substance or body, which is certainly the largest, and probably the most uniform body of which we have any knowledge.

This modern reincarnation of the ether had been proposed for the sole purpose of transporting light waves; an object moving through the ether would experience no mechanical resistance, so Earth's velocity through the ether could not be directly measured.

### The Galilean Transformations

In fact, *no* mechanical experiment is capable of determining the absolute velocity of an observer. It is impossible to tell whether you are at rest or in uniform motion (not accelerating). This general principle was recognized very early. Galileo described a laboratory completely enclosed below the deck of a smoothly sailing ship and argued that no experiment done in this uniformly moving laboratory could measure the ship's velocity. To see why, consider two *inertial reference frames*, $S$ and $S'$. As discussed in Section 2.2, an inertial reference frame may be thought of as a laboratory in which Newton's first law is valid: An object at rest will remain at rest, and an object in motion will remain in motion in a straight line at

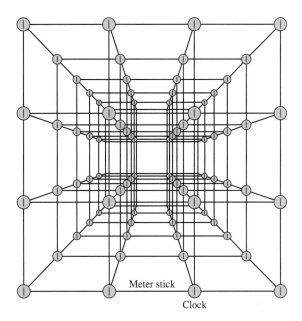

Meter stick

Clock

**FIGURE 4.1**   Inertial reference frame.

constant speed, unless acted upon by an external force. As shown in Fig. 4.1, the laboratory consists of (in principle) an infinite collection of meter sticks and synchronized clocks that can record the position and time of any event that occurs in the laboratory, *at the location of that event*; this removes the time delay involved in relaying information about an event to a distant recording device. With no loss of generality, the frame $S'$ can be taken as moving in the positive $x$-direction (relative to the frame $S$) with constant velocity $\mathbf{u}$, as shown in Fig. 4.2.[1] Furthermore, the clocks in the two frames can be started when the origins of the coordinate systems, $O$ and $O'$, coincide at time $t = t' = 0$.

Observers in the two frames $S$ and $S'$ measure the same moving object, recording its positions $(x, y, z)$ and $(x', y', z')$ at time $t$ and $t'$, respectively. An appeal to common sense and intuition leads to the conclusion that these measurements are related by the **Galilean transformation equations**:

$$x' = x - ut \tag{4.1}$$

$$y' = y \tag{4.2}$$

$$z' = z \tag{4.3}$$

$$t' = t. \tag{4.4}$$

---

[1]This does *not* imply that the frame $S$ is at rest and that $S'$ is moving. $S'$ could be at rest while $S$ moves in the negative $x'$-direction, or both frames could be moving. The point of the following argument is that there is *no way to tell*; only the *relative velocity* $\mathbf{u}$ is meaningful.

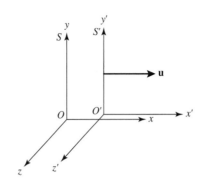

**FIGURE 4.2**   Inertial reference frames $S$ and $S'$.

Taking time derivatives with respect to either $t$ or $t'$ (since they are always equal) shows how the components of the object's velocity $\mathbf{v}$ and $\mathbf{v}'$ measured in the two frames are related:

$$v'_x = v_x - u$$

$$v'_y = v_y$$

$$v'_z = v_z,$$

or, in vector form,

$$\mathbf{v}' = \mathbf{v} - \mathbf{u}. \tag{4.5}$$

Since $\mathbf{u}$ is constant, another time derivative shows that the *same* acceleration is obtained for the object as measured in both reference frames:

$$\mathbf{a}' = \mathbf{a}.$$

Thus $\mathbf{F} = m\mathbf{a} = m\mathbf{a}'$ for the object of mass $m$; Newton's laws are obeyed in both reference frames. Whether a laboratory is located in the hold of Galileo's ship or anywhere else in the universe, no mechanical experiment can be done to measure the laboratory's absolute velocity.

### The Michelson–Morley Experiment

Maxwell's discovery that electromagnetic waves move through the ether with a speed of $c \simeq 3 \times 10^8$ m s$^{-1}$ seemed to open the possibility of detecting Earth's *absolute* motion through the ether by measuring the speed of light from Earth's frame of reference and comparing it with Maxwell's theoretical value of $c$. In 1887 two Americans, the physicist Albert A. Michelson (1852–1931) and his chemist colleague Edward W. Morley (1838–1923), performed a classic experiment that attempted this measurement of Earth's absolute velocity. Although Earth orbits the Sun at approximately 30 km s$^{-1}$, the results of the Michelson–Morley experiment were consistent with a velocity of Earth through the ether of *zero*![2] Furthermore, as Earth spins on its axis and orbits the Sun, a laboratory's speed through the ether should be constantly changing. The constantly shifting "ether wind" should easily be detected. However, all of the many physicists who have since repeated the Michelson–Morley experiment with increasing precision have reported the same null result. Everyone measures *exactly* the same value for the speed of light, regardless of the velocity of the laboratory on Earth or the velocity of the source of the light.

On the other hand, Eq. (4.5) implies that two observers moving with a relative velocity **u** should obtain *different* values for the speed of light. The contradiction between the commonsense expectation of Eq. (4.5) and the experimentally determined constancy of the speed of light means that this equation, and the equations from which it was derived (the Galilean transformation equations, 4.1–4.4), cannot be correct. Although the Galilean transformations adequately describe the familiar low-speed world of everyday life where $v/c \ll 1$, they are in sharp disagreement with the results of experiments involving velocities near the speed of light. A crisis in the Newtonian paradigm was developing.

## 4.2 ■ THE LORENTZ TRANSFORMATIONS

The young Albert Einstein (1875–1955; see Fig. 4.3) enjoyed discussing a puzzle with his friends: What would you see if you looked in a mirror while moving at the speed of light? Would you see your image in the mirror, or not? This was the beginning of Einstein's search for a simple, consistent picture of the universe, a quest that would culminate in his theories

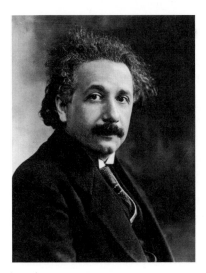

**FIGURE 4.3**  Albert Einstein (1875–1955). (Courtesy of Yerkes Observatory.)

---

[2]Strictly speaking, a laboratory on Earth is not in an inertial frame of reference, because Earth both spins on its axis and accelerates as it orbits the Sun. However, these noninertial effects are unimportant for the Michelson–Morley experiment.

of relativity. After much reflection, Einstein finally rejected the notion of an all-pervading ether.

### Einstein's Postulates

In 1905 Einstein introduced his two postulates of special relativity[3] in a remarkable paper, "On the Electrodynamics of Moving Bodies."

> The phenomena of electrodynamics as well as of mechanics possess no properties corresponding to the idea of absolute rest. They suggest rather that ... the same laws of electrodynamics and optics will be valid for all frames of reference for which the equations of mechanics hold good. We will raise this conjecture (the purport of which will hereafter be called the "Principle of Relativity") to the status of a postulate, and also introduce another postulate, which is only apparently irreconcilable to the former, namely, that light is always propagated in empty space with a definite speed $c$ which is independent of the state of motion of the emitting body.

In other words, **Einstein's postulates** are

> **The Principle of Relativity** The laws of physics are the same in all inertial reference frames.

> **The Constancy of the Speed of Light** Light moves through a vacuum at a constant speed $c$ that is independent of the motion of the light source.

### The Derivation of the Lorentz Transformations

Einstein then went on to derive the equations that lie at the heart of his theory of special relativity, the **Lorentz transformations**.[4] For the two inertial reference frames shown in Fig. 4.2, the most general set of linear transformation equations between the space and time coordinates $(x, y, z, t)$ and $(x', y', z', t')$ of the *same event* measured from $S$ and $S'$ are

$$x' = a_{11}x + a_{12}y + a_{13}z + a_{14}t \tag{4.6}$$

$$y' = a_{21}x + a_{22}y + a_{23}z + a_{24}t \tag{4.7}$$

$$z' = a_{31}x + a_{32}y + a_{33}z + a_{34}t \tag{4.8}$$

$$t' = a_{41}x + a_{42}y + a_{43}z + a_{44}t. \tag{4.9}$$

If the transformation equations were not linear, then the length of a moving object or the time interval between two events would depend on the choice of origin for the frames $S$ and $S'$. This is unacceptable, since the laws of physics cannot depend on the numerical coordinates of an arbitrarily chosen coordinate system.

---

[3]The theory of *special* relativity deals only with inertial reference frames, whereas the *general* theory includes accelerating frames.

[4]These equations were first derived by Hendrik A. Lorentz (1853–1928) of the Netherlands but were applied to a different situation involving a reference frame at absolute rest with respect to the ether.

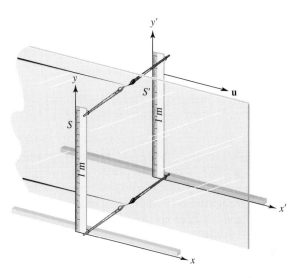

**FIGURE 4.4** Paint brush demonstration that $y' = y$.

The coefficients $a_{ij}$ can be determined by using Einstein's two postulates and some simple symmetry arguments. Einstein's first postulate, the Principle of Relativity, implies that lengths *perpendicular* to **u**, the velocity of frame $S$ relative to $S'$, are unchanged. To see this, imagine that each frame has a meter stick oriented along the $y$- and $y'$-axes, with one end of each meter stick located at the origin of its respective frame; see Fig. 4.4. Paint brushes are mounted perpendicular at both ends of each meter stick, and the frames are separated by a sheet of glass that extends to infinity in the $x$–$y$ plane. Each brush paints a line on the glass sheet as the two frames pass each other. Let's say that frame $S$ uses blue paint, and frame $S'$ uses red paint. If an observer in frame $S$ measures the meter stick in frame $S'$ to be shorter than his own meter stick, he will see the red lines painted *inside* his blue lines on the glass. But by the Principle of Relativity, an observer in frame $S'$ would measure the meter stick in frame $S$ as being shorter than her own meter stick and would see the blue lines painted *inside* her red lines. Both color lines cannot lie inside the other; the only conclusion is that blue and red lines must overlap. The lengths of the meter sticks, perpendicular to **u**, are unchanged. Thus $y' = y$ and $z' = z$, so that $a_{22} = a_{33} = 1$, whereas $a_{21}$, $a_{23}$, $a_{24}$, $a_{31}$, $a_{32}$, and $a_{34}$ are all zero.

Another simplification comes from requiring that Eq. (4.9) give the same result if $y$ is replaced by $-y$ or $z$ is replaced by $-z$. This must be true because rotational symmetry about the axis parallel to the relative velocity **u** implies that a time measurement cannot depend on the side of the $x$-axis on which an event occurs. Thus $a_{42} = a_{43} = 0$.

Finally, consider the motion of the origin $O'$ of frame $S'$. Since the frames' clocks are assumed to be synchronized at $t = t' = 0$ when the origins $O$ and $O'$ coincide, the $x$-coordinate of $O'$ is given by $x = ut$ in frame $S$ and by $x' = 0$ in frame $S'$. Thus Eq. (4.6) becomes

$$0 = a_{11}ut + a_{12}y + a_{13}z + a_{14}t,$$

which implies that $a_{12} = a_{13} = 0$ and $a_{11}u = -a_{14}$. Collecting the results found thus far reveals that Eqs. (4.6–4.9) have been reduced to

$$x' = a_{11}(x - ut) \tag{4.10}$$

$$y' = y \tag{4.11}$$

$$z' = z \tag{4.12}$$

$$t' = a_{41}x + a_{44}t. \tag{4.13}$$

At this point, these equations would be consistent with the commonsense Galilean transformation equations (4.1–4.4) if $a_{11} = a_{44} = 1$ and $a_{41} = 0$. However, only one of Einstein's postulates has been employed in the derivation thus far: the Principle of Relativity championed by Galileo himself.

Now the argument introduces the second of Einstein's postulates: Everyone measures exactly the same value for the speed of light. Suppose that when the origins $O$ and $O'$ coincide at time $t = t' = 0$, a flashbulb is set off at the common origins. At a later time $t$, an observer in frame $S$ will measure a spherical wavefront of light with radius $ct$, moving away from the origin $O$ with speed $c$ and satisfying

$$x^2 + y^2 + z^2 = (ct)^2. \tag{4.14}$$

Similarly, at a time $t'$, an observer in frame $S'$ will measure a spherical wavefront of light with radius $ct'$, moving away from the origin $O'$ with speed $c$ and satisfying

$$x'^2 + y'^2 + z'^2 = \left(ct'\right)^2. \tag{4.15}$$

Inserting Eqs. (4.10–4.13) into Eq. (4.15) and comparing the result with Eq. (4.14) reveal that $a_{11} = a_{44} = 1/\sqrt{1 - u^2/c^2}$ and $a_{41} = -ua_{11}/c^2$. Thus the Lorentz transformation equations linking the space and time coordinates $(x, y, z, t)$ and $(x', y', z', t')$ of the *same event* measured from $S$ and $S'$ are

$$x' = \frac{x - ut}{\sqrt{1 - u^2/c^2}} \tag{4.16}$$

$$y' = y \tag{4.17}$$

$$z' = z \tag{4.18}$$

$$t' = \frac{t - ux/c^2}{\sqrt{1 - u^2/c^2}}. \tag{4.19}$$

Whenever the Lorentz transformations are used, you should be certain that the situation is consistent with the geometry of Fig. 4.2, where the inertial reference frame $S'$ is moving in

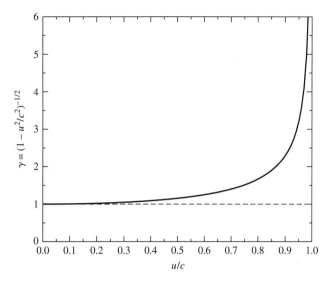

**FIGURE 4.5** The Lorentz factor $\gamma$.

the positive $x$-direction with velocity **u** relative to the frame $S$. The ubiquitous factor of

$$\gamma \equiv \frac{1}{\sqrt{1 - u^2/c^2}},\tag{4.20}$$

called the **Lorentz factor**, may be used to estimate the importance of relativistic effects. Roughly speaking, relativity differs from Newtonian mechanics by 1% ($\gamma = 1.01$) when $u/c \simeq 1/7$ and by 10% when $u/c \simeq 5/12$; see Fig. 4.5. In the low-speed Newtonian world, the Lorentz transformations reduce to the Galilean transformation equations (4.1–4.4). A similar requirement holds for all relativistic formulas; they must agree with the Newtonian equations in the low-speed limit of $u/c \to 0$.

The inverse Lorentz transformations can be derived algebraically, or they can be obtained more easily by switching primed and unprimed quantities and by replacing $u$ with $-u$. (Be sure you understand the physical basis for these substitutions.) Either way, the inverse transformations are found to be

$$x = \frac{x' + ut'}{\sqrt{1 - u^2/c^2}}\tag{4.21}$$

$$y = y'\tag{4.22}$$

$$z = z'\tag{4.23}$$

$$t = \frac{t' + ux'/c^2}{\sqrt{1 - u^2/c^2}}.\tag{4.24}$$

### Four-Dimensional Spacetime

The Lorentz transformation equations form the core of the theory of special relativity, and they have many surprising and unusual implications. The most obvious surprise is the intertwining roles of spatial and temporal coordinates in the transformations. In the words of Einstein's professor, Hermann Minkowski (1864–1909), "Henceforth space by itself, and time by itself, are doomed to fade away into mere shadows, and only a kind of union between the two will preserve an independent reality." The drama of the physical world unfolds on the stage of a four-dimensional **spacetime**, where events are identified by their spacetime coordinates $(x, y, z, t)$.

## 4.3 ■ TIME AND SPACE IN SPECIAL RELATIVITY

Suppose an observer in frame $S$ measures two flashbulbs going off at the *same time t* but at *different x*-coordinates $x_1$ and $x_2$. Then an observer in frame $S'$ would measure the time interval $t'_1 - t'_2$ between the flashbulbs going off to be (see Eq. 4.19)

$$t'_1 - t'_2 = \frac{(x_2 - x_1)\, u/c^2}{\sqrt{1 - u^2/c^2}}.$$  (4.25)

*According to the observer in frame S', if $x_1 \neq x_2$, then the flashbulbs do not go off at the same time!* Events that occur simultaneously in one inertial reference frame do not occur simultaneously in all other inertial reference frames. There is no such thing as two events that occur at different locations happening *absolutely* at the same time. Equation (4.25) shows that if $x_1 < x_2$, then $t'_1 - t'_2 > 0$ for positive $u$; flashbulb 1 is measured to go off *after* flashbulb 2. An observer moving at the same speed in the opposite direction ($u$ changed to $-u$) will come to the opposite conclusion: Flashbulb 2 goes off *after* flashbulb 1. The situation is symmetric; an observer in frame $S'$ will conclude that the flashbulb he or she passes first goes off *after* the other flashbulb. It is tempting to ask, "Which observer is *really* correct?" However, this question is meaningless and is equivalent to asking, "Which observer is *really* moving?" Neither question has an answer because "really" has no meaning in this situation. There is no absolute simultaneity, just as there is no absolute motion. Each observer's measurement is correct, as made from his or her own frame of reference.

The implications of this **downfall of universal simultaneity** are far-reaching. The absence of a universal simultaneity means that clocks in relative motion will not stay synchronized. Newton's idea of an absolute universal time that "of itself and from its own nature flows equably without regard to anything external" has been overthrown. Different observers in relative motion will measure *different* time intervals between the *same* two events!

### Proper Time and Time Dilation

Imagine that a strobe light located at rest relative to frame $S'$ produces a flash of light every $\Delta t'$ seconds; see Fig. 4.6. If one flash is emitted at time $t'_1$, then the next flash will be emitted at time $t'_2 = t'_1 + \Delta t'$, as measured by a clock in frame $S'$. Using Eq. (4.24) with $x'_1 = x'_2$,

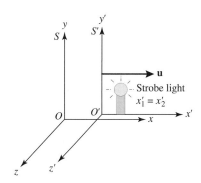

**FIGURE 4.6**   A strobe light at rest ($x' =$ constant) in frame $S'$.

the time interval $\Delta t \equiv t_2 - t_1$ between the *same* two flashes measured by a clock in frame $S$ is

$$t_2 - t_1 = \frac{(t_2' - t_1') + (x_2' - x_1')\,u/c^2}{\sqrt{1 - u^2/c^2}}$$

or

$$\Delta t = \frac{\Delta t'}{\sqrt{1 - u^2/c^2}}. \tag{4.26}$$

Because the clock in frame $S'$ is *at rest* relative to the strobe light, $\Delta t'$ will be called $\Delta t_{\text{rest}}$. Frame $S'$ is called the clock's **rest frame**. Similarly, because the clock in frame $S$ is *moving* relative to the strobe light, $\Delta t$ will be called $\Delta t_{\text{moving}}$. Thus Eq. (4.26) becomes

$$\Delta t_{\text{moving}} = \frac{\Delta t_{\text{rest}}}{\sqrt{1 - u^2/c^2}}. \tag{4.27}$$

This equation shows the effect of **time dilation** on a moving clock. It says that the time interval between two events is measured differently by different observers in relative motion. The *shortest time interval* is measured by a clock *at rest* relative to the two events. This clock measures the **proper time** between the two events. Any other clock moving relative to the two events will measure a longer time interval between them.

The effect of time dilation is often described by the phrase "moving clocks run slower" without explicitly identifying the two events involved. This easily leads to confusion, since the moving and rest subscripts in Eq. (4.27) mean "moving" or "at rest" *relative to the two events*. To gain insight into this phrase, imagine that you are holding clock $C$ while it ticks once each second and, at the same time, are measuring the ticks of an identical clock $C'$ moving relative to you. The two events to be measured are consecutive ticks of clock $C'$. Since clock $C'$ is at rest relative to itself, it measures a time $\Delta t_{\text{rest}} = 1$ s between its own ticks. However, using your clock $C$, you measure a time

$$\Delta t_{\text{moving}} = \frac{\Delta t_{\text{rest}}}{\sqrt{1 - u^2/c^2}} = \frac{1 \text{ s}}{\sqrt{1 - u^2/c^2}} > 1 \text{ s}$$

between the ticks of clock $C'$. Because you measure clock $C'$ to be ticking slower than once per second, you conclude that clock $C'$, which is moving relative to you, is running more slowly than your clock $C$. Very accurate atomic clocks have been flown around the world on jet airliners and have confirmed that moving clocks do indeed run slower, in agreement with relativity.[5]

## Proper Length and Length Contraction

Both time dilation and the downfall of simultaneity contradict Newton's belief in absolute time. Instead, the time measured between two events differs for different observers in relative motion. Newton also believed that "absolute space, in its own nature, without relation to anything external, remains always similar and immovable." However, the Lorentz transformation equations require that different observers in relative motion will measure space differently as well.

Imagine that a rod lies along the $x'$-axis of frame $S'$, at rest relative to that frame; $S'$ is the rod's rest frame (see Fig. 4.7). Let the left end of the rod have coordinate $x_1'$, and let the right end of the rod have coordinate $x_2'$. Then the length of the rod as measured in frame $S'$ is $L' = x_2' - x_1'$. What is the length of the rod measured from $S$? Because the rod is moving relative to $S$, care must be taken to measure the $x$-coordinates $x_1$ and $x_2$ of the ends of the rod *at the same time*. Then Eq. (4.16), with $t_1 = t_2$, shows that the length $L = x_2 - x_1$ measured in $S$ may be found from

$$x_2' - x_1' = \frac{(x_2 - x_1) - u(t_2 - t_1)}{\sqrt{1 - u^2/c^2}}$$

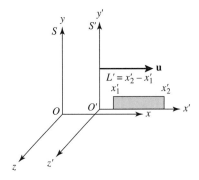

**FIGURE 4.7**    A rod at rest in frame $S'$.

---

[5]See Hafele and Keating (1972a, 1972b) for the details of this test of time dilation.

or

$$L' = \frac{L}{\sqrt{1 - u^2/c^2}}. \tag{4.28}$$

Because the rod is *at rest* relative to $S'$, $L'$ will be called $L_{rest}$. Similarly, because the rod is *moving* relative to $S$, $L$ will be called $L_{moving}$. Thus Eq. (4.28) becomes

$$\boxed{L_{moving} = L_{rest}\sqrt{1 - u^2/c^2}.} \tag{4.29}$$

This equation shows the effect of **length contraction** on a moving rod. It says that length or distance is measured differently by two observers in relative motion. If a rod is moving relative to an observer, that observer will measure a shorter rod than will an observer at rest relative to it. The *longest length*, called the rod's **proper length**, is measured in the rod's rest frame. Only lengths or distances *parallel* to the direction of the relative motion are affected by length contraction; distances perpendicular to the direction of the relative motion are unchanged (cf. Eqs. 4.17–4.18).

### Time Dilation and Length Contraction Are Complementary

Time dilation and length contraction are not independent effects of Einstien's new way of looking at the universe. Rather, they are complementary; the magnitude of either effect depends on the motion of the event being observed relative to the observer.

---

**Example 4.3.1.**   Cosmic rays from space collide with the nuclei of atoms in Earth's upper atmosphere, producing elementary particles called *muons*. Muons are unstable and decay after an average lifetime $\tau = 2.20 \ \mu s$, as measured in a laboratory where the muons are at rest. That is, the number of muons in a given sample should decrease with time according to $N(t) = N_0 \, e^{-t/\tau}$, where $N_0$ is the number of muons originally in the sample at time $t = 0$. At the top of Mt. Washington in New Hampshire, a detector counted 563 muons hr$^{-1}$ moving downward at a speed $u = 0.9952c$. At sea level, 1907 m below the first detector, another detector counted 408 muons hr$^{-1}$.[6]

The muons take $(1907 \text{ m})/(0.9952c) = 6.39 \ \mu s$ to travel from the top of Mt. Washington to sea level. Thus it might be expected that the number of muons detected per hour at sea level would have been

$$N = N_0 \, e^{-t/\tau} = (563 \text{ muons hr}^{-1}) \, e^{-(6.39 \ \mu s)/(2.20 \ \mu s)} = 31 \text{ muons hr}^{-1}.$$

This is much less than the 408 muons hr$^{-1}$ actually measured at sea level! How did the muons live long enough to reach the lower detector? The problem with the preceding calculation is that the lifetime of 2.20 $\mu s$ is measured in the muon's rest frame as $\Delta t_{rest}$, but the experimenter's clocks on Mt. Washington and below are moving relative to the muons.

*continued*

---

[6] Details of this experiment can be found in Frisch and Smith (1963).

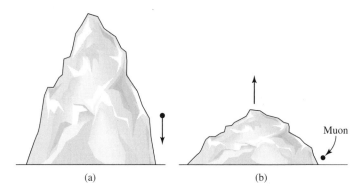

(a)                                    (b)

**FIGURE 4.8**    Muons moving downward past Mt. Washington. (a) Mountain frame. (b) Muon frame.

They measure the muon's lifetime to be

$$\Delta t_{\text{moving}} = \frac{\Delta t_{\text{rest}}}{\sqrt{1 - u^2/c^2}} = \frac{2.20 \ \mu\text{s}}{\sqrt{1 - (0.9952)^2}} = 22.5 \ \mu\text{s},$$

more than *ten* times a muon's lifetime when measured in its own rest frame. The moving muons' clocks run slower, so more of them survive long enough to reach sea level. Repeating the preceding calculation using the muon lifetime as measured by the experimenters gives

$$N = N_0 \, e^{-t/\tau} = (563 \text{ muons hr}^{-1}) \, e^{-(6.39 \ \mu\text{s})/(22.5 \ \mu\text{s})} = 424 \text{ muons hr}^{-1}.$$

When the effects of time dilation are included, the theoretical prediction is in excellent agreement with the experimental result.

From a muon's rest frame, its lifetime is only 2.20 $\mu$s. How would an observer riding along with the muons, as shown in Fig. 4.8, explain their ability to reach sea level? The observer would measure a severely length-contracted Mt. Washington (in the direction of the relative motion only). The distance traveled by the muons would not be $L_{\text{rest}} = 1907$ m but, rather, would be

$$L_{\text{moving}} = L_{\text{rest}}\sqrt{1 - u^2/c^2} = (1907 \text{ m})\sqrt{1 - (0.9952)^2} = 186.6 \text{ m}.$$

Thus it would take $(186.6 \text{ m})/(0.9952c) = 0.625 \ \mu$s for the muons to travel the length-contracted distance to the detector at sea level, as measured by an observer in the muons' rest frame. That observer would then calculate the number of muons reaching the lower detector to be

$$N = N_0 \, e^{-t/\tau} = (563 \text{ muons hr}^{-1}) \, e^{-(0.625 \ \mu\text{s})/(2.20 \ \mu\text{s})} = 424 \text{ muons hr}^{-1},$$

in agreement with the previous result. This shows that an effect due to time dilation as measured in one frame may instead be attributed to length contraction as measured in another frame.

The effects of time dilation and length contraction are both symmetric between two observers in relative motion. Imagine two identical spaceships that move in opposite directions, passing each other at some relativistic speed. Observers aboard each spaceship will measure the other ship's length as being shorter than their own, and the other ship's clocks as running slower. *Both observers are right*, having made correct measurements from their respective frames of reference.

You should not think of these effects as being due to some sort of "optical illusion" caused by light taking different amounts of time to reach an observer from different parts of a moving object. The language used in the preceding discussions has involved the *measurement* of an event's spacetime coordinates $(x, y, z, t)$ using meter sticks and clocks located *at that event*, so there is no time delay. Of course, no actual laboratory has an infinite collection of meter sticks and clocks, and the time delays caused by finite light-travel times must be taken into consideration. This will be important in determining the relativistic Doppler shift formula, which follows.

### The Relativistic Doppler Shift

In 1842 the Austrian physicist Christian Doppler showed that as a source of sound moves through a medium (such as air), the wavelength is compressed in the forward direction and expanded in the backward direction. This change in wavelength of any type of wave caused by the motion of the source or the observer is called a **Doppler shift**. Doppler deduced that the difference between the wavelength $\lambda_{obs}$ observed for a moving source of sound and the wavelength $\lambda_{rest}$ measured in the laboratory for a reference source at rest is related to the radial velocity $v_r$ (the component of the velocity directly toward or away from the observer; see Fig. 1.15) of the source through the medium by

$$\frac{\lambda_{obs} - \lambda_{rest}}{\lambda_{rest}} = \frac{\Delta\lambda}{\lambda_{rest}} = \frac{v_r}{v_s}, \tag{4.30}$$

where $v_s$ is the speed of sound in the medium. However, this expression cannot be precisely correct for light. Experimental results such as those of Michelson and Morley led Einstein to abandon the ether concept, and they demonstrated that no medium is involved in the propagation of light waves. The Doppler shift for light is a qualitatively different phenomenon from its counterpart for sound waves.

Consider a distant source of light that emits a light signal at time $t_{rest,1}$ and another signal at time $t_{rest,2} = t_{rest,1} + \Delta t_{rest}$ as measured by a clock *at rest* relative to the source. If this light source is moving relative to an observer with velocity **u**, as shown in Fig. 4.9, then the time between receiving the light signals at the observer's location will depend on both the effect of time dilation and the different distances traveled by the signals from the source to the observer. (The light source is assumed to be sufficiently far away that the signals travel along parallel paths to the observer.) Using Eq. (4.27), we find that the time between the emission of the light signals as measured in the observer's frame is $\Delta t_{rest}/\sqrt{1 - u^2/c^2}$. In this time, the observer determines that the distance to the light source has changed by an amount $u\,\Delta t_{rest}\cos\theta/\sqrt{1 - u^2/c^2}$. Thus the time interval $\Delta t_{obs}$ between the arrival of the

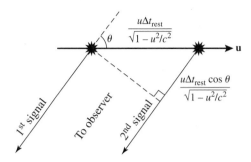

**FIGURE 4.9**   Relativistic Doppler shift.

two light signals at the observer's location is

$$\Delta t_{\text{obs}} = \frac{\Delta t_{\text{rest}}}{\sqrt{1 - u^2/c^2}} [1 + (u/c) \cos \theta]. \tag{4.31}$$

If $\Delta t_{\text{rest}}$ is taken to be the time between the emission of the light wave crests, and if $\Delta t_{\text{obs}}$ is the time between their arrival, then the frequencies of the light wave are $\nu_{\text{rest}} = 1/\Delta t_{\text{rest}}$ and $\nu_{\text{obs}} = 1/\Delta t_{\text{obs}}$. The equation describing the **relativistic Doppler shift** is thus

$$\nu_{\text{obs}} = \frac{\nu_{\text{rest}} \sqrt{1 - u^2/c^2}}{1 + (u/c) \cos \theta} = \frac{\nu_{\text{rest}} \sqrt{1 - u^2/c^2}}{1 + v_r/c}, \tag{4.32}$$

where $v_r = u \cos \theta$ is the *radial velocity* of the light source. If the light source is moving directly away from the observer ($\theta = 0°$, $v_r = u$) or toward the observer ($\theta = 180°$, $v_r = -u$), then the relativistic Doppler shift reduces to

$$\nu_{\text{obs}} = \nu_{\text{rest}} \sqrt{\frac{1 - v_r/c}{1 + v_r/c}} \qquad \text{(radial motion).} \tag{4.33}$$

There is also a **transverse Doppler shift** for motion perpendicular to the observer's line of sight ($\theta = 90°$, $v_r = 0$). This transverse shift is entirely due to the effect of time dilation. Note that, unlike formulas describing the Doppler shift for sound, Eqs. (4.32) and (4.33) do not distinguish between the velocity of the source and the velocity of the observer. Only the relative velocity is important.

When astronomers observe a star or galaxy moving away from or toward Earth, the wavelength of the light they receive is shifted toward longer or shorter wavelengths, respectively. If the source of light is moving *away* from the observer ($v_r > 0$), then $\lambda_{\text{obs}} > \lambda_{\text{rest}}$. This shift to a longer wavelength is called a **redshift**. Similarly, if the source is moving *toward* the observer ($v_r < 0$), then there is a shift to a shorter wavelength, a **blueshift**.[7] Because

---

[7]Doppler himself maintained that all stars would be white if they were at rest and that the different colors of the stars were due to their Doppler shifts. However, the stars move much too slowly for their Doppler shifts to significantly change their colors.

most of the objects in the universe outside of our own Milky Way Galaxy are moving away from us, redshifts are commonly measured by astronomers. A **redshift parameter** $z$ is used to describe the change in wavelength; it is defined as

$$z \equiv \frac{\lambda_{\text{obs}} - \lambda_{\text{rest}}}{\lambda_{\text{rest}}} = \frac{\Delta\lambda}{\lambda_{\text{rest}}}. \tag{4.34}$$

The observed wavelength $\lambda_{\text{obs}}$ is obtained from Eq. (4.33) and $c = \lambda\nu$,

$$\lambda_{\text{obs}} = \lambda_{\text{rest}}\sqrt{\frac{1 + v_r/c}{1 - v_r/c}} \qquad \text{(radial motion)}, \tag{4.35}$$

and the redshift parameter becomes

$$z = \sqrt{\frac{1 + v_r/c}{1 - v_r/c}} - 1 \qquad \text{(radial motion)}. \tag{4.36}$$

In general, Eq. (4.34), together with $\lambda = c/\nu$, shows that

$$z + 1 = \frac{\Delta t_{\text{obs}}}{\Delta t_{\text{rest}}}. \tag{4.37}$$

This expression indicates that if the luminosity of an astrophysical source with redshift parameter $z > 0$ (receding) is observed to vary during a time $\Delta t_{\text{obs}}$, then the change in luminosity occurred over a *shorter* time $\Delta t_{\text{rest}} = \Delta t_{\text{obs}}/(z + 1)$ in the rest frame of the source.

---

**Example 4.3.2.**    In its rest frame, the quasar SDSS 1030+0524 produces a hydrogen emission line of wavelength $\lambda_{\text{rest}} = 121.6$ nm. On Earth, this emission line is observed to have a wavelength of $\lambda_{\text{obs}} = 885.2$ nm. The redshift parameter for this quasar is thus

$$z = \frac{\lambda_{\text{obs}} - \lambda_{\text{rest}}}{\lambda_{\text{rest}}} = 6.28.$$

Using Eq. (4.36), we may calculate the speed of recession of the quasar:

$$z = \sqrt{\frac{1 + v_r/c}{1 - v_r/c}} - 1$$

$$\frac{v_r}{c} = \frac{(z + 1)^2 - 1}{(z + 1)^2 + 1} \tag{4.38}$$

$$= 0.963.$$

*continued*

Quasar SDSS 1030+0524 appears to be moving away from us at more than 96% of the speed of light! However, objects that are enormously distant from us, such as quasars, have large apparent recessional speeds due to the overall expansion of the universe. In these cases the increase in the observed wavelength is actually due to *the expansion of space itself* (which stretches the wavelength of light) rather than being due to the motion of the object through space! This *cosmological redshift* is a consequence of the Big Bang.

This quasar was discovered as a product of the massive Sloan Digital Sky Survey; see Becker, et al. (2001) for further information about this object.

---

Suppose the speed $u$ of the light source is small compared to that of light ($u/c \ll 1$). Using the expansion (to first order)

$$(1 + v_r/c)^{\pm 1/2} \simeq 1 \pm \frac{v_r}{2c},$$

together with Eqs. (4.34) and (4.35) for radial motion, then shows that for low speeds,

$$z = \frac{\Delta \lambda}{\lambda_{\text{rest}}} \simeq \frac{v_r}{c}, \tag{4.39}$$

where $v_r > 0$ for a receding source ($\Delta \lambda > 0$) and $v_r < 0$ for an approaching source ($\Delta \lambda < 0$). Although this equation is similar to Eq. (4.30), you should bear in mind that Eq. (4.39) is an *approximation*, valid only for low speeds. Misapplying this equation to the relativistic quasar SDSS 1030+0524 discussed in Example 4.3.2 would lead to the erroneous conclusion that the quasar is moving away from us at 6.28 times the speed of light!

### The Relativistic Velocity Transformation

Because space and time intervals are measured differently by different observers in relative motion, velocities must be transformed as well. The equations describing the relativistic transformation of velocities may be easily found from the Lorentz transformation equations (4.16–4.19) by writing them as differentials. Then dividing the $dx'$, $dy'$, and $dz'$ equations by the $dt'$ equation gives the **relativistic velocity transformations**:

$$v'_x = \frac{v_x - u}{1 - uv_x/c^2} \tag{4.40}$$

$$v'_y = \frac{v_y \sqrt{1 - u^2/c^2}}{1 - uv_x/c^2} \tag{4.41}$$

$$v'_z = \frac{v_z \sqrt{1 - u^2/c^2}}{1 - uv_x/c^2}. \tag{4.42}$$

As with the inverse Lorentz transformations, the inverse velocity transformations may be obtained by switching primed and unprimed quantities and by replacing $u$ with $-u$. It is left as an exercise to show that these equations do satisfy the second of Einstein's postulates: Light travels through a vacuum at a constant speed that is independent of the motion of the light source. From Eqs. (4.40–4.42), if **v** has a magnitude of $c$, so does **v**' (see Problem 4.12).

**Example 4.3.3.**   As measured in the reference frame $S'$, a light source is at rest and radiates light equally in all directions. In particular, half of the light is emitted into the forward (positive $x'$) hemisphere. Is this situation any different when viewed from frame $S$, which measures the light source traveling in the positive $x$-direction with a relativistic speed $u$?

Consider a light ray whose velocity components measured in $S'$ are $v'_x = 0$, $v'_y = c$, and $v'_z = 0$. This ray travels along the boundary between the forward and backward hemispheres of light as measured in $S'$. However, as measured in frame $S$, this light ray has the velocity components given by the inverse transformations of Eqs. (4.40–4.42):

$$v_x = \frac{v'_x + u}{1 + uv'_x/c^2} = u$$

$$v_y = \frac{v'_y\sqrt{1 - u^2/c^2}}{1 + uv'_x/c^2} = c\sqrt{1 - u^2/c^2}$$

$$v_z = \frac{v'_z\sqrt{1 - u^2/c^2}}{1 + uv'_x/c^2} = 0.$$

As measured in frame $S$, the light ray is not traveling perpendicular to the $x$-axis; see Fig. 4.10.

In fact, for $u/c$ close to 1, the angle $\theta$ measured between the light ray and the $x$-axis may be found from $\sin\theta = v_y/v$, where

$$v = \sqrt{v_x^2 + v_y^2 + v_z^2} = c$$

is the speed of the light ray measured in frame $S$. Thus

$$\boxed{\sin\theta = \frac{v_y}{v} = \sqrt{1 - u^2/c^2} = \gamma^{-1},} \tag{4.43}$$

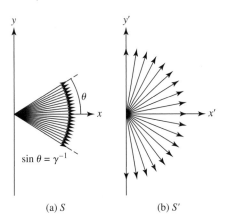

(a) $S$                                (b) $S'$

**FIGURE 4.10**   Relativistic headlight effect. (a) Frame $S$. (b) Frame $S'$.

*continued*

where $\gamma$ is the Lorentz factor defined by Eq. (4.20). For relativistic speeds $u \approx c$, implying that $\gamma$ is very large, so $\sin \theta$ (and hence $\theta$) becomes very small. All of the light emitted into the forward hemisphere, as measured in $S'$, is concentrated into a narrow cone in the direction of the light source's motion when measured in frame $S$. Called the **headlight effect**, this result plays an important role in many areas of astrophysics. For example, as relativistic electrons spiral around magnetic field lines, they emit light in the form of **synchrotron radiation**. The radiation is concentrated in the direction of the electron's motion and is strongly plane-polarized. Synchrotron radiation is an important electromagnetic radiation process in the Sun, Jupiter's magnetosphere, pulsars, and active galaxies.

## 4.4 ■ RELATIVISTIC MOMENTUM AND ENERGY

Up to this point, only relativistic kinematics has been considered. Einstein's theory of special relativity also requires new definitions for the concepts of momentum and energy. The ideas of conservation of linear momentum and energy are two of the cornerstones of physics. According to the Principle of Relativity, if momentum is conserved in one inertial frame of reference, then it must be conserved in all inertial frames. At the end of this section, it is shown that this requirement leads to a definition of the **relativistic momentum vector p**:

$$\mathbf{p} = \frac{m\mathbf{v}}{\sqrt{1 - v^2/c^2}} = \gamma m\mathbf{v}, \tag{4.44}$$

where $\gamma$ is the Lorentz factor defined by Eq. (4.20). *Warning:* Some authors prefer to separate the "$m$" and the "$\mathbf{v}$" in this formula by defining a "relativistic mass," $m/\sqrt{1 - v^2/c^2}$. There is no compelling reason for this separation, and it can be misleading. In this text, the mass $m$ of a particle is taken to be the same value in *all* inertial reference frames; it is **invariant** under a Lorentz transformation, and so there is no reason to qualify the term as a "rest mass." Thus the mass of a moving particle *does not* increase with increasing speed, although its *momentum* approaches infinity as $v \to c$. Also note that the "$v$" in the denominator is the magnitude of the particle's velocity relative to the observer, *not* the relative velocity $u$ between two arbitrary frames of reference.

### The Derivation of $E = mc^2$

Using Eq. (4.44) and the relation between kinetic energy and work from Section 2.2, we can derive an expression for the relativistic kinetic energy. The starting point is Newton's second law, $\mathbf{F} = d\mathbf{p}/dt$, applied to a particle of mass $m$ that is initially at rest.[8] Consider a force of magnitude $F$ that acts on the particle in the $x$-direction. The particle's final kinetic energy $K$ equals the total work done by the force on the particle as it travels from its initial

---

[8]It is left as an exercise to show that $\mathbf{F} = m\mathbf{a}$ is *not correct*, since at relativistic speeds the force and the acceleration need *not* be in the same direction!

position $x_i$ to its final position $x_f$:

$$K = \int_{x_i}^{x_f} F\,dx = \int_{x_i}^{x_f} \frac{dp}{dt}\,dx = \int_{p_i}^{p_f} \frac{dx}{dt}\,dp = \int_{p_i}^{p_f} v\,dp,$$

where $p_i$ and $p_f$ are the initial and final momenta of the particle, respectively. Integrating the last expression by parts and using the initial condition $p_i = 0$ give

$$K = p_f v_f - \int_0^{v_f} p\,dv$$

$$= \frac{m v_f^2}{\sqrt{1 - v_f^2/c^2}} - \int_0^{v_f} \frac{mv}{\sqrt{1 - v^2/c^2}}\,dv$$

$$= \frac{m v_f^2}{\sqrt{1 - v_f^2/c^2}} + mc^2\left(\sqrt{1 - v_f^2/c^2} - 1\right).$$

If we drop the $f$ subscript, the expression for the **relativistic kinetic energy** becomes

$$K = mc^2\left(\frac{1}{\sqrt{1 - v^2/c^2}} - 1\right) = mc^2(\gamma - 1). \qquad (4.45)$$

Although it is not apparent that this formula for the kinetic energy reduces to either of the familiar forms $K = \frac{1}{2}mv^2$ or $K = p^2/2m$ in the low-speed Newtonian limit, both forms must be true if Eq. (4.45) is to be correct. The proofs will be left as exercises.

The right-hand side of this expression for the kinetic energy consists of the difference between two energy terms. The first is identified as the **total relativistic energy** $E$,

$$E = \frac{mc^2}{\sqrt{1 - v^2/c^2}} = \gamma mc^2. \qquad (4.46)$$

The second term is an energy that does not depend on the speed of the particle; the particle has this energy even when it is at rest. The term $mc^2$ is called the **rest energy** of the particle:

$$E_{\text{rest}} = mc^2. \qquad (4.47)$$

The particle's kinetic energy is its total energy minus its rest energy. When the energy of a particle is given as (for example) 40 MeV, the implicit meaning is that the particle's *kinetic energy* is 40 MeV; the rest energy is not included. Finally, there is a very useful expression relating a particle's total energy $E$, the magnitude of its momentum $p$, and its rest energy $mc^2$. It states that

$$E^2 = p^2 c^2 + m^2 c^4. \qquad (4.48)$$

As we will discuss in Section 5.2, this equation is valid even for particles that have no mass, such as photons.

For a *system* of $n$ particles, the total energy $E_{\text{sys}}$ of the system is the sum of the total energies $E_i$ of the individual particles: $E_{\text{sys}} = \sum_{i=1}^{n} E_i$. Similarly, the vector momentum $\mathbf{p}_{\text{sys}}$ of the system is the sum of the momenta $\mathbf{p}_i$ of the individual particles: $\mathbf{p}_{\text{sys}} = \sum_{i=1}^{n} \mathbf{p}_i$. If the momentum of the system of particles is conserved, then the total energy is also conserved, *even for inelastic collisions* in which the kinetic energy of the system, $K_{\text{sys}} = \sum_{i=1}^{n} K_i$, is reduced. The kinetic energy lost in the inelastic collisions goes into increasing the rest energy, and hence the mass, of the particles. This increase in rest energy allows the total energy of the system to be conserved. Mass and energy are two sides of the same coin; one can be transformed into the other.

---

**Example 4.4.1.**   In a one-dimensional completely inelastic collision, two identical particles of mass $m$ and speed $v$ approach each other, collide head-on, and merge to form a single particle of mass $M$. The initial energy of the system of particles is

$$E_{\text{sys},i} = \frac{2mc^2}{\sqrt{1 - v^2/c^2}}.$$

Since the initial momenta of the particles are equal in magnitude and opposite in direction, the momentum of the system $\mathbf{p}_{\text{sys}} = 0$ before and after the collision. Thus after the collision, the particle is at rest and its final energy is

$$E_{\text{sys},f} = Mc^2.$$

Equating the initial and final energies of the system shows that the mass $M$ of the conglomerate particle is

$$M = \frac{2m}{\sqrt{1 - v^2/c^2}}.$$

Thus the particle mass has increased by an amount

$$\Delta m = M - 2m = \frac{2m}{\sqrt{1 - v^2/c^2}} - 2m = 2m \left( \frac{1}{\sqrt{1 - v^2/c^2}} - 1 \right).$$

The origin of this mass increase may be found by comparing the initial and final values of the kinetic energy. The initial kinetic energy of the system is

$$K_{\text{sys},i} = 2mc^2 \left( \frac{1}{\sqrt{1 - v^2/c^2}} - 1 \right)$$

and the final kinetic energy $K_{\text{sys},f} = 0$. Dividing the kinetic energy lost in this inelastic collision by $c^2$ equals the particle mass increase, $\Delta m$.

**The Derivation of Relativistic Momentum (Eq. 4.44)**

To justify Eq. (4.44) for the relativistic momentum, we will consider a glancing elastic collision between two identical particles of mass $m$. This collision will be observed from three carefully chosen inertial reference frames, as shown in Fig. 4.11. When measured in an inertial reference frame $S''$, the two particles $A$ and $B$ have velocities and momenta that are equal in magnitude and opposite in direction, both before and after the collision. As a result, the total momentum must be zero both before and after the collision; momentum is conserved. This collision can also be measured from two other reference frames, $S$ and $S'$. From Fig. 4.11, if $S$ moves in the negative $x''$-direction with a velocity equal to the $x''$-component of particle $A$ in $S''$, then as measured from frame $S$, the velocity of particle $A$ has only a $y$-component. Similarly, if $S'$ moves in the positive $x''$-direction with a velocity equal to the $x''$-component of particle $B$ in $S''$, then as measured from frame $S'$, the velocity of particle $B$ has only a $y$-component. Actually, the figures for frames $S$ and $S'$ would be *identical* if the figures for one of these frames were rotated by 180° and the $A$ and $B$ labels were reversed. This means that the change in the $y$-component of particle $A$'s momentum as measured in frame $S$ is the same as the change in the $y'$-component of particle $B$'s momentum as measured in the frame $S'$, except for a change in sign (due to the 180° rotation): $\Delta p_{A,y} = -\Delta p'_{B,y}$. On the other hand, momentum must be conserved in frames $S$ and $S'$, just as it is in frame $S''$. This means that, measured in frame $S'$, the sum of the changes in the $y'$-components of particle $A$'s and $B$'s momenta must be zero: $\Delta p'_{A,y} + \Delta p'_{B,y} = 0$. Combining these results gives

$$\Delta p'_{A,y} = \Delta p_{A,y}. \tag{4.49}$$

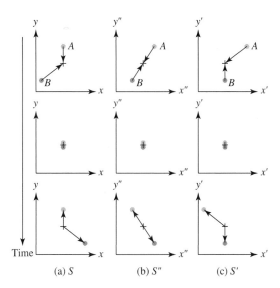

(a) $S$      (b) $S''$      (c) $S'$

**FIGURE 4.11** An elastic collision measured in frames (a) $S$, (b) $S''$, and (c) $S'$. As observed from frame $S''$, frame $S$ moves in the negative $x''$-direction, along with particle $A$, and frame $S'$ moves in the positive $x''$-direction, along with particle $B$. For each reference frame, a vertical sequence of three figures shows the situation before (top), during, and after the collision.

So far, the argument has been independent of a specific formula for the relativistic momentum vector **p**. Let's assume that the relativistic momentum vector has the form **p** $= fm$**v**, where $f$ is a relativistic factor that depends on the *magnitude* of the particle's velocity, but not its direction. As the particle's speed $v \to 0$, it is required that the factor $f \to 1$ to obtain agreement with the Newtonian result.[9]

A second assumption allows the relativistic factor $f$ to be determined: The $y$- and $y'$-components of each particle's velocity are chosen to be arbitrarily small compared to the speed of light. Thus the $y$- and $y'$-components of particle $A$'s velocity in frames $S$ and $S'$ are extremely small, and the $x'$-component of particle $A$'s velocity in frame $S'$ is taken to be relativistic. Since

$$v'_A = \sqrt{v'^2_{A,x} + v'^2_{A,y}} \approx c$$

in frame $S'$, the relativistic factor $f'$ for particle $A$ in frame $S'$ is not equal to 1, whereas in frame $S$, $f$ is arbitrarily close to unity. If $v_{A,y}$ is the final $y$-component of particle $A$'s velocity, and similarly for $v'_{A,y}$, then Eq. (4.49) becomes

$$2f' m v'_{A,y} = 2m v_{A,y}. \tag{4.50}$$

The relative velocity $u$ of frames $S$ and $S'$ is needed to relate $v'_{A,y}$ and $v_{A,y}$ using Eq. (4.41). Because $v_{A,x} = 0$ in frame $S$, Eq. (4.40) shows that $u = -v'_{A,x}$; that is, the relative velocity $u$ of frame $S'$ relative to frame $S$ is just the negative of the $x'$-component of particle $A$'s velocity in frame $S'$. Furthermore, because the $y'$-component of particle $A$'s velocity is arbitrarily small, we can set $v'_{A,x} = v'_A$, the magnitude of particle $A$'s velocity as measured in frame $S'$, and so use $u = -v'_A$. Inserting this into Eq. (4.41) with $v_{A,x} = 0$ gives

$$v'_{A,y} = v_{A,y}\sqrt{1 - v'^2_A/c^2}.$$

Finally, inserting this relation between $v'_{A,y}$ and $v_{A,y}$ into Eq. (4.50) and canceling terms reveals the relativistic factor $f$ to be

$$f = \frac{1}{\sqrt{1 - v'^2_A/c^2}},$$

as measured in frame $S'$. Dropping the prime superscript and the $A$ subscript (which merely identify the reference frame and particle involved) gives

$$f = \frac{1}{\sqrt{1 - v^2/c^2}}.$$

The formula for the **relativistic momentum vector p** $= fm$**v** is thus

$$\mathbf{p} = \frac{m\mathbf{v}}{\sqrt{1 - v^2/c^2}} = \gamma m\mathbf{v}.$$

[9]There is no requirement that relativistic formulas appear similar to their low-speed Newtonian counterparts (cf. Eq. 4.45). However, this simple argument produces the correct result.

## SUGGESTED READING

### General

French, A. P. (ed.), *Einstein: A Centenary Volume*, Harvard University Press, Cambridge, MA, 1979.

Gardner, Martin, *The Relativity Explosion*, Vintage Books, New York, 1976.

### Technical

Becker, Robert H. et al., "Evidence for Reionization at $Z \sim 6$: Detection of a Gunn–Peterson Trough in a $Z = 6.28$ Quasar," *The Astronomical Journal*, preprint, 2001.

Bregman, Joel N. et al., "Multifrequency Observations of the Optically Violent Variable Quasar 3C 446," *The Astrophysical Journal*, *331*, 746, 1988.

Frisch, David H., and Smith, James H., "Measurement of the Relativistic Time Dilation Using $\mu$-Mesons," *American Journal of Physics*, *31*, 342, 1963.

Hafele, J. C., and Keating, Richard E., "Around-the-World Atomic Clocks: Predicted Relativistic Time Gains," *Science*, *177*, 166, 1972a.

Hafele, J. C., and Keating, Richard E., "Around-the-World Atomic Clocks: Observed Relativistic Time Gains," *Science*, *177*, 168, 1972b.

McCarthy, Patrick J. et al., "Serendipitous Discovery of a Redshift 4.4 QSO," *The Astrophysical Journal Letters*, *328*, L29, 1988.

Sloan Digital Sky Survey, `http://www.sdss.org`

Resnick, Robert, and Halliday, David, *Basic Concepts in Relativity and Early Quantum Theory*, Second Edition, John Wiley and Sons, New York, 1985.

Taylor, Edwin F., and Wheeler, John A., *Spacetime Physics*, Second Edition, W. H. Freeman, San Francisco, 1992.

## PROBLEMS

**4.1** Use Eqs. (4.14) and (4.15) to derive the Lorentz transformation equations from Eqs. (4.10–4.13).

**4.2** Because there is no such thing as absolute simultaneity, two observers in relative motion may disagree on which of two events $A$ and $B$ occurred first. Suppose, however, that an observer in reference frame $S$ measures that event $A$ occurred first and *caused* event $B$. For example, event $A$ might be pushing a light switch, and event $B$ might be a light bulb turning on. Prove that an observer in another frame $S'$ cannot measure event $B$ (the effect) occurring before event $A$ (the cause). The temporal order of cause and effect is preserved by the Lorentz transformation equations. *Hint:* For event $A$ to cause event $B$, information must have traveled from $A$ to $B$, and the fastest that *anything* can travel is the speed of light.

**4.3** Consider the special *light clock* shown in Fig. 4.12. The light clock is at rest in frame $S'$ and consists of two perfectly reflecting mirrors separated by a vertical distance $d$. As measured by an observer in frame $S'$, a light pulse bounces vertically back and forth between the two mirrors; the time interval between the pulse leaving and subsequently returning to the bottom mirror is $\Delta t'$. However, an observer in frame $S$ sees a moving clock and determines that the time interval

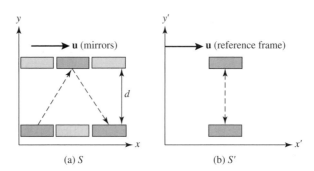

**FIGURE 4.12**   (a) A light clock that is moving in frame $S$, and (b) at rest in frame $S'$.

between the light pulse leaving and returning to the bottom mirror is $\Delta t$. Use the fact that both observers must measure that the light pulse moves with speed $c$, plus some simple geometry, to derive the time-dilation equation (4.27).

**4.4** A rod moving relative to an observer is measured to have its length $L_{\mathrm{moving}}$ contracted to one-half of its length when measured at rest. Find the value of $u/c$ for the rod's rest frame relative to the observer's frame of reference.

**4.5** An observer $P$ stands on a train station platform as a high-speed train passes by at $u/c = 0.8$. The observer $P$, who measures the platform to be 60 m long, notices that the front and back ends of the train line up exactly with the ends of the platform at the same time.

   **(a)** How long does it take the train to pass $P$ as he stands on the platform, as measured by his watch?

   **(b)** According to a rider $T$ on the train, how long is the train?

   **(c)** According to a rider $T$ on the train, what is the length of the train station platform?

   **(d)** According to a rider $T$ on the train, how much time does it take for the train to pass observer $P$ standing on the train station platform?

   **(e)** According to a rider $T$ on the train, the ends of the train will *not* simultaneously line up with the ends of the platform. What time interval does $T$ measure between when the front end of the train lines up with the front end of the platform, and when the back end of the train lines up with the back end of the platform?

**4.6** An astronaut in a starship travels to $\alpha$ Centauri, a distance of approximately 4 ly as measured from Earth, at a speed of $u/c = 0.8$.

   **(a)** How long does the trip to $\alpha$ Centauri take, as measured by a clock on Earth?

   **(b)** How long does the trip to $\alpha$ Centauri take, as measured by the starship pilot?

   **(c)** What is the distance between Earth and $\alpha$ Centauri, as measured by the starship pilot?

   **(d)** A radio signal is sent from Earth to the starship every 6 months, as measured by a clock on Earth. What is the time interval between reception of one of these signals and reception of the next signal aboard the starship?

   **(e)** A radio signal is sent from the starship to Earth every 6 months, as measured by a clock aboard the starship. What is the time interval between reception of one of these signals and reception of the next signal on Earth?

   **(f)** If the wavelength of the radio signal sent from Earth is $\lambda = 15$ cm, to what wavelength must the starship's receiver be tuned?

**4.7** Upon reaching $\alpha$ Centauri, the starship in Problem 4.6 immediately reverses direction and travels back to Earth at a speed of $u/c = 0.8$. (Assume that the turnaround itself takes *zero* time.) Both Earth and the starship continue to emit radio signals at 6-month intervals, as measured by their respective clocks. Make a table for the entire trip showing at what times Earth receives the signals from the starship. Do the same for the times when the starship receives the signals from Earth. Thus an Earth observer and the starship pilot will agree that the pilot has aged 4 years less than the Earth observer during the round-trip voyage to $\alpha$ Centauri.

**4.8** In its rest frame, quasar Q2203+29 produces a hydrogen emission line of wavelength 121.6 nm. Astronomers on Earth measure a wavelength of 656.8 nm for this line. Determine the redshift parameter and the apparent speed of recession for this quasar. (For more information about this quasar, see McCarthy et al. 1988.)

**4.9** Quasar 3C 446 is violently variable; its luminosity at optical wavelengths has been observed to change by a factor of 40 in as little as 10 days. Using the redshift parameter $z = 1.404$ measured for 3C 446, determine the time for the luminosity variation as measured in the quasar's rest frame. (For more details, see Bregman et al. 1988.)

**4.10** Use the Lorentz transformation equations (4.16–4.19) to derive the velocity transformation equations (4.40–4.42).

**4.11** The *spacetime interval*, $\Delta s$, between two events with coordinates

$$(x_1, y_1, z_1, t_1) \qquad \text{and} \qquad (x_2, y_2, z_2, t_2)$$

is defined by

$$(\Delta s)^2 \equiv (c\Delta t)^2 - (\Delta x)^2 - (\Delta y)^2 - (\Delta z)^2.$$

**(a)** Use the Lorentz transformation equations (4.16–4.19) to show that $\Delta s$ has the same value in all reference frames. The spacetime interval is said to be *invariant* under a Lorentz transformation.

**(b)** If $(\Delta s)^2 > 0$, then the interval is *timelike*. Show that in this case,

$$\Delta \tau \equiv \frac{\Delta s}{c}$$

is the proper time between the two events. Assuming that $t_1 < t_2$, could the first event possibly have caused the second event?

**(c)** If $(\Delta s)^2 = 0$, then the interval is *lightlike* or *null*. Show that only light could have traveled between the two events. Could the first event possibly have caused the second event?

**(d)** If $(\Delta s)^2 < 0$, then the interval is *spacelike*. What is the physical significance of $\sqrt{-(\Delta s)^2}$? Could the first event possibly have caused the second event?

The concept of a spacetime interval will play a key role in the discussion of general relativity in Chapter 17.

**4.12** General expressions for the components of a light ray's velocity as measured in reference frame $S$ are

$$v_x = c \sin \theta \cos \phi$$

$$v_y = c \sin \theta \sin \phi$$

$$v_z = c \cos \theta,$$

where $\theta$ and $\phi$ are the angular coordinates in a spherical coordinate system.

**(a)** Show that

$$v = \sqrt{v_x^2 + v_y^2 + v_z^2} = c.$$

**(b)** Use the velocity transformation equations to show that, as measured in reference frame $S'$,

$$v' = \sqrt{v_x'^2 + v_y'^2 + v_z'^2} = c,$$

and so confirm that the speed of light has the constant value $c$ in all frames of reference.

**4.13** Starship $A$ moves away from Earth with a speed of $v_A/c = 0.8$. Starship $B$ moves away from Earth in the opposite direction with a speed of $v_B/c = 0.6$. What is the speed of starship $A$ as measured by starship $B$? What is the speed of starship $B$ as measured by starship $A$?

**4.14** Use Newton's second law, $\mathbf{F} = d\mathbf{p}/dt$, and the formula for relativistic momentum, Eq. (4.44), to show that the acceleration vector $\mathbf{a} = d\mathbf{v}/dt$ produced by a force $\mathbf{F}$ acting on a particle of mass $m$ is

$$\mathbf{a} = \frac{\mathbf{F}}{\gamma m} - \frac{\mathbf{v}}{\gamma mc^2}\,(\mathbf{F} \cdot \mathbf{v}),$$

where $\mathbf{F} \cdot \mathbf{v}$ is the vector dot product between the force $\mathbf{F}$ and the particle velocity $\mathbf{v}$. Thus the acceleration depends on the particle's velocity and is not in general in the same direction as the force.

**4.15** Suppose a constant force of magnitude $F$ acts on a particle of mass $m$ initially at rest.

**(a)** Integrate the formula for the acceleration found in Problem 4.14 to show that the speed of the particle after time $t$ is given by

$$\frac{v}{c} = \frac{(F/m)t}{\sqrt{(F/m)^2 t^2 + c^2}}.$$

**(b)** Rearrange this equation to express the time $t$ as a function of $v/c$. If the particle's initial acceleration at time $t = 0$ is $a = g = 9.80$ m s$^{-2}$, how much time is required for the particle to reach a speed of $v/c = 0.9$? $v/c = 0.99$? $v/c = 0.999$? $v/c = 0.9999$? $v/c = 1$?

**4.16** Find the value of $v/c$ when a particle's kinetic energy equals its rest energy.

**4.17** Prove that in the low-speed Newtonian limit of $v/c \ll 1$, Eq. (4.45) does reduce to the familiar form $K = \frac{1}{2}mv^2$.

**4.18** Show that the relativistic kinetic energy of a particle can be written as

$$K = \frac{p^2}{(1 + \gamma)m},$$

where $p$ is the magnitude of the particle's relativistic momentum. This demonstrates that in the low-speed Newtonian limit of $v/c \ll 1$, $K = p^2/2m$ (as expected).

**4.19** Derive Eq. (4.48).

# 5

# The Interaction of Light and Matter

## 5.1 ■ SPECTRAL LINES

In 1835 a French philosopher, Auguste Comte (1798–1857), considered the limits of human knowledge. In his book *Positive Philosophy*, Comte wrote of the stars, "We see how we may determine their forms, their distances, their bulk, their motions, but we can never know anything of their chemical or mineralogical structure." Thirty-three years earlier, however, William Wollaston (1766–1828), like Newton before him, passed sunlight through a prism to produce a rainbow-like spectrum. He discovered that a number of dark **spectral lines** were superimposed on the continuous spectrum where the Sun's light had been absorbed at certain discrete wavelengths. By 1814, the German optician Joseph von Fraunhofer (1787–1826) had cataloged 475 of these dark lines (today called **Fraunhofer lines**) in the solar spectrum. While measuring the wavelengths of these lines, Fraunhofer made the first observation capable of proving Comte wrong. Fraunhofer determined that the wavelength of one prominent dark line in the Sun's spectrum corresponds to the wavelength of the yellow light emitted when salt is sprinkled in a flame. The new science of *spectroscopy* was born with the identification of this sodium line.

### Kirchhoff's Laws

The foundations of spectroscopy were established by Robert Bunsen (1811–1899), a German chemist, and by Gustav Kirchhoff (1824–1887), a Prussian theoretical physicist. Bunsen's burner produced a colorless flame that was ideal for studying the spectra of heated substances. He and Kirchhoff then designed a *spectroscope* that passed the light of a flame spectrum through a prism to be analyzed. The wavelengths of light absorbed and emitted by an element were found to be the same; Kirchhoff determined that 70 dark lines in the solar spectrum correspond to 70 bright lines emitted by iron vapor. In 1860 Kirchhoff and Bunsen published their classic work *Chemical Analysis by Spectral Observations*, in which they developed the idea that every element produces its own pattern of spectral lines and thus may be identified by its unique spectral line "fingerprint." Kirchhoff summarized the production of spectral lines in three laws, which are now known as **Kirchhoff's laws**:

- A hot, dense gas or hot solid object produces a continuous spectrum with no dark spectral lines.[1]

- A hot, diffuse gas produces bright spectral lines (**emission lines**).

- A cool, diffuse gas in front of a source of a continuous spectrum produces dark spectral lines (**absorption lines**) in the continuous spectrum.

### Applications of Stellar Spectra Data

An immediate application of these results was the identification of elements found in the Sun and other stars. A new element previously unknown on Earth, *helium*,[2] was discovered spectroscopically on the Sun in 1868; it was not found on Earth until 1895. Figure 5.1 shows the visible portion of the solar spectrum, and Table 5.1 lists some of the elements responsible for producing the dark absorption lines.

Another rich line of investigation was pursued by measuring the Doppler shifts of spectral lines. For individual stars, $v_r \ll c$, and so the *low-speed approximation* of Eq. (4.39),

$$\frac{\lambda_{\text{obs}} - \lambda_{\text{rest}}}{\lambda_{\text{rest}}} = \frac{\Delta\lambda}{\lambda_{\text{rest}}} = \frac{v_r}{c}, \tag{5.1}$$

can be utilized to determine their radial velocities. By 1887 the radial velocities of Sirius, Procyon, Rigel, and Arcturus had been measured with an accuracy of a few kilometers per second.

---

**Example 5.1.1.**   The rest wavelength $\lambda_{\text{rest}}$ for an important spectral line of hydrogen (known as H$\alpha$) is 656.281 nm when measured in air. However, the wavelength of the H$\alpha$ absorption line in the spectrum of the star Vega in the constellation Lyra is measured to be 656.251 nm at a ground-based telescope. Equation (5.1) shows that the radial velocity of Vega is

$$v_r = \frac{c\,(\lambda_{\text{obs}} - \lambda_{\text{rest}})}{\lambda_{\text{rest}}} = -13.9 \text{ km s}^{-1};$$

the minus sign means that Vega is approaching the Sun. Recall from Section 1.3, however, that stars also have a *proper motion*, $\mu$, perpendicular to the line of sight. Vega's angular position in the sky changes by $\mu = 0.35077''$ yr$^{-1}$. At a distance of $r = 7.76$ pc, this proper motion is related to the star's transverse velocity, $v_\theta$, by Eq. (1.5). Expressing $r$ in meters and $\mu$ in radians per second results in

$$v_\theta = r\mu = 12.9 \text{ km s}^{-1}.$$

---

[1] In the first of Kirchhoff's laws, "hot" actually means any temperature above 0 K. However, according to Wien's displacement law (Eq. 3.19), a temperature of several thousand degrees K is required for $\lambda_{\text{max}}$ to fall in the visible portion of the electromagnetic spectrum. As will be discussed in Chapter 9, it is the opacity or *optical depth* of the gas that is responsible for the continuous blackbody spectrum.

[2] The name *helium* comes from *Helios*, a Greek Sun god.

This transverse velocity is comparable to Vega's radial velocity. Vega's speed through space relative to the Sun is thus

$$v = \sqrt{v_r^2 + v_\theta^2} = 19.0 \text{ km s}^{-1}.$$

The average speed of stars in the solar neighborhood is about 25 km s$^{-1}$. In reality, the measurement of a star's radial velocity is complicated by the 29.8 km s$^{-1}$ motion of Earth around the Sun, which causes the observed wavelength $\lambda_{\text{obs}}$ of a spectral line to vary sinusoidally over the course of a year. This effect of Earth's speed may be easily compensated for by subtracting the component of Earth's orbital velocity along the line of sight from the star's measured radial velocity.

### Spectrographs

Modern methods can measure radial velocities with an accuracy of better than $\pm 3$ m s$^{-1}$! Today astronomers use *spectrographs* to measure the spectra of stars and galaxies; see Fig. 5.2.[3] After passing through a narrow slit, the starlight is collimated by a mirror and directed onto a *diffraction grating*. A diffraction grating is a piece of glass onto which narrow, closely spaced lines have been evenly ruled (typically several thousand lines per millimeter); the grating may be made to transmit the light (a *transmission grating*) or reflect the light (a *reflection grating*). In either case, the grating acts like a long series of neighboring double slits. Different wavelengths of light have their maxima occurring at different angles $\theta$ given by Eq. (3.11):

$$d \sin \theta = n\lambda \qquad (n = 0, 1, 2, \ldots),$$

where $d$ is the distance between adjacent lines of the grating, $n$ is the order of the spectrum, and $\theta$ is measured from the line normal (or perpendicular) to the grating. ($n = 0$ corresponds to $\theta = 0$ for all wavelengths, so the light is not dispersed into a spectrum in this case.) The spectrum is then focused onto a photographic plate or electronic detector for recording.

The ability of a spectrograph to resolve two closely spaced wavelengths separated by an amount $\Delta\lambda$ depends on the order of the spectrum, $n$, and the total number of lines of the grating that are illuminated, $N$. The smallest difference in wavelength that the grating can resolve is

$$\Delta\lambda = \frac{\lambda}{nN}, \tag{5.2}$$

where $\lambda$ is either of the closely spaced wavelengths being measured. The ratio $\lambda/\Delta\lambda$ is the **resolving power** of the grating.[4]

---

[3]As we will discuss in Chapter 7, measuring the radial velocities of stars in binary star systems allows the *masses* of the stars to be determined. The same methods have now been used to detect numerous extrasolar planets.

[4]In some cases, the resolving power of a spectrograph may be determined by other factors—for example, the slit width.

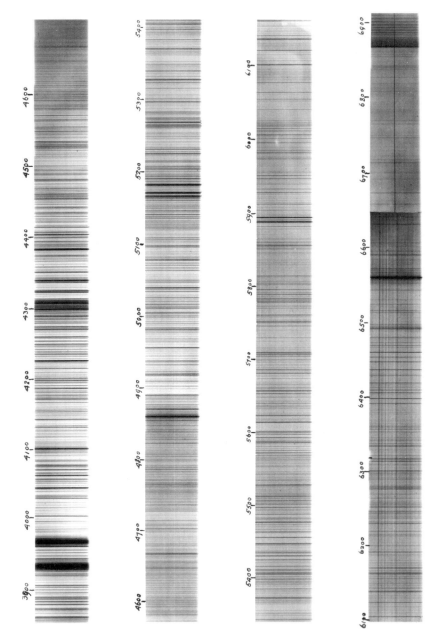

**FIGURE 5.1**  The solar spectrum with Fraunhofer lines. Note that the wavelengths are expressed in angstroms (1 Å = 0.1 nm), a commonly used wavelength unit in astronomy. Modern depictions of spectra are typically shown as plots of flux as a function of wavelength; see, for instance, Figs. 8.4 and 8.5. (Courtesy of The Observatories of the Carnegie Institution of Washington.)

**TABLE 5.1**   Wavelengths of some of the stronger Fraunhofer lines measured in air near sea level. The atomic notation is explained in Section 8.1, and the equivalent width of a spectral line is defined in Section 9.5. The difference in wavelengths of spectral lines when measured in air versus in vacuum are discussed in Example 5.3.1. (Data from Lang, *Astrophysical Formulae*, Third Edition, Springer, New York, 1999.)

| Wavelength (nm) | Name | Atom | Equivalent Width (nm) |
|---|---|---|---|
| 385.992 | | Fe I | 0.155 |
| 388.905 | | $H_8$ | 0.235 |
| 393.368 | K | Ca II | 2.025 |
| 396.849 | H | Ca II | 1.547 |
| 404.582 | | Fe I | 0.117 |
| 410.175 | h, H$\delta$ | H I | 0.313 |
| 422.674 | g | Ca I | 0.148 |
| 434.048 | G′, H$\gamma$ | H I | 0.286 |
| 438.356 | d | Fe I | 0.101 |
| 486.134 | F, H$\beta$ | H I | 0.368 |
| 516.733 | $b_4$ | Mg I | 0.065 |
| 517.270 | $b_2$ | Mg I | 0.126 |
| 518.362 | $b_1$ | Mg I | 0.158 |
| 588.997 | $D_2$ | Na I | 0.075 |
| 589.594 | $D_1$ | Na I | 0.056 |
| 656.281 | C, H$\alpha$ | H I | 0.402 |

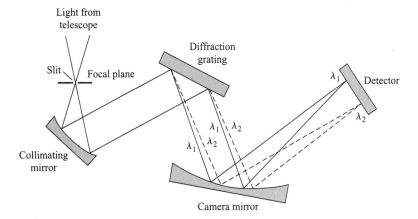

**FIGURE 5.2**   Spectrograph.

Astronomers recognized the great potential for uncovering the secrets of the stars in the empirical rules that had been obtained for the spectrum of light: Wien's law, the Stefan–Boltzmann equation, Kirchhoff's laws, and the new science of spectroscopy. By 1880 Gustav Wiedemann (1826–1899) found that a detailed investigation of the Fraunhofer lines could reveal the temperature, pressure, and density of the layer of the Sun's atmosphere that produces the lines. The splitting of spectral lines by a magnetic field was discovered by Pieter Zeeman (1865–1943) of the Netherlands in 1897, raising the possibility of measuring stellar magnetic fields. But a serious problem blocked further progress: However impressive, these results lacked the solid theoretical foundation required for the interpretation of stellar spectra. For example, the absorption lines produced by hydrogen are much stronger for Vega than for the Sun. Does this mean that Vega's composition contains significantly more hydrogen than the Sun's? The answer is no, but how can this information be gleaned from the dark absorption lines of a stellar spectrum recorded on a photographic plate? The answer required a new understanding of the nature of light itself.

## 5.2 ■ PHOTONS

Despite Heinrich Hertz's absolute certainty in the wave nature of light, the solution to the riddle of the continuous spectrum of blackbody radiation led to a complementary description, and ultimately to new conceptions of matter and energy. Planck's constant $h$ (see Section 3.5) is the basis of the modern description of matter and energy known as **quantum mechanics**. Today $h$ is recognized as a fundamental constant of nature, like the speed of light $c$ and the universal gravitational constant $G$. Although Planck himself was uncomfortable with the implications of his discovery of energy quantization, quantum theory was to develop into what is today a spectacularly successful description of the physical world. The next step forward was taken by Einstein, who convincingly demonstrated the reality of Planck's quantum bundles of energy.

### The Photoelectric Effect

When light shines on a metal surface, electrons are ejected from the surface, a result called the **photoelectric effect**. The electrons are emitted with a range of energies, but those originating closest to the surface have the maximum kinetic energy, $K_{max}$. A surprising feature of the photoelectric effect is that the value of $K_{max}$ does *not* depend on the brightness of the light shining on the metal. Increasing the intensity of a monochromatic light source will eject *more* electrons but will not increase their maximum kinetic energy. Instead, $K_{max}$ varies with the *frequency* of the light illuminating the metal surface. In fact, each metal has a characteristic *cutoff frequency* $\nu_c$ and a corresponding *cutoff wavelength* $\lambda_c = c/\nu_c$; electrons will be emitted only if the frequency $\nu$ of the light satisfies $\nu > \nu_c$ (or the wavelength satisfies $\lambda < \lambda_c$). This puzzling frequency dependence is nowhere to be found in Maxwell's classic description of electromagnetic waves. Equation (3.12) for the Poynting vector admits no role for the frequency in describing the energy carried by a light wave.

Einstein's bold solution was to take seriously Planck's assumption of the quantized energy of electromagnetic waves. According to Einstein's explanation of the photoelectric effect, the light striking the metal surface consists of a stream of massless particles called

**photons.**[5] The energy of a single photon of frequency $\nu$ and wavelength $\lambda$ is just the energy of Planck's quantum of energy:

$$E_{\text{photon}} = h\nu = \frac{hc}{\lambda}. \tag{5.3}$$

---

**Example 5.2.1.**   The energy of a single photon of visible light is small by everyday standards. For red light of wavelength $\lambda = 700$ nm, the energy of a single photon is

$$E_{\text{photon}} = \frac{hc}{\lambda} \simeq \frac{1240 \text{ eV nm}}{700 \text{ nm}} = 1.77 \text{ eV}.$$

Here, the product $hc$ has been expressed in the convenient units of (electron volts) $\times$ (nanometers); recall that $1 \text{ eV} = 1.602 \times 10^{-19}$ J. For a single photon of blue light with $\lambda = 400$ nm,

$$E_{\text{photon}} = \frac{hc}{\lambda} \simeq \frac{1240 \text{ eV nm}}{400 \text{ nm}} = 3.10 \text{ eV}.$$

How many visible photons ($\lambda = 500$ nm) are emitted each second by a 100-W light bulb (assuming that it is monochromatic)? The energy of each photon is

$$E_{\text{photon}} = \frac{hc}{\lambda} \simeq \frac{1240 \text{ eV nm}}{500 \text{ nm}} = 2.48 \text{ eV} = 3.97 \times 10^{-19} \text{ J}.$$

This means that the 100-W light bulb emits $2.52 \times 10^{20}$ photons per second. As this huge number illustrates, with so many photons nature does not appear "grainy." We see the world as a continuum of light, illuminated by a flood of photons.

---

Einstein reasoned that when a photon strikes the metal surface in the photoelectric effect, its energy may be absorbed by a single electron. The electron uses the photon's energy to overcome the binding energy of the metal and so escape from the surface. If the *minimum* binding energy of electrons in a metal (called the **work function** of the metal, usually a few eV) is $\phi$, then the maximum kinetic energy of the ejected electrons is

$$\boxed{K_{\text{max}} = E_{\text{photon}} - \phi = h\nu - \phi = \frac{hc}{\lambda} - \phi.} \tag{5.4}$$

Setting $K_{\text{max}} = 0$, the cutoff frequency and wavelength for a metal are seen to be $\nu_c = \phi/h$ and $\lambda_c = hc/\phi$, respectively.

The photoelectric effect established the reality of Planck's quanta. Albert Einstein was awarded the 1921 Nobel Prize, not for his theories of special and general relativity, but "for his services to theoretical physics, and especially for his discovery of the law of the

---

[5]Only a massless particle can move with the speed of light, since a massive particle would have infinite energy; see Eq. (4.45). The term *photon* was first used in 1926 by the physicist G. N. Lewis (1875–1946).

photoelectric effect."[6] Today astronomers take advantage of the quantum nature of light in various instruments and detectors, such as the CCDs (charge-coupled devices) that will be described in Section 6.2.

## The Compton Effect

In 1922, the American physicist Arthur Holly Compton (1892–1962) provided the most convincing evidence that light does in fact manifest its particle-like nature when interacting with matter. Compton measured the change in the wavelength of X-ray photons as they were scattered by free electrons. Because photons are massless particles that move at the speed of light, the relativistic energy equation, Eq. (4.48) (with mass $m = 0$ for photons), shows that the energy of a photon is related to its momentum $p$ by

$$E_{\text{photon}} = h\nu = \frac{hc}{\lambda} = pc. \tag{5.5}$$

Compton considered the "collision" between a photon and a free electron, initially at rest. As shown in Fig. 5.3, the electron is scattered in the direction $\phi$ and the photon is scattered by an angle $\theta$. Because the photon has lost energy to the electron, the wavelength of the photon has increased.

In this collision, both (relativistic) momentum and energy are conserved. It is left as an exercise to show that the final wavelength of the photon, $\lambda_f$, is greater than its initial

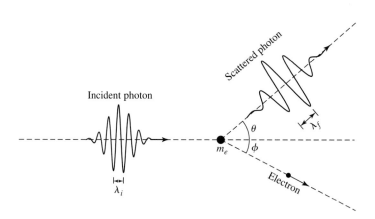

**FIGURE 5.3**    The Compton effect: The scattering of a photon by a free electron. $\theta$ and $\phi$ are the scattering angles of the photon and electron, respectively.

---

[6]Partly in recognition of his determination of an accurate value of Planck's constant $h$ using Eq. (5.4), the American physicist Robert A. Millikan (1868–1953) also received a Nobel Prize (1923) for his work on the photoelectric effect.

wavelength, $\lambda_i$, by an amount

$$\Delta\lambda = \lambda_f - \lambda_i = \frac{h}{m_e c}(1 - \cos\theta),\qquad(5.6)$$

where $m_e$ is the mass of the electron. Today, this change in wavelength is known as the **Compton effect**. The term $h/m_e c$ in Eq. (5.6), called the **Compton wavelength**, $\lambda_C$, is the characteristic change in the wavelength of the scattered photon and has the value $\lambda_C = 0.00243$ nm, 30 times smaller than the wavelength of the X-ray photons used by Compton. Compton's experimental verification of this formula provided convincing evidence that photons are indeed massless particles that nonetheless carry momentum, as described by Eq. (5.5). This is the physical basis for the force exerted by radiation upon matter, which we discussed in Section 3.3 in terms of radiation pressure.

## 5.3 ■ THE BOHR MODEL OF THE ATOM

The pioneering work of Planck, Einstein, and others at the beginning of the twentieth century revealed the **wave–particle duality** of light. Light exhibits its wave properties as it *propagates* through space, as demonstrated by its double-slit interference pattern. On the other hand, light manifests its particle nature when it *interacts* with matter, as in the photoelectric and Compton effects. Planck's formula describing the energy distribution of blackbody radiation explained many of the features of the continuous spectrum of light emitted by stars. But what physical process was responsible for the dark absorption lines scattered throughout the continuous spectrum of a star, or for the bright emission lines produced by a hot, diffuse gas in the laboratory?

### The Structure of the Atom

In the very last years of the nineteenth century, Joseph John Thomson (1856–1940) discovered the **electron** while working at Cambridge University's Cavendish Laboratory. Because bulk matter is electrically neutral, atoms were deduced to consist of negatively charged electrons and an equal positive charge of uncertain distribution. Ernest Rutherford (1871–1937) of New Zealand, working at England's University of Manchester, discovered in 1911 that an atom's positive charge was concentrated in a tiny, massive nucleus. Rutherford directed high-speed alpha particles (now known to be helium nuclei) onto thin metal foils. He was amazed to observe that a few of the alpha particles were bounced backward by the foils instead of plowing through them with only a slight deviation. Rutherford later wrote: "It was quite the most incredible event that has ever happened to me in my life. It was almost as incredible as if you fired a 15-inch shell at a piece of tissue paper and it came back and hit you." Such an event could occur only as the result of a single collision of the alpha particle with a minute, massive, positively charged nucleus. Rutherford calculated that the radius of the nucleus was 10,000 times smaller than the radius of the atom itself, showing that ordinary matter is mostly empty space! He established that an electrically neutral atom consists of $Z$ electrons (where $Z$ is an integer), with $Z$ positive elementary charges confined to the nucleus. Rutherford coined the term **proton** to refer to the nucleus of the hydrogen atom ($Z = 1$), 1836 times more massive than the electron. But how were these charges arranged?

### The Wavelengths of Hydrogen

The experimental data were abundant. The wavelengths of 14 spectral lines of hydrogen had been precisely determined. Those in the visible region of the electromagnetic spectrum are 656.3 nm (red, H$\alpha$), 486.1 nm (turquoise, H$\beta$), 434.0 nm (blue, H$\gamma$), and 410.2 nm (violet, H$\delta$). In 1885 a Swiss school teacher, Johann Balmer (1825–1898), had found, by trial and error, a formula to reproduce the wavelengths of these spectral lines of hydrogen, today called the **Balmer series** or **Balmer lines**:

$$\frac{1}{\lambda} = R_H \left( \frac{1}{4} - \frac{1}{n^2} \right), \tag{5.7}$$

where $n = 3, 4, 5, \ldots$, and $R_H = 1.09677583 \times 10^7 \pm 1.3 \text{ m}^{-1}$ is the experimentally determined Rydberg constant for hydrogen.[7] Balmer's formula was very accurate, to within a fraction of a percent. Inserting $n = 3$ gives the wavelength of the H$\alpha$ Balmer line, $n = 4$ gives H$\beta$, and so on. Furthermore, Balmer realized that since $2^2 = 4$, his formula could be generalized to

$$\frac{1}{\lambda} = R_H \left( \frac{1}{m^2} - \frac{1}{n^2} \right), \tag{5.8}$$

with $m < n$ (both integers). Many nonvisible spectral lines of hydrogen were found later, just as Balmer had predicted. Today, the lines corresponding to $m = 1$ are called *Lyman lines*. The Lyman series of lines is found in the ultraviolet region of the electromagnetic spectrum. Similarly, inserting $m = 3$ into Eq. (5.8) produces the wavelengths of the *Paschen* series of lines, which lie entirely in the infrared portion of the spectrum. The wavelengths of important selected hydrogen lines are given in Table 5.2.

Yet all of this was sheer numerology, with no foundation in the physics of the day. Physicists were frustrated by their inability to construct a model of even this simplest of atoms. A planetary model of the hydrogen atom, consisting of a central proton and one electron held together by their mutual electrical attraction, should have been most amenable to analysis. However, a model consisting of a single electron and proton moving around their common center of mass suffers from a basic instability. According to Maxwell's equations of electricity and magnetism, an accelerating electric charge emits electromagnetic radiation. The orbiting electron should thus lose energy by emitting light with a continuously increasing frequency (the orbital frequency) as it spirals down into the nucleus. This theoretical prediction of a continuous spectrum disagreed with the discrete emission lines actually observed. Even worse was the calculated timescale: The electron should plunge into the nucleus in only $10^{-8}$ s. Obviously, matter is stable over much longer periods of time!

---

[7]$R_H$ is named in honor of Johannes Rydberg (1854–1919), a Swedish spectroscopist.

**TABLE 5.2**   The wavelengths of selected hydrogen spectral lines in air. (Based on Cox, (ed.), *Allen's Astrophysical Quantities*, Fourth Edition, Springer, New York, 2000.)

| Series Name | Symbol | Transition | Wavelength (nm) |
|---|---|---|---|
| Lyman | Ly$\alpha$ | $2 \leftrightarrow 1$ | 121.567 |
| | Ly$\beta$ | $3 \leftrightarrow 1$ | 102.572 |
| | Ly$\gamma$ | $4 \leftrightarrow 1$ | 97.254 |
| | Ly$_{\text{limit}}$ | $\infty \leftrightarrow 1$ | 91.18 |
| Balmer | H$\alpha$ | $3 \leftrightarrow 2$ | 656.281 |
| | H$\beta$ | $4 \leftrightarrow 2$ | 486.134 |
| | H$\gamma$ | $5 \leftrightarrow 2$ | 434.048 |
| | H$\delta$ | $6 \leftrightarrow 2$ | 410.175 |
| | H$\epsilon$ | $7 \leftrightarrow 2$ | 397.007 |
| | H$_8$ | $8 \leftrightarrow 2$ | 388.905 |
| | H$_{\text{limit}}$ | $\infty \leftrightarrow 3$ | 364.6 |
| Paschen | Pa$\alpha$ | $4 \leftrightarrow 3$ | 1875.10 |
| | Pa$\beta$ | $5 \leftrightarrow 3$ | 1281.81 |
| | Pa$\gamma$ | $6 \leftrightarrow 3$ | 1093.81 |
| | Pa$_{\text{limit}}$ | $\infty \leftrightarrow 3$ | 820.4 |

### Bohr's Semiclassical Atom

Theoretical physicists hoped that the answer might be found among the new ideas of photons and quantized energy. A Danish physicist, Niels Bohr (1885–1962; see Fig. 5.4) came to the rescue in 1913 with a daring proposal. The dimensions of Planck's constant, J × s, are equivalent to kg × m s$^{-1}$ × m, the units of angular momentum. Perhaps the angular momentum of the orbiting electron was quantized. This quantization had been previously introduced into atomic models by the British astronomer J. W. Nicholson. Although Bohr knew that Nicholson's models were flawed, he recognized the possible significance of the quantization of angular momentum. Just as an electromagnetic wave of frequency $v$ could have the energy of only an integral number of quanta, $E = nh v$, suppose that the value of the angular momentum of the hydrogen atom could assume only integral multiples of Planck's constant divided by $2\pi$: $L = nh/2\pi = n\hbar$.[8] Bohr hypothesized that in orbits with precisely these allowed values of the angular momentum, the electron would be stable and would not radiate *in spite of its centripetal acceleration*. What would be the result of such a bold departure from classical physics?

   To analyze the mechanical motion of the atomic electron–proton system, we start with the mathematical description of their electrical attraction given by **Coulomb's law**. For two charges $q_1$ and $q_2$ separated by a distance $r$, the electric force on charge 2 due to charge 1 has the familiar form

$$\mathbf{F} = \frac{1}{4\pi\epsilon_0} \frac{q_1 q_2}{r^2} \hat{\mathbf{r}}, \qquad (5.9)$$

---

[8]The quantity $\hbar \equiv h/2\pi = 1.054571596 \times 10^{-34}$ J s and is pronounced "h-bar."

**FIGURE 5.4**   Niels Bohr (1885–1962). (Courtesy of The Niels Bohr Archive, Copenhagen.)

where $\epsilon_0 = 8.854187817\ldots \times 10^{-12}$ F m$^{-1}$ is the *permittivity of free space*[9] and $\hat{\mathbf{r}}$ is a unit vector directed from charge 1 toward charge 2.

Consider an electron of mass $m_e$ and charge $-e$ and a proton of mass $m_p$ and charge $+e$ in circular orbits around their common center of mass, under the influence of their mutual electrical attraction, $e$ being the *fundamental charge*, $e = 1.602176462 \times 10^{-19}$ C. Recall from Section 2.3 that this two-body problem may be treated as an equivalent one-body problem by using the reduced mass

$$\mu = \frac{m_e m_p}{m_e + m_p} = \frac{(m_e)(1836.15266\,m_e)}{m_e + 1836.15266\,m_e} = 0.999455679\,m_e$$

and the total mass

$$M = m_e + m_p = m_e + 1836.15266\,m_e = 1837.15266\,m_e = 1.0005446\,m_p$$

of the system. Since $M \simeq m_p$ and $\mu \simeq m_e$, the hydrogen atom may be thought of as being composed of a proton of mass $M$ that is at rest and an electron of mass $\mu$ that follows a circular orbit of radius $r$ around the proton; see Fig. 5.5. The electrical attraction between the electron and the proton produces the electron's centripetal acceleration $v^2/r$, as described by Newton's second law:

$$\mathbf{F} = \mu \mathbf{a},$$

implying

$$\frac{1}{4\pi\epsilon_0}\frac{q_1 q_2}{r^2}\,\hat{\mathbf{r}} = -\mu\,\frac{v^2}{r}\,\hat{\mathbf{r}},$$

---

[9]Formally, $\epsilon_0$ is *defined* as $\epsilon_0 \equiv 1/\mu_0 c^2$, where $\mu_0 \equiv 4\pi \times 10^{-7}$ N A$^{-2}$ is the *permeability of free space* and $c \equiv 2.99792458 \times 10^8$ m s$^{-1}$ is the defined speed of light.

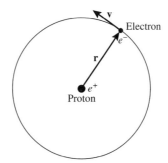

**FIGURE 5.5**   The Bohr model of the hydrogen atom.

or

$$-\frac{1}{4\pi\epsilon_0}\frac{e^2}{r^2}\hat{\mathbf{r}} = -\mu\frac{v^2}{r}\hat{\mathbf{r}}.$$

Canceling the minus sign and the unit vector $\hat{\mathbf{r}}$, this expression can be solved for the kinetic energy, $\frac{1}{2}\mu v^2$:

$$K = \frac{1}{2}\mu v^2 = \frac{1}{8\pi\epsilon_0}\frac{e^2}{r}. \tag{5.10}$$

Now the electrical potential energy $U$ of the Bohr atom is[10]

$$U = -\frac{1}{4\pi\epsilon_0}\frac{e^2}{r} = -2K.$$

Thus the total energy $E = K + U$ of the atom is

$$E = K + U = K - 2K = -K = -\frac{1}{8\pi\epsilon_0}\frac{e^2}{r}. \tag{5.11}$$

Note that the relation between the kinetic, potential, and total energies is in accordance with the virial theorem for an inverse-square force, as discussed for gravity in Section 2.4, $E = \frac{1}{2}U = -K$. Because the kinetic energy must be positive, the total energy $E$ is negative. This merely indicates that the electron and the proton are *bound*. To **ionize** the atom (that is, to remove the proton and electron to an infinite separation), an amount of energy of magnitude $|E|$ (or more) must be added to the atom.

Thus far the derivation has been completely classical in nature. At this point, however, we can use Bohr's quantization of angular momentum,

$$L = \mu v r = n\hbar, \tag{5.12}$$

---

[10]This is found from a derivation analogous to the one leading to the gravitational result, Eq. (2.14). The zero of potential energy is taken to be zero at $r = \infty$.

to rewrite the kinetic energy, Eq. (5.10).

$$\frac{1}{8\pi\epsilon_0}\frac{e^2}{r} = \frac{1}{2}\mu v^2 = \frac{1}{2}\frac{(\mu vr)^2}{\mu r^2} = \frac{1}{2}\frac{(n\hbar)^2}{\mu r^2}.$$

Solving this equation for the radius $r$ shows that the only values allowed by Bohr's quantization condition are

$$r_n = \frac{4\pi\epsilon_0\hbar^2}{\mu e^2}n^2 = a_0 n^2, \tag{5.13}$$

where $a_0 = 5.291772083 \times 10^{-11}$ m $= 0.0529$ nm is known as the **Bohr radius**. Thus the electron can orbit at a distance of $a_0, 4a_0, 9a_0, \ldots$ from the proton, but no other separations are allowed. According to Bohr's hypothesis, when the electron is in one of these orbits, the atom is stable and emits no radiation.

Inserting this expression for $r$ into Eq. (5.11) reveals that the allowed energies of the Bohr atom are

$$E_n = -\frac{\mu e^4}{32\pi^2\epsilon_0^2\hbar^2}\frac{1}{n^2} = -13.6 \text{ eV}\frac{1}{n^2}. \tag{5.14}$$

The integer $n$, known as the **principal quantum number**, completely determines the characteristics of each orbit of the Bohr atom. Thus, when the electron is in the lowest orbit (the *ground state*), with $n = 1$ and $r_1 = a_0$, its energy is $E_1 = -13.6$ eV. With the electron in the ground state, it would take at least 13.6 eV to ionize the atom. When the electron is in the *first excited state*, with $n = 2$ and $r_2 = 4a_0$, its energy is greater than it is in the ground state: $E_2 = -13.6/4$ eV $= -3.40$ eV.

If the electron does not radiate in any of its allowed orbits, then what is the origin of the spectral lines observed for hydrogen? Bohr proposed that a photon is emitted or absorbed when an electron makes a transition from one orbit to another. Consider an electron as it "falls" from a higher orbit, $n_{\text{high}}$, to a lower orbit, $n_{\text{low}}$, without stopping at any intermediate orbit. (This is *not* a fall in the classical sense; the electron is *never* observed between the two orbits.) The electron loses energy $\Delta E = E_{\text{high}} - E_{\text{low}}$, and this energy is carried away from the atom by a single photon. Equation (5.14) leads to an expression for the wavelength of the emitted photon,

$$E_{\text{photon}} = E_{\text{high}} - E_{\text{low}}$$

or

$$\frac{hc}{\lambda} = \left(-\frac{\mu e^4}{32\pi^2\epsilon_0^2\hbar^2}\frac{1}{n_{\text{high}}^2}\right) - \left(-\frac{\mu e^4}{32\pi^2\epsilon_0^2\hbar^2}\frac{1}{n_{\text{low}}^2}\right),$$

which gives

$$\frac{1}{\lambda} = \frac{\mu e^4}{64\pi^3 \epsilon_0^2 \hbar^3 c} \left( \frac{1}{n_{\text{low}}^2} - \frac{1}{n_{\text{high}}^2} \right). \tag{5.15}$$

Comparing this with Eqs. (5.7) and (5.8) reveals that Eq. (5.15) is just the generalized Balmer formula for the spectral lines of hydrogen, with $n_{\text{low}} = 2$ for the Balmer series. Inserting values into the combination of constants in front of the parentheses shows that this term is exactly the Rydberg constant for hydrogen:

$$R_H = \frac{\mu e^4}{64\pi^3 \epsilon_0^2 \hbar^3 c} = 10967758.3 \text{ m}^{-1}.$$

This value is in perfect agreement with the experimental value quoted following Eq. (5.7) for the hydrogen lines determined by Johann Balmer, and this agreement illustrates the great success of Bohr's model of the hydrogen atom.[11]

---

**Example 5.3.1.**   What is the wavelength of the photon emitted when an electron makes a transition from the $n = 3$ to the $n = 2$ orbit of the Bohr hydrogen atom? The energy lost by the electron is carried away by the photon, so

$$E_{\text{photon}} = E_{\text{high}} - E_{\text{low}}$$

$$\frac{hc}{\lambda} = -13.6 \text{ eV} \frac{1}{n_{\text{high}}^2} - \left( -13.6 \text{ eV} \frac{1}{n_{\text{low}}^2} \right)$$

$$= -13.6 \text{ eV} \left( \frac{1}{3^2} - \frac{1}{2^2} \right).$$

Solving for the wavelength gives $\lambda = 656.469$ nm in a vacuum. This result is within 0.03% of the measured value of the H$\alpha$ spectral line, as quoted in Example 5.1.1 and Table 5.2.

   The discrepancy between the calculated and the observed values is due to the measurements being made in air rather than in vacuum. Near sea level, the speed of light is slower than in vacuum by a factor of *approximately* 1.000297. Defining the **index of refraction** to be $n = c/v$, where $v$ is the measured speed of light in the medium, $n_{\text{air}} = 1.000297$. Given that $\lambda \nu = v$ for wave propagation, and since $\nu$ cannot be altered in moving from one medium to another without resulting in unphysical discontinuities in the electromagnetic field of the light wave, the measured wavelength must be proportional to the wave speed. Thus $\lambda_{\text{air}}/\lambda_{\text{vacuum}} = v_{\text{air}}/c = 1/n_{\text{air}}$. Solving for the measured wavelength of the H$\alpha$ line in air yields

$$\lambda_{\text{air}} = \lambda_{\text{vacuum}}/n_{\text{air}} = 656.469 \text{ nm}/1.000297 = 656.275 \text{ nm}.$$

*continued*

---

[11]The slightly different Rydberg constant, $R_\infty$, found in many texts assumes an infinitely heavy nucleus. The reduced mass, $\mu$, in the expression for $R_H$ is replaced by the electron mass, $m_e$, in $R_\infty$.

This result differs from the quoted value by only 0.0009%. The remainder of the discrepancy is due to the fact that the index of refraction is wavelength dependent. The index of refraction also depends on environmental conditions such as temperature, pressure, and humidity.[12]

Unless otherwise noted, throughout the remainder of this text, wavelengths will be assumed to be measured in air (from the ground).

---

The reverse process may also occur. If a photon has an energy equal to the *difference* in energy between two orbits (with the electron in the lower orbit), the photon may be absorbed by the atom. The electron uses the photon's energy to make an upward transition from the lower orbit to the higher orbit. The relation between the photon's wavelength and the quantum numbers of the two orbits is again given by Eq. (5.15).

After the quantum revolution, the physical processes responsible for Kirchhoff's laws (discussed in Section 5.1) finally became clear.

- A hot, dense gas or hot solid object produces a continuous spectrum with no dark spectral lines. This is the continuous spectrum of blackbody radiation emitted at any temperature above absolute zero and described by the Planck functions $B_\lambda(T)$ and $B_\nu(T)$. The wavelength $\lambda_{max}$ at which the Planck function $B_\lambda(T)$ obtains its maximum value is given by Wien's displacement law, Eq. (3.15).

- A hot, diffuse gas produces bright emission lines. Emission lines are produced when an electron makes a downward transition from a higher orbit to a lower orbit. The energy lost by the electron is carried away by a single photon. For example, the hydrogen Balmer emission lines are produced by electrons "falling" from higher orbits down to the $n = 2$ orbit; see Fig. 5.6(a).

- A cool, diffuse gas in front of a source of a continuous spectrum produces dark absorption lines in the continuous spectrum. Absorption lines are produced when an electron makes a transition from a lower orbit to a higher orbit. If an incident photon in the continuous spectrum has exactly the right amount of energy, equal to

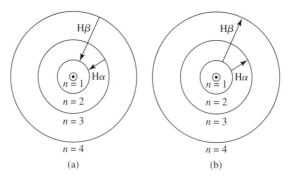

(a)                                      (b)

**FIGURE 5.6**   Balmer lines produced by the Bohr hydrogen atom. (a) Emission lines. (b) Absorption lines.

---

[12] See, for example, Lang, *Astrophysical Formulae*, 1999, page 185 for a fitting formula for $n(\lambda)$.

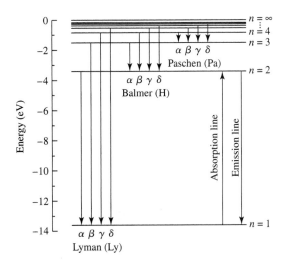

**FIGURE 5.7**   Energy level diagram for the hydrogen atom showing Lyman, Balmer, and Paschen lines (downward arrows indicate emission lines; upward arrow indicates absorption lines).

the difference in energy between a higher orbit and the electron's initial orbit, the photon is absorbed by the atom and the electron makes an upward transition to that higher orbit. For example, the hydrogen Balmer absorption lines are produced by atoms absorbing photons that cause electrons to make transitions from the $n = 2$ orbit to higher orbits; see Figs. 5.6(b) and 5.7.

Despite the spectacular successes of Bohr's model of the hydrogen atom, it is not quite correct. Although angular momentum is quantized, it *does not* have the values assigned by Bohr.[13] Bohr painted a *semiclassical* picture of the hydrogen atom, a miniature Solar System with an electron circling the proton in a classical circular orbit. In fact, the electron orbits are not circular. They are not even orbits at all, in the classical sense of an electron at a precise location moving with a precise velocity. Instead, on an atomic level, nature is "fuzzy," with an attendant uncertainty that cannot be avoided. It was fortunate that Bohr's model, with all of its faults, led to the correct values for the energies of the orbits and to a correct interpretation of the formation of spectral lines. This intuitive, easily imagined model of the atom is what most physicists and astronomers have in mind when they visualize atomic processes.

## 5.4 ■ QUANTUM MECHANICS AND WAVE–PARTICLE DUALITY

The last act of the quantum revolution began with the musings of a French prince, Louis de Broglie (1892–1987; see Fig. 5.8). Wondering about the recently discovered wave–particle duality for light, he posed a profound question: If light (classically thought to be a wave)

---

[13]As we will see in the next section, instead of $L = n\hbar$, the actual values of the orbital angular momentum are $L = \sqrt{\ell(\ell + 1)}\,\hbar$, where $\ell$, an integer, is a new quantum number.

**FIGURE 5.8**   Louis de Broglie (1892–1987). (Courtesy of AIP Niels Bohr Library.)

could exhibit the characteristics of particles, might not particles sometimes manifest the properties of waves?

### de Broglie's Wavelength and Frequency

In his 1927 Ph.D. thesis, de Broglie extended the wave–particle duality to all of nature. Photons carry both energy $E$ and momentum $p$, and these quantities are related to the frequency $\nu$ and wavelength $\lambda$ of the light wave by Eq. (5.5):

$$\nu = \frac{E}{h} \tag{5.16}$$

$$\lambda = \frac{h}{p}. \tag{5.17}$$

de Broglie proposed that these equations be used to define a frequency and a wavelength for *all* particles. The **de Broglie wavelength** and **frequency** describe not only massless photons but massive electrons, protons, neutrons, atoms, molecules, people, planets, stars, and galaxies as well. This seemingly outrageous proposal of matter waves has been confirmed in countless experiments. Figure 5.9 shows the interference pattern produced by *electrons* in a double-slit experiment. Just as Thomas Young's double-slit experiment established the wave properties of light, the electron double-slit experiment can be explained only by the wave-like behavior of electrons, with *each* electron propagating through *both* slits.[14] The wave–particle duality applies to everything in the physical world; everything exhibits its wave properties in its *propagation* and manifests its particle nature in its *interactions*.

[14]See Chapter 6 of Feynman (1965) for a fascinating description of the details and profound implications of the electron double-slit experiment.

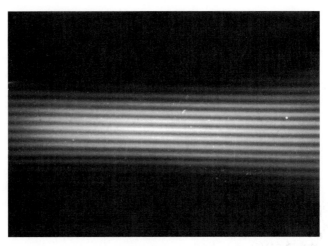

**FIGURE 5.9**   Interference pattern from an electron double-slit experiment. (Figure from Jönsson, *Zeitschrift für Physik, 161*, 454, 1961.)

---

**Example 5.4.1.**   Compare the wavelengths of a free electron moving at $3 \times 10^6$ m s$^{-1}$ and a 70-kg man jogging at 3 m s$^{-1}$. For the electron,

$$\lambda = \frac{h}{p} = \frac{h}{m_e v} = 0.242 \text{ nm},$$

which is about the size of an atom and much shorter than the wavelength of visible light. Electron microscopes utilize electrons with wavelengths one million times shorter than visible wavelengths to obtain a much higher resolution than is possible with optical microscopes.

The wavelength of the jogging man is

$$\lambda = \frac{h}{p} = \frac{h}{m_{\text{man}} v} = 3.16 \times 10^{-36} \text{ m},$$

which is completely negligible on the scale of the everyday world, and even on atomic or nuclear scales. Thus the jogging gentleman need not worry about diffracting when returning home through his doorway!

---

Just what are the waves that are involved in the wave–particle duality of nature? In a double-slit experiment, each photon or electron must pass through *both* slits, since the interference pattern is produced by the constructive and destructive interference of the two waves. Thus the wave cannot convey information about where the photon or electron is, but only about where it *may* be. The wave is one of *probability*, and its amplitude is denoted by the Greek letter $\Psi$ (psi). The square of the wave amplitude, $|\Psi|^2$, at a certain location describes the probability of finding the photon or electron at that location. In the double-slit experiment, photons or electrons are never found where the waves from slits 1 and 2 have destructively interfered—that is, where $|\Psi_1 + \Psi_2|^2 = 0$.

### Heisenberg's Uncertainty Principle

The wave attributes of matter lead to some unexpected conclusions of paramount importance for the science of astronomy. For example, consider Fig. 5.10(a). The probability wave, $\Psi$, is a sine wave, with a precise wavelength $\lambda$. Thus the momentum $p = h/\lambda$ of the particle described by this wave is known exactly. However, because $|\Psi|^2$ consists of a number of equally high peaks extending out to $x = \pm\infty$, the particle's location is perfectly uncertain. The particle's position can be narrowed down if several sine waves with different wavelengths are added together, so they destructively interfere with one another nearly everywhere. Figure 5.10(b) shows the resulting combination of waves, $\Psi$, is approximately zero everywhere except at one location. Now the particle's position may be determined with a greater certainty because $|\Psi|^2$ is large only for a narrow range of values of $x$. However, the value of the particle's momentum has become more uncertain because $\Psi$ is now a combination of waves of various wavelengths. This is nature's intrinsic trade-off: The uncertainty in a particle's position, $\Delta x$, and the uncertainty in its momentum, $\Delta p$, are inversely related. As one decreases, the other must increase. This fundamental inability of a particle to *simultaneously* have a well-defined position and a well-defined momentum is a direct result of the wave–particle duality of nature. A German physicist, Werner Heisenberg (1901–1976), placed this inherent "fuzziness" of the physical world in a firm theoretical framework. He demonstrated that the uncertainty in a particle's position multiplied by the uncertainty in its momentum must be *at least* as large as $\hbar/2$:

$$\Delta x \, \Delta p \geq \frac{1}{2} \hbar. \tag{5.18}$$

Today this is known as **Heisenberg's uncertainty principle**. The equality is rarely realized in nature, and the form often employed for making estimates is

$$\Delta x \, \Delta p \approx \hbar. \tag{5.19}$$

A similar statement relates the uncertainty of an energy measurement, $\Delta E$, and the time interval, $\Delta t$, over which the energy measurement is taken:

$$\Delta E \, \Delta t \approx \hbar. \tag{5.20}$$

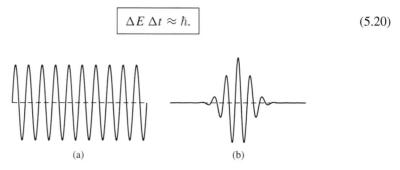

(a)          (b)

**FIGURE 5.10** Two examples of a probability wave, $\Psi$: (a) a single sine wave and (b) a pulse composed of many sine waves.

As the time available for an energy measurement increases, the inherent uncertainty in the result decreases. In Section 9.5, we will apply this version of the uncertainty principle to the sharpness of spectral lines.

---

**Example 5.4.2.**   Imagine an electron confined within a region of space the size of a hydrogen atom. We can estimate the minimum speed and kinetic energy of the electron using Heisenberg's uncertainty principle. Because we know only that the particle is within an atom-size region of space, we can take $\Delta x \approx a_0 = 5.29 \times 10^{-11}$ m. This implies that the uncertainty in the electron's momentum is roughly

$$\Delta p \approx \frac{\hbar}{\Delta x} = 1.98 \times 10^{-24} \text{ kg m s}^{-1}.$$

Thus, if the magnitude of the momentum of the electron were repeatedly measured, the resulting values would vary within a range $\pm \Delta p$ around some average (or *expected*) value. Since this expected value, as well as the individual measurements, must be $\geq 0$, the expected value must be at least as large as $\Delta p$. Thus we can equate the minimum expected value of the momentum with its uncertainty: $p_{min} \approx \Delta p$. Using $p_{min} = m_e v_{min}$, the minimum speed of the electron is estimated to be

$$v_{min} = \frac{p_{min}}{m_e} \approx \frac{\Delta p}{m_e} \approx 2.18 \times 10^6 \text{ m s}^{-1}.$$

The minimum kinetic energy of the (nonrelativistic) electron is approximately

$$K_{min} = \frac{1}{2} m_e v_{min}^2 \approx 2.16 \times 10^{-18} \text{ J} = 13.5 \text{ eV}.$$

This is in good agreement with the kinetic energy of the electron in the ground state of the hydrogen atom. An electron confined to such small region *must* move rapidly with at least this speed and this energy. In Chapter 16, we will see that this subtle quantum effect is responsible for supporting white dwarf and neutron stars against the tremendous inward pull of gravity.

---

## Quantum Mechanical Tunneling

When a ray of light attempts to travel from a glass prism into air, it may undergo *total internal reflection* if it strikes the surface at an angle greater than the critical angle $\theta_c$, where the critical angle is related to the indices of refraction of the glass and air by

$$\sin \theta_c = \frac{n_{air}}{n_{glass}}.$$

This familiar result is nonetheless surprising because, even though the ray of light is totally reflected, the index of refraction of the outside air appears in this formula. In fact, the electromagnetic wave does enter the air, but it ceases to be oscillatory and instead dies away exponentially. In general, when a classical wave such as a water or light wave enters a

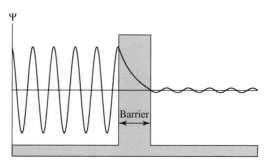

**FIGURE 5.11**   Quantum mechanical tunneling (barrier penetration) of a particle traveling to the right.

medium through which it cannot propagate, it becomes *evanescent* and its amplitude decays exponentially with distance.

This total internal reflection can in fact be frustrated by placing another prism next to the first prism so that their surfaces nearly (but not quite) touch. Then the evanescent wave in the air may enter the second prism before its amplitude has completely died away. The electromagnetic wave once again becomes oscillatory upon entering the glass, and so the ray of light has traveled from one prism to another without passing through the air gap between the prisms. In the language of particles, photons have **tunneled** from one prism to another without traveling in the space between them.

The wave–particle duality of nature implies that particles can also tunnel through a region of space (a barrier) in which they cannot exist classically, as illustrated in Fig. 5.11. The barrier must not be too wide (not more than a few particle wavelengths) if tunneling is to take place; otherwise, the amplitude of the evanescent wave will have declined to nearly zero. This is consistent with Heisenberg's uncertainty principle, which implies that a particle's location cannot be determined with an uncertainty that is less than its wavelength. Thus, if the barrier is only a few wavelengths wide, the particle may suddenly appear on the other side of the barrier. **Barrier penetration** is extremely important in radioactive decay, where alpha particles tunnel out of an atom's nucleus; in modern electronics, where it is the basis for the "tunnel diode"; and inside stars, where the rates of nuclear fusion reactions depend upon tunneling.

### Schrödinger's Equation and the Quantum Mechanical Atom

What are the implications for Bohr's model of the hydrogen atom? Heisenberg's uncertainty principle does not allow classical orbits, with their simultaneously precise values of the electron's position and momentum. Instead, the electron *orbitals* must be imagined as fuzzy clouds of probability, with the clouds being more "dense" in regions where the electron is more likely to be found (see Fig. 5.12). In 1925 a complete break from classical physics was imminent, one that would fully incorporate de Broglie's matter waves.

As described in Section 3.3, Maxwell's equations of electricity and magnetism can be manipulated to produce a wave equation for the electromagnetic waves that describe the propagation of photons. Similarly, a wave equation discovered in 1926 by Erwin Schrödinger (1877–1961), an Austrian physicist, led to a true **quantum mechanics**, the quantum analog of the classical mechanics that originated with Galileo and Newton. The Schrödinger

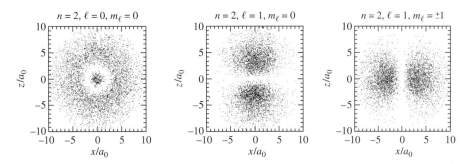

**FIGURE 5.12**   Electron orbitals of the hydrogen atom. Left: $2s$ orbital. Middle: $2p$ orbital with $m_\ell = 0$. Right: $2p$ orbital with $m_\ell = \pm 1$. The quantum numbers $n$, $\ell$, and $m_\ell$ are described in the text.

equation can be solved for the probability waves that describe the allowed values of a particle's energy, momentum, and so on, as well as the particle's propagation through space. In particular, the Schrödinger equation can be solved analytically for the hydrogen atom, giving exactly the same set of allowed energies as those obtained by Bohr (cf. Eq. 5.11). However, in addition to the principal quantum number $n$, Schrödinger found that two additional quantum numbers, $\ell$ and $m_\ell$, are required for a complete description of the electron orbitals. These additional numbers describe the angular momentum vector, **L**, of the atom. Instead of the quantization used by Bohr, $L = n\hbar$, the solution to the Schrödinger equation shows that the permitted values of the magnitude of the angular momentum $L$ are actually

$$L = \sqrt{\ell(\ell + 1)}\,\hbar, \tag{5.21}$$

where $\ell = 0, 1, 2, \ldots, n - 1$, and $n$ is the principal quantum number that determines the energy.

Note that it is common practice to refer to the angular momentum quantum numbers by their historical spectroscopic designations $s, p, d, f, g, h$, and so on, corresponding to $\ell = 0$, 1, 2, 3, 4, 5, etc. When the associated principle quantum number is used in combination with the angular momentum quantum number, the principle quantum number precedes the spectroscopic designation. For example, ($n = 2$, $\ell = 1$) corresponds to $2p$, and ($n = 3$, $\ell = 2$) is given as $3d$. This notation was used in the caption of Fig. 5.12 and is also used in Fig. 5.13.

The $z$-component of the angular momentum vector, $L_z$, can assume only the values $L_z = m_\ell \hbar$, with $m_\ell$ equal to any of the $2\ell + 1$ integers between $-\ell$ and $+\ell$ inclusive. Thus the angular momentum vector can point in $2\ell + 1$ different directions. For our purposes, the important point is that the values of the energy of an *isolated* hydrogen atom do not depend on $\ell$ and $m_\ell$. In the absence of a preferred direction in space, the direction of the angular momentum has no effect on the atom's energy. Different orbitals, labeled by different values of $\ell$ and $m_\ell$ (see Fig. 5.12), are said to be **degenerate** if they have the same value of the principal quantum number $n$ and so have the same energy. Electrons making a transition from a given orbital to one of several degenerate orbitals will produce the *same* spectral line, because they experience the same change in energy.

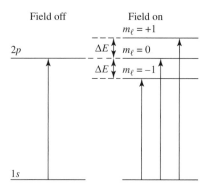

**FIGURE 5.13**   Splitting of absorption lines by the Zeeman effect.

However, the atom's surroundings may single out one spatial direction as being different from another. For example, an electron in an atom will feel the effect of an external magnetic field. The magnitude of this effect will depend on the $2\ell + 1$ possible orientations of the electron's motion, as given by $m_\ell$, and the magnetic field strength, $B$, where the units of $B$ are teslas (T).[15] As the electron moves through the magnetic field, the normally degenerate orbitals acquire slightly different energies. Electrons making a transition between these formerly degenerate orbitals will thus produce spectral lines with slightly different frequencies. The splitting of spectral lines in a weak magnetic field is called the **Zeeman effect** and is shown in Fig. 5.13. The three frequencies of the split lines in the simplest case (called the *normal Zeeman effect*) are

$$
\nu = \nu_0 \qquad \text{and} \qquad \nu_0 \pm \frac{eB}{4\pi\mu}, \tag{5.22}
$$

where $\nu_0$ is the frequency of the spectral line in the absence of a magnetic field and $\mu$ is the reduced mass. Although the energy levels are split into $2\ell + 1$ components, electron transitions involving these levels produce just three spectral lines with different polarizations.[16] Viewed from different directions, it may happen that not all three lines will be visible. For example, when looking parallel to the magnetic field (as when looking down on a sunspot), the unshifted line of frequency $\nu_0$ is absent.

Thus the Zeeman effect gives astronomers a probe of the magnetic fields observed around sunspots and on other stars. Even if the splitting of the spectral line is too small to be directly detected, the different polarizations across the closely spaced components can still be measured and the magnetic field strength deduced.

---

**Example 5.4.3.**   Interstellar clouds may contain very weak magnetic fields, as small as $B \approx 2 \times 10^{-10}$ T. Nevertheless, astronomers have been able to measure this magnetic field. Using radio telescopes, they detect the variation in polarization that occurs across the

---

[15]Another commonly used unit of magnetic field strength is gauss, where $1\,\mathrm{G} = 10^{-4}$ T. Earth's magnetic field is roughly 0.5 G, or $5 \times 10^{-5}$ T.

[16]See the discussion beginning on page 136 concerning selection rules.

blended Zeeman components of the absorption lines that are produced by these interstellar clouds of hydrogen gas. The change in frequency, $\Delta \nu$, produced by a magnetic field of this magnitude can be calculated from Eq. (5.22) by using the mass of the electron, $m_e$, for the reduced mass $\mu$:

$$\Delta \nu = \frac{eB}{4\pi m_e} = 2.8 \text{ Hz},$$

a minute change. The total difference in frequency from one side of this blended line to the other is twice this amount, or 5.6 Hz. For comparison, the frequency of the radio wave emitted by hydrogen with $\lambda = 21$ cm is $\nu = c/\lambda = 1.4 \times 10^9$ Hz, 250 million times larger!

---

### Spin and the Pauli Exclusion Principle

Attempts to understand more complicated patterns of magnetic field splitting (the *anomalous Zeeman effect*), usually involving an even number of unequally spaced spectral lines, led physicists in 1925 to discover a *fourth* quantum number. In addition to its orbital motion, the electron possesses a **spin**. This is *not* a classical top-like rotation but purely a quantum effect that endows the electron with a *spin angular momentum* **S**. **S** is a vector of constant magnitude

$$S = \sqrt{\frac{1}{2}\left(\frac{1}{2} + 1\right)}\,\hbar = \frac{\sqrt{3}}{2}\,\hbar,$$

with a $z$-component $S_z = m_s \hbar$. The only values of the fourth quantum number, $m_s$, are $\pm \frac{1}{2}$.

With each orbital, or *quantum state*, labeled by four quantum numbers, physicists wondered how many electrons in a multielectron atom could occupy the same quantum state. The answer was supplied in 1925 by an Austrian theoretical physicist, Wolfgang Pauli (1900–1958): No two electrons can occupy the same quantum state. The **Pauli exclusion principle**, that *no two electrons can share the same set of four quantum numbers*, explained the electronic structure of atoms, thereby providing an explanation of the properties of the periodic table of the elements, the well-known chart from any introductory chemistry text. Despite this success, Pauli was unhappy about the lack of a firm theoretical understanding of electron spin. Spin was stitched onto quantum theory in an ad hoc manner, and the seams showed. Pauli lamented this patchwork theory and asked, "How can one avoid despondency if one thinks of the anomalous Zeeman effect?"

The final synthesis arrived in 1928 from an unexpected source. A brilliant English theoretical physicist, Paul Adrien Maurice Dirac (1902–1984), was working at Cambridge to combine Schrödinger's wave equation with Einstein's theory of special relativity. When he finally succeeded in writing a relativistic wave equation for the electron, he was delighted to see that the mathematical solution automatically included the spin of the electron. It also explained and extended the Pauli exclusion principle by dividing the world of particles into two fundamental groups: fermions and bosons. **Fermions**[17] are particles such as electrons,

---

[17]The fermion is named after the Italian physicist Enrico Fermi (1901–1954).

protons, and neutrons[18] that have a spin of $\frac{1}{2}\hbar$ (or an odd integer times $\frac{1}{2}\hbar$, such as $\frac{3}{2}\hbar$, $\frac{5}{2}\hbar$, ...). Fermions obey the Pauli exclusion principle, so no two fermions of the same type can have the same set of quantum numbers. The exclusion principle for fermions, along with Heisenberg's uncertainty relation, explains the structure of white dwarfs and neutron stars, as will be discussed in Chapter 16. **Bosons**[19] are particles such as photons that have an integral spin of 0, $\hbar$, $2\hbar$, $3\hbar$, .... Bosons do not obey the Pauli exclusion principle, so any number of bosons can occupy the same quantum state.

As a final bonus, the Dirac equation predicted the existence of antiparticles. A particle and its antiparticle are identical except for their opposite electric charges and magnetic moments. Pairs of particles and antiparticles may be created from the energy of gamma-ray photons (according to $E = mc^2$). Conversely, particle–antiparticle pairs may annihilate each other, with their mass converted back into the energy of two gamma-ray photons. As we will see in Section 17.3, pair creation and annihilation play a major role in the evaporation of black holes.

### The Complex Spectra of Atoms

With the full list of four quantum numbers ($n$, $\ell$, $m_\ell$, and $m_s$) that describe the detailed state an each electron in an atom, the number of possible energy levels increases rapidly with the number of electrons. When we take into account the additional complications of external magnetic fields, and the electromagnetic interactions between the electrons themselves and between the electrons and the nucleus, the spectra can become very complicated indeed. Figure 5.14 shows some of the available energy levels for the two electrons in the neutral helium atom.[20] Imagine the complexity of the relatively abundant iron atom with its 26 electrons!

Although energy levels exist for electrons with various combinations of quantum numbers, it is not always easy for an electron to make a transition from one quantum state with a specific set of quantum numbers to another quantum state. In particular, Nature imposes a set of **selection rules** that restrict certain transitions. For example, a careful investigation of Fig. 5.14 will show that only transitions involving $\Delta\ell = \pm 1$ are shown (from $^1P$ to $^1S$, or from $^1F$ to $^1D$, for instance). These transitions are referred to as **allowed** transitions and can happen spontaneously on timescales of $10^{-8}$ s. On the other hand, transitions that do not satisfy the requirement that $\Delta\ell = \pm 1$ are known as **forbidden** transitions.

In the case of the Zeeman effect first discussed on page 134, it was pointed out that only three transitions could occur between the $1s$ and $2p$ energy levels (recall Fig. 5.13). This is because of another set of selection rules requiring that $\Delta m_\ell = 0$ or $\pm 1$ and forbidding transitions between orbitals if both orbitals have $m_\ell = 0$.

Although forbidden transitions may occur, they require much longer times if they are to occur with any significant probability. Since collisions between atoms trigger transitions and can compete with spontaneous transitions, very low gas densities are required for measurable intensities to be observed from forbidden transitions. Such environments do exist in astronomy, such as in the diffuse interstellar medium or in the outer atmospheres

---

[18]The neutron was not discovered until 1932 by James Chadwick (1891–1974), the same year that the positron (antimatter electron) was discovered by Carl Anderson (1905–1991).

[19]The boson is named in honor of the Indian physicist S. N. Bose (1894–1974).

[20]Figure 5.14 is known as a Grotrian diagram.

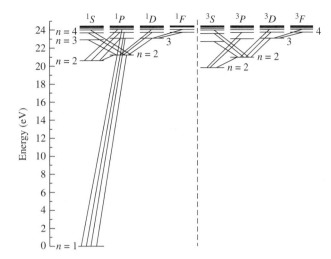

**FIGURE 5.14** Some of the electronic energy levels of the helium atom. A small number of possible allowed transitions are also indicated. (Data courtesy of the National Institute of Standards and Technology.)

of stars. (It is beyond the scope of this text to discuss the detailed physics that underlies the existence of selection rules.)

The revolution in physics started by Max Planck culminated in the quantum atom and gave astronomers their most powerful tool: a theory that would enable them to analyze the spectral lines observed for stars, galaxies, and nebulae.[21] Different atoms, and combinations of atoms in molecules, have orbitals of distinctly different energies; thus they can be identified by their spectral line "fingerprints." The specific spectral lines produced by an atom or molecule depend on which orbitals are occupied by electrons. This, in turn, depends on its surroundings: the temperature, density, and pressure of its environment. These and other factors, such as the strength of a surrounding magnetic field, may be determined by a careful examination of spectral lines. Much of Chapters 8 and 9 will be devoted to the practical application of the quantum atom to stellar atmospheres.

## SUGGESTED READING

### General

Feynman, Richard, *The Character of Physical Law*, The M.I.T. Press, Cambridge, MA, 1965.

French, A. P., and Kennedy, P. J. (eds.), *Niels Bohr: A Centenary Volume*, Harvard University Press, Cambridge, MA, 1985.

Hey, Tony, and Walters, Patrick, *The New Quantum Universe*, Cambridge University Press, Cambridge, 2003.

---

[21] Nearly all of the physicists mentioned in this chapter won the Nobel Prize for physics or chemistry in recognition of their work.

Pagels, Heinz R., *The Cosmic Code*, Simon and Schuster, New York, 1982.

Segre, Emilio, *From X-Rays to Quarks*, W. H. Freeman and Company, San Francisco, 1980.

### Technical

Cox, Arthur N. (ed.), *Allen's Astrophysical Quantities*, Fourth Edition, Springer, New York, 2000.

Harwit, Martin, *Astrophysical Concepts*, Third Edition, Springer, New York, 1998.

Lang, Kenneth R., *Astrophysical Formulae*, Third Edition, Springer, New York, 1999.

Marcy, Geoffrey W., et al, "Two Substellar Companions Orbiting HD 168443," *Astrophysical Journal*, *555*, 418, 2001.

Resnick, Robert, and Halliday, David, *Basic Concepts in Relativity and Early Quantum Theory*, Second Edition, John Wiley and Sons, New York, 1985.

Shu, Frank H., *The Physics of Astrophysics*, University Science Books, Mill Valley, CA, 1991.

## PROBLEMS

**5.1** Barnard's star, named after the American astronomer Edward E. Barnard (1857–1923), is an orange star in the constellation Ophiuchus. It has the largest known proper motion ($\mu = 10.3577''$ yr$^{-1}$) and the fourth-largest parallax angle ($p = 0.54901''$). Only the stars in the triple system $\alpha$ Centauri have larger parallax angles. In the spectrum of Barnard's star, the H$\alpha$ absorption line is observed to have a wavelength of 656.034 nm when measured from the ground.

(a) Determine the radial velocity of Barnard's star.

(b) Determine the transverse velocity of Barnard's star.

(c) Calculate the speed of Barnard's star through space.

**5.2** When salt is sprinkled on a flame, yellow light consisting of two closely spaced wavelengths, 588.997 nm and 589.594 nm, is produced. They are called the *sodium D lines* and were observed by Fraunhofer in the Sun's spectrum.

(a) If this light falls on a diffraction grating with 300 lines per millimeter, what is the angle between the second-order spectra of these two wavelengths?

(b) How many lines of this grating must be illuminated for the sodium D lines to just be resolved?

**5.3** Show that $hc \simeq 1240$ eV nm.

**5.4** The photoelectric effect can be an important heating mechanism for the grains of dust found in interstellar clouds (see Section 12.1). The ejection of an electron leaves the grain with a positive charge, which affects the rates at which other electrons and ions collide with and stick to the grain to produce the heating. This process is particularly effective for ultraviolet photons ($\lambda \approx 100$ nm) striking the smaller dust grains. If the average energy of the ejected electron is about 5 eV, estimate the work function of a typical dust grain.

**5.5** Use Eq. (5.5) for the momentum of a photon, plus the conservation of relativistic momentum and energy [Eqs. (4.44) and (4.48), respectively], to derive Eq. (5.6) for the change in wavelength of the scattered photon in the Compton effect.

**5.6** Consider the case of a "collision" between a photon and a free proton, initially at rest. What is the characteristic change in the wavelength of the scattered photon in units of nanometers? How does this compare with the Compton wavelength, $\lambda_C$?

**5.7** Verify that the units of Planck's constant are the units of angular momentum.

**5.8** A one-electron atom is an atom with $Z$ protons in the nucleus and with all but one of its electrons lost to ionization.

    **(a)** Starting with Coulomb's law, determine expressions for the orbital radii and energies for a Bohr model of the one-electron atom with $Z$ protons.

    **(b)** Find the radius of the ground-state orbit, the ground-state energy, and the ionization energy of singly ionized helium (He II).

    **(c)** Repeat part (b) for doubly ionized lithium (Li III).

**5.9** To demonstrate the relative strengths of the electrical and gravitational forces of attraction between the electron and the proton in the Bohr atom, suppose the hydrogen atom were held together *solely* by the force of gravity. Determine the radius of the ground-state orbit (in units of nm and AU) and the energy of the ground state (in eV).

**5.10** Calculate the energies and vacuum wavelengths of all possible photons that are emitted when the electron cascades from the $n = 3$ to the $n = 1$ orbit of the hydrogen atom.

**5.11** Find the shortest vacuum-wavelength photon emitted by a downward electron transition in the Lyman, Balmer, and Paschen series. These wavelengths are known as the *series limits*. In which regions of the electromagnetic spectrum are these wavelengths found?

**5.12** An electron in a television set reaches a speed of about $5 \times 10^7$ m s$^{-1}$ before it hits the screen. What is the wavelength of this electron?

**5.13** Consider the de Broglie wave of the electron in the Bohr atom. The circumference of the electron's orbit must be an integral number of wavelengths, $n\lambda$; see Fig. 5.15. Otherwise, the electron wave will find itself *out of phase* and suffer destructive interference. Show that this requirement leads to Bohr's condition for the quantization of angular momentum, Eq. (5.12).

**5.14** A white dwarf is a very dense star, with its ions and electrons packed extremely close together. Each electron may be considered to be located within a region of size $\Delta x \approx 1.5 \times 10^{-12}$ m. Use Heisenberg's uncertainty principle, Eq. (5.19), to estimate the minimum speed of the electron. Do you think that the effects of relativity will be important for these stars?

**5.15** An electron spends roughly $10^{-8}$ s in the first excited state of the hydrogen atom before making a spontaneous downward transition to the ground state.

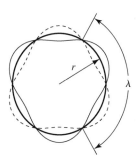

**FIGURE 5.15** Three de Broglie wavelengths spanning an electron's orbit in the Bohr atom.

(a) Use Heisenberg's uncertainty principle (Eq. 5.20) to determine the uncertainty $\Delta E$ in the energy of the first excited state.

(b) Calculate the uncertainty $\Delta \lambda$ in the wavelength of the photon involved in a transition (either upward or downward) between the ground and first excited states of the hydrogen atom. Why can you assume that $\Delta E = 0$ for the ground state?

This increase in the width of a spectral line is called *natural broadening*.

**5.16** Each quantum state of the hydrogen atom is labeled by a set of four quantum numbers: $\{n, \ell, m_\ell, m_s\}$.

(a) List the sets of quantum numbers for the hydrogen atom having $n = 1$, $n = 2$, and $n = 3$.

(b) Show that the degeneracy of energy level $n$ is $2n^2$.

**5.17** The members of a class of stars known as Ap stars are distinguished by their strong global magnetic fields (usually a few tenths of one tesla).[22] The star HD215441 has an unusually strong magnetic field of 3.4 T. Find the frequencies and wavelengths of the three components of the H$\alpha$ spectral line produced by the normal Zeeman effect for this magnetic field.

## COMPUTER PROBLEM

**5.18** One of the most important ideas of the physics of waves is that *any* complex waveform can be expressed as the sum of the harmonics of simple cosine and sine waves. That is, any wave function $f(x)$ can be written as

$$f(x) = a_0 + a_1 \cos x + a_2 \cos 2x + a_3 \cos 3x + a_4 \cos 4x + \cdots$$
$$+ b_1 \sin x + b_2 \sin 2x + b_3 \sin 3x + b_4 \sin 4x + \cdots .$$

The coefficients $a_n$ and $b_n$ tell how much of each harmonic goes into the recipe for $f(x)$. This series of cosine and sine terms is called the *Fourier series* for $f(x)$. In general, both cosine and sine terms are needed, but in this problem you will use only the sine terms; all of the $a_n \equiv 0$.

On page 130, the process of constructing a wave pulse by adding a series of sine waves was described. The Fourier sine series that you will use to construct your wave employs only the *odd* harmonics and is given by

$$\Psi = \frac{2}{N+1} (\sin x - \sin 3x + \sin 5x - \sin 7x + \cdots \pm \sin Nx) = \frac{2}{N+1} \sum_{\substack{n=1 \\ n \text{ odd}}}^{N} (-1)^{(n-1)/2} \sin nx,$$

where $N$ is an odd integer. The leading factor of $2/(N+1)$ does not change the shape of $\Psi$ but scales the wave for convenience so that its maximum value is equal to 1 for any choice of $N$.

(a) Graph $\Psi$ for $N = 5$, using values of $x$ (in radians) between 0 and $\pi$. What is the width, $\Delta x$, of the wave pulse?

(b) Repeat part (a) for $N = 11$.

(c) Repeat part (a) for $N = 21$.

(d) Repeat part (a) for $N = 41$.

(e) If $\Psi$ represents the probability wave of a particle, for which value of $N$ is the position of the particle known with the least uncertainty? For which value of $N$ is the momentum of the particle known with the least uncertainty?

---

[22]The letter A is the star's spectral type (to be discussed in Section 8.1), and the letter p stands for "peculiar."

# Telescopes

## 6.1 ■ BASIC OPTICS

From the beginning, astronomy has been an observational science. In comparison with what was previously possible with the naked eye, Galileo's use of the new optical device known as the telescope greatly improved our ability to observe the universe (see Section 2.2). Today we continue to enhance our ability to "see" faint objects and to resolve them in greater detail. As a result, modern observational astronomy continues to supply scientists with more clues to the physical nature of our universe.

Although observational astronomy now covers the entire range of the electromagnetic spectrum, along with many areas of particle physics, the most familiar part of the field remains in the optical regime of the human eye (approximately 400 nm to 700 nm). Consequently, telescopes and detectors designed to investigate optical-wavelength radiation will be discussed in some detail. Furthermore, much of what we learn in studying telescopes and detectors in the optical regime will apply to other wavelength regions as well.

### Refraction and Reflection

Galileo's telescope was a **refracting** telescope that made use of lenses through which light would pass, ultimately forming an image. Later, Newton designed and built a **reflecting** telescope that made use of mirrors as the principal optical component. Both refractors and reflectors remain in use today.

To understand the effects of an optical system on the light coming from an astronomical object, we will focus first on refracting telescopes. The path of a light ray through a lens can be understood using **Snell's law** of refraction. Recall that as a light ray travels from one transparent medium to another, its path is bent at the interface. The amount that the ray is bent depends on the ratio of the wavelength-dependent indices of refraction $n_\lambda \equiv c/v_\lambda$ of each material, where $v_\lambda$ represents the speed of light within the specific medium.[1] If $\theta_1$

---

[1] It is only in a vacuum that $v_\lambda \equiv c$, independent of wavelength. The speed of light is wavelength-dependent in other environments.

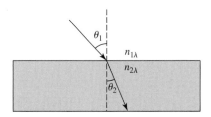

**FIGURE 6.1**   Snell's law of refraction.

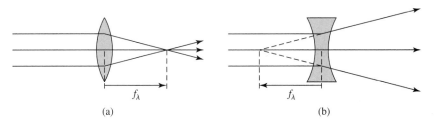

(a)                                          (b)

**FIGURE 6.2**   (a) A converging lens, $f_\lambda > 0$. (b) A diverging lens, $f_\lambda < 0$.

is the angle of incidence, measured with respect to the normal to the interface between the two media, and $\theta_2$ is the angle of refraction, also measured relative to the normal to the interface (see Fig. 6.1), then Snell's law is given by

$$n_{1\lambda} \sin \theta_1 = n_{2\lambda} \sin \theta_2. \tag{6.1}$$

If the surfaces of the lens are shaped properly, a beam of light rays of a given wavelength, originally traveling parallel to the axis of symmetry of the lens (called the **optical axis** of the system) can be brought to a focus at a point along that axis by a *converging* lens [Fig. 6.2(a)]. Alternatively, the light can be made to diverge by a *diverging* lens and the light rays will appear to originate from a single point along the axis [Fig. 6.2(b)]. The unique point in either case is referred to as the **focal point** of the lens, and the distance to that point from the center of the lens is known as the **focal length**, $f$. For a converging lens the focal length is taken to be positive, and for a diverging lens the focal length is negative.

The focal length of a given thin lens can be calculated directly from its index of refraction and geometry. If we assume that both surfaces of the lens are spheroidal, then it can be shown that the focal length $f_\lambda$ is given by the **lensmaker's formula**,

$$\frac{1}{f_\lambda} = (n_\lambda - 1) \left( \frac{1}{R_1} + \frac{1}{R_2} \right), \tag{6.2}$$

where $n_\lambda$ is the index of refraction of the lens and $R_1$ and $R_2$ are the radii of curvature of each surface, taken to be positive if the specific surface is convex and negative if it is concave (see Fig. 6.3).[2]

---

[2]It is worth noting that many authors choose to define the sign convention for the radii of curvature in terms of the direction of the incident light. This choice means that Eq. (6.2) must be expressed in terms of the *difference* in the reciprocals of the radii of curvature.

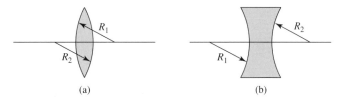

**FIGURE 6.3** The sign convention for the radii of curvature of a lens in the lensmaker's formula. (a) $R_1 > 0$, $R_2 > 0$. (b) $R_1 < 0$, $R_2 < 0$.

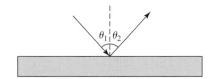

**FIGURE 6.4** The law of reflection, $\theta_1 = \theta_2$.

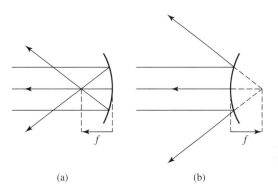

**FIGURE 6.5** (a) A converging mirror, $f > 0$. (b) A diverging mirror, $f < 0$.

For mirrors $f$ is wavelength-independent, since reflection depends only on the fact that the angle of incidence always equals the angle of reflection ($\theta_1 = \theta_2$; see Fig. 6.4). Furthermore, in the case of a spheroidal mirror (Fig. 6.5), the focal length becomes $f = R/2$, where $R$ is the radius of curvature of the mirror, either positive (converging) or negative (diverging), a fact that can be demonstrated by simple geometry. Converging mirrors are generally used as the main mirrors in reflecting telescopes, although either diverging or flat mirrors may be used in other parts of the optical system.

## The Focal Plane

For an extended object, the image will also necessarily be extended. If a photographic plate or some other detector is to record this image, the detector must be placed in the focal plane of the telescope. The **focal plane** is defined as the plane passing through the focal point and oriented perpendicular to the optical axis of the system. Since, for all practical purposes, any astronomical object can reasonably be assumed to be located infinitely far

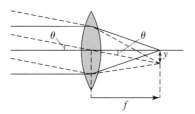

**FIGURE 6.6**   The plate scale, determined by the focal length of the optical system.

from the telescope,[3] all of the rays coming from that object are essentially parallel to one another, although not necessarily parallel to the optical axis. If the rays are not parallel to the optical axis, distortion of the image can result; this is just one of many forms of *aberration* discussed later.

The image separation of two point sources on the focal plane is related to the focal length of the lens being used. Figure 6.6 shows the rays of two point sources, the direction of one source being along the optical axis of a converging lens and the other being at an angle $\theta$ with respect to the optical axis. At the position of the focal plane, the rays from the on-axis source will converge at the focal point while the rays from the other will *approximately* meet at a distance $y$ from the focal point. Now, from simple geometry, $y$ is given by

$$y = f \tan \theta$$

(the wavelength dependence of $f$ is implicitly assumed). If it is assumed that the field of view of the telescope is small, then $\theta$ must also be small. Using the small-angle approximation, $\tan \theta \simeq \theta$, for $\theta$ expressed in radians, we find

$$y = f\theta. \tag{6.3}$$

This immediately leads to the differential relation known as the **plate scale**, $d\theta/dy$,

$$\boxed{\frac{d\theta}{dy} = \frac{1}{f},} \tag{6.4}$$

which connects the angular separation of the objects with the linear separation of their images at the focal plane. As the focal length of the lens is increased, the linear separation of the images of two point sources separated by an angle $\theta$ also increases.

### Resolution and the Rayleigh Criterion

Unfortunately, the ability to "see" two objects in space that have a small angular separation $\theta$ is not simply a matter of choosing a focal length sufficiently long to produce the necessary plate scale. A fundamental limit exists in our ability to *resolve* those objects. This limitation is due to diffraction produced by the advancing wavefronts of light coming from those objects.

---

[3]Technically, this implies that the distance to the astronomical object is much greater than the focal length of the telescope.

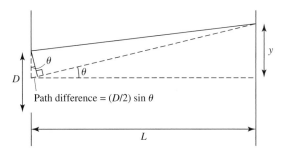

**FIGURE 6.7** For a minimum to occur, the path difference between paired rays must be a half-wavelength.

This phenomenon is closely related to the well-known single-slit diffraction pattern, which is similar to the Young double-slit interference pattern discussed in Section 3.3.

To understand single-slit diffraction, consider a slit of width $D$ (see Fig. 6.7). Assuming that the advancing wavefronts are coherent, any ray passing through the opening (or **aperture**) and arriving at a specific point in the focal plane can be thought of as being associated with another ray passing through the aperture exactly one-half of a slit width away and arriving at the same point. If the two rays are one-half wavelength ($\lambda/2$) out of phase, then destructive interference will occur. This leads to the relation

$$\frac{D}{2} \sin \theta = \frac{1}{2}\lambda,$$

or

$$\sin \theta = \frac{\lambda}{D}.$$

We can next consider dividing the aperture into four equal segments and pairing up a ray from the edge of the opening with one passing through a point one-quarter of a slit width away. For destructive interference to occur in this case, it is necessary that

$$\frac{D}{4} \sin \theta = \frac{1}{2}\lambda,$$

which gives

$$\sin \theta = 2\frac{\lambda}{D}.$$

This analysis may be continued by considering the aperture as being divided into six segments, then eight segments, then ten segments, and so on. We see, therefore, that the condition for minima to occur as a result of destructive interference from light passing through a single slit is given in general by

$$\boxed{\sin \theta = m\frac{\lambda}{D},} \tag{6.5}$$

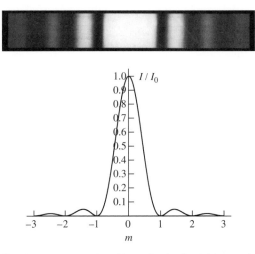

**FIGURE 6.8**   The diffraction pattern produced by a single slit. (Photograph from Cagnet, Francon, and Thrierr, *Atlas of Optical Phenomena*, Springer-Verlag, Berlin, 1962.)

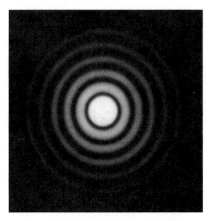

**FIGURE 6.9**   The circular aperture diffraction pattern of a point source. (Photograph from Cagnet, Francon, and Thrierr, *Atlas of Optical Phenomena*, Springer-Verlag, Berlin, 1962.)

where $m = 1, 2, 3, \ldots$ for dark fringes (as in Eq. 3.11). The intensity pattern produced by the light passing through a single slit is shown in Fig. 6.8.

The analysis for light passing through a circular aperture such as a telescope is similar, although somewhat more sophisticated. Due to the symmetry of the problem, the diffraction pattern appears as concentric rings (see Fig. 6.9). To evaluate this two-dimensional problem, it is necessary to perform a double integral over the aperture, considering the path differences of all possible pairs of rays passing through the aperture. The solution was first obtained in 1835 by Sir George Airy (1801–1892), Astronomer Royal of England; the central bright spot of the diffraction pattern is known as the **Airy disk**. Equation (6.5) remains appropriate for describing the locations of *both* the maxima and the minima, but $m$ is no longer an integer.

**TABLE 6.1** The locations and intensity maxima of the diffraction rings produced by a circular aperture.

| Ring | $m$ | $I/I_0$ |
|---|---|---|
| Central maximum | 0.000 | 1.00000 |
| First minimum | 1.220 | |
| Second maximum | 1.635 | 0.01750 |
| Second minimum | 2.233 | |
| Third maximum | 2.679 | 0.00416 |
| Third minimum | 3.238 | |

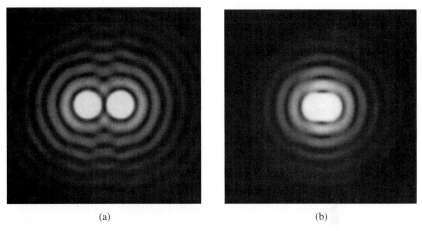

(a)  (b)

**FIGURE 6.10** The superimposed diffraction patterns from two point sources. (a) The sources are easily resolved. (b) The two sources are barely resolvable. (Photographs from Cagnet, Francon, and Thrierr, *Atlas of Optical Phenomena*, Springer-Verlag, Berlin, 1962.)

Table 6.1 lists the values of $m$, along with the relative intensities of the maxima, for the first three orders.

As can be seen in Fig. 6.10, when the diffraction patterns of two sources are sufficiently close together (e.g., there is a very small angular separation, $\theta_{min}$), the diffraction rings are no longer clearly distinguished and it becomes impossible to resolve the two sources. The two images are said to be unresolved when the central maximum of one pattern falls inside the location of the first minimum of the other. This *arbitrary* resolution condition is referred to as the **Rayleigh criterion**.[4] Assuming that $\theta_{min}$ is quite small, and invoking the small-angle approximation, $\sin \theta_{min} \simeq \theta_{min}$, where $\theta_{min}$ is expressed in radians, the Rayleigh

---

[4]By undertaking a careful analysis of the diffraction patterns of the sources, it is possible to resolve objects that are somewhat more closely spaced than allowed by the Rayleigh criterion.

criterion is given by

$$\boxed{\theta_{\min} = 1.22 \, \frac{\lambda}{D}}$$

(6.6)

for a circular aperture. Therefore, the resolution of a telescope improves with increasing aperture size and when shorter wavelengths are observed, just as expected for diffraction phenomena.

## Seeing

Unfortunately, despite the implications of Eq. (6.6), the resolution of ground-based optical telescopes does not improve without limit as the size of the primary lens or mirror is increased, unless certain complex, real-time adjustments are made to the optical system (see page 159). This consequence is due to the turbulent nature of Earth's atmosphere. Local changes in atmospheric temperature and density over distances ranging from centimeters to meters create regions where the light is refracted in nearly random directions, causing the image of a point source to become blurred. Since virtually all stars effectively appear as point sources, even when viewed through the largest telescopes, atmospheric turbulence produces the well-known "twinkling" of stellar images. The quality of the image of a stellar point source at a given observing location at a specific time is referred to as **seeing**. Some of the best seeing conditions found anywhere in the world are at **Mauna Kea Observatories** in Hawaii, located 4200 m (13,800 feet) above sea level, where the resolution is between 0.5″ and 0.6″ approximately 50% of the time, improving to 0.25″ on the best nights (Fig. 6.11). Other locations known for their excellent seeing are **Kitt Peak National Observatory** near Tucson, Arizona, **Tenerife** and **La Palma** of the Canary Islands, and several sites in the Chilean Andes Mountains [**Cerro Tololo Inter-American Observatory**, the **Cerro La Silla** and **Cerro Paranal** sites of the **European Southern Observatory**, and **Cerro Pachón**, location of Gemini South (Gemini North is on Mauna Kea)]. As a result, these sites have become locations where significant collections of optical telescopes and/or large-aperture optical telescopes have been built.

It is interesting to note that since the angular size of most planets is actually larger than the scale of atmospheric turbulence, distortions tend to be averaged out over the size of the image, and the "twinkling" effect is removed.

---

**Example 6.1.1.**   After many years of delays, the **Hubble Space Telescope** (HST) was finally placed in an orbit 610 km (380 miles) high by the Space Shuttle *Discovery* in April 1990 [see Fig. 6.12(a)]. At this altitude, HST is above the obscuring atmosphere of Earth yet still accessible for needed repairs, instrument upgrades or replacement, or a boost in its constantly decaying orbit.[5] HST is the most ambitious and, at a cost of approximately $2 billion, the most expensive scientific project ever completed.

*continued on page 150*

---

[5]Decaying orbits are caused by the drag produced by Earth's extended, residual atmosphere. The extent of the atmosphere is determined in part by the heating associated with the solar cycle; see Section 11.3.

(a)

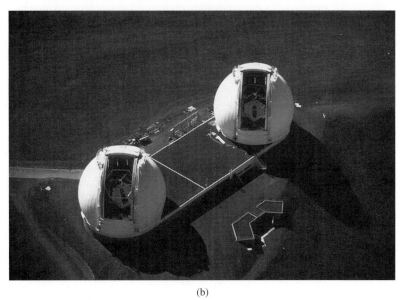

(b)

**FIGURE 6.11** (a) The Mauna Kea Observatories in Hawaii. Among the telescopes visible in this view are Gemini North (open silver dome left of center, 8.1 m, optical/IR, operated by a seven-country consortium), the Canada-France-Hawaii Telescope (front center, 3.6 m optical), twin W. M. Keck Telescopes (back right, two 10 m, optical, Caltech and University of California, United States), and Japan's Subaru Telescope (left of Keck I and Keck II, 8.2 m, optical/IR). (Copyright 1998, Richard Wainscoat.) (b) Keck I and Keck II. These telescopes can be operated as an optical interferometer. (Copyright 1998, Richard Wainscoat.)

HST has a 2.4-m (94-inch) primary mirror. When we observe at the ultraviolet wavelength of the hydrogen Lyman alpha (Ly$\alpha$) line, 121.6 nm, the Rayleigh criterion implies a resolution limit of

$$\theta = 1.22 \left( \frac{121.6 \text{ nm}}{2.4 \text{ m}} \right) = 6.18 \times 10^{-8} \text{ rad} = 0.0127''.$$

This is roughly the equivalent of the angle subtended by a quarter from 400 km away! It was projected that HST would not quite be "diffraction-limited" in the ultraviolet region due to *extremely* small imperfections in the surfaces of the mirrors. Since resolution is proportional to wavelength and mirror defects become less significant as the wavelength increases, HST should have been nearly diffraction-limited at the red end of the visible spectrum. Unfortunately, because of an error in the grinding of the primary mirror, an optimal shape was not obtained. Consequently, those initial expectations were not realized until corrective optics packages were installed during a repair mission in December 1993 [Fig. 6.12(b)].

---

## Aberrations

Both lens and mirror systems suffer from inherent image distortions known as **aberrations**. Often these aberrations are common to both types of systems, but **chromatic aberration** is unique to refracting telescopes. The problem stems from the fact that the focal length of a lens is wavelength-dependent. Equation (6.1) shows that since the index of refraction varies with wavelength, the angle of refraction at the interface between two different media must also depend on wavelength. This translates into a wavelength-dependent focal length (Eq. 6.2) and, as a result, a focal point for blue light that differs from that for red light. The problem of chromatic aberration can be diminished somewhat by the addition of correcting lenses. The demonstration of this procedure is left as an exercise.

Several aberrations result from the shape of the reflecting or refracting surface(s). Although it is easier, and therefore cheaper, to grind lenses and mirrors into spheroids, not all areas of these surfaces will focus a parallel set of light rays to a single point. This effect, known as **spherical aberration**, can be overcome by producing carefully designed optical surfaces (paraboloids).

The cause of HST's initial imaging problems is a classic case of spherical aberration. A mistake that was made while grinding the primary left the center of the mirror too shallow by approximately two microns. The result of this minute error was that light reflected from near the edge of the mirror came to a focus almost 4 cm behind light reflected from the central portion. When the best possible compromise focal plane was used, the image of a point source (such as a distant star) had a definable central core and an extended, diffuse halo. Although the central core was quite small (0.1″ radius), unfortunately it contained only 15% of the energy. The halo included more than half of the total energy and had a diameter of about 1.5″ (typical of traditionally designed ground-based telescopes). The remainder of the energy (approximately 30%) was spread out over an even larger area. Some of HST's original spherical aberration was compensated for by the use of computer programs designed to analyze the images produced by the flawed optical system and mathematically

(a) (b)

**FIGURE 6.12** (a) The 1990 launch of the Hubble Space Telescope aboard the Space Shuttle *Discovery*. (b) HST and the Space Shuttle *Endeavour* during the December 1993 repair mission to install optical systems to compensate for a misshapen primary mirror. (Courtesy of NASA.)

create corrected versions. In addition, during the repair mission in 1993, special corrective optics packages were installed in the telescope. Today the spherical aberration problem of HST is only a bad memory of what can go wrong.

Even when paraboloids are used, mirrors are not necessarily free from aberrations. **Coma** produces elongated images of point sources that lie off the optical axis, because the focal lengths of paraboloids are a function of $\theta$, the angle between the direction of an incoming light ray and the optical axis. **Astigmatism** is a defect that derives from having different parts of a lens or mirror converge an image at slightly different locations on the focal plane. When a lens or mirror is designed to correct for astigmatism, **curvature of field** can then be a problem. Curvature of field is due to the focusing of images on a curve rather than on a plane. Yet another potential difficulty occurs when the plate scale (Eq. 6.4) depends on the distance from the optical axis; this effect is referred to as **distortion of field**.

### The Brightness of an Image

In addition to resolution and aberration issues, telescope design must also consider the desired **brightness** of an image. It might be assumed that the brightness of an extended (resolved) image would increase with the area of the telescope lens, since more photons are collected as the aperture size increases; however, this assumption is not necessarily correct. To understand the brightness of an image, we begin by considering the **intensity** of the radiation. Some of the energy radiated from an infinitesimal portion of the surface of the source of area $d\sigma$ [shown in Fig. 6.13(a)] will enter a cone of differential **solid angle** $d\Omega \equiv dA_\perp / r^2$, where $dA_\perp$ is an infinitesimal amount of surface area that is located a distance $r$ from $d\sigma$ and oriented perpendicular to the position vector $\mathbf{r}$ [Fig. 6.13(b)].[6] The

---

[6]The unit of solid angle is the steradian, sr. It is left as an exercise to show that $\Omega_{\text{tot}} = \oint d\Omega = 4\pi$ sr; the total solid angle about a point $P$, resulting from an integration over a closed surface containing that point, is $4\pi$ sr.

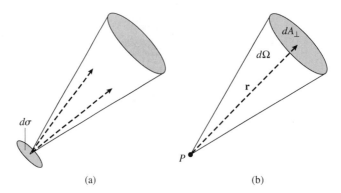

**FIGURE 6.13**    (a) The geometry of intensity. (b) The definition of solid angle.

intensity is given by the amount of energy per unit time interval $dt$, and per unit wavelength interval $d\lambda$, radiated from $d\sigma$ into a differential solid angle $d\Omega$; the units of intensity are $\text{W m}^{-2} \text{ nm}^{-1} \text{ sr}^{-1}$.

Consider an object located at a distance $r$ far from a telescope of focal length $f$. Assuming that the object is effectively infinitely far away (i.e., $r \gg f$), the image intensity $I_i$ may be determined from geometry. If an infinitesimal amount of surface area, $dA_0$, of the object has a surface intensity given by $I_0$, then the amount of energy per second per unit wavelength interval radiated into the solid angle defined by the telescope's aperture, $d\Omega_{T,0}$, is given by

$$I_0 \, d\Omega_{T,0} \, dA_0 = I_0 \left( \frac{A_T}{r^2} \right) dA_0,$$

where $A_T$ is the area of the telescope's aperture [see Fig. 6.14(a)]. Since an image will form from the photons emitted by the object, all of the photons coming from $dA_0$ within the solid angle $d\Omega_{T,0}$ must strike an area $dA_i$ on the focal plane.[7] Therefore,

$$I_0 \, d\Omega_{T,0} \, dA_0 = I_i \, d\Omega_{T,i} \, dA_i,$$

where $d\Omega_{T,i}$ is the solid angle defined by the telescope's aperture as seen from the image, or

$$I_0 \left( \frac{A_T}{r^2} \right) dA_0 = I_i \left( \frac{A_T}{f^2} \right) dA_i.$$

Solving for the image intensity gives

$$I_i = I_0 \left( \frac{dA_0/r^2}{dA_i/f^2} \right).$$

However, as can be seen in Fig. 6.14(b), the solid angle $d\Omega_{0,T}$ containing the entire object as seen from the center of the telescope's aperture must equal the solid angle $d\Omega_{i,T}$ of the

[7]Assuming, of course, that no photons are absorbed or scattered out of the beam in transit.

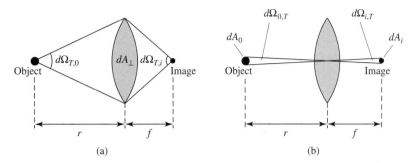

**FIGURE 6.14**   The effect of telescopes on image intensity $(r \gg f)$. (a) The solid angles subtended by the telescope, as measured from the object and the image. (b) The solid angle subtended by the object and the image, as measured from the center of the telescope.

entire image, also seen from the telescope center, or $d\Omega_{0,T} = d\Omega_{i,T}$. This implies that

$$\frac{dA_0}{r^2} = \frac{dA_i}{f^2}.$$

Substituting into the expression for the image intensity gives the result that

$$I_i = I_0;$$

the image intensity is identical to the object intensity, independent of the area of the aperture. This result is completely analogous to the simple observation that a wall does not appear to get brighter when the observer walks toward it.

The concept that describes the effect of the light-gathering power of telescopes is the **illumination** $J$, the amount of light energy per second focused onto a unit area of the resolved image. Since the amount of light collected from the source is proportional to the area of the aperture, the illumination $J \propto \pi(D/2)^2 = \pi D^2/4$, where $D$ is the diameter of the aperture. We have also shown that the linear size of the image is proportional to the focal length of the lens (Eq. 6.3); therefore, the image area must be proportional to $f^2$, and correspondingly, the illumination must be inversely proportional to $f^2$. Combining these results, the illumination must be proportional to the square of the ratio of the aperture diameter to the focal length. The inverse of this ratio is often referred to as the **focal ratio**,

$$\boxed{F \equiv \frac{f}{D}.} \tag{6.7}$$

Thus the illumination is related to the focal ratio by

$$J \propto \frac{1}{F^2}. \tag{6.8}$$

Since the number of photons per second striking a unit area of photographic plate or some other detector is described by the illumination, the illumination indicates the amount of time required to collect the photons needed to form a sufficiently bright image for analysis.

---

**Example 6.1.2.**    The twin multimirror telescopes of the **Keck Observatory** at Mauna Kea have primary mirrors 10 m in diameter with focal lengths of 17.5 m. The focal ratios of these mirrors are

$$F = \frac{f}{D} = 1.75.$$

It is standard to express focal ratios in the form $f/F$, where $f/$ signifies that the focal ratio is being referenced. Using this notation, the Keck telescopes have 10-m, $f/1.75$ primary mirrors.

---

We now see that the size of the aperture of a telescope is critical for two reasons: A larger aperture both improves resolution and increases the illumination. On the other hand, a longer focal length increases the linear size of the image but decreases the illumination. For a *fixed* focal ratio, increasing the diameter of the telescope results in greater spatial resolution, but the illumination remains constant. The proper design of a telescope must take into account the principal applications that are intended for the instrument.

## 6.2 ■ OPTICAL TELESCOPES

In the last section we studied some of the fundamental aspects of optics in the context of astronomical observing. We now build on those concepts to consider design features of optical telescopes.

### Refracting Telescopes

The major optical component of a refracting telescope is the primary or *objective* lens of focal length $f_{\text{obj}}$. The purpose of the objective lens is to collect as much light as possible and with the greatest possible resolution, bringing the light to a focus at the focal plane. A photographic plate or other detector may be placed at the focal plane to record the image, or the image may be viewed with an eyepiece, which serves as a magnifying glass. The eyepiece would be placed at a distance from the focal plane equal to its focal length, $f_{\text{eye}}$, causing the light rays to be refocused at infinity. Figure 6.15 shows the path of rays coming from a point source lying off the optical axis at an angle $\theta$. The rays ultimately emerge from the eyepiece at an angle $\phi$ from the optical axis. The **angular magnification** produced by this arrangement of lenses can be shown to be (Problem 6.5)

$$m = \frac{f_{\text{obj}}}{f_{\text{eye}}}. \tag{6.9}$$

Clearly, eyepieces of different focal lengths can produce different angular magnifications. Viewing a large image requires a long objective focal length, in combination with a short focal length for the eyepiece.

Recall, however, that the illumination decreases with the square of the objective's focal length (Eq. 6.8). To compensate for the diminished illumination, a larger-diameter objective is needed. Unfortunately, significant practical limitations exist for the size of the objective

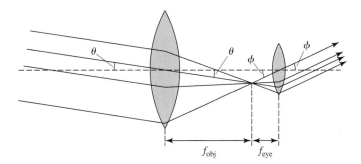

**FIGURE 6.15**    A refracting telescope is composed of an objective lens and an eyepiece.

lens of a refracting telescope. Because light must pass through the objective lens, it is possible to support the lens only from its edges. As a result, when the size and weight of the lens are increased, deformation in its shape occurs because of gravity. The specific form of the deformation depends on the position of the objective, which changes as the orientation of the telescope changes.

Another problem related to size is the difficulty in constructing a lens that is sufficiently free of defects. Since light must pass through the lens, its entire *volume* must be nearly optically perfect. Furthermore, *both* surfaces of the lens must be ground with great precision. Specifically, any defects in the material from which the lens is made and any deviations from the desired shape of the surface must be kept to less than some small fraction of the wavelength, typically $\lambda/20$. When observing at 500 nm, this implies that any defects must be smaller than approximately 25 nm. (Recall that the diameter of an atom is on the order of 0.1 nm.)

Yet another difficulty with a large objective lens occurs because of its slow thermal response. When the dome is opened, the temperature of the telescope must adjust to its new surroundings. This produces thermally driven air currents around the telescope, significantly affecting seeing. The shape of the telescope will also change as a consequence of thermal expansion, making it advantageous to minimize the "thermal mass" of the telescope as much as possible.

A mechanical problem also arises with long focal-length refractors. Due to the long lever arm involved, placing a massive detector on the end of the telescope will create a large amount of torque that requires compensation.

We have already discussed the unique problem of chromatic aberration in lenses, a complication not shared by mirrors. Considering all of the challenges inherent in the design and construction of refracting telescopes, the vast majority of all large modern telescopes are reflectors. The largest refracting telescope in use today is at the **Yerkes Observatory** in Williams Bay, Wisconsin (Fig. 6.16 on the following page). It was built in 1897 and has a 40-in (1.02-m) objective with a focal length of 19.36 m.

**Reflecting Telescopes**

With the exception of chromatic aberration, most of the basic optical principles already discussed apply equally well to reflectors and refractors. A reflecting telescope is designed by

**FIGURE 6.16**   The 40-in (1.02-m) telescope at Yerkes Observatory was built in 1897 and is the largest refractor in the world. (Courtesy of Yerkes Observatory.)

replacing the objective lens with a mirror, significantly reducing or completely eliminating many of the problems already discussed. Because the light does not pass through a mirror, only the *one* reflecting surface needs to be ground with precision. Also, the weight of the mirror can be minimized by creating a honeycomb structure behind the reflecting surface, removing a large amount of unnecessary mass. In fact, because the mirror is supported from behind rather than along its edges, it is possible to design an active system of pressure pads that can help to eliminate distortions in the mirror's shape produced by thermal effects and the changes in the gravitational force on the mirror as the telescope moves (a process known as **active optics**).

Reflecting telescopes are not completely free of drawbacks, however. Since the objective mirror reflects light back along the direction from which it came, the focal point of the mirror, known as the **prime focus**, is in the path of the incoming light [see Fig. 6.17(a)]. An observer or a detector can be placed at this position, but then some of the incident light is cut off (see Fig. 6.18). If the detector is too large, a substantial amount of light will be lost.

Isaac Newton first found a solution to the problem by placing a small, flat mirror in the reflected light's path, changing the location of the focal point; this arrangement is depicted in Fig. 6.17(b). Of course, the presence of this secondary mirror does block some of the incoming light from the primary, but if the ratio of the areas of the primary and secondary is sufficiently large, the effect of the lost light can be minimized. A **Newtonian** telescope design suffers from the drawback that the eyepiece (or detector) must be placed at a significant distance from the center of mass of the telescope. If a massive detector were used, it would exert a significant torque on the telescope.

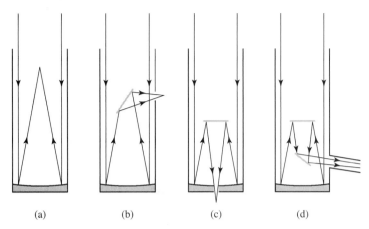

**FIGURE 6.17** Schematic drawings of various telescope optical systems: (a) Prime focus, (b) Newtonian, (c) Cassegrain, (d) coudé.

**FIGURE 6.18** Edwin Hubble (1889–1953) working at the prime focus of the Hale reflecting telescope on Mount Palomar. (Courtesy of Palomar/Caltech.)

Since the region of the primary mirror located behind the secondary is effectively useless anyway, it is possible to bore a hole in the primary and use the secondary to reflect the light back through the hole. This **Cassegrain** design [Fig. 6.17(c)] makes it possible to place heavy instrument packages near the center of mass of the telescope and permits an observer to stay near the bottom of the telescope, rather than near the top, as is the case for Newtonians. In this type of design the secondary mirror is usually convex, effectively increasing the focal length of the system.

The classical Cassegrain design uses a parabolic primary mirror. However, an important modification to the Cassegrain design, known as a **Ritchey–Chrétien** design, uses a hyperbolic primary mirror rather than a parabolic one.

If the instrument package is too massive, it is often more effective to bring the light directly to a special laboratory in which the detector is located. A **coudé telescope** [Fig. 6.17(d)] uses a series of mirrors to reflect the light down the telescope's mount to a *coudé room* located below the telescope. Because of the extended optical path, it is possible to create a very long focal length with a coudé telescope. This can be particularly useful in high-resolution work or in high-dispersion spectral line studies (see Section 5.1).

A unique instrument is the **Schmidt** telescope, specifically designed to provide a wide-angle field of view with low distortion. Schmidt telescopes are generally used as cameras, with the photographic plate located at the prime focus. To minimize coma, a spheroidal primary mirror is used, combined with a "correcting" lens to help remove spherical aberration. Whereas a large Cassegrain telescope may have a field of view of a few arc minutes across, a Schmidt camera has a field of view of several degrees. These instruments provide important survey studies of large regions of the sky. For example, sky survey plates from the Palomar and UK Schmidt telescopes have been scanned to produce the Guide Star Catalogue II containing 998,402,801 objects as faint as 19.5 magnitudes.[8] The stellar data in this catalog are being used to supply the reference (or *guide*) stars needed to orient the Hubble Space Telescope.

## Telescope Mounts

Producing high-resolution, deep-sky images of faint objects requires that the telescope be pointed at a fixed region of the sky for an extended period of time. This is necessary so that enough photons will be collected to ensure that the desired object can be seen. Such *time integration* requires careful *guiding* (or positional control) of the telescope while compensating for the rotation of Earth.

In order to account for Earth's rotation, perhaps the most common type of telescope mount (especially for smaller telescopes) is the **equatorial mount**. It incorporates a polar axis that is aligned to the north celestial pole, and the telescope simply rotates about that axis to compensate for the changing altitude and azimuth of the object of interest. With an equatorial mount, it is a simple matter to adjust the position of the telescope in both right ascension and declination. Unfortunately, for a massive telescope an equatorial mount can be extremely expensive and difficult to build. An alternative, more easily constructed mount for large telescopes, the **altitude–azimuth mount**, permits motion both parallel and perpendicular to the horizon. In this case, however, the tracking of a celestial object requires the continuous calculation of its altitude and azimuth from knowledge of the object's right ascension and declination, combined with knowledge of the local sidereal time and latitude of the telescope. A second difficulty with altitude–azimuth mounts is the effect of the continuous rotation of image fields. Without proper adjustment, this can create complications when guiding the telescope during an extended exposure or when a spectrum is obtained by passing the light through a long slit. Fortunately, rapid computer calculations can compensate for all of these effects.

---

[8]The count of objects in GSC II quoted here was effective in May of 2006.

**TABLE 6.2** Optical and/or near-infrared telescopes with apertures of 8 meters or more.

| Name | Size (m) | Site | First Light |
|---|---|---|---|
| Gemini North | 8.1 | Mauna Kea, Hawaii | 1999 |
| Gemini South | 8.1 | Cerro Pachón, Chile | 2002 |
| Subaru | 8.2 | Mauna Kea, Hawaii | 1999 |
| Very Large Telescope (VLT–Antu)[a] | 8.2 | Cerro Paranal, Chile | 1998 |
| Very Large Telescope (VLT)–Kueyen[a] | 8.2 | Cerro Paranal, Chile | 1999 |
| Very Large Telescope (VLT)–Melipal[a] | 8.2 | Cerro Paranal, Chile | 2000 |
| Very Large Telescope (VLT)–Yepun[a] | 8.2 | Cerro Paranal, Chile | 2000 |
| Large Binocular Telescope (LBT)[b] | 8.4 × 2 | Mt. Graham, Arizona | 2005 |
| Hobby–Eberly Telescope (HET)[c] | 9.2 | McDonald Observatory, Texas | 1999 |
| Keck I[d] | 10 | Mauna Kea, Hawaii | 1993 |
| Keck II[d] | 10 | Mauna Kea, Hawaii | 1996 |
| Gran Telescopio Canarias (GTC) | 10.4 | La Palma, Canary Islands | 2005 |
| Southern African Large Telescope (SALT)[e] | 11 | Sutherland, South Africa | 2005 |

[a] The four 8.2-m VLT telescopes, together with three 1.8-m auxiliary telescopes, can serve as an optical/IR interferometer.

[b] The two 8.4-m mirrors will sit on a single mount, with an effective collecting area of an 11.8-m aperture.

[c] Mounted with a fixed altitude angle of 55°. The mirror measures 11.1 m by 9.8 m with an effective aperture of 9.2 m.

[d] The two 10-m Keck telescopes, together with four 1.8-m outrigger telescopes, can serve as an optical/IR interferometer.

[e] Mounted with a fixed altitude angle of 37°.

## Large-Aperture Telescopes

In addition to long integration times, large aperture sizes play an important role in obtaining a sufficient number of photons to study a faint source (recall that the illumination is proportional to the diameter of the primary mirror of the telescope, Eq. 6.8). With tremendous improvements in telescope design, and aided by the development of high-speed computers, it has become possible to build very large-aperture telescopes. Table 6.2 contains a list of optical and/or near-infrared telescopes with apertures of greater than 8 m that are currently in operation. A number of much larger-aperture ground-based telescopes are also currently being considered, with effective mirror diameters ranging from 20 m to 100 m.

## Adaptive Optics

While large-aperture ground-based telescopes are able to gather many more photons than smaller telescopes over the same time interval, they are generally unable to resolve the object any more effectively without significant effort. In fact, even a ground-based 10-m telescope located at a site with exceptional seeing (e.g. the Keck telescopes at Mauna Kea) cannot resolve a source any better than an amateur's 20-cm backyard telescope can without the aide of *active optics* to correct distortions in the telescope's mirrors (page 156) and **adaptive optics** to compensate for atmospheric turbulence. In the later case, a small, deformable ("rubber") mirror is employed that has tens or perhaps hundreds of piezoelectric crystals attached to the back that act like tiny actuators. In order to counteract changes in the shape of the wavefronts coming from the source due to Earth's atmosphere, these crystals make micrometer-size adjustments to the shape of the mirror several hundred of times per second. In order to determine the changes that need to be applied, the telescope automatically monitors a guide star that is very near the target object.[9] Fluctuations in the guide star determine the adjustments that must be made to the deformable mirror. This process is somewhat easier in the near-infrared simply because of the longer wavelengths involved. As a result, adaptive optics systems have been successful in providing near-diffraction-limited images at near-infrared wavelengths.

**Space-Based Observatories**

In another effort to overcome the inherent imaging problems imposed by Earth's atmosphere, observational astronomy is also carried out in space. The **Hubble Space Telescope** (Fig. 6.12; named for Edwin Hubble) has a 2.4-m, $f/24$, primary that is the smoothest mirror ever constructed, with no surface imperfection larger than $1/50$ of the 632.8-nm test wavelength. Long exposures of 150 hours or more allow the telescope to "see" objects at least as faint as 30th magnitude. The optical system used by HST operates from 120 nm to 1 $\mu$m (ultraviolet to infrared, respectively) and is of the Ritchey–Chrétien type.

As HST approaches the end of its operational lifetime, plans are under way to replace it with the **James Webb Space Telescope** (JWST). The design specifications call for a telescope that will operate in the wavelength range between 600 nm and 28 $\mu$m, and it will have a primary mirror with a 6-m-diameter aperture. Unlike HST's low-Earth orbit, JWST will orbit about a gravitationally stable point that is located along the line connecting Earth and the Sun, but in the direction away from the Sun. This point, known as the second Lagrange point (L2), represents a balance between the gravitational forces of the Sun and Earth, and the centrifugal force due to its motion around the Sun as seen in a noninertial reference frame.[10] This location was chosen for the spacecraft in order to minimize thermal emissions that could otherwise affect its infrared detectors.

**Electronic Detectors**

Although the human eye and photographic plates have traditionally been the tools of astronomers to record images and spectra, other, more efficient devices are typically used in modern astronomy today. In particular, the semiconductor detector known as the charge-coupled device (CCD) has revolutionized the way in which photons are counted. Whereas the human eye has a very low *quantum efficiency* of approximately 1% (one photon in one hundred is detected), and photographic plates do only slightly better, CCDs are able to detect nearly 100% of the incident photons. Moreover, CCDs are able to detect a very wide range of wavelengths. From soft (low-energy) X-rays to the infrared, they have a linear response: Ten times as many photons produce a signal ten times stronger. CCDs also have

---

[9]In most cases, a sufficiently bright guide star does not exist close enough to the target. At a small number of observatories, an artificial laser guide star may be used in these circumstances. This is accomplished by firing a very powerful and carefully tuned laser into the sky in order to excite sodium atoms at an altitude of approximately 90 km.

[10]For more information about the Lagrange points, see Figure 18.3 and the discussion in Section 18.1.

a wide dynamic range and so can differentiate between very bright and very dim objects that are viewed simultaneously.

A CCD works by collecting electrons that are excited into higher energy states (conduction bands) when the detector is struck by a photon (a process similar to the photoelectric effect). The number of electrons collected in each pixel is then proportional to the brightness of the image at that location. The $2\frac{1}{2}$ million pixels of HST's second-generation Wide Field and Planetary Camera (WF/PC 2) are the individual elements of four $800 \times 800$ pixel CCD cameras, with each pixel capable of holding up to 70,000 electrons. HST's Advanced Camera for Surveys (ACS) contains an array of $4144 \times 4136$ (or 17,139,584) pixels for high-resolution survey work.

Given the rapid improvement in both ground-based and orbital telescopes, along with the tremendous advancements in detector technologies, it is clear that the future of optical astronomy is indeed a bright one.

## 6.3 ■ RADIO TELESCOPES

In 1931 Karl Jansky (1905–1950) was conducting experiments for Bell Laboratories related to the production of radio-wavelength static from thunderstorms. During the course of his investigations Jansky discovered that some of the static in his receiver was of "extraterrestrial origin." By 1935 he had correctly concluded that much of the signal he was measuring originated in the plane of the Milky Way, with the strongest emission coming from the constellation Sagittarius, which lies in the direction of the center of our Galaxy. Jansky's pioneering work represented the birth of *radio astronomy*, a whole new field of observational study.

Today radio astronomy plays an important role in our investigation of the electromagnetic spectrum. Radio waves are produced by a variety of mechanisms related to a range of physical processes, such as the interactions of charged particles with magnetic fields. This window on the universe provides astronomers and physicists with valuable clues to the inner workings of some of nature's most spectacular phenomena.

### Spectral Flux Density

Since radio waves interact with matter differently than visible light does, the devices used to detect and measure it are necessarily very different from optical telescopes. The parabolic dish of a typical radio telescope reflects the radio energy of the source to an antenna. The signal is then amplified and processed to produce a *radio map* of the sky at a particular wavelength, like the one shown in Fig. 6.19.

The strength of a radio source is measured in terms of the **spectral flux density**, $S(\nu)$, the amount of energy per second, per unit frequency interval striking a unit area of the telescope. To determine the total amount of energy per second (the power) collected by the receiver, the spectral flux must be integrated over the telescope's collecting area and over the frequency interval for which the detector is sensitive, referred to as the *bandwidth*. If $f_\nu$ is a function describing the efficiency of the detector at the frequency $\nu$, then the amount

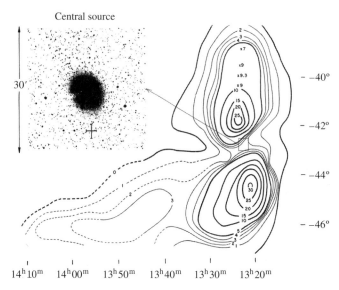

**FIGURE 6.19** A radio map of Centaurus A, together with an optical image of the same region. The contours show lines of constant radio power. (Figure from Matthews, Morgan, and Schmidt, *Ap. J.*, *140*, 35, 1964.)

of energy detected per second becomes[11]

$$P = \int_A \int_\nu S(\nu) f_\nu \, d\nu \, dA. \tag{6.10}$$

If the detector is 100% efficient over a frequency interval $\Delta\nu$ (i.e., $f_\nu = 1$), and if $S(\nu)$ can be considered to be constant over that interval, then the integral simplifies to give

$$P = SA\,\Delta\nu,$$

where $A$ is the effective area of the aperture.

A typical radio source has a spectral flux density $S(\nu)$ on the order of one jansky (Jy), where 1 Jy = $10^{-26}$ W m$^{-2}$ Hz$^{-1}$. Spectral flux density measurements of several mJy are not uncommon. With such weak sources, a large aperture is needed to collect enough photons to be measurable.

---

**Example 6.3.1.** The third strongest radio source in the sky, after the Sun and Cassiopeia A (a nearby supernova remnant), is the galaxy Cygnus A (see Fig. 6.20). At 400 MHz (a wavelength of 75 cm), its spectral flux density is 4500 Jy. Assuming that a 25-m-diameter

---

[11]A similar expression applies to optical telescopes since filters and detectors (including the human eye) are frequency dependent; see Section 3.6.

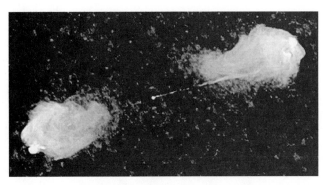

**FIGURE 6.20** A Very Large Array (VLA; see page 166) radio-wavelength image of the relativistic jets coming from the core of the galaxy Cygnus A. (Courtesy of the National Radio Astronomy Observatory, ©NRAO/AUI.)

radio telescope is 100% efficient and is used to collect the radio energy of this source over a frequency bandwidth of 5 MHz, the total power detected by the receiver would be

$$P = S(\nu)\pi \left(\frac{D}{2}\right)^2 \Delta\nu = 1.1 \times 10^{-13} \text{ W}.$$

### Improving Resolution: Large Apertures and Interferometry

One problem that radio telescopes share with optical telescopes is the need for greater resolution. Rayleigh's criterion (Eq. 6.6) applies to radio telescopes just as it does in the visible regime, except that radio wavelengths are much longer than those involved in optical work. Therefore, to obtain a level of resolution comparable to what is reached in the visible, much larger diameters are needed.

**Example 6.3.2.** To obtain a resolution of $1''$ at a wavelength of 21 cm using a single aperture, the dish diameter must be

$$D = 1.22 \frac{\lambda}{\theta} = 1.22 \left(\frac{21 \text{ cm}}{4.85 \times 10^{-6} \text{ rad}}\right) = 52.8 \text{ km}.$$

For comparison, the largest single-dish radio telescope in the world is the fixed dish 300 m (1000 ft) in diameter at **Arecibo Observatory**, Puerto Rico (see Fig. 6.21).

One advantage of working at such long wavelengths is that small deviations from an ideal parabolic shape are not nearly as crucial. Since the relevant criterion is to be within some small fraction of a wavelength (say $\lambda/20$) of what is considered a perfect shape, variations of 1 cm are tolerable when observing at 21 cm.

Although it is impractical to build individual dishes of sufficient size to produce the resolution at radio wavelengths that is anything like what is obtainable from the ground

**FIGURE 6.21** The 300-m radio telescope at Arecibo Observatory, Puerto Rico. (Courtesy of the NAIC–Arecibo Observatory, which is operated by Cornell University for the National Science Foundation.)

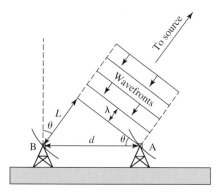

**FIGURE 6.22** The technique of radio interferometry.

in the visible regime, astronomers have nevertheless been able to resolve radio images to better than $0.001''$. This remarkable resolution is accomplished through a process not unlike the interference technique used in the Young double-slit experiment.

Figure 6.22 shows two radio telescopes separated by a baseline of distance $d$. Since the distance from telescope B to the source is greater than the distance from telescope A to the source by an amount $L$, a specific wavefront will arrive at B after it has reached A. The two signals will be *in phase* and their superposition will result in a maximum if $L$ is equal to an integral number of wavelengths ($L = n\lambda$, where $n = 0, 1, 2, \ldots$ for constructive interference). Similarly, if $L$ is an odd integral number of half-wavelengths, then the signals will be exactly *out of phase* and a superposition of signals will result in a minimum in the signal strength [$L = (n - \frac{1}{2})\lambda$, where $n = 1, 2, \ldots$ for destructive interference]. Since the pointing angle $\theta$ is related to $d$ and $L$ by

$$\sin \theta = \frac{L}{d}, \tag{6.11}$$

it is possible to accurately determine the position of the source by using the interference pattern produced by combining the signals of the two antennas. Equation (6.11) is completely analogous to Eq. (3.11) describing the Young double-slit experiment.

Clearly the ability to resolve an image improves with a longer baseline $d$. **Very long baseline interferometry** (VLBI) is possible over the size of a continent or even between continents. In such cases the data can be recorded on site and delivered to a central location for processing at a later time. It is only necessary that the observations be simultaneous and that the exact time of data acquisition be recorded.

Although a single antenna has its greatest level of sensitivity in the direction in which it is pointing, the antenna can also be sensitive to radio sources at angles far from the direction desired. Figure 6.23 shows a typical **antenna pattern** for a single radio telescope. It is a polar coordinate plot describing the direction of the antenna pattern along with the relative sensitivity in each direction; the longer the *lobe*, the more sensitive the telescope is in that direction. Two characteristics are immediately noticeable: First, the main lobe is not infinitesimally thin (the directionality of the beam is not perfect), and second, side lobes exist that can result in the accidental detection of unwanted sources that are indistinguishable from the desired source.

The narrowness of the main lobe is described by specifying its angular width at half its length, referred to as the **half-power beam width** (HPBW). This width can be decreased, and the effect of the side lobes can be significantly reduced, by the addition of other telescopes to produce the desired diffraction pattern. This property is analogous to the increase in sharpness of a grating diffraction pattern as the number of grating lines is increased; see Eq. (5.2).

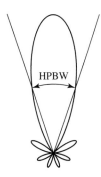

**FIGURE 6.23**   A typical antenna pattern for a single radio telescope. The width of the main lobe is describable by the half-power beam width (HPBW).

**FIGURE 6.24**    The Very Large Array (VLA) near Socorro, New Mexico. (Courtesy of the National Radio Astronomy Observatory, ©NRAO/AUI.)

The **Very Large Array** (VLA) located near Socorro, New Mexico, consists of 27 radio telescopes in a movable **Y** configuration with a maximum configuration diameter of 27 km. Each individual dish has a diameter of 25 m and uses receivers sensitive at a variety of frequencies (see Fig. 6.24). The signal from each of the separate telescopes is combined with all of the others and analyzed by computer to produce a high-resolution map of the sky; Fig. 6.20 is an example of an image produced by the VLA. Of course, along with the resolution gain, the 27 telescopes combine to produce an effective collecting area that is 27 times greater than that of an individual telescope.

The National Radio Astronomy Observatory (NRAO) plans to modernize and significantly expand the capabilities of the VLA. During Phase I, the **Expanded Very Large Array** (EVLA) will receive new, more sensitive receivers, extensive fiber-optic connections between the telescopes and the control facility, and greatly enhanced software and computational capabilities. Phase II of the expansion plan calls for the addition of approximately 8 new telescopes located throughout New Mexico that will augment the currently existing 27 telescopes. With baselines of up to 350 km, these new telescopes will greatly increase the resolution capability of the EVLA. The present VLA has a point source sensitivity of 10 $\mu$Jy, a highest-frequency resolution of 381 Hz, and a spatial resolution (at 5 GHz) of 0.4″. After completion of Phase II, the EVLA will have a point source sensitivity of 0.6 $\mu$Jy, a highest-frequency resolution of 0.12 Hz, and a spatial resolution (at 5 GHz) of 0.04″, which are improvements over the existing facility by one to two orders of magnitude.

NRAO also operates the **Very Long Baseline Array** (VLBA), composed of a series of 10 radio telescopes scattered throughout the continental United States, Hawaii, and St. Croix in the U.S. Virgin Islands. With a maximum baseline of 8600 km (5000 miles), the VLBA can achieve resolutions of better than 0.001″.

In addition to these and other radio observatories around the world, a major international effort is under way to construct the **Atacama Large Millimeter Array** (ALMA). ALMA will be composed of 50 12-m diameter antennas with baselines up to 12 km in length. Situated at an altitude of 5000 m (16,400 ft) in the Atacama desert region of Llano de Chajnantor in northern Chile, ALMA will be ideally located to work in the wavelength region from 10 mm to 350 $\mu$m (900 GHz to 70 GHz). At those wavelengths ALMA will be able to probe deeply into dusty regions of space where stars and planets are believed to be forming, as well as to study the earliest stages of galaxy formation—all critical problems in modern astrophysics. It is anticipated that ALMA will begin early scientific operations with a partial array in 2007. The full array is expected to be in operation by 2012.

## 6.4 ■ INFRARED, ULTRAVIOLET, X-RAY, AND GAMMA-RAY ASTRONOMY

Given the enormous amount of data supplied by optical and radio observations, it is natural to consider studies in other wavelength regions as well. Unfortunately, such observations are either difficult or impossible to perform from the ground because Earth's atmosphere is opaque to most wavelength regions outside of the visible and radio bands.

### Atmospheric Windows in the Electromagnetic Spectrum

Figure 6.25 shows the transparency of the atmosphere as a function of wavelength. Long-wavelength ultraviolet radiation and some regions in the infrared are able to traverse the atmosphere with limited success, but other wavelength regions are completely blocked. For this reason, special measures must be taken to gather information at many photon energies.

The primary contributor to infrared absorption is water vapor. As a result, if an observatory can be placed above most of the atmospheric water vapor, some observations can be made from the ground. To this end, both NASA and the United Kingdom operate infrared telescopes (3 m and 3.8 m, respectively) on Mauna Kea where the humidity is quite low. However, even at an altitude of 4200 m, the problem is not completely solved. To get above more of the atmosphere, balloon and aircraft observations have also been used.

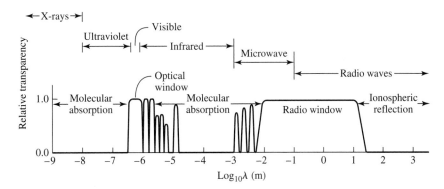

**FIGURE 6.25** The transparency of Earth's atmosphere as a function of wavelength.

Besides atmospheric absorption, the situation in the infrared is complicated still further because steps must be taken to cool the detector, if not the entire telescope. Using Wien's displacement law (Eq. 3.15), the peak wavelength of a blackbody of temperature 300 K is found to be nearly 10 $\mu$m. Thus the telescope and its detectors can produce radiation in just the wavelength region the observer might be interested in. Of course, the atmosphere itself can radiate in the infrared as well, including the production of molecular IR emission lines.

## Observing Above the Atmosphere

In 1983 the **Infrared Astronomy Satellite** (IRAS) was placed in an orbit 900 km (560 miles) high, well above Earth's obscuring atmosphere. The 0.6-m imaging telescope was cooled to liquid helium temperatures, and its detectors were designed to observe at a variety of wavelengths from 12 $\mu$m to 100 $\mu$m. Before its coolant was exhausted, IRAS proved to be very successful. Among its many accomplishments was the detection of dust in orbit around young stars, possibly indicating the formation of planetary systems. IRAS was also responsible for many important observations concerning the nature of galaxies.

Based upon the success of IRAS, the European Space Agency (in collaboration with Japan and the United States) launched the 0.6-m **Infrared Space Observatory** (ISO) in 1995. The observatory was cooled, just as IRAS was, but to obtain nearly 1000 times the resolution of IRAS, ISO was able to point toward a target for a much longer period of time, which enabled it to collect a greater number of photons. ISO ceased operations in 1998, after depletion of its liquid helium coolant.

The most recent and largest infrared observatory ever launched is the **Spitzer Space Telescope** [Fig. 6.26; named for Lyman Spitzer, Jr. (1914–1997)]. After many years of delay, this telescope was successfully placed into orbit on August 25, 2003. Trailing Earth in a heliocentric orbit, Spitzer is able to observe the heavens in the wavelength range from 3 $\mu$m to 180 $\mu$m. With a 0.85-m, f/12 mirror constructed of light-weight beryllium and cooled to less than 5.5 K, the telescope is able to provide diffraction-limited performance at wavelengths of 6.5 $\mu$m and longer. It is anticipated that Spitzer will have an operational lifetime of at least 2.5 years.[12]

Designed to investigate the electromagnetic spectrum at the longer wavelengths of the microwave regime, the **COsmic Background Explorer** (COBE) was launched in 1989 and finally switched off in 1993. COBE made a number of important observations, including very precise measurements of the 2.7 K blackbody spectrum believed to be the remnant fireball of the Big Bang.

As in other wavelength regions, a number of challenges exist when observing in the ultraviolet portion of the electromagnetic spectrum. In this case, because of the short wavelengths involved (as compared to optical observations), great care must be taken to provide a very precise reflecting surface. As has already been mentioned, even the HST primary mirror has imperfections that prohibit shorter UV wavelengths from being observed at the theoretical resolution limit.

A second UV observing problem stems from the fact that glass is opaque to these short-wavelength photons (as it is for much of the infrared). Consequently, glass lenses cannot

---

[12]The Spitzer Space Telescope was the last of NASA's four great orbiting observatories to be placed in orbit. The others were the Hubble Space Telescope (Fig. 6.12), the Compton Gamma Ray Observatory [Fig. 6.27(a)], and the Chandra X-ray Observatory [Fig. 6.27(b)].

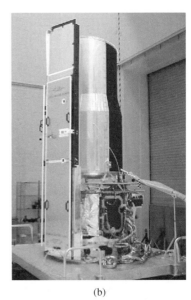

(a)                                                    (b)

**FIGURE 6.26**    (a) The Spitzer Infrared Telescope during construction. The 0.85-m beryllium primary mirror is visible. (Courtesy of NASA/JPL.) (b) The almost fully assembled Spitzer Space Telescope in the laboratory. (Courtesy of NASA/JPL.)

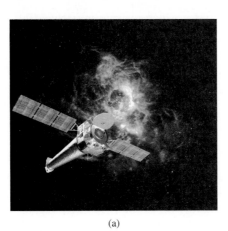

(a)                                                    (b)

**FIGURE 6.27**    (a) An artist's conception of the Chandra X-ray Observatory. (Illustration: NASA/MSFC.) (b) The Compton Gamma Ray Observatory being deployed by the Space Shuttle *Atlantis* in 1991.

be used in the optical system of a telescope designed to observe in the ultraviolet. Lenses made of crystal provide an appropriate substitute, however.

A real workhorse of ultraviolet astronomy was the **International Ultraviolet Explorer**. Launched in 1978 and operational until 1996, the IUE proved to be a remarkably productive and durable instrument. Today, HST, with its sensitivity down to 120 nm, provides another important window on the ultraviolet universe. At even shorter wavelengths, the

**Extreme Ultraviolet Explorer**, launched in 1992 and switched off more than eight years later, made observations between 7 nm and 76 nm. The data from these telescopes have given astronomers important information concerning a vast array of astrophysical processes, including mass loss from hot stars, cataclysmic variable stars, and compact objects such as white dwarfs and pulsars.

At even shorter wavelengths, X-ray and gamma-ray astronomy yields information about very energetic phenomena, such as nuclear reaction processes and the environments around black holes. As a result of the very high photon energies involved, X-ray and gamma-ray observations require techniques that differ markedly from those at longer wavelengths. For instance, traditional glass mirrors are useless for forming images in this regime because of the great penetrating power of these photons. However, it is still possible to image sources by using grazing-incidence reflections (incident angle close to 90°). X-ray spectra can also be obtained using techniques such as Bragg scattering, an interference phenomenon produced by photon reflections from atoms in a regular crystal lattice. The distance between the atoms corresponds to the separation between slits in an optical diffraction grating.

In 1970 **UHURU** (also known as the Small Astronomy Satellite–1, SAS 1) made the first comprehensive survey of the X-ray sky. In the late 1970s the three **High Energy Astrophysical Observatories**, including the **Einstein Observatory**, discovered thousands of X-ray and gamma-ray sources. Between 1990 and 1999, the X-ray observatory ROSAT (the **Roentgen Satellite**), a German–American–British satellite consisting of two detectors and an imaging telescope operating in the range of 0.51 nm to 12.4 nm, investigated the hot coronas of stars, supernova remnants, and quasars. Japan's **Advanced Satellite for Cosmology and Astrophysics**, which began its mission in 1993, also made valuable X-ray observations of the heavens before attitude control was lost as a result of a geomagnetic storm July 14, 2000.

Launched in 1999 and named for the Nobel Prize–winning astrophysicist Subrahmanyan Chandrasekhar (1910–1995), the **Chandra X-ray Observatory** [Fig. 6.27(a)] operates in the energy range from 0.2 keV to 10 keV (6.2 nm to 0.1 nm, respectively) with an angular resolution of approximately $0.5''$. Because X-rays cannot be focused in the same way that longer wavelengths can, grazing incidence mirrors are used to achieve the outstanding resolving power of Chandra.

The European Space Agency operates another X-ray telescope also launched in 1999, the **X-ray Multi-Mirror Newton Observatory** (XMM-Newton). Complementing Chandra's sensitivity range, XMM-Newton operates between 0.01 nm and 1.2 nm.

The **Compton Gamma Ray Observatory** [CGRO; Fig. 6.27(b)] observed the heavens at wavelengths shorter than those measured by the X-ray telescopes. Placed into orbit by the Space Shuttle *Atlantis* in 1991, the observatory was deorbited into the Pacific Ocean in June 2000.

### 6.5 ■ ALL-SKY SURVEYS AND VIRTUAL OBSERVATORIES

Our ability to probe the heavens at wavelengths spanning the electromagnetic spectrum has provided an enormous amount of information not previously available from ground-based observations made exclusively in the visible wavelength regime. For example, Fig. 6.28 illustrates the change in the appearance of the sky when different wavelength regions are

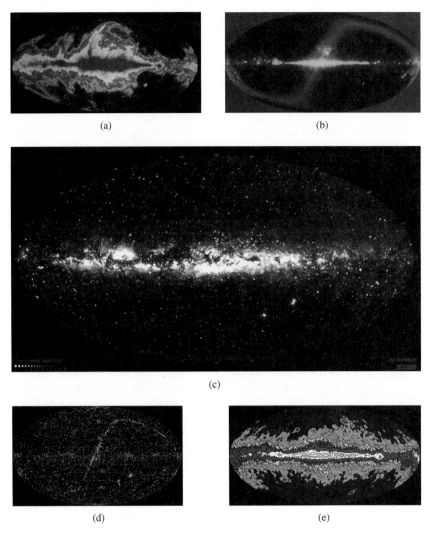

**FIGURE 6.28** Observations of the entire sky as seen in various wavelengths. The plane of the Milky Way Galaxy is clearly evident running horizontally across each image. Also evident in several of the images is the plane of our Solar System, running diagonally from lower left to upper right. (a) Radio (courtesy of the Max–Planck Institut für Radioastronomie), (b) infrared (courtesy of the COBE Science Working Group and NASA's Goddard Space Flight Center), (c) visible (courtesy of Lund Observatory), (d) ultraviolet (courtesy of NASA's Goddard Space Flight Center), and (e) gamma-ray wavelengths (courtesy of NASA).

explored (radio, infrared, visible, ultraviolet, and gamma-rays). Notice that the plane of our Milky Way galaxy is clearly evident in each of the wavelength bands but that other features are not necessarily present in each image.

The ground-based and space-based observatories described in this chapter by no means constitute a complete list. For instance, along with the orbital telescopes discussed in Section 6.4 are many others designed for general observing, or to carry out specialized studies, such as solar observations [e.g., the **Solar and Heliospheric Observatory** (SOHO) and the **Transition Region and Coronal Explorer** (TRACE)] or the determination of highly accurate positions of, and distances to, celestial objects[13] [the **Hipparcos Space Astrometry Mission** (completed, ESA), the **SIM PlanetQuest Mission** (SIM, anticipated launch 2011, NASA), and **Gaia** (anticipated launch 2011, ESA)].

Moreover, from the ground, large-scale, automated surveys have been, or are being, conducted in various wavelength regimes. For instance, the visible **Sloan Digital Sky Survey** (SDSS) and the near infrared **Two-Micron All Sky Survey** (2MASS) will produce tremendous volumes of data that need to be analyzed. SDSS alone will result in 15 terabytes of data (comparable to all of the information contained in the Library of Congress). Petabyte-sized data sets are also being envisioned in the not-too-distant future.

Given the enormous volumes of data that have already been and will be produced by ground-based and space-based astronomical observatories, together with the tremendous amount of information that already exists in on-line journals and databases, great attention is being given to developing web-based virtual observatories. The goal of these projects is to create user interfaces that give astronomers access to already-existing observational data. For instance, an astronomer could query a virtual observatory database for all of the observations that have ever been made in a specified region of the sky over any wavelength band. That data would then be downloaded to the astronomer's desktop computer or mainframe for study. In order to accomplish this task, common data formats must be created, and data analysis and visualization tools must be developed to aide in this challenging project in information technology. At the time this text was written, several prototype virtual observatories had been developed, such as *Skyview*[14] hosted by NASA's Goddard Space Flight Center, or the *Guide Star Catalogs* and the *Digitized Sky Survey*[15] maintained by the Space Telescope Science Institute. On-line access to a large number of currently existing databases are also available at the National Space Science Data Center (NSSDC).[16] In addition, several initiatives are under way to integrate and standardize the efforts. In the United States, the National Science Foundation (NSF) has funded the **National Virtual Observatory** project;[17] in Europe, the **Astrophysical Virtual Observatory** project is under way; the United Kingdom is pursuing **Astrogrid**; and Australia is working on the **Australian Virtual Observatory**. It is hoped that all of these efforts will ultimately be combined to create an **International Virtual Observatory**.

With the past successes of ground-based and orbital observatories, astronomers have been able to make great strides in our understanding of the universe. Given the current advances in detectors, observational techniques, new observational facilities, and virtual observatories, the future holds tremendous promise for providing significantly improved studies of known objects in the heavens. However, perhaps the most exciting implications of these observational advances are to be found in as yet undiscovered and unanticipated phenomena.

---

[13]**Astrometry** is the subdiscipline in astronomy that determines positional information of celestial objects.

[14]Skyview is available at `http://skyview.gsfc.nasa.gov`.

[15]The Guide Star Catalogs and the Digitized Sky Survey can both be consulted at `http://www-gsss.stsci.edu`.

[16]The NSSDC website is located at `http://nssdc.gsfc.nasa.gov/`.

[17]The National Virtual Observatory website is located at `http://www.us-vo.org`.

## SUGGESTED READING

### General

Colless, Matthew, "The Great Cosmic Map: The 2dF Galaxy Redshift Survey," *Mercury*, March/April 2003.

Frieman, Joshua A., and SubbaRao, Mark, "Charting the Heavens: The Sloan Digital Sky Survey," *Mercury*, March/April 2003.

Fugate, Robert Q., and Wild, Walter J., "Untwinkling the Stars—Part I," *Sky and Telescope*, May 1994.

Martinez, Patrick (ed.), *The Observer's Guide to Astronomy: Volume 1*, Cambridge University Press, Cambridge, 1994.

O'Dell, C. Robert, "Building the Hubble Space Telescope," *Sky and Telescope*, July 1989.

Schilling, Govert, "Adaptive Optics," *Sky and Telescope*, October 2001.

Schilling, Govert, "The Ultimate Telescope," *Mercury*, May/June 2002.

Sherrod, P. Clay, *A Complete Manual of Amateur Astronomy: Tools and Techniques for Astronomical Observations*, Prentice-Hall, Englewood Cliffs, NJ, 1981.

Stephens, Sally, " 'We Nailed It!' A First Look at the New and Improved Hubble Space Telescope," *Mercury*, January/February 1994.

Van Dyk, Schuyler, "The Ultimate Infrared Sky Survey: The 2MASS Survey," *Mercury*, March/April 2003.

White, James C. II, "Seeing the Sky in a Whole New Way," *Mercury*, March/April 2003.

Wild, Walter J., and Fugate, Robert Q., "Untwinkling the Stars—Part II," *Sky and Telescope*, June 1994.

### Technical

Beckers, Jacques M., "Adaptive Optics for Astronomy: Principles, Performance, and Applications," *Annual Review of Astronomy and Astrophysics*, *31*, 1993.

Culhane, J. Leonard, and Sanford, Peter W., *X-ray Astronomy*, Faber and Faber, London, 1981.

Jenkins, Francis A., and White, Harvey E., *Fundamentals of Optics*, Fourth Edition, McGraw-Hill, New York, 1976.

Kellermann, K. I., and Moran, J. M., "The Development of High-Resoution Imaging in Radio Astronomy," *Annual Review of Astronomy and Astrophysics*, *39*, 2001.

Kraus, John D., *Radio Astronomy*, Second Edition, Cygnus-Quasar Books, Powell, Ohio, 1986.

Quirrenbach, Andreas, "Optical Interferometry," *Annual Review of Astronomy and Astrophysics*, *39*, 2001.

Thompson, A. R., Moran, J. M., and Swenson, G. W., *Interferometry and Synthesis in Radio Astronomy*, Second Edition, Wiley, New York, 2001.

## PROBLEMS

**6.1** For some point $P$ in space, show that for any arbitrary closed surface surrounding $P$, the integral over a solid angle about $P$ gives

$$\Omega_{\text{tot}} = \oint d\Omega = 4\pi.$$

**6.2** The light rays coming from an object do not, in general, travel parallel to the optical axis of a lens or mirror system. Consider an arrow to be the object, located a distance $p$ from the center of a simple converging lens of focal length $f$, such that $p > f$. Assume that the arrow is perpendicular to the optical axis of the system with the tail of the arrow located on the axis. To locate the image, draw two light rays coming from the tip of the arrow:

**(i)** One ray should follow a path *parallel* to the optical axis until it strikes the lens. It then bends toward the focal point of the side of the lens opposite the object.

**(ii)** A second ray should pass directly through the center of the lens undeflected. (This assumes that the lens is sufficiently thin.)

The intersection of the two rays is the location of the tip of the image arrow. All other rays coming from the tip of the object that pass through the lens will also pass through the image tip. The tail of the image is located on the optical axis, a distance $q$ from the center of the lens. The image should also be oriented perpendicular to the optical axis.

**(a)** Using similar triangles, prove the relation

$$\frac{1}{p} + \frac{1}{q} = \frac{1}{f}.$$

**(b)** Show that if the distance of the object is much larger than the focal length of the lens ($p \gg f$), then the image is effectively located on the focal plane. This is essentially always the situation for astronomical observations.

The analysis of a diverging lens or a mirror (either converging or diverging) is similar and leads to the same relation between object distance, image distance, and focal length.

**6.3** Show that if two lenses of focal lengths $f_1$ and $f_2$ can be considered to have zero physical separation, then the effective focal length of the combination of lenses is

$$\frac{1}{f_{\text{eff}}} = \frac{1}{f_1} + \frac{1}{f_2}.$$

*Note:* Assuming that the actual physical separation of the lenses is $x$, this approximation is strictly valid only when $f_1 \gg x$ and $f_2 \gg x$.

**6.4 (a)** Using the result of Problem 6.3, show that a compound lens system can be constructed from two lenses of different indices of refraction, $n_{1\lambda}$ and $n_{2\lambda}$, having the property that the resultant focal lengths of the compound lens at two specific wavelengths $\lambda_1$ and $\lambda_2$, respectively, can be made equal, or

$$f_{\text{eff},\lambda_1} = f_{\text{eff},\lambda_2}.$$

**(b)** Argue qualitatively that this condition does not guarantee that the focal length will be constant for all wavelengths.

**6.5** Prove that the angular magnification of a telescope having an objective focal length of $f_{obj}$ and an eyepiece focal length of $f_{eye}$ is given by Eq. (6.9) when the objective and the eyepiece are separated by the sum of their focal lengths, $f_{obj} + f_{eye}$.

**6.6** The diffraction pattern for a single slit (Figs. 6.7 and 6.8) is given by

$$I(\theta) = I_0 \left[ \frac{\sin(\beta/2)}{\beta/2} \right]^2,$$

where $\beta \equiv 2\pi D \sin\theta/\lambda$.

**(a)** Using l'Hôpital's rule, prove that the intensity at $\theta = 0$ is given by $I(0) = I_0$.

**(b)** If the slit has an aperture of 1.0 $\mu$m, what angle $\theta$ corresponds to the first minimum if the wavelength of the light is 500 nm? Express your answer in degrees.

**6.7 (a)** Using the Rayleigh criterion, estimate the angular resolution limit of the human eye at 550 nm. Assume that the diameter of the pupil is 5 mm.

**(b)** Compare your answer in part (a) to the angular diameters of the Moon and Jupiter. You may find the data in Appendix C helpful.

**(c)** What can you conclude about the ability to resolve the Moon's disk and Jupiter's disk with the unaided eye?

**6.8 (a)** Using the Rayleigh criterion, estimate the theoretical diffraction limit for the angular resolution of a typical 20-cm (8-in) amateur telescope at 550 nm. Express your answer in arcseconds.

**(b)** Using the information in Appendix C, estimate the minimum size of a crater on the Moon that can be resolved by a 20-cm (8-in) telescope.

**(c)** Is this resolution limit likely to be achieved? Why or why not?

**6.9** The New Technology Telescope (NTT) is operated by the European Southern Observatory at Cerro La Silla. This telescope was used as a testbed for evaluating the adaptive optics technology used in the VLT. The NTT has a 3.58-m primary mirror with a focal ratio of $f/2.2$.

**(a)** Calculate the focal length of the primary mirror of the New Technology Telescope.

**(b)** What is the value of the plate scale of the NTT?

**(c)** $\epsilon$ Bootes is a double star system whose components are separated by 2.9″. Calculate the linear separation of the images on the primary mirror focal plane of the NTT.

**6.10** When operated in "planetary" mode, HST's WF/PC 2 has a focal ratio of $f/28.3$ with a plate scale of 0.0455″ pixel$^{-1}$. Estimate the angular size of the field of view of one CCD in the planetary mode.

**6.11** Suppose that a radio telescope receiver has a bandwidth of 50 MHz centered at 1.430 GHz (1 GHz = 1000 MHz). Assume that, rather than being a perfect detector over the entire bandwidth, the receiver's frequency dependence is triangular, meaning that the sensitivity of the detector is 0% at the edges of the band and 100% at its center. This filter function can be expressed as

$$f_\nu = \begin{cases} \dfrac{\nu}{\nu_m - \nu_\ell} - \dfrac{\nu_\ell}{\nu_m - \nu_\ell} & \text{if } \nu_\ell \le \nu \le \nu_m \\[2ex] -\dfrac{\nu}{\nu_u - \nu_m} + \dfrac{\nu_u}{\nu_u - \nu_m} & \text{if } \nu_m \le \nu \le \nu_u \\[2ex] 0 & \text{elsewhere.} \end{cases}$$

(a) Find the values of $\nu_\ell$, $\nu_m$, and $\nu_u$.

(b) Assume that the radio dish is a 100% efficient reflector over the receiver's bandwidth and has a diameter of 100 m. Assume also that the source NGC 2558 (a spiral galaxy with an apparent visual magnitude of 13.8) has a constant spectral flux density of $S = 2.5$ mJy over the detector bandwidth. Calculate the total power *measured* at the receiver.

(c) Estimate the power emitted at the source in this frequency range if $d = 100$ Mpc. Assume that the source emits the signal isotropically.

**6.12** What would the diameter of a single radio dish need to be to have a collecting area equivalent to that of the 27 telescopes of the VLA?

**6.13** How much must the pointing angle of a two-element radio interferometer be changed in order to move from one interference maximum to the next? Assume that the two telescopes are separated by the diameter of Earth and that the observation is being made at a wavelength of 21 cm. Express your answer in arcseconds.

**6.14** Assuming that ALMA is completed with the currently envisioned 50 antennas, how many unique baselines will exist within the array?

**6.15** The technical specifications for the planned SIM PlanetQuest mission call for the ability to resolve two point sources with an accuracy of better than $0.000004''$ for objects as faint as 20th magnitude in visible light. This will be accomplished through the use of optical interferometry.

(a) Assuming that grass grows at the rate of 2 cm per week, and assuming that SIM could observe a blade of grass from a distance of 10 km, how long would it take for SIM to detect a measurable change in the length of the blade of grass?

(b) Using a baseline of the diameter of Earth's orbit, how far away will SIM be able to determine distances using trigonometric parallax, assuming the source is bright enough? (For reference, the distance from the Sun to the center of the Milky Way Galaxy is approximately 8 kpc.)

(c) From your answer to part (b), what would the apparent magnitude of the Sun be from that distance?

(d) The star Betelgeuse (in Orion) has an absolute magnitude of $-5.14$. How far could Betelgeuse be from SIM and still be detected? (Neglect any effects of dust and gas between the star and the spacecraft.)

**6.16** (a) Using data available in the text or on observatory websites, list the wavelength ranges (in m) and photon energy ranges (in eV) covered by the following telescopes: VLA, ALMA, Spitzer, JWST, VLT/VLTI, Keck/Keck Interferometer, HST, IUE, EUVE, Chandra, CGRO.

(b) Graphically illustrate the wavelength coverage of each of the telescopes listed in part (a) by drawing a horizontal bar over a horizontal axis like the one shown in Fig. 6.25.

(c) Using photon energies rather than wavelengths, create a graphic similar to the one in part (b).

## COMPUTER PROBLEM

**6.17** Suppose that two identical slits are situated next to each other in such a way that the axes of the slits are parallel and oriented vertically. Assume also that the two slits are the same distance from a flat screen. Different light sources of identical intensity are placed behind each slit so that the two sources are incoherent, which means that double-slit interference effects can be neglected.

(a) If the two slits are separated by a distance such that the central maximum of the diffraction pattern corresponding to the first slit is located at the second minimum of the second slit's diffraction pattern, plot the resulting superposition of intensities (i.e., the total intensity at each location). Include at least two minima to the left of the central maximum of the leftmost slit and at least two minima to the right of the central maximum of the rightmost slit. *Hint:* Refer to the equation given in Problem 6.6 and plot your results as a function of $\beta$.

**(b)** Repeat your calculations for the case when the two slits are separated by a distance such that the central maximum of one slit falls at the location of the first minimum of the second (the Rayleigh criterion for single slits).

**(c)** What can you conclude about the ability to resolve two individual sources (the slits) as the sources are brought progressively closer together?

# PART

# II

# The Nature of Stars

# 7

# Binary Systems and Stellar Parameters

## 7.1 ■ THE CLASSIFICATION OF BINARY STARS

A detailed understanding of the structure and evolution of stars (the goal of Part II) requires knowledge of their physical characteristics. We have seen that knowledge of blackbody radiation curves, spectra, and parallax enables us to determine a star's effective temperature, luminosity, radius, composition, and other parameters. However, the only direct way to determine the mass of a star is by studying its gravitational interaction with other objects.

In Chapter 2 Kepler's laws were used to calculate the masses of members of our Solar System. However, the universality of the gravitational force allows Kepler's laws to be generalized to include the orbits of stars about one another and even the orbital interactions of galaxies, as long as proper care is taken to refer all orbits to the center of mass of the system.

Fortunately, nature has provided ample opportunity for astronomers to observe binary star systems. At least half of all "stars" in the sky are actually multiple systems, two or more stars in orbit about a common center of mass. Analysis of the orbital parameters of these systems provides vital information about a variety of stellar characteristics, including mass.

The methods used to analyze the orbital data vary somewhat depending on the geometry of the system, its distance from the observer, and the relative masses and luminosities of the components. Consequently, binary star systems are classified according to their specific observational characteristics.

- **Optical double**. These systems are not actually binaries at all but simply two stars that lie along the same line of sight (i.e., they have similar right ascensions and declinations). As a consequence of their large physical separations, the stars are not gravitationally bound, and hence the system is not useful in determining stellar masses.

- **Visual binary**. Both stars in the binary can be resolved independently, and if the orbital period is not prohibitively long, it is possible to monitor the motion of each member of the system. These systems provide important information about the angular separation of the stars from their mutual center of mass. If the distance to the binary is also known, the linear separations of the stars can then be calculated.

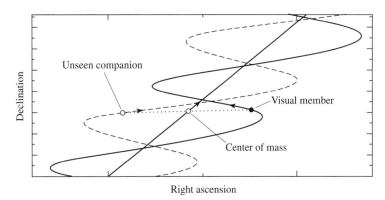

**FIGURE 7.1**   An astrometric binary, which contains one visible member. The unseen component is implied by the oscillatory motion of the observable star in the system. The proper motion of the entire system is reflected in the straight-line motion of the center of mass.

- **Astrometric binary**. If one member of a binary is significantly brighter than the other, it may not be possible to observe both members directly. In such a case the existence of the unseen member may be deduced by observing the oscillatory motion of the visible component. Since Newton's first law requires that a constant velocity be maintained by a mass unless a force is acting upon it, such an oscillatory behavior requires that another mass be present (see Fig. 7.1).

- **Eclipsing binary**.   For binaries that have orbital planes oriented approximately along the line of sight of the observer, one star may periodically pass in front of the other, blocking the light of the eclipsed component (see Fig. 7.2). Such a system is recognizable by regular variations in the amount of light received at the telescope. Not only do observations of these *light curves* reveal the presence of two stars, but the data can also provide information about relative effective temperatures and radii of each star based on the depths of the light curve minima and the lengths of the eclipses. Details of such an analysis will be discussed in Section 7.3.

- **Spectrum binary**.   A spectrum binary is a system with two superimposed, independent, discernible spectra. The Doppler effect (Eq. 4.35) causes the spectral lines of a star to be shifted from their rest frame wavelengths if that star has a nonzero radial velocity. Since the stars in a binary system are constantly in motion about their mutual center of mass, there must necessarily be periodic shifts in the wavelength of every spectral line of each star (unless the orbital plane is exactly perpendicular to the line of sight, of course). It is also apparent that when the lines of one star are blueshifted, the lines of the other must be redshifted relative to the wavelengths that would be produced if the stars were moving with the constant velocity of the center of mass. However, it may be that the orbital period is so long that the time dependence of the spectral wavelengths is not readily apparent. In any case, if one star is not overwhelmingly more luminous than its companion and if it is not possible to resolve each star separately, it may still be possible to recognize the object as a binary system by observing the superimposed and oppositely Doppler-shifted spectra.

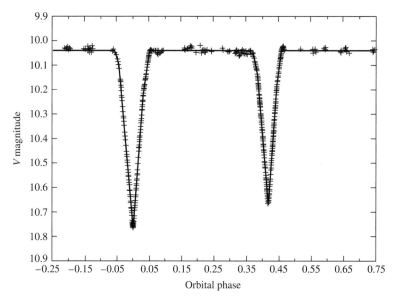

**FIGURE 7.2**    The $V$ magnitude light curve of YY Sagittarii, an eclipsing binary star. The data from many orbital periods have been plotted on this light curve as a function of phase, where the phase is defined to be 0.0 at the primary minimum. This system has an orbital period $P = 2.6284734$ d, an eccentricity $e = 0.1573$, and orbital inclination $i = 88.89°$ (see Section 7.2). (Figure adopted from Lacy, C. H. S., *Astron. J.*, *105*, 637, 1993.)

Even if the Doppler shifts are not significant (if the orbital plane is perpendicular to the line of sight, for instance), it may still be possible to detect two sets of super-imposed spectra if they originate from stars that have significantly different spectral features (see the discussion of spectral classes in Section 8.1).

- **Spectroscopic binary**.  If the period of a binary system is not prohibitively long and if the orbital motion has a component along the line of sight, a periodic shift in spectral lines will be observable. Assuming that the luminosities of the stars are comparable, both spectra will be observable. However, if one star is much more luminous than the other, then the spectrum of the less luminous companion will be overwhelmed and only a single set of periodically varying spectral lines will be seen. In either situation, the existence of a binary star system is revealed. Figure 7.3 shows the relationship between spectra and orbital phase for a spectroscopic binary star system.

These specific classifications are not mutually exclusive. For instance, an unresolved system could be both an eclipsing and a spectroscopic binary. It is also true that some systems can be significantly more useful than others in providing information about stellar characteristics. Three types of systems can provide us with mass determinations: visual binaries combined with parallax information; visual binaries for which radial velocities are available over a complete orbit; and eclipsing, double-line, spectroscopic binaries.

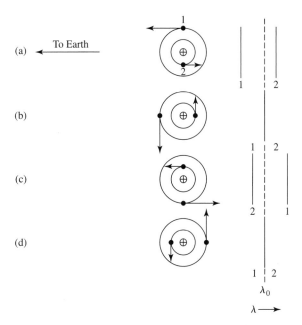

**FIGURE 7.3** The periodic shift in spectral features of a double-line spectroscopic binary. The relative wavelengths of the spectra of Stars 1 and 2 are shown at four different phases during the orbit: (a) Star 1 is moving toward the observer while Star 2 is moving away. (b) Both stars have velocities perpendicular to the line of sight. (c) Star 1 is receding from the observer while Star 2 is approaching. (d) Again both stars have velocities perpendicular to the line of sight. $\lambda_0$ represents the wavelength of the observed line Doppler-shifted by the velocity of the center of mass of the system.

## 7.2 ■ MASS DETERMINATION USING VISUAL BINARIES

When the angular separation between components of a binary system is greater than the resolution limit imposed by local seeing conditions and the fundamental diffraction limitation of the Rayleigh criterion, it becomes possible to analyze the orbital characteristics of the individual stars. From the orbital data, the orientation of the orbits and the system's center of mass can be determined, providing knowledge of the ratio of the stars' masses. If the distance to the system is also known, from trigonometric parallax for instance, the linear separation of the stars can be determined, leading to the individual masses of the stars in the system.

To see how a visual binary can yield mass information, consider two stars in orbit about their mutual center of mass. Assuming that the orbital plane is perpendicular to the observer's line of sight, we see from the discussion of Section 2.3 that the ratio of masses may be found from the ratio of the angular separations of the stars from the center of mass. Using Eq. (2.19) and considering only the lengths of the vectors $\mathbf{r}_1$ and $\mathbf{r}_2$, we find that

$$\frac{m_1}{m_2} = \frac{r_2}{r_1} = \frac{a_2}{a_1}, \tag{7.1}$$

where $a_1$ and $a_2$ are the semimajor axes of the ellipses. If the distance from the observer to the binary star system is $d$, then the angles subtended by the semimajor axes are

$$\alpha_1 = \frac{a_1}{d} \quad \text{and} \quad \alpha_2 = \frac{a_2}{d},$$

where $\alpha_1$ and $\alpha_2$ are measured in radians. Substituting, we find that the mass ratio simply becomes

$$\frac{m_1}{m_2} = \frac{\alpha_2}{\alpha_1}. \tag{7.2}$$

Even if the distance to the star system is not known, the mass ratio may still be determined. Note that since only the ratio of the subtended angles is needed, $\alpha_1$ and $\alpha_2$ may be expressed in arcseconds, the unit typically used for angular measure in astronomy.

The general form of Kepler's third law (Eq. 2.37),

$$P^2 = \frac{4\pi^2}{G(m_1 + m_2)} a^3,$$

gives the sum of the masses of the stars, provided that the semimajor axis ($a$) of the orbit of the reduced mass is known. Since $a = a_1 + a_2$ (the proof of this is left as an exercise), the semimajor axis can be determined directly only if the distance to the system has been determined. Assuming that $d$ is known, $m_1 + m_2$ may be combined with $m_1/m_2$ to give each mass separately.

This process is complicated somewhat by the proper motion of the center of mass[1] (see Fig. 7.1) and by the fact that most orbits are not conveniently oriented with their planes perpendicular to the line of sight of the observer. Removing the proper motion of the center of mass from the observations is a relatively simple process since the center of mass must move at a constant velocity. Fortunately, estimating the orientation of the orbits is also possible and can be taken into consideration.

Let $i$ be the **angle of inclination** between the plane of an orbit and the plane of the sky, as shown in Fig. 7.4; note that the orbits of both stars are necessarily in the same plane. As a special case, assume that the orbital plane and the plane of the sky (defined as being perpendicular to the line of sight) intersect along a line parallel to the minor axis, forming a **line of nodes**. The observer will not measure the actual angles subtended by the semimajor axes $\alpha_1$ and $\alpha_2$ but their projections onto the plane of the sky, $\tilde{\alpha}_1 = \alpha_1 \cos i$ and $\tilde{\alpha}_2 = \alpha_2 \cos i$. This geometrical effect plays no role in calculating the mass ratios since the $\cos i$ term will simply cancel in Eq. (7.2):

$$\frac{m_1}{m_2} = \frac{\alpha_2}{\alpha_1} = \frac{\alpha_2 \cos i}{\alpha_1 \cos i} = \frac{\tilde{\alpha}_2}{\tilde{\alpha}_1}.$$

However, this projection effect can make a significant difference when we are using Kepler's third law. Since $\alpha = a/d$ ($\alpha$ in radians), Kepler's third law may be solved for the sum of

---

[1]The annual wobble of stellar positions due to trigonometric parallax must also be considered, when significant.

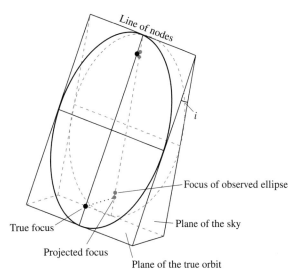

**FIGURE 7.4** An elliptical orbit projected onto the plane of the sky produces an observable elliptical orbit. The foci of the original ellipse do not project onto the foci of the observed ellipse, however.

the masses to give

$$m_1 + m_2 = \frac{4\pi^2}{G} \frac{(\alpha d)^3}{P^2} = \frac{4\pi^2}{G} \left( \frac{d}{\cos i} \right)^3 \frac{\tilde{\alpha}^3}{P^2}, \qquad (7.3)$$

where $\tilde{\alpha} = \tilde{\alpha}_1 + \tilde{\alpha}_2$.

To evaluate the sum of the masses properly, we must deduce the angle of inclination. This can be accomplished by carefully noting the apparent position of the center of mass of the system. As illustrated in Fig. 7.4, the projection of an ellipse tilted at an angle $i$ with respect to the plane of the sky will result in an observed ellipse with a different eccentricity. However, the center of mass will not be located at one of the foci of the projection—a result that is inconsistent with Kepler's first law. Thus the geometry of the true ellipse may be determined by comparing the observed stellar positions with mathematical projections of various ellipses onto the plane of the sky.

Of course, the problem of projection has been simplified here. Not only can the angle of inclination $i$ be nonzero, but the ellipse may be tilted about its major axis and rotated about the line of sight to produce any possible orientation. However, the general principles already mentioned still apply, making it possible to deduce the true shapes of the stars' elliptical orbits, as well as their masses.

It is also possible to determine the individual masses of members of visual binaries, even if the distance is not known. In this situation, detailed radial velocity data are needed. The projection of velocity vectors onto the line of sight, combined with information about the stars' positions and the orientation of their orbits, provides a means for determining the semimajor axes of the ellipses, as required by Kepler's third law.

## 7.3 ■ ECLIPSING, SPECTROSCOPIC BINARIES

A wealth of information is available from a binary system even if it is not possible to resolve each of its stars individually. This is particularly true for a double-line, eclipsing, spectroscopic binary star system. In such a system, not only is it possible to determine the individual masses of the stars, but astronomers may be able to deduce other parameters as well, such as the stars' radii and the ratio of their fluxes, and hence the ratio of their effective temperatures. (Of course, eclipsing systems are not restricted to spectroscopic binaries but may occur in other types of binaries as well, such as visual binaries.)

### The Effect of Eccentricity on Radial Velocity Measurements

Consider a spectroscopic binary star system for which the spectra of both stars may be seen (a double-line, spectroscopic binary). Since the individual members of the system cannot be resolved, the techniques used to determine the orientation and eccentricity of the orbits of visual binaries are not applicable. Also, the inclination angle $i$ clearly plays a role in the solution obtained for the stars' masses because it directly influences the measured radial velocities. If $v_1$ is the velocity of the star of mass $m_1$ and $v_2$ is the velocity of the star of mass $m_2$ at some instant, then, referring to Fig. 7.4, the observed radial velocities cannot exceed $v_{1r}^{\max} = v_1 \sin i$ and $v_{2r}^{\max} = v_2 \sin i$, respectively. Therefore, the actual measured radial velocities depend upon the positions of the stars at that instant. As a special case, if the directions of motion of the stars happen to be perpendicular to the line of sight, then the observed radial velocities will be zero.

For a star system having circular orbits, the speed of each star will be constant. If the plane of their orbits lies in the line of sight of the observer ($i = 90°$), then the measured radial velocities will produce sinusoidal *velocity curves*, as in Fig. 7.5. Changing the orbital inclination does not alter the shape of the velocity curves; it merely changes their amplitudes

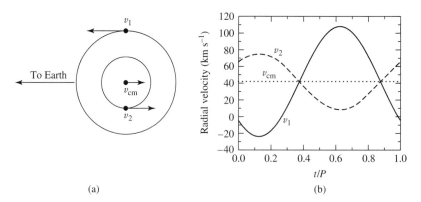

(a)                                                            (b)

**FIGURE 7.5**  The orbital paths and radial velocities of two stars in circular orbits ($e = 0$). In this example, $M_1 = 1 \, M_\odot$, $M_2 = 2 \, M_\odot$, the orbital period is $P = 30$ d, and the radial velocity of the center of mass is $v_{cm} = 42$ km s$^{-1}$. $v_1$, $v_2$, and $v_{cm}$ are the velocities of Star 1, Star 2, and the center of mass, respectively. (a) The plane of the circular orbits lies along the line of sight of the observer. (b) The observed radial velocity curves.

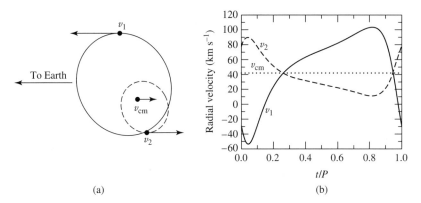

(a)     (b)

**FIGURE 7.6** The orbital paths and radial velocities of two stars in elliptical orbits ($e = 0.4$). As in Fig. 7.5, $M_1 = 1\ M_\odot$, $M_2 = 2\ M_\odot$, the orbital period is $P = 30$ d, and the radial velocity of the center of mass is $v_{cm} = 42$ km s$^{-1}$. In addition, the orientation of periastron is $45°$. $v_1$, $v_2$, and $v_{cm}$ are the velocities of Star 1, Star 2, and the center of mass, respectively. (a) The plane of the orbits lies along the line of sight of the observer. (b) The observed radial velocity curves.

by the factor $\sin i$. To estimate $i$ and the actual orbital velocities, therefore, other information about the system is necessary.

When the eccentricity, $e$, of the orbits is not zero, the observed velocity curves become skewed, as shown in Fig. 7.6. The exact shapes of the curves also depend strongly on the orientation of the orbits with respect to the observer, even for a given inclination angle.

In reality, many spectroscopic binaries possess nearly circular orbits, simplifying the analysis of the system somewhat. This occurs because close binaries tend to circularize their orbits due to tidal interactions over timescales that are short compared to the lifetimes of the stars involved.

### The Mass Function and the Mass–Luminosity Relation

If we assume that the orbital eccentricity is very small ($e \ll 1$), then the speeds of the stars are essentially constant and given by $v_1 = 2\pi a_1/P$ and $v_2 = 2\pi a_2/P$ for stars of mass $m_1$ and $m_2$, respectively, where $a_1$ and $a_2$ are the radii (semimajor axes) and $P$ is the period of the orbits. Solving for $a_1$ and $a_2$ and substituting into Eq. (7.1), we find that the ratio of the masses of the two stars becomes

$$\frac{m_1}{m_2} = \frac{v_2}{v_1}. \tag{7.4}$$

Because $v_{1r} = v_1 \sin i$ and $v_{2r} = v_2 \sin i$, Eq. (7.4) can be written in terms of the observed radial velocities rather than actual orbital velocities:

$$\boxed{\frac{m_1}{m_2} = \frac{v_{2r}/\sin i}{v_{1r}/\sin i} = \frac{v_{2r}}{v_{1r}}.} \tag{7.5}$$

As is the situation with visual binaries, we can determine the ratio of the stellar masses without knowing the angle of inclination.

However, as is also the case with visual binaries, finding the sum of the masses does require knowledge of the angle of inclination. Replacing $a$ with

$$a = a_1 + a_2 = \frac{P}{2\pi} (v_1 + v_2)$$

in Kepler's third law (Eq. 2.37) and solving for the sum of the masses, we have

$$m_1 + m_2 = \frac{P}{2\pi G} (v_1 + v_2)^3 .$$

Writing the actual radial velocities in terms of the observed values, we can express the sum of the masses as

$$m_1 + m_2 = \frac{P}{2\pi G} \frac{(v_{1r} + v_{2r})^3}{\sin^3 i} . \tag{7.6}$$

It is clear from Eq. (7.6) that the sum of the masses can be obtained only if both $v_{1r}$ and $v_{2r}$ are measurable. Unfortunately, this is not always the case. If one star is much brighter than its companion, the spectrum of the dimmer member will be overwhelmed. Such a system is referred to as a *single-line spectroscopic binary*. If the spectrum of Star 1 is observable but the spectrum of Star 2 is not, Eq. (7.5) allows $v_{2r}$ to be replaced by the ratio of the stellar masses, giving a quantity that is dependent on both of the system masses and the angle of inclination. If we substitute, Eq. (7.6) becomes

$$m_1 + m_2 = \frac{P}{2\pi G} \frac{v_{1r}^3}{\sin^3 i} \left( 1 + \frac{m_1}{m_2} \right)^3 .$$

Rearranging terms gives

$$\frac{m_2^3}{(m_1 + m_2)^2} \sin^3 i = \frac{P}{2\pi G} v_{1r}^3 . \tag{7.7}$$

The right-hand side of this expression, known as the **mass function**, depends only on the readily observable quantities, period and radial velocity. Since the spectrum of only one star is available, Eq. (7.5) cannot provide any information about mass ratios. As a result, the mass function is useful only for statistical studies or if an estimate of the mass of at least one component of the system already exists by some indirect means. If either $m_1$ or $\sin i$ is unknown, the mass function sets a lower limit for $m_2$, since the left-hand side is always less than $m_2$.

Even if both radial velocities are measurable, it is not possible to get exact values for $m_1$ and $m_2$ without knowing $i$. However, since stars can be grouped according to their effective temperatures and luminosities (see Section 8.2), and assuming that there is a relationship between these quantities and mass, then a statistical mass estimate for each class may be found by choosing an appropriately averaged value for $\sin^3 i$. An integral average of $\sin^3 i$ ($\langle \sin^3 i \rangle$) evaluated between $0°$ and $90°$ has a value $3\pi/16 \simeq 0.589$.[2] However, since no

---

[2]The proof is left as an exercise.

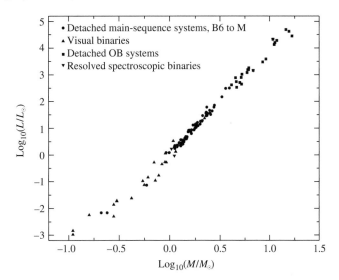

**FIGURE 7.7** The mass–luminosity relation. (Data from Popper, *Annu. Rev. Astron. Astrophys.*, *18*, 115, 1980.)

Doppler shift will be noticeable if the inclination angle is very small, it is more likely that a spectroscopic binary star system will be discovered if $i$ differs significantly from $0°$. This *selection effect* associated with detecting binary systems suggests that a larger value of $\langle \sin^3 i \rangle \simeq 2/3$ is more representative.

Evaluating masses of binaries has shown the existence of a well-defined **mass–luminosity relation** for the large majority of stars in the sky (see Fig. 7.7). One of the goals of the next several chapters is to investigate the origin of this relation in terms of fundamental physical principles.

### Using Eclipses to Determine Radii and Ratios of Temperatures

A good estimate of $i$ is possible in the special situation that a spectroscopic binary star system is observed to be an eclipsing system as well. Unless the distance of separation between the components of the binary is not much larger than the sum of the radii of the stars involved, an eclipsing system implies that $i$ must be close to $90°$, as suggested in Fig. 7.8. Even if it were assumed that $i = 90°$ while the actual value was closer to $75°$, an error of only 10% would result in the calculation of $\sin^3 i$ and in the determination of $m_1 + m_2$.

From the light curves produced by eclipsing binaries, it is possible to improve the estimate of $i$ still further. Figure 7.9 indicates that if the smaller star is completely eclipsed by the larger one, a nearly constant minimum will occur in the measured brightness of the system during the period of occultation. Similarly, even though the larger star will not be fully hidden from view when the smaller companion passes in front of it, a constant amount of area will still be obscured for a time, and again a nearly constant, though diminished amount of light will be observed. When one star is not completely eclipsed by its companion (Fig. 7.10), the minima are no longer constant, implying that $i$ must be less than $90°$.

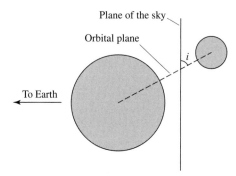

**FIGURE 7.8** The geometry of an eclipsing, spectroscopic binary requires that the angle of inclination $i$ be close to $90°$.

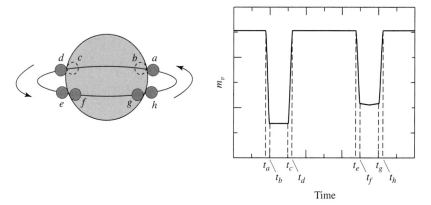

**FIGURE 7.9** The light curve of an eclipsing binary for which $i = 90°$. The times indicated on the light curve correspond to the positions of the smaller star relative to its larger companion. It is assumed in this example that the smaller star is hotter than the larger one.

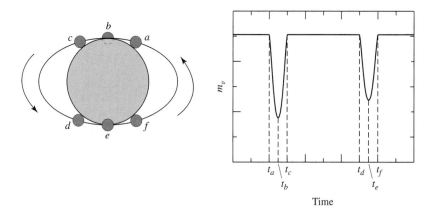

**FIGURE 7.10** The light curve of a partially eclipsing binary. It is assumed in this example that the smaller star is hotter than its companion.

Using measurements of the duration of eclipses, it is also possible to find the radii of each member of an eclipsing, spectroscopic binary. Referring again to Fig. 7.9, if we assume that $i \simeq 90°$, the amount of time between *first contact* ($t_a$) and minimum light ($t_b$), combined with the velocities of the stars, leads directly to the calculation of the radius of the smaller component. For example, if the semimajor axis of the smaller star's orbit is sufficiently large compared to either star's radius, and if the orbit is nearly circular, we can assume that the smaller object is moving approximately perpendicular to the line of sight of the observer during the duration of the eclipse. In this case the radius of the smaller star is simply

$$r_s = \frac{v}{2}(t_b - t_a),\tag{7.8}$$

where $v = v_s + v_\ell$ is the *relative* velocity of the two stars ($v_s$ and $v_\ell$ are the velocities of the small and large stars, respectively). Similarly, if we consider the amount of time between $t_b$ and $t_c$, the size of the larger member can also be determined. It can be quickly shown that the radius of the larger star is just

$$r_\ell = \frac{v}{2}(t_c - t_a) = r_s + \frac{v}{2}(t_c - t_b).\tag{7.9}$$

---

**Example 7.3.1.** An analysis of the spectrum of an eclipsing, double-line, spectroscopic binary having a period of $P = 8.6$ yr shows that the maximum Doppler shift of the hydrogen Balmer Hα (656.281 nm) line is $\Delta\lambda_s = 0.072$ nm for the smaller member and only $\Delta\lambda_\ell = 0.0068$ nm for its companion. From the sinusoidal shapes of the velocity curves, it is also apparent that the orbits are nearly circular. Using Eqs. (4.39) and (7.5), we find that the mass ratio of the two stars must be

$$\frac{m_\ell}{m_s} = \frac{v_{rs}}{v_{r\ell}} = \frac{\Delta\lambda_s}{\Delta\lambda_\ell} = 10.6.$$

Assuming that the orbital inclination is $i = 90°$, the Doppler shift of the smaller star implies that the maximum measured radial velocity is

$$v_{rs} = \frac{\Delta\lambda_s}{\lambda}c = 33 \text{ km s}^{-1}$$

and the radius of its orbit must be

$$a_s = \frac{v_{rs}P}{2\pi} = 1.42 \times 10^{12} \text{ m} = 9.5 \text{ AU}.$$

In the same manner, the orbital velocity and radius of the other star are $v_{r\ell} = 3.1$ km s$^{-1}$ and $a_\ell = 0.90$ AU, respectively. Therefore, the semimajor axis of the reduced mass becomes $a = a_s + a_\ell = 10.4$ AU.

*continued*

The sum of the masses can now be determined from Kepler's third law. If Eq. (2.37) is written in units of solar masses, astronomical units, and years, we have

$$m_s + m_\ell = a^3/P^2 = 15.2 \ M_\odot.$$

Solving for the masses independently yields $m_s = 1.3 \ M_\odot$ and $m_\ell = 13.9 \ M_\odot$.

Furthermore, from the light curve for this system, it is found that $t_b - t_a = 11.7$ hours and $t_c - t_b = 164$ days. Using Eq. (7.8) reveals that the radius of the smaller star is

$$r_s = \frac{(v_{rs} + v_{r\ell})}{2} (t_b - t_a) = 7.6 \times 10^8 \ \text{m} = 1.1 \ R_\odot,$$

where one solar radius is $1 \ R_\odot = 6.96 \times 10^8$ m. Equation (7.9) now gives the radius of the larger star, which is found to be $r_\ell = 369 \ R_\odot$.

In this particular system, the masses and radii of the stars are found to differ significantly.

---

The ratio of the effective temperatures of the two stars can also be obtained from the light curve of an eclipsing binary. This is accomplished by considering the objects as blackbody radiators and comparing the amount of light received during an eclipse with the amount received when both members are fully visible.

Referring once more to the sample binary system depicted in Fig. 7.9, it can be seen that the dip in the light curve is deeper when the smaller, hotter star is passing behind its companion. To understand this effect, recall that the radiative surface flux is given by Eq. (3.18),

$$F_r = F_{\text{surf}} = \sigma T_e^4.$$

Regardless of whether the smaller star passes behind or in front of the larger one, the same total cross-sectional area is eclipsed. Assuming for simplicity that the observed flux is constant across the disks,[3] the amount of light detected from the binary when both stars are fully visible is given by

$$B_0 = k \left( \pi r_\ell^2 F_{r\ell} + \pi r_s^2 F_{rs} \right),$$

where $k$ is a constant that depends on the distance to the system, the amount of intervening material between the system and the detector, and the nature of the detector. The deeper, or *primary*, minimum occurs when the hotter star passes behind the cooler one. If, as in the last example, the smaller star is hotter and therefore has the larger surface flux, and the smaller star is entirely eclipsed, the amount of light detected during the primary minimum may be expressed as

$$B_p = k\pi r_\ell^2 F_{r\ell}$$

while the brightness of the *secondary* minimum is

$$B_s = k \left( \pi r_\ell^2 - \pi r_s^2 \right) F_{r\ell} + k\pi r_s^2 F_{rs}.$$

---

[3]Stars often appear darker near the edges of their disks, a phenomenon referred to as *limb darkening*. This effect will be discussed in Section 9.3.

Since it is generally not possible to determine $k$ exactly, ratios are employed. Consider the ratio of the depth of the primary to the depth of the secondary. Using the expressions for $B_0$, $B_p$, and $B_s$, we find immediately that

$$\frac{B_0 - B_p}{B_0 - B_s} = \frac{F_{rs}}{F_{r\ell}} \tag{7.10}$$

or, from Eq. (3.18),

$$\boxed{\frac{B_0 - B_p}{B_0 - B_s} = \left(\frac{T_s}{T_\ell}\right)^4.} \tag{7.11}$$

---

**Example 7.3.2.** Further examination of the light curve of the binary system discussed in Example 7.3.1 provides information on the relative temperatures of the two stars. Photometric observations show that at maximum light the bolometric magnitude is $m_{\text{bol},0} = 6.3$, at the primary minimum $m_{\text{bol},p} = 9.6$, and at the secondary minimum $m_{\text{bol},s} = 6.6$. From Eq. (3.3), the ratio of brightnesses between the primary minimum and maximum light is

$$\frac{B_p}{B_0} = 100^{(m_{\text{bol},0} - m_{\text{bol},p})/5} = 0.048.$$

Similarly, the ratio of brightnesses between the secondary minimum and maximum light is

$$\frac{B_s}{B_0} = 100^{(m_{\text{bol},0} - m_{\text{bol},s})/5} = 0.76.$$

Now, by rewriting Eq. (7.10), we find that the ratio of the radiative fluxes is

$$\frac{F_{rs}}{F_{r\ell}} = \frac{1 - B_p/B_0}{1 - B_s/B_0} = 3.97.$$

Finally, from Eq. (3.18),

$$\frac{T_s}{T_\ell} = \left(\frac{F_{rs}}{F_{r\ell}}\right)^{1/4} = 1.41.$$

---

### A Computer Modeling Approach

The modern approach to analyzing the data from binary star systems involves computing detailed models that can yield important information about a variety of physical parameters. Not only can masses, radii, and effective temperatures be determined, but for many systems other details can be described as well. For instance, gravitational forces, combined with the effects of rotation and orbital motion, alter the stars' shapes; they are no longer simply spherical objects but may become elongated (these effects will be discussed in more detail in Section 18.1). The models may also incorporate information about the nonuniform

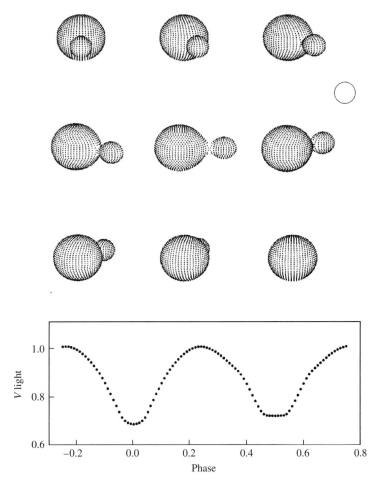

**FIGURE 7.11**   A synthetic light curve of RR Centauri, an eclipsing binary star system for which the two components are in close contact. The open circle represents the size of the Sun. The orbital and physical characteristics of the RR Cen system are $P = 0.6057$ d, $e = 0.0$, $M_1 = 1.8$ M$_\odot$, $M_2 = 0.37$ M$_\odot$. The spectral classification of the primary is F0V (see Section 8.1 for a discussion of stellar spectral classifications). (Figure adapted from R. E. Wilson, *Publ. Astron. Soc. Pac.*, *106*, 921, 1994; ©Astronomical Society of the Pacific.)

distribution of flux across the observed disks of the stars, variations in surface temperatures, and so on. Once the shapes of the gravitational equipotential surfaces and other parameters are determined, *synthetic* (theoretical) light curves can be computed for various wavelength bands ($U$, $B$, $V$, etc.), which are then compared to the observational data. Adjustments in the model parameters are made until the light curves agree with the observations. One such model for the binary system RR Centauri is shown in Fig. 7.11. In this system the two stars are actually in contact with each other, producing interesting and subtle effects in the light curve.

In order to introduce you to the process of modeling binary systems, the simple code TwoStars is described in Appendix K and available on the companion website. TwoStars makes the simplifying assumption that the stars are perfectly spherically symmetric. Thus TwoStars is capable of generating light curves, radial velocity curves, and astrometric data for systems in which the two stars are well separated. The simplifying assumptions imply that TwoStars is incapable of modeling the details of more complicated systems such as RR Cen, however.[4]

The study of binary star systems provides valuable information about the observable characteristics of stars. These results are then employed in developing a theory of stellar structure and evolution.

## 7.4 ■ THE SEARCH FOR EXTRASOLAR PLANETS

For hundreds of years, people have looked up at the night sky and wondered if planets might exist around other stars.[5] However, it wasn't until October 1995 that Michel Mayor and Didier Queloz of the Geneva Observatory announced the discovery of a planet around the solar-type star 51 Pegasi. This discovery represented the first detection of an extrasolar planet around a typical star.[6] Within one month of the announced discovery of 51 Peg, Geoffery W. Marcy and R. Paul Butler of the University of California, Berkeley, and the Carnegie Institution of Washington, respectively, announced that they had detected planets around two other Sun-like stars, 70 Vir and 47 UMa. By May 2006, just over ten years after the original announcements, 189 extrasolar planets had been discovered orbiting 163 stars that are similar to our own Sun.

This modern discovery of extrasolar planets at such a prodigious rate was made possible by dramatic advances in detector technology, the availability of large-aperture telescopes, and diligent, long-term observing campaigns. Given the huge disparity between the luminosity of the parent star and any orbiting planets, direct observation of a planet has proved very difficult; the planet's reflected light is simply overwhelmed by the luminosity of the star.[7] As a result, more indirect methods are usually required to detect extrasolar planets. Three techniques that have all been used successfully are based on ideas discussed in this chapter: radial velocity measurements, astrometric wobbles, and eclipses.[8] The first method, the detection of radial velocity variations in parent stars induced by the gravitational tug of the orbiting planets has been by far the most prolific method at the time this text was written.

---

[4]More sophisticated binary star modeling codes are available for download on the Internet or may be purchased. Examples include WD95, originally written by Wilson and Devinney and later modified by Kallrath, et al., and Binary Maker by Bradstreet and Steelman.

[5]In fact, it is thought that Giordano Bruno (1548–1600), a one-time Dominican monk, was executed for his belief in a Copernican universe filled with an infinite number of inhabited worlds around other stars; recall Section 1.2.

[6]In 1992, Alexander Wolszczan, of the Arecibo Radio Observatory in Puerto Rico, and Dale Frail, of the National Radio Astronomy Observatory, detected three Earth- and Moon-sized planets around a pulsar (PSR 1257+12), an extremely compact collapsed star that was produced following a supernova explosion; see Section 16.7. This discovery was made by noting variations in the extremely regular radio emission coming from the collapsed star.

[7]In April 2004, G. Chauvin and colleagues used the VLT/NACO of the European Southern Observatory to obtain an infrared image of a giant extrasolar planet of spectral type between L5 and L9.5 orbiting the brown dwarf 2MASSWJ1207334–393254. HST/NICMOS was also able to observe the brown dwarf's planetary companion.

[8]Another technique has also been employed in the search for extrasolar planets; it is based on the gravitational lensing of light; see Chapter 17.

---

**Example 7.4.1.** The so-called *reflex motion* of the parent star is extremely small. For example, consider the motion of Jupiter around the Sun. Jupiter's orbital period is 11.86 yr, the semimajor axis of its orbit is 5.2 AU, and its mass is only 0.000955 $M_\odot$. Assuming that the orbit of Jupiter is essentially circular (its actual eccentricity is just $e = 0.0489$), the planet's orbital velocity is approximately

$$v_J = 2\pi a/P = 13.1 \text{ km s}^{-1}.$$

According to Eq. (7.5), the Sun's orbital velocity about their mutual center of mass is only

$$v_\odot = \frac{m_J}{M_\odot} v_J = 12.5 \text{ m s}^{-1}.$$

This is similar to the top speed of a world-class sprinter from Earth.

---

Incredibly, today it is possible to measure radial velocity variations as small as 3 m s$^{-1}$, a slow jog in the park. Marcy, Butler, and their research-team colleagues accomplish this level of detection by passing starlight through an iodine vapor. The imprinted absorption lines from the iodine are used as zero-velocity reference lines in the high-resolution spectrum of the star. By comparing the absorption and emission-line wavelengths of the star to the iodine reference wavelengths, it is possible to determine very precise radial velocities. The high-resolution spectrographs used by the team were designed and built by another team member, Steve Vogt of the University of California, Santa Cruz.

The analysis of the radial velocities requires much more work before the true reflex motion radial velocity variations of the star can be deduced, however. In order to determine the source of the variations, it is first necessary to eliminate all other sources of radial velocities superimposed on the observed spectra. These include the rotation and wobble of Earth, the orbital velocity of Earth around the Sun, and the gravitational effects of the other planets in our Solar System on Earth and our Sun. After all of these corrections have been made, the radial velocity of the target star can be referenced to the true center of mass of our Solar System.

In addition to the motions in our Solar System, motions of the target star itself must be taken into account. For instance, if the target star is rotating, radial velocities due to the approaching and receding edges of its apparent disk will blur the absorption lines used to measure radial velocity. Pulsations of the surface of the star (see Chapter 14), surface convection (Section 10.4), and the movement of surface features such as star spots (e.g., Section 11.3), can also confuse the measurements and degrade the velocity resolution limit.

All of the planets discovered by the radial velocity technique are quite close to their parent star and very massive. For instance, the lower limit for the mass of the planet orbiting 51 Peg is 0.45 $M_J$ (where $M_J$ is the mass of Jupiter), it has an orbital period of just 4.23077 d, and the semimajor axis of its orbit is only 0.051 AU. The lower limit on the mass of the planet orbiting HD 168443c is 16.96 $M_J$, its orbital period is 1770 d, and the semimajor axis of its orbit is 2.87 AU. As the length of time that stars are observed increases, longer orbital-period planets will continue to be discovered, as will lower-mass planets.

Careful analysis of the radial velocity curves of one star, HD 209458, led researchers to predict and then detect transits of an extrasolar planet across the star's disk in 1999. The dimming of the light due to the transits is completely analogous to the eclipsing,

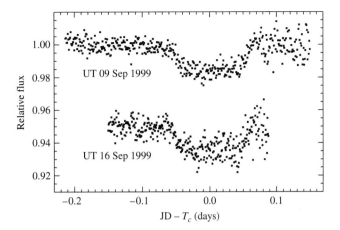

**FIGURE 7.12** The photometric detection of two transits of an extrasolar planet across the disk of HD 2094589 in September 1999. The September 16 transit was artificially offset by −0.05 relative to the transit of September 9 in order to avoid overlap of the data. $T_c$ designates the midpoint of the transit, and JD represents the Julian Date (time) of the particular measurement. (Figure adapted from Charbonneau, Brown, Latham, and Mayor, *Ap. J., 529*, L45, 2000.)

spectroscopic binary star systems discussed in Section 7.3. Given the very small size of the planet relative to HD 209458, the dimming of the light was only about 4 mmag (milli-magnitudes); see Fig. 7.12. Based on the additional information provided by the light curves during the transits, Charbonneau, Brown, Latham, and Mayor were able to determine that the transiting planet has a radius of approximately 1.27 $R_J$ (Jupiter radii) and that the orbital inclination $i = 87.1° \pm 0.2°$. Having restricted the value of the inclination angle, it then became possible to largely remove the uncertainty of the $\sin i$ term, resulting in a mass determination for the planet of 0.63 $M_J$. From the radial velocity data, the mass and radius of HD 209458 are 1.1 $M_\odot$ and 1.1 $R_\odot$, respectively. The orbital period of the planet is $P = 3.5250 \pm 0.003$ d, and the semimajor axis of its orbit is $a = 0.0467$ AU.

To date a number of planets have been detected by the dimming of starlight resulting from their transits of the disks of their parent stars. However, OGLE-TR-56b was the first system for which a planet was detected before it was found by the radial velocity technique. The orbital period of the planet is only 29 h and it orbits just 4.5 stellar radii (0.023 AU) from its parent star. The measurement was made by detecting a drop of slightly more than 0.01 magnitudes in the brightness of the star. An advantage of this technique is the ability to detect relatively distant systems; OGLE-TR-56b is approximately 1500 pc from Earth. In addition, from the transit time, the radius of the planet can also be determined, enabling an estimate of the density of the planet. The mass of the planet orbiting OGLE-TR-56b is estimated to be 0.9 $M_J$, as confirmed by follow-up radial velocity measurements, and its radius is just slightly larger than Jupiter's.

The reflex motion of a star due to the pull of a planet was detected for the first time in 2002. The Hubble Space Telescope's Fine Guidance Sensors were used to measure 0.5-milliarcsecond wobbles in Gliese 876, a tenth-magnitude star located 4.7 pc from Earth. By adding the third dimension of projection onto the plane of the sky, the mass of the planet previously obtained by the radial velocity technique was refined to give a value between

1.9 $M_J$ and 2.4 $M_J$. When future astrometric missions are launched (see Section 6.5), it is likely that this technique will result in detection of many more planets.

Although Earth-sized planets have yet to be discovered around solar-type stars, with missions such as NASA's Terrestrial Planet Finder being planned, and exquisitely sensitive astrometric missions such as SIM PlanetQuest and Gaia, it seems likely that such discoveries will occur soon.

## SUGGESTED READING

### General

Burnham, Robert Jr., *Burnham's Celestial Handbook: An Observer's Guide to the Universe Beyond the Solar System*, Revised and Enlarged Edition, Dover Publications, New York, 1978.

Jones, Kenneth Glyn (ed.), *Webb Society Deep-Sky Observer's Handbook*, Second Edition, Enslow Publishers, Hillside, NJ, 1986.

Marcy, Geoffrey, and Butler, R. Paul, "New Worlds: The Diversity of Planetary Systems," *Sky and Telescope*, March 1998.

Marcy, Geoffrey, et al., *California and Carnegie Planet Search Web Site*, `http://exoplanets.org`.

NASA, *Planet Quest: The Search for Another Earth Web Site*, `http://planetquest.jpl.nasa.gov`.

Pasachoff, Jay M., *Field Guide to the Stars and Planets*, Fourth Edition, Houghton Mifflin, Boston, 2000.

### Technical

Batten, Alan H., Fletcher, J. Murray, and MacCarthy, D. G., "Eighth Catalogue of the Orbital Elements of Spectroscopic Binary Systems," *Publications of the Dominion Astrophysical Observatory*, *17*, 1989.

Böhm-Vitense, Erika, *Introduction to Stellar Astrophysics: Basic Stellar Observations and Data*, Volume 1, Cambridge University Press, Cambridge, 1989.

Bradstreet, D. H., and Steelman, D. P., "Binary Maker 3.0—An Interactive Graphics-Based Light Curve Synthesis Program Written in Java," *Bulletin of the American Astronomical Society*, January 2003.

Charbonneau, David, Brown, Timothy M., Latham, David W., and Mayor, Michel, "Detection of Planetary Transits Across a Sun-Like Star," *The Astrophysical Journal*, *529*, L45, 2000.

Eggen, O. J., "Masses of Visual Binary Stars," *Annual Review of Astronomy and Astrophysics*, *5*, 105, 1967.

Kallrath, Josef, and Milone, Eugene F., *Eclipsing Binary Stars: Modeling and Analysis*, Springer–Verlag, New York, 1999.

Kallrath, J, Milone, E. F., Terrell, D., and Young, A. T., "Recent Improvements to a Version of the Wilson–Devinney Program," *The Astrophysical Journal Supplement Series*, *508*, 308, 1998.

Kitchin, C. R., *Astrophysical Techniques*, Third Edition, Institute of Physics Publications, Philadelphia, 1998.

Marcy, Geoffrey W., and Butler, R. Paul, " Detection of Extrasolar Giant Planets," *Annual Review of Astronomy and Astrophysics, 36*, 57, 1998.

Mayor, M., and Queloz, D., "A Jupiter-Mass Companion to a Solar-Type Star," *Nature, 378*, 355, 1995.

Popper, Daniel M., "Determination of Masses of Eclipsing Binary Stars," *Annual Review of Astronomy and Astrophysics, 5*, 85, 1967.

Popper, Daniel M., "Stellar Masses," *Annual Review of Astronomy and Astrophysics, 18*, 115, 1980.

Wilson, R. E., "Binary-Star Light-Curve Models," *Publications of the Astronomical Society of the Pacific, 106*, 921, 1994.

## PROBLEMS

**7.1** Consider two stars in orbit about a mutual center of mass. If $a_1$ is the semimajor axis of the orbit of star of mass $m_1$ and $a_2$ is the semimajor axis of the orbit of star of mass $m_2$, prove that the semimajor axis of the orbit of the *reduced mass* is given by $a = a_1 + a_2$. *Hint:* Review the discussion of Section 2.3 and recall that $\mathbf{r} = \mathbf{r}_2 - \mathbf{r}_1$.

**7.2** In Problem 2.9 we discussed integral averages that implicitly assumed a probability distribution (or *weighting function*) that was constant throughout the interval over which the integral was applied. When a *normalized* weighting function $w(\tau)$ is considered, such that

$$\int_0^\tau w(\tau)\, d\tau = 1,$$

then the integral average of $f(\tau)$ becomes

$$\langle f(\tau) \rangle = \int_0^\tau f(\tau)\, w(\tau)\, d\tau.$$

Comparison with Problem 2.9 reveals that the weighting function implicitly used in that case was $w(\tau) = 1/\tau$ over the interval 0 to $\tau$.

In evaluating $\langle \sin^3 i \rangle$ between 0 rad and $\pi/2$ rad (0° and 90°, respectively) as discussed on page 188, it is more likely that the radial velocity variations will be detected if the plane of the orbit is oriented along the line of sight. The weighting function should therefore take into consideration the projection of the plane of the orbital velocity onto the line of sight.

**(a)** Select an appropriate weighting function and show that your weighting function is normalized over the interval $i = 0$ to $\pi/2$.

**(b)** Prove that $\langle \sin^3 i \rangle = 3\pi/16$.

**7.3** Assume that two stars are in circular orbits about a mutual center of mass and are separated by a distance $a$. Assume also that the angle of inclination is $i$ and their stellar radii are $r_1$ and $r_2$.

**(a)** Find an expression for the smallest angle of inclination that will just barely produce an eclipse. *Hint:* Refer to Fig. 7.8.

**(b)** If $a = 2$ AU, $r_1 = 10\,R_\odot$, and $r_2 = 1\,R_\odot$, what minimum value of $i$ will result in an eclipse?

**7.4** Sirius is a visual binary with a period of 49.94 yr. Its measured trigonometric parallax is 0.37921″ ± 0.00158″ and, assuming that the plane of the orbit is in the plane of the sky, the true angular extent of the semimajor axis of the reduced mass is 7.61″. The ratio of the distances of Sirius A and Sirius B from the center of mass is $a_A/a_B = 0.466$.

   **(a)** Find the mass of each member of the system.

   **(b)** The absolute bolometric magnitude of Sirius A is 1.36, and Sirius B has an absolute bolometric magnitude of 8.79. Determine their luminosities. Express your answers in terms of the luminosity of the Sun.

   **(c)** The effective temperature of Sirius B is approximately 24,790 K ±100 K. Estimate its radius, and compare your answer to the radii of the Sun and Earth.

**7.5** ζ Phe is a 1.67-day spectroscopic binary with nearly circular orbits. The maximum measured Doppler shifts of the brighter and fainter components of the system are 121.4 km s$^{-1}$ and 247 km s$^{-1}$, respectively.

   **(a)** Determine the quantity $m \sin^3 i$ for each star.

   **(b)** Using a statistically chosen value for $\sin^3 i$ that takes into consideration the Doppler-shift selection effect, estimate the individual masses of the components of ζ Phe.

**7.6** From the light and velocity curves of an eclipsing, spectroscopic binary star system, it is determined that the orbital period is 6.31 yr, and the maximum radial velocities of Stars A and B are 5.4 km s$^{-1}$ and 22.4 km s$^{-1}$, respectively. Furthermore, the time period between first contact and minimum light ($t_b - t_a$) is 0.58 d, the length of the primary minimum ($t_c - t_b$) is 0.64 d, and the apparent bolometric magnitudes of maximum, primary minimum, and secondary minimum are 5.40 magnitudes, 9.20 magnitudes, and 5.44 magnitudes, respectively. From this information, and assuming circular orbits, find the

   **(a)** Ratio of stellar masses.

   **(b)** Sum of the masses (assume $i \simeq 90°$).

   **(c)** Individual masses.

   **(d)** Individual radii (assume that the orbits are circular).

   **(e)** Ratio of the effective temperatures of the two stars.

**7.7** The $V$-band light curve of YY Sgr is shown in Fig. 7.2. Neglecting bolometric corrections, estimate the ratio of the temperatures of the two stars in the system.

**7.8** Refer to the synthetic light curve and model of RR Centauri shown in Fig. 7.11.

   **(a)** Indicate the approximate points on the light curve (as a function of phase) that correspond to the orientations depicted.

   **(b)** Explain qualitatively the shape of the light curve.

**7.9** Data from binary star systems were used to illustrate the mass–luminosity relation in Fig. 7.7. A strong correlation also exists between mass and the effective temperatures of stars. Use the data provided in Popper, *Annu. Rev. Astron. Astrophys.*, *18*, 115, 1980 to create a graph of $\log_{10} T_e$ as a function of $\log_{10}(M/M_\odot)$. Use the data from Popper's Table 2, Table 4, Table 7 (excluding the α Aur system), and Table 8 (include only those stars with spectral types in the Sp column that end with the Roman numeral V). The stars that are excluded in Tables 7 and 8 are evolved stars with structures significantly different from the main sequence stars.[9] The article by Popper may be available in your library or it can be downloaded from the NASA Astrophysics Data System (NASA ADS) at http://adswww.harvard.edu.

---

[9]Spectral types and different classes of stars will be discussed in detail in Chapter 8.

**7.10** Give two reasons why the radial velocity technique for detecting planets around other stars favors massive planets (*Jupiters*) with relatively short orbital periods.

**7.11** Explain why radial velocity detections of extrasolar planets yield only lower limits on the masses of the orbiting planets. What value is actually measured, and what unknown orbital parameter is involved?

**7.12** From the data given in the text, determine the masses of the following stars (in solar masses):

(**a**) 51 Peg

(**b**) HD 168443c

**7.13** Suppose that you are an astronomer on a planet orbiting another star. While you are observing our Sun, Jupiter passes in front of it. Estimate the fractional decrease in the brightness of the star, assuming that you are observing a flat disk of constant flux, with a temperature of $T_e = 5777$ K. *Hint:* Neglect Jupiter's contribution to the total brightness of the system.

**7.14** From the data given in the text, combined with the information in Fig. 7.12, make a rough estimate of the radius of the orbiting planet, and compare your result with the quoted value. Be sure to explain each step used in computing your estimate.

## COMPUTER PROBLEMS

**7.15** (**a**) Use the computer program TwoStars, described in Appendix K and available on the companion website, to generate orbital radial velocity data similar to Fig. 7.6 for any choice of eccentricity. Assume that $M_1 = 0.5$ M$_\odot$, $R_1 = 1.8$ R$_\odot$, $T_{e_1} = 8190$ K, $M_2 = 2.0$ M$_\odot$, $R_2 = 0.63$ R$_\odot$, $T_{e_2} = 3840$ K, $P = 1.8$ yr, and $i = 30°$. Plot your results for $e = 0, 0.2, 0.4$, and 0.5. (You may assume that the center-of-mass velocity is zero and that the orientation of the major axis is perpendicular to the line of sight.)

(**b**) Verify your results for $e = 0$ by using the equations developed in Section 7.3.

(**c**) Explain how you might determine the eccentricity of an orbital system.

**7.16** The code TwoStars (Appendix K) can be used to analyze the apparent motions of binary stars across the plane of the sky. If fact, TwoStars was used to generate the data for Fig. 7.1. Assume that the binary system used in Problem 7.15 is located 3.2 pc from Earth and that its center of mass is moving through space with the vector components $(v'_x, v'_y, v'_z) = (30$ km s$^{-1}$, 42 km s$^{-1}$, $-15.3$ km s$^{-1}$). From the position data generated by TwoStars, plot the apparent positions of the stars in milliarcseconds for the case where $e = 0.4$.

**7.17** Figure 7.2 shows the light curve of the eclipsing binary YY Sgr. The code TwoStars, described in Appendix K and available on the companion website, can be used to roughly model this system. Use the data provided in the caption, and assume that the masses, radii, and effective temperatures of the two stars are $M_1 = 5.9$ M$_\odot$, $R_1 = 3.2$ R$_\odot$, $T_{e1} = 15,200$ K, and $M_2 = 5.6$ M$_\odot$, $R_2 = 2.9$ R$_\odot$, $T_{e2} = 13,700$ K. Also assume that the periastron angle is 214.6° and that the center of mass is at rest relative to the observer.

(**a**) Using TwoStars, create a synthetic light curve for the system.

(**b**) Using TwoStars, plot the radial velocities of the two stars.

**7.18** Using the data given in the text, and assuming that the orbital inclination is 90°, use TwoStars (Appendix K) to generate data that model the light curve of OGLE-TR-56b. You may assume that the radius of the planet is approximately the radius of Jupiter ($7 \times 10^7$ m) and its temperature is roughly 1000 K. Take the temperature of the star to be 3000 K. You may also assume that the planet's orbit is perfectly circular.

# CHAPTER

# 8

# The Classification of Stellar Spectra

## 8.1 ■ THE FORMATION OF SPECTRAL LINES

With the invention of photometry and spectroscopy, the new science of *astrophysics* progressed rapidly. As early as 1817, Joseph Fraunhofer had determined that different stars have different spectra. Stellar spectra were classified according to several schemes, the earliest of which recognized just three types of spectra. As instruments improved, increasingly subtle distinctions became possible.

### The Spectral Types of Stars

A spectral taxonomy developed at Harvard by Edward C. Pickering (1846–1919) and his assistant Williamina P. Fleming (1857–1911) in the 1890s labeled spectra with capital letters according to the strength of their hydrogen absorption lines, beginning with the letter A for the broadest lines. At about the same time, Antonia Maury (1866–1952), another of Pickering's assistants and a colleague of Fleming's, was developing a somewhat different classification scheme that she was using to study the widths of spectral lines. In her work Maury rearranged her classes in a way that would have been equivalent to placing Pickering's and Fleming's B class before the A stars. Then, in 1901, Annie Jump Cannon[1] (1863–1941; see Fig. 8.1), also employed by Pickering, and using the scheme of Pickering and Fleming while following the suggestion of Maury, rearranged the sequence of spectra by placing O and B before A, added decimal subdivisions (e.g., A0–A9), and consolidated many of the classes. With these changes, the Harvard classification scheme of "O B A F G K M" became a *temperature* sequence, running from the hottest blue O stars to the coolest red M stars. Generations of astronomy students have remembered this string of **spectral types** by memorizing the phrase "Oh Be A Fine Girl/Guy, Kiss Me." Stars nearer the beginning of this sequence are referred to as **early-type** stars, and those closer to the end are called **late-type** stars. These labels also distinguish the stars within the spectral subdivisions, so astronomers may speak of a K0 star as an "early K star" or refer to a B9 star as a "late B star." Cannon classified some 200,000 spectra between 1911 and 1914, and the results were

---

[1]The Annie J. Cannon Award is bestowed annually by the American Association of University Women and the American Astronomical Society for distinguished contributions to astronomy by a woman.

**FIGURE 8.1**    Annie Jump Cannon (1863–1941). (Courtesy of Harvard College Observatory.)

collected into the **Henry Draper Catalogue**.[2] Today, many stars are referred to by their HD numbers; Betelgeuse is HD 39801.

The physical basis of the Harvard spectral classification scheme remained obscure, however. Vega (spectral type A0) displays very strong hydrogen absorption lines, much stronger than the faint lines observed for the Sun (spectral type G2). On the other hand, the Sun's calcium absorption lines are much more intense than those of Vega. Is this a result of a variation in the *composition* of the two stars? Or are the different surface temperatures of Vega ($T_e = 9500$ K) and the Sun ($T_e = 5777$ K) responsible for the relative strengths of the absorption lines?

The theoretical understanding of the quantum atom achieved early in the twentieth century gave astronomers the key to the secrets of stellar spectra. As discussed in Section 5.3, absorption lines are created when an atom absorbs a photon with exactly the energy required for an electron to make an upward transition from a lower to a higher orbital. Emission lines are formed in the inverse process, when an electron makes a downward transition from a higher to a lower orbital and a single photon carries away the energy lost by the electron. The wavelength of the photon thus depends on the energies of the atomic orbitals involved in these transitions. For example, the Balmer absorption lines of hydrogen are caused by electrons making upward transitions from the $n = 2$ orbital to higher-energy orbitals, and Balmer emission lines are produced when electrons make downward transitions from higher-energy orbitals to the $n = 2$ orbital.

The distinctions between the spectra of stars with different temperatures are due to electrons occupying different atomic orbitals in the atmospheres of these stars. The details of spectral line formation can be quite complicated because electrons can be found in any of an atom's orbitals. Furthermore, the atom can be in any one of various stages of ionization and has a unique set of orbitals at each stage. An atom's stage of ionization is denoted by a

[2]In 1872 Henry Draper took the first photograph of a stellar spectrum. The catalog bearing his name was financed from his estate.

Roman numeral following the symbol for the atom. For example, H I and He I are neutral (not ionized) hydrogen and helium, respectively; He II is singly ionized helium, and Si III and Si IV refer to a silicon atom that has lost two and three electrons, respectively.

In the Harvard system devised by Cannon, the Balmer lines reach their maximum intensity in the spectra of stars of type A0, which have an effective temperature of $T_e = 9520$ K (recall Eq. 3.17). The visible spectral lines of neutral helium (He I) are strongest for B2 stars ($T_e = 22{,}000$ K), and the visible spectral lines of singly ionized calcium (Ca II) are most intense for K0 stars ($T_e = 5250$ K).[3]

Table 8.1 lists some of the defining criteria for various spectral types. In the table the term **metal** is used to indicate any element heavier than helium, a convention commonly adopted by astronomers because by far the most abundant elements in the universe are hydrogen and helium.

In addition to the traditional spectral types of the Harvard classification scheme (OBAFGKM), Table 8.1 also includes recently defined spectral types of very cool stars and **brown dwarfs**. Brown dwarfs are objects with too little mass to allow nuclear reactions to occur in their interiors in any substantial way, so they are not considered stars in the usual sense (see Section 10.6). The necessity of introducing these new spectral types came from all-sky surveys that detected a large number of objects with very low effective temperatures (1300 K to 2500 K for L spectral types and less than 1300 K for T spectral types).[4] In order to remember the new, cooler spectral types, one might consider extending the popular mnemonic by: "Oh Be A Fine Girl/Guy, Kiss Me—Less Talk!"

Figures 8.2 and 8.3 display some sample photographic spectra for various spectral types. You will note that hydrogen lines [e.g., H$\gamma$ (434.4 nm) and H$\delta$ (410.1 nm)] increase in width (strength) from O9 to A0, then decrease in width from A0 through F5, and nearly vanish by late K. Helium (He) lines are discernible in the spectra of early-type stars (O and early B) but begin to disappear in cooler stars.

Figures 8.4 and 8.5 also depict stellar spectra in a graphical format typical of modern digital detectors. Readily apparent is the shifting to longer wavelengths of the peak of the superimposed blackbody spectrum as the temperature of the star decreases (later spectral types). Also apparent are the H$\alpha$, H$\beta$, H$\gamma$, and H$\delta$ Balmer lines at 656.2 nm, 486.1 nm, 434.0 nm, and 410.2 nm, respectively. Note how these hydrogen absorption lines grow in strength from O to A and then decrease in strength for spectral types later than A. For later spectral types, the messy spectra are indicative of metal lines, with molecular lines appearing in the spectra of the coolest stars.

### The Maxwell–Boltzmann Velocity Distribution

To uncover the physical foundation of this classification system, two basic questions must be answered: In what orbitals are electrons most likely to be found? What are the relative numbers of atoms in various stages of ionization?

---

[3]The two prominent spectral lines of Ca II are usually referred to as the H ($\lambda = 396.8$ nm) and K ($\lambda = 393.3$ nm) lines of calcium. The nomenclature for the H line was devised by Fraunhofer; the K line was named by E. Mascart (1837–1908) in the 1860s.

[4]The surveys that discovered large numbers of these objects are the Sloan Digital Sky Survey (SDSS) and the 2-Micron All-Sky Survey (2MASS).

**TABLE 8.1** Harvard Spectral Classification.

| Spectral Type | Characteristics |
| --- | --- |
| O | Hottest blue-white stars with few lines<br>Strong He II absorption (sometimes emission) lines.<br>He I absorption lines becoming stronger. |
| B | Hot blue-white<br>He I absorption lines strongest at B2.<br>H I (Balmer) absorption lines becoming stronger. |
| A | White<br>Balmer absorption lines strongest at A0, becoming weaker later.<br>Ca II absorption lines becoming stronger. |
| F | Yellow-white<br>Ca II lines continue to strengthen as Balmer lines continue to weaken.<br>Neutral metal absorption lines (Fe I, Cr I). |
| G | Yellow<br>Solar-type spectra.<br>Ca II lines continue becoming stronger.<br>Fe I, other neutral metal lines becoming stronger. |
| K | Cool orange<br>Ca II H and K lines strongest at K0, becoming weaker later.<br>Spectra dominated by metal absorption lines. |
| M | Cool red<br>Spectra dominated by molecular absorption bands,<br>   especially titanium oxide (TiO) and vanadium oxide (VO).<br>Neutral metal absorption lines remain strong. |
| L | Very cool, dark red<br>Stronger in infrared than visible.<br>Strong molecular absorption bands of metal hydrides (CrH, FeH), water<br>   ($H_2O$), carbon monoxide (CO), and alkali metals (Na, K, Rb, Cs).<br>TiO and VO are weakening. |
| T | Coolest, Infrared<br>Strong methane ($CH_4$) bands but weakening CO bands. |

S and C spectral types for evolved giant stars are discussed on page 466.

The answers to both questions are found in an area of physics known as **statistical mechanics**. This branch of physics studies the statistical properties of a system composed of many members. For example, a gas can contain a huge number of particles with a large range of speeds and energies. Although in practice it would be impossible to calculate the detailed behavior of any single particle, the gas as a whole does have certain well-defined properties, such as its temperature, pressure, and density. For such a gas in thermal

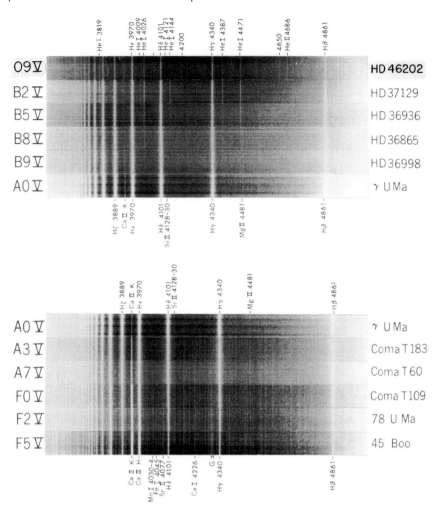

**FIGURE 8.2**    Stellar spectra for main-sequence classes O9–F5. Note that these spectra are displayed as negatives; absorption lines appear bright. Wavelengths are given in angstroms. (Figure from Abt, et al., *An Atlas of Low-Dispersion Grating Stellar Spectra*, Kitt Peak National Observatory, Tucson, AZ, 1968.)

equilibrium (the gas is not rapidly increasing or decreasing in temperature, for instance), the **Maxwell–Boltzmann velocity distribution function**[5] describes the fraction of particles having a given range of speeds. The number of gas particles per unit volume having speeds between $v$ and $v + dv$ is given by

$$n_v \, dv = n \left( \frac{m}{2\pi kT} \right)^{3/2} e^{-mv^2/2kT} \, 4\pi v^2 \, dv, \tag{8.1}$$

[5]This name honors James Clerk Maxwell and Ludwig Boltzmann (1844–1906), the latter of whom is considered the founder of statistical mechanics.

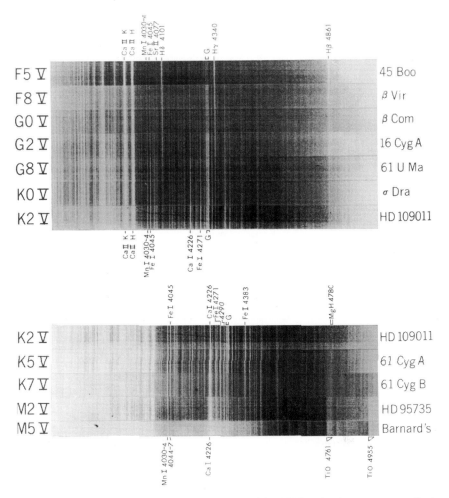

**FIGURE 8.3** Stellar spectra for main-sequence classes F5–M5. Note that these spectra are displayed as negatives; absorption lines appear bright. Wavelengths are given in angstroms. (Figure from Abt, et al., *An Atlas of Low-Dispersion Grating Stellar Spectra*, Kitt Peak National Observatory, Tucson, AZ, 1968.)

where $n$ is the total number density (number of particles per unit volume), $n_v \equiv \partial n / \partial v$, $m$ is a particle's mass, $k$ is Boltzmann's constant, and $T$ is the temperature of the gas *in kelvins*. Figure 8.6 shows the Maxwell–Boltzmann distribution of molecular speeds in terms of the *fraction* of molecules having a speed between $v$ and $v + dv$. The exponent of the distribution function is the ratio of a gas particle's kinetic energy, $\frac{1}{2}mv^2$, to the characteristic thermal energy, $kT$. It is difficult for a significant number of particles to have an energy much greater or less than the thermal energy; the distribution peaks when these energies are equal, at a **most probable speed** of

$$v_{\mathrm{mp}} = \sqrt{\frac{2kT}{m}}. \tag{8.2}$$

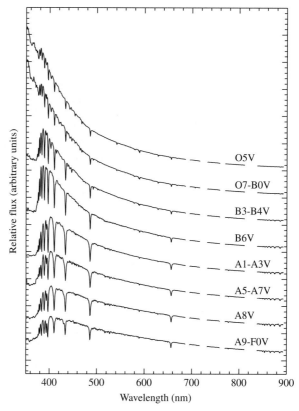

**FIGURE 8.4** Digitized spectra of main sequence classes O5–F0 displayed in terms of relative flux as a function of wavelength. Modern spectra obtained by digital detectors (as opposed to photographic plates) are generally displayed graphically. (Data from Silva and Cornell, *Ap. J. Suppl.*, *81*, 865, 1992.)

The high-speed exponential "tail" of the distribution function results in a somewhat higher (average) **root-mean-square speed**[6] of

$$v_{\text{rms}} = \sqrt{\frac{3kT}{m}}. \tag{8.3}$$

---

**Example 8.1.1.** The area under the curve between two speeds is equal to the fraction of gas particles in that range of speeds. In order to determine the fraction of hydrogen atoms in a gas of $T = 10{,}000$ K having speeds between $v_1 = 2 \times 10^4$ m s$^{-1}$ and $v_2 = 2.5 \times 10^4$ m s$^{-1}$, it is necessary to integrate the Maxwell–Boltzmann distribution between these two limits,

---

[6]The root-mean-square speed is the square root of the average (mean) value of $v^2$: $v_{\text{rms}} = \sqrt{\overline{v^2}}$.

or

$$N/N_{\text{total}} = \frac{1}{n} \int_{v_1}^{v_2} n_v \, dv$$

$$= \left(\frac{m}{2\pi kT}\right)^{3/2} \int_{v_1}^{v_2} e^{-mv^2/2kT} \, 4\pi v^2 \, dv. \qquad (8.4)$$

Although Eq. (8.4) has a closed-form solution when $v_1 = 0$ and $v_2 \to \infty$, it must be evaluated numerically in other cases. This can be accomplished crudely by evaluating the integrand using an average value of the velocity over the interval, multiplied by the width of the interval, or

$$N/N_{\text{total}} = \frac{1}{n} \int_{v_1}^{v_2} n_v(v) \, dv \simeq \frac{1}{n} n_v(\bar{v})\,(v_2 - v_1),$$

where $\bar{v} \equiv (v_1 + v_2)/2$. Substituting, we find

$$N/N_{\text{total}} \simeq \left(\frac{m}{2\pi kT}\right)^{3/2} e^{-m\bar{v}^2/2kT} \, 4\pi \bar{v}^2 \,(v_2 - v_1)$$

$$\simeq 0.125.$$

Approximately 12.5% of the hydrogen atoms in a gas at 10,000 K have speeds between $2 \times 10^4$ m s$^{-1}$ and $2.5 \times 10^4$ m s$^{-1}$. A more careful numerical integration over the range gives 12.76%.

---

## The Boltzmann Equation

The atoms of a gas gain and lose energy as they collide. As a result, the distribution in the speeds of the impacting atoms, given by Eq. (8.1), produces a definite distribution of the electrons among the atomic orbitals. This distribution of electrons is governed by a fundamental result of statistical mechanics: Orbitals of higher energy are less likely to be occupied by electrons.

Let $s_a$ stand for the specific set of quantum numbers that identifies a state of energy $E_a$ for a system of particles. Similarly, let $s_b$ stand for the set of quantum numbers that identifies a state of energy $E_b$. For example, $E_a = -13.6$ eV for the lowest orbit of the hydrogen atom, with $s_a = \{n = 1, \ \ell = 0, \ m_\ell = 0, \ m_s = +1/2\}$ identifying a specific state with that energy (recall Section 5.4 for a discussion of quantum numbers). Then the ratio of the probability $P(s_b)$ that the system is in state $s_b$ to the probability $P(s_a)$ that the system is in state $s_a$ is given by

$$\frac{P(s_b)}{P(s_a)} = \frac{e^{-E_b/kT}}{e^{-E_a/kT}} = e^{-(E_b - E_a)/kT}, \qquad (8.5)$$

where $T$ is the common temperature of the two systems. The term $e^{-E/kT}$ is called the *Boltzmann factor*.[7]

---

[7]The energies encountered in this context are usually given in units of electron volts (eV), so it is useful to remember that at a room temperature of 300 K, the product $kT$ is approximately 1/40 eV.

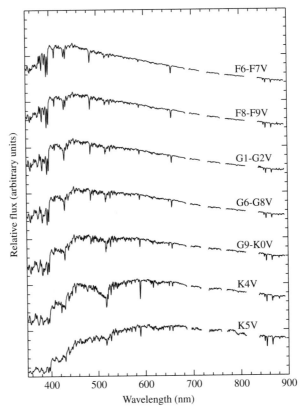

**FIGURE 8.5**   Digitized spectra of main sequence classes F6–K5 displayed in terms of relative flux as a function of wavelength. (Data from Silva and Cornell, *Ap. J. Suppl.*, *81*, 865, 1992.)

The Boltzmann factor plays such a fundamental role in the study of statistical mechanics that Eq. (8.5) merits further consideration. Suppose, for example, that $E_b > E_a$; the energy of state $s_b$ is greater than the energy of state $s_a$. Notice that as the thermal energy $kT$ decreases toward zero (i.e., $T \rightarrow 0$), the quantity $-(E_b - E_a)/kT \rightarrow -\infty$, and so $P(s_b)/P(s_a) \rightarrow 0$. This is just what is to be expected if there isn't any thermal energy available to raise the energy of an atom to a higher level. On the other hand, if there is a great deal of thermal energy available (i.e., $T \rightarrow \infty$), then $-(E_b - E_a)/kT \rightarrow 0$ and $P(s_b)/P(s_a) \rightarrow 1$. Again this is what would be expected since with an unlimited reservoir of thermal energy, all available energy levels of the atom should be accessible with equal probability. You can quickly verify that if we had assumed instead that $E_b < E_a$, the expected results would again be obtained in the limits of $T \rightarrow 0$ and $T \rightarrow \infty$.

It is often the case that the energy levels of the system may be **degenerate**, with more than one quantum state having the same energy. That is, if states $s_a$ and $s_b$ are degenerate, then $E_a = E_b$ but $s_a \neq s_b$. When taking averages, we must count each of the degenerate states separately. To account properly for the number of states that have a given energy, define $g_a$ to be the number of states with energy $E_a$. Similarly, define $g_b$ to be the number of states with energy $E_b$. These are called the **statistical weights** of the energy levels.

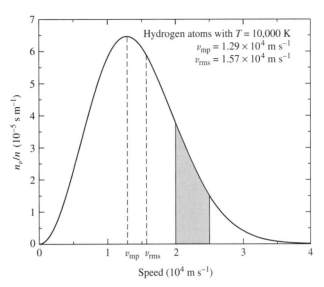

**FIGURE 8.6** Maxwell–Boltzmann distribution function, $n_v/n$, for hydrogen atoms at a temperature of 10,000 K. The fraction of hydrogen atoms in the gas having velocities between $2 \times 10^4$ m s$^{-1}$ and $2.5 \times 10^4$ m s$^{-1}$ is the shaded area under the curve between those two velocities; see Example 8.1.1.

---

**Example 8.1.2.**  The ground state of the hydrogen atom is twofold degenerate. In fact, although "ground state" is the standard terminology, the plural "ground states" would be more precise because these are *two* quantum states that have the same energy of $-13.6$ eV (for $m_s = \pm 1/2$).[8] In the same manner, the "first excited state" actually consists of *eight* degenerate quantum states with the same energy of $-3.40$ eV.

Table 8.2 shows the set of quantum numbers $\{n, \ell, m_\ell, m_s\}$ that identifies each state; it also shows each state's energy. Notice that there are $g_1 = 2$ ground states with the energy $E_1 = -13.6$ eV, and $g_2 = 8$ first excited states with the energy $E_2 = -3.40$ eV. This result agrees with that of Problem 5.16, which demonstrated that the degeneracy of energy level $n$ of the hydrogen atom is $2n^2$.

---

The ratio of the probability $P(E_b)$ that the system will be found in *any* of the $g_b$ degenerate states with energy $E_b$ to the probability $P(E_a)$ that the system is in *any* of the $g_a$ degenerate states with energy $E_a$ is given by

$$\frac{P(E_b)}{P(E_a)} = \frac{g_b \, e^{-E_b/kT}}{g_a \, e^{-E_a/kT}} = \frac{g_b}{g_a} e^{-(E_b - E_a)/kT}.$$

Stellar atmospheres contain a vast number of atoms, so the ratio of probabilities is indistinguishable from the ratio of the number of atoms. Thus, for the atoms of a given element in a specified state of ionization, the ratio of the number of atoms $N_b$ with energy $E_b$ to

---

[8] In reality, the two "ground states" of the hydrogen atom are not precisely degenerate. As explained in Section 12.1, the two states actually have slightly different energies, enabling the hydrogen atom to emit 21-cm radio waves, an important signature of hydrogen gas in interstellar space.

**TABLE 8.2**   Quantum Numbers and Energies for the Hydrogen Atom.

| | Ground States $s_1$ | | | Energy $E_1$ (eV) |
|---|---|---|---|---|
| $n$ | $\ell$ | $m_\ell$ | $m_s$ | |
| 1 | 0 | 0 | +1/2 | −13.6 |
| 1 | 0 | 0 | −1/2 | −13.6 |

| | First Excited States $s_2$ | | | Energy $E_2$ (eV) |
|---|---|---|---|---|
| $n$ | $\ell$ | $m_\ell$ | $m_s$ | |
| 2 | 0 | 0 | +1/2 | −3.40 |
| 2 | 0 | 0 | −1/2 | −3.40 |
| 2 | 1 | 1 | +1/2 | −3.40 |
| 2 | 1 | 1 | −1/2 | −3.40 |
| 2 | 1 | 0 | +1/2 | −3.40 |
| 2 | 1 | 0 | −1/2 | −3.40 |
| 2 | 1 | −1 | +1/2 | −3.40 |
| 2 | 1 | −1 | −1/2 | −3.40 |

the number of atoms $N_a$ with energy $E_a$ *in different states of excitation* is given by the **Boltzmann equation**,

$$\frac{N_b}{N_a} = \frac{g_b \, e^{-E_b/kT}}{g_a \, e^{-E_a/kT}} = \frac{g_b}{g_a} e^{-(E_b - E_a)/kT}. \tag{8.6}$$

**Example 8.1.3.**   For a gas of neutral hydrogen atoms, at what temperature will equal numbers of atoms have electrons in the ground state ($n = 1$) and in the first excited state ($n = 2$)?[9] Recall from Example 8.1.2 that the degeneracy of the $n$th energy level of the hydrogen atom is $g_n = 2n^2$. Associating state $a$ with the ground state and state $b$ with the first excited state, setting $N_2 = N_1$ on the left-hand side of Eq. (8.6), and using Eq. (5.14) for the energy levels lead to

$$1 = \frac{2(2)^2}{2(1)^2} e^{-[(-13.6 \text{ eV}/2^2) - (-13.6 \text{ eV}/1^2)]/kT},$$

or

$$\frac{10.2 \text{ eV}}{kT} = \ln(4).$$

[9]We have reverted to the standard practice of referring to the two degenerate states of lowest energy as the "ground state" and to the eight degenerate states of next-lowest energy as the "first excited state."

Solving for the temperature yields[10]

$$T = \frac{10.2 \text{ eV}}{k \ln (4)} = 8.54 \times 10^4 \text{ K}.$$

High temperatures are required for a significant number of hydrogen atoms to have electrons in the first excited state. Figure 8.7 shows the relative occupancy of the ground and first excited states, $N_2/(N_1 + N_2)$, as a function of temperature.[11] This result is somewhat puzzling, however. Recall that the Balmer absorption lines are produced by electrons in hydrogen atoms making an upward transition from the $n = 2$ orbital. If, as shown in Example 8.1.3, temperatures on the order of 85,000 K are needed to provide electrons in the first excited state, then why do the Balmer lines reach their maximum intensity at a much lower temperature of 9520 K? Clearly, according to Eq. (8.6), at temperatures higher than 9520 K an even greater proportion of the electrons will be in the first excited state rather than in the ground state. If this is the case, then what is responsible for the diminishing strength of the Balmer lines at higher temperatures?

### The Saha Equation

The answer lies in also considering the relative number of atoms *in different stages of ionization*. Let $\chi_i$ be the ionization energy needed to remove an electron from an atom (or

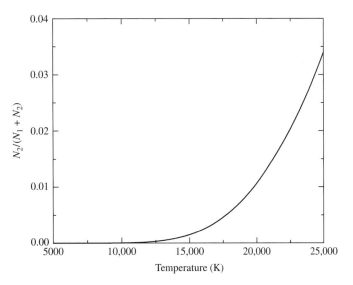

**FIGURE 8.7**  $N_2/(N_1 + N_2)$ for the hydrogen atom obtained via the Boltzmann equation.

---

[10]When we are working with electron volts, the Boltzmann constant can be expressed in the convenient form $k = 8.6173423 \times 10^{-5}$ eV K$^{-1}$.

[11]For the remainder of this section, we will use $a = 1$ for the ground state energy and $b = 2$ for the energy of the first excited state.

ion) in the ground state, thus taking it from ionization stage $i$ to stage $(i + 1)$. For example, the ionization energy of hydrogen, the energy needed to convert it from H I to H II, is $\chi_I = 13.6$ eV. However, it may be that the initial and final ions are not in the ground state. An average must be taken over the orbital energies to allow for the possible partitioning of the atom's electrons among its orbitals. This procedure involves calculating the **partition functions**, $Z$, for the initial and final atoms. The partition function is simply the weighted sum of the number of ways the atom can arrange its electrons with the same energy, with more energetic (and therefore less likely) configurations receiving less weight from the Boltzmann factor when the sum is taken. If $E_j$ is the energy of the $j$th energy level and $g_j$ is the degeneracy of that level, then the partition function $Z$ is defined as

$$Z = \sum_{j=1}^{\infty} g_j \, e^{-(E_j - E_1)/kT}. \tag{8.7}$$

If we use the partition functions $Z_i$ and $Z_{i+1}$ for the atom in its initial and final stages of ionization, the ratio of the number of atoms in stage $(i + 1)$ to the number of atoms in stage $i$ is

$$\frac{N_{i+1}}{N_i} = \frac{2 Z_{i+1}}{n_e Z_i} \left( \frac{2\pi m_e kT}{h^2} \right)^{3/2} e^{-\chi_i/kT}. \tag{8.8}$$

This equation is known as the **Saha equation**, after the Indian astrophysicist Meghnad Saha (1894–1956), who first derived it in 1920. Because a free electron is produced in the ionization process, it is not surprising to find the number density of free electrons (number of free electrons per unit volume), $n_e$, on the right-hand side of the Saha equation. Note that as the number density of free electrons increases, the number of atoms in the higher stage of ionization decreases, since there are more electrons with which the ion may recombine. The factor of 2 in front of the partition function $Z_{i+1}$ reflects the two possible spins of the free electron, with $m_s = \pm 1/2$. The term in parentheses is also related to the free electron, with $m_e$ being the electron mass.[12] Sometimes the pressure of the free electrons, $P_e$, is used in place of the electron number density; the two are related by the ideal gas law written in the form

$$P_e = n_e kT.$$

Then the Saha equation takes the alternative form

$$\frac{N_{i+1}}{N_i} = \frac{2kT \, Z_{i+1}}{P_e Z_i} \left( \frac{2\pi m_e kT}{h^2} \right)^{3/2} e^{-\chi_i/kT}. \tag{8.9}$$

[12]The term in parentheses is the number density of electrons for which the quantum energy (such as discussed in Example 5.4.2) is roughly equal to the characteristic thermal energy $kT$. For the classical conditions encountered in stellar atmospheres, this term is much greater than $n_e$.

The electron pressure ranges from $0.1$ N m$^{-2}$ in the atmospheres of cooler stars to $100$ N m$^{-2}$ for hotter stars. In Section 9.5, we will describe how the electron pressure is determined for stellar atmospheres.

### Combining the Boltzmann and Saha Equations

We are now finally ready to consider the combined effects of the Boltzmann and Saha equations and how they influence the stellar spectra that we observe.

---

**Example 8.1.4.**   Consider the degree of ionization in a stellar atmosphere that is assumed to be composed of pure hydrogen. Assume for simplicity that the electron pressure is a constant $P_e = 20$ N m$^{-2}$.

The Saha equation (8.9) will be used to calculate the fraction of atoms that are ionized, $N_{II}/N_{total} = N_{II}/(N_I + N_{II})$, as the temperature $T$ varies between 5000 K and 25,000 K. However, the partition functions $Z_I$ and $Z_{II}$ must be determined first. A hydrogen ion is just a proton and so has no degeneracy; thus $Z_{II} = 1$. The energy of the first excited state of hydrogen is $E_2 - E_1 = 10.2$ eV above the ground state energy. Because $10.2$ eV $\gg kT$ for the temperature regime under consideration, the Boltzmann factor $e^{-(E_2-E_1)/kT} \ll 1$. Nearly all of the H I atoms are therefore in the ground state (recall the previous example), so Eq. (8.7) for the partition function simplifies to $Z_I \simeq g_1 = 2(1)^2 = 2$.

Inserting these values into the Saha equation with $\chi_I = 13.6$ eV gives the ratio of ionized to neutral hydrogen, $N_{II}/N_I$. This ratio is then used to find the fraction of ionized hydrogen, $N_{II}/N_{total}$, by writing

$$\frac{N_{II}}{N_{total}} = \frac{N_{II}}{N_I + N_{II}} = \frac{N_{II}/N_I}{1 + N_{II}/N_I};$$

the results are displayed in Fig. 8.8. This figure shows that when $T = 5000$ K, essentially none of the hydrogen atoms are ionized. At about 8300 K, 5% of the atoms have become ionized. Half of the hydrogen is ionized at a temperature of 9600 K, and when $T$ has risen to 11,300 K, all but 5% of the hydrogen is in the form of H II. Thus the ionization of hydrogen takes place within a temperature interval of approximately 3000 K. This range of temperatures is quite limited compared to the temperatures of tens of millions of degrees routinely encountered inside stars. The narrow region inside a star where hydrogen is partially ionized is called a hydrogen **partial ionization zone** and has a characteristic temperature of approximately 10,000 K for a wide range of stellar parameters.

Now we can see why the Balmer lines are observed to attain their maximum intensity at a temperature of 9520 K, instead of at the much higher characteristic temperatures (on the order of 85,000 K) required to excite electrons to the $n = 2$ energy level of hydrogen. The strength of the Balmer lines depends on $N_2/N_{total}$, the fraction of *all* hydrogen atoms that are in the first excited state. This is found by combining the results of the Boltzmann and Saha equations. Because virtually all of the neutral hydrogen atoms are in either the ground state or the first excited state, we can employ the approximation $N_1 + N_2 \simeq N_I$ and write

$$\frac{N_2}{N_{total}} = \left(\frac{N_2}{N_1 + N_2}\right)\left(\frac{N_I}{N_{total}}\right) = \left(\frac{N_2/N_1}{1 + N_2/N_1}\right)\left(\frac{1}{1 + N_{II}/N_I}\right).$$

*continued*

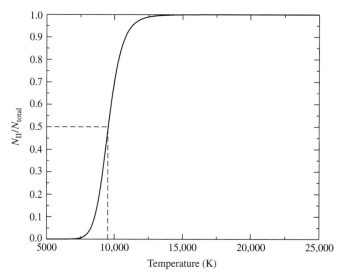

**FIGURE 8.8**   $N_{II}/N_{total}$ for hydrogen from the Saha equation when $P_e = 20$ N m$^{-2}$. Fifty percent ionization occurs at $T \simeq 9600$ K.

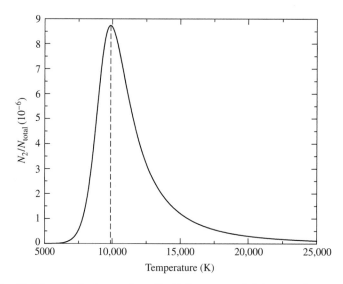

**FIGURE 8.9**   $N_2/N_{total}$ for hydrogen from the Boltzmann and Saha equations, assuming $P_e = 20$ N m$^{-2}$. The peak occurs at approximately 9900 K.

Figure 8.9 shows that in this example, the hydrogen gas would produce the most intense Balmer lines at a temperature of 9900 K, in good agreement with the observations. *The diminishing strength of the Balmer lines at higher temperatures is due to the rapid ionization of hydrogen above 10,000 K.* Figure 8.10 summarizes this situation.

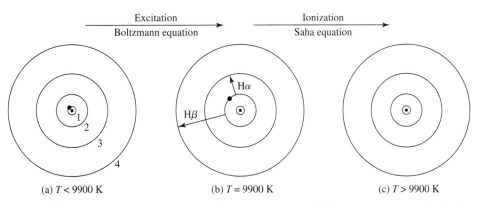

**FIGURE 8.10**   The electron's position in the hydrogen atom at different temperatures. In (a), the electron is in the ground state. Balmer absorption lines are produced only when the electron is initially in the first excited state, as shown in (b). In (c), the atom has been ionized.

Of course, stellar atmospheres are not composed of pure hydrogen, and the results obtained in Example 8.1.4 depended on an appropriate value for the electron pressure. In stellar atmospheres, there is typically one helium atom for every ten hydrogen atoms. The presence of ionized helium provides more electrons with which the hydrogen ions can recombine. Thus, when helium is added, it takes a *higher* temperature to achieve the same degree of hydrogen ionization.

It should also be emphasized that the Saha equation can be applied only to a gas in *thermodynamic equilibrium*, so that the Maxwell–Boltzmann velocity distribution is obeyed.[13] Furthermore, the density of the gas must not be too great (less than roughly 1 kg m$^{-3}$ for stellar material), or the presence of neighboring ions will distort an atom's orbitals and lower its ionization energy.

**Example 8.1.5.**   The Sun's "surface" is a thin layer of the solar atmosphere called the *photosphere*; see Section 11.2. The characteristic temperature of the photosphere is $T = T_e = 5777$ K, and it has about 500,000 hydrogen atoms for each calcium atom with an electron pressure of about 1.5 N m$^{-2}$.[14] From this information and knowledge of the appropriate statistical weights and partition functions, the Saha and Boltzmann equations can be used to estimate the relative strengths of the absorption lines due to hydrogen (the Balmer lines) and those due to calcium (the Ca II H and K lines).

We must compare the number of neutral hydrogen atoms with electrons in the first excited state (which produce the Balmer lines) to the number of singly ionized calcium atoms with electrons in the ground state (which produce the Ca II H and K lines). As in Example 8.1.4, we will use the Saha equation to determine the degree of ionization and will use the Boltzmann equation to reveal the distribution of electrons between the ground and first excited states.

*continued*

---

[13]Thermodynamic equilibrium will be discussed in detail in Section 9.2.
[14]See Cox (2000), page 348 for a model solar photosphere.

Let's consider hydrogen first. If we substitute the partition functions found in Example 8.1.4 into the Saha equation (8.9), the ratio of ionized to neutral hydrogen is

$$\left[\frac{N_{II}}{N_I}\right]_H = \frac{2kT\,Z_{i+1}}{P_e\,Z_i}\left(\frac{2\pi m_e kT}{h^2}\right)^{3/2} e^{-\chi_i/kT} = 7.70 \times 10^{-5} \simeq \frac{1}{13{,}000}.$$

Thus there is only one hydrogen ion (H II) for every 13,000 neutral hydrogen atoms (H I) at the Sun's surface. Almost none of the hydrogen is ionized.

The Boltzmann equation (8.6) reveals how many of these neutral hydrogen atoms are in the first excited state. Using $g_n = 2n^2$ for hydrogen (implying $g_1 = 2$ and $g_2 = 8$), we have

$$\left[\frac{N_2}{N_1}\right]_{H\,I} = \frac{g_2}{g_1} e^{-(E_2-E_1)/kT} = 5.06 \times 10^{-9} \simeq \frac{1}{198{,}000{,}000}.$$

The result is that only one of every 200 million hydrogen atoms is in the first excited state and capable of producing Balmer absorption lines:

$$\frac{N_2}{N_{\text{total}}} = \left(\frac{N_2}{N_1 + N_2}\right)\left(\frac{N_1}{N_{\text{total}}}\right) = 5.06 \times 10^{-9}.$$

We now turn to the calcium atoms. The ionization energy $\chi_I$ of Ca I is 6.11 eV, about half of the 13.6 eV ionization energy of hydrogen. We will soon see, however, that this small difference has a great effect on the ionization state of the atoms. Note that the Saha equation is very sensitive to the ionization energy because $\chi/kT$ appears as an *exponent* and $kT \approx 0.5$ eV $\ll \chi$. Thus a difference of several electron volts in the ionization energy produces a change of many powers of $e$ in the Saha equation.

Evaluating the partition functions $Z_I$ and $Z_{II}$ for calcium is a bit more complicated than for hydrogen, and the results have been tabulated elsewhere:[15] $Z_I = 1.32$ and $Z_{II} = 2.30$. Thus the ratio of ionized to un-ionized calcium is

$$\left[\frac{N_{II}}{N_I}\right]_{Ca} = \frac{2kT\,Z_{II}}{P_e\,Z_I}\left(\frac{2\pi m_e kT}{h^2}\right)^{3/2} e^{-\chi_I/kT} = 918.$$

Practically all of the calcium atoms are in the form of Ca II; only one atom out of 900 remains neutral. Now we can use the Boltzmann equation to estimate how many of these calcium ions are in the ground state, capable of forming the Ca II H and K absorption lines. The next calculation will consider the K ($\lambda = 393.3$ nm) line; the results for the H ($\lambda = 396.8$ nm) line are similar. The first excited state of Ca II is $E_2 - E_1 = 3.12$ eV above the ground state. The degeneracies for these states are $g_1 = 2$ and $g_2 = 4$. Thus the ratio of the number of Ca II ions in the first excited state to those in the ground state is

$$\left[\frac{N_2}{N_1}\right]_{Ca\,II} = \frac{g_2}{g_1} e^{-(E_2-E_1)/kT} = 3.79 \times 10^{-3} = \frac{1}{264}.$$

Out of every 265 Ca II ions, all but one are in the ground state and are capable of producing the Ca II K line. This implies that nearly *all* of the calcium atoms in the Sun's photosphere

---

[15]The values of the partition functions used here are from Aller (1963); see also Cox (2000), page 32.

are singly ionized and in the ground state,[16] so that almost all of the calcium atoms are available for forming the H and K lines of calcium:

$$\left[\frac{N_1}{N_{\text{total}}}\right]_{\text{Ca II}} \simeq \left[\frac{N_1}{N_1 + N_2}\right]_{\text{Ca II}} \left[\frac{N_{\text{II}}}{N_{\text{total}}}\right]_{\text{Ca}}$$

$$= \left(\frac{1}{1 + [N_2/N_1]_{\text{Ca II}}}\right)\left(\frac{[N_{\text{II}}/N_{\text{I}}]_{\text{Ca}}}{1 + [N_{\text{II}}/N_{\text{I}}]_{\text{Ca}}}\right)$$

$$= \left(\frac{1}{1 + 3.79 \times 10^{-3}}\right)\left(\frac{918}{1 + 918}\right)$$

$$= 0.995.$$

Now it becomes clear why the Ca II H and K lines are so much stronger in the Sun's spectrum than are the Balmer lines. There are 500,000 hydrogen atoms for every calcium atom in the solar photosphere, but only an extremely small fraction, $5.06 \times 10^{-9}$, of these hydrogen atoms are un-ionized and in the first excited state, capable of producing a Balmer line. Multiplying these two factors,

$$(500,000) \times (5.06 \times 10^{-9}) \approx 0.00253 = \frac{1}{395},$$

reveals that there are approximately 400 times more Ca II ions with electrons in the ground state (to produce the Ca II H and K lines) than there are neutral hydrogen atoms with electrons in the first excited state (to produce the Balmer lines). The strength of the H and K lines is *not* due to a greater abundance of calcium in the Sun. Rather, the strength of these Ca II lines reflects the sensitive temperature dependence of the atomic states of excitation and ionization.

Figure 8.11 shows how the strength of various spectral lines varies with spectral type and temperature. As the temperature changes, a smooth variation from one spectral type to the next occurs, indicating that there are only minor differences in the composition of stars, as inferred from their spectra. The first person to determine the composition of the stars and discover the dominant role of hydrogen in the universe was Cecilia Payne (1900–1979). Her 1925 Ph.D. thesis, in which she calculated the relative abundances of 18 elements in stellar atmospheres, is among the most brilliant ever done in astronomy. In Section 9.5, we will see just how the relative abundances of atoms and molecules in stellar atmospheres are measured.

## 8.2 ■ THE HERTZSPRUNG–RUSSELL DIAGRAM

Early in the twentieth century, as astronomers accumulated data for an increasingly large sample of stars, they became aware of the wide range of stellar luminosities and absolute magnitudes. The O stars at one end of the Harvard sequence tended to be both brighter and

---

[16]It is left as an exercise to show that only a very small fraction of calcium atoms are doubly ionized (Ca III).

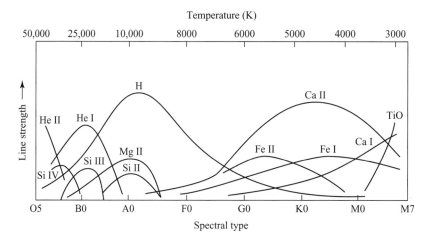

**FIGURE 8.11**    The dependence of spectral line strengths on temperature.

hotter than the M stars at the other end. In addition, the empirical mass–luminosity relation (see Fig. 7.7), deduced from the study of binary stars, showed that O stars are more massive than M stars. These regularities led to a theory of stellar evolution[17] that described how stars might cool off as they age. This theory (no longer accepted) held that stars begin their lives as young, hot, bright blue O stars. It was suggested that as they age, stars become less massive as they exhaust more and more of their "fuel" and that they then gradually become cooler and fainter until they fade away as old, dim red M stars. Although incorrect, a vestige of this idea remains in the terms *early* and *late* spectral types.

**An Enormous Range in Stellar Radii**

If this idea of stellar cooling were correct, then there should be a relation between a star's absolute magnitude and its spectral type. A Danish engineer and amateur astronomer, Ejnar Hertzsprung (1873–1967), analyzed stars whose absolute magnitudes and spectral types had been accurately determined. In 1905 he published a paper confirming the expected correlation between these quantities. However, he was puzzled by his discovery that stars of type G or later had a range of magnitudes, despite having the same spectral classification. Hertzsprung termed the brighter stars **giants**. This nomenclature was natural, since the Stefan–Boltzmann law (Eq. 3.17) shows that

$$R = \frac{1}{T_e^2} \sqrt{\frac{L}{4\pi\sigma}}.$$

(8.10)

If two stars have the same temperature (as inferred for stars having the same spectral type), then the more luminous star must be larger.

---

[17]Stellar evolution describes the change in the structure and composition of an individual star as it ages. This usage of the term *evolution* differs from that in biology, where it describes the changes that occur over generations, rather than during the lifetime of a single individual.

Hertzsprung presented his results in tabular form only. Meanwhile, at Princeton University, Henry Norris Russell (1877–1957) independently came to the same conclusions as Hertzsprung. Russell used the same term, *giant*, to describe the luminous stars of late spectral type and the term **dwarf** stars for their dim counterparts. In 1913 Russell published the diagram shown in Fig. 8.12. It records a star's observed properties: absolute magnitude on the vertical axis (with brightness increasing upward) and spectral type running horizontally (so temperature increases to the *left*). This first "Russell diagram" shows most of the features of its modern successor, the **Hertzsprung–Russell (H–R) diagram**.[18] More than 200 stars were plotted, most within a band reaching from the upper left-hand corner, home of the hot, bright O stars, to the lower right-hand corner, where the cool, dim M stars reside. This band, called the **main sequence**, contains between 80% and 90% of all stars in the H–R diagram. In the upper right-hand corner are the giant stars. A single **white dwarf**, 40 Eridani B, sits at the lower left.[19] The vertical bands of stars in Russell's diagram are a result of the discrete classification of spectral types. A more recent version of an observational H–R diagram is shown in Fig. 8.13 with the absolute visual magnitude of each star plotted versus its color index and spectral type.[20]

Figure 8.14 shows another version of the H–R diagram. Based on the average properties of main-sequence stars as listed in Appendix G, this diagram has a theorist's orientation: The luminosity and effective temperature are plotted for each star, rather than the observationally determined quantities of absolute magnitude and color index or spectral type. The uneven nature of the main sequence is an artifact of the slight differences among the references used to compile the tables in this appendix. The Sun (G2) is found on the main sequence, as is Vega (A0). Both axes are scaled logarithmically to accommodate the huge span of stellar luminosities, ranging from about $5 \times 10^{-4}$ $L_\odot$ to nearly $10^6$ $L_\odot$.[21] Actually, the main sequence is not a line but, rather, has a finite width, as shown in Figs. 8.12 and 8.13, owing to the changes in a star's temperature and luminosity that occur while it is on the main sequence and to slight differences in the compositions of stars. The giant stars occupy the region above the lower main sequence, with the **supergiants**, such as Betelgeuse, in the extreme upper right-hand corner. The white dwarfs (which, despite their name, are often not white at all) lie well below the main sequence.

The radius of a star can be easily determined from its position on the H–R diagram. The Stefan–Boltzmann law in the form of Eq. (8.10) shows that if two stars have the same surface temperature, but one star is 100 times more luminous than the other, then the

---

[18]The names of Hertzsprung and Russell were forever joined by another Danish astronomer, Bengt Strömgren (1908–1987), who suggested that the diagram be named after its two inventors. Strömgren's suggestion that star clusters be studied led to a clarification of the ideas of stellar evolution.

[19]Russell merely considered this star to be an extremely underluminous binary companion of the star 40 Eridani A; the extraordinary nature of white dwarfs was yet to be discovered. Note that the term *dwarf* refers to the stars on the main sequence and should not be confused with the *white dwarf* designation for stars lying well below the main sequence.

[20]Note that Fig. 8.13 suggests that a correlation exists between color index and spectral type, both of which are reflections of the effective temperature of the star. Recall that color index is closely related to the blackbody spectrum of a star (Section 3.6).

[21]Extremely late and early spectral types are not included in Fig. 8.14. The dimmest main-sequence stars are difficult to find, and the brightest have very short lifetimes, making their detection unlikely. As a result, only a handful of stars belonging to these classifications are known—too few to establish their average properties.

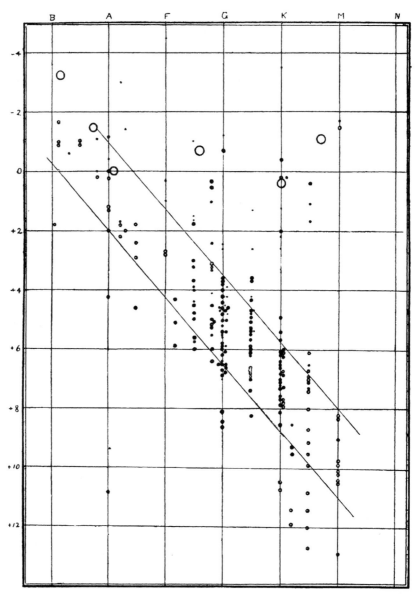

**FIGURE 8.12**    Henry Norris Russell's first diagram, with spectral types listed along the top and absolute magnitudes on the left-hand side. (Figure from Russell, *Nature*, *93*, 252, 1914.)

radius of the more luminous star is $\sqrt{100} = 10$ times larger. On a logarithmically plotted H–R diagram, the locations of stars having the same radii fall along diagonal lines that run roughly parallel to the main sequence (lines of constant radius are also shown in Fig. 8.14). The main-sequence stars show some variation in their sizes, ranging from roughly 20 $R_\odot$ at the extreme upper left end of the main sequence down to 0.1 $R_\odot$ at the lower right end. The giant stars fall between roughly 10 $R_\odot$ and 100 $R_\odot$. For example, Aldebaran ($\alpha$ Tauri),

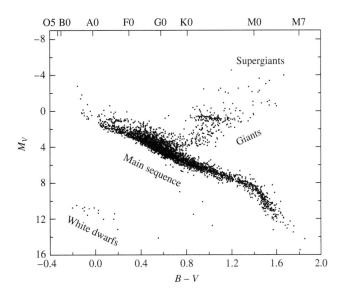

**FIGURE 8.13** An observer's H–R diagram. The data are from the Hipparcos catalog. More than 3700 stars are included here with parallax measurements determined to better than 20%. (Data courtesy of the European Space Agency.)

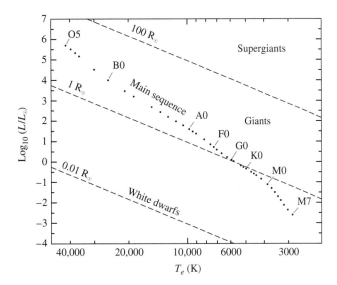

**FIGURE 8.14** The theorist's Hertzsprung–Russell diagram. The dashed lines indicate lines of constant radius.

the gleaming "eye" of the constellation Taurus (the Bull), is an orange giant star that is 45 times larger than the Sun.

The supergiant stars are even larger. Betelgeuse, a pulsating variable star, contracts and expands roughly between 700 and 1000 times the radius of the Sun with a period of approximately 2070 days. If Betelgeuse were located at the Sun's position, its surface would at times extend past the orbit of Jupiter. The star $\mu$ Cephei in the constellation of Cepheus (a king of Ethiopia) is even larger and would swallow Saturn.[22]

The existence of such a simple relation between luminosity and temperature for main-sequence stars is a valuable clue that the position of a star on the main sequence is governed by a single factor. This factor is the star's *mass*.[23] The masses of stars along the main sequence are listed in Appendix G. The most massive O stars listed in that table are observed to have masses of 60 $M_\odot$,[24] and the lower end of the main sequence is bounded by M stars having at least 0.08 $M_\odot$.[25] Combining the radii and masses known for main-sequence stars, we can calculate the average density of the stars. The result, perhaps surprising, is that main-sequence stars have roughly the same density as water. Moving up the main sequence, we find that the larger, more massive, early-type stars have a *lower* average density.

---

**Example 8.2.1.**   The Sun, a G2 main-sequence star, has a mass of $M_\odot = 1.9891 \times 10^{30}$ kg and a radius of $R_\odot = 6.95508 \times 10^8$ m. Its average density is thus

$$\overline{\rho}_\odot = \frac{M_\odot}{\frac{4}{3}\pi R_\odot^3} = 1410 \text{ kg m}^{-3}.$$

Sirius, the brightest-appearing star in the sky, is classified as an A1 main sequence star with a mass of 2.2 $M_\odot$ and a radius of 1.6 $R_\odot$. The average density of Sirius is

$$\overline{\rho} = \frac{2.2\,M_\odot}{\frac{4}{3}\pi(1.6\,R_\odot)^3} = 760 \text{ kg m}^{-3} = 0.54\,\overline{\rho}_\odot,$$

which is about 76 percent of the density of water. However, this is enormously dense compared to a giant or supergiant star. The mass of Betelgeuse is estimated to lie between 10 and 15 $M_\odot$; we will adopt 10 $M_\odot$ here. For illustration, if we take the maximum radius of this pulsating star to be about 1000 $R_\odot$, then the average density of Betelgeuse (at maximum size) is roughly

$$\overline{\rho} = \frac{10\,M_\odot}{\frac{4}{3}\pi(1000\,R_\odot)^3} = 10^{-8}\,\overline{\rho}_\odot!$$

Thus Betelgeuse is a tenuous, ghostly object—a hundred thousand times less dense than the air we breathe. It is difficult even to define what is meant by the "surface" of such a wraith-like star.

---

[22] $\mu$ Cephei is a pulsating variable like Betelgeuse and has a period of 730 days. One of the reddest stars visible in the night sky, $\mu$ Cephei, is known as the *Garnet Star*.

[23] In Chapter 10 we will see how mass determines the location of a star on the main sequence.

[24] Theoretical calculations indicate that main-sequence stars as massive as 90 $M_\odot$ may exist, and recent observations have been made of a few stars with masses estimated near 100 $M_\odot$; see Section 15.3.

[25] Stars less massive than 0.08 $M_\odot$ have insufficient temperatures in their cores to support significant nuclear burning (see Chapter 10).

## Morgan–Keenan Luminosity Classes

Hertzsprung wondered whether there might be some difference in the spectra of giant and main-sequence stars of the same spectral type (or same effective temperature). He found just such a variation in spectra among the stars cataloged by Antonia Maury. In her classification scheme she had noted line width variations that she referred to as a *c*-characteristic. The subtle differences in the relative strengths of spectral lines for stars of similar effective temperatures and different luminosities are depicted in Fig. 8.15. The work begun by Hertzsprung and Maury, and further developed by other astronomers, culminated in the 1943 publication of the *Atlas of Stellar Spectra* by William W. Morgan (1906–1994) and Phillip C. Keenan (1908–2000) of Yerkes Observatory. Their atlas consists of 55 prints of spectra that clearly display the effect of temperature and luminosity on stellar spectra and includes the criteria for the classification of each spectrum. The *MKK Atlas* established the *two-dimensional* Morgan–Keenan (M–K) system of spectral classification.[26] A **luminosity class**, designated by a Roman numeral, is appended to a star's Harvard spectral type. The numeral "I" (subdivided into classes Ia and Ib) is reserved for the supergiant stars, and "V" denotes a main-sequence star. The ratio of the strengths of two closely spaced lines is often employed to place a star in the appropriate luminosity class. In general, for stars of the same spectral type, narrower lines are usually produced by more luminous stars.[27] The Sun is a G2 V star, and Betelgeuse is classified as M2 Ia.[28] The series of Roman numerals extends below the main sequence; the subdwarfs (class VI or "sd") reside slightly to the left of the main sequence because they are deficient in metals. The M–K system does not extend to the white dwarfs, which are classified by the letter D. Figure 8.16 shows the corresponding divisions on the H–R diagram and the locations of a selection of specific stars, and Table 8.3 lists the luminosity classes.

The two-dimensional M–K classification scheme enables astronomers to locate a star's position on the Hertzsprung–Russell diagram based *entirely* on the appearance of its spectrum. Once the star's absolute magnitude, $M$, has been read from the vertical axis of the H–R diagram, the *distance* to the star can be calculated from its apparent magnitude, $m$, via Eq. (3.5),

$$d = 10^{(m-M+5)/5},$$

where $d$ is in units of parsecs. This method of distance determination, called **spectroscopic parallax**, is responsible for many of the distances measured for stars,[29] but its accuracy is limited because there is not a perfect correlation between stellar absolute magnitudes and luminosity classes. The intrinsic scatter of roughly $\pm 1$ magnitude for a specific luminosity class renders $d$ uncertain by a factor of about $10^{1/5} = 1.6$.

---

[26]Edith Kellman of Yerkes printed the 55 spectra and was a co-author of the atlas; hence the additional "K" in *MKK Atlas*.

[27]In Section 9.5, we will find that because the atmospheres of more luminous stars are less dense, there are fewer collisions between atoms. Collisions can distort the energies of atomic orbitals, leading to broadening of the spectral lines.

[28]Betelgeuse, a pulsating variable star, is sometimes given the intermediate classification M2 Iab.

[29]Since the technique of parallax is not involved, the term *spectroscopic parallax* is a misnomer, although the name does at least imply a distance determination.

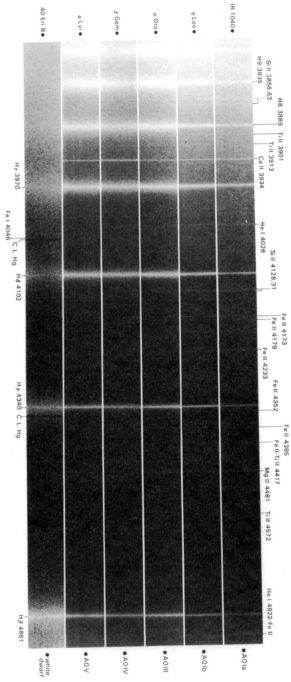

**FIGURE 8.15** A comparison of the strengths of the hydrogen Balmer lines in types A0 Ia, A0 Ib, A0 III, A0 IV, A0 V, and a white dwarf, showing the narrower lines found in supergiants. These spectra are displayed as negatives, so absorption lines appear bright. (Figure from Yamashita, Nariai, and Norimoto, *An Atlas of Representative Stellar Spectra*, University of Tokyo Press, Tokyo, 1978.)

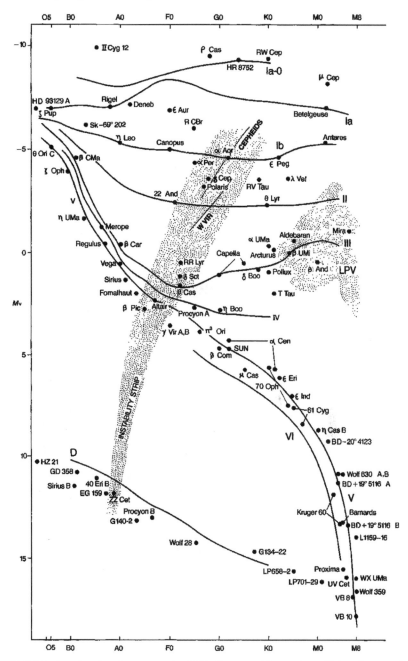

**FIGURE 8.16** Luminosity classes on the H–R diagram. (Figure from Kaler, *Stars and Stellar Spectra*, © Cambridge University Press 1989. Reprinted with the permission of Cambridge University Press.)

**TABLE 8.3**    Morgan–Keenan Luminosity Classes.

| Class | Type of Star |
|-------|--------------|
| Ia-O | Extreme, luminous supergiants |
| Ia | Luminous supergiants |
| Ib | Less luminous supergiants |
| II | Bright giants |
| III | Normal giants |
| IV | Subgiants |
| V | Main-sequence (dwarf) stars |
| VI, sd | Subdwarfs |
| D | White dwarfs |

## SUGGESTED READING

### General

Aller, Lawrence H., *Atoms, Stars, and Nebulae*, Third Edition, Cambridge University Press, New York, 1991.

Dobson, Andrea K., and Bracher, Katherine, "A Historical Introduction to Women in Astronomy," *Mercury*, January/February 1992.

Hearnshaw, J. B., *The Analysis of Starlight*, Cambridge University Press, Cambridge, 1986.

Herrmann, Dieter B., *The History of Astronomy from Hershel to Hertzsprung*, Cambridge University Press, Cambridge, 1984.

Hoffleit, Dorrit, "Reminiscenses on Antonia Maury and the *c*-Characteristic," *The MK Process at 50 Years*, Corbally, C. J., Gray, R. O., and Garrison, R. F. (editors), *ASP Conference Series*, *60*, 215, 1994.

Kaler, James B., *Stars and Their Spectra*, Cambridge University Press, Cambridge, 1997.

### Technical

Aller, Lawrence H., *The Atmospheres of the Sun and Stars*, Ronald Press, New York, 1963.

Böhm-Vitense, Erika, *Stellar Astrophysics, Volume 2: Stellar Atmospheres*, Cambridge University Press, Cambridge, 1989.

Cox, Arthur N. (editor), *Allen's Astrophysical Quantities*, Fourth Edition, AIP Press, New York, 2000.

Geballe, T. R., et al., "Toward Spectral Classification of L and T Dwarfs: Infrared and Optical Spectroscopy and Analysis," *The Astrophysical Journal*, *564*, 466, 2002.

Kirkpatrick, J. Davy, et al., "Dwarfs Cooler Than "M": The Definition of Spectral Type "L" Using Discoveries From the 2-Micron All-Sky Survey (2MASS)," *The Astrophysical Journal*, *519*, 802, 1999.

Mihalas, Dimitri, *Stellar Atmospheres*, Second Edition, W.H. Freeman, San Francisco, 1978.

Novotny, Eva, *Introduction to Stellar Atmospheres and Interiors*, Oxford University Press, New York, 1973.

Padmanabhan, T., *Theoretical Astrophysics*, Cambridge University Press, Cambridge, 2000.

## PROBLEMS

**8.1** Show that at room temperature, the thermal energy $kT \approx 1/40$ eV. At what temperature is $kT$ equal to 1 eV? to 13.6 eV?

**8.2** Verify that Boltzmann's constant can be expressed in terms of electron volts rather than joules as $k = 8.6173423 \times 10^{-5}$ eV K$^{-1}$ (see Appendix A).

**8.3** Use Fig. 8.6, the graph of the Maxwell–Boltzmann distribution for hydrogen gas at 10,000 K, to estimate the fraction of hydrogen atoms with a speed within 1 km s$^{-1}$ of the most probable speed, $v_{mp}$.

**8.4** Show that the most probable speed of the Maxwell–Boltzmann distribution of molecular speeds (Eq. 8.1) is given by Eq. (8.2).

**8.5** For a gas of neutral hydrogen atoms, at what temperature is the number of atoms in the first excited state only 1% of the number of atoms in the ground state? At what temperature is the number of atoms in the first excited state 10% of the number of atoms in the ground state?

**8.6** Consider a gas of neutral hydrogen atoms, as in Example 8.1.3.

   **(a)** At what temperature will equal numbers of atoms have electrons in the ground state and in the second excited state ($n = 3$)?

   **(b)** At a temperature of 85,400 K, when equal numbers ($N$) of atoms are in the ground state and in the first excited state, how many atoms are in the second excited state ($n = 3$)? Express your answer in terms of $N$.

   **(c)** As the temperature $T \rightarrow \infty$, how will the electrons in the hydrogen atoms be distributed, according to the Boltzmann equation? That is, what will be the relative numbers of electrons in the $n = 1, 2, 3, \ldots$ orbitals? Will this in fact be the distribution that actually occurs? Why or why not?

**8.7** In Example 8.1.4, the statement was made that "nearly all of the H I atoms are in the ground state, so Eq. (8.7) for the partition function simplifies to $Z_1 \simeq g_1 = 2(1)^2 = 2$." Verify that this statement is correct for a temperature of 10,000 K by evaluating the first three terms in Eq. (8.7) for the partition function.

**8.8** Equation (8.7) for the partition function actually diverges as $n \rightarrow \infty$. Why can we ignore these large-$n$ terms?

**8.9** Consider a box of electrically neutral hydrogen gas that is maintained at a constant volume $V$. In this simple situation, the number of free electrons must equal the number of H II ions: $n_e V = N_{II}$. Also, the total number of hydrogen atoms (both neutral and ionized), $N_t$, is related to the density of the gas by $N_t = \rho V/(m_p + m_e) \simeq \rho V/m_p$, where $m_p$ is the mass of the proton. (The tiny mass of the electron may be safely ignored in this expression for $N_t$.) Let the density of the gas be $10^{-6}$ kg m$^{-3}$, typical of the photosphere of an A0 star.

   **(a)** Make these substitutions into Eq. (8.8) to derive a quadratic equation for the fraction of ionized atoms:

$$\left(\frac{N_{II}}{N_t}\right)^2 + \left(\frac{N_{II}}{N_t}\right)\left(\frac{m_p}{\rho}\right)\left(\frac{2\pi m_e kT}{h^2}\right)^{3/2} e^{-\chi_1/kT} - \left(\frac{m_p}{\rho}\right)\left(\frac{2\pi m_e kT}{h^2}\right)^{3/2} e^{-\chi_1/kT} = 0.$$

   **(b)** Solve the quadratic equation in part (a) for the fraction of ionized hydrogen, $N_{II}/N_t$, for a range of temperatures between 5000 K and 25,000 K. Make a graph of your results, and compare it with Fig. 8.8.

**8.10** In this problem, you will follow a procedure similar to that of Example 8.1.4 for the case of a stellar atmosphere composed of pure helium to find the temperature at the middle of the He I partial ionization zone, where half of the He I atoms have been ionized. (Such an atmosphere would be found on a white dwarf of spectral type DB; see Section 16.1.) The ionization energies of neutral helium and singly ionized helium are $\chi_I = 24.6$ eV and $\chi_{II} = 54.4$ eV, respectively. The partition functions are $Z_I = 1$, $Z_{II} = 2$, and $Z_{III} = 1$ (as expected for any completely ionized atom). Use $P_e = 20$ N m$^{-2}$ for the electron pressure.

(a) Use Eq. (8.9) to find $N_{II}/N_I$ and $N_{III}/N_{II}$ for temperatures of 5000 K, 15,000 K, and 25,000 K. How do they compare?

(b) Show that $N_{II}/N_{total} = N_{II}/(N_I + N_{II} + N_{III})$ can be expressed in terms of the ratios $N_{II}/N_I$ and $N_{III}/N_{II}$.

(c) Make a graph of $N_{II}/N_{total}$ similar to Fig. 8.8 for a range of temperatures from 5000 K to 25,000 K. What is the temperature at the middle of the He I partial ionization zone? Because the temperatures of the middle of the hydrogen and He I partial ionization zones are so similar, they are sometimes considered to be a single partial ionization zone with a characteristic temperature of $1–1.5 \times 10^4$ K.

**8.11** Follow the procedure of Problem 8.10 to find the temperature at the middle of the He II partial ionization zone, where half of the He II atoms have been ionized. This ionization zone is found at a greater depth in the star, and so the electron pressure is larger—use a value of $P_e = 1000$ N m$^{-2}$. Let your temperatures range from 10,000 K to 60,000 K. This particular ionization zone plays a crucial role in pulsating stars, as will be discussed in Section 14.2.

**8.12** Use the Saha equation to determine the fraction of hydrogen atoms that are ionized, $N_{II}/N_{total}$, at the center of the Sun. Here the temperature is 15.7 million K and the number density of electrons is about $n_e = 6.1 \times 10^{31}$ m$^{-3}$. (Use $Z_I = 2$.) Does your result agree with the fact that practically *all* of the Sun's hydrogen is ionized at the Sun's center? What is the reason for any discrepancy?

**8.13** Use the information in Example 8.1.5 to calculate the ratio of doubly to singly ionized calcium atoms (Ca III/Ca II) in the Sun's photosphere. The ionization energy of Ca II is $\chi_{II} = 11.9$ eV. Use $Z_{III} = 1$ for the partition function of Ca III. Is your result consistent with the statement in Example 8.1.5 that in the solar photosphere, "nearly all of the calcium atoms are available for forming the H and K lines of calcium"?

**8.14** Consider a giant star and a main-sequence star of the *same spectral type*. Appendix G shows that the giant star, which has a lower atmospheric density, has a slightly lower temperature than the main-sequence star. Use the Saha equation to explain why this is so. Note that this means that there is not a perfect correspondence between temperature and spectral type!

**8.15** Figure 8.14 shows that a white dwarf star typically has a radius that is only 1% of the Sun's. Determine the average density of a 1-$M_\odot$ white dwarf.

**8.16** The blue-white star Fomalhaut ("the fish's mouth" in Arabic) is in the southern constellation of Pisces Austrinus. Fomalhaut has an apparent visual magnitude of $V = 1.19$. Use the H–R diagram in Fig. 8.16 to determine the distance to this star.

# CHAPTER

# 9

# Stellar Atmospheres

## 9.1 ■ THE DESCRIPTION OF THE RADIATION FIELD

The light that astronomers receive from a star comes from the star's atmosphere, the layers of gas overlying the opaque interior. A flood of photons pours from these layers, releasing the energy produced by the thermonuclear reactions, gravitational contraction, and cooling in the star's center. The temperature, density, and composition of the atmospheric layers from which these photons escape determine the features of the star's spectrum. To interpret the observed spectral lines properly, we must describe how light travels through the gas that makes up a star.

### The Specific and Mean Intensities

Figure 9.1 shows a ray of light with a wavelength between $\lambda$ and $\lambda + d\lambda$ passing through a surface of area $dA$ at an angle $\theta$ into a cone of solid angle $d\Omega$.[1] The angle $\theta$ is measured from the direction perpendicular to the surface, so $dA \cos\theta$ is the area $dA$ projected onto a plane perpendicular to the direction in which the radiation is traveling. Defining

$$E_\lambda \equiv \frac{\partial E}{\partial \lambda},$$

$E_\lambda \, d\lambda$ is assumed to be the amount of energy that these rays carry into the cone in a time interval $dt$. Then the **specific intensity** of the rays is defined as

$$I_\lambda \equiv \frac{\partial I}{\partial \lambda} \equiv \frac{E_\lambda \, d\lambda}{d\lambda \, dt \, dA \, \cos\theta \, d\Omega}. \tag{9.1}$$

---

[1]The surface is a mathematical location in space and is not necessarily a real physical surface. The concept of a solid angle and its units of steradians (sr) was discussed in Section 6.1.

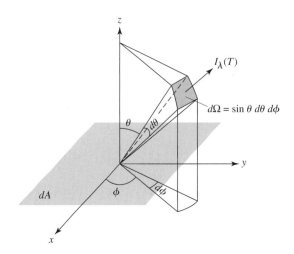

**FIGURE 9.1**    Intensity $I_\lambda$.

Although the energy $E_\lambda \, d\lambda$ in the numerator is vanishingly small, the differentials in the denominator are also vanishingly small, so the ratio approaches a limiting value of $I_\lambda$. The specific intensity is usually referred to simply as the *intensity*. Thus, in spherical coordinates,

$$E_\lambda \, d\lambda = I_\lambda \, d\lambda \, dt \, dA \, \cos\theta \, d\Omega = I_\lambda \, d\lambda \, dt \, dA \, \cos\theta \, \sin\theta \, d\theta \, d\phi \qquad (9.2)$$

is the amount of electromagnetic radiation energy having a wavelength between $\lambda$ and $\lambda + d\lambda$ that passes in time $dt$ through the area $dA$ into a solid angle $d\Omega = \sin\theta \, d\theta \, d\phi$. The specific intensity therefore has units of W m$^{-3}$sr$^{-1}$.[2] The Planck function $B_\lambda$, Eq. (3.22), is an example of the specific intensity for the special case of blackbody radiation. In general, however, the energy of the light need not vary with wavelength in the same way as it does for blackbody radiation. Later we will see under what circumstances we may set $I_\lambda = B_\lambda$.

Imagine a light ray of intensity $I_\lambda$ as it propagates through a vacuum. Because $I_\lambda$ is defined in the limit $d\Omega \to 0$, the energy of the ray *does not spread out (or diverge)*. The intensity is therefore *constant* along any ray traveling through empty space.

In general, the specific intensity $I_\lambda$ does vary with direction, however. The **mean intensity** of the radiation is found by integrating the specific intensity over all directions and dividing the result by $4\pi$ sr, the solid angle enclosed by a sphere, to obtain an average value of $I_\lambda$. In spherical coordinates, this average value is[3]

$$\langle I_\lambda \rangle \equiv \frac{1}{4\pi} \int I_\lambda \, d\Omega = \frac{1}{4\pi} \int_{\phi=0}^{2\pi} \int_{\theta=0}^{\pi} I_\lambda \, \sin\theta \, d\theta \, d\phi. \qquad (9.3)$$

For an isotropic radiation field (one with the same intensity in all directions), $\langle I_\lambda \rangle = I_\lambda$. Blackbody radiation is isotropic and has $\langle I_\lambda \rangle = B_\lambda$.

---

[2]Recall from Section 3.5 that W m$^{-3}$ indicates an energy per second per unit area *per unit wavelength interval*, W m$^{-2}$m$^{-1}$, *not* an energy per second per unit volume.

[3]Many texts refer to the average intensity as $J_\lambda$ instead of $\langle I_\lambda \rangle$. However, in this text the notation $\langle I_\lambda \rangle$ has been selected to explicitly illustrate the average nature of the quantity.

## The Specific Energy Density

To determine how much energy is contained within the radiation field, we can use a "trap" consisting of a small cylinder of length $dL$, open at both ends, with perfectly reflecting walls inside; see Fig. 9.2. Light entering the trap at one end travels and (possibly) bounces back and forth until it exits the other end of the trap. The energy inside the trap is the same as what would be present at that location if the trap were removed. The radiation that enters the trap at an angle $\theta$ travels through the trap in a time $dt = dL/(c\cos\theta)$. Thus the amount of energy inside the trap with a wavelength between $\lambda$ and $\lambda + d\lambda$ that is due to the radiation that enters at angle $\theta$ is

$$E_\lambda \, d\lambda = I_\lambda \, d\lambda \, dt \, dA \, \cos\theta \, d\Omega = I_\lambda \, d\lambda \, dA \, d\Omega \, \frac{dL}{c}.$$

The quantity $dA \, dL$ is just the volume of the trap, so the **specific energy density** (energy per unit volume having a wavelength between $\lambda$ and $\lambda + d\lambda$) is found by dividing $E_\lambda \, d\lambda$ by $dL \, dA$, integrating over all solid angles, and using Eq. (9.3):

$$
\begin{aligned}
u_\lambda \, d\lambda &= \frac{1}{c} \int I_\lambda \, d\lambda \, d\Omega \\
&= \frac{1}{c} \int_{\phi=0}^{2\pi} \int_{\theta=0}^{\pi} I_\lambda \, d\lambda \, \sin\theta \, d\theta \, d\phi \\
&= \frac{4\pi}{c} \langle I_\lambda \rangle \, d\lambda.
\end{aligned}
\tag{9.4}
$$

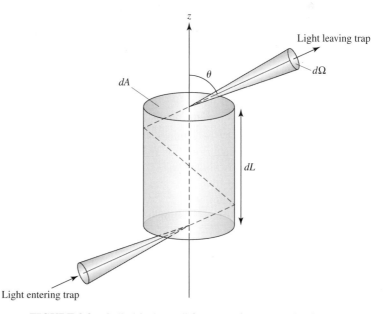

**FIGURE 9.2**   Cylindrical "trap" for measuring energy density $u_\lambda$.

For an isotropic radiation field, $u_\lambda \, d\lambda = (4\pi/c)I_\lambda \, d\lambda$, and for blackbody radiation,

$$u_\lambda \, d\lambda = \frac{4\pi}{c} B_\lambda \, d\lambda = \frac{8\pi hc/\lambda^5}{e^{hc/\lambda kT} - 1} \, d\lambda. \tag{9.5}$$

At times it may be more useful to express the blackbody energy density in terms of the frequency, $\nu$, of the light by employing Eq. (3.24):

$$u_\nu \, d\nu = \frac{4\pi}{c} B_\nu \, d\nu = \frac{8\pi h\nu^3/c^3}{e^{h\nu/kT} - 1} \, d\nu. \tag{9.6}$$

Thus $u_\nu \, d\nu$ is the energy per unit volume with a frequency between $\nu$ and $\nu + d\nu$.

The total energy density, $u$, is found by integrating over all wavelengths or over all frequencies:

$$u = \int_0^\infty u_\lambda \, d\lambda = \int_0^\infty u_\nu \, d\nu.$$

For blackbody radiation ($I_\lambda = B_\lambda$), Eq. (3.28) shows that

$$u = \frac{4\pi}{c} \int_0^\infty B_\lambda(T) \, d\lambda = \frac{4\sigma T^4}{c} = aT^4, \tag{9.7}$$

where $a \equiv 4\sigma/c$ is known as the *radiation constant* and has the value

$$a = 7.565767 \times 10^{-16} \text{ J m}^{-3}\text{K}^{-4}.$$

### The Specific Radiative Flux

Another quantity of interest is $F_\lambda$, the **specific radiative flux**. $F_\lambda \, d\lambda$ is the *net* energy having a wavelength between $\lambda$ and $\lambda + d\lambda$ that passes each second through a unit area in the direction of the $z$-axis:

$$F_\lambda \, d\lambda = \int I_\lambda \, d\lambda \, \cos\theta \, d\Omega = \int_{\phi=0}^{2\pi} \int_{\theta=0}^{\pi} I_\lambda \, d\lambda \, \cos\theta \, \sin\theta \, d\theta \, d\phi. \tag{9.8}$$

The factor of $\cos\theta$ determines the $z$-component of a light ray and allows the cancelation of oppositely directed rays. For an isotropic radiation field there is no net transport of energy, and so $F_\lambda = 0$.

Both the radiative flux and the specific intensity describe the light received from a celestial source, and you may wonder which of these quantities is actually measured by a telescope's photometer, pointed at the source of light. The answer depends on whether the source is resolved by the telescope. Figure 9.3(a) shows a source of light, uniform over its entire surface,[4] that is resolved by the telescope; the angle $\theta$ subtended by the source as a whole is much larger than $\theta_{\text{min}}$, the smallest angle resolvable according to Rayleigh's

---

[4]The assumption of a uniform light source precludes dimming effects such as limb darkening, which will be discussed later.

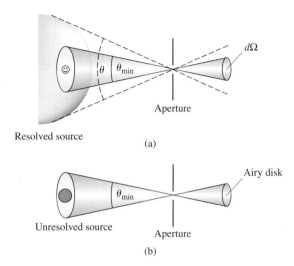

**FIGURE 9.3**   The measurement of (a) the specific intensity for a resolved source and (b) the radiative flux for an unresolved source. Note that any object with an angular resolution smaller than $\theta_{min}$ on the surface of the resolved source (such as a surface feature on a planet) remains unresolved.

criterion. In this case, what is being measured is the *specific intensity*, the amount of energy per second passing through the aperture area into the solid angle $\Omega_{min}$ defined by $\theta_{min}$. For example, at a wavelength of 501 nm, the measured value of the specific intensity at the center of the Sun's disk is

$$I_{501} = 4.03 \times 10^{13} \text{ W m}^{-3}\text{sr}^{-1}.$$

Now imagine that the source is moved twice as far away. According to the inverse square law for light, Eq. (3.2), there will be only $(1/2)^2 = 1/4$ as much energy received from each square meter of the source. If the source is still resolved, however, then the amount of source area that contributes energy to the solid angle $\Omega_{min}$ has increased by a factor of 4, resulting in the same amount of energy reaching each square meter of the detector. The specific intensity of light rays from the source is thus measured to be constant.[5]

However, it is the *radiative flux* that is measured for an unresolved source. As the source recedes farther and farther, it will eventually subtend an angle $\theta$ smaller than $\theta_{min}$, and it can no longer be resolved by the telescope. When $\theta < \theta_{min}$, the energy received from the entire source will disperse throughout the diffraction pattern (the Airy disk and rings; recall Section 6.1) determined by the telescope's aperture. Because the light arriving at the detector leaves the surface of the source at all angles [see Fig. 9.3(b)], the detector is effectively integrating the specific intensity over all directions. This is just the definition of the radiative flux, Eq. (9.8). As the distance $r$ to the source increases further, the amount of energy falling within the Airy disk (and consequently the value of the radiative flux) decreases as $1/r^2$, as expected.

---

[5]We encountered this argument in the statement in Section 6.1 that the image and object intensities of a resolved object are the same.

## Radiation Pressure

Because a photon possesses an energy $E$, Einstein's relativistic energy equation (Eq. 4.48) tells us that even though it is massless, a photon also carries a momentum of $p = E/c$ and thus can exert a **radiation pressure**. This radiation pressure can be derived in the same way that gas pressure is found for molecules bouncing off a wall. Figure 9.4 shows photons reflected at an angle $\theta$ from a perfectly reflecting surface of area $dA$ into a solid angle $d\Omega$. Because the angle of incidence equals the angle of reflection, the solid angles shown for the incident and reflected photons are the same size and inclined by the same angle $\theta$ on opposing sides of the $z$-axis. The change in the $z$-component of the momentum of photons with wavelengths between $\lambda$ and $\lambda + d\lambda$ that are reflected from the area $dA$ in a time interval $dt$ is

$$
\begin{aligned}
dp_\lambda \, d\lambda &= \left[ (p_\lambda)_{\text{final},z} - (p_\lambda)_{\text{initial},z} \right] d\lambda \\
&= \left[ \frac{E_\lambda \cos\theta}{c} - \left( -\frac{E_\lambda \cos\theta}{c} \right) \right] d\lambda \\
&= \frac{2 \, E_\lambda \cos\theta}{c} \, d\lambda \\
&= \frac{2}{c} \, I_\lambda \, d\lambda \, dt \, dA \, \cos^2\theta \, d\Omega,
\end{aligned}
$$

where the last expression was obtained from Eq. (9.2). Dividing $dp_\lambda$ by $dt$ and $dA$ gives $(dp_\lambda/dt)/dA$. But from Newton's second and third laws, $-dp_\lambda/dt$ is the force exerted by the photons on the area $dA$, although we will ignore the minus sign, which merely says that the force is in the $-z$-direction. Thus the radiation pressure is the force per unit area, $(dp_\lambda/dt)/dA$, produced by the photons within the solid angle $d\Omega$. Integrating over the hemisphere of all incident directions results in $P_{\text{rad},\lambda} \, d\lambda$, the radiation pressure exerted by

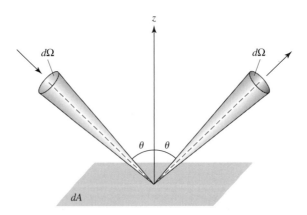

**FIGURE 9.4**   Radiation pressure produced by incident photons from the solid angle $d\Omega$.

those photons having a wavelength between $\lambda$ and $\lambda + d\lambda$:

$$P_{\text{rad},\lambda}\, d\lambda = \frac{2}{c} \int_{\text{hemisphere}} I_\lambda\, d\lambda\, \cos^2\theta\, d\Omega \quad \text{(reflection)}$$

$$= \frac{2}{c} \int_{\phi=0}^{2\pi} \int_{\theta=0}^{\pi/2} I_\lambda\, d\lambda\, \cos^2\theta\, \sin\theta\, d\theta\, d\phi.$$

Just as the pressure of a gas exists throughout the volume of the gas and not just at the container walls, the radiation pressure of a "photon gas" exists everywhere in the radiation field. Imagine removing the reflecting surface $dA$ in Fig. 9.4 and replacing it with a mathematical surface. The incident photons will now keep on going through $dA$; instead of reflected photons, photons will be streaming through $dA$ from the other side. Thus, for an *isotropic radiation field*, there will be no change in the expression for the radiation pressure if the leading factor of 2 (which originated in the change in momentum upon reflection of the photons) is removed and the angular integration is extended over all solid angles:

$$P_{\text{rad},\lambda}\, d\lambda = \frac{1}{c} \int_{\text{sphere}} I_\lambda\, d\lambda\, \cos^2\theta\, d\Omega \quad \text{(transmission)} \tag{9.9}$$

$$= \frac{1}{c} \int_{\phi=0}^{2\pi} \int_{\theta=0}^{\pi} I_\lambda\, d\lambda\, \cos^2\theta\, \sin\theta\, d\theta\, d\phi$$

$$= \frac{4\pi}{3c} I_\lambda\, d\lambda \quad \text{(isotropic radiation field).} \tag{9.10}$$

However, it may be that the radiation field is *not* isotropic. In that case, Eq. (9.9) for the radiation pressure is still valid but the pressure depends on the orientation of the mathematical surface $dA$.

The total radiation pressure produced by photons of all wavelengths is found by integrating Eq. (9.10):

$$P_{\text{rad}} = \int_0^\infty P_{\text{rad},\lambda}\, d\lambda.$$

For blackbody radiation, it is left as a problem to show that

$$P_{\text{rad}} = \frac{4\pi}{3c} \int_0^\infty B_\lambda(T)\, d\lambda = \frac{4\sigma T^4}{3c} = \frac{1}{3} a T^4 = \frac{1}{3} u. \tag{9.11}$$

Thus the **blackbody radiation pressure** is one-third of the energy density. (For comparison, the pressure of an ideal monatomic gas is two-thirds of its energy density.)

## 9.2 ■ STELLAR OPACITY

The classification of stellar spectra is an ongoing process. Even the most basic task, such as finding the "surface"[6] temperature of a particular star, is complicated by the fact that stars are not actually blackbodies. The Stefan–Boltzmann relation, in the form of Eq. (3.17), defines a star's effective temperature, but some effort is required to obtain a more accurate value of the "surface" temperature.[7] Figure 9.5 shows that the Sun's spectrum deviates substantially from the shape of the blackbody Planck function, $B_\lambda$, because solar absorption lines remove light from the Sun's continuous spectrum at certain wavelengths. The decrease in intensity produced by the dense series of metallic absorption lines in the solar spectrum is especially effective; this effect is called **line blanketing**. In other wavelength regimes (e.g., X-ray and UV), emission lines may augment the intensity of the continuous spectrum.

### Temperature and Local Thermodynamic Equilibrium

Although we often think in terms of *the temperature* at a particular location, there are actually many different measures of temperature within a star, defined according to the physical process being described:

- The **effective temperature**, which is obtained from the Stefan–Boltzmann law (Eq. 3.17), is uniquely defined for a specific level within a star and is an important global descriptor of that star.

- The **excitation temperature** is defined by the Boltzmann equation (8.6).

- The **ionization temperature** is defined by the Saha equation (8.8).

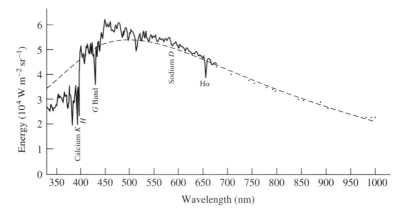

**FIGURE 9.5**   The spectrum of the Sun in 2 nm wavelength intervals. The dashed line is the curve of an ideal blackbody having the Sun's effective temperature. (Figure adapted from Aller, *Atoms, Stars, and Nebulae*, Third Edition, Cambridge University Press, New York, 1991.)

---

[6]The "surface" of a star is defined as the region where the emergent visual continuum forms, namely the *photosphere* (see Section 11.2).

[7]See Böhm-Vitense (1981) for more details concerning the determination of temperatures.

- The **kinetic temperature** is contained in the Maxwell–Boltzmann distribution, Eq. (8.1).

- The **color temperature** is obtained by fitting the shape of a star's continuous spectrum to the Planck function, Eq. (3.22).

With the exception of the effective temperature, the remaining temperatures apply to any location within the star and vary according to the conditions of the gas. Although defined differently, the excitation temperature, the ionization temperature, the kinetic temperature, and the color temperature are the same for the simple case of a gas confined within an "ideal box." The confined gas particles and blackbody radiation will come into equilibrium, individually and with each other, and can be described by a single well-defined temperature. In such a steady-state condition, no net flow of energy through the box or between the matter and the radiation occurs. Every process (e.g., the absorption of a photon) occurs at the same rate as its inverse process (e.g., the emission of a photon). This condition is called **thermodynamic equilibrium**.

However, a star cannot be in perfect thermodynamic equilibrium. A net outward flow of energy occurs through the star, and the temperature, however it is defined, varies with location. Gas particles and photons at one position in the star may have arrived there from other regions, either hotter or cooler (in other words, there is no "ideal box"). The distribution in particle speeds and photon energies thus reflects a range of temperatures. As the gas particles collide with one another and interact with the radiation field by absorbing and emitting photons, the description of the processes of excitation and ionization becomes quite complex. However, the idealized case of a single temperature can still be employed if the distance over which the temperature changes significantly is large compared with the distances traveled by the particles and photons between collisions (their *mean free paths*). In this case, referred to as **local thermodynamic equilibrium** (LTE), the particles and photons cannot escape the local environment and so are effectively confined to a limited volume (an approximated "box") of nearly constant temperature.

---

**Example 9.2.1.** The photosphere is the surface layer of the Sun's atmosphere where the photons can escape into space (see page 253 and Section 11.2). According to a model solar atmosphere (see Cox, page 348), the temperature in one region of the photosphere varies from 5580 K to 5790 K over a distance of 25.0 km. The characteristic distance over which the temperature varies, called the *temperature scale height*, $H_T$, is given by

$$H_T \equiv \frac{T}{|dT/dr|} = \frac{5685 \text{ K}}{(5790 \text{ K} - 5580 \text{ K})/(25.0 \text{ km})} = 677 \text{ km},$$

where the average temperature has been used for the value of $T$.

How does the temperature scale height of 677 km compare with the average distance traveled by an atom before hitting another atom? The density of the photosphere at that level is about $\rho = 2.1 \times 10^{-4} \text{ kg m}^{-3}$, consisting primarily of neutral hydrogen atoms in the ground state. Assuming a pure hydrogen gas for convenience, the number of hydrogen

*continued*

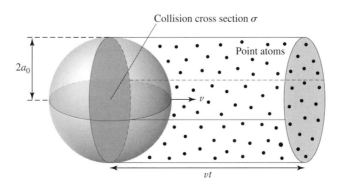

**FIGURE 9.6**    Mean free path, $\ell$, of a hydrogen atom.

atoms per cubic meter is roughly

$$n = \frac{\rho}{m_H} = 1.25 \times 10^{23} \text{ m}^{-3},$$

where $m_H$ is the mass of a hydrogen atom. In an approximate sense, two of these atoms will "collide" if their centers pass within two Bohr radii, $2a_0$, of each other.[8] As shown in Fig. 9.6, we may consider the equivalent problem of a single atom of radius $2a_0$ moving with speed $v$ through a collection of stationary points that represent the centers of the other atoms. In an amount of time $t$, this atom has moved a distance $vt$ and has swept out a cylindrical volume $V = \pi(2a_0)^2 vt = \sigma vt$, where $\sigma \equiv \pi(2a_0)^2$ is the collision **cross section** of the atom in this classical approximation.[9] Within this volume $V$ are $nV = n\sigma vt$ point atoms with which the moving atom has collided. Thus the average distance traveled between collisions is

$$\ell = \frac{vt}{n\sigma vt} = \frac{1}{n\sigma}. \tag{9.12}$$

The distance $\ell$ is the **mean free path** between collisions.[10] For a hydrogen atom,

$$\sigma = \pi(2a_0)^2 = 3.52 \times 10^{-20} \text{ m}^2.$$

Thus the mean free path in this situation is

$$\ell = \frac{1}{n\sigma} = 2.27 \times 10^{-4} \text{ m}.$$

The mean free path is several billion times smaller than the temperature scale height. As a result, the atoms in the gas see an essentially constant kinetic temperature between collisions. They are effectively confined within a limited volume of space in the photosphere. Of course this cannot be true for the photons as well, since the Sun's photosphere is the visible layer

[8]This treats the atoms as solid spheres, a classical approximation to the quantum atom.

[9]The concept of *cross section*, which will be discussed in more detail in Section 10.3, actually represents a probability of particle interactions but has units of cross-sectional area.

[10]A more careful calculation, using a Maxwellian velocity distribution for all of the atoms, results in a mean free path that is smaller by a factor of $\sqrt{2}$.

of the solar surface that we observe from Earth. Thus, by the very definition of photosphere, the photons must be able to escape freely into space. To say more about the photon mean free path and the concept of LTE, and to better understand the solar spectrum shown in Fig. 9.5, we must examine the interaction of particles and photons in some detail.

### The Definition of Opacity

We now turn to a consideration of a beam of parallel light rays traveling through a gas. Any process that removes photons from a beam of light will be collectively termed **absorption**. In this sense then, absorption includes the *scattering* of photons (such as Compton scattering, discussed in Section 5.2) as well as the true absorption of photons by atomic electrons making upward transitions. In sufficiently cool gases, molecular energy-level transitions may also occur and must be included.

The change in the intensity, $dI_\lambda$, of a ray of wavelength $\lambda$ as it travels through a gas is proportional to its intensity, $I_\lambda$, the distance traveled, $ds$, and the density of the gas, $\rho$. That is,

$$dI_\lambda = -\kappa_\lambda \rho I_\lambda \, ds. \tag{9.13}$$

The distance $s$ is measured along the path traveled by the beam and increases in the direction that the beam travels; the minus sign in Eq. (9.13) shows that the intensity *decreases* with distance due to the absorption of photons. The quantity $\kappa_\lambda$ is called the **absorption coefficient**, or **opacity**, with the $\lambda$ subscript implicitly indicating that the opacity is wavelength-dependent ($\kappa_\lambda$ is sometimes referred to as a *monochromatic opacity*). The opacity is the cross section for absorbing photons of wavelength $\lambda$ per unit mass of stellar material and has units of $m^2 \, kg^{-1}$. In general, the opacity of a gas is a function of its composition, density, and temperature.[11]

---

**Example 9.2.2.**   Consider a beam of light traveling through a gas with initial intensity $I_{\lambda,0}$ at $s = 0$. The final intensity $I_{\lambda,f}$ after the light has traveled a distance $s$ may be found by integrating Eq. (9.13):

$$\int_{I_{\lambda,0}}^{I_{\lambda,f}} \frac{dI_\lambda}{I_\lambda} = -\int_0^s \kappa_\lambda \rho \, ds.$$

This leads to

$$I_\lambda = I_{\lambda,0} e^{-\int_0^s \kappa_\lambda \rho \, ds}, \tag{9.14}$$

where the $f$ subscript has been dropped. For the specific case of a uniform gas of constant opacity and density,

$$I_\lambda = I_{\lambda,0} e^{-\kappa_\lambda \rho s}.$$

*continued*

---

[11]Note that there is some inconsistency in the terminology; some authors refer to *opacity* as the inverse of the mean free path of the photons.

For pure absorption (with emission processes neglected), there is no way of replenishing the photons lost from the beam. The intensity declines exponentially, falling by a factor of $e^{-1}$ over a characteristic distance of $\ell = 1/\kappa_\lambda \rho$. In the solar photosphere where the density is approximately $\rho = 2.1 \times 10^{-4}$ kg m$^{-3}$, the opacity (at a wavelength of 500 nm) is $\kappa_{500} = 0.03$ m$^2$ kg$^{-1}$. Thus the characteristic distance traveled by a photon before being removed from the beam at this level in the photosphere is

$$\ell = \frac{1}{\kappa_{500}\rho} = 160 \text{ km.}$$

Recalling Example 9.2.1, this distance is comparable to the temperature scale height $H_T = 677$ km. This implies that the photospheric photons do not see a constant temperature, and so local thermodynamic equilibrium (LTE) is not strictly valid in the photosphere. The temperature of the regions from which the photons have traveled will be somewhat different from the local kinetic temperature of the gas. Although LTE is a commonly invoked assumption in stellar atmospheres, it must be used with caution.

## Optical Depth

For scattered photons, the characteristic distance $\ell$ is in fact the mean free path of the photons. From Eq. (9.12),

$$\ell = \frac{1}{\kappa_\lambda \rho} = \frac{1}{n\sigma_\lambda}.$$

Both $\kappa_\lambda \rho$ and $n\sigma_\lambda$ can be thought of as the fraction of photons scattered per meter of distance travelled. Note that the mean free path is different for photons of different wavelengths.

It is convenient to define an **optical depth**, $\tau_\lambda$, back along a light ray by

$$\boxed{d\tau_\lambda = -\kappa_\lambda \rho \, ds,} \tag{9.15}$$

where $s$ is the distance measured along the photon's path in its direction of motion (when observing the light from a star, we are *looking back* along the path traveled by the photon; see Fig. 9.7). The difference in optical depth between a light ray's initial position ($s = 0$) and its final position after traveling a distance $s$ is

$$\Delta \tau_\lambda = \tau_{\lambda,f} - \tau_{\lambda,0} = -\int_0^s \kappa_\lambda \rho \, ds. \tag{9.16}$$

Note that $\Delta \tau_\lambda < 0$; as the light approaches an observer, it is traveling through material with diminishing optical depth. The outermost layers of a star may be taken to be at $\tau_\lambda = 0$ for all wavelengths, after which the light travels unimpeded to observers on Earth. With this definition of $\tau_\lambda = 0$, Eq. (9.16) gives the initial optical depth, $\tau_{\lambda,0}$, of a ray of light that

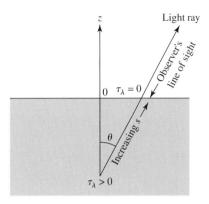

**FIGURE 9.7** Optical depth $\tau_\lambda$ measured back along a ray's path.

traveled a distance $s$ to reach the top of the atmosphere:

$$0 - \tau_{\lambda,0} = -\int_0^s \kappa_\lambda \rho \, ds$$

$$\tau_\lambda = \int_0^s \kappa_\lambda \rho \, ds. \tag{9.17}$$

The "0" subscript has been dropped with the understanding that $\tau_\lambda$ is the optical depth of the ray's initial position, a distance $s$ $(s > 0)$ from the top of the atmosphere.

Combining Eq. (9.17) with Eq. (9.14) of Example 9.2.2 for the case of pure absorption, we find that the decline in the intensity of a ray that travels through a gas from an optical depth $\tau_\lambda$ to reach the observer is given by

$$I_\lambda = I_{\lambda,0} e^{-\tau_\lambda}. \tag{9.18}$$

Thus, if the optical depth of the ray's starting point is $\tau_\lambda = 1$, the intensity of the ray will decline by a factor of $e^{-1}$ before escaping from the star. *The optical depth may be thought of as the number of mean free paths from the original position to the surface, as measured along the ray's path.* As a result, we typically see no deeper into an atmosphere at a given wavelength than $\tau_\lambda \approx 1$. Of course, for pure absorption the intensity of the ray declines exponentially regardless of its direction of travel through the gas. But we can observe only those rays traveling toward us, and this is reflected in our choice of $\tau_\lambda = 0$ at the top of the atmosphere. Other choices of where $\tau_\lambda = 0$ may be more useful in some situations.

If $\tau_\lambda \gg 1$ for a light ray passing through a volume of gas, the gas is said to be *optically thick*; if $\tau_\lambda \ll 1$, the gas is *optically thin*. Because the optical depth varies with wavelength, a gas may be optically thick at one wavelength and optically thin at another. For example, Earth's atmosphere is optically thin at visible wavelengths (we can see the stars), but optically thick at X-ray wavelengths; e.g., recall Fig. 6.25.

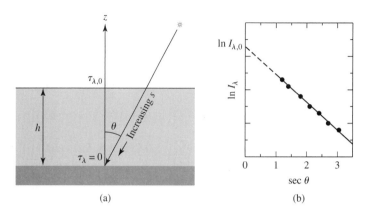

**FIGURE 9.8**    (a) A light ray entering Earth's atmosphere at an angle $\theta$. (b) $\ln I_\lambda$ vs. $\sec \theta$.

---

**Example 9.2.3.**    In Section 5.2, we stated that measurements of a star's radiative flux and apparent magnitude are routinely corrected for the absorption of light by Earth's atmosphere. Figure 9.8(a) shows a ray of intensity $I_{\lambda,0}$ entering Earth's atmosphere at an angle $\theta$ and traveling to a telescope on the ground. The intensity of the light detected at the telescope is $I_\lambda$; the problem is to determine the value of $I_{\lambda,0}$. If we take $\tau_\lambda = 0$ at the telescope and $h$ to be the height of the atmosphere, then the optical depth of the light ray's path through the atmosphere may be found from Eq. (9.17). Using $ds = -dz/\cos\theta = -\sec\theta\,dz$ yields

$$\tau_\lambda = \int_0^s \kappa_\lambda \rho \, ds = -\int_h^0 \kappa_\lambda \rho \, \frac{dz}{\cos\theta} = \sec\theta \int_0^h \kappa_\lambda \rho \, dz = \tau_{\lambda,0} \sec\theta,$$

where $\tau_{\lambda,0}$ is the optical depth for a vertically traveling photon ($\theta = 0$). Substituting into Eq. (9.18), the intensity of the light received at the telescope is therefore given by

$$I_\lambda = I_{\lambda,0} e^{-\tau_{\lambda,0} \sec\theta}. \tag{9.19}$$

There are two unknowns in this equation, $I_{\lambda,0}$ and $\tau_{\lambda,0}$; neither can be determined by a single observation. However, as time passes and as Earth rotates on its axis, the angle $\theta$ will change, and a semilog graph of several measurements of the received intensity $I_\lambda$ as a function of $\sec\theta$ can be made. As shown in Fig. 9.8(b), the *slope* of the best-fitting straight line is $-\tau_{\lambda,0}$. Extrapolating the best-fitting line to $\sec\theta = 0$ provides the value of $I_{\lambda,0}$ at the point where the line intercepts the $I_\lambda$-axis.[12] In this way, measurements of the specific intensity or radiative flux can be corrected for absorption by Earth's atmosphere.

---

## General Sources of Opacity

The opacity of the stellar material is determined by the details of how photons interact with particles (atoms, ions, and free electrons). If the photon passes within $\sigma_\lambda$ of the particle,

---

[12]Note that since $\sec\theta \geq 1$, the best-fitting straight line must be extrapolated to the mathematically unavailable value of 0.

where $\sigma_\lambda$ is the particle's cross-sectional area (or effective target area), the photon may be either absorbed or scattered. In an absorption process, the photon ceases to exist and its energy is given up to the thermal energy of the gas. In a scattering process the photon continues on in a different direction. Both absorption and scattering can remove photons from a beam of light, and so contribute to the opacity, $\kappa_\lambda$, of the stellar material. If the opacity varies slowly with wavelength, it determines the star's continuous spectrum (or *continuum*). The dark absorption lines superimposed on the continuum are the result of a rapid variation in the opacity with wavelength (recall, for example, Figs. 8.2 and 8.3).

In general, there are four primary sources of opacity available for removing stellar photons from a beam. Each involves a change in the quantum state of an electron, and the terms *bound* and *free* are used to describe whether the electron is bound to an atom or ion in its initial and final states.

1. **Bound–bound transitions** (excitations and de-excitations) occur when an electron in an atom or ion makes a transition from one orbital to another. An electron can make an upward transition from a lower- to a higher-energy orbital when a photon of the appropriate energy is absorbed. Thus $\kappa_{\lambda,\mathrm{bb}}$, the bound–bound opacity, is small except at those discrete wavelengths capable of producing an upward atomic transition. It is $\kappa_{\lambda,\mathrm{bb}}$ that is responsible for forming the absorption lines in stellar spectra. The reverse process, emission, occurs when the electron makes a downward transition from a higher- to a lower-energy orbital.

   If an electron absorbs a photon and then returns directly to its initial orbital (where it was before absorbing the photon), then a single photon is emitted in a random direction. The net result of this absorption–emission sequence is essentially a scattered photon. Otherwise, if the electron makes a transition to an orbital other than its initial one, the original photon is not recovered and the process is one of true absorption. If, while in its excited state, the atom or ion collides with a neighboring particle, collisional de-excitation may result. When this occurs, the energy lost by the atom or ion becomes a part of the thermal energy of the gas.

   An important by-product of this absorption process is degrading of the average energy of the photons in the radiation field. For example, if one photon is absorbed but two photons are emitted as the electron cascades down to its initial orbital, then the average photon energy has been reduced by half. There is no simple equation for bound–bound transitions that describes all of the contributions to the opacity by individual spectral lines.

2. **Bound–free absorption**, also known as *photoionization*, occurs when an incident photon has enough energy to ionize an atom. The resulting free electron can have any energy, so any photon with a wavelength $\lambda \leq hc/\chi_n$, where $\chi_n$ is the ionization energy of the $n$th orbital, can remove an electron from an atom. Thus $\kappa_{\lambda,\mathrm{bf}}$, the bound–free opacity, is one source of the continuum opacity. The cross section for the photoionization of a hydrogen atom in quantum state $n$ by a photon of wavelength $\lambda$ is

$$\sigma_{\mathrm{bf}} = 1.31 \times 10^{-19} \frac{1}{n^5} \left(\frac{\lambda}{500 \text{ nm}}\right)^3 \text{ m}^2,$$

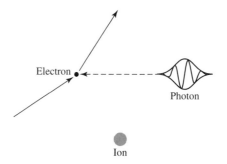

**FIGURE 9.9**    Free–free absorption of a photon.

which is comparable to the collision cross section for hydrogen found in Example 9.2.1. The inverse process of free–bound emission occurs when a free electron recombines with an ion, emitting one or more photons in random directions. As with bound–bound emission, this also contributes to reducing the average energy of the photons in the radiation field.

3. **Free–free absorption** is a scattering process, shown in Fig. 9.9, that takes place when a free electron in the vicinity of an ion absorbs a photon, causing the speed of the electron to increase. In this process the nearby ion is necessary in order to conserve both energy and momentum. (It is left as an exercise to show that an isolated free electron cannot absorb a photon.) Since this mechanism can occur for a continuous range of wavelengths, free–free opacity, $\kappa_{\lambda,\mathrm{ff}}$, is another contributor to the continuum opacity. It may also happen that as it passes near an ion, the electron loses energy by emitting a photon, which causes the electron to slow down. This process of free–free emission is also known as *bremsstrahlung*, which means "braking radiation" in German.

4. **Electron scattering** is as advertised: A photon is scattered (*not absorbed*) by a free electron through the process of *Thomson scattering*. In this process, the electron can be thought of as being made to oscillate in the electromagnetic field of the photon. However, because the electron is tiny, it makes a poor target for an incident photon, resulting in a small cross section. The cross section for Thomson scattering has the same value for photons of all wavelengths:

$$\sigma_T = \frac{1}{6\pi\epsilon_0^2}\left(\frac{e^2}{m_e c^2}\right)^2 = 6.65 \times 10^{-29}\ \mathrm{m}^2. \qquad (9.20)$$

This is typically two billion times smaller than the hydrogen cross section for photoionization, $\sigma_{\mathrm{bf}}$. The small size of the Thomson cross section means that electron scattering is most effective as a source of opacity when the electron density is very high, which requires high temperature. In the atmospheres of the hottest stars (and in the interiors of all stars), where most of the gas is completely ionized, other sources of opacity that involve bound electrons are eliminated. In this high-temperature regime, the opacity due to electron scattering, $\kappa_{\mathrm{es}}$, dominates the continuum opacity.

A photon may also be scattered by an electron that is loosely bound to an atomic nucleus. This result is called *Compton scattering* if the photon's wavelength is much smaller than the atom or *Rayleigh scattering* if the photon's wavelength is much larger. In Compton scattering, the change in the wavelength and energy of the scattered photon is very small (recall the discussion of the Compton wavelength on page 119), so Compton scattering is usually lumped together with Thomson scattering. The cross section for Rayleigh scattering from a loosely bound electron is smaller than the Thomson cross section; it is proportional to $1/\lambda^4$ and so decreases with increasing photon wavelength. Rayleigh scattering can be neglected in most atmospheres, but it is important in the UV for the extended envelopes of supergiant stars, and in cool main-sequence stars.[13] The scattering of photons from small particles is also responsible for the reddening of starlight as it passes through interstellar dust; see Section 12.1.

---

**Example 9.2.4.**   The energy of an electron in the $n = 2$ orbit of a hydrogen atom is given by Eq. (5.14):

$$E_2 = -\frac{13.6}{2^2} \text{ eV} = -3.40 \text{ eV}.$$

A photon must have an energy of at least $\chi_2 = 3.40$ eV to eject this electron from the atom. Thus any photon with a wavelength

$$\lambda \leq \frac{hc}{\chi_2} = 364.7 \text{ nm}$$

is capable of ionizing a hydrogen atom in the first excited state ($n = 2$). The opacity of the stellar material suddenly increases at wavelengths $\lambda \leq 364.7$ nm, and the radiative flux measured for the star accordingly decreases. The abrupt drop in the continuous spectrum of a star at this wavelength, called the **Balmer jump**, is evident in the Sun's spectrum (Fig. 9.5). The size of the Balmer jump in hot stars depends on the fraction of hydrogen atoms that are in the first excited state. This fraction is determined by the temperature via the Boltzmann equation (Eq. 8.6), so a measurement of the size of the Balmer jump can be used to determine the temperature of the atmosphere. For cooler or very hot stars with other significant sources of opacity, the analysis is more complicated, but the size of the Balmer jump can still be used as a probe of atmospheric temperatures.

The wavelength 364.7 nm is right in the middle of the bandwidth of the ultraviolet ($U$) filter in the *UBV* system, described on page 75. As a result, the Balmer jump will tend to decrease the amount of light received in the bandwidth of the $U$ filter and so *increase* both the ultraviolet magnitude $U$ and the color index ($U - B$) observed for a star. This effect will be strongest when $N_2/N_{\text{total}}$, the fraction of all hydrogen atoms that are in the first

*continued*

---

[13]Rayleigh scattering is also important in planetary atmospheres and is responsible for Earth's blue sky, for instance.

excited state, is a maximum. From Example 8.1.4, this occurs at a temperature of 9600 K, about the temperature of an A0 star on the main sequence. A careful examination of the color–color diagram in Fig. 3.11 reveals that this is indeed the spectral type at which the value of $U - B$ differs most from its blackbody value. The effect of line blanketing affects the measured color indices, making the star appear more red than a model blackbody star of the same effective temperature, and thus increasing the values of both $U - B$ and $B - V$.

## Continuum Opacity and the $H^-$ Ion

The primary source of the continuum opacity in the atmospheres of stars later than F0 is the photoionization of $H^-$ ions. An $H^-$ ion is a hydrogen atom that possesses an extra electron. Because of the partial shielding that the nucleus provides, a second electron can be loosely bound to the atom on the side of the ion opposite that of the first electron. In this position the second electron is closer to the positively charged nucleus than it is to the negatively charged electron. Therefore, according to Coulomb's law (Eq. 5.9), the net force on the extra electron is attractive.

The binding energy of the $H^-$ ion is only 0.754 eV, compared with the 13.6 eV required to ionize the ground state hydrogen atom. As a result, any photon with energy in excess of the ionization energy can be absorbed by an $H^-$ ion, liberating the extra electron; the remaining energy becomes kinetic energy. Conversely, an electron captured by a hydrogen atom to form $H^-$ will release a photon corresponding to the kinetic energy lost by the electron together with the ion's binding energy,

$$H + e^- \rightleftharpoons H^- + \gamma.$$

Since 0.754 eV corresponds to a photon with a wavelength of 1640 nm, any photon with a wavelength less than that value can remove an electron from the ion (bound–free opacity). At longer wavelengths, $H^-$ can also contribute to the opacity through free–free absorption. Consequently, $H^-$ ions are an important source of continuum opacity for stars cooler than F0. However, the $H^-$ ions become increasingly ionized at higher temperatures and therefore make less of a contribution to the continuum opacity. For stars of spectral types B and A, the photoionization of hydrogen atoms and free–free absorption are the main sources of the continuum opacity. At the even higher temperatures encountered for O stars, the ionization of atomic hydrogen means that electron scattering becomes more and more important, with the photoionization of helium also contributing to the opacity.

Molecules can survive in cooler stellar atmospheres and contribute to the bound–bound and bound–free opacities; the large number of discrete molecular absorption lines is an efficient impediment to the flow of photons. Molecules can also be broken apart into their constituent atoms by the absorption of photons in the process of *photodissociation*, which plays an important role in planetary atmospheres.

The total opacity is the sum of the opacities due to all of the preceding sources:

$$\kappa_\lambda = \kappa_{\lambda,\text{bb}} + \kappa_{\lambda,\text{bf}} + \kappa_{\lambda,\text{ff}} + \kappa_{\text{es}} + \kappa_{H^-}$$

(the $H^-$ opacity is explicitly included because of its unique and critical contribution to the opacity in many stellar atmospheres, including our Sun). The total opacity depends not only on the wavelength of the light being absorbed but also on the composition, density, and temperature of the stellar material.[14]

## The Rosseland Mean Opacity

It is often useful to employ an opacity that has been averaged over all wavelengths (or frequencies) to produce a function that depends only on the composition, density, and temperature. Although a variety of different schemes have been developed to compute a wavelength-independent opacity, by far the most commonly used is the **Rosseland mean opacity**, often simply referred to as the **Rosseland mean**.[15] This *harmonic mean* gives the greatest contribution to the lowest values of opacity. In addition, the Rosseland mean incorporates a weighting function that depends on the rate at which the blackbody spectrum varies with temperature (recall Eq. 3.24).[16] Formally, the Rosseland mean opacity is defined as

$$
\frac{1}{\bar{\kappa}} \equiv \frac{\displaystyle\int_0^\infty \frac{1}{\kappa_\nu} \frac{\partial B_\nu(T)}{\partial T}\, d\nu}{\displaystyle\int_0^\infty \frac{\partial B_\nu(T)}{\partial T}\, d\nu}. \tag{9.21}
$$

Unfortunately, there is no simple equation that is capable of describing all of the complex contributions to the opacity by individual spectral lines in bound–bound transitions, and so an analytic expression for the Rosseland mean cannot be given for these processes. However, approximation formulae have been developed for both the average bound–free and free–free opacities:

$$
\bar{\kappa}_{bf} = 4.34 \times 10^{21} \frac{g_{bf}}{t} Z(1+X) \frac{\rho}{T^{3.5}} \ \mathrm{m^2\ kg^{-1}} \tag{9.22}
$$

$$
\bar{\kappa}_{ff} = 3.68 \times 10^{18} g_{ff}\, (1-Z)(1+X) \frac{\rho}{T^{3.5}} \ \mathrm{m^2\ kg^{-1}}, \tag{9.23}
$$

where $\rho$ is the density (in $\mathrm{kg\ m^{-3}}$) and $T$ is the temperature (in kelvins). $X$ and $Z$ are the **mass fractions**, or fractional abundances (by mass), of hydrogen and metals, respectively.[17]

---

[14]The additional dependencies of the opacity on the electron number density, states of excitation and ionization of the atoms and ions, and other factors can all be calculated from the composition, density, and temperature.

[15]This wavelength-averaged opacity was introduced in 1924 by the Norwegian astronomer Svein Rosseland (1894–1985).

[16]You may also wish to refer to Problem 7.2 for a discussion of the role of weighting functions in integral averages.

[17]As we noted on page 204, because the primary components of most stellar gases are hydrogen and helium, all other constituents are frequently lumped together and referred to as *metals*. In certain applications, however, it is necessary to specify the composition in greater detail. In these cases, each species is represented by its own mass fraction.

Together with the mass fraction of helium, $Y$, their formal definitions are

$$X \equiv \frac{\text{total mass of hydrogen}}{\text{total mass of gas}} \tag{9.24}$$

$$Y \equiv \frac{\text{total mass of helium}}{\text{total mass of gas}} \tag{9.25}$$

$$Z \equiv \frac{\text{total mass of metals}}{\text{total mass of gas}}. \tag{9.26}$$

Clearly, $X + Y + Z = 1$.

The **Gaunt factors**, $g_{bf}$ and $g_{ff}$, are quantum-mechanical correction terms first calculated by J. A. Gaunt. These Gaunt factors are both $\approx 1$ for the visible and ultraviolet wavelengths of interest in stellar atmospheres. The additional correction factor, $t$, in the equation for the bound–free opacity is called the **guillotine factor** and describes the cutoff of an atom's contribution to the opacity after it has been ionized. Typical values of $t$ lie between 1 and 100.

Both of these formulae have the functional form $\overline{\kappa} = \kappa_0 \rho / T^{3.5}$, where $\kappa_0$ is approximately constant for a given composition. The first forms of these expressions were derived by H. A. Kramers (1894–1952) in 1923 using classical physics and the Rosseland mean. Any opacity having this density and temperature dependence is referred to as a **Kramers opacity law**.

Because the cross section for electron scattering is independent of wavelength, the Rosseland mean for this case has the particularly simple form

$$\overline{\kappa}_{es} = 0.02(1 + X) \text{ m}^2 \text{ kg}^{-1}. \tag{9.27}$$

An estimate of the contribution to the mean opacity provided by the $H^-$ ion may also be included over the temperature range 3000 K $\leq T \leq$ 6000 K and for densities between $10^{-7}$ kg m$^{-3}$ $\leq \rho \leq 10^{-2}$ kg m$^{-3}$ when $X \sim 0.7$ and $0.001 < Z < 0.03$ (the values of $X$ and $Z$ are typical of main-sequence stars). Specifically,

$$\overline{\kappa}_{H^-} \approx 7.9 \times 10^{-34} (Z/0.02) \rho^{1/2} T^9 \text{ m}^2 \text{ kg}^{-1}. \tag{9.28}$$

The total Rosseland mean opacity, $\overline{\kappa}$, is the average of the sum of the individual contributors to the opacity:

$$\overline{\kappa} = \overline{\kappa_{bb} + \kappa_{bf} + \kappa_{ff} + \kappa_{es} + \kappa_{H^-}}.$$

Figure 9.10 shows the results of an extensive computer calculation of the Rosseland mean opacity from first principles using detailed quantum physics. The calculation was carried out by Carlos Iglesias and Forrest Rogers for a composition with $X = 0.70$ and $Z = 0.02$.[18] The values of $\overline{\kappa}$ are plotted as a function of the temperature for several densities.

---

[18]A specific mixture of elements known as the Anders–Grevesse abundances were used to calculate the opacities shown.

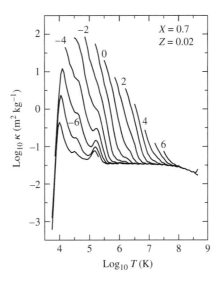

**FIGURE 9.10** Rosseland mean opacity for a composition that is 70% hydrogen, 28% helium, and 2% metals by mass. The curves are labeled by the logarithmic value of the density ($\log_{10} \rho$ in kg m$^{-3}$). (Data from Iglesias and Rogers, *Ap. J.*, *464*, 943, 1996.)

Considering the details of Fig. 9.10, first notice that the opacity increases with increasing density for a given temperature. Next, starting at the left-hand side of the figure, follow a constant-density plot as it rises steeply with increasing temperature. This reflects the increase in the number of free electrons produced by the ionization of hydrogen and helium. (Recall from Example 8.1.4 that the hydrogen partial ionization zone has a characteristic temperature of 10,000 K, and neutral helium is ionized at about the same temperature.) The decline of the plot after the peak in the opacity roughly follows a Kramers law, $\bar{\kappa} \propto T^{-3.5}$, and is due primarily to the bound–free and free–free absorption of photons. The He II ion loses its remaining electron at a characteristic temperature of 40,000 K for a wide range of stellar parameters; the slight increase in the number of free electrons produces a small "bump" seen near that temperature. Another bump, evident above $10^5$ K, is the result of the ionization of certain metals, most notably iron. Finally, the plot reaches a flat floor at the right-hand side of the figure. Electron scattering dominates at the highest temperatures, when nearly all of the stellar material is ionized and there are few bound electrons available for bound–bound and bound–free processes. The form of Eq. (9.27) for electron scattering, with no density or temperature dependence, requires that all of the curves in Fig. 9.10 converge to the same constant value in the high-temperature limit.

## 9.3 ■ RADIATIVE TRANSFER

In an equilibrium, steady-state star, there can be no change in the total energy contained within any layer of the stellar atmosphere or interior.[19] In other words, the mechanisms

---

[19]This is not the case for a star that is *not* in equilibrium. For instance, pulsating stars (to be discussed in Chapter 14), periodically absorb or "dam up" the outward flow of energy, driving the oscillations.

involved in absorbing and emitting energy must be precisely in balance throughout the star. In this section, the competition between the absorption and emission processes will be described, first in qualitative terms and later in more quantitative detail.

## Photon Emission Processes

Any process that adds photons to a beam of light will be called **emission**. Thus emission includes the scattering of photons into the beam, as well as the true emission of photons by electrons making downward atomic transitions. Each of the four primary sources of opacity listed in Section 9.2 has an inverse emission process: bound–bound and free–bound emission, free–free emission (bremsstrahlung), and electron scattering. The simultaneous and complementary processes of absorption and emission hinder the flow of photons through the star by redirecting the paths of the photons and redistributing their energy. Thus in a star there is not a direct flow of photons streaming toward the surface, carrying energy outward at the speed of light. Instead, the individual photons travel only temporarily with the beam as they are repeatedly scattered in random directions following their encounters with gas particles.

## The Random Walk

As the photons diffuse upward through the stellar material, they follow a haphazard path called a **random walk**. Figure 9.11 shows a photon that undergoes a net vector displacement $\mathbf{d}$ as the result of making a large number $N$ of randomly directed steps, each of length $\ell$ (the mean free path):

$$\mathbf{d} = \boldsymbol{\ell}_1 + \boldsymbol{\ell}_2 + \boldsymbol{\ell}_3 + \cdots + \boldsymbol{\ell}_N.$$

Taking the vector dot product of $\mathbf{d}$ with itself gives

$$
\begin{aligned}
\mathbf{d} \cdot \mathbf{d} = {} & \boldsymbol{\ell}_1 \cdot \boldsymbol{\ell}_1 + \boldsymbol{\ell}_1 \cdot \boldsymbol{\ell}_2 + \cdots + \boldsymbol{\ell}_1 \cdot \boldsymbol{\ell}_N \\
& + \boldsymbol{\ell}_2 \cdot \boldsymbol{\ell}_1 + \boldsymbol{\ell}_2 \cdot \boldsymbol{\ell}_2 + \cdots + \boldsymbol{\ell}_2 \cdot \boldsymbol{\ell}_N \\
& + \cdots + \boldsymbol{\ell}_N \cdot \boldsymbol{\ell}_1 + \boldsymbol{\ell}_N \cdot \boldsymbol{\ell}_2 + \cdots + \boldsymbol{\ell}_N \cdot \boldsymbol{\ell}_N \\
= {} & \sum_{i=1}^{N} \sum_{j=1}^{N} \boldsymbol{\ell}_i \cdot \boldsymbol{\ell}_j,
\end{aligned}
$$

or

$$
\begin{aligned}
d^2 = {} & N\ell^2 + \ell^2 [\cos\theta_{12} + \cos\theta_{13} + \cdots + \cos\theta_{1N} \\
& + \cos\theta_{21} + \cos\theta_{23} + \cdots + \cos\theta_{2N} \\
& + \cdots + \cos\theta_{N1} + \cos\theta_{N2} + \cdots + \cos\theta_{N(N-1)}] \\
= {} & N\ell^2 + \ell^2 \sum_{i=1}^{N} \sum_{\substack{j=1 \\ j\neq i}}^{N} \cos\theta_{ij},
\end{aligned}
$$

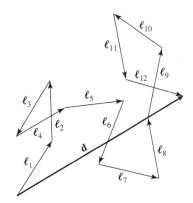

**FIGURE 9.11** Displacement **d** of a random-walking photon.

where $\theta_{ij}$ is the angle between the vectors $\boldsymbol{\ell}_i$ and $\boldsymbol{\ell}_j$. For a large number of randomly directed steps, the sum of all the cosine terms approaches zero. As a result, for a random walk, the displacement $d$ is related to the size of each step, $\ell$, by

$$d = \ell\sqrt{N}. \tag{9.29}$$

Thus the transport of energy through a star by radiation may be extremely inefficient. As a photon follows its tortuous path to the surface of a star,[20] it takes 100 steps to travel a distance of $10\ell$; 10,000 steps to travel $100\ell$; and one million steps to travel $1000\ell$.[21] Because the optical depth at a point is roughly the number of photon mean free paths from that point to the surface (as measured along a light ray's straight path), Eq. (9.29) implies that the distance to the surface is $d = \tau_\lambda \ell = \ell\sqrt{N}$. The average number of steps needed for a photon to travel the distance $d$ before leaving the surface is then

$$N = \tau_\lambda^2, \tag{9.30}$$

for $\tau_\lambda \gg 1$. As might be expected, when $\tau_\lambda \approx 1$, a photon may escape from that level of the star. A more careful analysis (performed in Section 9.4) shows that the average level in the atmosphere from which photons of wavelength $\lambda$ escape is at a characteristic optical depth of about $\tau_\lambda = 2/3$. *Looking into a star at any angle, we always look back to an optical depth of about $\tau_\lambda = 2/3$, as measured straight back along the line of sight.* In fact, a star's photosphere is defined as the layer from which its visible light originates—that is, where $\tau_\lambda \approx 2/3$ for wavelengths in the star's continuum.

The realization that an observer looking vertically down on the surface of a star sees photons from $\tau_\lambda \approx 2/3$ offers an important insight into the formation of spectral lines.

---

[20]Strictly speaking, an individual photon does not make the entire journey, but rather, along with being scattered, photons may be absorbed and re-emitted during the "collisions."

[21]As will be discussed in Section 10.4, the process of transporting energy by radiation is sometimes so inefficient that another transport process, *convection*, must take over.

Recalling the definition of optical depth, Eq. (9.17),

$$\tau_\lambda = \int_0^s \kappa_\lambda \rho \, ds,$$

we see that if the opacity $\kappa_\lambda$ increases at some wavelength, then the actual distance back along the ray to the level where $\tau_\lambda = 2/3$ decreases for that wavelength. One cannot see as far into murky material, so an observer will not see as deeply into the star at wavelengths where the opacity is greater than average (i.e., greater than the continuum opacity). This implies that if the temperature of the stellar atmosphere decreases outward, then these higher regions of the atmosphere will be cooler. As a result, the intensity of the radiation at $\tau_\lambda \approx 2/3$ will decline the most for those wavelengths at which the opacity is greatest, resulting in absorption lines in the continuous spectrum. Therefore, the temperature *must* decrease outward for the formation of absorption lines. This is the analog for stellar atmospheres of Kirchhoff's law that a cool, diffuse gas in front of a source of a continuous spectrum produces dark spectral lines in the continuous spectrum.

**Limb Darkening**

Another implication of receiving radiation from an optical depth of about two-thirds is shown in Fig. 9.12. The line of sight of an observer on Earth viewing the Sun is vertically downward at the center of the Sun's disk but makes an increasingly larger angle $\theta$ with the vertical near the edge, or *limb*, of the Sun. Looking near the limb, the observer will not see as deeply into the solar atmosphere and will therefore see a lower temperature at an optical depth of two-thirds (compared to looking at the center of the disk). As a result, the limb of

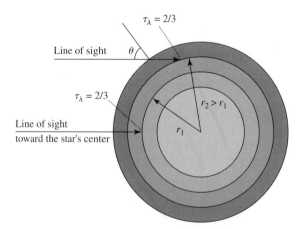

**FIGURE 9.12**    Limb darkening. The distance traversed within the atmosphere of the star to reach a specified radial distance $r$ from the star's center increases along the line of sight of the observer as $\theta$ increases. This implies that to reach a specified optical depth (e.g., $\tau_\lambda = 2/3$), the line of sight terminates at greater distances (and cooler temperatures) from the star's center as $\theta$ increases. Note that the physical scale of the photosphere has been greatly exaggerated for illustration purposes. The thickness of a typical photosphere is on the order of 0.1% of the stellar radius.

the Sun appears darker than its center. This **limb darkening** is clearly seen in Fig. 11.11 for the Sun and has also been observed in the light curves of some eclipsing binaries. More detailed information on limb darkening may be found later in this section.

### The Radiation Pressure Gradient

Considering the meandering nature of a photon's journey to the surface, it may seem surprising that the energy from the deep interior of the star ever manages to escape into space. At great depth in the interior of the star, the photon's mean free path is only a fraction of a centimeter. After a few scattering encounters, the photon is traveling in a nearly random direction, hundreds of millions of meters from the surface. This situation is analogous to the motions of air molecules in a closed room. An individual molecule moves about with a speed of nearly 500 m s$^{-1}$, and it collides with other air molecules several billion times per second. As a result, the molecules are moving in random directions. Because there is no overall migration of the molecules in a closed room, a person standing in the room feels no wind. However, opening a window may cause a breeze if a pressure difference is established between one side of the room and the other. The air in the room responds to this pressure gradient, producing a net flux of molecules toward the area of lower pressure.

In a star the same mechanism causes a "breeze" of photons to move toward the surface of the star. Because the temperature in a star decreases outward, the radiation pressure is smaller at greater distances from the center (cf., Eq. 9.11 for the blackbody radiation pressure). This gradient in the radiation pressure produces the slight net movement of photons toward the surface that carries the radiative flux. As we will discover later in this section, this process is described by

$$\frac{dP_{\text{rad}}}{dr} = -\frac{\overline{\kappa}\rho}{c} F_{\text{rad}}. \tag{9.31}$$

Thus the transfer of energy by radiation is a subtle process involving the slow upward diffusion of randomly walking photons, drifting toward the surface in response to minute differences in the radiation pressure. The description of a "beam" or a "ray" of light is only a convenient fiction, used to define the direction of motion momentarily shared by the photons that are continually absorbed and scattered into and out of the beam. Nevertheless, we will continue to use the language of photons traveling in a beam or ray of light, realizing that a specific photon is in the beam for only an instant.

## 9.4 ■ THE TRANSFER EQUATION

In this section, we will focus on a more thorough examination of the flow of radiation through a stellar atmosphere.[22] We will develop and solve the basic equation of radiative transfer using several standard assumptions. In addition, we will derive the variation of temperature with optical depth in a simple model atmosphere before applying it to obtain a quantitative description of limb darkening.

---

[22]Although the focus of this discussion is on stellar atmospheres, much of the discussion is applicable to other environments as well, such as light traversing an interstellar gas cloud.

**The Emission Coefficient**

In the following discussions of beams and light rays, the primary consideration is the net flow of energy in a given direction, not the specific path taken by individual photons. First, we will examine the emission process that increases the intensity of a ray of wavelength $\lambda$ as it travels through a gas. The increase in intensity $dI_\lambda$ is proportional to both $ds$, the distance traveled in the direction of the ray, and $\rho$, the density of the gas. For pure emission (no absorption of the radiation),

$$dI_\lambda = j_\lambda \rho \, ds, \tag{9.32}$$

where $j_\lambda$ is the **emission coefficient** of the gas. The emission coefficient, which has units of m s$^{-3}$ sr$^{-1}$, varies with the wavelength of the light.

As a beam of light moves through the gas in a star, its specific intensity, $I_\lambda$, changes as photons traveling with the beam are removed by absorption or scattering out of the beam, and are replaced by photons emitted from the surrounding stellar material, or scattered into the beam. Combining Eq. (9.13) for the decrease in intensity due to the absorption of radiation with Eq. (9.32) for the increase produced by emission gives the general result

$$dI_\lambda = -\kappa_\lambda \rho I_\lambda \, ds + j_\lambda \rho \, ds. \tag{9.33}$$

The ratio of the rates at which the competing processes of emission and absorption occur determines how rapidly the intensity of the beam changes. This is similar to describing the flow of traffic on an interstate highway. Imagine following a group of cars as they leave Los Angeles, traveling north on I-15. Initially, nearly all of the cars on the road have California license plates. Driving north, the number of cars on the road declines as more individuals exit than enter the highway. Eventually approaching Las Vegas, the number of cars on the road increases again, but now the surrounding cars bear Nevada license plates. Continuing onward, the traffic fluctuates as the license plates eventually change to those of Utah, Idaho, and Montana. Most of the cars have the plates of the state they are in, with a few cars from neighboring states and even fewer from more distant locales. At any point along the way, the number of cars on the road reflects the local population density. Of course, this is to be expected; the surrounding area is the source of the traffic entering the highway, and the rate at which the traffic changes is determined by the ratio of the number of entering to exiting automobiles. This ratio determines how rapidly the cars on the road from elsewhere are replaced by the cars belonging to the local population. Thus the traffic constantly changes, always tending to resemble the number and types of automobiles driven by the people living nearby.

**The Source Function and the Transfer Equation**

In a stellar atmosphere or interior, the same considerations describe the competition between the rates at which photons are plucked out of a beam of light by absorption, and introduced into the beam by emission processes. The ratio of the rates of emission and absorption determines how rapidly the intensity of the beam of light changes and describes the tendency of the population of photons in the beam to resemble the local source of photons in the surrounding stellar material. To introduce the ratio of emission to absorption, we divide

Eq. (9.33) by $-\kappa_\lambda \rho\, ds$:

$$-\frac{1}{\kappa_\lambda \rho}\frac{dI_\lambda}{ds} = I_\lambda - \frac{j_\lambda}{\kappa_\lambda}.$$

The ratio of the emission coefficient to the absorption coefficient is called the **source function**, $S_\lambda \equiv j_\lambda/\kappa_\lambda$. It describes how photons originally traveling with the beam are removed and replaced by photons from the surrounding gas.[23] The source function, $S_\lambda$, has the same units as the intensity, W m$^{-3}$ sr$^{-1}$. Therefore, in terms of the source function,

$$\boxed{-\frac{1}{\kappa_\lambda \rho}\frac{dI_\lambda}{ds} = I_\lambda - S_\lambda.}\tag{9.34}$$

This is one form of the **equation of radiative transfer** (usually referred to as the **transfer equation**).[24] According to the transfer equation, if the intensity of the light does not vary (so that the left-hand side of the equation is zero), then the intensity is equal to the source function, $I_\lambda = S_\lambda$. If the intensity of the light is *greater* than the source function (the right-hand side of the transfer equation is greater than 0), then $dI_\lambda/ds$ is less than 0, and the intensity *decreases* with distance. On the other hand, if the intensity is *less* than the source function, the intensity *increases* with distance. This is merely a mathematical restatement of the tendency of the photons found in the beam to resemble the local source of photons in the surrounding gas. Thus *the intensity of the light tends to become equal to the local value of the source function*, although the source function itself may vary too rapidly with distance for an equality to be attained.

### The Special Case of Blackbody Radiation

The source function for the special case of blackbody radiation can be found by considering a box of optically thick gas maintained at a constant temperature $T$. The confined particles and blackbody radiation are in thermodynamic equilibrium, with no net flow of energy through the box or between the gas particles and the radiation. With the particles and photons in equilibrium, individually and with each other, every process of absorption is balanced by an inverse process of emission. The intensity of the radiation is described by the Planck function, $I_\lambda = B_\lambda$. Furthermore, because the intensity is constant throughout the box, $dI_\lambda/ds = 0$, and so $I_\lambda = S_\lambda$. *For the case of thermodynamic equilibrium, the source function is equal to the Planck function, $S_\lambda = B_\lambda$.*

As mentioned in Section 9.2, a star cannot be in perfect thermodynamic equilibrium; there is a net flow of energy from the center to the surface. Deep in the atmosphere, where $\tau_\lambda \gg 1$ as measured along a vertical ray, a random-walking photon will take at least $\tau_\lambda^2$ steps to reach the surface (recall Eq. 9.30) and so will suffer many scattering events before escaping from the star. Thus, at a depth at which the photon mean free path is small compared to the

---

[23]As a ratio involving the inverse processes of absorption and emission, the source function is less sensitive to the detailed properties of the stellar material than are $j_\lambda$ and $\kappa_\lambda$ individually.

[24]It is assumed that the atmosphere is in a steady state, not changing with time. Otherwise, a time-derivative term would have to be included in the transfer equation.

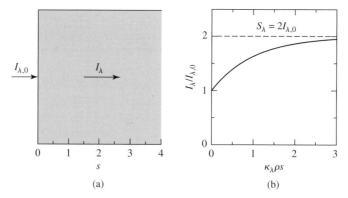

**FIGURE 9.13**    Transformation of the intensity of a light ray traveling through a volume of gas. (a) A light ray entering a volume of gas. (b) Intensity of the light ray. The horizontal axis has units of $\kappa_\lambda \rho s$, the number of optical depths traveled into the gas.

temperature scale height, the photons are effectively confined to a limited volume, a region of nearly constant temperature. The conditions for local thermodynamic equilibrium (LTE) are satisfied, and so, as already seen, the source function is equal to the Planck function, $S_\lambda = B_\lambda$. Making the assumption of LTE in a problem means setting $S_\lambda = B_\lambda$. However, even in LTE, the intensity of the radiation, $I_\lambda$, will not necessarily be equal to $B_\lambda$ unless $\tau_\lambda \gg 1$. In summary, saying that $I_\lambda = B_\lambda$ is a statement that the radiation field is described by the Planck function, while $S_\lambda = B_\lambda$ describes the physical source of the radiation, $j_\lambda / \kappa_\lambda$, as one that produces blackbody radiation.

---

**Example 9.4.1.**    To see how the intensity of a light ray tends to become equal to the local value of the source function, imagine a beam of light of initial intensity $I_{\lambda,0}$ at $s = 0$ entering a volume of gas of constant density, $\rho$, that has a *constant* opacity, $\kappa_\lambda$, and a *constant* source function, $S_\lambda$. Then it is left as an exercise to show that the transfer equation (Eq. 9.34) may be easily solved for the intensity of the light as a function of the distance $s$ traveled into the gas:

$$I_\lambda(s) = I_{\lambda,0}\, e^{-\kappa_\lambda \rho s} + S_\lambda (1 - e^{-\kappa_\lambda \rho s}). \tag{9.35}$$

As shown in Fig. 9.13 for the case of $S_\lambda = 2I_{\lambda,0}$, this solution describes the transformation of the intensity of the light ray from its initial value of $I_{\lambda,0}$ to $S_\lambda$, the value of the source function. The characteristic distance for this change to occur is $s = 1/\kappa_\lambda \rho$, which is one photon mean free path (recall Example 9.2.2), or one optical depth into the gas.

---

### The Assumption of a Plane-Parallel Atmosphere

Although the transfer equation is the basic tool that describes the passage of light through a star's atmosphere, a reader seeing it for the first time may be prone to despair. In this troublesome equation, the intensity of the light must depend on the direction of travel to account for the net outward flow of energy. And although absorption and emission coefficients are the same for light traveling in all directions (implying that the source function is independent of direction), the absorption and emission coefficients depend on the temperature and density in a rather complicated way.

However, if astronomers are to learn anything about the physical conditions in stellar atmospheres, such as temperature or density, they must know where (at what depth) a spectral line is formed. A vast amount of effort has therefore been devoted to solving and understanding the implications of the transfer equation, and several powerful techniques have been developed that simplify the analysis considerably.

We will begin by rewriting Eq. (9.34) in terms of the optical depth $\tau_\lambda$, defined by Eq. (9.15), resulting in

$$\frac{dI_\lambda}{d\tau_\lambda} = I_\lambda - S_\lambda. \tag{9.36}$$

Unfortunately, because the optical depth is measured along the path of the light ray, neither the optical depth nor the distance $s$ in Eq. (9.34) corresponds to a unique geometric depth in the atmosphere. Consequently, the optical depth must be replaced by a meaningful measure of position.

To find a suitable replacement, we introduce the first of several standard approximations. The atmospheres of stars near the main sequence are physically thin compared with the size of the star, analogous to the skin of an onion. The atmosphere's radius of curvature is thus much larger than its thickness, and we may consider the atmosphere as a *plane-parallel slab*. As shown in Fig. 9.14, the $z$-axis is assumed to be in the vertical direction, with $z = 0$ at the top of this **plane-parallel atmosphere**.

Next, a **vertical optical depth**, $\tau_{\lambda,v}(z)$, is defined as

$$\tau_{\lambda,v}(z) \equiv \int_z^0 \kappa_\lambda \rho \, dz. \tag{9.37}$$

Comparison with Eq. (9.17) reveals that this is just the initial optical depth of a ray traveling

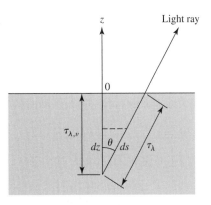

**FIGURE 9.14** Plane-parallel stellar atmosphere.

vertically upward from an initial position ($z < 0$) to the surface ($z = 0$) where $\tau_{\lambda,v} = 0$.[25] However, a ray that travels upward at an angle $\theta$ from the same initial position $z$ has farther to go through the same layers of the atmosphere in order to reach the surface. Therefore, the optical depth measured along this ray's path to the surface, $\tau_\lambda$, is *greater* than the vertical optical depth, $\tau_{\lambda,v}(z)$. Since $dz = ds\cos\theta$, the two optical depths are related by

$$\tau_\lambda = \frac{\tau_{\lambda,v}}{\cos\theta} = \tau_{\lambda,v}\sec\theta. \tag{9.38}$$

The vertical optical depth is a true vertical coordinate, analogous to $z$, that increases in the $-z$-direction. Its value does not depend on the direction of travel of a light ray, and so it can be used as a meaningful position coordinate in the transfer equation. Replacing $\tau_\lambda$ by $\tau_{\lambda,v}$ in Eq. (9.36) results in

$$\cos\theta \frac{dI_\lambda}{d\tau_{\lambda,v}} = I_\lambda - S_\lambda. \tag{9.39}$$

This form of the transfer equation is usually employed when dealing with the approximation of a plane-parallel atmosphere.

Of course, the value of the vertical optical depth at a level $z$ is wavelength-dependent because of the wavelength-dependent opacity in Eq. (9.37). In order to simplify the following analysis, and to permit the identification of an atmospheric level with a unique value of $\tau_v$, the opacity is assumed to be *independent of wavelength* (we usually take it to be equal to the Rosseland mean opacity, $\overline{\kappa}$). A model stellar atmosphere, for which the simplifying assumption is made that the opacity is independent of wavelength, is called a *gray atmosphere*, reflecting its indifference to the spectrum of wavelengths. If we write $\overline{\kappa}$ instead of $\kappa_\lambda$ in Eq. (9.37), the vertical optical depth no longer depends on wavelength; we can therefore write $\tau_v$ instead of $\tau_{\lambda,v}$ in the transfer equation (Eq. 9.39). The remaining wavelength dependencies may be removed by integrating the transfer equation over all wavelengths, using

$$I = \int_0^\infty I_\lambda\, d\lambda \qquad \text{and} \qquad S = \int_0^\infty S_\lambda\, d\lambda.$$

With the preceding changes, the transfer equation appropriate for a plane-parallel gray atmosphere is

$$\cos\theta \frac{dI}{d\tau_v} = I - S. \tag{9.40}$$

This equation leads to two particularly useful relations between the various quantities describing the radiation field. First, integrating over all solid angles, and recalling that $S$ depends only on the local conditions of the gas, independent of direction, we get

$$\frac{d}{d\tau_v}\int I\cos\theta\, d\Omega = \int I\, d\Omega - S\int d\Omega. \tag{9.41}$$

---

[25]Recall that as the light approaches the surface (and the observer on Earth), it is traveling through smaller and smaller values of the optical depth.

Using $\int d\Omega = 4\pi$ together with the definitions of the radiative flux $F_{\text{rad}}$ (Eq. 9.8) and the mean intensity $\langle I \rangle$ (Eq. 9.3), both integrated over all wavelengths, we find

$$\frac{dF_{\text{rad}}}{d\tau_v} = 4\pi(\langle I \rangle - S).$$

The second relation is found by first multiplying the transfer equation (9.40) by $\cos\theta$ and again integrating over all solid angles:

$$\frac{d}{d\tau_v} \int I \cos^2\theta \, d\Omega = \int I \cos\theta \, d\Omega - S \int \cos\theta \, d\Omega.$$

The term on the left is the radiation pressure multiplied by the speed of light (recall Eq. 9.9). The first term on the right-hand side is the radiative flux. In spherical coordinates, the second integral on the right-hand side evaluates to

$$\int \cos\theta \, d\Omega = \int_{\phi=0}^{2\pi} \int_{\theta=0}^{\pi} \cos\theta \sin\theta \, d\theta \, d\phi = 0.$$

Thus

$$\frac{dP_{\text{rad}}}{d\tau_v} = \frac{1}{c} F_{\text{rad}}. \tag{9.42}$$

In Problem 9.16, you will find that in a spherical coordinate system with its origin at the center of the star, this equation is

$$\frac{dP_{\text{rad}}}{dr} = -\frac{\overline{\kappa}\rho}{c} F_{\text{rad}},$$

which is just Eq. (9.31). As mentioned previously, this result can be interpreted as saying that the net radiative flux is driven by differences in the radiation pressure, with a "photon wind" blowing from high to low $P_{\text{rad}}$. Equation (9.31) will be employed in Chapter 10 to determine the temperature structure in the interior of a star.

In an equilibrium stellar atmosphere, every process of absorption is balanced by an inverse process of emission; no net energy is subtracted from or added to the radiation field. In a plane-parallel atmosphere, this means that the radiative flux must have the same value at every level of the atmosphere, including its surface. From Eq. (3.18),

$$F_{\text{rad}} = \text{constant} = F_{\text{surf}} = \sigma T_e^4. \tag{9.43}$$

Because the flux is a constant, $dF_{\text{rad}}/d\tau_v = 0$, which implies that the mean intensity must be equal to the source function,

$$\langle I \rangle = S. \tag{9.44}$$

Equation (9.42) may now be integrated to find the radiation pressure as a function of the vertical optical depth:

$$P_{\text{rad}} = \frac{1}{c} F_{\text{rad}} \tau_v + C, \tag{9.45}$$

where $C$ is the constant of integration.

### The Eddington Approximation

If we knew how the radiation pressure varied with temperature for the general case (and not just for blackbody radiation), we could use Eq. (9.45) to determine the temperature structure of our plane-parallel gray atmosphere. We would have to *assume* a description of the angular distribution of the intensity. In an approximation that we owe to the brilliant English physicist Sir Arthur Stanley Eddington (1882–1944), the intensity of the radiation field is assigned one value, $I_{out}$, in the $+z$-direction (outward) and another value, $I_{in}$, in the $-z$-direction (inward); see Fig. 9.15. Both $I_{out}$ and $I_{in}$ vary with depth in the atmosphere, and in particular, $I_{in} = 0$ at the top of the atmosphere, where $\tau_v = 0$. It is left as an exercise to show that with this **Eddington approximation**,[26] the mean intensity, radiative flux, and radiation pressure are given by

$$\langle I \rangle = \frac{1}{2} \left( I_{out} + I_{in} \right) \tag{9.46}$$

$$F_{rad} = \pi \left( I_{out} - I_{in} \right) \tag{9.47}$$

$$P_{rad} = \frac{2\pi}{3c} \left( I_{out} + I_{in} \right) = \frac{4\pi}{3c} \langle I \rangle. \tag{9.48}$$

[Note that because the flux is a constant, Eq. (9.47) shows that there is a constant difference between $I_{out}$ and $I_{in}$ at any level of the atmosphere.]

Inserting the last relation for the radiation pressure into Eq. (9.45), we find that

$$\frac{4\pi}{3c} \langle I \rangle = \frac{1}{c} F_{rad} \tau_v + C. \tag{9.49}$$

The constant $C$ can be determined by evaluating Eqs. (9.46) and (9.47) at the top of the atmosphere, where $\tau_v = 0$ and $I_{in} = 0$. The result is that $\langle I(\tau_v = 0) \rangle = F_{rad}/2\pi$. Inserting

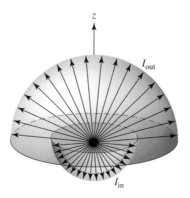

**FIGURE 9.15**   The Eddington approximation.

[26]Actually, there are several more mathematical ways of implementing the Eddington approximation, but they are all equivalent.

this into Eq. (9.49) with $\tau_v = 0$ shows that

$$C = \frac{2}{3c} F_{\text{rad}}.$$

With this value of $C$, Eq. (9.49) becomes

$$\frac{4\pi}{3} \langle I \rangle = F_{\text{rad}} \left( \tau_v + \frac{2}{3} \right). \tag{9.50}$$

Of course, we already know that the radiative flux is a constant, given by Eq. (9.43). Using this results in an expression for the mean intensity as a function of the vertical optical depth:

$$\langle I \rangle = \frac{3\sigma}{4\pi} T_e^4 \left( \tau_v + \frac{2}{3} \right). \tag{9.51}$$

We may now derive the final approximation to determine the temperature structure of our model atmosphere. If the atmosphere is assumed to be in local thermodynamic equilibrium, another expression for the mean intensity can be found and combined with Eq. (9.51). By the definition of LTE, the source function is equal to the Planck function, $S_\lambda = B_\lambda$. Integrating $B_\lambda$ over all wavelengths (see Eq. 3.28) shows that for LTE,

$$S = B = \frac{\sigma T^4}{\pi},$$

and so, from Eq. (9.44),

$$\langle I \rangle = \frac{\sigma T^4}{\pi}. \tag{9.52}$$

Equating expressions (9.51) and (9.52) finally results in the variation of the temperature with vertical optical depth in a plane-parallel gray atmosphere in LTE, assuming the Eddington approximation:[27]

$$\boxed{T^4 = \frac{3}{4} T_e^4 \left( \tau_v + \frac{2}{3} \right).} \tag{9.53}$$

This relation is well worth the effort of its derivation, because it reveals some important aspects of real stellar atmospheres. First, notice that $T = T_e$ at $\tau_v = 2/3$, *not* at $\tau_v = 0$. Thus the "surface" of a star, which by definition has temperature $T_e$ [recall the Stefan–Boltzmann equation, Eq. (3.17)], is *not* at the top of the atmosphere, where $\tau_v = 0$, but deeper down, where $\tau_v = 2/3$. This result may be thought of as the average point of origin of the observed photons. Although this result came at the end of a string of assumptions, it can be generalized to the statement that *when looking at a star, we see down to a vertical optical depth of $\tau_v \approx 2/3$, averaged over the disk of the star*. The importance of this for the formation and interpretation of spectral lines was discussed on page 254.

[27] You are encouraged to refer to Mihalas, Chapter 3, for a more detailed discussion of the gray atmosphere, including a more sophisticated development of the relation $T^4 = \frac{3}{4} T_e^4 [\tau_v + q(\tau_v)]$, where the Eddington approximation $[q(\tau_v) \equiv \frac{2}{3}]$ is a special case.

**Limb Darkening Revisited**

We now move on to take a closer look at limb darkening (recall Fig. 9.12). A comparison of theory and observations of limb darkening can provide valuable information about how the source function varies with depth in a star's atmosphere. To see how this is done, we first solve the general form of the transfer equation (Eq. 9.36),

$$\frac{dI_\lambda}{d\tau_\lambda} = I_\lambda - S_\lambda,$$

at least formally, rather than by making assumptions. (The inevitable assumptions will be required soon enough.) Multiplying both sides by $e^{-\tau_\lambda}$, we have

$$\frac{dI_\lambda}{d\tau_\lambda}\, e^{-\tau_\lambda} - I_\lambda e^{-\tau_\lambda} = -S_\lambda\, e^{-\tau_\lambda}$$

$$\frac{d}{d\tau_\lambda}(e^{-\tau_\lambda} I_\lambda) = -S_\lambda\, e^{-\tau_\lambda}$$

$$d(e^{-\tau_\lambda} I_\lambda) = -S_\lambda\, e^{-\tau_\lambda}\, d\tau_\lambda.$$

If we integrate from the initial position of the ray, at optical depth $\tau_{\lambda,0}$ where $I_\lambda = I_{\lambda,0}$, to the top of the atmosphere, at optical depth $\tau_\lambda = 0$ where $I_\lambda = I_\lambda(0)$, the result for the emergent intensity at the top of the atmosphere, $I_\lambda(0)$, is

$$I_\lambda(0) = I_{\lambda,0} e^{-\tau_{\lambda,0}} - \int_{\tau_{\lambda,0}}^{0} S_\lambda e^{-\tau_\lambda}\, d\tau_\lambda. \tag{9.54}$$

This equation has a very straightforward interpretation. The emergent intensity on the left is the sum of two positive contributions. The first term on the right is the initial intensity of the ray, reduced by the effects of absorption along the path to the surface. The second term, also positive,[28] represents the emission at every point along the path, attenuated by the absorption between the point of emission and the surface.

We now return to the geometry of a plane-parallel atmosphere and the vertical optical depth, $\tau_v$. However, we do *not* assume a gray atmosphere, LTE, or make the Eddington approximation. As shown in Fig. 9.16, the problem of limb darkening amounts to determining the emergent intensity $I_\lambda(0)$ as a function of the angle $\theta$. Equation (9.54), the formal solution to the transfer equation, is easily converted to this situation by using Eq. (9.38) to replace $\tau_\lambda$ with $\tau_{\lambda,v} \sec\theta$ (the vertical optical depth) to get

$$I(0) = I_0 e^{-\tau_{v,0}\sec\theta} - \int_{\tau_{v,0}\sec\theta}^{0} S \sec\theta\, e^{-\tau_v \sec\theta}\, d\tau_v.$$

Although both $I$ and $\tau_v$ depend on wavelength, the $\lambda$ subscript has been dropped to simplify the notation; the approximation of a gray atmosphere has *not* been made. To include the contributions to the emergent intensity from all layers of the atmosphere, we take the value

---

[28]Remember that the optical depth, measured along the ray's path, decreases in the direction of travel, so $d\tau_\lambda$ is negative.

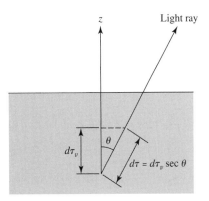

**FIGURE 9.16**  Finding $I(0)$ as a function of $\theta$ for limb darkening in plane-parallel geometry.

of the initial position of the rays to be at $\tau_{v,0} = \infty$. Then the first term on the right-hand side vanishes, leaving

$$I(0) = \int_0^\infty S \sec\theta \, e^{-\tau_v \sec\theta} \, d\tau_v. \tag{9.55}$$

If we knew how the source function depends on the vertical optical depth, this equation could be integrated to find the emergent intensity as a function of the direction of travel, $\theta$, of the ray. Although the form of the source function is not known, a good guess will be enough to estimate $I(0)$. Suppose that the source function has the form

$$S = a + b\tau_v, \tag{9.56}$$

where $a$ and $b$ are wavelength-dependent numbers to be determined. Inserting this into Eq. (9.55) and integrating (the details are left as an exercise) show that the emergent intensity for this source function is

$$I_\lambda(0) = a_\lambda + b_\lambda \cos\theta, \tag{9.57}$$

where the $\lambda$ subscripts have been restored to the appropriate quantities to emphasize their wavelength dependence. By making careful measurements of the variation in the specific intensity across the disk of the Sun, the values of $a_\lambda$ and $b_\lambda$ for the solar source function can be determined for a range of wavelengths. For example, for a wavelength of 501 nm, Böhm-Vitense (1989) supplies values of $a_{501} = 1.04 \times 10^{13}$ W m$^{-3}$ sr$^{-1}$ and $b_{501} = 3.52 \times 10^{13}$ W m$^{-3}$ sr$^{-1}$.

---

**Example 9.4.2.** Solar limb darkening provides an opportunity to test the accuracy of our "plane-parallel gray atmosphere in LTE using the Eddington approximation." In the preceding discussion of an equilibrium gray atmosphere, it was found that the mean intensity is equal to the source function,

$$\langle I \rangle = S$$

*continued*

(Eq. 9.44). Then, with the additional assumptions of the Eddington approximation and LTE, Eqs. (9.52) and (9.53) can be used to determine the mean intensity and thus the source function:

$$S = \langle I \rangle = \frac{\sigma T^4}{\pi} = \frac{3\sigma}{4\pi} T_e^4 \left( \tau_v + \frac{2}{3} \right).$$

Taking the source function to have the form of Eq. (9.56), $S = a + b\tau_v$, as used earlier for limb darkening (*after integrating over all wavelengths*), the values of the coefficients are

$$a = \frac{\sigma}{2\pi} T_e^4 \qquad \text{and} \qquad b = \frac{3\sigma}{4\pi} T_e^4.$$

The emergent intensity then will have the form of Eq. (9.57), $I(0) = a + b\cos\theta$ (again after integrating over all wavelengths). The ratio of the emergent intensity at angle $\theta$, $I(\theta)$, to that at the center of the star, $I(\theta = 0)$, is thus

$$\frac{I(\theta)}{I(\theta = 0)} = \frac{a + b\cos\theta}{a + b} = \frac{2}{5} + \frac{3}{5}\cos\theta. \tag{9.58}$$

We can compare the results of this calculation with observations of solar limb darkening in integrated light (made by summing over all wavelengths). Figure 9.17 shows both the observed values of $I(\theta)/I(\theta = 0)$ and the values from Eq. (9.58). The agreement is remarkably good, despite our numerous approximations. However, be forewarned that the agreement is much worse for observations made at a given wavelength (see Böhm-Vitense, 1989) as a consequence of wavelength-dependent opacity effects such as line blanketing.

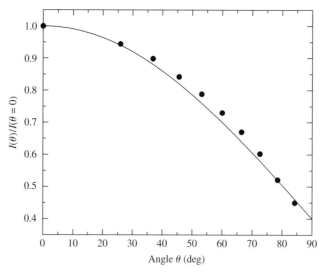

**FIGURE 9.17** A theoretical Eddington approximation of solar limb darkening for light integrated over all wavelengths. The dots are observational data for the Sun. Although a good fit, the Eddington approximation is not perfect, which implies that a more detailed model must be developed; see, for example, Problem 9.29.

## 9.5 ∎ THE PROFILES OF SPECTRAL LINES

We now have a formidable theoretical arsenal to bring to bear on the analysis of spectral lines. The shape of an individual spectral line contains a wealth of information about the environment in which it was formed.

### Equivalent Widths

Figure 9.18 shows a graph of the radiant flux, $F_\lambda$, as a function of wavelength for a typical absorption line. In the figure, $F_\lambda$ is expressed as a fraction of $F_c$, the value of the flux from the continuous spectrum outside the spectral line. Near the central wavelength, $\lambda_0$, is the *core* of the line, and the sides sweeping upward to the continuum are the line's *wings*. Individual lines may be narrow or broad, shallow or deep. The quantity $(F_c - F_\lambda)/F_c$ is referred to as the *depth* of the line. The strength of a spectral line is measured in terms of its **equivalent width**. The equivalent width $W$ of a spectral line is defined as the width of a box (shaded in Fig. 9.18) reaching up to the continuum that has the same area as the spectral line. That is,

$$W = \int \frac{F_c - F_\lambda}{F_c}\, d\lambda, \tag{9.59}$$

where the integral is taken from one side of the line to the other. The equivalent width of a line in the visible spectrum, shaded in Fig. 9.18, is usually on the order of 0.01 nm. Another measure of the width of a spectral line is the change in wavelength from one side of the line to the other, where its depth $(F_c - F_\lambda)/(F_c - F_{\lambda_0}) = 1/2$; this is called the *full width at half-maximum* and will be denoted by $(\Delta\lambda)_{1/2}$.

The spectral line shown in Fig. 9.18 is termed **optically thin** because there is no wavelength at which the radiant flux has been completely blocked. The opacity $\kappa_\lambda$ of the stellar

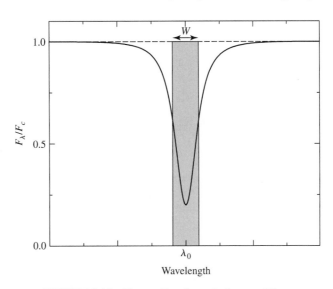

**FIGURE 9.18** The profile of a typical spectral line.

material is greatest at the wavelength $\lambda_0$ at the line's center and decreases moving into the wings. From the discussion on page 254, this means that the center of the line is formed at higher (and cooler) regions of the stellar atmosphere. Moving into the wings from $\lambda_0$, the line formation occurs at progressively deeper (and hotter) layers of the atmosphere, until it merges with the continuum-producing region at an optical depth of 2/3. In Section 11.2 this idea will be applied to the absorption lines produced in the solar photosphere.

### Processes That Broaden Spectral Lines

Three main processes are responsible for the broadening of spectral lines. Each of these mechanisms produces its own distinctive line shape or *line profile*.

1. **Natural broadening.** Spectral lines cannot be infinitely sharp, even for motionless, isolated atoms. According to Heisenberg's uncertainty principle (recall Eq. 5.20), as the time available for an energy measurement decreases, the inherent uncertainty of the result increases. Because an electron in an excited state occupies its orbital for only a brief instant, $\Delta t$, the orbital's energy, $E$, cannot have a precise value. Thus the uncertainty in the energy, $\Delta E$, of the orbital is approximately

$$\Delta E \approx \frac{\hbar}{\Delta t}.$$

(The electron's lifetime in the ground state may be taken as infinite, so in that case $\Delta E = 0$.) Electrons can make transitions from and to anywhere within these "fuzzy" energy levels, producing an uncertainty in the wavelength of the photon absorbed or emitted in a transition. Using Eq. (5.3) for the energy of a photon, $E_{\text{photon}} = hc/\lambda$, we find that the uncertainty in the photon's wavelength has a magnitude of roughly

$$\Delta\lambda \approx \frac{\lambda^2}{2\pi c}\left(\frac{1}{\Delta t_i} + \frac{1}{\Delta t_f}\right), \tag{9.60}$$

where $\Delta t_i$ is the lifetime of the electron in its initial state and $\Delta t_f$ is the lifetime in the final state. (The proof is left as a problem.)

---

**Example 9.5.1.**   The lifetime of an electron in the first and second excited states of hydrogen is about $\Delta t = 10^{-8}$ s. The natural broadening of the H$\alpha$ line of hydrogen, $\lambda = 656.3$ nm, is then

$$\Delta\lambda \approx 4.57 \times 10^{-14}\text{ m} = 4.57 \times 10^{-5}\text{ nm}.$$

---

A more involved calculation shows that the full width at half-maximum of the line profile for natural broadening is

$$(\Delta\lambda)_{1/2} = \frac{\lambda^2}{\pi c}\frac{1}{\Delta t_0}, \tag{9.61}$$

where $\Delta t_0$ is the average waiting time for a specific transition to occur. This results in a typical value of

$$(\Delta\lambda)_{1/2} \simeq 2.4 \times 10^{-5} \text{ nm},$$

in good agreement with the preceding estimate.

2. **Doppler broadening.** In thermal equilibrium, the atoms in a gas, each of mass $m$, are moving randomly about with a distribution of speeds that is described by the Maxwell–Boltzmann distribution function (Eq. 8.1), with the most probable speed given by Eq. (8.2), $v_{mp} = \sqrt{2kT/m}$. The wavelengths of the light absorbed or emitted by the atoms in the gas are Doppler-shifted according to (nonrelativistic) Eq. (4.30), $\Delta\lambda/\lambda = \pm|v_r|/c$. Thus the width of a spectral line due to Doppler broadening should be approximately

$$\Delta\lambda \approx \frac{2\lambda}{c}\sqrt{\frac{2kT}{m}}.$$

---

**Example 9.5.2.**   For hydrogen atoms in the Sun's photosphere ($T = 5777$ K), the Doppler broadening of the H$\alpha$ line should be about

$$\Delta\lambda \approx 0.0427 \text{ nm},$$

roughly 1000 times greater than for natural broadening.

---

A more in-depth analysis, taking into account the different directions of the atoms' motions with respect to one another and to the line of sight of the observer, shows that the full width at half-maximum of the line profile for Doppler broadening is

$$(\Delta\lambda)_{1/2} = \frac{2\lambda}{c}\sqrt{\frac{2kT\ln 2}{m}}. \tag{9.62}$$

Although the line profile for Doppler broadening is much wider at half-maximum than for natural broadening, the line depth for Doppler broadening decreases *exponentially* as the wavelength moves away from the central wavelength $\lambda_0$. This rapid decline is due to the high-speed exponential "tail" of the Maxwell–Boltzmann velocity distribution and is a much faster falloff in strength than for natural broadening.

Doppler shifts caused by the large-scale turbulent motion of large masses of gas (as opposed to the random motion of the individual atoms) can also be accommodated by Eq. (9.62) if the distribution of turbulent velocities follows the Maxwell–Boltzmann distribution. In that case,

$$(\Delta\lambda)_{1/2} = \frac{2\lambda}{c}\sqrt{\left(\frac{2kT}{m} + v_{\text{turb}}^2\right)\ln 2}, \tag{9.63}$$

where $v_{\text{turb}}$ is the most probable turbulent speed. The effect of turbulence on line profiles is particularly important in the atmospheres of giant and supergiant stars. In fact, the existence of turbulence in the atmospheres of these stars was first deduced from the inordinately large effect of Doppler broadening on their spectra.

Other sources of Doppler broadening involve orderly, coherent mass motions, such as stellar rotation, pulsation, and mass loss. These phenomena can have a substantial effect on the shape and width of the line profiles but cannot be combined with the results of Doppler broadening produced by random thermal motions obeying the Maxwell–Boltzmann distribution. For example, the characteristic P Cygni profile associated with mass loss will be discussed in Section 12.3 (see Fig. 12.17).

3. **Pressure (and collisional) broadening.** The orbitals of an atom can be perturbed in a collision with a neutral atom or by a close encounter involving the electric field of an ion. The results of individual collisions are called *collisional broadening*, and the statistical effects of the electric fields of large numbers of closely passing ions is termed *pressure broadening*; however, in the following discussion, both of these effects will be collectively referred to as pressure broadening. In either case, the outcome depends on the average time between collisions or encounters with other atoms and ions.

Calculating the precise width and shape of a pressure-broadened line is quite complicated. Atoms and ions of the same or different elements, as well as free electrons, are involved in these collisions and close encounters. The general shape of the line, however, is like that found for natural broadening, Eq. (9.61), and the line profile shared by natural and pressure broadening is sometimes referred to as a *damping profile* (also known as a *Lorentz profile*), so named because the shape is characteristic of the spectrum of radiation emitted by an electric charge undergoing damped simple harmonic motion. The values of the full width at half-maximum for natural and pressure broadening usually prove to be comparable, although the pressure profile can at times be more than an order of magnitude wider.

An estimate of pressure broadening due to collisions with atoms of a single element can be obtained by taking the value of $\Delta t_0$ in Eq. (9.61) to be the average time between collisions. This time is approximately equal to the mean free path between collisions divided by the average speed of the atoms. Using Eq. (9.12) for the mean free path and Eq. (8.2) for the speed, we find that

$$\Delta t_0 \approx \frac{\ell}{v} = \frac{1}{n\sigma\sqrt{2kT/m}},$$

where $m$ is the mass of an atom, $\sigma$ is its collision cross section, and $n$ is the number density of the atoms. Thus the width of the spectral line due to pressure broadening is on the order of

$$\Delta\lambda = \frac{\lambda^2}{c}\frac{1}{\pi\,\Delta t_0} \approx \frac{\lambda^2}{c}\frac{n\sigma}{\pi}\sqrt{\frac{2kT}{m}}. \tag{9.64}$$

Note that the width of the line is proportional to the number density $n$ of the atoms.

The physical reason for the Morgan–Keenan luminosity classes is now clear. The narrower lines observed for the more luminous giant and supergiant stars are due to the lower number densities in their extended atmospheres. Pressure broadening (with the width of the line profile proportional to $n$) broadens the lines formed in the denser atmospheres of main-sequence stars, where collisions occur more frequently.

---

**Example 9.5.3.**   Again, consider the hydrogen atoms in the Sun's photosphere, where the temperature is 5777 K and the number density of hydrogen atoms is about $1.5 \times 10^{23}$ m$^{-3}$. Then the pressure broadening of the H$\alpha$ line should be roughly

$$\Delta\lambda \approx 2.36 \times 10^{-5} \text{ nm},$$

which is comparable to the result for natural broadening found earlier. However, if the number density of the atoms in the atmosphere of a star is larger, the line width will be larger as well—more than an order of magnitude larger in some cases.

---

**The Voigt Profile**

The total line profile, called a **Voigt profile**, is due to the contributions of both the Doppler and damping profiles. The wider line profile for Doppler broadening dominates near the central wavelength $\lambda_0$. Farther from $\lambda_0$, however, the exponential decrease in the line depth for Doppler broadening means that there is a transition to a damping profile in the wings at a distance of about 1.8 times the Doppler value of $(\Delta\lambda)_{1/2}$ from the center of the line. Thus line profiles tend to have *Doppler cores* and *damping wings*. Figure 9.19 schematically shows the Doppler and damping line profiles.

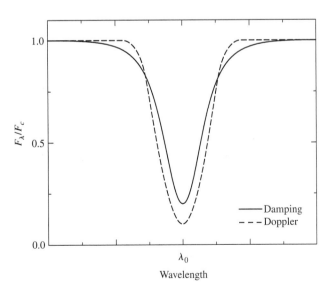

**FIGURE 9.19**   Schematic damping and Doppler line profiles, scaled so they have the same equivalent width.

---

**Example 9.5.4.**   As a review of the ideas of spectral line formation discussed here and in Chapter 8, consider the subdwarfs of luminosity class VI or "sd," which reside to the left of the main sequence (see Fig. 8.16). The spectra of these subdwarfs show that they are deficient in the atoms of metals (elements heavier than helium). Because ionized metals are an important source of electrons in stellar atmospheres, the electron number density is reduced. As mentioned in Section 8.1, fewer electrons with which ions may recombine means that a higher degree of ionization for all atoms can be achieved at the same temperature. Specifically, this reduces the number of $H^-$ ions in the atmosphere by ionizing them, thereby diluting this dominant source of continuum opacity. As a consequence of a lower opacity, we can see longer distances into these stars before reaching an optical depth of $\tau_\lambda = 2/3$. The forest of metallic lines (which are already weakened by the low metal abundance of the subdwarfs) appears even weaker against the brighter continuum. Thus, as a result of an under-abundance of metals, the spectrum of a subdwarf appears to be that of a hotter and brighter star of earlier spectral type with less prominent metal lines (see Table 8.1). This is why it is more accurate to say that these stars are displaced to the *left* of the main sequence, toward higher temperatures, rather than one magnitude below the main sequence.

---

The simplest model used for calculating a line profile assumes that the star's photosphere acts as a source of blackbody radiation and that the atoms above the photosphere remove photons from this continuous spectrum to form absorption lines. Although this **Schuster–Schwarzschild model** is inconsistent with the idea that photons of wavelength λ originate at an optical depth of $\tau_\lambda = 2/3$, it is still a useful approximation. In order to carry out the calculation, values for the temperature, density, and composition must be adopted for the region above the photosphere where the line is formed. The temperature and density determine the importance of Doppler and pressure broadening and are also used in the Boltzmann and Saha equations.

The calculation of a spectral line depends not only on the abundance of the element forming the line but also on the quantum-mechanical details of how atoms absorb photons. Let $N$ be the number of atoms of a certain element lying *above a unit area* of the photosphere. $N$ is a **column density** and has units of $m^{-2}$. (In other words, suppose a hollow tube with a cross section of 1 $m^2$ was stretched from the observer to the photosphere; the tube would then contain $N$ atoms of the specified type.) To find the number of absorbing atoms per unit area, $N_a$, that have electrons in the proper orbital for absorbing a photon at the wavelength of the spectral line, the temperature and density are used in the Boltzmann and Saha equations to calculate the atomic states of excitation and ionization. Our goal is to determine the value of $N_a$ by comparing the calculated and observed line profiles.

This task is complicated by the fact that not all transitions between atomic orbitals are equally likely. For example, an electron initially in the $n = 2$ orbital of hydrogen is about five times more likely to absorb an Hα photon and make a transition to the $n = 3$ orbital than it is to absorb an Hβ photon and jump to the $n = 4$ orbital. The relative probabilities of an electron making a transition from the same initial orbital are given by the *f-values* or *oscillator strengths* for that orbital. For hydrogen, $f = 0.637$ for the Hα transition and $f = 0.119$ for Hβ. The oscillator strengths may be calculated numerically or measured in the laboratory, and they are defined so that the $f$-values for transitions from the same initial

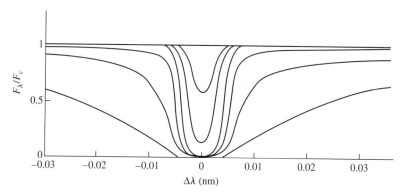

**FIGURE 9.20** Voigt profiles of the K line of Ca II. The shallowest line is produced by $N_a = 3.4 \times 10^{15}$ ions m$^{-2}$, and the ions are ten times more abundant for each successively broader line. (Adapted from Novotny, *Introduction to Stellar Atmospheres and Interiors*, Oxford University Press, New York, 1973.)

orbital add up to the number of electrons in the atom or ion. Thus the oscillator strength is the effective number of electrons per atom participating in a transition, and so multiplying the number of absorbing atoms per unit area by the $f$-value gives the number of atoms lying above each square meter of the photosphere that are actively involved in producing a given spectral line, $f N_a$. Figure 9.20 shows the Voigt profiles of the K line of Ca II ($\lambda = 393.3$ nm) for various values of the number of absorbing calcium ions.

### The Curve of Growth

The **curve of growth** is an important tool that astronomers use to determine the value of $N_a$ and thus the abundances of elements in stellar atmospheres. As seen in Fig. 9.20, the equivalent width, $W$, of the line varies with $N_a$. A curve of growth, shown in Fig. 9.21, is a logarithmic graph of the equivalent width, $W$, as a function of the number of absorbing atoms, $N_a$. To begin with, imagine that a specific element is not present in a stellar atmosphere. As some of that element is introduced, a weak absorption line appears that is initially optically thin. If the number of the absorbing atoms is doubled, twice as much light is removed, and the equivalent width of the line is twice as great. So $W \propto N_a$, and the curve of growth is initially linear with $\ln N_a$. As the number of absorbing atoms continues to increase, the center of the line becomes optically thick as the maximum amount of flux at the line's center is absorbed.[29] With the addition of still more atoms, the line bottoms out and becomes saturated. The wings of the line, which are still optically thin, continue to deepen. This occurs with relatively little change in the line's equivalent width and produces a flattening on the curve of growth where $W \propto \sqrt{\ln N_a}$. Increasing the number of absorbing atoms still further increases the width of the pressure-broadening profile [recall Eq. (9.64)],

---

[29]The zero flux at the center of the line shown in Fig. 9.20 is a peculiarity of the Schuster–Schwarzschild model. Actually, there is always *some* flux received at the central wavelength, $\lambda_0$, even for very strong, optically thick lines. As a rule, the flux at any wavelength cannot fall below $F_\lambda = \pi S_\lambda(\tau_\lambda = 2/3)$, the value of the source function at an optical depth of $2/3$; see Problem 9.20.

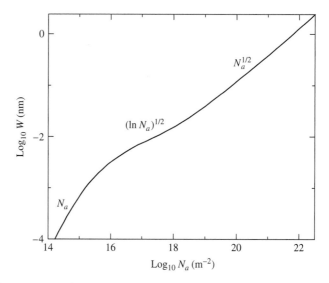

**FIGURE 9.21**    The curve of growth for the K line of Ca II. As $N_a$ increases, the functional dependence of the equivalent width ($W$) changes. At various positions along the curve of growth, $W$ is proportional to the functional forms indicated. (Figure adapted from Aller, *The Atmospheres of the Sun and Stars*, Ronald Press, New York, 1963.)

enabling it to contribute to the wings of the line. The equivalent width grows more rapidly, although not as steeply as at first, with approximately $W \propto \sqrt{N_a}$ for the total line profile. Using the curve of growth and a measured equivalent width, we can obtain the number of absorbing atoms. The Boltzmann and Saha equations are then used to convert this value into the total number of atoms of that element lying above the photosphere.

To reduce the errors involved in using a single spectral line, it is advantageous to locate, on a single curve of growth, the positions of the equivalent widths of several lines formed by transitions from the same initial orbital.[30] This can be accomplished by plotting $\log_{10}(W/\lambda)$ on the vertical axis and $\log_{10}[f N_a(\lambda/500 \text{ nm})]$ on the horizontal axis. This scaling results in a general curve of growth that can be used for several lines. Figure 9.22 shows a general curve of growth for the Sun. The use of such a curve of growth is best illustrated by an example.

---

**Example 9.5.5.**    We will use Fig. 9.22 to find the number of sodium atoms above each square meter of the Sun's photosphere from measurements of the 330.238-nm and 588.997-nm absorption lines of sodium (Table 9.1). Values of $T = 5800$ K and $P_e = 1$ N m$^{-2}$ were used for the temperature and electron pressure, respectively, to construct this curve of growth and will be adopted in the calculations that follow.

Both of these lines are produced when an electron makes an upward transition from the ground state orbital of the neutral Na I atom, and so these lines have the same value of $N_a$,

---

[30]This is just one of several possible ways of scaling the curve of growth. The assumptions used to obtain such a scaling are not valid for all broad lines (such as hydrogen) and may lead to inaccurate results.

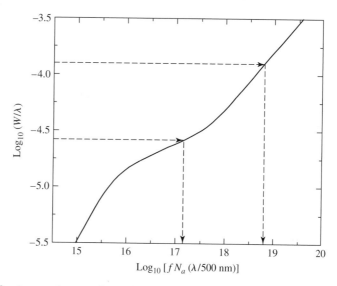

FIGURE 9.22 A general curve of growth for the Sun. The arrows refer to the data used in Example 9.5.5. (Figure adapted from Aller, *Atoms, Stars, and Nebulae*, Revised Edition, Harvard University Press, Cambridge, MA, 1971.)

TABLE 9.1 Data for Solar Sodium Lines. (From Aller, *Atoms, Stars, and Nebulae*, Revised Edition, Harvard University Press, Cambridge, MA, 1971.)

| $\lambda$ (nm) | $W$ (nm) | $f$ | $\log_{10}(W/\lambda)$ | $\log_{10}[f(\lambda/500 \text{ nm})]$ |
|---|---|---|---|---|
| 330.238 | 0.0088 | 0.0214 | $-4.58$ | $-1.85$ |
| 588.997 | 0.0730 | 0.645 | $-3.90$ | $-0.12$ |

the number of absorbing sodium atoms per unit area above the continuum-forming layer of the photosphere. This number can be found using the values of $\log_{10}(W/\lambda)$ with the general curve of growth, Fig. 9.22, to obtain a value of $\log_{10}[f N_a (\lambda/500 \text{ nm})]$ for each line. The results are

$$\log_{10}\left(\frac{f N_a \lambda}{500 \text{ nm}}\right) = 17.20 \quad \text{for the 330.238 nm line}$$

$$= 18.83 \quad \text{for the 588.997 nm line.}$$

To obtain the value of the number of absorbing atoms per unit area, $N_a$, we use the measured values of $\log_{10}[f(\lambda/500 \text{ nm})]$ together with

$$\log_{10} N_a = \log_{10}\left(\frac{f N_a \lambda}{500 \text{ nm}}\right) - \log_{10}\left(\frac{f\lambda}{500 \text{ nm}}\right),$$

to find

$$\log_{10} N_a = 17.15 - (-1.85) = 19.00 \quad \text{for the 330.238 nm line}$$

*continued*

and

$$\log_{10} N_a = 18.80 - (-0.12) = 18.92 \quad \text{for the 588.997 nm line.}$$

The average value of $\log_{10} N_a$ is 18.96; thus there are about $10^{19}$ Na I atoms in the ground state per square meter of the photosphere.

To find the total number of sodium atoms, the Boltzmann and Saha equations must be used; Eqs. (8.6) and (8.9), respectively. The difference in energy between the final and initial states $[E_b - E_a$ in Eq. (8.6)] is just the energy of the emitted photon. Using Eq. (5.3), the exponential term in the Boltzmann equation is

$$e^{-(E_b - E_a)/kT} = e^{-hc/\lambda kT}$$

$$= 5.45 \times 10^{-4} \quad \text{for the 330.238 nm line}$$

$$= 1.48 \times 10^{-2} \quad \text{for the 588.997 nm line,}$$

so nearly all of the neutral Na I atoms are in the ground state.

All that remains is to determine the total number of sodium atoms per unit area in all stages of ionization. If there are $N_{\mathrm{I}} = 10^{19}$ neutral sodium atoms per square meter, then the number of singly ionized atoms, $N_{\mathrm{II}}$, comes from the Saha equation:

$$\frac{N_{\mathrm{II}}}{N_{\mathrm{I}}} = \frac{2kT Z_{\mathrm{II}}}{P_e Z_{\mathrm{I}}} \left( \frac{2\pi m_e kT}{h^2} \right)^{3/2} e^{-\chi_1/kT}.$$

Using $Z_{\mathrm{I}} = 2.4$ and $Z_{\mathrm{II}} = 1.0$ for the partition functions and $\chi_{\mathrm{I}} = 5.14\,\text{eV}$ for the ionization energy of neutral sodium leads to $N_{\mathrm{II}}/N_{\mathrm{I}} = 2.43 \times 10^3$. There are about 2430 singly ionized sodium atoms for every neutral sodium atom in the Sun's photosphere,[31] so the total number of sodium atoms per unit area above the photosphere is about

$$N = 2430 N_{\mathrm{I}} = 2.43 \times 10^{22} \text{ m}^{-2}.$$

The mass of a sodium atom is $3.82 \times 10^{-26}$ kg, so the mass of sodium atoms above each square meter of the photosphere is roughly $9.3 \times 10^{-4}$ kg m$^{-2}$. (A more detailed analysis leads to a slightly lower value of $5.4 \times 10^{-4}$ kg m$^{-2}$.) For comparison, the mass of hydrogen atoms per unit area is about 11 kg m$^{-2}$.

---

Thus the number of absorbing atoms can be determined by comparing the equivalent widths measured for different absorption lines produced by atoms or ions initially in the same state (and so having the same column density in the stellar atmosphere) with a theoretical curve of growth. A curve-of-growth analysis can also be applied to lines originating from atoms or ions in different initial states; then applying the Boltzmann equation to the relative numbers of atoms and ions in these different states of excitation allows the excitation temperature to be calculated. Similarly, it is possible to use the Saha equation to find either the electron pressure or the ionization temperature (if the other is known) in the atmosphere from the relative numbers of atoms at various stages of ionization.

[31]The ionization energy for Na II is 47.3 eV. This is sufficiently large to guarantee that $N_{\mathrm{III}} \ll N_{\mathrm{II}}$, so higher states of ionization can be neglected.

### Computer Modeling of Stellar Atmospheres

The ultimate refinement in the analysis of stellar atmospheres is the construction of a *model atmosphere* on a computer. Each atmospheric layer is involved in the formation of line profiles and contributes to the spectrum observed for the star. All of the ingredients of the preceding discussion, plus the equations of hydrostatic equilibrium, thermodynamics, statistical and quantum mechanics, and the transport of energy by radiation and convection, are combined with extensive libraries of opacities to calculate how the temperature, pressure, and density vary with depth below the surface.[32] These models not only provide details regarding line profiles; they also provide information about such fundamental properties as the effective temperature and surface gravity of the star. Only when the variables of the model have been "fine-tuned" to obtain good agreement with the observations can astronomers finally claim to have decoded the vast amount of information carried in the light from a star.

This basic procedure has led astronomers to an understanding of the abundances of the elements in the Sun (see Table 9.2) and other stars. Hydrogen and helium are by far the most common elements, followed by oxygen, carbon, and nitrogen; for every $10^{12}$ atoms of hydrogen, there are $10^{11}$ atoms of helium and about $10^9$ atoms of oxygen. These figures are in very good agreement with abundances obtained from meteorites, giving astronomers

**TABLE 9.2** The Most Abundant Elements in the Solar Photosphere. The relative abundance of an element is given by $\log_{10}(N_{el}/N_H) + 12$. (Data from Grevesse and Sauval, *Space Science Reviews*, *85*, 161, 1998.)

| Element | Atomic Number | Log Relative Abundance |
|---|---|---|
| Hydrogen | 1 | 12.00 |
| Helium | 2 | $10.93 \pm 0.004$ |
| Oxygen | 8 | $8.83 \pm 0.06$ |
| Carbon | 6 | $8.52 \pm 0.06$ |
| Neon | 10 | $8.08 \pm 0.06$ |
| Nitrogen | 7 | $7.92 \pm 0.06$ |
| Magnesium | 12 | $7.58 \pm 0.05$ |
| Silicon | 14 | $7.55 \pm 0.05$ |
| Iron | 26 | $7.50 \pm 0.05$ |
| Sulfur | 16 | $7.33 \pm 0.11$ |
| Aluminum | 13 | $6.47 \pm 0.07$ |
| Argon | 18 | $6.40 \pm 0.06$ |
| Calcium | 20 | $6.36 \pm 0.02$ |
| Sodium | 11 | $6.33 \pm 0.03$ |
| Nickel | 28 | $6.25 \pm 0.04$ |

---

[32]Details of the construction of a model star will be deferred to Chapter 10.

confidence in their results.[33] This knowledge of the basic ingredients of the universe provides invaluable observational tests and constraints for some of the most fundamental theories in astronomy: the nucleosynthesis of light elements as a result of stellar evolution, the production of heavier elements by supernovae, and the Big Bang that produced the primordial hydrogen and helium that started it all.

## SUGGESTED READING

### General

Hearnshaw, J. B., *The Analysis of Starlight*, Cambridge University Press, Cambridge, 1986.

Kaler, James B., *Stars and Their Spectra*, Cambridge University Press, Cambridge, 1997.

### Technical

Aller, Lawrence H., *The Atmospheres of the Sun and Stars*, Ronald Press, New York, 1963.

Aller, Lawrence H., *Atoms, Stars, and Nebulae*, Third Edition, Cambridge University Press, New York, 1991.

Böhm-Vitense, Erika, "The Effective Temperature Scale," *Annual Review of Astronomy and Astrophysics*, *19*, 295, 1981.

Böhm-Vitense, Erika, *Stellar Astrophysics, Volume 2: Stellar Atmospheres*, Cambridge University Press, Cambridge, 1989.

Cox, Arthur N. (editor), *Allen's Astrophysical Quantities*, Fourth Edition, AIP Press, New York, 2000.

Gray, David F., *The Observation and Analysis of Stellar Photospheres*, Third Edition, Cambridge University Press, Cambridge, 2005.

Grevesse, N., and Sauval, A. J., "Standard Solar Composition," *Space Science Reviews*, *85*, 161, 1998.

Iglesias, Carlos J., and Rogers, Forrest J., "Updated OPAL Opacities," *The Astrophysical Journal*, *464*, 943, 1996.

Mihalas, Dimitri, *Stellar Atmospheres*, Second Edition, W.H. Freeman, San Francisco, 1978.

Mihalas, Dimitri, and Weibel-Mihalas, Barbara, *Foundations of Radiation Hydrodynamics*, Dover Publications, Inc., Mineola, NY, 1999.

Novotny, Eva, *Introduction to Stellar Atmospheres and Interiors*, Oxford University Press, New York, 1973.

Rogers, Forrest, and Iglesias, Carlos, "The OPAL Opacity Code,"
http://www-phys.llnl.gov/Research/OPAL/opal.html.

Rybicki, George B., and Lightman, Alan P., *Radiative Processes in Astrophysics*, John Wiley and Sons, New York, 1979.

---

[33]A notable exception is lithium, whose solar relative abundance of $10^{1.16}$ is significantly less than the value of $10^{3.31}$ obtained from meteorites. The efficient depletion of the Sun's lithium, sparing only one of every 140 lithium atoms, is probably due to its destruction by nuclear reaction processes when the lithium is transported into the hot interior of the star by convection.

## PROBLEMS

**9.1** Evaluate the energy of the blackbody photons inside your eye. Compare this with the visible energy inside your eye while looking at a 100-W light bulb that is 1 m away. You can assume that the light bulb is 100% efficient, although in reality it converts only a few percent of its 100 watts into visible photons. Take your eye to be a hollow sphere of radius 1.5 cm at a temperature of 37°C. The area of the eye's pupil is about 0.1 cm$^2$. Why is it dark when you close your eyes?

**9.2 (a)** Find an expression for $n_\lambda \, d\lambda$, the number density of blackbody photons (the number of blackbody photons per m$^3$) with a wavelength between $\lambda$ and $\lambda + d\lambda$.

  **(b)** Find the total number of photons inside a kitchen oven set at 400°F (477 K), assuming a volume of 0.5 m$^3$.

**9.3 (a)** Use the results of Problem 9.2 to find the total number density, $n$, of blackbody photons of all wavelengths. Also show that the average energy per photon, $u/n$, is

$$\frac{u}{n} = \frac{\pi^4 kT}{15(2.404)} = 2.70kT. \tag{9.65}$$

  **(b)** Find the average energy per blackbody photon at the center of the Sun, where $T = 1.57 \times 10^7$ K, and in the solar photosphere, where $T = 5777$ K. Express your answers in units of electron volts (eV).

**9.4** Derive Eq. (9.11) for the blackbody radiation pressure.

**9.5** Consider a spherical blackbody of radius $R$ and temperature $T$. By integrating Eq. (9.8) for the radiative flux with $I_\lambda = B_\lambda$ over all outward directions, derive the Stefan–Boltzmann equation in the form of Eq. (3.17). (You will also have to integrate over all wavelengths and over the surface area of the sphere.)

**9.6** Using the root-mean-square speed, $v_{\text{rms}}$, estimate the mean free path of the nitrogen molecules in your classroom at room temperature (300 K). What is the average time between collisions? Take the radius of a nitrogen molecule to be 0.1 nm and the density of air to be 1.2 kg m$^{-3}$. A nitrogen molecule contains 28 nucleons (protons and neutrons).

**9.7** Calculate how far you could see through Earth's atmosphere if it had the opacity of the solar photosphere. Use the value for the Sun's opacity from Example 9.2.2 and 1.2 kg m$^{-3}$ for the density of Earth's atmosphere.

**9.8** In Example 9.2.3, suppose that only two measurements of the specific intensity, $I_1$ and $I_2$, are available, made at angles $\theta_1$ and $\theta_2$. Determine expressions for the intensity $I_{\lambda,0}$ of the light above Earth's atmosphere and for the vertical optical depth of the atmosphere, $\tau_{\lambda,0}$, in terms of these two measurements.

**9.9** Use the laws of conservation of relativistic energy and momentum to prove that an isolated electron cannot absorb a photon.

**9.10** By measuring the slope of the curves in Fig. 9.10, verify that the decline of the curves after the peak in the opacity follows a Kramers law, $\overline{\kappa} \propto T^{-n}$, where $n \approx 3.5$.

**9.11** According to one model of the Sun, the central density is $1.53 \times 10^5$ kg m$^{-3}$ and the Rosseland mean opacity at the center is 0.217 m$^2$ kg$^{-1}$.

  **(a)** Calculate the mean free path of a photon at the center of the Sun.

(b) Calculate the average time it would take for the photon to escape from the Sun if this mean free path remained constant for the photon's journey to the surface. (Ignore the fact that identifiable photons are constantly destroyed and created through absorption, scattering, and emission.)

**9.12** If the temperature of a star's atmosphere is *increasing* outward, what type of spectral lines would you expect to find in the star's spectrum at those wavelengths where the opacity is greatest?

**9.13** Consider a large hollow spherical shell of hot gas surrounding a star. Under what circumstances would you see the shell as a glowing *ring* around the star? What can you say about the optical thickness of the shell?

**9.14** Verify that the emission coefficient, $j_\lambda$, has units of m s$^{-3}$ sr$^{-1}$.

**9.15** Derive Eq. (9.35) in Example 9.4.1, which shows how the intensity of a light ray is converted from its initial intensity $I_\lambda$ to the value $S_\lambda$ of the source function.

**9.16** The transfer equation, Eq. (9.34), is written in terms of the distance, $s$, measured along the path of a light ray. In different coordinate systems, the transfer equation will look slightly different, and care must be taken to include all of the necessary terms.

(a) Show that in a spherical coordinate system, with the center of the star at the origin, the transfer equation has the form

$$-\frac{\cos\theta'}{\kappa_\lambda \rho}\frac{dI_\lambda}{dr} = I_\lambda - S_\lambda,$$

where $\theta'$ is the angle between the ray and the outward radial direction. Note that you cannot simply replace $s$ with $r$!

(b) Use this form of the transfer equation to derive Eq. (9.31).

**9.17** For a plane-parallel atmosphere, show that the Eddington approximation leads to expressions for the mean intensity, radiative flux, and radiation pressure given by Eqs. (9.46–9.48).

**9.18** Using the Eddington approximation for a plane-parallel atmosphere, determine the values of $I_{\text{in}}$ and $I_{\text{out}}$ as functions of the vertical optical depth. At what depth is the radiation isotropic to within 1%?

**9.19** Using the results for the plane-parallel gray atmosphere in LTE, determine the ratio of the effective temperature of a star to its temperature at the top of the atmosphere. If $T_e = 5777$ K, what is the temperature at the top of the atmosphere?

**9.20** Show that for a plane-parallel gray atmosphere in LTE, the (constant) value of the radiative flux is equal to $\pi$ times the source function evaluated at an optical depth of 2/3:

$$F_{\text{rad}} = \pi S(\tau_v = 2/3).$$

This function, called the **Eddington–Barbier relation**, says that the radiative flux received from the surface of the star is determined by the value of the source function at $\tau_v = 2/3$.

**9.21** Consider a horizontal plane-parallel slab of gas of thickness $L$ that is maintained at a constant temperature $T$. Assume that the gas has optical depth $\tau_{\lambda,0}$, with $\tau_\lambda = 0$ at the top surface of the slab. Assume further that no radiation enters the gas from outside. Use the general solution of the transfer equation (9.54) to show that when looking at the slab from above, you see blackbody radiation if $\tau_{\lambda,0} \gg 1$ and emission lines (where $j_\lambda$ is large) if $\tau_{\lambda,0} \ll 1$. You may assume that the source function, $S_\lambda$, does not vary with position inside the gas. You may also assume thermodynamic equilibrium when $\tau_{\lambda,0} \gg 1$.

**9.22** Consider a horizontal plane-parallel slab of gas of thickness $L$ that is maintained at a constant temperature $T$. Assume that the gas has optical depth $\tau_{\lambda,0}$, with $\tau_\lambda = 0$ at the top surface of the slab. Assume further that incident radiation of intensity $I_{\lambda,0}$ enters the bottom of the slab from outside. Use the general solution of the transfer equation (9.54) to show that when looking at the slab from above, you see blackbody radiation if $\tau_{\lambda,0} \gg 1$. If $\tau_{\lambda,0} \ll 1$, show that you see absorption lines superimposed on the spectrum of the incident radiation if $I_{\lambda,0} > S_\lambda$ and emission lines superimposed on the spectrum of the incident radiation if $I_{\lambda,0} < S_\lambda$. (These latter two cases correspond to the spectral lines formed in the Sun's photosphere and chromosphere, respectively; see Section 11.2.) You may assume that the source function, $S_\lambda$, does not vary with position inside the gas. You may also assume thermodynamic equilibrium when $\tau_{\lambda,0} \gg 1$.

**9.23** Verify that if the source function is $S_\lambda = a_\lambda + b_\lambda \tau_{\lambda,v}$, then the emergent intensity is given by Eq. (9.57), $I_\lambda(0) = a_\lambda + b_\lambda \cos\theta$.

**9.24** Suppose that the shape of a spectral line is fit with one-half of an ellipse, such that the semimajor axis $a$ is equal to the maximum depth of the line (let $F_\lambda = 0$) and the minor axis $2b$ is equal to the maximum width of the line (where it joins the continuum). What is the equivalent width of this line? *Hint:* You may find Eq. (2.4) useful.

**9.25** Derive Eq. (9.60) for the uncertainty in the wavelength of a spectral line due to Heisenberg's uncertainty principle.

**9.26** The two solar absorption lines given in Table 9.3 are produced when an electron makes an upward transition from the ground state orbital of the neutral Na I atom.

   **(a)** Using the general curve of growth for the Sun, Fig. 9.22, repeat the procedure of Example 9.5.5 to find $N_a$, the number of absorbing sodium atoms per unit area of the photosphere.

   **(b)** Combine your results with those of Example 9.5.5 to find an average value of $N_a$. Use this value to plot the positions of the four sodium absorption lines on Fig. 9.22, and confirm that they do all lie on the curve of growth.

**9.27** Pressure broadening (due to the presence of the electric fields of nearby ions) is unusually effective for the spectral lines of hydrogen. Using the general curve of growth for the Sun with these broad hydrogen absorption lines will result in an overestimate of the amount of hydrogen present. The following calculation nevertheless demonstrates just how abundant hydrogen is in the Sun.

   The two solar absorption lines given in Table 9.4 belong to the Paschen series, produced when an electron makes an upward transition from the $n = 3$ orbital of the hydrogen atom.

   **(a)** Using the general curve of growth for the Sun, Fig. 9.22, repeat the procedure of Example 9.5.5 to find $N_a$, the number of absorbing hydrogen atoms per unit area of the photosphere (those with electrons initially in the $n = 3$ orbital).

   **(b)** Use the Boltzmann and Saha equations to calculate the total number of hydrogen atoms above each square meter of the Sun's photosphere.

**TABLE 9.3** Data for Solar Sodium Lines for Problem 9.26. (Data from Aller, *Atoms, Stars, and Nebulae*, Revised Edition, Harvard University Press, Cambridge, MA, 1971.)

| $\lambda$ (nm) | $W$ (nm) | $f$ |
|---|---|---|
| 330.298 | 0.0067 | 0.0049 |
| 589.594 | 0.0560 | 0.325 |

**TABLE 9.4**    Data for Solar Hydrogen Lines for Problem 9.27. (Data from Aller, *Atoms, Stars, and Nebulae*, Revised Edition, Harvard University Press, Cambridge, MA, 1971.)

| $\lambda$ (nm) | $W$ (nm) | $f$ |
|---|---|---|
| 1093.8 (Pa$\gamma$) | 0.22 | 0.0554 |
| 1004.9 (Pa$\delta$) | 0.16 | 0.0269 |

## COMPUTER PROBLEMS

**9.28** In this problem, you will use the values of the density and opacity at various points near the surface of the star to calculate the optical depth of these points. The data in Table 9.5 were obtained from the stellar model building program StatStar, described in Section 10.5 and Appendix L. The first point listed is at the surface of the stellar model.

(a) Find the optical depth at each point by numerically integrating Eq. (9.15). Use a simple trapezoidal rule such that

$$d\tau = -\kappa\rho\,ds$$

becomes

$$\tau_{i+1} - \tau_i = -\left(\frac{\kappa_i\rho_i + \kappa_{i+1}\rho_{i+1}}{2}\right)(r_{i+1} - r_i),$$

where $i$ and $i+1$ designate adjacent zones in the model. Note that because $s$ is measured along the path traveled by the photons, $ds = dr$.

(b) Make a graph of the temperature (vertical axis) vs. the optical depth (horizontal axis).

(c) For each value of the optical depth, use Eq. (9.53) to calculate the temperature for a plane-parallel gray atmosphere in LTE. Plot these values of $T$ on the same graph.

(d) The StatStar program utilizes a simplifying assumption that the surface temperature is zero (see Appendix L). Comment on the validity of the surface value of $T$ that you found.

**9.29** The binary star code TwoStars, discussed in Section 7.3 and Appendix K, makes use of an empirical limb darkening formula developed by W. Van Hamme (*Astronomical Journal, 106,* 1096, 1993):

$$\frac{I(\theta)}{I(\theta = 0)} = 1 - x(1 - \cos\theta) - y\cos\theta\,\log_{10}(\cos\theta),$$

where $x = 0.648$ and $y = 0.207$ for solar-type stars (other coefficients are provided for other types of stars).

(a) Plot Van Hamme's formula for limb darkening over the range $0 \le \theta \le 90°$. (Be sure to correctly treat the singularity in the function at $\theta = 90°$.)

(b) Plot Eq. (9.58), which is based on the Eddington approximation, on the same graph.

(c) Where is the difference between the two formulae the greatest?

(d) Compare the two curves to the observational data shown in Fig. 9.17. Which curve best represents the solar data?

**TABLE 9.5** A 1 $M_\odot$ StatStar Model for Problem 9.28. $T_e = 5504$ K.

| $i$ | $r$ (m) | $T$ (K) | $\rho \ (\text{kg m}^{-3})$ | $\kappa \ (\text{m}^2 \ \text{kg}^{-1})$ |
|---|---|---|---|---|
| 0 | 7.100764E+08 | 0.000000E+00 | 0.000000E+00 | 0.000000E+00 |
| 1 | 7.093244E+08 | 3.379636E+03 | 2.163524E−08 | 2.480119E+01 |
| 2 | 7.092541E+08 | 3.573309E+03 | 3.028525E−08 | 2.672381E+01 |
| 3 | 7.091783E+08 | 3.826212E+03 | 4.206871E−08 | 2.737703E+01 |
| 4 | 7.090959E+08 | 4.133144E+03 | 5.814973E−08 | 2.708765E+01 |
| 5 | 7.090062E+08 | 4.488020E+03 | 8.015188E−08 | 2.625565E+01 |
| 6 | 7.089085E+08 | 4.887027E+03 | 1.103146E−07 | 2.517004E+01 |
| 7 | 7.088019E+08 | 5.329075E+03 | 1.517126E−07 | 2.399474E+01 |
| 8 | 7.086856E+08 | 5.815187E+03 | 2.085648E−07 | 2.281158E+01 |
| 9 | 7.085588E+08 | 6.347784E+03 | 2.866621E−07 | 2.165611E+01 |
| 10 | 7.084205E+08 | 6.930293E+03 | 3.939580E−07 | 2.054686E+01 |
| 11 | 7.082697E+08 | 7.566856E+03 | 5.413734E−07 | 1.948823E+01 |
| 12 | 7.081052E+08 | 8.262201E+03 | 7.439096E−07 | 1.848131E+01 |
| 13 | 7.079259E+08 | 9.021603E+03 | 1.022171E−06 | 1.752513E+01 |
| 14 | 7.077303E+08 | 9.850881E+03 | 1.404459E−06 | 1.661785E+01 |
| 15 | 7.075169E+08 | 1.075642E+04 | 1.929644E−06 | 1.575731E+01 |
| 16 | 7.072843E+08 | 1.174520E+04 | 2.651111E−06 | 1.494128E+01 |
| 17 | 7.070306E+08 | 1.282486E+04 | 3.642174E−06 | 1.416754E+01 |
| 18 | 7.067540E+08 | 1.400375E+04 | 5.003513E−06 | 1.343396E+01 |
| 19 | 7.064524E+08 | 1.529096E+04 | 6.873380E−06 | 1.273849E+01 |
| 20 | 7.061235E+08 | 1.669643E+04 | 9.441600E−06 | 1.207917E+01 |
| 21 | 7.057649E+08 | 1.823102E+04 | 1.296880E−05 | 1.145414E+01 |
| 22 | 7.053741E+08 | 1.990656E+04 | 1.781279E−05 | 1.086165E+01 |
| 23 | 7.049480E+08 | 2.173599E+04 | 2.446473E−05 | 1.030001E+01 |
| 24 | 7.044836E+08 | 2.373341E+04 | 3.359882E−05 | 9.767631E+00 |
| 25 | 7.039774E+08 | 2.591421E+04 | 4.614038E−05 | 9.263005E+00 |
| 26 | 7.034259E+08 | 2.829519E+04 | 6.335925E−05 | 8.784696E+00 |
| 27 | 7.028250E+08 | 3.089468E+04 | 8.699788E−05 | 8.331344E+00 |
| 28 | 7.021704E+08 | 3.373266E+04 | 1.194469E−04 | 7.901659E+00 |
| 29 | 7.014574E+08 | 3.683096E+04 | 1.639859E−04 | 7.494416E+00 |
| 30 | 7.006810E+08 | 4.021337E+04 | 2.251132E−04 | 7.108452E+00 |
| 31 | 6.998356E+08 | 4.390583E+04 | 3.089976E−04 | 6.742665E+00 |
| 32 | 6.989155E+08 | 4.793666E+04 | 4.240980E−04 | 6.396010E+00 |
| 33 | 6.979141E+08 | 5.233670E+04 | 5.820105E−04 | 6.067495E+00 |
| 34 | 6.968247E+08 | 5.713961E+04 | 7.986295E−04 | 5.756179E+00 |
| 35 | 6.956399E+08 | 6.238205E+04 | 1.095736E−03 | 5.461170E+00 |
| 36 | 6.943518E+08 | 6.810401E+04 | 1.503169E−03 | 5.181621E+00 |
| 37 | 6.929517E+08 | 7.434904E+04 | 2.061803E−03 | 4.916730E+00 |
| 38 | 6.914307E+08 | 8.116461E+04 | 2.827602E−03 | 4.665735E+00 |
| 39 | 6.897790E+08 | 8.860239E+04 | 3.877181E−03 | 4.427914E+00 |
| 40 | 6.879861E+08 | 9.671869E+04 | 5.315384E−03 | 4.202584E+00 |
| 41 | 6.860411E+08 | 1.055748E+05 | 7.285639E−03 | 3.989094E+00 |

# 10

# The Interiors of Stars

## 10.1 ■ HYDROSTATIC EQUILIBRIUM

In the last two chapters, many of the observational details of stellar spectra were discussed, along with the basic physical principles behind the production of the observed lines. Analysis of that light, collected by ground-based and space-based telescopes, enables astronomers to determine a variety of quantities related to the outer layers of stars, such as effective temperature, luminosity, and composition. However, with the exceptions of the ongoing detection of neutrinos from the Sun (which will be discussed later in this chapter and in Chapter 11) and the one-time detection from Supernova 1987A (Section 15.3), no direct way exists to observe the central regions of stars.

### Determining the Internal Structures of Stars

To deduce the detailed internal structure of stars requires the generation of computer models that are consistent with all known physical laws and that ultimately agree with observable surface features. Although much of the theoretical foundation of stellar structure was understood by the first half of the twentieth century, it wasn't until the 1960s that sufficiently fast computing machines became available to carry out all of the necessary calculations. Arguably one of the greatest successes of theoretical astrophysics has been the detailed computer modeling of stellar structure and evolution. However, despite all of the successes of such calculations, numerous questions remain unanswered. The solution to many of these problems requires a more detailed theoretical understanding of the physical processes in operation in the interiors of stars, combined with even greater computational power.

The theoretical study of stellar structure, coupled with observational data, clearly shows that stars are dynamic objects, usually changing at an imperceptibly slow rate by human standards, although they can sometimes change in very rapid and dramatic ways, such as during a supernova explosion. That such changes must occur can be seen by simply considering the observed energy output of a star. In the Sun, $3.839 \times 10^{26}$ J of energy is emitted every second. This rate of energy output would be sufficient to melt a 0°C block of ice measuring 1 AU × 1 mile × 1 mile in only 0.3 s, assuming that the absorption of the energy was 100% efficient. Because stars do not have infinite supplies of energy, they must eventually use up their reserves and die. *Stellar evolution is the result of a constant fight against the relentless pull of gravity.*

## The Derivation of the Hydrostatic Equilibrium Equation

The gravitational force is always attractive, implying that an opposing force must exist if a star is to avoid collapse. This force is provided by pressure. To calculate how the pressure must vary with depth, consider a cylinder of mass $dm$ whose base is located a distance $r$ from the center of a spherical star (see Fig. 10.1). The areas of the top and bottom of the cylinder are each $A$ and the cylinder's height is $dr$. Furthermore, assume that the only forces acting on the cylinder are gravity and the pressure force, which is always normal to the surface and may vary with distance from the center of the star. Using Newton's second law $\mathbf{F} = m\mathbf{a}$, we have the net force on the cylinder:

$$dm \frac{d^2 r}{dt^2} = F_g + F_{P,t} + F_{P,b},$$

where $F_g < 0$ is the gravitational force directed inward and $F_{P,t}$ and $F_{P,b}$ are the pressure forces on the top and bottom of the cylinder, respectively. Note that since the pressure forces on the side of the cylinder will cancel, they have been explicitly excluded from the expression. Because the pressure force is always normal to the surface, the force exerted on the top of the cylinder must necessarily be directed toward the center of the star ($F_{P,t} < 0$),

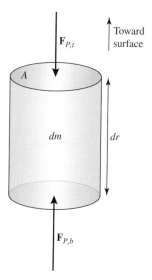

**FIGURE 10.1**   In a static star the gravitational force on a mass element is exactly canceled by the outward force due to a pressure gradient in the star. A cylinder of mass $dm$ is located at a distance $r$ from the center of the star. The height of the cylinder is $dr$, and the areas of the top and bottom are both $A$. The density of the gas is assumed to be $\rho$ at that position.

whereas the force on the bottom is directed outward ($F_{P,b} > 0$). Writing $F_{P,t}$ in terms of $F_{P,b}$ and a correction term $dF_P$ that accounts for the change in force due to a change in $r$ results in

$$F_{P,t} = -\left(F_{P,b} + dF_P\right).$$

Substitution into the previous expression gives

$$dm\,\frac{d^2r}{dt^2} = F_g - dF_P. \qquad (10.1)$$

As we noted in Example 2.2.1, the gravitational force on a small mass $dm$ located at a distance $r$ from the center of a spherically symmetric mass is

$$F_g = -G\frac{M_r\,dm}{r^2}, \qquad (10.2)$$

where $M_r$ is the mass inside the sphere of radius $r$, often referred to as the *interior mass*. The contribution to the gravitational force by spherically symmetric mass shells located outside $r$ is zero (the proof of this is left to Problem 10.2).

Pressure is defined as the amount of force per unit area exerted on a surface, or

$$P \equiv \frac{F}{A}.$$

Allowing for a difference in pressures $dP$ between the top of the cylinder and the bottom due to the different forces exerted on each surface, the differential force may be expressed as

$$dF_P = A\,dP. \qquad (10.3)$$

Substituting Eqs. (10.2) and (10.3) into Eq. (10.1) gives

$$dm\frac{d^2r}{dt^2} = -G\frac{M_r\,dm}{r^2} - A\,dP. \qquad (10.4)$$

If the density of the gas in the cylinder is $\rho$, its mass is just

$$dm = \rho A\,dr,$$

where $A\,dr$ is the cylinder's volume. Using this expression in Eq. (10.4) yields

$$\rho A\,dr\frac{d^2r}{dt^2} = -G\frac{M_r\rho A\,dr}{r^2} - A\,dP.$$

Finally, dividing through by the volume of the cylinder, we have

$$\rho\frac{d^2r}{dt^2} = -G\frac{M_r\rho}{r^2} - \frac{dP}{dr}. \qquad (10.5)$$

This is the equation for the radial motion of the cylinder, assuming spherical symmetry.

If we assume further that the star is static, then the acceleration must be zero. In this case Eq. (10.5) reduces to

$$\frac{dP}{dr} = -G\frac{M_r \rho}{r^2} = -\rho g, \tag{10.6}$$

where $g \equiv GM_r/r^2$ is the local acceleration of gravity at radius $r$. Equation (10.6), the condition of **hydrostatic equilibrium**, represents one of the fundamental equations of stellar structure for spherically symmetric objects under the assumption that accelerations are negligible. Equation (10.6) clearly indicates that in order for a star to be static, a *pressure gradient dP/dr* must exist to counteract the force of gravity. It is not the pressure that supports a star, but the change in pressure with radius. Furthermore, the pressure must *decrease* with increasing radius; the pressure is necessarily larger in the interior than it is near the surface.

---

**Example 10.1.1.**  To obtain a very crude estimate of the pressure at the center of the Sun, assume that $M_r = 1\ M_\odot$, $r = 1\ R_\odot$, and $\rho = \overline{\rho}_\odot = 1410\ \mathrm{kg\ m^{-3}}$ is the *average* solar density (see Example 8.2.1). Assume also that the surface pressure is exactly zero. Then, converting the differential equation to a difference equation, the left hand side of Eq. (10.6) becomes

$$\frac{dP}{dr} \sim \frac{P_s - P_c}{R_s - 0} \sim -\frac{P_c}{R_\odot},$$

where $P_c$ is the central pressure, and $P_s$ and $R_s$ are the surface pressure and radius, respectively. Substituting into the equation of hydrostatic equilibrium and solving for the central pressure, we find

$$P_c \sim G\frac{M_\odot \overline{\rho}_\odot}{R_\odot} \sim 2.7 \times 10^{14}\ \mathrm{N\ m^{-2}}.$$

To obtain a more accurate value, we need to *integrate* the hydrostatic equilibrium equation from the surface to the center, taking into consideration the change in the interior mass $M_r$ at each point, together with the variation of density with radius $\rho_r \equiv \rho(r)$, giving

$$\int_{P_s}^{P_c} dP = P_c = -\int_{R_s}^{R_c} \frac{GM_r\rho}{r^2}\, dr.$$

Actually carrying out the integration requires functional forms of $M_r$ and $\rho$. Unfortunately, such explicit expressions are not available, implying that further relationships between such quantities must be developed.

From a more rigorous calculation, a solar model gives a central pressure of nearly $2.34 \times 10^{16}\ \mathrm{N\ m^{-2}}$. This value is much larger than the one obtained from our crude estimate because of the increased density near the center of the Sun. As a reference, one atmosphere of pressure is 1 atm $= 1.013 \times 10^5\ \mathrm{N\ m^{-2}}$; therefore, the more realistic model predicts a central pressure of $2.3 \times 10^{11}$ atm!

---

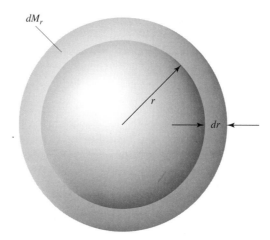

**FIGURE 10.2**    A spherically symmetric shell of mass $dM_r$ having a thickness $dr$ and located a distance $r$ from the center of the star. The local density of the shell is $\rho$.

### The Equation of Mass Conservation

A second relationship involving mass, radius, and density also exists. Again, for a spherically symmetric star, consider a *shell* of mass $dM_r$ and thickness $dr$, located a distance $r$ from the center, as in Fig. 10.2. Assuming that the shell is sufficiently thin (i.e., $dr \ll r$), the volume of the shell is approximately $dV = 4\pi r^2 \, dr$. If the local density of the gas is $\rho$, the shell's mass is given by

$$dM_r = \rho(4\pi r^2 dr).$$

Rewriting, we arrive at the **mass conservation equation**,

$$\boxed{\frac{dM_r}{dr} = 4\pi r^2 \rho,} \tag{10.7}$$

which dictates how the interior mass of a star must change with distance from the center. Equation (10.7) is the second of the fundamental equations of stellar structure.

### 10.2 ■ PRESSURE EQUATION OF STATE

Up to this point no information has been provided about the origin of the pressure term required by Eq. (10.6). To describe this macroscopic manifestation of particle interactions, it is necessary to derive a pressure **equation of state** of the material. Such an equation of state relates the dependence of pressure on other fundamental parameters of the material. One well-known example of a pressure equation of state is the **ideal gas law**, often expressed as

$$PV = NkT,$$

where $V$ is the volume of the gas, $N$ is the number of particles, $T$ is the temperature, and $k$ is Boltzmann's constant.

Although this expression was first determined experimentally, it is informative to derive it from fundamental physical principles. The approach used here will also provide a general method for considering environments where the assumptions of the ideal gas law do not apply, a situation frequently encountered in astrophysical problems.

### The Derivation of the Pressure Integral

Consider a cylinder of gas of length $\Delta x$ and cross-sectional area $A$, as in Fig. 10.3. The gas contained in the cylinder is assumed to be composed of point particles, each of mass $m$, that interact through perfectly elastic collisions only—in other words, as an ideal gas. To determine the pressure exerted on one of the ends of the container, examine the result of an impact on the right wall by an individual particle. Since, for a perfectly elastic collision, the angle of reflection from the wall must be equal to the angle of incidence, the change in momentum of the particle is necessarily entirely in the $x$-direction, normal to the surface. From Newton's second law[1] ($\mathbf{f} = m\mathbf{a} = d\mathbf{p}/dt$) and third law, the *impulse* $\mathbf{f}\,\Delta t$ delivered to the wall is just the negative of the change in momentum of the particle, or

$$\mathbf{f}\,\Delta t = -\Delta\mathbf{p} = 2p_x\hat{\mathbf{i}},$$

where $p_x$ is the component of the particle's initial momentum in the $x$-direction. Now the average force exerted by the particle over a period of time can be determined by evaluating the time interval between collisions with the right wall. Since the particle must traverse the length of the container twice before returning for a second reflection, the time interval between collisions with the same wall by the same particle is given by

$$\Delta t = 2\frac{\Delta x}{v_x},$$

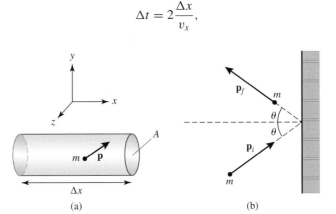

(a)                                                    (b)

**FIGURE 10.3**   (a) A cylinder of gas of length $\Delta x$ and cross-sectional area $A$. Assume that the gas contained in the cylinder is an ideal gas. (b) The collision of an individual point mass with one of the ends of the cylinder. For a perfectly elastic collision, the angle of reflection must equal the angle of incidence.

---

[1]Note that a lowercase $\mathbf{f}$ is used here to indicate that the force is due to a *single* particle.

so that the average force exerted on the wall by a single particle over that time period is given by

$$f = \frac{2p_x}{\Delta t} = \frac{p_x v_x}{\Delta x},$$

where it is assumed that the direction of the force vector is normal to the surface.

Now, because $p_x \propto v_x$, the numerator is proportional to $v_x^2$. To evaluate this, recall that the magnitude of the velocity vector is given by $v^2 = v_x^2 + v_y^2 + v_z^2$. For a sufficiently large collection of particles in random motion, the likelihood of motion in each of the three directions is the same, or $\overline{v_x^2} = \overline{v_y^2} = \overline{v_z^2} = v^2/3$. Substituting $\frac{1}{3} pv$ for $p_x v_x$, the average force per particle having momentum $p$ is

$$f(p) = \frac{1}{3} \frac{pv}{\Delta x}.$$

It is usually the case that the particles have a range of momenta. If the number of particles with momenta between $p$ and $p + dp$ is given by the expression $N_p \, dp$, then the total number of particles in the container is

$$N = \int_0^\infty N_p \, dp.$$

The contribution to the total force, $dF(p)$, by *all* particles in that momentum range is given by

$$dF(p) = f(p)N_p \, dp = \frac{1}{3} \frac{N_p}{\Delta x} pv \, dp.$$

Integrating over all possible values of the momentum, the total force exerted by particle collisions is

$$F = \frac{1}{3} \int_0^\infty \frac{N_p}{\Delta x} pv \, dp.$$

Dividing both sides of the expression by the surface area of the wall $A$ gives the pressure on the surface as $P = F/A$. Noting that $\Delta V = A \, \Delta x$ is just the volume of the cylinder and defining $n_p \, dp$ to be the number of particles *per unit volume* having momenta between $p$ and $p + dp$, or

$$n_p \, dp \equiv \frac{N_p}{\Delta V} \, dp,$$

we find that the pressure exerted on the wall is

$$P = \frac{1}{3} \int_0^\infty n_p pv \, dp. \tag{10.8}$$

This expression, which is sometimes called the **pressure integral**, makes it possible to compute the pressure, given some *distribution function, $n_p \, dp$*.

### The Ideal Gas Law in Terms of the Mean Molecular Weight

Equation (10.8) is valid for both massive and massless particles (such as photons) traveling at any speed. For the special case of massive, nonrelativistic particles, we may use $p = mv$ to write the pressure integral as

$$P = \frac{1}{3} \int_0^\infty m n_v v^2 \, dv, \tag{10.9}$$

where $n_v \, dv = n_p \, dp$ is the number of particles per unit volume having speeds between $v$ and $v + dv$.

The function $n_v \, dv$ is dependent on the physical nature of the system being described. In the case of an ideal gas, $n_v \, dv$ is the Maxwell–Boltzmann velocity distribution described in Chapter 8 (Eq. 8.1),

$$n_v \, dv = n \left( \frac{m}{2\pi kT} \right)^{3/2} e^{-mv^2/2kT} 4\pi v^2 \, dv,$$

where $n = \int_0^\infty n_v \, dv$ is the particle number density. Substituting into the pressure integral finally gives

$$P_g = nkT \tag{10.10}$$

(the proof is left as an exercise in Problem 10.5). Since $n \equiv N/V$, Eq. (10.10) is just the familiar ideal gas law.

In astrophysical applications it is often convenient to express the ideal gas law in an alternative form. Since $n$ is the particle number density, it is clear that it must be related to the mass density of the gas. Allowing for a variety of particles of different masses, it is then possible to express $n$ as

$$n = \frac{\rho}{\overline{m}},$$

where $\overline{m}$ is the average mass of a gas particle. Substituting, the ideal gas law becomes

$$P_g = \frac{\rho kT}{\overline{m}}.$$

We now define a new quantity, the **mean molecular weight**, as

$$\mu \equiv \frac{\overline{m}}{m_H},$$

where $m_H = 1.673532499 \times 10^{-27}$ kg is the mass of the hydrogen atom. *The mean molecular weight is just the average mass of a free particle in the gas, in units of the mass of hydrogen.* The ideal gas law can now be written in terms of the mean molecular weight as

$$\boxed{P_g = \frac{\rho kT}{\mu m_H}.} \tag{10.11}$$

The mean molecular weight depends on the composition of the gas as well as on the state of ionization of each species. The level of ionization enters because free electrons must be included in the average mass per particle $\overline{m}$. This implies that a detailed analysis of the Saha equation (8.8) is necessary to calculate the relative numbers of ionization states. When the gas is either completely neutral or completely ionized, the calculation simplifies significantly, however.

For a completely neutral gas,

$$\overline{m}_n = \frac{\displaystyle\sum_j N_j m_j}{\displaystyle\sum_j N_j}, \qquad (10.12)$$

where $m_j$ and $N_j$ are, respectively, the mass and the total number of atoms of type $j$ that are present in the gas, and the sums are assumed to be carried out over all types of atoms. Dividing by $m_H$ yields

$$\mu_n = \frac{\displaystyle\sum_j N_j A_j}{\displaystyle\sum_j N_j},$$

where $A_j \equiv m_j/m_H$. Similarly, for a completely ionized gas,

$$\mu_i \simeq \frac{\displaystyle\sum_j N_j A_j}{\displaystyle\sum_j N_j(1+z_j)},$$

where $1 + z_j$ accounts for the nucleus plus the number of free electrons that result from completely ionizing an atom of type $j$. (Do not confuse $z_j$ with $Z$, the mass fraction of metals.)

By inverting the expression for $\overline{m}$, it is possible to write alternative equations for $\mu$ in terms of mass fractions.[2] Recalling that $\overline{m} = \mu m_H$, Eq. (10.12) for a neutral gas gives

$$\frac{1}{\mu_n m_H} = \frac{\displaystyle\sum_j N_j}{\displaystyle\sum_j N_j m_j}$$

$$= \frac{\text{total number of particles}}{\text{total mass of gas}}$$

---

[2]Recall the definitions of mass fractions on page 250.

$$= \sum_j \frac{\text{number of particles from } j}{\text{mass of particles from } j} \cdot \frac{\text{mass of particles from } j}{\text{total mass of gas}}$$

$$= \sum_j \frac{N_j}{N_j A_j m_H} X_j$$

$$= \sum_j \frac{1}{A_j m_H} X_j,$$

where $X_j$ is the mass fraction of atoms of type $j$. Solving for $1/\mu_n$, we have

$$\frac{1}{\mu_n} = \sum_j \frac{1}{A_j} X_j. \tag{10.13}$$

Thus, for a neutral gas,

$$\frac{1}{\mu_n} \simeq X + \frac{1}{4} Y + \left\langle \frac{1}{A} \right\rangle_n Z. \tag{10.14}$$

$\langle 1/A \rangle_n$ is a weighted average of all elements in the gas heavier than helium. For solar abundances, $\langle 1/A \rangle_n \sim 1/15.5$.

The mean molecular weight of a completely ionized gas may be determined in a similar way. It is necessary only to include the *total* number of particles contained in the sample, both nuclei and electrons. For instance, each hydrogen atom contributes one free electron, together with its nucleus, to the total number of particles. Similarly, one helium atom contributes two free electrons plus its nucleus. Therefore, for a completely ionized gas, Eq. (10.13) becomes

$$\frac{1}{\mu_i} = \sum_j \frac{1 + z_j}{A_j} X_j. \tag{10.15}$$

Including hydrogen and helium explicitly, we have

$$\frac{1}{\mu_i} \simeq 2X + \frac{3}{4} Y + \left\langle \frac{1 + z}{A} \right\rangle_i Z. \tag{10.16}$$

For elements much heavier than helium, $1 + z_j \simeq z_j$, where $z_j \gg 1$ represents the number of protons (or electrons) in an atom of type $j$. It also holds that $A_j \simeq 2z_j$, the relation being based on the facts that sufficiently massive atoms have approximately the same number of protons and neutrons in their nuclei and that protons and neutrons have very similar masses (see page 299). Thus

$$\left\langle \frac{1 + z}{A} \right\rangle_i \simeq \frac{1}{2}.$$

If we assume that $X = 0.70$, $Y = 0.28$, and $Z = 0.02$, a composition typical of younger stars, then with these expressions for the mean molecular weight, $\mu_n = 1.30$ and $\mu_i = 0.62$.

### The Average Kinetic Energy Per Particle

Further investigation of the ideal gas law shows that it is also possible to combine Eq. (10.10) with the pressure integral (Eq. 10.9) to find the average kinetic energy per particle. Equating, we see that

$$nkT = \frac{1}{3} \int_0^\infty m n_v v^2 \, dv.$$

This expression can be rewritten to give

$$\frac{1}{n} \int_0^\infty n_v v^2 \, dv = \frac{3kT}{m}.$$

However, the left-hand side of this expression is just the integral average of $v^2$ weighted by the Maxwell–Boltzmann distribution function. Thus

$$\overline{v^2} = \frac{3kT}{m},$$

or

$$\frac{1}{2} m \overline{v^2} = \frac{3}{2} kT. \tag{10.17}$$

It is worth noting that the factor of 3 arose from averaging particle velocities over the three coordinate directions (or *degrees of freedom*), which was performed on page 290. Thus the average kinetic energy of a particle is $\frac{1}{2}kT$ per degree of freedom.

### Fermi–Dirac and Bose–Einstein Statistics

As has already been mentioned, there are stellar environments where the assumptions of the ideal gas law do not hold even approximately. For instance, in the pressure integral it was assumed that the upper limit of integration for velocity was infinity. Of course, this cannot be the case since, from Einstein's theory of special relativity, the maximum possible value of velocity is $c$, the speed of light. Furthermore, the effects of quantum mechanics were also neglected in the derivation of the ideal gas law. When the Heisenberg uncertainty principle and the Pauli exclusion principle are considered, a distribution function different from the Maxwell–Boltzmann distribution results. The **Fermi–Dirac** distribution function considers these important principles and leads to a very different pressure equation of state when applied to extremely dense matter such as that found in white dwarf stars and neutron stars. These exotic objects will be discussed in detail in Chapter 16. As mentioned in Section 5.4, particles such as electrons, protons, and neutrons that obey **Fermi–Dirac statistics** are called **fermions**.

Another statistical distribution function is obtained if it is assumed that the presence of some particles in a particular state enhances the likelihood of others being in the same state, an effect somewhat opposite to that of the Pauli exclusion principle. **Bose–Einstein statistics** has a variety of applications, including understanding the behavior of photons. Particles that obey Bose–Einstein statistics are known as **bosons**.

Just as special relativity and quantum mechanics must give classical results in the appropriate limits, Fermi–Dirac and Bose–Einstein statistics also approach the classical regime at sufficiently low densities and velocities. In these limits both distribution functions become indistinguishable from the classical Maxwell–Boltzmann distribution function.

### The Contribution Due to Radiation Pressure

Because photons possess momentum $p_\gamma = h\nu/c$ (Eq. 5.5), they are capable of delivering an impulse to other particles during absorption or reflection. Consequently, electromagnetic radiation results in another form of pressure. It is instructive to rederive the expression for radiation pressure found in Chapter 9 by making use of the pressure integral. Substituting the speed of light for the velocity $v$, using the expression for photon momentum, and using an identity for the distribution function, $n_p \, dp = n_\nu \, d\nu$, the general pressure integral, Eq. (10.8), now describes the effect of radiation, giving

$$P_{\text{rad}} = \frac{1}{3} \int_0^\infty h\nu n_\nu \, d\nu.$$

At this point, the problem again reduces to finding an appropriate expression for $n_\nu \, d\nu$. Since photons are bosons, the Bose–Einstein distribution function would apply. However, the problem may also be solved by realizing that $n_\nu \, d\nu$ represents the number density of photons having frequencies lying in the range between $\nu$ and $\nu + d\nu$. Multiplying by the energy of each photon in that range would then give the *energy density* over the frequency interval, or

$$P_{\text{rad}} = \frac{1}{3} \int_0^\infty u_\nu \, d\nu, \tag{10.18}$$

where $u_\nu \, d\nu = h\nu n_\nu \, d\nu$. But the energy density distribution function is found from the Planck function for blackbody radiation, Eq. (9.6). Substituting into Eq. (10.18) and performing the integration lead to

$$\boxed{P_{\text{rad}} = \frac{1}{3} a T^4,} \tag{10.19}$$

where $a$ is the radiation constant found previously in Eq. (9.7).

In many astrophysical situations the pressure due to photons can actually exceed by a significant amount the pressure produced by the gas. In fact it is possible that the magnitude of the force due to radiation pressure can become sufficiently great that it surpasses the gravitational force, resulting in an overall expansion of the system.

Combining both the ideal gas and radiation pressure terms, the total pressure becomes

$$P_t = \frac{\rho k T}{\mu m_H} + \frac{1}{3} a T^4. \tag{10.20}$$

---

**Example 10.2.1.** Using the results of Example 10.1.1, we can estimate the central temperature of the Sun. Neglecting the radiation pressure term, the central temperature is found

*continued*

from the ideal gas law equation of state to be

$$T_c = \frac{P_c \mu m_H}{\rho k}.$$

Using $\overline{\rho}_\odot$, a value of $\mu_i = 0.62$ appropriate for complete ionization,[3] and the estimated value for the central pressure, we find that

$$T_c \sim 1.44 \times 10^7 \text{ K}$$

which is in reasonable agreement with more detailed calculations. One solar model gives a central temperature of $1.57 \times 10^7$ K. At this temperature, the pressure due to radiation is only $1.53 \times 10^{13}$ N m$^{-2}$, 0.065% of the gas pressure.

## 10.3 ■ STELLAR ENERGY SOURCES

As we have already seen, the rate of energy output of stars (their luminosities) is very large. However, the question of the source of that energy has not yet been addressed. Clearly, one measure of the lifetime of a star must be related to how long it can sustain its power output.

### Gravitation and the Kelvin–Helmholtz Timescale

One likely source of stellar energy is gravitational potential energy. Recall that the gravitational potential energy of a system of two particles is given by Eq. (2.14),

$$U = -G\frac{Mm}{r}.$$

As the distance between $M$ and $m$ diminishes, the gravitational potential energy becomes *more negative*, implying that energy must have been converted to other forms, such as kinetic energy. If a star can manage to convert its gravitational potential energy into heat and then radiate that heat into space, the star may be able to shine for a significant period of time. However, we must also remember that by the virial theorem (Eq. 2.47) the total energy of a system of particles in equilibrium is one-half of the system's potential energy. Therefore, only one-half of the change in gravitational potential energy of a star is actually available to be radiated away; the remaining potential energy supplies the thermal energy that heats the star.

Calculating the gravitational potential energy of a star requires consideration of the interaction between every possible pair of particles. This is not as difficult as it might first seem. The gravitational force on a point mass $dm_i$ located outside of a spherically symmetric mass $M_r$ is

$$dF_{g,i} = G\frac{M_r \, dm_i}{r^2}$$

---

[3]Since, as we will see in the next chapter, the Sun has already converted a significant amount of its core hydrogen into helium via nuclear reactions, the actual value of $\mu_i$ is closer to 0.84.

and is directed toward the center of the sphere. This is just the same force that would exist if all of the mass of the sphere were located at its center, a distance $r$ from the point mass. This immediately implies that the gravitational potential energy of the point mass is

$$dU_{g,i} = -G\frac{M_r \, dm_i}{r}.$$

If, rather than considering an individual point mass, we assume that point masses are distributed uniformly within a shell of thickness $dr$ and mass $dm$ (where $dm$ is the sum of all the point masses $dm_i$), then

$$dm = 4\pi r^2 \rho \, dr,$$

where $\rho$ is the mass density of the shell and $4\pi r^2 \, dr$ is its volume. Thus

$$dU_g = -G\frac{M_r 4\pi r^2 \rho}{r} \, dr.$$

Integrating over all mass shells from the center of the star to the surface, its total gravitational potential energy becomes

$$U_g = -4\pi G \int_0^R M_r \rho r \, dr, \tag{10.21}$$

where $R$ is the radius of the star.

An exact calculation of $U_g$ requires knowledge of how $\rho$, and consequently $M_r$, depend on $r$. Nevertheless, an approximate value can be obtained by assuming that $\rho$ is constant and equal to its average value, or

$$\rho \sim \overline{\rho} = \frac{M}{\frac{4}{3}\pi R^3},$$

$M$ being the total mass of the star. Now we may also approximate $M_r$ as

$$M_r \sim \frac{4}{3}\pi r^3 \overline{\rho}.$$

If we substitute into Eq. (10.21), the total gravitational potential energy becomes

$$U_g \sim -\frac{16\pi^2}{15} G\overline{\rho}^2 R^5 \sim -\frac{3}{5}\frac{GM^2}{R}. \tag{10.22}$$

Lastly, applying the virial theorem, the total mechanical energy of the star is

$$E \sim -\frac{3}{10}\frac{GM^2}{R}. \tag{10.23}$$

---

**Example 10.3.1.** If the Sun were originally much larger than it is today, how much energy would have been liberated in its gravitational collapse? Assuming that its original radius

*continued*

was $R_i$, where $R_i \gg 1 \, R_\odot$, then the energy radiated away during collapse would be

$$\Delta E_g = -\left(E_f - E_i\right) \simeq -E_f \simeq \frac{3}{10} \frac{GM_\odot^2}{R_\odot} \simeq 1.1 \times 10^{41} \text{ J}.$$

Assuming also that the luminosity of the Sun has been roughly constant throughout its lifetime, it could emit energy at that rate for approximately

$$t_{KH} = \frac{\Delta E_g}{L_\odot} \tag{10.24}$$

$$\sim 10^7 \text{ yr}.$$

$t_{KH}$ is known as the **Kelvin–Helmholtz timescale**. Based on radioactive dating techniques, however, the estimated age of rocks on the Moon's surface is over $4 \times 10^9$ yr. It seems unlikely that the age of the Sun is less than the age of the Moon! Therefore, gravitational potential energy alone cannot account for the Sun's luminosity throughout its entire lifetime. As we shall see in later chapters, however, gravitational energy can play an important role during some phases of the evolution of stars.

---

Another possible energy source involves chemical processes. However, since chemical reactions are based on the interactions of orbital electrons in atoms, the amount of energy available to be released per atom is not likely to be more than 1–10 electron volts, typical of the atomic energy levels in hydrogen and helium (see Section 5.3). Given the number of atoms present in a star, the amount of chemical energy available is also far too low to account for the Sun's luminosity over a reasonable period of time (Problem 10.3).

### The Nuclear Timescale

The nuclei of atoms may also be considered as sources of energy. Whereas electron orbits involve energies in the electron volt (eV) range, nuclear processes generally involve energies millions of times larger (MeV). Just as chemical reactions can result in the transformation of atoms into molecules or one kind of molecule into another, nuclear reactions change one type of nucleus into another.

The nucleus of a particular **element** is specified by the number of protons, $Z$, it contains (not to be confused with the mass fraction of metals), with each proton carrying a charge of $+e$. Obviously, in a neutral atom the number of protons must exactly equal the number of orbital electrons. An **isotope** of a given element is identified by the number of neutrons, $N$, in the nucleus, with neutrons being electrically neutral, as the name implies. (All isotopes of a given element have the same number of protons.) Collectively, protons and neutrons are referred to as **nucleons**, the number of nucleons in a particular isotope being $A = Z + N$. Since protons and neutrons have very similar masses and greatly exceed the mass of electrons, $A$ is a good indication of the mass of the isotope and is often referred to as the *mass number.* [4] The masses of the proton, neutron, and electron are, respectively,

$$m_p = 1.67262158 \times 10^{-27} \text{ kg} = 1.00727646688 \text{ u}$$

$$m_n = 1.67492716 \times 10^{-27} \text{ kg} = 1.00866491578 \text{ u}$$

$$m_e = 9.10938188 \times 10^{-31} \text{ kg} = 0.0005485799110 \text{ u}.$$

It is often convenient to express the masses of nuclei in terms of *atomic mass units*; 1 u $= 1.66053873 \times 10^{-27}$ kg, exactly one-twelfth the mass of the isotope carbon-12. The masses of nuclear particles are also frequently expressed in terms of their rest mass energies, in units of MeV. Using Einstein's $E = mc^2$, we find 1 u $= 931.494013$ MeV/$c^2$. When masses are expressed simply in terms of rest mass energies, as is often the case, the factor $c^2$ is implicitly assumed.

The simplest isotope of hydrogen is composed of one proton and one electron and has a mass of $m_H = 1.00782503214$ u. This mass is actually very slightly *less* than the combined masses of the proton and electron taken separately. In fact, if the atom is in its ground state, the exact mass difference is 13.6 eV, which is just its ionization potential. Since mass is equivalent to a corresponding amount of energy, and the total mass–energy of the system must be conserved, any loss in energy when the electron and proton combine to form an atom must come at the expense of a loss in total mass.

Similarly, energy is also released with an accompanying loss in mass when nucleons are combined to form atomic nuclei. A helium nucleus, composed of two protons and two neutrons, can be formed by a series of nuclear reactions originally involving four hydrogen nuclei (i.e., 4H $\rightarrow$ He + low mass remnants). Such reactions are known as **fusion** reactions, since lighter particles are "fused" together to form a heavier particle. (Conversely, a **fission** reaction occurs when a massive nucleus is split into smaller fragments.) The total mass of the four hydrogen atoms is 4.03130013 u, whereas the mass of one helium atom is $m_{He} = 4.002603$ u. Neglecting the contribution of low-mass remnants such as neutrinos, the combined mass of the hydrogen atoms *exceeds* the mass of the helium atom by $\Delta m = 0.028697$ u, or 0.7%. Therefore, the total amount of energy released in forming the helium nucleus is $E_b = \Delta mc^2 = 26.731$ MeV. This is known as the **binding energy** of the helium nucleus. If the nucleus were to be broken apart into its constituent protons and neutrons, the amount of energy required to accomplish the task would be 26.731 MeV.

---

**Example 10.3.2.**   Is this source of nuclear energy sufficient to power the Sun during its lifetime? For simplicity, assume also that the Sun was originally 100% hydrogen and that only the inner 10% of the Sun's mass becomes hot enough to convert hydrogen into helium.

Since 0.7% of the mass of hydrogen would be converted to energy in forming a helium nucleus, the amount of nuclear energy available in the Sun would be

$$E_{\text{nuclear}} = 0.1 \times 0.007 \times M_\odot c^2 = 1.3 \times 10^{44} \text{ J}.$$

*continued*

---

[4]The quantity $A_j$ defined on page 292 is approximately equal to the mass number.

This gives a **nuclear timescale** of approximately

$$t_{\text{nuclear}} = \frac{E_{\text{nuclear}}}{L_\odot}$$

(10.25)

$$\sim 10^{10} \text{ yr},$$

more than enough time to account for the age of Moon rocks.

### Quantum Mechanical Tunneling

Apparently, sufficient energy is available in the nuclei of atoms to provide a source for stellar luminosities, but can nuclear reactions actually occur in the interiors of stars? For a reaction to occur, the nuclei of atoms must collide, forming new nuclei in the process. However, all nuclei are positively charged, meaning that a Coulomb potential energy barrier must be overcome before contact can occur. Figure 10.4 shows the characteristic shape of the potential energy curve that an atomic nucleus would experience when approaching another nucleus. The curve is composed of two parts: The portion outside of the nucleus is the potential energy that exists between two positively charged nuclei, and the portion inside the nucleus forms a *potential well* governed by the **strong nuclear force** that binds the nucleus together. The strong nuclear force is a very short-range force that acts between all nucleons within the atom. It is an attractive force that dominates the Coulomb repulsion between protons. Clearly, if such a force did not exist, a nucleus would immediately fly apart.

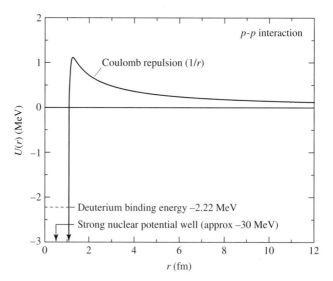

**FIGURE 10.4** The potential energy curve characteristic of nuclear reactions. The Coulomb repulsion between positive nuclei results in a barrier that is inversely proportional to the separation between nuclei and is proportional to the product of their charges. The nuclear potential well inside the nucleus is due to the attractive strong nuclear force.

If we assume that the energy required to overcome the Coulomb barrier is provided by the thermal energy of the gas, and that all nuclei are moving nonrelativistically, then the temperature $T_{classical}$ required to overcome the barrier can be estimated. Since all of the particles in the gas are in random motion, it is appropriate to refer to the relative velocity $v$ between two nuclei and their *reduced mass*, $\mu_m$, given by Eq. (2.22) (note that we are not referring here to the mean molecular weight, $\mu$). Equating the initial kinetic energy of the reduced mass to the potential energy of the barrier gives the position of the classical "turn-around point." Now, using Eq. (10.17) yields

$$\frac{1}{2}\mu_m \overline{v^2} = \frac{3}{2}kT_{classical} = \frac{1}{4\pi\epsilon_0}\frac{Z_1 Z_2 e^2}{r},$$

where $T_{classical}$ denotes the temperature required for an average particle to overcome the barrier, $Z_1$ and $Z_2$ are the numbers of protons in each nucleus, and $r$ is their distance of separation. Assuming that the radius of a typical nucleus is on the order of 1 fm $= 10^{-15}$ m, the temperature needed to overcome the Coulomb potential energy barrier is approximately

$$T_{classical} = \frac{Z_1 Z_2 e^2}{6\pi\epsilon_0 kr} \tag{10.26}$$

$$\sim 10^{10} \text{ K}$$

for a collision between two protons ($Z_1 = Z_2 = 1$). However, the central temperature of the Sun is only $1.57 \times 10^7$ K, much lower than required here. Even taking into consideration the fact that the Maxwell–Boltzmann distribution indicates that a significant number of particles have speeds well in excess of the average speed of particles in the gas, classical physics is unable to explain how a sufficient number of particles can overcome the Coulomb barrier to produce the Sun's observed luminosity.

As was mentioned in Section 5.4, quantum mechanics tells us that it is never possible to know both the position and the momentum of a particle to unlimited accuracy. The Heisenberg uncertainty principle states that the uncertainties in position and momentum are related by

$$\Delta x \Delta p_x \geq \frac{\hbar}{2}.$$

The uncertainty in the position of one proton colliding with another may be so large that even though the kinetic energy of the collision is insufficient to overcome the classical Coulomb barrier, one proton might nevertheless find itself within the central potential well defined by the strong force of the other. This quantum mechanical tunneling has no classical counterpart (recall the discussion in Section 5.4). Of course, the greater the ratio of the potential energy barrier height to the particle's kinetic energy or the wider the barrier, the less likely tunneling becomes.

As a crude estimate of the effect of tunneling on the temperature necessary to sustain nuclear reactions, assume that a proton must be within approximately one de Broglie wavelength of its target in order to tunnel through the Coulomb barrier. Recalling that the wavelength of a massive particle is given by $\lambda = h/p$ (Eq. 5.17), rewriting the kinetic energy in terms of momentum,

$$\frac{1}{2}\mu_m v^2 = \frac{p^2}{2\mu_m},$$

and setting the distance of closest approach equal to one wavelength (where the potential energy barrier height is equal to the original kinetic energy) give

$$\frac{1}{4\pi\epsilon_0}\frac{Z_1 Z_2 e^2}{\lambda} = \frac{p^2}{2\mu_m} = \frac{(h/\lambda)^2}{2\mu_m}.$$

Solving for $\lambda$ and substituting $r = \lambda$ into Eq. (10.26), we find the quantum mechanical estimate of the temperature required for a reaction to occur:

$$T_{\text{quantum}} = \frac{Z_1^2 Z_2^2 e^4 \mu_m}{12\pi^2 \epsilon_0^2 h^2 k}. \tag{10.27}$$

Again assuming the collision of two protons, $\mu_m = m_p/2$ and $Z_1 = Z_2 = 1$. Substituting, we find that $T_{\text{quantum}} \sim 10^7$ K. In this case, if we assume the effects of quantum mechanics, the temperature required for nuclear reactions is consistent with the estimated central temperature of the Sun.

## Nuclear Reaction Rates and the Gamow Peak

Now that the possibility of a nuclear energy source has been established, we need a more detailed description of nuclear *reaction rates* in order to apply them to the development of stellar models. For instance, not all particles in a gas of temperature $T$ will have sufficient kinetic energy and the necessary wavelength to tunnel through the Coulomb barrier successfully. Consequently, the reaction rate per energy interval must be described in terms of the number density of particles having energies within a specific range, combined with the probability that those particles can actually tunnel through the Coulomb barrier of the target nucleus. The total nuclear reaction rate is then integrated over all possible energies.

First consider the number density of nuclei within a specified energy interval. As we have seen, the Maxwell–Boltzmann distribution (Eq. 8.1) relates the number density of particles with velocities between $v$ and $v + dv$ to the temperature of the gas. Assuming that particles are initially sufficiently far apart that the potential energy may be neglected, the nonrelativistic[5] kinetic energy relation describes the total energy of the particles, or $K = E = \mu_m v^2/2$. Solving for the velocity and substituting, we can write the Maxwell–Boltzmann distribution in terms of the number of particles with kinetic energies between $E$ and $E + dE$ as

$$n_E\, dE = \frac{2n}{\pi^{1/2}}\frac{1}{(kT)^{3/2}} E^{1/2} e^{-E/kT}\, dE \tag{10.28}$$

(the proof of this relation is left to Problem 10.6).

---

[5] In astrophysical processes, nuclei are usually nonrelativistic, except in the extreme environment of neutron stars. Because of the much smaller masses of electrons, it cannot be assumed that they are also nonrelativistic, however.

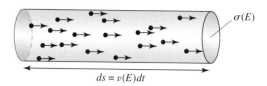

**FIGURE 10.5**   The number of reactions per unit time between particles of type $i$ and a target $x$ of cross section $\sigma(E)$ may be thought of in terms of the number of particles in a cylinder of cross-sectional area $\sigma(E)$ and length $ds = v(E)\, dt$ that will reach the target in a time interval $dt$.

Equation (10.28) gives the number of particles per unit volume that have energies in the range $dE$, but it does not describe the probability that particles will actually interact. To account for this factor, the idea of a **cross section** is re-introduced.[6] Define the cross section $\sigma(E)$ to be the number of reactions per target nucleus per unit time, divided by the flux of incident particles, or

$$\sigma(E) \equiv \frac{\text{number of reactions/nucleus/time}}{\text{number of incident particles/area/time}}.$$

Although $\sigma(E)$ is strictly a measure of probability, as mentioned in Chapter 9, it can be thought of as *roughly* the cross-sectional area of the target particle; any incoming particle that strikes within that area, centered on the target, will result in a nuclear reaction.

To find the reaction rate in units of reactions volume$^{-1}$ time$^{-1}$, consider the number of particles that will hit a target of cross-sectional area $\sigma(E)$, assuming that all of the incident particles are moving in one direction. Let $x$ denote a target particle and $i$ denote an incident particle. If the number of incident particles per unit volume having energies between $E$ and $E + dE$ is $n_{iE}\, dE$, then the number of reactions, $dN_E$, is the number of particles that can strike $x$ in a time interval $dt$ with a velocity $v(E) = \sqrt{2E/\mu_m}$.

The number of incident particles is just the number contained within a cylinder of volume $\sigma(E)v(E)\, dt$ (see Fig. 10.5), or

$$dN_E = \sigma(E)v(E)n_{iE}\, dE\, dt.$$

Now, the number of incident particles per unit volume with the appropriate velocity (or kinetic energy) is some fraction of the total number of particles in the sample,

$$n_{iE}\, dE = \frac{n_i}{n}\, n_E\, dE,$$

where $n_i = \int_0^\infty n_{iE}\, dE$, $n = \int_0^\infty n_E\, dE$, and $n_E\, dE$ is given by Eq. (10.28). Therefore, the number of reactions per target nucleus per time interval $dt$ having energies between $E$ and $E + dE$ is

$$\frac{\text{reactions per nucleus}}{\text{time interval}} = \frac{dN_E}{dt} = \sigma(E)v(E)\frac{n_i}{n}\, n_E\, dE.$$

[6]The concept of a cross section was first discussed in the context of determining the mean free path between collisions in Section 9.2.

Finally, if there are $n_x$ targets per unit volume, the total number of reactions per unit volume per unit time, integrated over all possible energies, is

$$r_{ix} = \int_0^\infty n_x n_i \sigma(E) v(E) \frac{n_E}{n} \, dE. \tag{10.29}$$

To evaluate Eq. (10.29) we must know the functional form of $\sigma(E)$. Unfortunately, $\sigma(E)$ changes rapidly with energy, and its functional form is complicated. It is also important to compare $\sigma(E)$ with experimental data. However, stellar thermal energies are quite low compared to energies found in laboratory experimentation, and significant extrapolation is usually required to obtain comparison data for stellar nuclear reaction rates.

The process of determining $\sigma(E)$ can be improved somewhat if the terms most strongly dependent on energy are factored out first. We have already suggested that the cross section can be roughly thought of as being a physical area. Moreover, the size of a nucleus, measured in terms of its ability to "touch" target nuclei, is approximately one de Broglie wavelength in radius ($r \sim \lambda$). Combining these ideas, the cross section of the nucleus $\sigma(E)$ should be proportional to

$$\sigma(E) \propto \pi \lambda^2 \propto \pi \left( \frac{h}{p} \right)^2 \propto \frac{1}{E}.$$

To obtain the last expression, we have again used the nonrelativistic relation, $K = E = \mu_m v^2 / 2 = p^2 / 2\mu_m$.

We have also mentioned previously that the ability to tunnel through the Coulomb barrier is related to the ratio of the barrier height to the initial kinetic energy of the incoming nucleus, a factor that must be considered in the cross section. If the barrier height $U_c$ is zero, the probability of successfully penetrating it necessarily equals one (100%). As the barrier height increases relative to the initial kinetic energy of the incoming nucleus, the probability of penetration must decrease, asymptotically approaching zero as the potential energy barrier height goes to infinity. In fact, the tunneling probability is exponential in nature. Since $\sigma(E)$ must be related to the tunneling probability, we have

$$\sigma(E) \propto e^{-2\pi^2 U_c / E}. \tag{10.30}$$

The factor of $2\pi^2$ arises from the strict quantum mechanical treatment of the problem. Again assuming that $r \sim \lambda = h/p$, taking the ratio of the barrier potential height $U_c$ to particle kinetic energy $E$ gives

$$\frac{U_c}{E} = \frac{Z_1 Z_2 e^2 / 4\pi \epsilon_0 r}{\mu_m v^2 / 2} = \frac{Z_1 Z_2 e^2}{2\pi \epsilon_0 h v}.$$

After some manipulation, we find that

$$\sigma(E) \propto e^{-bE^{-1/2}}, \tag{10.31}$$

where

$$b \equiv \frac{\pi \mu_m^{1/2} Z_1 Z_2 e^2}{2^{1/2} \epsilon_0 h}.$$

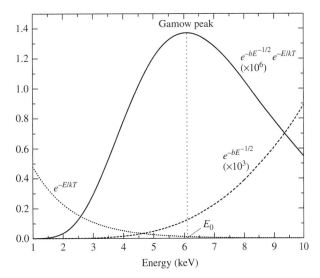

**FIGURE 10.6**  The likelihood that a nuclear reaction will occur is a function of the kinetic energy of the collision. The Gamow peak arises from the contribution of the $e^{-E/kT}$ Maxwell–Boltzmann high-energy tail and the $e^{-bE^{-1/2}}$ Coulomb barrier penetration term. This particular example represents the collision of two protons at the central temperature of the Sun. (Note that $e^{-bE^{-1/2}}$ and $e^{-bE^{-1/2}}e^{-E/kT}$ have been multiplied by $10^3$ and $10^6$, respectively, to more readily illustrate the functional dependence on energy.)

Clearly, $b$ depends on the masses and electric charges of the two nuclei involved in the interaction.

Combining the previous results and defining $S(E)$ to be some (we hope) slowly varying function of energy, we may now express the cross section as[7]

$$\sigma(E) = \frac{S(E)}{E} e^{-bE^{-1/2}}. \tag{10.32}$$

Substituting Eqs. (10.28) and (10.32) into Eq. (10.29) and simplifying, the reaction rate integral becomes

$$r_{ix} = \left(\frac{2}{kT}\right)^{3/2} \frac{n_i n_x}{(\mu_m \pi)^{1/2}} \int_0^\infty S(E)\, e^{-bE^{-1/2}} e^{-E/kT}\, dE. \tag{10.33}$$

In Eq. (10.33), the term $e^{-E/kT}$ represents the high-energy wing of the Maxwell–Boltzmann distribution, and the term $e^{-bE^{-1/2}}$ comes from the penetration probability. As can be seen in Fig. 10.6, the product of these two factors produces a strongly peaked curve, known as the **Gamow peak** after George Gamow (1904–1968), the physicist who first

---

[7]The angular momentum of the interacting particles also plays a role in nuclear reaction rates, but it is generally a minor component for reactions of astrophysical significance.

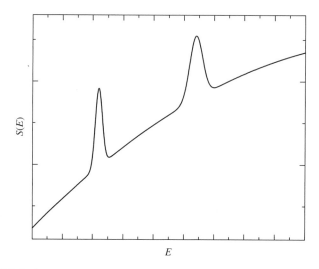

**FIGURE 10.7**    A hypothetical example of the effect of resonance on $S(E)$.

investigated Coulomb barrier penetration. The top of the curve occurs at the energy

$$E_0 = \left( \frac{bkT}{2} \right)^{2/3}.$$

(10.34)

As a consequence of the Gamow peak, the greatest contribution to the reaction rate integral comes in a fairly narrow energy band that depends on the temperature of the gas, together with the charges and masses of the constituents of the reaction.

Assuming that $S(E)$ is indeed slowly varying across the Gamow peak, it may be approximated by its value at $E_0$ [$S(E) \simeq S(E_0) =$ constant] and removed from inside of the integral. Also, it is generally much easier to extrapolate laboratory results if they are expressed in terms of $S(E)$.

### Resonance

In some cases, however, $S(E)$ can vary quite rapidly, peaking at specific energies, as illustrated schematically in Fig. 10.7. These energies correspond to energy levels within the nucleus, analogous to the orbital energy levels of electrons. It is a **resonance** between the energy of the incoming particle and differences in energy levels within the nucleus that accounts for these strong peaks. A detailed discussion of these resonance peaks is beyond the scope of this book.[8]

### Electron Screening

Yet another factor influencing reaction rates is **electron screening**. On average, the electrons liberated when atoms are ionized at the high temperatures of stellar interiors produce a

---

[8] See Clayton (1983) or Arnett (1996) for excellent and detailed discussions of resonance peaks.

"sea" of negative charge that partially hides the target nucleus, reducing its *effective* positive charge. The result of this reduced positive charge is a lower Coulomb barrier to the incoming nucleus and an enhanced reaction rate. By including electron screening, the effective Coulomb potential becomes

$$U_{\text{eff}} = \frac{1}{4\pi\epsilon_0} \frac{Z_1 Z_2 e^2}{r} + U_s(r),$$

where $U_s(r) < 0$ is the electron screening contribution. Electron screening can be significant, sometimes enhancing the helium-producing reactions by 10% to 50%.

### Representing Nuclear Reaction Rates Using Power Laws

It is often illuminating to write the complicated reaction rate equations in the form of a power law centered at a particular temperature. Neglecting the screening factor, in the case of a two-particle interaction, the reaction rate would become

$$r_{ix} \simeq r_0 X_i X_x \rho^{\alpha'} T^{\beta},$$

where $r_0$ is a constant, $X_i$ and $X_x$ are the mass fractions of the two particles, and $\alpha'$ and $\beta$ are determined from the power law expansion of the reaction rate equations. Usually $\alpha' = 2$ for a two-body collision, and $\beta$ can range from near unity to 40 or more.

   By combining the reaction rate equation with the amount of energy released per reaction, we can calculate the amount of energy released per second in each kilogram of stellar material. If $\mathcal{E}_0$ is the amount of energy released per reaction, the amount of energy liberated per kilogram of material per second becomes

$$\epsilon_{ix} = \left(\frac{\mathcal{E}_0}{\rho}\right) r_{ix},$$

or, in the form of a power law,

$$\epsilon_{ix} = \epsilon_0' X_i X_x \rho^{\alpha} T^{\beta}, \tag{10.35}$$

where $\alpha = \alpha' - 1$. $\epsilon_{ix}$ has units of W kg$^{-1}$ and the sum of $\epsilon_{ix}$ for all reactions is the total nuclear energy generation rate. This form of the nuclear energy generation rate will be used later to show the dependence of energy production on temperature and density for several reaction sequences typically operating in stellar interiors.

### The Luminosity Gradient Equation

To determine the luminosity of a star, we must now consider all of the energy generated by stellar material. The contribution to the total luminosity due to an infinitesimal mass $dm$ is simply

$$dL = \epsilon \, dm,$$

where $\epsilon$ is the *total* energy released per kilogram per second by all nuclear reactions and by gravity, or $\epsilon = \epsilon_{\text{nuclear}} + \epsilon_{\text{gravity}}$. It is worth noting that $\epsilon_{\text{gravity}}$ could be negative if the star is expanding, a point to be discussed later. For a spherically symmetric star, the mass of a thin shell of thickness $dr$ is just $dm = dM_r = \rho \, dV = 4\pi r^2 \rho \, dr$ (recall Fig. 10.2). Substituting and dividing by the shell thickness, we have

$$\frac{dL_r}{dr} = 4\pi r^2 \rho \epsilon, \tag{10.36}$$

where $L_r$ is the *interior luminosity* due to all of the energy generated within the star's interior out to the radius $r$. Equation (10.36) is another of the fundamental stellar structure equations.

## Stellar Nucleosynthesis and Conservation Laws

The remaining problem in understanding nuclear reactions is the exact sequence of steps by which one element is converted into another, a process known as **nucleosynthesis**. Our estimate of the nuclear timescale for the Sun was based on the assumption that four hydrogen nuclei are converted into helium. However, it is highly unlikely that this could occur via a four-body collision (i.e., all nuclei hitting simultaneously). For the process to occur, the final product must be created by a chain of reactions, each involving much more probable two-body interactions. In fact, we derived the reaction rate equation under the assumption that only two nuclei would collide at any one time.

The process by which a chain of nuclear reactions leads to the final product cannot happen in a completely arbitrary way, however; a series of particle *conservation laws* must be obeyed. In particular, during every reaction it is necessary to conserve electric charge, the number of nucleons, and the number of leptons. The term **lepton** means a "light thing" and includes electrons, positrons, neutrinos, and antineutrinos.

Although antimatter is extremely rare in comparison with matter, it plays an important role in subatomic physics, including nuclear reactions. Antimatter particles are identical to their matter counterparts but have opposite attributes, such as electric charge. Antimatter also has the characteristic (often used in science fiction) that a collision with its matter counterpart results in complete annihilation of both particles, accompanied by the production of energetic photons. For instance,

$$e^- + e^+ \rightarrow 2\gamma,$$

where $e^-$, $e^+$, and $\gamma$ denote an electron, positron, and photon, respectively. Note that two photons are required to conserve both momentum and energy simultaneously.

Neutrinos and antineutrinos (symbolized by $\nu$ and $\overline{\nu}$, respectively) are an interesting class of particles in their own right and will be discussed often in the remainder of this text.[9] Neutrinos are electrically neutral and have a very small but non-zero mass ($m_\nu < 2.2 \, \text{eV}/c^2$). One of the interesting characteristics of a neutrino is its extremely small cross section for interactions with other matter, making it very difficult to detect. Typically

---

[9]These particles were originally proposed by Wolfgang Pauli in 1930, in order that energy and momentum might be conserved in certain reaction processes. In 1934, they were given the name *neutrinos* ("little neutral ones") by Italian physicist Enrico Fermi (1901–1954).

$\sigma_\nu \sim 10^{-48}$ m$^2$, implying that at densities common to stellar interiors, a neutrino's mean free path is on the order of $10^{18}$ m $\sim 10$ pc, or nearly $10^9$ R$_\odot$! After being produced in the deep interior, neutrinos almost always successfully escape from the star. One exception to this transparency of stellar material to neutrinos occurs with important consequences during a supernova explosion, as will be discussed in Chapter 15.

Since electrons and positrons have charges equal in magnitude to that of a proton, these leptons will contribute to the charge conservation requirement while their total **lepton numbers** must also be conserved. Note that in counting the number of leptons involved in a nuclear reaction, we treat matter and antimatter differently. Specifically, the total number of matter leptons *minus* the total number of antimatter leptons must remain constant.

To assist in counting the number of nucleons and the total electric charge, nuclei will be represented in this text by the symbol

$$ {}_Z^A\text{X}, $$

where X is the chemical symbol of the element (H for hydrogen, He for helium, etc.), $Z$ is the number of protons (the total positive charge, in units of $e$), and $A$ is the mass number (the total number of nucleons, protons plus neutrons).[10]

### The Proton–Proton Chains

Applying the conservation laws, one chain of reactions that can convert hydrogen into helium is the first **proton–proton chain** (PP I). It involves a reaction sequence that ultimately results in

$$ 4\,{}_1^1\text{H} \rightarrow {}_2^4\text{He} + 2e^+ + 2\nu_e + 2\gamma $$

through the intermediate production of deuterium (${}_1^2\text{H}$) and helium-3 (${}_2^3\text{He}$). The entire **PP I** reaction chain is[11]

$$ {}_1^1\text{H} + {}_1^1\text{H} \rightarrow {}_1^2\text{H} + e^+ + \nu_e \tag{10.37} $$

$$ {}_1^2\text{H} + {}_1^1\text{H} \rightarrow {}_2^3\text{He} + \gamma \tag{10.38} $$

$$ {}_2^3\text{He} + {}_2^3\text{He} \rightarrow {}_2^4\text{He} + 2\,{}_1^1\text{H}. \tag{10.39} $$

Each step of the PP I chain has its own reaction rate, since different Coulomb barriers and cross sections are involved. The slowest step in the sequence is the initial one, because it involves the *decay* of a proton into a neutron via $p^+ \rightarrow n + e^+ + \nu_e$. Such a decay involves the **weak force**, another of the four known forces.[12]

---

[10]Since an element is uniquely determined by the number of protons ($Z$) in the nucleus, specifying both X and $Z$ is redundant. As a result, some texts use the less cumbersome notation $^A$X. However, this notation makes it more difficult to keep track of the electric charge in a nuclear reaction.

[11]Approximately 0.4% of the time, the first reaction step is accomplished by the so-called pep reaction: ${}_1^1\text{H} + e^- + {}_1^1\text{H} \rightarrow {}_1^2\text{H} + \nu_e$.

[12]Each of the four forces has now been mentioned: the gravitational force, which involves all particles with mass–energy; the electromagnetic force, associated with photons and electric charge; the strong force that binds nuclei together; and the weak force of radioactive beta (electron/positron) decay.

The production of helium-3 nuclei in the PP I chain also provides for the possibility of their interaction directly with helium-4 nuclei, resulting in a second branch of the proton–proton chain. In an environment characteristic of the center of the Sun, 69% of the time a helium-3 interacts with another helium-3 in the PP I chain, whereas 31% of the time the **PP II** chain occurs:

$$\ce{^3_2He} + \ce{^4_2He} \rightarrow \ce{^7_4Be} + \gamma \tag{10.40}$$

$$\ce{^7_4Be} + e^- \rightarrow \ce{^7_3Li} + \nu_e \tag{10.41}$$

$$\ce{^7_3Li} + \ce{^1_1H} \rightarrow 2\,\ce{^4_2He}. \tag{10.42}$$

Yet another branch, the **PP III** chain, is possible because the capture of an electron by the beryllium-7 nucleus in the PP II chain competes with the capture of a proton (a proton is captured only 0.3% of the time in the center of the Sun):

$$\ce{^7_4Be} + \ce{^1_1H} \rightarrow \ce{^8_5B} + \gamma \tag{10.43}$$

$$\ce{^8_5B} \rightarrow \ce{^8_4Be} + e^+ + \nu_e \tag{10.44}$$

$$\ce{^8_4Be} \rightarrow 2\,\ce{^4_2He}. \tag{10.45}$$

The three branches of the proton–proton (pp) chain, along with their branching ratios, are summarized in Fig. 10.8.

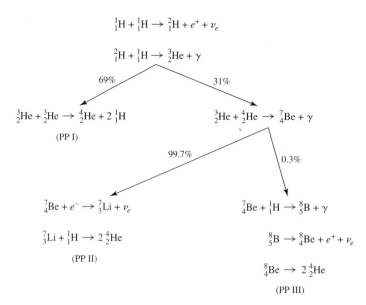

**FIGURE 10.8**    The three branches of the pp chain, along with the branching ratios appropriate for conditions in the core of the Sun.

Beginning with Eq. (10.33), the nuclear energy generation rate for the combined pp chain is calculated to be

$$\epsilon_{pp} = 0.241 \rho X^2 f_{pp} \psi_{pp} C_{pp} T_6^{-2/3} e^{-33.80 T_6^{-1/3}} \ \text{W kg}^{-1}, \qquad (10.46)$$

where $T_6$ is a dimensionless expression of temperature in units of $10^6$ K (or $T_6 \equiv T/10^6$ K). $f_{pp} = f_{pp}(X, Y, \rho, T) \simeq 1$ is the pp chain screening factor, $\psi_{pp} = \psi_{pp}(X, Y, T) \simeq 1$ is a correction factor that accounts for the simultaneous occurrence of PP I, PP II, and PP III, and $C_{pp} \simeq 1$ involves higher-order correction terms.[13]

When written as a power law (e.g., Eq. 10.35) near $T = 1.5 \times 10^7$ K, the energy generation rate has the form

$$\epsilon_{pp} \simeq \epsilon'_{0,pp} \rho X^2 f_{pp} \psi_{pp} C_{pp} T_6^4, \qquad (10.47)$$

where $\epsilon'_{0,pp} = 1.08 \times 10^{-12}$ W m$^3$ kg$^{-2}$. The power law form of the energy generation rate demonstrates a relatively modest temperature dependence of $T^4$ near $T_6 = 15$.

## The CNO Cycle

A second, independent cycle also exists for the production of helium-4 from hydrogen. This cycle was proposed by Hans Bethe (1906–2005) in 1938, just six years after the discovery of the neutron. In the **CNO cycle**, carbon, nitrogen, and oxygen are used as catalysts, being consumed and then regenerated during the process. Just as with the pp chain, the CNO cycle has competing branches. The first branch culminates with the production of carbon-12 and helium-4:

$$^{12}_{6}\text{C} + {}^{1}_{1}\text{H} \rightarrow {}^{13}_{7}\text{N} + \gamma \qquad (10.48)$$

$$^{13}_{7}\text{N} \rightarrow {}^{13}_{6}\text{C} + e^+ + \nu_e \qquad (10.49)$$

$$^{13}_{6}\text{C} + {}^{1}_{1}\text{H} \rightarrow {}^{14}_{7}\text{N} + \gamma \qquad (10.50)$$

$$^{14}_{7}\text{N} + {}^{1}_{1}\text{H} \rightarrow {}^{15}_{8}\text{O} + \gamma \qquad (10.51)$$

$$^{15}_{8}\text{O} \rightarrow {}^{15}_{7}\text{N} + e^+ + \nu_e \qquad (10.52)$$

$$^{15}_{7}\text{N} + {}^{1}_{1}\text{H} \rightarrow {}^{12}_{6}\text{C} + {}^{4}_{2}\text{He}. \qquad (10.53)$$

The second branch occurs only about 0.04% of the time and arises when the last reaction (Eq. 10.53) produces oxygen-16 and a photon, rather than carbon-12 and helium-4:

$$^{15}_{7}\text{N} + {}^{1}_{1}\text{H} \rightarrow {}^{16}_{8}\text{O} + \gamma \qquad (10.54)$$

$$^{16}_{8}\text{O} + {}^{1}_{1}\text{H} \rightarrow {}^{17}_{9}\text{F} + \gamma \qquad (10.55)$$

$$^{17}_{9}\text{F} \rightarrow {}^{17}_{8}\text{O} + e^+ + \nu_e \qquad (10.56)$$

$$^{17}_{8}\text{O} + {}^{1}_{1}\text{H} \rightarrow {}^{14}_{7}\text{N} + {}^{4}_{2}\text{He}. \qquad (10.57)$$

[13]Expressions for the various correction terms are given in the stellar structure code StatStar, described in Appendix L.

The energy generation rate for the CNO cycle is given by

$$\epsilon_{CNO} = 8.67 \times 10^{20} \rho X X_{CNO} C_{CNO} T_6^{-2/3} e^{-152.28 T_6^{-1/3}} \text{ W kg}^{-1}, \tag{10.58}$$

where $X_{CNO}$ is the total mass fraction of carbon, nitrogen, and oxygen, and $C_{CNO}$ is a higher-order correction term. When written as a power law centered about $T = 1.5 \times 10^7$ K (see Eq. 10.35), this energy equation becomes

$$\epsilon_{CNO} \simeq \epsilon'_{0,CNO} \rho X X_{CNO} T_6^{19.9}, \tag{10.59}$$

where $\epsilon'_{0,CNO} = 8.24 \times 10^{-31}$ W m$^3$ kg$^{-2}$. As shown by the power law dependence, the CNO cycle is much more strongly temperature-dependent than is the pp chain. This property implies that low-mass stars, which have smaller central temperatures, are dominated by the pp chains during their "hydrogen burning" evolution, whereas more massive stars, with their higher central temperatures, convert hydrogen to helium by the CNO cycle. The transition in stellar mass between stars dominated by the pp chain and those dominated by the CNO cycle occurs for stars slightly more massive than our Sun. This difference in nuclear reaction processes plays an important role in the structure of stellar interiors, as will be seen in the next section.

When hydrogen is converted into helium by either the pp chain or the CNO cycle, the mean molecular weight $\mu$ of the gas increases. If neither the temperature nor the density of the gas changes, the ideal gas law predicts that the central pressure will necessarily decrease. As a result, the star would no longer be in hydrostatic equilibrium and would begin to collapse. This collapse has the effect of actually raising both the temperature and the density to compensate for the increase in $\mu$ (recall the virial theorem, Eq. 2.46). When the temperature and density become sufficiently high, helium nuclei can overcome their Coulomb repulsion and begin to "burn."

### The Triple Alpha Process of Helium Burning

The reaction sequence by which helium is converted into carbon is known as the **triple alpha process**. The process takes its name from the historical result that the mysterious alpha particles detected in some types of radioactive decay were shown by Rutherford to be helium-4 ($^4_2$He) nuclei. The triple alpha process is

$$^4_2\text{He} + {}^4_2\text{He} \rightleftharpoons {}^8_4\text{Be} \tag{10.60}$$

$$^8_4\text{Be} + {}^4_2\text{He} \rightarrow {}^{12}_6\text{C} + \gamma. \tag{10.61}$$

In the triple alpha process, the first step produces an unstable beryllium nucleus that will rapidly decay back into two separate helium nuclei if not immediately struck by another alpha particle. As a result, this reaction may be thought of as a three-body interaction, and therefore, the reaction rate depends on $(\rho Y)^3$. The nuclear energy generation rate is given by

$$\epsilon_{3\alpha} = 50.9 \rho^2 Y^3 T_8^{-3} f_{3\alpha} e^{-44.027 T_8^{-1}} \text{ W kg}^{-1}, \tag{10.62}$$

where $T_8 \equiv T/10^8$ K and $f_{3\alpha}$ is the screening factor for the triple alpha process. Written as a power law centered on $T = 10^8$ K (see Eq. 10.35), it demonstrates a very dramatic temperature dependence:

$$\epsilon_{3\alpha} \simeq \epsilon'_{0,3\alpha} \rho^2 Y^3 f_{3\alpha} T_8^{41.0}. \tag{10.63}$$

With such a strong dependence, even a small increase in temperature will produce a large increase in the amount of energy generated per second. For instance, an increase of only 10% in temperature raises the energy output rate by more than 50 times!

### Carbon and Oxygen Burning

In the high-temperature environment of helium burning, other competing processes are also at work. After sufficient carbon has been generated by the triple alpha process, it becomes possible for carbon nuclei to capture alpha particles, producing oxygen. Some of the oxygen in turn can capture alpha particles to produce neon.

$$^{12}_{6}\text{C} + ^{4}_{2}\text{He} \rightarrow ^{16}_{8}\text{O} + \gamma \tag{10.64}$$

$$^{16}_{8}\text{O} + ^{4}_{2}\text{He} \rightarrow ^{20}_{10}\text{Ne} + \gamma \tag{10.65}$$

At helium-burning temperatures, the continued capture of alpha particles leading to progressively more massive nuclei quickly becomes prohibitive due to the ever higher Coulomb barrier.

If a star is sufficiently massive, still higher central temperatures can be obtained and many other nuclear products become possible. Examples of available reactions include carbon burning reactions near $6 \times 10^8$ K,

$$^{12}_{6}\text{C} + ^{12}_{6}\text{C} \rightarrow
\begin{cases}
^{16}_{8}\text{O} + 2\,^{4}_{2}\text{He} & *** \\
^{20}_{10}\text{Ne} + ^{4}_{2}\text{He} \\
^{23}_{11}\text{Na} + p^+ \\
^{23}_{12}\text{Mg} + n & *** \\
^{24}_{12}\text{Mg} + \gamma
\end{cases}
\tag{10.66}$$

and oxygen burning reactions near $10^9$ K,

$$^{16}_{8}\text{O} + ^{16}_{8}\text{O} \rightarrow
\begin{cases}
^{24}_{12}\text{Mg} + 2\,^{4}_{2}\text{He} & *** \\
^{28}_{14}\text{Si} + ^{4}_{2}\text{He} \\
^{31}_{15}\text{P} + p^+ \\
^{31}_{16}\text{S} + n \\
^{32}_{16}\text{S} + \gamma
\end{cases}
\tag{10.67}$$

Reactions marked by *** are ones for which energy is absorbed rather than released and are referred to as being **endothermic**; energy-releasing reactions are **exothermic**. In endothermic reactions the product nucleus actually possesses more energy per nucleon than did the nuclei from which it formed. Such reactions occur at the expense of the energy released by exothermic reactions or by gravitational collapse (the virial theorem). In general, endothermic reactions are much less likely to occur than exothermic reactions under conditions that normally prevail in stellar interiors.

### The Binding Energy Per Nucleon

A useful quantity in understanding the energy release in nuclear reactions is the binding energy per nucleon, $E_b/A$, where

$$E_b = \Delta mc^2 = \left[ Zm_p + (A - Z)m_n - m_{\text{nucleus}} \right] c^2.$$

Figure 10.9 shows $E_b/A$ versus the mass number. It is apparent that for relatively small values of $A$ (less than 56), several nuclei have abnormally high values of $E_b/A$ relative to others of similar mass. Among these unusually stable nuclei are $^4_2$He and $^{16}_8$O, which, along with $^1_1$H, are the most abundant nuclei in the universe. This unusual stability arises from an inherent shell structure of the nucleus, analogous to the shell structure of atomic energy levels that accounts for the chemical nature of elements. These unusually stable nuclei are called *magic nuclei*.

It is believed that shortly after the Big Bang the early universe was composed primarily of hydrogen and helium, with no heavy elements. Today, Earth and its inhabitants contain an abundance of heavier metals. The study of stellar nucleosynthesis strongly suggests that

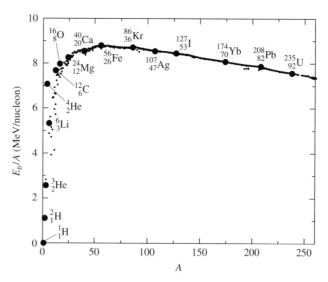

**FIGURE 10.9**   The binding energy per nucleon, $E_b/A$, as a function of mass number, $A$. Notice that several nuclei, most notably $^4_2$He (see also $^{12}_6$C and $^{16}_8$O), lie well above the general trend of the other nuclei, indicating unusual stability. At the peak of the curve is $^{56}_{26}$Fe, the most stable of all nuclei.

these heavier nuclei were generated in the interiors of stars. It can be said that we are all "star dust," the product of heavy element generation within previous generations of stars.

Another important feature of Fig. 10.9 is the broad peak around $A = 56$. At the top of the peak is an isotope of iron, $^{56}_{26}\text{Fe}$, the most stable of all nuclei. As successively more massive nuclei are created in stellar interiors, the iron peak is approached from the left. These fusion reactions result in the liberation of energy.[14] Consequently, the ultimate result of successive chains of nuclear reactions within stars is the production of iron, assuming sufficient energy is available to overcome the Coulomb barrier. If a star is massive enough to create the central temperatures and densities necessary to produce iron, the results are spectacular, as we will see in Chapter 15.

Considering what we have learned in this section about stellar nucleosynthesis, it should come as no surprise that the most abundant nuclear species in the cosmos are, in order, $^1_1\text{H}$, $^4_2\text{He}$, $^{16}_8\text{O}$, $^{12}_6\text{C}$, $^{20}_{10}\text{Ne}$, $^{14}_7\text{N}$, $^{24}_{12}\text{Mg}$, $^{28}_{14}\text{Si}$, and $^{56}_{26}\text{Fe}$ (recall Table 9.2).[15] The abundances are the result of the dominant nuclear reaction processes that occur in stars, together with the nuclear configurations that result in the most stable nuclei. Reactions capable of producing the diversity of other isotopes are discussed in Section 15.3.

## 10.4 ■ ENERGY TRANSPORT AND THERMODYNAMICS

One stellar structure equation remains to be developed. We have already related the fundamental quantities $P$, $M$, and $L$ to the independent variable $r$ through differential equations that describe hydrostatic equilibrium, mass conservation, and energy generation, respectively; see Eqs. (10.6), (10.7), and (10.36). However, we have not yet found a differential equation relating the basic parameter of temperature, $T$, to $r$. Moreover, we have not explicitly developed equations that describe the processes by which heat generated via nuclear reactions or gravitational contraction is carried from the deep interior to the surface of the star.

### Three Energy Transport Mechanisms

Three different energy transport mechanisms operate in stellar interiors. **Radiation** allows the energy produced by nuclear reactions and gravitation to be carried to the surface via photons, the photons being absorbed and re-emitted in nearly random directions as they encounter matter (recall the discussion in Section 9.3). This suggests that the opacity of the material must play an important role, as one would expect. **Convection** can be a very efficient transport mechanism in many regions of a star, with hot, buoyant mass elements carrying excess energy outward while cool elements fall inward. Finally, **conduction** transports heat via collisions between particles. Although conduction can play an important role in some stellar environments, it is generally insignificant in most stars throughout the majority of their lifetimes and will not be discussed further here.

---

[14]Energy is also released when the peak is approached from the right via fission reactions that produce nuclei of smaller mass, again resulting in more stable nuclei. This type of reaction process is important in the fission reactors of nuclear power plants.

[15]The relative abundances of the elements will be discussed further in Section 15.3; see Fig. 15.16.

## The Radiative Temperature Gradient

First consider radiation transport. In Chapter 9 we found that the radiation pressure gradient is given by Eq. (9.31),

$$\frac{dP_{rad}}{dr} = -\frac{\overline{\kappa}\rho}{c} F_{rad},$$

where $F_{rad}$ is the outward radiative flux. However, from Eq. (10.19), the radiation pressure gradient may also be expressed as

$$\frac{dP_{rad}}{dr} = \frac{4}{3} a T^3 \frac{dT}{dr}.$$

Equating the two expressions, we have

$$\frac{dT}{dr} = -\frac{3}{4ac} \frac{\overline{\kappa}\rho}{T^3} F_{rad}.$$

Finally, if we use the expression for the radiative flux (Eq. 3.2), written in terms of the local radiative luminosity of the star at radius $r$,

$$F_{rad} = \frac{L_r}{4\pi r^2},$$

the temperature gradient for radiative transport becomes

$$\boxed{\frac{dT}{dr} = -\frac{3}{4ac} \frac{\overline{\kappa}\rho}{T^3} \frac{L_r}{4\pi r^2}.} \tag{10.68}$$

As either the flux or the opacity increases, the temperature gradient must become steeper (more negative) if radiation is to transport all of the required luminosity outward. The same situation holds as the density increases or the temperature decreases.

## The Pressure Scale Height

If the temperature gradient becomes too steep, convection can begin to play an important role in the transport of energy. Physically, convection involves mass motions: hot parcels of matter move upward as cooler, denser parcels sink. Unfortunately, convection is a much more complex phenomenon than radiation at the macroscopic level. In fact, no truly satisfactory prescription yet exists to describe it adequately in stellar environments. *Fluid mechanics*, the field of physics describing the motion of gases and liquids, relies on a complicated set of three-dimensional equations known as the *Navier–Stokes equations*. However, at present, due in large part to the current limitations in computing power,[16] most stellar structure codes are one-dimensional (i.e., depend on $r$ only). It becomes necessary, therefore, to approximate an explicitly three-dimensional process by a one-dimensional phenomenological

---

[16]This limitation is being overcome to some extent with the development of ever faster computers with more memory, and through the implementation of more sophisticated numerical techniques.

theory. To complicate the situation even more, when convection is present in a star, it is generally quite turbulent, requiring a detailed understanding of the amount of viscosity (fluid friction) and heat dissipation involved. Also, a characteristic length scale for convection, typically referred to in terms of the **pressure scale height**, is often comparable to the size of the star. Lastly, the timescale for convection, taken to be the amount of time required for a convective element to travel a characteristic distance, is in some cases approximately equal to the timescale for changes in the structure of the star, implying that convection is strongly coupled to the star's dynamic behavior. The impact of these complications on the behavior of the star is not yet fundamentally understood.

The situation is not completely hopeless, however. Despite the difficulties encountered in attempting to treat stellar convection exactly, approximate (and even reasonable) results can usually be obtained. To estimate the size of a convective region in a star, consider the pressure scale height, $H_P$, defined as

$$\frac{1}{H_P} \equiv -\frac{1}{P}\frac{dP}{dr}. \tag{10.69}$$

If we assume for the moment that $H_P$ is a constant, we can solve for the variation of pressure with radius, giving

$$P = P_0 e^{-r/H_P}.$$

Obviously, if $r = H_P$, then $P = P_0 e^{-1}$, so that $H_P$ is the distance over which the gas pressure decreases by a factor of $e$. To find a convenient general expression for $H_P$, recall that from the equation for hydrostatic equilibrium (Eq. 10.6), $dP/dr = -\rho g$, where $g = GM_r/r^2$ is the local acceleration of gravity. Substituting into Eq. (10.69), the pressure scale height is simply

$$H_P = \frac{P}{\rho g}. \tag{10.70}$$

---

**Example 10.4.1.**   To estimate a typical value for the pressure scale height in the Sun, assume that $\overline{P} = P_c/2$, where $P_c$ is the central pressure, $\overline{\rho}_\odot$ is the average solar density, and

$$\overline{g} = \frac{G(M_\odot/2)}{(R_\odot/2)^2} = 550 \text{ m s}^{-2}.$$

Then we have

$$H_P \simeq 1.8 \times 10^8 \text{ m} \sim R_\odot/4.$$

Detailed calculations show that $H_P \sim R_\odot/10$ is more typical.

---

### Internal Energy and the First Law of Thermodynamics

Understanding convective heat transport in stars, even in an approximate way, begins with some knowledge of thermodynamics. In the study of heat transport, conservation of energy is expressed by **the first law of thermodynamics,**

$$dU = dQ - dW,$$ (10.71)

where the change in the internal energy of a mass element $dU$ is given by the amount of heat *added* to that element, $dQ$, minus the work done *by* that element on its surroundings, $dW$. Throughout our discussion we will assume that these energy changes are measured *per unit mass*.

The internal energy of a system $U$ is a **state function**, meaning that its value depends only on the present conditions of the gas, not on the history of any changes leading to its current state. Consequently, $dU$ is independent of the actual process involved in the change. On the other hand, neither heat nor work is a state function. The amount of heat added to a system or the amount of work done by a system depends on the ways in which the processes are carried out. $dQ$ and $dW$ are referred to as *inexact differentials*, reflecting their path dependence.

Consider an ideal monatomic gas, a gas composed of single particles with no ionization. The total internal energy per unit mass is given by

$$U = (\text{average energy/particle}) \times (\text{number of particles/mass})$$

$$= \overline{K} \times \frac{1}{\overline{m}}$$

where $\overline{m} = \mu m_H$ is the average mass of a single particle in the gas. For an ideal gas, $\overline{K} = 3kT/2$ and the internal energy is given by

$$U = \frac{3}{2}\left(\frac{k}{\mu m_H}\right)T = \frac{3}{2}nRT,$$ (10.72)

where $n$ is the number of moles[17] *per unit mass*, $R = 8.314472\,\text{J mole}^{-1}\,\text{K}^{-1}$ is the universal gas constant,[18] and

$$nR = \frac{k}{\mu m_H}.$$

Clearly $U = U(\mu, T)$ is a function of the composition of the gas and its temperature. In this case of an ideal monatomic gas, the internal energy is just the kinetic energy per unit mass.

## Specific Heats

The change in heat of the mass element $dQ$ is generally expressed in terms of the **specific heat** $C$ of the gas. The specific heat is defined as the amount of heat required to raise the

---

[17] 1 mole $= N_A$ particles, where $N_A = 6.02214199 \times 10^{23}$ is Avogadro's number, defined as the number of $^{12}_{6}\text{C}$ atoms required to produce exactly 12 grams of a pure sample.
[18] $R = N_A k$.

temperature of a unit mass of a material by a unit temperature interval, or

$$C_P \equiv \left. \frac{\partial Q}{\partial T} \right|_P \qquad \text{and} \qquad C_V \equiv \left. \frac{\partial Q}{\partial T} \right|_V ,$$

where $C_P$ and $C_V$ are the specific heats at constant pressure and volume, respectively.

Consider next the amount of work per unit mass, $dW$, done by the gas on its surroundings. Suppose that a cylinder of cross-sectional area $A$ is filled with a gas of mass $m$ and pressure $P$. The gas then exerts a force $F = PA$ on an end of the cylinder. If the end of the cylinder is a piston that moves through a distance $dr$, the work *per unit mass* performed by the gas may be expressed as

$$dW = \left( \frac{F}{m} \right) dr = \left( \frac{PA}{m} \right) dr = P \, dV,$$

$V$ being defined as the **specific volume**, the volume *per unit mass*, or $V \equiv 1/\rho$. The first law of thermodynamics may now be expressed in the useful form

$$dU = dQ - P \, dV. \tag{10.73}$$

At constant volume, $dV = 0$, which gives $dU = dU|_V = dQ|_V$, or

$$dU = \left. \frac{\partial Q}{\partial T} \right|_V dT = C_V \, dT. \tag{10.74}$$

[It is important to note that because $dU$ is independent of any specific process, the second equality of Eq. (10.74) is always valid, regardless of the type of thermodynamic process involved.] But from Eq. (10.72), $dU = (3nR/2) \, dT$ for a monatomic gas. Thus

$$C_V = \frac{3}{2} n R. \tag{10.75}$$

To find $C_P$ for a monatomic gas, note that

$$dU = \left. \frac{\partial Q}{\partial T} \right|_P dT - P \left. \frac{\partial V}{\partial T} \right|_P dT. \tag{10.76}$$

In addition, from Eq. (10.11), the ideal gas law can be written as

$$PV = nRT. \tag{10.77}$$

Considering all possible differential changes in quantities in Eq. (10.77), we find that

$$P \, dV + V \, dP = RT \, dn + nR \, dT \tag{10.78}$$

(recall that $R$ is a constant). For constant $P$ and $n$, Eq. (10.78) implies that $P \, dV/dT = nR$. Substituting this result into Eq. (10.76) along with $dU = C_V \, dT$ and the definition of $C_P$, we arrive at

$$C_P = C_V + nR. \tag{10.79}$$

Equation (10.79) is valid for all situations for which the ideal gas law applies.

Define the parameter $\gamma$ to be the ratio of specific heats, or

$$\gamma \equiv \frac{C_P}{C_V}. \tag{10.80}$$

For a monatomic gas, we see that $\gamma = 5/3$. If ionization occurs, some of the heat that would normally go into increasing the average kinetic energy of the particles must go into ionizing the atoms instead. Therefore the temperature of the gas, which is a measure of its internal energy, will not rise as rapidly, implying larger values for the specific heats in a partial ionization zone. As both $C_P$ and $C_V$ increase, $\gamma$ approaches unity.[19]

### The Adiabatic Gas Law

Since the change in internal energy is independent of the process involved, consider the special case of an **adiabatic process** ($dQ = 0$) for which no heat flows into or out of the mass element. Then the first law of thermodynamics (Eq. 10.73) becomes

$$dU = -P\,dV.$$

However, from Eq. (10.78) with constant $n$,

$$P\,dV + V\,dP = nR\,dT.$$

Also, since $dU = C_V\,dT$, we have

$$dT = \frac{dU}{C_V} = -\frac{P\,dV}{C_V}.$$

Combining these results yields

$$P\,dV + V\,dP = -\left(\frac{nR}{C_V}\right)P\,dV,$$

which may be rewritten by using Eqs. (10.79) and (10.80), to give

$$\gamma \frac{dV}{V} = -\frac{dP}{P}. \tag{10.81}$$

Solving this differential equation leads to the adiabatic gas law,

$$PV^{\gamma} = K, \tag{10.82}$$

where $K$ is a constant. Using the ideal gas law, a second adiabatic relation may be obtained:

$$P = K'T^{\gamma/(\gamma-1)}, \tag{10.83}$$

where $K'$ is another constant. Because of its special role in Eqs. (10.82) and (10.83), $\gamma$ is often referred to as the "adiabatic gamma," specifying a particularly simple equation of state.

---

[19]The variation of $\gamma$ also plays an important role in the dynamic stability of stars. This factor will be discussed further in Section 14.3.

## The Adiabatic Sound Speed

Using the results obtained thus far, we can now calculate a sound speed through the material. The sound speed is related to the compressibility of the gas and to its inertia (represented by density) and is given by

$$v_s = \sqrt{B/\rho},$$

where $B \equiv -V(\partial P/\partial V)_{\text{ad}}$ is the bulk modulus of the gas.[20] The bulk modulus describes how much the volume of the gas will change with changing pressure. From Eq. (10.81), the adiabatic sound speed becomes

$$v_s = \sqrt{\gamma P/\rho}. \tag{10.84}$$

---

**Example 10.4.2.** Assuming a monatomic gas, a typical adiabatic sound speed for the Sun is

$$\overline{v}_s \simeq \left( \frac{5}{3} \frac{\overline{P}}{\rho_\odot} \right)^{1/2} \simeq 4 \times 10^5 \text{ m s}^{-1},$$

where $\overline{P} \sim P_c/2$ was assumed. The amount of time needed for a sound wave to traverse the radius of the Sun would then be

$$t \simeq R_\odot/\overline{v}_s \simeq 29 \text{ minutes}.$$

---

## The Adiabatic Temperature Gradient

Returning now to the specific problem of describing convection, we first consider the situation where a hot convective bubble of gas rises and expands *adiabatically*, meaning that the bubble does not exchange heat with its surroundings. After it has traveled some distance, it finally *thermalizes*, giving up any excess heat as it loses its identity and dissolves into the surrounding gas. Differentiating the ideal gas law (Eq. 10.11) yields an expression involving the bubble's *temperature gradient* (how the bubble's temperature changes with position):

$$\frac{dP}{dr} = -\frac{P}{\mu}\frac{d\mu}{dr} + \frac{P}{\rho}\frac{d\rho}{dr} + \frac{P}{T}\frac{dT}{dr}. \tag{10.85}$$

Using the adiabatic relationship between pressure and density (Eq. 10.82), and recalling that $V \equiv 1/\rho$ is the specific volume, we have

$$P = K\rho^\gamma. \tag{10.86}$$

---

[20]Formally, the bulk modulus, and therefore the sound speed, must be defined in terms of a process by which pressure varies with volume. Since sound waves typically propagate through a medium too quickly for a significant amount of heat to enter or leave a mass element in the gas, we usually assume that the process is adiabatic.

Differentiating and rewriting, we obtain

$$\frac{dP}{dr} = \gamma \frac{P}{\rho} \frac{d\rho}{dr}. \tag{10.87}$$

If we assume for simplicity that $\mu$ is a constant, Eqs. (10.85) and (10.87) may be combined to give *the adiabatic temperature gradient* (designated by the subscript ad)

$$\left. \frac{dT}{dr} \right|_{ad} = \left( 1 - \frac{1}{\gamma} \right) \frac{T}{P} \frac{dP}{dr}. \tag{10.88}$$

Using Eq. (10.6) and the ideal gas law, we finally obtain

$$\boxed{ \left. \frac{dT}{dr} \right|_{ad} = - \left( 1 - \frac{1}{\gamma} \right) \frac{\mu m_H}{k} \frac{G M_r}{r^2}. } \tag{10.89}$$

It is sometimes helpful to express Eq. (10.89) in another, equivalent form. Recalling that $g = G M_r / r^2$, $k / \mu m_H = nR$, $\gamma = C_P / C_V$, and $C_P - C_V = nR$, and that $n$, $C_P$, and $C_V$ are per unit mass, we have

$$\left. \frac{dT}{dr} \right|_{ad} = - \frac{g}{C_P}. \tag{10.90}$$

This result describes how the temperature of the gas *inside* the bubble changes as the bubble rises and expands adiabatically.

If the star's *actual temperature gradient* (designated by the subscript act) is *steeper* than the adiabatic temperature gradient given in Eq. (10.89), or

$$\left| \frac{dT}{dr} \right|_{act} > \left| \frac{dT}{dr} \right|_{ad},$$

the temperature gradient is said to be **superadiabatic** (recall that $dT/dr < 0$). It will be shown that in the deep interior of a star, if $|dT/dr|_{act}$ is just *slightly* larger than $|dT/dr|_{ad}$, this may be sufficient to carry nearly all of the luminosity by convection. Consequently, it is often the case that either radiation or convection dominates the energy transport in the deep interiors of stars, while the other energy transport mechanism contributes very little to the total energy outflow. The particular mechanism in operation is determined by the temperature gradient. However, near the surface of the star the situation is much more complicated: Both radiation and convection can carry significant amounts of energy simultaneously.

## A Criterion for Stellar Convection

Just what condition must be met if convection is to dominate over radiation in the deep interior? When will a hot bubble of gas continue to rise rather than sink back down after being displaced upward? Figure 10.10 shows a convective bubble traveling a distance $dr$ through the surrounding medium. According to Archimedes's principle, if the initial density

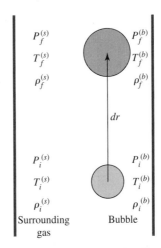

**FIGURE 10.10** A convective bubble traveling outward a distance $dr$. The initial conditions of the bubble are given by $P_i^{(b)}$, $T_i^{(b)}$, and $\rho_i^{(b)}$, for the pressure, temperature, and density, respectively, while the initial conditions for the surrounding gas at the same level are designated by $P_i^{(s)}$, $T_i^{(s)}$, and $\rho_i^{(s)}$, respectively. Final conditions for either the bubble or the surrounding gas are indicated by an $f$ subscript.

of the bubble is less than that of its surroundings ($\rho_i^{(b)} < \rho_i^{(s)}$), it will begin to rise. Now, the buoyant force *per unit volume* exerted on a bubble that is totally submersed in a fluid of density $\rho_i^{(s)}$ is given by

$$f_B = \rho_i^{(s)} g.$$

If we subtract the downward gravitational force per unit volume on the bubble, given by

$$f_g = \rho_i^{(b)} g,$$

the net force per unit volume on the bubble becomes

$$f_{\text{net}} = -g\,\delta\rho, \tag{10.91}$$

where $\delta\rho \equiv \rho_i^{(b)} - \rho_i^{(s)} < 0$ initially. If, after traveling an infinitesimal distance $dr$, the bubble now has a greater density than the surrounding material ($\rho_f^{(b)} > \rho_f^{(s)}$), it will sink again and convection will be prohibited. On the other hand, if $\rho_f^{(b)} < \rho_f^{(s)}$, the bubble will continue to rise and convection will result.

To express this condition in terms of temperature gradients, assume that the gas is initially *very nearly in thermal equilibrium*, with $T_i^{(b)} \simeq T_i^{(s)}$ and $\rho_i^{(b)} \simeq \rho_i^{(s)}$. Also assume that the bubble expands adiabatically and that the bubble and surrounding gas pressures are equal at all times, $P_f^{(b)} = P_f^{(s)}$. Now, since it is assumed that the bubble has moved an infinitesimal distance, it is possible to express the final quantities in terms of the initial quantities and their gradients by using a Taylor expansion. To first order,

$$\rho_f^{(b)} \simeq \rho_i^{(b)} + \left.\frac{d\rho}{dr}\right|^{(b)} dr \quad \text{and} \quad \rho_f^{(s)} \simeq \rho_i^{(s)} + \left.\frac{d\rho}{dr}\right|^{(s)} dr.$$

If the densities inside and outside the bubble remain nearly equal (as is usually the case except near the surfaces of some stars), substituting these results into the convection condition, $\rho_f^{(b)} < \rho_f^{(s)}$, gives

$$\left.\frac{d\rho}{dr}\right|^{(b)} < \left.\frac{d\rho}{dr}\right|^{(s)}. \tag{10.92}$$

We now want to express this solely in terms of quantities for the surroundings. Using Eq. (10.87) for the adiabatically rising bubble to rewrite the left-hand side of (10.92) and using Eq. (10.85) to rewrite the right-hand side (again assuming $d\mu/dr = 0$), we find

$$\frac{1}{\gamma} \frac{\rho_i^{(b)}}{P_i^{(b)}} \left.\frac{dP}{dr}\right|^{(b)} < \frac{\rho_i^{(s)}}{P_i^{(s)}} \left[ \left.\frac{dP}{dr}\right|^{(s)} - \frac{P_i^{(s)}}{T_i^{(s)}} \left.\frac{dT}{dr}\right|^{(s)} \right].$$

Recalling that $P^{(b)} = P^{(s)}$ at all times, it is necessary that

$$\left.\frac{dP}{dr}\right|^{(b)} = \left.\frac{dP}{dr}\right|^{(s)} = \frac{dP}{dr},$$

where the superscripts on the pressure gradient have been shown to be redundant. Substituting, and canceling equivalent initial conditions,

$$\frac{1}{\gamma} \frac{dP}{dr} < \frac{dP}{dr} - \frac{P_i^{(s)}}{T_i^{(s)}} \left.\frac{dT}{dr}\right|^{(s)}.$$

Dropping subscripts for initial conditions and superscripts designating the surrounding material, we arrive at the requirement

$$\left(\frac{1}{\gamma} - 1\right) \frac{dP}{dr} < -\frac{P}{T} \left.\frac{dT}{dr}\right|_{\text{act}}, \tag{10.93}$$

where the temperature gradient is the *actual* temperature gradient of the surrounding gas. Multiplying by the negative quantity $-T/P$ requires that the direction of the inequality be reversed, giving

$$\left(1 - \frac{1}{\gamma}\right) \frac{T}{P} \frac{dP}{dr} > \left.\frac{dT}{dr}\right|_{\text{act}}.$$

But from Eq. (10.88), we see that the left-hand side of the inequality is just the adiabatic temperature gradient. Thus

$$\left.\frac{dT}{dr}\right|_{\text{ad}} > \left.\frac{dT}{dr}\right|_{\text{act}}$$

is the condition for the gas bubble to keep rising. Finally, since $dT/dr < 0$ (the temperature decreases as the stellar radius increases), taking the absolute value of the equation again requires that the direction of the inequality be reversed, or

$$\left| \frac{dT}{dr} \right|_{\text{act}} > \left| \frac{dT}{dr} \right|_{\text{ad}}. \tag{10.94}$$

If the actual temperature gradient is superadiabatic, convection will result, assuming that $\mu$ does not vary.

Equation (10.93) may be used to find another useful, and equivalent, condition for convection. Since $dT/dr < 0$ and $1/\gamma - 1 < 0$ (recall that $\gamma > 1$),

$$\frac{T}{P} \left( \frac{dT}{dr} \right)^{-1} \frac{dP}{dr} < -\frac{1}{\gamma^{-1} - 1},$$

which may be simplified to give

$$\frac{T}{P} \frac{dP}{dT} < \frac{\gamma}{\gamma - 1},$$

or, for convection to occur,

$$\frac{d \ln P}{d \ln T} < \frac{\gamma}{\gamma - 1}. \tag{10.95}$$

For an ideal monatomic gas, $\gamma = 5/3$ and convection will occur in some region of a star when $d \ln P/d \ln T < 2.5$. In that case the temperature gradient $(dT/dr)$ is given approximately by Eq. (10.89). When $d \ln P/d \ln T > 2.5$, the region is stable against convection and $dT/dr$ is given by Eq. (10.68).

By comparing Eq. (10.68) for the radiative temperature gradient with either Eq. (10.89) or Eq. (10.90), together with the condition for convection written in terms of the temperature gradient, Eq. (10.94), it is possible to develop some understanding of which conditions are likely to lead to convection over radiation. In general, convection will occur when (1) the stellar opacity is large, implying that an unachievably steep temperature gradient $(|dT/dr|_{\text{act}})$ would be necessary for radiative transport, (2) a region exists where ionization is occurring, causing a large specific heat and a low adiabatic temperature gradient $(|dT/dr|_{\text{ad}})$, and (3) the temperature dependence of the nuclear energy generation rate is large, causing a steep radiative flux gradient and a large temperature gradient. In the atmospheres of many stars, the first two conditions can occur simultaneously, whereas the third condition would occur only deep in stellar interiors. In particular, the third condition can occur when the highly temperature-dependent CNO cycle or triple alpha processes are occurring.

## The Mixing-Length Theory of Superadiabatic Convection

It has already been suggested that the temperature gradient must be only slightly superadiabatic in the deep interior in order for convection to carry most of the energy. We will now justify that assertion.

We begin by returning to the fundamental criterion for convection, $\rho_f^{(b)} < \rho_f^{(s)}$. Since the pressure of the bubble and that of its surroundings are always equal, the ideal gas law implies that $T_f^{(b)} > T_f^{(s)}$, assuming thermal equilibrium initially. Therefore, the temperature of the surrounding gas must decrease more rapidly with radius, so

$$\left. \frac{dT}{dr} \right|^{(s)} - \left. \frac{dT}{dr} \right|^{(b)} > 0$$

is required for convection. Since the temperature gradients are negative, we have

$$\left. \frac{dT}{dr} \right|^{(b)} - \left. \frac{dT}{dr} \right|^{(s)} > 0.$$

Assuming that the bubble moves essentially adiabatically, and designating the temperature gradient of the surroundings as the actual average temperature gradient of the star, let

$$\left. \frac{dT}{dr} \right|^{(b)} = \left. \frac{dT}{dr} \right|_{ad} \quad \text{and} \quad \left. \frac{dT}{dr} \right|^{(s)} = \left. \frac{dT}{dr} \right|_{act}.$$

After the bubble travels a distance $dr$, its temperature will exceed the temperature of the surrounding gas by[21]

$$\delta T = \left( \left. \frac{dT}{dr} \right|_{ad} - \left. \frac{dT}{dr} \right|_{act} \right) dr = \delta \left( \frac{dT}{dr} \right) dr. \qquad (10.96)$$

We use $\delta$ here to indicate the difference between the value of a quantity associated with the bubble and the same quantity associated with the surroundings, both determined at a specified radius $r$, just as was done for Eq. (10.91).

Now assume that a hot, rising bubble travels some distance

$$\ell = \alpha H_P$$

before dissipating, at which point it thermalizes with its surroundings, giving up its excess heat at constant pressure (since $P^{(b)} = P^{(s)}$ at all times). The distance $\ell$ is called the **mixing length**, $H_P$ is the pressure scale height (see Eq. 10.70), and

$$\alpha \equiv \ell / H_P,$$

*the ratio of mixing length to pressure scale height* is an adjustable parameter, or *free parameter*, generally assumed to be of order unity. (From comparisons of numerical stellar models with observations, values of $0.5 < \alpha < 3$ are typical.)

After the bubble travels one mixing length, the excess heat flow *per unit volume* from the bubble into its surroundings is just

$$\delta q = (C_P \, \delta T) \rho,$$

---

[21] In some texts, $\delta \left( \frac{dT}{dr} \right) \equiv \Delta \nabla T$.

where $\delta T$ is calculated from Eq. (10.96) by substituting $\ell$ for $dr$. Multiplying by the average velocity $\overline{v}_c$ of the convective bubble, we obtain the convective flux (the amount of energy per unit area per unit time carried by a bubble):

$$F_c = \delta q\,\overline{v}_c = (C_P\,\delta T)\rho\overline{v}_c. \tag{10.97}$$

Note that $\rho\overline{v}$ is a *mass flux*, or the amount of mass per second that crosses a unit area oriented perpendicular to the direction of the flow. Mass flux is a quantity that is often encountered in fluid mechanics.

The average velocity $\overline{v}$ may be found from the net force *per unit volume*, $f_{net}$, acting on the bubble. Using the ideal gas law and assuming constant $\mu$, we can write

$$\delta P = \frac{P}{\rho}\,\delta\rho + \frac{P}{T}\,\delta T.$$

Since the pressure is always equal between the bubble and its surroundings, $\delta P \equiv P^{(b)} - P^{(s)} = 0$. Thus

$$\delta\rho = -\frac{\rho}{T}\,\delta T.$$

From Eq. (10.91),

$$f_{net} = \frac{\rho g}{T}\,\delta T.$$

However, we assumed that the initial temperature difference between the bubble and its surroundings is essentially zero, or $\delta T_i \approx 0$. Consequently the buoyant force must also be quite close to zero initially. Since $f_{net}$ increases linearly with $\delta T$, we may take an average over the distance $\ell$ between the initial and final positions, or

$$\langle f_{net}\rangle = \frac{1}{2}\frac{\rho g}{T}\,\delta T_f.$$

Neglecting viscous forces, the work done per unit volume by the buoyant force over the distance $\ell$ goes into the kinetic energy of the bubble, or

$$\frac{1}{2}\rho v_f^2 = \langle f_{net}\rangle\,\ell.$$

Choosing an *average* kinetic energy over one mixing length leads to some average value of $v^2$, namely $\beta v^2$, where $\beta$ has a value in the range $0 < \beta < 1$. Now the average convective bubble velocity becomes

$$\overline{v}_c = \left(\frac{2\beta\,\langle f_{net}\rangle\,\ell}{\rho}\right)^{1/2}.$$

Substituting the net force per unit volume, using Eq. (10.96) with $dr = \ell$, and rearranging, we have

$$\bar{v}_c = \left(\frac{\beta g}{T}\right)^{1/2} \left[\delta\left(\frac{dT}{dr}\right)\right]^{1/2} \ell$$

$$= \beta^{1/2} \left(\frac{T}{g}\right)^{1/2} \left(\frac{k}{\mu m_H}\right) \left[\delta\left(\frac{dT}{dr}\right)\right]^{1/2} \alpha, \tag{10.98}$$

where we obtained the last equation by replacing the mixing length with $\alpha H_P$ and using Eq. (10.70) together with the ideal gas law.

After some manipulation, Eqs. (10.97) and (10.98) finally yield an expression for the convective flux:

$$F_c = \rho C_P \left(\frac{k}{\mu m_H}\right)^2 \left(\frac{T}{g}\right)^{3/2} \beta^{1/2} \left[\delta\left(\frac{dT}{dr}\right)\right]^{3/2} \alpha^2. \tag{10.99}$$

Fortunately, $F_c$ is not very sensitive to $\beta$, but it does depend strongly on $\alpha$ and $\delta(dT/dr)$.

The derivation leading to the prescription for the convective flux given by Eq. (10.99) is known as the **mixing-length theory**. Although basically a phenomenological theory containing arbitrary constants, the mixing-length theory is generally quite successful in predicting the results of observations.

To evaluate $F_c$, we still need to know the difference between the temperature gradients of the bubble and its surroundings. Suppose, for simplicity, that *all* of the flux is carried by convection, so that

$$F_c = \frac{L_r}{4\pi r^2},$$

where $L_r$ is the interior luminosity. This will allow us to estimate the difference in temperature gradients needed for this special case. Solving Eq. (10.99) for the temperature gradient difference gives

$$\delta\left(\frac{dT}{dr}\right) = \left[\frac{L_r}{4\pi r^2} \frac{1}{\rho C_P \alpha^2} \left(\frac{\mu m_H}{k}\right)^2 \left(\frac{g}{T}\right)^{3/2} \beta^{-1/2}\right]^{2/3}. \tag{10.100}$$

Dividing Eq. (10.100) by Eq. (10.90) for the adiabatic temperature gradient gives an estimate of how superadiabatic the actual temperature gradient must be to carry all of the flux by convection alone:

$$\frac{\delta(dT/dr)}{|dT/dr|_{ad}} = \left(\frac{L_r}{4\pi r^2}\right)^{2/3} C_P^{1/3} \rho^{-2/3} \alpha^{-4/3} \left(\frac{\mu m_H}{k}\right)^{4/3} \frac{1}{T} \beta^{-1/3}.$$

---

**Example 10.4.3.**   Using values typical of the base of the Sun's convection zone, assuming a monatomic gas throughout, and assuming $\alpha = 1$ and $\beta = 1/2$, we can estimate a characteristic adiabatic temperature gradient, the degree to which the actual gradient is superadiabatic, and the convective bubble velocity.

Assume that $M_r = 0.976 \, M_\odot$, $L_r = 1 \, L_\odot$, $r = 0.714 \, R_\odot$, $g = GM_r/r^2 = 525 \, \text{m s}^{-2}$, $C_P = 5nR/2$, $P = 5.59 \times 10^{12} \, \text{N m}^{-2}$, $\rho = 187 \, \text{kg m}^{-3}$, $\mu = 0.606$, and $T = 2.18 \times 10^6 \, \text{K}$. Then, from Eq. (10.90),

$$\left. \frac{dT}{dr} \right|_{\text{ad}} \sim 0.015 \, \text{K m}^{-1},$$

and from Eq. (10.100),

$$\delta \left( \frac{dT}{dr} \right) \sim 6.7 \times 10^{-9} \, \text{K m}^{-1}.$$

The relative amount by which the actual temperature gradient is superadiabatic is then

$$\frac{\delta(dT/dr)}{|dT/dr|_{\text{ad}}} \sim 4.4 \times 10^{-7}.$$

For parameters appropriate for the deep interior, convection is certainly adequately approximated by the adiabatic temperature gradient.

The convective velocity needed to carry all of the convective flux is found from Eq. (10.98),

$$\bar{v}_c \sim 50 \, \text{m s}^{-1} \sim 10^{-4} \, v_s,$$

where $v_s$ is the local solar sound speed (see Eq. 10.84).

---

Near the surface of a star, where the presence of ionization results in a larger value for $C_P$ and where $\rho$ and $T$ get much smaller, the ratio of the superadiabatic excess to the adiabatic gradient can become significantly larger, with the convective velocity possibly approaching the sound speed. In this situation, a detailed study of the relative amounts of convective and radiative flux must be considered. This will not be discussed further here.

Although the mixing length theory is adequate for many problems, it is incomplete. For instance, $\alpha$ and $\beta$ are *free parameters* that must be chosen for a particular problem; they may even vary throughout the star. There are also stellar conditions for which the time-independent mixing length theory is inherently unsatisfactory. As one example, consider stellar pulsations; during a star's pulsation cycle the outer layers of the star are oscillating with periods comparable to the timescale for convection, given by $t_c = \ell/\bar{v}_c$. In such cases, rapid changes in the physical conditions in the star directly couple to the driving of the convective bubbles, which in turn alters the structure of the star. Although much effort (and some progress) has been made in developing a full, time-dependent convection theory for stellar interiors, at present no theory exists that completely describes this complex behavior. Much work remains to be done in understanding the important details of stellar convection.

## 10.5 ■ STELLAR MODEL BUILDING

We have now derived all of the fundamental differential equations of stellar structure. These equations, together with a set of relations describing the physical properties of the stellar material, may be solved to obtain a theoretical stellar model.

### A Summary of the Equations of Stellar Structure

For convenience, the basic *time-independent* (static) stellar structure equations are summarized:

$$\frac{dP}{dr} = -G\frac{M_r \rho}{r^2} \tag{10.6}$$

$$\frac{dM_r}{dr} = 4\pi r^2 \rho \tag{10.7}$$

$$\frac{dL_r}{dr} = 4\pi r^2 \rho \epsilon \tag{10.36}$$

$$\frac{dT}{dr} = -\frac{3}{4ac}\frac{\bar{\kappa}\rho}{T^3}\frac{L_r}{4\pi r^2} \qquad \text{(radiation)} \tag{10.68}$$

$$= -\left(1 - \frac{1}{\gamma}\right)\frac{\mu m_H}{k}\frac{GM_r}{r^2} \qquad \text{(adiabatic convection)} \tag{10.89}$$

The last equation assumes that the convective temperature gradient is purely adiabatic and is applied when

$$\frac{d\ln P}{d\ln T} < \frac{\gamma}{\gamma - 1}. \tag{10.95}$$

If the star is static, as assumed above, then $\epsilon = \epsilon_{\text{nuclear}}$. However, if the structure of the stellar model is changing over time, we must include the energy contribution due to gravity, $\epsilon = \epsilon_{\text{nuclear}} + \epsilon_{\text{gravity}}$. The introduction of the gravitational energy term adds an explicit time dependence to the equations that is not present in the purely static case. This can be seen by realizing that the virial theorem requires that one-half of the gravitational potential energy that is lost must be converted into heat. The *rate* of energy production (per unit mass) by gravity is then $dQ/dt$. Therefore $\epsilon_{\text{gravity}} = -dQ/dt$, the minus sign indicating that heat is liberated *from* the material.

### Entropy

As a note of interest, it is often useful to express the gravitational energy generation rate in terms of the change in the **entropy** per unit mass (the specific entropy), defined by[22]

$$\boxed{dS \equiv \frac{dQ}{T}.} \tag{10.101}$$

Then the energy generation rate is seen to be due to the change in entropy of the material, or

$$\epsilon_{\text{gravity}} = -T\frac{dS}{dt}. \tag{10.102}$$

---

[22]Although $dQ$ is an inexact differential, it can be shown that the entropy is a state function.

If the star is collapsing, $\epsilon_{gravity}$ will be positive; if it is expanding, $\epsilon_{gravity}$ will be negative. Thus, as the star contracts, its entropy decreases. This is not a violation of the second law of thermodynamics, which states that the entropy of a *closed* system must always remain the same (reversible process) or increase (irreversible process). Since a star is not a closed system, its entropy may decrease locally while the entropy of the remainder of the universe increases by a greater amount. The entropy is carried out of the star by photons and neutrinos.

When changes in the structure of the star are sufficiently rapid that accelerations can no longer be neglected, Eq. (10.6) must be replaced by the exact expression, Eq. (10.5). Such a situation can occur during a supernova explosion or during stellar pulsations. The effect of the acceleration term in stellar pulsations will be discussed in Chapter 14.

### The Constitutive Relations

The basic stellar structure equations [(10.6), (10.7), (10.36), (10.68), and (10.89)] require information concerning the physical properties of the matter from which the star is made. The required conditions are the equations of state of the material and are collectively referred to as **constitutive relations**. Specifically, we need relationships for the pressure, the opacity, and the energy generation rate, in terms of fundamental characteristics of the material: the density, temperature, and composition. In general,

$$P = P(\rho, T, \text{composition}) \tag{10.103}$$

$$\overline{\kappa} = \overline{\kappa}(\rho, T, \text{composition}) \tag{10.104}$$

$$\epsilon = \epsilon(\rho, T, \text{composition}) \tag{10.105}$$

The pressure equation of state can be quite complex in the deep interiors of certain classes of stars, where the density and temperature can become extremely high. However, in most situations, the ideal gas law, combined with the expression for radiation pressure, is a good first approximation, particularly when the variation in the mean molecular weight with composition and ionization is properly calculated. The pressure equation of state developed earlier (Eq. 10.20) includes both the ideal gas law and radiation pressure.

The opacity of the stellar material cannot be expressed exactly by a single formula. Instead, it is calculated explicitly for various compositions at specific densities and temperatures and presented in tabular form. Stellar structure codes either interpolate in a density–temperature grid to obtain the opacity for the specified conditions, or, alternatively, use a "fitting function", based on the tabulated values. A similar situation also occurs for accurate calculations of the pressure equation of state. Although no accurate fitting function can be constructed to account for bound–bound opacities, approximate expressions for bound–free, free–free, electron scattering, and $H^-$ ion opacities were presented in Section 9.2; see Eqs. (9.22), (9.23), (9.27), and (9.28), respectively.

To calculate the nuclear energy generation rate, we can use formulas such as those presented in Section 10.3 for the pp chain (Eq. 10.46) and the CNO cycle (Eq. 10.58). In more sophisticated calculations, *reaction networks* are employed that yield individual reaction rates for each step of a process and equilibrium abundances for each isotope in the mixture.

### Boundary Conditions

The actual solution of the stellar structure equations, including the constitutive relations, requires appropriate **boundary conditions** that specify physical constraints to the mathematical equations. Boundary conditions play the essential role of defining the limits of integration. The central boundary conditions are fairly obvious—namely that the interior mass and luminosity must approach zero at the center of the star, or

$$\left.\begin{array}{ccc} M_r & \rightarrow & 0 \\ L_r & \rightarrow & 0 \end{array}\right\} \quad \text{as } r \rightarrow 0. \tag{10.106}$$

This simply means that the star is physically realistic and does not contain a hole, a core of negative luminosity, or central points of infinite $\rho$ or $\epsilon$!

A second set of boundary conditions is required at the surface of the star. The simplest set of assumptions is that the temperature, pressure, and density all approach zero at some surface value for the star's radius, $R_\star$, or

$$\left.\begin{array}{ccc} T & \rightarrow & 0 \\ P & \rightarrow & 0 \\ \rho & \rightarrow & 0 \end{array}\right\} \quad \text{as } r \rightarrow R_\star. \tag{10.107}$$

Strictly, the conditions of Eqs. (10.107) will never be obtained in a real star (as is obviously the case for the temperature). Therefore, it is often necessary to use more sophisticated surface boundary conditions, such as when the star being modeled has an extended atmosphere or is losing mass, as most stars do.

### The Vogt–Russell Theorem

Given the basic stellar structure equations, constitutive relations, and boundary conditions, we can now specify the type of star to be modeled. As can be seen by examination of Eq. (10.6), the pressure gradient at a given radius is dependent on the interior mass and the density. Similarly, the radiative temperature gradient (Eq. 10.36) depends on the local temperature, density, opacity, and interior luminosity, while the luminosity gradient is a function of the density and energy generation rate. The pressure, opacity, and energy generation rate in turn depend explicitly on the density, temperature, and composition at that location. If the interior mass at the surface of the star (i.e., the entire stellar mass) is specified, along with the composition, surface radius, and luminosity, application of the surface boundary conditions allows for a determination of the pressure, interior mass, temperature, and interior luminosity at an infinitesimal distance $dr$ below the surface of the star.[23] Continuing this *numerical integration* of the stellar structure equations to the center of the star must result in agreement with the central boundary conditions (Eqs. 10.106). Since the values

---

[23] It is also necessary to specify the average density over that distance. Since $\rho$ is assumed to be zero at the surface, and since it depends explicitly on the pressure and temperature, which are also assumed to be zero at the surface and are initially unknown below the surface, an immediate difficulty arises; the right-hand sides of Eqs. (10.6) and (10.68) are *zero*, so $P$ and $T$ never increase from their surface values! One solution to this initial-step problem is outlined in Appendix L. More sophisticated solutions require an iterative procedure, continually correcting previous estimates until a self-consistent answer is obtained to within some specified level of accuracy.

of the various gradients are directly related to the composition of the star, it is not possible to specify any arbitrary combination of surface radius and luminosity after the mass and composition have been selected. This set of constraints is known as the **Vogt–Russell theorem**:

> The mass and the composition structure throughout a star uniquely determine its radius, luminosity, and internal structure, as well as its subsequent evolution.

*The dependence of a star's evolution on mass and composition is a consequence of the change in composition due to nuclear burning.*[24] The statement of the Vogt–Russell "theorem" given here is somewhat misleading since there are other parameters that can influence stellar interiors, such as magnetic fields and rotation. However, these parameters are assumed to have little effect in most stars and will not be discussed further.[25]

### Numerical Modeling of the Stellar Structure Equations

With the exception of a special family of approximate solutions to the stellar structure equations known as *polytropes* (to be discussed on page 334), the system of differential equations, along with their constitutive relations, cannot be solved analytically. Instead, as already mentioned, it is necessary to integrate the system of equations numerically. This is accomplished by approximating the differential equations by *difference equations*—by replacing $dP/dr$ by $\Delta P/\Delta r$, for instance. The star is then imagined to be constructed of spherically symmetric **shells**, as in Fig. 10.11, and the "integration" is carried out from some initial radius in finite steps by specifying some increment $\delta r$.[26] It is then possible to increment each of the fundamental physical parameters through successive applications of the difference equations. For instance, if the pressure in zone $i$ is given by $P_i$, then the pressure in the next deepest zone, $P_{i+1}$, is found from

$$P_{i+1} = P_i + \frac{\Delta P}{\Delta r}\,\delta r,$$

where $\delta r$ is negative.

The numerical integration of the stellar structure equations may be carried out from the surface toward the center, from the center toward the surface, or, as is often done, in both directions simultaneously. If the integration is carried out in both directions, the solutions will meet at some *fitting point* where the variables must vary smoothly from one solution to the other. This last approach is frequently taken because the most important physical processes in the outer layers of stars generally differ from those in the deep interiors. The transfer of radiation through optically thin zones and the ionization of hydrogen and helium

---

[24]In this sense, Eq. (10.36) does contain an implicit time dependence due to stellar nucleosynthesis.

[25]Even without the complications of magnetic fields and rotation, the Vogt–Russell "theorem" can be violated in certain special circumstances. However, an actual star (as opposed to a theoretical model) would probably adopt one unique structure as a consequence of its evolutionary history. In this sense, the Vogt–Russell "theorem" should be considered a general rule rather than a rigorous law.

[26]Codes that treat the radius as an independent variable are called **Eulerian** codes. **Lagrangian** codes treat the mass as an independent variable. In the Lagrangian formulation, the differential equations are rewritten using Eq. (10.7); the hydrostatic equilibrium equation can be written in the form $dP/dM$, for instance.

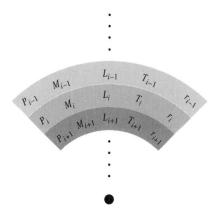

**FIGURE 10.11** Zoning in a numerical stellar model. The star is assumed to be constructed of spherically symmetric mass shells, with the physical parameters associated with each zone being specified by the stellar structure equations, the constitutive relations, the boundary conditions, and the star's mass and composition. In research-quality codes some quantities are specified in the middle of mass shells (e.g., $P$ and $T$), whereas others are associated with the interfaces between shells (e.g., $r$, $M_r$, and $L_r$).

occur close to the surface, while nuclear reactions occur near the center. By integrating in both directions, it is possible to decouple these processes somewhat, simplifying the problem.

Simultaneously matching the surface and central boundary conditions for a desired stellar model usually requires several iterations before a satisfactory solution is obtained. If the surface-to-center and center-to-surface integrations do not agree at the fitting point, the starting conditions must be changed. This is accomplished in a series of attempts, called *iterations*, where the initial conditions of the next integration are estimated from the outcome of the previous integration. A process of successive iterations is also necessary if the star is integrated from the surface to the center or from the center to the surface; in these cases the fitting points are simply the center and surface, respectively.

A very simple stellar structure code (called StatStar) is presented in Appendix L. StatStar integrates the stellar structure equations developed in this chapter in their time-independent form from the outside of the star to the center using the appropriate constitutive relations; it also assumes a constant (or homogeneous) composition throughout. Many of the sophisticated numerical techniques present in research codes have been neglected so that the basic elements of stellar model building can be more easily understood, as have the detailed calculations of the pressure equation of state and the opacity. The complex formalism of the mixing-length theory has also been left out in favor of the simplifying assumption of adiabatic convection. Despite these approximations, very reasonable models may be obtained for stars lying on the main sequence of the H–R diagram.

### Polytropic Models and the Lane–Emden Equation

As we mentioned previously, it is not generally possible to solve the system of stellar structure equations and their associated constitutive relations analytically; we must employ numerical solutions to "build" stellar models. However, under very special and restrictive situations, it is possible to find analytic solutions to a subset of the equations. The first work in this area was carried out by J. Homer Lane (1819–1880), who wrote a paper on the equilibrium of stellar configurations in the *American Journal of Science* in 1869. That work was later extended significantly by Robert Emden (1862–1940). Today, the famous equation that helps us describe analytical stellar models is referred to as the *Lane–Emden equation*.

To understand the motivation of developing the Lane–Emden equation, note that careful inspection of the stellar structure equations shows that the mechanical equations of stellar structure (Eqs. 10.6 and 10.7) could be solved simultaneously without reference to the energy equations (10.36, and either 10.68 or 10.89) if only a simple relationship existed between pressure and density. Of course, as we have seen, such a simple relationship does not generally exist; normally, temperature and composition must also enter into the pressure equation of state, often in a complicated way. However, under certain circumstances, such as for an adiabatic gas (see Eq. 10.86), the pressure can be written explicitly in terms of the density alone. Hypothetical stellar models in which the pressure depends on density in the form $P = K\rho^{\gamma}$ are known as **polytropes**. The development of polytropic models is well worth the effort since their relative simplicity allows us to gain some insight into stellar structure without all of the complications inherent in full-blown numerical models.

To derive the Lane–Emden equation, we begin with the equation for hydrostatic equilibrium, Eq. (10.6). Rewriting the equation and taking the radial derivative of both sides gives

$$\frac{d}{dr}\left(\frac{r^2}{\rho}\frac{dP}{dr}\right) = -G\frac{dM_r}{dr}.$$

We immediately see that Eq. (10.7) can be used to eliminate the mass gradient. Substituting, we get

$$\frac{d}{dr}\left(\frac{r^2}{\rho}\frac{dP}{dr}\right) = -G(4\pi r^2 \rho)$$

or

$$\frac{1}{r^2}\frac{d}{dr}\left(\frac{r^2}{\rho}\frac{dP}{dr}\right) = -4\pi G\rho. \tag{10.108}$$

As an aside, it is worth pointing out here that Eq. (10.108) is actually a slightly camouflaged form of a very well-studied differential equation known as *Poisson's equation*. It is left as an exercise to show that Eq. (10.108) can be rewritten in the form

$$\frac{1}{r^2}\frac{d}{dr}\left(r^2\frac{d\Phi_g}{dr}\right) = 4\pi G\rho, \tag{10.109}$$

which is the spherically symmetric form of Poisson's equation for the gravitational potential energy per unit mass, $\Phi_g \equiv U_g/m$.[27]

---

[27]Poisson's equation shows up frequently in physics. For example, Gauss' Law, one of Maxwell's equations of electromagnetic theory, can be reformulated into Poisson's equation by replacing the electric field vector with the negative of the gradient of the electrostatic potential.

To solve Eq. (10.108), we now employ the relationship $P(\rho) = K\rho^\gamma$, where $K$ and $\gamma > 0$ are constants. This functional form of the pressure equation is known generally as a polytropic equation of state. Substituting, taking the appropriate derivative, and simplifying, we have

$$\frac{\gamma K}{r^2} \frac{d}{dr}\left[r^2 \rho^{\gamma-2} \frac{d\rho}{dr}\right] = -4\pi G\rho.$$

It is customary to rewrite the expression slightly by letting $\gamma \equiv (n+1)/n$, where $n$ is historically referred to as the *polytropic index*. Then

$$\left(\frac{n+1}{n}\right) \frac{K}{r^2} \frac{d}{dr}\left[r^2 \rho^{(1-n)/n} \frac{d\rho}{dr}\right] = -4\pi G\rho.$$

In order to simplify the last expression somewhat, it is now useful to rewrite the equation in a dimensionless form. Expressing the density in terms of a scaling factor and a dimensionless function $D(r)$, let

$$\rho(r) \equiv \rho_c[D_n(r)]^n, \quad \text{where } 0 \le D_n \le 1.$$

(As you might suspect, $\rho_c$ will turn out to be the central density of the polytropic stellar model.) Again substituting and simplifying, we arrive at

$$\left[(n+1)\left(\frac{K\rho_c^{(1-n)/n}}{4\pi G}\right)\right] \frac{1}{r^2} \frac{d}{dr}\left[r^2 \frac{dD_n}{dr}\right] = -D_n^n.$$

Careful study of our last equation reveals that the collective constant in square brackets has the units of distance squared. Defining

$$\lambda_n \equiv \left[(n+1)\left(\frac{K\rho_c^{(1-n)/n}}{4\pi G}\right)\right]^{1/2}$$

and introducing the dimensionless independent variable $\xi$ via

$$r \equiv \lambda_n \xi,$$

we finally arrive at

$$\boxed{\frac{1}{\xi^2} \frac{d}{d\xi}\left[\xi^2 \frac{dD_n}{d\xi}\right] = -D_n^n,} \tag{10.110}$$

which is the famous **Lane–Emden equation**.

Solving Eq. (10.110) for the dimensionless function $D_n(\xi)$ in terms of $\xi$ for a specific polytropic index $n$ leads directly to the profile of density with radius $\rho_n(r)$. The polytropic equation of state $P_n(r) = K\rho_n^{(n+1)/n}$ provides the pressure profile. In addition, if the ideal gas law and radiation pressure are assumed for constant composition (Eq. 10.20), then the temperature profile, $T(r)$, is also obtained.

In order to actually solve this second-order differential equation, it is necessary to impose two boundary conditions (which effectively specify the two constants of integration). Assuming that the "surface" of the star is that location where the pressure goes to zero (and correspondingly the density of the gas also goes to zero), then

$$D_n(\xi_1) = 0 \text{ specifies the surface at } \xi = \xi_1,$$

where $\xi_1$ is the location of the first zero of the solution.

Next consider the center of the star. If $r = \delta$ represents a distance infinitesimally close to the center of the star, then the mass contained within a volume of radius $\delta$ is given by

$$M_r = \frac{4\pi}{3}\bar{\rho}\,\delta^3$$

where $\bar{\rho}$ is the average density of the gas within the radius $\delta$. Substituting into the equation for hydrostatic equilibrium, Eq. (10.6), we have

$$\frac{dP}{dr} = -G\frac{M_r\rho}{r^2} = -\frac{4\pi}{3}G\bar{\rho}^2\,\delta \to 0 \text{ as } \delta \to 0.$$

Since $P = K\rho^{(n+1)/n}$, this implies that

$$\frac{d\rho}{dr} \to 0 \text{ as } r \to 0,$$

which immediately leads to the central boundary condition

$$\frac{dD_n}{d\xi} = 0 \text{ at } \xi = 0.$$

In addition, in order for $\rho_c$ to represent the central density of the star, it is also necessary that $D_n(0) = 1$ (this condition isn't strictly a boundary condition, it simply *normalizes* the density scaling function, $D_n$).

With the boundary conditions specified, it is now possible to compute the total mass of a star of a specific polytropic index. From Eq. (10.7),

$$M = 4\pi \int_0^R r^2\rho\,dr,$$

where $R = \lambda_n\xi_1$ represents the radius of the star. Rewriting in terms of the dimensionless quantities yields

$$M = 4\pi \int_0^{\xi_1} \lambda_n^2\xi^2\rho_c D_n^n\,d(\lambda_n\xi),$$

or

$$M = 4\pi \lambda_n^3 \rho_c \int_0^{\xi_1} \xi^2 D_n^n \, d\xi.$$

Although this expression could be integrated directly with knowledge of $D_n(\xi)$, it can also be rewritten directly by noting, from the Lane–Emden equation and the central boundary condition, that

$$\xi^2 D_n^n = -\frac{d}{d\xi}\left[\xi^2 \frac{dD_n}{d\xi}\right]$$

gives

$$M = -4\pi \lambda_n^3 \rho_c \xi_1^2 \left.\frac{dD_n}{d\xi}\right|_{\xi_1},$$

where $(dD_n/d\xi)|_{\xi_1}$ means that the derivative of $D_n$ is evaluated at the surface.

Although the Lane–Emden equation is compact and elegant, it is important to bear in mind its many limitations. Recall that Eq. (10.110) contains no information about either energy transport or energy generation within a star; the equation only describes hydrostatic equilibrium and mass conservation, and then only within the highly idealized class of polytropic equations of state. Nevertheless, the Lane–Emden equation is capable of giving us some important insights into the structures of stars.

There are only three analytic solutions to the Lane–Emden equation, namely $n = 0$, 1, and 5. The $n = 0$ solution is given by

$$D_0(\xi) = 1 - \frac{\xi^2}{6}, \quad \text{with } \xi_1 = \sqrt{6}.$$

It is left as an exercise for you to derive the $n = 0$ solution. The solution for $n = 1$ is the well-known "sinc" function

$$D_1(\xi) = \frac{\sin \xi}{\xi}, \quad \text{with } \xi_1 = \pi,$$

and the $n = 5$ solution is given by

$$D_5(\xi) = [1 + \xi^2/3]^{-1/2}, \quad \text{with } \xi_1 \to \infty.$$

In the latter case you are asked to verify that although the radius of the star is infinite, the total mass of the star is actually finite. This is not the case for values of $n > 5$. Thus, the physical limits of $n$ are constrained to the range $0 \leq n \leq 5$. Graphical representations of $D_0$, $D_1$, and $D_5$ are shown in Fig. 10.12.

This discussion of polytropes was originally motivated by the equation of state of an adiabatic gas. For the case of an ideal, monatomic gas, $\gamma = 5/3$, which implies that $n = 1.5$. In addition, as we shall see later, in Chapter 16 (see Eq. 16.12), certain extremely compressed stars in their final stage of evolution known as white dwarfs can also be described

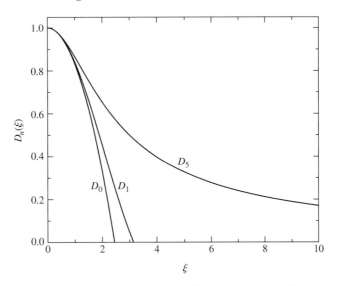

**FIGURE 10.12** The analytic solutions to the Lane–Emden equation: $D_0(\xi)$, $D_1(\xi)$, and $D_5(\xi)$.

by polytropes of index 1.5 (technically these are non-relativistic, completely degenerate stars). Although the important $n = 1.5$ case cannot be solved analytically, it can be solved numerically.

Another important polytropic index is the $n = 3$ "Eddington standard model" associated with a star in radiative equilibrium. To see how this model corresponds to radiative equilibrium, consider a polytrope that is supported by both an ideal gas and radiation pressure (see Eq. 10.20). If the total pressure at a certain location in the star is represented by $P$, and the contribution to that pressure due to an ideal gas is given by

$$P_g = \frac{\rho k T}{\mu m_H} = \beta P, \tag{10.111}$$

where $0 \leq \beta \leq 1$, then the contribution due to radiation pressure is

$$P_r = \frac{1}{3} a T^4 = (1 - \beta) P. \tag{10.112}$$

Since we are looking for a polytropic equation of state that can be expressed independent of temperature, we can combine the last two expressions to eliminate $T$. Solving for $T$ in Eq. (10.111) and substituting into Eq. (10.112), we obtain

$$\frac{1}{3} a \left( \frac{\beta P \mu m_H}{\rho k} \right)^4 = (1 - \beta) P.$$

This leads immediately to an expression for the total pressure in terms of the density, namely

$$P = K \rho^{4/3} \tag{10.113}$$

where

$$K \equiv \left[ \frac{3(1-\beta)}{a} \right]^{1/3} \left( \frac{k}{\beta \mu m_H} \right)^{4/3}.$$

Since $\gamma = 4/3$, this implies that $n = 3$.[28]

Certainly the two most physically significant polytropic models correspond to $n = 1.5$ and $n = 3$. Although neither model can be solved analytically, the use of computers and numerical integration algorithms allow us to explore their structure and behavior relatively easily. Careful study of these polytropes can yield important insights into the structures of more realistic, although significantly more complex stellar models.

## 10.6 ■ THE MAIN SEQUENCE

The analysis of stellar spectra tells us that the atmospheres of the vast majority of all stars are composed primarily of hydrogen, usually about 70% by mass ($X \sim 0.7$), whereas the mass fraction of metals varies from near zero to approximately 3% ($0 < Z < 0.03$). Assuming that the initial composition of a star is **homogeneous** (meaning that the composition is the same throughout), the first set of nuclear fusion reactions ought to be those that convert hydrogen into helium (the pp chains and/or the CNO cycle). Recall that these reactions occur at the lowest temperatures because the associated Coulomb barrier is lower than that for the burning of more massive nuclei. Consequently, the structure of a homogeneous, hydrogen-rich star ought to be strongly influenced by hydrogen nuclear burning deep within its interior.

Because of the predominance of hydrogen that initially exists in the core, and since hydrogen burning is a relatively slow process, the interior composition and structure of the star will change slowly. As we saw in Example 10.3.2, a rough estimate of the hydrogen-burning lifetime of the Sun is 10 billion years. Of course, the surface conditions will not be completely static. By the Vogt–Russell theorem, any change in composition or mass requires a readjustment of the effective temperature and luminosity; *the observational characteristic of the star must change as a consequence of the central nuclear reactions.* As long as changes in the core are slow, so are the evolutionary changes in the observed surface features.[29]

Since most stars have similar compositions, the structures of stars ought to vary smoothly with mass. Recall from Examples 10.1.1 and 10.2.1 that as the mass increases, the central pressure and the central temperature should increase. Therefore, for stars of low mass, the pp chain will dominate since less energy is required to initiate these reactions than the reactions of the CNO cycle. For high-mass stars, the CNO cycle will likely dominate because of its very strong temperature dependence.

---

[28]We will learn in Chapter 16 that stars supported solely by a fully relativistic, completely degenerate gas can also be described by a polytropic index of 3; see Eq. (16.15).

[29]Some short-period surface changes can occur that are essentially decoupled from the long-term variations in the core. Stellar pulsations require specific conditions to exist, but their timescales are usually much shorter than the nuclear timescale. These oscillations will be discussed in Chapter 14.

At some point, as progressively less massive stars are considered, the central temperature will diminish to the point where nuclear reactions are no longer able to stabilize a star against gravitational contraction. This has been shown to occur at approximately 0.072 $M_\odot$ for solar composition (the lower limit is slightly higher, 0.09 $M_\odot$, for stars with virtually no metal content, $Z \simeq 0$). At the other extreme, stars with masses greater than approximately 90 $M_\odot$ become subject to thermal oscillations in their centers that may produce significant variations in the nuclear energy generation rates over timescales as short as 8 hours.[30]

### The Eddington Luminosity Limit

Along with thermal oscillations, the stability of very massive stars is directly affected by their extremely high luminosities. As can be seen by Eq. (10.20), if the temperature is sufficiently high and the gas density is low enough, it is possible for radiation pressure to dominate over the gas pressure in certain regions of the star, a situation that can occur in the outer layers of very massive stars. In this case the pressure gradient is approximately given by Eq. (9.31). Combined with the relationship between radiant flux and luminosity (Eq. 3.2), the pressure gradient near the surface may be written as

$$\frac{dP}{dr} \simeq -\frac{\overline{\kappa}\rho}{c}\frac{L}{4\pi r^2}.$$

But hydrostatic equilibrium (Eq. 10.6) demands that the pressure gradient near the star's surface must also be given by

$$\frac{dP}{dr} = -G\frac{M\rho}{r^2},$$

where $M$ is the star's mass. Combining, and solving for the luminosity, we have

$$\boxed{L_{\text{Ed}} = \frac{4\pi G c}{\overline{\kappa}}M.} \tag{10.114}$$

$L_{\text{Ed}}$ is the maximum radiative luminosity that a star can have and still remain in hydrostatic equilibrium. If the luminosity exceeds $L_{\text{Ed}}$, mass loss must occur, driven by radiation pressure. This luminosity maximum, known as the **Eddington limit**, appears in a number of areas of astrophysics, including late stages of stellar evolution, novae, and the structure of accretion disks.

For our purposes, it is possible to make an estimate of the Eddington luminosity for stars on the upper end of the main sequence. The effective temperatures of these massive stars are in the range of 50,000 K, high enough that most of the hydrogen is ionized in their photospheres. Therefore, the major contribution to the opacity is from electron scattering, and we can replace $\overline{\kappa}$ by Eq. (9.27). For $X = 0.7$, Eq. (10.114) becomes

$$L_{\text{Ed}} \simeq 1.5 \times 10^{31}\,\frac{M}{M_\odot}\,\text{W} \qquad \text{or} \qquad \frac{L_{\text{Ed}}}{L_\odot} \simeq 3.8 \times 10^4\,\frac{M}{M_\odot}.$$

---

[30]This so-called $\epsilon$ *pulsation mechanism* will be discussed in Section 14.2.

For a 90 $M_\odot$ star, $L_{Ed} \simeq 3.5 \times 10^6$ $L_\odot$, roughly three times the expected main-sequence value.

The fairly close correspondence between the theoretical and Eddington luminosities implies that the envelopes of massive main-sequence stars are loosely bound at best. In fact, observations of the few stars with masses estimated to be near 100 $M_\odot$ indicate that they are suffering from large amounts of mass loss and exhibit variability in their luminosities.

### Variations of Main-Sequence Stellar Parameters with Mass

From theoretical models that are computed in the mass range of hydrogen burning, it is possible to obtain a numerical relationship between $M$ and $L$ that agrees well with the observational mass–luminosity relation shown in Fig. 7.7. It is also possible to locate each of the models on a theoretical H–R diagram (see Fig. 10.13). By comparison with Fig. 8.13, it can be seen that stars undergoing hydrogen burning in their cores lie along the observational main sequence!

The range in main-sequence luminosities is from near $5 \times 10^{-4}$ $L_\odot$ to approximately $1 \times 10^6$ $L_\odot$, a variation of over nine orders of magnitude, while the masses change by only three orders of magnitude. Because of the enormous rate of energy output from upper main-sequence stars, they consume their core hydrogen in a much shorter period of time than do stars on the lower end of the main sequence. As a result, *main-sequence lifetimes decrease with increasing luminosity.* Estimates of the range of main-sequence lifetimes are left as an exercise.

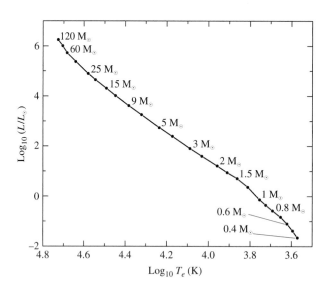

**FIGURE 10.13** The locations of stellar models on a theoretical H–R diagram. The models were computed using the stellar structure equations and constitutive relations. (Data from Schaller, et al., *Astron. Astrophys. Suppl.*, *96*, 269, 1992, and Charbonnel, et al., *Astron. Astrophys. Suppl.*, *135*, 405, 1999.)

Effective temperatures are much less dependent on stellar mass. From approximately 1700 K for 0.072 $M_\odot$ stars to near 53,000 K for 90 $M_\odot$ stars, the increase in effective temperature amounts to a factor of only about 20. However, this variation is still large enough to dramatically change the stellar spectrum, since the dissociation energies of molecules and the ionization potentials of most elements lie within this range, as we demonstrated in Chapter 8. Consequently, by comparison with theoretical models, it is possible to correlate main-sequence masses with observed spectra.

The interior structure of stars along the main sequence also varies with mass, primarily in the location of convection zones. In the upper portion of the main sequence, where energy generation is due to the strongly temperature-dependent CNO cycle, convection is dominant in the core. This occurs because the rate of energy generation changes quickly with radius, and radiation is not efficient enough to transport all of the energy being released in nuclear reactions. Outside of the hydrogen-burning core, radiation is again capable of handling the flux, and convection ceases. As the stellar mass decreases, so does the central temperature and the energy output of the CNO cycle until, near 1.2 $M_\odot$, the pp chain begins to dominate and the core becomes radiative. Meanwhile, near the surface of the star, as the effective temperature decreases with decreasing mass, the opacity increases, in part because of the location of the zone of hydrogen ionization (recall Fig. 9.10). The increase in opacity makes convection more efficient than radiation near the surfaces of stars with masses less than approximately 1.3 $M_\odot$. This has the effect of creating convection zones near the surfaces of these stars. As we continue to move down the main sequence, the bottom of the surface convection zone lowers until the entire star becomes convective near 0.3 $M_\odot$.

Through the use of the fundamental physical principles developed thus far in this text, we have been able to build realistic models of main-sequence stars and develop an understanding of their interiors. However, other stars remain on the observational H–R diagram that do not lie along the main sequence (see Fig. 8.13). By considering the changes in stellar structure that occur because of changes in composition due to nuclear burning (the Vogt–Russell theorem), it will become possible to explain their existence as well. The evolution of stars is discussed in Chapters 12 and 13.

## SUGGESTED READING

### General

Kippenhahn, Rudolf, *100 Billion Suns*, Basic Books, New York, 1983.

### Technical

Arnett, David, *Supernovae and Nucleosynthesis*, Princeton University Press, Princeton, 1996.

Bahcall, John N., *Neutrino Astrophysics*, Cambridge University Press, Cambridge, 1989.

Bahcall, John N., Pinsonneault, M. H., and Basu, Sarbani, "Solar Models: Current Epoch and Time Dependences, Neutrinos, and Helioseismological Properties," *The Astrophysical Journal*, *555*, 990, 2001.

Barnes, C. A., Clayton, D. D., and Schramm, D. N. (eds.), *Essays in Nuclear Astrophysics*, Cambridge University Press, Cambridge, 1982.

Bowers, Richard L., and Deeming, Terry, *Astrophysics I: Stars*, Jones and Bartlett, Publishers, Boston, 1984.

Chabrier, Gilles, and Baraffe, Isabelle, "Theory of Low-Mass Stars and Substellar Objects," *Annual Review of Astronomy and Astrophysics*, *38*, 337, 2000.

Chandrasekhar, S., *An Introduction to the Study of Stellar Structure*, Dover Publications, Inc., New York, 1967.

Clayton, Donald D., *Principles of Stellar Evolution and Nucleosynthesis*, University of Chicago Press, Chicago, 1983.

Cox, J. P., and Giuli, R. T., *Principles of Stellar Structure*, Gordon and Breach, New York, 1968.

Fowler, William A., Caughlan, Georgeanne R., and Zimmerman, Barbara A., "Thermonuclear Reaction Rates, I," *Annual Review of Astronomy and Astrophysics*, *5*, 525, 1967.

Fowler, William A., Caughlan, Georgeanne R., and Zimmerman, Barbara A., "Thermonuclear Reaction Rates, II," *Annual Review of Astronomy and Astrophysics*, *13*, 69, 1975.

Hansen, Carl J., Kawaler, Steven D., and Trimble, Virginia *Stellar Interiors: Physical Principles, Structure, and Evolution*, Second Edition, Springer-Verlag, New York, 2004.

Harris, Michael J., Fowler, William A., Caughlan, Georgeanne R., and Zimmerman, Barbara A., "Thermonuclear Reaction Rates, III," *Annual Review of Astronomy and Astrophysics*, *21*, 165, 1983.

Iben, Icko, Jr., "Stellar Evolution Within and Off the Main Sequence," *Annual Review of Astronomy and Astrophysics*, *5*, 571, 1967.

Iglesias, Carlos A, and Rogers, Forrest J., "Updated Opal Opacities," *The Astrophysical Journal*, *464*, 943, 1996.

Kippenhahn, Rudolf, and Weigert, Alfred, *Stellar Structure and Evolution*, Springer-Verlag, Berlin, 1990.

Liebert, James, and Probst, Ronald G., "Very Low Mass Stars," *Annual Review of Astronomy and Astrophysics*, *25*, 473, 1987.

Padmanabhan, T., *Theoretical Astrophysics*, Cambridge University Press, Cambridge, 2002.

Prialnik, Dina, *An Introduction to the Theory of Stellar Structure and Evolution*, Cambridge University Press, Cambridge, 2000.

Novotny, Eva, *Introduction to Stellar Atmospheres and Interiors*, Oxford University Press, New York, 1973.

Shore, Steven N., *The Tapestry of Modern Astrophysics*, John Wiley and Sons, Hoboken, 2003.

## PROBLEMS

**10.1** Show that the equation for hydrostatic equilibrium, Eq. (10.6), can also be written in terms of the optical depth $\tau$, as

$$\frac{dP}{d\tau} = \frac{g}{\kappa}.$$

This form of the equation is often useful in building model stellar atmospheres.

**10.2** Prove that the gravitational force on a point mass located anywhere inside a hollow, spherically symmetric shell is zero. Assume that the mass of the shell is $M$ and has a constant density $\rho$. Assume also that the radius of the inside surface of the shell is $r_1$ and that the radius of the outside surface is $r_2$. The mass of the point is $m$.

**10.3** Assuming that 10 eV could be released by every atom in the Sun through chemical reactions, estimate how long the Sun could shine at its current rate through chemical processes alone. For simplicity, assume that the Sun is composed entirely of hydrogen. Is it possible that the Sun's energy is entirely chemical? Why or why not?

**10.4** **(a)** Taking into consideration the Maxwell–Boltzmann velocity distribution, what temperature would be required for two protons to collide if quantum mechanical tunneling is neglected? Assume that nuclei having velocities ten times the root-mean-square (rms) value for the Maxwell–Boltzmann distribution can overcome the Coulomb barrier. Compare your answer with the estimated central temperature of the Sun.

**(b)** Using Eq. (8.1), calculate the ratio of the number of protons having velocities ten times the rms value to those moving at the rms velocity.

**(c)** Assuming (incorrectly) that the Sun is pure hydrogen, estimate the number of hydrogen nuclei in the Sun. Could there be enough protons moving with a speed ten times the rms value to account for the Sun's luminosity?

**10.5** Derive the ideal gas law, Eq. (10.10). Begin with the pressure integral (Eq. 10.9) and the Maxwell–Boltzmann velocity distribution function (Eq. 8.1).

**10.6** Derive Eq. (10.28) from Eq. (8.1).

**10.7** By invoking the virial theorem (Eq. 2.46), make a crude estimate of an "average" temperature for the Sun. Is your result consistent with other estimates obtained in Chapter 10? Why or why not?

**10.8** Show that the form of the Coulomb potential barrier penetration probability given by Eq. (10.31) follows directly from Eq. (10.30).

**10.9** Prove that the energy corresponding to the Gamow peak is given by Eq. (10.34).

**10.10** Calculate the ratio of the energy generation rate for the pp chain to the energy generation rate for the CNO cycle given conditions characteristic of the center of the present-day (evolved) Sun, namely $T = 1.5696 \times 10^7$ K, $\rho = 1.527 \times 10^5$ kg m$^{-3}$, $X = 0.3397$, and $X_{CNO} = 0.0141$.[31] Assume that the pp chain screening factor is unity ($f_{pp} = 1$) and that the pp chain branching factor is unity ($\psi_{pp} = 1$).

---

[31] The interior values assumed here are taken from the solar model of Bahcall, Pinsonneault, and Basu, *Ap. J.*, *555*, 990, 2001.

**10.11**  Beginning with Eq. (10.62) and writing the energy generation rate in the form

$$\epsilon(T) = \epsilon'' T_8^\alpha,$$

show that the temperature dependence for the triple alpha process, given by Eq. (10.63), is correct. $\epsilon''$ is a function that is independent of temperature.

*Hint:* First take the natural logarithm of both sides of Eq. (10.62) and then differentiate with respect to $\ln T_8$. Follow the same procedure with your power law form of the equation and compare the results. You may want to make use of the relation

$$\frac{d\ln\epsilon}{d\ln T_8} = \frac{d\ln\epsilon}{\frac{1}{T_8}dT_8} = T_8\frac{d\ln\epsilon}{dT_8}.$$

**10.12**  The $Q$ value of a reaction is the amount of energy released (or absorbed) during the reaction. Calculate the $Q$ value for each step of the PP I reaction chain (Eqs. 10.37–10.39). Express your answers in MeV. The masses of $_1^2\text{H}$ and $_2^3\text{He}$ are 2.0141 u and 3.0160 u, respectively.

**10.13**  Calculate the amount of energy released or absorbed in the following reactions (express your answers in MeV):

(a) $_6^{12}\text{C} + _6^{12}\text{C} \rightarrow _{12}^{24}\text{Mg} + \gamma$

(b) $_6^{12}\text{C} + _6^{12}\text{C} \rightarrow _8^{16}\text{O} + 2\,_2^4\text{He}$

(c) $_9^{19}\text{F} + _1^1\text{H} \rightarrow _8^{16}\text{O} + _2^4\text{He}$

The mass of $_6^{12}\text{C}$ is 12.0000 u, by definition, and the masses of $_8^{16}\text{O}$, $_9^{19}\text{F}$, and $_{12}^{24}\text{Mg}$ are 15.99491 u, 18.99840 u, and 23.98504 u, respectively. Are these reactions exothermic or endothermic?

**10.14**  Complete the following reaction sequences. Be sure to include any necessary leptons.

(a) $_{14}^{27}\text{Si} \rightarrow _{13}^{?}\text{Al} + e^+ + \underline{?}$

(b) $_{13}^{?}\text{Al} + _1^1\text{H} \rightarrow _{12}^{24}\text{Mg} + _?^4\underline{?}$

(c) $_{17}^{35}\text{Cl} + _1^1\text{H} \rightarrow _{18}^{36}\text{Ar} + \underline{?}$

**10.15**  Prove that Eq. (10.83) follows from Eq. (10.82).

**10.16**  Show that Eq. (10.109) can be obtained from Eq. (10.108).

**10.17**  Starting with the Lane–Emden equation and imposing the necessary boundary conditions, prove that the $n = 0$ polytrope has a solution given by

$$D_0(\xi) = 1 - \frac{\xi^2}{6}, \quad \text{with } \xi_1 = \sqrt{6}.$$

**10.18**  Describe the density structure associated with an $n = 0$ polytrope.

**10.19**  Derive an expression for the total mass of an $n = 5$ polytrope, and show that although $\xi_1 \rightarrow \infty$, the mass is finite.

**10.20**  (a) On the same graph, plot the density structure of stars of polytropic indices $n = 0$, $n = 1$, and $n = 5$. *Hint:* You will want to plot $\rho_n/\rho_c$ vs. $r/\lambda_n$.

(b) What can you conclude about the concentration of density with radius for increasing polytropic index?

(c) From the trend that you observe for the analytic solutions to the Lane–Emden equation, what would you expect regarding the density concentration of an adiabatically convective stellar model compared to a model in radiative equilibrium?

(d) Explain your conclusion in part (c) in terms of the physical processes of convection and radiation.

**10.21** Estimate the hydrogen-burning lifetimes of stars near the lower and upper ends of the main sequence. The lower end of the main sequence[32] occurs near 0.072 $M_\odot$, with $\log_{10} T_e = 3.23$ and $\log_{10}(L/L_\odot) = -4.3$. On the other hand, an 85 $M_\odot$ star[33] near the upper end of the main sequence has an effective temperature and luminosity of $\log_{10} T_e = 4.705$ and $\log_{10}(L/L_\odot) = 6.006$, respectively. Assume that the 0.072 $M_\odot$ star is entirely convective so that, through convective mixing, all of its hydrogen, rather than just the inner 10%, becomes available for burning.

**10.22** Using the information given in Problem 10.21, calculate the radii of a 0.072 $M_\odot$ star and a 85 $M_\odot$ star. What is the ratio of their radii?

**10.23** (a) Estimate the Eddington luminosity of a 0.072 $M_\odot$ star and compare your answer to the main-sequence luminosity given in Problem 10.21. Assume $\bar{\kappa} = 0.001$ m$^2$ kg$^{-1}$. Is radiation pressure likely to be significant in the stability of a low-mass main-sequence star?

(b) If a 120 $M_\odot$ star forms with $\log_{10} T_e = 4.727$ and $\log_{10}(L/L_\odot) = 6.252$, estimate its Eddington luminosity, assuming the opacity is due to electron scattering. Compare your answer with the actual luminosity of the star.

## COMPUTER PROBLEMS

**10.24** (a) Use a numerical integration algorithm such as the Euler method to compute the density profile for the $n = 1.5$ and $n = 3$ polytropes. Be sure to correctly incorporate the boundary conditions in your integrations.

(b) Plot your results and compare them with the $n = 0$, $n = 1$, and $n = 5$ analytic models determined in Problem 10.20.

**10.25** Verify that the basic equations of stellar structure [Eqs. (10.6), (10.7), (10.36), (10.68)] are satisfied by the 1 $M_\odot$ StatStar model available for download from the companion website; see Appendix L. This may be done by selecting two adjacent zones and numerically computing the derivatives on the left-hand sides of the equations, for example

$$\frac{dP}{dr} \simeq \frac{P_{i+1} - P_i}{r_{i+1} - r_i},$$

and comparing your results with results obtained from the right-hand sides using average values of quantities for the two zones [e.g., $M_r = (M_i + M_{i+1})/2$].

Perform your calculations for two adjacent shells at temperatures near $5 \times 10^6$ K, and then compare your results for the left- and right-hand sides of each equation by determining relative errors. Note that the model assumes complete ionization everywhere and has the uniform composition $X = 0.7$, $Y = 0.292$, $Z = 0.008$. Your results on the left- and right-hand sides of the stellar structure equations will not agree exactly because StatStar uses a Runge–Kutta numerical algorithm that carries out intermediate steps not shown in the output file.

**10.26** The companion website contains an example of a theoretical 1.0 $M_\odot$ main-sequence star produced by the stellar structure code StatStar, described in Appendix L. Using StatStar, build

---

[32] Data from Chabrier, et al., *Ap. J.*, *542*, 464, 2000.
[33] Data from Schaller, et al., *Astron. Astrophys. Suppl. Ser.*, *96*, 269, 1992.

a second main-sequence star with a mass of 0.75 $M_\odot$ that has a homogeneous composition of $X = 0.7$, $Y = 0.292$, and $Z = 0.008$. For these values, the model's luminosity and effective temperature are 0.189 $L_\odot$ and 3788.5 K, respectively. Compare the central temperatures, pressures, densities, and energy generation rates between the 1.0 $M_\odot$ and 0.75 $M_\odot$ models. Explain the differences in the central conditions of the two models.

**10.27** Use the stellar structure code StatStar described in Appendix L, together with the theoretical StatStar H–R diagram and mass–effective temperature data provided on the companion website, to calculate a homogeneous, main-sequence model having the composition $X = 0.7$, $Y = 0.292$, and $Z = 0.008$. (*Note:* It may be more illustrative to assign each student in the class a different mass for this problem so that the results can be compared.)

(a) After obtaining a satisfactory model, plot $P$ versus $r$, $M_r$ versus $r$, $L_r$ versus $r$, and $T$ versus $r$.

(b) At what temperature has $L_r$ reached approximately 99% of its surface value? 50% of its surface value? Is the temperature associated with 50% of the total luminosity consistent with the rough estimate found in Eq. (10.27)? Why or why not?

(c) What are the values of $M_r/M_\star$ for the two temperatures found in part (b)? $M_\star$ is the total mass of the stellar model.

(d) If each student in the class calculated a different mass, compare the changes in the following quantities with mass:

(i)    The central temperature.
(ii)   The central density.
(iii)  The central energy generation rate.
(iv)   The extent of the central convection zone with mass fraction and radius.
(v)    The effective temperature.
(vi)   The radius of the star.

(e) If each student in the class calculated a different mass:

(i)    Plot each model on a graph of luminosity versus mass (i.e., plot $L_\star/L_\odot$ versus $M_\star/M_\odot$).

(ii)   Plot $\log_{10}(L_\star/L_\odot)$ versus $\log_{10}(M_\star/M_\odot)$ for each stellar model.

(iii)  Using an approximate power law relation of the form

$$L_\star/L_\odot = (M_\star/M_\odot)^\alpha,$$

find an appropriate value for $\alpha$. $\alpha$ may differ for different compositions or vary somewhat with mass. This is known as the mass–luminosity relation (see Fig. 7.7).

**10.28** Repeat Problem 10.27 using the same mass but a different composition; assume $X = 0.7$, $Y = 0.290$, $Z = 0.010$.

(a) For a given mass, which model ($Z = 0.008$ or $Z = 0.010$) has the higher central temperature? the greater central density?

(b) Referring to the appropriate stellar structure equations and constitutive relations, explain your results in part (a).

(c) Which model has the largest energy generation rate at the center? Why?

(d) How do you account for the differences in effective temperature and luminosity between your two models?

# CHAPTER

# 11

# The Sun

## 11.1 ■ THE SOLAR INTERIOR

Over the last few chapters we have investigated the theoretical foundations of stellar structure, treating the star as being composed of an atmosphere and an interior. The distinction between the two regions is fairly nebulous. Loosely, the atmosphere is considered to be that region where the optical depth is less than unity and the simple approximation of photons *diffusing* through optically thick material is not justified (see Eq. 9.31). Instead, atomic line absorption and emission must be considered in detail in the stellar atmosphere. On the other hand, nuclear reaction processes deep in the stellar interior plays a crucial role in the star's energy output and its inevitable evolution.

Due to its proximity to us, the star for which we have the greatest amount of observational data is our Sun. From ground-based and space-based observatories, we are able to measure with high precision the composition of our Sun's surface; its luminosity, effective temperature, radius, magnetic fields, and rotation rates; the oscillation frequencies (vibrations) throughout its interior;[1] and the rate at which neutrinos are produced via nuclear reactions in its core. This tremendous wealth of information provides us with rigorous tests of our understanding of the physical processes operating within stellar atmospheres and interiors.

### The Evolutionary History of the Sun

Based on its observed luminosity and effective temperature, our Sun is classified as a typical main-sequence star of spectral class G2 with a surface composition of $X = 0.74$, $Y = 0.24$, and $Z = 0.02$ (the mass fractions of hydrogen, helium, and metals, respectively). To understand how it has evolved to this point, recall that according to the Vogt–Russell theorem the mass and composition of a star dictate its internal structure. Our Sun has been converting hydrogen to helium via the pp chain during most of its lifetime, thereby changing its composition and its structure. By comparing the results of radioactive dating tests of Moon rocks and meteorites with stellar evolution calculations and the present-day

---

[1] Helioseismology, the study of solar oscillations, is discussed in Chapter 14.

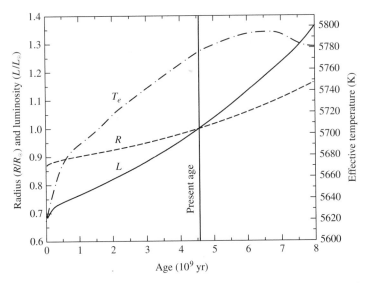

**FIGURE 11.1**    The evolution of the Sun on the main sequence. As a result of changes in its internal composition, the Sun has become larger and brighter. The solid line indicates its luminosity, the dashed line its radius, and the dash-dot line its effective temperature. The luminosity and radius curves are relative to present-day values. (Data from Bahcall, Pinsonneault, and Basu, *Ap. J.*, *555*, 990, 2001.)

observable Sun, the current age of the Sun is determined to be approximately $4.57 \times 10^9$ yr.[2] Furthermore, as depicted in Fig. 11.1, since becoming a main-sequence star, the Sun's luminosity has increased nearly 48% (from 0.677 $L_\odot$) while its radius has increased 15% from an initial value of 0.869 $R_\odot$.[3] The Sun's effective temperature has also increased from 5620 K to its present-day value of 5777 K (recall Example 3.4.2).

You may be wondering what impact this evolution has had on Earth. Interestingly, from a theoretical standpoint it is not at all clear how this change in solar energy output altered our planet during its history, primarily because of uncertainties in the behavior of the terrestrial environment. Understanding the complex interaction between the Sun and Earth involves the detailed calculation of convection in Earth's atmosphere, as well as the effects of the atmosphere's time-varying composition and the nature of the continually changing reflectivity, or **albedo**,[4] of Earth's surface.

### The Present-Day Interior Structure of the Sun

Consistent with the current age of the Sun, a **solar model** may be constructed for the present-day Sun using the physical principles discussed in preceding chapters. Table 11.1 gives the values of the central temperature, pressure, density, and composition for one such

---

[2]Radioactive dating of the oldest known objects in the Solar System, calcium-aluminum-rich inclusions (CAIs) in meteorites, leads to a determination of the age of the Solar System of $4.5672 \pm 0.0006$ Gyr.

[3]The data quoted here and in the following discussion are from the solar model of Bahcall, Pinsonneault, and Basu, *Ap. J.*, *555*, 990, 2001.

[4]Earth's albedo, the ratio of reflected to incident sunlight, is affected by the amount of surface water and ice.

**TABLE 11.1**   Central Conditions in the Sun. (Data from Bahcall, Pinsonneault, and Basu, *Ap. J.,* *555*, 990, 2001.)

| | |
|---|---|
| Temperature | $1.570 \times 10^7$ K |
| Pressure | $2.342 \times 10^{16}$ N m$^{-2}$ |
| Density | $1.527 \times 10^5$ kg m$^{-3}$ |
| $X$ | 0.3397 |
| $Y$ | 0.6405 |

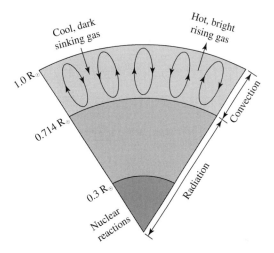

**FIGURE 11.2**   A schematic diagram of the Sun's interior.

solar model, and a schematic diagram of the model is shown in Fig. 11.2. According to the evolutionary sequence leading to this model, during its lifetime the mass fraction of hydrogen ($X$) in the Sun's center has decreased from its initial value of 0.71 to 0.34, while the central mass fraction of helium ($Y$) has increased from 0.27 to 0.64. In addition, due to diffusive settling of elements heavier than hydrogen, the mass fraction of hydrogen near the surface has increased by approximately 0.03, while the mass fraction of helium has decreased by 0.03.

Because of the Sun's past evolution, its composition is no longer homogeneous but instead shows the influence of ongoing nucleosynthesis, surface convection, and elemental diffusion (settling of heavier elements). The composition structure of the Sun is shown in Fig. 11.3 for $^1_1$H, $^3_2$He, and $^4_2$He. Since the Sun's primary energy production mechanism is the pp chain, $^3_2$He is an intermediate species in the reaction sequence. During the conversion of hydrogen to helium, $^3_2$He is produced and then destroyed again (see Fig. 10.8). At the top of the hydrogen-burning region where the temperature is lower, $^3_2$He is relatively more abundant because it is produced more easily than it is destroyed.[5] At greater depths, the higher temperatures allow the $^3_2$He–$^3_2$He interaction to proceed more rapidly, and the $^3_2$He

---

[5]Recall that much higher temperatures are required for helium–helium interactions than proton–proton interactions.

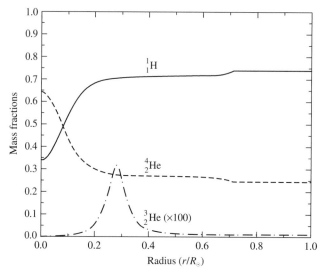

**FIGURE 11.3**    The abundances of $^1_1$H, $^3_2$He, and $^4_2$He as a function of radius for the Sun. Note that the abundance of $^3_2$He is multiplied by a factor of 100. (Data from Bahcall, Pinsonneault, and Basu, *Ap. J., 555*, 990, 2001.)

abundance again decreases (the temperature profile of the Sun is shown in Fig. 11.4). The slight ramp in the $^1_1$H and $^4_2$He curves near 0.7 $R_\odot$ reflects evolutionary changes in the position of the base of the surface convection zone, combined with the effects of elemental diffusion. Within the convection zone, turbulence results in essentially complete mixing and a homogeneous composition. The base of the present-day convection zone is at 0.714 $R_\odot$.

The largest contribution to the energy production in the Sun occurs at approximately one-tenth of the solar radius, as can be seen in the Sun's interior luminosity profile and the curve of its derivative with respect to radius (Fig. 11.5). If this result seems unexpected, consider that the mass conservation equation (Eq. 10.7),

$$\frac{dM_r}{dr} = 4\pi r^2 \rho,$$

gives

$$dM_r = 4\pi r^2 \rho \, dr = \rho \, dV, \tag{11.1}$$

indicating that the amount of mass within a certain radius interval increases with radius simply because the volume of a spherical shell, $dV = 4\pi r^2 \, dr$, increases with $r$ for a fixed choice of $dr$. Of course, the mass contained in the shell also depends on the density of the gas. Consequently, even if the amount of energy liberated per kilogram of material ($\epsilon$) decreases steadily from the center outward, the largest contribution to the total luminosity will occur, not at the center, but in a shell that contains a significant amount of mass. In the case of the middle-aged Sun, the decrease in the amount of available hydrogen fuel at its center will also influence the location of the peak in the energy production region (see Eq. 10.46).

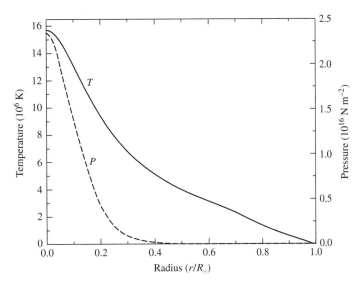

**FIGURE 11.4** The temperature and pressure profiles in the solar interior. (Data from Bahcall, Pinsonneault, and Basu, *Ap. J.*, *555*, 990, 2001.)

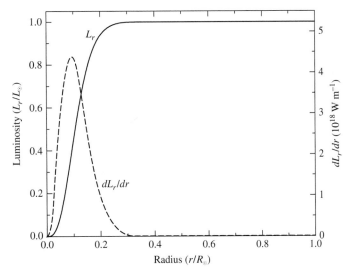

**FIGURE 11.5** The interior luminosity profile of the Sun and the derivative of the interior luminosity as a function of radius. (Data from Bahcall, Pinsonneault, and Basu, *Ap. J.*, *555*, 990, 2001.)

Figures 11.4 and 11.6 show just how rapidly the pressure and density change with radius in the Sun. These variations are forced on the solar structure by the condition of hydrostatic equilibrium (Eq. 10.6), the ideal gas law (Eq. 10.11), and the composition structure of the star. Of course, boundary conditions applied to the stellar structure equations require that both $\rho$ and $P$ become negligible at the surface (Eq. 10.107).

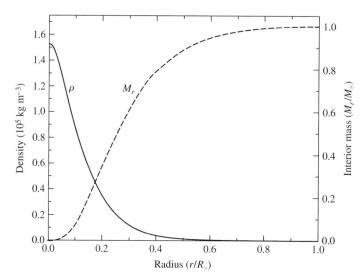

**FIGURE 11.6**    The density profile and the interior mass of the Sun as a function of radius. (Data from Bahcall, Pinsonneault, and Basu, *Ap. J.*, *555*, 990, 2001.)

Figure 11.6 also shows the interior mass ($M_r$) as a function of radius. Notice that 90% of the mass of the star is located within roughly one-half of its radius. This should not come as a complete surprise since the density increases significantly as the center of the Sun is approached. Integration of the density function over the volume of the star from the center outward (i.e., the integration of Eq. 11.1) yields the interior mass function.

The question remains as to how the energy generated in the interior is transported outward. Recall that in Chapter 10 we determined a criterion for the onset of convection in stellar interiors, namely that the temperature gradient become superadiabatic (Eq. 10.94),

$$\left|\frac{dT}{dr}\right|_{act} > \left|\frac{dT}{dr}\right|_{ad},$$

where the "act" and "ad" subscripts designate the actual and adiabatic temperature gradients, respectively. Under the simplifying assumption of an ideal monatomic gas, this condition becomes (Eq. 10.95),

$$\frac{d \ln P}{d \ln T} < 2.5.$$

$d \ln P / d \ln T$ is plotted versus $r/R_\odot$ in Fig. 11.7. As can be seen, the Sun is purely radiative below $r/R_\odot = 0.714$ and becomes convective above that point. Physically this occurs because the opacity in the outer portion of the Sun becomes large enough to inhibit the transport of energy by radiation; recall that the radiative temperature gradient is proportional to the opacity (see Eq. 10.68). When the temperature gradient becomes too large, convection becomes the more efficient means of energy transport. Throughout most of the region of convective energy transport, $d \ln P / d \ln T \simeq 2.5$, which is characteristic of the nearly adiabatic temperature gradient of most convection zones. The rapid rise in $d \ln P / d \ln T$

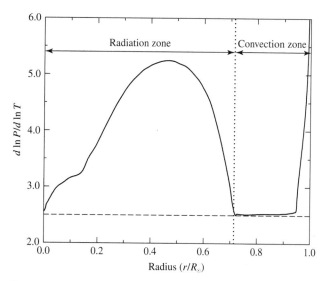

**FIGURE 11.7** The convection condition $d \ln P / d \ln T$ plotted versus $r / R_\odot$. The dashed horizontal line represents the boundary between adiabatic convection and radiation for an ideal monatomic gas. The onset of convection does not exactly agree with the ideal adiabatic case because of the incorporation of a sophisticated equation of state and a more detailed treatment of convection physics. The rapid rise in $d \ln P / d \ln T$ near the surface is associated with the highly superadiabatic nature of convection in that region (i.e., the adiabatic approximation that convection occurs when $d \ln P / d \ln T < 2.5$ is invalid near the surface of the Sun). [$d \ln P / d \ln T$ was computed using data from Bahcall, Pinsonneault, and Basu, *Ap. J.*, *555*, 990, 2001. The data for the zones above 0.95 $R_\odot$ are from Cox, Arthur N. (editor), *Allen's Astrophysical Quantities*, Fourth Edition, AIP Press, New York, 2000.]

above 0.95 $R_\odot$ is due to the significant departure of the actual temperature gradient from an adiabatic one. In this case convection must be described by a more detailed treatment, such as the mixing-length theory discussed at the end of Section 10.4.

Notice that $d \ln P / d \ln T$ also decreases to almost 2.5 at the center of the Sun. Although the Sun remains purely radiative in the center, the large amount of energy that must be transported outward pushes the temperature gradient in the direction of becoming superadiabatic. We will see in Chapter 13 that stars only slightly more massive than the Sun are convective in their centers because of the stronger temperature dependence of the CNO cycle as compared to the pp chain.

Clearly, an enormous amount of information is available regarding the solar interior, as derived from the direct and careful application of the stellar structure equations and the fundamental physical principles described in the last three chapters. A very complete and reasonable model of the Sun can be produced that is consistent with evolutionary timescales and fits the global characteristics of the star, specifically its mass, luminosity, radius, effective temperature, and surface composition; precise measurements of oscillation frequencies (see Chapter 14); and, as we will see in the next section, its observed surface convection zone.

One aspect of the observed Sun that is not yet fully consistent with the current solar model is the abundance of lithium. The observed lithium abundance at the Sun's surface is actually somewhat less than expected and may imply some need for adjustments in the model through refined treatments of convection, rotation, and/or mass loss. The lithium problem will be discussed further in Chapter 13.

### The Solar Neutrino Problem: A Detective Story Solved

Another significant discrepancy had existed between observations and the solar model for several decades, the resolution of which led to an important new understanding of fundamental physics. The **solar neutrino problem** was first noticed when Raymond Davis began measuring the neutrino flux from the Sun in 1970 using a detector located almost one mile below ground in the Homestake Gold Mine in Lead, South Dakota (Fig. 11.8). Because of the very low cross section of neutrino interactions with other matter, neutrinos can easily travel completely through Earth while other particles originating from space cannot. As a result, the underground detector was assured of measuring what it was designed to measure—neutrinos created eight minutes earlier in the solar core.

The Davis neutrino detector contained 615,000 kg of cleaning fluid, $C_2Cl_4$ (tetrachlorethylene) in a volume of 377,000 liters (100,000 gallons). One isotope of chlorine ($^{37}_{17}Cl$) is capable of interacting with neutrinos of sufficient energy to produce a radioactive isotope of argon that has a half-life of 35 days,

$$^{37}_{17}Cl + \nu_e \rightleftharpoons \,^{37}_{18}Ar + e^-.$$

The threshold energy for this reaction, 0.814 MeV, is less than the energies of the neutrinos produced in every step of the pp chain except the crucial first one, $^1_1H + \,^1_1H \rightarrow \,^2_1H + e^+ + \nu_e$ (recall the reaction sequences depicted in Fig. 10.8). However, the reaction that accounted

**FIGURE 11.8**  Raymond Davis's solar neutrino detector. The tank was located 1478 m (4850 ft) below ground in the Homestake Gold Mine in Lead, South Dakota, and was filled with 615,000 kg of $C_2Cl_4$ in a volume of 377,000 liters (100,000 gallons). (Courtesy of Brookhaven National Laboratory.)

for 77% of the neutrinos detected in the Davis experiment is the decay of $^8_5$B in the PP III chain,

$$^8_5\text{B} \rightarrow {}^8_4\text{Be} + e^+ + \nu_e.$$

Unfortunately, this reaction is very rare, producing only for one pp chain termination in 5000.

John Bahcall (1935–2005), a colleague of Davis, was able to compute the anticipated rate at which solar neutrinos should have been detected by the chlorine experiment (the capture rate). The complex calculation was based on the rate of neutrino production by $^8_5$B decay in the PP III chain as computed from the solar model, combined with the probability that a solar neutrino will interact with a chlorine atom in the Homestake experiment.

Once every few months Davis and his collaborators carefully purged the accumulated argon from the tank and determined the number of argon atoms produced. The capture rate was measured in terms of the **solar neutrino unit**, or SNU (1 SNU $\equiv 10^{-36}$ reactions per target atom per second). With approximately $2.2 \times 10^{30}$ atoms of $^{37}_{17}$Cl atoms in the tank, if only one argon atom was produced each day, this rate would have corresponded to 5.35 SNU.

Results of 108 extractions from the Davis experiment between 1970 and 1994 are shown in Fig. 11.9. Bahcall predicted that the experiment should have yielded a capture rate of 7.9 SNU while the actual data gave an average of $2.56 \pm 0.16$ SNU; only one argon atom was produced every two days in that 100,000 gallon tank!

Other neutrino experiments, fundamentally different from the $^{37}_{17}$Cl experiment, have confirmed the discrepancy between the prediction of the solar model and observed neutrino counts. Japan's underground Super-Kamiokande observatory (Fig. 11.10) detects the

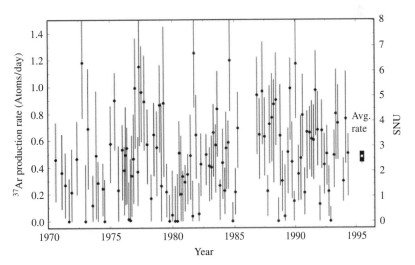

**FIGURE 11.9**   Results of the Davis solar neutrino experiment from 1970 to 1994. The uncertainties in the experimental data are shown by vertical error bars associated with each run. The predicted solar neutrino capture rate for the $^{37}_{17}$Cl detector was 7.9 SNU based on solar models without neutrino oscillations. (Figure adapted from Cleveland, et al., *Ap. J., 496*, 505, 1998.)

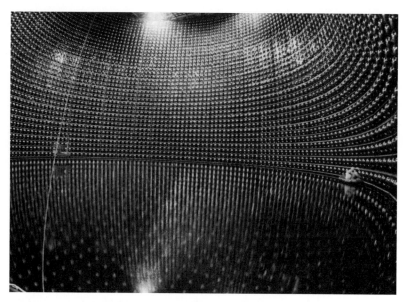

**FIGURE 11.10**    Super-Kamiokande neutrino observatory in Japan contains $4.5 \times 10^7$ kg (50,000 tons) of pure water. As neutrinos pass through the water, they scatter electrons at speeds greater than the speed of light through water. The pale blue Cerenkov light that is produced is detected by the 11,200 inwardly-directed photomultiplier tubes, signaling the presence of the passing neutrino. [Photo courtesy of Kamioka Observatory, ICRR (Institute for Cosmic Ray Research), The University of Tokyo.]

**Cerenkov light** that is produced when neutrinos scatter electrons, causing the electrons to move at speeds greater than the speed of light in water.[6] The number of neutrinos detected by Super-Kamiokande (and Kamiokande II before it) are less than half the number expected from solar models. The Soviet–American Gallium Experiment (SAGE), located at the Baksan Neutrino Laboratory (inside a mountain in the Caucasus), and GALLEX (at the Gran Sasso underground laboratory in Italy) measure the low-energy pp chain neutrinos that dominate the Sun's neutrino flux. SAGE and GALLEX make their detections via a reaction that converts gallium into germanium,

$$\nu_e + {}^{71}_{31}\text{Ga} \rightarrow {}^{71}_{32}\text{Ge} + e^-.$$

After considering the expected number of background counts from sources other than the Sun, both experiments also confirm the deficit of neutrinos first established by the Davis detector.

The search for a theoretical resolution to the solar neutrino problem considered two general approaches: Either some fundamental physical process operating in the solar model is incorrect, or something happens to the neutrinos on their way from the Sun's core to Earth. The first of these possibilities inspired an intense reexamination of a host of features of the

---

[6]Note that this *does not* violate Einstein's special theory of relativity since the special theory applies to the speed of light *in a vacuum*. The speed of light in any other medium is always less than the speed of light in a vacuum.

solar model, including nuclear reaction rates, the opacity of stellar material, the evolution of the Sun up to its present state, variations in the composition of the solar interior, and several exotic suggestions (including *dark matter* in the Sun's core). However, none of these suggested solutions was able to satisfy all of the observational constraints simultaneously, particularly neutrino counts and solar oscillation frequencies.

An elegant solution to the solar neutrino problem proposed that the solar model is essentially correct but that the neutrinos produced in the Sun's core actually change before they reach Earth. The **Mikheyev–Smirnov–Wolfenstein** (or **MSW**) **effect** involves the transformation of neutrinos from one type to another. This idea is an extension of the *electroweak theory* of particle physics that combines the electromagnetic theory with the theory of weak interactions governing some types of radioactive decay. The neutrinos produced in the various branches of the pp chain are all electron neutrinos ($\nu_e$); however, two other *flavors* of neutrinos also exist—the muon neutrino ($\nu_\mu$) and the tau neutrino ($\nu_\tau$). The MSW effect suggests that neutrinos oscillate among flavors, being electron neutrinos, muon neutrinos, and/or tau neutrinos during their passage through the Sun. The neutrino oscillations are caused by interactions with electrons as the neutrinos travel toward the surface. Because the chlorine (Davis), water (Kamiokande and Super-Kamiokande), and gallium detectors (SAGE and GALLEX) have different threshold energies and they are sensitive only to the electron neutrino, their results were determined to be consistent with the MSW theory.

One testable consequence of the MSW effect is that if neutrinos oscillate between flavors, they must necessarily have mass. This is because a change of neutrino flavor can occur only between neutrinos having different masses. The required mass difference needed for the MSW solution to the solar neutrino problem is much less than the current experimentally established upper limit on the mass of the electron neutrino of approximately 2.2 eV. Even though the standard electroweak theory does not predict masses for the neutrinos, many reasonable extensions of this theory do allow for masses in the right range. These extended theories, known as **grand unified theories** (**GUT**s), are currently the focus of intense research by high-energy (particle) physicists.

Confirmation of neutrino oscillations came in 1998 when Super-Kamiokande was used to detect *atmospheric neutrinos* that are produced when high-energy cosmic rays (charged particles from space) collide with Earth's upper atmosphere. Cosmic rays are capable of creating both electron and muon neutrinos, but not tau neutrinos. The Super-Kamiokande group was able to determine that the number of muon neutrinos traveling upward after having traversed the diameter of Earth was significantly reduced relative to the number traveling downward. The difference in numbers is in excellent agreement with the theory of neutrino mixing (neutrinos oscillating among the three flavors), demonstrating for the first time that neutrinos are not massless particles.

Thus, after several decades of study, the solar neutrino problem was resolved by a profound advance in our understanding of particle physics and the nature of the fundamental forces. As a result of their contributions to this important scientific detective story, Raymond Davis and Masatoshi Koshiba, director of the Kamiokande research group that confirmed the neutrino detections, were two of the recipients of the 2002 Nobel Prize in physics.[7]

---

[7]The third recipient of the 2002 Nobel Prize, Riccardo Giacconi, used a rocket experiment to detect X-rays in space. Giacconi later designed the Uhuru and Einstein X-ray observatories and also served as the first director of the Space Telescope Science Institute.

In 2004, John Bahcall wrote of the efforts to solve the solar neutrino problem:

> I am astonished when I look back on what has been accomplished in the field of solar neutrino research over the past four decades. Working together, an international community of thousands of physicists, chemists, astronomers, and engineers has shown that counting radioactive atoms in a swimming pool full of cleaning fluid in a deep mine on Earth can tell us important things about the center of the Sun and about the properties of exotic fundamental particles called neutrinos. If I had not lived through the solar neutrino saga, I would not have believed it was possible.[8]

## 11.2 ■ THE SOLAR ATMOSPHERE

When we observe the Sun visually, it appears as though there is a very abrupt and clear edge to this hot, gaseous ball (Fig. 11.11). Of course, an actual "surface" does not exist; rather, what we are seeing is a region where the solar atmosphere is *optically thin* and photons originating from that level travel unimpeded through space. Even this region is not clearly defined, however, since some photons can always escape when the optical depth is somewhat greater than unity while others may be absorbed when the optical depth is less than unity, but the odds of a photon leaving the solar atmosphere diminish rapidly as the optical depth increases. Consequently, the Sun's atmosphere changes from being optically thin to optically thick in only about 600 km. This relatively small distance (about 0.09% of the Sun's radius) is what gives the "edge" of the Sun its sharp appearance.

### The Photosphere

The region where the observed optical photons originate is known as the solar **photosphere**. Defining the base of the photosphere is somewhat arbitrary since some photons can originate from an optical depth significantly greater than unity. For instance, if 1% of the photons originating from a layer reach us, the optical depth would be approximately 4.5 at that level ($e^{-4.5} \sim 0.01$); if 0.1% reach us, the optical depth would be about 6.9. Of course, since the opacity and optical depth are wavelength dependent, the base of the photosphere is also wavelength dependent if it is defined in terms of the optical depth. Given the arbitrariness of the definition, the base of the photosphere for the Sun is sometimes simply defined to be 100 km below the level where the optical depth at a wavelength of 500 nm is unity. At this depth, $\tau_{500} \simeq 23.6$ and the temperature is approximately 9400 K.

Moving upward through the solar photosphere, the temperature of the gas decreases from its base value to a minimum of 4400 K about 525 km above the $\tau_{500} = 1$ level. It is this temperature minimum that defines the top of the photosphere. Above this point, the temperature begins to rise again. The approximate thicknesses of the various components of the Sun's atmosphere to be discussed in this section are depicted in Fig. 11.12.

As was discussed in Section 9.4, on average the solar flux is emitted from an optical depth of $\tau = 2/3$ (the Eddington approximation). This leads to the identification of the effective temperature with the temperature of the gas at this depth, or $T_e = T_{\tau=2/3} = 5777$ K.

---

[8]"Solving the Mystery of the Missing Neutrinos," John N. Bahcall (2004), Nobel *e*-Museum, http://nobelprize.org/physics/articles/bahcall/.

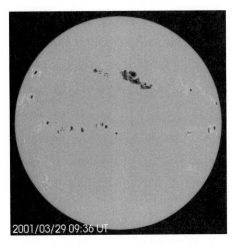

**FIGURE 11.11**  The solar disk appears sharp because of the rapid increase in optical depth with distance through the photosphere. Sunspots are visible on the surface of the disk in this image taken by SOHO/MDI on March 29, 2001. [SOHO (ESA & NASA)]

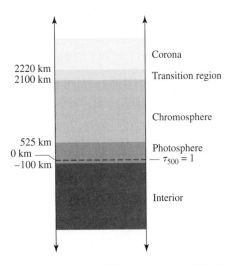

**FIGURE 11.12**  The thicknesses of the components of the Sun's atmosphere.

Recall that the Sun radiates predominantly as a blackbody in the visible and infrared portions of the spectrum (note the relatively smooth features of the solar spectrum depicted in Fig. 9.5). This observation suggests that there exists a source of opacity that is basically continuous across wavelength. The continuum opacity is due in part to the presence of the $H^-$ ions in the photosphere (recall the discussion on page 248).

Using the Saha equation (Eq. 8.8), we can determine the ratio of the number of $H^-$ ions to neutral hydrogen atoms. It is left as an exercise to show that in the Sun's photosphere, only about one in $10^7$ hydrogen atoms actually forms an $H^-$ ion. The importance of $H^-$ in the Sun is due to the fact that even though the abundance of the ion is quite low, neutral hydrogen is not capable of contributing significantly to the continuum.

Of course, optical depth is a function not only of the distance that a photon must travel to the surface of the Sun, but also of the wavelength-dependent opacity of the solar material (Eq. 9.17). Consequently, photons can originate from or be absorbed at different physical depths in the atmosphere, depending upon their wavelengths. Since a spectral line is not infinitesimally thin, but actually covers a range of wavelengths (recall Section 9.5), even different parts of the same line are formed at different levels of the atmosphere. Thus solar observations with high-wavelength resolution can be used to probe the atmosphere at various depths, providing a wealth of information about its structure.

Absorption lines, including Fraunhofer lines, are produced in the photosphere (see Section 5.1). According to Kirchhoff's laws, the absorption lines must be produced where the gas is cooler than the bulk of the continuum-forming region. Line formation must also occur between the observer and the region where much of the continuum is produced. In reality, the Fraunhofer lines are formed in the same layers where $H^-$ produces the continuum. However, the darkest part of the line (its center) originates from regions higher in the photosphere, where the gas is cooler. This is because the opacity is greatest in the center of the line, making it more difficult to see deeper into the photosphere. Moving away from the central wavelength toward the wing of the line implies that absorption is occurring at progressively deeper levels. At wavelengths sufficiently far from the central peak, the edge of the line merges with the continuum being produced at the base of the photosphere. This effect is illustrated in Fig. 11.13 (recall the discussion at the beginning of Section 9.5).

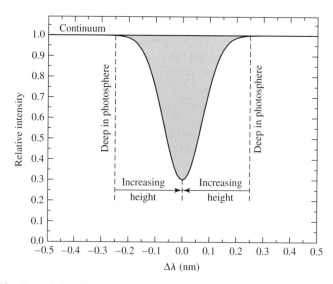

**FIGURE 11.13**  The relationship between absorption line strength and depth in the photosphere for a typical spectral line. The wings of the line are formed deeper in the photosphere than is the center of the line.

### Solar Granulation

When the base of the photosphere is observed (see Fig. 11.14), it appears as a patchwork of bright and dark regions that are constantly changing, with individual regions appearing and then disappearing. With a spatial extent of roughly 700 km, the characteristic lifetime for one of these regions is five to ten minutes. This patchwork structure is known as **granulation** and is the top of the convection zone protruding into the base of the photosphere.

Figure 11.15 shows a high-resolution spectrum of solar granulation spanning a number of convection cells. The appearance of wiggles in the absorption lines occurs because some parts of the region are Doppler blueshifted while others are redshifted. Using Eq. (4.39), we

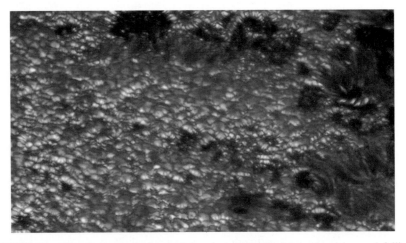

**FIGURE 11.14** Granulation at the base of the photosphere is due to the rising and falling gas bubbles produced by the underlying convection zone. (This three-dimensional image is from the Swedish 1-m Solar Telescope, operated on the island of La Palma by the Institute for Solar Physics of the Royal Swedish Academy of Sciences in the Spanish Observatorio del Roque de los Muchachos of the Instituto de Astrofísica de Canarias.)

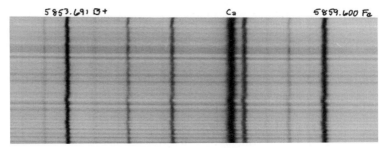

**FIGURE 11.15** A spectrum of a portion of the photospheric granulation showing absorption lines that indicate the presence of radial motions. Wiggles to the left are toward shorter wavelengths and are blueshifted while wiggles to the right are redshifted. The wavelengths shown at the top of the image are given in angstroms. (Courtesy of W. Livingston and the National Optical Astronomy Observatories.)

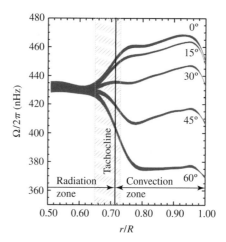

**FIGURE 11.16**   The rotation period of the Sun varies with latitude and depth. $\Omega$, the angular frequency, has units of radians per second. (Adapted from a figure courtesy of NSF's National Solar Observatory.)

find that radial velocities of $0.4\,\mathrm{km\ s^{-1}}$ are common; brighter regions produce the blueshifted sections of the lines while darker regions produce the redshifted sections. Thus the bright cells are the vertically rising hot convective bubbles carrying energy from the solar interior. When those bubbles reach the optically thin photosphere, the energy is released via photons and the resulting cooler, darker gas sinks back into the interior. The lifetime of a typical granule is the amount of time needed for a convective eddy to rise and fall the distance of one mixing length. Solar granulation provides us with a visual verification of the results of the stellar structure equations applied to our Sun.

### Differential Rotation

Photospheric absorption lines may also be used to measure the rotation rate of the Sun. By measuring Doppler shifts at the solar limb, we find that the Sun rotates *differentially* (i.e., the rate of rotation depends on the latitude being observed). At the equator the rotation period is approximately 25 days, increasing to 36 days at the poles.

Observations of solar oscillations have revealed that the Sun's rotation also varies with radius; see Fig. 11.16. Near the base of the convection zone, the differing rotation rates with latitude converge in a region known as the **tachocline**. The strong shear that is set up in this region is believed to result in electric currents in the highly conducting plasma, which in turn generate the Sun's magnetic field. Thus the tachocline is probably the source of the Sun's magnetic field. (The complex manifestations of the Sun's dynamic magnetic field will be discussed extensively in Section 11.3.)

### The Chromosphere

The **chromosphere**, with an intensity that is only about $10^{-4}$ of the value for the photosphere, is that portion of the solar atmosphere that lies just above the photosphere and extends upward for approximately 1600 km (2100 km above $\tau_{500} = 1$). Analysis of the light produced in the chromosphere indicates that the gas density drops by more than a factor of $10^4$ and that the temperature begins to *increase* with increasing altitude, from 4400 K to about 10,000 K.

Reference to the Boltzmann and Saha equations (8.6 and 8.8, respectively) shows that lines that are not produced at the lower temperatures and higher densities of the photosphere can form in the environment of the chromosphere. For instance, along with the hydrogen Balmer lines, the lines of He II, Fe II, Si II, Cr II, and Ca II (in particular, the Ca II H and K lines, 396.8 nm and 393.3 nm, respectively) can appear in the spectrum.

Although certain Fraunhofer lines appear as absorption lines in the visible and near ultraviolet portions of the spectrum, others begin to appear as emission lines at shorter (and much longer) wavelengths. Again Kirchhoff's laws offer an explanation, suggesting that a hot, low-density gas must be responsible. Because the interior of the Sun is optically thick below the base of the photosphere, the area of emission line production must occur elsewhere. With the peak of the blackbody spectrum near 500 nm, the strength of the continuum decreases rapidly at shorter and longer wavelengths (Fig. 3.8). As a result, emission lines produced outside of the visible portion of the spectrum are not overwhelmed by the blackbody radiation.

Visible wavelength emission lines are not normally seen against the bright solar disk, but they can be observed near the limb of the Sun for a few seconds at the beginning and end of a total eclipse of the Sun; this phenomenon is referred to as a **flash spectrum**. During this period, the portion of the Sun that is still visible takes on a reddish hue because of the dominance of the Balmer H$\alpha$ emission line, a line that is normally observed only as an absorption line in the Sun's atmosphere.

Using filters that restrict observations to the wavelengths of the emission lines produced in the chromosphere (particularly H$\alpha$), it is possible to see a great deal of structure in this portion of the atmosphere. **Supergranulation** becomes evident on scales of 30,000 km, showing the continued effects of the underlying convection zone. Doppler studies again reveal convective velocities on the order of 0.4 km s$^{-1}$, with gas rising in the centers of the supergranules and sinking at their edges. Also present are vertical filaments of gas, known as **spicules**, extending upward from the chromosphere for 10,000 km (Fig. 11.17). An individual spicule may have a lifetime of only 15 minutes, but at any given moment spicules cover several percent of the surface of the Sun. Doppler studies show that mass motions are present in spicules, with material moving outward at approximately 15 km s$^{-1}$.

### The Transition Region

Above the chromosphere, the temperature rises very rapidly within approximately 100 km (see Fig. 11.18), reaching more than $10^5$ K before the temperature gradient flattens somewhat. The temperature then continues to rise more slowly, eventually exceeding $10^6$ K. This **transition region** may be selectively observed at various altitudes in the ultraviolet and extreme ultraviolet parts of the electromagnetic spectrum. For instance, the 121.6-nm Lyman-alpha (Ly$\alpha$) emission line of hydrogen ($n = 2 \rightarrow n = 1$) is produced at the top of the chromosphere at 20,000 K, the C III 97.7-nm line originates at a level where the temperature is 90,000 K, the 103.2-nm line of O VI occurs at 300,000 K, and Mg X creates a

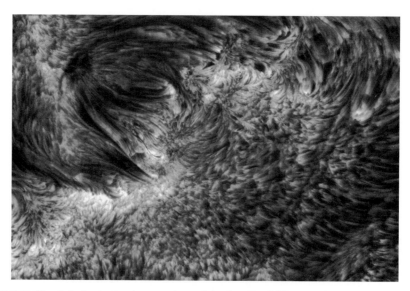

**FIGURE 11.17**   Spicules in the chromosphere of the Sun. In addition, small sunspots are visible in the upper left quadrant of the image, and brighter areas known as plage regions are also visible. The observations were made using the H$\alpha$ emission line. Features as small as 130 km are evident in this image. (Courtesy of the Royal Swedish Academy of Sciences.)

62.5-nm line at $1.4 \times 10^6$ K. Figure 11.19 shows images of the Sun at various wavelengths and heights above the base of the photosphere.

### The Corona

When the Moon fully occults the photosphere during a total solar eclipse, the radiation from the faint **corona** becomes visible (Fig. 11.20). The corona, located above the transition region, extends out into space without a well-defined outer boundary and has an energy output that is nearly $10^6$ times less intense than that of the photosphere. The number density of particles at the base of the corona is typically $10^{15}$ particles m$^{-3}$, whereas in the vicinity of Earth, the number density of particles originating from the Sun (solar wind particles) have a characteristic value of $10^7$ particles m$^{-3}$ (this can be compared with $10^{25}$ particles m$^{-3}$ at sea level in Earth's atmosphere). Because the density of the corona is so low, it is essentially transparent to most electromagnetic radiation (except long radio wavelengths) and is not in local thermodynamic equilibrium (LTE). For gases that are not in LTE, a unique temperature is not strictly definable (see Section 9.2). However, the temperatures obtained by considering thermal motions, ionization levels, and radio emissions do give reasonably consistent results. For instance, the presence of Fe XIV lines indicates temperatures in excess of $2 \times 10^6$ K, as do line widths produced by thermal Doppler broadening.

Based on the radiation coming from the corona, three distinct structural components can be identified:

- The **K corona** (from *Kontinuierlich*, the German word for "continuous") produces the continuous white light emission that results from photospheric radiation scattered

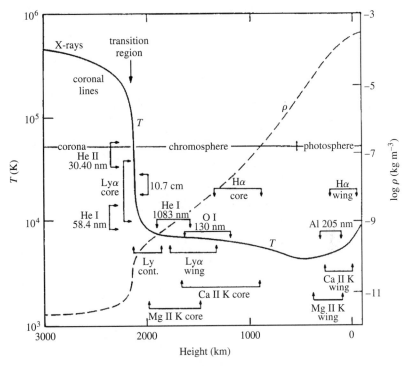

**FIGURE 11.18**   Logarithmic plots of the temperature structure (solid line) and mass density structure (dashed line) of the upper atmosphere of the Sun. The altitudes observed at various wavelengths are also depicted. [Figure adapted from Avrett, in *Encyclopedia of Astronomy and Astrophysics*, Paul Murdin (ed.), Institute of Physics Publishing, Bristol, 2001, page 2480.]

by free electrons. Contributions to the coronal light due to the K corona primarily occur between 1 and 2.3 $R_\odot$ from the center of the Sun. The spectral lines evident in the photosphere are essentially blended by the large Doppler shifts that are caused by the high thermal velocities of the electrons.

- The **F corona** (for Fraunhofer) comes from the scattering of photospheric light by dust grains that are located beyond 2.3 $R_\odot$. Because dust grains are much more massive and slower than electrons, Doppler broadening is minimal and the Fraunhofer lines are still detectable. The F corona actually merges with the **zodiacal light**, the faint glow found along the ecliptic that is a reflection of the Sun's light from interplanetary dust.

- The **E corona** is the source of the emission lines that are produced by the highly ionized atoms located throughout the corona; the E corona overlaps the K and F coronas. Since the temperatures are extremely high in the corona, the exponential term in the Saha equation encourages ionization because thermal energies are comparable to ionization potentials. The very low number densities also encourage ionization since the chance of recombination is greatly reduced.

**FIGURE 11.19**   Visible features of the Sun at various wavelengths. The central image is a three-color composite of the corona obtained by TRACE at 17.1 nm, 19.5 nm, and 28.4 nm. Clockwise starting from the top are a SOHO/MDI magnetic map, white light, TRACE 170 nm continuum, TRACE Lyα, TRACE 17.1 nm, TRACE 19.5 nm, TRACE 28.4 nm, and a Yohkoh/SXT X-ray image. [The Transition Region and Coronal Explorer, TRACE, is a mission of the Stanford-Lockheed Institute for Space Research (a joint program of the Lockheed-Martin Advanced Technology Center's Solar and Astrophysics Laboratory and Stanford's Solar Observatories Group) and part of the NASA Small Explorer program.]

The low number densities allow forbidden transitions to occur, producing spectral lines that are generally seen only in astrophysical environments where gases are extremely thin (recall the discussion of selection rules in Section 5.4). Forbidden transitions occur from atomic energy levels that are **metastable**; electrons do not readily make transitions from metastable states to lower energy states without assistance. Whereas allowed transitions occur on timescales on the order of $10^{-8}$ s, spontaneous forbidden transitions may require one second or longer. In gases at higher densities, electrons are able to escape from metastable states through collisions with other atoms or ions, but in the corona these collisions are rare. Consequently, given enough time, some electrons will be able to make spontaneous transitions from metastable states to lower energy states, accompanied by the emission of photons.

(a) (b)

**FIGURE 11.20** (a) The quiet solar corona seen during a total solar eclipse in 1954. The shape of the corona is elongated along the Sun's equator. (Courtesy of J. D. R. Bahng and K. L. Hallam.) (b) The active corona tends to have a very complex structure. This image of the July 11, 1991, eclipse is a composite of five photographs that was processed electronically. (Courtesy of S. Albers.)

Since the blackbody continuum emission from the Sun decreases like $\lambda^{-2}$ for sufficiently long wavelengths (see Eq. 3.22), the amount of photospheric radio emission is negligible. The solar corona, however, is a source of radio-wavelength radiation that is not associated with the blackbody continuum. Some radio emission arises from free–free transitions of electrons that pass near atoms and ions. During these close encounters, photons may be emitted as the electrons' energies are decreased slightly. From the conservation of energy, the greater the change in the energy of an electron, the more energetic the resulting photon and the shorter its wavelength. Clearly, the closer an electron comes to an ion, the more likely it is that the electron's energy will change appreciably. Since more frequent and closer encounters are expected if the number density is larger, shorter-wavelength radio emissions should be observed nearer the Sun. Radio wavelengths of 1 to 20 cm are observed from the chromosphere through the lower corona, while longer wavelength radiation originates from the outer corona. It is important to note that synchrotron radiation by relativistic electrons also contributes to the observed radio emission from the solar corona (recall the discussion of the headlight effect in Example 4.3.3).

Photospheric emissions are negligible in the X-ray wavelength range as well. In this case the blackbody continuum decreases very rapidly, dropping off like $\lambda^{-5}e^{-hc/\lambda kT}$. Consequently, any emission in X-ray wavelengths from the corona will completely overwhelm the output from the photosphere. In fact, because of the high temperatures of the corona, its X-ray spectrum is very rich in emission lines. This is due to the high degree of ionization that exists for all of the elements present, together with the ability of the corona to excite a large number of atomic transitions. Given the many electrons that are present in heavy elements such as iron and the vast number of available energy levels, each such element is capable of producing an extensive emission spectrum. Figure 11.21 shows a section of the X-ray emission spectrum of the solor corona. It displays a sample of the lines that are observed in one portion of the X-ray wavelength band, along with the ions responsible for their production.

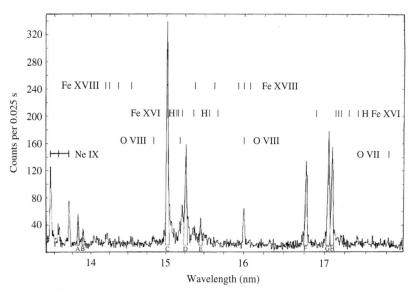

**FIGURE 11.21**   A section of the X-ray emission spectrum of the solar corona. (Figure adapted from Parkinson, *Astron. Astrophys.*, *24*, 215, 1973.)

## Coronal Holes and the Solar Wind

An image of the X-ray Sun is shown in Fig. 11.22. This fascinating picture indicates that X-ray emission is not uniform. Active (bright and hot) regions exist, along with darker, cooler regions known as **coronal holes**. Moreover, even in the coronal holes, localized bright spots of enhanced X-ray emission appear and disappear on a timescale of several hours. Smaller features are also apparent within the regions of generally bright X-ray emission.

The weaker X-ray emission coming from coronal holes is characteristic of the lower densities and temperatures that exist in those regions, as compared to the rest of the corona. The explanation for the existence of coronal holes is tied to the Sun's magnetic field and the generation of the *fast* **solar wind**, a continuous stream of ions and electrons escaping from the Sun and moving through interplanetary space at speeds of approximately 750 km s$^{-1}$. A gusty, *slow solar wind*, with speeds of roughly one-half those of the fast wind appears to be produced by streamers in the corona associated with closed magnetic fields.

Just like the magnetic field that is produced by a current loop, the magnetic field of the Sun is generally that of a dipole, at least on a global scale (Fig. 11.23). Although its value can differ significantly in localized regions (as we will see in the next section), the strength of the field is typically a few times $10^{-4}$ T near the surface.[9] Coronal holes correspond to those parts of the magnetic field where the field lines are *open*, while the X-ray bright regions are associated with *closed* field lines; open field lines extend out to great distances from the Sun, while closed lines form loops that return to the Sun.

---

[9]The magnetic field near the surface of Earth is approximately $6 \times 10^{-5}$ T.

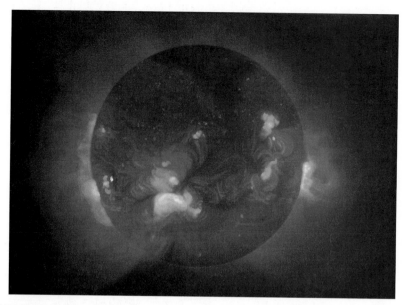

**FIGURE 11.22**   An X-ray image of the Sun obtained by the Soft X-Ray Telescope on the Yohkoh Solar Observatory, May 8, 1992. Bright regions are hotter X-ray regions and darker regions are cooler. A dark coronal hole is evident at the top of the image. (From the Yohkoh mission of ISAS, Japan. The X-ray telescope was prepared by the Lockheed Palo Alto Research Laboratory, the National Astronomical Observatory of Japan, and the University of Tokyo with the support of NASA and ISAS.)

The Lorentz force equation,

$$\mathbf{F} = q\,(\mathbf{E} + \mathbf{v} \times \mathbf{B}),$$

(11.2)

describing the force exerted on a charged particle of velocity $\mathbf{v}$ in an electric field $\mathbf{E}$ and a magnetic field $\mathbf{B}$ states that the force due to the magnetic field is always mutually perpendicular to both the direction of the velocity vector and the field (the cross product). Providing that electric fields are negligible, charged particles are forced to spiral around magnetic field lines and cannot actually cross them except by collisions (Fig. 11.24). This implies that closed magnetic field lines tend to trap charged particles, not allowing them to escape. In regions of open field lines, however, particles can actually follow the lines out away from the Sun. Consequently, the solar wind originates from the regions of open magnetic field lines, namely the coronal holes. The details observed in the X-ray-bright regions, as well as the localized bright spots in the coronal holes, are due to the higher densities of the electrons and ions that are trapped in large and small magnetic field loops.

The existence of ongoing mass loss from the Sun was deduced long before it was ever detected directly, as evidenced by the tails of comets. The tails are generally composed of two parts, a curved dust tail and a straight ion tail, both of which are always pointed away from the Sun (Fig. 11.25). The force exerted on dust grains by photons (radiation pressure)

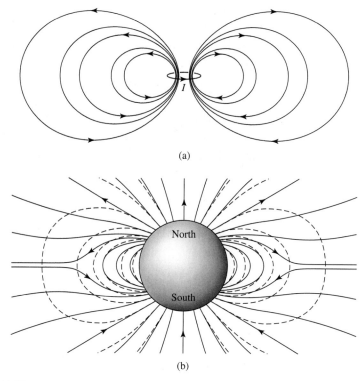

(a)

(b)

**FIGURE 11.23** (a) The characteristic dipole magnetic field of a current loop. (b) A generalized depiction of the global magnetic field of the Sun. The dashed lines show the field of a perfect magnetic dipole.

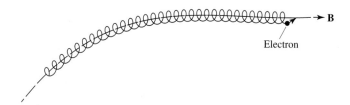

**FIGURE 11.24** A charged particle is forced to spiral around a magnetic field line because the Lorentz force is mutually perpendicular to both the velocity of the particle and the direction of the magnetic field.

is sufficient to push the dust tail back; the curvature of the tail is due to the different orbital speeds of the individual dust grains, which, according to Kepler's third law, are a function of their varying distances from the Sun. However, the ion tail cannot be explained by radiation pressure; the interaction between photons and the ions is not efficient enough. Rather, it is the electric force between the ions of the solar wind and the ions in the comet that counts for the direction of the ion tail. This interaction allows momentum to be transferred to the cometary ions, driving them straight away from the Sun.

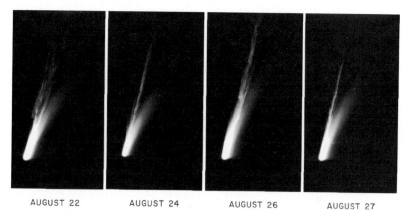

AUGUST 22        AUGUST 24        AUGUST 26        AUGUST 27

**FIGURE 11.25** Comet Mrkos in 1957. The dust tail of a comet is curved and its ion tail is straight. (Courtesy of Palomar/Caltech.)

**FIGURE 11.26** Aurora australis seen over the South Pole. (NASA)

The **aurora borealis** and the **aurora australis** (the northern and southern lights, respectively) are also products of the solar wind (see Fig. 11.26). As the ions from the Sun interact with Earth's magnetic field, they become trapped in it. Bouncing back and forth between the north magnetic pole and the south magnetic pole, these ions form the **Van Allen radiation belts**. Ions that are sufficiently energetic will collide with the atoms in Earth's upper atmosphere near the magnetic poles, causing the atmospheric atoms to become excited or ionized. The resulting de-excitations or recombinations emit the photons that produce the spectacular light displays observed from high northern and southern latitudes.

Using rockets and satellites, characteristics of the two solar winds can be measured as they pass near Earth. In addition, the Ulysses spacecraft, placed in a polar orbit around the Sun, was able to detect the wind well out of the plane of Earth's orbit. At a distance of 1 AU from the Sun, the solar wind velocity ranges from approximately 200 km s$^{-1}$ to 750 km s$^{-1}$, with a typical density of $7 \times 10^6$ ions m$^{-3}$ and characteristic kinetic temperatures of $4 \times 10^4$ K for protons and $10^5$ K for electrons. Although the winds are composed primarily of protons and electrons, heavier ions are present as well.

---

**Example 11.2.1.**   The mass loss rate of the Sun may be estimated from the data given above. We know that all of the mass leaving the Sun must also pass through a sphere of radius 1 AU centered on the Sun; otherwise it would collect at some location in space. If we further assume (for simplicity) that the mass loss rate is spherically symmetric, then the amount of mass crossing a spherical surface of radius $r$ in an amount of time $t$ is just the mass density of the gas multiplied by the volume of the shell of gas that can travel across the sphere during that time interval, or

$$dM = \rho \, dV = (nm_H)(4\pi r^2 \, v \, dt),$$

where $n$ is the number density of ions (mostly hydrogen), $m_H$ is approximately the mass of a hydrogen ion, $v$ is the ion velocity, and $dV = A \, dr \simeq 4\pi r^2 \, v \, dt$ is the volume of a shell that crosses a spherical surface in an amount of time $dt$. Dividing both sides by $dt$, we obtain the mass loss rate,

$$\frac{dM}{dt} = 4\pi r^2 \, nm_H v = 4\pi r^2 \, \rho v. \tag{11.3}$$

By convention, stellar mass loss rates are generally given in *solar masses per year* and symbolized by $\dot{M} \equiv dM/dt$. Using $v = 500$ km s$^{-1}$, $r = 1$ AU, and $n = 7 \times 10^6$ protons m$^{-3}$, we find that

$$\dot{M}_\odot \simeq 3 \times 10^{-14} \, M_\odot \; \text{yr}^{-1}.$$

At this rate it would require more than $10^{13}$ yr before the entire mass of the Sun is dissipated. However, the interior structure of the Sun is changing much more rapidly than this, so the effect of the present-day solar wind on the evolution of the Sun is minimal.

As an interesting aside, in 1992 both *Voyagers I* and *II* detected radio noise at frequencies of 1.8 to 3.5 kHz originating from the outer reaches of the Solar System. It is believed that the noise is produced where particles from the solar wind collide with the interstellar medium, producing a termination shock. (The interstellar medium is the dust and gas located between the stars; see Section 12.1). The 1992 observations represented the first detection of the **heliopause**, the outer limit of the Sun's electromagnetic influence. In 2005, when Voyager I was about 95 AU from Earth and traveling at 3.6 AU per year, it passed through the termination shock into the region known as the **heliosheath**. The strongest evidence that Voyager I did in fact cross the termination shock comes from the measurement of a sudden significant increase in the strength of the magnetic field that is carried by the solar wind. This increased magnetic field strength is due to the slowing of the solar wind particles and the resulting increase in particle density.

---

**The Parker Wind Model**

We now consider how the expansion of the solar corona produces the solar wind. This is a result of the corona's high temperature, together with the high thermal conductivity of the ionized gas, referred to as a **plasma**. The ability of the plasma to conduct heat implies that the corona is almost isothermal (recall Fig. 11.18).

In 1958 Eugene Parker developed an approximately isothermal model of the solar wind that has been successful in describing many of its basic features. To see why the solar wind is inevitable, begin by considering the condition of hydrostatic equilibrium, Eq. (10.6). If the mass of the corona is insignificant compared to the total mass of the Sun, then $M_r \simeq M_\odot$ in that region and the hydrostatic equilibrium equation becomes

$$\frac{dP}{dr} = -\frac{GM_\odot \rho}{r^2}.$$
(11.4)

Next, assuming for simplicity that the gas is completely ionized and composed entirely of hydrogen, the number density of protons is given by

$$n \simeq \frac{\rho}{m_p}$$

since $m_p \simeq m_H$. From the ideal gas law (Eq. 10.11), the pressure of the gas may be written as

$$P = 2nkT,$$

where $\mu = 1/2$ for ionized hydrogen and $m_H \simeq m_p$. Substituting expressions for the pressure and density into Eq. (11.4), the hydrostatic equilibrium equation becomes

$$\frac{d}{dr}(2nkT) = -\frac{GM_\odot n m_p}{r^2}.$$
(11.5)

Making the assumption that the gas is isothermal, Eq. (11.5) can be integrated directly to give an expression for the number density (and therefore the pressure) as a function of radius. It is left as an exercise to show that

$$n(r) = n_0 e^{-\lambda(1 - r_0/r)},$$
(11.6)

where

$$\lambda \equiv \frac{GM_\odot m_p}{2kT r_0}$$

and $n = n_0$ at some radius $r = r_0$. Note that $\lambda$ is approximately the ratio of a proton's gravitational potential energy and its thermal kinetic energy at a distance $r_0$ from the center of the Sun. We now see that the pressure structure is just

$$P(r) = P_0 e^{-\lambda(1 - r_0/r)},$$

where $P_0 = 2n_0 kT$.

An immediate consequence of Eq. (11.2) is that in our isothermal approximation the pressure *does not* approach zero as $r$ goes to infinity. To estimate the limiting values of $n(r)$ and $P(r)$, let $T = 1.5 \times 10^6$ K and $n_0 = 3 \times 10^{13}$ m$^{-3}$ at about $r_0 = 1.4\,R_\odot$, values typical of the inner corona. Then $\lambda \simeq 5.5$, $n(\infty) \simeq 1.2 \times 10^{11}$ m$^{-3}$, and $P(\infty) \simeq 5 \times 10^{-6}$ N m$^{-2}$. However, as we will see in Section 12.1, with the exception of localized clouds of material, the actual densities and pressures of interstellar dust and gas are much lower than those just derived.

Given the inconsistency that exists between the isothermal, hydrostatic solution to the structure of the corona and the conditions in interstellar space, at least one of the assumptions made in the derivation must be incorrect. Although the assumption that the corona is approximately isothermal is not completely valid, it is roughly consistent with observations. Recall that near Earth ($r \sim 215\,R_\odot$), the solar wind is characterized by temperatures on the order of $10^5$ K, indicating that the temperature of the gas is not decreasing rapidly with distance. It can be shown that solutions that allow for a realistically varying temperature structure still do not eliminate the problem of a predicted gas pressure significantly in excess of the interstellar value. Apparently, it is the assumption that the corona is in hydrostatic equilibrium that is wrong. Since $P(\infty)$ greatly exceeds the pressures in interstellar space, material must be expanding outward from the Sun, implying the existence of the solar wind.

### The Hydrodynamic Nature of the Upper Solar Atmosphere

If we are to develop an understanding of the structure of the solar atmosphere, the simple approximation of hydrostatic equilibrium must be replaced by a set of **hydrodynamic equations** that describe the flow. In particular, when we write

$$\frac{d^2 r}{dt^2} = \frac{dv}{dt} = \frac{dv}{dr}\frac{dr}{dt} = v\frac{dv}{dr},$$

Eq. (10.5) becomes

$$\rho v \frac{dv}{dr} = -\frac{dP}{dr} - G\frac{M_r \rho}{r^2}, \tag{11.7}$$

where $v$ is the velocity of the flow. With the introduction of a new variable (velocity), another expression that describes the conservation of mass flow across boundaries must also be included, specifically

$$4\pi r^2 \rho v = \text{constant},$$

which is just the relationship that was used in Example 11.2.1 to estimate the Sun's mass loss rate. This expression immediately implies that

$$\frac{d(\rho v r^2)}{dr} = 0.$$

At the top of the convection zone, the motion of the hot, rising gas and the return flow of the cool gas sets up longitudinal waves (pressure waves) that propagate outward through the photosphere and into the chromosphere. The outward flux of wave energy, $F_E$, is governed by the expression

$$F_E = \frac{1}{2}\rho v_w^2 v_s, \tag{11.8}$$

where $v_s$ is the local sound speed and $v_w$ is the velocity amplitude of the oscillatory wave motion for individual particles being driven about their equilibrium positions by the "piston" of the convection zone.

From Eq. (10.84), the sound speed is given by

$$v_s = \sqrt{\gamma P/\rho}.$$

Since, according to the ideal gas law, $P = \rho k T/\mu m_H$, the sound speed may also be written as

$$v_s = \sqrt{\frac{\gamma k T}{\mu m_H}} \propto \sqrt{T}$$

for fixed $\gamma$ and $\mu$.

When the wave is first generated at the top of the convection zone, $v_w < v_s$. However, the density of the gas that these waves travel through decreases significantly with altitude, dropping by four orders of magnitude in approximately 1000 km. If we assume that very little mechanical energy is lost in moving through the photosphere (i.e., $4\pi r^2 F_E$ is approximately constant) and that $v_s$ remains essentially unchanged since the temperature varies by only about a factor of two across the photosphere and chromosphere, the rapid decrease in density means that $v_w$ must increase significantly (approximately two orders of magnitude). As a result, the wave motion quickly becomes *supersonic* ($v_w > v_s$) as particles in the wave try to travel through the medium faster than the local speed of sound. The result is that the wave develops into a **shock wave**, much like the shock waves that produce sonic booms behind supersonic aircraft.

A shock wave is characterized by a very steep density change over a short distance, called the **shock front**.[10] As a shock moves through a gas, it produces a great deal of heating via collisions, leaving the gas behind the shock highly ionized. This heating comes at the expense of the mechanical energy of the shock, and the shock quickly dissipates. Thus the gas in the chromosphere and above is effectively heated by the mass motions created in the convection zone.

### Magnetohydrodynamics and Alfvén Waves

It should be noted that our discussion of the hydrodynamic equations has failed to account for the influence of the Sun's magnetic field. It is believed that the temperature structure throughout the outer solar atmosphere, including the very steep *positive* temperature gradient in the transition region, is due at least in part to the presence of the magnetic field, coupled with mass motions produced by the convection zone. **Magnetohydrodynamics**

---

[10]Recall the discussion of the heliopause at the end of Example 11.2.1.

(usually mercifully shortened to **MHD**) is the study of the interactions between magnetic fields and plasmas. Owing to the great complexity of the problem, a complete solution to the set of MHD equations applied to the outer atmosphere of the Sun does not yet exist. However, some aspects of the solution can be described.

The presence of the magnetic field allows for the generation of a second kind of wave motion. These waves may be thought of as transverse waves that propagate along the magnetic field lines as a consequence of the restoring force of tension associated with the magnetic field lines. To understand the origin of this restoring force, recall that establishing a magnetic field (which is always generated by moving electric charges, or currents) requires that energy be expended. The energy used to establish the field can be thought of as being stored within the magnetic field itself; thus the space containing the magnetic field also contains a magnetic energy density. The value of the magnetic energy density is given by

$$u_m = \frac{B^2}{2\mu_0}. \qquad (11.9)$$

If a volume $V$ of plasma containing a number of magnetic field lines is compressed in a direction perpendicular to the lines, the density of field lines necessarily increases.[11] But the density of field lines is just a description of the strength of the magnetic field itself, so the energy density of the magnetic field also increases during compression. An amount of mechanical work must therefore have been done in compressing the field lines in the gas. Since work is given by $W = \int P\,dV$, the compression of the plasma must imply the existence of a magnetic pressure. It can be shown that the magnetic pressure is numerically equal to the magnetic energy density, or

$$P_m = \frac{B^2}{2\mu_0}. \qquad (11.10)$$

When a magnetic field line gets displaced by some amount perpendicular to the direction of the line, a *magnetic pressure gradient* becomes established; the pressure in the direction of the displacement increases as indicated by an increase in the number density of field lines, while at the same time the pressure in the opposite direction decreases. This pressure change then tends to push the line back again, restoring the original density of field lines. This process may be thought of as analogous to the oscillations that occur in a string when a portion of the string is displaced; it is the tension in the string that pulls it back when it is plucked. The "tension" that restores the position of the magnetic field line is just the magnetic pressure gradient.

As with the traveling motion of a wave on a string, a disturbance in the magnetic field line can also propagate down the line. This transverse MHD wave is called an **Alfvén wave**.[12]

---

[11] Recall that if the electric field is negligible, charged particles must spiral around field lines. This implies that if the charged particles are pushed, they drag the field lines with them; the field lines are said to be "frozen in" the plasma.

[12] Alfvén waves are named for Hannes Olof Gösta Alfvén, (1908–1995), who was awarded the Nobel Prize in 1970 for his fundamental studies in magnetohydrodynamics.

The speed of propagation of the Alfvén wave may be estimated by making a comparison with the sound speed in a gas. Since the adiabatic sound speed is given by

$$v_s = \sqrt{\frac{\gamma P_g}{\rho}},$$

where $\gamma$ is of order unity, by analogy the *Alfvén speed* should be approximately

$$v_m \sim \sqrt{\frac{P_m}{\rho}} = \frac{B}{\sqrt{2\mu_0 \rho}}.$$

A more careful treatment gives the result

$$v_m = \frac{B}{\sqrt{\mu_0 \rho}}. \tag{11.11}$$

---

**Example 11.2.2.** Using Eqs. (10.84) and (11.11), the sound speed and Alfvén speed may be compared for the photosphere. The gas pressure at the top of the photosphere is roughly $140\ \mathrm{N\ m^{-2}}$, with a density of $4.9 \times 10^{-6}\ \mathrm{kg\ m^{-3}}$. Assuming an ideal monatomic gas for which $\gamma = 5/3$,

$$v_s \simeq 7000\ \mathrm{m\ s^{-1}}.$$

Notice that this speed is significantly lower than we found using global solar values in Example 10.4.2; apparently, the sound speed is much larger in the Sun's interior.

Taking a typical surface magnetic field strength to be $2 \times 10^{-4}$ T, the magnetic pressure is (from Eq. 11.10)

$$P_m \simeq 0.02\ \mathrm{N\ m^{-2}},$$

and the Alfvén speed is

$$v_m \simeq 10\ \mathrm{m\ s^{-1}}.$$

The magnetic pressure may generally be neglected in photospheric hydrostatic considerations since it is smaller than the gas pressure by roughly four orders of magnitude. However, we will see in the next section that much larger magnetic field strengths can exist in localized regions on the Sun's surface.

---

Since Alfvén waves can propagate along magnetic field lines, they may also transport energy outward. According to Maxwell's equations, a time-varying magnetic field produces an electric field, which in turn creates electrical currents in the highly conductive plasma. This implies that some resistive Joule heating will occur in the ionized gas, causing the temperature to rise. Thus MHD waves can also contribute to the temperature structure of the upper solar atmosphere.

**FIGURE 11.27**    The Sun's rotation creates a spiral pattern in the solar magnetic field in interplanetary space, known as the Parker spiral. The drag produced by the spiraling magnetic field causes angular momentum to be transferred away from the Sun. This diagram shows the heliospheric current sheet that separates regions of space where the magnetic field points toward or away from the Sun. The orbits of the planets out to Jupiter are depicted. (Courtesy of Prof. John M. Wilcox and NASA artist Werner Heil.)

Because of the Sun's rotation, its open magnetic field lines are dragged along through interplanetary space (Fig. 11.27). Since the solar wind is forced to move with the field lines, a torque is produced that actually slows the Sun's rotation. Said another way, the solar wind is transferring angular momentum away from the Sun. As a result, the Sun's rotation rate will decrease significantly over its lifetime. Interestingly, the differential rotation present in the photosphere is not manifested in the corona. Apparently, the magnetic field, which so strongly influences the structure of the corona, does not exhibit differential rotation at this height.

### The Outer Atmospheres of Other Stars

Although this chapter is devoted to our Sun, the most thoroughly studied of all stars, the outer atmospheres of other stars can be investigated as well. For instance, observations indicate that the rotation rates of solar-type stars seem to decrease with age. Furthermore, late main-sequence stars, with their convective envelopes, generally have much slower rotation rates than stars on the upper end of the main sequence. Perhaps winds are transferring angular momentum away from these lower-mass stars as well.

A host of satellites such as EUVE, FUSE, ROSAT, ASCA, XMM-Newton, and the Chandra X-Ray Observatory have also provided us with valuable UV and X-ray observations of other stars. It appears that stars along the main sequence that are cooler than spectral class F have emission lines in the ultraviolet that are similar to those observed coming from the Sun's chromosphere and transition region. In addition, X-ray observations indicate corona-like emissions. These stars are also those for which stellar structure calculations indicate that surface convection zones should exist. Apparently, the same mechanisms that are heating the outer atmosphere of our Sun are also in operation in other stars.

## 11.3 ■ THE SOLAR CYCLE

Some of the most fascinating and complex features of the solar atmosphere are transient in nature. However, as we will learn in this section, many observational features of the solar atmosphere are also cyclic.

### Sunspots

It was Galileo who made the first telescopic observations of **sunspots** (recall Fig. 11.11). Sunspots are even visible occasionally with the unaided eye, but making such observations is ***strongly discouraged*** because of the potential for eye damage.

Reliable observations made over the past two centuries indicate that the number of sunspots is approximately periodic, going from minimum to maximum and back to minimum again nearly every 11 years (Fig. 11.28). The average latitude of sunspot formation is also periodic, again over an 11-year cycle. A plot of sunspot location as a function of time is shown in Fig. 11.29, along with a plot of the percentage of the solar surface covered by sunspots. Because of its wing-like appearance, the top portion of Fig. 11.29 has come to be known as the **butterfly diagram**. Individual sunspots are short-lived features, typically surviving no more than a month or so. During its lifetime, a sunspot will remain at a constant latitude, although succeeding sunspots tend to form at progressively lower latitudes. As the last sunspots of one cycle vanish near the Sun's equator, a new cycle begins near ±40° (north and south) of the equator. The largest number of spots (sunspot maximum) typically occurs at intermediate latitudes.

The key to understanding sunspots lies in their strong magnetic fields. A typical sunspot is shown in Fig. 11.30. The darkest portion of the sunspot is known as the **umbra** and

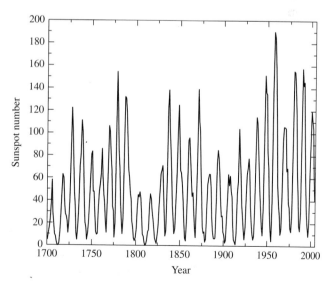

**FIGURE 11.28** The number of sunspots between 1700 and 2005 indicates an 11-year periodicity. (Data from the World Data Center for the Sunspot Index at the Royal Observatory of Belgium.)

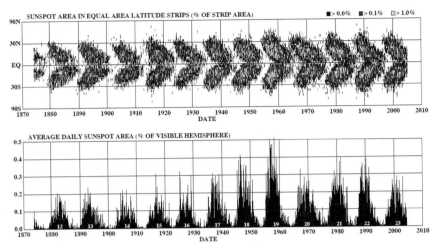

**FIGURE 11.29**    The upper figure depicts the butterfly diagram, showing sunspot latitudes with time. The lower figure shows the percentage of the Sun's surface covered by sunspots as a function of time. (Courtesy of Dr. David H. Hathaway, NASA/Marshall Space Flight Center.)

may measure as much as 30,000 km in diameter. (For reference, the diameter of Earth is 12,756 km.) The umbra is usually surrounded by a filament-like structure, called the **penumbra**, whose mere appearance suggests the presence of magnetic lines of force. The existence of a strong magnetic field can be verified by observing individual spectral lines produced within the spot. As was discussed in Section 5.4, the strength and polarity of magnetic fields can be measured by observing the Zeeman effect, the splitting of spectral lines that results from removing the degeneracy inherent in atomic energy levels. The amount of splitting is proportional to the strength of the magnetic field, whereas the polarization of the light corresponds to the direction of the field. Figure 11.31 shows an example of the splitting of a spectral line measured across a sunspot. Magnetic field strengths of several tenths of a tesla and greater have been measured in the centers of umbral regions, with field strengths decreasing across penumbral regions. Furthermore, polarization measurements indicate that the direction of a typical umbral magnetic field is vertical, becoming horizontal across the penumbra.

Sunspots are generally located in groups. Typically, a dominant sunspot leads in the direction of rotation, and one or more sunspots follow. During an 11-year cycle, the lead sunspot will always have the same polarity in one hemisphere—say, a north pole in the geographic northern hemisphere—while the lead sunspot in the other hemisphere will have the opposite polarity (e.g., a south pole in the geographic southern hemisphere); trailing sunspots have the opposite polarity. Even when a large collection of trailing spots exist, resulting from a tangled magnetic field pattern, a basically *bipolar* field is present. During the next 11-year cycle, polarities will be reversed; the sunspot with a magnetic south polarity will lead in the northern hemisphere, and vice versa in the southern hemisphere. Accompanying this local polarity reversal is a global polarity reversal: the overall dipole field of the Sun

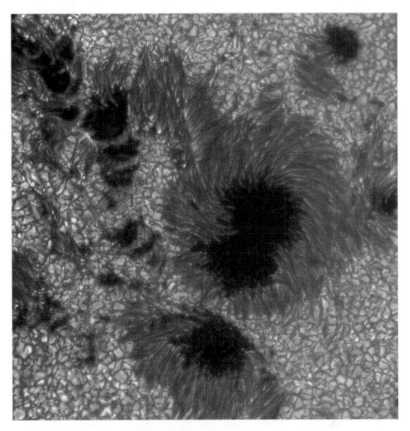

**FIGURE 11.30**  A typical sunspot group. The dark umbra of the central sunspot is clearly evident, as is the filamentary structure of its penumbra. (Courtesy of the Royal Swedish Academy of Sciences.)

will change so that the magnetic north pole of the Sun will switch from the geographic north pole to the geographic south pole. Polarity reversal always occurs during sunspot minimum, when the first sunspots are beginning to form at the highest latitudes. When the polarity reversal is considered, the Sun is said to have a **22-year cycle**. This important magnetic behavior is illustrated in Fig. 11.32.

The dark appearance of sunspots is due to their significantly lower temperatures. In the central portion of the umbra the temperature may be as low as 3900 K, compared with the Sun's effective temperature of 5777 K. From Eq. (3.18), this implies a surface bolometric flux that is a factor of $(5777/3900)^4 = 4.8$ lower than that of the surrounding photosphere.[13] Observations obtained from the Solar Maximum Mission satellite (SMM) have shown that this decrease in surface flux affects the overall energy output of the Sun. When a number of large sunspots exist, the solar luminosity is depressed by roughly 0.1%. Since convection is the principal energy transport mechanism just below the photosphere, and since strong

---

[13]A 3900-K blackbody is very bright, of course. However, when seen through a filter dark enough to make viewing the rest of the 5777-K photosphere comfortable, the sunspot appears dark.

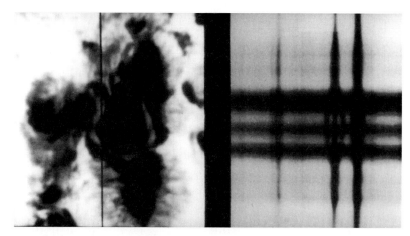

**FIGURE 11.31**    The Zeeman splitting of the Fe 525.02-nm spectral line due to the presence of a strong magnetic field in a sunspot. The spectrograph slit was aligned vertically across a sunspot, resulting in a wavelength dependence that runs from left to right in the image. The slit extended beyond the image of the sunspot. (Courtesy of the National Optical Astronomy Observatories/National Solar Observatory.)

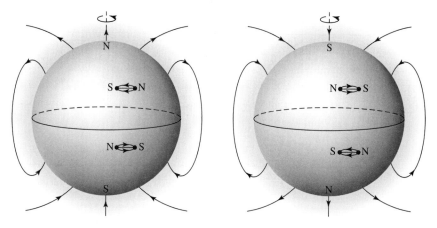

**FIGURE 11.32**    The global magnetic field orientation of the Sun, along with the magnetic polarity of sunspots during successive 11-year periods.

magnetic fields inhibit motion through the "freezing in" of field lines in a plasma, it is likely that the mass motion of convective bubbles is inhibited in sunspots, thereby decreasing the flow of energy through the sunspots.

Along with luminosity variations on a timescale of months (the typical lifetime of an individual sunspot), the Sun's luminosity seems to experience variability on a much longer timescale, as does the number of sunspots. For instance, very few sunspots were observed between 1645 and 1715; this time interval has come to be called the **Maunder minimum**

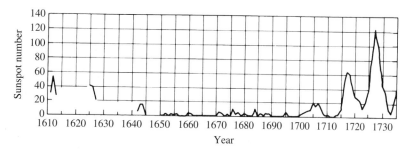

**FIGURE 11.33** An unusually small number of sunspots were observed between 1645 and 1715 (the Maunder minimum). (Adapted from a figure courtesy of J. A. Eddy, High Altitude Observatory.)

(see Fig. 11.33).[14] Surprisingly, during this period the average temperature in Europe was significantly lower, consistent with the solar luminosity being a few tenths of a percent less than it is today. John Eddy has proposed that there is a very long-term periodicity on which the solar cycle is superimposed. This long-period variation goes through grand sunspot maxima and minima that may last for centuries. Evidence in support of this suggestion is found on Earth in the relative numbers of atmospheric carbon dioxide molecules that contain radioactive carbon atoms ($^{14}_{6}C$), as preserved in the 7000-year-long record of tree rings. The importance of $^{14}_{6}C$ in long-term sunspot studies lies in an inverse correlation between sunspots and the amount of $^{14}_{6}C$ present in Earth's atmosphere. $^{14}_{6}C$ is a radioactive isotope of carbon that is produced when extremely energetic charged particles from space, called **cosmic rays**, collide with atmospheric nitrogen. Cosmic rays are affected by the magnetic field of the Sun, which in turn is affected by solar activity. During the Maunder minimum, the amount of atmospheric $^{14}_{6}C$ increased significantly and was incorporated into the rings of living trees. The amount of $^{14}_{6}C$ also seems to correlate well with the advance and retreat of glaciers over the past 5000 years.

With lower temperatures in sunspots, the gas pressure is necessarily lower than in the surrounding material (see Eq. 10.11). However, the gravitational force is essentially the same. From these considerations alone, it seems as though the gas within a sunspot ought to sink into the interior of the star, an effect that is not observed. Without the benefit of a sufficiently large gas pressure gradient to support a sunspot, another component to the pressure must exist. As we have already seen in the last section, a magnetic field is accompanied by a pressure term. It is this extra magnetic pressure that provides the support necessary to keep a sunspot from sinking or being compressed by the surrounding gas pressure.

### Plages

A variety of other phenomena are also associated with sunspot activity. **Plages** (from the French word for *beach*) are chromospheric regions of bright $H\alpha$ emission located near

---

[14]With the development and continual improvement of the telescope beginning during the early phase of the Maunder minimum (recall that Galileo died in 1642 and Newton was born in the same year), the Maunder minimum was not a manifestation of poor observations.

active sunspots (recall Fig. 11.17). They usually form before the sunspots appear and usually disappear after the sunspots vanish from a particular area. Plages have higher densities than the surrounding gas and are products of the magnetic fields. Apparently the cause of the decreased brightness of sunspots does not play an important role in plages.

### Solar Flares

**Solar flares** are eruptive events that are known to release from $10^{17}$ J of energy at the lower detection limit to as much as $10^{25}$ J of energy over time intervals ranging from milliseconds to more than an hour.[15] The physical dimensions of a flare are enormous, with a large flare reaching 100,000 km in length (see Fig. 11.34a). During an eruption, the hydrogen Balmer line, H$\alpha$, appears locally in emission rather than in absorption, as is usually the case, implying that photon production occurs above much of the absorbing material. When observed in H$\alpha$, a flare is often seen on the disk as two ribbons of light (Fig. 11.34b). Along with H$\alpha$, other types of electromagnetic radiation are produced that can range from kilometer-wavelength *nonthermal* radio waves due to synchrotron radiation (see Section 4.3) to very short-wavelength hard X-ray and gamma-ray emission lines.

Charged particles are also ejected outward at high speeds, many escaping into interplanetary space as **solar cosmic rays**. In the largest flares the ejected charged particles, mostly protons and helium nuclei, may reach Earth in 30 minutes, disrupting some communications and posing a very serious threat to any unprotected astronauts. Shock waves are also generated and can occasionally propagate several astronomical units before dissipating.

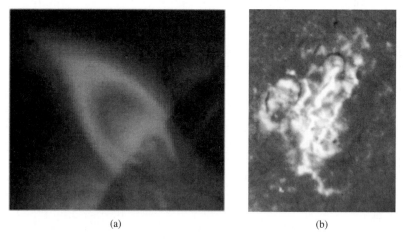

(a)                                    (b)

**FIGURE 11.34**   (a) A solar flare seen at the limb of the Sun, observed by the Yohkoh Soft X-ray Telescope, March 18, 1999, 16:40 UT. (From the Yohkoh mission of ISAS, Japan. The X-ray telescope was prepared by the Lockheed Palo Alto Research Laboratory, the National Astronomical Observatory of Japan, and the University of Tokyo with the support of NASA and ISAS.) (b) A two-ribbon flare seen in H$\alpha$ on October 19, 1989. (Courtesy of the National Optical Astronomy Observatories.)

---

[15]For comparison, a one-megaton bomb releases approximately $10^{16}$ J.

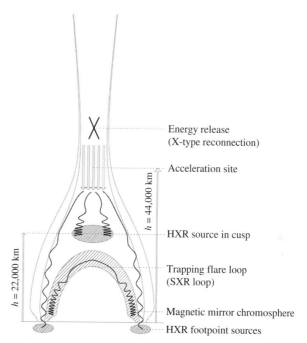

Energy release
(X-type reconnection)

Acceleration site

$h = 44,000$ km

HXR source in cusp

Trapping flare loop
(SXR loop)

$h = 22,000$ km

Magnetic mirror chromosphere

HXR footpoint sources

**FIGURE 11.35**   A model of the January 13, 1992, Masuda solar flare. Note the two hard X-ray (HXR) footpoint sources associated with Hα flare ribbons [see Fig. 11.34(b)]. Electrons are accelerated downward along the magnetic field lines until they collide with the chromosphere. The soft X-ray (SXR) loop may be compared to Fig. 11.34(a). (Figure adapted from Aschwanden, et al., *Ap. J.*, *464*, 985, 1996.)

The answer to the question of what powers solar flares lies in the location of the flare eruption. Flares develop in regions where the magnetic field intensity is great, namely in sunspot groups. From the discussion of the previous section, the creation of magnetic fields results in energy being stored in those magnetic fields (Eq. 11.9). If a magnetic field disturbance could quickly release the stored energy, a flare might develop. It is left as an exercise to show that both the amount of energy stored in the magnetic field and the timescale involved in perturbing it via Alfvén waves are consistent with the creation of a solar flare. However, details of the energy conversion, such as particle acceleration, are still a matter of active research.

A model of a solar flare is illustrated in Fig. 11.35. The general mechanism of a solar flare involves the **reconnection** of magnetic field lines. A disturbance in magnetic field loops (perhaps due to the Sun's convection zone) causes the creation of a sheet of current in the highly conducting plasma (recall Lenz's law). The finite resistance in the plasma results in Joule heating of the gas, causing temperatures to reach $10^7$ K. Particles accelerated away from the reconnection point and away from the Sun may escape entirely, producing solar cosmic rays. Radio-wavelength radiation is generated by the synchrotron process of charged particles spiraling around the magnetic field lines. Soft X-ray emission results from the high temperatures in the loop below the acceleration (reconnection) point. Hα emission at the base of the magnetic field lines (the two Hα ribbons) is produced by recombining electrons and protons that are accelerated away from the reconnection point, toward the chromosphere.

In addition, high-energy particles accelerated toward the chromosphere produce hard X-rays and gamma rays due to surface nuclear reactions. Examples of important nuclear reactions associated with solar flares are **spallation reactions** that break heavier nuclei into lighter nuclei, such as

$$\,^1_1\text{H} + \,^{16}_{8}\text{O} \rightarrow \,^{12}_{6}\text{C}^* + \,^4_2\text{He} + \,^1_1\text{H},$$

where C* represents a carbon nucleus in an excited state, followed by the de-excitation reaction

$$\,^{12}_{6}\text{C}^* \rightarrow \,^{12}_{6}\text{C} + \gamma,$$

with $E_\gamma = 4.438$ MeV, or

$$\,^1_1\text{H} + \,^{20}_{10}\text{Ne} \rightarrow \,^{16}_{8}\text{O}^* + \,^4_2\text{He} + \,^1_1\text{H},$$

followed by the de-excitation reaction

$$\,^{16}_{8}\text{O}^* \rightarrow \,^{16}_{8}\text{O} + \gamma,$$

with $E_\gamma = 6.129$ MeV. Other examples of reactions produced by flares on the Sun's surface include electron–positron annihilation,

$$e^- + e^+ \rightarrow \gamma + \gamma$$

where $E_\gamma = 0.511$ MeV, and the production of deuterium by

$$\,^1_1\text{H} + \text{n} \rightarrow \,^2_1\text{H}^* \rightarrow \,^2_1\text{H} + \gamma,$$

where $E_\gamma = 2.223$ MeV.

### Solar Prominences

**Solar prominences** are also related to the Sun's magnetic field. **Quiescent prominences** are curtains of ionized gas that reach well into the corona and can remain stable for weeks or months. The material in the prominence has collected along the magnetic field lines of an active region, with the result that the gas is cooler (with a typical temperature of 8000 K) and more dense than the surrounding coronal gas. This causes the gas to "rain" back down into the chromosphere. When viewed in H$\alpha$ at the limb of the Sun, quiescent prominences appear as bright structures against the thin corona. However, when viewed in the continuum against the solar disk, a quiescent prominence appears as a dark **filament**, absorbing the light emitted from below. An example of a quiescent prominence is shown in Fig. 11.36(a).

An **eruptive** (or **active**) **prominence** (Fig. 11.36b) may exist for only a few hours and may abruptly develop from a quiescent prominence. It appears as though a relatively stable magnetic field configuration can suddenly become unstable, causing the prominence to lift away from the Sun. Although the mechanism is related to that of a solar flare, the outcome is somewhat different; rather than most of the energy going into electromagnetic radiation, the energy of an eruptive prominence is converted into mass motions as gas is ejected from the Sun.

(a)

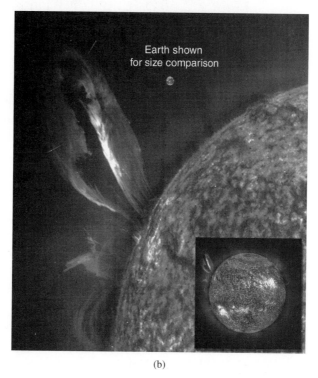

Earth shown
for size comparison

(b)

**FIGURE 11.36**  (a) A quiescent hedgerow prominence. (Courtesy of Big Bear Solar Observatory, California Institute of Technology.) (b) An eruptive prominence observed by the SOHO Extreme Ultraviolet Imaging Telescope (EIT) on July 24, 1999. [SOHO (ESA & NASA)]

### Coronal Mass Ejections

Even more spectacular is a **coronal mass ejection** (CME). CMEs have been observed since the early 1970s using spacecraft such as NASA's seventh Orbiting Solar Observatory (OSO 7) and Skylab. Most recently, CMEs have been observed routinely by SOHO's Large Angle Spectrometric COronograph (LASCO); see Fig. 11.37. LASCO uses an occulting disk to create an artificial solar eclipse, allowing it to observe the white-light corona from

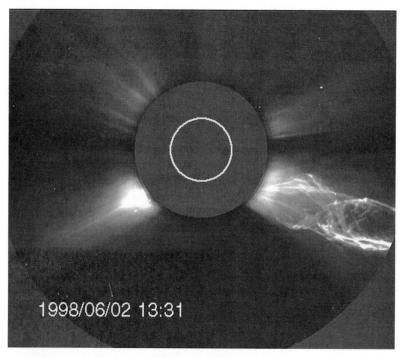

**FIGURE 11.37** A coronal mass ejection observed by the SOHO LASCO instrument on June 2, 1998. Note the intricacy of the magnetic field lines within the CME. The white circle on the occulting disk represents the size of the Sun out to the photosphere. [SOHO (ESA & NASA)]

a few solar radii out to 30 $R_\odot$. With the detection of thousands of CMEs it appears that there is about one CME per day when averaged over the 11-year sunspot cycle. When the Sun is more active (i.e., near sunspot maximum) the frequency may be about 3.5 events per day, and during sunspot minimum the number of events may decrease to roughly one every five days. During a CME event, between $5 \times 10^{12}$ kg and $5 \times 10^{13}$ kg of material may be ejected from the Sun at speeds ranging from 400 km s$^{-1}$ to over 1000 km s$^{-1}$. CMEs appear to be associated with eruptive prominences approximately 70% of the time, and with flares only about 40% of the time. One can think of a CME as a magnetic bubble lifting off of the Sun's surface after a magnetic reconnection event, carrying a significant fraction of the mass of the solar corona with it.

### The Time-Dependent Shape of the Corona

Yet another feature of the solar cycle involves the shape of the corona itself. During a period of little solar activity, when there are few sunspots and few, if any, flares or prominences, the **quiet corona** is generally more extended at the equator than at the poles, consistent with a nearly dipole magnetic field. Near sunspot maximum, the **active corona** is more complex in shape, as is the structure of the magnetic field. Examples of the shape of the corona during sunspot minimum and maximum are seen in Figs. 11.20(a) and 11.20(b), respectively. Evidently, the changing shape of the corona, like other solar activity, is due to the dynamic structure of the Sun's magnetic field.

### The Magnetic Dynamo Theory

A **magnetic dynamo** model describing many of the components of the solar cycle was first proposed by Horace Babcock in 1961. Despite its general success in describing the major features of the solar cycle, the model is not yet able to provide adequate explanations of many of the important details of solar activity. Any complete picture of the solar cycle will require a full treatment of the MHD equations in the solar environment, including differing rotation rates with latitude and depth in the Sun, convection, solar oscillations, heating of the upper atmosphere, and mass loss. Of course, not all of these processes are likely to play equally important roles in the study of the solar cycle, but it is important to understand the degree to which each of them contributes to the particular phenomenon under investigation.

As depicted in Fig. 11.38, because the magnetic field lines are "frozen into" the gas, the differential rotation of the Sun drags the lines along, converting a *poloidal* field (essentially

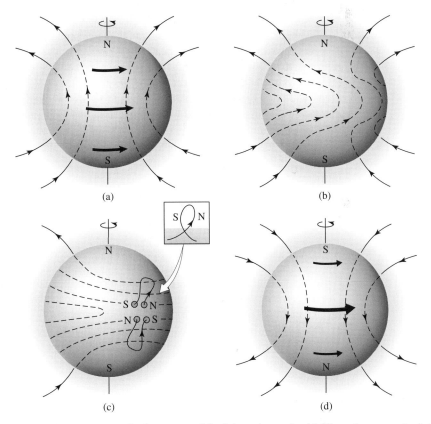

**FIGURE 11.38** The magnetic dynamo model of the solar cycle. (a) The solar magnetic field is initially a poloidal field. (b) Differential rotation drags the "frozen-in" magnetic field lines around the Sun, converting the poloidal field into a toroidal field. (c) Turbulent convection twists the field lines into magnetic ropes, causing them to rise to the surface as sunspots, the polarity of the lead spots corresponding to the original polarity of the poloidal field. (d) As the cycle progresses, successive sunspot groups migrate toward the equator where magnetic field reconnection reestablishes the poloidal field, but with the original polarity reversed.

a simple magnetic dipole) to one that has a significant *toroidal* component (field lines that are wrapped around the Sun). The turbulent convection zone then has the effect of twisting the lines, creating regions of intense magnetic fields, called magnetic *ropes*. The buoyancy produced by magnetic pressure (Eq. 11.10) causes the ropes to rise to the surface, appearing as sunspot groups. The polarity of the sunspots is due to the direction of the magnetic field along the ropes; consequently, every lead spot in one hemisphere will have the same polarity while the lead spots in the other hemisphere will have the opposite polarity.

Initially, the little twisting that does develop occurs at higher latitudes; during sunspot minimum. As the differential rotation continues to drag the field lines along and convective turbulence ties them in knots, more sunspots develop at intermediate latitudes, producing a sunspot maximum. It would seem that ultimately the greatest amount of twisting and the largest number of sunspots should develop near the equator. However, sunspots from the two hemispheres tend to cancel out near the equator since the polarities of their leading spots are opposed. As a result, the number of sunspots appearing near the equator is small. Finally, the cancelation of magnetic fields near the equator causes the poloidal field to be reestablished, but with its original polarity reversed. This process takes approximately 11 years. The entire procedure repeats continuously, with the polarity of the magnetic field returning to its original orientation every other cycle. Hence, the entire solar cycle is actually 22 years long when magnetic field polarities are considered.

As we have already seen, details related to specific phenomena, such as the cause of the decreased flux coming from sunspots or the exact process of flare generation, are not yet well understood. The same situation also holds for the more fundamental magnetic dynamo itself. Although the preceding discussion describes the behavior of the solar cycle in an approximate way, even such basic results as the timescales involved have not yet been accurately modeled. A successful magnetic dynamo model must not only produce the general location and numbers of sunspots and flares, but it must also do so with the observed 22-year periodicity. Moreover, the dynamo model must replicate the much slower variation that appears to be responsible for the Maunder minimum.

### Evidence of Magnetic Activity in Other Stars

Fortunately, some evidence does exist that the basic ideas behind the solar cycle are correct. Observations of other cool main-sequence stars indicate that they possess activity cycles much like the solar cycle. It was pointed out in the last section that late main-sequence stars exhibit observational characteristics consistent with the existence of hot coronae. It was also mentioned that angular momentum is apparently lost via stellar winds. Both phenomena agree with the theoretical onset of surface convection in low-mass stars, a major component of the dynamo theory.

Other forms of magnetic activity have also been seen in some stars. Observations indicate the existence of **flare stars**, main-sequence stars of class M that demonstrate occasional, rapid fluctuations in brightness. If flares the size of those on the Sun were to occur on the much dimmer M stars, the flares would contribute significantly to the total luminosity of those stars, producing the short-term changes that are observed. Much larger flares may also be generated by other stars as well: On April 24, 2004, the star GJ 3685A released a flare that was roughly one million times more energetic than a large solar flare. The event was detected serendipitously by NASA's Galaxy Evolution Explorer.

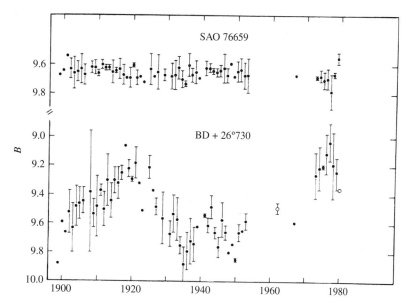

**FIGURE 11.39**   The light curve of BD + 26°730, a BY Dra star. SAO 76659 is a nearby reference star. (Figure from Hartmann et al., *Ap. J.*, *249*, 662, 1981.)

**Starspots** are also observed to exist on stars other than the Sun. Starspots are revealed by their effect on the luminosity of a star, which can be measured at a level of 1%. Two classes of stars, RS Canum Venaticorum and BY Draconis stars,[16] show significant long-term variations that are attributed to starspots covering appreciable fractions of their surfaces. For example, Fig. 11.39 shows a variation of over 0.6 magnitude in the B band for the BY Draconis star, BD + 26°730. Starspots can even be used to measure stellar rotation.

Magnetic fields have also been detected directly on several cool main-sequence stars by measuring Zeeman-broadened spectral lines. Analysis of the data indicates field strengths of several tenths of a tesla over significant fractions of the stellar surfaces. The existence of the strong fields correlates with their observed luminosity variations.

From our discussion in this chapter, it should be clear that astrophysics has had a great deal of success in explaining many of the features of our Sun. The stellar structure equations describe the major aspects of the solar interior, and much of the Sun's complex atmosphere is also understood. But many other important issues remain to be resolved, such as the surface abundance of lithium, the intricate details of the solar cycle, and the interaction between the Sun and Earth's climate. Much exciting and challenging work remains to be done before we can feel confident that we fully understand the star that is closest to us.

---

[16]Classes of stars that show light variations, *variable stars*, are usually named after the first star discovered that exhibits the specific characteristics. RS CVn and BY Dra are main-sequence stars of spectral classes F–G and K–M, respectively. The letters RS and BY indicate that these are variable stars; Canum Venaticorum and Draconis are the constellations in which the stars are located.

## SUGGESTED READING

### General

Bahcall, John N., "Solving the Mystery of the Missing Neutrinos," Nobel *e*-Museum, `http://www.nobel.se/physics/articles/bahcall/`, 2004.

Golub, Leon, and Pasachoff, Jay M., *Nearest Star: The Surprising Science of Our Sun*, Harvard University Press, Cambridge, MA, 2001.

Lang, Kenneth R., *The Cambridge Encyclopedia of the Sun*, Cambridge University Press, Cambridge, 2001.

Semeniuk, Ivan, "Astronomy and the New Neutrino," *Sky and Telescope*, September, 2004.

The Solar and Heliospheric Observatory (SOHO), `http://sohowww.nascom.nasa.gov/`.

Transition Region and Coronal Explorer (TRACE), `http://vestige.lmsal.com/TRACE/`.

Yohkoh Solar Observatory, `http://www.lmsal.com/SXT/`.

Zirker, Jack B., *Journey from the Center of the Sun*, Princeton University Press, Princeton, 2002.

### Technical

Aschwanden, Markus J., Poland, Arthur I., and Rabin, Douglas, M., "The New Solar Corona," *Annual Review of Astronomy and Astrophysics*, *39*, 175, 2001.

Aschwanden, Markus J., *Physics of the Solar Corona: An Introduction*, Springer, Berlin, 2004.

Bahcall, John N., *Neutrino Astrophysics*, Cambridge University Press, Cambridge, 1989.

Bahcall, John N., and Ulrich, Roger K., "Solar Models, Neutrino Experiments, and Helioseismology," *Reviews of Modern Physics*, *60*, 297, 1988.

Bahcall, John N., Pinsonneault, M. H., and Basu, Sarbani, "Solar Models: Current Epoch and Time Dependences, Neutrinos, and Helioseismological Properties," *The Astrophysical Journal*, *555*, 990, 2001.

Bhattacharjee, A., "Impulsive Magnetic Reconnection in the Earth's Magnetotail and the Solar Corona," *Annual Review of Astronomy and Astrophysics*, *27*, 421, 1989.

Bai, T., and Sturrock, P. A., "Classification of Solar Flares," *Annual Review of Astronomy and Astrophysics*, *42*, 365, 2004.

Böhm-Vitense, Erika, *Introduction to Stellar Astrophysics, Volume I: Basic Stellar Observations and Data*, Cambridge University Press, Cambridge, 1989.

Böhm-Vitense, Erika, *Introduction to Stellar Astrophysics, Volume II: Stellar Atmospheres*, Cambridge University Press, Cambridge, 1989.

Cleveland, Bruce T., et al., "Measurement of the Solar Electron Neutrino Flux with the Homestake Chlorine Detector," *The Astrophysical Journal*, *496*, 505, 1998.

Cox, A. N., Livingston, W. C., and Matthews, M. S. (eds.), *Solar Interior and Atmosphere*, University of Arizona Press, Tucson, 1991.

Foukal, Peter V., *Solar Astrophysics*, John Wiley and Sons, New York, 1990.

Griffiths, David J., *Introduction to Electrodynamics*, Third Edition, Prentice-Hall, Upper Saddle River, NJ, 1999.

Kivelson, Margaret G., and Russell, Christopher T. (eds.), *Introduction to Space Physics*, Cambridge University Press, Cambridge, 1995.

Lang, Kenneth R., *The Sun from Space*, Springer, Berlin, 2000.

Parker, E. N., "Dynamics of Interplanetary Gas and Magnetic Fields," *The Astrophysical Journal*, *128*, 664, 1958.

Thompson, Michael J., Christensen-Dalsgaard, Jørgen, Miesch, Mark S., and Toomre, Juri, "The Internal Rotation of the Sun," *Annual Review of Astronomy and Astrophysics*, *41*, 599, 2003.

## PROBLEMS

**11.1** Using Fig. 11.1, verify that the change in the Sun's effective temperature over the past 4.57 billion years is consistent with the variations in its radius and luminosity.

**11.2** **(a)** At what rate is the Sun's mass decreasing due to nuclear reactions? Express your answer in solar masses per year.

**(b)** Compare your answer to part (a) with the mass loss rate due to the solar wind.

**(c)** Assuming that the solar wind mass loss rate remains constant, would either mass loss process significantly affect the total mass of the Sun over its entire main-sequence lifetime?

**11.3** Using the Saha equation, calculate the ratio of the number of $H^-$ ions to neutral hydrogen atoms in the Sun's photosphere. Take the temperature of the gas to be the effective temperature, and assume that the electron pressure is $1.5 \ \mathrm{N \ m^{-2}}$. Note that the Pauli exclusion principle requires that only one state can exist for the ion because its two electrons must have opposite spins.

**11.4** The Paschen series of hydrogen ($n = 3$) can contribute to the visible continuum for the Sun since the series limit occurs at 820.8 nm. However, it is the contribution from the $H^-$ ion that dominates the formation of the continuum. Using the results of Problem 11.3, along with the Boltzmann equation, estimate the ratio of the number of $H^-$ ions to hydrogen atoms in the $n = 3$ state.

**11.5** **(a)** Using Eq. (9.63) and neglecting turbulence, estimate the full width at half-maximum of the hydrogen $H\alpha$ absorption line due to random thermal motions in the Sun's photosphere. Assume that the temperature is the Sun's effective temperature.

**(b)** Using $H\alpha$ redshift data for solar granulation, estimate the full width at half-maximum when convective turbulent motions are included with thermal motions.

**(c)** What is the ratio of $v_{\mathrm{turb}}^2$ to $2kT/m$?

**(d)** Determine the relative change in the full width at half-maximum due to Doppler broadening when turbulence is included. Does turbulence make a significant contribution to $(\Delta\lambda)_{1/2}$ in the solar photosphere?

**11.6** Estimate the thermally Doppler-broadened line widths for the hydrogen Ly$\alpha$, C III, O VI, and Mg X lines given on page 365; use the temperatures provided. Take the masses of H, C, O, and Mg to be 1 u, 12 u, 16 u, and 24 u, respectively.

**11.7** **(a)** Using Eq. (3.22), show that in the Sun's photosphere,

$$\ln\left(B_a/B_b\right) \approx 11.5 + \frac{hc}{kT}\left(\frac{1}{\lambda_b} - \frac{1}{\lambda_a}\right)$$

where $B_a/B_b$ is the ratio of the amount of blackbody radiation emitted at $\lambda_a = 10$ nm to the amount emitted at $\lambda_b = 100$ nm, centered in a wavelength band 0.1 nm wide.

**(b)** What is the value of this expression for the case where the temperature is taken to be the effective temperature of the Sun?

**(c)** Writing the ratio in the form $B_a/B_b = 10^x$, determine the value of $x$.

**11.8** The gas pressure at the base of the photosphere is approximately $2 \times 10^4$ N m$^{-2}$ and the mass density is $3.2 \times 10^{-4}$ kg m$^{-3}$. Estimate the sound speed at the base of the photosphere, and compare your answer with the values at the top of the photosphere and averaged throughout the Sun.

**11.9** Suppose that you are attempting to make observations through an optically thick gas that has a constant density and temperature. Assume that the density and temperature of the gas are $2.2 \times 10^{-4}$ kg m$^{-3}$ and 5777 K, respectively, typical of the values found in the Sun's photosphere. If the opacity of the gas at one wavelength ($\lambda_1$) is $\kappa_{\lambda 1} = 0.026$ m$^2$ kg$^{-1}$ and the opacity at another wavelength ($\lambda_2$) is $\kappa_{\lambda 2} = 0.030$ m$^2$ kg$^{-1}$, calculate the distance into the gas where the optical depth equals 2/3 for each wavelength. At which wavelength can you see farther into the gas? How much farther? This effect allows astronomers to probe the Sun's atmosphere at different depths (recall Fig. 11.13).

**11.10** **(a)** Using the data given in Example 11.2.2, estimate the pressure scale height at the base of the photosphere.

**(b)** Assuming that the ratio of the mixing length to the pressure scale height is 2.2, use the measured Doppler velocity of solar granulation to estimate the amount of time required for a convective bubble to travel one mixing length. Compare this value to the characteristic lifetime of a granule.

**11.11** Show that Eq. (11.6) follows directly from Eq. (11.5).

**11.12** Calculate the magnetic pressure in the center of the umbra of a large sunspot. Assume that the magnetic field strength is 0.2 T. Compare your answer with a typical value of $2 \times 10^4$ N m$^{-2}$ for the gas pressure at the base of the photosphere.

**11.13** Assume that a large solar flare erupts in a region where the magnetic field strength is 0.03 T and that it releases $10^{25}$ J in one hour.

**(a)** What was the magnetic energy density in that region before the eruption began?

**(b)** What minimum volume would be required to supply the magnetic energy necessary to fuel the flare?

**(c)** Assuming for simplicity that the volume involved in supplying the energy for the flare eruption was a cube, compare the length of one side of the cube with the typical size of a large flare.

**(d)** How long would it take an Alfvén wave to travel the length of the flare?

(e) What can you conclude about the assumption that magnetic energy is the source of solar flares, given the physical dimensions and timescales involved?

**11.14** Assuming that an average of one coronal mass ejection occurs per day and that a typical CME ejects $10^{13}$ kg of material, estimate the annual mass loss from CMEs and compare your answer with the annual mass loss from the solar wind. Express your answer as a percentage of CME mass loss to solar wind mass loss.

**11.15** Assume that the velocity of a CME directed toward Earth is 400 km s$^{-1}$ and that the mass of the CME is $10^{13}$ kg.

(a) Estimate the kinetic energy contained in the CME, and compare your answer to the energy released in a large flare. Express your answer as a percentage of the energy of the flare.

(b) Estimate the transit time for the CME to reach Earth.

(c) Briefly explain how astronomers are able to "predict" the occurrence of aurorae in advance of magnetic storms on Earth.

**11.16** (a) Calculate the frequency shift produced by the normal Zeeman effect in the center of a sunspot that has a magnetic field strength of 0.3 T.

(b) By what fraction would the wavelength of one component of the 630.25-nm Fe I spectral line change as a consequence of a magnetic field of 0.3 T?

**11.17** From the data given in Fig. 11.16, estimate the rotation period of the solar interior at the base of the tachocline.

**11.18** Argue from Eq. (11.9) and the work integral that magnetic pressure is given by Eq. (11.10).

# 12

# The Interstellar Medium and Star Formation

## 12.1 ■ INTERSTELLAR DUST AND GAS

When we look into the heavens, it appears as though the stars are unchanging, point-like sources of light that shine steadily. On casual inspection, even our own Sun appears constant. But, as we have seen in the last chapter, this is not the case; sunspots come and go, flares erupt, significant amounts of matter are launched into space via coronal mass ejections, the corona itself changes shape, and even the Sun's luminosity appears to be fluctuating over human timescales, as evidenced by the Maunder minimum. Of course, over the 4.57-Gyr lifetime of our Sun, the luminosity, effective temperature, and radius have all changed substantially (recall Fig. 11.1).

In fact all stars change. Usually the changes are so gradual and over such long time intervals when measured in human terms that we do not notice them without very careful telescopic observation. Occasionally, however, the changes are extremely rapid and dramatic, as in the case of a supernova explosion. By invoking the understanding we developed thus far of the physics of stellar interiors and atmospheres, we can now begin to examine the processes governing how stars evolve during their lives.

### The Interstellar Medium

In some sense the evolution of stars is a cyclic process. A star is born out of gas and dust that exists between the stars, known as the **interstellar medium** (ISM). During its lifetime, depending on the star's total mass, much of that material may be returned to the ISM through stellar winds and explosive events. Subsequent generations of stars can then form from this processed material. As a result, to understand the evolution of a star, it is important to study the nature of the ISM.

Understanding the interstellar medium is critical for more than its role in stellar evolution, however. The ISM is of profound importance in describing the structure, dynamics, and evolution of our Milky Way Galaxy, as well as galaxies throughout the universe. In addition, it impacts our observations of everything from relatively nearby stars to the most remote galaxies and quasars.

More fundamentally, the ISM is an enormous and complex environment that provides an important laboratory for testing our understanding of astrophysics at many levels. The dynamics of the ISM involve turbulent gas motions, shocks, and galactic magnetic fields that lace through interstellar space. Thus, modeling the ISM ultimately requires detailed solutions to the equations of magnetohydrodynamics. The dust, molecules, atoms, ions, and free electrons that permeate the ISM challenge our understanding of radiative transfer, thermodynamics, and quantum mechanics. Moreover, the production and destruction of dust grains and complex molecules requires a detailed understanding of chemistry in an environment not reproducible in a terrestrial laboratory.

As an introduction to astrophysical processes, this text is unable to explore all of the fascinating aspects of the interstellar medium. Consequently, the present section serves only as a brief introduction to general aspects of the ISM.

### Interstellar Extinction

On a dark night some of the **dust clouds** that populate our Milky Way Galaxy can be seen in the band of stars that is the disk of the Galaxy (see Fig. 12.1). It is not that these dark regions are devoid of stars, but rather that the stars located behind intervening dust clouds are obscured. This obscuration, referred to as **interstellar extinction**, is due to the summative effects of scattering and absorption of starlight (as depicted in Fig. 12.2).

Given the effect that extinction can have on the apparent magnitude of a star, the distance modulus equation (Eq. 3.6) must be modified appropriately. In a given wavelength band centered on $\lambda$, we now have

$$m_\lambda = M_\lambda + 5\log_{10} d - 5 + A_\lambda, \tag{12.1}$$

where $d$ is the distance in pc and $A_\lambda > 0$ represents the number of magnitudes of interstellar extinction present along the line of sight. If $A_\lambda$ is large enough, a star that would otherwise be visible to the naked eye or through a telescope could no longer be detected. This is the reason for the dark bands running through the Milky Way.

**FIGURE 12.1**   Dust clouds obscure the stars located behind them in the disk of the Milky Way. (Courtesy of Palomar/Caltech.)

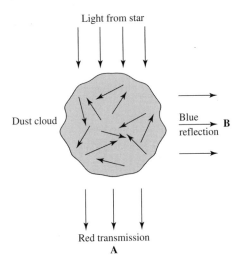

**FIGURE 12.2**  An interstellar cloud containing significant amounts of dust along with the gas (a dust cloud) can both scatter and absorb light that passes through it. The amount of scattering and absorption depends on the number density of dust grains, the wavelength of the light, and the thickness of the cloud. Since shorter wavelengths are affected more significantly than longer ones, a star lying behind the cloud appears reddened to observer A. Observer B sees the scattered shorter wavelengths as a blue reflection nebula.

Clearly $A_\lambda$ must be related to the optical depth of the material, measured back along the line of sight. From Eq. (9.18), the fractional change in the intensity of the light is given by

$$I_\lambda/I_{\lambda,0} = e^{-\tau_\lambda},$$

where $I_{\lambda,0}$ is the intensity in the absence of interstellar extinction. Combining this with Eq. (3.4), we can now relate the optical depth to the change in apparent magnitude due to extinction, giving

$$m_\lambda - m_{\lambda,0} = -2.5 \log_{10}\left(e^{-\tau_\lambda}\right) = 2.5\tau_\lambda \log_{10} e = 1.086\tau_\lambda.$$

But the change in apparent magnitude is just $A_\lambda$, so

$$A_\lambda = 1.086\tau_\lambda. \tag{12.2}$$

*The change in magnitude due to extinction is approximately equal to the optical depth along the line of sight.*

From the expression for the optical depth given by Eq. (9.17) and the discussion at the end of Example 9.2.2, the optical depth through the cloud is given by

$$\tau_\lambda = \int_0^s n_d(s')\,\sigma_\lambda\,ds', \tag{12.3}$$

where $n_d(s')$ is the number density of scattering dust grains and $\sigma_\lambda$ is the scattering cross section. If $\sigma_\lambda$ is constant along the line of sight, then

$$\tau_\lambda = \sigma_\lambda \int_0^s n_d(s')\,ds' = \sigma_\lambda N_d, \qquad (12.4)$$

where $N_d$, the dust grain *column density*, is the number of scattering dust particles in a thin cylinder with a cross section of 1 m$^2$ stretching from the observer to the star. Thus we see that the amount of extinction depends on the amount of interstellar dust that the light passes through, as one would expect.

### The Mie Theory

If we assume for simplicity, as was first done by Gustav Mie (1868–1957) in 1908, that dust particles are spherical and each has a radius $a$, then the geometrical cross section that a particle presents to a passing photon is just $\sigma_g = \pi a^2$. We may now define the dimensionless **extinction coefficient** $Q_\lambda$ to be

$$Q_\lambda \equiv \frac{\sigma_\lambda}{\sigma_g},$$

where $Q_\lambda$ depends on the composition of the dust grains.

Mie was able to show that when the wavelength of the light is on the order of the size of the dust grains, then $Q_\lambda \sim a/\lambda$, implying that

$$\sigma_\lambda \propto \frac{a^3}{\lambda} \qquad (\lambda \gtrsim a). \qquad (12.5)$$

In the limit that $\lambda$ becomes very large relative to $a$, $Q_\lambda$ goes to zero. On the other hand, if $\lambda$ becomes very small relative to $a$, it can be shown that $Q_\lambda$ approaches a constant, independent of $\lambda$ so that

$$\sigma_\lambda \propto a^2 \qquad (\lambda \ll a). \qquad (12.6)$$

These limiting behaviors can be understood by analogy to waves on the surface of a lake. If the wavelength of the waves is much larger than an object in their way, such as a grain of sand, the waves pass by almost completely unaffected ($\sigma_\lambda \sim 0$). On the other hand, if the waves are much smaller than the obstructing object—for instance, an island—they are simply blocked; the only waves that continue on are those that miss the island altogether. Similarly, at sufficiently short wavelengths, the only light we detect passing through the dust cloud is the light that travels between the particles.

Combining the ideas already discussed, it is clear that the amount of extinction, as measured by $A_\lambda$, must be wavelength-dependent. Since the longer wavelengths of red light are not scattered as strongly as blue light, the starlight passing through intervening dust clouds becomes reddened as the blue light is removed. This **interstellar reddening** causes stars to appear redder than their effective temperatures would otherwise imply. Fortunately, it is possible to detect this change by carefully analyzing the absorption and emission lines in the star's spectrum.

Much of the incident blue light is scattered out of its original path and can leave the cloud in virtually any direction. As a result, looking at the cloud in a direction other than along the line of sight to a bright star behind the cloud, an observer will see a blue **reflection nebula** (recall Fig. 12.2) such as the Pleiades (Fig. 13.16b). This process is analogous to Rayleigh scattering, which produces a blue sky on Earth. The difference between Mie scattering and Rayleigh scattering is that the sizes of the scattering molecules associated with Rayleigh scattering are much smaller than the wavelength of visible light, leading to $\sigma_\lambda \propto \lambda^{-4}$.

---

**Example 12.1.1.**   A certain star, located 0.8 kpc from Earth, is found to be dimmer than expected at 550 nm by $A_V = 1.1$ magnitudes, where $A_V$ is the amount of extinction as measured through the *visual wavelength* filter (see Section 3.6). If $Q_{550} = 1.5$ and the dust grains are assumed to be spherical with radii of 0.2 $\mu$m, estimate the average density $(\overline{n})$ of material between the star and Earth.

From Eq. (12.2), the optical depth along the line of sight is nearly equal to the amount of extinction in magnitudes, or $\tau_{550} \simeq 1$. Also,

$$\sigma_{550} = \pi a^2 Q_{550} \simeq 2 \times 10^{-13} \text{ m}^2.$$

Now the column density of the dust along the line of sight is given by Eq. (12.4),

$$N_d = \frac{\tau_{550}}{\sigma_{550}} \simeq 5 \times 10^{12} \text{ m}^{-2}.$$

Finally, since $N_d = \int_0^s n(s')\, ds' = \overline{n} \times 0.8$ kpc, we have

$$\overline{n} = \frac{N_d}{0.8 \text{ kpc}} = 2 \times 10^{-7} \text{ m}^{-3}.$$

Number densities of this magnitude are typical of the plane of the Milky Way Galaxy.

---

### Molecular Contributions to Interstellar Extinction Curves

Predictions of the Mie theory work well for longer wavelengths, typically from the infrared into the visible wavelength region. However, at ultraviolet wavelengths significant deviations become apparent, as can be seen by considering the ratio of $A_\lambda$, the extinction in a wavelength band centered at $\lambda$, to the extinction in some reference wavelength band, such as $A_V$. This ratio is often plotted versus reciprocal wavelength $\lambda^{-1}$, as in Fig. 12.3. Alternatively, **color excesses** are sometimes plotted instead, such as $(A_\lambda - A_V)/(A_B - A_V)$ or $E(B - V) \equiv (B - V)_{\text{intrinsic}} - (B - V)_{\text{observed}}$.

At longer wavelengths (the left side of the graph) the data agree well with the Mie theory. For wavelengths shorter than the blue wavelength band $(B)$, however, the curves begin to diverge significantly, deviating from the expected relation, $A_\lambda/A_V \propto \lambda^{-1}$. Particularly evident is the "bump" in the ultraviolet at 217.5 nm or 4.6 $\mu$m$^{-1}$. At even shorter wavelengths, the extinction curve tends to rise sharply as the wavelength decreases.

The existence of the "bump" in Fig. 12.3 gives us some hint of the composition of the dust. Graphite, a well-ordered form of carbon, interacts strongly with light near 217.5 nm.

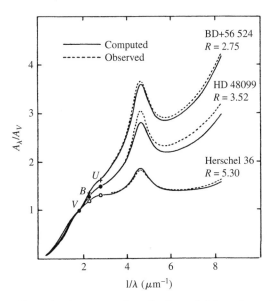

**FIGURE 12.3** Interstellar extinction curves along the lines of sight to three stars. The dashed lines represent the observational data, and the solid lines are theoretical fits. The $U$, $B$, and $V$ wavelength bands are indicated for reference. (Figure adapted from Mathis, *Annu. Rev. Astron. Astrophys.*, *28*, 37, 1990. Reproduced with permission from the *Annual Review of Astronomy and Astrophysics*, Volume 28, ©1990 by Annual Reviews Inc.)

Although it is uncertain how carbon can organize into large graphite particles in the interstellar medium, the strength of the "bump," the abundance of carbon, and the existence of the 217.5-nm resonance have led most researchers to suggest that graphite may be a major component of interstellar dust.

Another possible source of the 217.5-nm feature may be **polycyclic aromatic hydrocarbons** (PAHs; see Fig. 12.4). These are complex organic planar molecules with multiple benzene ring-like structures that are probably responsible for a series of molecular bands that have been observed in emission in the light from diffuse dust clouds.[1] The so-called *unidentified infrared emission bands* exist in the wavelength range between 3.3 $\mu$m and 12 $\mu$m; they appear to be due to vibrations in the C-C and C-H bonds common in PAHs. Just as transitions between atomic energy levels are quantized, so are the energies associated with molecular bonds. In the case of molecular bonds, however, the energy levels tend to be grouped in closely spaced bands, producing characteristic broad features in the spectrum of the light. The vibration, rotation, and bending of molecular bonds are all quantized, yielding complex spectra that may be difficult to identify in large molecules.

Interstellar dust is composed of other particles as well, as evidenced by the existence of dark absorption bands at wavelengths of 9.7 $\mu$m and 18 $\mu$m in the near-infrared. These features are believed to be the result of the stretching of the Si-O molecular bond and the

---

[1]The fact that molecules as complex as PAHs can exist in space has also been confirmed by their presence in certain types of meteorites found on Earth, known as **carbonaceous meteorites**.

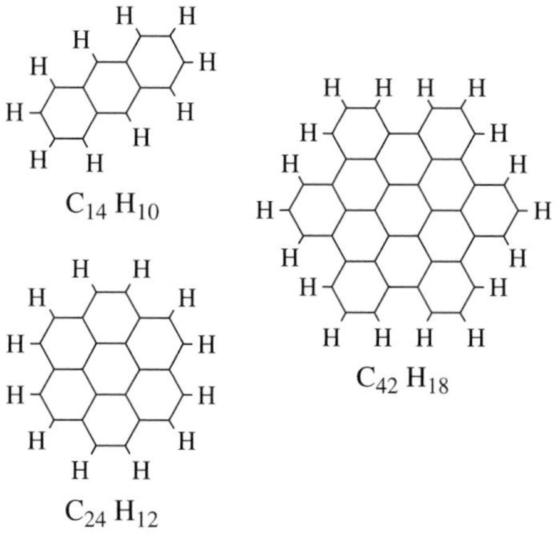

**FIGURE 12.4**   The structures of several polycyclic aromatic hydrocarbons: $C_{14}H_{10}$ (anthracene), $C_{24}H_{12}$ (coronene), $C_{42}H_{18}$ (hexabenzocoronene). The hexagonal structures are shorthand for indicating the presence of a carbon atom at each corner of the hexagon.

bending of Si-O-Si bonds in silicates, respectively. The existence of these absorption bands involving silicon indicates that silicate grains are also present in the dust clouds and the diffuse dust of the ISM.

An important characteristic of the light scattered from interstellar dust is that it tends to be slightly polarized. The amount of polarization is typically a few percent and depends on wavelength. This necessarily implies that the dust grains cannot be perfectly spherical. Furthermore, they must be at least somewhat aligned along a unique direction since the electric field vectors of the radiation are preferentially oriented in a particular direction. The most likely way to establish such an alignment is for the grains to interact with a weak magnetic field. Because less energy is required, the particles tend to rotate with their long axes perpendicular to the direction of the magnetic field.

All of these observations give us some clues to the nature of the dust in the ISM. Apparently the dust in the ISM is composed of both graphite and silicate grains ranging in size from several microns down to fractions of a nanometer, the characteristic size of the smaller PAHs. It appears that many of the features of the interstellar extinction curve can be reproduced by combining the contributions from all of these components.

### Hydrogen as the Dominant Component of the ISM

Although dust produces most of the obscuration that is readily noticeable, the dominant component of the ISM is hydrogen gas in its various forms: neutral hydrogen (H I), ionized hydrogen (H II), and molecular hydrogen ($H_2$). Hydrogen comprises approximately 70% of the mass of matter in the ISM, and helium makes up most of the remaining mass; metals, such as carbon and silicon, account for only a few percent of the total.

Most hydrogen in *diffuse* interstellar hydrogen clouds is in the form of H I in the ground state. As a result, the H I is generally incapable of producing emission lines by downward transitions of electrons from one orbit to another. It is also difficult to observe H I in absorption, since UV-wavelength photons are required to lift the electrons out of the ground state. However, in certain unique circumstances, orbiting observatories have detected absorption lines produced by cold clouds of H I when there are strong UV sources lying behind them.

### 21-cm Radiation of Hydrogen

Fortunately, it is still generally possible to identify neutral hydrogen in the diffuse ISM. This is done by detecting the unique radio-wavelength **21-cm line**. The 21-cm line is produced by the reversal of the spin of the electron relative to the proton in the atom's nucleus. Recall (Section 5.4) that both electrons and protons possess an inherent spin angular momentum, with the $z$-component of the spin angular momentum vector having one of two possible orientations, corresponding to the two allowed values of the spin quantum number, $m_s = \pm\frac{1}{2}$. Because these particles are also electrically charged, their intrinsic spins endow them with dipole magnetic fields, much like those of bar magnets. If the spins of the electron and proton are aligned (e.g., both spin axes are in the same direction), the atom has slightly more energy than if they are anti-aligned (see Fig. 12.5). As a result, if the electron's spin "flips" from being aligned with the proton to being anti-aligned, energy must be lost from the atom. If the spin flip is not due to a collision with another atom, then a photon is emitted. Of course, a photon can also be absorbed, exciting a hydrogen atom into aligning its electron and proton spins. The wavelength of the photon associated with the spin flip is 21.1 cm, corresponding to a frequency of 1420 MHz.

The emission of a 21-cm photon from an individual hydrogen atom is extremely rare. Once in the excited state, several million years can pass on average before that atom will emit a photon. Competing with this spontaneous emission are collisions between hydrogen atoms that may result in either excitation or de-excitation. In the low-density environment of the diffuse ISM, collisions occur on timescales of hundreds of years. Although this is far shorter than the spontaneous emission timescale, statistically some atoms are still able to make the necessary spontaneous transition. In contrast, the best vacuums produced in

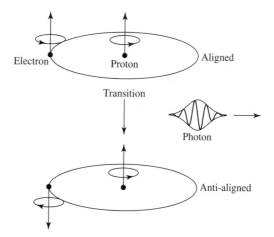

**FIGURE 12.5** When the spins of the electron and proton in a hydrogen atom go from being aligned to being anti-aligned, a 21-cm-wavelength photon is emitted.

Earth-based laboratories have densities much greater than those found in the ISM, meaning that collision rates are significantly higher in laboratory environments and virtually all of the atoms in the laboratory are de-excited before they can emit 21-cm radiation. The existence of 21-cm radiation was predicted in the early 1940s and first detected in 1951. Since then it has become an important tool in mapping the location and density of H I, measuring radial velocities using the Doppler effect, and estimating magnetic fields using the Zeeman effect. 21-cm radiation is particularly valuable in determining the structure and kinematic properties of galaxies, including our own.

Although H I is quite abundant, the rarity of 21-cm emission (or absorption) from individual atoms means that the center of this line can remain optically thin over large interstellar distances. Assuming that the line profile is a Gaussian, like the shape of the Doppler line profile shown in Fig. 9.19, the optical depth of the line center is given by

$$\tau_H = 5.2 \times 10^{-23} \frac{N_H}{T \, \Delta v}, \tag{12.7}$$

where $N_H$ is the column density of H I (in units of m$^{-2}$), $T$ is the temperature of the gas (in kelvins), and $\Delta v$ is the full width of the line at half maximum (in km s$^{-1}$). [Note that since the line width is due primarily to the Doppler effect (Eq. 4.30), $\Delta v$ is expressed in units of velocity, rather than in wavelength units; typically $\Delta v \sim 10$ km s$^{-1}$.]

As long as the 21-cm hydrogen line is optically thin (i.e., on the linear part of the curve of growth; Fig. 9.21), the optical depth is proportional to the neutral hydrogen column density. Studies of **diffuse H I clouds** indicate temperatures of 30 to 80 K, number densities in the range of $1 \times 10^8$ m$^{-3}$ to $8 \times 10^8$ m$^{-3}$, and masses on the order of 1–100 M$_\odot$.

Comparing $\tau_H$ with $A_V$ along the same line of sight shows that $N_H$ is generally proportional to $N_d$ (the column density of dust) when $A_V < 1$. This observation suggests that dust and gas are distributed together throughout the ISM. However, when $A_V > 1$, this correlation breaks down; the column density of H I no longer increases as rapidly as the column density of dust. Apparently, other physical processes are involved when the dust becomes optically thick.

Optically thick dust clouds shield hydrogen from sources of ultraviolet radiation. One consequence of this shielding is that molecular hydrogen can exist without the threat of undergoing dissociation by UV photon absorption. Dust can also enhance the $H_2$ formation rate beyond what would be expected by random collisions of hydrogen atoms. This enhancement occurs for two reasons: (1) A dust grain can provide a *site* on the surface of the grain where the hydrogen atoms can meet, rather than requiring chance encounters in the ISM, and (2) the dust provides a *sink* for the binding energy that must be liberated if a stable molecule is to form. The liberated energy goes into heating the grain and ejecting the $H_2$ molecule from the formation site. If the column density of atomic hydrogen is sufficiently large ($N_H$ on the order of $10^{25}$ m$^{-2}$), it can also shield $H_2$ from UV photodissociation. Consequently, **molecular clouds** are surrounded by shells of H I.

## Molecular Tracers of $H_2$

Since the structure of $H_2$ differs greatly from that of atomic hydrogen, the $H_2$ molecule does not emit 21-cm radiation. This explains why $N_H$ and $A_V$ are poorly correlated in molecular clouds when $A_V > 1$; the number density of atomic hydrogen decreases significantly as the hydrogen becomes locked up in its molecular form.

Unfortunately, $H_2$ is very difficult to observe directly because the molecule does not have any emission or absorption lines in the visible or radio portions of the electromagnetic spectrum at the cool temperatures typical of the ISM. In special circumstances when $T > 2000$ K, it is possible to detect rotational and vibrational bands (known collectively as *rovibrational bands*) associated with the molecular bond. However, in most instances it becomes necessary to use other molecules as *tracers* of $H_2$ by making the assumption that their abundances are proportional to the abundance of $H_2$. Because of its relatively high abundance (approximately $10^{-4}$ that of $H_2$), the most commonly investigated tracer is carbon monoxide, CO, although other molecules have also been used, including CH, OH, CS, $C_3H_2$, $HCO^+$, and $N_2H^+$. It is also possible to use **isotopomers** of the molecules, such as $^{13}CO$ or $C^{18}O$, to further refine studies of molecular clouds. Given that molecules have moments of inertia that affect their spectra, different isotopes in molecules result in different spectral wavelengths; see, for example, Problem 12.7. (Note that when the specific isotope is not indicated, it is assumed that the most abundant isotope is implied; thus CO implies $^{12}C^{16}O$.)

During collisions the tracer molecules become excited (or de-excited) and spontaneous transitions from excited states result in the emission of photons in wavelength regions that are more easily observed than those associated with $H_2$, such as the 2.6-mm transition of CO. Since collision rates depend on both the gas temperature (or thermal kinetic energy) and the number densities of the species, molecular tracers can provide information about the environment within a molecular cloud. In fact, an estimate of atomic and molecular collision rates can be made in a way completely analogous to the approach used to obtain the nuclear reaction rate equation (Eq. 10.29).

### The Classification of Interstellar Clouds

The results of these studies show that conditions within molecular clouds can vary widely. Consequently, any effort to specify a discrete classification scheme is destined to fail because the delineation between types is blurred at best. However, even with that caveat, a broad classification scheme is still useful for distinguishing the general characteristics of specific environments.

In clouds where the hydrogen gas is primarily atomic and the interstellar extinction is roughly $1 < A_V < 5$, molecular hydrogen may be found in regions of higher column density. Such clouds are sometimes referred to as **diffuse molecular clouds**, or alternatively as **translucent molecular clouds**. Conditions in diffuse molecular clouds are typical of diffuse H I clouds but with somewhat higher masses; they have temperatures of 15 to 50 K, $n \sim 5 \times 10^8$ to $5 \times 10^9$ m$^{-3}$, $M \sim 3$ to $100$ M$_\odot$, and they measure several parsecs across. Both H I clouds and diffuse molecular clouds tend to be irregularly shaped.

**Giant molecular clouds** (GMCs) are enormous complexes of dust and gas where temperatures are typically $T \sim 15$ K, number densities are in the range $n \sim 1 \times 10^8$ to $3 \times 10^8$ m$^{-3}$, masses are typically $10^5$ M$_\odot$ but may reach $10^6$ M$_\odot$, and typical sizes are on the order of 50 pc across. The famous Horsehead Nebula, also known as Barnard 33 (B33),

**FIGURE 12.6**    The Horsehead Nebula is part of the Orion giant molecular cloud complex. The "horsehead" appearance is due to dust protruding into an H II (ionized hydrogen) environment. (European Southern Observatory)

is shown in Fig. 12.6. The Horsehead Nebula is a portion of the Orion giant molecular cloud complex. Thousands of GMCs are known to exist in our Galaxy, mostly in its spiral arms.

Overall, the structure of GMCs tend to be clumpy with local regions of significantly greater density. **Dark cloud complexes** of roughly $10^4 \, M_\odot$ have $A_V \sim 5, n \sim 5 \times 10^8 \, \text{m}^{-3}$, diameters on the order of 10 pc, and characteristic temperatures of 10 K. Smaller, individual **clumps** may be even more dense, with $A_V \sim 10$, $n \sim 10^9 \, \text{m}^{-3}$, diameters of a couple of parsecs, temperatures of 10 K or so, and masses of 30 $M_\odot$. At even smaller scales are **dense cores** with masses on the order of 10 $M_\odot$, $A_V > 10$, $n \sim 10^{10} \, \text{m}^{-3}$, characteristic diameters of 0.1 pc, and temperatures of 10 K. Finally, in some localized regions of GMCs, observations reveal **hot cores** with characteristic sizes of 0.05 to 0.1 pc, where $A_V \sim 50$ to 1000, $T \sim 100$ to 300 K, $n \sim 10^{13}$ to $10^{15} \, \text{m}^{-3}$, and $M \sim 10$ to 3000 $M_\odot$. Based on observations from infrared telescopes such as NASA's Spitzer Space Telescope and the European Space Agency's Infrared Space Observatory, hot cores appear to have massive, young O and B stars embedded within them, suggesting strongly that these are regions of recent star formation.

Located outside of larger molecular complexes are the almost spherical clouds known as **Bok globules** (see, for example, Fig. 12.7).[2] These globules are characterized by large visual extinctions ($A_V \sim 10$), low temperatures ($T \sim 10$ K), relatively large number densities ($n > 10^{10} \, \text{m}^{-3}$), low masses ($M \sim 1$ to 1000 $M_\odot$), and small sizes of typically less than 1 pc. Infrared surveys of Bok globules have revealed that many, perhaps most, of these objects harbor young low-luminosity stars in their centers, implying that Bok globules are also sites of active star formation. In fact, Bok globules appear to be dense cores that have

---

[2]Bok globules are named after Bart Bok (1906–1983), who first studied these objects in the 1940s.

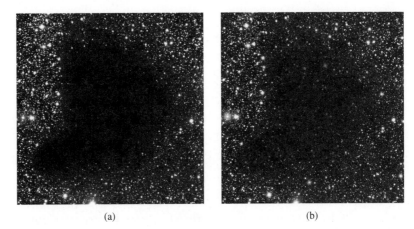

**FIGURE 12.7**   The Bok globule, Barnard 68 (B68), observed in visible light [(a) composite of BVI bands] and in infrared light [(b) composite of BIK bands]. The visible image was obtained by one of the 8-m telescopes of the European Southern Observatory's Very Large Telescope at Paranal. The infrared image was obtained by ESO's 3.58-m New Technology Telescope at La Silla. Notice that significantly reddened stars can be seen through the globule in the infrared. (Interstellar reddening is the result of scattering photons off of dust grains.) (European Southern Observatory)

been stripped of their surrounding molecular gas by nearby hot, massive stars. The process by which stars form out of the ISM will be considered in the next section.

### Interstellar Chemistry

Along with the molecules and dust grains already discussed, the ISM is rich in other molecules as well. As of June 2005, radio observations have resulted in the positive identification of 125 molecules (not including isotopomers), ranging in complexity from diatomic molecules such as $H_2$ and CO, and triatomic molecules such as $H_2O$ and $H_3^+$, to fairly long organic strings, including $HC_{11}N$.

Given the complex nature of the molecules present in the interstellar medium, it is evident that the chemistry of the ISM is also quite complex. The specific processes in operation in a given molecular cloud depend on the density and temperature of the gas, as well as its composition and the presence of dust grains. We noted on page 406 that dust grains must be present for the formation of molecular hydrogen, $H_2$, the dominant constituent in molecular clouds. It is also likely that dust grains can help facilitate the formation of numerous other molecules as well, including CH, NH, OH, $CH_2$, CO, $CO_2$, and $H_2O$. In fact, in sufficiently dense clouds, the formation of molecules on the surfaces of grains can actually lead to the development of icy mantles on the grains. Absorption signatures of solid CO, $CO_2$, $H_2O$, $CH_4$, $CH_3OH$, $NH_3$, and other ices have been measured in combination with the infrared spectra of silicate dust grains.

In addition to the chemistry that can occur on grain surfaces, it is also possible for molecules to form in the gas phase. For example, the hydroxyl molecule (OH) can form through a series of reactions involving atomic and molecular ions, including the ionic water molecule, $H_2O^+$:

$$H^+ + O \rightarrow O^+ + H$$

$$O^+ + H_2 \rightarrow OH^+ + H$$

$$OH^+ + H_2 \rightarrow H_2O^+ + H$$

$$H_2O^+ + e^- \rightarrow OH + H. \tag{12.8}$$

Eq. (12.8) competes with another reaction involving molecular hydrogen,

$$H_2O^+ + H_2 \rightarrow H_3O^+ + H,$$

leading to the production of either a hydroxyl molecule (75% of the time) or a water molecule via

$$H_3O^+ + e^- \rightarrow \begin{cases} OH + H_2 \\ H_2O + H. \end{cases} \tag{12.9}$$

### The Heating and Cooling of the ISM

Not only are molecules and dust grains critical in understanding the chemistry of the ISM, but they also play important roles in the heating and cooling of the material between the stars. You may have noticed that diffuse molecular clouds have higher gas temperatures than giant molecular clouds, and the dense cores of GMCs are even cooler yet. On the other hand, the hot cores of GMCs have significantly greater temperatures. What are the physical causes of these observational trends?

Much of the heating of the interstellar medium comes from **cosmic rays**, charged particles that travel through space with sometimes astonishing amounts of energy. A single proton may possess an energy ranging anywhere from 10 to $10^{14}$ MeV.[3] The highest energy cosmic rays are extremely rare, but energies in the range $10^3$ to $10^8$ MeV are common. The sources of cosmic rays include stellar flares and supernova explosions and will be discussed in Section 15.5.

Heating by cosmic rays comes primarily through the ionization of hydrogen atoms and molecules as a result of collisions with cosmic ray protons;

$$p^+ + H \rightarrow H^+ + e^- + p^+$$

$$p^+ + H_2 \rightarrow H_2^+ + e^- + p^+.$$

When an atom or molecule is ionized, an electron is ejected that carries some of the original kinetic energy of the proton with it. It is this ejected electron that interacts with the ISM to increase the average kinetic energy of the ISM's constituents via collisions with molecules (see, for example, Eqs. 12.8 and 12.9). Those molecules then collide with other molecules

---

[3] $10^{14}$ MeV is roughly the kinetic energy of a tennis ball of mass 0.057 kg traveling at 100 km h$^{-1}$ (approximately 60 mph).

in the gas, distributing thermal kinetic energy throughout the cloud, thereby raising the temperature of the cloud.

Other sources of heating in molecular clouds include the ionization of carbon atoms by ultraviolet starlight resulting in ejected electrons, the photoelectric ejection of electrons from dust grains by ultraviolet starlight, the absorption of light energy into the lattice of dust grains, and the ionization of hydrogen by stellar X-rays. Shocks from supernovae or strong stellar winds can also produce some heating of molecular clouds in special cases.

To balance the heating processes, cooling mechanisms must also be in operation. The primary mechanism for cooling is based on the emission of infrared photons. Recalling Mie scattering (Eq. 12.5), when photon wavelengths are on the order of, or longer than, the size of dust grains, they are less likely to be scattered. IR photons can pass more easily through the molecular cloud than can shorter-wavelength photons, allowing the IR photons to transport energy out of the cloud.

IR photons are produced in molecular clouds through collisions between ions, atoms, molecules, and dust grains. Typically a collision between ions, atoms, or molecules results in one of the species being left in an excited state; the energy of the excited state comes from the kinetic energy of the collision. The species in the excited state then decays back to the ground state through the emission of an IR photon. For example,

$$O + H \rightarrow O^* + H \tag{12.10}$$

$$O^* \rightarrow O + \gamma. \tag{12.11}$$

Here $O^*$ represents an excited state of the oxygen atom. The collisional kinetic energy (thermal energy) is thus transformed into an IR photon that escapes the cloud. Collisional excitations of $C^+$ and CO by H and $H_2$, respectively, are also significant contributors to cooling of molecular clouds.

Collisions involving dust grains can also result in cooling of molecular clouds. This process is similar to ionic, atomic, and molecular collisions in that the lattice of a dust grain can be left with excess thermal energy after the collision. The grain then emits infrared energy that is able to escape from the cloud.

### The Sources of Dust Grains

It is apparent that even though dust grains make up only about one percent of the mass of a molecular cloud, they are important constituents in determining its chemistry and physics. The question of the source of these grains then naturally arises. Although observations indicate that dust grains can be formed in the envelopes of very cool stars, aided by the enhanced density in those environments relative to molecular clouds, grains can also be easily destroyed by UV and X-ray photons. Dust grains are also formed as a product of supernova explosions and stellar winds. However, none of these sources appear to be able to provide the abundance of massive grains found in molecular clouds. Rather, it appears that grains probably grow by a process of coagulation within the molecular clouds themselves. Dust grain formation represents just one of many areas of active research into the nature of the ISM.

## 12.2 ■ THE FORMATION OF PROTOSTARS

Our understanding of stellar evolution has developed significantly since the 1960s, reaching the point where much of the life history of a star is well determined. This success has been due to advances in observational techniques, improvements in our knowledge of the physical processes important in stars, and increases in computational power. In the remainder of this chapter and in Chapter 13 we will present an overview of the lives of stars, leaving detailed discussions of some special phases of evolution until later, specifically stellar pulsation, supernovae, and compact objects (stellar corpses).

### The Jeans Criterion

Despite many successes, important questions remain concerning how stars change during their lifetimes. One area where the picture is far from complete is in the earliest stage of evolution, the formation of pre-nuclear-burning objects known as **protostars** from interstellar molecular clouds.

If globules and cores in molecular clouds are the sites of star formation, what conditions must exist for collapse to occur? Sir James Jeans (1877–1946) first investigated this problem in 1902 by considering the effects of small deviations from hydrostatic equilibrium. Although several simplifying assumptions are made in the analysis, such as neglecting effects due to rotation, turbulence, and galactic magnetic fields, it provides important insights into the development of protostars.

The virial theorem (Eq. 2.46),

$$2K + U = 0,$$

describes the condition of equilibrium for a stable, gravitationally bound system.[4] We have already seen that the virial theorem arises naturally in the discussion of orbital motion, and we have also invoked it in estimating the amount of gravitational energy contained within a star (Eq. 10.22). The virial theorem may also be used to estimate the conditions necessary for protostellar collapse.

If twice the total internal kinetic energy of a molecular cloud ($2K$) exceeds the absolute value of the gravitational potential energy ($|U|$), the force due to the gas pressure will dominate the force of gravity and the cloud will expand. On the other hand, if the internal kinetic energy is too low, the cloud will collapse. The boundary between these two cases describes the critical condition for stability when rotation, turbulence, and magnetic fields are neglected.

Assuming a spherical cloud of constant density, the gravitational potential energy is approximately (Eq. 10.22)

$$U \sim -\frac{3}{5}\frac{GM_c^2}{R_c},$$

where $M_c$ and $R_c$ are the mass and radius of the cloud, respectively. We may also estimate

---

[4]We have implicitly assumed that the kinetic and potential energy terms are averaged over time, as in Section 2.4.

the cloud's internal kinetic energy, given by

$$K = \frac{3}{2} NkT,$$

where $N$ is the total number of particles. But $N$ is just

$$N = \frac{M_c}{\mu m_H},$$

where $\mu$ is the mean molecular weight. Now, by the virial theorem, the condition for collapse ($2K < |U|$) becomes

$$\frac{3 M_c kT}{\mu m_H} < \frac{3}{5} \frac{G M_c^2}{R_c}. \tag{12.12}$$

The radius may be replaced by using the initial mass density of the cloud, $\rho_0$, assumed here to be constant throughout the cloud,

$$R_c = \left( \frac{3 M_c}{4 \pi \rho_0} \right)^{1/3}. \tag{12.13}$$

After substitution into Eq. (12.12), we may solve for the minimum mass necessary to initiate the spontaneous collapse of the cloud. This condition is known as the **Jeans criterion**:

$$M_c > M_J,$$

where

$$M_J \simeq \left( \frac{5kT}{G \mu m_H} \right)^{3/2} \left( \frac{3}{4 \pi \rho_0} \right)^{1/2} \tag{12.14}$$

is called the **Jeans mass**. Using Eq. (12.13), the Jeans criterion may also be expressed in terms of the minimum radius necessary to collapse a cloud of density $\rho_0$:

$$R_c > R_J, \tag{12.15}$$

where

$$R_J \simeq \left( \frac{15kT}{4 \pi G \mu m_H \rho_0} \right)^{1/2} \tag{12.16}$$

is the **Jeans length**.

The Jeans mass derivation given above neglected the important fact that there must exist an external pressure on the cloud due to the surrounding interstellar medium (such as the encompassing GMC in the case of an embedded dense core). Although we will not derive the expression here, the critical mass required for gravitational collapse in the presence of an external gas pressure of $P_0$ is given by the **Bonnor–Ebert mass**,

$$M_{BE} = \frac{c_{BE} v_T^4}{P_0^{1/2} G^{3/2}},\tag{12.17}$$

where

$$v_T \equiv \sqrt{kT/\mu m_H}\tag{12.18}$$

is the *isothermal sound speed* ($\gamma = 1$ in Eq. 10.84), and the dimensionless constant $c_{BE}$ is given by

$$c_{BE} \simeq 1.18.$$

It is shown in Problem 12.11 that the Jeans mass (Eq. 12.14) can be written in the form of Eq. (12.17) with $c_J \simeq 5.46$ replacing $c_{BE}$. The smaller constant for the Bonnor–Ebert mass is to be expected since an external compression force due to $P_0$ is being exerted on the cloud.[5]

---

**Example 12.2.1.**   For a typical diffuse hydrogen cloud, $T = 50$ K and $n = 5 \times 10^8$ m$^{-3}$. If we assume that the cloud is entirely composed of H I, $\rho_0 = m_H n_H = 8.4 \times 10^{-19}$ kg m$^{-3}$. Taking $\mu = 1$ and using Eq. (12.14), the minimum mass necessary to cause the cloud to collapse spontaneously is approximately $M_J \sim 1500\,M_\odot$. However, this value significantly exceeds the estimated 1 to 100 $M_\odot$ believed to be contained in H I clouds. Hence diffuse hydrogen clouds are stable against gravitational collapse.

On the other hand, for a dense core of a giant molecular cloud, typical temperatures and number densities are $T = 10$ K and $n_{H_2} = 10^{10}$ m$^{-3}$. Since dense clouds are predominantly molecular hydrogen, $\rho_0 = 2m_H n_{H_2} = 3 \times 10^{-17}$ kg m$^{-3}$ and $\mu \simeq 2$. In this case the Jeans mass is $M_J \sim 8\,M_\odot$, characteristic of the masses of dense cores being on the order of 10 $M_\odot$. Apparently the dense cores of GMCs are unstable to gravitational collapse, consistent with being sites of star formation.

If the Bonnor–Ebert mass (Eq. 12.17) is used as the critical collapse condition, then the required mass reduces to approximately 2 $M_\odot$.

---

### Homologous Collapse

In the case that the criterion for gravitational collapse has been satisfied in the absence of rotation, turbulence, or magnetic fields, the molecular cloud will collapse. If we make the simplifying (and possibly unrealistic) assumption that any existing pressure gradients are too small to influence the motion appreciably, then the cloud is essentially in free-fall during the first part of its evolution. Furthermore, throughout the free-fall phase, the temperature

---

[5]You may be interested to know that the derivation of Eq. (12.17) involves the *isothermal Lane–Emden equation*, analogous to Eq. (10.110) used in deriving polytropes in Chapter 10.

of the gas remains nearly constant (i.e., the collapse is said to be *isothermal*). This is true as long as the cloud remains optically thin and the gravitational potential energy released during the collapse can be efficiently radiated away. In this case the spherically symmetric hydrodynamic equation (Eq. 10.5) can be used to describe the contraction if we assume that $|dP/dr| \ll GM_r\rho/r^2$. After canceling the density on both sides of the expression, we have

$$\frac{d^2r}{dt^2} = -G\frac{M_r}{r^2}. \tag{12.19}$$

Of course, the right-hand side of Eq. (12.19) is just the local acceleration of gravity at a distance $r$ from the center of a spherical cloud. As usual, the mass of the sphere interior to the radius $r$ is denoted by $M_r$.

To describe the behavior of the surface of a sphere of radius $r$ within the collapsing cloud as a function of time, Eq. (12.19) must be integrated over time. Since we are interested only in the surface that encloses $M_r$, the mass interior to $r$ will remain a constant during that collapse. As a result, we may replace $M_r$ by the product of the initial density $\rho_0$ and the initial spherical volume, $4\pi r_0^3/3$. Then, if we multiply both sides of Eq. (12.19) by the velocity of the surface of the sphere, we arrive at the expression

$$\frac{dr}{dt}\frac{d^2r}{dt^2} = -\left(\frac{4\pi}{3}G\rho_0 r_0^3\right)\frac{1}{r^2}\frac{dr}{dt},$$

which can be integrated once with respect to time to give

$$\frac{1}{2}\left(\frac{dr}{dt}\right)^2 = \left(\frac{4\pi}{3}G\rho_0 r_0^3\right)\frac{1}{r} + C_1.$$

The integration constant, $C_1$, can be evaluated by requiring that the velocity of the sphere's surface be zero at the beginning of the collapse, or $dr/dt = 0$ when $r = r_0$. This gives

$$C_1 = -\frac{4\pi}{3}G\rho_0 r_0^2.$$

Substituting and solving for the velocity at the surface, we have

$$\frac{dr}{dt} = -\left[\frac{8\pi}{3}G\rho_0 r_0^2\left(\frac{r_0}{r} - 1\right)\right]^{1/2}. \tag{12.20}$$

Note that the negative root was chosen because the cloud is collapsing.

To integrate Eq. (12.20) so that we can obtain an expression for the position as a function of time, we make the substitutions

$$\theta \equiv \frac{r}{r_0}$$

and

$$\chi \equiv \left(\frac{8\pi}{3}G\rho_0\right)^{1/2},$$

which leads to the differential equation

$$\frac{d\theta}{dt} = -\chi \left( \frac{1}{\theta} - 1 \right)^{1/2}. \tag{12.21}$$

Making yet another substitution,

$$\theta \equiv \cos^2 \xi, \tag{12.22}$$

and after some manipulation, Eq. (12.21) becomes

$$\cos^2 \xi \, \frac{d\xi}{dt} = \frac{\chi}{2}. \tag{12.23}$$

Equation (12.23) may now be integrated directly with respect to $t$ to yield

$$\frac{\xi}{2} + \frac{1}{4} \sin 2\xi = \frac{\chi}{2} t + C_2. \tag{12.24}$$

Lastly, the integration constant, $C_2$, must be evaluated. Doing so requires that $r = r_0$ when $t = 0$, which implies that $\theta = 1$, or $\xi = 0$ at the beginning of the collapse. Therefore, $C_2 = 0$.

We have finally arrived at the equation of motion for the gravitational collapse of the cloud, given in parameterized form by

$$\xi + \frac{1}{2} \sin 2\xi = \chi t. \tag{12.25}$$

Our task now is to extract the behavior of the collapsing cloud from this equation. From Eq. (12.25), it is possible to calculate the **free-fall timescale** for a cloud that has satisfied the Jeans criterion. Let $t = t_{\text{ff}}$ when the radius of the collapsing sphere reaches zero ($\theta = 0$, $\xi = \pi/2$).[6] Then

$$t_{\text{ff}} = \frac{\pi}{2\chi}.$$

Substituting the value for $\chi$, we have

$$\boxed{t_{\text{ff}} = \left( \frac{3\pi}{32} \frac{1}{G\rho_0} \right)^{1/2}.} \tag{12.26}$$

You should notice that the free-fall time is actually independent of the initial radius of the sphere. Consequently, as long as the original density of the spherical molecular cloud was uniform, all parts of the cloud will take the same amount of time to collapse, and the density will increase at the same rate everywhere. This behavior is known as a **homologous collapse**.

---

[6]This is obviously an unphysical final condition, since it implies infinite density. If $r_0 \gg r_{\text{final}}$, however, then $r_{\text{final}} \simeq 0$ is a reasonable approximation for our purposes here.

However, if the cloud is somewhat centrally condensed when the collapse begins, the free-fall time will be shorter for material near the center than for material farther out. Thus, as the collapse progresses, the density will increase more rapidly near the center than in other regions. In this case the collapse is referred to as an **inside-out collapse**.

---

**Example 12.2.2.** Using data given in Example 12.2.1 for a dense core of a giant molecular cloud, we may estimate the amount of time required for the collapse. Assuming a density of $\rho_0 = 3 \times 10^{-17}$ kg m$^{-3}$ that is constant throughout the core, Eq. (12.26) gives

$$t_{ff} = 3.8 \times 10^5 \text{ yr.}$$

To investigate the actual behavior of the collapse in our simplified model, we must first solve Eq. (12.25) for $\xi$, given a value for $t$, and then use Eq. (12.22) to find $\theta = r/r_0$. However, Eq. (12.25) cannot be solved explicitly, so numerical techniques must be employed. The numerical solution of the homologous collapse of the molecular cloud is shown in Fig. 12.8. Notice that the collapse is quite slow initially and accelerates quickly as $t_{ff}$ is approached. At the same time, the density increases very rapidly during the final stages of collapse.

---

**The Fragmentation of Collapsing Clouds**

Since the masses of fairly large molecular clouds could exceed the Jeans limit, from Eq. (12.14) our simple analysis seems to imply that stars can form with very large masses, possibly up to the initial mass of the cloud. However, observations show that this does not happen. Furthermore, it appears that stars frequently (perhaps even preferentially) tend to form in groups, ranging from binary star systems to clusters that contain hundreds of thousands of members (see Section 13.3).

The process of **fragmentation** that segments a collapsing cloud is an aspect of star formation that is under significant investigation. To see that fragmentation must occur by

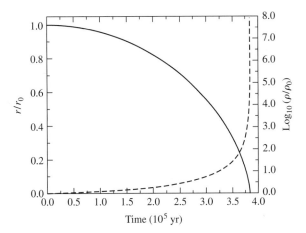

**FIGURE 12.8** The homologous collapse of a molecular cloud, as discussed in Example 12.2.2. $r/r_0$ is shown as the solid line and $\log_{10}(\rho/\rho_0)$ is shown as the dashed line. The initial density of the cloud was $\rho_0 = 3 \times 10^{-17}$ kg m$^{-3}$ and the free-fall time is $3.8 \times 10^5$ yr.

some mechanism(s), refer again to the equation for the Jeans mass (Eq. 12.14). An important consequence of the collapse of a molecular cloud is that the density of the cloud increases by many orders of magnitude during free-fall (Fig. 12.8). Consequently, since $T$ remains nearly constant throughout much of the collapse, it appears that the Jeans mass must decrease. After collapse has begun, any initial inhomogeneities in density will cause individual sections of the cloud to satisfy the Jeans mass limit independently and begin to collapse locally, producing smaller features within the original cloud. This cascading collapse could lead to the formation of large numbers of smaller objects.

It is important to point out that one challenge with the overly simplified scenario described here is that the process implies that far too many stars would be produced. It is likely that only about 1% of the cloud actually forms stars.

What is it that stops the fragmentation process? Since we observe a galaxy filled with stars that have masses on the order of the mass of the Sun, the cascading fragmentation of the cloud cannot proceed without interruption. The answer to the question lies in our implicit assumption that the collapse is isothermal, which in turn implies that the only term that changes in Eq. (12.14) is the density. Clearly this cannot be the case since stars have temperatures much higher than 10 to 100 K. If the energy that is released during a gravitational collapse is radiated away efficiently, the temperature can remain nearly constant. At the other extreme, if the energy cannot be transported out of the cloud at all (an *adiabatic* collapse), then the temperature must rise. Of course, the real situation must be somewhere between these two limits, but by considering each of these special cases carefully, we can begin to understand some of the important features of the problem.

If the collapse changes from being essentially isothermal to adiabatic, the associated temperature rise would begin to affect the value of the Jeans mass. In Chapter 10 we saw that for an adiabatic process the pressure of the gas is related to its density by $\gamma$, the ratio of specific heats (Eq. 10.86). Using the ideal gas law (Eq. 10.11), an adiabatic relation between density and temperature can be obtained,

$$T = K'' \rho^{\gamma - 1}, \tag{12.27}$$

where $K''$ is a constant. Substituting this expression into Eq. (12.14), we find that for an adiabatic collapse, the dependence of the Jeans mass on density becomes

$$M_J \propto \rho^{(3\gamma - 4)/2}.$$

For atomic hydrogen $\gamma = 5/3$, giving $M_J \propto \rho^{1/2}$; the Jeans mass *increases* with increasing density for a perfectly adiabatic collapse of a cloud. This behavior means that the collapse results in a minimum value for the mass of the fragments produced. The minimum mass depends on the point when the collapse goes from being predominantly isothermal to adiabatic.

Of course, this transition is not instantaneous or even complete. However, it is possible to make a crude order-of-magnitude estimate of the lower mass limit of the fragments. As we have already mentioned, according to the virial theorem, energy must be liberated during the collapse of the cloud. From Eq. (10.22) and the discussion contained in Example 10.3.1, the energy released is roughly

$$\Delta E_g \simeq \frac{3}{10} \frac{G M_J^2}{R_J}$$

for a spherical cloud just satisfying the Jeans criterion at some point during the collapse. Averaged over the free-fall time, the luminosity due to gravity is given by

$$L_{ff} \simeq \frac{\Delta E_g}{t_{ff}} \sim G^{3/2} \left( \frac{M_J}{R_J} \right)^{5/2},$$

where we have made use of Eq. (12.26) and have neglected terms of order unity.

If the cloud were optically thick and in thermodynamic equilibrium, the energy would be emitted as blackbody radiation. However, during collapse the process of releasing the energy is less efficient than for an ideal blackbody. Following Eq. (3.17), we may express the radiated luminosity as

$$L_{rad} = 4\pi R^2 e\sigma T^4,$$

where an efficiency factor, $0 < e < 1$, has been introduced to indicate the deviation from thermodynamic equilibrium. If the collapse is perfectly isothermal and escaping radiation does not interact at all with overlying infalling material, $e \sim 0$. If, on the other hand, energy emitted by some parts of the cloud is absorbed and then re-emitted by other parts of the cloud, thermodynamic equilibrium would more nearly apply and $e$ would be closer to unity.

Equating the two expressions for the cloud's luminosity,

$$L_{ff} = L_{rad},$$

and rearranging, we have

$$M_J^{5/2} = \frac{4\pi}{G^{3/2}} R_J^{9/2} e\sigma T^4.$$

Making use of Eq. (12.13) to eliminate the radius, and then using Eq. (12.14) to write the density in terms of the Jeans mass, we arrive at an estimate of when adiabatic effects become important, expressed in terms of the minimum obtainable Jeans mass:

$$M_{J_{min}} = 0.03 \left( \frac{T^{1/4}}{e^{1/2}\mu^{9/4}} \right) \, \text{M}_\odot, \tag{12.28}$$

where $T$ is expressed in kelvins. If we take $\mu \sim 1$, $e \sim 0.1$, and $T \sim 1000$ K at the time when adiabatic effects may start to become significant, $M_J \sim 0.5 \, \text{M}_\odot$; fragmentation ceases when the segments of the original cloud begin to reach the range of solar mass objects. The estimate is relatively insensitive to other reasonable choices for $T$, $e$, and $\mu$. For instance, if $e \sim 1$ then $M_J \sim 0.2 \, \text{M}_\odot$.

### Additional Physical Processes in Protostellar Star Formation

We have, of course, left out a number of important features in our calculations. For instance, we have freely used the Jeans criterion during each point in the collapse of the cloud to discuss the process of fragmentation. This cannot be correct, since our estimate of the Jeans criterion was based on a perturbation of a static cloud; no consideration was made of the initial velocity of the cloud's outer layers. We have also neglected the details

of radiation transport through the cloud, as well as vaporization of the dust grains, dissociation of molecules, and ionization of the atoms. Nevertheless, it is worth noting that as
unsophisticated as the preceding analysis was, it did illustrate important aspects of the fundamental problem and left us with a result that is reasonable. Such preliminary approaches
to understanding complex physical systems are powerful tools in our study of nature.[7] More
sophisticated estimates of the complex process of cessation of fragmentation place the limit
an order of magnitude lower than determined above, at about 0.01 $M_\odot$.

Perhaps just as important to the problem of the collapse process are the possible effects of
rotation (angular momentum), the deviation from spherical symmetry, turbulent motions in
the gas, and the presence of magnetic fields. For example, an appreciable amount of angular
momentum present in the original cloud is likely to result in a disk-like structure for at least
a part of the original material, since collapse will proceed at a more rapid rate along the
axis of rotation relative to collapse along the equator (aspects of the angular momentum of
the collapsing cloud will be explored in Problem 12.18).

It is also apparent from careful investigations of molecular clouds that magnetic fields
must also play a crucial role and, in fact, are likely to control the onset of collapse. That
mechanisms other than gravity must be involved becomes clear in simply considering the
free-fall time of the dense core discussed in Example 12.2.2. From that calculation, the
collapse of the dense core should occur on a timescale on the order of $10^5$ yr. While this
may seem long by human standards, it is quite short on stellar evolution timescales. This
would imply that almost as soon as a dense core forms, it begins producing stars. This would
also imply that dense cores should be very rare; however, many dense cores are observable
throughout our Galaxy.

Zeeman measurements of various molecular clouds indicate the presence of magnetic
fields with strengths typically on the order of magnitude of 1 to 100 nT. If the magnetic
field of a cloud is "frozen in," and the cloud is compressed, the magnetic field strength will
increase, leading to an increase in the magnetic pressure and resistance to the compression.
In fact, if the cloud is stable to collapse because of magnetic pressure, it will remain so as
long as the magnetic field does not decay (recall the discussion of magnetic fields associated
with sunspots in Chapter 11).

During the derivation of the Jeans criterion, the virial theorem was invoked using a
balance between gravitational potential energy and the cloud's internal (thermal) kinetic
energy. Absent from that calculation was the inclusion of energy due to the presence of
magnetic fields. When magnetic fields are included, the critical mass can be expressed as

$$M_B = c_B \frac{\pi R^2 B}{G^{1/2}},$$ 
(12.29)

where $c_B = 380 \, \text{N}^{1/2} \, \text{m}^{-1} \, \text{T}^{-1}$ for a magnetic field permeating a spherical, uniform cloud.
If $B$ is expressed in nT and $R$ in units of pc, then Eq. (12.29) can be written in the more
illustrative form

$$M_B \simeq 70 \, M_\odot \left( \frac{B}{1 \, \text{nT}} \right) \left( \frac{R}{1 \, \text{pc}} \right)^2.$$ 
(12.30)

---

[7]This type of approach is sometimes called a "back-of-the-envelope" calculation because of the relatively small
space required to carry out the estimate. Extensive use of "back-of-the-envelope" calculations is made throughout
this text to illustrate the effects of key physical processes.

If the mass of the cloud is less than $M_B$, the cloud is said to be **magnetically subcritical** and stable against collapse, but if the mass of the cloud exceeds $M_B$, the cloud is **magnetically supercritical** and the force due to gravity will overwhelm the ability of the magnetic field to resist collapse.

---

**Example 12.2.3.**   For the dense core considered in Examples 12.2.1 and 12.2.2, if the dense core has a magnetic field of 100 nT threading through it, and if it has a radius of 0.1 pc, the magnetic critical mass would be $M_B \simeq 70 \ M_\odot$, implying that a dense core of mass 10 $M_\odot$ would be stable against collapse. However, if $B = 1$ nT, then $M_B \simeq 0.7 \ M_\odot$ and collapse would occur.

---

**Ambipolar Diffusion**

The last example hints at another possibility for triggering the collapse of a dense core. If a core that was originally subcritical were to become supercritical, collapse could ensue. This could happen in one of two ways: a group of subcritical clouds could combine to form a supercritical cloud, or the magnetic field could be rearranged so that the field strength is lessened in a portion of the cloud. It appears that both processes may occur, although the latter process seems to dominate the pre-collapse evolution of most molecular clouds.

Recall that only charged particles such as electrons or ions are tied to magnetic field lines; neutrals are not affected directly. Given that dense molecular cores are dominated by neutrals, how can magnetic fields have any substantial effect on the collapse? The answer lies in the collisions between neutrals and the ions (electrons do not significantly affect neutral atoms or molecules through collisions). As neutrals try to drift across magnetic field lines, they collide with the "frozen-in" ions, and the motions of the neutrals are inhibited. However, if there is a net defined direction for the motion of neutrals due to gravitational forces, they will still tend to migrate slowly in that direction. This slow migration process is known as **ambipolar diffusion**.

To determine the relative impact of ambipolar diffusion, we need to estimate a characteristic timescale for the diffusion process. This is done by comparing the size of the molecular cloud to the time it takes for a neutral to drift across the cloud. It can be shown that the timescale for ambipolar diffusion is approximately

$$t_{AD} \simeq \frac{2R}{v_{\text{drift}}} \simeq 10 \ \text{Gyr} \ \left( \frac{n_{H_2}}{10^{10} \ \text{m}^{-3}} \right) \left( \frac{B}{1 \ \text{nT}} \right)^{-2} \left( \frac{R}{1 \ \text{pc}} \right)^2 . \qquad (12.31)$$

Once collapse begins, magnetic fields can be further altered by undergoing reconnection events similar to those of solar flares.

---

**Example 12.2.4.**   Returning to the dense core we used in previous examples, if $B = 1$ nT and $R = 0.1$ pc, we find from Eq. (12.31) that the timescale for ambipolar diffusion is 100 Myr. This is several hundred times longer than the free-fall timescale determined in Example 12.2.2. Clearly the ambipolar diffusion process can control the evolution of a dense core for a long time before free-fall collapse begins.

---

### Numerical Simulations of Protostellar Evolution

To investigate the nature of the gravitational collapse of a cloud in detail, we must solve the magnetohydrodynamic equations numerically. Unfortunately, limits in computing power and numerical methods still necessitate making numerous and significant simplifying assumptions. These numerical models do exhibit many of the characteristics that were illustrated by our crude analytical studies, but other important aspects of the collapse become apparent that were not contained in the physics that has already been discussed.[8]

Consider a spherical cloud of approximately 1 $M_\odot$ and solar composition that is supercritical. Initially the early stages of the free-fall collapse are nearly isothermal because light near the center of the collapse can travel significant distances before being absorbed by dust. Owing to an initial slight increase in density toward the center of the cloud, the free-fall timescale is shorter near the center and the density increases more rapidly there (inside-out collapse). When the density of the material near the center of the collapse region reaches approximately $10^{-10}$ kg m$^{-3}$, the region becomes optically thick and the collapse becomes more adiabatic. The opacity of the cloud at this point is primarily due to the presence of dust.

The increased pressure that occurs when the collapse becomes adiabatic substantially slows the rate of collapse near the core. At this point the central region is nearly in hydrostatic equilibrium with a radius of approximately 5 AU. It is this central object that is referred to as a protostar.

One observable consequence of the cloud becoming optically thick is that the gravitational potential energy being released during the collapse is converted into heat and then radiated away in the infrared as blackbody radiation. By computing the rate of energy release (the luminosity) and the radius of the cloud where the optical depth is $\tau = 2/3$, the effective temperature may be determined using Eq. (3.17). (At this point in its evolution, the optical depth is determined by the dust, and so the photosphere is a dust photosphere.)

With the identification of a photosphere, it becomes possible to plot the location of the simulated cloud on the H–R diagram as a function of time. Curves that depict the life histories of stars on the H–R diagram are known as **evolutionary tracks**. Figure 12.9 shows theoretical evolutionary tracks of 0.05, 0.1, 0.5, 1, 2, and 10 $M_\odot$ clouds computed by one research group through the protostar phase. As the collapse continues to accelerate during the early stages, the luminosity of the protostar increases along with its effective temperature.

Above the developing protostellar core, material is still in free-fall. When the infalling material meets the nearly hydrostatic core, a shock wave develops where the speed of the material exceeds the local sound speed (the material is supersonic). It is at this shock front that the infalling material loses a significant fraction of its kinetic energy in the form of heat that "powers" the cloud and produces much of its luminosity.

When the temperature reaches approximately 1000 K, the dust within the developing protostar begins to vaporize and the opacity drops. This means that the radius where $\tau = 2/3$ is substantially reduced, approaching the surface of the hydrostatic core. Since the luminosity remains high during this phase, a corresponding increase in the effective temperature must occur.

---

[8]Some of the first calculations of protostellar collapse were performed by Richard Larson in 1969. His pioneering work neglected the complicated physics associated with rotation, turbulence, and magnetic fields but did include thermodynamics, radiative transfer, and other important physical processes.

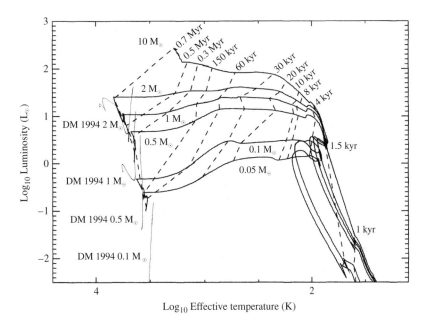

**FIGURE 12.9** Theoretical evolutionary tracks of the gravitational collapse of 0.05, 0.1, 0.5, 1, 2, and 10 $M_\odot$ clouds through the protostar phase (solid lines). The dashed lines show the times since collapse began. The light dotted lines are pre-main-sequence evolutionary tracks of 0.1, 0.5, 1, and 2 $M_\odot$ stars from D'Antona and Mazzitelli, *Ap. J. Suppl.*, *90*, 457, 1994. Note that the horizontal axis is plotted with effective temperature increasing to the left, as is characteristic of all H–R diagrams. (Figure adapted from Wuchterl and Tscharnuter, *Astron. Astrophys.*, *398*, 1081, 2003.)

As the overlying material continues to fall onto the hydrostatic core, the temperature of the core slowly increases. Eventually the temperature becomes high enough (approximately 2000 K) to cause the molecular hydrogen to dissociate into individual atoms. This process absorbs energy that would otherwise provide a pressure gradient sufficient to maintain hydrostatic equilibrium. As a result, the core becomes dynamically unstable and a second collapse occurs.[9] After the core radius has decreased to a value about 30% larger than the present size of the Sun, hydrostatic equilibrium is re-established. At this point, the core mass is still much less than its final value, implying that **accretion** is still ongoing.

After the core collapse, a second shock front is established as the envelope continues to accrete infalling material. When the nearly flat, roughly constant luminosity part of the evolutionary track is reached in Fig. 12.9, accretion has settled into a quasi-steady main accretion phase. At about the same time, temperatures in the deep interior of the protostar have increased enough that deuterium ($^{2}_{1}$H) begins to burn (Eq. 10.38), producing up to 60% of the luminosity of the 1 $M_\odot$ protostar. Note that this reaction is favored over the first step in the PP I chain because it has a fairly large cross section, $\sigma(E)$, at low temperatures.

With only a finite amount of mass available from the original cloud, and with only a limited amount of deuterium available to burn, the luminosity must eventually decrease. When deuterium burn-out occurs, the evolutionary track bends sharply downward and the

---

[9]Dynamical instabilities will be discussed further in Chapter 14 in connection with pulsating stars.

effective temperature decreases slightly. The evolution has now reached a quasi-static pre-main-sequence phase that will be discussed in the next section.

The theoretical scenario just described leads to the possibility of observational verification. Since it is expected that the collapse should occur deep within a molecular cloud, the protostar itself would likely be shielded from direct view by a cocoon of dust. Consequently, any observational evidence of the collapse would be in the form of small infrared sources embedded within dense cores or Bok globules. The detection of protostellar collapse is made more difficult by the relatively small value for the free-fall time, meaning that protostars are fairly short-lived objects.

The search for protostars is under way in infrared and millimeter wavelengths, and a number of strong candidates have been identified, including B335, a Bok globule in the constellation of Aquila, L1527 in Taurus, and numerous objects in the Orion Nebula. B335 is probably the best-studied case and is almost a perfect test of the theory of protostellar collapse since it seems to have very little turbulence or rotation.

Some astronomers believe that by studying the details of the infrared spectra of these sources, they have been able to identify possible spectral signatures of infalling dust and gas around the embedded infrared objects. These tell-tale features involve Doppler-shifted sub-structures in the profiles of spectral lines. For an optically thick line, a central absorption feature is often visible (see Fig. 12.10). The source of the absorption feature is cool material between the observer and the source of the line (the hotter central region). The broad wings of the line result from Doppler-shifted light coming from infalling gas. The blueshifted wing is from infalling gas on the far side of the cloud (therefore moving toward the observer), and the redshifted wing is from infalling gas on the near side of the cloud. Infall has been identified in starless dense cores as well.[10]

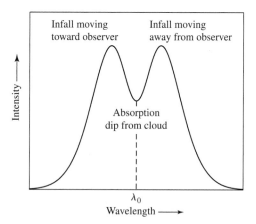

**FIGURE 12.10**  A line profile of a spherical, infalling cloud. The wings are Doppler shifted due to infalling material. The central absorption is produced by intervening material far from the central collapse. The redshifted wing arises from material in front of the central region moving away from the observer, and the blueshifted wing is due to material in the back of the cloud moving toward the observer.

---

[10]Another example of line profile signatures indicating mass motions is discussed in the next section; see Fig. 12.17.

## 12.3 ■ PRE-MAIN-SEQUENCE EVOLUTION

As we discovered in the last section, once the collapse of a molecular cloud has begun, it is characterized by the free-fall timescale given by Eq. (12.26). With the formation of a quasi-static protostar, the rate of evolution becomes controlled by the rate at which the star can thermally adjust to the collapse. This is just the Kelvin–Helmholtz timescale discussed in Example 10.3.1 (see Eq. 10.24); the gravitational potential energy liberated by the collapse is released over time and is the source of the object's luminosity. Since $t_{KH} \gg t_{ff}$, protostellar evolution proceeds at a much slower rate than free-fall collapse. For instance, a 1 $M_\odot$ star requires almost 40 Myr to contract quasi-statically to its main-sequence structure.

### The Hayashi Track

With the steadily increasing effective temperature of the protostar, the opacity of the outer layers becomes dominated by the $H^-$ ion, the extra electrons coming from the partial ionization of some of the heavier elements in the gas that have lower ionization potentials. As with the envelope of the main-sequence Sun, this large opacity contribution causes the envelope of a contracting protostar to become convective. In fact, in some cases the convection zone extends all the way to the center of the star. In 1961, C. Hayashi demonstrated that because of the constraints convection puts on the structure of a star, a deep convective envelope limits its quasi-static evolutionary path to a line that is nearly vertical in the H–R diagram. Consequently, as the protostar collapse slows, its luminosity decreases while its effective temperature increases slightly. It is this evolution along the **Hayashi track** that appears as the downward turn at the end of the evolutionary tracks shown in Fig. 12.9.

The Hayashi track actually represents a boundary between "allowed" hydrostatic stellar models and those that are "forbidden." To the right of the Hayashi track, there is no mechanism that can adequately transport the luminosity out of the star at those low effective temperatures; hence no stable stars can exist there. To the left of the Hayashi track, convection and/or radiation is responsible for the necessary energy transport. Note that this distinction between allowed and forbidden models is not in conflict with the free-fall evolution of collapsing gas clouds found to the right of the Hayashi track since those objects are far from being in hydrostatic equilibrium.

### Classical Calculations of Pre-Main-Sequence Evolution

In 1965, before detailed protostellar collapse calculations were performed, Icko Iben, Jr. computed the final stages of collapse onto the main sequence for stars of various masses. In each case he started his models on the Hayashi track. All of those models neglected the effects of rotation, magnetic fields, and mass loss. Since that time, significant improvements have been made in our understanding of the physical processes involved in stellar structure and evolution, including refined nuclear reaction rates, new opacities, and the inclusion of mass loss or accretion. Some modern evolutionary calculations have also included the effects of rotation.[11] The **pre-main-sequence evolutionary tracks** for a sequence of masses

---

[11] Some calculations have also begun considering the effects of magnetic fields, but the results presented in this text do not include those recent preliminary results.

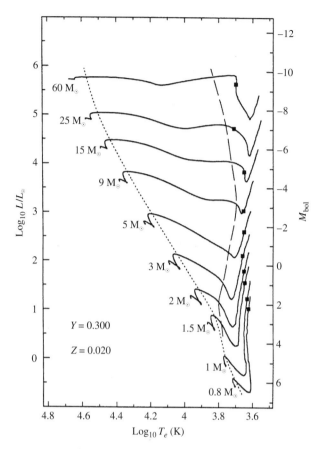

**FIGURE 12.11** Classical pre-main-sequence evolutionary tracks computed for stars of various masses with the composition $X = 0.68$, $Y = 0.30$, and $Z = 0.02$. The direction of evolution on each track is generally from low effective temperature to high effective temperature (right to left). The mass of each model is indicated beside its evolutionary track. The square on each track indicates the onset of deuterium burning in these calculations. The long-dash line represents the point on each track where convection in the envelope stops and the envelope becomes purely radiative. The short-dash line marks the onset of convection in the core of the star. Contraction times for each track are given in Table 12.1. (Figure adapted from Bernasconi and Maeder, *Astron. Astrophys.*, *307*, 829, 1996.)

computed with state-of-the-art physics are shown in Fig. 12.11, and the total time for each evolutionary track is given in Table 12.1.

Consider the pre-main-sequence evolution of a 1 $M_\odot$ star, beginning on the Hayashi track. With the high $H^-$ opacity near the surface, the star is completely convective during approximately the first one million years of the collapse. In these models, deuterium burning also occurs during this early period of collapse, beginning at the square indicated on the evolutionary tracks in Fig. 12.11.[12] However, since $^2_1H$ is not very abundant, the nuclear

[12]Note that since these calculations did not include the formation of the protostar from the direct collapse of the cloud as was done for the tracks in Fig. 12.9, there is a fundamental inconsistency between when deuterium burning occurs in the two sets of calculations.

**TABLE 12.1** Pre-main-sequence contraction times for the classical models presented in Fig. 12.11. (Data from Bernasconi and Maeder, *Astron. Astrophys.*, *307*, 829, 1996.)

| Initial Mass ($M_\odot$) | Contraction Time (Myr) |
|---|---|
| 60 | 0.0282 |
| 25 | 0.0708 |
| 15 | 0.117 |
| 9 | 0.288 |
| 5 | 1.15 |
| 3 | 7.24 |
| 2 | 23.4 |
| 1.5 | 35.4 |
| 1 | 38.9 |
| 0.8 | 68.4 |

reactions have little effect on the overall collapse; they simply slow the rate of collapse slightly.

As the central temperature continues to rise, increasing levels of ionization decrease the opacity in that region (see Fig. 9.10) and a radiative core develops, progressively encompassing more and more of the star's mass. At the point of minimum luminosity in the tracks following the descent along the Hayashi track, the existence of the radiative core allows energy to escape into the convective envelope more readily, causing the luminosity of the star to increase again. Also, as required by Eq. (3.17), the effective temperature continues to increase, since the star is still shrinking.

At about the time that the luminosity begins to increase again, the temperature near the center has become high enough for nuclear reactions to begin in earnest, although not yet at their equilibrium rates. Initially, the first two steps of the PP I chain [the conversion of $^1_1H$ to $^3_2He$; Eqs. (10.37) and (10.38)] and the CNO reactions that turn $^{12}_6C$ into $^{14}_7N$ [Eqs. (10.48–10.50)] dominate the nuclear energy production. With time, these reactions provide an increasingly larger fraction of the luminosity, while the energy production due to gravitational collapse makes less of a contribution to $L$.

Due to the onset of the highly temperature-dependent CNO reactions, a steep temperature gradient is established in the core, and some convection again develops in that region. At the local maximum in the luminosity on the H–R diagram near the short dashed line, the rate of nuclear energy production has become so great that the central core is forced to expand somewhat, causing the gravitational energy term in Eq. (10.36) to become negative [recall that $\epsilon = \epsilon_{nuclear} + \epsilon_{gravity}$; see Eq. (10.102)]. This effect is apparent at the surface as the total luminosity decreases toward its main-sequence value, accompanied by a decrease in the effective temperature.

When the $^{12}_6C$ is finally exhausted, the core completes its readjustment to nuclear burning, reaching a sufficiently high temperature for the remainder of the PP I chain to become important. At the same time, with the establishment of a stable energy source, the gravitational energy term becomes insignificant and the star finally settles onto the main sequence. It is worth noting that the time required for a 1 $M_\odot$ star to reach the main sequence, according to the detailed numerical model just described, is not very different from the crude estimate of the Kelvin–Helmholtz timescale performed in Example 10.3.1.

For stars with masses lower than our Sun's, the evolution is somewhat different. For stars with masses $M \lesssim 0.5\ M_\odot$ (not shown in Fig. 12.11), the upward branch is missing just before the main sequence. This happens because the central temperature never gets hot enough to burn $^{12}_{6}C$ efficiently (recall that our estimates of the central pressure and temperature of the Sun in Examples 10.1.1 and 10.2.1 were roughly proportional to the mass of the star). In fact, as was mentioned in Section 10.6, if the mass of the collapsing protostar is less than approximately $0.072\ M_\odot$, the core never gets hot enough to generate sufficient energy by nuclear reactions to stabilize the star against gravitational collapse. As a result, the stable hydrogen-burning main sequence is never obtained. This explains the lower end of the main sequence.

Another important difference exists between solar-mass stars and stars of lower mass that can reach the main sequence: Temperatures remain cool enough and the opacity stays sufficiently high in low-mass stars that a radiative core never develops. Consequently, these stars remain fully convective all the way to the main sequence.

### The Formation of Brown Dwarfs

Below about $0.072\ M_\odot$, some nuclear burning will still occur, but not at a rate necessary to form a main-sequence star. Above about $0.06\ M_\odot$ the core temperature of the star is great enough to burn lithium, and above a mass of approximately $0.013\ M_\odot$ deuterium burning occurs ($0.013\ M_\odot$ is roughly thirteen times the mass of Jupiter). This last value is also in agreement with the cessation of fragmentation discussed on page 420. The objects in the range between about $0.013\ M_\odot$ and $0.072\ M_\odot$ are known as **brown dwarfs** and have spectral types of L and T (recall Table 8.1). The first confirmed discovery of a brown dwarf, Gliese 229B, was announced in 1995. Since that time hundreds of brown dwarfs have been detected thanks to near-infrared all-sky surveys, such as the Two Micron All Sky Survey (2MASS) and the Sloan Digital Sky Survey (SDSS). Given their very low luminosities and difficulty of detection, the number of objects found to date suggest that brown dwarfs are prevalent throughout the Milky Way Galaxy.

### Massive Star Formation

For massive stars, the central temperature quickly becomes high enough to burn $^{12}_{6}C$ as well as convert $^{1}_{1}H$ into $^{3}_{2}He$. This means that these stars leave the Hayashi track at higher luminosities and evolve nearly horizontally across the H–R diagram. Because of the much larger central temperatures, the full CNO cycle becomes the dominant mechanism for hydrogen burning in these main-sequence stars. Since the CNO cycle is so strongly temperature-dependent, the core remains convective even after the main sequence is reached.

### Possible Modifications to the Classical Models

The general pre-main-sequence evolutionary track calculations described above contain numerous approximations, as already discussed. It is likely that rotation plays an important role, along with turbulence and magnetic fields. It is also likely that the initial environments contain inhomogeneities in cloud densities, strong stellar winds, and ionizing radiation from nearby, massive stars.

These classical models also assume initial structures that are very large, with radii that are effectively infinitely greater than their final values. Given that dense cores have dimensions on the order of 0.1 pc, the initial radii of clouds undergoing protostellar collapse must be much smaller than traditionally assumed. In addition, the assumption of pressure-free protostellar collapse may also be a poor one; more realistic calculations probably require an initial contraction that is quasi-static (after all, the dark cores are roughly in hydrostatic equilibrium).

To complicate matters further, the more massive stars also interact with infalling material in such a way that a feedback loop may develop, limiting the amount of mass that they can accrete via the classical process discussed to this point; recall the discussion of the Eddington limit (e.g. Eq. 10.6).

In light of these various complications, some astronomers have suggested that significant modifications to the classical pre-main-sequence evolutionary tracks may be required. Theoretical evolutionary sequences beginning with smaller initial radii lead to a **birth line** where protostars first become visible. This birth line places an upper limit on the observed luminosities of protostars.

In addition, some observations suggest that stars with masses greater than about 10 $M_\odot$ or so may not form at all by the classical pre-main-sequence process described above. This apparent effect could be due to limiting feedback mechanisms, such as the high luminosity of ionizing radiation associated with high effective temperatures. Instead of the collapse of single protostellar clouds, the more massive stars may form by mergers of smaller stars in dense protostellar environments. On the other hand, some researchers have argued that the need for mergers can be avoided because rotation implies that most of the infalling mass collapses to an **accretion disk**[13] that forms around the star. The accretion disk then feeds the growing massive star, minimizing the impact of high amounts of ionizing radiation on the infalling gas and dust.

### The Zero-Age Main Sequence (ZAMS)

The diagonal line in the H–R diagram where stars of various masses first reach the main sequence and begin equilibrium hydrogen burning is known as the **zero-age main sequence** (ZAMS). Inspection of the classical results given in Table 12.1 shows that the amount of time required for stars to collapse onto the ZAMS is inversely related to mass; a 0.8 $M_\odot$ star takes over 68 Myr to reach the ZAMS, whereas a 60 $M_\odot$ star makes it to the ZAMS in only 28,000 years!

This inverse relationship between star-formation time and stellar mass may also signal a problem with classical pre-main-sequence evolutionary models. The reason is that if the most massive stars do indeed form first in a cluster of stars, the intense radiation that they produce would likely disperse the cloud before their low-mass siblings would ever have a chance to develop.

Clearly much work remains before we can say that pre-main-sequence stellar evolution is understood.

---

[13]Accretion disks will be discussed in detail in Chapter 18.

**The Initial Mass Function (IMF)**

From observational studies it is apparent that more low-mass than high-mass stars form when an interstellar cloud fragments. This implies that the number of stars that form per mass interval per unit volume (or per unit area in the Milky Way's disk) is strongly mass-dependent. This functional dependence is known as the **initial mass function** (IMF). One theoretical estimate of the IMF is shown in Fig. 12.12. However, a particular IMF depends on a variety of factors, including the local environment in which a cluster of stars forms from a given cloud complex in the ISM.

As a consequence of the process of fragmentation, most stars form with relatively low mass. Given the disparity in the numbers of stars formed in different mass ranges, combined with the very different rates of evolution, it is not surprising that massive stars are extremely rare, while low-mass stars are found in abundance. Observations also suggest that although the IMF is quite uncertain below about 0.1 $M_\odot$, rather than falling off sharply as indicated in Fig. 12.12, the curve may be fairly flat, resulting in large numbers of low-mass stars and brown dwarfs.

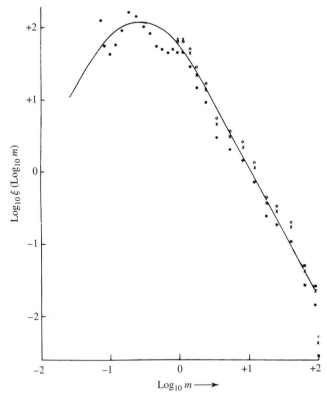

**FIGURE 12.12**    The initial mass function, $\xi$, shows the number of stars per unit area of the Milky Way's disk per unit interval of logarithmic mass that is produced in different mass intervals. The individual points represent observational data and the solid line is a theoretical estimate. Masses are in solar units. (Figure adapted from Rana, *Astron. Astrophys.*, *184*, 104, 1987.)

## H II Regions

When hot, massive stars reach the ZAMS with O or B spectral types, they do so shrouded in a cloak of gas and dust. The bulk of their radiation is emitted in the ultraviolet portion of the electromagnetic spectrum. Those photons that are produced with energies in excess of 13.6 eV can ionize the ground-state hydrogen gas (H I) in the ISM that still surrounds the newly formed star. Of course, if these **H II regions** are in equilibrium, the rate of ionization must equal the rate of recombination; photons must be absorbed and ions must be produced at the same rate that free electrons and protons recombine to form neutral hydrogen atoms. When recombination occurs, the electron does not necessarily fall directly to the ground state but can cascade downward, producing a number of lower-energy photons, many of which will be in the visible portion of the spectrum. The dominant visible wavelength photon produced in this way results from the transition between $n = 3$ and $n = 2$, the red line of the Balmer series (H$\alpha$). Consequently, because of this energy cascade, H II regions appear to fluoresce in red light.

These **emission nebulae** are considered by some to be among the most beautiful objects in the night sky. One of the more famous H II regions is the Orion nebula (M42),[14] found in the sword of the Orion constellation. M42 is part of the Orion A complex (see Fig. 12.13), which also contains a giant molecular cloud (OMC 1) and a very young cluster of stars (the Trapezium cluster). The first protostar candidates were discovered in this region as well.

The size of an H II region can be estimated by considering the requirement of equilibrium. Let $N$ be the number of photons *per second* produced by the O or B star with sufficient energy to ionize hydrogen from the ground state ($\lambda < 91.2$ nm). Assuming that all of the

**FIGURE 12.13** The H II region in Orion A is associated with a young OB association, the Trapezium cluster, and a giant molecular cloud. The Orion complex is 450 pc away. (Courtesy of the National Optical Astronomy Observatories.)

---

[14]M42 is the entry number in the well-known Messier catalog, a popular collection of observing objects for amateur astronomers. The Messier catalog is listed in Appendix H.

energetic photons are ultimately absorbed by the hydrogen in the H II region, the rate of photon creation must equal the rate of recombination. If this equilibrium condition did not develop, the size of the region would continue to grow as the photons traveled ever farther before encountering un-ionized gas.

Next, let $\alpha n_e n_H$ be the number of recombinations *per unit volume per second*, where $\alpha$ is a quantum-mechanical recombination coefficient that describes the likelihood that an electron and a proton can form a hydrogen atom, given their number densities (obviously, the more electrons and protons that are present, the greater the chance of recombination; hence the product $n_e n_H$).[15] At about 8000 K, a temperature characteristic of H II regions, $\alpha = 3.1 \times 10^{-19}$ m$^3$ s$^{-1}$. If we assume that the gas is composed entirely of hydrogen and is electrically neutral, then for every ion produced, one electron must have been liberated, or $n_e = n_H$. With this equality, the expression for the recombination rate can be multiplied by the volume of the H II region, assumed here to be spherical, and then set equal to the number of ionizing photons produced per second. Finally, solving for the radius of the H II region gives

$$r_S \simeq \left( \frac{3N}{4\pi\alpha} \right)^{1/3} n_H^{-2/3}. \tag{12.32}$$

$r_S$ is called the **Strömgren radius**, after Bengt Strömgren (1908–1987), the astrophysicist who first carried out the analysis in the late 1930s.

---

**Example 12.3.1.** From Appendix G, the effective temperature and luminosity of an O6 star are $T_e \simeq 45{,}000$ K and $L \simeq 1.3 \times 10^5$ L$_\odot$, respectively. According to Wien's law (Eq. 3.15), the peak wavelength of the blackbody spectrum is given by

$$\lambda_{\max} = \frac{0.0029 \text{ m K}}{T_e} = 64 \text{ nm}.$$

Since this is significantly shorter than the 91.2-nm limit necessary to produce ionization from the hydrogen ground state, it can be assumed that most of the photons created by an O6 star are capable of causing ionization.

The energy of one 64-nm photon can be calculated from Eq. (5.3), giving

$$E_\gamma = \frac{hc}{\lambda} = 19 \text{ eV}.$$

Now, assuming for simplicity that all of the emitted photons have the same (peak) wavelength, the total number of photons produced by the star per second is just

$$N \simeq L/E_\gamma \simeq 1.6 \times 10^{49} \text{ photons s}^{-1}.$$

Lastly, taking $n_H \sim 10^8$ m$^{-3}$ to be a typical value an H II region, we find

$$r_S \simeq 3.5 \text{ pc}.$$

Values of $r_S$ range from less than 0.1 pc to greater than 100 pc.

---

[15]Note that this expression is somewhat analogous to Eq. (10.29), the generalized nuclear reaction rate equation.

## The Effects of Massive Stars on Gas Clouds

As a massive star forms, the protostar will initially appear as an infrared source embedded inside the molecular cloud. With the rising temperature, first the dust will vaporize, then the molecules will dissociate, and finally, as the star reaches the main sequence, the gas immediately surrounding it will ionize, resulting in the creation of an H II region inside of an existing H I region.

Now, because of the star's high luminosity, radiation pressure will begin to drive significant amounts of mass loss, which then tends to disperse the remainder of the cloud. If several O and B stars form at the same time, it may be that much of the mass that has not yet become gravitationally bound to more slowly forming low-mass protostars will be driven away, halting any further star formation. Moreover, if the cloud was originally marginally bound (near the limit of criticality), the loss of mass will diminish the potential energy term in the virial theorem, with the result that the newly formed cluster of stars and protostars will become unbound (i.e., the stars will tend to drift apart). Figure 12.14 shows such a process under way in the Carina Nebula, located approximately 3000 pc from Earth. Another famous example of the effects of ionizing radiation of nearby massive stars is the production of the pillars in M16, the Eagle Nebula (Fig. 12.15).

## OB Associations

Groups of stars that are dominated by O and B main-sequence stars are referred to as **OB associations**. Studies of their individual kinematic velocities and masses generally lead to the conclusion that they cannot remain gravitationally bound to one another as permanent stellar clusters. One such example is the Trapezium cluster in the Orion A complex, believed to be less than 10 million years old. It is currently densely populated with stars

(a)                                    (b)

**FIGURE 12.14**  (a) An infrared image of a portion of the Carina Nebula. Eta Carina, a very young and marginally stable star of more than 100 $M_\odot$ is located above the image. The strong winds and intense ultraviolet radiation from Eta Carina and other massive stars in the region are shredding the nebula. Other, lower-mass newborn stars, such as those just above the pillar to the right of center in the image, are being inhibited from growing larger because of the destruction of the nebula by their much more massive siblings. [NASA/JPL-Caltech/N. Smith (University of Colorado at Boulder)] (b) The same region observed in visible light. Much less detail is observable because of the obscuration due to dust in the cloud. (NOAO)

**FIGURE 12.15**    The giant gas pillars of the Eagle Nebula (M16). The left most pillar is more than 1 pc long from base to top. Ionizing radiation from massive newborn stars off the top edge of the image are causing the gas in the cloud to photoevaporate. [Courtesy of NASA, ESA, STScI, J. Hester and P. Scowen (Arizona State University).]

$(> 2 \times 10^3 \text{ pc}^{-3})$, most of which have masses in the range of 0.5 to 2.0 $M_\odot$. Doppler shift measurements of the radial velocities of $^{13}$CO show that the gas in the vicinity is very turbulent. Apparently, the nearby O and B stars are dispersing the gas, and the cluster is becoming unbound.

### T Tauri Stars

**T Tauri stars** are an important class of low-mass pre-main-sequence objects that represent a transition between stars that are still shrouded in dust (IR sources) and main-sequence stars. T Tauri stars, named after the first star of their class to be identified (located in the constellation of Taurus), are characterized by unusual spectral features and by large and fairly rapid irregular variations in luminosity, with timescales on the order of days. The positions of T Tauri stars on the H–R diagram are shown in Fig. 12.16; theoretical pre-main-sequence evolutionary tracks are also included. The masses of T Tauri stars range from 0.5 to about 2 $M_\odot$.

Many T Tauri stars exhibit strong emission lines from hydrogen (the Balmer series), from Ca II (the H and K lines), and from iron, as well as absorption lines of lithium. Interestingly, forbidden lines of [O I] and [S II] are also present in the spectra of many T Tauri stars.[16] The

---

[16]Recall the discussion of allowed and forbidden transitions in the Sun's corona (Section 11.2).

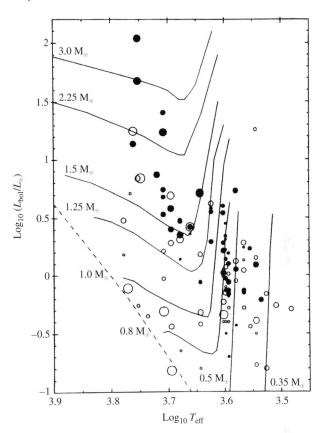

**FIGURE 12.16**  The positions of T Tauri stars on the H–R diagram. The sizes of the circles indicate the rate of rotation. Stars with strong emission lines are indicated by filled circles, and weak emission line stars are represented by open circles. Theoretical pre-main-sequence evolutionary tracks are also included. (Figure adapted from Bertout, *Annu. Rev. Astron. Astrophys.*, *27*, 351, 1989. Reproduced with permission from the *Annual Review of Astronomy and Astrophysics*, Volume 27, ©1989 by Annual Reviews Inc.)

existence of forbidden lines in a spectrum is an indication of extremely low gas densities. (Note that, to distinguish them from "allowed" lines, forbidden lines are usually indicated by square brackets, e.g., [O I].)

Not only can information be gleaned from spectra by determining which lines are present and with what strengths, but information is also contained in the *shapes* of those lines as a function of wavelength.[17] An important example is found in the shapes of some of the lines in T Tauri stars. The H$\alpha$ line often exhibits the characteristic shape shown in Fig. 12.17(a). Superimposed on a rather broad emission peak is an absorption trough at the short-wavelength edge of the line. This unique line shape is known as a **P Cygni profile**, after the first star observed to have emission lines with blueshifted absorption components.

---

[17]Line profiles were first discussed in Section 9.5; recall also Fig. 12.10.

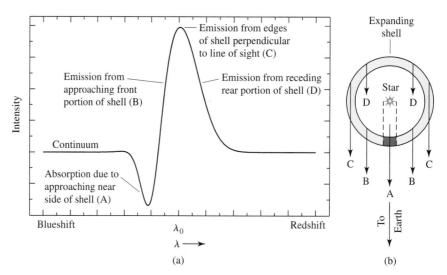

**FIGURE 12.17**   (a) A spectral line exhibiting a P Cygni profile is characterized by a broad emission peak with a superimposed blueshifted absorption trough. (b) A P Cygni profile is produced by an expanding mass shell. The emission peak is due to the outward movement of material perpendicular to the line of sight, whereas the blueshifted absorption feature is caused by the approaching matter in the shaded region, intercepting photons coming from the central star.

The interpretation given for the existence of P Cygni profiles in a star's spectrum is that the star is experiencing significant mass loss. Recall from Kirchhoff's laws (Section 5.1) that emission lines are produced by a hot, diffuse gas when there is little intervening material between the source and the observer. In this case the emission source is that portion of the expanding shell of the T Tauri star that is moving nearly perpendicular to the line of sight, as illustrated by the geometry shown in Fig. 12.17(b). Absorption lines are the result of light passing through a cooler, diffuse gas; the shaded portion of the expanding shell absorbs the photons emitted by the hotter star behind it. Since the shaded part of the shell (A) is moving toward the observer, the absorption is blueshifted relative to the emission component (typically by 80 km s$^{-1}$ for T Tauri stars). The mass loss rates of T Tauri stars average about $\dot{M} = 10^{-8}$ M$_\odot$ yr$^{-1}$.[18]

In some extreme cases, line profiles of T Tauri stars have gone from P Cygni profiles to *inverse* P Cygni profiles (redshifted absorption) on timescales of days, indicating mass accretion rather than mass loss. Mass accretion rates appear to be on the same order as mass loss rates. Apparently the environment around a T Tauri star is very unstable.

### FU Orionis Stars

In some instances, it appears that T Tauri stars have gone through very significant increases in mass accretion rates, reaching values on the order of $\dot{M} = 10^{-4}$ M$_\odot$ yr$^{-1}$. At the same time the luminosities of the stars increase by four magnitudes or more, with the increases lasting for decades. The first star observed to undergo this abrupt increase in accretion

---

[18]This value is much higher than the Sun's current rate of mass loss ($10^{-14}$ M$_\odot$ yr$^{-1}$; see Example 11.2.1).

was FU Orionis, for which the **FU Orionis stars** are named. Apparently, instabilities in a circumstellar accretion disk around an FU Orionis star can result in on the order of 0.01 $M_\odot$ being dumped onto the central star over the century or so duration of the outburst. During that time the inner disk can outshine the central star by a factor of 100 to 1000, while strong, high-velocity winds in excess of 300 km s$^{-1}$ occur. It has been suggested that T Tauri stars may go through several FU Orionis events during their lifetimes.

### Herbig Ae/Be Stars

Closely related to the T-Tauri stars are **Herbig Ae/Be stars**, named for George Herbig. These pre-main-sequence stars are of spectral types A or B and have strong emission lines (hence the Ae/Be designations). Their masses range from 2 to 10 $M_\odot$ and they tend to be enveloped in some remaining dust and gas. He Ae/Be stars are not as thoroughly studied as T-Tauri stars, in large part because of their much shorter lifetimes (recall Table 12.1) and in part because fewer intermediate-mass than lower-mass stars form from a cloud (Fig. 12.12).

### Herbig–Haro Objects

Along with expanding shells, mass loss during pre-main-sequence evolution can also occur from **jets** of gas that are ejected in narrow beams in opposite directions.[19] **Herbig–Haro objects**, first discovered in the vicinity of the Orion nebula in the early 1950s by George Herbig and Guillermo Haro (1913–1988), are apparently associated with the jets produced by young protostars, such as T Tauri stars. As the jets expand supersonically into the interstellar medium, collisions excite the gas, resulting in bright objects with emission-line spectra. Figure 12.18(a) shows a Hubble Space Telescope image of the Herbig–Haro objects HH 1 and HH 2, which were created by material ejected at speeds of several hundred kilometers per second from a star shrouded in a cocoon of dust. The jets associated with another Herbig–Haro object, HH 47, are shown in Fig. 12.18(b).

Continuous emission is also observed in some protostellar objects and is due to the reflection of light from the parent star. A circumstellar accretion disk is apparent in Fig. 12.19 around HH 30. The surfaces of the disk are illuminated by the central star, which is again hidden from view behind the dust in the disk. Also apparent are jets originating from deep within the accretion disk, possibly from the central star itself. These accretion disks seem to be responsible for many of the characteristics associated with the protostellar objects, including emission lines, mass loss, jets, and perhaps even some of the luminosity variations. Unfortunately, details concerning the physical processes involved are not fully understood. An early model of the production of Herbig–Haro objects like HH 1 and HH 2 is shown in Fig. 12.20.

### Young Stars with Circumstellar Disks

Observations have revealed that other young stars also possess circumstellar disks of material orbiting them. Two well-known examples are Vega and $\beta$ Pictoris. An infrared image of $\beta$ Pic and its disk is shown in Fig. 12.21. $\beta$ Pic has also been observed in the ultraviolet lines of Fe II by the Hubble Space Telescope. It appears that clumps of material are falling from

---

[19]As we shall see in later chapters, astrophysical jets occur in a variety of phenomena over enormous ranges of energy and physical size.

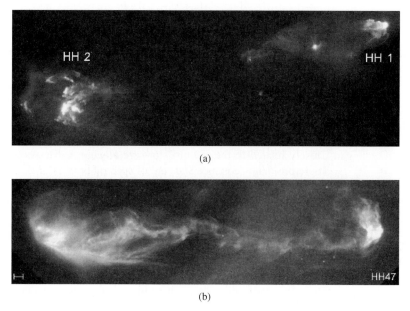

(a)

(b)

**FIGURE 12.18**   (a) The Herbig–Haro objects HH 1 and HH 2 are located just south of the Orion nebula and are moving away from a young protostar hidden inside a dust cloud near the center of the image. [Courtesy of J. Hester (Arizona State University), the WF/PC 2 Investigation Definition Team, and NASA.] (b) A jet associated with HH 47. The scale at the lower left is 1000 AU. (Courtesy of J. Morse/STScI, and NASA.)

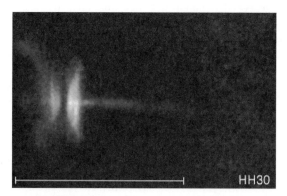

**FIGURE 12.19**   The circumstellar disk and jets of the protostellar object, HH 30. The central star is obscured by dust in the plane of the disk. The scale at the lower left is 1000 AU. [Courtesy of C. Burrows (STScI and ESA), the WF/PC 2 Investigation Definition Team, and NASA.]

the disk into the star at the rate of two or three per week. Larger objects may be forming in the disk as well, possibly protoplanets. It has been suggested that these disks may in fact be **debris disks** rather than accretion disks, meaning that the observed material is due to collisions between objects already formed in the disks. An artist's conception of the $\beta$ Pic system is shown in Fig. 12.22.

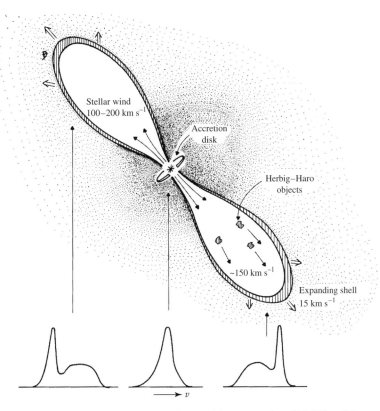

**FIGURE 12.20**    An early model of a T Tauri star with an accretion disk. The disk powers and collimates jets that expand into the interstellar medium, producing Herbig–Haro objects. (Figure adapted from Snell, Loren, and Plambeck, *Ap. J. Lett.*, *239*, L17, 1980.)

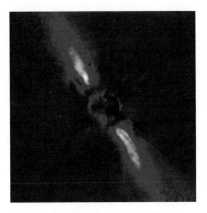

**FIGURE 12.21**    An infrared image of $\beta$ Pictoris, showing its circumstellar debris disk. (European Southern Observatory)

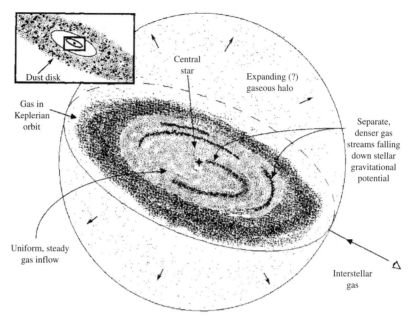

**FIGURE 12.22**   An artist's conception of the $\beta$ Pictoris system. Clumps of material appear to be falling into the star at the rate of two or three clumps per week. Some matter may also be leaving the system as an expanding halo. (Figure adapted from Boggess et al., *Ap. J. Lett.*, *377*, L49, 1991.)

## Proplyds

Shortly after the December 1993 refurbishment mission of the Hubble Space Telescope, HST made observations of the Orion Nebula. The images in Fig. 12.23 were obtained using the emission lines of H$\alpha$, [N II], and [O III]. Analysis of the data has revealed that 56 of the 110 stars brighter than $V = 21$ mag are surrounded by disks of circumstellar dust and gas. The circumstellar disks, termed **proplyds**, appear to be protoplanetary disks associated with young stars that are less than 1 million years old. Based on observations of the ionized material in the proplyds, the disks seem to have masses much greater than $2 \times 10^{25}$ kg (for reference, the mass of Earth is $5.974 \times 10^{24}$ kg).

## Circumstellar Disk Formation

Apparently, disk formation is fairly common during the collapse of protostellar clouds. Undoubtedly this is due to the spin-up of the cloud as required by the conservation of angular momentum. As the radius of the protostar decreases, so does its moment of inertia. This implies that in the absence of external torques, the protostar's angular velocity must increase. It is left as an exercise (Problem 12.18) to show that by including a centripetal acceleration term in Eq. (12.19) and requiring conservation of angular momentum, the collapse perpendicular to the axis of rotation can be halted before the collapse along the axis, resulting in disk formation.

A problem immediately arises when the effect of angular momentum is included in the collapse. Conservation of angular momentum arguments lead us to expect that all main-

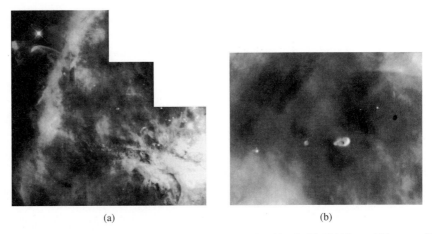

(a)                                                        (b)

**FIGURE 12.23**   Images of the Orion Nebula (M42) obtained by the Hubble Space Telescope. Note that (b) is an enlarged view of the central region of (a). Numerous proplyds are visible in the field of view of the camera. (Courtesy of C. Robert O'Dell/Vanderbilt University, NASA, and ESA.)

sequence stars ought to be rotating very rapidly, at rates close to breakup. However, observations show that this is not generally the case. Apparently the angular momentum is transferred away from the collapsing star. One suggestion (discussed in Section 11.2) is that magnetic fields, anchored to convection zones within the stars and coupled to ionized stellar winds, slow the rotation by applying torques. Evidence in support of this idea exists in the form of apparent solar-like coronal activity in the outer atmospheres of many T Tauri stars.

Along with the problems associated with rotation and magnetic fields, mass loss may also play an important role in the evolution of pre-main-sequence stars. Although these problems are being investigated, much work remains to be done before we can hope to understand all of the details of protostellar collapse and pre-main-sequence evolution.

## SUGGESTED READING

### General

Knapp, Gillian, "The Stuff Between the Stars," *Sky and Telescope*, May 1995.

Nadis, Steve, "Searching for the Molecules of Life in Space," *Sky and Telescope*, January 2002.

Renyolds, Ronald J., "The Gas Between the Stars," *Scientific American*, January 2002.

### Technical

Aller, Lawrence H., *Atoms, Stars, and Nebulae*, Third Edition, Cambridge University Press, Cambridge, 1991.

Dickey, John M., and Lockman, Felix J., "H I in the Galaxy," *Annual Review of Astronomy and Astrophysics*, *28*, 215, 1990.

Draine, B. T., "Interstellar Dust Grains," *Annual Review of Astronomy and Astrophysics*, *41*, 241, 2003.

Dopita, Michael A., and Sutherland, Ralph S., *Astrophysics of the Diffuse Universe*, Springer, Berlin, 2003.

Dyson, J. E., and Williams, D. A., *Physics of the Interstellar Medium*, Second Edition, Institute of Physics Publishing, Bristol, 1997.

Evans, Neal J. II, "Physical Conditions in Regions of Star Formation," *Annual Review of Astronomy and Astrophysics*, *37*, 311, 1999.

Iben, Icko Jr., "Stellar Evolution. I. The Approach to the Main Sequence," *The Astrophysical Journal*, *141*, 993, 1965.

Krügel, Endrik, *The Physics of Interstellar Dust*, Institute of Physics Publishing, Bristol, 2003.

Larson, Richard B., "Numerical Calculations of the Dynamics of a Collapsing Protostar," *Monthly Notices of the Royal Astronomical Society*, *145*, 271, 1969.

Larson, Richard B., "The Physics of Star Formation," *Reports of Progress in Physics*, *66*, 1651, 2003.

Lequeux, James, *The Interstellar Medium*, Springer, Berlin, 2003.

Mannings, Vincent, Boss, Alan P., and Russell, Sara S. (eds.), *Protostars and Planets IV*, The University of Arizona Press, Tucson, 2000.

O'Dell, C. R., and Wen, Zheng, "Postrefurbishment Mission Hubble Space Telescope Images of the Core of the Orion Nebula: Proplyds, Herbig-Haro Objects, and Measurements of a Circumstellar Disk," *The Astrophysical Journal*, *436*, 194, 1994.

Osterbrock, Donald E., *Astrophysics of Gaseous Nebulae and Active Galactic Nuclei*, Second Edition, University Science Books, Sausalito, CA, 2006.

Reipurth, Bo, and Bally, John, "Herbig-Haro Flows: Probes of Early Stellar Evolution," *Annual Review of Astronomy and Astrophysics*, *39*, 403, 2001.

Shu, Frank H., Adams, Fred C., and Lizano, Susana, "Star Formation in Molecular Clouds: Observation and Theory," *Annual Review of Astronomy and Astrophysics*, *25*, 23, 1987.

Stahler, Steven W., "Pre-Main-Sequence Stars," *The Encyclopedia of Astronomy and Astrophysics*, Institute of Physics Publishing, 2000.

Stahler, Steven W., and Palla, Francesco, *The Formation of Stars*, Wiley-VCH, Weinheim, 2004.

## PROBLEMS

**12.1** In a certain part of the North American Nebula, the amount of interstellar extinction in the visual wavelength band is 1.1 magnitudes. The thickness of the nebula is estimated to be 20 pc, and it is located 700 pc from Earth. Suppose that a B spectral class main-sequence star is observed in the direction of the nebula and that the absolute visual magnitude of the star is known to be $M_V = -1.1$ from spectroscopic data. Neglect any other sources of extinction between the observer and the nebula.

(a) Find the apparent visual magnitude of the star if it is lying just in front of the nebula.

(b) Find the apparent visual magnitude of the star if it is lying just behind the nebula.

(c) Without taking the existence of the nebula into consideration, based on its apparent magnitude, how far away does the star in part (b) appear to be? What would be the percentage error in determining the distance if interstellar extinction were neglected?

**12.2** Estimate the temperature of a dust grain that is located 100 AU from a newly formed F0 main-sequence star. *Hint:* Assume that the dust grain is in thermal equilibrium—meaning that the amount of energy absorbed by the grain in a given time interval must equal the amount of energy radiated away during the same interval of time. Assume also that the dust grain is spherically symmetric and emits and absorbs radiation as a perfect blackbody. You may want to refer to Appendix G for the effective temperature and radius of an F0 main-sequence star.

**12.3** The Boltzmann factor, $e^{-(E_2 - E_1)/kT}$, helps determine the relative populations of energy levels (see Section 8.1). Using the Boltzmann factor, estimate the temperature required for a hydrogen atom's electron and proton to go from being anti-aligned to being aligned. Are the temperatures in H I clouds sufficient to produce this low-energy excited state?

**12.4** An H I cloud produces a 21-cm line with an optical depth at its center of $\tau_H = 0.5$ (the line is optically thin). The temperature of the gas is 100 K, the line's full width at half-maximum is 10 km s$^{-1}$, and the average atomic number density of the cloud is estimated to be $10^7$ m$^{-3}$. From this information and Eq. (12.7), find the thickness of the cloud. Express your answer in pc.

**12.5** Using an approach analogous to the development of Eq. (10.29) for nuclear reaction rates, make a crude estimate of the number of random collisions per cubic meter per second between CO and H$_2$ molecules in a giant molecular cloud that has a temperature of 15 K and a number density of $n_{H_2} = 10^8$ m$^{-3}$. Assume (incorrectly) that the molecules are spherical in shape with radii of approximately 0.1 nm, the characteristic size of an atom.

**12.6** Explain why astronomers would use the isotopomers $^{13}$CO or C$^{18}$O, rather than the more common CO molecule, to probe deeply into a giant molecular cloud.

**12.7** The rotational kinetic energy of a molecule is given by

$$E_{\text{rot}} = \frac{1}{2} I \omega^2 = \frac{L^2}{2I},$$

where $L$ is the molecule's angular momentum and $I$ is its moment of inertia. The angular momentum is restricted by quantum mechanics to the discrete values

$$L = \sqrt{\ell(\ell+1)}\hbar$$

where $\ell = 0, 1, 2, \ldots$.

(a) For a diatomic molecule,

$$I = m_1 r_1^2 + m_2 r_2^2,$$

where $m_1$ and $m_2$ are the masses of the individual atoms and $r_1$ and $r_2$ are their separations from the center of mass of the molecule. Using the ideas developed in Section 2.3, show that $I$ may be written as

$$I = \mu r^2,$$

where $\mu$ is the reduced mass and $r$ is the separation between the atoms in the molecule.

(b) The separation between the carbon and oxygen atoms in CO is approximately 0.12 nm, and the atomic masses of $^{12}$C, $^{13}$C, and $^{16}$O are 12.000 u, 13.003 u, and 15.995 u, respectively. Calculate the moments of inertia for $^{12}$CO and $^{13}$CO.

**(c)** What is the wavelength of the photon that is emitted by $^{12}$CO during a transition between two rotational angular momentum states $\ell = 3$ and $\ell = 2$? To which part of the electromagnetic spectrum does this correspond?

**(d)** Repeat part (c) for $^{13}$CO. How do astronomers distinguish among different isotopes in the interstellar medium?

**12.8 (a)** Equations (12.10) and (12.11) illustrate a cooling mechanism for a molecular cloud accomplished through the excitation of oxygen atoms. Explain why the excitation of hydrogen rather than oxygen is not an effective cooling mechanism.

**(b)** Why are the temperatures of hot cores significantly greater than dense cores?

**12.9** In light of the cooling mechanisms discussed for molecular clouds, explain why dense cores are generally cooler than the surrounding giant molecular clouds, and why GMCs are cooler than diffuse molecular clouds.

**12.10** Calculate the Jeans length for the dense core of the giant molecular cloud in Example 12.2.1.

**12.11** Show that the Jeans mass (Eq. 12.14) can also be written in the form

$$ M_J = \frac{c_J v_T^4}{P_0^{1/2} G^{3/2}} \tag{12.33} $$

where the isothermal sound speed, $v_T$, is given by Eq. (12.18), $P_0$ is the pressure associated with the density $\rho_0$ and temperature $T$, and $c_J \simeq 5.46$ is a dimensionless constant.

**12.12** By invoking the requirements of hydrostatic equilibrium, explain why the assumption of a constant gas pressure $P_0$ in Eq. (12.33) cannot be correct for a static cloud without magnetic fields. What does that imply about the assumptions of constant mass density in an isothermal molecular cloud having a constant composition throughout?

**12.13 (a)** By using the ideal gas law, calculate $|dP/dr| \approx |\Delta P/\Delta r| \sim P_c/R_J$ at the beginning of the collapse of the dense core of a giant molecular cloud, where $P_c$ is an approximate value for the central pressure of the cloud. Assume that $P = 0$ at the edge of the molecular cloud and take its mass and radius to be the Jeans values found in Example 12.2.1 and in Problem 12.10. You should also assume the cloud temperature and density given in Example 12.2.1.

**(b)** Show that, given the accuracy of our crude estimates, $|dP/dr|$ found in part (a) is much smaller than $GM_r\rho/r^2$. What does this say about the core's dynamics?

**(c)** Show that as long as the collapse remains isothermal, the contribution of $dP/dr$ in Eq. (10.5) continues to decrease relative to $GM_r\rho/r^2$, supporting the assumption made in Eq. (12.19) that $dP/dr$ can be neglected once free-fall collapse begins.

**12.14** Assuming that the free-fall acceleration of the surface of a collapsing cloud remains constant during the entire collapse, derive an expression for the free-fall time. Show that your answer differs from Eq. (12.26) only by a term of order unity.

**12.15** Using Eq. (10.84), estimate the adiabatic sound speed of the dense core of the giant molecular cloud discussed in Examples 12.2.1 and 12.2.2. Use this speed to find the amount of time required for a sound wave to cross the cloud, $t_s = 2R_J/v_s$, and compare your answer to the estimate of the free-fall time found in Example 12.2.2. Explain your result.

**12.16** Using the information contained in the text, derive Eq. (12.28).

**12.17** Estimate the gravitational energy per unit volume in the dense core of the giant molecular cloud in Example 12.2.1, and compare that with the magnetic energy density that would be contained in the cloud if it had a magnetic field of uniform strength, $B = 1$ nT. [*Hint:* Refer to Eq. (11.9).] Could magnetic fields play a significant role in the collapse of a cloud?

**12.18 (a)** Beginning with Eq. (12.19), adding a centripetal acceleration term, and using conservation of angular momentum, show that the collapse of a cloud will stop in the plane perpendicular to its axis of rotation when the radius reaches

$$r_f = \frac{\omega_0^2 r_0^4}{2 G M_r}$$

where $M_r$ is the interior mass, and $\omega_0$ and $r_0$ are the original angular velocity and radius of the surface of the cloud, respectively. Assume that the initial radial velocity of the cloud is zero and that $r_f \ll r_0$. You may also assume (incorrectly) that the cloud rotates as a rigid body during the entire collapse. *Hint:* Recall from the discussion leading to Eq. (11.7) that $d^2 r/dt^2 = v_r \, dv_r/dr$. (Since no centripetal acceleration term exists for collapse along the rotation axis, disk formation is a consequence of the original angular momentum of the cloud.)

**(b)** Assume that the original cloud had a mass of 1 $M_\odot$ and an initial radius of 0.5 pc. If collapse is halted at approximately 100 AU, find the initial angular velocity of the cloud.

**(c)** What was the original rotational velocity (in m s$^{-1}$) of the edge of the cloud?

**(d)** Assuming that the moment of inertia is approximately that of a uniform solid sphere, $I_{\text{sphere}} = \frac{2}{5} M r^2$, when the collapse begins and that of a uniform disk, $I_{\text{disk}} = \frac{1}{2} M r^2$, when it stops, determine the rotational velocity at 100 AU.

**(e)** Calculate the time required, after the collapse has stopped, for a piece of mass to make one complete revolution around the central protostar. Compare your answer with the orbital period at 100 AU expected from Kepler's third law. Why would you not expect the two periods to be identical?

**12.19** Assuming a mass loss rate of $10^{-7}$ $M_\odot$ yr$^{-1}$ and a stellar wind velocity of 80 km s$^{-1}$ from a T Tauri star, estimate the mass density of the wind at a distance of 100 AU from the star. (*Hint:* Refer to Example 11.2.1.) Compare your answer with the density of the giant molecular cloud in Example 12.2.1.

# Main Sequence and Post-Main-Sequence Stellar Evolution

**13.1** *Evolution on the Main Sequence*
**13.2** *Late Stages of Stellar Evolution*
**13.3** *Stellar Clusters*

## 13.1 ■ EVOLUTION ON THE MAIN SEQUENCE

In Section 10.6 we learned that the existence of the main sequence is due to the nuclear reactions that convert hydrogen into helium in the cores of stars. The evolutionary process of protostellar collapse to the zero-age main sequence was discussed in Chapter 12. In this chapter we will follow the lives of stars as they age, beginning on the main sequence. This evolutionary process is an inevitable consequence of the relentless force of gravity and the change in chemical composition due to nuclear reactions.

### Stellar Evolution Timescales

To maintain their luminosities, stars must tap sources of energy contained within, either nuclear or gravitational.[1] Pre-main-sequence evolution is characterized by two basic time-scales: the free-fall timescale (Eq. 12.26) and the thermal Kelvin–Helmholtz timescale (Eq. 10.24). Main-sequence and post-main-sequence evolution are also governed by a third timescale, the timescale of nuclear reactions (Eq. 10.25). As we saw in Example 10.3.2, the nuclear timescale is on the order of $10^{10}$ years for the Sun, much longer than the Kelvin–Helmholtz timescale of roughly $10^7$ years, estimated in Example 10.3.1. It is the difference in timescales for the various phases of evolution of individual stars that explains why approximately 80% to 90% of all stars in the solar neighborhood are observed to be main-sequence stars (see Section 8.2); we are more likely to find stars on the main sequence simply because that stage of evolution requires the most time; later stages of evolution proceed more rapidly. However, as a star switches from one nuclear source to the next, gravitational energy can play a major role and the Kelvin–Helmholtz timescale will again become important.

---

[1] We have already seen in Problem 10.3 that chemical energy cannot play a significant role in the energy budgets of stars.

## Width of the Main Sequence

Careful study of the main sequence of an observational H–R diagram such as Fig. 8.13 or the observational mass–luminosity relation (Fig. 7.7) reveals that these curves are not simply thin lines but have finite widths. The widths of the main sequence and the mass–luminosity relation are due to a number of factors, including observational errors, differing chemical compositions of the individual stars in the study, and varying stages of evolution on the main sequence.

## Low-Mass Main-Sequence Evolution

In this section, we will consider the evolution of stars on the main sequence. Although all stars on the main sequence are converting hydrogen into helium and, as a result, share similar evolutionary characteristics, differences do exist. For instance, as was mentioned in Section 10.6, zero-age main-sequence (ZAMS) stars with masses greater than about $1.2 \, M_\odot$ have convective cores due to the highly temperature-dependent CNO cycle. On the other hand, ZAMS stars with masses less than $1.2 \, M_\odot$ are dominated by the less temperature-dependent pp chain. This implies that ZAMS stars in the range $0.3 \, M_\odot$ to $1.2 \, M_\odot$ possess radiative cores. However, the lowest-mass ZAMS stars again have convective cores because their high surface opacities drive surface convection zones deep into the interior, making the entire star convective.

First consider a typical low-mass main-sequence star such as the Sun. As we noted in the discussion of Fig. 11.1, the Sun's luminosity, radius, and temperature have all increased steadily since it reached the ZAMS 4.57 Gyr ago. This evolution occurs because, as the pp chain converts hydrogen into helium, the mean molecular weight $\mu$ of the core increases (Eq. 10.16). According to the ideal gas law (Eq. 10.11), unless the density and/or temperature of the core also increases, there will be insufficient gas pressure to support the overlying layers of the star. As a result, the core must be compressed. While the density of the core increases, gravitational potential energy is released, and, as required by the virial theorem (Section 2.4), half of the energy is radiated away and half of the energy goes into increasing the thermal energy and hence the temperature of the gas. One consequence of this temperature increase is that the region of the star that is hot enough to undergo nuclear reactions increases slightly during the main-sequence phase of evolution. In addition, since the pp chain nuclear reaction rate goes as $\rho X^2 T_6^4$ (see Eq. 10.47), the increased temperature and density more than offset the decrease in the mass fraction of hydrogen, and the luminosity of the star slowly increases, along with its radius and effective temperature.

Main-sequence and post-main-sequence evolutionary tracks of stars of various masses were first computed in a pioneering study by Icko Iben, Jr., and published in the mid-1960s. Modern calculations of theoretical evolutionary tracks that include the effects of convective overshooting as well as mass lost from stars during their lifetimes are shown in Fig. 13.1.[2] According to the calculations, the amount of time required to evolve from the zero-age main sequence to points indicated in Fig. 13.1 are as given in Table 13.1. The locus of points

---

[2]Convective overshooting takes into consideration the inertia of a convective bubble, which causes it to travel some distance into an otherwise radiative region of the star.

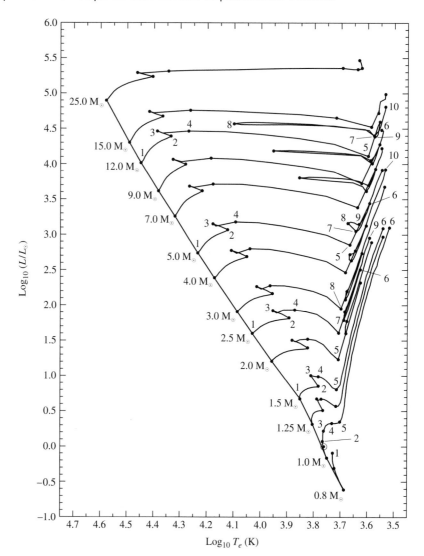

**FIGURE 13.1**   Main-sequence and post-main-sequence evolutionary tracks of stars with an initial composition of $X = 0.68$, $Y = 0.30$, and $Z = 0.02$. The location of the present-day Sun (see Fig. 13.2) is depicted by the solar symbol ($\odot$) between points 1 and 2 on the 1 $M_\odot$ track. The elapsed times to points indicated on the diagram are given in Table 13.1. To enhance readability, only the points on the evolutionary tracks for 0.8, 1.0, 1.5, 2.5, 5.0, and 12.0 $M_\odot$ are labeled. The model calculations include mass loss and convective overshooting. The diagonal line connecting the locus of points 1 is the zero-age main sequence. For complete, and annotated, evolutionary tracks of 1 $M_\odot$ and 5 $M_\odot$ stars, see Figs. 13.4 and 13.5, respectively. (Data from Schaller et al., *Astron. Astrophys. Suppl.*, *96*, 269, 1992.)

**TABLE 13.1** The elapsed times since reaching the zero-age main sequence to the indicated points in Fig. 13.1, measured in millions of years (Myr). (Data from Schaller et al., *Astron. Astrophys. Suppl.*, *96*, 269, 1992.)

| Initial Mass | 1 | 2 | 3 | 4 | 5 |
| (M$_\odot$) | 6 | 7 | 8 | 9 | 10 |
| --- | --- | --- | --- | --- | --- |
| 25 | 0 | 6.33044 | 6.40774 | 6.41337 | 6.43767 |
|  | 6.51783 | 7.04971 | 7.0591 |  |  |
| 15 | 0 | 11.4099 | 11.5842 | 11.5986 | 11.6118 |
|  | 11.6135 | 11.6991 | 12.7554 |  |  |
| 12 | 0 | 15.7149 | 16.0176 | 16.0337 | 16.0555 |
|  | 16.1150 | 16.4230 | 16.7120 | 17.5847 | 17.6749 |
| 9 | 0 | 25.9376 | 26.3886 | 26.4198 | 26.4580 |
|  | 26.5019 | 27.6446 | 28.1330 | 28.9618 | 29.2294 |
| 7 | 0 | 42.4607 | 43.1880 | 43.2291 | 43.3388 |
|  | 43.4304 | 45.3175 | 46.1810 | 47.9727 | 48.3916 |
| 5 | 0 | 92.9357 | 94.4591 | 94.5735 | 94.9218 |
|  | 95.2108 | 99.3835 | 100.888 | 107.208 | 108.454 |
| 4 | 0 | 162.043 | 164.734 | 164.916 | 165.701 |
|  | 166.362 | 172.38 | 185.435 | 192.198 | 194.284 |
| 3 | 0 | 346.240 | 352.503 | 352.792 | 355.018 |
|  | 357.310 | 366.880 | 420.502 | 440.536 |  |
| 2.5 | 0 | 574.337 | 584.916 | 586.165 | 589.786 |
|  | 595.476 | 607.356 | 710.235 | 757.056 |  |
| 2 | 0 | 1094.08 | 1115.94 | 1117.74 | 1129.12 |
|  | 1148.10 | 1160.96 | 1379.94 | 1411.25 |  |
| 1.5 | 0 | 2632.52 | 2690.39 | 2699.52 | 2756.73 |
|  | 2910.76 |  |  |  |  |
| 1.25 | 0 | 4703.20 | 4910.11 | 4933.83 | 5114.83 |
|  | 5588.92 |  |  |  |  |
| 1 | 0 | 7048.40 | 9844.57 | 11386.0 | 11635.8 |
|  | 12269.8 |  |  |  |  |
| 0.8 | 0 | 18828.9 | 25027.9 |  |  |

labeled 1 represents the theoretical ZAMS, with the present-day Sun located between points 1 and 2 on the 1 $M_\odot$ track.

We discussed solar models in some detail in Section 11.1, where Figs. 11.3–11.7 showed the internal structure of one such model as a function of radius. In particular, Fig. 11.3 illustrated the partial depletion of hydrogen in the core, together with the accompanying increase in the amount of helium. The internal structure of the present-day Sun is also shown in Fig. 13.2, this time as a function of interior mass. Along with radius, density, temperature, pressure, and luminosity, the figure illustrates the mass fractions of the species $_1^1$H, $_2^3$He, $_6^{12}$C, $_7^{14}$N, and $_8^{16}$O. As the star's evolution on the main sequence continues, eventually the hydrogen at its center will be completely depleted. Such a situation is illustrated in Fig. 13.3 for a 1 $M_\odot$ star approximately 9.8 Gyr after arriving on the ZAMS; this model roughly corresponds to point 3 in Fig. 13.1.

With the depletion of hydrogen in the core, the generation of energy via the pp chain must stop. However, by now the core temperature has increased to the point that nuclear fusion continues to generate energy in a thick hydrogen-burning shell around a small, predominantly helium core. This effect can be seen in the luminosity curve in Fig. 13.3. Note that the luminosity remains close to zero throughout the inner 3% of the star's mass. At the same time, the temperature is nearly constant over the same region. That the helium core must be isothermal when the luminosity gradient is zero can be seen from the radiative temperature gradient, given by Eq. (10.68). Since $L_r \simeq 0$ over a finite region, $dT/dr \simeq 0$

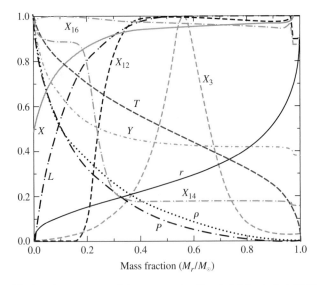

**FIGURE 13.2**    The interior structure of the present-day Sun (a 1 $M_\odot$ star), 4.57 Gyr after reaching the ZAMS. The model is located between points 1 and 2 in Fig. 13.1. The maximum ordinate values of the parameters are $r = 1.0$ $R_\odot$, $L = 1.0$ $L_\odot$, $T = 15.69 \times 10^6$ K, $\rho = 1.527 \times 10^5$ kg m$^{-3}$, $P = 2.342 \times 10^{16}$ N m$^{-2}$, $X = 0.73925$, $Y = 0.64046$, $X_3 = 3.19 \times 10^{-3}$, $X_{12} = 3.21 \times 10^{-3}$, $X_{14} = 5.45 \times 10^{-3}$, and $X_{16} = 9.08 \times 10^{-3}$. (Data from Bahcall, Pinsonneault, and Basu, *Ap. J.*, **555**, 990, 2001.)

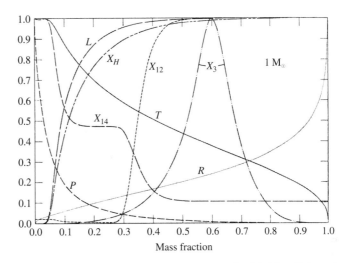

**FIGURE 13.3**  The interior structure of a 1 $M_\odot$ star near point 3 in Fig. 13.1, as described by the pioneering calculations of Icko Iben. Although specific values of quantities in modern models differ somewhat from those given here, state-of-the-art models do not significantly differ qualitatively from these calculations. The maximum ordinate values of the parameters for the Iben model are $R = 1.2681\ R_\odot$, $P = 1.3146 \times 10^{17}$ N m$^{-2}$, $T = 19.097 \times 10^6$ K, $L = 2.1283\ L_\odot$, $X_H = 0.708$, $X_3 = 5.15 \times 10^{-3}$, $X_{12} = 3.61 \times 10^{-3}$, and $X_{14} = 1.15 \times 10^{-2}$. The radius of the star is 1.3526 $R_\odot$. (Figure adapted from Iben, *Ap. J.*, *47*, 624, 1967.)

and $T$ is nearly constant. For an isothermal core to support the material above it in hydrostatic equilibrium, the required pressure gradient must be the result of a continuous increase in density as the center of the star is approached.

At this point, the luminosity being generated in the thick shell actually exceeds what was produced by the core during the phase of core hydrogen burning. As a result, the evolutionary track continues to rise beyond point 3 in Fig. 13.1, although not all of the energy generated reaches the surface; some of it goes into a slow expansion of the envelope. Consequently, the effective temperature begins to decrease slightly and the evolutionary track bends to the right. As the hydrogen-burning shell continues to consume its nuclear fuel, the ash from nuclear burning causes the isothermal helium core to grow in mass while the star moves farther to the red in the H–R diagram.

### The Schönberg–Chandrasekhar Limit

This phase of evolution ends when the mass of the isothermal core has become too great and the core is no longer capable of supporting the material above it. The maximum fraction of a star's mass that can exist in an isothermal core and still support the overlying layers was first estimated by M. Schönberg and Chandrasekhar in 1942; it is given by

$$\left( \frac{M_{ic}}{M} \right)_{SC} \simeq 0.37 \left( \frac{\mu_{env}}{\mu_{ic}} \right)^2,$$

(13.1)

where $\mu_{env}$ and $\mu_{ic}$ are the mean molecular weights of the overlying envelope and the isothermal core, respectively. The **Schönberg–Chandrasekhar limit** is another consequence of the virial theorem. Based on the physical tools we have developed so far, an approximate form of this result can be obtained. The analysis is presented beginning on the facing page.

The maximum fraction of the mass of a star that can be contained in an isothermal core and still maintain hydrostatic equilibrium is a function of the mean molecular weights of the core and the envelope. When the mass of the isothermal helium core exceeds this limit, the core collapses on a Kelvin–Helmholtz timescale, and the star evolves very rapidly relative to the nuclear timescale of main-sequence evolution. This occurs at the points labeled 4 in Fig. 13.1. For stars below about 1.2 $M_\odot$, this defines the end of the main-sequence phase. What happens next is the subject of Section 13.2.

---

**Example 13.1.1.**   If a star is formed with the initial composition $X = 0.68$, $Y = 0.30$, and $Z = 0.02$, and if complete ionization is assumed at the core–envelope boundary, we find from Eq. (10.16) that $\mu_{env} \simeq 0.63$. Assuming that all of the hydrogen has been converted into helium in the isothermal core, $\mu_{ic} \simeq 1.34$. Therefore, from Eq. (13.1), the Schönberg–Chandrasekhar limit is

$$\left( \frac{M_{ic}}{M} \right)_{SC} \simeq 0.08.$$

The isothermal core will collapse if its mass exceeds 8% of the star's total mass.

---

### The Degenerate Electron Gas

The mass of an isothermal core may exceed the Schönberg–Chandrasekhar limit if an additional source of pressure can be found to supplement the ideal gas pressure. This can occur if the electrons in the gas start to become **degenerate**. When the density of a gas becomes sufficiently high, the electrons in the gas are forced to occupy the lowest available energy levels. Since electrons are fermions and obey the Pauli exclusion principle (see Section 5.4), they cannot all occupy the same quantum state. Consequently, the electrons are stacked into progressively higher energy states, beginning with the ground state. In the case of *complete degeneracy*, the pressure of the gas is due entirely to the resultant nonthermal motions of the electrons, and therefore it becomes independent of the temperature of the gas.

If the electrons are nonrelativistic, the pressure of a completely degenerate electron gas is given by

$$P_e = K\rho^{5/3}, \tag{13.2}$$

where $K$ is a constant.[3] If the degeneracy is only partial, some temperature dependence remains.[4] The isothermal core of a 1 $M_\odot$ star between points 3 and 4 in Fig. 13.1 is partially degenerate; consequently, the core mass can reach approximately 13% of the entire mass

---

[3]Note that Eq. (13.2) is a polytropic equation of state with a polytropic index of $n = 1.5$; see page 334ff.
[4]The physics of degenerate gases will be discussed in more detail in Section 16.3.

of the star before it begins to collapse. Less massive stars exhibit even higher levels of degeneracy on the main sequence and may not exceed the Schönberg–Chandrasekhar limit at all before the next stage of nuclear burning commences.

## Main-Sequence Evolution of Massive Stars

The evolution of more massive stars on the main sequence is similar to that of their lower-mass cousins with one important difference: the existence of a convective core. The convection zone continually mixes the material, keeping the core composition nearly homogeneous. This is because the timescale for convection, defined by the amount of time it takes a convective element to travel one mixing length (see Section 10.4), is much shorter than the nuclear timescale. For a 5 $M_\odot$ star, the central convection zone decreases somewhat in mass during core hydrogen burning, leaving behind a slight composition gradient. Moving up the main sequence, as the star evolves the convection zone in the core retreats more rapidly with increasing stellar mass, disappearing entirely before the hydrogen is exhausted for those stars with masses greater than about 10 $M_\odot$.

When the mass fraction of hydrogen reaches about $X = 0.05$ in the core of a 5 $M_\odot$ star (point 2 in Fig. 13.1), the entire star begins to contract. With the release of some gravitational potential energy, the luminosity increases slightly. Since the radius decreases, the effective temperature must also increase. For stars with masses greater than 1.2 $M_\odot$, this stage of overall contraction is defined to be the end of the main-sequence phase of evolution.

## A Derivation of the Schönberg–Chandrasekhar limit

To estimate the Schönberg–Chandrasekhar limit, begin by dividing the equation of hydrostatic equilibrium (Eq. 10.6) by the equation of mass conservation (Eq. 10.7). This gives

$$\frac{dP}{dM_r} = -\frac{GM_r}{4\pi r^4},$$ (13.3)

which is just the condition of hydrostatic equilibrium, written with the interior mass as the independent variable.[5] Rewriting, Eq. (13.3) may be expressed as

$$4\pi r^3 \frac{dP}{dM_r} = -\frac{GM_r}{r}.$$ (13.4)

The left-hand side is just

$$4\pi r^3 \frac{dP}{dM_r} = \frac{d(4\pi r^3 P)}{dM_r} - 12\pi r^2 P \frac{dr}{dM_r} = \frac{d(4\pi r^3 P)}{dM_r} - \frac{3P}{\rho},$$

where Eq. (10.7) was used to obtain the last expression. Substituting back into Eq. (13.4) and integrating over the mass ($M_{ic}$) of the isothermal core, we have

$$\int_0^{M_{ic}} \frac{d(4\pi r^3 P)}{dM_r} dM_r - \int_0^{M_{ic}} \frac{3P}{\rho} dM_r = -\int_0^{M_{ic}} \frac{GM_r}{r} dM_r.$$ (13.5)

[5]This is called the Lagrangian form of the condition for hydrostatic equilibrium.

To evaluate Eq. (13.5), we will consider each term separately. The first term on the left-hand side is just

$$\int_0^{M_{ic}} \frac{d(4\pi r^3 P)}{dM_r} dM_r = 4\pi R_{ic}^3 P_{ic},$$

where $R_{ic}$ and $P_{ic}$ are the radius and the gas pressure at the *surface* of the isothermal core, respectively (note that $r = 0$ at $M_r = 0$).

The second term on the left-hand side of Eq. (13.5) can also be evaluated quickly by realizing that, from the ideal gas law,

$$\frac{P}{\rho} = \frac{kT_{ic}}{\mu_{ic} m_H},$$

where $T_{ic}$ and $\mu_{ic}$ are the temperature and mean molecular weight throughout the isothermal core, respectively.[6] Thus

$$\int_0^{M_{ic}} \frac{3P}{\rho} dM_r = \frac{3M_{ic}kT_{ic}}{\mu_{ic} m_H} = 3N_{ic}kT_{ic} = 2K_{ic},$$

where

$$N_{ic} \equiv \frac{M_{ic}}{\mu_{ic} m_H}$$

is the number of gas particles in the core and

$$K_{ic} = \frac{3}{2} N_{ic} k T_{ic}$$

is the total thermal energy of the core, assuming an ideal monatomic gas.

The right-hand side of Eq. (13.5) is simply the gravitational potential energy of the core, or

$$-\int_0^{M_{ic}} \frac{GM_r}{r} dM_r = U_{ic}.$$

Substituting each term into Eq. (13.5), we find

$$4\pi R_{ic}^3 P_{ic} - 2K_{ic} = U_{ic}. \tag{13.6}$$

This expression should be compared with the form of the virial theorem developed in Section 2.4; see Eq. (2.46). If we had integrated from the center of the star to the surface, where $P \simeq 0$, we would have obtained our original version of the theorem. The difference lies in the nonzero pressure boundary condition. Thus Eq. (13.6) is a generalized form of the virial theorem for stellar interiors in hydrostatic equilibrium.

---

[6]The core is actually supported in part by electron degeneracy pressure, meaning that the ideal gas law is not strictly valid. For our purposes here, however, the assumption of an ideal gas gives reasonable results.

Next, from Eq. (10.22), the gravitational potential energy of the core may be approximated as

$$U_{ic} \sim -\frac{3}{5} \frac{G M_{ic}^2}{R_{ic}}.$$

Furthermore, the internal thermal energy of the core is just

$$K_{ic} = \frac{3 M_{ic} k T_{ic}}{2 \mu_{ic} m_H}.$$

Introducing these expressions into Eq. (13.6) and solving for the pressure at the surface of the isothermal core, we have

$$P_{ic} = \frac{3}{4\pi R_{ic}^3} \left( \frac{M_{ic} k T_{ic}}{\mu_{ic} m_H} - \frac{1}{5} \frac{G M_{ic}^2}{R_{ic}} \right). \tag{13.7}$$

Notice that there are two competing terms in Eq. (13.7); the first term is due to the thermal energy in the core and the second is due to gravitational effects. For specific values of $T_{ic}$ and $R_{ic}$, as the core mass increases, the thermal energy tends to increase the pressure at the surface of the core while the gravitational term tends to decrease it. For some value of $M_{ic}$, $P_{ic}$ is maximized, meaning that there exists an upper limit on how much pressure the isothermal core can exert in order to support the overlying envelope.

To determine when $P_{ic}$ is a maximum, we must differentiate Eq. (13.7) with respect to $M_{ic}$ and set the derivative equal to zero. It is left as an exercise to show that the radius of the isothermal core for which $P_{ic}$ is a maximum is given by

$$R_{ic} = \frac{2}{5} \frac{G M_{ic} \mu_{ic} m_H}{k T_{ic}} \tag{13.8}$$

and that the maximum value of the surface pressure that can be produced by an isothermal core is given by

$$P_{ic,\max} = \frac{375}{64\pi} \frac{1}{G^3 M_{ic}^2} \left( \frac{k T_{ic}}{\mu_{ic} m_H} \right)^4. \tag{13.9}$$

The important feature of Eq. (13.9) is that, as the core mass increases, the maximum pressure at the surface of the core *decreases*. At some point, it may no longer be possible for the core to support the overlying layers of the star's envelope. Clearly this critical condition must be related to the mass contained in the envelope and therefore to the total mass of the star.

To estimate the mass that can be supported by the isothermal core, we need to determine the pressure exerted on the core by the overlying envelope. In hydrostatic equilibrium, this pressure must not exceed the maximum possible pressure that may be supported by the isothermal core. To estimate the envelope pressure, we will start again with Eq. (13.3), this time integrating from the surface of the star to the surface of the isothermal core. Assuming for simplicity that the pressure at the surface of the star is zero,

$$P_{ic,\text{env}} = \int_0^{P_{ic,\text{env}}} dP$$

$$= -\int_M^{M_{ic}} \frac{GM_r}{4\pi r^4} \, dM_r$$

$$\simeq -\frac{G}{8\pi \langle r^4 \rangle} \left( M_{ic}^2 - M^2 \right),$$

where $M$ is the total mass of the star and $\langle r^4 \rangle$ is some average value of $r^4$ between the surface of the star of radius $R$ and the surface of the core. Assuming that $M_{ic}^2 \ll M^2$, and making the crude approximation that $\langle r^4 \rangle \sim R^4/2$, we have

$$P_{ic,\text{env}} \sim \frac{G}{4\pi} \frac{M^2}{R^4} \tag{13.10}$$

for the pressure at the core's surface due to the weight of the envelope.

The quantity $R^4$ can be written in terms of the mass of the star and the temperature of the isothermal core through the use of the ideal gas law,

$$T_{ic} = \frac{P_{ic,\text{env}} \mu_{\text{env}} m_H}{\rho_{ic,\text{env}} k}, \tag{13.11}$$

where $\mu_{\text{env}}$ is the mean molecular weight of the envelope and $\rho_{ic,\text{env}}$ is the gas density at the core–envelope interface. Making the rough estimate that

$$\rho_{ic,\text{env}} \sim \frac{M}{4\pi R^3/3}, \tag{13.12}$$

using Eq. (13.10), and solving for $R$, Eq. (13.11) gives

$$R \sim \frac{1}{3} \frac{GM}{T_{ic}} \frac{\mu_{\text{env}} m_H}{k}.$$

Substituting our solution for the radius of the envelope back into Eq. (13.10), we arrive at an expression for the pressure at the core–envelope interface due to the overlying envelope:

$$P_{ic,\text{env}} \sim \frac{81}{4\pi} \frac{1}{G^3 M^2} \left( \frac{kT_{ic}}{\mu_{\text{env}} m_H} \right)^4.$$

Note that $P_{ic,\text{env}}$ is independent of the mass of the isothermal core.

Finally, to estimate the Schönberg–Chandrasekhar limit, we set the maximum pressure of the isothermal core (Eq. 13.9) equal to the pressure needed to support the overlying envelope (Eq. 13.1). This immediately simplifies to give

$$\frac{M_{ic}}{M} \sim 0.54 \left( \frac{\mu_{\text{env}}}{\mu_{ic}} \right)^2.$$

Our result is only slightly larger than the one obtained originally by Schönberg and Chandrasekhar (Eq. 13.1).

## 13.2 ■ LATE STAGES OF STELLAR EVOLUTION

Following the completion of the main-sequence phase of stellar evolution, a complicated sequence of evolutionary stages occurs that may involve nuclear burning in the cores of stars together with nuclear burning in concentric mass shells. At various times, core burning and/or nuclear burning in a mass shell may cease, accompanied by a readjustment of the structure of the star. This readjustment may involve expansion or contraction of the core or envelope and the development of extended convection zones. As the final stages of evolution are approached, extensive mass loss from the surface also plays a critical role in determining the star's ultimate fate.

As examples of post-main-sequence stellar evolution, we will continue to explore changes in the structures over time of a *low-mass star* of 1 $M_\odot$ and an *intermediate-mass star* of 5 $M_\odot$. Detailed depictions of their evolutionary tracks in the H–R diagram are shown in Figs. 13.4 and 13.5, respectively.

### Evolution Off the Main Sequence

As mentioned in Section 13.1, the end of the main-sequence phase of evolution occurs when hydrogen burning ceases in the core of the star (in Fig. 13.1 this corresponds to point 3 for the 1 $M_\odot$ star and point 2 for the 5 $M_\odot$ star). In the case of the 1 $M_\odot$ star, the core begins to contract while a thick hydrogen-burning shell continues to consume available fuel. With the rising temperature in the shell due to core contraction, the shell actually produces more energy than the core did on the main sequence, causing the luminosity to increase, the envelope to expand slightly, and the effective temperature to decrease.

The situation is somewhat different for the 5 $M_\odot$ star, however. Rather than a thick hydrogen-burning shell immediately producing energy with the cessation of hydrogen burning in the core, the entire star participates in an overall contraction on a Kelvin–Helmholtz timescale. This contraction phase releases gravitational potential energy, causing the luminosity to increase slightly, the radius of the star to decrease, and the effective temperature to increase (corresponding to the evolution between points 2 and 3 in Fig. 13.1). Eventually the temperature outside the helium core increases sufficiently to cause a thick shell of hydrogen to burn (point 3 in Fig. 13.1). At this point the interior chemical composition of the 5 $M_\odot$ star resembles that of Fig. 13.6. Because the ignition of the shell is quite rapid, the overlying envelope is forced to expand slightly, absorbing some of the energy released by the shell. As a result, the luminosity decreases momentarily and the effective temperature drops, as can be seen in both Figs. 13.1 and 13.5. A sketch of the star's structure at this point is given in Fig. 13.7.

### The Subgiant Branch

For both low- and intermediate-mass stars, as the shell continues to consume the hydrogen that is available at the base of the star's envelope, the helium core steadily increases in mass

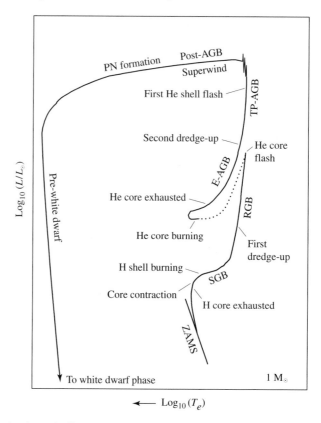

**FIGURE 13.4** A schematic diagram of the evolution of a low-mass star of 1 M$_\odot$ from the zero-age main sequence to the formation of a white dwarf star (see Section 16.1). The dotted phase of evolution represents rapid evolution following the helium core flash. The various phases of evolution are labeled as follows: Zero-Age-Main-Sequence (ZAMS), Sub-Giant Branch (SGB), Red Giant Branch (RGB), Early Asymptotic Giant Branch (E-AGB), Thermal Pulse Asymptotic Giant Branch (TP-AGB), Post-Asymptotic Giant Branch (Post-AGB), Planetary Nebula formation (PN formation), and Pre-white dwarf phase leading to white dwarf phase.

and becomes nearly isothermal. At points 4 in Fig. 13.1, the Schönberg–Chandrasekhar limit is reached and the core begins to contract rapidly, causing the evolution to proceed on the much faster Kelvin–Helmholtz timescale. The gravitational energy released by the rapidly contracting core again causes the envelope of the star to expand and the effective temperature cools, resulting in redward evolution on the H–R diagram. This phase of evolution is known as the **subgiant branch** (SGB).

As the core contracts, a nonzero temperature gradient is soon re-established because of the release of gravitational potential energy. At the same time, the temperature and density of the hydrogen-burning shell increase, and, although the shell begins to narrow significantly, the rate at which energy is generated by the shell increases rapidly. Once again the stellar envelope expands, absorbing some of the energy produced by the shell

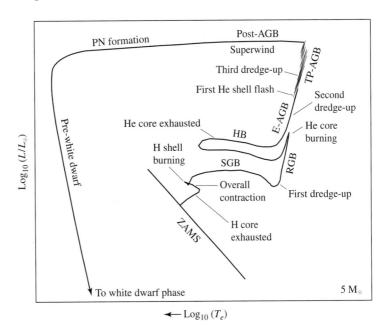

**FIGURE 13.5**   A schematic diagram of the evolution of an intermediate-mass star of 5 M$_\odot$ from the zero-age main sequence to the formation of a white dwarf star (see Section 16.1). The diagram is labeled according to Fig. 13.4 with the addition of the Horizontal Branch (HB).

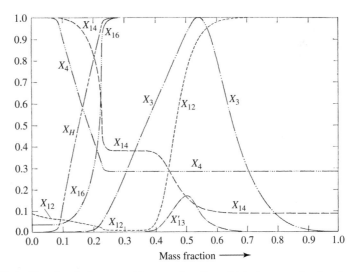

**FIGURE 13.6**   The chemical composition as a function of interior mass fraction for a 5 M$_\odot$ star during the phase of overall contraction, following the main-sequence phase of core hydrogen burning. The maximum mass fractions of the indicated species are $X_H = 0.708$, $X_3 = 1.296 \times 10^{-4}$ ($^3_2$He), $X_4 = 0.9762$ ($^4_2$He), $X_{12} = 3.61 \times 10^{-3}$ ($^{12}_6$C), $X'_{13} = 3.61 \times 10^{-3}$ ($^{13}_6$C), $X_{14} = 0.0145$ ($^{14}_7$N), and $X_{16} = 0.01080$ ($^{16}_8$O). (Figure adapted from Iben, *Ap. J.*, *143*, 483, 1966.)

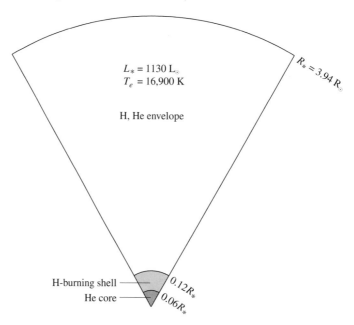

**FIGURE 13.7**   A 5 $M_\odot$ star with a helium core and a hydrogen-burning shell shortly after shell ignition (point 3 in Fig. 13.1). (Data from Iben, *Ap. J.*, *143*, 483, 1966.)

before the energy reaches the surface. For the 5 $M_\odot$ star, in a situation analogous to thick hydrogen shell burning immediately following overall contraction, the expanding envelope actually absorbs enough energy for a time to cause the luminosity to decrease slightly before recovering (point 5 in Fig. 13.1).

### The Red Giant Branch

With the expansion of the stellar envelope and the decrease in effective temperature, the photospheric opacity increases due to the additional contribution of the $H^-$ ion. The result is that a convection zone develops near the surface for both low- and intermediate-mass stars. As the evolution continues toward points 5 in Fig. 13.1, the base of the convection zone extends deep into the interior of the star. With the nearly adiabatic temperature gradient associated with convection throughout much of the stellar interior, and the efficiency with which the energy is transported to the surface, the star begins to rise rapidly upward along the **red giant branch** (RGB) of the H-R diagram. This path is essentially the same one followed by pre-main-sequence stars descending the Hayashi track prior to the onset of core hydrogen burning.

As the star climbs the RGB, its convection zone deepens until the base reaches down into regions where the chemical composition has been modified by nuclear processes. In particular, because of its rather large nuclear reaction cross section, lithium burns via collisions with protons at relatively cool temperatures (greater than about $2.7 \times 10^6$ K). This means that because of the evolution of the star to this point, lithium has become nearly depleted over most of the interior of the star (the inner 98% of the mass for the 5 $M_\odot$ star).[7] At the same time, nuclear processing has increased the mass fraction of $^3_2$He over the middle third of the star (see, for example, Figs. 13.3 and 13.6) as well as altered the abundance ratios of the various species in the CNO cycle. When the surface convection zone encounters this chemically modified region, the processed material becomes mixed with the material above it. The effect is observable changes in the composition of the photosphere; the amount of lithium at the surface will decrease and the amount of $^3_2$He will increase. At the same time, convection transports $^{12}_6$C inward and $^{14}_7$N outward, decreasing the observable ratio of $X_{12}/X_{14}$. Other abundance ratios such as $X'_{13}/X_{12}$ will also be modified. This transport of materials from the deep interior to the surface is referred to as the **first dredge-up** phase. Nature has provided us with an opportunity to directly observe the products of nuclear reactions deep within stellar interiors. These observable changes in surface composition provide an important test of the predictions of stellar evolution theory.

### The Red Giant Tip

At the tip of the RGB (points 6 in Fig. 13.1) the central temperature and density ($1.3 \times 10^8$ K and $7.7 \times 10^6$ kg m$^{-3}$ for the 5 $M_\odot$ star) have finally become high enough that quantum-mechanical tunneling becomes effective through the Coulomb barrier acting between $^4_2$He nuclei, allowing the triple alpha process to begin. Some of the resulting $^{12}_6$C is further processed into $^{16}_8$O as well.

With the onset of a new and strongly temperature-dependent source of energy (see Eqs. 10.60 and 10.61), the core expands. Although the hydrogen-burning shell remains the dominant source of the star's luminosity, the expansion of the core pushes the hydrogen-burning shell outward, cooling it and causing the rate of energy output of the shell to decrease somewhat. The result is an abrupt decrease in the luminosity of the star. At the same time, the envelope contracts and the effective temperature begins to increase again.

### The Helium Core Flash

An interesting difference arises at this point between the evolution of stars with masses greater than about 1.8 $M_\odot$ and those that have masses less than 1.8 $M_\odot$. For stars of lower mass, as the helium core continues to collapse during evolution up to the tip of the red giant branch, the core becomes strongly electron-degenerate. Furthermore, significant neutrino losses from the core of the star prior to reaching the tip of the RGB result in a negative temperature gradient near the center (i.e., a temperature inversion develops); the core is actually refrigerated somewhat because of the energy that is carried away by the easily escaping neutrinos! When the temperature and density become high enough to initiate the triple alpha process (approximately $10^8$ K and $10^7$ kg m$^{-3}$, respectively), the ensuing energy release is almost explosive. The ignition of helium burning occurs initially in a shell around the center of the star, but the entire core quickly becomes involved and the temperature inversion is lifted. The luminosity generated by the helium-burning core reaches $10^{11}$ $L_\odot$, comparable to that of an entire galaxy! However, this tremendous energy release lasts for

---

[7]Recall from Section 11.1 that the surface abundance of lithium is also lower than expected in the present-day Sun.

only a few seconds, and most of the energy never even reaches the surface. Instead, it is absorbed by the overlying layers of the envelope, possibly causing some mass to be lost from the surface of the star. This short-lived phase of evolution of low-mass stars is referred to as the **helium core flash**. The origin of the explosive energy release is in the very weak temperature dependence of electron degeneracy pressure and the strong temperature dependence of the triple alpha process. The energy generated must first go into "lifting" the degeneracy. Only after this occurs can the energy go into thermal (kinetic) energy required to expand the core, which decreases the density, lowers the temperature, and slows the reaction rate.

It is because of the very rapid pace of the helium core flash that stellar evolution calculations of low-mass stars are often terminated at that point. Given the dramatic changes occuring deep in the interior of the star, it is very difficult to follow the evolution adequately; very small time steps are required to model the evolution, meaning that a great deal of computer time is needed to follow a star through the helium core flash (in fact, the star evolves *much faster* than the computer can model it). This is why the evolutionary tracks of stars with masses of 1, 1.25, and 1.5 $M_\odot$ are not followed past points 6 in Fig. 13.1. This is also why the annotated evolutionary track in Fig. 13.4 immediately following the red giant tip is indicated by a dotted line; the evolution is extremely rapid and the computations are resumed when quiescent helium core burning and hydrogen shell burning are established in the star.

### The Horizontal Branch

For both low- and intermediate-mass stars, as the envelope of the model contracts following the red giant tip, the increasing compression of the hydrogen-burning shell eventually causes the energy output of the shell, and the overall energy output of the stars, to begin to rise again. With the associated increase in effective temperature, the deep convection zone in the envelope rises toward the surface, while at the same time, a convective core develops. The appearance of a convective core is due to the high temperature sensitivity of the triple alpha process (just as the convective core of an upper-main-sequence star arises because of the temperature dependence of the CNO cycle). This generally horizontal evolution is the blueward portion of the **horizontal branch** (HB) loop. The blueward portion of the HB is essentially the helium-burning analog of the hydrogen-burning main sequence, but with a much shorter timescale.

When the evolution of the star reaches its most blueward point (point 8 in Fig. 13.1 for the 5 $M_\odot$ star), the mean molecular weight of the core has increased to the point that the core begins to contract, accompanied by the expansion and cooling of the star's envelope. Shortly after beginning the redward portion of the HB loop, the core helium is exhausted, having been converted to carbon and oxygen. Again the redward evolution proceeds rapidly as the inert CO core contracts, much like the rapid evolution across the SGB following the extinction of core hydrogen burning.

During their passage along the horizontal branch, many stars develop instabilities in their outer envelopes, leading to periodic pulsations that are readily observable as variations in luminosity, temperature, radius, and surface radial velocity. Since these oscillations depend sensitively on the internal structure of the star, stellar pulsations provide yet another test of stellar structure theory.[8]

With the increase in core temperature associated with its contraction, a thick helium-burning shell develops outside the CO core. As the core continues to contract, the helium-burning shell narrows and strengthens, forcing the material above the shell to expand and cool. This results in a temporary turn-off of the hydrogen-burning shell.

Along with the contraction of the helium-exhausted core, neutrino production increases to the point that the core cools a bit. As a consequence of the increasing central density and decreasing temperature, electron degeneracy pressure becomes an important component of the total pressure in the carbon–oxygen core.

### The Early Asymptotic Giant Branch

The next phase of the evolution illustrated in Figs. 13.4 and 13.5 is very similar to the evolution following the exhaustion of the hydrogen-burning cores. When the redward evolution reaches the Hayashi track, the evolutionary track bends upward along a path referred to as the **asymptotic giant branch** (AGB). (The AGB is so named because the evolutionary track approaches the line of the RGB asymptotically from the left. The AGB may be thought of as the helium-burning-shell analog to the hydrogen-burning-shell RGB.) At this point in its evolution the core temperature of the 5 $M_\odot$ star is approximately $2 \times 10^8$ K, and its density is on the order of $10^9$ kg m$^{-3}$. A diagram of an *early* AGB (E-AGB) star with two shell sources is shown in Fig. 13.8. Although two shell sources are depicted, it is the helium-burning shell that dominates the energy output during the E-AGB; the hydrogen-burning shell is nearly inactive at this point. Note that the diagram is not to scale; in order to visualize the structure from the hydrogen-burning shell inward, that region was enlarged by a factor of 100 relative to the surface of the star.

The expanding envelope initially absorbs much of the energy produced by the helium-burning shell. As the effective temperature continues to decrease, the convective envelope deepens again, this time extending downward to the chemical discontinuity between the hydrogen-rich outer layer and the helium-rich region above the helium-burning shell. The mixing that results during this **second dredge-up** phase increases the helium and nitrogen content of the envelope. (The increase in nitrogen is due to the previous conversion of carbon and oxygen into nitrogen in the intershell region.)

### The Thermal-Pulse Asymptotic Giant Branch

Near the upper portion of the AGB (labeled TP-AGB in Figs. 13.4 and 13.5 for *thermal-pulse* AGB), the dormant hydrogen-burning shell eventually reignites and again dominates the energy output of the star. However, during this phase of evolution, the narrowing helium-burning shell begins to turn on and off quasi-periodically. These intermittent **helium shell flashes** occur because the hydrogen-burning shell is dumping helium ash onto the helium layer below. As the mass of the helium layer increases, its base becomes slightly degenerate. Then, when the temperature at the base of the helium shell increases sufficiently, a helium

---

[8]Pulsating variable stars will be discussed in greater detail in Chapter 14.

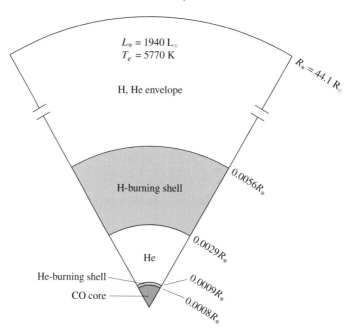

**FIGURE 13.8**   A 5 $M_\odot$ star on the early asymptotic giant branch with a carbon–oxygen core and hydrogen- and helium-burning shells. Note that relative to the surface radius, the scale of the shells and core has been increased by a factor of 100 for clarity. (Data from Iben, *Ap. J.*, *143*, 483, 1966.)

shell flash occurs, analogous to the earlier helium core flashes of low-mass stars (although much less energetic). This drives the hydrogen-burning shell outward, causing it to cool and turn off for a time. Eventually the burning in the helium shell diminishes, the hydrogen-burning shell recovers, and the process repeats. The period between pulses is a function of the mass of the star, ranging from thousands of years for stars near 5 $M_\odot$ to hundreds of thousands of years for low-mass stars (0.6 $M_\odot$), with the pulse amplitude growing with each successive event; see Fig. 13.9. This phase of periodic activity in the deep interior of the star is evident in abrupt changes in luminosity at the surface (see the TP-AGB phases in Figs. 13.4 and 13.5).

Details of thermal pulses for a 7 $M_\odot$ star are shown in Fig. 13.10. Following a helium shell flash, the luminosity arising from the hydrogen-burning shell drops appreciably while the energy output from the helium-burning shell increases. This is because the hydrogen-burning shell is pushed outward, causing it to cool. Since the hydrogen-burning shell is responsible for most of the energy output of the star, the star's luminosity abruptly decreases when a helium shell flash occurs. At the same time, the radius of the surface of the star also decreases and the star's effective temperature increases. After a period of time, the energy output of the helium shell diminishes when the degeneracy is lifted, the hydrogen-burning shell moves deeper into the star, and the hydrogen-burning shell once again dominates the star's total energy output. As a result, the surface radius, luminosity, and effective temperature relax back to near their pre-flash values. It is important to note, however, that

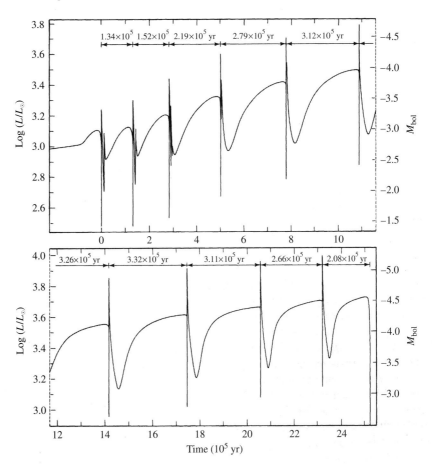

**FIGURE 13.9**  The surface luminosity as a function of time for a 0.6 M$_\odot$ stellar model that is undergoing helium shell flashes on the TP-AGB. (Figure adapted from Iben, *Ap. J.*, *260*, 821, 1982.)

the overall evolutionary track of the star is toward greater luminosity and lower effective temperature throughout the TP-AGB.

A class of pulsating variable stars known as **long-period variables** (LPVs) are AGB stars. (LPVs have pulsation periods of 100 to 700 days and include the subclass of **Mira variable stars**.) It has been suggested that the structural changes arising from shell flashes could cause observable changes in the periods of some of these stars, providing another possible test of stellar evolution theory. In fact, several Miras (e.g., W Dra, R Aql, and R Hya) have been observed to be undergoing relatively rapid period changes.

### Third Dredge-Up and Carbon Stars

Because of the sudden increase in energy flux from the helium-burning shell during a flash episode, a convection zone is established between the helium-burning shell and the hydrogen-burning shell. At the same time, the depth of the envelope convection zone

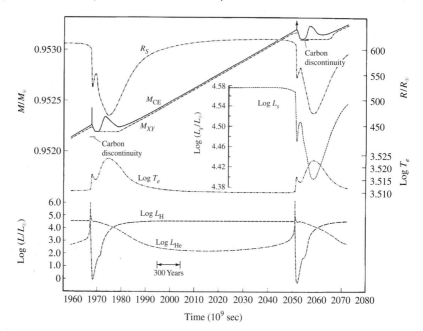

**FIGURE 13.10**   Time-dependent changes in the properties of a 7 $M_\odot$ AGB star produced by helium shell flashes on the TP-AGB. The quantities shown are the surface radius ($R_S$), the interior mass fractions of the base of the convective envelope ($M_{CE}$) and the hydrogen–helium discontinuity ($M_{XY}$), the star's luminosity and effective temperature ($L_s$ and $T_e$, respectively), and the luminosities of the hydrogen- and helium-burning shells ($L_H$ and $L_{He}$, respectively). (Figure adapted from Iben, *Ap. J.*, *196*, 525, 1975.)

increases with the pulse strength of the flashes. For stars that are massive enough ($M >$ 2 $M_\odot$), the convection zones will merge and eventually extend down into regions where carbon has been synthesized. In the region between the hydrogen- and helium-burning shells, the abundance of carbon exceeds that of oxygen by a factor of five to ten. This is in sharp contrast to the general excess of oxygen over carbon in the atmospheres of most stars. During this **third dredge-up** phase, the carbon-rich material is brought to the surface, decreasing the ratio of oxygen to carbon. If there are multiple third dredge-up events arising from repeated helium shell flashes, the oxygen-rich spectrum of a star will transform over time to a carbon-rich spectrum. This appears to explain the difference, observed spectroscopically, between oxygen-rich giants where the number density of oxygen atoms in the atmosphere exceeds the number density of carbon atoms ($N_O > N_C$) and carbon-rich giants ($N_C > N_O$) called **carbon stars**.

Carbon stars are designated with a special **C spectral type** (overlapping the traditional K and M types). These stars are distinguished by an abundance of carbon-rich molecules in their atmospheres, such as SiC, rather than SiO of typical M stars. This occurs because carbon monoxide (CO) is a very tightly bound molecule. If the atmosphere of the star contains more oxygen than carbon, the carbon is almost completely tied up in CO, leaving oxygen to form additional molecules. Conversely, if the atmosphere contains more carbon than oxygen, the oxygen is tied up in CO, allowing carbon to form in molecules.

Intermediate between the M and C spectral types are the **S spectral type** stars. These stars show ZrO lines in their atmospheres, having replaced the TiO lines of M stars. S stars have almost identical abundances of carbon and oxygen in their atmospheres. S and C spectral types represent additions to the spectral classification scheme presented in Table 8.1.

Of particular interest in the atmospheres of evolved TP-AGB stars is the presence of technetium (Tc), an element with no stable isotopes. In particular, $^{99}_{43}$Tc is the most abundant isotope of technetium found in the atmospheres of TP-AGB stars, yet it has a half-life of only 200,000 yr. The existence of technetium in S and C stars strongly suggests that the isotope must have been formed very recently in the star's history and dredged up to the surface from the deep interior.

### s-Process Nucleosynthesis

Technetium-99 is one of a host of isotopes formed by the slow capture of neutrons by existing nuclei. Numerous nuclear reactions, such as carbon burning (Eq. 10.66) and oxygen burning (Eq. 10.67) release neutrons. Since neutrons do not have an electric charge, they can easily collide with nuclei (there is no Coulomb barrier to tunnel through). If the flux of neutrons is not too great, radioactive nuclei produced by the absorption of stray neutrons have time to decay into other nuclei before they absorb another neutron. $^{99}_{43}$Tc is one product of this *slow s-process* nucleosynthesis.

### Mass Loss and AGB Evolution

AGB stars are known to lose mass at a rapid rate, sometimes as high as $\dot{M} \sim 10^{-4} \, M_\odot \, \text{yr}^{-1}$. The effective temperatures of these stars are also quite cool (around 3000 K). As a result, dust grains form in the expelled matter. Since silicate grains tend to form in an environment rich in oxygen, and graphite grains will form in a carbon-rich environment, the composition of the ISM may be related to the relative numbers of carbon- and oxygen-rich stars. Observations of ultraviolet extinction curves in the Milky Way and the Large and Small Magellanic Clouds[9] support the idea that mass loss from these stars does, in fact, help enrich the ISM.

As evolution up the AGB continues, what happens next is strongly dependent on the original mass of the star and the amount of mass loss experienced by that star during its lifetime. It appears that the final evolutionary behavior of stars can be separated into two basic groups: those with ZAMS masses above about 8 $M_\odot$ and those with masses below this value. The distinction between the two mass regimes is based on whether or not the core of the star will undergo significant further nuclear burning. In the remainder of this section, we will consider the final evolution of stars with initial masses less than 8 $M_\odot$, leaving the ultimate evolution of more massive stars to Chapter 15.

As stars with initial masses below 8 $M_\odot$ continue to evolve up the AGB, the helium-burning shell converts more and more of the helium into carbon and then into oxygen, increasing the mass of the carbon–oxygen core. At the same time, the core continues to contract slowly, causing its central density to increase. Depending on the star's mass, neutrino energy losses may decrease the central temperature somewhat during this phase. In any event, the densities in the core become large enough that electron degeneracy pressure

---

[9]The LMC and the SMC are small satellite galaxies of the Milky Way, visible in the southern hemisphere.

begins to dominate. This situation is very similar to the development of an electron-degenerate helium core in a low-mass star during its rise up the red giant branch.

For stars with ZAMS masses less than about 4 $M_\odot$, the carbon–oxygen core will never become large enough and hot enough to ignite nuclear burning. On the other hand, *if the important contribution of mass loss is ignored* for stars between 4 $M_\odot$ and 8 $M_\odot$, theory suggests that the C–O core would reach a sufficiently large mass that it could no longer remain in hydrostatic equilibrium, even with the assistance of pressure from the degenerate electron gas. The outcome of this situation is catastrophic core collapse. The maximum value of 1.4 $M_\odot$ for a completely degenerate core is known as the **Chandrasekhar limit**.[10]

However, as has already been mentioned, observations of AGB stars do show enormous mass loss rates. When these mass loss rates are included in evolution calculations on the AGB, the situation described in the last paragraph does not actually occur. Instead, for stars between 4 $M_\odot$ and 8 $M_\odot$, mass loss prevents catastrophic core collapse. Instead of collapse, these stars experience additional nucleosynthesis in their cores, leading to core compositions of oxygen, neon, and magnesium (ONeMg cores) and masses remaining below the Chandrasekhar limit of 1.4 $M_\odot$. Nuclear reactions capable of producing this composition were given in Eqs. (10.66) and (10.67). In addition, reactions such as

$$^{22}_{10}\text{Ne} + {}^{4}_{2}\text{He} \rightarrow {}^{25}_{12}\text{Mg} + n$$

$$^{22}_{10}\text{Ne} + {}^{4}_{2}\text{He} \rightarrow {}^{26}_{12}\text{Mg} + \gamma$$

also affect the composition of these cores.

Unfortunately, our understanding of the mechanism(s) that cause this mass loss is poor. Some astronomers have suggested that the mass loss may be linked to the helium shell flashes or perhaps to the periodic envelope pulsations of LPVs. Other proposed mechanisms stem from the high luminosities and low surface gravities of these stars, coupled with radiation pressure on the dust grains, dragging the gas with them. Whatever the cause, its influence on the evolution of AGB stars is significant.

As one might expect, the rate of mass loss accelerates with time because the luminosity and radius are increasing while the mass is decreasing during continued evolution up the AGB. The decreasing mass and increasing radius of the star imply that the surface gravity is also decreasing, and the surface material is becoming progressively less tightly bound. Consequently, mass loss becomes increasingly more important as AGB evolution continues.

In the latest stages of evolution on the AGB, a **superwind** develops with $\dot{M} \sim 10^{-4}$ $M_\odot$ yr$^{-1}$. Whether shell flashes, envelope pulsations, or some other mechanism is the reason, the observed high mass loss rates seem to be responsible for the existence of a class of objects known as **OH/IR sources**. These objects appear to be stars shrouded in optically thick dust clouds that radiate their energy primarily in the infrared part of the electromagnetic spectrum.

The OH part of the OH/IR designation is due to the detection of OH molecules, which can be seen via their **maser emission**.[11] A maser is the molecular analog of a laser; electrons

---

[10]The Chandrasekhar limit plays a critical role in the formation of the final products of stellar evolution, namely white dwarfs, neutron stars, and black holes. The physics of the Chandrasekhar limit will be discussed in some detail in Section 16.4.

[11]The term *maser* is an acronym for *m*icrowave *a*mplification by *s*timulated *e*mission of *r*adiation.

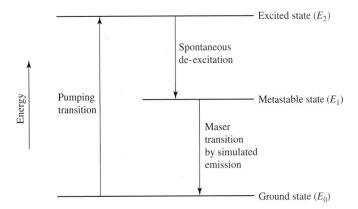

**FIGURE 13.11**    A schematic diagram of a hypothetical three-level maser. The intermediate energy level is a relatively long-lived metastable state. A transition from the metastable state to the lowest energy level can occur through stimulated emission by a photon of energy equal to the energy difference between the two states ($E_\gamma = h\nu = E_1 - E_0$).

are "pumped up" from a lower energy level into a higher, long-lived metastable energy state. The electron then makes a downward transition back to a lower state when it is stimulated by a photon with an energy equal to the difference in energies between the two states. The original photon and the emitted photon will travel in the same direction and will be in phase with each other; hence the amplification of radiation. A schematic energy level diagram of a hypothetical three-level maser is depicted in Fig. 13.11.

### Post-Asymptotic Giant Branch

As the cloud around the OH/IR source continues to expand, it eventually becomes optically thin, exposing the central star, which characteristically exhibits the spectrum of an F or G supergiant. At this point in the evolution of our 1 and 5 $M_\odot$ stars (Figs. 13.4 and 13.5, respectively), the evolutionary tracks have turned blueward, leaving the TP-AGB and moving nearly horizontally across the H–R diagram as **post-AGB** stars. During the ensuing final phase of mass loss, the remainder of the star's envelope is expelled, revealing the cinders produced by its long history of nuclear reactions. With only a very thin layer of material remaining above them, the hydrogen- and helium-burning shells are extinguished, and the luminosity of the star drops rapidly. The hot central object, now revealed, will cool to become a **white dwarf star**, which is essentially the old red giant's degenerate C–O core (or ONeMg core in the case of the more massive stars), surrounded by a thin layer of residual hydrogen and helium. This important class of stars, the end products of the evolution of stars with initial main-sequence masses less than 8 $M_\odot$, will be discussed in Chapter 16.

The last stages of evolution of a 0.6 $M_\odot$ model are depicted in Fig. 13.12. The position of the star on the H–R diagram at the start of each flash episode is indicated by a number next to the evolutionary track (eleven pulses in all), with the resulting excursions in luminosity and effective temperature indicated for pulses 7, 9, and 10. It is after the tenth pulse that the star leaves the AGB, ejecting its envelope during the nearly constant luminosity path

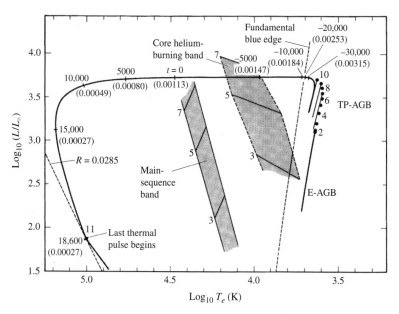

**FIGURE 13.12**    The AGB and post-AGB evolution of a 0.6 $M_\odot$ star undergoing mass loss. The initial composition of the model is $X = 0.749$, $Y = 0.25$, and $Z = 0.001$. The main-sequence and horizontal branches of 3, 5, and 7 $M_\odot$ stars are shown for reference. Details of the figure are discussed in the body of the text. (Figure adapted from Iben, *Ap. J.*, *260*, 821, 1982.)

across the H–R diagram. The amount of mass remaining in the hydrogen-rich envelope is indicated in parentheses along the evolutionary track (in $M_\odot$). Also indicated is the amount of time before (negative) or after (positive) the point when the star's effective temperature was 30,000 K (the time is measured in years). Following the eleventh helium shell flash, the star finally loses the last remnants of its envelope and becomes a white dwarf of radius 0.0285 $R_\odot$.[12]

### Planetary Nebulae

The expanding shell of gas around a white dwarf progenitor is called a **planetary nebula**. Examples of planetary nebulae are shown in Figs. 13.13–13.15. These beautiful, glowing clouds of gas were given this name in the nineteenth century because, when viewed through a small telescope, they look somewhat like giant gaseous planets.

A planetary nebula owes its appearance to the ultraviolet light emitted by the hot, condensed central star. The ultraviolet photons are absorbed by the gas in the nebula, causing the atoms to become excited or ionized. When the electrons cascade back down to lower

---

[12]The line labeled "Fundamental blue edge" corresponds to the high-temperature limit for fundamental mode pulsations of a class of variable stars known as **RR Lyraes**. This important class of objects will be discussed extensively in Chapter 14.

(a)

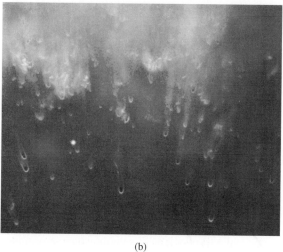

(b)

**FIGURE 13.13** (a) The Helix nebula (NGC 7293) is one of the closest planetary nebulae to Earth, 213 pc away in the constellation of Aquarius. Its angular diameter in the sky is about 16 arcmin, roughly one-half the angular size of the full moon. The pre-white dwarf star is visible at the center of the nebula. [Credit: NASA, ESA, C.R. O'Dell (Vanderbilt University), M. Meixner, and P. McCullough.] (b) A close-up of "cometary knots" in the Helix nebula. The central star is located beyond the bottom of the picture. [Credit: NASA, NOAO, ESA, the Hubble Helix Nebula Team, M. Meixner (STScI), and T. A. Rector (NRAO).]

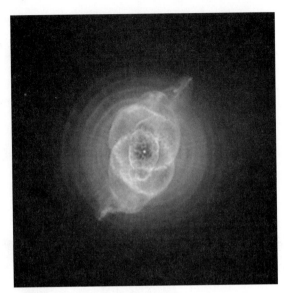

**FIGURE 13.14** NGC 6543 (the "Cat's Eye") is a planetary nebula in Draco, 900 pc away. The complex structure may be due to high-speed jets and the presence of a companion star, making NGC 6543 part of a binary star system. The jets are clearly visible in the upper right-hand and lower left-hand portions of the image. Note the central star in the image. [Credit: NASA, ESA, HEIC, and the Hubble Heritage Team (STScI/AURA). Acknowledgment: R. Corradi (Isaac Newton Group of Telescopes, Spain) and Z. Tsvetanov (NASA).]

energy levels, photons are emitted whose wavelengths are in the visible portion of the electromagnetic spectrum. As a result, the cloud appears to glow in visible light.[13]

The bluish-green coloration of many planetary nebulae is due to the 500.68-nm and 495.89-nm forbidden lines of [O III] (forbidden lines of [O II] and [Ne III] are also common), and the reddish coloration comes from ionized hydrogen and nitrogen. Characteristic temperatures of these objects are in the range of the ionization temperature of hydrogen, $10^4$ K.

With the advent of high-resolution images of planetary nebulae obtained by telescopes such as the Hubble Space Telescope, astronomers have come to realize that the morphologies of planetary nebulae are often much more complex than might have been expected of a spherically symmetric parent TP-AGB star. Some planetaries, like the Helix nebula in Fig. 13.13(a), look as though they have a ringlike structure. This is because gas is ejected preferentially along the equator of the star due to the presence of angular momentum, and our viewing angle is down the star's rotation axis. Suggestions for the surprising array of structures include varying viewing angles, multiple ejections of material from the stellar surface, the presence of one or more companion stars, and magnetic fields.

---

[13]This process is reminiscent of the creation of H II regions around newly formed O and B main-sequence stars, discussed in Section 12.3.

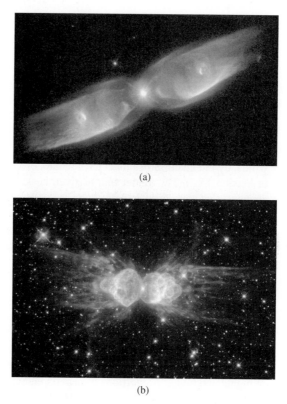

(a)

(b)

**FIGURE 13.15** Examples of two "butterfly" planetary nebulae. (a) M2-9 is a bipolar planetary nebula 800 pc distant in Ophiucus. [Credit: Bruce Balick (University of Washington), Vincent Icke (Leiden University, The Netherlands), Garrelt Mellema (Stockholm University), and NASA.] (b) Menzel 3 (Mz 3), in Norma, is also known as the Ant nebula. Outflow velocities of 1000 km s$^{-1}$ are much greater than for any other similar object. [Credit: NASA, ESA, and the Hubble Heritage Team (STScI/AURA). Acknowledgment: R. Sahai (Jet Propulsion Lab) and B. Balick (University of Washington).]

Significant detail is also evident at smaller scales. Figure 13.13(b) shows so-called cometary knots that are pointed radially away from the central star in the Helix nebula. These clumps of material have dark cores with luminous cusps on the sides facing the star.

The expansion velocities of planetaries, as measured by Doppler-shifted spectral lines, show that the gas is typically moving away from the central stars with speeds of between 10 and 30 km s$^{-1}$, although much greater speeds have been measured, as in the case of Mz 3 [Fig. 13.15(b)]. Combined with characteristic length scales of around 0.3 pc, their estimated ages are on the order of 10,000 years. After only about 50,000 years, a planetary nebula will dissipate into the ISM. Compared with the entire lifetime of a star, the phase of planetary nebula ejection is fleeting indeed.

Despite their short lifetimes, roughly 1500 planetary nebulae are known to exist in the Milky Way Galaxy. Given the fact that we are unable to observe the entire galaxy from

Earth, it is estimated that the number of planetaries is probably close to 15,000. If, on average, each planetary contains about 0.5 $M_\odot$ of material, the ISM is being enriched at the rate of roughly one solar mass per year through this process.

## 13.3 ■ STELLAR CLUSTERS

Over the past two chapters we have seen a story develop that depicts the lives of stars. They are formed from the ISM, only to return most of that material to the ISM through stellar winds, by the ejection of planetary nebulae, or via supernova explosions (to be discussed in Section 15.3). The matter that is given back, however, has been enriched with heavier elements that were produced through the various sequences of nuclear reactions governing a star's life. As a result, when the next generation of stars is formed, it possesses higher concentrations of these heavy elements than did its ancestors. This cyclic process of star formation, death, and rebirth is evident in the variations in composition between stars.

### Population I, II, and III Stars

The universe began with the Big Bang 13.7 billion years ago. At that time hydrogen and helium were essentially the only elements produced by the nucleosynthesis that occurred during the initial fireball. Consequently, the first stars to form did so with virtually no metal content; $Z = 0$. The next generation of stars that formed were extremely **metal-poor**, having very low but non-zero values of $Z$. Each succeeding generation of star production resulted in higher and higher proportions of heavier elements, leading to metal-rich stars for which $Z$ may reach values as high as 0.03. The (thus far hypothetical) original stars that formed immediately after the Big Bang are referred to as **Population III** stars, metal-poor stars with $Z \gtrsim 0$ are referred to as **Population II**, and metal-rich stars are called **Population I**.

The classifications of Population II and Population I are due originally to their identifications with kinematically distinct groups of stars within our Galaxy. Population I stars have velocities relative to the Sun that are low compared to Population II stars. Furthermore, Population I stars are found predominantly in the disk of the Milky Way, while Population II stars can be found well above or below the disk. It was only later that astronomers realized that these two groups of stars differed chemically as well. Not only do populations tell us something about evolution, but the kinematic characteristics, positions, and compositions of Population I and Population II stars also provide us with a great deal of information about the formation and evolution of the Milky Way Galaxy.

### Globular Clusters and Galactic (Open) Clusters

Recall from Section 12.2 that during the collapse of a molecular cloud, **stellar clusters** can form, ranging in size from tens of stars to hundreds of thousands of stars. Every member of a given cluster formed from the same cloud, they all formed with essentially identical compositions, and they all formed within a relatively short period of time. Thus, excluding such effects as rotation, magnetic fields, and membership in a binary star system, the Vogt–Russell theorem suggests that the differences in evolutionary states between the various stars in the cluster are due solely to their initial masses.

(a)                                            (b)

**FIGURE 13.16**   (a) M13, the great globular cluster in Hercules, is located approximately 7000 pc from Earth. (From the Digitized Sky Survey at STScI. Courtesy of Palomar/Caltech, the National Geographic Society, and the Space Telescope Science Institute.) (b) The Pleiades is a galactic cluster found in the constellation of Taurus, at a distance of 130 pc. (Courtesy of the National Optical Astronomical Observatories.)

Extreme Population II clusters formed when the Galaxy was very young, making them some of the oldest objects in the Milky Way. They also contain the largest number of members. Figure 13.16(a) shows M13, one such **globular cluster**, located in the constellation of Hercules. Population I clusters, such as the Pleiades [Fig. 13.16(b)], tend to be smaller and younger. These smaller clusters are called alternately **galactic clusters** or **open clusters**.

### Spectroscopic Parallax

The H–R diagrams of clusters can be constructed in a self-consistent way without knowledge of the exact distances to them. Since the dimensions of a typical cluster are small relative to its distance from Earth, little error is introduced by assuming that each member of the cluster has the same distance modulus. As a result, plotting the apparent magnitude rather than the absolute magnitude only amounts to shifting the position of each star in the diagram vertically by the same amount. By matching the observational main sequence of the cluster to a main sequence calibrated in absolute magnitude, the distance modulus of the cluster can be determined, giving the cluster's distance from the observer. This method of distance determination is known as **spectroscopic parallax** (the method is also often referred to as **main-sequence fitting**).

### Color–Magnitude Diagrams

Rather than attempting to determine the effective temperature of every member of a cluster by undertaking a detailed spectral line analysis of each star (which would be a major project for a globular cluster, even assuming that the stars were bright enough to get good spectra), it is much faster to determine their color indices $(B - V)$. With knowledge of the apparent magnitude and the color index of each star, a **color–magnitude diagram** can be constructed. Color–magnitude diagrams for M3 (a globular cluster) and h and $\chi$ Persei (a *double* galactic cluster) are shown in Figs. 13.17 and 13.18, respectively.

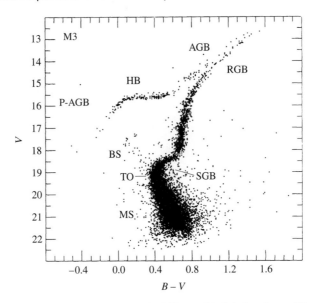

**FIGURE 13.17**   A color–magnitude diagram for M3, an old globular cluster. The major phases of stellar evolution are indicated: main sequence (MS); blue stragglers (BS); the main-sequence turn-off point (TO); the subgiant branch of hydrogen shell burning (SGB); the red giant branch along the Hayashi track, prior to helium core burning (RGB); the horizontal branch during helium core burning (HB); the asymptotic giant branch during hydrogen and helium shell burning (AGB); post-AGB evolution proceeding to the white dwarf phase (P-AGB). (Figure adapted from Renzini and Fusi Pecci, *Annu. Rev. Astron. Astrophys.*, *26*, 199, 1988. Reproduced with permission from the *Annual Review of Astronomy and Astrophysics*, Volume 26, ©1988 by Annual Reviews Inc.)

## Isochrones and Cluster Ages

Clusters, and their associated color–magnitude diagrams, offer nearly ideal tests of many aspects of stellar evolution theory. By computing the evolutionary tracks of stars of various masses, all having the same composition as the cluster, it is possible to plot the position of each evolving model on the H–R diagram when the model reaches the age of the cluster. (The curve connecting these positions is known as an **isochrone**.) The relative number of stars at each location on the isochrone depends on the number of stars in each mass range within the cluster (the initial mass function; see Fig. 12.12), combined with the different rates of evolution during each phase. Therefore, star counts in a color–magnitude diagram can shed light on the timescales involved in stellar evolution.

As the cluster ages, beginning with the initial collapse of the molecular cloud, the most massive and least abundant stars will arrive on the main sequence first, evolving rapidly. Before the lowest-mass stars have even reached the main sequence, the most massive ones have already evolved into the red giant region, perhaps even undergoing supernova explosions. These disparate rates of evolution can be seen by comparing Figs. 12.11 and 13.1 for pre-main-sequence and post-main-sequence evolution, respectively, together with their associated tables.

Since core hydrogen-burning lifetimes are inversely related to mass, continued evolution of the cluster means that the main-sequence **turn-off point**, defined as the point where stars

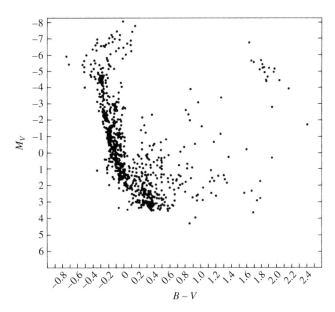

**FIGURE 13.18**   A color–magnitude diagram for the young double galactic cluster, h and $\chi$ Persei. Note that the most massive stars are pulling away from the main sequence while the low-mass stars in the middle of the diagram are still contracting onto the main sequence. Red giants are present in the upper right-hand corner of the diagram. (Figure adapted from Wildey, *Ap. J. Suppl.*, *8*, 439, 1964.)

in the cluster are currently leaving the main sequence, becomes redder and less luminous with time. Consequently, it is possible to estimate the age of a cluster by the location of the uppermost point of its main sequence. This fundamental technique is an important tool for determining ages of stars, clusters, our Milky Way Galaxy, and other galaxies with observable clusters, and even for establishing a lower limit on the age of the universe itself. A composite color–magnitude diagram of a number of clusters is shown in Fig. 13.19. Labeled vertically on the right-hand side is the age of the cluster corresponding to the location of the main-sequence turn-off point.

### The Hertzsprung Gap

Another consequence of varying timescales can be seen in the color–magnitude diagram of h and $\chi$ Persei (Fig. 13.18). Apparent are red giants, together with low-mass pre-main-sequence stars. Also evident in the diagram is the complete absence of stars between the massive ones that are just leaving the main sequence and the few in the red giant region. It is unlikely that this represents an incomplete survey, since these stars are the brightest members of the cluster. Rather, it points out the very rapid evolution that occurs just after leaving the main sequence. This feature, known as the **Hertzsprung gap**, is a common characteristic of the color–magnitude diagrams of young, galactic clusters. The existence of the Hertzsprung gap is due to evolution on a Kelvin–Helmholtz timescale across the SGB, following the point when the hydrogen-depleted core exceeds the Schönberg–Chandrasekhar limit.

Notice in Fig. 13.19 that the cluster M67 does not show the existence of the Hertzsprung gap; the same can be said of M3 (Fig. 13.17). Recall that below about 1.25 $M_\odot$, the rapid

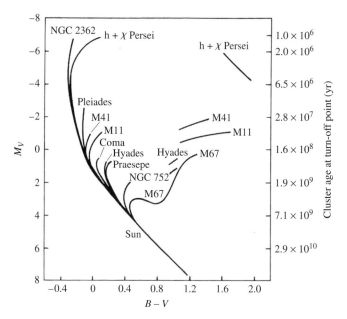

**FIGURE 13.19**    A composite color–magnitude diagram for a set of Population I galactic clusters. The absolute visual magnitude is indicated on the left-hand vertical axis, and the age of the cluster, based on the location of its turn-off point, is labeled on the right-hand side. (Figure adapted from an original diagram by A. Sandage.)

contraction phase related to the Schönberg–Chandrasekhar limit is much less pronounced. As a result, color–magnitude diagrams of old globular clusters with turn-off points near or less than 1 $M_\odot$ have continuous distributions of stars leading to the red giant region.

### Relatively Few AGB and Post-AGB Stars

Close inspection of Fig. 13.17 also shows that a relatively small number of stars exist on the asymptotic giant branch and only a few stars are to be found in the region labeled P-AGB (post-asymptotic giant branch). This is just a consequence of the very rapid pace of evolution during this phase of heavy mass loss that leads directly to the formation of white dwarfs.

### Blue Stragglers

It should be pointed out that a group of stars, known as **blue stragglers**, can be found above the turn-off point of M3. Although our understanding of these stars is incomplete, it appears that their tardiness in leaving the main sequence is due to some unusual aspect of their evolution. The most likely scenarios appear to be mass exchange with a binary star companion,[14] or collisions between two stars, extending the star's main-sequence lifetime.

---

[14] Mass exchange between close binaries is the subject of Chapter 18.

## A Work in Progress

The successful comparisons between theory and observation that are provided by stellar clusters give strong support to the idea that our picture of stellar evolution is fairly complete, although perhaps in need of some fine-tuning. Continued refinements in stellar opacities, revisions in nuclear reaction cross sections, and much-needed improvements in the treatment of convection will probably lead to even better agreement with observations. However, much fundamental work remains to be done as well, such as developing a better understanding of the effects of mass loss, rotation, magnetic fields, and the presence of a close companion.

## SUGGESTED READING

### General

Balick, B., et al., "The Shaping of Planetary Nebulae," *Sky and Telescope*, February 1987.

Harpaz, Amos, "The Formation of a Planetary Nebula," *The Physics Teacher*, May 1991.

Kwok, Sun, *Cosmic Butterflies: The Colorful Mysteries of Planetary Nebulae*, Cambridge University Press, Cambridge, 2001.

*The Space Telescope Science Institute*, http://www.stsci.edu

### Technical

Aller, Lawrence H., *Atoms, Stars, and Nebulae*, Third Edition, Cambridge University Press, Cambridge, 1991.

Ashman, Keith M., and Zepf, Stephen E., *Globular Cluster Systems*, Cambridge University Press, Cambridge, 1998.

Busso, M., Gallino, R., and Wasserburg, G. J., "Nucleosynthesis in Asymptotic Giant Branch Stars: Relevance for Galactic Enrichment and Solar System Formation," *Annual Review of Astronomy and Astrophysics*, *37*, 239, 1999.

Carney, Bruce W., and Harris, William E., *Star Clusters*, Springer-Verlag, Berlin, 2001.

Hansen, Carl J., Kawaler, Steven D., and Trimble, Virginia *Stellar Interiors: Physical Principles, Structure, and Evolution*, Second Edition, Springer-Verlag, New York, 2004.

Herwig, Falk, "Evolution of Asymptotic Giant Branch Stars," *Annual Review of Astronomy and Astrophysics*, *43*, 435, 2005.

Iben, Icko Jr., "Stellar Evolution Within and Off the Main Sequence," *Annual Review of Astronomy and Astrophysics*, *5*, 571, 1967.

Iben, Icko Jr., and Renzini, Alvio, "Asymptotic Giant Branch Evolution and Beyond," *Annual Review of Astronomy and Astrophysics*, *21*, 271, 1983.

Kippenhahn, Rudolf, and Weigert, Alfred, *Stellar Structure and Evolution*, Springer-Verlag, Berlin, 1990.

Kwok, Sun, *The Origin and Evolution of Planetary Nebulae*, Cambridge University Press, Cambridge, 2000.

Padmanabhan, T., *Theoretical Astrophysics*, Cambridge University Press, Cambridge, 2001.

Prialnik, Dina, *An Introduction to the Theory of Stellar Structure and Evolution*, Cambridge University Press, Cambridge, 2000.

Schaller, G., et al., "New grids of stellar models from 0.8 to 120 solar masses at $Z = 0.020$ and $Z = 0.001$," *Astronomy and Astrophysics Supplement Series*, *96*, 269, 1992.

Willson, Lee Anne, "Mass Loss from Cool Stars: Impact on the Evolution of Stars and Stellar Populations," *Annual Review of Astronomy and Astrophysics*, *38*, 573, 2000.

## PROBLEMS

**13.1** (a) For a 5 $M_\odot$ star, use the data in Table 13.1 associated with Fig. 13.1 to construct a table that expresses the evolutionary times between points 2 and 3, points 3 and 4, and so on, as a percentage of the lifetime of the star on the main sequence between points 1 and 2.

   (b) How long does it take a 5 $M_\odot$ star to cross the Hertzsprung gap relative to its main-sequence lifetime?

   (c) How long does the 5 $M_\odot$ star spend on the blueward portion of the horizontal branch relative to its main-sequence lifetime?

   (d) How long does the 5 $M_\odot$ star spend on the redward portion of the horizontal branch relative to its main-sequence lifetime?

**13.2** Estimate the Kelvin–Helmholtz timescale for a 5 $M_\odot$ star on the subgiant branch and compare your result with the amount of time the star spends between points 4 and 5 in Fig. 13.1.

**13.3** (a) Beginning with Eq. (13.7), show that the radius of the isothermal core for which the gas pressure is a maximum is given by Eq. (13.8). Recall that this solution assumes that the gas in the core is ideal and monatomic.

   (b) From your results in part (a), show that the maximum pressure at the surface of the isothermal core is given by Eq. (13.9).

**13.4** During the first dredge-up phase of a 5 $M_\odot$ star, would you expect the composition ratio $X'_{13}/X_{12}$ to increase or decrease? Explain your reasoning. *Hint:* You may find Fig. 13.6 helpful.

**13.5** Use Eq. (10.27) to show that the ignition of the triple alpha process at the tip of the red giant branch ought to occur at more than $10^8$ K.

**13.6** In an attempt to identify the important components of AGB mass loss, various researchers have proposed parameterizations of the mass loss rate that are based on fitting observed rates for a specified set of stars with some general equation that includes measurable quantities associated with the stars in the sample. One of the most popular, developed by D. Reimers, is given by

$$\dot{M} = -4 \times 10^{-13} \eta \frac{L}{gR} \ M_\odot \ \text{yr}^{-1}, \qquad (13.13)$$

where $L$, $g$, and $R$ are the luminosity, surface gravity, and radius of the star, respectively (all in solar units; $g_\odot = 274$ m s$^{-2}$). $\eta$ is a *free parameter* whose value is expected to be near unity. Note that the minus sign has been explicitly included here, indicating that the mass of the star is decreasing.

   (a) Explain qualitatively why $L$, $g$, and $R$ enter Eq. (13.13) in the way they do.

   (b) Estimate the mass loss rate of a 1 $M_\odot$ AGB star that has a luminosity of 7000 $L_\odot$ and a temperature of 3000 K.

**13.7** **(a)** Show that the Reimers mass loss rate, given by Eq. (13.13) in Problem 13.6, can also be written in the form

$$\dot{M} = -4 \times 10^{-13} \eta \frac{LR}{M} \; M_\odot \; \mathrm{yr}^{-1},$$

where $L$, $R$, and $M$ are all in solar units.

**(b)** Assuming (incorrectly) that $L$, $R$, and $\eta$ do not change with time, derive an expression for the mass of the star as a function of time. Let $M = M_0$ when the mass loss phase begins.

**(c)** Using $L = 7000 \; L_\odot$, $R = 310 \; R_\odot$, $M_0 = 1 \; M_\odot$, and $\eta = 1$, make a graph of the star's mass as a function of time.

**(d)** How long would it take for a star with an initial mass of $1 \; M_\odot$ to be reduced to the mass of the degenerate carbon–oxygen core ($0.6 \; M_\odot$)?

**13.8** The Helix nebula is a planetary nebula with an angular diameter of $16'$ that is located approximately 213 pc from Earth.

**(a)** Calculate the diameter of the nebula.

**(b)** Assuming that the nebula is expanding away from the central star at a constant velocity of $20 \; \mathrm{km \; s^{-1}}$, estimate its age.

**13.9** An old version of stellar evolution, popular at the beginning of the twentieth century, maintained that stars begin their lives as large, cool spheres of gas, like the giant stars on the H–R diagram. They then contract and heat up under the pull of their own gravity to become hot, bright blue O stars. For the remainder of their lives they lose energy, becoming dimmer and redder with age. As they slowly move down the main sequence, they eventually end up as cool, dim red M stars. Explain how observations of stellar clusters, plotted on an H–R diagram, contradict this idea.

**13.10** **(a)** Using data available in Tables 12.1 and 13.1, compare the pre-main sequence evolutionary time of a $0.8 M_\odot$ star with the lifetime on the main sequence for a $15 \; M_\odot$ star. How does this information help to explain the appearance of a color–magnitude diagram such as Fig. 13.18?

**(b)** Estimate the mass of a star that would have a main-sequence lifetime comparable to the pre-main-sequence evolutionary time of a $0.8 \; M_\odot$ star.

**13.11** **(a)** The age of the universe is 13.7 Gyr. Compare this value to the main-sequence lifetime of a $0.8 \; M_\odot$ star. Why isn't it useful to compute the detailed post-main-sequence evolution of stars with masses much lower than the mass of the Sun?

**(b)** Would you expect to find globular clusters with main-sequence turn-off points below $0.8 \; M_\odot$? Explain your answer.

**13.12** **(a)** Show that $\log_{10}(L_V/L_B)$ + constant is, to within a multiplicative constant, equivalent to the color index, $B - V$.

**(b)** Estimating best-fit curves through the data given in Fig. 13.20, trace the two color–magnitude diagrams, placing them on a single graph. Note that the abscissas have been normalized so that the lowest-luminosity stars of both clusters are located at the same positions on their respective diagrams.

**(c)** Given that 47 Tuc is relatively metal-rich for a globular cluster ($Z/Z_\odot = 0.17$, where $Z_\odot$ is the solar value) and M15 is metal-poor ($Z/Z_\odot = 0.0060$), explain the difference in colors between the two clusters. *Hint:* You may wish to refer back to the discussion in Example 9.5.4 (Section 9.5).

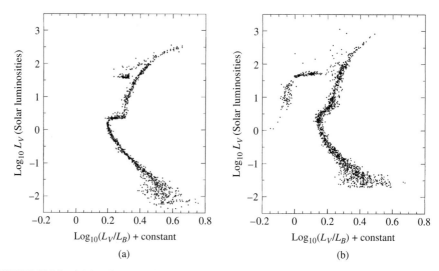

**FIGURE 13.20**     (a) A color–magnitude diagram for 47 Tuc, a relatively metal-rich globular cluster with $Z/Z_\odot = 0.17$. (Data from Hesser et al., *Publ. Astron. Soc. Pac.*, *99*, 739, 1987; figure courtesy of William E. Harris.) (b) A color–magnitude diagram for M15, a metal-poor globular cluster with $Z/Z_\odot = 0.0060$. (Data from Durrell and Harris, *Astron. J.*, *105*, 1420, 1993; figure courtesy of William E. Harris.)

**13.13** Using the technique of main-sequence fitting, estimate the distance to M3; refer to Figs. 13.17 and 13.19.

# CHAPTER
# 14

# Stellar Pulsation

## 14.1 ■ OBSERVATIONS OF PULSATING STARS

In August of 1595, a Lutheran pastor and amateur astronomer named David Fabricius (1564–1617) observed the star *o* Ceti. As he watched over a period of months, the brightness of this second-magnitude star in the constellation Cetus (the Sea Monster) slowly faded. By October, the star had vanished from the sky. Several more months passed as the star eventually recovered and returned to its former brilliance. In honor of this miraculous event, *o* Ceti was named Mira, meaning "wonderful."

Mira continued its rhythmic dimming and brightening, and by 1660 the 11-month period of its cycle was established. The regular changes in brightness were mistakenly attributed to dark "blotches" on the surface of a rotating star. Supposedly, Mira would appear fainter when these dark areas were turned toward Earth.

Figure 14.1 shows the *light curve* of Mira for a 51-year interval. Today astronomers recognize that the changes in Mira's brightness are due not to dark spots on its surface but to the fact that Mira is a **pulsating star**, a star that dims and brightens as its surface expands and contracts. Mira is the prototype of the **long-period variables**, stars that have somewhat irregular light curves and pulsation periods between 100 and 700 days.

Nearly two centuries elapsed before another pulsating star was discovered. In 1784 John Goodricke (1764–1786) of York, England, found that the brightness of the star δ Cephei varies regularly with a period of 5 days, 8 hours, 48 minutes. This discovery cost Goodricke his life; he contracted pneumonia while observing δ Cephei and died at the age of 21. The light curve of δ Cephei, shown in Fig. 14.2, is less spectacular than that of *o* Ceti. It varies by less than one magnitude in brightness and never fades from view. Nevertheless, pulsating stars similar to δ Cephei, called **classical Cepheids**, are vitally important to astronomy.

### The Period–Luminosity Relation

By 2005, nearly 40,000 pulsating stars had been cataloged by astronomers. One woman, Henrietta Swan Leavitt (1868–1921; see Fig. 14.3), discovered more than 5% of these stars

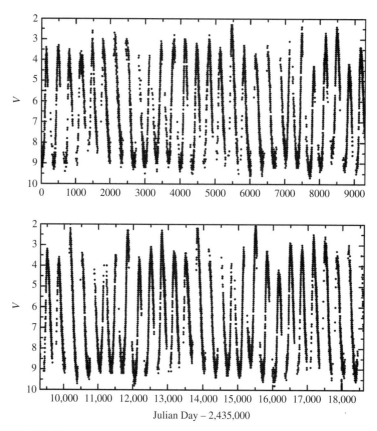

**FIGURE 14.1**   The light curve of Mira from September 14, 1954 (JD 2,435,000) through September 2005. Recall that magnitudes dimmer than 6 are undetectable to the unaided eye. (We acknowledge with thanks the variable-star observations from the AAVSO International Database contributed by observers worldwide.)

while working as a "computer" for Edward Charles Pickering (1846–1919) at Harvard University. Her tedious task was to compare two photographs of the same field of stars taken at different times and detect any star that varied in brightness. Eventually she discovered 2400 classical Cepheids with periods between 1 and 50 days, most of them located in the Small Magellanic Cloud (SMC). Leavitt took advantage of this opportunity to investigate the nature of the classical Cepheids in the SMC. Noticing that the more luminous Cepheids took longer to go through their pulsation cycles, she plotted the apparent magnitudes of these SMC stars against their pulsation periods. The resulting graph, shown in Fig. 14.4, demonstrated that the apparent magnitudes of classical Cepheids are closely correlated with their periods, with an uncertainty of only $\Delta m \approx \pm 0.5$ at a given period.

Because all of the stars in the Small Magellanic Cloud are roughly the same distance from us (about 61 kpc), the differences in their apparent magnitudes must be the same as the differences in their absolute magnitudes [cf. Eq. (3.6) for the distance modulus]. Thus the

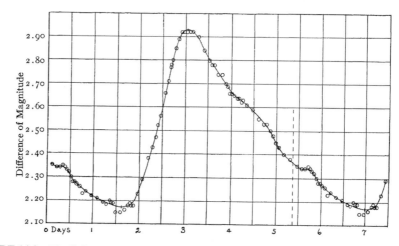

**FIGURE 14.2**   The light curve of δ Cephei. Its pulsation period is 5.37 days. (Figure from Stebbins, Joel, *Ap. J.*, *27*, 188, 1908.)

**FIGURE 14.3**   Henrietta Swan Leavitt (1868–1921). (Courtesy of Harvard College Observatory.)

observed differences in these stars' apparent brightnesses must reflect intrinsic differences in their luminosities. Astronomers were excited at the prospect of determining the absolute magnitude or luminosity of a distant Cepheid simply by timing its pulsation, because knowing both a star's apparent and absolute magnitudes allows the distance of the star to be easily determined from the distance modulus, Eq. (3.6). This would permit the measurement of large distances in the universe, far beyond the limited range of parallax techniques. The only stumbling block was the calibration of Leavitt's relation. An independent distance to a

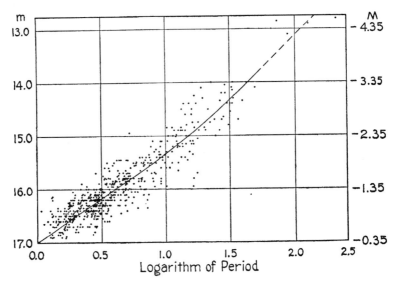

**FIGURE 14.4**    Classical Cepheids in the Small Magellanic Cloud, with the period in units of days. (Figure from Shapley, *Galaxies*, Harvard University Press, Cambridge, MA, 1961.)

single Cepheid had to be obtained to measure its absolute magnitude and luminosity. Once this difficult chore was accomplished, the resulting **period–luminosity relation** could be used to measure the distance to any Cepheid.

The nearest classical Cepheid is Polaris, some 200 pc away. In the early twentieth century, this distance was too great to be reliably measured by stellar parallax. However, in 1913, Ejnar Hertzsprung succeeded in using the longer baseline provided by the Sun's motion through space, together with statistical methods, to find the distances to Cepheids having a specified period. (The measurement of the absolute magnitude of a Cepheid is also complicated by the dimming effect of interstellar extinction; recall Eq. 12.1.)

The calibrated period–luminosity relation depicted in Fig. 14.5 for the $V$ band is described by

$$M_{\langle V \rangle} = -2.81 \log_{10} P_{\rm d} - 1.43, \tag{14.1}$$

where $M_{\langle V \rangle}$ is the average absolute $V$ magnitude and $P_{\rm d}$ is the pulsation period in units of days. In terms of the average luminosity of the star, the relation is given by

$$\log_{10} \frac{\langle L \rangle}{L_\odot} = 1.15 \log_{10} P_{\rm d} + 2.47. \tag{14.2}$$

Astronomers can substantially decrease the scatter in the period–luminosity relation by making observations in infrared wavelengths where interstellar extinction is less of a problem. One such fit, made using magnitudes measured in the infrared $H$ band (centered at 1.654 $\mu$m), is illustrated in Fig. 14.6(a). The data are for 92 Cepheids in the Large

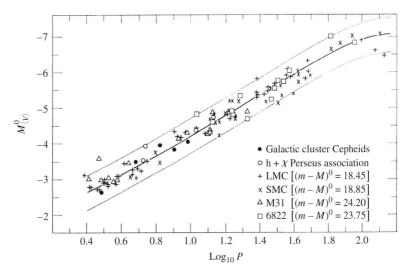

**FIGURE 14.5**   The period–luminosity relation for classical Cepheids. (Figure adapted from Sandage and Tammann, *Ap. J.*, *151*, 531, 1968.)

Magellanic Cloud. The infrared period–luminosity fit is given by

$$H = -3.234 \log_{10} P_{\mathrm{d}} + 16.079. \qquad (14.3)$$

The scatter can be further reduced by adding a color term to the fit. Using the infrared color index $J - K_s$, Fig. 14.6(b) shows that the fit is indeed somewhat tighter ($J$ and $K_s$ are centered at 1.215 $\mu$m and 2.157 $\mu$m, respectively). The fit for this **period–luminosity–color relation** is given by

$$H = -3.428 \log_{10} P_{\mathrm{d}} + 1.54\langle J - K_s \rangle + 15.637. \qquad (14.4)$$

Classical Cepheids provide astronomy with its third dimension and supply the foundation for the measurement of extragalactic distances. Because Cepheids are supergiant stars (luminosity class Ib), about fifty times the Sun's size and thousands of times more luminous, they can be seen over intergalactic distances. They serve as "standard candles," beacons scattered throughout the night sky that serve as mileposts for astronomical surveys of the universe.

### The Pulsation Hypothesis for Brightness Variations

The important use of Cepheids as cosmic distance indicators does not require an understanding of the physical reasons for their light variations. In fact, the observed changes in brightness were once thought to be caused by tidal effects in the atmospheres of binary stars. However, in 1914 the American astronomer Harlow Shapley (1885–1972) argued that the binary theory was fatally flawed because the size of the star would exceed the size of the

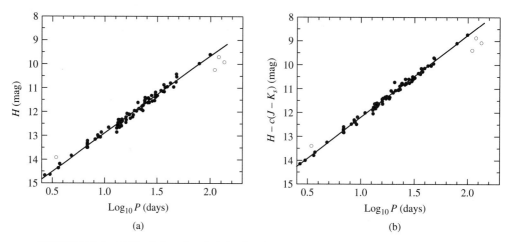

**FIGURE 14.6**    (a) The infrared period–luminosity relation for 92 Cepheids in the Large Magellanic Cloud. The infrared $H$ band was used for the observations. The open circles represent four Cepheids that were excluded from the least-squares linear fit. (b) The period–luminosity–color relation for the same Cepheids. (Data from Persson, S. E., et al., *Astron. J.*, *128*, 2239, 2004.)

orbit for some variables. Shapley advanced an alternative idea: that the observed variations in the brightness and temperature of classical Cepheids were caused by the radial pulsation of single stars. He proposed that these stars were rhythmically "breathing" in and out, becoming alternately brighter and dimmer in the process. Four years later Sir Arthur Stanley Eddington provided a firm theoretical framework for the pulsation hypothesis, which received strong support from the observed correlations among the variations in brightness, temperature, and surface velocity throughout the pulsation cycle. Figure 14.7 shows the measured changes in magnitude, temperature, radius, and surface velocity for $\delta$ Cephei. The change in brightness is primarily due to the roughly 1000 K variation in $\delta$ Cephei's surface temperature; the accompanying change in size makes a lesser contribution to the luminosity. Although the total excursion of $\delta$ Cephei's surface from its equilibrium radius is large in absolute terms (a bit more than the diameter of the Sun), it is still only about 5% to 10% of the size of this supergiant star. The spectral type of $\delta$ Cephei changes continuously throughout the cycle, varying between F5 (hottest) and G2 (coolest). A careful examination of Fig. 14.7 reveals that the magnitude and surface velocity curves are nearly identical in shape. Thus, the star is brightest when its surface is expanding outward most rapidly, *after* it has passed through its minimum radius. Later in this chapter we will see that the explanation of this **phase lag** of maximum luminosity behind minimum radius has its origin in the mechanism that maintains the oscillations.

### The Instability Strip

The Milky Way Galaxy is estimated to contain several million pulsating stars. Considering that our Galaxy consists of several hundred billion stars, this implies that stellar pulsation must be a transient phenomenon. The positions of the pulsating variables on the H–R diagram (see Figs. 14.8 and 8.16) confirm this conclusion. Rather than being located on the main sequence, where stars spend most of their lives, the majority of pulsating stars occupy

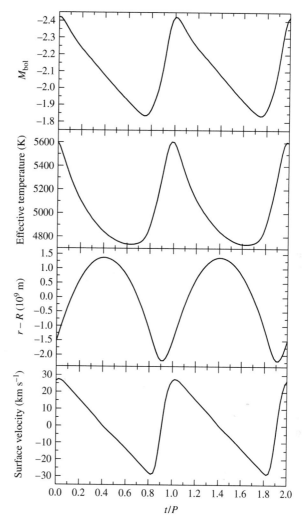

**FIGURE 14.7** Observed pulsation properties of δ Cephei, a typical classical Cepheid. (Data from Schwarzschild, *Harvard College Observatory Circular, 431*, 1938.)

a narrow (about 600–1100 K wide), nearly vertical **instability strip** on the right-hand side of the H–R diagram. Theoretical evolutionary tracks for stars of various masses are also shown in Fig. 14.8. As stars evolve along these tracks, they begin to pulsate as they enter the instability strip and cease their oscillations upon leaving. Of course, evolutionary timescales are far too long for us to observe the onset and cessation of a single star's oscillations, but several stars have been caught in the final phase of their pulsational history.

### Some Classes of Pulsating Stars

Astronomers have divided pulsating stars into several classes. Some of these are listed in Table 14.1. The W Virginis stars are metal-deficient (Population II) Cepheids and are

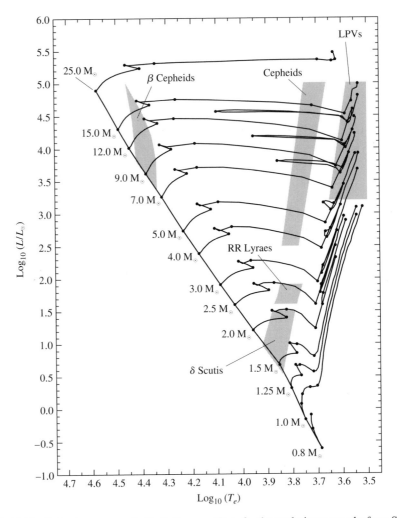

**FIGURE 14.8**  Pulsating stars on the H–R diagram. (Data for the evolutionary tracks from Schaller, et al., *Astron. Astrophys. Suppl.*, *96*, 269, 1992.)

about four times less luminous than classical Cepheids with the same period. Their period–luminosity relation is thus lower than and parallel to the one shown for the classical Cepheids in Fig. 14.5. RR Lyrae stars, also Population II, are horizontal-branch stars found in globular clusters. Because all RR Lyrae stars have nearly the same luminosity, they are also useful yardsticks for distance measurements. The $\delta$ Scuti variables are evolved F stars found near the main sequence of the H–R diagram. They exhibit both radial and nonradial oscillations; the latter is a more complicated motion that will be discussed in Section 14.4. Below the main sequence (not shown in Fig. 14.8; however, see Fig. 16.4) are the pulsating white dwarfs, called ZZ Ceti stars.

All of the types of stars listed thus far lie within the instability strip, and they share a common mechanism that drives the oscillations. The long-period variables such as Mira

**TABLE 14.1**   Pulsating Stars. (Adopted from Cox, *The Theory of Stellar Pulsation*, Princeton University Press, Princeton, NJ, 1980.)

| Type | Range of Periods | Population Type | Radial or Nonradial |
|------|------------------|-----------------|---------------------|
| Long-Period Variables | 100–700 days | I,II | R |
| Classical Cepheids | 1–50 days | I | R |
| W Virginis stars | 2–45 days | II | R |
| RR Lyrae stars | 1.5–24 hours | II | R |
| $\delta$ Scuti stars | 1–3 hours | I | R,NR |
| $\beta$ Cephei stars | 3–7 hours | I | R,NR |
| ZZ Ceti stars | 100–1000 seconds | I | NR |

and the $\beta$ Cephei stars are located outside of the instability strip occupied by the classical Cepheids and RR Lyrae stars. Their unusual positions on the H–R diagram will be discussed in the next section.

## 14.2 ■ THE PHYSICS OF STELLAR PULSATION

Geologists and geophysicists have obtained a wealth of information about Earth's interior from their study of the seismic waves produced by earthquakes and other sources. In the same manner, astrophysicists model the pulsational properties of stars to understand better their internal structure. By numerically calculating an evolutionary sequence of stellar models and then comparing the pulsational characteristics (periods, amplitudes, and details of the light and radial velocity curves) of the models with those actually observed, astronomers can further test their theories of stellar structure and evolution and obtain a detailed view of the interior of a star.[1]

### The Period–Density Relation

The radial oscillations of a pulsating star are the result of sound waves resonating in the star's interior. A rough estimate of the pulsation period,[2] $\Pi$, may be easily obtained by considering how long it would take a sound wave to cross the diameter of a model star of radius $R$ and constant density $\rho$. The adiabatic sound speed is given by Eq. (10.84),

$$v_s = \sqrt{\frac{\gamma P}{\rho}}.$$

The pressure may be found from Eq. (10.6) for hydrostatic equilibrium, using the

---

[1] Several other ways of testing the ideas of stellar structure and evolution were discussed in Chapter 13.

[2] Throughout the following discussion, $\Pi$ will be used to designate the pulsation period so that it is not confused with the pressure, $P$. $\Pi$ is commonly used for the pulsation period in stellar pulsation theory studies. ($T$, another symbol commonly used for period, would lead to confusion with temperature.)

(unrealistic) assumption of constant density. Thus

$$\frac{dP}{dr} = -\frac{GM_r\rho}{r^2} = -\frac{G\left(\frac{4}{3}\pi r^3 \rho\right)\rho}{r^2} = -\frac{4}{3}\pi G\rho^2 r.$$

This is readily integrated using the boundary condition that $P = 0$ at the surface to obtain the pressure as a function of $r$,

$$P(r) = \frac{2}{3}\pi G\rho^2 \left(R^2 - r^2\right). \tag{14.5}$$

Thus the pulsation period is roughly

$$\Pi \approx 2 \int_0^R \frac{dr}{v_s} \approx 2 \int_0^R \frac{dr}{\sqrt{\frac{2}{3}\gamma \pi G\rho \left(R^2 - r^2\right)}},$$

or

$$\boxed{\Pi \approx \sqrt{\frac{3\pi}{2\gamma G\rho}}.} \tag{14.6}$$

Qualitatively, this shows that the pulsation period of a star is inversely proportional to the square root of its mean density. Referring to Fig. 14.8 and Table 14.1, this **period–mean density relation** explains why the pulsation period decreases as we move down the instability strip from the very tenuous supergiants to the very dense white dwarfs.[3] The tight period–luminosity relation discovered by Leavitt exists because the instability strip is roughly parallel to the luminosity axis of the H–R diagram (the finite width of the instability strip is reflected in the $\pm 0.5$ magnitude uncertainty in the period–luminosity relation). The quantitative agreement of Eq. (14.6) with the observed periods of Cepheids is not too bad, considering its crude derivation. If we take $M = 5\,M_\odot$ and $R = 50\,R_\odot$ for a typical Cepheid, then $\Pi \approx 10$ days. This falls nicely within the range of periods measured for the classical Cepheids.

### Radial Modes of Pulsation

The sound waves involved in the **radial modes** of stellar pulsation are essentially *standing waves*, similar to the standing waves that occur in an organ pipe that is open at one end; see Fig. 14.9. Both the star and the organ pipe can sustain several modes of oscillation. The standing wave for each mode has a *node* at one end (the star's center, the pipe's closed end), where the gases do not move, and an *antinode* at the other end (the star's surface, the pipe's open end). For the **fundamental mode**, the gases move in the same direction at every point in the star or pipe. There is a single node between the center and the surface for the **first overtone** mode,[4] with the gases moving in opposite directions on either side of

---

[3]Pulsating white dwarfs exhibit nonradial oscillations, and their periods are longer than predicted by the period–mean density relation.

[4]Some texts use the unfortunate term *first harmonic* for the first overtone.

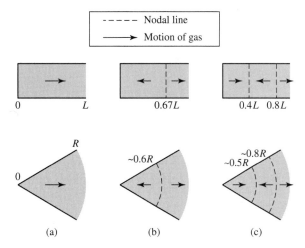

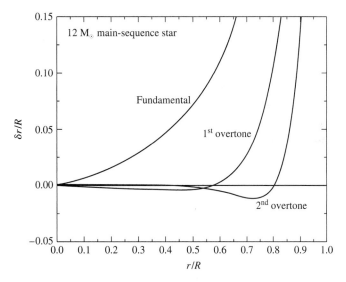

**FIGURE 14.9** Standing sound waves in an organ pipe and in a star for (a) the fundamental mode, (b) the first overtone, and (c) the second overtone.

**FIGURE 14.10** Radial modes for a pulsating star. The waveform for each mode has been arbitrarily scaled so that $\delta r / R = 1$ at the surface of the star. Actually, the maximum surface ratio of $\delta r / R$ is approximately 0.05 to 0.10 for a classical Cepheid.

the node, and two nodes for the **second overtone** mode. Figure 14.10 shows the fractional displacement, $\delta r / R$, of the stellar material from its equilibrium position for several radial modes of a 12 $M_\odot$ main-sequence star model. Note that $\delta r / R$ has been arbitrarily scaled to unity at the stellar surface.

For radial modes, the motion of the stellar material occurs primarily in the surface regions, but there is some oscillation deep inside the star. This effect is most prominent for the fundamental mode, where non-negligible amplitudes exist. For the stellar model used in Fig. 14.10, at $r = 0.5R$, $\delta r/R$ is about 7% of its surface value. For the first overtone at the same location, $\delta r/R$ is less than 1% of its surface value and is in the opposite direction, and for the second overtone, the oscillation is nearly zero amplitude ($r = 0.5R$ is close to a node for the second overtone).

The vast majority of the classical Cepheids and W Virginis stars pulsate in the fundamental mode. The RR Lyrae variables pulsate in either the fundamental or the first overtone mode, with a few oscillating in both modes simultaneously. The long-period variables, such as Mira, may also oscillate in either the fundamental mode or the first overtone, although this is still not entirely clear.

### Eddington's Thermodynamic Heat Engine

To explain the mechanism that powers these standing sound waves, Eddington proposed that pulsating stars are thermodynamic heat engines. The gases comprising the layers of the star do $P \, dV$ work as they expand and contract throughout the pulsation cycle. If the integral $\oint P \, dV > 0$ for the cycle, a layer does net positive work on its surroundings and contributes to driving the oscillations; if $\oint P \, dV < 0$, the net work done by the layer is negative and tends to dampen the oscillations. Figures 14.11 and 14.12 show $P$–$V$ diagrams for a driving layer and a damping layer, respectively, in a numerical calculation of the oscillation of an RR Lyrae star. If the total work (found by adding up the contributions of all the layers of the star) is positive, the oscillations will grow in amplitude. The oscillations will decay if the total work is negative. These changes in the pulsation amplitude continue until an equilibrium value is reached, when the total work done by all the layers is zero.

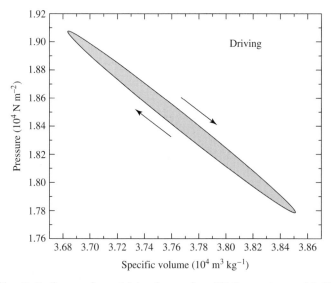

**FIGURE 14.11**    $P$–$V$ diagram for a driving layer of an RR Lyrae star model. You may recall the analogous use of $P$–$V$ diagrams in discussing heat engines in introductory physics courses. A clockwise path in a $P$–$V$ diagram corresponds with net driving.

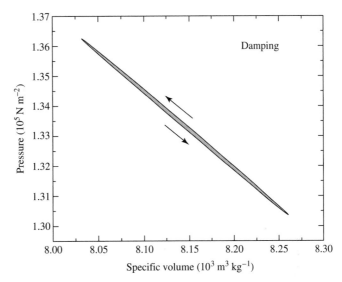

**FIGURE 14.12** *P–V* diagram for a damping layer of an RR Lyrae star model. A counterclockwise path in a *P–V* diagram corresponds to net damping.

As for any heat engine, the net work done by each layer of the star during one cycle is the difference between the heat flowing into the gas and the heat leaving the gas. For driving, the heat must enter the layer during the high-temperature part of the cycle and leave during the low-temperature part. Just as the spark plug of an automobile engine fires at the end of the compression stroke, the driving layers of a pulsating star must absorb heat around the time of their maximum compression. In this case the maximum pressure will occur *after* maximum compression, and the oscillations will be amplified.

### The Nuclear $\epsilon$ Mechanism

In what region of the star can this driving take place? An obvious possibility was first considered by Eddington: When the center of the star is compressed, its temperature and density rise, increasing the rate at which thermonuclear energy is generated. However, recall from Fig. 14.10 that the displacement $\delta r/R$ has a node at the center of the star. The pulsation amplitude is very small near the center. Although this energy mechanism (called the $\epsilon$-**mechanism**) does in fact operate in the core of a star, it is usually not enough to drive the star's pulsation. However, as mentioned in Section 10.6, variations in the nuclear energy generation rate ($\epsilon$) produce oscillations that may contribute to preventing the formation of stars with masses greater than approximately 90 $M_\odot$.

### Eddington's Valve

Eddington then suggested an alternative, a *valve mechanism*. If a layer of the star became more opaque upon compression, it could "dam up" the energy flowing toward the surface and push the surface layers upward. Then, as this expanding layer became more transparent, the trapped heat could escape and the layer would fall back down to begin the cycle anew. In Eddington's own words, "To apply this method we must make the star more heat-tight when compressed than when expanded; in other words, *the opacity must increase with compression.*"

In most regions of the star, however, the opacity actually *decreases* with compression. Recall from Section 9.2 that for a Kramers law, the opacity $\kappa$ depends on the density and temperature of the stellar material as $\kappa \propto \rho/T^{3.5}$. As the layers of a star are compressed, their density and temperature both increase. But because the opacity is more sensitive to the temperature than to the density, the opacity of the gases usually decreases upon compression. It takes special circumstances to overcome the damping effect of most stellar layers, which explains why stellar pulsation is observed for only one of every $10^5$ stars.

### Opacity Effects and the $\kappa$ and $\gamma$ Mechanisms

The conditions responsible for exciting and maintaining the stellar oscillations were first identified by the Russian astronomer S. A. Zhevakin and then verified in detailed calculations by a German and two Americans, Rudolph Kippenhahn, Norman Baker, and John P. Cox (1926–1984). They found that the regions of a star where Eddington's valve mechanism can successfully operate are its *partial ionization zones* (cf. Section 8.1). In these layers of the star where the gases are partially ionized, part of the work done on the gases as they are compressed produces further ionization rather than raising the temperature of the gas.[5] With a smaller temperature rise, the increase in density with compression produces a corresponding increase in the Kramers opacity; see Fig. 14.13. Similarly, during expansion, the temperature does not decrease as much as expected since the ions now recombine with electrons and release energy. Again, the density term in the Kramers law dominates, and

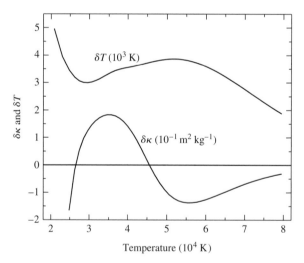

**FIGURE 14.13** Variations in the temperature and opacity throughout an RR Lyrae star model at the time of maximum compression. In the He II partial ionization zone ($T \approx 40{,}000$ K), $\delta\kappa > 0$ and $\delta T$ is reduced. These are the $\kappa$- and $\gamma$-mechanisms that drive the star's oscillations.

---

[5]As discussed in Section 10.4, this causes the specific heats $C_P$ and $C_V$ to have larger values in a partial ionization zone.

the opacity decreases with decreasing density during the expansion. This layer of the star can thus absorb heat during compression, be pushed outward to release the heat during expansion, and fall back down again to begin another cycle. Astronomers refer to this opacity mechanism as the **$\kappa$-mechanism**.

In a partial ionization zone, the $\kappa$-mechanism is reinforced by the tendency of heat to flow into the zone during compression simply because its temperature has increased less than the adjacent stellar layers. This effect is called the **$\gamma$-mechanism**, after the smaller ratio of specific heats caused by the increased values of $C_P$ and $C_V$. Partial ionization zones are the pistons that drive the oscillations of stars; they modulate the flow of energy through the layers of the star and are the direct cause of stellar pulsation.

### The Hydrogen and Helium Partial Ionization Zones

In most stars there are two main ionization zones. The first is a broad zone where both the ionization of neutral hydrogen (H I→H II) and the first ionization of helium (He I→He II) occur in layers with a characteristic temperature of 1 to 1.5 $\times 10^4$ K. These layers are collectively referred to as the **hydrogen partial ionization zone**. The second, deeper zone involves the second ionization of helium (He II→He III), which occurs at a characteristic temperature of $4 \times 10^4$ K and is called the **He II partial ionization zone**.

The location of these ionization zones within the star determines its pulsational properties. As shown in Fig. 14.14, if the star is too hot (7500 K), the ionization zones will be located very near the surface. At this position, the density is quite low, and there is not enough mass available to drive the oscillations effectively. This accounts for the hot **blue edge** of the instability strip on the H–R diagram. In a cooler star (6500 K), the characteristic

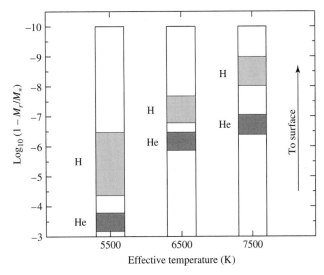

**FIGURE 14.14**   Hydrogen and helium ionization zones in stars of different temperatures. For each point in the star, the vertical axis displays the logarithm of the fraction of the star's mass that lies *above* that point.

temperatures of the ionization zones are found deeper in the star. There is more mass for the ionization zone "piston" to push around, and the first overtone mode may be excited. (Whether a mode is actually excited depends on whether the positive work generated within the ionization zones is sufficient to overcome the damping due to negative work of the other layers of the star.) In a still cooler star (5500 K), the ionization zones occur deep enough to drive the fundamental mode of pulsation. However, if a star's surface temperature is too low, the onset of efficient convection in its outer layers may dampen the oscillations. Because the transport of energy by convection is more effective when the star is compressed, the convecting stellar material may lose heat at minimum radius. This could overcome the damming up of heat by the ionization zones—and so quench the pulsation of the star. The cool **red edge** of the instability strip is the result of the damping effect of convection.[6]

Detailed numerical calculations of the pulsation of model stars produce an instability strip that is in good agreement with its observed location on the H–R diagram. These computations show that it is the He II partial ionization zone that is primarily responsible for driving the oscillations of stars within the instability strip. If the effect of the helium ionization zone is artificially removed, the model stars will not pulsate.

The hydrogen ionization zone plays a more subtle role. As a star pulsates, the hydrogen ionization zone moves toward or away from the surface as the zone expands and contracts in response to the changing temperature of the stellar gases. It happens that the star is brightest when the *least mass* lies between the hydrogen ionization zone and the surface. As a star oscillates, the location of an ionization zone changes with respect to both its radial position, $r$, and its mass interior to $r$, $M_r$. The luminosity incident on the *bottom* of the hydrogen ionization zone is indeed a maximum at minimum radius, but this merely propels the zone outward (through mass) most rapidly at that instant. The emergent luminosity is thus greatest *after* minimum radius, when the zone is nearest the surface. This delaying action of the hydrogen partial ionization zone produces the phase lag observed for classical Cepheids and RR Lyrae stars.

The mechanisms responsible for the pulsation of stars outside the instability strip are not always as well understood. The long-period variables are red supergiants (AGB stars) with huge, diffuse convective envelopes surrounding a compact core. Their spectra are dominated by molecular absorption lines and emission lines that reveal the existence of atmospheric shock waves and significant mass loss. While we understand that the *hydrogen* partial ionization zone drives the pulsation of a long-period variable star, many details remain to be explained, such as how its oscillations interact with its outer atmosphere.[7]

### $\beta$ Cephei Stars and the Iron Opacity "Bump"

The $\beta$ Cephei stars pose another interesting challenge. Being situated in the upper left-hand side of the H–R diagram, these stars are very hot and luminous. $\beta$ Cepheis are early B stars with effective temperatures in the range of 20,000 to 30,000 K and typically with luminosity classes of III, IV, and V. Given their high effective temperatures, hydrogen is completely

---

[6]Much work remains to be done on the effect of convection on stellar pulsation, although some results have been obtained for RR Lyrae and ZZ Ceti stars. Progress has been hampered by the present lack of a fundamental theory of time-dependent convection.

[7]In Section 16.2, we will see that the ZZ Ceti stars are also driven by the hydrogen partial ionization zone.

ionized, and the helium ionization zone is too near the surface to effectively drive pulsations in these stars. After years of investigation it was realized that the $\kappa$ and $\gamma$ mechanisms are still active in $\beta$ Cephei stars, but the element responsible for the driving is iron. Although the abundance of iron is low in all stars (recall, for example, Table 9.2), the large number of absorption lines in the spectrum of iron implies that iron contributes significantly to stellar opacities at temperatures near 100,000 K. This effect can be seen in the "iron bump" above 100,000 K in the plot of opacity vs. temperature shown in Fig. 9.10. The depth of this iron ionization region is sufficient to produce net positive pulsational driving in these stars.

## 14.3 ■ MODELING STELLAR PULSATION

In Section 10.5, the construction of a stellar model in hydrostatic equilibrium was described. The star was considered to be divided into a number of concentric mass shells. The differential equations of static stellar structure were then converted into difference equations and applied to each mass shell, and the system of equations was solved on a computer subject to certain boundary conditions at the center and surface of the stellar model.

### Nonlinear Hydrodynamic Models

Because a pulsating star is not in hydrostatic equilibrium, the stellar structure equations collected at the beginning of Section 10.5 cannot be used in their present form. Instead, a more general set of equations is employed that takes the oscillation of the mass shells into account. For example, Newton's second law (Eq. 10.5),

$$\rho \frac{d^2 r}{dt^2} = -G \frac{M_r \rho}{r^2} - \frac{dP}{dr}, \tag{14.7}$$

must be used instead of Eq. (10.6) for hydrostatic equilibrium. Once the differential equations describing the nonequilibrium mechanical and thermal behavior of a star have been assembled, along with the appropriate constitutive relations, they may be replaced by difference equations as described in Section 10.5 and solved numerically. In essence, the model star is mathematically displaced from its equilibrium configuration and then "released" to begin its oscillation. The mass shells expand and contract, pushing against each other as they move. If conditions are right, the ionization zones in the model star will drive the oscillations, and the pulsation amplitude will slowly increase; otherwise the amplitude will decay away. Computer programs that carry out these calculations have been quite successful at modeling the details observed in the light and radial velocity curves of Cepheid variables.

The main advantage of the preceding approach is that it is a **nonlinear** calculation, capable in principle of modeling the complexities of large pulsation amplitudes and reproducing the nonsinusoidal shape of actual light curves. One disadvantage lies in the computer resources required: This process requires a significant amount of CPU time and memory. Many (sometimes thousands of) oscillations must be calculated before the model settles down into a well-behaved periodic motion, and even more periods may be required for the model to reach its *limit cycle*, when the pulsation amplitude has reached its final value. In fact, in some cases the computer simulations of certain classes of pulsating stars may never attain a truly periodic solution but exhibit chaotic behavior instead, as observed in some real stars.

A second disadvantage of nonlinear calculations lies in the challenges involved in accurately converging models at each time step. Numerical instabilities in the nonlinear equations can cause calculations to misbehave and lead to unphysical solutions. This is particularly true when theories of time-dependent convection are required for red giants and supergiant stars.

## Linearizing the Hydrodynamic Equations

An alternative to the nonlinear approach is to **linearize** the differential equations by considering only small-amplitude oscillations. This is done by writing every variable in the differential equations as an equilibrium value (found in the static model of the star) plus a small change due to the pulsation. For example, the pressure $P$ would be written as $P = P_0 + \delta P$, where $P_0$ is the value of the pressure in a mass shell of the equilibrium model, and $\delta P$ is the small change in pressure that occurs as that mass shell moves in the oscillating model star. Thus $\delta P$ is a function of time, but $P_0$ is constant. When the variables written in this manner are inserted into the differential equations, the terms containing only equilibrium quantities cancel, and terms that involve powers of the deltas higher than the first, such as $(\delta P)^2$, may be discarded because they are negligibly small. The resulting linearized differential equations and their associated boundary conditions, also linearized, are similar to the equations for a wave on a string or in an organ pipe. Only certain standing waves with specific periods are permitted, and so the pulsation modes of the star are cleanly identified. The equations are still sufficiently complicated that a computer solution is required, but the time involved is much less than that required for a nonlinear calculation. The penalties for adopting the linearized approach are that the motion of the star is forced to be sinusoidal (as it must be for small amplitudes of oscillation), and the limiting value of the pulsation amplitude cannot be determined. Modeling the complexities of the full nonlinear behavior of the stellar model is thus sacrificed.

---

**Example 14.3.1.**   In this example, we consider an unrealistic, but very instructive, model of a pulsating star called a **one-zone model**; see Fig. 14.15. It consists of a central point mass equal to the entire mass of the star, $M$, surrounded by a single thin, spherical shell of mass $m$ and radius $R$ that represents the surface layer of the star. The interior of the shell is filled with a massless gas of pressure $P$ whose sole function is to support the shell against the gravitational pull of the central mass $M$. Newton's second law (Eq. 14.7) applied to this shell is

$$m \frac{d^2 R}{dt^2} = -\frac{GMm}{R^2} + 4\pi R^2 P. \tag{14.8}$$

For the equilibrium model, the left-hand side of this equation is zero, so

$$\frac{GMm}{R_0^2} = 4\pi R_0^2 P_0. \tag{14.9}$$

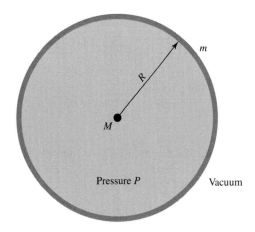

**FIGURE 14.15** One-zone model of a pulsating star.

The linearization is accomplished by writing the star's radius and pressure as

$$R = R_0 + \delta R \qquad \text{and} \qquad P = P_0 + \delta P$$

and inserting these expressions into Eq. (14.8), giving

$$m \frac{d^2(R_0 + \delta R)}{dt^2} = -\frac{GMm}{(R_0 + \delta R)^2} + 4\pi (R_0 + \delta R)^2 (P_0 + \delta P).$$

Using the first-order approximation

$$\frac{1}{(R_0 + \delta R)^2} \approx \frac{1}{R_0^2} \left(1 - 2\frac{\delta R}{R_0}\right)$$

and keeping only those terms involving the first powers of the deltas results in

$$m \frac{d^2(\delta R)}{dt^2} = -\frac{GMm}{R_0^2} + \frac{2GMm}{R_0^3} \delta R + 4\pi R_0^2 P_0 + 8\pi R_0 P_0 \delta R + 4\pi R_0^2 \delta P,$$

where $d^2 R_0 / dt^2 = 0$ has been used for the equilibrium model. The first and third terms on the right-hand side cancel (see Eq. 14.9), leaving

$$m \frac{d^2(\delta R)}{dt^2} = \frac{2GMm}{R_0^3} \delta R + 8\pi R_0 P_0 \delta R + 4\pi R_0^2 \delta P. \tag{14.10}$$

This is the linearized version of Newton's second law for our one-zone model.

To reduce the two variables $\delta R$ and $\delta P$ to one, we now assume that the oscillations are *adiabatic*. In this case, the pressure and volume of the model are related by the adiabatic

*continued*

relation $PV^\gamma = $ constant, where $\gamma$ is the ratio of specific heats of the gas. Since the volume of the one-zone model is just $\frac{4}{3}\pi R^3$, the adiabatic relation says that $PR^{3\gamma} = $ constant. It is left as a problem to show that the linearized version of this expression is

$$\frac{\delta P}{P_0} = -3\gamma \frac{\delta R}{R_0}. \tag{14.11}$$

Using this equation, $\delta P$ can be eliminated from Eq. (14.10). In addition, $8\pi R_0 P_0$ can be replaced by $2GMm/R_0^3$ through the use of Eq. (14.9). As a result, the mass $m$ of the shell cancels, leaving the linearized equation for $\delta R$:

$$\frac{d^2(\delta R)}{dt^2} = -(3\gamma - 4)\frac{GM}{R_0^3}\delta R. \tag{14.12}$$

If $\gamma > 4/3$ (so the right-hand side of the equation is negative), this is just the familiar equation for simple harmonic motion. It has the solution $\delta R = A\sin(\omega t)$, where $A$ is the pulsation amplitude and $\omega$ is the angular pulsation frequency. Inserting this expression for $\delta R$ into Eq. (14.12) results in

$$\omega^2 = (3\gamma - 4)\frac{GM}{R_0^3}. \tag{14.13}$$

Finally, the pulsation period of the one-zone model is just $\Pi = 2\pi/\omega$, or

$$\Pi = \frac{2\pi}{\sqrt{\frac{4}{3}\pi G\rho_0(3\gamma - 4)}}, \tag{14.14}$$

where $\rho_0 = M/\frac{4}{3}\pi R_0^3$ is the average density of the equilibrium model. For an ideal monatomic gas (appropriate for hot stellar gases), $\gamma = 5/3$. Except for factors of order unity, this is the same as our earlier period estimate (Eq. 14.6) obtained by considering the time required for a sound wave to cross the diameter of a star.

---

In Example 14.3.1, the approximations that the pulsation of the one-zone model was *linear* and *adiabatic* were used to simplify the calculation. Note that the pulsation amplitude, $A$, canceled in this example. The inability to calculate the amplitude of the oscillations is an inherent drawback of the linearized approach to pulsation.

### Nonlinear and Nonadiabatic Calculations

Because no heat is allowed to enter or leave the layers of a stellar model in an adiabatic analysis, the amplitude (whatever it may be) of the oscillation remains constant. However, astronomers need to know which modes will grow and which will decay away. This calculation must include the physics involved in Eddington's valve mechanism. The equations describing the transfer of heat and radiation through the stellar layers (similar to those discussed in Section 10.4) must be incorporated in such a *nonadiabatic* computation. These nonadiabatic expressions may also be linearized and solved to obtain the periods and growth rates of the individual modes. However, a more sophisticated and costly *nonlinear*, *non*adiabatic calculation is needed to reproduce the complicated light and radial velocity curves that are observed for some variable stars. The computer problem at the end of this chapter asks you to carry out a nonlinear (but still adiabatic) calculation of the pulsation of this one-zone model.

### Dynamical Stability

Equation (14.12) provides a very important insight into the **dynamical stability** of a star. If $\gamma < 4/3$, then the right-hand side of Eq. (14.12) is positive. The solution is now $\delta R = A e^{-\kappa t}$, where $\kappa^2$ is the same as $\omega^2$ in Eq. (14.13). Instead of pulsating, the star *collapses* if $\gamma < 4/3$. The increase in gas pressure is not enough to overcome the inward pull of gravity and push the mass shell back out again, resulting in a *dynamically unstable* model. This instability caused by a reduction in the value of $\gamma$ will be seen again in Section 16.4, where the effect of relativity on white dwarf stars is described.

For the case of *nonadiabatic* oscillations, the time dependence of the pulsation is usually taken to be the real part of $e^{i\sigma t}$, where $\sigma$ is the complex frequency $\sigma = \omega + i\kappa$. In this expression, $\omega$ is the usual pulsation frequency, while $\kappa$ is a *stability coefficient*. The pulsation amplitude is then proportional to $e^{-\kappa t}$, and $1/\kappa$ is the characteristic time for the growth or decay of the oscillations.

## 14.4 ■ NONRADIAL STELLAR PULSATION

As some types of stars pulsate, their surfaces do not move uniformly in and out in a simple "breathing" motion. Instead, such a star executes a more complicated type of **nonradial** motion in which some regions of its surface expand while other areas contract.

### Nonradial Oscillations and Spherical Harmonic Functions

Figure 14.16 shows the angular patterns for several nonradial modes. If the stellar surface is moving outward within the lighter regions, then it is moving inward within the shaded areas. Scalar quantities such as the change in pressure ($\delta P$) follow the same pattern, having positive values in some areas and negative values in others. Formally, these patterns are described by the real parts of the spherical harmonic functions, $Y_\ell^m(\theta, \phi)$, where $\ell$ is a non-negative integer and $m$ is equal to any of the $2\ell + 1$ integers between $-\ell$ and $+\ell$.[8] There are $\ell$ *nodal circles* (where $\delta r = 0$), with $|m|$ of these circles passing through the poles of the star and the remaining $\ell - |m|$ nodal circles being parallel to the star's equator. If $\ell = m = 0$, then the pulsation is purely radial.

A few examples of $Y_\ell^m(\theta, \phi)$ functions are

$$Y_0^0(\theta, \phi) = K_0^0$$

$$Y_1^0(\theta, \phi) = K_1^0 \cos\theta$$

$$Y_1^{\pm 1}(\theta, \phi) = K_1^{\pm 1} \sin\theta \, e^{\pm i\phi}$$

[8] Spherical harmonics are often encountered in physics when spherical symmetry is employed. A common example in the undergraduate physics curriculum is the use of spherical harmonics to describe the quantum mechanical wave functions of a hydrogen atom. Recall the discussion in Section 5.4 and the orbitals depicted in Fig. 5.12.

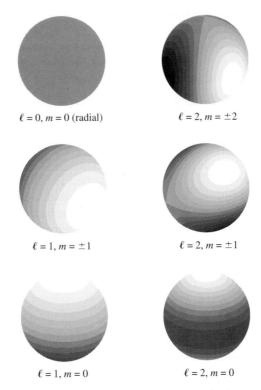

$\ell = 0, m = 0$ (radial)            $\ell = 2, m = \pm 2$

$\ell = 1, m = \pm 1$            $\ell = 2, m = \pm 1$

$\ell = 1, m = 0$            $\ell = 2, m = 0$

**FIGURE 14.16**    Nonradial pulsation patterns. The modes of pulsation are represented by the real parts of the spherical harmonic functions, $Y_\ell^m(\theta, \phi)$.

$$Y_2^0(\theta, \phi) = K_2^0(3\cos^2\theta - 1)$$
$$Y_2^{\pm 1}(\theta, \phi) = K_2^{\pm 1}\sin\theta\cos\theta e^{\pm i\phi}$$
$$Y_2^{\pm 2}(\theta, \phi) = K_2^{\pm 2}\cos^2\theta e^{\pm 2i\phi}$$

where the $K_\ell^m$s are "normalization" constants and $i$ is the imaginary number $i \equiv \sqrt{-1}$. Recall from Euler's formula that $e^{\pm mi\phi} = \cos(m\phi) \pm i\sin(m\phi)$. Thus, the real part of $e^{\pm mi\phi}$ is just $\cos(m\phi)$.

The patterns for nonzero $m$ represent *traveling waves* that move across the star parallel to its equator. (Imagine these patterns on a beach ball, with the ball slowly spinning about the vertical axis.) The time required for the waves to travel around the star is $|m|$ times the star's pulsation period. However, it is important to note that the star itself may not be rotating at all. Just as water waves may travel across the surface of a lake without the water itself making the trip, these traveling waves are disturbances that pass through the stellar gases.[9]

---

[9]Observations of nonradially pulsating stars are considered in Problem 14.10.

### The p and f Modes

In Section 14.2, the radial pulsation of stars was attributed to standing sound waves in the stellar interior. For the case of nonradial oscillations, the sound waves can propagate horizontally as well as radially to produce waves that travel around the star. Because *pressure* provides the restoring force for sound waves, these nonradial oscillations are called **p-modes**. A complete description of a p-mode requires specification of its radial and angular nodes. For example, a $p_2$ mode may be thought of as the nonradial analog of a radial second overtone mode. The $p_2$ mode with $\ell = 4$ and $m = -3$ has two radial nodes between the center and the surface, and its angular pattern has four nodal lines, three through the poles and one parallel to the equator. Figure 14.17 shows two p-modes for a $12 \, M_\odot$ main-sequence star model; you may note the similarities between this figure and Fig. 14.10, with most of the motion occurring near the stellar surface. Also shown is the **f-mode**, which can be thought of as a surface gravity wave (note the rapid rise in amplitude with radius). The frequency of the f-mode is intermediate between the p-modes and the g-modes (discussed later). There is no radial analog for the f-mode.

### The Acoustic Frequency

An estimate of the angular frequency of a p-mode may be obtained from the time for a sound wave to travel one horizontal wavelength, from one angular nodal line to the next. This horizontal wavelength is given by the expression

$$\lambda_h = \frac{2\pi r}{\sqrt{\ell(\ell + 1)}}, \tag{14.15}$$

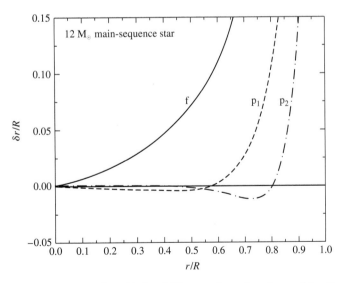

**FIGURE 14.17** Nonradial p-modes with $\ell = 2$. The waveforms have been arbitrarily scaled so that $\delta r / R = 1$ at the star's surface. The f-mode is also shown.

where $r$ is the radial distance from the center of the star. The **acoustic frequency** at this depth in the star is then defined as

$$S_\ell = \frac{2\pi}{\text{time for sound to travel } \lambda_h},$$

which can be written as

$$S_\ell = 2\pi \left[ \frac{v_s}{2\pi r / \sqrt{\ell(\ell+1)}} \right]$$

$$= \sqrt{\frac{\gamma P}{\rho}} \frac{\sqrt{\ell(\ell+1)}}{r}, \tag{14.16}$$

where $v_s$ is the adiabatic sound speed given by Eq. (10.84). Because the speed of sound is proportional to the square root of the temperature [recall from the ideal gas law, Eq. (10.11), that $P/\rho \propto T$], the acoustic frequency is large in the deep interior of the star and decreases with increasing $r$. The frequency of a p-mode is determined by the average value of $S_\ell$, with the largest contributions to the average coming from the regions of the star where the oscillations are most energetic.

In the absence of rotation, the pulsation period depends only on the number of radial nodes and the integer $\ell$. The period is independent of $m$ because with no rotation there are no well-defined poles or equator; thus $m$ has no physical significance. On the other hand, if the star is rotating, the rotation itself defines the poles and equator, and the pulsation frequencies for modes with different values of $m$ become separated or *split* as the traveling waves move either with or against the rotation (the sign of $m$ determines the direction in which the waves move around the star). The amount by which the pulsation frequencies are split depends on the angular rotation frequency, $\Omega$, of the star, with the rotationally produced shift in frequency proportional to the product $m\Omega$ for the simple case of uniform rotation. As we will discuss later, this frequency splitting provides a powerful probe for measuring the rotation of the Sun's interior.

### The g Modes

Just as pressure supplies the restoring force for the compression and expansion of the p-mode sound waves, *gravity* is the source of the restoring force for another class of nonradial oscillations called **g-modes**. The g-modes are produced by *internal gravity waves*. These waves involve a "sloshing" back and forth of the stellar gases, which is ultimately connected to the *buoyancy* of stellar material. Because "sloshing" cannot occur for purely radial motion, there are no radial analogs for the g-modes.

### The Brunt–Väisälä (Buoyancy) Frequency

To gain a better understanding of this oscillatory motion for g-modes, consider a small bubble of stellar material that is displaced upward from its equilibrium position in the star by an amount $dr$, as shown in Fig. 10.10.[10] We will assume that this motion occurs

1. slowly enough that the pressure within the bubble, $P^{(b)}$, is always equal to the pressure of its surroundings, $P^{(s)}$; and
2. rapidly enough that there is no heat exchanged between the bubble and its surroundings.

The second assumption means that the expansion and compression of the gas bubble are *adiabatic*. If the density of the displaced bubble is greater than the density of its new surroundings, the bubble will fall back to its original position. The net restoring force *per unit volume* on the bubble in its final position is the difference between the upward buoyant force (given by Archimedes's law) and the downward gravitational force:

$$f_{\text{net}} = \left( \rho_f^{(s)} - \rho_f^{(b)} \right) g,$$

where $g = GM_r/r^2$ is the local value of the gravitational acceleration. Using a Taylor expansion for the densities about their initial positions results in

$$f_{\text{net}} = \left[ \left( \rho_i^{(s)} + \frac{d\rho^{(s)}}{dr} dr \right) - \left( \rho_i^{(b)} + \frac{d\rho^{(b)}}{dr} dr \right) \right] g.$$

The initial densities of the bubble and its surroundings are the same, so these terms cancel, leaving

$$f_{\text{net}} = \left( \frac{d\rho^{(s)}}{dr} - \frac{d\rho^{(b)}}{dr} \right) g\, dr.$$

Because the motion of the bubble is adiabatic, Eq. (10.87) can be used to replace $d\rho^{(b)}/dr$:

$$f_{\text{net}} = \left( \frac{d\rho^{(s)}}{dr} - \frac{\rho_i^{(b)}}{\gamma P_i^{(b)}} \frac{dP^{(b)}}{dr} \right) g\, dr.$$

Looking at this equation, all of the "$b$" superscripts may be changed to "$s$" because the initial densities are equal, and according to the first assumption given, the pressures inside and outside the bubble are *always* the same. Thus all quantities in this equation refer to the stellar material surrounding the bubble. With that understanding, the subscripts may be dropped completely, resulting in

$$f_{\text{net}} = \left( \frac{1}{\rho} \frac{d\rho}{dr} - \frac{1}{\gamma P} \frac{dP}{dr} \right) \rho g\, dr.$$

For convenience, the term in parentheses is defined as

$$A \equiv \frac{1}{\rho} \frac{d\rho}{dr} - \frac{1}{\gamma P} \frac{dP}{dr}. \tag{14.17}$$

---

[10]The following discussion is just a reexamination of the problem of convection, last seen in Section 10.4, from another perspective.

Thus the net force per unit volume acting on the bubble is

$$f_{net} = \rho A g \, dr. \tag{14.18}$$

If $A > 0$, the net force on the displaced bubble has the same sign as $dr$, and so the bubble will continue to move away from its equilibrium position. This is the condition necessary for *convection* to occur, and it is equivalent to the other requirements previously found for convective instability, such as Eq. (10.94). However, if $A < 0$, then the net force on the bubble will be in a direction opposite to the displacement, and so the bubble will be pushed back toward its equilibrium position. In this case, Eq. (14.18) has the form of Hooke's law, with the restoring force proportional to the displacement. Thus if $A < 0$, the bubble will oscillate about its equilibrium position with simple harmonic motion.

Dividing the force per unit volume, $f_{net}$, by the mass per unit volume, $\rho$, gives the force per unit mass, or acceleration: $a = f_{net}/\rho = A g \, dr$. Because the acceleration is simply related to the displacement for simple harmonic motion,[11] we have

$$a = -N^2 \, dr = A g \, dr,$$

where $N$ is the angular frequency of the bubble about its equilibrium position, called the **Brunt–Väisälä frequency** or the **buoyancy frequency**,

$$N = \sqrt{-Ag} = \sqrt{\left( \frac{1}{\gamma P} \frac{dP}{dr} - \frac{1}{\rho} \frac{d\rho}{dr} \right) g}. \tag{14.19}$$

The buoyancy frequency is zero at the center of the star (where $g = 0$) and at the edges of convection zones (where $A = 0$). Recall that $A < 0$ where there is no convection, so $N$ is larger in regions that are more stable against convection. Inside a convection zone, where $A > 0$, the buoyancy frequency is not defined.

### The g and p Modes as Probes of Stellar Structure

The "sloshing" effect of neighboring regions of the star produces the internal gravity waves that are responsible for the g-modes of a nonradially pulsating star. The frequency of a g-mode is determined by the value of $N$ averaged across the star. Figure 14.18 shows several g-modes for the same stellar model that was used for Fig. 14.17. A comparison of these two figures reveals significant differences between these classes of modes, making them very useful to astronomers attempting to study the interior of the Sun and other stars. Most important, notice the difference in the vertical scales of the two figures. The g-modes involve significant movement of the stellar material deep within the star, while the p-mode's motions are confined near the stellar surface. Thus g-modes provide a view into the very heart of a star, while p-modes allow a diagnosis of the conditions in its surface layers.

---

[11]For example, recall that $F = ma = -kx$ for a spring. The acceleration is $a = -\omega^2 x$, where $\omega = \sqrt{k/m}$ is the angular frequency of the spring's motion.

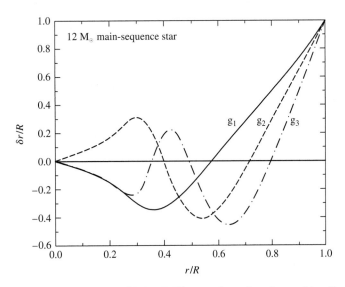

**FIGURE 14.18** Nonradial g-modes with $\ell = 2$. The waveforms have been arbitrarily scaled so that $\delta r/R = 1$ at the star's surface.

## 14.5 ■ HELIOSEISMOLOGY AND ASTEROSEISMOLOGY

All of the ideas of nonradial pulsation come into play in the science of **helioseismology**, the study of the oscillations of the Sun first observed in 1962 by American astronomers Robert Leighton (1919–1997), Robert Noyes, and George Simon. A typical solar oscillation mode has a very low amplitude, with a surface velocity of only 0.10 m s$^{-1}$ or less,[12] and a luminosity variation $\delta L/L_\odot$ of only $10^{-6}$. With an incoherent superposition of roughly *ten million* modes rippling through its surface and interior, our star is "ringing" like a bell.

### The Five-Minute Solar Oscillations

The oscillations observed on the Sun have modes with periods between three and eight minutes and very short horizontal wavelengths ($\ell$ ranging from 0 to 1000 or more). These so-called *five-minute oscillations* have been identified as p-modes. The five-minute p-modes are concentrated below the photosphere within the Sun's convection zone; Fig. 14.19 shows a typical p-mode. g-modes are located deep in the solar interior, below the convection zone. By studying these p-mode oscillations, astronomers have been able to gain new insights into the structure of the Sun in these regions.[13]

---

[12]These incredibly precise velocity measurements are made by carefully observing the Doppler shifts of spectral absorption lines such as Fe I (557.6099 nm) through a narrow slit that follows the rotating solar surface.

[13]A series of 160-minute "g-modes" were believed to have been observed as well. However, continuous observations over 690 days using the GOLF instrument onboard the SOHO spacecraft were unable to detect any evidence of the controversial mode. It is believed that ground-based observations that indicated a 160-minute mode were due to harmonic effects associated with Earth's atmosphere; note that 160 minutes is exactly 1/9 of the 24-hour solar day.

**FIGURE 14.19**    Five-minute $p_{15}$ mode with $\ell = 20$ and $m = 16$. The solar convection zone is the stippled region, where the p-modes are found. (Courtesy of National Optical Astronomy Observatories.)

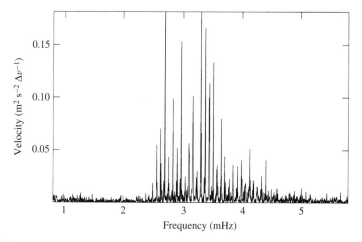

**FIGURE 14.20**    Relative power of solar p-modes; a period of five minutes corresponds to a frequency of 3.33 mHz. (Figure adapted from Grec, Fossat, and Pomerantz, *Nature*, *288*, 541, 1980.)

Figure 14.20 shows the relative power contained in the solar p-modes. This information can also be plotted in another manner, as shown in Fig. 14.21, with $\ell$ on the horizontal axis and the pulsation frequency on the vertical axis. Circles show the observed frequencies, and each continuous ridge corresponds to a different p-mode ($p_1$, $p_2$, $p_3$, etc.). The superimposed lines are the *theoretical* frequencies calculated for a solar model. All of the observed five-minute modes have been identified in this way. The fit is certainly impressive but not quite exact. A solar model must be carefully tuned to obtain the best agreement between the theoretical and observed p-mode frequencies. This procedure can reveal much about the depth of the solar convection zone and about the rotation and composition of the outer layers of the Sun.

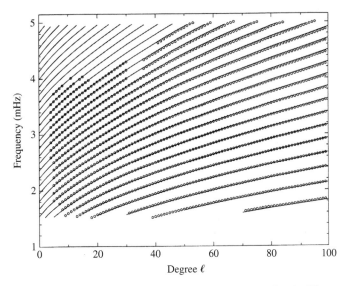

**FIGURE 14.21** Solar p-modes: observations (circles) and theory (lines). (Figure adapted from Libbrecht, *Space Sci. Rev.*, *47*, 275, 1988.)

### Differential Rotation and the Solar Convection Zone

As mentioned in Chapter 11, based on helioseismology studies, combined with detailed stellar evolution calculations, the base of the solar convection zone is known to be located at 0.714 $R_\odot$, with a temperature of about $2.18 \times 10^6$ K. The rotational splitting observed for p-mode frequencies indicates that differential rotation observed at the Sun's surface decreases slightly down through the convection zone (recall Fig. 11.16). Those p-modes with shorter horizontal wavelengths (larger $\ell$) penetrate less deeply into the convection zone, so the difference in rotational frequency splitting with $\ell$ reveals the depth dependence of the rotation. The measurement of the variation in rotation with the distance from the solar equator comes from the dependence of the rotational frequency splitting on $m$. Below the convection zone, the equatorial and polar rotation rates converge to a single value at $r/R_\odot \approx 0.65$. Because a change in the rotation rate with depth is needed to convert the Sun's magnetic field from a poloidal to a toroidal geometry (as discussed in Section 11.3), these results indicate that the Sun's magnetic dynamo is probably seated in the tachocline at the interface between the radiation zone and the convection zone.

### Tests of Composition

The abundance of helium in the outer layers of the Sun can also be inferred from a comparison of the observed and theoretical p-mode ridges in Fig. 14.21. The results are consistent with a value of $Y = 0.2437$ for the mass fraction of helium at the Sun's surface.

### Probing the Deep Interior

Astronomers have experienced more difficulty in their attempts to use the solar g-modes as a probe of the Sun's interior. Because the g-modes dwell beneath the convection zone, their amplitudes are significantly diminished at the Sun's surface. To date, no definite identification of g-modes has been made. Nevertheless, the potential rewards of using these oscillations to learn more about the core of the Sun compel astronomers to apply their observational ingenuity to these g-modes.

### Driving Solar Oscillations

The question of the mechanism responsible for driving the solar oscillations has not yet been conclusively answered. Our main-sequence Sun is not a normal pulsating star. It lies far beyond the red edge of the instability strip on the H–R diagram (see Fig. 14.8) where turbulent convection overcomes the tendency of the ionization zones to absorb heat at maximum compression. Eddington's valve mechanism thus cannot be responsible for the solar oscillations. However, the timescale for convection near the top of the convection zone is a few minutes, and it is strongly suspected that the p-modes are driven by tapping into the turbulent energy of the convection zone itself, where the p-modes are confined.

### $\delta$ Scuti Stars and Rapidly Oscillating Ap Stars

The techniques of helioseismology can be applied to other stars as well. **Asteroseismology** is the study of the pulsation modes of stars in order to investigate their internal structures, chemical composition, rotation, and magnetic fields.

$\delta$ Scuti stars are Population I main-sequence stars and giant stars in the spectral class range A to F. They tend to pulsate in low-overtone radial modes, as well as in low-order p-modes (and possibly g-modes). The amplitudes of $\delta$ Scutis are fairly small, ranging from a few mmag to roughly 0.8 mag. Population II subgiants also exhibit radial and nonradial oscillations and are known as SX Phoenicis stars.

Another interesting class of pulsating stars are the **rapidly oscillating Ap stars** (roAp), found in the same portion of the H–R diagram as the $\delta$ Scuti stars. These stars have peculiar surface chemical compositions (hence the "p" designation), are rotating, and have strong magnetic fields. The unusual chemical composition is likely due to settling of heavier elements, similar to the elemental diffusion that has occurred near the surface of the Sun. Some elements may also have been elevated in the atmosphere if they have a significant number of absorption lines near the peak of the star's blackbody spectrum. These atoms preferentially absorb photons that impart a net upward momentum. If the atmosphere is sufficiently stable against turbulent motions, some of these atoms will tend to drift upward.

roAp stars have very small pulsation amplitudes of less than 0.016 mag. It appears that they primarily pulsate in higher-order p-modes and that the axis for the pulsation is aligned with the magnetic field axis, which is tilted somewhat to the rotation axis (an oblique rotator model). roAp stars are among the most well-studied of main-sequence stars other than the Sun, but the pulsation driving mechanism still remains in question.

### SUGGESTED READING

### General

*The American Association of Variable Star Observers*, http://www.aavso.org/.

Giovanelli, Ronald, *Secrets of the Sun*, Cambridge University Press, Cambridge, 1984.

Kaler, James B., *Stars and Their Spectra*, Cambridge University Press, Cambridge, 1997.

Leibacher, John W., et al., "Helioseismology," *Scientific American*, September 1985.

Zirker, Jack B., *Sunquakes: Probing the Interior of the Sun*, Johns Hopkins University Press, Baltimore, 2003.

## Technical

Aller, Lawrence H., *Atoms, Stars, and Nebulae*, Third Edition, Cambridge University Press, Cambridge, 1991.

Brown, Timothy M., et al., "Inferring the Sun's Internal Angular Velocity from Observed p-Mode Frequency Splittings," *The Astrophysical Journal, 343*, 526, 1989.

Clayton, Donald D., *Principles of Stellar Evolution and Nucleosynthesis*, University of Chicago Press, Chicago, 1983.

Cox, John P., *The Theory of Stellar Pulsation*, Princeton University Press, Princeton, NJ, 1980.

Freedman, Wendy L., et al., "Distance to the Virgo Cluster Galaxy M100 from Hubble Space Telescope Observations of Cepheids," *Nature, 371*, 757, 1994.

*General Catalogue of Variable Stars*, Sternberg Astronomical Institute, Moscow, Russia, `http://www.sai.msu.su/groups/cluster/gcvs/gcvs/`.

Hansen, Carl J., Kawaler, Steven D., and Trimble, Virginia *Stellar Interiors: Physical Principles, Structure, and Evolution*, Second Edition, Springer-Verlag, New York, 2004.

Perrson, S. E., et al., "New Cepheid Period–Luminosity Relations for the Large Magellanic Cloud: 92 Near-Infrared Light Curves," *The Astronomical Journal, 128*, 2239, 2004.

Svestka, Zdenek, and Harvey, John W. (eds.), *Helioseismic Diagnostics of Solar Convection and Activity*, Kluwer Academic Publishers, Dordrecht, 2000.

## PROBLEMS

**14.1** Use the light curve for Mira, Fig. 14.1, to estimate the ratio of Mira's luminosity at visible wavelengths, when it is brightest to when it is dimmest. For what fraction of its pulsation cycle is Mira visible to the naked eye?

**14.2** If the intrinsic uncertainty in the period–luminosity relation shown in Fig. 14.5 is $\Delta M \approx 0.5$ magnitude, find the resulting fractional uncertainty in the calculated distance to a classical Cepheid.

**14.3** Several remote classical Cepheids were discovered in 1994 by the Hubble Space Telescope in the galaxy denoted M100. (M100 is a member of the Virgo cluster, a rich cluster of galaxies.) Figure 14.22 shows the period–luminosity relation for these Cepheids. Use the two Cepheids nearest the figure's best-fit line to estimate the distance to M100. The mean visual extinction is $A_V = 0.15 \pm 0.17$ magnitudes for the M100 Cepheids. Compare your result to the distance of $17.1 \pm 1.8$ Mpc obtained by Wendy Freedman and her colleagues. You are referred to Freedman et al. (1994) for more information on the discovery and importance of these remote pulsating stars.

**14.4** Make a graph similar to Fig. 14.5 showing the period–luminosity relation for both the classical Cepheids and W Virginis stars.

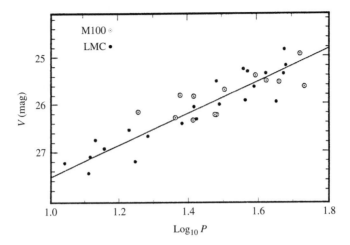

**FIGURE 14.22**    A composite period–luminosity relation for Problem 14.3. The white circles denote Cepheids in M100, and the black circles show nearby Cepheids found in the Large Magellanic Cloud (a small galaxy that neighbors our Milky Way Galaxy). The average visual magnitudes of the LMC Cepheids have been increased by the same amount to match those of the M100 variables. The required increase in $V$ for a best fit is then used to find the relative distances to the LMC and M100. (Adapted from Freedman et al., *Nature*, *371*, 757, 1994.)

**14.5** Assuming (incorrectly) that the oscillations of $\delta$ Cephei are sinusoidal, calculate the greatest excursion of its surface from its equilibrium position.

**14.6** Use Eq. (14.6) to estimate the pulsation period that the Sun would have if it were to oscillate radially.

**14.7** Derive Eq. (14.11) by linearizing the adiabatic relation

$$PV^{\gamma} = \text{constant.}$$

**14.8** **(a)** Linearize the Stefan–Boltzmann equation in the form of Eq. (3.17) to show that

$$\frac{\delta L}{L_0} = 2\,\frac{\delta R}{R_0} + 4\,\frac{\delta T}{T_0}.$$

  **(b)** Linearize the adiabatic relation $TV^{\gamma-1} = \text{constant}$, and so find a relation between $\delta L/L_0$ and $\delta R/R_0$ for a spherical blackbody model star composed of an ideal monatomic gas.

**14.9** Consider a general potential energy function, $U(r)$, for a force $\mathbf{F} = -(dU/dr)\hat{\mathbf{r}}$ on a particle of mass $m$. Assume that the origin ($r = 0$) is a point of stable equilibrium. By expanding $U(r)$ in a Taylor series about the origin, show that if a particle is displaced slightly from the origin and then released, it will undergo simple harmonic motion about the origin. This explains why the linearization procedure of Section 14.3 is guaranteed to result in sinusoidal oscillations.

**14.10** Figure 14.23 shows a view of a hypothetical nonradially pulsating ($\ell = 2, m = -2$), rotating star from above the star's north pole. From the vantage point of Earth, astronomers view the star along its equatorial plane. Assuming that a spectral absorption line appears as in Fig. 9.18 when the bottom of Fig. 14.23 is facing Earth, sketch the changes in the appearance of the

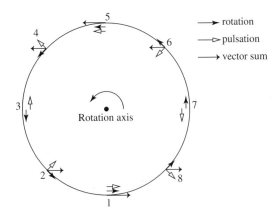

**FIGURE 14.23**  Surface velocities for a rotating, pulsating star ($\ell = 2, m = -2$) for Problem 14.10. The arrows indicate the surface velocities due to rotation alone, pulsation alone, and their vector sum.

line profile due to Doppler shifts caused by the total surface velocity as the star rotates. (Don't worry about the timing; just sketch the spectral line as seen from the eight different points of view shown that are directly over the star's equator.) Assume that the equivalent width of the line does not change. You may wish to compare your line profiles with those actually observed for a nonradially pulsating star such as the $\beta$ Cephei star 12 Lacertae; see Smith, *Ap. J.*, *240*, 149, 1980. For convenience, the magnitudes of the rotation and pulsation velocities are assumed to be equal.

**14.11**  Show that Eq. (10.94), the condition for convection to occur, is the same as the requirement that $A > 0$, where $A$ is given by Eq. (14.17). Assume that the mean molecular weight, $\mu$, does not vary.

**14.12**  In a convection zone, the timescale for convection (see Section 10.4) is related to the value of $A$ (Eq. 14.17) by

$$t_c \simeq 2\sqrt{2/Ag}.$$

Table 14.2 shows the values of the pressure and density at two points near the top of the Sun's convection zone as described by a solar model. Use these values and $\gamma = 5/3$ to obtain an estimate of the timescale for convection near the top of the Sun's convection zone. How does your answer compare with the range of periods observed for the Sun's p-modes?

**TABLE 14.2**  Data from a Solar Model for Problem 14.12. (Data from Joyce Guzik, private communication.)

| $r$ (m) | $P$ (N m$^{-2}$) | $\rho$ (kg m$^{-3}$) |
|---|---|---|
| $6.959318 \times 10^8$ | 9286.0 | $2.2291 \times 10^{-4}$ |
| $6.959366 \times 10^8$ | 8995.7 | $2.1925 \times 10^{-4}$ |

## COMPUTER PROBLEM

**14.13** In this problem you will carry out a nonlinear calculation of the radial pulsation of the one-zone model described in Example 14.3.1. The equations that describe the oscillation of this model star are Newton's second law for the forces on the shell,

$$m \frac{dv}{dt} = -\frac{GMm}{R^2} + 4\pi R^2 P, \qquad (14.20)$$

and the definition of the velocity, $v$, of the mass shell,

$$v = \frac{dR}{dt}. \qquad (14.21)$$

As in Example 14.3.1, we assume that the expansion and contraction of the gas are adiabatic:

$$P_i V_i^\gamma = P_f V_f^\gamma, \qquad (14.22)$$

where the "initial" and "final" subscripts refer to any two instants during the pulsation cycle.

(a) Explain in words the meaning of each term in Eq. (14.20).

(b) Use Eq. (14.22) to show that

$$P_i R_i^{3\gamma} = P_f R_f^{3\gamma}. \qquad (14.23)$$

(c) You will not be taking derivatives. Instead, you will take the difference between the initial and final values of the radius $R$ and radial velocity $v$ of the shell divided by the time interval $\Delta t$ separating the initial and final values. That is, you will use $(v_f - v_i)/\Delta t$ instead of $dv/dt$, and $(R_f - R_i)/\Delta t$ instead of $dR/dt$ in Eqs. (14.20) and (14.21). A careful analysis shows that you should use $R = R_i$ and $P = P_i$ on the right-hand side of Eq. (14.20), and use $v = v_f$ on the left-hand side of Eq. (14.21). Make these substitutions in Eqs. (14.20) and (14.21), and show that you can write

$$v_f = v_i + \left( \frac{4\pi R_i^2 P_i}{m} - \frac{GM}{R_i^2} \right) \Delta t \qquad (14.24)$$

and

$$R_f = R_i + v_f \Delta t. \qquad (14.25)$$

(d) Now you are ready to calculate the oscillation of the model star. The mass of a typical classical Cepheid is $M = 1 \times 10^{31}$ kg (5 $M_\odot$), and the mass of the surface layers may be arbitrarily assigned $m = 1 \times 10^{26}$ kg. For starting values at time $t = 0$, take

$$R_i = 1.7 \times 10^{10} \text{ m}$$

$$v_i = 0 \text{ m s}^{-1}$$

$$P_i = 5.6 \times 10^4 \text{ N m}^{-2}$$

and use a time interval of $\Delta t = 10^4$ s. Take the ratio of specific heats to be $\gamma = 5/3$ for an ideal monatomic gas. Use Eq. (14.24) to calculate the final velocity $v_f$ at the end of one time interval (at time $t = 1 \times 10^4$ s); then use Eq. (14.25) to calculate the final radius $R_f$ and Eq. (14.23) to calculate the final pressure $P_f$. Now take these final values to be your new initial values, and find new values for $R$, $v$, and $P$ after two time intervals (at time $t = 2 \times 10^4$ s). Continue to find $R$, $v$, and $P$ for 150 time intervals, until $t = 1.5 \times 10^6$ s. Make three graphs of your results: $R$ vs. $t$, $v$ vs. $t$, and $P$ vs. $t$. Plot the time on the horizontal axis.

(e) From your graphs, measure the period $\Pi$ of the oscillation (both in seconds and in days) and the equilibrium radius, $R_0$, of the model star. Compare this value of the period with that obtained from Eq. (14.14). Also compare your results with the period and radial velocity observed for $\delta$ Cephei.

# CHAPTER

# 15

# The Fate of Massive Stars

## 15.1 ■ POST-MAIN-SEQUENCE EVOLUTION OF MASSIVE STARS

Astronomers have been observing the fascinating Southern Hemisphere star $\eta$ Carinae ($\alpha = 10^h\ 45^m\ 03.59^s$, $\delta = -59°\ 41'\ 04.26''$) since at least 1600. Between 1600 and about 1830, observers had reported the star as being about second magnitude, although it was sometimes reported to be a fourth-magnitude star. Then, in 1820 or 1830, it may have started becoming more active. However, in 1837 $\eta$ Car suddenly brightened significantly, fluctuating between zero and first magnitude for about twenty years. At one point, its magnitude reached about $-1$, making it the second brightest extra-Solar-System object in the sky (only Sirius was brighter). During this period John Herschel (1792–1871) described $\eta$ Carinae as being "fitfully variable."

$\eta$ Car's remarkable brightness achieved in 1837 is all the more impressive given that the star is approximately 2300 pc from Earth (by comparison, Sirius is only 2.64 pc from Earth). After 1856, this mysterious star began to fade again, dropping to about eighth magnitude by 1870. With the exception of a lesser brightening event that took place between 1887 and 1895, $\eta$ Car has been relatively quiet since the "Great Eruption" between 1837 and 1856. Over the past century and a half, $\eta$ Car has brightened slightly, and it is currently at a visual magnitude of about six.

Another star in the Milky Way Galaxy has behaved similarly. P Cygni was apparently too faint to be seen by the naked eye prior to 1600 but then suddenly appeared, reaching third magnitude. Following this early eruption, P Cyg faded from view, only to reappear in 1655, becoming nearly as bright as it was in 1600. P Cyg has been a roughly constant fifth-magnitude star since 1700, although it may have brightened slightly over the past several centuries. [You may recall that spectral line profiles known as P Cygni profiles (see Fig. 12.17), named for this episodic star, are indicative of mass loss.]

### Luminous Blue Variables

A small number of other stars are known to behave similarly to $\eta$ Carinae and P Cygni, some in our galaxy and some farther away. S Doradus, located in the Large Magellanic Cloud (LMC), a satellite galaxy of the Milky Way, is perhaps the best-known extragalactic example. Similar stars were discovered in nearby galaxies by Edwin Hubble and Allan Sandage. This class of stars is referred to by several different names, including **S Doradus variables**, **Hubble–Sandage variables**, and **luminous blue variables** (LBVs). In this text, we will adopt the designation of LBV for this class of stars.

While perhaps an extreme example of an LBV, $\eta$ Car is certainly the best-studied representative of the class. A Hubble Space Telescope image of $\eta$ Car is shown in Fig. 15.1. Its bipolar structure, known as the "homunculus," is clearly evident, as is its equatorial disk. From Doppler measurements, the lobes are expanding outward at about 650 km s$^{-1}$, although a number of different velocities can be recorded along any particular line of sight. The expanding lobes are largely hollow, but the material in the shells contains molecules of $H_2$, CH, and OH. However, it appears that the homunculus is significantly depleted in C and O, while being enriched in He and N. This would suggest that the ejected material has undergone nuclear processing by the CNO cycle (see Eqs. 10.48–10.57). The present rate of mass loss from $\eta$ Car is on the order of $10^{-3}$ M$_\odot$ yr$^{-1}$, but it probably ejected one to three solar masses of material during the twenty years of the Great Eruption.

During the Great Eruption, $\eta$ Car's luminosity may have been about $2 \times 10^7$ L$_\odot$, whereas its present quiescent luminosity is near $5 \times 10^6$ L$_\odot$. It is also estimated that the central star's effective temperature is roughly 30,000 K. Although at the time this text was written, $m_V \sim 6$ for $\eta$ Car, most of its luminosity is initially emitted in the ultraviolet wavelength region, owing to the high effective temperature. Much of the UV radiation is scattered, absorbed, and re-emitted by dust grains in the infrared portion of the electromagnetic

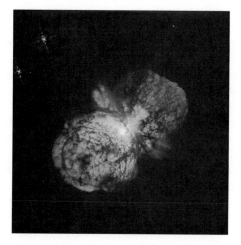

**FIGURE 15.1** $\eta$ Carinae is a luminous blue variable that is estimated to have a mass of 120 M$_\odot$ and is rapidly losing mass. Each lobe has a diameter of approximately 0.1 pc. [Courtesy of Jon Morse (University of Colorado) and NASA.]

spectrum at temperatures ranging from 200 K to 1000 K. The visual magnitude alone leads to an underestimate of the total luminosity of the object today.

As a class, LBVs tend to have high effective temperatures of between 15,000 K and 30,000 K, with luminosities in excess of $10^6$ $L_\odot$. This places LBVs in the upper-left-hand portion of the H–R diagram. Given the composition of their atmospheres and ejecta, LBVs are clearly evolved, post-main-sequence stars. It also appears that LBVs cluster in an instability region of the H–R diagram, suggesting that their behavior is transient, turning on shortly after they leave the main sequence and ceasing after some period of time.

A variety of mechanisms have been proposed to explain the behavior of LBVs, including their variability and dramatic mass loss. As discussed in Section 10.6, the upper end of the main sequence is very near the Eddington luminosity limit where the force due to radiation pressure may equal or exceed the force of gravity on surface layers of the star. The derived expression for the Eddington luminosity limit, Eq. (10.114), is a function of the Rosseland mean opacity of the surface layers,

$$L_{\text{Ed}} = \frac{4\pi Gc}{\overline{\kappa}} M.$$

The "classical" Eddington limit assumes that the opacity is due entirely to scattering from free electrons (Eq. 9.27), which is constant for a completely ionized gas. A "modified" Eddington limit has been proposed by Roberta M. Humphreys and Kris Davidson, in which some temperature-dependent component of opacity, perhaps due to iron lines, modifies the opacity term as the star evolves to the right in the H–R diagram. As the temperature decreases and the opacity increases, the Eddington luminosity would drop below the actual luminosity of the star, implying that radiation pressure dominates gravity, driving mass loss from the envelope.

A second suggestion is that atmospheric pulsation instabilities may develop, much like those in Cepheids, RR Lyraes, and long-period variables. Some preliminary nonlinear pulsation studies have suggested that large-amplitude oscillations can develop in LBVs, which could conceivably drive mass loss as outwardly moving mass shells are lifted off the surface during the pulsation cycle. Moreover, such pulsations are likely to be very irregular in such a weakly bound atmosphere. Unfortunately, these models are very sensitive to the treatment of time-dependent convection, which is poorly understood in the context of stellar pulsations.

Also intriguing is the apparent high rotation velocity of at least some LBVs. Rapid rotation would result in decreasing the "effective" gravity at the equator of these stars due to centrifugal effects, making the gases in the atmospheres of the equatorial regions easier to drive away from the surface. It has been suggested that the equatorial disk around $\eta$ Car could have formed from just such an effect during the lesser eruption between 1887 and 1895.

The possibility that LBVs are members of binary star systems has also been suggested as influencing the behavior of these stars. Interestingly, $\eta$ Car exhibits a 5.54-yr periodicity in the equivalent widths of some of its spectral lines, hinting at the presence of a binary companion, although it is unclear how the presence of a companion can cause the effects observed.

It may turn out that more than one of the mechanisms discussed above could influence the behavior of LBVs, or perhaps the principal mechanism is one that has not yet been identified.

### Wolf–Rayet Stars

Closely related to the LBVs are the **Wolf–Rayet stars** (WR). The first WRs were discovered by C. J. E. Wolf and G. Rayet while working at the Paris Observatory in 1867. Using a visual-wavelength spectrometer to conduct a survey of stars in Cygnus, they observed three stars all within one degree of each other that exhibited unusually strong, very broad emission lines, rather than the absorption lines usually seen in other stars. Today, more than 220 WR stars have been identified in the Milky Way Galaxy, although the total number of WRs in the Galaxy is estimated to be between 1000 and 2000 on the basis of sampling statistics (see Fig. 15.2).

Along with the strong emission lines, WR stars are very hot, with effective temperatures of 25,000 K to 100,000 K. WRs are also losing mass at rates in excess of $10^{-5}$ $M_\odot$ $yr^{-1}$ with wind speeds ranging from 800 km $s^{-1}$ to more than 3000 km $s^{-1}$. In addition, there is strong evidence that many, and perhaps all, WR stars are rapidly rotating, with equatorial rotation speeds of typically 300 km $s^{-1}$.

Whereas LBVs are all very massive stars of 85 $M_\odot$ or more, WRs can have progenitor masses as low as 20 $M_\odot$. WRs also do not demonstrate the dramatic variability that is characteristic of LBVs.

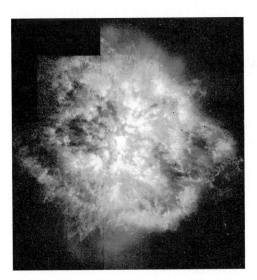

**FIGURE 15.2** The nebula M1-67 around the Wolf–Rayet star WR 124. The surface temperature of the star is about 50,000 K. Clumpiness is clearly evident in the nebula, and the mass of each blob is about 30 $M_\oplus$. WR 124 is at a distance of 4600 pc in Sagittarius. [Courtesy of Yves Grosdidier (University of Montreal and Observatoire de Strasbourg), Anthony Moffat (Université de Montréal), Gilles Joncas (Universite Laval), Agnes Acker (Observatoire de Strasbourg), and NASA.]

What really sets Wolf–Rayet stars apart from other stars is their unusual spectra. Not only are the spectra dominated by broad emission lines, but they also reveal a composition that is decidedly atypical. Today we recognize three classes of WR stars: WN, WC, and WO. The spectra of WNs are dominated by emission lines of helium and nitrogen, although emission from carbon, oxygen, and hydrogen is detectable in some WN stars. WC stars exhibit emission lines of helium and carbon, with a distinct absence of nitrogen and hydrogen lines. Finally, the WO stars, which are much rarer than either WNs or WCs, have spectra containing prominent oxygen lines, with some contribution from highly ionized species.

The literature further sub-classifies WN and WC stars based on the degree of ionization of species in the atmosphere. For example, WN2 stars show spectral lines of He II, N IV, and O VI, and WN9 stars contain spectra of low-ionization species such as He I and N III. "Early" (E) and "late" (L) types are also mentioned; WNE stars are Wolf–Rayet stars of ionization classes, WN2 to WN5, and WNL stars are of ionization classes WN6 through WN11. Similarly WC4 stars have higher ionization levels (He II, O IV, C VI), and WC9 stars exhibit lower ionization levels (e.g., He I and C II). WCEs range for WC4 through WC6, and WCLs include WC7 through WC9.

This strange trend in composition from WN to WC to WO was eventually recognized to be a direct consequence of the mass loss of these stars. WNs have lost virtually all of their hydrogen-dominated envelopes, revealing material synthesized by nuclear reactions in the core. Convection in the core of the star has brought equilibrium CNO cycle-processed material to the surface. Further mass loss results in the ejection of the CNO processed material, exposing helium-burning material generated by the triple alpha process (Eqs. 10.60 and 10.61). Then, if the star survives long enough, mass loss will eventually strip away all but the oxygen component of the triple-alpha ash.

In addition to LBVs and WRs, the upper portion of the H–R diagram also contains **blue supergiant stars** (BSG), **red supergiant stars** (RSG), and **Of stars** (O supergiants with pronounced emission lines).

## A General Evolutionary Scheme for Massive Stars

In a scheme originally suggested by Peter Conti in 1976 and subsequently modified, a general evolutionary path for massive stars has been outlined. In each case the star ends its life in a **supernova** (SN) explosion, to be discussed in detail in Section 15.3. (The masses listed below are only approximate.)[1]

$$M > 85\,M_\odot : O \rightarrow Of \rightarrow LBV \rightarrow WN \rightarrow WC \rightarrow SN$$

$$40\,M_\odot < M < 85\,M_\odot : O \rightarrow Of \rightarrow WN \rightarrow WC \rightarrow SN$$

$$25\,M_\odot < M < 40\,M_\odot : O \rightarrow RSG \rightarrow WN \rightarrow WC \rightarrow SN$$

$$20\,M_\odot < M < 25\,M_\odot : O \rightarrow RSG \rightarrow WN \rightarrow SN$$

$$10\,M_\odot < M < 20\,M_\odot : O \rightarrow RSG \rightarrow BSG \rightarrow SN$$

This qualitative evolutionary scheme has been supported by detailed numerical evolutionary models of massive star formation. Evolutionary tracks for stars of solar composition

[1]This version has been adopted from Massey, *Annu. Rev. Astron. Astrophys.*, *41*, 15, 2003.

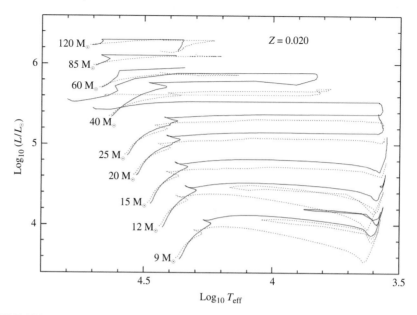

**FIGURE 15.3** The evolution of massive stars with $Z = 0.02$. The solid lines are evolutionary tracks computed with initial rotation velocities of 300 km s$^{-1}$, and the dotted lines are evolutionary tracks for stars without rotation. Mass loss has been included in the models and significantly impacts the evolution of these stars. (Figure from Meynet and Maeder, *Astron. Astrophys.*, *404*, 975, 2003.)

ranging in mass from 9 M$_\odot$ to 120 M$_\odot$ are shown in Fig. 15.3. These models of Georges Meynet and Andrés Maeder include mass loss typical of massive stars. The models are also computed with and without rotation; when rotation is included, the equatorial rotation speed is taken to be 300 km s$^{-1}$. Meynet and Maeder point out that rotation can have an appreciable affect on stellar evolution, including driving internal mixing and enhancing mass loss.

### The Humphreys–Davidson Luminosity Limit

These massive-star evolutionary tracks indicate that the most massive stars never evolve to the red supergiant portion of the H–R diagram. This is in agreement with the qualitative evolutionary scenario presented above, and it is also consistent with observations. Humphreys and Davidson were the first to point out that there is an upper-luminosity cut-off in the H–R diagram that includes a diagonal component running from highest luminosities and effective temperatures to lower values in both parameters. At that point, when full redward evolutionary tracks develop for stars below about 40 M$_\odot$, the **Humphreys–Davidson luminosity limit** continues at constant luminosity.

Although very massive stars are extremely rare (only one 100 M$_\odot$ star exists for every one million 1 M$_\odot$ stars), they play a major role in the dynamics and chemical evolution of the interstellar medium. The tremendous amount of kinetic energy deposited in the ISM through the stellar winds of massive stars has a significant impact on the kinematics of the ISM. In fact, when very massive stars form, they have the ability to quench star formation in their regions. The ultraviolet light from massive stars also ionizes gas clouds in their region. And, in addition, the highly enriched gases of massive stellar winds increase the metal content of the ISM, resulting in the formation of increasingly metal-rich stars. Besides being spectacular and exotic objects, massive stars are critically important to the evolution of the galaxies in which they reside.

## 15.2 ■ THE CLASSIFICATION OF SUPERNOVAE

In A.D. 1006 an extremely bright star suddenly appeared in Lupis. Reaching an estimated apparent visual magnitude of $m_V = -9$, it was reportedly bright enough to read by at night. This event was recorded by astrologers in Europe, China, Japan, Egypt, and Iraq. Based on their writings, it is likely that **Supernova 1006** (SN 1006) appeared about April 30, 1006, and faded from view roughly one year later.

Other, similar events have been seen throughout human history, although only rarely. Perhaps the most famous celestial event of its kind occurred on July 4, 1054, only 48 years after the A.D. 1006 event, when a "guest star" appeared in the night sky in the constellation of Taurus. Yang Wei-T'e, a court astrologer during China's Sung dynasty, recorded the remarkable event, noting that "after more than a year it gradually became invisible." In addition to having been carefully documented in the official records of the Sung dynasty, the star was noted by the Japanese and Koreans and was also recorded in an Arabic medical textbook. There is evidence, although this is the subject of some debate, that Europeans may have witnessed the event as well. As with the A.D. 1006 event, this amazing star was visible during daylight. With the development of powerful telescopes, modern astronomers have identified a rapidly expanding cloud, known as the **Crab supernova remnant** at the reported location of this ancient "guest star" (see Fig. 15.4).

It was five hundred years before another star suddenly appeared in the heavens in such dramatic fashion. Tycho Brahe [the most famous astronomer of his day; recall Fig. 2.1(a)] and others witnessed a supernova in A.D. 1572. This strange occurrence was clearly in contrast to the widely held belief in the Western world at the time that the heavens were unchanging. Not to be outdone, his student Johannes Kepler [Fig. 2.1(b)] also witnessed a supernova explosion in A.D. 1604. These two events are now known as **Tycho's supernova** and **Kepler's supernova**, respectively.

Unfortunately, Kepler's supernova was the last supernova observed to occur in the Milky Way Galaxy. However, on February 24.23 UT, 1987, Ian Shelton, using a 10-inch astrograph at the Las Campanas Observatory in Chile, detected **SN 1987A** just southwest of a massive molecular cloud region in the LMC known as 30 Doradus; the supernova is shown in Fig. 15.5. It was the first time since the development of modern instruments that a supernova had been seen so close to Earth (the distance to the LMC is 50 kpc). The excitement of the astronomical community worldwide was immediate and intense. It was quickly realized that the progenitor of this spectacular supernova was a blue supergiant star. Nicholas Sanduleak (1933–1990) had cataloged the star Sk −69 202 while investigating hot stars in the Magellanic Clouds.[2] The chance to observe a supernova from such a close vantage point using

---

[2]Sk −69 202 gets its name from being the 202nd entry in the −69° declination band of the Sanduleak catalog of stars in the Magellanic Clouds.

**FIGURE 15.4** The Crab supernova remnant, located 2000 pc away in the constellation of Taurus. The remnant is the result of a Type II supernova that was observed for the first time on July 4, 1054. [Courtesy of NASA, ESA, J. Hester and A. Loll (Arizona State University).]

**FIGURE 15.5** A portion of the Large Magellanic Cloud showing SN 1987A on the lower right-hand side of the photograph. 30 Doradus (also known as the Tarantula Nebula) is an immense H II region that is clearly evident on the left-hand side of the photograph. (Courtesy of the European Southern Observatory, ©ESO.)

the arsenal of tools available to modern astrophysics provided an ideal opportunity to test our theory of the fate of massive stars.

## Classes of Supernovae

Today astronomers are able to routinely observe supernovae in other galaxies (hence the "A" in SN 1987A representing the first supernova reported that year). However, supernovae are exceedingly rare events, typically occurring about once every one hundred years or so in any one galaxy. As the spectra and light curves of supernovae have been carefully studied, it has been realized that there are several distinct classes of supernovae with different classes of underlying progenitors and mechanisms.

**Type I** supernovae were identified first as those supernovae that do not exhibit any hydrogen lines in their spectra. Given that hydrogen is the most abundant element in the universe, this fact alone suggests something unusual about these objects. Conversely, the spectra of **Type II** supernovae contain strong hydrogen lines.

Type I supernovae can be further subdivided according to their spectra. Those Type I spectra that show a strong Si II line at 615 nm are called **Type Ia**. The others are designated **Type Ib** or **Type Ic**, based on the presence (Ib) or absence (Ic) of strong helium lines.

Figure 15.6 shows examples of the spectra of each of the four types of supernovae (SNe) discussed here.

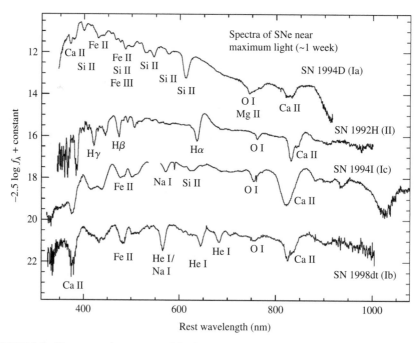

**FIGURE 15.6**    Representative spectra of the four types of supernovae; Type Ia, Ib, Ic, and II. Note that although SN 1994I (Type Ic) does exhibit a weak Si II absorption line, it is much less prominent than the Si lines in a Type Ia. Brightness is in arbitrary flux units. (Courtesy of Thomas Matheson, National Optical Astronomy Observatory.)

The lack of hydrogen lines in Type I supernovae indicates that the stars involved have been stripped of their hydrogen envelopes. The differences in the spectral signatures between Type Ia and Types Ib and Ic indicate that different physical mechanisms are at work. This is reflected in the different environments observed for these outbursts. Type Ia supernovae are found in all types of galaxies, including ellipticals that show very little evidence of recent star formation. On the other hand, Types Ib and Ic have been seen only in spiral galaxies, near sites of recent star formation (H II regions). This implies that short-lived massive stars are probably involved with Types Ib and Ic, but not with Type Ia.

Figure 15.7 shows a composite light curve at blue ($B$) wavelengths for Type I's. The typical peak brightness of a Type Ia is $M_B = -18.4$, while the light curves of Types Ib and Ic supernovae are fainter by 1.5 to 2 magnitudes in blue light but are otherwise similar. All Type I supernovae show similar rates of decline of their brightness after maximum, about 0.065 ($\pm 0.007$) magnitude per day at 20 days. After about 50 days, the rate of dimming slows and becomes constant, with Type Ia's declining 50% faster than the others (0.015 mag d$^{-1}$ vs. 0.010 mag d$^{-1}$). It is believed that SN 1006 and the supernovae detected by Tycho (SN 1572) and Kepler (SN 1604) were Type I's.

Observationally, Type II supernovae are characterized by a rapid rise in luminosity, reaching a maximum brightness that is typically 1.5 mag dimmer than Type Ia's. The peak light output is followed by a steady decrease, dropping six to eight magnitudes in a year. Their spectra also exhibit lines associated with hydrogen and heavier elements. Furthermore, P Cygni profiles are common in many lines (indicating rapid expansion; see Section 12.3). The Crab supernova (SN 1054) and SN 1987A were Type II's.

The light curves of Type II supernovae can be classified as either **Type II-P** (plateau) or **Type II-L** (linear). Composite $B$ magnitude light curves of each type are shown in Fig. 15.8. A temporary but clear plateau exists between about 30 and 80 days after maximum light for Type II-P supernovae; no such detectable plateau exists for Type II-L objects. Type II-P supernovae also occur approximately ten times as often as Type II-L's.

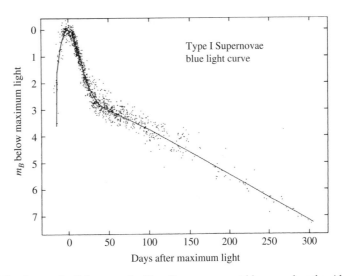

**FIGURE 15.7** Composite light curve for Type I supernovae at blue wavelengths. All magnitudes are relative to $m_B$ at maximum. (Figure adapted from Doggett and Branch, *Astron. J.*, *90*, 2303, 1985.)

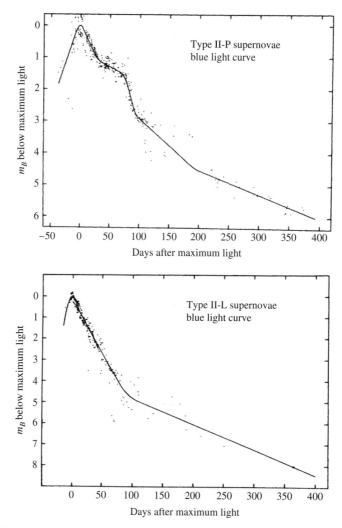

**FIGURE 15.8**    The characteristic shapes of Type II-P and Type II-L light curves. These are composite light curves, based on the observations of many supernovae. (Figures adapted from Doggett and Branch, *Astron. J.*, *90*, 2303, 1985.)

A summary decision tree of supernova classification is given in Fig. 15.9.

Of course nature loves to confound our clean classification schemes. SN 1993J in the spiral galaxy M81 in Ursa Major initially displayed strong hydrogen emission lines (i.e., Type II), but within a month the hydrogen lines were replaced by helium and its appearance changed to that of a Type Ib. This provides some indication that at least Type Ib's and Type II's are related in some way. We will learn that Type Ic's are also closely related.

We know today that Type Ia's are fundamentally different events from other supernovae. In Section 15.3 we will discuss the physics involved in Types Ib, Ic, and II supernovae, postponing a detailed discussion of Type Ia's until Section 18.5.

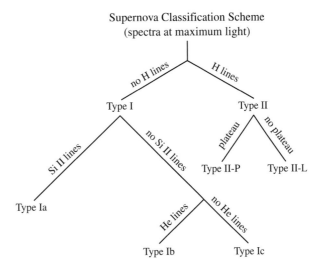

**FIGURE 15.9** The classification of supernovae based on their spectra at maximum light and the existence or absence of a plateau in the Type II light curve.

## 15.3 ■ CORE-COLLAPSE SUPERNOVAE

The goal of understanding the physical processes involved in generating supernovae has been a long-standing challenge. The sheer amount of energy released in a supernova event is staggering. A typical Type II releases $10^{46}$ J of energy, with about 1% of that appearing as kinetic energy of the ejected material and less than 0.01% being released as the photons that produce the spectacular visual display. As we will see later, the remainder of the energy is radiated in the form of neutrinos. Similar values are obtained for Type Ib and Type Ic supernovae.

---

**Example 15.3.1.** To illustrate how much energy is involved in a Type II event, consider the equivalent amount of rest mass, and how much iron could ultimately be produced by releasing that much nuclear binding energy.

From $E = mc^2$, the energy released by a Type II supernova corresponds to a rest mass of

$$m = E/c^2 = 10^{46} \text{ J}/c^2 = 1 \times 10^{29} \text{ kg} = 0.06 \text{ M}_\odot.$$

The binding energy of an iron-56 nucleus ($^{56}_{26}$Fe), the most stable of all nuclei, is 492.26 MeV, and the mass of the nucleus is 55.934939 u (recall Fig. 10.9). In order to release $10^{46}$ J of energy through the formation of iron nuclei from protons and neutrons, it would be necessary to form

$$N = \left( \frac{1 \times 10^{46} \text{ J}}{492.26 \text{ MeV/nucleus}} \right) \left( \frac{1 \text{ MeV}}{1.6 \times 10^{-13} \text{ J}} \right) = 1.3 \times 10^{56} \text{ nuclei},$$

*continued*

corresponding to a mass of iron of

$$m = N(55.93 \text{ u})(1.66 \times 10^{-27} \text{ kg/u}) = 1.2 \times 10^{31} \text{ kg} = 5.9 \text{ M}_\odot.$$

Of course this would require that 5.9 M$_\odot$ of iron be produced all at once in an explosive event if this is the source of the energy for a Type II supernova! On the other hand, if this much iron were broken down to the original protons and neutrons, the equivalent amount of energy would need to be absorbed.

As we will see later, *iron is not formed* as a result of releasing the energy involved in a supernova explosion; in fact, the energy source is not nuclear. However, iron is critically involved in the process in a perhaps unexpected way.

### Core-Collapse Supernova Mechanism

The post-main-sequence evolution of stars more massive than about 8 M$_\odot$ is decidedly different from what was described in Chapter 13. Although hydrogen is converted into helium on the main sequence, followed by helium burning leading to a carbon–oxygen core, the very high temperature in the core of a massive star means that carbon and oxygen can burn as well. The end result is that rather than the star ending its life through the formation of a planetary nebula, a catastrophic supernova explosion occurs instead.

What follows is a discussion of that evolutionary process. Although all of the details have yet to be worked out at the time of writing, the story of how Type Ib, Type Ic, and Type II supernovae are produced is becoming clearer. The three types are all closely related and all involve the collapse of a massive, evolved stellar core. Hence, collectively Types Ib, Ic, and II are known as **core-collapse supernovae**.

As the helium-burning shell continues to add ash to the carbon–oxygen core, and as the core continues to contract, it eventually ignites in carbon burning, generating a variety of by-products, such as $^{16}_{8}$O, $^{20}_{10}$Ne, $^{23}_{11}$Na, $^{23}_{12}$Mg, and $^{24}_{12}$Mg (see Eqs. 10.64–10.66). This leads to a succession of nuclear reaction sequences, the exact details of which depend sensitively on the mass of the star.

Assuming that each reaction sequence reaches equilibrium, an "onion-like" shell structure develops in the interior of the star. Following carbon burning, the oxygen in the resulting neon–oxygen core will ignite (Eq. 10.67), producing a new core composition dominated by $^{28}_{14}$Si. Finally, at temperatures near $3 \times 10^9$ K, silicon burning can commence through a series of reactions such as

$$^{28}_{14}\text{Si} + {}^{4}_{2}\text{He} \rightleftharpoons {}^{32}_{16}\text{S} + \gamma \tag{15.1}$$

$$^{32}_{16}\text{S} + {}^{4}_{2}\text{He} \rightleftharpoons {}^{36}_{18}\text{Ar} + \gamma \tag{15.2}$$

$$\vdots$$

$$^{52}_{24}\text{Cr} + {}^{4}_{2}\text{He} \rightleftharpoons {}^{56}_{28}\text{Ni} + \gamma. \tag{15.3}$$

Silicon burning produces a host of nuclei centered near the $^{56}_{26}$Fe peak of the binding energy per nucleon curve shown in Fig. 10.9, the most abundant of which are probably $^{54}_{26}$Fe, $^{56}_{26}$Fe, and $^{56}_{28}$Ni. Any further reactions that produce nuclei more massive than $^{56}_{26}$Fe are endothermic

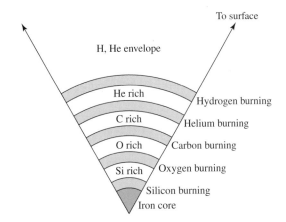

**FIGURE 15.10** The onion-like interior of a massive star that has evolved through core silicon burning. Inert regions of processed material are sandwiched between the nuclear burning shells. The inert regions exist because the temperature and density are not sufficient to cause nuclear reactions to occur with that composition. (This drawing is not to scale.)

and cannot contribute to the luminosity of the star. Grouping all of the products together, silicon burning is said to produce an *iron core*. A sketch of the onion-like interior structure of a massive star following silicon burning is given in Fig. 15.10.

Because carbon, oxygen, and silicon burning produce nuclei with masses progressively nearer the *iron peak* of the binding energy curve, less and less energy is generated per unit mass of fuel. As a result, the timescale for each succeeding reaction sequence becomes shorter (recall Example 10.3.1). For example, for a 20 $M_\odot$ star, the main-sequence lifetime (core hydrogen burning) is roughly $10^7$ years, core helium burning requires $10^6$ years, carbon burning lasts 300 years, oxygen burning takes roughly 200 days, and silicon burning is completed in only two days!

At the very high temperatures now present in the core, the photons possess enough energy to destroy heavy nuclei (note the reverse arrows in the silicon-burning sequence), a process known as **photodisintegration**. Particularly important are the photodisintegration of $^{56}_{26}\text{Fe}$ and $^4_2\text{He}$:

$$^{56}_{26}\text{Fe} + \gamma \rightarrow 13\,^4_2\text{He} + 4n \tag{15.4}$$

$$^4_2\text{He} + \gamma \rightarrow 2p^+ + 2n. \tag{15.5}$$

When the mass of the contracting iron core has become large enough and the temperature sufficiently high, photodisintegration can, in a very short period of time, undo what the star has been trying to do its entire life, namely produce elements more massive than hydrogen and helium. Of course, this process of stripping iron down to individual protons and neutrons is highly endothermic, as suggested in Example 15.3.1; thermal energy is removed from the gas that would otherwise have resulted in the pressure necessary to support the core of the star. The core masses for which this process occurs vary from 1.3 $M_\odot$ for a 10 $M_\odot$ ZAMS star to 2.5 $M_\odot$ for a 50 $M_\odot$ star.

Under the extreme conditions that now exist (e.g., $T_c \sim 8 \times 10^9$ K and $\rho_c \sim 10^{13}$ kg m$^{-3}$ for a 15 M$_\odot$ star), the free electrons that had assisted in supporting the star through degeneracy pressure are captured by heavy nuclei and by the protons that were produced through photodisintegration; for instance,

$$p^+ + e^- \to n + \nu_e. \tag{15.6}$$

The amount of energy that escapes the star in the form of neutrinos becomes enormous; during silicon burning the photon luminosity of a 20 M$_\odot$ stellar model is $4.4 \times 10^{31}$ W while the neutrino luminosity is $3.1 \times 10^{38}$ W.

Through the photodisintegration of iron, combined with electron capture by protons and heavy nuclei, most of the core's support in the form of electron degeneracy pressure is suddenly gone and the core begins to collapse extremely rapidly. In the inner portion of the core, the collapse is homologous, and the velocity of the collapse is proportional to the distance away from the center of the star (recall Eq. 12.26 and the discussion of a homologous free-fall collapse during the formation of a protostar). At the radius where the velocity exceeds the local sound speed, the collapse can no longer remain homologous and the inner core decouples from the now supersonic outer core, which is left behind and nearly in free-fall. During the collapse, speeds can reach almost 70,000 km s$^{-1}$ in the outer core, and within about one second a volume the size of Earth has been compressed down to a radius of 50 km!

---

**Example 15.3.2.**   If a mass with the radius of Earth ($R_\oplus$) collapses to a radius of only 50 km, a tremendous amount of gravitational potential energy would be released. Can this energy release be responsible for the energy of a core-collapse supernova?

Assume for simplicity that we can use Newtonian physics to estimate the amount of energy released during the collapse. From the virial theorem (see Eq. 10.23), the potential energy of a spherically symmetric star of constant density is

$$U = -\frac{3}{10}\frac{GM^2}{R}.$$

Equating the energy of a Type II supernova, $E_{II} = 10^{46}$ J, to the gravitational energy released during the collapse, and given that $R_f = 50$ km $\ll R_\oplus$, the amount of mass required to produce the supernova would be

$$M \simeq \sqrt{\frac{10}{3}\frac{E_{II}R_f}{G}} \simeq 5 \times 10^{30} \text{ kg} \simeq 2.5 \text{ M}_\odot.$$

This value is characteristic of the core masses mentioned earlier.

---

Since mechanical information will propagate through the star only at the speed of sound and because the core collapse proceeds so quickly, there is not enough time for the outer layers to learn about what has happened inside. The outer layers, including the oxygen, carbon, and helium shells, as well as the outer envelope, are left in the precarious position of being almost suspended above the catastrophically collapsing core.

The homologous collapse of the inner core continues until the density there exceeds about $8 \times 10^{17}$ kg m$^{-3}$, roughly three times the density of an atomic nucleus. At that point, the nuclear material that now makes up the inner core stiffens because the strong force (usually attractive) suddenly becomes repulsive. This is a consequence of the Pauli exclusion principle applied to neutrons.[3] The result is that the inner core rebounds somewhat, sending pressure waves outward into the infalling material from the outer core. When the velocity of the pressure waves reach the sound speed, they build into a shock wave that begins to move outward.

As the shock wave encounters the infalling outer iron core, the high temperatures that result cause further photodisintegration, robbing the shock of much of its energy. For every 0.1 M$_\odot$ of iron that is broken down into protons and neutrons, the shock loses $1.7 \times 10^{44}$ J.

Computer simulations indicate that at this point the shock stalls, becoming nearly stationary, with infalling material accreting onto it. In other words, the shock has become an **accretion shock**, somewhat akin to the situation during protostellar collapse, discussed in Section 12.2. However, below the shock, a **neutrinosphere** develops from the processes of photodisintegration and electron capture. Since the overlying material is now so dense that even neutrinos cannot easily penetrate it, some of the neutrino energy ($\sim 5\%$) will be deposited in the matter just behind the shock. This additional energy heats the material and allows the shock to resume its march toward the surface. If this does not happen quickly enough, the initially outflowing material will fall back onto the core, meaning that an explosion doesn't occur.

The success of core-collapse supernova models seems to hinge very sensitively on the details of three-dimensional simulations, which allow hot, rising plumes of gas to mix with colder, infalling gas. The challenges lie in the details of convection, the need to treat the neutrino physics properly (including electron, muon, and tau neutrinos, and their antiparticles), and the very high resolution required for the calculations [up to $10^9$ mesh points (locations) in the computational grid]. It may also be necessary ultimately to include a proper treatment of sound waves, differential rotation, and magnetic fields in order to describe all of the observed details of a supernova explosion. This level of computational sophistication challenges even the world's most powerful supercomputers.

Assuming that the scenario just described is essentially correct, and that the shock is able to resume its march to the surface, the shock will drive the envelope and the remainder of the nuclear-processed matter in front of it. The total kinetic energy in the expanding material is on the order of $10^{44}$ J, roughly 1% of the energy liberated in neutrinos. Finally, when the material becomes optically thin at a radius of about $10^{13}$ m, or roughly 100 AU, a tremendous optical display results, releasing approximately $10^{42}$ J of energy in the form of photons, with a peak luminosity of nearly $10^{36}$ W, or roughly $10^9$ L$_\odot$, which is capable of competing with the brightness of an entire galaxy.

The events just described—the catastrophic collapse of an iron core, the generation of a shock wave, and the ensuing ejection of the star's envelope—are believed to be the general mechanism that creates a core-collapse supernova. The details that result in a Type II rather than a Type Ib or Type Ic supernova have to do with the composition and mass of the envelope at the time of the core collapse and the amount of radioactive material synthesized in the ejecta.

---

[3]Neutrons, along with electrons and protons, are fermions.

Type II supernovae, which are more common than either Type Ib or Type Ic, are usually red supergiant stars in the extreme upper-right-hand corner of the H–R diagram at the time they undergo catastrophic core collapse. Type Ib's and Type Ic's have lost various amounts of their envelopes prior to detonation. It is now believed that these are the products of exploded Wolf–Rayet stars. Type Ib's and Type Ic's may correspond to the detonation of WN and WC Wolf–Rayets, respectively; recall the Conti scenario for the evolution of massive stars described on page 522.

## Stellar Remnants of a Core-Collapse Supernova

If the initial mass of the star on the main sequence was not too large (perhaps $M_{ZAMS} <$ 25 $M_\odot$), the remnant in the inner core will stabilize and become a **neutron star** (essentially a gigantic nucleus), supported by degenerate neutron pressure.[4] However, if the initial stellar mass is much larger, even the pressure of neutron degeneracy cannot support the remnant against the pull of gravity, and the final collapse will be complete, producing a **black hole** (an object whose mass has collapsed to a *singularity* of infinite density).[5] In either case, the creation of these exotic objects is accompanied by a tremendous production of neutrinos, the majority of which escape into space with a total energy on the order of the binding energy of a neutron star, approximately $3 \times 10^{46}$ J. This represents roughly 100 times more energy than the Sun will produce over its entire main-sequence lifetime!

## The Light Curves and the Radioactive Decay of the Ejecta

A Type II-P supernova is the most common type of core-collapse supernova. The source of the plateau in Type II-P light curves is due largely to the energy deposited by the shock into the hydrogen-rich envelope. The gas, which was ionized by the shock, enters a stage of prolonged recombination, releasing the energy at a nearly constant temperature of about 5000 K.

The plateau may be supported further by the energy deposited in the envelope by the **radioactive decay** of $^{56}_{28}$Ni that was produced by the shock front during its march through the star (the half-life of $^{56}_{28}$Ni is $\tau_{1/2} = 6.1$ days). It is expected that the explosive nucleosynthesis of the supernova shock should have produced significant amounts of other radioactive isotopes as well, such as $^{57}_{27}$Co ($\tau_{1/2} = 271$ days), $^{22}_{11}$Na ($\tau_{1/2} = 2.6$ yr), and $^{44}_{22}$Ti ($\tau_{1/2} \simeq$ 47 yr). If the isotopes are present in sufficient quantities, each in turn may contribute to the overall light curve, causing the slope of the curve to change.

The $^{56}_{28}$Ni is transformed into $^{56}_{27}$Co through the *beta-decay* reaction[6]

$$^{56}_{28}\text{Ni} \rightarrow \,^{56}_{27}\text{Co} + e^+ + \nu_e + \gamma. \qquad (15.7)$$

The energy released by the decay is deposited into the optically thick expanding shell, which is then radiated away from the supernova remnant's photosphere. This "holds up" the light

---

[4]The upper mass limit of the progenitor that results in the formation of a neutron star depends on how metal-rich the original star was. A star sufficiently metal-rich may form a neutron star even if its initial mass is much greater than 25 $M_\odot$.

[5]We leave the detailed discussion of neutron stars and black holes to Chapters 16 and 17, respectively.

[6]Electrons and positrons are also known as $\beta$ particles.

curve for a time, extending the observed plateau. Eventually the expanding gas cloud will become optically thin, exposing the central product of the explosion, the neutron star or black hole.

$^{56}_{27}$Co, the product of the radioactive decay of $^{56}_{28}$Ni, is itself radioactive, with a longer half-life of 77.7 days:

$$^{56}_{27}\text{Co} \rightarrow {}^{56}_{26}\text{Fe} + e^+ + \nu_e + \gamma. \tag{15.8}$$

This implies that as the luminosity of the supernova diminishes over time, it should be possible to detect the contribution to the light being made by $^{56}_{27}$Co. Type II-L supernovae appear to have had progenitor stars with significantly reduced hydrogen envelopes, implying that the signature of the radioactive decay becomes evident almost immediately after the event.

Since radioactive decay is a statistical process, the rate of decay must be proportional to the number of atoms remaining in the sample, or

$$\frac{dN}{dt} = -\lambda N, \tag{15.9}$$

where $\lambda$ is a constant. It is left as an exercise to show that Eq. (15.9) can be integrated to give

$$\boxed{N(t) = N_0 e^{-\lambda t},} \tag{15.10}$$

where $N_0$ is the original number of radioactive atoms in the sample (see Fig. 15.11), and

$$\lambda = \frac{\ln 2}{\tau_{1/2}}.$$

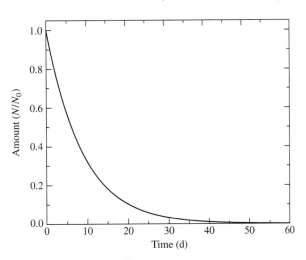

**FIGURE 15.11** The radioactive decay of $^{56}_{28}$Ni, with a half-life of $\tau_{1/2} = 6.1$ days. There is a 50% chance that any given $^{56}_{28}$Ni atom will decay during a time interval of 6.1 days. If the original sample is entirely composed of $^{56}_{28}$Ni, after $n$ successive half-lives the fraction of Ni atoms remaining is $2^{-n}$.

Since the rate at which decay energy is being deposited into the supernova remnant must be proportional to $dN/dt$, the slope of the bolometric light curve is given by

$$\frac{d \log_{10} L}{dt} = -0.434\lambda \tag{15.11}$$

or

$$\frac{dM_{\text{bol}}}{dt} = 1.086\lambda. \tag{15.12}$$

Therefore, by measuring the slope of the light curve, we can determine $\lambda$ and verify the presence of large quantities of a specific radioactive isotope, like $^{56}_{27}\text{Co}$.

Given its proximity to Earth, the most carefully studied supernova to date has been SN 1987A. However, almost as soon as it was discovered, astronomers realized that SN 1987A was unusual when compared with other, more distant Type II's that had been observed. This was most evident in the rather slow rise to maximum light (taking 80 days), which peaked only at an absolute bolometric magnitude of $-15.5$, whereas a typical Type II reaches $M_{\text{bol}} = -18$. The light curve through day 1444 after the outburst is shown in Fig. 15.12.

The decay of the 0.075 $M_\odot$ of $^{56}_{28}\text{Ni}$ that was produced by the shock occurred while the timescale required for energy to be radiated away was still quite long. Consequently, the

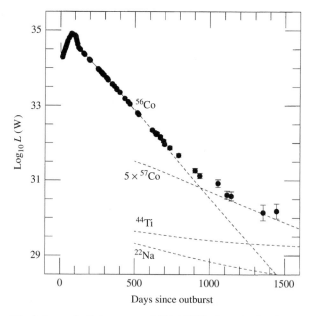

**FIGURE 15.12** The bolometric light curve of SN 1987A through the first 1444 days after the explosion. The dashed lines show the contributions expected from the radioactive isotopes produced by the shock wave. The initial masses are estimated to be $^{56}_{28}\text{Ni}$ (and later $^{56}_{27}\text{Co}$), 0.075 $M_\odot$; $^{57}_{27}\text{Co}$, 0.009 $M_\odot$ (five times the solar abundance); $^{44}_{22}\text{Ti}$, $1 \times 10^{-4}$ $M_\odot$; and $^{22}_{11}\text{Na}$, $2 \times 10^{-6}$ $M_\odot$. (Figure adapted from Suntzeff et al., *Ap. J. Lett.*, *384*, L33, 1992.)

added decay energy produced a bump on the light curve near maximum light, rather than forming a plateau. By the time the resulting $^{56}_{27}$Co began to decay, this diffusion timescale had become sufficiently short that the decrease in the luminosity of the remnant began to track closely the rate of decay of cobalt-56. Subsequently, the next important radioactive isotope, $^{57}_{27}$Co, began to play an important role in the development of the light curve. The expected contributions of the various radioactive isotopes to the light curve of SN 1987A are shown in Fig. 15.12, the slopes in the light curve being related to the half-lives of the isotopes through Eq. (15.11).

SN 1987A also allowed astronomers, for the first time, to directly measure the X-ray and gamma-ray emission lines produced by radioactive decay. In particular, the 847-keV and 1238-keV lines of $^{56}_{27}$Co were detected by a number of experiments, confirming the presence of this isotope. Doppler shift measurements indicate that the heavier isotopes in the remnant are expanding at several thousand kilometers per second.

### The Subluminous Nature of SN 1987A

The mystery of the subluminous nature of SN 1987A was solved when the identity of its progenitor was established. The star that blew up was the 12th-magnitude *blue* supergiant (spectral class B3 I), Sk −69 202. Since what exploded was a much smaller blue supergiant, rather than a red supergiant (as is usually assumed to be the case), the star was more dense. As a result, before the thermal energy produced by the shock could diffuse out and escape as light, it was converted into the mechanical energy required to lift the envelope of the star out of the deeper potential well of a blue supergiant. Measurements of H$\alpha$ lines indicate that some of the outer hydrogen envelope was ejected at speeds near 30,000 km s$^{-1}$, or 0.1$c$!

The available observations of Sk −69 202, together with theoretical evolutionary models, suggest that the progenitor of SN 1987A had a mass of roughly 20 $M_\odot$ when it was on the main sequence and that it lost perhaps a few solar masses before its iron core collapsed (estimated to be between 1.4 and 1.6 $M_\odot$). Although it was apparently a red supergiant for between several hundred thousand and one million years, it evolved to the blue just 40,000 years before the explosion (refer to the evolutionary tracks in Fig. 15.3; the blueward loop of a 20 $M_\odot$ star is not shown, but the blueward loop of a 25 $M_\odot$ star is present on the diagram.). Supporting this hypothesis is the observation that hydrogen was more abundant in the envelope of Sk −69 202 than was helium, suggesting that the star had not suffered extensive amounts of mass loss. Whether and when a massive star evolves from being a red supergiant to a blue supergiant before exploding depends sensitively on the mass of the star (it cannot be much more than about 20 $M_\odot$), its composition (it must be metal poor, as are the stars of the LMC), the rate of mass loss (which must be low), and the treatment of convection (always a major uncertainty in theoretical stellar models).

### Supernova Remnants

There are now many examples of **supernova remnants** (SNR), including the Crab Nebula, located in the constellation of Taurus (recall Fig. 15.4). Today, nearly 1000 years since the SN 1054 explosion, the Crab is still expanding at a rate of almost 1450 km s$^{-1}$ and it has a luminosity of $8 \times 10^4$ $L_\odot$. Much of the radiation being emitted is in the form of highly polarized synchrotron radiation (see Section 4.3), indicating the presence of relativistic

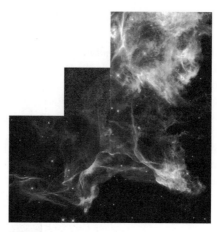

**FIGURE 15.13**   An HST WF/PC 2 image of a portion of the Cygnus Loop, 800 pc away. (Courtesy of J. Hester/Arizona State University and NASA.)

electrons that are spiraling around magnetic field lines. The ongoing source of the electrons and the continued high luminosity so long after the explosion remained major puzzles in astronomy until the discovery of a **pulsar** (a rapidly spinning neutron star) at the center of the Crab SNR. Pulsars will be discussed extensively in Chapter 16.

A second example of a supernova remnant is shown in Fig. 15.13. The image is of a small portion of the 15,000-year-old Cygnus Loop nebula, located 800 pc from Earth in the constellation of Cygnus. The remnant is expanding from left to right in the image, producing shock fronts several astronomical units wide as the debris from the supernova explosion encounters material in the interstellar medium. The shocks excite and ionize the ISM, causing the observed emission.

Although mass loss prior to the explosion could not have been excessive, the progenitor of SN 1987A did lose some mass, resulting in a very unusual structure around the expanding supernova remnant. The Hubble Space Telescope has recorded three rings around SN 1987A (Fig. 15.14). The innermost ring measures 0.42 pc in diameter and lies in a plane that contains the center of the supernova explosion. It glows in visible light as a consequence of emissions from O III energized by radiation from the supernova and appears elongated because it is inclined relative to our line of sight. The material making up the central ring was ejected by stellar winds 20,000 years before the explosion of SN 1987A.

The two larger rings are not in planes containing the central explosion but lie in front of and behind the star. One explanation for these fascinating and unexpected features is that Sk −69 202 resided near a companion star, possibly a neutron star or a black hole (see Chapters 16 and 17, respectively). As this companion source wobbles, narrow jets of radiation from the source "paint" the rings on an hourglass-shaped, bipolar distribution of mass that was ejected from Sk −69 202. It is in the denser equatorial plane of the bipolar mass distribution that the central ring is located. In support of this hypothesis, researchers believe that they may have identified the source of these beams of radiation about 0.1 pc from the center of the supernova explosion, consistent with the fact that the larger rings appear to be offset from the explosion's center. Opponents of this model believe that the explanation is too complicated: It requires two sources of high-energy radiation, one to

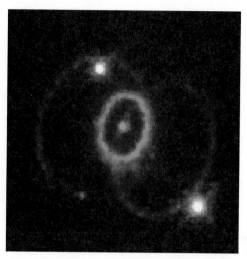

**FIGURE 15.14** Rings around SN 1987A, detected by the Hubble Space Telescope in 1994. The diameter of the inner ring is 0.42 pc. (Courtesy of Dr. Christopher Burrows, ESA/STScI and NASA.)

explain the central ring and another to explain the larger ones. Alternatively, the larger rings could be the product of a hot, fast stellar wind from the blue supergiant progenitor overtaking the slower, cooler wind given off by the star when it was still a red supergiant.

In the summer of 1990, fluctuating radio emissions were finally detected from the supernova. Although radio wavelength energy was detected during the first days following the explosion, SN 1987A had remained radio quiet since that time. Apparently the shock wave, still propagating outward at a speed close to $0.1c$, collided with clumps of material lost from Sk $-69\ 202$ prior to the supernova event.

The shock front from the expanding supernova remnant of SN 1987A began to collide with the slower-moving stellar wind comprising the inner ring in 1996. The result was a brilliant display of bright clumps in the inner ring that developed over the next several years. The expanding shock front and the inner ring are shown in successive images in Fig. 15.15.

### The Detection of Neutrinos from SN 1987A

Arguably the most exciting early observations of SN 1987A were based on its neutrinos, representing the first time that neutrinos had been detected from an astronomical source other than the Sun. The measurement of the neutrino burst confirmed the basic theory of core-collapse supernovae and amounts to our "seeing" the formation of a neutron star out of the collapsed iron core.

The arrival of the neutrino burst was recorded over a period of $12\frac{1}{2}$ seconds, beginning at February 23.316 UT, 1987, three hours *before* the arrival of the photons at February 23.443 UT. Twelve events were recorded at Japan's Kamiokande II Cerenkov detector, and at the same time, eight events were detected by the underground IMB[7] Cerenkov detector near

---

[7]IMB stands for the consortium that operates the observatory: University of California at Irvine, University of Michigan, and Brookhaven National Laboratory.

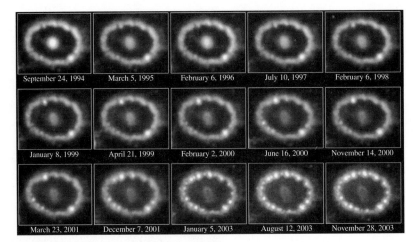

**FIGURE 15.15**    The expanding nebula from SN 1987A is seen in the center of this series of images. The inner ring shown in Fig. 15.14 is being overtaken by the shock front, causing clumps of gas to glow. The bright spot visible in the lower-right-hand portion of the ring in every image is another star that happens to be situated in the line of sight of the ring. [NASA and R. Kirshner (Harvard–Smithsonian Center for Astrophysics).]

Fairport, Ohio.[8] Assuming that the exploding star became optically thin to neutrinos before the shock wave reached the surface, and assuming further that the neutrinos traveled faster than the shock while still inside the star, the neutrinos began their trip to Earth ahead of the photons. Given the fact that the neutrinos arrived ahead of the light, their velocity through space must have been very near the speed of light (within one part in $10^8$). This observation, together with the absence of any significant dispersion in the arrival time of neutrinos of different energies (i.e., higher-energy neutrinos did not arrive any earlier than lower-energy ones), suggests that the rest mass of electron neutrinos must be quite small. The upper limit on the electron neutrino, based on data from SN 1987A, is $m_e \leq 16$ eV, consistent with the results of laboratory experiments that place the upper limit at 2.2 eV.

### The Search for a Compact Remnant of SN 1987A

Interestingly, as of 2006, neutrinos have been the only direct evidence of the formation of a compact object at the center of SN 1987A. All attempts to detect a remnant in optical, ultraviolet, or X-ray wavelengths have failed. In addition, efforts to find any evidence of a surviving binary companion have also been unsuccessful. The upper limit on the luminosity in the optical portion of the electromagnetic spectrum is currently less than $8 \times 10^{26}$ W, equivalent to the optical energy output of an F6 main-sequence star. Ultaviolet spectra lead to an upper limit of $L_{\rm UV} \leq 1.7 \times 10^{27}$ W, and Chandra has set an upper limit on the X-ray luminosity of $L_X \leq 5.5 \times 10^{26}$ W in the energy band between 2 and 10 keV.

---

[8]The famous Davis solar neutrino detector (discussed extensively in Section 11.1) did not measure any neutrinos from SN 1987A; the solar neutrino background was much *larger* than the neutrino count from the supernova in the energy range of the detector!

These limits are sufficiently stringent that models for the form of the compact companion and its environment (thin accretion disk, thick accretion disk, spherical accretion, etc.) are becoming seriously restricted. Perhaps future infrared observations with Spitzer will finally detect the elusive compact remnant.

Although SN 1987A has presented some interesting twists in our study of stellar evolution, it has also confirmed or clarified important aspects of the theory.

### Chemical Abundance Ratios in the Universe

Following the discussion of core-collapse supernovae, and in anticipation of results from our discussion of Type Ia supernovae coming up in Section 18.5, it is worth revisiting the chemical compositions and observed abundance ratios in the universe. A critical component in determining the success of current stellar evolution theory is the ability to explain the observed abundance ratios of the elements.

The chemical composition of the Sun's photosphere is shown in Fig. 15.16, with all values normalized to $10^{12}$ for hydrogen; see also Table 9.2. By far the most abundant element in the universe is hydrogen, with helium being less abundant by about a factor of 10. It is believed that hydrogen is primordial, having been synthesized immediately following the Big Bang that began the universe. Much of the present-day helium was also produced directly from the Big Bang, while the remainder was generated from hydrogen burning in stellar interiors.

Relative to hydrogen and helium, lithium, beryllium, and boron are very under-abundant. There are two reasons for this: They are not prominent end products of nuclear reaction chains, and they can be destroyed by collisions with protons. For lithium this occurs at

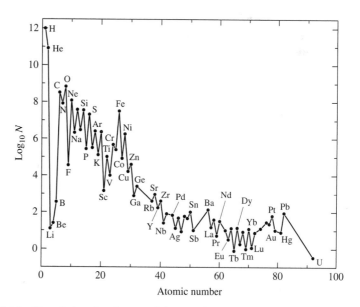

**FIGURE 15.16**   The relative abundances of elements in the Sun's photosphere. All abundances are normalized relative to $10^{12}$ hydrogen atoms. (Data from Grevesse and Sauval, *Space Sci. Rev., 85,* 161, 1998.)

temperatures greater than about $2.7 \times 10^6$ K, while for beryllium the required temperature is $3.5 \times 10^6$ K. It is the Sun's surface convection zone that is responsible for transporting the surface lithium, beryllium, and boron into the interior. When the present-day solar composition is compared with the abundances of meteorites (which should be similar to the Sun's primordial composition), we find that the relative abundances of beryllium are comparable but that the Sun's surface composition of lithium is smaller than the meteorites' lithium abundance by a factor of about 100. This suggests that lithium has been destroyed in the Sun since the star's formation but that beryllium has not been appreciably depleted. Apparently the base of the solar convection zone extends down sufficiently far to burn lithium but not far enough to burn beryllium. However, combining stellar structure theory, including the mixing-length theory of convection, with the analysis of solar oscillations (see Section 14.5) indicates that the base of the convection zone extends down to $2.3 \times 10^6$ K, not far enough to burn lithium adequately. The disagreement of standard models with the observations is known as the **solar lithium problem**.[9]

Peaks occur in Fig. 15.16 for elements such as carbon, nitrogen, oxygen, neon, and so on because they are created as a consequence of a star's evolutionary trek toward the iron peak and because they are relatively stable $\alpha$-particle-rich nuclei (recall Fig. 10.9).

Core-collapse supernovae are also responsible for the generation of significant quantities of oxygen, and Type Ia supernovae are responsible for the creation of most of the iron observed in the cosmos.

### s-Process and r-Process Nucleosynthesis

When nuclei having progressively higher values of $Z$ (the number of protons) form via stellar nucleosynthesis, it becomes increasingly difficult for other charged particles, such as protons, alpha particles, and so on, to react with them. The cause is the existence of a high Coulomb potential barrier. However, the same limitation does not exist when neutrons collide with these nuclei. Consequently, nuclear reactions involving neutrons can occur even at relatively low temperatures, assuming, of course, that free neutrons are present in the gas. The reactions with neutrons

$$^A_Z X + n \rightarrow {}^{A+1}_Z X + \gamma$$

result in more massive nuclei that are either stable or unstable against the beta-decay reaction,

$$^{A+1}_Z X \rightarrow {}^{A+1}_{Z+1} X + e^- + \overline{\nu}_e + \gamma.$$

If the beta-decay half-life is short compared to the timescale for neutron capture, the neutron-capture reaction is said to be a *slow process* or an **s-process** reaction (recall the discussion on page 467). s-Process reactions tend to yield stable nuclei, either directly or secondarily via beta decay. On the other hand, if the half-life for the beta-decay reaction is long compared with the time scale for neutron capture, the neutron-capture reaction is termed a *rapid*

---

[9]Perhaps because of their momenta, descending convective bubbles *overshoot* the bottom of the zone of convective instability, causing lithium to be transported deeper than the standard models suggest. Additional effects may also derive from diffusion and the interaction of convection with rotation.

*process* or ***r*-process** and results in neutron-rich nuclei. *s*-Process reactions tend to occur in normal phases of stellar evolution, whereas *r*-processes can occur during a supernova when a large flux of neutrinos exists. Although neither process plays significant roles in energy production, they do account for the abundance ratios of nuclei with $A > 60$.

## 15.4 ■ GAMMA-RAY BURSTS

One of the many great detective stories of modern astrophysics began in the 1960s with the launch of the Vela series of military satellites. The Vela spacecraft were designed to monitor compliance of the former Soviet Union with the 1963 nuclear test ban treaty by looking for sudden bursts of gamma rays of terrestrial origin. By 1967 it was clear that the **gamma-ray bursts** (GRB) that were being detected were coming from above rather than below, but it was not until 1973 that this information was released to the public.

About once per day, at some random location in the sky, a shower of gamma-ray photons with energies ranging from about 1 keV to many GeV appears. (Although the lower end of this range includes X-ray photons, most of the energy is in gamma rays.) The bursts last from $10^{-2}$ to $10^3$ s, and they have rise times as fast as $10^{-4}$ s, followed by an exponential decay. The bursts are usually multiply peaked and complex, although there is no typical burst profile. Two examples of gamma-ray bursts recorded by the Burst and Transient Source Experiment (BATSE) onboard the Compton Gamma-Ray Observatory (CGRO) are shown in Fig. 15.17.

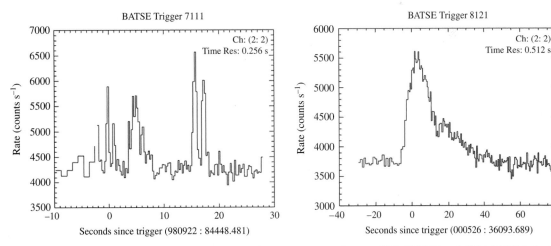

**FIGURE 15.17**   Light curves of two gamma-ray bursts, GRB 980922 and GRB 000526, in the energy range between 50 keV and 100 keV. The data were obtained by BATSE onboard the Compton Gamma-Ray Observatory. The dates of the two events are recorded in their designations; GRB 980922 occurred on September 22, 1998, and GRB 000526 occurred on May 26, 2000. GRB 0000526 was the last gamma-ray burst recorded by BATSE before the Compton Gamma-Ray Observatory was deorbited. (Courtesy of the BATSE Team – NASA.)

**Are the Sources of GRBs Galactic or Extragalactic?**

Until the late 1990s much of the mystery of GRBs was associated with their distances and the resulting implied energy of the events. It was not clear whether gamma-ray bursts originated in our Solar System (perhaps in the Oort cloud of comets), within our Galaxy, or the farthest reaches of the universe.

Without knowing the distance to the sources of the gamma rays, we must give their energy output in terms of the total energy received per unit area of detector surface during the burst (the energy flux integrated over the duration of the burst). This quantity, called the **fluence**, $S$, can be as small as $10^{-12}$ J m$^{-2}$ or as great as $10^{-7}$ J m$^{-2}$. The most energetic bursts recorded by the CGRO saturated its detectors. For example, the "Super Bowl burst," named because it occurred on Super Bowl Sunday (January 31) in 1993, lasted only a second, but an afterglow of energetic photons (up to 1 GeV each) persisted for about 100 seconds. Then, on December 15, 1994, the CGRO measured a burst that lasted for 90 minutes, with a peak photon energy of 18 GeV!

---

**Example 15.4.1.**   Suppose that the fluence of a particular GRB was determined to be $10^{-7}$ J m$^{-2}$. Assuming that the source of the burst was located 50,000 AU away in the Oort cloud of comets within our Solar System, and assuming further that the emission of the energy from the source was isotropic, the energy of the burst would necessarily have been

$$E = \left(4\pi r^2\right) S = 4\pi (50{,}000 \text{ AU})^2 \left(10^{-7} \text{ J m}^{-2}\right) = 7 \times 10^{25} \text{ J.}$$

On the other hand, if the source of the GRB were located 1 Gpc away in a distant galaxy, then the amount of energy involved in the burst (again assuming isotropy) would have been

$$E = \left(4\pi r^2\right) S = 4\pi (1 \text{ Gpc})^2 \left(10^{-7} \text{ J m}^{-2}\right) = 1 \times 10^{45} \text{ J,}$$

comparable to the energy released in a Type II supernova, including its neutrino emission.

The difference between the two estimates is almost 20 orders of magnitude! Clearly it is critical to understand the distance to these objects in order to begin to understand the underlying process.

---

Even before the CGRO was released by the Space Shuttle *Atlantis* on April 5, 1991, most astronomers agreed that the gamma-ray burst mechanism involved the presence of a neutron star. The short rise times of the bursts, when multiplied by the speed of light, result in a characteristic length comparable to the size of a neutron star ($ct_{\text{rise}}$ as small as 30 km). Observations of emission lines corresponding to photons with energies of roughly 350 to 500 keV are thought to be due to the 511-keV photons that are created when an electron and a positron annihilate each other and produce two gamma-ray photons near the surface of a neutron star. The energy of the photons is reduced by as much as 25% as they climb out of the severe potential well of the neutron star. Other spectral features with photons in the range of 20–60 keV have been identified as cyclotron lines due to a $10^8$ T magnetic field, typical of the fields of pulsars. The consensus was that the gamma-ray bursts were produced by neutron stars in the thick disk of the Galaxy at distances of a few hundred parsecs; the GRBs occurred either by tapping their internal energy (as in a pulsar glitch) or through accretion in a close binary system.

2704 BATSE Gamma-Ray Bursts

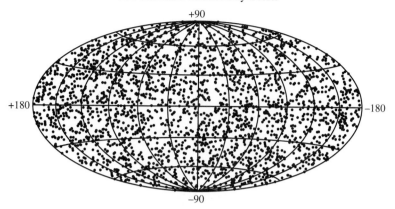

**FIGURE 15.18**   The isotropic angular distribution of 2704 gamma-ray bursts seen by the BATSE detector onboard the CGRO. (Courtesy of the BATSE Team – NASA.)

There were problems with this scenario. If the bursts were due to accretion in a binary system, why didn't they repeat? Furthermore, the bursts were distributed uniformly over the celestial sphere, instead of being concentrated in the plane of the Milky Way like most pulsars and X-ray binaries. It was thought that this could be due to the relative insensitivity of the gamma-ray detectors employed in the pre-Compton era. These instruments were unable to see sources beyond the disk of stars in our part of the Galaxy, and so the distribution of bursts appeared to be isotropic. (In the same manner, if we could see only the nearest stars, they would be uniformly scattered over the night sky instead of being concentrated along the Milky Way.) It was thought that this situation would surely change with the launch of the much more sensitive Compton Gamma Ray Observatory satellite. On average, one burst was observed every 25 hours by the spacecraft. Figure 15.18 shows the distribution of 2704 gamma-ray bursts observed by BATSE; there is no statistically significant deviation from an isotropic distribution.

Equally interesting was the finding that although the sources are spread evenly across the sky, they do not appear to be distributed homogeneously throughout space.[10] A classic argument was employed to determine whether there is an edge to the distribution. Let $E$ be the energy of a gamma-ray burst, located at a distance $r$ from Earth. Then the fluence is

$$S = \frac{E}{4\pi r^2}, \tag{15.13}$$

assuming an isotropic burst. Solving this expression for $r$, we have

$$r(S) = \left( \frac{E}{4\pi S} \right)^{1/2}.$$

---

[10]A homogeneous distribution would have the same number density of sources everywhere, independent of distance or direction.

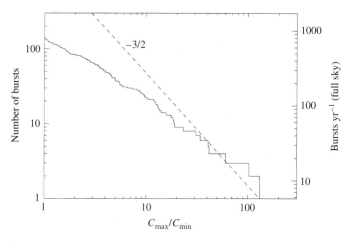

**FIGURE 15.19**   Violation of the proportionality $N \propto S^{-3/2}$, indicating an edge to the distribution of gamma-ray burst sources. The maximum gamma-ray count rate, $C_{max}$, is plotted instead of the fluence, $S$; $C_{min}$ is the weakest burst the CGRO can confidently detect. (Figure adapted from Meegan et al., *Nature*, *355*, 143, 1992.)

Assuming that all burst sources have the same intrinsic energy, $E$, then for a specific value of $S$ (say, $S_0$), all of the sources within a sphere of radius $r(S_0)$ will be observed to have a fluence $S \geq S_0$.[11] If there are $n$ burst sources per unit volume, then the number of sources with a fluence equal to or greater than $S_0$ is

$$N(S) = \frac{4}{3}\pi n r^3(S) = \frac{4}{3}\pi n \left(\frac{E}{4\pi S}\right)^{3/2},$$  (15.14)

where the "0" subscript has been dropped. Thus, if the sources of the gamma-ray bursts are distributed uniformly throughout space, the number of bursts observed with a fluence greater than some value $S$ would be proportional to $S^{-3/2}$. The Compton results show that this proportionality is violated when $S$ is small enough to include the more distant, fainter, sources; see Fig. 15.19.[12] This implies that there is an edge to the distribution; the burst sources do not extend outward without limit (but of course the limit could be the edge of the observable universe). When this result is combined with the fact that the distribution is isotropic, the implication is that Earth is near the center of a spherically symmetric distribution of gamma-ray burst sources.

The resolution to the distance question finally came with the detection of GRB 970228 on February 28, 1997, by the BeppoSAX spacecraft. (BeppoSAX is the product of a collaboration between the Italian Space Agency and the Netherlands Agency for Aerospace Programs.) The Gamma-Ray Burst Monitor on BeppoSAX first noticed the event, allowing

---

[11] It may be that there are different populations of burst sources with different characteristic values of $E$. However, the following argument is still valid if each population is homogeneously distributed.

[12] Note that in Fig. 15.19, the maximum count rate is plotted instead of the fluence. It is the maximum count rate that determines whether a burst is detected, making it more appropriate to use in the statistics of burst counts.

its wide-field X-ray camera to localize the region of the sky to within 3'. Within a few hours, the observatory's narrow-field X-ray telescopes were able to further localize the source. With rapid knowledge of the position of the GRB, it became possible for other ground-based and orbiting observatories to investigate that area of the sky as well. Even after the gamma-ray signature had vanished, fading X-ray and optical counterparts were detected. Deep images of the region, obtained using the Keck Observatory and the Hubble Space Telescope, revealed that the GRB had occurred in a distant galaxy, indicating that GRB 970228 originated at a cosmological (extragalactic) distance.

Additional cosmological discoveries of GRBs from BeppoSAX and other rapid-response gamma-ray telescopes [including NASA's High Energy Transient Explorer (HETE-2) and Swift missions] were quickly followed up with optical identifications that have verified the cosmological distances involved in GRB events. Numerous GRBs have since been associated with fading X-ray, optical, and radio counterparts in distant galaxies. This implies that GRBs are among the most energetic phenomena in the universe, comparable to the staggering energy release of core-collapse supernovae.

### Two Classes of GRBs

Now that the true distances to gamma-ray bursts can be determined and the scale of the energy output confirmed, it becomes possible to evaluate proposed mechanisms for generating GRBs. After the study of thousands of events, it is clear that there are two basic classes of gamma-ray bursts. Those events that last longer than 2 seconds are referred to as **long–soft** GRBs, while those that are shorter than 2 seconds are **short–hard** events. "Soft" and "hard" refer to having more of the event energy at lower energies or higher energies, respectively.

Just as there are two fundamentally different types of supernovae (Type Ia and core-collapse), it appears that there are two fundamentally different types of gamma-ray bursts. The short–hard bursts seem to be associated with neutron star–neutron star or neutron star–black hole mergers, whereas long–soft bursts may be connected with supernovae. As with Type Ia supernovae, we will defer the discussion of short–hard GRBs until later (Section 18.6), after we have had an opportunity to study the physics of compact objects more thoroughly.

### Core-Collapse Supernovae and Long–Soft GRBs

A direct link between a supernova and a long–soft GRB was established with the detection of GRB 980425. At a distance of 40 Mpc, GRB 980425 was determined to be about five orders of magnitude less energetic ($8 \times 10^{40}$ J) than a typical GRB. What was particularly important, however, was the detection of a supernova, SN 1998bw, at the same location. It seems that SN 1998bw was a particularly energetic Type Ib or Ic supernova with a total energy output of between 2 and $6 \times 10^{45}$ J (about 30 times greater than a typical Type Ib/c). It is likely that the remnant core that collapsed was 3 $M_\odot$, resulting in a black hole.

A second GRB–supernova connection has also been identified, this one between GRB 030329 and SN 2003dh. In this case the energy of the gamma-ray burst is more typical of other GRBs.

### Models of Long–Soft GRBs

Several models have been proposed to explain long–soft gamma-ray bursts, but one common ingredient in the models involves the *beaming of highly relativistic matter*. The prodigious amount of energy that is apparently produced in a GRB may be reduced by replacing the assumption of isotropic emission (recall Example 15.4.1) with **relativistic jets**, illustrated schematically in Fig. 15.20. According to Eq. (4.43), radiation is emitted in a cone having an opening angle of half-width $\theta \sim 1/\gamma$ for $\gamma \gg 1$ (see Fig. 4.10), where

$$\gamma \equiv \frac{1}{\sqrt{1 - u^2/c^2}}$$

is the Lorentz factor (see Eq. 4.20). If the emitted radiation is beamed forward (rather than isotropically) as the jet of material advances at speeds near the speed of light, and if Earth lies within the beam, the amount of energy produced by the GRB will seem to be greater than it actually is. Models suggest that Lorentz factors of up to 100 or more may be possible in gamma-ray bursts, implying that the actual energy produced and emitted into the solid angle of the jet may be smaller by a factor of $\gamma^2$—perhaps 10,000 times less than the isotropic assumption would suggest.

One challenge with the concept of a highly relativistic jet is that expanding material could encounter baryon-rich material, causing the jet to slow down. This would happen after the jet material had swept up matter with a total rest-mass energy on the order of $\gamma mc^2$, where $m$ is the mass of the jet's material.

The first viable model proposed to explain long–soft bursts is the **collapsar** model of Stan Woosley (sometimes also referred to as a *hypernova* model). As we will learn in Section 16.6, neutron stars have an upper mass limit based on the ability of neutron degeneracy to support the extremely compact star (similar to electron degeneracy pressure). Using sophisticated equations of state for neutrons at very high densities ($\rho \sim 10^{18}$ kg m$^{-3}$), investigators have estimated that the maximum mass of a nonrotating neutron star is about 2.2 M$_\odot$. When a core-collapse supernova occurs, either a neutron star or a black hole will form, depending on the mass, metallicity, and rotation of the progenitor star. Woosley's models have suggested that for a progenitor star with sufficiently great mass (possibly a Wolf–Rayet star), the central object to form will be a black hole with a debris disk surrounding it. The collimating effect of the debris disk and associated magnetic fields would lead to a jet emanating from the center of the supernova. Since the jet material will be highly relativistic, it will appear to be further collimated. The jet will plow its way through the overlying material of the infalling stellar envelope producing bursts of gamma-rays. One version of Woosley's collapsar model is shown in Fig. 15.21.

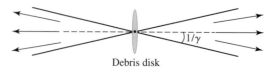

Debris disk

**FIGURE 15.20**   A relativistic jet of material will appear to the observer as having a cone opening of half-width $\theta \sim 1/\gamma$.

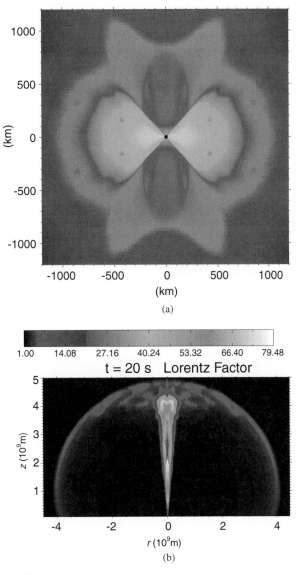

**FIGURE 15.21** A collapsar model of the formation of a gamma-ray burst event. (a) The central region of the star following the formation of a black hole and debris disk. (b) The emerging relativistic jet. (Figures courtesy of Weiqun Zhang and Stan Woosley.)

An alternative to the collapsar model is the **supranova** model. Rather than the black hole forming immediately during the core-collapse supernova, a delay in the mechanism occurs. Although the upper mass limit of a static neutron star is believed to be 2.2 $M_\odot$, a rapidly rotating neutron star can be supported with a mass of up to 2.9 $M_\odot$. Neutron stars are also likely to have very strong magnetic fields associated with them that could lead to a slowing of the rotation rate.

The supranova model suggests that a supermassive rotating neutron star may form with $M > 2.2\,M_\odot$ from a core-collapse supernova that would then slow down over the course of weeks or months until it was no longer stable against further gravitational collapse. A catastrophic collapse to a black hole would then result. If a debris disk were to form around the black hole, the disk, in combination with the magnetic field, could produce a relativistic jet and a gamma-ray burst. One advantage of this model is that it would naturally explain the absence of baryons that would slow the jet since the envelope of the star would have been swept up in the supernova explosion that happened earlier.

## 15.5 ■ COSMIC RAYS

On August 7, 1912, Victor F. Hess (1883–1964) and two colleagues ascended in a balloon to an altitude of 5 km. During that six-hour flight, Hess made careful readings from three electroscopes that he used to measure the intensity of radiation.[13] As the balloon rose, Hess determined that the level of radiation increased with altitude. From his experiment he concluded, "The results of these observations seem best explained by a radiation of great penetrating power entering our atmosphere from above…" This event marked the birth of the study of **cosmic rays**.[14]

### Charged Particles from Space

Although referred to as "rays," this penetrating radiation is actually composed of charged particles. A wide range of masses and classes of particles have been identified in cosmic rays, from electrons, positrons, protons, and muons to a host of nuclei including, but not limited to, carbon, oxygen, neon, magnesium, silicon, iron, and nickel (products of stellar nucleosynthesis).

Particularly striking is the wide range of energies involved, from less than $10^7$ eV to at least $3 \times 10^{20}$ eV. Near the low end of the energy spectrum, cosmic rays impinge on the atmosphere with fluxes of more than 1 particle m$^{-2}$ s$^{-1}$, whereas the highest-energy cosmic rays have fluxes that are very low: less than 1 particle km$^{-2}$ century$^{-1}$. The flux as a function of energy is shown in Fig. 15.22.

### Sources of Cosmic Rays

The question naturally arises as to what the source (or sources) of these particles may be. One obvious answer is our Sun. As we saw in Chapter 11, the solar wind, flares, and coronal mass ejections emit charged particles into space on a routine basis. Although abundant, these **solar cosmic rays** (also known as **solar energetic particles**) are relatively low-energy particles as cosmic rays go. Recall that the fast solar wind has ion speeds on the order of 750 km s$^{-1}$. For protons, this corresponds to an energy of $E \sim 3$ keV. Even in

---

[13]An electroscope is a device used to detect electric charges. Simple electroscopes are often used as lecture demonstrations in introductory physics courses.

[14]Hess was a recipient of the Nobel Prize in 1936 for his discovery of cosmic rays. Hess shared the prize with Carl David Anderson, who discovered positrons; he identified them amid the myriad of cosmic-ray particles.

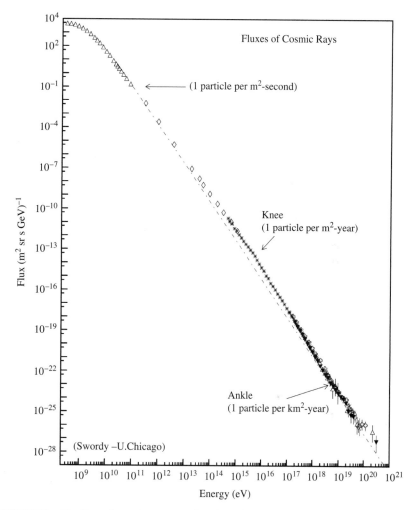

**FIGURE 15.22**   The flux of cosmic rays as a function of energy. (Ref: J. Cronin, T. K. Gaisser, and S. P. Swordy, *Sci. Amer.*, *276*, 44, 1997.)

energetic coronal mass ejections, when protons may travel at $v \sim 0.1c$, the energies are only on the order of 10 MeV, corresponding to the lowest-energy particles shown in the nearly flat portion of the curve in Fig. 15.22.

The sources of higher-energy cosmic rays have been identified with supernovae ($E \lesssim 10^{16}$ eV; notice the "knee" in Fig. 15.22 at between $10^{15}$ and $10^{16}$ eV). At such high energies the speed of the particle is essentially $c$ and the rest energy is negligible (for a proton, $mc^2 \simeq 937$ MeV $\sim 1$ GeV).

Consider the radius of the "orbit" of cosmic-ray particles about the magnetic field lines in the region of space in which they find themselves. From the Lorentz force equation (Eq. 11.2), if we neglect any contribution from electric fields, the force on a charged particle in a magnetic field is given by

$$F_B = qvB$$

for the special case when the velocity of the particle is perpendicular to the magnetic field. Since the Lorentz force is always perpendicular to the direction of motion, the force is centripetal and results in a circular path of the particle around the magnetic field. This implies that

$$\frac{\gamma m v^2}{r} = qvB,$$

where $\gamma$ is the Lorentz factor. Solving for $r$, the **Larmor radius** (or the **gyroradius**) of the orbit is given by

$$r = \frac{\gamma m v}{q B}. \tag{15.15}$$

Taking $v \sim c$, we find

$$r = \frac{\gamma m c^2}{qcB} = \frac{E}{qcB}. \tag{15.16}$$

---

**Example 15.5.1.**   If the Larmor radius of the "orbit" significantly exceeds the size scale for the magnetic field, the particle cannot be considered to be bound to the associated system. In interstellar space, magnetic field strengths of $10^{-10}$ T are typical. For a proton with an energy of $10^{15}$ eV, the Larmor radius is

$$r = 3 \times 10^{16} \text{ m} = 1 \text{ pc}.$$

This radius is characteristic of the size of a supernova remnant, suggesting that for energies much larger than $10^{15}$ eV, cosmic-ray particles are not likely to be bound to a supernova remnant.

---

Example 15.5.1 indicates that cosmic-ray particles with energies below about $10^{15}$ eV are possibly associated with supernova remnants, but once their energies exceed that limit, they escape from the remnant. It has long been suggested that the shock waves associated with supernovae could be sites of acceleration of cosmic-ray particles. Enrico Fermi was the first to propose a mechanism by which supernovae could accelerate charged particles to ultra-relativistic energies. He suggested that charged particles trapped in magnetic fields can be accelerated to very high energies through successive collisions with the advancing shock wave. After absorbing energy from the shock, a particle is accelerated forward in the direction of the shock motion (think of an elastic collision of a particle with an advancing wall). However, being tied to the magnetic field in the shock's vicinity, the particle is forced to return, only to collide with the shock again, receiving additional energy. The process is repeated many times until the particle possesses sufficient energy to escape the bonds of the

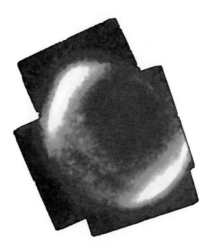

**FIGURE 15.23**   An X-ray image of SN 1006 obtained by ASCA. (Photo Credit: Dr. Eric V. Gotthelf, Columbia University.)

supernova's magnetic field. The power-law nature of the cosmic-ray spectrum ($F \propto E^{-\alpha}$) in Fig. 15.22 is characteristic of this nonthermal source of energy for these particles.

An X-ray image of SN 1006 (Fig. 15.23) obtained by Japan's Advanced Satellite for Cosmology and Astrophysics (ASCA) suggests strongly that supernova remnant shock fronts are indeed the source of acceleration of cosmic-ray particles with energies below about $10^{15}$ eV.

It remains unclear where the highest-energy cosmic rays come from. From the existence of the "knee" above $10^{15}$ eV in the cosmic-ray spectrum, it appears that the source of higher-energy cosmic rays is fundamentally different from the supernova source below $10^{15}$ eV. In addition, the "ankle" near $10^{19}$ eV suggests yet another source for the most extreme cosmic rays. It has been proposed that the cosmic rays with energies between the "knee" and the "ankle" may be due to acceleration in the vicinity of neutron stars or black holes. On the other hand, the most energetic cosmic rays may originate from outside our Galaxy, possible from collisions involving intergalactic shocks, or perhaps from the active regions in the centers of most galaxies where supermassive black holes are believed to reside.

## SUGGESTED READING

### General

Cronin, James W., Gaisser, Thomas K., and Swordy, Simon P., "Cosmic Rays at the Energy Frontier," *Scientific American, 276*, 44, 1997.

Friedlander, Michael W., *A Thin Cosmic Rain: Particles from Outer Space*, Harvard University Press, Cambridge, MA, 2000.

Hurley, Kevin, "Probing the Gamma-Ray Sky," *Sky and Telescope*, December 1992.

Lattimer, J., and Burrows, A., "Neutrinos from Supernova 1987A," *Sky and Telescope*, October 1988.

Marschall, Laurence A., *The Supernova Story*, Princeton University Press, Princeton, 1994.

Wheeler, J. Craig, *Cosmic Catastrophes: Supernovae, Gamma-Ray Bursts, and Adventures in Hyperspace*, Cambridge University Press, Cambridge, 2000.

Woosley, S., and Weaver, T., "The Great Supernova of 1987," *Scientific American*, August 1989.

## Technical

Arnett, David, *Supernovae and Nucleosynthesis: An Investigation of the History of Matter, from the Big Bang to the Present*, Princeton University Press, Princeton, 1996.

Arnett, W. David, Bahcall, John N., Kirshner, Robert P., and Woosley, Stanford E., "Supernova 1987A," *Annual Review of Astronomy and Astrophysics*, 27, 629, 1989.

Blaes, Omer M., "Theories of Gamma-Ray Bursts," *The Astrophysical Journal Supplement*, 92, 643, 1994.

Davidson, Kris, and Humphreys, Roberta M., "Eta Carinae and Its Environment," *Annual Review of Astronomy and Astrophysics*, 35, 1, 1997.

Fenimore, E. E., and Galassi, M. (eds.), *Gamma-Ray Bursts: 30 Years of Discovery*, AIP Conference Proceedings, 727, 2004.

Galama, T., et al., "The Decay of Optical Emission from the Gamma-Ray Burst GRB 970228," *Nature*, 387, 479, 1997.

Hansen, Carl J., Kawaler, Steven D., and Trimble, Virginia, *Stellar Interiors: Physical Principles, Structure, and Evolution*, Second Edition, Springer-Verlag, New York, 2004.

Heger, A., et al., "How Massive Stars End Their Life," *The Astrophysical Journal*, 591, 288, 2003.

Humphreys, Roberta M., and Davidson, Kris, "The Luminous Blue Variables: Astrophysical Geysers," *Publications of the Astronomical Society of the Pacific*, 106, 1025, 1994.

Massey, Philip, "Massive Stars in the Local Group: Implications for Stellar Evolution and Star Formation," *Annual Review of Astronomy and Astrophysics*, 41, 15, 2003.

Mészáros, P., "Theories of Gamma-Ray Bursts," *Annual Review of Astronomy and Astrophysics*, 40, 137, 2002.

Meynet, G., and Maeder, A., "Stellar Evolution with Rotation X. Wolf–Rayet Star Populations at Solar Metallicity," *Astronomy and Astrophysics*, 404, 975, 2003.

Petschek, Albert G. (ed.), *Supernovae*, Springer-Verlag, New York, 1990.

Piran, Tsvi, "The Physics of Gamma-Ray Bursts," *Reviews of Modern Physics*, 76, 1143, 2004.

Schlickeiser, Reinhard, *Cosmic Ray Astrophysics*, Springer-Verlag, Berlin, 2002.

Shore, Steven N., *The Tapestry of Modern Astrophysics*, John Wiley & Sons, Inc., Hoboken, 2003.

Stahl, O., et al., "A Spectroscopic Event of $\eta$ Car Viewed from Different Directions: The Data and First Results," *Astronomy and Astrophysics*, 435, 303, 2005.

Woosley, S. E., Zhang, Weiqun, and Heger, A., "The Central Engines of Gamma-Ray Bursts," *Gamma-Ray Burst and Afterglow Astronomy 2001: A Workshop Celebrating the First Year of the HETE Mission*, AIP Conference Proceedings, 662, 185, 2003.

## PROBLEMS

**15.1** Estimate the Eddington limit for $\eta$ Car and compare your answer with the luminosity of that star. Is your answer consistent with its behavior? Why or why not?

**15.2** During the Great Eruption of $\eta$ Car, the apparent visual magnitude reached a characteristic value of $m_V \sim 0$. Assume that the interstellar extinction to $\eta$ Car is 1.7 magnitudes and that the bolometric correction is essentially zero.

(a) Estimate the luminosity of $\eta$ Car during the Great Eruption.

(b) Determine the total amount of photon energy liberated during the twenty years of the Great Eruption.

(c) If 3 $M_\odot$ of material was ejected at a speed of 650 km s$^{-1}$, how much energy went into the kinetic energy of the ejecta?

**15.3** The angular extent of one of the lobes of $\eta$ Car is approximately 8.5″. Assuming a constant expansion of the lobes of 650 km s$^{-1}$, estimate how long it has been since the Great Eruption that produced the lobes. Is this likely to be an overestimate or an underestimate? Justify your answer.

**15.4** (a) Show that the amount of radioactive material remaining in an initially pure sample is given by Eq. (15.10).

(b) Prove that

$$\lambda = \frac{\ln 2}{\tau_{1/2}}.$$

**15.5** Assume that the 1 $M_\odot$ core of a 10 $M_\odot$ star collapses to produce a Type II supernova. Assume further that 100% of the energy released by the collapsing core is converted to neutrinos and that 1% of the neutrinos are absorbed by the overlying envelope to power the ejection of the supernova remnant. Estimate the final radius of the stellar remnant if sufficient energy is to be liberated to just barely eject the remaining 9 $M_\odot$ to infinity. Be sure to state clearly any additional assumptions you make in determining your estimate of the final radius of the remnant.

**15.6** (a) The angular size of the Crab SNR is $4' \times 2'$ and its distance from Earth is approximately 2000 pc (see Fig. 15.4). Estimate the linear dimensions of the nebula.

(b) Using the measured expansion rate of the Crab and ignoring any accelerations since the time of the supernova explosion, estimate the age of the nebula.

**15.7** Taking the distance to the Crab to be 2000 pc, and assuming that the absolute bolometric magnitude at maximum brightness was characteristic of a Type II supernova, estimate its peak apparent magnitude. Compare this to the maximum brightness of the planet Venus ($m \simeq -4$), which is sometimes visible in the daytime.

**15.8** Using Eq. (12.26), make a crude estimate of the amount of time required for the homologous collapse of the inner portion of the iron core of a massive star, marking the beginning of a core-collapse supernova.

**15.9** (a) Assuming that the light curve of a supernova is dominated by the energy released in the radioactive decay of an isotope that has a decay constant of $\lambda$, show that the slope of the light curve is given by Eq. (15.11).

(b) Prove that Eq. (15.12) follows from Eq. (15.11).

**15.10** If the linear decline of a supernova light curve is powered by the radioactive decay of the ejecta, find the rate of decline (in mag d$^{-1}$) produced by the decay of $^{56}_{27}$Co $\rightarrow$ $^{56}_{26}$Fe, with a half-life of 77.7 days.

**15.11** The energy released during the decay of one $^{56}_{27}$Co atom is 3.72 MeV. If 0.075 M$_\odot$ of cobalt was produced by the decay of $^{56}_{28}$Ni in SN 1987A, estimate the amount of energy released per second through the radioactive decay of cobalt:

(a) just after the formation of the cobalt.

(b) one year after the explosion.

(c) Compare your answers with the light curve of SN 1987A given in Fig. 15.12.

**15.12** The neutrino flux from SN 1987A was estimated to be $1.3 \times 10^{14}$ m$^{-2}$ at the location of Earth. If the average energy per neutrino was approximately 4.2 MeV, estimate the amount of energy released via neutrinos during the supernova explosion.

**15.13** Using Eq. (10.22), estimate the gravitational binding energy of a neutron star with a mass 1.4 M$_\odot$ and a radius of 10 km. Compare your answer with the amount of energy released in neutrinos during the collapse of the iron core of Sk $-69$ 202 (the progenitor of SN 1987A).

**15.14** It is estimated that there are approximately 100,000 neutron stars in the Milky Way Galaxy. Show that if the observed gamma-ray bursts are associated with neutron stars in our Galaxy, then each source *must* repeat. If you make the extreme assumption that each neutron star produces bursts, what would be the average time between bursts?

**15.15** Consider an electron and positron that annihilate each other at the surface of a neutron star ($M = 1.4$ M$_\odot$, $R = 10$ km), producing two gamma-ray photons of the same energy. Show that each gamma ray has an energy of at least 511 keV.

**15.16** Suppose there are two populations of gamma-ray burst sources with energies $E_1$ and $E_2$. Show that if the sources are distributed homogeneously throughout the universe with number densities $n_1$ and $n_2$, respectively, then the total number of bursts observed to have a fluence $\geq S$ is proportional to $S^{-3/2}$.

**15.17** The highest-energy cosmic-ray particle that had been recorded at the time this text was written was measured by the Fly's Eye HiRes experiment in the Utah desert in 1991. The energy of the particle was $3 \times 10^{20}$ eV.

(a) Convert the energy of the particle to joules.

(b) If the particle was a baseball of mass 0.143 kg, calculate the speed of the ball.

(c) Convert your answer to miles per hour and compare your answer to the speed of a fast ball of the fastest major league pitchers (approximately 100 mph, or 45 m s$^{-1}$).

**15.18** Using Eq. (15.16), show that cosmic-ray particles with energies of greater than $10^{19}$ eV are not likely to be bound to the Milky Way Galaxy. (A characteristic size scale for the Galaxy is about 30 kpc.) What about particles with energies in the range of $10^{16}$ eV to $10^{19}$ eV?

**15.19** Nonthermal spectra are often represented by power laws of the form

$$F = CE^{-\alpha}.$$

Figure 15.22 shows a power-law spectrum for cosmic rays. Determine the value of $\alpha$ in the region $10^{11}$ eV to the "knee," and from the "knee" to the "ankle."

**15.20** Calculate the Lorentz factor for a proton with an energy of $10^{20}$ eV.

# CHAPTER

# 16 The Degenerate Remnants of Stars

## 16.1 ■ THE DISCOVERY OF SIRIUS B

In 1838 Friedrich Wilhelm Bessel (1784–1846) used the technique of stellar parallax to find the distance to the star 61 Cygni. Following this first successful measurement of a stellar distance, Bessel applied his talents to another likely candidate: Sirius, the brightest appearing star in the sky. Its parallax angle of $p'' = 0.379''$ corresponds to a distance of only 2.64 pc, or 8.61 ly (see Appendix E). Sirius's brilliance in the night sky is due in part to its proximity to Earth. As he followed the star's path through the heavens, Bessel found that it deviated slightly from a straight line. After ten years of precise observations, Bessel concluded in 1844 that Sirius is actually a binary star system. Although unable to detect the companion of the brighter star, he deduced that its orbital period was about 50 years (the modern value is 49.9 years) and predicted its position. The search was on for the unseen "Pup," the faint companion of the luminous "Dog Star."

The telescopes of Bessel's time were incapable of finding the Pup so close to the glare of its bright counterpart, and following Bessel's death in 1846 the enthusiasm for the quest waned. Finally, in 1862, Alvan Graham Clark (1832–1897), son of the prominent American lensmaker Alvan Clark (1804–1887), tested his father's new 18-inch refractor (3 inches larger than any previous instrument) on Sirius, and he promptly discovered the Pup at its predicted position. The dominant Sirius A was found to be nearly one thousand times brighter than the Pup, now called Sirius B; see Fig. 16.1. The details of their orbits about their center of mass (see Fig. 16.2 and Problem 7.4) revealed that Sirius A and Sirius B have masses of about 2.3 $M_\odot$ and 1.0 $M_\odot$, respectively. A more recent determination for the mass of Sirius B is $1.053 \pm 0.028$ $M_\odot$, and it is this value that we will use.

Clark's discovery of Sirius B was made near the opportune time of apastron, when the two stars were most widely separated (by just 10''). The great difference in their luminosities ($L_A = 23.5$ $L_\odot$ and $L_B = 0.03$ $L_\odot$) makes observations at other times much more difficult.

557

**FIGURE 16.1**    The white dwarf, Sirius B, beside the overexposed image of Sirius A. (Courtesy of Lick Observatory.)

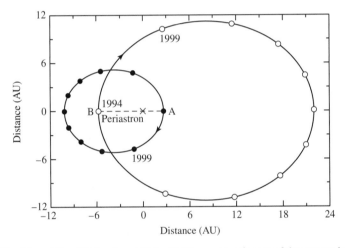

**FIGURE 16.2**    The orbits of Sirius A and Sirius B. The center of mass of the system is marked with an "×."

When the next apastron arrived 50 years later, spectroscopists had developed the tools to measure the stars' surface temperatures. From the Pup's faint appearance, astronomers expected it to be cool and red. They were startled when Walter Adams (1876–1956), working at Mt. Wilson Observatory in 1915, discovered that, to the contrary, Sirius B is a hot, blue-white star that emits much of its energy in the ultraviolet. A modern value of the temperature of Sirius B is 27,000 K, much hotter than Sirius A's 9910 K.

The implications for the star's physical characteristics were astounding. Using the Stefan–Boltzmann law, Eq. (3.17), to calculate the size of Sirius B results in a radius of only $5.5 \times 10^6$ m $\approx 0.008$ $R_\odot$. Sirius B has the mass of the Sun confined within a volume smaller than Earth! The average density of Sirius B is $3.0 \times 10^9$ kg m$^{-3}$, and the acceleration due to gravity at its surface is about $4.6 \times 10^6$ m s$^{-2}$. On Earth, the pull of gravity on a teaspoon of white-dwarf material would be $1.45 \times 10^5$ N (over 16 tons), and on the surface of the white dwarf it would weigh 470,000 times more. This fierce gravity reveals itself in the spectrum of Sirius B; it produces an immense pressure near the surface that results in very broad hydrogen absorption lines; see Fig. 8.15.[1] Apart from these lines, its spectrum is a featureless continuum.

Astronomers first reacted to the discovery of Sirius B by dismissing the results, calling them "absurd." However, the calculations are so simple and straightforward that this attitude soon changed to the one expressed by Eddington in 1922: "Strange objects, which persist in showing a type of spectrum entirely out of keeping with their luminosity, may ultimately teach us more than a host which radiate according to rule." Like all sciences, astronomy advances most rapidly when confronted with exceptions to its theories.

## 16.2 ■ WHITE DWARFS

Obviously Sirius B is not a normal star. It is a **white dwarf**, a class of stars that have approximately the mass of the Sun and the size of Earth. Although as many as one-quarter of the stars in the vicinity of the Sun may be white dwarfs, the average characteristics of these faint stars have been difficult to determine because a complete sample has been obtained only within 10 pc of the Sun.

### Classes of White Dwarf Stars

Figures 8.14 and 8.16 show that the white dwarfs occupy a narrow sliver of the H–R diagram that is roughly parallel to and below the main sequence. Although white dwarfs are typically whiter than normal stars, the name itself is something of a misnomer since they come in all colors, with surface temperatures ranging from less than 5000 K to more than 80,000 K. Their spectral type, D (for "dwarf"), has several subdivisions. The largest group (about two-thirds of the total number, including Sirius B), called **DA white dwarfs**, display only pressure-broadened hydrogen absorption lines in their spectra. Hydrogen lines are absent from the **DB white dwarfs** (8%), which show only helium absorption lines, and the **DC white dwarfs** (14%) show no lines at all—only a continuum devoid of features. The remaining types include **DQ white dwarfs**, which exhibit carbon features in their spectra, and **DZ white dwarfs** with evidence of metal lines.

### Central Conditions in White Dwarfs

It is instructive to estimate the conditions at the center of a white dwarf of mass $M_{wd}$ and radius $R_{wd}$, using the values for Sirius B given in the preceding section. Equation (14.5) with $r = 0$ shows that the central pressure is roughly[2]

$$P_c \approx \frac{2}{3} \pi G \rho^2 R_{wd}^2 \approx 3.8 \times 10^{22} \text{ N m}^{-2}, \tag{16.1}$$

---

[1] Recall the discussion of pressure broadening in Section 9.5.
[2] Remember that Eq. (14.5) was obtained for the unrealistic assumption of constant density.

about 1.5 million times larger than the pressure at the center of the Sun. A crude estimate of the central temperature may be obtained from Eq. (10.68) for the radiative temperature gradient,[3]

$$\frac{dT}{dr} = -\frac{3}{4ac} \frac{\overline{\kappa}\rho}{T^3} \frac{L_r}{4\pi r^2}$$

or

$$\frac{T_{\text{wd}} - T_c}{R_{\text{wd}} - 0} = -\frac{3}{4ac} \frac{\overline{\kappa}\rho}{T_c^3} \frac{L_{\text{wd}}}{4\pi R_{\text{wd}}^2}.$$

Assuming that the surface temperature, $T_{\text{wd}}$, is much smaller than the central temperature and using $\overline{\kappa} = 0.02 \text{ m}^2 \text{ kg}^{-1}$ for electron scattering (Eq. 9.27 with $X = 0$) give

$$T_c \approx \left[ \frac{3\overline{\kappa}\rho}{4ac} \frac{L_{\text{wd}}}{4\pi R_{\text{wd}}} \right]^{1/4} \approx 7.6 \times 10^7 \text{ K}.$$

Thus the central temperature of a white dwarf is several times $10^7$ K.

These estimated values for a white dwarf lead directly to a surprising conclusion. Although hydrogen makes up roughly 70% of the visible mass of the universe, it cannot be present in appreciable amounts below the surface layers of a white dwarf. Otherwise, the dependence of the nuclear energy generation rates on density and temperature (see Eq. 10.46 for the pp chain and Eq. 10.58 for the CNO cycle) would produce white dwarf luminosities several orders of magnitude larger than those actually observed. Similar reasoning applied to other reaction sequences implies that thermonuclear reactions are not involved in producing the energy radiated by white dwarfs and that their centers must therefore consist of particles that are incapable of fusion at these densities and temperatures.

As was discussed in Section 13.2, white dwarfs are manufactured in the cores of low- and intermediate-mass stars (those with an initial mass below 8 or 9 $M_\odot$ on the main sequence) near the end of their lives on the asymptotic giant branch of the H–R diagram. Because any star with a helium core mass exceeding about 0.5 $M_\odot$ will undergo fusion, most white dwarfs consist primarily of completely ionized carbon and oxygen nuclei.[4] As the aging giant expels its surface layers as a planetary nebula, the core is exposed as a white dwarf progenitor. The distribution of DA white dwarf masses is sharply peaked at 0.56 $M_\odot$, with some 80% lying between 0.42 $M_\odot$ and 0.70 $M_\odot$; see Fig. 16.3. The much larger main-sequence masses quoted earlier imply that significant amounts of mass loss occurred while on the asymptotic giant branch, involving thermal pulses and a superwind.

## Spectra and Surface Composition

The exceptionally strong pull of the white dwarf's gravity is responsible for the characteristic hydrogen spectrum of DA white dwarfs. Heavier nuclei are pulled below the surface while

---

[3]As we will discuss later in Section 16.5, the assumption of a radiative temperature gradient is incorrect because the energy is actually carried outward by electron conduction. However, Eq. (10.68) is sufficient for the purpose of this estimation.

[4]Low-mass helium white dwarfs may also exist, and rare oxygen–neon–magnesium white dwarfs have been detected in a few novae.

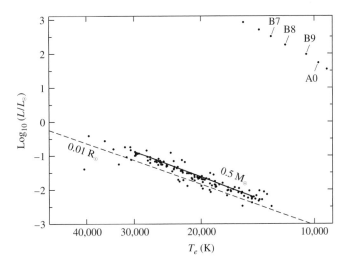

**FIGURE 16.3**   DA white dwarfs on an H–R diagram. A line marks the location of the 0.50 $M_\odot$ white dwarfs, and a portion of the main sequence is at the upper right. (Data from Bergeron, Saffer, and Liebert, *Ap. J.*, *394*, 228, 1992.)

the lighter hydrogen rises to the top, resulting in a thin outer layer of hydrogen covering a layer of helium on top of the carbon–oxygen core.[5] This vertical stratification of nuclei according to their mass takes only 100 years or so in the hot atmosphere of the star. The origin of the non-DA (e.g., DB and DC) white dwarfs is not yet clear. Efficient mass-loss may occur on the asymptotic giant branch associated with the thermal pulse or superwind phases, stripping the white dwarf of nearly all of its hydrogen. Alternatively, a single white dwarf may be transformed between the DA and non-DA spectral types by convective mixing in its surface layers.[6] For example, the helium convection zone's penetration into a thin hydrogen layer above could change a DA into a DB white dwarf by diluting the hydrogen with additional helium.

**Pulsating White Dwarfs**

White dwarfs with surface temperatures of $T_e \approx 12,000$ K lie within the instability strip of the H–R diagram and pulsate with periods between 100 and 1000 s; see Fig. 8.16 and Table 14.1. These **ZZ Ceti** variables, named after the prototype discovered in 1968 by Arlo Landolt, are variable DA white dwarfs; hence they are also known as **DAV stars**. The pulsation periods correspond to nonradial g-modes that resonate within the white dwarf's surface layers of hydrogen and helium.[7] Because these g-modes involve almost perfectly

---

[5]Estimates of the relative masses of the hydrogen and helium layers range from $m(\text{H})/m(\text{He}) \approx 10^{-2}$ to $10^{-11}$ for DA white dwarfs.

[6]As we will see in Section 16.5, steep temperature gradients produce convection zones in the white dwarf's surface layers.

[7]The nonradial pulsation of stars was discussed in Section 14.4. Unlike the g-modes of normal stars, shown in Fig. 14.18, the g-modes of white dwarfs are confined to their surface layers.

horizontal displacements, the radii of these compact pulsators hardly change. Their brightness variations (typically a few tenths of a magnitude) are due to temperature variations on the stars' surfaces. Since most stars will end their lives as white dwarfs, these must be the most common type of variable star in the universe, although only about seventy had been detected at the time this text was written.

Successful numerical calculations of pulsating white dwarf models were carried out by American astronomer Don Winget and others. They were able to demonstrate that it is the *hydrogen* partial ionization zone that is responsible for driving the oscillations of the ZZ Ceti stars, as mentioned in Section 14.2. These computations also confirmed the elemental stratification of white dwarf envelopes. Winget and his colleagues went on to predict that hotter DB white dwarfs should also exhibit g-mode oscillations driven by the *helium* partial ionization zone. Within a year's time, this prediction was confirmed when the first **DBV** star ($T_e \approx 27{,}000$ K) was discovered by Winget and his collaborators.[8] The location of the DAV and DBV stars on the H–R diagram is shown in Fig. 16.4, along with the very hot DOV and PNNV ($T_e \approx 10^5$ K) variables that are associated with the birth of white dwarfs. ("PNN" stands for planetary nebula nuclei and the DO spectral type marks the

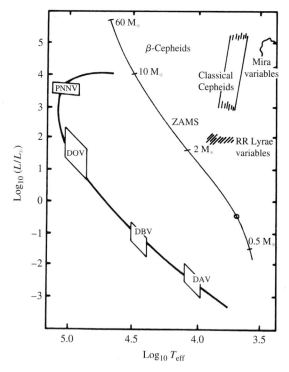

**FIGURE 16.4**    Compact pulsators on the H–R diagram. (Figure adapted from Winget, *Advances in Helio- and Asteroseismology*, Christensen-Dalsgaard and Frandsen (eds.), Reidel, Dordrecht, 1988.)

---

[8] Readers interested in this unique prediction and in the subsequent discovery of a new type of star are referred to Winget et al. (1982a,b).

transition to the white dwarf stage.) All of these stars have multiple periods, simultaneously displaying at least 3, and as many as 125, different frequencies. Astronomers are deciphering the data to obtain a detailed look at the structure of white dwarfs.

## 16.3 ■ THE PHYSICS OF DEGENERATE MATTER

We now delve below the surface to ask, What can support a white dwarf against the relentless pull of its gravity? It is easy to show (Problem 16.4) that normal gas and radiation pressure are completely inadequate. The answer was discovered in 1926 by the British physicist Sir Ralph Howard Fowler (1889–1944), who applied the new idea of the Pauli exclusion principle (recall Section 5.4) to the electrons within the white dwarf. The qualitative argument that follows elucidates the fundamental physics of the **electron degeneracy pressure** described by Fowler.

### The Pauli Exclusion Principle and Electron Degeneracy

Any system—whether an atom of hydrogen, an oven filled with blackbody photons, or a box filled with gas particles—consists of quantum states that are identified by a set of quantum numbers. Just as the oven is filled with standing waves of electromagnetic radiation that are described by three quantum numbers (specifying the number of photons of wavelength $\lambda$ traveling in the $x$-, $y$-, and $z$-directions), a box of gas particles is filled with standing de Broglie waves that are also described by three quantum numbers (specifying the particle's component of momentum in each of three directions). If the gas particles are fermions (such as electrons or neutrons), then the Pauli exclusion principle allows at most one fermion in each quantum state because no two fermions can have the same set of quantum numbers.

In an everyday gas at standard temperature and pressure, only one of every $10^7$ quantum states is occupied by a gas particle, and the limitations imposed by the Pauli exclusion principle become insignificant. Ordinary gas has a *thermal* pressure that is related to its temperature by the ideal gas law. However, as energy is removed from the gas and its temperature falls, an increasingly large fraction of the particles are forced into the lower energy states. If the gas particles are fermions, only one particle is allowed in each state; thus all the particles cannot crowd into the ground state. Instead, as the temperature of the gas is lowered, the fermions will fill up the lowest available unoccupied states, starting with the ground state, and then successively occupy the excited states with the lowest energy. Even in the limit $T \rightarrow 0$ K, the vigorous motion of the fermions in excited states produces a pressure in the fermion gas. At zero temperature, *all* of the lower energy states and *none* of the higher energy states are occupied. Such a fermion gas is said to be completely **degenerate**.

### The Fermi Energy

The maximum energy ($\varepsilon_F$) of any electron in a completely degenerate gas at $T = 0$ K is known as the **Fermi energy**; see Fig. 16.5. To determine this limiting energy, imagine a three-dimensional box of length $L$ on each side. Thinking of the electrons as being standing

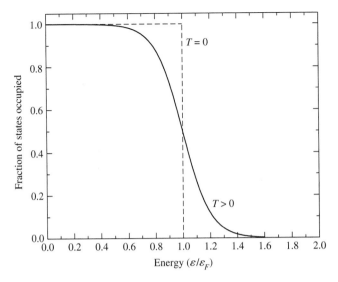

**FIGURE 16.5** Fraction of states of energy $\varepsilon$ occupied by fermions. For $T = 0$, all fermions have $\varepsilon \leq \varepsilon_F$, but for $T > 0$, some fermions have energies in excess of the Fermi energy.

waves in the box, we note that their wavelengths in each dimension are given by

$$\lambda_x = \frac{2L}{N_x}, \qquad \lambda_y = \frac{2L}{N_y}, \qquad \lambda_z = \frac{2L}{N_z},$$

where $N_x$, $N_y$, and $N_z$ are integer quantum numbers associated with each dimension. Recalling that the de Broglie wavelength is related to momentum (Eq. 5.17),

$$p_x = \frac{hN_x}{2L}, \qquad p_y = \frac{hN_y}{2L}, \qquad p_y = \frac{hN_x}{2L}.$$

Now, the total kinetic energy of a particle can be written as

$$\varepsilon = \frac{p^2}{2m},$$

where $p^2 = p_x^2 + p_y^2 + p_z^2$. Thus,

$$\varepsilon = \frac{h^2}{8mL^2}(N_x^2 + N_y^2 + N_z^2) = \frac{h^2N^2}{8mL^2}, \qquad (16.2)$$

where $N^2 \equiv N_x^2 + N_y^2 + N_z^2$, analogous to the "distance" from the origin in "$N$-space" to the point $(N_x, N_y, N_z)$.

The total number of electrons in the gas corresponds to the total number of unique quantum numbers, $N_x$, $N_y$, and $N_z$ times two. The factor of two arises from the fact that electrons are spin $\frac{1}{2}$ particles, so $m_s = \pm 1/2$ implies that two electrons can have the same combination of $N_x$, $N_y$, and $N_z$ and still posses a unique set of *four* quantum numbers (including spin). Now, each integer coordinate in $N$-space (e.g., $N_x = 1, N_y = 3, N_z = 1$) corresponds to the quantum state of two electrons. With a large enough sample of electrons, they can be thought of as occupying each integer coordinate out to a radius of $N = \sqrt{N_x^2 + N_y^2 + N_z^2}$, but only for the positive octant of $N$-space where $N_x > 0$, $N_y > 0$, and $N_z > 0$. This means that the total number of electrons will be

$$N_e = 2 \left(\frac{1}{8}\right) \left(\frac{4}{3}\pi N^3\right).$$

Solving for $N$ yields

$$N = \left(\frac{3N_e}{\pi}\right)^{1/3}.$$

Substituting into Eq. (16.2) and simplifying, we find that the Fermi energy is given by

$$\varepsilon_F = \frac{\hbar^2}{2m} \left(3\pi^2 n\right)^{2/3}, \tag{16.3}$$

where $m$ is the mass of the electron and $n \equiv N_e/L^3$ is the number of electrons per unit volume. The average energy per electron at zero temperature is $\frac{3}{5}\varepsilon_F$. (Of course the derivation above applies for any fermion, not just electrons.)

**The Condition for Degeneracy**

At any temperature above absolute zero, some of the states with an energy less than $\varepsilon_F$ will become vacant as fermions use their thermal energy to occupy other, more energetic states. Although the degeneracy will not be precisely complete when $T > 0$ K, the assumption of complete degeneracy is a good approximation at the densities encountered in the interior of a white dwarf. All but the most energetic particles will have an energy less than the Fermi energy. To understand how the degree of degeneracy depends on both the temperature and the density of the white dwarf, we first express the Fermi energy in terms of the density of the electron gas. For full ionization, the number of electrons per unit volume is

$$n_e = \left(\frac{\text{\# electrons}}{\text{nucleon}}\right) \left(\frac{\text{\# nucleons}}{\text{volume}}\right) = \left(\frac{Z}{A}\right) \frac{\rho}{m_H}, \tag{16.4}$$

where $Z$ and $A$ are the number of protons and nucleons, respectively, in the white dwarf's nuclei, and $m_H$ is the mass of a hydrogen atom.[9] Thus the Fermi energy is proportional to the 2/3 power of the density,

$$\varepsilon_F = \frac{\hbar^2}{2m_e} \left[3\pi^2 \left(\frac{Z}{A}\right) \frac{\rho}{m_H}\right]^{2/3}. \tag{16.5}$$

---

[9]The hydrogen mass is adopted as a representative mass of the proton and neutron.

Now compare the Fermi energy with the average thermal energy of an electron, $\frac{3}{2}kT$ (where $k$ is Boltzmann's constant; see Eq. 10.17). In rough terms, if $\frac{3}{2}kT < \varepsilon_F$, then an average electron will be unable to make a transition to an unoccupied state, and the electron gas will be degenerate. That is, for a degenerate gas,

$$\frac{3}{2}kT < \frac{\hbar^2}{2m_e}\left[3\pi^2\left(\frac{Z}{A}\right)\frac{\rho}{m_H}\right]^{2/3},$$

or

$$\frac{T}{\rho^{2/3}} < \frac{\hbar^2}{3m_e k}\left[\frac{3\pi^2}{m_H}\left(\frac{Z}{A}\right)\right]^{2/3} = 1261 \text{ K m}^2 \text{ kg}^{-2/3}$$

for $Z/A = 0.5$. Defining

$$\mathcal{D} \equiv 1261 \text{ K m}^2 \text{ kg}^{-2/3},$$

the condition for degeneracy may be written as

$$\boxed{\frac{T}{\rho^{2/3}} < \mathcal{D}.} \tag{16.6}$$

The smaller the value of $T/\rho^{2/3}$, the more degenerate the gas.

---

**Example 16.3.1.**   How important is electron degeneracy at the centers of the Sun and Sirius B? At the center of the standard solar model (see Table 11.1), $T_c = 1.570 \times 10^7$ K and $\rho_c = 1.527 \times 10^5$ kg m$^{-3}$. Then

$$\frac{T_c}{\rho_c^{2/3}} = 5500 \text{ K m}^2 \text{ kg}^{-2/3} > \mathcal{D}.$$

In the Sun, electron degeneracy is quite weak and plays a very minor role, supplying only a few tenths of a percent of the central pressure. However, as the Sun continues to evolve, electron degeneracy will become increasingly important (Fig. 16.6). As described in Section 13.2, the Sun will develop a degenerate helium core while on the red giant branch of the H–R diagram, leading eventually to a core helium flash. Later, on the asymptotic giant branch, the progenitor of a carbon–oxygen white dwarf will form in the core to be revealed when the Sun's surface layers are ejected as a planetary nebula.

For Sirius B, the values of the density and central temperature estimated above lead to

$$\frac{T_c}{\rho_c^{2/3}} = 37 \text{ K m}^2 \text{ kg}^{-2/3} \ll \mathcal{D},$$

so complete degeneracy is a valid assumption for Sirius B.

---

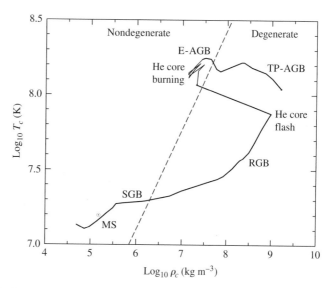

**FIGURE 16.6** Degeneracy in the Sun's center as it evolves. (Data from Mazzitelli and D'Antona, *Ap. J.*, *311*, 762, 1986.)

### Electron Degeneracy Pressure

We now estimate the electron degeneracy pressure by combining two key ideas of quantum mechanics:

1. The Pauli exclusion principle, which allows at most one electron in each quantum state; and

2. Heisenberg's uncertainty principle in the form of Eq. (5.19),

$$\Delta x \, \Delta p_x \approx \hbar,$$

which requires that an electron confined to a small volume of space have a correspondingly high uncertainty in its momentum. Because the minimum value of the electron's momentum, $p_{min}$, is approximately $\Delta p$, more closely confined electrons will have greater momenta.

When we make the unrealistic assumption that all of the electrons have the same momentum, $p$, Eq. (10.8) for the pressure integral becomes

$$P \approx \frac{1}{3} n_e p v, \tag{16.7}$$

where $n_e$ is the total electron number density.

In a completely degenerate electron gas, the electrons are packed as tightly as possible, and for a uniform number density of $n_e$, the separation between neighboring electrons is about $n_e^{-1/3}$. However, to satisfy the Pauli exclusion principle, the electrons must maintain their identities as different particles. That is, the uncertainty in their positions cannot be larger than their physical separation. Identifying $\Delta x \approx n_e^{-1/3}$ for the limiting case of complete degeneracy, we can use Heisenberg's uncertainty relation to estimate the momentum of an electron. In one coordinate direction,

$$p_x \approx \Delta p_x \approx \frac{\hbar}{\Delta x} \approx \hbar n_e^{1/3} \tag{16.8}$$

(see Example 5.4.2). However, in a three-dimensional gas each of the directions is equally likely, implying that

$$p_x^2 = p_y^2 = p_z^2,$$

which is just a statement of the equipartition of energy among all the coordinate directions. Therefore,

$$p^2 = p_x^2 + p_y^2 + p_z^2 = 3p_x^2,$$

or

$$p = \sqrt{3}p_x.$$

Using Eq. (16.4) for the electron number density with full ionization gives

$$p \approx \sqrt{3}\hbar \left[ \left( \frac{Z}{A} \right) \frac{\rho}{m_H} \right]^{1/3}.$$

For nonrelativistic electrons, the speed is

$$v = \frac{p}{m_e}$$

$$\approx \frac{\sqrt{3}\hbar}{m_e} n_e^{1/3} \tag{16.9}$$

$$\approx \frac{\sqrt{3}\hbar}{m_e} \left[ \left( \frac{Z}{A} \right) \frac{\rho}{m_H} \right]^{1/3}. \tag{16.10}$$

Inserting Eqs. (16.4), (16.8), and (16.10) into Eq. (16.7) for the electron degeneracy pressure results in

$$P \approx \frac{\hbar^2}{m_e} \left[ \left( \frac{Z}{A} \right) \frac{\rho}{m_H} \right]^{5/3}. \tag{16.11}$$

This is roughly a factor of two smaller than the exact expression for the pressure due to a completely degenerate, nonrelativistic electron gas $P$,

$$P = \frac{(3\pi^2)^{2/3}}{5} \frac{\hbar^2}{m_e} n_e^{5/3},$$

or

$$P = \frac{(3\pi^2)^{2/3}}{5} \frac{\hbar^2}{m_e} \left[ \left( \frac{Z}{A} \right) \frac{\rho}{m_H} \right]^{5/3}. \tag{16.12}$$

Using $Z/A = 0.5$ for a carbon–oxygen white dwarf, Eq. (16.12) shows that the electron degeneracy pressure available to support a white dwarf such as Sirius B is about $1.9 \times 10^{22}$ N m$^{-2}$, within a factor of two of the estimate of the central pressure made previously (Eq. 16.1). *Electron degeneracy pressure is responsible for maintaining hydrostatic equilibrium in a white dwarf.*

You may have noticed that Eq. (16.12) is the polytropic equation of state, $P = K\rho^{5/3}$, corresponding to $n = 1.5$. This implies that the extensive tools associated with the Lane–Emden equation (Eq. 10.110), developed beginning on page 334, can be used to study these objects. Of course, to understand them in detail requires careful numerical calculations involving the details of the complex equation of state of partially degenerate gases, nonzero temperatures, and changing compositions.

### 16.4 ∎ THE CHANDRASEKHAR LIMIT

The requirement that degenerate electron pressure must support a white dwarf star has profound implications. In 1931, at the age of 21, the Indian physicist Subrahmanyan Chandrasekhar announced his discovery that *there is a maximum mass for white dwarfs*. In this section we will consider the physics that leads to this amazing conclusion.

#### The Mass–Volume Relation

The relation between the radius, $R_{wd}$, of a white dwarf and its mass, $M_{wd}$, may be found by setting the estimate of the central pressure, Eq. (16.1), equal to the electron degeneracy pressure, Eq. (16.12):

$$\frac{2}{3} \pi G \rho^2 R_{wd}^2 = \frac{(3\pi^2)^{2/3}}{5} \frac{\hbar^2}{m_e} \left[ \left( \frac{Z}{A} \right) \frac{\rho}{m_H} \right]^{5/3}.$$

Using $\rho = M_{wd}/\frac{4}{3}\pi R_{wd}^3$ (assuming constant density), this leads to an estimate of the radius of the white dwarf,

$$R_{wd} \approx \frac{(18\pi)^{2/3}}{10} \frac{\hbar^2}{Gm_e M_{wd}^{1/3}} \left[ \left( \frac{Z}{A} \right) \frac{1}{m_H} \right]^{5/3}. \tag{16.13}$$

For a 1 $M_{\odot}$ carbon–oxygen white dwarf, $R \approx 2.9 \times 10^6$ m, too small by roughly a factor of two but an acceptable estimate. More important is the surprising implication that $M_{wd} R_{wd}^3 =$ constant, or

$$M_{wd} V_{wd} = \text{constant.} \tag{16.14}$$

The volume of a white dwarf is inversely proportional to its mass, so more massive white dwarfs are actually *smaller*. This **mass–volume relation** is a result of the star deriving its support from electron degeneracy pressure. The electrons must be more closely confined to generate the larger degeneracy pressure required to support a more massive star. In fact, the mass–volume relation implies that $\rho \propto M_{wd}^2$.

According to the mass–volume relation, piling more and more mass onto a white dwarf would eventually result in shrinking the star down to zero volume as its mass becomes infinite. However, if the density exceeds about $10^9$ kg m$^{-3}$, there is a departure from this relation. To see why this is so, use Eq. (16.10) to estimate the speed of the electrons in Sirius B:

$$v \approx \frac{\hbar}{m_e} \left[ \left( \frac{Z}{A} \right) \frac{\rho}{m_H} \right]^{1/3} = 1.1 \times 10^8 \text{ m s}^{-1},$$

over one-third the speed of light! If the mass–volume relation were correct, white dwarfs a bit more massive than Sirius B would be so small and dense that their electrons would exceed the limiting value of the speed of light. This impossibility points out the dangers of ignoring the effects of relativity in our expressions for the electron speed (Eq. 16.10) and pressure (Eq. 16.11).[10] Because the electrons are moving more slowly than the nonrelativistic Eq. (16.10) would indicate, there is less electron pressure available to support the star. Thus a massive white dwarf is *smaller* than predicted by the mass–volume relation. Indeed, zero volume occurs for a finite value of the mass; in other words, there is a limit to the amount of matter that can be supported by electron degeneracy pressure.

**Dynamical Instability**

To appreciate the effect of relativity on the stability of a white dwarf, recall that Eq. (16.12) (which is valid only for approximately $\rho < 10^9$ kg m$^{-3}$) is of the polytropic form $P = K\rho^{5/3}$, where $K$ is a constant. Comparing this with Eq. (10.86) shows that the value of the ratio of specific heats is $\gamma = 5/3$ in the nonrelativistic limit. As we discussed in Section 14.3, this means that the white dwarf is dynamically stable. If it suffers a small perturbation, it will return to its equilibrium structure instead of collapsing. However, in the extreme relativistic limit, the electron speed $v = c$ must be used instead of Eq. (16.10) to find the electron degeneracy pressure. The result is

$$P = \frac{(3\pi^2)^{1/3}}{4} \hbar c \left[ \left( \frac{Z}{A} \right) \frac{\rho}{m_H} \right]^{4/3} \tag{16.15}$$

(see, for example, Problem 16.6). In this limit $\gamma = 4/3$, which corresponds to *dynamical instability*. The smallest departure from equilibrium will cause the white dwarf to collapse as electron degeneracy pressure fails.[11] As was explained in Section 15.3, approaching this

---

[10]It is left as an exercise to show that relativistic effects must be included for densities greater than $10^9$ kg m$^{-3}$.

[11]In fact, the strong gravity of the white dwarf, as described by Einstein's general theory of relativity (see Section 17.1), acts to raise the critical value of $\gamma$ for dynamical instability slightly above 4/3.

limiting case leads to the collapse of the degenerate core in an aging supergiant, resulting in a core-collapse supernova. (Note that Eq. 16.15 is a polytropic equation of state, $P = K\rho^{4/3}$, with a polytropic index of $n = 3$.)

### Estimating the Chandrasekhar Limit

An approximate value for the maximum white-dwarf mass may be obtained by setting the estimate of the central pressure, Eq. (16.1) with $\rho = M_{wd}/\frac{4}{3}\pi R_{wd}^3$, equal to Eq. (16.15) with $Z/A = 0.5$. The radius of the white dwarf cancels, leaving

$$M_{Ch} \sim \frac{3\sqrt{2\pi}}{8}\left(\frac{\hbar c}{G}\right)^{3/2}\left[\left(\frac{Z}{A}\right)\frac{1}{m_H}\right]^2 = 0.44\ M_\odot \qquad (16.16)$$

for the greatest possible mass. Note that Eq. (16.16) contains three fundamental constants—$\hbar$, $c$, and $G$—representing the combined effects of quantum mechanics, relativity, and Newtonian gravitation on the structure of a white dwarf. A precise derivation with $Z/A = 0.5$ results in a value of $M_{Ch} = 1.44\ M_\odot$, called the **Chandrasekhar limit**. Figure 16.7 shows the mass–radius relation for white dwarfs.[12] No white dwarf has been discovered with a mass exceeding the Chandrasekhar limit.[13]

It is important to emphasize that neither the nonrelativistic nor the relativistic formula for the electron degeneracy pressure developed here (Eqs. 16.12 and 16.15, respectively) contains the temperature. Unlike the gas pressure of the ideal gas law and the expression for

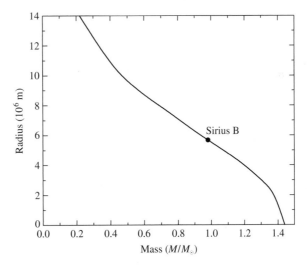

**FIGURE 16.7** Radii of white dwarfs of $M_{wd} \leq M_{Ch}$ at $T = 0$ K.

[12]Figure 16.7 does not include complications such as the electrostatic attraction between the nuclei and electrons in a white dwarf, thus tending to reduce the radius slightly.

[13]It is natural to wonder about the outcome of sneaking up on the Chandrasekhar limit by adding just a bit more mass to white dwarf with very nearly 1.44 $M_\odot$. This will be considered in Section 18.5, where Type Ia supernovae are discussed.

radiation pressure, the pressure of a completely degenerate electron gas is independent of its temperature. This has the effect of decoupling the mechanical structure of the star from its thermal properties. However, the decoupling is never perfect since $T > 0$. As a result, the correct expression for the pressure involves treating the gas as partially degenerate and relativistic, but with $v < c$. This is a challenging equation of state to deal with properly.[14]

We have already seen one implication of this decoupling in Section 13.2, where the helium core flash was described as the result of the independence of the mechanical and thermal behavior of the degenerate helium core of a low-mass star. When helium burning begins in the core, it proceeds without an accompanying increase in pressure that would normally expand the core and therefore restrain the rising temperature. The resulting rapid rise in temperature leads to a runaway production of nuclear energy—the helium flash—which lasts until the temperature becomes sufficiently high to remove the degeneracy of the core, allowing it to expand. On the other hand, a star may have so little mass that its core temperature never becomes high enough to initiate helium burning. The result in this case is the formation of a helium white dwarf.

## 16.5 ■ THE COOLING OF WHITE DWARFS

Most stars end their lives as white dwarfs. These glowing embers scattered throughout space are a galaxy's memory of its past glory. Because no fusion occurs in their interiors, white dwarfs simply cool off at an essentially constant radius as they slowly deplete their supply of thermal energy (recall Fig. 16.3). Much effort has been directed at understanding the rate at which a white dwarf cools so its lifetime and the time of its birth may be calculated. Just as paleontologists can read the history of Earth's life in the fossil record, astronomers may be able to recover the history of star formation in our Galaxy by studying the statistics of white-dwarf temperatures. This section will be devoted to a discussion of the principles involved in this stellar archaeology.

### Energy Transport

First we must ask how energy is transported outward from the interior of a white dwarf. In an ordinary star, photons travel much farther than atoms do before suffering a collision that robs them of energy (recall Examples 9.2.1 and 9.2.2). As a result, photons are normally more efficient carriers of energy to the stellar surface. In a white dwarf, however, the degenerate electrons can travel long distances before losing energy in a collision with a nucleus, since the vast majority of the lower-energy electron states are already occupied. Thus, in a white dwarf, energy is carried by **electron conduction** rather than by radiation. This is so efficient that the interior of a white dwarf is nearly isothermal, with the temperature dropping significantly only in the nondegenerate surface layers. Figure 16.8 shows that a white dwarf consists of a nearly constant-temperature interior surrounded by a thin nondegenerate envelope that transfers heat less efficiently, causing the energy to leak out

---

[14]You are referred to Clayton (1983) or Hansen, Kawaler, and Trimble (2004) for a discussion of partial electron degeneracy.

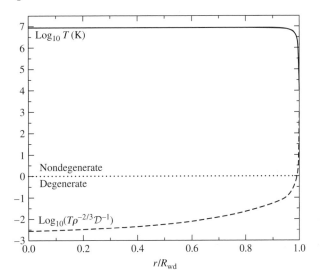

**FIGURE 16.8** Temperature and degree of degeneracy in the interior of a white dwarf model. The horizontal dotted line marks the boundary between degeneracy and nondegeneracy as described by Eq. (16.6).

slowly. The steep temperature gradient near the surface creates convection zones that may alter the appearance of the white dwarf's spectrum as it cools (as described in Section 16.2).

The structure of the nondegenerate surface layers of a star is described at the beginning of Appendix L. For a white dwarf of surface luminosity $L_{wd}$ and mass $M_{wd}$, Eq. (L.1) for the pressure $P$ as a function of the temperature $T$ in the envelope is[15]

$$P = \left( \frac{4}{17} \frac{16\pi ac}{3} \frac{GM_{wd}}{L_{wd}} \frac{k}{\kappa_0 \mu m_H} \right)^{1/2} T^{17/4}, \tag{16.17}$$

where $\kappa_0$ (called "$A$" in Eq. L.1) is the coefficient of the bound–free Kramers opacity law in Eq. (9.22),

$$\kappa_0 = 4.34 \times 10^{21} \, Z(1 + X) \, \text{m}^2 \, \text{kg}^{-1}.$$

Using the ideal gas law (Eq. 10.11) to replace the pressure results in a relation between the density and the temperature,

$$\rho = \left( \frac{4}{17} \frac{16\pi ac}{3} \frac{GM_{wd}}{L_{wd}} \frac{\mu m_H}{\kappa_0 k} \right)^{1/2} T^{13/4}. \tag{16.18}$$

The transition between the nondegenerate surface layers of the star and its isothermal, degenerate interior of temperature $T_c$ is described by setting the two sides of Eq. (16.6) equal

---

[15]Equation (16.17) assumes that the envelope is in radiative equilibrium, with the energy carried outward by photons. Even when convection occurs in the surface layers of a white dwarf, it is not expected to have a large effect on the cooling.

to each other. Using this to replace the density results in an expression for the luminosity at the white dwarf's surface in terms of its interior temperature,

$$L_{\text{wd}} = \frac{4\mathcal{D}^3}{17} \frac{16\pi ac}{3} \frac{Gm_H}{\kappa_0 k} \mu M_{\text{wd}} T_c^{7/2}$$

$$= CT_c^{7/2}, \tag{16.19}$$

where

$$C \equiv \frac{4\mathcal{D}^3}{17} \frac{16\pi ac}{3} \frac{Gm_H}{\kappa_0 k} \mu M_{\text{wd}}$$

$$= 6.65 \times 10^{-3} \left( \frac{M_{\text{wd}}}{M_\odot} \right) \frac{\mu}{Z(1+X)}.$$

Note that the luminosity is proportional to $T_c^{7/2}$ (the *interior* temperature) and that it varies as the fourth power of the *effective* temperature according to the Stefan–Boltzmann law, Eq. (3.17). Thus the surface of a white dwarf cools more slowly than its isothermal interior as the star's thermal energy leaks into space.

---

**Example 16.5.1.**   Equation (16.19) can be used to estimate the interior temperature of a 1 $M_\odot$ white dwarf with $L_{\text{wd}} = 0.03\ L_\odot$. Arbitrarily assuming values of $X = 0$, $Y = 0.9$, $Z = 0.1$ for the nondegenerate envelope (so $\mu \simeq 1.4$) results in[16]

$$T_c = \left[ \frac{L_{\text{wd}}}{6.65 \times 10^{-3}} \left( \frac{M_\odot}{M_{\text{wd}}} \right) \frac{Z(1+X)}{\mu} \right]^{2/7} = 2.8 \times 10^7\ \text{K}.$$

Equating the two sides of the degeneracy condition, Eq. (16.6), shows that the density at the base of the nondegenerate envelope is about

$$\rho = \left( \frac{T_c}{\mathcal{D}} \right)^{3/2} = 3.4 \times 10^6\ \text{kg m}^{-3}.$$

This result is several orders of magnitude less than the average density of a 1 $M_\odot$ white dwarf such as Sirius B and confirms that the envelope is indeed thin, contributing very little to the star's total mass.

---

### The Cooling Timescale

A white dwarf's thermal energy resides primarily in the kinetic energy of its nuclei; the degenerate electrons cannot give up a significant amount of energy because nearly all of the lower energy states are already occupied. If we assume for simplicity that the composition is uniform, then the total number of nuclei in the white dwarf is equal to the star's mass, $M_{\text{wd}}$,

---

[16]Because the amount of hydrogen is quite small even in a DA white dwarf, this composition is a reasonable choice for both type DA and type DB.

divided by the mass of a nucleus, $Am_H$. Furthermore, since the average thermal energy of a nucleus is $\frac{3}{2}kT$, the thermal energy available for radiation is

$$U = \frac{M_{wd}}{Am_H} \frac{3}{2} kT_c. \tag{16.20}$$

If we use the value of $T_c$ from Example 16.5.1 and $A = 12$ for carbon, Eq. (16.20) gives approximately $6.0 \times 10^{40}$ J. A crude estimate of the characteristic timescale for cooling, $\tau_{cool}$, can be obtained simply by dividing the thermal energy by the luminosity. Thus

$$\tau_{cool} = \frac{U}{L_{wd}} = \frac{3}{2} \frac{M_{wd}k}{Am_H C T_c^{5/2}}, \tag{16.21}$$

which is about $5.2 \times 10^{15}$ s $\approx 170$ million years. This is an underestimate, because the cooling timescale increases as $T_c$ decreases. The more detailed calculation that follows shows that a white dwarf spends most of its life cooling slowly with a low temperature and luminosity.

### The Change in Luminosity with Time

The depletion of the internal energy provides the luminosity, so Eqs. (16.19) and (16.20) give

$$-\frac{dU}{dt} = L_{wd}$$

or

$$-\frac{d}{dt}\left(\frac{M_{wd}}{Am_H} \frac{3}{2} kT_c\right) = CT_c^{7/2}.$$

If the initial temperature of the interior is $T_0$ when $t = 0$, then this expression may be integrated to obtain the core temperature as a function of time:

$$T_c(t) = T_0 \left(1 + \frac{5}{3} \frac{Am_H C T_0^{5/2}}{M_{wd}k} t\right)^{-2/5} = T_0 \left(1 + \frac{5}{2} \frac{t}{\tau_0}\right)^{-2/5}, \tag{16.22}$$

where $\tau_0$ is the timescale for cooling at the initial temperature of $T_0$; that is, $\tau_0 = \tau_{cool}$ at time $t_0$. Inserting this into Eq. (16.19) shows that the luminosity of the white dwarf first declines sharply from its initial value of $L_0 = CT_0^{7/2}$ and then dims much more gradually as time passes:

$$L_{wd} = L_0 \left(1 + \frac{5}{3} \frac{Am_H C^{2/7} L_0^{5/7}}{M_{wd}k} t\right)^{-7/5} = L_0 \left(1 + \frac{5}{2} \frac{t}{\tau_0}\right)^{-7/5}. \tag{16.23}$$

The solid line in Fig. 16.9 shows the decline in the luminosity of a pure carbon 0.6 $M_\odot$ white dwarf calculated from Eq. (16.23). The dashed line is a curve obtained for a sequence

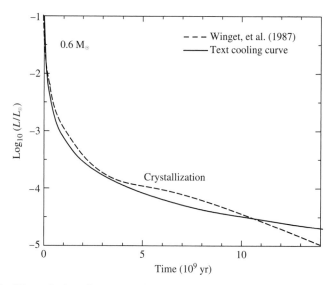

**FIGURE 16.9**    Theoretical cooling curves for 0.6 $M_\odot$ white-dwarf models. [The solid line is from Eq. (16.23), and the dashed line is from Winget et al., *Ap. J. Lett.*, *315*, L77, 1987.]

of more realistic white-dwarf models[17] that include thin surface layers of hydrogen and helium overlying the carbon core. The insulating effect of these layers slows the cooling by about 15%. Also included are some of the intriguing phenomena that occur as the white dwarf's internal temperature drops.

### Crystallization

As a white dwarf cools, it crystallizes in a gradual process that starts at the center and moves outward. The upturned "knee" in the dashed curve in Fig. 16.9 at about $L_{wd}/L_\odot \approx 10^{-4}$ occurs when the cooling nuclei begin settling into a crystalline lattice. The regular crystal structure is maintained by the mutual electrostatic repulsion of the nuclei; it minimizes their energy as they vibrate about their average position in the lattice. As the nuclei undergo this phase change, they release their latent heat (about $kT$ per nucleus), slowing the star's cooling and producing the knee in the cooling curve. Later, as the white dwarf's temperature continues to drop, the crystalline lattice actually accelerates the cooling as the coherent vibration of the regularly spaced nuclei promotes further energy loss. This is reflected in the subsequent downturn in the cooling curve. Thus the ultimate monument to the lives of most stars will be a "diamond in the sky," a cold, dark, Earth-size sphere of crystallized carbon and oxygen floating through the depths of space.[18]

---

[17]You are referred to Winget et al. (1987) for details of this and other cooling curves.

[18]Unlike a terrestrial diamond, the white dwarf's nuclei are arrayed in a body-centered cubic lattice like that of metallic sodium.

**Comparing Theory with Observations**

Despite the large uncertainties in the measurement of surface temperatures resulting from high surface gravities and broad spectral features,[19] it is possible to observe the cooling of a pulsating white dwarf. As the star's temperature declines, its period $P$ slowly changes according to $dP/dt \propto T^{-1}$ (approximately). Extremely precise measurements of a rapidly cooling DOV star yield a period derivative of $P/|dP/dt| = 1.4 \times 10^6$ years, in excellent agreement with the theoretical value. Measuring period changes for the more slowly cooling DBV and DAV stars are even more difficult.

This interest in an accurate calculation of the decline in a white dwarf's temperature reflects the hope of using these fossil stars as a tool for uncovering the history of star formation in our Galaxy. Figure 16.10, from Winget et al. (1987), illustrates how this might be accomplished. Each circle (both open and filled) in the figure is the observed number of white dwarfs per cubic parsec with the absolute visual magnitude given at the top of the figure. The dramatically sudden drop in the population of white dwarfs with $L_{\rm wd}/L_\odot < -4.5$ is inconsistent with the assumption that stars have been forming in our Galaxy throughout the infinite past. Instead, this decline can best be explained if the first white dwarfs were formed and began cooling $9.0 \pm 1.8$ billion years ago. Figure 16.10 shows the theoretically expected distribution of white dwarf luminosities based on this cooling time, calculated using theoretical cooling curves similar to the one shown in Fig. 16.9

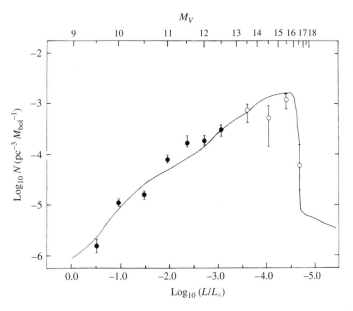

**FIGURE 16.10**   Observed and theoretical distribution of white-dwarf luminosities. (Figure adapted from Winget et al., *Ap. J. Lett.*, *315*, L77, 1987.)

[19]For Sirius B, effective temperatures ranging from 27,000 K to 32,000 K are often quoted.

together with the observed distribution of white-dwarf masses. Furthermore, adding the time spent in the pre-white-dwarf stages of stellar evolution implies that star formation in the disk of our Galaxy began about $9.3 \pm 2.0$ billion years ago.[20] This time is about 3 billion years shorter than the age determined for the Milky Way's globular clusters, which formed at an earlier epoch.

## 16.6 ■ NEUTRON STARS

Two years after James Chadwick (1891–1974) discovered the neutron in 1932, a German astronomer and a Swiss astrophysicist, Walter Baade (1893–1960) and Fritz Zwicky (1898–1974) of Mount Wilson Observatory, proposed the existence of **neutron stars**. These two astronomers, who also coined the term *supernova*, went on to suggest that "supernovae represent the transitions from ordinary stars into neutron stars, which in their final stages consist of extremely closely packed neutrons."

### Neutron Degeneracy

Because neutron stars are formed when the degenerate core of an aging supergiant star nears the Chandrasekhar limit and collapses, we take $M_{Ch}$ (rounded to two figures) for a typical neutron star mass. A 1.4-solar-mass neutron star would consist of $1.4 \, M_\odot / m_n \approx 10^{57}$ neutrons—in effect, a huge nucleus with a mass number of $A \approx 10^{57}$ that is held together by gravity and supported by **neutron degeneracy pressure**.[21] It is left as an exercise to show that

$$R_{ns} \approx \frac{(18\pi)^{2/3}}{10} \frac{\hbar^2}{G M_{ns}^{1/3}} \left( \frac{1}{m_H} \right)^{8/3} \tag{16.24}$$

is the expression for the estimated neutron star radius, analogous to Eq. (16.13) for a white dwarf. For $M_{ns} = 1.4 \, M_\odot$, this yields a value of 4400 m. As we found with Eq. (16.13) for white dwarfs, this estimate is too small by a factor of about 3. That is, the actual radius of a $1.4 \, M_\odot$ neutron star lies roughly between 10 and 15 km; we will adopt a value of 10 km for the radius. As will be seen, there are many uncertainties involved in the construction of a model neutron star.

### The Density of a Neutron Star

This incredibly compact stellar remnant would have an average density of $6.65 \times 10^{17}$ kg m$^{-3}$, greater than the typical density of an atomic nucleus, $\rho_{nuc} \approx 2.3 \times 10^{17}$ kg m$^{-3}$. In some sense, the neutrons in a neutron star must be "touching" one another. At the density of a neutron star, all of Earth's human inhabitants could be crowded into a cube 1.5 cm on each side.[22]

---

[20]Other, more recent studies have obtained similar results for the age of the *thin disk* of our Galaxy based on white-dwarf cooling times; age estimates range from 9 Gyr to 11 Gyr.

[21]Like electrons, neutrons are fermions and so are subject to the Pauli exclusion principle.

[22]Astronomer Frank Shu has commented that this shows "how much of humanity is empty space"!

The pull of gravity at the surface of a neutron star is fierce. For a 1.4 $M_\odot$ neutron star with a radius of 10 km, $g = 1.86 \times 10^{12}$ m s$^{-2}$, 190 billion times stronger than the acceleration of gravity at Earth's surface. An object dropped from a height of one meter would arrive at the star's surface with a speed of $1.93 \times 10^6$ m s$^{-1}$ (about 4.3 million mph).

---

**Example 16.6.1.**   The inadequacy of using Newtonian mechanics to describe neutron stars can be demonstrated by calculating the escape velocity at the surface. Using Eq. (2.17), we find

$$v_{esc} = \sqrt{2G M_{ns}/R_{ns}} = 1.93 \times 10^8 \text{ m s}^{-1} = 0.643c.$$

This can also be seen by considering the ratio of the Newtonian gravitational potential energy to the rest energy of an object of mass $m$ at the star's surface:

$$\frac{G M_{ns}m/R_{ns}}{mc^2} = 0.207.$$

Clearly, the effects of relativity must be included for an accurate description of a neutron star. This applies not only to Einstein's theory of special relativity, described in Chapter 4, but also to his theory of gravity, called the *general theory of relativity*, which will be considered in Section 17.1. Nevertheless, we will use both relativistic formulas and the more familiar Newtonian physics to reach qualitatively correct conclusions about neutron stars.

---

## The Equation of State

To appreciate the exotic nature of the material constituting a neutron star and the difficulties involved in calculating the equation of state, imagine compressing the mixture of iron nuclei and degenerate electrons that make up an iron white dwarf at the center of a massive supergiant star.[23] Specifically, we are interested in the equilibrium configuration of $10^{57}$ nucleons (protons and neutrons), together with enough free electrons to provide zero net charge. The equilibrium arrangement is the one that involves the least energy.

Initially, at low densities the nucleons are found in iron nuclei. This is the outcome of the minimum-energy compromise between the repulsive Coulomb force between the protons and the attractive nuclear force between all of the nucleons. However, as mentioned in the discussion of the Chandrasekhar limit (Section 16.4), when $\rho \approx 10^9$ kg m$^{-3}$ the electrons become relativistic. Soon thereafter, the minimum-energy arrangement of protons and neutrons changes because the energetic electrons can convert protons in the iron nuclei into neutrons by the process of electron capture (Eq. 15.6),

$$p^+ + e^- \rightarrow n + \nu_e.$$

Because the neutron mass is slightly greater than the sum of the proton and electron masses, and the neutrino's rest-mass energy is negligible, the electron must supply the kinetic energy to make up the difference in energy; $m_n c^2 - m_p c^2 - m_e c^2 = 0.78$ MeV.

---

[23]Because the mechanical and thermal properties of degenerate matter are independent of one another, we will assume for convenience that $T = 0$ K. The iron nuclei are then arranged in a crystalline lattice.

**Example 16.6.2.** We will obtain an estimate of the density at which the process of electron capture begins for a simple mixture of hydrogen nuclei (protons) and relativistic degenerate electrons,

$$p^+ + e^- \rightarrow n + \nu_e.$$

In the limiting case when the neutrino carries away no energy, we can equate the relativistic expression for the electron kinetic energy, Eq. (4.45), to the difference between the neutron rest energy and combined proton and electron rest energies and write

$$m_e c^2 \left( \frac{1}{\sqrt{1 - v^2/c^2}} - 1 \right) = (m_n - m_p - m_e)c^2,$$

or

$$\left( \frac{m_e}{m_n - m_p} \right)^2 = 1 - \frac{v^2}{c^2}.$$

Although Eq. (16.10) for the electron speed is strictly valid only for nonrelativistic electrons, it is accurate enough to be used in this estimate. Inserting this expression for $v$ leads to

$$\left( \frac{m_e}{m_n - m_p} \right)^2 \approx 1 - \frac{\hbar^2}{m_e^2 c^2} \left[ \left( \frac{Z}{A} \right) \frac{\rho}{m_H} \right]^{2/3}.$$

Solving for $\rho$ shows that the density at which electron capture begins is approximately

$$\rho \approx \frac{A m_H}{Z} \left( \frac{m_e c}{\hbar} \right)^3 \left[ 1 - \left( \frac{m_e}{m_n - m_p} \right)^2 \right]^{3/2} \approx 2.3 \times 10^{10} \text{ kg m}^{-3},$$

using $A/Z = 1$ for hydrogen. This is in reasonable agreement with the actual value of $\rho = 1.2 \times 10^{10} \text{ kg m}^{-3}$.

We considered free protons in Example 16.6.2 to avoid the complications that arise when they are bound in heavy nuclei. A careful calculation that takes into account the surrounding nuclei and relativistic degenerate electrons, as well as the complexities of nuclear physics, reveals that the density must exceed $10^{12}$ kg m$^{-3}$ for the protons in $^{56}_{26}$Fe nuclei to capture electrons. At still higher densities, the most stable arrangement of nucleons is one where the neutrons and protons are found in a lattice of increasingly neutron-rich nuclei so as to decrease the energy due to the Coulomb repulsion between protons. This process is known as **neutronization** and produces a sequence of nuclei such as $^{56}_{26}$Fe, $^{62}_{28}$Ni, $^{64}_{28}$Ni, $^{66}_{28}$Ni, $^{86}_{36}$Kr, ..., $^{118}_{36}$Kr. Ordinarily, these supernumerary neutrons would revert to protons via the standard $\beta$-decay process,

$$n \rightarrow p^+ + e^- + \overline{\nu}_e.$$

However, under the conditions of complete electron degeneracy, there are no vacant states available for an emitted electron to occupy, so the neutrons cannot decay back into protons.[24]

When the density reaches about $4 \times 10^{14}$ kg m$^{-3}$, the minimum-energy arrangement is one in which some of the neutrons are found *outside* the nuclei. The appearance of these free neutrons is called **neutron drip** and marks the start of a three-component mixture of a lattice of neutron-rich nuclei, nonrelativistic degenerate free neutrons, and relativistic degenerate electrons.

The fluid of free neutrons has the striking property that it has no viscosity. This occurs because a spontaneous pairing of the degenerate neutrons has taken place. The resulting combination of two fermions (the neutrons) is a boson (recall Section 5.4) and so is not subject to the restrictions of the Pauli exclusion principle. Because degenerate bosons can *all* crowd into the lowest energy state, the fluid of paired neutrons can lose no energy. It is a **superfluid** that flows without resistance. Any whirlpools or vortices in the fluid will continue to spin forever without stopping.

As the density increases further, the number of free neutrons increases as the number of electrons declines. The neutron degeneracy pressure exceeds the electron degeneracy pressure when the density reaches roughly $4 \times 10^{15}$ kg m$^{-3}$. As the density approaches $\rho_{\text{nuc}}$, the nuclei effectively dissolve as the distinction between neutrons inside and outside of nuclei becomes meaningless. This results in a fluid mixture of free neutrons, protons, and electrons dominated by neutron degeneracy pressure, with both the neutrons and protons paired to form superfluids. The fluid of pairs of positively charged protons is also **superconducting**, with zero electrical resistance. As the density increases further, the ratio of neutrons:protons:electrons approaches a limiting value of 8:1:1, as determined by the balance between the competing processes of electron capture and $\beta$-decay inhibited by the presence of degenerate electrons.

The properties of the neutron star material when $\rho > \rho_{\text{nuc}}$ are still poorly understood. A complete theoretical description of the behavior of a sea of free neutrons interacting via the strong nuclear force in the presence of protons and electrons is not yet available, and there is little experimental data on the behavior of matter in this density range. A further complication is the appearance of sub-nuclear particles such as *pions* $(\pi)$ produced by the decay of a neutron into a proton and a negatively charged pion, $n \rightarrow p^+ + \pi^-$, which occurs spontaneously in neutron stars when $\rho > 2\rho_{\text{nuc}}$.[25] Nevertheless, these are the values of the density encountered in the interiors of neutron stars, and the difficulties mentioned are the primary reasons for the uncertainty in the structure calculated for model neutron stars.

### Neutron Star Models

Table 16.1 summarizes the composition of the neutron star material at various densities. After an equation of state that relates the density and pressure has been obtained, a model of the star can be calculated by numerically integrating general-relativistic versions of the

---

[24]An *isolated* neutron decays into a proton in about 10.2 minutes, the half-life for that process.

[25]The $\pi^-$ is a negatively charged particle that is 273 times more massive than the electron. It mediates the strong nuclear force that holds an atomic nucleus together. (The strong force between nucleons was described in Section 10.3.) Pions have been produced and studied in high-energy accelerator laboratories.

**TABLE 16.1**    Composition of Neutron Star Material.

| Transition density $(\text{kg m}^{-3})$ | Composition | Degeneracy pressure |
|---|---|---|
| | iron nuclei, nonrelativistic free electrons | electron |
| $\approx 1 \times 10^9$ | electrons become relativistic | |
| | iron nuclei, relativistic free electrons | electron |
| $\approx 1 \times 10^{12}$ | neutronization | |
| | neutron-rich nuclei, relativistic free electrons | electron |
| $\approx 4 \times 10^{14}$ | neutron drip | |
| | neutron-rich nuclei, free neutrons, relativistic free electrons | electron |
| $\approx 4 \times 10^{15}$ | neutron degeneracy pressure dominates | |
| | neutron-rich nuclei, superfluid free neutrons, relativistic free electrons | neutron |
| $\approx 2 \times 10^{17}$ | nuclei dissolve | |
| | superfluid free neutrons, superconducting free protons, relativistic free electrons | neutron |
| $\approx 4 \times 10^{17}$ | pion production | |
| | superfluid free neutrons, superconducting free protons, relativistic free electrons, other elementary particles (pions, ...?) | neutron |

stellar structure equations collected at the beginning of Section 10.5. The first quantitative model of a neutron star was calculated by J. Robert Oppenheimer (1904–1967) and G. M. Volkoff (1914–2000) at Berkeley in 1939. Figure 16.11 shows the result of a recent calculation of a 1.4 $M_\odot$ neutron star model. Although the details are sensitive to the equation of state used, this model displays some typical features.

1. The outer crust consists of heavy nuclei, in the form of either a fluid "ocean" or a solid lattice, and relativistic degenerate electrons. Nearest the surface, the nuclei are probably $^{56}_{26}$Fe. At greater depth and density, increasingly neutron-rich nuclei are encountered until neutron drip begins at the bottom of the outer crust (where $\rho \approx 4 \times 10^{14} \text{ kg m}^{-3}$).

2. The inner crust consists of a three-part mixture of a lattice of nuclei such as $^{118}_{36}$Kr, a superfluid of free neutrons, and relativistic degenerate electrons. The bottom of the inner crust occurs where $\rho \approx \rho_{\text{nuc}}$, and the nuclei dissolve.

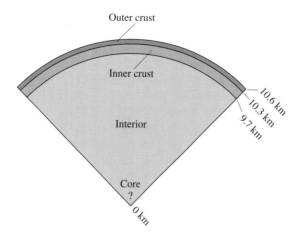

**FIGURE 16.11**   A 1.4 $M_\odot$ neutron star model.

3. The interior of the neutron star consists primarily of superfluid neutrons, with a smaller number of superfluid, superconducting protons and relativistic degenerate electrons.

4. There may or may not be a solid core consisting of pions or other sub-nuclear particles. The density at the center of a 1.4 $M_\odot$ neutron star is about $10^{18}$ kg m$^{-3}$.

### The Chandrasekhar Limit for Neutron Stars

Like white dwarfs, neutron stars obey a mass–volume relation,

$$M_{\text{ns}} V_{\text{ns}} = \text{constant}, \tag{16.25}$$

so neutron stars become smaller and more dense with increasing mass. However, this mass–volume relation fails for more massive neutron stars because there is a point beyond which neutron degeneracy pressure can no longer support the star. Hence, there is a maximum mass for neutron stars, analogous to the Chandrasekhar mass for white dwarfs. As might be expected, the value of this maximum mass is different for different choices of the equation of state. However, detailed computer modeling of neutron stars, along with a very general argument involving the general theory of relativity, shows that the maximum mass possible for a neutron star cannot exceed about 2.2 $M_\odot$ if it is static, and 2.9 $M_\odot$ if it is rotating rapidly.[26] If a neutron star is to remain dynamically stable and resist collapsing, it must be able to respond to a small disturbance in its structure by rapidly adjusting its pressure to compensate. However, there is a limit to how quickly such an adjustment can be made because these changes are conveyed by sound waves that must move more slowly than light. If a neutron star's mass exceeds 2.2 $M_\odot$ in the static case or 2.9 $M_\odot$ in the rapidly rotating case, it cannot generate pressure quickly enough to avoid collapsing. The result is a black hole (as will be discussed in Section 17.3).

---

[26]Recall from Section 15.4 that centrifugal effects provide additional support to a rapidly rotating neutron star.

## Rapid Rotation and Conservation of Angular Momentum

Several properties of neutron stars were anticipated before they were observed. For example, neutron stars must rotate very rapidly. If the iron core of the pre-supernova supergiant star were rotating even slowly, the decrease in radius would be so great that the conservation of angular momentum would guarantee the formation of a rapidly rotating neutron star.

The scale of the collapse can be found from Eqs. (16.13) and (16.24) for the estimated radii of a white dwarf and neutron star if we assume that the progenitor core is characteristic of a white dwarf composed entirely of iron. Although the leading constants in both expressions are spurious (a by-product of the approximations made), the *ratio* of the radii is more accurate:

$$\frac{R_{\text{core}}}{R_{\text{ns}}} \approx \frac{m_n}{m_e} \left(\frac{Z}{A}\right)^{5/3} = 512,$$

where $Z/A = 26/56$ for iron has been used. Now apply the conservation of angular momentum to the collapsing core (which is assumed here for simplicity to lose no mass, so $M_{\text{core}} = M_{\text{wd}} = M_{\text{ns}}$). Treating each star as a sphere with a moment of inertia of the form $I = CMR^2$, we have[27]

$$I_i \omega_i = I_f \omega_f$$

$$C M_i R_i^2 \omega_i = C M_f R_f^2 \omega_f$$

$$\omega_f = \omega_i \left(\frac{R_i}{R_f}\right)^2.$$

In terms of the rotation period $P$, this is

$$P_f = P_i \left(\frac{R_f}{R_i}\right)^2. \tag{16.26}$$

For the specific case of an iron core collapsing to form a neutron star, Eq. (16.6) shows that

$$P_{\text{ns}} \approx 3.8 \times 10^{-6} \, P_{\text{core}}. \tag{16.27}$$

The question of how fast the progenitor core may be rotating is difficult to answer. As a star evolves, its contracting core is not completely isolated from the surrounding envelope, so one cannot use the simple approach to conservation of angular momentum described above.[28] For purposes of estimation, we will take $P_{\text{core}} = 1350$ s, the rotation period observed for the white dwarf 40 Eridani B (shown in the H–R diagrams of Figs. 8.12 and 8.16). Inserting this into Eq. (16.27) results in a rotation period of about $5 \times 10^{-3}$ s. Thus neutron stars will be rotating very rapidly when they are formed, with rotation periods on the order of a few milliseconds.

---

[27]The constant $C$ is determined by the distribution of mass inside the star. For example, $C = 2/5$ for a uniform sphere. We assume that the progenitor core and neutron star have about the same value of $C$.

[28]The core and envelope may exchange angular momentum by magnetic fields or rotational mixing via the very slow *meridional currents* that generally circulate upward at the poles and downward at the equator of a rotating star.

### "Freezing In" Magnetic Field Lines

Another property predicted for neutron stars is that they should have extremely strong magnetic fields. The "freezing in" of magnetic field lines in a conducting fluid or gas (mentioned in Section 11.3 in connection with sunspots) implies that the *magnetic flux* through the surface of a white dwarf will be conserved as it collapses to form a neutron star. The flux of a magnetic field through a surface $\mathcal{S}$ is defined as the surface integral

$$\Phi \equiv \int_{\mathcal{S}} \mathbf{B} \cdot d\mathbf{A},$$

where $\mathbf{B}$ is the magnetic field vector (see Fig. 16.12). In approximate terms, if we ignore the geometry of the magnetic field, this means that the product of the magnetic field strength and the area of the star's surface remains constant. Thus

$$B_i 4\pi R_i^2 = B_f 4\pi R_f^2. \tag{16.28}$$

In order to use Eq. (16.28) to estimate the magnetic field of a neutron star, we must first know what the strength of the magnetic field is for the iron core of a pre-supernova star. Although this is not at all clear, we can use the largest observed white-dwarf magnetic field of $B \approx 5 \times 10^4$ T as an extreme case, which is large compared to a typical white-dwarf magnetic field of perhaps 10 T, and huge compared with the Sun's global field of about $2 \times 10^{-4}$ T. Then, using Eq. (16.6), the magnetic field of the neutron star would be

$$B_{\text{ns}} \approx B_{\text{wd}} \left( \frac{R_{\text{wd}}}{R_{\text{ns}}} \right)^2 = 1.3 \times 10^{10} \text{ T}.$$

This shows that neutron stars could be formed with extremely strong magnetic fields, although smaller values such as $10^8$ T or less are more typical.

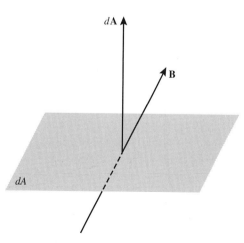

**FIGURE 16.12** Magnetic flux, $d\Phi = \mathbf{B} \cdot d\mathbf{A}$, through an element of surface area $d\mathbf{A}$.

**Neutron Star Temperatures**

The final property of neutron stars is the most obvious. They were extremely hot when they were forged in the "fires" of a supernova, with $T \sim 10^{11}$ K. During the first day, the neutron star cools by emitting neutrinos via the so-called **URCA process**,[29]

$$n \rightarrow p^+ + e^- + \overline{\nu}_e$$

$$p^+ + e^- \rightarrow n + \nu_e.$$

As the nucleons shuttle between being neutrons and being protons, large numbers of neutrinos and antineutrinos are produced that fly unhindered into space, carrying away energy and thus cooling the neutron star. This process can continue only as long as the nucleons are not degenerate, and it is suppressed after the protons and neutrons settle into the lowest unoccupied energy states. This degeneracy occurs about one day after the formation of the neutron star, when its internal temperature has dropped to about $10^9$ K. Other neutrino-emitting processes continue to dominate the cooling for approximately the first thousand years, after which photons emitted from the star's surface take over. The neutron star is a few hundred years old when its internal temperature has declined to $10^8$ K, with a surface temperature of several million K. By now the cooling has slowed considerably, and the surface temperature will hover around $10^6$ K for the next ten thousand years or so as the neutron star cools at an essentially constant radius.

It is interesting to calculate the blackbody luminosity of a 1.4 M$_\odot$ neutron star with a surface temperature of $T = 10^6$ K. From the Stefan–Boltzmann law, Eq. (3.17),

$$L = 4\pi R^2 \sigma T_e^4 = 7.13 \times 10^{25} \text{ W}.$$

Although this is comparable to the luminosity of the Sun, the radiation is primarily in the form of X-rays since, according to Wien's displacement law, Eq. (3.19),

$$\lambda_{\text{max}} = \frac{(500 \text{ nm})(5800 \text{ K})}{T} = 2.9 \text{ nm}.$$

Prior to the advent of X-ray observatories such as ROSAT, ASCA, and Chandra, astronomers held little hope of ever observing such an exotic object, barely the size of San Diego, California.

## 16.7 ■ PULSARS

Jocelyn Bell spent two years setting up a forest of 2048 radio dipole antennae over four and a half acres of English countryside. She and her Ph.D. thesis advisor, Anthony Hewish, were using this radio telescope, tuned to a frequency of 81.5 MHz, to study the scintillation ("flickering") that is observed when the radio waves from distant sources known as quasars

---

[29]The URCA process, which efficiently removes energy from a hot neutron star, is named for the Casino de URCA in Rio de Janeiro, in remembrance of the efficiency with which it removed money from an unlucky physicist. The casino was closed by Brazil in 1955.

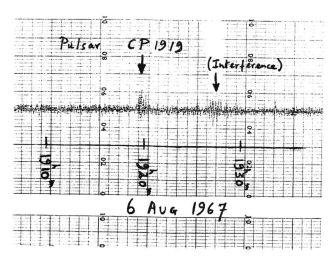

**FIGURE 16.13**   Discovery of the first pulsar, PSR 1919+21 ("CP" stands for Cambridge Pulsar). (Figure from Lyne and Graham-Smith, *Pulsar Astronomy*, ©Cambridge University Press, New York, 1990. Reprinted with the permission of Cambridge University Press.)

pass through the solar wind. In July 1967, Bell was puzzled to find a bit of "scruff" that reappeared every 400 feet or so on the rolls of her strip chart recorder; see Fig. 16.13. Careful measurements showed that this quarter inch of ink reappeared every 23 hours and 56 minutes, indicating that its source passed over her fixed array of antennae once every sidereal day. Bell concluded that the source was out among the stars rather than within the Solar System. To better resolve the signal, she used a faster recorder and discovered that the scruff consisted of a series of regularly spaced radio pulses 1.337 s apart (the pulse **period**, $P$). Such a precise celestial clock was unheard of, and Bell and Hewish considered the possibility that these might be signals from an extraterrestrial civilization. If this were true, she felt annoyed that the aliens had chosen such an inconvenient time to make contact. She recalled, "I was now two and a half years through a three year studentship and here was some silly lot of Little Green Men using *my* telescope and *my* frequency to signal to planet Earth." When Bell found another bit of scruff, coming from another part of the sky, her relief was palpable. She wrote, "It was highly unlikely that two lots of Little Green Men could choose the same unusual frequency and unlikely technique to signal to the same inconspicuous planet Earth!"

Hewish, Bell, and their colleagues announced the discovery of these mysterious **pulsars**,[30] and several more were quickly found by other radio observatories. At the time this text was written, more than 1500 pulsars were known, and each is designated by a "PSR"

[30]The term *pulsar* was coined by the science correspondent for the London *Daily Telegraph*. See Hewish et al. (1968) for details of the discovery of pulsars. In 1974 Hewish was awarded a share of the Nobel Prize, along with Martin Ryle (1918–1984), for their work in radio astronomy. Fred Hoyle (1915–2001) and others have argued that Jocelyn Bell should have shared the prize as well; Hewish had designed the radio array and observational technique, but Bell was the first to notice the pulsar signal. This controversial omission has inspired references to the award as the "no-Bell" prize.

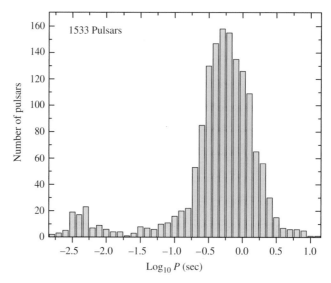

**FIGURE 16.14**   The distribution of periods for 1533 pulsars. The millisecond pulsars are clearly evident on the left. The average period is about 0.795 s. (Data from Manchester, R. N., Hobbs, G. B., Teoh, A., and Hobbs, M., *A. J.*, *129*, 1993, 2005. Data available at http://www.atnf.csiro.au/research/pulsar/psrcat.)

prefix (for *P*ulsating *S*ource of *R*adio) followed by its right ascension ($\alpha$) and declination ($\delta$). For example, the source of Bell's scruff is PSR 1919+21, identifying its position as $\alpha = 19^{\mathrm{h}}19^{\mathrm{m}}$ and $\delta = +21°$.

### General Characteristics

All known pulsars share the following characteristics, which are crucial clues to their physical nature:

- Most pulsars have periods between 0.25 s and 2 s, with an average time between pulses of about 0.795 s (see Fig. 16.14). The pulsar with the longest known period is PSR 1841-0456 ($P = 11.8$ s); Terzan 5ad (PSR J1748-2446ad) is the fastest known pulsar ($P = 0.00139$ s).

- Pulsars have extremely well-defined pulse periods and would make exceptionally accurate clocks. For example, the period of PSR 1937+214 has been determined to be $P = 0.00155780644887275$ s, a measurement that challenges the accuracy of the best atomic clocks. (Such precise determinations are possible because of the enormous number of pulsar measurements that can be made, given their very short periods.)

- The periods of all pulsars increase very gradually as the pulses slow down, the rate of increase being given by the period derivative $\dot{P} \equiv dP/dt$.[31] Typically, $\dot{P} \approx 10^{-15}$,

---

[31]Note that $\dot{P}$ is measured in terms of seconds of period change per second and so is unitless.

and the *characteristic lifetime* (the time it would take the pulses to cease if $\dot{P}$ were constant) is $P/\dot{P} \approx$ a few $10^7$ years. The value of $\dot{P}$ for PSR 1937+214 is unusually small, $\dot{P} = 1.051054 \times 10^{-19}$. This corresponds to a characteristic lifetime of $P/\dot{P} = 1.48 \times 10^{16}$ s, or about 470 million years.

## Possible Pulsar Models

These characteristics enabled astronomers to deduce the basic components of pulsars. In the paper announcing their discovery, Hewish, Bell, and their co-authors suggested that an oscillating neutron star might be involved, but American astronomer Thomas Gold (1920–2004) quickly and convincingly argued instead that pulsars are rapidly rotating neutron stars.

There are three obvious ways of obtaining rapid regular pulses in astronomy:

1. **Binary stars**. If the orbital periods of a binary star system are to fall in the range of the observed pulsar periods, then extremely compact stars must be involved—either white dwarfs or neutron stars. The general form of Kepler's third law, Eq. (2.37), shows that if two 1 $M_\odot$ stars were to orbit each other every 0.79 s (the average pulsar period), then their separation would be only $1.6 \times 10^6$ m. This is much less than the $5.5 \times 10^6$ m radius of Sirius B, and the separation would be even smaller for more rapid pulsars. This eliminates even the smallest, most massive white dwarfs from consideration.

    Neutron stars are so small that two of them could orbit each other with a period in agreement with those observed for pulsars. However, this possibility is ruled out by Einstein's general theory of relativity. As the two neutron stars rapidly move through space and time, gravitational waves are generated that carry energy away from the binary system. As the neutron stars slowly spiral closer together, their orbital period *decreases*, according to Kepler's third law. This contradicts the observed *increase* in the periods of the pulsars and so eliminates binary neutron stars as a source of the radio pulses.[32]

2. **Pulsating stars**. As we noted in Section 16.2, white dwarfs oscillate with periods between 100 and 1000 s. The periods of these nonradial g-modes are much longer than the observed pulsar periods. Of course, it might be imagined that a radial oscillation is involved with the pulsars. However, the period for the radial fundamental mode is a few seconds, too long to explain the faster pulses.

    A similar argument eliminates neutron star oscillations. Neutron stars are about $10^8$ times more dense than white dwarfs. According to the period–mean density relation for stellar pulsation (recall Section 14.2), the period of oscillation is proportional to $1/\sqrt{\rho}$. This implies that neutron stars should vibrate approximately $10^4$ times more rapidly than white dwarfs, with a radial fundamental mode period around $10^{-4}$ s and nonradial g-modes between $10^{-2}$ s and $10^{-1}$ s. These periods are much too short for the slower pulsars.

---

[32]Gravitational waves will be described in more detail in Section 18.6, as will the binary system of two neutron stars in which these waves have been indirectly detected.

3. **Rotating stars**. The enormous angular momentum of a rapidly rotating compact star would guarantee its precise clock-like behavior. But how fast can a star spin? Its angular velocity, $\omega$, is limited by the ability of gravity to supply the centripetal force that keeps the star from flying apart. This constraint is most severe at the star's equator, where the stellar material moves most rapidly. Ignore the inevitable equatorial bulging caused by rotation and assume that the star remains circular with radius $R$ and mass $M$. Then the maximum angular velocity may be found by equating the centripetal and gravitational accelerations at the equator,

$$\omega_{max}^2 R = G \frac{M}{R^2},$$

so that the minimum rotation period is $P_{min} = 2\pi/\omega_{max}$, or

$$P_{min} = 2\pi \sqrt{\frac{R^3}{GM}}. \tag{16.29}$$

For Sirius B, $P_{min} \approx 7$ s, which is much too long. However, for a $1.4 \, M_\odot$ neutron star, $P_{min} \approx 5 \times 10^{-4}$ s. Because this is a *minimum* rotation time, it can accommodate the complete range of periods observed for pulsars.

### Pulsars as Rapidly Rotating Neutron Stars

Only one alternative has emerged unscathed from this process of elimination, namely, that pulsars are rapidly rotating neutron stars. This conclusion was strengthened by the discovery in 1968 of pulsars associated with the Vela and Crab supernovae remnants. (Today dozens of pulsars are known to be associated with supernova remnants.) In addition, the Crab pulsar PSR 0531-21 has a very short pulse period of only 0.0333 s. No white dwarf could rotate 30 times per second without disintegrating, and the last doubts about the identity of pulsars were laid to rest. Until the discovery of the **millisecond pulsars** ($P \approx 10$ ms or less) in 1982, the Crab pulsar held the title of the fastest known pulsar (see Fig. 16.14).[33] The Vela and Crab pulsars not only produce radio bursts but also pulse in other regions of the electromagnetic spectrum ranging from radio to gamma rays, including visible flashes as shown in Fig. 16.15. These young pulsars (and a few others) also display **glitches** when their periods abruptly *decrease* by a tiny amount ($|\Delta P|/P \approx 10^{-6}$ to $10^{-8}$); see Fig. 16.16.[34] These sudden spinups are separated by uneven intervals of several years.

### Geminga

The nearest pulsar yet detected is only some 90 pc away. PSR 0633+1746, nicknamed Geminga, was well known as a strong source of gamma rays for 17 years before its identity as a pulsar was established in 1992.[35] With a period of 0.237 s, Geminga pulses in both

---

[33]It is likely that the millisecond pulsars have rapid rotation periods that are a consequence of their membership in close binary systems; more than half of the known millisecond pulsars belong to binaries. For this reason, millisecond pulsars will be discussed in more detail in Section 18.6.

[34]See page 602 for a discussion of possible glitch mechanisms.

[35]*Geminga* means "does not exist" in Milanese dialect, accurately reflecting its long-mysterious nature.

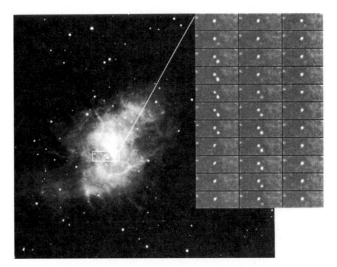

**FIGURE 16.15**   A sequence of images showing the flashes at visible wavelengths from the Crab pulsar, located at the center of the Crab Nebula (left). A foreground star can be seen as the constant point of light above and to the left of the Crab pulsar. (Courtesy of National Optical Astronomy Observatories.)

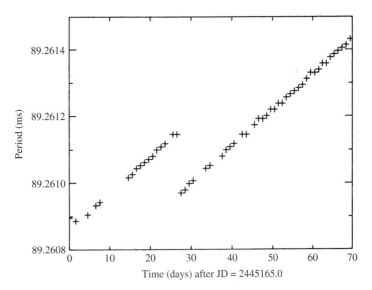

**FIGURE 16.16**   A glitch in the Vela pulsar. (Figure adapted from McCulloch et al., *Aust. J. Phys.*, *40*, 725, 1987.)

gamma and X-rays (but not at radio wavelengths) and may display glitches. In visible light, its absolute magnitude is fainter than +23.

### Evidence for a Core-Collapse Supernova Origin

Although at least one-half of all stars in the sky are known to be members of multiple-star systems, only a few percent of pulsars are known to belong to binary systems. Pulsars also move much faster through space than do normal stars, sometimes with speeds in excess of 1000 km s$^{-1}$. Both of these observations are consistent with a supernova origin for pulsars. This is because it is highly likely that a core-collapse supernova explosion is not perfectly spherically symmetric, so the forming pulsar could receive a kick, possibly ejecting it from any binary system that it may have been a part of initially. One hypothesis is that the pulsar is formed with an associated asymmetric jet and that, like a jet engine, the pulsar jet could launch the pulsar at high speed away from its formation point.

### Synchrotron and Curvature Radiation

Observations of the Crab Nebula, the remnant of the A.D. 1054 supernova, clearly reveal its intimate connection with the pulsar at its center. As shown in Fig. 16.15, the expanding nebula produces a ghostly glow surrounding gaseous filaments that wind throughout it. Interestingly, if the present rate of expansion is extrapolated backward in time, the nebula converges to a point about 90 years *after* the supernova explosion was observed. Obviously the nebula must have been expanding more slowly in the past than it is now, which implies that the expansion is actually accelerating.

In 1953, the Russian astronomer I. Shklovsky (1916–1985) proposed that the white light is **synchrotron radiation** produced when relativistic electrons spiral along magnetic field lines. From the equation for the magnetic force on a moving charge $q$,

$$\mathbf{F}_m = q\,(\mathbf{v} \times \mathbf{B}),$$

the component of an electron's velocity $\mathbf{v}$ perpendicular to the field lines produces a circular motion around the lines, while the component of the velocity along the lines is not affected; see Fig. 16.17. As they follow the curved field lines, the relativistic electrons accelerate and emit electromagnetic radiation. It is called synchrotron radiation if the circular motion around the field lines dominates or **curvature radiation** if the motion is primarily along the field lines. In both cases, the shape of the continuous spectrum produced depends on the energy distribution of the emitting electrons and so is easily distinguished from the spectrum of blackbody radiation.[36] The radiation is strongly linearly polarized in the plane of the circular motion for synchrotron radiation and is strongly linearly polarized in the plane of the curving magnetic field line for curvature radiation. As a test of his theory, Shklovsky predicted that the white light from the Crab Nebula would be found to be strongly linearly

---

[36]Both synchrotron and curvature radiation are sometimes called *nonthermal* to distinguish them from the thermal origin of blackbody radiation.

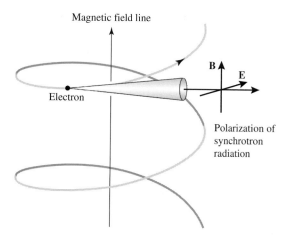

Magnetic field line

B

E

Electron

Polarization of
synchrotron
radiation

**FIGURE 16.17** Synchrotron radiation emitted by a relativistic electron as it spirals around a magnetic field line.

polarized. His prediction was subsequently confirmed as the light from some emitting regions of the nebula was measured to be 60% linearly polarized.

### The Energy Source for the Crab's Synchrotron Radiation

The identification of the white glow as synchrotron radiation raised new questions. It implied that magnetic fields of $10^{-7}$ T must permeate the Crab Nebula. This was puzzling because, according to theoretical estimates, long ago the expansion of the nebula should have weakened the magnetic field far below this value. Furthermore, the electrons should have radiated away all of their energy after only 100 years. It is clear that the production of synchrotron radiation today requires both a replenishment of the magnetic field and a continuous injection of new energetic electrons. The total power needed for the accelerating expansion of the nebula, the relativistic electrons, and the magnetic field is calculated to be about $5 \times 10^{31}$ W, or more than $10^5$ $L_\odot$.

The energy source is the rotating neutron star at the heart of the Crab Nebula. It acts as a huge flywheel and stores an immense amount of rotational kinetic energy. As the star slows down, its energy supply decreases.

To calculate the rate of energy loss, write the rotational kinetic energy in terms of the period and moment of inertia of the neutron star:

$$K = \frac{1}{2}I\omega^2 = \frac{2\pi^2 I}{P^2}.$$

Then the rate at which the rotating neutron star is losing energy is

$$\frac{dK}{dt} = -\frac{4\pi^2 I \dot{P}}{P^3}. \tag{16.30}$$

**Example 16.7.1.** Assuming that the neutron star is a uniform sphere with $R = 10\,\text{km}$ and $M = 1.4\,\text{M}_\odot$, its moment of inertia is approximately

$$I = \frac{2}{5}MR^2 = 1.1 \times 10^{38}\,\text{kg m}^2.$$

Inserting $P = 0.0333\,\text{s}$ and $\dot{P} = 4.21 \times 10^{-13}$ for the Crab pulsar gives $dK/dt \approx 5.0 \times 10^{31}\,\text{W}$. Remarkably, this is exactly the energy required to power the Crab Nebula. The slowing down of the neutron star flywheel has enabled the nebula to continue shining and expanding for nearly 1000 years.

It is important to realize that this energy is not transported to the nebula by the pulse itself. The radio luminosity of the Crab's pulse is about $10^{24}\,\text{W}$, 200 million times smaller than the rate at which energy is delivered to the nebula. (For older pulsars, the radio pulse luminosity is typically $10^{-5}$ of the spin-down rate of energy loss.) Thus the pulse process, whatever it may be, is a minor component of the total energy-loss mechanism.

Figure 16.18 shows an HST view of the immediate environment of the Crab pulsar. The ring-like halo seen on the west side of the pulsar is a glowing torus of gas; it may be the result of a polar jet from the pulsar forcing its way through the surrounding nebula. Just to the east of the pulsar, about 1500 AU away, is a bright knot of emission from shocked material in the jet, perhaps due to an instability in the jet itself. Another knot is seen at

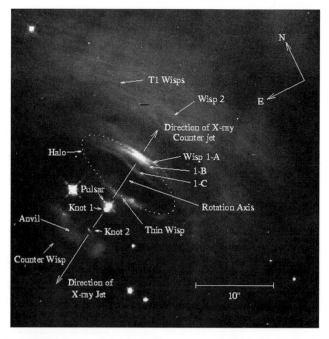

**FIGURE 16.18** An HST image of the immediate surroundings of the Crab pulsar. (Figure from Hester et al., *Ap. J., 448*, 240, 1995.)

a distance of 9060 AU. Low time-resolution "movies" of the central region of the Crab supernova remnant obtained by long-term observations by HST and Chandra are actually able to show the expansion and evolution of that portion of the nebula. Some of the wisps appear to moving outward at between $0.35c$ and $0.5c$.[37]

### The Structure of the Pulses

Before describing the details of a model pulsar, it is worth taking a closer look at the pulses themselves. As can be seen in Fig. 16.19, the pulses are brief and are received over a small fraction of the pulse period (typically from 1% to 5%). Generally, they are received at radio wave frequencies between roughly 20 MHz and 10 GHz.

As the pulses travel through interstellar space, the time-varying electric field of the radio waves causes the electrons that are encountered along the way to vibrate. This process slows the radio waves below the speed of light in a vacuum, $c$, with a greater retardation at lower frequencies. Thus a sharp pulse emitted at the neutron star, with all frequencies peaking at the same time, is gradually drawn out or *dispersed* as it travels to Earth (see Fig. 16.20). Because more distant pulsars exhibit a greater pulse dispersion, these time delays can be used to measure the distances to pulsars. The results show that the known pulsars are concentrated within the plane of our Milky Way Galaxy (Fig. 16.21) at typical distances of hundreds to thousands of parsecs.

Figure 16.22 shows that there is a substantial variation in the shape of the individual pulses received from a given pulsar. Although a typical pulse consists of a number of brief *subpulses*, the *integrated pulse profile*, an average built up by adding together a train of 100 or more pulses, is remarkably stable. Some pulsars have more than one average pulse profile and abruptly switch back and forth between them (Fig. 16.23). The subpulses may appear at random times in the "window" of the main pulse, or they may march across in a phenomenon known as *drifting subpulses*, as shown in Fig. 16.24. For about 30% of all known pulsars, the individual pulses may simply disappear or *null*, only to reappear up to 100 periods later. Drifting subpulses may even emerge from a nulling event in step with those that entered the null. Finally, the radio waves of many pulsars are strongly linearly polarized (up to 100%), a feature that indicates the presence of a strong magnetic field.

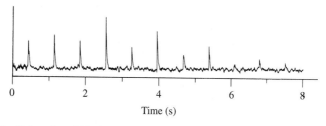

Time (s)

**FIGURE 16.19** Pulses from PSR 0329+54 with a period of 0.714 s. (Figure adapted from Manchester and Taylor, *Pulsars*, W. H. Freeman and Co., New York, 1977.)

---

[37] See Hester, et al., *Ap. J.*, *577*, L49, 2002. The movies are at
http://chandra.harvard.edu/photo/2002/0052/movies.html.

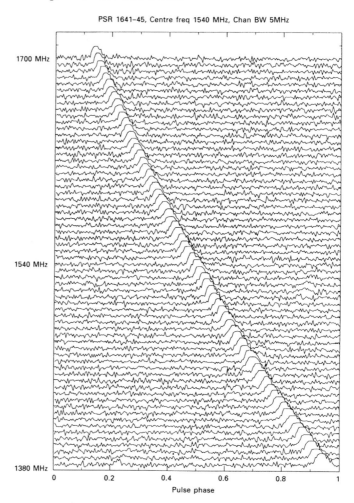

FIGURE 16.20   Dispersion of the pulse from PSR 1641-45. (Figure from Lyne and Graham-Smith, *Pulsar Astronomy*, ©Cambridge University Press, New York, 1990. Reprinted with the permission of Cambridge University Press.)

## The Basic Pulsar Model

The basic pulsar model, shown in Fig. 16.25, consists of a rapidly rotating neutron star with a strong dipole magnetic field (two poles, north and south) that is inclined to the rotation axis at an angle $\theta$. As explained in the previous section, the rapid rotation and the strong dipole field both arise naturally following the collapse of the core of a supergiant star.

First, we need to obtain a measure of the strength of the pulsar's magnetic field. As the pulsar rotates, the magnetic field at any point in space will change rapidly. According to Faraday's law, this will induce an electric field at that point. Far from the star (near the **light cylinder** defined in Fig. 16.26) the time-varying electric and magnetic fields form an electromagnetic wave that carries energy away from the star. For this particular situation,

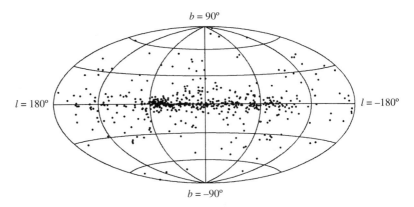

**FIGURE 16.21** Distribution of 558 pulsars in galactic coordinates, with the center of the Milky Way in the middle. The clump of pulsars at $\ell = 60°$ is a selection effect due to the fixed orientation of the Arecibo radio telescope. (Figure from Taylor, Manchester, and Lyne, *Ap. J. Suppl.*, *88*, 529, 1993.)

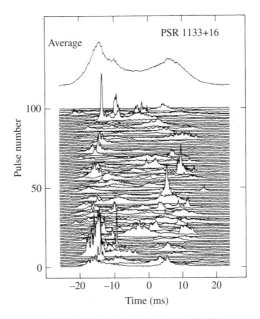

**FIGURE 16.22** The average of 500 pulses (top) and a series of 100 consecutive pulses (below) for PSR 1133+16. (Figure adapted from Cordes, *Space Sci. Review*, *24*, 567, 1979.)

the radiation is called **magnetic dipole radiation**. Although it is beyond the scope of this book to consider the model in detail, we note that the energy per second emitted by the rotating magnetic dipole is

$$\frac{dE}{dt} = -\frac{32\pi^5 B^2 R^6 \sin^2 \theta}{3\mu_0 c^3 P^4},$$

(16.31)

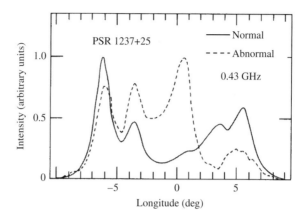

**FIGURE 16.23**   Changes in the integrated pulse profile of PSR 1237+25 due to mode switching. This pulsar displays five distinct subpulses. (Figure adapted from Bartel et al., *Ap. J.*, *258*, 776, 1982.)

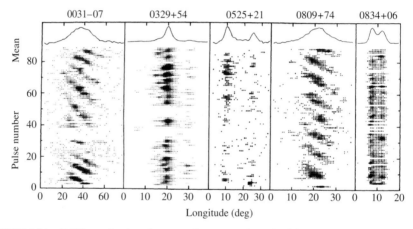

**FIGURE 16.24**   Drifting subpulses for two pulsars; note that PSR 0031-07 also nulls. (Figure from Taylor et al., *Ap. J.*, *195*, 513, 1975.)

where $B$ is the field strength at the magnetic pole of the star of radius $R$. The minus sign indicates that the neutron star is drained of energy, causing its rotation period, $P$, to increase. Note that the factor of $1/P^4$ means that the neutron star will lose energy much more quickly at smaller periods. Since the average pulsar period is 0.79 s, most pulsars are born spinning considerably faster than their current rates, with typical initial periods of a few milliseconds.

Assuming that all of the rotational kinetic energy lost by the star is carried away by magnetic dipole radiation, $dE/dt = dK/dt$. Using Eqs. (16.30) and (16.31), this is

$$-\frac{32\pi^5 B^2 R^6 \sin^2 \theta}{3\mu_0 c^3 P^4} = -\frac{4\pi^2 I \dot{P}}{P^3}. \tag{16.32}$$

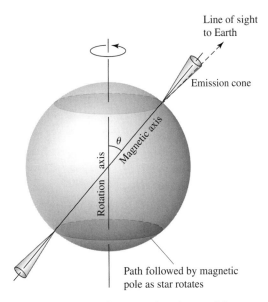

**FIGURE 16.25**   A basic pulsar model.

This can be easily solved for the magnetic field at the pole of the neutron star,

$$B = \frac{1}{2\pi R^3 \sin \theta} \sqrt{\frac{3\mu_0 c^3 I P \dot{P}}{2\pi}}. \qquad (16.33)$$

---

**Example 16.7.2.**   We will estimate the magnetic field strength at the poles of the Crab pulsar (PSR 0531-21), with $P = 0.0333$ s and $\dot{P} = 4.21 \times 10^{-13}$. Assuming that $\theta = 90°$, Eq. (16.33) then gives a value of $8.0 \times 10^8$ T. As we have seen, the Crab pulsar is interacting with the dust and gas in the surrounding nebula, so there are other torques that contribute to slowing down the pulsar's spin. This value of $B$ is therefore an overestimate; the accepted value of the Crab pulsar's magnetic field is $4 \times 10^8$ T.[38] Values of $B$ around $10^8$ T are typical for most pulsars.

However, repeating the calculation for PSR 1937+214 with $P = 0.00156$ s, $\dot{P} = 1.05 \times 10^{-19}$, and assuming the same value for the moment of inertia, we find the magnetic field strength to be only $B = 8.6 \times 10^4$ T. This much smaller value distinguishes the millisecond pulsars and provides another hint that these fastest pulsars may have a different origin or environment.

---

### Correlation Between Period Derivatives and Pulsar Classes

Figure 16.27 shows the distribution of period derivatives for pulsars as a function of pulsar period. Although the vast majority of pulsars fall into a large grouping in the middle of

---

[38]The suggestion that the Crab Nebula is powered by the magnetic dipole radiation from a rotating neutron star was made by the Italian astronomer Franco Pacini in 1967, a year *before* the discovery of pulsars!

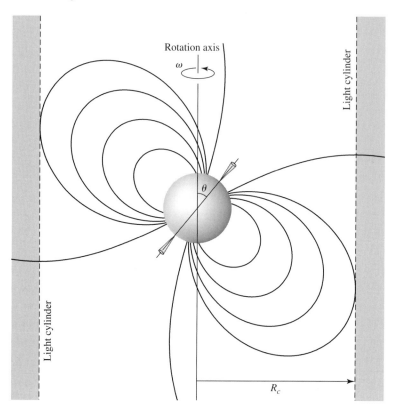

**FIGURE 16.26**   The light cylinder around a rotating neutron star. The cylinder's radius $R_c$ is where a point co-rotating with the neutron star would move at the speed of light: $R_c = c/\omega = cP/2\pi$.

the plot, the millisecond pulsars show a clear correlation with pulsars known to exist in binary systems. Other classes of pulsars are also evident: Pulsars known to emit energy at X-ray wavelengths have the longest periods and have the largest period derivatives, whereas high-energy pulsars that emit energies from radio frequencies through the infrared or higher frequencies tend to have larger values of $\dot{P}$ but otherwise typical periods. Note that although nearly all of the pulsars represented in Fig. 16.27 have positive values of $\dot{P}$, some of them, primarily the binary pulsars, actually have values of $\dot{P} < 0$, meaning that their periods are decreasing (they are speeding up!). Figure 16.27 may be compared with the histogram of pulsar periods shown in Fig. 16.14.

### Toward a Model of Pulsar Emission

Developing a detailed model of the pulsar's emission mechanism has been an exercise in frustration because almost every observation is open to more than one interpretation. The emission of radiation is the most poorly understood aspect of pulsars, and at present there is agreement only on the most general features of how a neutron star manages to produce radio waves. The following discussion summarizes a popular model of the pulse process.

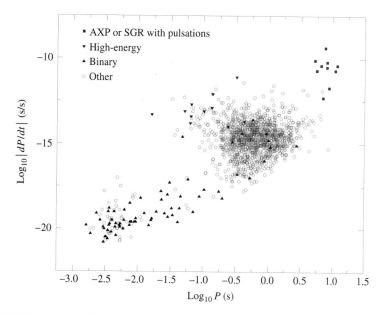

**FIGURE 16.27** The absolute value of the time derivative of period ($|\dot{P}|$) versus period ($P$) for all pulsars for which $\dot{P}$ has been determined. Special classes of pulsars are depicted separately: Anomalous X-ray pulsars (AXP) or Soft Gamma Repeaters (SGR) with pulsations, high-energy pulsars with emitted frequencies between radio and infrared or higher, and binary pulsars (with one or more known binary companions) are depicted separately. All remaining pulsars are indicated as "other." Note the abundance of known binary pulsars among the millisecond pulsars. (Data from Manchester, Hobbs, Teoh, and Hobbs, *A. J.*, *129*, 1993, 2005. Data available at http://www.atnf.csiro.au/research/pulsar/psrcat.)

You should keep in mind, however, that there is as yet no general consensus on whether the object being discussed actually occurs in nature or only in the minds of astrophysicists!

It is at least certain that the rapidly changing magnetic field near the rotating pulsar induces a huge electric field at the surface. The electric field of about $6.3 \times 10^{10}$ V m$^{-1}$ easily overcomes the pull of gravity on charged particles in the neutron star's crust. For example, the electric force on a proton is about 300 million times stronger than the force of gravity, and the ratio of the electric force on an electron to the gravitational force is even more overwhelming. Depending on the direction of the electric field, either negatively charged electrons or positively charged ions will be continuously ripped from the neutron star's polar regions. This creates a **magnetosphere** of charged particles surrounding the pulsar that is dragged around with the pulsar's rotation. However, the speed of the co-rotating particles cannot exceed the speed of light, so at the light cylinder the charged particles are spun away, carrying the magnetic field with them in a pulsar "wind." Such a wind may be responsible for the replenishment of the Crab Nebula's magnetic field and the continual delivery of relativistic particles needed to keep the nebula shining.

The charged particles ejected from the vicinity of the pulsar's magnetic poles are quickly accelerated to relativistic speeds by the induced electric field. As the electrons follow the curved magnetic field lines, they emit curvature radiation in the form of energetic gamma-ray photons. This radiation is emitted in a narrow beam in the instantaneous direction of motion of the electron, a consequence of the relativistic headlight effect discussed in Section 4.3. Each gamma-ray photon has so much energy that it can spontaneously convert this energy into an electron–positron pair via Einstein's relation $E = mc^2$. (This process, described by $\gamma \rightarrow e^- + e^+$, is just the inverse of the annihilation process mentioned in Section 10.3 for the Sun's interior.) The electrons and positrons are accelerated and in turn emit their own gamma rays, which create more electron–positron pairs, and so on. A cascade of pair production is thus initiated near the magnetic poles of the neutron star. Coherent beams of curvature radiation emitted by bunches of these particles may be responsible for the individual subpulses that contribute to the integrated pulse profile.

As these particles continue to curve along the magnetic field lines, they emit a continuous spectrum of curvature radiation in the forward direction, producing a narrow cone of radio waves radiating from the magnetic polar regions.[39] As the neutron star rotates, these radio waves sweep through space in a way reminiscent of the light from a rotating lighthouse beacon. If the beam happens to fall on a radio telescope on a blue-green planet in a distant Solar System, the astronomers there will detect a regular series of brief radio pulses.

As the pulsar ages and slows down, the structure of the underlying neutron star must adapt to the reduced rotational stresses. As a consequence, perhaps the crust settles a fraction of a millimeter and the star spins faster as a result of its decreased moment of inertia, or perhaps the superfluid vortices in the neutron star's core become momentarily "unpinned" from the underside of the solid crust where they are normally attached, giving the crust a sudden jolt. Either possibility could produce a small but abrupt increase in the rotation speed, and the astronomers on Earth would record a glitch for the pulsar (recall Fig. 16.16).

The question of a pulsar's final fate, as its period increases beyond several seconds, has several possible answers. It may be that the neutron star's magnetic field, originally produced by the collapse of the pre-supernova star's degenerate stellar core, decays with a characteristic time of 9 million years or so. Then, at some future time when the pulsar's period has been reduced to several seconds, its magnetic field may no longer be strong enough to sustain the pulse mechanism, and the pulsar turns off. On the other hand, it may be that the magnetic field does not decay appreciably but is maintained by a dynamo-like mechanism involving the differential rotation of the crust and core of the neutron star. However, rotation itself is an essential ingredient of any pulsar emission mechanism. As a pulsar ages and slows down, its beam will become weaker even if the magnetic field does not decay. In this case, the radio pulses may become too faint to be detected as the pulsar simply fades below the sensitivity of radio telescopes. The timescale for the decay of a neutron star's magnetic field is a matter of considerable debate, and both scenarios are consistent with the observations.

## Magnetars and Soft Gamma Repeaters

The preceding sketch reflects the current state of uncertainty about the true nature of pulsars. There are few objects in astronomy that offer such a wealth of intriguing observational detail

---

[39]The visible, X-ray, and gamma-ray pulses received from the Crab, Vela, Circinus, and Geminga pulsars may originate farther out in the pulsar's magnetosphere.

and yet are so lacking in a consistent theoretical description. Regardless of whether the basic picture outlined is vindicated or is supplanted by another view (perhaps involving a disk of material surrounding the neutron star), pulsar theorists will continue to take advantage of this unique natural laboratory for studying matter under the most extreme conditions.

To complicate the picture further, it is now believed that a class of extremely magnetic neutron stars known as **magnetars** exists. Magnetars have magnetic field strengths that are on the order of $10^{11}$ T, several orders of magnitude greater than typical pulsars. They also have relatively slow rotation periods of 5 to 8 seconds. Magnetars were first proposed to explain the **soft gamma repeaters** (SGRs), objects that emit bursts of hard X-rays and soft gamma-rays with energies of up to 100 keV (recall Fig. 16.27). Only a few SGRs are known to exist in the Milky Way Galaxy, and one has been detected in the Large Magellanic Cloud. Each of the SGRs is also known to correlate with supernova remnants of fairly young age ($\sim 10^4$ y). This would suggest that magnetars, if they are the source of the SGRs, are short-lived phenomena. Perhaps the Galaxy has many "extinct," or low-energy, magnetars scattered through it.

The emission mechanism of intense X-rays from SGRs is thought to be associated with stresses in the magnetic fields of magnetars that cause the surface of the neutron star to crack. The resulting readjustment of the surface produces a *super-Eddington* release of energy (roughly $10^3$ to $10^4$ times the Eddington luminosity limit in X-rays). In order to obtain such high luminosities, it is believed that the radiation must be confined; hence the need for very high magnetic field strengths.

Magnetars are distinguished from ordinary pulsars by the fact that the energy of the magnetar's field plays the major role in the energetics of the system, rather than rotation, as is the case for pulsars. Clearly much remains to be learned about the exotic environment of rapidly rotating, degenerate spheres with radii on the order of 10 km and densities exceeding the density of the nucleus of an atom.

## SUGGESTED READING

### General

Burnell, Jocelyn Bell, "The Discovery of Pulsars," *Serendipitous Discoveries in Radio Astronomy*, National Radio Astronomy Observatory, Green Bank, WV, 1983.

Graham-Smith, F., "Pulsars Today," *Sky and Telescope*, September 1990.

Kawaler, Stephen D., and Winget, Donald E., "White Dwarfs: Fossil Stars," *Sky and Telescope*, August 1987.

Nather, R. Edward, and Winget, Donald E., "Taking the Pulse of White Dwarfs," *Sky and Telescope*, April 1992.

Trimble, Virginia, "White Dwarfs: The Once and Future Suns," *Sky and Telescope*, October 1986.

### Technical

Clayton, Donald D., *Principles of Stellar Evolution and Nucleosynthesis*, University of Chicago Press, Chicago, 1983.

D'Antona, Francesca, and Mazzitelli, Italo, "Cooling of White Dwarfs," *Annual Review of Astronomy and Astrophysics*, *28*, 139, 1990.

Gold, T., "Rotating Neutron Stars as the Origin of the Pulsating Radio Sources," *Nature*, *218*, 731, 1968.

Hansen, Brad M. S., and Liebert, James, "Cool White Dwarfs," *Annual Review of Astronomy and Astrophysics*, *41*, 465, 2003.

Hansen, Carl J., Kawaler, Steven D., and Trimble, Virginia, *Stellar Interiors: Physical Principles, Structure, and Evolution*, Second Edition, Springer-Verlag, New York, 2004.

Hewish, A., et al., "Observations of a Rapidly Pulsating Radio Source," *Nature*, *217*, 709, 1968.

Kalogera, Vassiliki, and Baym, Gordon, "The Maximum Mass of a Neutron Star," *The Astrophysical Journal*, *470*, L61, 1996.

Liebert, James, "White Dwarf Stars," *Annual Review of Astronomy and Astrophysics*, *18*, 363, 1980.

Lyne, Andrew G., and Graham-Smith, F., *Pulsar Astronomy*, Third Edition, Cambridge University Press, Cambridge, 2006.

Manchester, Joseph H., and Taylor, Richard N., *Pulsars*, W. H. Freeman and Company, San Francisco, CA, 1977.

Michel, F. Curtis, *Theory of Neutron Star Magnetospheres*, The University of Chicago Press, Chicago, 1991.

Pacini, F., "Energy Emission from a Neutron Star," *Nature*, *216*, 567, 1967.

Salaris, Maurizio, et al., "The Cooling of CO White Dwarfs: Influence of the Internal Chemical Distribution," *The Astrophysical Journal*, *486*, 413, 1997.

Shapiro, Stuart L., and Teukolsky, Saul A., *Black Holes, White Dwarfs, and Neutron Stars*, John Wiley and Sons, New York, 1983.

Thompson, Christopher, and Duncan, Robert C., "The Soft Gamma Repeaters as Very Strongly Magnetized Neutron Stars – I. Radiative Mechanism for Outbursts," *Monthly Notices of the Royal Astronomical Society*, **275**, 255, 1995.

Winget, D. E., et al., "An Independent Method for Determining the Age of the Universe," *The Astrophysical Journal Letters*, *315*, L77, 1987.

Winget, D. E., et al., "Hydrogen-Driving and the Blue Edge of Compositionally Stratified ZZ Ceti Star Models," *The Astrophysical Journal Letters*, *252*, L65, 1982a.

Winget, Donald E., et al., "Photometric Observations of GD 358: DB White Dwarfs Do Pulsate," *The Astrophysical Journal Letters*, *262*, L11, 1982b.

## PROBLEMS

**16.1** The most easily observed white dwarf in the sky is in the constellation of Eridanus (the River Eridanus). Three stars make up the 40 Eridani system: 40 Eri A is a 4th-magnitude star similar to the Sun; 40 Eri B is a 10th-magnitude white dwarf; and 40 Eri C is an 11th-magnitude red M5 star. This problem deals only with the latter two stars, which are separated from 40 Eri A by 400 AU.

(a) The period of the 40 Eri B and C system is 247.9 years. The system's measured trigono-metric parallax is $0.201''$ and the true angular extent of the semimajor axis of the reduced mass is $6.89''$. The ratio of the distances of 40 Eri B and C from the center of mass is $a_B/a_C = 0.37$. Find the mass of 40 Eri B and C in terms of the mass of the Sun.

(b) The absolute bolometric magnitude of 40 Eri B is 9.6. Determine its luminosity in terms of the luminosity of the Sun.

(c) The effective temperature of 40 Eri B is 16,900 K. Calculate its radius, and compare your answer to the radii of the Sun, Earth, and Sirius B.

(d) Calculate the average density of 40 Eri B, and compare your result with the average density of Sirius B. Which is more dense, and why?

(e) Calculate the product of the mass and volume of both 40 Eri B and Sirius B. Is there a departure from the mass–volume relation? What might be the cause?

**16.2** The helium absorption lines seen in the spectra of DB white dwarfs are formed by excited He I atoms with one electron in the lowest ($n = 1$) orbital and the other in an $n = 2$ orbital. White dwarfs of spectral type DB are not observed with temperatures below about 11,000 K. Using what you know about spectral line formation, give a *qualitative* explanation why the helium lines would not be seen at lower temperatures. As a DB white dwarf cools below 12,000 K, into what spectral type does it change?

**16.3** Deduce a rough upper limit for $X$, the mass fraction of hydrogen, in the interior of a white dwarf. *Hint:* Use the mass and average density for Sirius B in the equations for the nuclear energy generation rate, and take $T = 10^7$ K for the central temperature. Set $\psi_{pp}$ and $f_{pp} = 1$ in Eq. (10.47) for the pp chain, and $X_{\text{CNO}} = 1$ in Eq. (10.59) for the CNO cycle.

**16.4** Estimate the ideal gas pressure and the radiation pressure at the center of Sirius B, using $3 \times 10^7$ K for the central temperature. Compare these values with the estimated central pressure, Eq. (16.1).

**16.5** By equating the pressure of an ideal gas of electrons to the pressure of a degenerate electron gas, determine a condition for the electrons to be degenerate, and compare it with the condition of Eq. (16.6). Use the exact expression (Eq. 16.12) for the electron degeneracy pressure.

**16.6** In the extreme relativistic limit, the electron speed $v = c$ must be used instead of Eq. (16.10) to find the electron degeneracy pressure. Use this to repeat the derivation of Eq. (16.11) and find

$$ P \approx \frac{\hbar c}{\sqrt{3}} \left[ \left( \frac{Z}{A} \right) \frac{\rho}{m_H} \right]^{4/3} . $$

**16.7** (a) At what speed do relativistic effects become important at a level of 10%? In other words, for what value of $v$ does the Lorentz factor, $\gamma$, become equal to 1.1?

(b) Estimate the density of the white dwarf for which the speed of a degenerate electron is equal to the value found in part (a).

(c) Use the mass–volume relation to find the approximate mass of a white dwarf with this average density. This is roughly the mass where white dwarfs depart from the mass–volume relation.

**16.8** Crystallization will occur in a cooling white dwarf when the electrostatic potential energy between neighboring nuclei, $Z^2 e^2 / 4\pi \epsilon_0 r$, dominates the characteristic thermal energy $kT$.

The ratio of the two is defined to be $\Gamma$,

$$\Gamma = \frac{Z^2 e^2}{4\pi \epsilon_0 r k T}.$$

In this expression, the distance $r$ between neighboring nuclei is customarily (and somewhat awkwardly) defined to be the radius of a sphere whose volume is equal to the volume per nucleus. Specifically, since the average volume per nucleus is $A m_H / \rho$, $r$ is found from

$$\frac{4}{3} \pi r^3 = \frac{A m_H}{\rho}.$$

(a) Calculate the value of the average separation $r$ for a 0.6 $M_\odot$ pure carbon white dwarf of radius 0.012 $R_\odot$.

(b) Much effort has been spent on precise numerical calculations of $\Gamma$ to obtain increasingly realistic cooling curves. The results indicate a value of about $\Gamma = 160$ for the onset of crystallization. Estimate the interior temperature, $T_c$, at which this occurs.

(c) Estimate the luminosity of a pure carbon white dwarf with this interior temperature. Assume a composition like that of Example 16.5.1 for the nondegenerate envelope.

(d) For roughly how many years could the white dwarf sustain the luminosity found in part (c), using just the latent heat of $kT$ per nucleus released upon crystallization? Compare this amount of time (when the white dwarf cools more slowly) with Fig. 16.9.

16.9 In the *liquid-drop model* of an atomic nucleus, a nucleus with mass number $A$ has a radius of $r_0 A^{1/3}$, where $r_0 = 1.2 \times 10^{-15}$ m. Find the density of this nuclear model.

16.10 If our Moon were as dense as a neutron star, what would its diameter be?

16.11 (a) Consider two point masses, each having mass $m$, that are separated vertically by a distance of 1 cm just above the surface of a neutron star of radius $R$ and mass $M$. Using Newton's law of gravity (Eq. 2.11), find an expression for the ratio of the gravitational force on the lower mass to that on the upper mass, and evaluate this expression for $R = 10$ km, $M = 1.4\,M_\odot$, and $m = 1$ g.

(b) An iron cube 1 cm on each side is held just above the surface of the neutron star described in part (a). The density of iron is 7860 kg m$^{-3}$. If iron experiences a stress (force per cross-sectional area) of $4.2 \times 10^7$ N m$^{-2}$, it will be permanently stretched; if the stress reaches $1.5 \times 10^8$ N m$^{-2}$, the iron will rupture. What will happen to the iron cube? (*Hint:* Imagine concentrating half of the cube's mass on each of its top and bottom surfaces.) What would happen to an iron meteoroid falling toward the surface of a neutron star?

16.12 Estimate the neutron degeneracy pressure at the center of a 1.4 $M_\odot$ neutron star (take the central density to be $1.5 \times 10^{18}$ kg m$^{-3}$), and compare this with the estimated pressure at the center of Sirius B.

16.13 (a) Assume that at a density just below neutron drip, all of the neutrons are in heavy neutron-rich nuclei such as $^{118}_{36}$Kr. Estimate the pressure due to relativistic degenerate electrons.

(b) Assume (*wrongly!*) that at a density just above neutron drip, all of the neutrons are free (and not in nuclei). Estimate the speed of the degenerate neutrons and the pressure they would produce.

16.14 Suppose that the Sun were to collapse down to the size of a neutron star (10-km radius).

(a) Assuming that no mass is lost in the collapse, find the rotation period of the neutron star.

**(b)** Find the magnetic field strength of the neutron star.

Even though our Sun will not end its life as a neutron star, this shows that the conservation of angular momentum and magnetic flux can easily produce pulsar-like rotation speeds and magnetic fields.

**16.15 (a)** Use Eq. (14.14) with $\gamma = 5/3$ to calculate the fundamental radial pulsation period for a one-zone model of a pulsating white dwarf (use the values for Sirius B) and a 1.4 $M_\odot$ neutron star. Compare these to the observed range of pulsar periods.

**(b)** Use Eq. (16.29) to calculate the minimum rotation period for the same stars, and compare them to the range of pulsar periods.

**(c)** Give an explanation for the similarity of your results.

**16.16 (a)** Determine the minimum rotation period for a 1.4 $M_\odot$ neutron star (the fastest it can spin without flying apart). For convenience, assume that the star remains spherical with a radius of 10 km.

**(b)** Newton studied the equatorial bulge of a homogeneous fluid body of mass $M$ that is *slowly* rotating with angular velocity $\Omega$. He proved that the difference between its equatorial radius ($E$) and its polar radius ($P$) is related to its average radius ($R$) by

$$\frac{E - P}{R} = \frac{5\Omega^2 R^3}{4GM}.$$

Use this to estimate the equatorial and polar radii for a 1.4 $M_\odot$ neutron star rotating with twice the minimum rotation period you found in part (a).

**16.17** If you measured the period of PRS 1937+214 and obtained the value on page 588, about how long would you have to wait before the last digit changed from a "5" to a "6"?

**16.18** Consider a pulsar that has a period $P_0$ and period derivative $\dot{P}_0$ at $t = 0$. Assume that the product $P\dot{P}$ remains constant for the pulsar (cf. Eq. 16.32).

**(a)** Integrate to obtain an expression for the pulsar's period $P$ at time $t$.

**(b)** Imagine that you have constructed a clock that would keep time by counting the radio pulses received from this pulsar. Suppose you also have a *perfect* clock ($\dot{P} = 0$) that is initially synchronized with the pulsar clock when they both read zero. Show that when the perfect clock displays the characteristic lifetime $P_0/\dot{P}_0$, the time displayed by the pulsar clock is $(\sqrt{3} - 1)P_0/\dot{P}_0$.

**16.19** During a glitch, the period of the Crab pulsar decreased by $|\Delta P| \approx 10^{-8}P$. If the increased rotation was due to an overall contraction of the neutron star, find the change in the star's radius. Assume that the pulsar is a rotating sphere of uniform density with an initial radius of 10 km.

**16.20** The Geminga pulsar has a period of $P = 0.237$ s and a period derivative of $\dot{P} = 1.1 \times 10^{-14}$. Assuming that $\theta = 90°$, estimate the magnetic field strength at the pulsar's poles.

**16.21 (a)** Find the radii of the light cylinders for the Crab pulsar and for the slowest pulsar, PSR 1841-0456. Compare these values to the radius of a 1.4 $M_\odot$ neutron star.

**(b)** The strength of a magnetic dipole is proportional to $1/r^3$. Determine the ratio of the magnetic field strengths at the light cylinder for the Crab pulsar and for PSR 1841-0456.

**16.22 (a)** Integrate Eq. (16.32) to obtain an expression for a pulsar's period $P$ at time $t$ if its initial period was $P_0$ at time $t = 0$.

**(b)** Assuming that the pulsar has had time to slow down enough that $P_0 \ll P$, show that the age $t$ of the pulsar is given approximately by

$$t = \frac{P}{2\dot{P}},$$

where $\dot{P}$ is the period derivative at time $t$.

**(c)** Evaluate this age for the case of the Crab pulsar, using the values found in Example 16.7.1. Compare your answer with the known age.

**16.23** One way of qualitatively understanding the flow of charged particles into a pulsar's magnetosphere is to imagine a charged particle of mass $m$ and charge $e$ (the fundamental unit of charge) at the equator of the neutron star. Assume for convenience that the star's rotation carries the charge perpendicular to the pulsar's magnetic field. The moving charge experiences a magnetic Lorentz force of $F_m = evB$ and a gravitational force, $F_g$. Show that the ratio of these forces is

$$\frac{F_m}{F_g} = \frac{2\pi e B R}{Pmg},$$

where $R$ is the star's radius and $g$ is the acceleration due to gravity at the surface. Evaluate this ratio for the case of a proton at the surface of the Crab pulsar, using a magnetic field strength of $10^8$ T.

**16.24** Find the minimum photon energy required for the creation of an electron–positron pair via the pair-production process $\gamma \rightarrow e^- + e^+$. What is the wavelength of this photon? In what region of the electromagnetic spectrum is this wavelength found?

**16.25** A subpulse involves a very narrow radio beam with a width between $1°$ and $3°$. Use Eq. (4.43) for the headlight effect to calculate the minimum speed of the electrons responsible for a $1°$ subpulse.

# CHAPTER

# 17

# General Relativity and Black Holes

## 17.1 ■ THE GENERAL THEORY OF RELATIVITY

Gravity, the weakest of the four forces of nature, plays a fundamental role in sculpting the universe on the largest scale. Newton's law of universal gravitation,

$$F = G\frac{Mm}{r^2}, \tag{17.1}$$

remained an unquestioned cornerstone of astronomers' understanding of heavenly motions until the beginning of the twentieth century. Its application had explained the motions of the known planets and had accurately predicted the existence and position of the planet Neptune in 1846. The sole blemish on Newtonian gravitation was the inexplicably large rate of shift in the orientation of Mercury's orbit.

The gravitational influences of the other planets cause the major axis of Mercury's elliptical orbit to slowly swing around the Sun in a counterclockwise direction relative to the fixed stars; see Fig. 17.1. The angular position at which perihelion occurs shifts at a rate of 574″ per century.[1] However, Newton's law of gravity was unable to explain 43″ per century of this shift, an inconsistency that led some mid-nineteenth century physicists to suggest that Eq. (17.1) should be modified from an exact inverse-square law. Others thought that an unseen planet, nicknamed Vulcan, might occupy an orbit inside Mercury's.

### The Curvature of Spacetime

Between the years 1907 and 1915, Albert Einstein developed a new theory of gravity, his **general theory of relativity**. In addition to resolving the mystery of Mercury's orbit, it predicted many new phenomena that were later confirmed by experiment. In this and the next section we will describe just enough of the physical content of general relativity to provide the background needed for future discussions of black holes and cosmology.

---

[1] The value of 1.5° per century encountered in some texts includes the very large effect of the precession of Earth's rotation axis on the celestial coordinate system, described in Section 1.3.

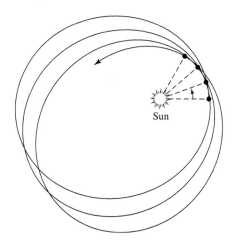

**FIGURE 17.1**    The perihelion shift of Mercury's orbit. Both the eccentricity of the orbit and the amount of shift in the location of perihelion in successive orbits have been exaggerated to better show the effect.

Einstein's view of the universe provides an exhilarating challenge to the imaginations of all students of astrophysics. But, before embarking on our study of general relativity, it will be helpful to take an advanced look at this new gravitational landscape.

The general theory of relativity is fundamentally a geometric description of how distances (intervals) in spacetime are measured in the presence of mass. For the moment, the effects on space and time will be considered separately, although you should always keep in mind that relativity deals with a unified spacetime. Near an object, both space and time must be described in a new way.

Distances between points in the space surrounding a massive object are altered in a way that can be interpreted as space becoming *curved* through a fourth spatial dimension perpendicular to *all* of the usual three spatial directions. The human mind balks at picturing this situation, but an analogy is easily found. Imagine four people holding the corners of a rubber sheet, stretching it tight and flat. This represents the flatness of empty space that exists in the absence of mass. Also imagine that a polar coordinate system has been painted on the sheet, with evenly spaced concentric circles spreading out from its center. Now lay a heavy bowling ball (representing the Sun) at the center of the sheet, and watch the indentation of the sheet as it curves down and stretches in response to the ball's weight, as pictured in Fig. 17.2. Closer to the ball, the sheet's curvature increases and the distance between points on the circles is stretched more. Just as the sheet curves in a third direction perpendicular to its original flat two-dimensional plane, the space surrounding a massive object may be thought of as curving in a fourth spatial dimension perpendicular to the usual three of "flat space."[2] The fact that mass has an effect on the surrounding space is the first essential element of general relativity. The curvature of space is just one aspect of the effect

---

[2]It is important to note that this fourth spatial dimension has nothing at all to do with the role played by time as a fourth nonspatial *coordinate* in the theory of relativity.

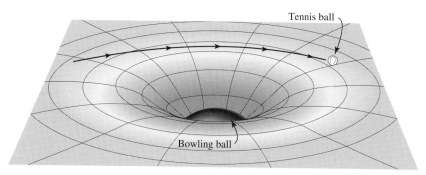

**FIGURE 17.2**    Rubber sheet analogy for curved space around the Sun. It is assumed that the rubber sheet is much larger than the area of curvature, so that the edges of the sheet have no effect on the curvature produced by the central mass.

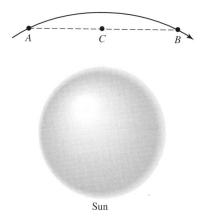

**FIGURE 17.3**    A photon's path around the Sun is shown by the solid line. The bend in the photon's trajectory is greatly exaggerated.

of mass on spacetime. In the language of unified spacetime, *mass acts on spacetime, telling it how to curve.*

Now imagine rolling a tennis ball, representing a planet, across the sheet. As it passes near the bowling ball, the tennis ball's path is curved. If the ball were rolled in just the right way under ideal conditions, it could even "orbit" the more massive bowling ball. In a similar manner, a planet orbits the Sun as it responds to the curved spacetime around it. Thus *curved spacetime acts on mass, telling it how to move.*

The passage of a ray of light near the Sun can be represented by rolling a ping-pong ball very rapidly past the bowling ball. Although the analogy with a massless photon is strained, it is reasonable to expect that as the photon moves through the curved space surrounding the Sun, its path will be deflected from a straight line. The bend of the photon's trajectory is small because the photon's speed carries it quickly through the curved space; see Fig. 17.3. In general relativity, gravity is the result of objects moving through curved spacetime, and everything that passes through, even massless particles such as photons, is affected.

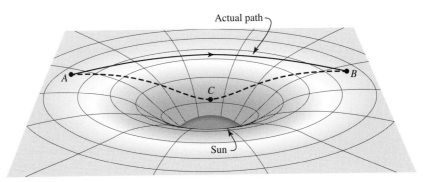

**FIGURE 17.4**   Comparison of two photon paths through curved space between points *A* and *B*. The projection of the path *ACB* onto the plane is the straight line depicted in Fig. 17.3.

Figure 17.3 hints at another aspect of general relativity. Since nothing can move between two points in space faster than light, light must always follow the quickest route between any two points.[3] In flat, empty space, this path is a straight line, but what is the quickest route through curved space? Suppose we use a series of mirrors to force the light beam to travel between points *A* and *B* by the apparent "shortcut" indicated by the dashed lines in Figs. 17.3 and 17.4. Would the light taking the dashed path outrace the beam free to follow its natural route through curved space? The answer is no—the curved beam would win the race. This result seems to imply that the beam following the dashed line would slow down along the way. However, this inference can't be correct because, according to the postulates of relativity (page 88), every observer, including one at point *C*, measures the same value for the speed of light. There are just two possible answers. The distance along the dashed line might actually be *longer* than the light beam's natural path, and/or time might run more slowly along the dashed path; either would retard the beam's passage. In fact, according to general relativity, these effects contribute equally to delaying the light beam's trip from *A* to *B* along the dashed line. The curving light beam actually does travel the *shorter* path. If two space travelers were to lay meter sticks end-to-end along the two paths, the dashed path would require a greater number of meter sticks because it penetrates farther into curved space, as shown in Fig. 17.4. In addition, the curvature of space involves a concomitant slowing down of time, so clocks placed along the dashed path would actually *run more slowly*. This is the final essential feature of general relativity: *Time runs more slowly in curved spacetime.*

It is important to note that all of the foregoing ideas have been tested experimentally many times, and in every case the results agree with general relativity. As soon as Einstein completed his theory, he applied it to the problem of Mercury's unexplained residual perihelion shift of 43" per century. Einstein wrote that his heart raced when his calculations exactly explained the discrepancy in terms of the planet's passage through the curved space near the Sun, saying that, "For a few days, I was beside myself with joyous excitement." Another triumph came in 1919 when the curving path of starlight passing near the Sun was first measured, by Arthur Stanley Eddington, during a total solar eclipse. As shown

---

[3]Throughout this chapter, light is assumed to be traveling in a vacuum.

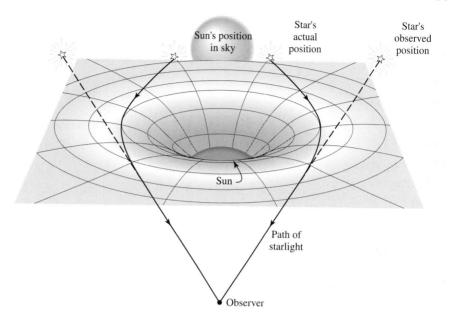

**FIGURE 17.5**   Bending of starlight measured during a solar eclipse.

in Fig. 17.5, the apparent positions of stars close to the Sun's eclipsed edge were shifted from their actual positions by a small angle. Einstein's theory predicted that this angular deflection would be 1.75″, in good agreement with Eddington's observations. General relativity has been tested continuously ever since. For instance, the superior conjunction of Mars that occurred in 1976 led to a spectacular confirmation of Einstein's theory. Radio signals beamed to Earth from the Viking spacecraft on Mars's surface were delayed as they traveled deep into the curved space surrounding the Sun. The time delay agreed with the predictions of general relativity to within 0.1%.

### The Principle of Equivalence

It is now time to retrace our steps and discover how Einstein came to his revolutionary understanding of gravity as geometry. One of the postulates of special relativity states that the laws of physics are the same in all inertial reference frames. Accelerating frames of reference are not inertial frames, because they introduce fictitious forces that depend on the acceleration. For example, an apple at rest on the seat of a car will not remain at rest if the car suddenly brakes to a halt. However, the acceleration produced by the force of gravity has a unique aspect. This may be clearly seen by noting a fundamental difference between Newton's law of gravity and Coulomb's law for the electrical force (Eq. 5.9).

Consider two objects separated by a distance $r$, one of mass $m$ and charge $q$, and the other of mass $M$ and charge $Q$. The magnitude of the acceleration ($a_g$) of mass $m$ *due to the gravitational force* is found from

$$ma_g = G\frac{mM}{r^2}, \tag{17.2}$$

while the magnitude of the acceleration ($a_e$) *due to the electrical force* is found from

$$ma_e = \frac{qQ}{4\pi\epsilon_0 r^2}. \tag{17.3}$$

The mass $m$ on the left-hand sides is an *inertial mass* and measures the object's resistance to being accelerated (its inertia). On the right-hand sides, the masses $m$ and $M$ and charges $q$ and $Q$ are numbers that couple the masses or charges to their respective forces and determine the strength of these forces. The mystery is the appearance of $m$ on both sides of the gravitational formula.

Why should a quantity that measures an object's inertia (which exists even in the complete absence of gravity) be the same as the "gravitational charge" that determines the force of gravity? The answer is that the notation in Eq. (17.2) is flawed, and the expression should properly be written as

$$m_i a_g = G \frac{m_g M_g}{r^2}$$

or

$$a_g = G \frac{M_g}{r^2} \frac{m_g}{m_i} \tag{17.4}$$

to clearly distinguish between the inertial and gravitational mass of each object. Similarly, for Eq. (17.3),

$$a_e = \frac{1}{4\pi\epsilon_0} \frac{qQ}{r^2} \frac{1}{m_i}.$$

In this case the only mass that enters the expression is the inertial mass.

It is an experimental fact, tested to a precision of 1 part in $10^{12}$, that $m_g/m_i$ in Eq. (17.4) is a constant. For convenience, this constant is chosen to be unity so the two types of mass will be numerically equal; if the gravitational mass were chosen to be *twice* the inertial mass, for example, the laws of physics would be unchanged except the gravitational constant $G$ would be assigned a new value only one-fourth as large. The proportionality of the inertial and gravitational masses means that at a given location, all objects experience the *same* gravitational acceleration. The constancy of $m_g/m_i$ is sometimes referred to as the **weak equivalence principle**.

This distinctive aspect of gravity, that every object falls with the same acceleration, has been known since the time of Galileo. It presented Einstein with both a problem and an opportunity to extend his theory of special relativity. He realized that if an entire laboratory were in free-fall, with all of its contents falling together, there would then be no way to detect its acceleration. In such a freely falling laboratory, it would be impossible to experimentally determine whether the laboratory was floating in space, far from any massive object, or falling freely in a gravitational field. Similarly, an observer watching an apple falling with an acceleration $g$ toward the floor of a laboratory would be unable to tell whether the laboratory was on Earth or far out in space, accelerating at a rate $g$ in the direction of the ceiling, as illustrated in Fig. 17.6. This posed a serious problem for the theory of special

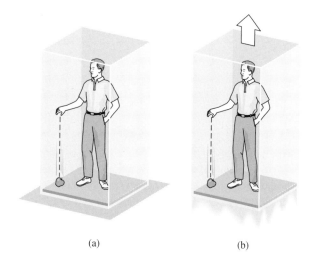

(a)                                                    (b)

**FIGURE 17.6**   Gravity is equivalent to an accelerating laboratory: (a) a laboratory on Earth, and (b) a laboratory accelerating in space.

relativity, which requires that inertial reference frames have a constant velocity. Because gravity is equivalent to an accelerating laboratory, an inertial reference frame cannot even be defined in the presence of gravity. Einstein had to find a way to remove gravity from the laboratory.

In 1907, Einstein had "the happiest thought of my life."

> I was sitting in a chair in the patent office at Bern when all of a sudden a thought occurred to me: "If a person falls freely he will not feel his own weight." I was startled. This simple thought made a deep impression on me. It impelled me toward a theory of gravitation.

The way to eliminate gravity in a laboratory is to surrender to it by entering into a state of free-fall; see Fig. 17.7.[4] However, there was an obstacle to applying this to special relativity because its inertial reference frames are *infinite* collections of meter sticks and synchronized clocks (recall Section 4.1). It would be impossible to eliminate gravity everywhere in an infinite, freely falling reference frame, because different points would have to be falling at different rates in different directions (toward the center of Earth, for example). Einstein realized that he would have to use *local* reference frames, just small enough that the acceleration due to gravity would be essentially constant in both magnitude and direction everywhere inside the reference frame (see Fig. 17.8). Gravity would then be abolished inside a local, freely falling reference frame.

---

[4]*Free-fall* means that there are no nongravitational forces accelerating the laboratory. In his meditation on general relativity, *A Journey into Gravity and Spacetime* (see Suggested Readings), John A. Wheeler prefers the term *free-float*. Since gravity has been abolished, why should falling even be mentioned? You are also urged to browse through the pages of *Gravitation* by Misner, Thorne, and Wheeler (1973) for additional insights into general relativity.

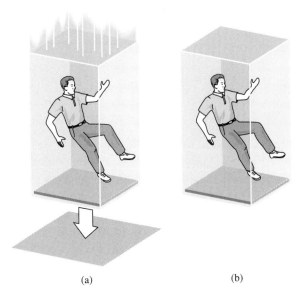

(a)                                              (b)

**FIGURE 17.7**    Gravity abolished in a freely falling laboratory: (a) a laboratory in free-fall, and (b) a laboratory floating in space.

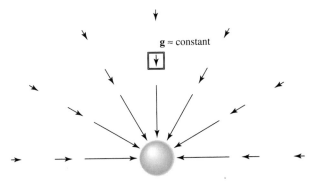

**FIGURE 17.8**    A local inertial reference frame, with $\mathbf{g} \approx$ constant inside. The arrows denote the gravitational acceleration vectors at those points around the mass.

In 1907 Einstein adopted this as the cornerstone of his theory of gravity, calling it the principle of equivalence.

> **The Principle of Equivalence:** All local, freely falling, nonrotating laboratories are fully equivalent for the performance of all physical experiments.

The restriction to nonrotating labs is necessary to eliminate the fictitious forces associated with rotation, such as the Coriolis and centrifugal forces. We will call these local, freely falling, nonrotating laboratories **local inertial reference frames**.

Note that special relativity is incorporated into the principle of equivalence. For example, measurements made from two local inertial frames in relative motion are related by the

Lorentz transformations (Eqs. 4.16–4.19) using the *instantaneous* value of the relative velocity between the two frames. Thus general relativity is in fact an extension of the theory of special relativity.

### The Bending of Light

We now move on to two simple thought experiments involving the equivalence principle that demonstrate the curvature of spacetime. For the first experiment, imagine a laboratory suspended above the ground by a cable [see Fig. 17.9(a)]. Let a photon of light leave a horizontal flashlight at the same instant the cable holding the lab is severed [Fig. 17.9(b)]. Gravity has been abolished from this freely falling lab, so it is now a local inertial reference frame. According to the equivalence principle, an observer falling with the lab will measure the light's path across the room as a straight horizontal line, in agreement with all of the laws of physics. But another observer on the ground sees a lab that is falling under the influence of gravity. Because the photon maintains a constant height above the lab's floor, the ground observer must measure a photon that falls with the lab, following a curved path. This displays the spacetime curvature represented by the rubber sheet analogy. The curved path taken by the photon is the quickest route possible through the curved spacetime surrounding Earth.

The angle of deflection, $\phi$, of the photon is very slight, as the following bit of geometry shows. Although the photon does not follow a circular path, we will use the *best-fitting circle* of radius $r_c$ to the actual path measured by the ground observer. Referring to Fig. 17.10, the center of the best-fitting circle is at point $O$, and the arc of the circle subtends an angle $\phi$ (exaggerated in the figure) between the radii $OA$ and $OB$. If the width of the lab is $\ell$, then the photon crosses the lab in time $t = \ell/c$. (The difference between the length of the arc and the width of the lab is negligible.) In this amount of time, the lab falls a distance

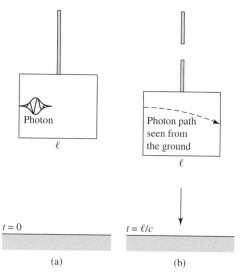

(a)                 (b)

**FIGURE 17.9** The equivalence principle for a horizontally traveling photon. The photon (a) leaves the left wall at $t = 0$, and (b) arrives at the right wall at $t = \ell/c$.

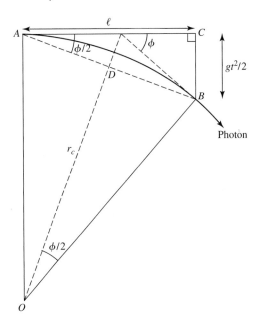

**FIGURE 17.10**    Geometry for the radius of curvature, $r_c$, and angular deflection, $\phi$.

$d = \frac{1}{2}gt^2$. Because triangles $ABC$ and $OBD$ are similar (each containing a right angle and another angle $\phi/2$),

$$\overline{BC}/\overline{AC} = \overline{BD}/\overline{OD}$$

$$\left(\frac{1}{2}gt^2\right) \bigg/ \ell = \left[\frac{\ell}{2\cos(\phi/2)}\right] \bigg/ \overline{OD}.$$

In fact, $\phi$ is so small that we can set $\cos(\phi/2) \simeq 1$ and the distance $\overline{OD} \simeq r_c$. Then, using $t = \ell/c$ and $g = 9.8$ m s$^{-2}$ for the acceleration of gravity near the surface of Earth, we find

$$r_c = \frac{c^2}{g} = 9.17 \times 10^{15} \text{ m,} \tag{17.5}$$

for the radius of curvature of the photon's path, which is nearly a light-year!

Of course, the angular deflection $\phi$ depends on the width $\ell$ of the lab. For example, if $\ell = 10$ m, then

$$\phi = \frac{\ell}{r_c} = 1.09 \times 10^{-15} \text{ rad,}$$

or only $2.25 \times 10^{-10}$ arcsecond. The large radius of the photon's path indicates that spacetime near Earth is only slightly curved. Nonetheless, the curvature is great enough to produce the circular orbits of satellites, which move slowly through the curved spacetime (slowly, that is, compared to the speed of light).

### Gravitational Redshift and Time Dilation

Our second thought experiment also begins with the laboratory suspended above the ground by a cable. This time, monochromatic light of frequency $\nu_0$ leaves a vertical flashlight on the floor at the same instant the cable holding the lab is severed. The freely falling lab is again a local inertial frame where gravity has been abolished, and so the equivalence principle requires that a frequency meter in the lab's ceiling record the *same* frequency, $\nu_0$, for the light that it receives. But an observer on the ground sees a lab that is falling under the influence of gravity. As shown in Fig. 17.11, if the light has traveled upward a height $h$ toward the meter in time $t = h/c$, then the meter has gained a downward speed toward the light of $v = gt = gh/c$ since the cable was released. Accordingly, we would expect that from the point of view of the ground observer, the meter should have measured a blueshifted frequency greater than $\nu_0$ by an amount given by Eq. (4.33). For the slow free-fall speeds involved here, this expected *increase* in frequency is

$$\frac{\Delta \nu}{\nu_0} = \frac{v}{c} = \frac{gh}{c^2}.$$

But in fact, the meter recorded *no change* in frequency. Therefore there must be another effect of the light's upward journey through the curved spacetime around Earth that exactly compensates for this blueshift. This is a **gravitational redshift** that tends to *decrease* the frequency of the light as it travels upward a distance $h$, given by

$$\frac{\Delta \nu}{\nu_0} = -\frac{v}{c} = -\frac{gh}{c^2}. \tag{17.6}$$

An outside observer, not in free-fall inside the lab, would measure only this gravitational

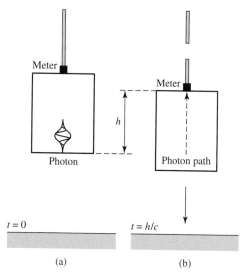

**FIGURE 17.11**   Equivalence principle for a vertically traveling light. The photon (a) leaves the floor at $t = 0$, and (b) arrives at the ceiling at $t = h/c$.

redshift. If the light were traveling downward, a corresponding blueshift would be measured. It is left as an exercise to show that this formula remains valid even if the light is traveling at an angle to the vertical, as long as $h$ is taken to be the *vertical* distance covered by the light.

---

**Example 17.1.1.**    In 1960, a test of the gravitational redshift formula was carried out at Harvard University. A gamma ray was emitted by an unstable isotope of iron, $^{57}_{26}$Fe, at the bottom of a tower 22.6 m tall, and received at the top of the tower. Using this value for $h$, the expected decrease in frequency of the gamma ray due to the gravitational redshift is

$$\frac{\Delta \nu}{\nu_0} = -\frac{gh}{c^2} = -2.46 \times 10^{-15}, \tag{17.7}$$

in excellent agreement with the experimental result of $\Delta \nu / \nu_0 = -(2.57 \pm 0.26) \times 10^{-15}$. More precise experiments carried out since that time have obtained agreement to within 0.007%.

   In actuality, the experiment was performed with both upward- and downward-traveling gamma rays, providing tests of both the gravitational redshift and blueshift.

---

An approximate expression for the total gravitational redshift for a beam of light that escapes out to infinity can be calculated by integrating Eq. (17.6) from an initial position $r_0$ to infinity, using $g = GM/r^2$ (Newtonian gravity) and setting $h$ equal to the differential radial element, $dr$ for a spherical mass, $M$, located at the origin. Some care must be taken when carrying out the integration, because Eq. (17.7) was derived using a *local* inertial reference frame. By integrating, we are really adding up the redshifts obtained for a chain of *different* frames. The radial coordinate $r$ can be used to measure distances for these frames only if spacetime is nearly flat [that is, if the radius of curvature given by Eq. (17.5) is very large compared with $r_0$]. In this case, the "stretching" of distances seen previously in the rubber sheet analogy is not too severe, and we can integrate

$$\int_{\nu_0}^{\nu_\infty} \frac{d\nu}{\nu} \simeq -\int_{r_0}^{\infty} \frac{GM}{r^2 c^2} \, dr,$$

where $\nu_0$ and $\nu_\infty$ are the frequencies at $r_0$ and infinity, respectively. The result is

$$\ln \left( \frac{\nu_\infty}{\nu_0} \right) \simeq -\frac{GM}{r_0 c^2},$$

which is valid when gravity is weak ($r_0/r_c = GM/r_0 c^2 \ll 1$). This can be rewritten as

$$\frac{\nu_\infty}{\nu_0} \simeq e^{-GM/r_0 c^2}. \tag{17.8}$$

Because the exponent is $\ll 1$, we use $e^{-x} \simeq 1 - x$ to get

$$\frac{\nu_\infty}{\nu_0} \simeq 1 - \frac{GM}{r_0 c^2}. \tag{17.9}$$

This approximation shows the first-order correction to the frequency of the photon.

The exact result for the gravitational redshift, valid even for a strong gravitational field, is

$$\frac{v_\infty}{v_0} = \left(1 - \frac{2GM}{r_0 c^2}\right)^{1/2}.$$

(17.10)

When gravity is weak and the exponent in Eq. (17.8) is $\ll 1$, we use $(1-x)^{1/2} \simeq 1 - x/2$ to recover Eq. (17.9).

The gravitational redshift can be incorporated into the redshift parameter defined by Eq. (4.34), giving

$$z = \frac{\lambda_\infty - \lambda_0}{\lambda_0} = \frac{v_0}{v_\infty} - 1$$

$$= \left(1 - \frac{2GM}{r_0 c^2}\right)^{-1/2} - 1$$

(17.11)

$$\simeq \frac{GM}{r_0 c^2},$$

(17.12)

where Eq. (17.12) is valid only for a weak gravitational field.

To understand the origin of the gravitational redshift, imagine a clock that is constructed to tick once with each vibration of a monochromatic light wave. The time between ticks is then equal to the period of the oscillation of the wave, $\Delta t = 1/v$. Then according to Eq. (17.10), as seen from an infinite distance, the gravitational redshift implies that the clock at $r_0$ will be observed to run more slowly than an identical clock at $r = \infty$. If an amount of time $\Delta t_0$ passes at position $r_0$ outside a spherical mass, $M$, then the time $\Delta t_\infty$ at $r = \infty$ is

$$\frac{\Delta t_0}{\Delta t_\infty} = \frac{v_\infty}{v_0} = \left(1 - \frac{2GM}{r_0 c^2}\right)^{1/2}.$$

(17.13)

For a weak field,

$$\frac{\Delta t_0}{\Delta t_\infty} \simeq 1 - \frac{GM}{r_0 c^2}.$$

(17.14)

We must conclude that *time passes more slowly as the surrounding spacetime becomes more curved*, an effect called **gravitational time dilation**. The gravitational redshift is therefore a consequence of time running at a slower rate near a massive object.

In other words, suppose two perfect, identical clocks are initially standing side by side, equally distant from a spherical mass. They are synchronized, and then one is slowly lowered below the other and then raised back to its original level. All observers will agree that when the clocks are again side by side, the clock that was lowered will be running behind the other because time in its vicinity passed more slowly while it was deeper in the mass's gravitational field.

---

**Example 17.1.2.** The white dwarf Sirius B has a radius of $R = 5.5 \times 10^6$ m and a mass of $M = 2.1 \times 10^{30}$ kg. The radius of curvature of the path of a horizontally traveling light beam near the surface of Sirius B is given by Eq. (17.5),

$$r_c = \frac{c^2}{g} = \frac{R^2 c^2}{GM} = 1.9 \times 10^{10} \text{ m.}$$

The fact that $GM/Rc^2 = R/r_c \ll 1$ indicates that the curvature of spacetime is not severe. Even at the surface of a white dwarf, gravity is considered relatively weak in terms of its effect on the curvature of spacetime.

From Eq. (17.12), the gravitational redshift suffered by a photon emitted at the star's surface is

$$z \simeq \frac{GM}{Rc^2} = 2.8 \times 10^{-4}.$$

This is in excellent agreement with the measured gravitational redshift for Sirius B of $(3.0 \pm 0.5) \times 10^{-4}$.

To compare the rate at which time passes at the surface of Sirius B with the rate at a great distance, suppose that exactly one hour is measured by a distant clock. The time recorded by a clock at the surface of Sirius B would be *less* than one hour by an amount found using Eq. (17.14):

$$\Delta t_\infty - \Delta t_0 = \Delta t_\infty \left( 1 - \frac{\Delta t_0}{\Delta t_\infty} \right) \simeq (3600 \text{ s}) \left( \frac{GM}{Rc^2} \right) = 1.0 \text{ s.}$$

The clock at the surface of Sirius B runs more slowly by about one second per hour compared to an identical clock far out in space.

---

The preceding experimental results (results obtained from tests of the equivalence principle) confirm the curvature of spacetime. In Section 17.2, we will learn that a freely falling particle takes the straightest possible path through curved spacetime.

## 17.2 ■ INTERVALS AND GEODESICS

We now consider the united concepts of space and time as expressed in *spacetime*, with four coordinates $(x, y, z, t)$ specifying each *event*.[5] Einstein's crowning achievement was the deduction of his *field equations* for calculating the geometry of spacetime produced by a given distribution of mass and energy. His equations have the form

$$\boxed{\mathcal{G} = -\frac{8\pi G}{c^4} \mathcal{T}.} \tag{17.15}$$

[5]Nothing special (in fact, nothing at all) need happen at an event. Recall from Section 4.1 that an event is simply a location in spacetime identified by $(x, y, z, t)$.

On the right is the *stress–energy tensor*, $\mathcal{T}$, which evaluates the effect of a given distribution of mass and energy on the curvature of spacetime, as described mathematically by the Einstein tensor, $\mathcal{G}$ (for $\mathcal{G}$ravity), on the left.[6] The appearance of Newton's gravitational constant, $G$, and the speed of light symbolizes the extension of relativity theory to include gravity. It is far beyond the scope of this book to delve further into this fascinating equation. We will be content merely to describe the curvature of spacetime around a spherical object of mass $M$ and radius $R$, then demonstrate how an object moves through the curved spacetime it encounters.

### Worldlines and Light Cones

Figure 17.12 shows three examples of some paths traced out in spacetime. In these *spacetime diagrams*, time is represented on the vertical axis, while space is depicted by the horizontal $x$–$y$ plane. The third spatial dimension, $z$, cannot be shown, so this figure deals only with motion that occurs in a plane. The path followed by an object as it moves through spacetime is called its **worldline**. Our task will be to calculate the worldline of a freely falling object in response to the local curvature of spacetime. The spatial components of such a worldline describe the trajectory of a baseball arcing toward an outfielder, a planet orbiting the Sun, or a photon attempting to escape from a black hole.

The worldlines of photons in flat spacetime point the way to an understanding of the geometry of spacetime. Suppose a flashbulb is set off at the origin at time $t = 0$; call this event $A$. As shown in Fig. 17.13, the worldlines of photons traveling in the $x$–$y$ plane form a **light cone** that represents a widening series of horizontal circular slices through the expanding spherical wavefront of light. The graph's axes are scaled so that the straight worldlines of light rays make $45°$ angles with the time axis.

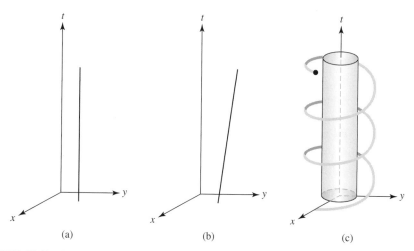

    (a)                               (b)                              (c)

**FIGURE 17.12** Worldlines for (a) a man at rest, (b) a woman running with constant velocity, and (c) a satellite orbiting Earth.

[6]Note that $E_{rest} = mc^2$ implies that both mass and energy contribute to the curvature of spacetime.

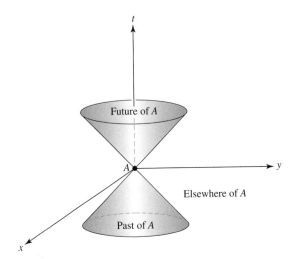

**FIGURE 17.13** Light cones generated by horizontally traveling photons leaving the origin at time $t = 0$.

A massive object initially at event $A$ must travel slower than light, so the angle between its worldline and the time axis must be less than $45°$. Therefore the region inside the light cone represents the possible *future* of event $A$. It consists of all of the events that can possibly be reached by a traveler initially at event $A$—and therefore all of the events that the traveler could ever influence in a causal way.

Extending the diverging photon worldlines back through the origin generates a lower light cone. Within this lower light cone is the possible *past* of event $A$, the collection of all events from which a traveler could have arrived just as the bulb flashed. In other words, the possible past consists of the locations in space and time of every event that could possibly have caused the flashbulb to go off.

Outside the future and past light cones is an unknowable *elsewhere*, that part of spacetime of which a traveler at event $A$ can have no knowledge and over which he or she can have no influence. It may come as a surprise to realize that vast regions of spacetime are hidden from us. You just can't get there from here.

In principle, every event in spacetime has a pair of light cones extending from it. The light cone divides spacetime into that event's future, past, and elsewhere. For any event in the past to have possibly influenced you, that event must lie within your past light cone, just as any event that you can ever possibly affect must lie within your future light cone. Your entire future worldline, your *destiny*, must therefore lie within your future light cone at every instant. Light cones act as spacetime horizons, separating the knowable from the unknowable.

### Spacetime Intervals, Proper Time, and Proper Distance

Measuring the progress of an object as it moves along its worldline involves defining a "distance" for spacetime. Consider the familiar case of purely spatial distances. If two points have Cartesian coordinates

$$(x_1, y_1, z_1) \qquad \text{and} \qquad (x_2, y_2, z_2),$$

then the distance $\Delta \ell$ measured along the straight line between the two points in flat space is defined by

$$(\Delta \ell)^2 = (x_2 - x_1)^2 + (y_2 - y_1)^2 + (z_2 - z_1)^2.$$

The analogous measure of "distance" in spacetime is called the **spacetime interval** (or simply *interval* for short), first encountered in Problem 4.11. Let two events $A$ and $B$ have spacetime coordinates

$$(x_A, y_A, z_A, t_A) \qquad \text{and} \qquad (x_B, y_B, z_B, t_B),$$

measured by an observer in an inertial reference frame, $S$. Then the interval $\Delta s$ measured along the straight worldline between the two events in flat spacetime is defined by

$$(\Delta s)^2 = [c(t_B - t_A)]^2 - (x_B - x_A)^2 - (y_B - y_A)^2 - (z_B - z_A)^2. \qquad (17.16)$$

In words,

$$(\text{interval})^2 = (\text{distance traveled by light in time } |t_B - t_A|)^2$$

$$- (\text{distance between events } A \text{ and } B)^2.$$

This definition of the interval is very useful because, as shown in Problem 4.11, $(\Delta s)^2$ is *invariant* under a Lorentz transformation (Eqs. 4.16–4.19). An observer in another inertial reference frame, $S'$, will measure the same value for the interval between events $A$ and $B$; that is, $\Delta s = \Delta s'$.

Note that $(\Delta s)^2$ may be positive, negative, or zero. The sign tells us whether light has enough time to travel between the two events. If $(\Delta s)^2 > 0$, then the interval is *timelike* and light has more than enough time to travel between events $A$ and $B$. An inertial reference frame $S$ can therefore be chosen that moves along the straight worldline connecting events $A$ and $B$ so that the two events happen at the *same location* in $S$ (at the origin, for example); see Fig. 17.14. Because the two events occur at the same place in $S$, the time measured between the two events is $\Delta s / c$. By definition, the time between two events that occur at the same location is the **proper time**, $\Delta \tau$, where

$$\boxed{\Delta \tau \equiv \frac{\Delta s}{c}} \qquad (17.17)$$

(recall Section 4.3). The proper time is just the elapsed time recorded by a watch moving along the worldline from $A$ to $B$. An observer in any inertial reference frame can use the interval to calculate the proper time between two events that are separated by a timelike interval.

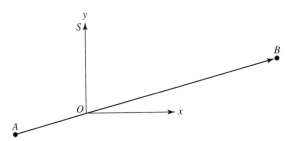

**FIGURE 17.14** An inertial reference frame $S$ moving along the timelike worldline connecting events $A$ and $B$. Both events occur at the origin of $S$.

If $(\Delta s)^2 = 0$, then the interval is *lightlike* or *null*. In this case, light has exactly enough time to travel between events $A$ and $B$. Only light can make the journey from one event to the other, and the proper time measured along a null interval is zero.

Finally, if $(\Delta s)^2 < 0$, then the interval is *spacelike*; light does not have enough time to travel between events $A$ and $B$. No observer could travel between the two events because speeds greater than $c$ would be required. The lack of absolute simultaneity in this situation, however, means that there are inertial reference frames in which the two events occur in the opposite temporal order, or even at the *same time*. By definition, the distance measured between two events $A$ and $B$ in a reference frame for which they occur simultaneously $(t_A = t_B)$ is the **proper distance** separating them,[7]

$$\Delta \mathcal{L} = \sqrt{-(\Delta s)^2}. \tag{17.18}$$

If a straight rod were connected between the locations of the two events, this would be the *rest length* of the rod. An observer in any inertial reference frame can use this to calculate the proper distance between two events that are separated by a spacelike interval.[8]

The interval is clearly related to the light cones discussed in the foregoing paragraphs. Let event $A$ be a flashbulb set off at the origin at time $t = 0$. The surfaces of the light cones, where the photons are at any time $t$, are the locations of all events $B$ that are connected to $A$ by a null interval. The events within the future and past light cones are connected to $A$ by a timelike interval, and the events that occur elsewhere are connected to $A$ by a spacelike interval.

### The Metric for Flat Spacetime

Returning to three-dimensional space for a moment, it is obvious that a path connecting two points in space doesn't have to be straight. Two points can be connected by infinitely many curved lines. To measure the distance along a curved path, $\mathcal{P}$, from one point to the

---

[7]In Section 4.3, when the emphasis was on length rather than distance, this was called the *proper length*. The terms may be used interchangeably, depending on the context.

[8]For both proper time and proper distance, the term *proper* has the connotation of "measured by an observer who is right there, moving along with the clock or the rod."

other, we use a differential distance formula called a **metric**,

$$(d\ell)^2 = (dx)^2 + (dy)^2 + (dz)^2.$$

Then $d\ell$ may be integrated along the path $\mathcal{P}$ (a *line integral*) to calculate the total distance between the two points,

$$\Delta\ell = \int_1^2 \sqrt{(d\ell)^2} = \int_1^2 \sqrt{(dx)^2 + (dy)^2 + (dz)^2} \quad \text{(along } \mathcal{P}\text{)}.$$

The distance between two points thus depends on the path connecting them. Of course, the *shortest* distance between two points in flat space is measured along a straight line. In fact, we can *define* the "straightest possible line" between two points as the path for which $\Delta\ell$ is a *minimum*.

Similarly, a worldline between two events in spacetime is not required to be straight; the two events can be connected by infinitely many curved worldlines. To measure the interval along a curved worldline, $\mathcal{W}$, connecting two events in spacetime with no mass present, we use the **metric for flat spacetime**,

$$(ds)^2 = (c\,dt)^2 - (d\ell)^2 = (c\,dt)^2 - (dx)^2 - (dy)^2 - (dz)^2. \tag{17.19}$$

Then $ds$ is integrated to determine the total interval along the worldline $\mathcal{W}$,

$$\Delta s = \int_A^B \sqrt{(ds)^2} = \int_A^B \sqrt{(c\,dt)^2 - (dx)^2 - (dy)^2 - (dz)^2} \quad \text{(along } \mathcal{W}\text{)}.$$

The interval is still related to the proper time measured along the worldline by Eq. (17.17). *The interval measured along any timelike worldline divided by the speed of light is always the proper time measured by a watch moving along that worldline.* The proper time is zero along a null worldline and is undefined for a spacelike worldline.

In flat spacetime, the interval measured along a straight timelike worldline between two events is a *maximum*. Any other worldline between the same two events will not be straight and will have a smaller interval. For a massless particle such as a photon, all worldlines have a null interval (so $\int \sqrt{(ds)^2} = 0$).

The maximal character of the interval of a straight worldline in flat spacetime is easily demonstrated. Figure 17.15 is a spacetime diagram showing two events, $A$ and $B$, that occur at times $t_A$ and $t_B$. The events are observed from an inertial reference frame, $S$, that moves from $A$ to $B$, chosen such that the two events occur at the origin of $S$. The interval measured along the straight worldline connecting $A$ and $B$ is

$$\Delta s(A \rightarrow B) = \int_A^B \sqrt{(ds)^2}$$

$$= \int_A^B \sqrt{(c\,dt)^2 - (dx)^2 - (dy)^2 - (dz)^2}$$

$$= \int_{t_A}^{t_B} c\,dt = c(t_B - t_A).$$

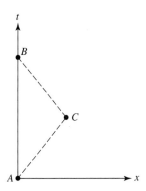

**FIGURE 17.15**    Worldlines connecting events $A$ and $B$.

Now consider the interval measured along another worldline connecting $A$ and $B$ that includes event $C$, which occurs at $(x, y, z, t) = (x_C, 0, 0, t_C)$. In this case,

$$\Delta s(A \rightarrow C \rightarrow B) = \int_A^C \sqrt{(ds)^2} + \int_C^B \sqrt{(ds)^2}$$

$$= \int_A^C \sqrt{(c\,dt)^2 - (dx)^2 - (dy)^2 - (dz)^2}$$

$$+ \int_C^B \sqrt{(c\,dt)^2 - (dx)^2 - (dy)^2 - (dz)^2}.$$

Using $dx/dt = v_{AC}$ for the constant velocity along worldline $A \rightarrow C$ in the first integral, and $dx/dt = v_{CB}$ for the constant velocity along $C \rightarrow B$ in the second integral, leads to

$$\Delta s(A \rightarrow C \rightarrow B) = (t_C - t_A)\sqrt{c^2 - v_{AC}^2} + (t_B - t_C)\sqrt{c^2 - v_{CB}^2}$$

$$< c(t_C - t_A) + c(t_B - t_C)$$

$$< \Delta s(A \rightarrow B).$$

Thus the straight worldline has the longer interval. Any worldline connecting event $A$ and $B$ can be represented as a series of short segments, so we can conclude that the interval $\Delta s$ is indeed a maximum for the straight worldline.

### Curved Spacetime and the Schwarzschild Metric

In a spacetime that is curved by the presence of mass, the situation is slightly more complicated. Even the "straightest possible worldline" will be curved. These straightest possible worldlines are called **geodesics**. In flat spacetime a geodesic is a straight worldline.

In curved spacetime, a timelike geodesic between two events has either a *maximum* or a *minimum* interval. In other words, the value of $\Delta s$ along a timelike geodesic is an *extremum*, either a maximum or a minimum, when compared with the intervals of nearby worldlines between the same two events.[9] In the situations we will encounter in this chapter, the intervals of timelike geodesics will be maxima. A massless particle such as a photon follows a *null geodesic*, with $\int \sqrt{(ds)^2} = 0$.[10] Einstein's key realization was that *the paths followed by freely falling objects through spacetime are geodesics.*

We are now prepared to deal with the effect of mass on the geometry of spacetime, based on the three fundamental features of general relativity:

- Mass acts on spacetime, telling it how to curve.

- Spacetime in turn acts on mass, telling it how to move.

- Any freely falling particle (including a photon) follows the straightest possible worldline, a geodesic, through spacetime. For a massive particle, the geodesic has a *maximum* or a *minimum* interval, while for light, the geodesic has a *null* interval.

These components of the theory will allow us to describe the curvature of spacetime around a massive spherical object and to determine how another object will move in response, whether it is a satellite orbiting Earth or a photon orbiting a black hole. For situations with spherical symmetry, it will be more convenient to use the familiar spherical coordinates $(r, \theta, \phi)$ instead of Cartesian coordinates. The metric between two nearby points in flat space is then

$$(d\ell)^2 = (dr)^2 + (r\,d\theta)^2 + (r\sin\theta\,d\phi)^2, \tag{17.20}$$

and the corresponding expression for the flat spacetime metric is

$$\boxed{(ds)^2 = (c\,dt)^2 - (dr)^2 - (r\,d\theta)^2 - (r\sin\theta\,d\phi)^2.} \tag{17.21}$$

Of course, spacetime will not be flat in the vicinity of a massive object. The specific situation to be investigated here is the motion of a particle through the curved spacetime produced by a massive sphere. It could be a planet, a star, or a black hole. The first task is to calculate how this massive object acts on spacetime, telling it how to curve. This requires a description of the metric for this curved spacetime that will replace Eq. (17.21) for a flat spacetime.

Before presenting this metric, we must emphasize that the variables $r$, $\theta$, $\phi$, and $t$ that appear in the expression for the metric are the *coordinates* used by an observer at rest a great ($\simeq$ infinite) distance from the origin. In the absence of a central mass at the origin, $r$ would be the distance from the origin, and differences in $r$ would measure the distance between points on a radial line. The time $t$ measured by clocks scattered throughout the coordinate system would remain synchronized, advancing everywhere at the same rate.

[9]In fact, a calculation of the intervals of nearby worldlines would show that the interval of a timelike geodesic corresponds to a maximum, a minimum, or an inflection point. You are referred to Section 13.4 of Misner, Thorne, and Wheeler (1973) for an interesting discussion of geodesics as worldlines of extremal proper time.

[10]The extremal principle for intervals cannot be directly applied to find the straightest possible worldline for a photon, since its interval is always null. However, the straightest possible worldline for a massless particle is the same as that for a massive particle in the limit of a vanishingly small mass as its velocity $v \to c$.

Now we place a sphere of mass $M$ and radius $R$ (which will be called a "planet") at the origin of our coordinate system. Some care must be taken in laying out the radial coordinate. The origin (which is inside the sphere) should not be used as a point of reference, and so we will avoid defining $r$ as "the distance from the origin." Instead, imagine a series of nested concentric spheres centered at the origin. The surface area of a sphere can be measured without approaching the origin, so the coordinate $r$ will be defined by the surface of that sphere having an area $4\pi r^2$. With this careful approach, we will find that these coordinates can be used with the metric for curved spacetime to measure distances in space and the passage of time near this massive sphere. As an object moves through this curved spacetime, its **coordinate speed** is just the rate at which its spatial coordinates change.

At a large distance ($r \simeq \infty$) from the planet, spacetime is essentially flat, and the gravitational time dilation of a photon received from the planet is given by Eq. (17.13). From this, it might be expected that $\sqrt{1 - 2GM/rc^2}$ would play a role in the metric for the spacetime surrounding the planet. Furthermore, recall from Section 17.1 that the stretching of space and the slowing down of time contribute equally to delaying a light beam's passage through curved spacetime. This provides a hint that the same factor will be involved in the metric's radial term. The angular terms are the same as those in Eq. (17.21) for flat spacetime.

These effects are indeed present in the metric that describes the curved spacetime surrounding a spherical mass, $M$. In 1916, just two months after Einstein published his general theory of relativity, the German astronomer Karl Schwarzschild (1873–1916) solved Einstein's field equations to obtain what is now called the **Schwarzschild metric**:

$$(ds)^2 = \left(c\,dt\sqrt{1 - 2GM/rc^2}\right)^2 - \left(\frac{dr}{\sqrt{1 - 2GM/rc^2}}\right)^2$$
$$- (r\,d\theta)^2 - (r\sin\theta\,d\phi)^2. \tag{17.22}$$

There is no other, easier way to obtain the Schwarzschild metric, so we must be content with the foregoing heuristic description of its terms.

It is important to realize that the Schwarzschild metric is the spherically symmetric *vacuum solution* of Einstein's field equations. That is, it is valid only in the empty space *outside* the object. The mathematical form of the metric is different in the object's interior, which is occupied by matter.

The Schwarzschild metric contains all of the effects considered in the last section. The "curvature of space" resides in the radial term. The radial distance measured simultaneously ($dt = 0$) between two nearby points on the same radial line ($d\theta = d\phi = 0$) is just the proper distance, Eq. (17.18),

$$d\mathcal{L} = \sqrt{-(ds)^2} = \frac{dr}{\sqrt{1 - 2GM/rc^2}}. \tag{17.23}$$

Thus the spatial distance $d\mathcal{L}$ between two points on the same radial line is *greater* than the coordinate difference $dr$. This is precisely what is represented by the stretched grid lines in the rubber sheet analogy of the previous section. The factor of $1/\sqrt{1 - 2GM/rc^2}$ must be included in any calculation of spatial distances. This is analogous to using a topographic

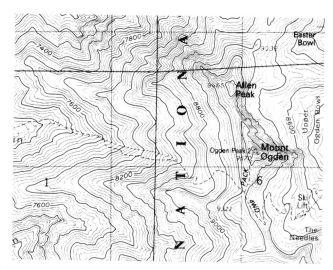

**FIGURE 17.16**   Topographical map with elevation contour lines. The shortest distance between two points on the map may not be a straight line. (Courtesy of USGS.)

map when planning a hike up a steep trail. The additional information provided by the map's elevation contour lines must be included in any calculation of the actual hiking distance, which is always greater than the difference in map coordinates; see Fig. 17.16.

The Schwarzschild metric also incorporates time dilation and the gravitational redshift (two aspects of the same effect). If a clock is at rest at the radial coordinate $r$, then the proper time $d\tau$ it records (Eq. 17.17) is related to the time $dt$ that elapses at an infinite distance by

$$d\tau = \frac{ds}{c} = dt\sqrt{1 - \frac{2GM}{rc^2}}, \tag{17.24}$$

which is, of course, just Eq. (17.13). Since $d\tau < dt$, this shows that time passes more slowly closer to the planet.

### The Orbit of a Satellite

Having finally learned how a spherical object of any mass acts on spacetime, telling it how to curve, we are now ready to calculate how curved spacetime acts on a particle, telling it how to move. The rest of this section will be devoted to using general relativity to find the motion of a satellite about the planet. All we need is the rule that it will follow the straightest possible worldline, the worldline with an *extremal* interval.[11]

At this point, you may be fondly recalling the simplicity of Newtonian gravity. According to Newton, the motion of a satellite in a circular orbit around Earth is found by simply equating the centripetal and gravitational accelerations. That is,

$$\frac{v^2}{r} = \frac{GM}{r^2},$$

[11] It is assumed that the satellite's mass $m$ is small enough that its effect on the surrounding spacetime is negligible.

where $v$ is the orbital speed. This immediately results in

$$v = \sqrt{\frac{GM}{r}}.$$

Einstein and Newton must agree in the limiting case of weak gravity, so this result must be concealed within the Schwarzschild metric for curved spacetime.[12] It can be found by using the Schwarzschild metric to find the straightest possible worldline for the satellite's circular orbit.

Powerful tools are available for calculating the worldline with the maximum interval between two fixed events. If we employed such an approach, the orbit of the satellite would emerge along with the laws of the conservation of energy, momentum, and angular momentum because they are built into Einstein's field equations. However, we will use a simpler strategy and assume from the beginning that the satellite travels above Earth's equator ($\theta = 90°$) in a circular orbit with a specified angular speed $\omega = v/r$. Inserting these choices, along with $dr = 0$, $d\theta = 0$, and $d\phi = \omega \, dt$, into the Schwarzschild metric gives

$$(ds)^2 = \left[ \left( c\sqrt{1 - 2GM/rc^2} \right)^2 - r^2\omega^2 \right] dt^2 = \left( c^2 - \frac{2GM}{r} - r^2\omega^2 \right) dt^2.$$

Integrating, the spacetime interval for one orbit is just

$$\Delta s = \int_0^{2\pi/\omega} \sqrt{c^2 - \frac{2GM}{r} - r^2\omega^2} \, dt. \tag{17.25}$$

When finding the value of $r$ for which the interval is an extremum, we must be certain that the endpoints of the satellite's worldline remain fixed. That is, the satellite's orbit must always begin and end at the same position, $r_0$, for all of the worldlines. To accommodate orbits of different radii, consider the "orbit" shown in Fig. 17.17. We start the satellite at $r_0$ and then move it (at nearly the speed of light) radially outward to the radius $r$ of its actual orbit. At the end of the orbit, the satellite returns just as rapidly to its starting point at $r_0$. Fortunately, the quick radial excursions at the beginning and the end of the orbit can be made with negligible contribution to the integral for the spacetime interval. (At almost the speed of light, the contribution is nearly null.) The net effect is a purely circular motion, so Eq. (17.25) can be used to evaluate the interval.

In Eq. (17.25), the limits of integration are constant and the only variable is $r$. The value of the radial coordinate $r$ for the orbit actually followed by the satellite must be the one for which $\Delta s$ is an extremum. This value may be found by taking the derivative of $\Delta s$ with respect to $r$ and setting it equal to zero:

$$\frac{d}{dr}(\Delta s) = \frac{d}{dr} \left( \int_0^{2\pi/\omega} \sqrt{c^2 - \frac{2GM}{r} - r^2\omega^2} \, dt \right) = 0.$$

[12] To avoid succumbing to Newtonian nostalgia, you should remember that when Einstein and Newton disagree, nature sides with Einstein.

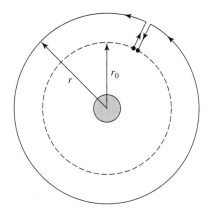

**FIGURE 17.17**   The "orbit" of a satellite, showing the radial motions used to keep the endpoints of the satellite's worldline fixed. The net effect is a circular orbit.

The derivative may be taken inside the integral to obtain

$$\frac{d}{dr}\sqrt{c^2 - \frac{2GM}{r} - r^2\omega^2} = 0,$$

implying

$$\frac{2GM}{r^2} - 2r\omega^2 = 0.$$

Thus, as promised,

$$v = r\omega = \sqrt{\frac{GM}{r}} \tag{17.26}$$

is the coordinate speed of the satellite for a circular orbit. [By coordinate speed, we simply mean that $v = r\,d\phi/dt$ is speed of the satellite measured in the $(r, \theta, \phi, t)$ coordinate system used by a distant observer.] Figure 17.18 illustrates how this straightest possible worldline through curved spacetime is projected onto the orbital plane, resulting in the satellite's circular orbit around Earth. In fact, this result is valid even for the very large spacetime curvature encountered around a black hole.

## 17.3 ■ BLACK HOLES

In 1783 John Michell (1724–1793), an English clergyman and amateur astronomer, considered the implications of Newton's corpuscular theory of light. If light were indeed a stream of particles, then it should be influenced by gravity. In particular, he conjectured that the gravity of a star 500 times larger than the Sun, but with the Sun's average density, would be sufficiently strong that even light could not escape from it. As you may verify using Eq. (2.17), the escape velocity of Michell's star would be the speed of light. Naively

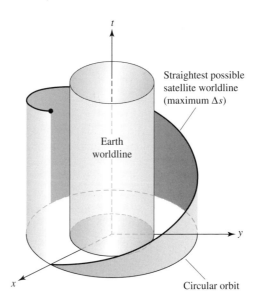

**FIGURE 17.18**    The straightest possible worldline through curved spacetime and its projection onto the orbital plane of the satellite.

setting the Newtonian formula for the escape velocity equal to $c$ shows that $R = 2GM/c^2$ is the radius of a star whose escape velocity equals the speed of light. In terms of the mass of the Sun, $R = 2.95(M/M_\odot)$ km. Even if this Newtonian derivation were correct, the resulting radius of such a star seemed unrealistically small, and so it held little interest for astronomers until the middle of the twentieth century.

In 1939 American physicists J. Robert Oppenheimer and Hartland Snyder (1913–1962) described the ultimate gravitational collapse of a massive star that had exhausted its sources of nuclear fusion. It was earlier that year that Oppenheimer and Volkoff had calculated the first models of neutron stars (see Section 16.6). We have seen that a neutron star cannot be more massive than about 3 $M_\odot$.[13] Oppenheimer and Snyder pursued the question of the fate of a degenerate star that might exceed this limit and surrender completely to the force of gravity.

### The Schwarzschild Radius

For the simplest case of a nonrotating star, the answer lies in the Schwarzschild metric, Eq. (17.22):

$$(ds)^2 = \left( c\, dt \sqrt{1 - 2GM/rc^2} \right)^2 - \left( \frac{dr}{\sqrt{1 - 2GM/rc^2}} \right)^2 - (r\, d\theta)^2 - (r \sin\theta\, d\phi)^2.$$

[13]As previously discussed, the upper mass limit of a neutron star is between 2.2 $M_\odot$ and 2.9 $M_\odot$ depending on the amount of rotation; recall Sections 15.4 and 16.6. We will adopt an approximate value of 3 $M_\odot$ for the purposes of this discussion.

When the radial coordinate of the star's surface has collapsed to

$$R_S = 2GM/c^2, \tag{17.27}$$

called the **Schwarzschild radius**, the square roots in the metric go to zero. The resulting behavior of space and time at $r = R_S$ is remarkable. For example, according to Eq. (17.17), the proper time measured by a clock at the Schwarzschild radius is $d\tau = 0$. Time has slowed to a complete stop, as measured from a vantage point that is at rest a great distance away.[14] From this viewpoint, *nothing ever happens at the Schwarzschild radius!*

This behavior is quite curious; does it imply that even light is frozen in time? The speed of light measured by an observer suspended above the collapsed star must always be $c$. But from far away, we can determine that light is delayed as it moves through curved spacetime. (Recall the time delay of radio signals from the Viking lander on Mars described in Section 17.1.) The apparent speed of light, the rate at which the spatial coordinates of a photon change, is called the *coordinate speed of light*. Starting with the Schwarzschild metric with $ds = 0$ for light,

$$0 = \left(c\, dt\sqrt{1 - 2GM/rc^2}\right)^2 - \left(\frac{dr}{\sqrt{1 - 2GM/rc^2}}\right)^2 - (r\, d\theta)^2 - (r\sin\theta\, d\phi)^2,$$

we can calculate the coordinate speed of a vertically traveling photon. Inserting $d\theta = d\phi = 0$ shows that, in general, the coordinate speed of light in the radial direction is

$$\frac{dr}{dt} = c\left(1 - \frac{2GM}{rc^2}\right) = c\left(1 - \frac{R_S}{r}\right). \tag{17.28}$$

When $r \gg R_S$, $dr/dt \simeq c$, as expected in flat spacetime. However, at $r = R_S$, $dr/dt = 0$ (see Fig. 17.19). Light is indeed frozen in time at the Schwarzschild radius. The spherical surface at $r = R_S$ acts as a barrier and prevents our receiving any information from within. For this reason, a star that has collapsed down within the Schwarzschild radius is called a **black hole**.[15] It is enclosed by the **event horizon**, the spherical surface at $r = R_S$. Note that the event horizon is a mathematical surface and need not coincide with any physical surface.

Although the interior of a black hole, inside the event horizon, is a region that is forever hidden from us on the outside, its properties may still be calculated. A nonrotating black hole has a particularly simple structure. At the center is the **singularity**, a point of zero volume and infinite density where all of the black hole's mass is located. Spacetime is infinitely curved at the singularity.[16] Cloaking the central singularity is the event horizon,

---

[14]You should recall that the spacetime coordinates $(r, \theta, \phi, t)$ in the Schwarzschild metric were established for use by an observer at rest at $r \simeq \infty$.

[15]The term *black hole* is the 1968 invention of the American theoretical physicist John A. Wheeler.

[16]The black hole's singularity is a real physical entity. It is not a mathematical artifact, as is the *mathematical* singularity exhibited by the Schwarzschild metric at the event horizon (where $1/\sqrt{1 - 2GM/rc^2} \to \infty$). Choosing another coordinate system would remove the divergence at the event horizon, so that divergence has no physical significance.

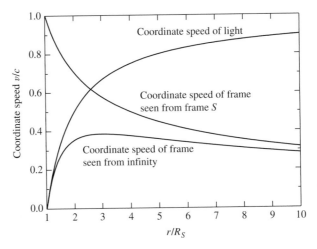

**FIGURE 17.19**    Coordinate speed of light, and coordinate speeds of a freely falling frame $S$ seen by an observer at rest at infinity and by an observer in the frame $S$. The radial coordinates are in terms of $R_S$ for a 10 $M_\odot$ black hole having a Schwarzschild radius of $\approx$ 30 km.

so the singularity can never be observed. In fact, there is a hypothesis dubbed the "Law of Cosmic Censorship" that forbids a *naked singularity* from appearing unclothed (without an associated event horizon).

### A Trip into a Black Hole

An object as bizarre as a black hole deserves closer scrutiny. Imagine an attempt to investigate the black hole by starting at a safe distance and reflecting a radio wave from an object at the event horizon. How much time will it take for a radio photon (or any photon) to reach the event horizon from a radial coordinate $r \gg R_S$ and then return? Since the round trip is symmetric, it is necessary only to find the time for either the journey in or out and then double the answer. It is easiest to integrate the coordinate speed of light in the radial direction, Eq. (17.28), between two arbitrary values of $r_1$ and $r_2$ to obtain the general answer,

$$\Delta t = \int_{r_1}^{r_2} \frac{dr}{dr/dt} = \int_{r_1}^{r_2} \frac{dr}{c(1 - R_S/r)} = \frac{r_2 - r_1}{c} + \frac{R_S}{c} \ln\left(\frac{r_2 - R_S}{r_1 - R_S}\right),$$

assuming that $r_1 < r_2$. Inserting $r_1 = R_S$ for the photon's original position, we find that $\Delta t = \infty$. Now, since the trip is symmetric, the same result applies if the photon started at $R_S$. According to the distant observer, the radio photon will *never* reach the event horizon. Instead, according to gravitational time dilation, the photon's coordinate velocity will slow down until it finally stops at the event horizon in the infinite future. In fact, any object falling toward the event horizon will suffer the same fate. Seen from the outside, even the surface of the star that collapsed to form the event horizon would be frozen, and so a black hole is in this sense a *frozen star*.

A brave (and *indestructible*) astronomer decides to test this remarkable conclusion. Starting from rest at a great distance, she volunteers to fall freely toward a 10 $M_\odot$ black hole

($R_S \simeq 30$ km). We remain behind to watch her local inertial frame $S$ as it falls with coordinate speed $dr/dt$ all the way to the event horizon. She gradually accelerates as she monitors her watch and shines a monochromatic flashlight back in our direction once every second. As her fall progresses, the light signals arrive farther and farther apart for several reasons: Subsequent signals must travel a longer distance as she accelerates, and her proper time $\tau$ is running more slowly than our coordinate time $t$ due to her location (gravitational time dilation) and her motion (special relativity time dilation). Furthermore, the coordinate speed of light becomes slower as she approaches the black hole, so the signals travel back to us more slowly. The frequency of the light waves we receive is also increasingly redshifted. This is caused by both her acceleration away from us and the gravitational redshift. The light becomes dimmer as well, as the rate at which her flashlight emits photons decreases (seen from our vantage point) and the energy per photon ($hc/\lambda$) also declines. Then when she is about $2R_S$ from the event horizon, the time between her signals begins to increase without limit as the strength of the signals decreases. The light is redshifted and dimmed into invisibility as time dilation brings her coordinate speed to zero (see Figs. 17.19 and 17.20). She is frozen in time, held for eternity like a fly caught in amber. Our successors could watch for millennia while stars were born, evolved, and died without receiving a single photon from her.

How does all of this appear to the brave astronomer, freely falling toward the black hole? Because gravity has been abolished in her local inertial frame, initially she does not notice her approach to the black hole. She monitors her watch (which displays her proper time, $\tau$), and she turns on her flashlight once per second. However, as she draws closer, she begins to feel as though she is being stretched in the radial direction and compressed in the perpendicular directions; see Fig. 17.21. The gravitational pull on her feet (nearer the black hole) is stronger than on her head, and the variation in the direction of gravity

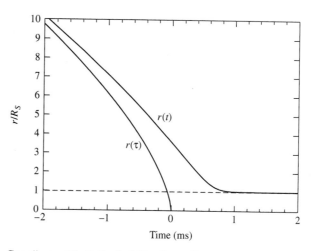

**FIGURE 17.20** Coordinate $r(t)$ of a freely falling frame $S$ according to an observer at rest at infinity, and $r(\tau)$ according to an observer in the frame $S$. The radial coordinates are in terms of $R_S$ for a $10\,M_\odot$ black hole.

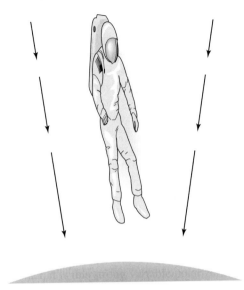

**FIGURE 17.21**   Tidal forces near a black hole.

from side to side produces a compression that is even more severe. These differential *tidal forces* increase in strength as she falls. In other words, the size of her local inertial frame (where gravity has been abolished) becomes increasing smaller as the spatial variation in the gravitational acceleration vector, **g**, increases. Were she not indestructible, our astronomer would be torn apart by the tidal force while still several hundred kilometers from the black hole.[17]

In just two milliseconds (proper time), she falls the final few hundred kilometers to the event horizon and crosses it. Her proper time continues normally, and she encounters no frozen stellar surface since it has fallen through long ago.[18] However, once inside the event horizon, her fate is sealed. It is impossible for any particle to be at rest when $r < R_S$, as can be seen from the Schwarzschild metric (Eq. 17.22). Using $dr = d\theta = d\phi = 0$ for an object at rest, the interval is given by

$$(ds)^2 = (c\,dt)^2 \left(1 - \frac{R_S}{r}\right) < 0$$

when $r < R_S$. This is a *spacelike* interval, which is not permitted for particles. Therefore it is impossible to remain at rest where $r < R_S$. Within the event horizon of a nonrotating black hole, all worldlines converge at the singularity. Even photons are pulled in toward the center. This means that the astronomer never has an opportunity to glimpse the singularity because no photons can reach her from there. She can, however, see the light that falls in behind her from events in the outside universe, but she does not see the entire history of

[17]You may recall Problem 16.11 concerning the stretching of an iron cube near a neutron star.
[18]The presence or absence of a frozen stellar surface at the event horizon makes no real difference; the Schwarzschild metric specifies the same spacetime curvature outside.

the universe as it unfolds. Although the elapsed coordinate time in the outside world does become infinite, the light from all of these events does not have time to reach the astronomer. Instead, these events occur in her "elsewhere." Just $6.6 \times 10^{-5}$ s of proper time after passing the event horizon, she is inexorably drawn to the singularity.[19]

### Mass Ranges of Black Holes

Black holes appear to exist with a range of masses. **Stellar-mass black holes**, with masses in the range of 3 to 15 $M_\odot$ or so, may form directly or indirectly as a consequence of the core-collapse of a sufficiently massive supergiant star, as discussed in Chapter 15. The direct collapse of the core to a black hole may be responsible for the production of collapsar, whereas a delayed collapse of a rapidly rotating neutron star may result in a supranova. It is also possible that a neutron star in a close binary system may gravitationally strip enough mass from its companion that the neutron star's self-gravity exceeds the ability of the degeneracy pressure to support it, again resulting in a black hole.

**Intermediate-mass black holes** (IMBHs) may exist that range in mass from roughly 100 $M_\odot$ to in excess of 1000 $M_\odot$ (or perhaps even greater than $10^4$ $M_\odot$). Evidence for them exists in the detection of sources known as **ultraluminous X-ray sources** (ULXs) that have been discovered by satellites such as Chandra and XMM-Newton. It is not entirely clear how these objects might form, although the correlation of IMBHs with the cores of globular clusters and low-mass galaxies suggests that they may develop in these dense stellar environments either by the mergers of stars to form a supermassive star that then core-collapses, or by the merger of stellar-mass black holes.

**Supermassive black holes** (SMBH) are known to exist at the centers of many (and probably most) galaxies. These enormous black holes range in mass from $10^5$ $M_\odot$ to $10^9$ $M_\odot$ (our own Milky Way Galaxy has a central black hole of mass $M = 3.7 \pm 0.2 \times 10^6$ $M_\odot$). How these behemoths formed remains an open question. One popular suggestion is that they formed from collisions between galaxies; another is that they formed as an extension of the formation process of IMBHs. Whatever the process, SMBHs appear to be closely linked with some bulk properties of galaxies, implying an important connection between galaxy formation and the formation of SMBHs.

Black holes may have also been manufactured in the earliest instants of the universe. Presumably, these **primordial black holes** would have been formed with a wide range of masses, from $10^{-8}$ kg to $10^5$ $M_\odot$. The only criterion for a black hole is that its entire mass must lie within the Schwarzschild radius, so the Schwarzschild metric is valid at the event horizon.

---

**Example 17.3.1.**   If Earth could somehow (miraculously) be compressed sufficiently to become a black hole, its radius would only be $R_S = 2GM_\oplus/c^2 = 0.009$ m. Although a primordial black hole could be this size, it is almost impossible to imagine packing Earth's entire mass into so small a ball.[20]

---

[19]A thorough description of the final view of the falling astronomer may be found in Rothman et al. (1985).
[20]You are reminded that the Schwarzschild metric is valid only outside matter. It does not describe the spacetime inside Earth.

### Black Holes Have No Hair!

Whatever the formation processes of black holes, they are certain to be very complicated. For example, the core-collapse of a star is almost certainly not symmetrical. Detailed calculations have demonstrated, however, that any irregularities are radiated away by gravitational waves (see Section 18.6). As a result, once the surface of the collapsing star reaches the event horizon, the exterior spacetime horizon is spherically symmetric and described by the Schwarzschild metric.

Another complication is the fact that all stars rotate, and therefore so will the resulting black hole. Remarkably, however, any black hole can be completely described by just three numbers: its mass, angular momentum, and electric charge.[21] Black holes have no other attributes or adornments, a condition commonly expressed by saying that "a black hole has no hair."[22]

There is a firm upper limit for a rotating black hole's angular momentum given by

$$L_{\max} = \frac{GM^2}{c}. \tag{17.29}$$

If the angular momentum of a rotating black hole were to exceed this limit, there would be no event horizon and a naked singularity would appear, in violation of the Law of Cosmic Censorship.

---

**Example 17.3.2.**    The maximum angular momentum for a solar-mass black hole is

$$L_{\max} = \frac{GM_\odot^2}{c} = 8.81 \times 10^{41} \text{ kg m}^2 \text{ s}^{-1}.$$

By comparison, the angular momentum of the Sun (assuming uniform rotation) is $1.63 \times 10^{41}$ kg m$^2$ s$^{-1}$, about 18% of $L_{\max}$. We should expect that many stars will have angular momenta that are comparable to $L_{\max}$, and so vigorous (if not maximal) rotation ought to be common for stellar-mass black holes.

---

### Spacetime Frame Dragging

The structure of a maximally rotating black hole is shown in Fig. 17.22.[23] The rotation has distorted the central singularity from a point into a flat ring, and the event horizon has assumed the shape of an ellipsoid. The figure also shows additional features caused by the rotation. As a massive object spins, it induces a rotation in the surrounding spacetime, a phenomenon known as **frame dragging**. To gain some insight into this effect, recall the behavior of a pendulum swinging at the north pole of Earth. As Earth rotates, the plane of the pendulum's swing remains fixed with respect to the distant stars. The stars define

---

[21] If magnetic monopoles exist, the "magnetic charge" would also be required for a complete specification. However, both magnetic and electric charge can be safely ignored because stars should be very nearly neutral.

[22] The "no hair" theorem actually applies only to the universe outside the event horizon. Inside, the spacetime geometry is complicated by the mass distribution of the collapsed star.

[23] The *Kerr metric* for a rotating black hole was derived from Einstein's field equations by a New Zealand mathematician, Roy Kerr, in 1963.

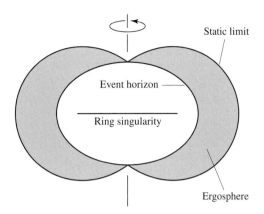

**FIGURE 17.22** The structure of a maximally rotating black hole, with the ring singularity seen edge-on. The location of the event horizon at the equator is $r = \frac{1}{2} R_S = GM/c^2$.

a nonrotating frame of reference for the universe, and it is relative to this frame that the pendulum's swing remains planar. However, the rotating spacetime close to a massive spinning object produces a local deviation from the nonrotating frame that describes the universe at large. Near a rotating black hole, frame dragging is so severe that there is a nonspherical region outside the event horizon called the **ergosphere** where any particle *must* move in the same direction that the black hole rotates. Spacetime within the ergosphere is rotating so rapidly that a particle would have to travel faster than the speed of light to remain at the same angular coordinate (e.g., at the same value of $\phi$ in the coordinate system used by a distant observer). The outer boundary of the ergosphere is called the **static limit**, so named because once beyond this boundary, a particle can remain at the same coordinate as the effect of frame dragging diminishes.

Even Earth's rotation produces very weak frame dragging. Detecting the effect of frame dragging was the mission of the Stanford Gravity Probe B experiment. The polar-orbit spacecraft was launched in April 2004 and ended data collection in October 2005 when the cryogenic liquid helium was used up. The experiment employed four superconducting gyroscopes made of precisely shaped spheres of fused quartz 3.8 cm in diameter. The gyroscopes were so nearly freely rotating that they formed an almost perfect spacetime reference frame. Although the predicted precession rate of the gyroscopes was only $0.042''$ yr$^{-1}$, the effect of frame dragging is cumulative. It is anticipated that frame dragging will be measurable by comparing the changes that occurred in the gyroscopes' different initial orientations.[24]

At this point, you should be warned that the previous descriptions of a black hole's structure inside the event horizon, such as Fig. 17.22, are based on vacuum solutions to Einstein's field equations. These solutions were obtained by ignoring the effects of the mass of the collapsing star, so the vacuum solutions do not describe the interior of a real black hole. Furthermore, the present laws of physics, including general relativity, break down under

---

[24]At the time the text was written, the year-long, painstaking process of data analysis was under way. Results of the experiment are expected to be announced in late 2006 (for updates, visit the Stanford Gravity Probe B website at http://einstein.stanford.edu/).

the extreme conditions found very near the center. The details of the singularity cannot be fully described until a theory of quantum gravity is found. The presence of a singularity seems assured, however. In 1965 an English mathematician, Roger Penrose, proved that *every* complete gravitational collapse must form a singularity.

### Tunnels in Spacetime

The possibility of using a black hole as a tunnel connecting one location in spacetime with another (perhaps in a different universe) has inspired both physicists and science fiction writers. Most conjectures of spacetime tunnels are based on vacuum solutions to Einstein's field equations and as such don't apply to the interiors of real black holes. Still, they have become part of the popular culture, and we will consider them briefly here. Figure 17.23 depicts a spacetime tunnel called a **Schwarzschild throat** (also known as an *Einstein–Rosen bridge*), which uses the Schwarzschild geometry of a nonrotating black hole to connect two regions of spacetime. The width of the throat is a minimum at the event horizon, and the "mouths" may be interpreted as opening onto two different locations in spacetime. It is tempting to imagine this as a tunnel, and writers of speculative fiction have dreamed of *white holes* pouring out mass or serving as passageways for starships. However, it appears that any attempt to send a tiny amount of matter or energy (even a stray photon) through the throat would cause it to collapse. For a real nonrotating black hole, all worldlines end at the inescapable singularity, where spacetime is infinitely curved. There is simply no way to bypass the singularity.

The story is somewhat different for a rotating black hole. Although spacetime is still infinitely curved at the ring singularity, all worldlines need not converge there. In fact, it is difficult for an infalling object to hit the singularity in a rotating black hole. Theorists have calculated worldlines for vacuum solutions that miss the singularity and emerge in the spacetime of another universe. But just as for nonrotating black holes, any attempt to pass

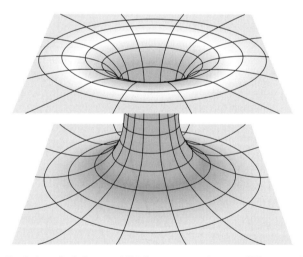

**FIGURE 17.23**   Depiction of a Schwarzschild throat connecting two different regions of spacetime. Any attempted passage of matter or energy through the throat would cause it to collapse.

the smallest amount of matter or energy along such a route would cause the passageway to collapse, thereby pinching it off. In summary, it seems extremely unlikely that black holes can provide a stable passageway for any matter or energy, even for idealized cases. For more realistic situations, any voyager attempting a trip through a black hole would end up being torn apart by the singularity.

Another possibility is that of a **wormhole**, a hypothetical tunnel between two points in spacetime separated by an arbitrarily great distance.[25] We will briefly consider nonrotating, spherically symmetric wormholes. They are described by *nonvacuum* solutions to Einstein's field equation. In other words, a wormhole must be threaded by some sort of *exotic material* whose tension prevents the collapse of the wormhole. There is no known mechanism that would allow a wormhole to arise naturally; it would have to be constructed by an incredibly advanced civilization. However, the theoretical possibilities alone are fascinating. These solutions to Einstein's field equations have no event horizon (permitting two-way trips through the wormhole) and involve survivable tidal forces. Journey times from one end through to the other can be less than one year (traveler's proper time), although the ends of the wormhole may be separated by interstellar or intergalactic distances.

The catch, of course, is the problematic existence of the exotic material needed to stabilize the wormhole. The unusual nature of the exotic material becomes apparent if we consider two light rays that converge on the wormhole and enter it, only to diverge when they exit the other end. This implies that the exotic material must be capable of gravitationally defocusing light, an "antigravity" effect involving the gravitational repulsion of the light by the material through which the rays pass. Exotic material meeting this requirement would have a negative energy density ($\rho c^2 < 0$), at least as experienced by the light rays. Although a negative energy density arises in certain quantum situations, it may or may not be allowed physically on macroscopic scales. We will leave wormholes as a fascinating possibility and abandon the discussion at this point, recalling Einstein's remark that "all our thinking is of the nature of a free play with concepts."[26]

## Stellar-Mass Black Hole Candidates

You may feel as though much of this section was borrowed from the pages of a science fiction novel. Extraordinary claims require extraordinary proof, and proof of the mere existence of black holes has been difficult to obtain. The problem lies in detecting an object only a few tens of kilometers across that emits no radiation directly. The best hope of astronomers has been to find a black hole in a close binary system. If the black hole in such a system is able to pull gas from the envelope of the normal companion star, the angular momentum of their orbital motion would cause a disk of gas to form around the black hole (see Fig. 17.24).[27] As the gas spirals down toward the event horizon, it is compressed and heated to millions of kelvins and emits X-rays. Only the gravity of a neutron star or a black hole can produce X-rays in a close binary system, and in fact the compact object in most X-ray binaries is

[25]The term *wormhole* recalls the holes eaten in some ancient map books by worms, providing a symbolic shortcut between the distant locations portrayed on the maps.

[26]You are referred to Morris and Thorne (1988) and Thorne (1994) for more details and further speculations concerning wormholes.

[27]Accretion disks in a close binary system will be discussed in more detail in Section 18.2.

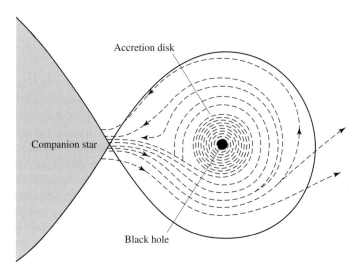

**FIGURE 17.24**   Gas pulled from a companion star forms an X-ray emitting disk around a black hole.

believed to be a neutron star. However, if an X-ray binary can be found in which the mass of the compact object exceeds 3 $M_\odot$, then a strong case can be made that the compact object is a black hole.

The first black hole to be tentatively identified in this way is Cygnus X-1, near the bright star $\eta$ Cygni in the middle of the swan's neck. Another promising candidate is LMC X-3, an X-ray binary in the Large Magellanic Cloud. Yet another compelling case involves the X-ray binary A0620−00, also known as V616 Mon.[28] The orbital velocities along the line of sight of *both* components of A0620−00 have been determined using Doppler-shifted spectral lines. A simple application of Kepler's laws shows that the mass of the compact object must be *at least* $3.82 \pm 0.24$ $M_\odot$, well above the 3 $M_\odot$ upper limit for a neutron star.

However, perhaps the strongest candidate is V404 Cygni. Originally determined to be a recurrent nova (see Section 18.4), V404 Cyg underwent an X-ray outburst in 1989 that was detected by the Ginga satellite. Examination with ground-based optical telescopes revealed a companion star of type K0 IV with a radial velocity amplitude of $211 \pm 4$ km s$^{-1}$ and an orbital period of $6.473 \pm 0.001$ d. The best estimate of the mass of the unseen companion to the K0 IV star is $12 \pm 2$ $M_\odot$. As more evidence accumulates, it seems that astronomers have finally found the extraordinary proof required for the existence of a black hole.

### Hawking Radiation

The black holes of classical general relativity last forever. A very general result derived by Stephen Hawking states that the surface area of a black hole's event horizon can never decrease. If a black hole coalesces with any other object, the result is an even larger black

---

[28]A0620−00 (V616 Mon) is on the border of the constellations Monoceros and Orion, about one-third of the way along a line from Betelgeuse to Sirius.

hole. In 1974, however, Hawking discovered a loophole in this law when he combined quantum mechanics with the theory of black holes and found that black holes can slowly *evaporate*. The key to this process is pair production, the formation of a particle–antiparticle pair just outside the event horizon of a black hole. Ordinarily the particles quickly recombine and disappear, but if one of the particles falls into the event horizon while its partner escapes, as shown in Fig. 17.25, this disappearing act may be thwarted. The black hole's gravitational energy was used to produce the two particles, and so the escaping particle has carried away some of the black hole's mass. The net effect as seen by an observer at a great distance is the emission of particles by the black hole, known as **Hawking radiation**, accompanied by a reduction in the black hole's mass.

The rate at which energy is carried away by particles in this manner is inversely proportional to the square of the black hole's mass, or $1/M^2$. For stellar-mass black holes, the emitted particles are photons and the rate of emission is minuscule. As the black hole's mass declines, however, the rate of emission increases. The final stage of a black hole's evaporation proceeds extremely rapidly, releasing a burst of all types of elementary particles. This tremendous explosion probably leaves behind only an empty region of flat spacetime.

The lifetime of a primordial black hole prior to its evaporation, $t_{\text{evap}}$, is quite long,

$$t_{\text{evap}} = 2560\pi^2 \left( \frac{2GM}{c^2} \right)^2 \left( \frac{M}{h} \right) \tag{17.30}$$

$$\approx 2 \times 10^{67} \left( \frac{M}{M_\odot} \right)^3 \text{ yr.}$$

Since the age of the universe is 13.7 billion years, this process is of no consequence for black holes formed by a collapsing star. However, a primordial black hole with a mass of

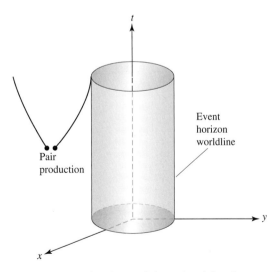

**FIGURE 17.25**   Spacetime diagram showing particle–antiparticle pairs created near the event horizon of a black hole.

roughly $1.7 \times 10^{11}$ kg would evaporate in about 13 billion years. Thus primordial black holes with this mass should be in the final, explosive stage of evaporation right now and could possibly be detected. The final burst of Hawking radiation is thought to release high-energy ($\approx 100$ MeV) gamma rays at a rate of $10^{13}$ W, together with electrons, positrons, and many other particles. The subsequent decay of these particles should produce additional gamma rays that would be observable by Earth-orbiting satellites. To date, measurements of the cosmic gamma-ray background at this energy have not detected anything that can be identified with the demise of a nearby primordial black hole. Although there is as yet no positive evidence that primordial black holes exist, this negative result is still important. It implies that on average there cannot be more than 200 primordial black holes with this mass in every cubic light-year of space.

## SUGGESTED READING

### General

Begelman, Mitchell, and Rees, Martin, *Gravity's Fatal Attraction: Black Holes in the Universe*, Scientific American Library, New York, 1996.

Charles, Philip A., and Wagner, R. Mark, "Black Holes in Binary Stars: Weighing the Evidence," *Sky and Telescope*, May 1996.

Davies, Paul, "Wormholes and Time Machines," *Sky and Telescope*, January 1992.

Ferguson, Kitty, *Prisons of Light—Black Holes*, Cambridge University Press, Cambridge, 1996.

Ford, Lawrence H., and Roman, Thomas A., "Negative Energy, Worm Holes, and Warp Drive," *Scientific American*, January, 2000.

Hawking, Stephen W., *A Brief History of Time*, Updated and Expanded Tenth Anniversary Edition, Bantam Books, New York, 1998.

Hawking, Stephen W., "The Quantum Mechanics of Black Holes," *Scientific American*, January 1977.

Lasota, Jean-Pierre, "Unmasking Black Holes," *Scientific American*, May 1999.

Luminet, Jean-Pierre, *Black Holes*, Cambridge University Press, Cambridge, 1992.

Morris, Michael S., and Thorne, Kip S., "Wormholes in Spacetime and Their Use for Interstellar Travel: A Tool for Teaching General Relativity," *American Journal of Physics*, *56*, 395, 1988.

Rothman, Tony, et al., *Frontiers of Modern Physics*, Dover Publications, New York, 1985.

Thorne, Kip S., *Black Holes and Time Warps: Einstein's Outrageous Legacy*, W. W. Norton and Co., New York, 1994.

Wheeler, John A., *A Journey into Gravity and Spacetime*, Scientific American Library, New York, 1990.

Will, Clifford, *Was Einstein Right?* Basic Books, New York, 1986.

### Technical

Bekenstein, Jacob D., "Black Hole Thermodynamics," *Physics Today*, January 1980.

Berry, Michael, *Principles of Cosmology and Gravitation*, Institute of Physics Publishing, Bristol, 1989.

Casares, J., Charles, P. A., and Naylor, T., "A 6.5-day Periodicity in the Recurrent Nova V404 Cygni Implying the Presence of a Black Hole," *Nature*, *335*, 614, 1992.

Charles, Philip A., "Black-Hole Candidates in X-Ray Binaries," *Encyclopedia of Astronomy and Astrophysics*, P. Murdin (ed.), Institute of Physics Publishing, Bristol, 2000.

Fré, P., Gorini, V., Magli, G., and Moschella, U. (eds.), *Classical and Quantum Black Holes*, Institute of Physics Publishing, 1999.

Hartle, James B., *Gravity: An Introduction to Einstein's General Relativity*, Addison-Wesley, San Francisco, 2003.

Misner, Charles W., Thorne, Kip S., and Wheeler, John A., *Gravitation*, W. H. Freeman and Co., San Francisco, 1973.

Ruffini, Remo, and Wheeler, John A., "Introducing the Black Hole," *Physics Today*, January 1991.

Shapiro, Stuart L., and Teukolsky, Saul A., *Black Holes, White Dwarfs, and Neutron Stars: The Physics of Compact Objects*, John Wiley and Sons, New York, 1983.

Taylor, Edwin F., and Wheeler, John A., *Exploring Black Holes: Introduction to General Relativity*, Addison Wesley Longman, San Francisco, 2000.

Wald, Robert M. (ed.), *Black Holes and Relativistic Stars*, University of Chicago Press, Chicago, 1998.

## PROBLEMS

**17.1** In the rubber sheet analogy of Section 17.1, a keen eye would notice that the tennis ball also depresses the sheet slightly, and so the bowling ball constantly tilts slightly toward the tennis ball as they orbit each other. Qualitatively compare this with the motion of two stars in a binary orbit.

**17.2** Show that Eq. (17.6) for the gravitational redshift remains valid even if the light travels upward at an angle $\theta$ measured from the vertical as long as $h$ is taken to be the *vertical distance* traveled by the light pulse.

**17.3** A photon near the surface of Earth travels a horizontal distance of 1 km. How far does the photon "fall" in this time?

**17.4** Leadville, Colorado, is at an altitude of 3.1 km above sea level. If a person there lives for 75 years (as measured by an observer at a great distance from Earth), how much longer would gravitational time dilation have allowed that person to live if he or she had moved at birth from Leadville to a city at sea level?

**17.5** **(a)** Estimate the radius of curvature of a horizontally traveling photon at the surface of a $1.4\,M_\odot$ neutron star, and compare the result with the 10-km radius of the star. Can general relativity be neglected when studying neutron stars?

**(b)** If one hour passes at the surface of the neutron star, how much time passes at a great distance? Compare the times obtained from the exact and approximate expressions, Eqs. (17.13) and (17.14), respectively.

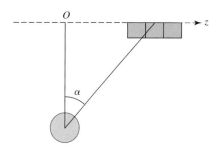

**FIGURE 17.26**    Local inertial frames for measuring the deflection of light near the Sun.

**17.6**  Imagine a series of rectangular local inertial reference frames suspended by cables in a line near the Sun's surface, as shown in Fig. 17.26. The frames are carefully lined up so that the tops and sides of neighboring frames are parallel, and the tops of the frames lie along the $z$-axis. A photon travels unhindered through the frames. As the photon enters each frame, the frame is released from rest and falls freely toward the center of the Sun.

(a) Show that as it passes through the frame located at angle $\alpha$ (shown in the figure), the angular deflection of the photon's path is

$$d\phi = \frac{g_0 \cos^3 \alpha}{c^2} \, dz,$$

where $dz$ is the width of the reference frame and $g_0$ is the Newtonian gravitational acceleration at the point of closest approach, $O$. The angular deflection is small, so assume that the photon is initially traveling in the $z$-direction as it enters the frame. (*Hint:* The width of the frame in the $z$-direction is $dz$, so the time for the photon to cross the frame can be taken to be $dz/c$.)

(b) Integrate the result you found in part (a) from $\alpha = -\pi/2$ to $+\pi/2$ and so find the total angular deflection of the photon as it passes through the curved spacetime near the Sun.

(c) Your answer (which is also the answer obtained by Einstein in 1911 before he arrived at his field equations) is only half the correct value of $1.75''$. Can you *qualitatively* account for the missing factor of two?

**17.7**  Assume that you are at the origin of a laboratory reference system at time $t = 0$ when you start your clock (event $A$). Determine whether the following events are within the future light cone of event $A$, within the past light cone of event $A$, or elsewhere.

(a) A flashbulb goes off 7 m away at time $t = 0$.

(b) A flashbulb goes off 7 m away at time $t = 2$ s.

(c) A flashbulb goes off 70 km away at time $t = 2$ s.

(d) A flashbulb goes off 700,000 km away at time $t = 2$ s.

(e) A supernova explodes 180,000 ly away at time $t = -5.7 \times 10^{12}$ s.

(f) A supernova explodes 180,000 ly away at time $t = 5.7 \times 10^{12}$ s.

(g) A supernova explodes 180,000 ly away at time $t = -5.6 \times 10^{12}$ s.

(h) A supernova explodes 180,000 ly away at time $t = 5.6 \times 10^{12}$ s.

For items (e) and (g), could an observer in another reference frame moving relative to yours measure that the supernova exploded *after* event $A$? For items (f) and (h), could an observer in another frame measure that the supernova exploded *before* event $A$?

**17.8** $\tau$ Ceti is the closest single star that is similar to the Sun. At time $t = 0$, Alice leaves Earth in her starship and travels at a speed of $0.95c$ to $\tau$ Ceti, 11.7 ly away as measured by astronomers on Earth. Her twin brother, Bob, remains at home, at $x = 0$.

(a) According to Bob, what is the interval between Alice's leaving Earth and arriving at $\tau$ Ceti?

(b) According to Alice, what is the interval between her leaving Earth and arriving at $\tau$ Ceti?

(c) Upon arriving at $\tau$ Ceti, Alice immediately turns around and returns to Earth at a speed of $0.95c$. (Assume that the actual turnaround takes negligible time.) What was the proper time for Alice during her round trip to $\tau$ Ceti?

(d) When she and Bob meet on her return to Earth, how much younger will Alice be than her brother?

**17.9** Consider a spherical blackbody of constant temperature and mass $M$ whose surface lies at radial coordinate $r = R$. An observer located at the surface of the sphere and a distant observer both measure the blackbody radiation given off by the sphere.

(a) If the observer at the surface of the sphere measures the luminosity of the blackbody to be $L$, use the gravitational time dilation formula, Eq. (17.13), to show that the observer at infinity measures

$$L_\infty = L \left( 1 - \frac{2GM}{Rc^2} \right). \tag{17.31}$$

(b) Both observers use Wien's law, Eq. (3.15), to determine the blackbody's temperature. Show that

$$T_\infty = T \sqrt{1 - \frac{2GM}{Rc^2}}. \tag{17.32}$$

(c) Both observers use the Stefan–Boltzmann law, Eq. (3.17), to determine the radius of the spherical blackbody. Show that

$$R_\infty = \frac{R}{\sqrt{1 - 2GM/Rc^2}}. \tag{17.33}$$

Thus, using the Stefan–Boltzmann law without including the effects of general relativity will lead to an *overestimate* of the size of a compact blackbody.

**17.10** In 1792 the French mathematician Simon-Pierre de Laplace (1749–1827) wrote that a hypothetical star, "of the same density as Earth, and whose diameter would be two hundred and fifty times larger than the Sun, would not, in consequence of its attraction, allow any of its rays to arrive at us." Use Newtonian mechanics to calculate the escape velocity of Laplace's star.

**17.11** Qualitatively describe the effects on the orbits of the planets if the Sun were suddenly to become a black hole.

**17.12** Consider four black holes with masses of $10^{12}$ kg, 10 $M_\odot$, $10^5$ $M_\odot$, and $10^9$ $M_\odot$.

(a) Calculate the Schwarzschild radius for each.

(b) Calculate the average density, defined by $\rho = M/(\frac{4}{3}\pi R_S^3)$, for each.

**17.13** (a) Show that the proper distance from the event horizon to a radial coordinate $r$ is given by

$$\Delta \mathcal{L} = r \sqrt{1 - \frac{R_S}{r}} + \frac{R_S}{2} \ln \left( \frac{1 + \sqrt{1 - R_S/r}}{1 - \sqrt{1 - R_S/r}} \right).$$

This illustrates the danger of interpreting $r$ as a distance instead of a coordinate. *Hint:* Integrate Eq. (17.23).

**(b)** Make a graph of $\Delta \mathcal{L}$ as a function of $r$ for values of $r$ between $r = R_S$ and $r = 10R_S$.

**(c)** Show that, for large values of $r$,

$$\Delta \mathcal{L} \simeq r.$$

Thus, far from the black hole, the radial coordinate $r$ can be treated as a distance.

**17.14** Verify that the area of the event horizon of a black hole is $4\pi R_S^2$. (*Hint:* Remember that the radial coordinate $r$ is *not* the distance to the center. Use the Schwarzschild metric as your starting point.)

**17.15** Equation (17.26) describes the coordinate speed of a massive particle orbiting a nonrotating black hole. However, it can be shown that the orbit is not stable unless $r \geq 3R_S$; any disturbance will cause a particle in a smaller orbit to spiral down to the event horizon.

**(a)** Find the coordinate speed of a particle in the smallest stable orbit around a 10 $M_\odot$ black hole.

**(b)** Find the orbital period (in coordinate time $t$) for this smallest stable orbit around a 10 $M_\odot$ black hole.

**17.16** **(a)** Find an expression for the coordinate speed of light in the $\phi$-direction.

**(b)** Consider Eq. (17.26) in the limit that the particle's mass goes to zero and its speed approaches that of light. Use your result for part (a) to show that $r = 1.5R_S$ for the circular orbit of a photon around a black hole.

**(c)** Find the orbital period (in coordinate time $t$) for this orbit around a 10 $M_\odot$ black hole.

**(d)** If a flashlight were beamed in the $\phi$-direction at $r = 1.5R_S$, what would happen? (The surface at $r = 1.5R_S$ is called the **photon sphere**.)

**17.17** To obtain a crude estimate for the maximum angular momentum of a rotating black hole, imagine (obviously incorrectly!) that the black hole's mass is distributed uniformly throughout a solid sphere of radius $R_S$ (the Schwarzschild radius). From basic physics, the moment of inertia of a uniform, rotating sphere is $I = \frac{2}{5}MR^2$, and the angular momentum of the sphere is $L = I\omega$, where $\omega$ is the sphere's angular velocity. From this classical approach, estimate the maximum angular momentum of the solid sphere, and compare your answer with Eq. (17.29). (Be sure to specify any additional assumptions you have made.) What is the percentage error in your estimate compared to the exact result?

**17.18** Use Eq. (17.29) to compare the maximum angular momentum of a 1.4 $M_\odot$ black hole with the angular momentum of the fastest known pulsar, which rotates with a period of 0.00139 s. Assume that the pulsar is a 1.4 $M_\odot$ uniform sphere of radius 10 km.

**17.19** An electron is a point-like particle of zero radius, so it is natural to wonder whether an electron could be a black hole. However, a black hole of mass $M$ cannot have an arbitrary amount of angular momentum $L$ and charge $Q$. These values must satisfy an inequality,

$$\left( \frac{GM}{c} \right)^2 \geq G \left( \frac{Q}{c} \right)^2 + \left( \frac{L}{M} \right)^2.$$

If this inequality were violated, the singularity would be found *outside* the event horizon, in violation of the Law of Cosmic Censorship. Use $\hbar/2$ for the electron's angular momentum to determine whether or not an electron is a black hole.

**17.20** **(a)** The angular rotation rate, $\Omega$, at which spacetime is dragged around a rotating mass must be proportional to its angular momentum $L$. The expression for $\Omega$ must also contain the constants $G$ and $c$, together with the radial coordinate $r$. Show on purely dimensional grounds that

$$\Omega = \text{constant} \times \frac{GL}{r^3 c^2},$$

where the constant (which you need not determine) is of order unity.

**(b)** Evaluate this for Earth, assuming that it is a uniformly rotating sphere. Set the leading constant equal to one, and express your answer in arcseconds per year. How much time would it take for a pendulum at the north pole to rotate *once* relative to the distant stars because of frame dragging?

**(c)** Repeat part (b) for the fastest known pulsar, expressing $\Omega$ in revolutions per second.

**17.21** **(a)** Use dimensional arguments to combine the fundamental constants $\hbar$, $c$, and $G$ into an expression that has units of mass. Evaluate your result, which is an estimate of the least massive primordial black hole formed in the first instant after the Big Bang. What is the mass in kilograms?

**(b)** What is the Schwarzschild radius for such a black hole?

**(c)** How long would it take light to travel this distance?

**(d)** What is the lifetime of this black hole before its evaporation?

**17.22** By combining gravitation ($G$), thermodynamics ($k$), and quantum mechanics ($\hbar$), Stephen Hawking calculated the temperature, $T$, of a nonrotating black hole to be

$$kT = \frac{\hbar c^3}{8\pi G M} = \frac{\hbar c}{4\pi R_S}, \tag{17.34}$$

where $R_S$ is the Schwarzschild radius.

**(a)** Verify that the expression has the right units.

**(b)** It was mentioned in the text that if a primoridal black hole formed at the beginning of the universe 13.7 Gyr ago with a mass of $1.7 \times 10^{11}$ kg, it would be reaching the end of its life now. Compute the temperature of a primordial black hole having a mass of $1.7 \times 10^{11}$ kg.

**(c)** Approximately what portion of the electromagnetic spectrum would this blackbody temperature correspond to?

**(d)** What would the radius of a sphere having the density of water be if it had a mass of $1.7 \times 10^{11}$ kg?

**(e)** Compute the temperature of a 10 $M_\odot$ black hole.

**17.23** In this problem we will derive the expression for the lifetime of a nonrotating, evaporating black hole (Eq. 17.30).

**(a)** Consider a black hole to be a perfectly radiating blackbody of temperature $T$, given by Eq. (17.34) in Problem 17.22. Assuming that the surface area of the black hole is given by $4\pi R_S^2$, where $R_S$ is the Schwarzschild radius, show that the luminosity of the black hole due to Hawking radiation is

$$L = \frac{\hbar c^6}{15360\pi G^2 M^2} = \frac{\hbar c^2}{3840\pi R_S^2}.$$

**(b)** The luminosity of the black hole must originate from a loss in the black hole's internal energy. Assuming that the energy of the black hole is given by $E = Mc^2$ and that $L = dE/dt$, show that the time required for the black hole to lose all of its mass to Hawking radiation is given by Eq. (17.30).

**17.24** In the X-ray binary system A0620$-$00, the radial orbital velocities for the normal star and the compact object are $v_{s,r} = 457$ km s$^{-1}$ and $v_{c,r} = 43$ km s$^{-1}$, respectively. The orbital period is 0.3226 day.

**(a)** Calculate the mass function (the right-hand side of Eq. 7.7),

$$\frac{m_c^3}{(m_s + m_c)^2} \sin^3 i,$$

where $m_s$ is the mass of the normal star, $m_c$ is the mass of its compact companion, and $i$ is the angle of inclination of the orbit. What does this result say about the mass of the compact object? (Note that the value of $v_{c,r}$ was not needed to obtain this result.)

**(b)** Now use the value of the orbital radial velocity of the compact object to determine its mass, assuming $i = 90°$. What does this result say about the mass of the compact object?

**(c)** The X-rays are not eclipsed in this system, so the angle of inclination must be less than approximately 85°. Suppose that the angle of inclination were 45°. What would the mass of the compact object be then?

**17.25** From the data in the text, make an estimate of the lower mass limit of the compact companion in V404 Cyg by setting the orbital inclination $i = 90°$. Refer to Appendix G in order to make a rough determination of the mass of the K0 IV companion. Why does your calculation result in a lower limit? What further information would be required to make a more precise determination?

# CHAPTER

# 18

# Close Binary Star Systems

## 18.1 ■ GRAVITY IN A CLOSE BINARY STAR SYSTEM

As explained in Chapter 7, at least half of all "stars" in the sky are actually multiple systems, consisting of two (or more) stars in orbit about their common center of mass. In most of these systems the stars are sufficiently far apart that they have a negligible impact on one another. They evolve essentially independently, living out their lives in isolation except for the gentle grip of gravity that binds them together.

If the stars are very close, with a separation roughly equal to the diameter of the larger star, then one or both stars may have their outer layers gravitationally deformed into a teardrop shape. As a star rotates through the tidal bulge raised by its partner's gravitational pull, it is forced to pulsate. These oscillations are damped by the mechanisms discussed in Section 14.2. Orbital and rotational energy is dissipated in this way until the system reaches the state of minimum energy for its (constant) angular momentum, resulting in synchronous rotation and circular orbits. Thereafter the same side of each star always faces the other as the system rotates rigidly in space, and no further energy can be lost by tidally driven oscillations.[1] The distorted star may even lose some of its photospheric gases to its companion. The spilling of gas from one star onto another can lead to some spectacular celestial fireworks, the subject of this chapter.

### Lagrangian Points and Equipotential Surfaces

To understand how gravity operates in a close binary star system, consider two stars in a circular orbit in the $x$–$y$ plane with angular velocity $\omega = v_1/r_1 = v_2/r_2$. Here, $v_1$ and $r_1$ are the orbital speed of Star 1 and its distance from the center of mass of the system, and similarly for Star 2. It is useful to choose a corotating coordinate system that follows the

---

[1] If one of the stars is a compact object such as a white dwarf or a neutron star, its spin may not be synchronized.

rotation of the two stars about their center of mass. If the center of mass is at the origin, then the stars will be at rest in this rotating reference frame, with their mutual gravitational attraction balanced by the outward "push" of a centrifugal force.[2] The centrifugal force vector on a mass $m$ in this frame a distance $r$ from the origin is then

$$\mathbf{F}_c = m\omega^2 r\,\hat{\mathbf{r}}, \tag{18.1}$$

in the outward radial direction.

It is usually easier to work with the gravitational potential energy, given by Eq. (2.14),

$$U_g = -G\frac{Mm}{r},$$

instead of with the gravitational force.[3] To do this in a rotating coordinate system, we must include a fictitious "centrifugal potential energy" in the potential energy term through the use of Eq. (2.13):

$$U_f - U_i = \Delta U_c = -\int_{\mathbf{r}_i}^{\mathbf{r}_f} \mathbf{F}_c \cdot d\mathbf{r}.$$

Here, $\mathbf{F}_c$ is the centrifugal force vector, $\mathbf{r}_i$ and $\mathbf{r}_f$ are the initial and final position vectors, respectively, and $d\mathbf{r}$ is the infinitesimal change in the position vector (recall Fig. 2.9). The change in centrifugal potential energy is thus

$$\Delta U_c = -\int_{r_i}^{r_f} m\omega^2 r\,dr = -\frac{1}{2}m\omega^2\left(r_f^2 - r_i^2\right).$$

Realizing that only *changes* in potential energy are physically meaningful, we can arbitrarily choose $U_c = 0$ at $r = 0$ to give the final result for the centrifugal potential energy,

$$U_c = -\frac{1}{2}m\omega^2 r^2. \tag{18.2}$$

Figure 18.1 shows a corotating coordinate system in which two stars with masses $M_1$ and $M_2$ are separated by a distance $a$. The stars are located on the $x$-axis at distances $r_1$ and $r_2$, respectively, from the center of mass, which is placed at the origin. Thus

$$r_1 + r_2 = a \quad \text{and} \quad M_1 r_1 = M_2 r_2. \tag{18.3}$$

Including the centrifugal term, the effective potential energy for a small test mass $m$ located in the plane of the orbit (the $x$–$y$ plane) is

$$U = -G\left(\frac{M_1 m}{s_1} + \frac{M_2 m}{s_2}\right) - \frac{1}{2}m\omega^2 r^2.$$

---

[2]The centrifugal force is an *inertial force* (as opposed to a physical force) that must be included when describing motion in a rotating coordinate system. There is another inertial force, called the *Coriolis force*, that will be neglected in what follows.

[3]Most stars can be treated as point masses in what follows because the mass is concentrated at their centers, allowing their teardrop shapes to be neglected.

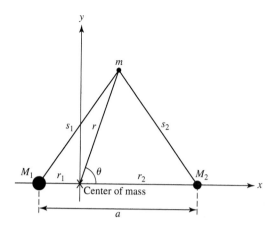

**FIGURE 18.1** Corotating coordinates for a binary star system.

For convenience, the effective potential energy can be divided by $m$ to obtain the **effective gravitational potential**, $\Phi$:

$$\Phi = -G\left(\frac{M_1}{s_1} + \frac{M_2}{s_2}\right) - \frac{1}{2}\omega^2 r^2. \tag{18.4}$$

This is just the effective potential energy *per unit mass*. From the law of cosines, the distances $s_1$ and $s_2$ are given by

$$s_1^2 = r_1^2 + r^2 + 2r_1 r \cos\theta \tag{18.5}$$

$$s_2^2 = r_2^2 + r^2 - 2r_2 r \cos\theta. \tag{18.6}$$

The angular frequency of the orbit, $\omega$, comes from Kepler's third law for the orbital period, $P$, Eq. (2.37),

$$\omega^2 = \left(\frac{2\pi}{P}\right)^2 = \frac{G(M_1 + M_2)}{a^3}. \tag{18.7}$$

Equations (18.3) and (18.4–18.7) can be used to evaluate the effective gravitational potential $\Phi$ at every point in the orbital plane of a binary star system. For example, Fig. 18.2 shows the value of $\Phi$ along the $x$-axis. The significance of this graph becomes clear when the $x$-component of the force on a small test mass $m$, initially at rest on the $x$-axis, is written as

$$F_x = -\frac{dU}{dx} = -m\frac{d\Phi}{dx} \tag{18.8}$$

(recall Eq. 2.15). The three "hilltops" labeled $L_1$, $L_2$, and $L_3$ are **Lagrangian points**, where there is no force on the test mass ($d\Phi/dx = 0$). At these three equilibrium points, the

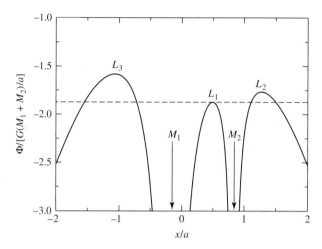

**FIGURE 18.2** The effective gravitational potential $\Phi$ for two stars of mass $M_1 = 0.85\ M_\odot$, $M_2 = 0.17\ M_\odot$ on the $x$-axis. The stars are separated by a distance $a = 5 \times 10^8$ m $= 0.718\ R_\odot$, with their center of mass located at the origin. The $x$-axis is in units of $a$, and $\Phi$ is expressed in units of $G(M_1 + M_2)/a = 2.71 \times 10^{11}$ J kg$^{-1}$. (In fact, the figure is the same for any $M_2/M_1 = 0.2$.) The dashed line is the value of $\Phi$ at the inner Lagrangian point. If the total energy per unit mass of a particle exceeds this value of $\Phi$, it can flow through the inner Lagrangian point between the two stars.

gravitational forces on $m$ due to $M_1$ and $M_2$ are balanced by the centrifugal force.[4] These equilibrium points are *unstable* because they are *local maxima* of $\Phi$; if the test mass is displaced slightly, the minus sign in Eq. (18.8) indicates that it will accelerate "downhill," away from its equilibrium position. The *inner Lagrangian point*, $L_1$, plays a central role in close binary systems. Approximate expressions for the distances from $L_1$ to $M_1$ and $M_2$, denoted respectively by $\ell_1$ and $\ell_2$, are

$$\ell_1 = a \left[ 0.500 - 0.227 \log_{10} \left( \frac{M_2}{M_1} \right) \right] \tag{18.9}$$

$$\ell_2 = a \left[ 0.500 + 0.227 \log_{10} \left( \frac{M_2}{M_1} \right) \right]. \tag{18.10}$$

Points in space that share the same value of $\Phi$ form an **equipotential surface**. Figure 18.3 shows *equipotential contours* that outline the intersection of several equipotential surfaces with the plane of the orbit. Very close to either of the masses $M_1$ or $M_2$, the equipotential surfaces are nearly spherical and centered on each mass. Farther away, the combined gravitational influence of $M_1$ and $M_2$ distorts the equipotential surfaces into teardrop shapes until

---

[4]From an inertial (nonrotating) frame of reference, this motion would be described by saying that gravitational forces of $M_1$ and $M_2$ produce the inward centripetal acceleration of the test mass as it orbits the center of mass of the system.

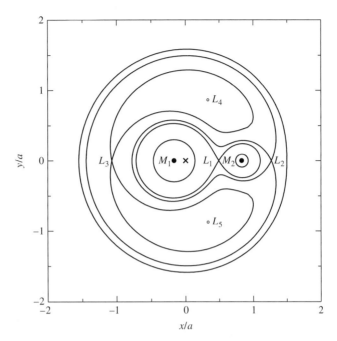

**FIGURE 18.3**   Equipotentials for $M_1 = 0.85\,\mathrm{M_\odot}$, $M_2 = 0.17\,\mathrm{M_\odot}$, and $a = 5 \times 10^8\,\mathrm{m} = 0.718\,\mathrm{R_\odot}$. The axes are in units of $a$, with the system's center of mass (the "$\times$") at the origin. Starting at the top of the figure and moving down toward the center of mass, the values of $\Phi$ in units of $G(M_1 + M_2)/a = 2.71 \times 10^{11}\,\mathrm{J\,kg^{-1}}$ for the equipotential curves are $\Phi = -1.875, -1.768, -1.583, -1.583, -1.768$ (the "dumbbell"), $-1.875$ (the Roche lobe), and $-3$ (the spheres). $L_4$ and $L_5$ are local maxima, with $\Phi = -1.431$.

they finally touch at the inner Lagrangian point. At even greater distances, the equipotential surfaces assume a "dumbbell" shape surrounding both masses.[5]

   These equipotential surfaces are *level surfaces* for binary stars. In a binary system, as one of the stars evolves, it will expand to fill successively larger equipotential surfaces (somewhat like inflating a balloon). To see this, consider that the effective gravity at each point is always *perpendicular* to the equipotential surface there.[6] Hydrostatic equilibrium guarantees that the pressure is constant along a surface of constant $\Phi$; there is no component of gravity parallel to an equipotential surface, and so a pressure difference in that direction

---

[5]We will not be concerned with the other equipotential contours on Fig. 18.3 that pass through the Lagrangian points $L_3$, $L_4$, and $L_5$. However, the *Trojan asteroids* that accumulate at two locations along Jupiter's orbit are collections of interplanetary rubble found at Lagrangian points $L_4$ and $L_5$. (If $M_1 > 24.96 M_2$, as for the Sun and Jupiter, then the Coriolis force is strong enough to cause $L_4$ and $L_5$ to be *stable* equilibrium points.) Each of the Lagrangian points $L_4$ and $L_5$ forms an equilateral triangle with masses $M_1$ and $M_2$ in Fig. 18.3, so the Trojan asteroids are found at about $60°$ ahead of and behind Jupiter in its orbit, with a spread due to the finite width of the potential well.

[6]The mathematical statement of this is $\mathbf{F} = -m\nabla\Phi$. This is analogous to an electric field vector being oriented perpendicular to an electrical equipotential surface, pointing from higher to lower voltage.

cannot be balanced and maintained. And because the pressure is due to the weight of the overlying layers of the star, the density must also be the same along each equipotential surface in order to produce a constant pressure there.

### Classes of Binary Star Systems

The appearance of a binary star system depends on which equipotential surfaces are filled by the stars. Binary stars with radii much less than their separation are nearly spherical (as shown by the small circles in Fig. 18.3). This situation describes a **detached binary** in which the stars evolve nearly independently. Detached binary systems have already been described in Chapter 7 as a primary source of astronomical information about the basic properties of stars.

If one star expands enough to fill the "figure-eight" contour in Fig. 18.3, then its atmospheric gases can escape through the inner Lagrangian point $L_1$ to be drawn toward its companion. The teardrop-shaped regions of space bounded by this particular equipotential surface are called **Roche lobes**.[7] The transfer of mass from one star to the other can begin when one of the stars has expanded beyond its Roche lobe. Such a system is called a **semidetached binary**. The star that fills its Roche lobe and loses mass is usually called the **secondary star**, with mass $M_2$, and its companion the **primary star** has mass $M_1$. The primary star may be either more or less massive than the secondary star.

It may happen that *both stars* fill, or even expand beyond, their Roche lobes. In this case, the two stars share a common atmosphere bounded by a dumbbell-shaped equipotential surface, such as the one passing through the Lagrangian point $L_2$. Such a system is called a **contact binary**. Figure 18.4 illustrates the three classes of binary stars.

### Mass Transfer Rate

A crude estimate of the rate at which mass is transferred in a semidetached binary may be obtained for the case of two stars of equal mass. Let the radius of the star that has expanded beyond its Roche lobe be $R$. The equipotential surface at the radius of this star will be modeled by two spheres of radius $R$ that overlap slightly by a distance $d$, as shown in Fig. 18.5. We will assume that stellar gas will escape from the filled lobe through the circular opening of radius $x$. If the density of the stellar material at the opening is $\rho$ and its speed toward the opening of area $A = \pi x^2$ is $v$, then it is left as an exercise (Problem 18.3) to show that the rate at which mass leaves the filled lobe, the **mass transfer rate**, is

$$\dot{M} = \rho v A. \tag{18.11}$$

A bit of geometry shows that

$$x = \sqrt{Rd} \tag{18.12}$$

when $d \ll R$. Using Eq. (8.3) for the thermal velocity of the gas particles results in the

---

[7]The term *Roche lobe* was chosen in honor of the nineteenth-century French mathematician Edouard Roche (1820–1883).

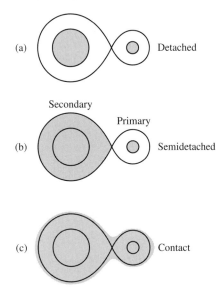

**FIGURE 18.4** The classification of binary star systems. (a) A detached system. (b) A semidetached system in which the secondary star has expanded to fill its Roche lobe. (c) A contact binary.

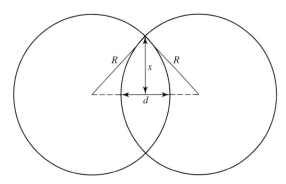

**FIGURE 18.5** Intersecting spheres used to estimate the mass transfer rate, $\dot{M}$.

estimate

$$\dot{M} \approx \rho v_{\text{rms}} \pi x^2 \tag{18.13}$$

or

$$\dot{M} \approx \pi R \, d\rho \sqrt{\frac{3kT}{m_H}}, \tag{18.14}$$

assuming a gas of hydrogen atoms. As the overfill distance $d$ becomes larger, the values of the density and temperature increase at the opening. In Problem 18.31, you will be asked to show that the mass transfer rate increases rapidly with $d$; a more detailed calculation results in $\dot{M} \propto d^3$.

---

**Example 18.1.1.**   Suppose a star like the Sun is in a semidetached binary system with a companion of equal mass, and it slightly overfills its Roche lobe to a point just below its photosphere. Using zone $i = 9$ for the StatStar model in Table 9.5, we have $d = r_0 - r_9 = 1.52 \times 10^6$ m, $T = 6348$ K, and $\rho = 2.87 \times 10^{-7}$ kg m$^{-3}$. Using $R = 7.10 \times 10^8$ m at the outermost point of the model, the rate at which this Sun-like star would lose its atmospheric gases would be roughly

$$\dot{M} \approx \pi R \, d\rho \sqrt{\frac{3kT}{m_H}} = 1.2 \times 10^{13} \text{ kg s}^{-1} = 1.9 \times 10^{-10} \text{ M}_\odot \text{ yr}^{-1}.$$

This is typical of the mass transfer rates for semidetached binary systems. The values of $\dot{M}$ inferred from observations of various systems range from $10^{-11}$ to $10^{-7}$ M$_\odot$ yr$^{-1}$. For comparison, recall from Example 11.2.1 that the solar wind transports mass away from the Sun at a much smaller rate, approximately $3 \times 10^{-14}$ M$_\odot$ yr$^{-1}$.

---

Before moving on to consider the consequences of the transfer of mass in semidetached binaries, it is worthwhile to consider the enormous energy that can be released when matter falls onto a star, especially onto a compact object such as a white dwarf or a neutron star.

---

**Example 18.1.2.**   Consider a mass $m = 1$ kg that starts at rest infinitely far from a star of mass $M$ and radius $R$. The initial total mechanical energy of the mass $m$ is

$$E = K + U = 0.$$

Using conservation of energy, we find that the kinetic energy of the mass when it arrives at the star's surface is

$$K = -U = G \frac{Mm}{R}.$$

This kinetic energy will be converted into heat and light upon impact with the star. If the star is a white dwarf with $M = 0.85$ M$_\odot$ and $R = 6.6 \times 10^6$ m $= 0.0095$ R$_\odot$, then the energy released by one kilogram of infalling matter is

$$G \frac{Mm}{R} = 1.71 \times 10^{13} \text{ J}.$$

This is 0.019% of the rest energy ($mc^2$) of one kilogram of material. For comparison, the amount of energy released by the thermonuclear fusion of one kilogram of hydrogen is

$$0.007mc^2 = 6.29 \times 10^{14} \text{ J}$$

(recall Example 10.3.2).

If the star is a neutron star with mass $M = 1.4 \, M_\odot$ and radius $R = 10$ km, then the energy released is much greater:

$$G\frac{Mm}{R} = 1.86 \times 10^{16} \text{ J}.$$

This is 21% of the rest energy of one kilogram, nearly 30 times greater than the energy that hydrogen fusion could provide! The calculations show that infalling matter is capable of generating immense amounts of energy.

Observations of celestial X-ray sources have revealed objects with a steady X-ray luminosity of approximately $10^{30}$ W. If this radiation were produced by gases pulled from a companion star that then fell onto a neutron star's surface, the amount of mass per second transferred between the two stars that would be needed to account for the observed luminosity is

$$\dot{M} = \frac{10^{30} \text{ W}}{1.86 \times 10^{16} \text{ J kg}^{-1}} = 5.38 \times 10^{13} \text{ kg s}^{-1},$$

which is only about $10^{-9} \, M_\odot \text{ yr}^{-1}$. This is similar to the mass transfer rate found in the previous example, a fortuitous agreement because $\dot{M}$ for semidetached systems can vary by several orders of magnitude.

---

## 18.2 ■ ACCRETION DISKS

The orbital motion of a semidetached binary can prevent the mass that escapes from the swollen secondary star from falling directly onto the primary star. The primary's movement is often enough to keep it out of the path of the gases that spill through the inner Lagrangian point. If the radius of the primary star is less than about 5% of the binary separation $a$, the mass stream will miss striking the primary's surface. Instead, the mass stream goes into orbit around the primary to form a thin **accretion disk** of hot gas in the orbital plane, as shown in Figs. 17.24 and 18.6.[8] **Viscosity**, an internal friction that converts the directed kinetic energy of bulk mass motion into random thermal motion, causes the orbiting gases to lose energy and slowly spiral inward toward the primary. The physical mechanism responsible for the viscosity in accretion disks is as yet poorly understood. The familiar molecular viscosity due to interparticle forces is far too weak to be effective. Other possibilities involve random motions of the gas, such as turbulence in the disk material caused by thermal convection or by a magnetohydrodynamic instability in the magnetic fields that interact with the differentially rotating disk (cf., Section 11.2). Whatever the mechanism, the gas is heated throughout its descent to increasingly higher temperatures as the lost orbital energy is converted into thermal energy. Finally, the plunging gas ends its journey at the star's surface.

---

[8]Astronomers refer to the process of accumulating mass from an outside source as *accretion*.

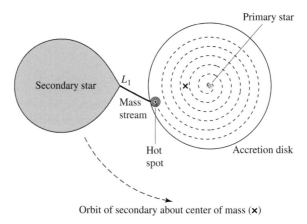

**FIGURE 18.6**   A semidetached binary showing the accretion disk around the primary star and the hot spot where the mass streaming through the inner Lagrangian point impacts the disk. This system's parameters correspond to those of Z Chamaeleontis, described in Example 18.4.1.

## The Temperature Profile and Luminosity

Just as a star may be treated as a blackbody in a rough first approximation, the assumption of an optically thick accretion disk radiating as a blackbody provides a simple, useful model. At each radial distance, an optically thick disk emits blackbody radiation with a continuous spectrum corresponding to the local disk temperature at that distance.

To estimate the temperature of a model accretion disk at a distance $r$ from the center of the primary star of mass $M_1$ and radius $R_1$, let's assume that the inward radial velocity of the disk gases is small compared with their orbital velocity. Then, to a good approximation, the gases follow circular Keplerian orbits, and the details of the viscous forces acting within the disk may be neglected. Furthermore, since the mass of the disk is very small compared with that of the primary, the orbiting material feels only the gravity of the central primary star. The *total energy* (kinetic plus potential) of a mass $m$ of orbiting gas is given by Eq. (2.35),

$$E = -G\frac{M_1 m}{2r}.$$

As the gas spirals inward, its total energy $E$ becomes more negative. The lost energy maintains the disk's temperature and is ultimately emitted in the form of blackbody radiation.

Now consider an annular ring of radius $r$ and width $dr$ within the disk, as shown in Fig. 18.7. If the rate at which mass is transferred from the secondary to the primary star is a constant $\dot{M}$, then in time $t$ the amount of mass that passes through the outer boundary of the circular ring shown in Fig. 18.7 is $\dot{M}t$. Assuming a *steady-state disk* that does not change with time, no mass is allowed to build up within the ring. Therefore during this time an amount of mass $\dot{M}t$ must also leave through the ring's inner boundary.

Conservation of energy requires that the energy $dE$ radiated by the ring in time $t$ be equal to the difference in the energy that passes through the ring's outer and inner boundaries:

$$dE = \frac{dE}{dr}\, dr = \frac{d}{dr}\left(-G\frac{M_1 m}{2r}\right)dr = G\frac{M_1 \dot{M}t}{2r^2}\, dr,$$

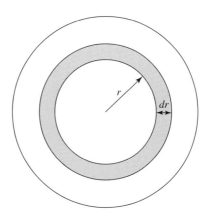

**FIGURE 18.7**   One of the (imaginary) annular rings constituting the accretion disk.

where $m = \dot{M}t$ has been used for the orbiting mass entering and leaving the ring. If the luminosity of the ring is $dL_{\text{ring}}$, then the energy radiated by the ring in time $t$ is related to $dL_{\text{ring}}$ by

$$dL_{\text{ring}}t = dE = G\,\frac{M_1\dot{M}t}{2r^2}\,dr.$$

Canceling the $t$'s and using the Stefan–Boltzmann law in the form of Eq. (3.16) with $A = 2(2\pi r\,dr)$ for the surface area of the ring (both sides) gives

$$dL_{\text{ring}} = 4\pi r\sigma T^4\,dr \tag{18.15}$$

$$= G\,\frac{M_1\dot{M}}{2r^2}\,dr \tag{18.16}$$

for the luminosity of the ring. Solving for $T$, the disk temperature at radius $r$, results in

$$T = \left(\frac{GM\dot{M}}{8\pi\sigma R^3}\right)^{1/4}\left(\frac{R}{r}\right)^{3/4}. \tag{18.17}$$

The "1" subscript has been dropped, with the understanding that $M$ and $R$ are the mass and radius of the primary star, and that $\dot{M}$ is the mass transfer rate for the semidetached binary system.

A more thorough analysis would take into account the thin turbulent boundary layer that must be produced when the rapidly orbiting disk gases encounter the surface of the primary star. This results in a better estimate of the disk temperature:

$$T = \left(\frac{3GM\dot{M}}{8\pi\sigma R^3}\right)^{1/4}\left(\frac{R}{r}\right)^{3/4}\left(1 - \sqrt{R/r}\right)^{1/4} \tag{18.18}$$

$$= T_{\text{disk}}\left(\frac{R}{r}\right)^{3/4}\left(1 - \sqrt{R/r}\right)^{1/4}, \tag{18.19}$$

where

$$T_{\text{disk}} \equiv \left( \frac{3GM\dot{M}}{8\pi\sigma R^3} \right)^{1/4} \tag{18.20}$$

is a characteristic temperature of the disk. Actually, $T_{\text{disk}}$ is roughly twice the maximum disk temperature (Problem 18.4),

$$T_{\text{max}} = 0.488 \left( \frac{3GM\dot{M}}{8\pi\sigma R^3} \right)^{1/4} = 0.488 T_{\text{disk}}, \tag{18.21}$$

which occurs at $r = (49/36)R$; see Fig. 18.13.[9] When $r \gg R$, the last term on the right-hand side of Eq. (18.19) may be neglected, leaving

$$T = \left( \frac{3GM\dot{M}}{8\pi\sigma R^3} \right)^{1/4} \left( \frac{R}{r} \right)^{3/4} = T_{\text{disk}} \left( \frac{R}{r} \right)^{3/4} \qquad (r \gg R). \tag{18.22}$$

This differs from our simple estimate, Eq. (18.17), by a factor of $3^{1/4} = 1.32$.

Integrating Eq. (18.16) for the luminosity of each ring from $r = R$ to $r = \infty$ results in an expression for the disk luminosity,

$$L_{\text{disk}} = G \frac{M\dot{M}}{2R}. \tag{18.23}$$

However, recall from Example 18.1.2 that without an accretion disk, the accretion luminosity (the rate at which falling matter delivers kinetic energy to the primary star) is twice as great:

$$L_{\text{acc}} = G \frac{M\dot{M}}{R}. \tag{18.24}$$

Thus, if half of the available accretion energy is radiated away as the gases spiral down through the disk, then the remaining half must be deposited at the surface of the star (or in the turbulent boundary layer between the rapidly rotating disk and the more slowly rotating primary star).[10]

---

**Example 18.2.1.** The maximum disk temperature, $T_{\text{max}}$, and the value of the disk luminosity for the white dwarf and neutron star used in Example 18.1.2 can now be evaluated. For a white dwarf with $M = 0.85$ $M_{\odot}$, $R = 0.0095$ $R_{\odot}$, and $\dot{M} = 10^{13}$ kg s$^{-1}$ ($1.6 \times 10^{-10}$ $M_{\odot}$ yr$^{-1}$), Eq. (18.21) is

$$T_{\text{max}} = 0.488 \left( \frac{3GM\dot{M}}{8\pi\sigma R^3} \right)^{1/4} = 2.62 \times 10^4 \text{ K}.$$

[9]Including the boundary layer results in $T = 0$ where the disk meets the star's surface, an unrealistic artifact of the assumptions of the model.
[10]This result is just another consequence of the virial theorem.

According to Wien's displacement law, Eq. (3.19), at this temperature the blackbody spectrum peaks at a wavelength of

$$\lambda_{max} = \frac{(500 \text{ nm})(5800 \text{ K})}{26{,}200 \text{ K}} = 111 \text{ nm},$$

which is in the ultraviolet region of the electromagnetic spectrum (Table 3.1). From Eq. (18.23), the luminosity of the accretion disk is

$$L_{disk} = G \frac{M \dot{M}}{2R} = 8.55 \times 10^{25} \text{ W},$$

or about 0.22 $L_\odot$.

Turning now to a neutron star with $M = 1.4 \text{ M}_\odot$, $R = 10$ km, and $\dot{M} = 10^{14}$ kg s$^{-1}$ ($1.6 \times 10^{-9}$ M$_\odot$ yr$^{-1}$), the maximum disk temperature is

$$T_{max} = 0.488 \left( \frac{3GM\dot{M}}{8\pi \sigma R^3} \right)^{1/4} = 6.86 \times 10^6 \text{ K}.$$

Its blackbody spectrum peaks at a wavelength of

$$\lambda_{max} = \frac{(500 \text{ nm})(5800 \text{ K})}{686{,}000 \text{ K}} = 0.423 \text{ nm},$$

which is in the X-ray region of the electromagnetic spectrum. The luminosity of the neutron star's accretion disk is

$$L_{disk} = G \frac{M \dot{M}}{2R} = 9.29 \times 10^{29} \text{ W},$$

over 2400 $L_\odot$. Thus the inner regions of accretion disks around white dwarfs should shine in the ultraviolet, whereas those around neutron stars will be strong X-ray sources.[11]

## The Radial Extent of an Accretion Disk

The radial extent of the accretion disk can be estimated by finding the value of $r = r_{circ}$ where a continuous stream of mass that passes through $L_1$ will settle into a circular orbit around the primary star. This may be done by considering the angular momentum of a parcel of mass $m$ about the primary star; see Fig. 18.8. Assuming that the motion of the mass at the inner Lagrangian point is due solely to the orbital motion of the binary system, the angular momentum, $L$, of the mass located there is

$$L = m\omega \ell_1^2 = m\ell_1^2 \sqrt{\frac{G(M_1 + M_2)}{a^3}},$$

where Eq. (18.7) has been used for the angular frequency of the orbit, and $\ell_1$ is given by Eq. (18.9).

---

[11]Actually, as we will see in Section 18.6, the accretion disk around a white dwarf or neutron star may be disrupted by the star's magnetic field and so may not extend down to its surface. Such systems are strong sources of X-rays.

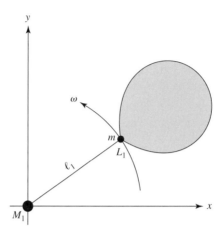

**FIGURE 18.8**    A parcel of mass $m$ passing through the inner Lagrangian point $L_1$, seen from a frame of reference with the primary star at rest at the origin.

The mass $m$ does not immediately enter into a circular orbit. Instead, the stream of mass to which $m$ belongs flows around the primary star and collides with itself after one orbit. The orbits of the mass parcels are made circular around the primary by the collisions as energy is lost while angular momentum is conserved. When the parcel of mass has settled into a circular orbit of radius $r_{circ}$ around $M_1$, its angular momentum is

$$L = m\sqrt{GM_1 r_{circ}},$$

where Eq. (2.30) was used for a circular orbit with $\mu = mM_1/(m + M_1) \simeq m$. Equating these two expressions for the angular momentum results in

$$r_{circ} = a \left(\frac{\ell_1}{a}\right)^4 \left(1 + \frac{M_2}{M_1}\right)$$

$$= a \left[0.500 - 0.227 \log_{10}\left(\frac{M_2}{M_1}\right)\right]^4 \left(1 + \frac{M_2}{M_1}\right). \tag{18.25}$$

Since the total angular momentum must be conserved when only internal and central forces act, you may wonder what happens to the angular momentum lost by the infalling material as it spirals through the accretion disk. As shown in Pringle (1981), orbiting material that is initially in the form of a narrow ring at $r = r_{circ}$ will spread, moving both inward and outward. The time for this migration of the disk material probably ranges from a few days to a few weeks. While most of the matter spirals inward, a small amount of the mass carries the "missing" angular momentum to the outer edge of the disk. From there, the angular momentum may be carried away from the system by wind-driven mass loss. If the accretion disk extends 80% to 90% of the way out to the inner Lagrangian point, angular momentum may also be returned to the orbital motion of the two stars by tides raised in the disk by the secondary star. Because of this outward migration of mass, we will adopt

$$R_{\text{disk}} \approx 2r_{\text{circ}} \tag{18.26}$$

as a rough estimate of the outer radius of the accretion disk.

### Eclipsing, Semidetached Binary Systems

It is comforting to know that there is evidence, obtained from observing *eclipsing* semidetached binary systems, that the objects described above actually exist. Observations of light curves for eclipsing semidetached binaries, such as shown in Fig. 18.9, indicate the presence of a hot spot where the mass transfer stream collides with the outer edge of the accretion disk. The light curve can be interpreted as the result of observing consecutive "slices" of the disk as they disappear and then reappear from behind the primary star. In fact, Fig. 18.9 can be used to re-create an image of the disk itself, shown in Fig. 18.10.[12] Because the hot spot is on the trailing side of the disk during the eclipse (see Fig. 18.12), more light is received from the disk near the beginning of the eclipse (when the hot spot is still visible) than near the end (when the hot spot is still hidden). This produces the deficit in intensity on the right-hand side of the light curve in Fig. 18.9.

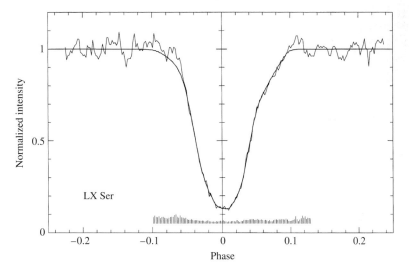

**FIGURE 18.9** The light curve of the eclipse of the accretion disk in the LX Serpentis binary system. The jagged line is the observed light curve, and the smooth line shows the fit calculated from a reconstructed image of the accretion disk, shown in Fig. 18.10. (Figure adapted from Rutten, van Paradijs, and Tinbergen, *Astron. Astrophys.*, *260*, 213, 1992.)

[12]Using slices of the emerging disk to reconstruct an image of the accretion disk is somewhat analogous to using a CAT scan (computerized axial tomography) in a hospital to mathematically reassemble X-ray slices of the human body. Because there is more than one model disk that will reproduce a given light curve, a technique called *maximum entropy* is used to choose the smoothest possible model for the final disk image.

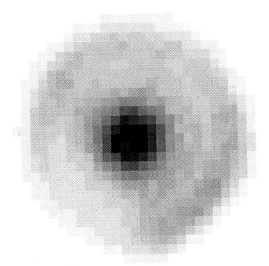

**FIGURE 18.10**   A reconstructed negative image of the accretion disk in the LX Serpentis binary system. The hot spot on the edge of the lower right portion of the disk is smeared out in the azimuthal direction and has the appearance of a partial ring. (Figure from Rutten, van Paradijs, and Tinbergen, *Astron. Astrophys.*, *260*, 213, 1992.)

## 18.3 ■ A SURVEY OF INTERACTING BINARY SYSTEMS

The life history of a close binary system is quite complicated, with many possible variations depending on the initial masses and separation of the two stars involved. As mass passes from one star to the other, the mass ratio $M_2/M_1$ will change. The resulting redistribution of angular momentum affects the orbital period of the system as well as the separation of the two stars. The extent of the Roche lobes, given by Eqs. (18.9) and (18.10), depends on both the separation and the mass ratio of the stars, so it too will vary accordingly.

### The Effects of Mass Transfer

The effects of mass transfer can be illustrated by considering the total angular momentum of the system. The contribution of the stars' rotation to the total angular momentum is small and may be neglected. The orbital angular momentum is given by Eq. (2.30) with an eccentricity of $e = 0$ for a circular orbit,

$$L = \mu\sqrt{GMa}.$$

In this expression, $\mu$ is the reduced mass (Eq. 2.22),

$$\mu = \frac{M_1 M_2}{M_1 + M_2},$$

and $M = M_1 + M_2$ is the total mass of the two stars. Assuming (to a first approximation) that no mass or angular momentum is removed from the system via stellar winds or gravitational radiation, both the total mass and the angular momentum of the system remain constant as mass is transferred between the two stars.[13] That is, $dM/dt = 0$ and $dL/dt = 0$.

Some useful insights concerning the effect of the transfer of mass on the separation of the two stars can be gained by taking a time derivative of the expression for the angular momentum:

$$\frac{dL}{dt} = \frac{d}{dt}\left(\mu\sqrt{GMa}\right)$$

$$0 = \sqrt{GM}\left(\frac{d\mu}{dt}\sqrt{a} + \frac{\mu}{2\sqrt{a}}\frac{da}{dt}\right)$$

$$\frac{1}{a}\frac{da}{dt} = -\frac{2}{\mu}\frac{d\mu}{dt}. \tag{18.27}$$

Remembering that the total mass, $M$, remains constant, we find that the time derivative of the reduced mass is

$$\frac{d\mu}{dt} = \frac{1}{M}\left(\frac{dM_1}{dt}M_2 + M_1\frac{dM_2}{dt}\right).$$

The mass lost by one star is gained by the other. Writing $\dot{M} \equiv dM/dt$, this means that $\dot{M}_1 = -\dot{M}_2$, and so

$$\frac{d\mu}{dt} = \frac{\dot{M}_1}{M}(M_2 - M_1).$$

Inserting this into Eq. (18.27) achieves our result,

$$\frac{1}{a}\frac{da}{dt} = 2\dot{M}_1\frac{M_1 - M_2}{M_1 M_2}. \tag{18.28}$$

Equation (18.28) describes the consequence of mass transfer on the separation of the binary system. The angular frequency of the orbit will also be affected, as shown by using Kepler's third law in the form of Eq. (18.7). Since $M_1 + M_2 = $ constant, Kepler's third law states that $\omega \propto a^{-3/2}$ so that

$$\frac{1}{\omega}\frac{d\omega}{dt} = -\frac{3}{2}\frac{1}{a}\frac{da}{dt}. \tag{18.29}$$

As the orbital separation decreases, the angular frequency increases.

## The Evolution of a Binary System

The following description illustrates the probable evolution of a binary system that is destined to become a cataclysmic variable. The starting point is a widely separated binary system with main-sequence stars having an initial orbital period ranging from a few months

[13]In fact, gravitational radiation, which will be discussed in Section 18.6, is primarily responsible for the loss of angular momentum in some short-period binary systems ($P < 14$ hours).

to a few years. At the start, suppose that Star 1 is more massive than Star 2, so $M_1 - M_2 > 0$. Star 1 therefore evolves more rapidly and, depending on its mass, may become a red giant or supergiant before it begins to overflow its Roche lobe. This initiates the transfer of mass from Star 1 to Star 2 (so $\dot{M}_1 < 0$). According to Eqs. (18.28) and (18.29), in this situation $da/dt$ is negative and $d\omega/dt$ is positive; the stars spiral closer together with an increasingly shorter period.

Now, from Eq. (18.9), as $a$ decreases and $M_2/M_1$ increases, the Roche lobe around Star 1 shrinks, as measured by the distance of Star 1 from the inner Lagrangian point. The mass transfer rate accelerates under the positive feedback of a shrinking Roche lobe, eventually producing an extended atmosphere around both stars, as shown in Fig. 18.4(c). The system is now a contact binary, with the degenerate core of Star 1 and the main-sequence Star 2 sharing a common gaseous envelope. The two stars transfer angular momentum to this envelope as they slowly spiral inward to a much smaller separation and shorter period. If the cores of the two stars merge, the result will be single star, which may explain the observations of blue stragglers in stellar clusters, previously discussed in Section 13.3. Alternatively, the envelope surrounding the stars may be ejected. In fact, several systems have been observed in which a binary is found at the center of a planetary nebula, possibly the result of the ejection of a common envelope. (For the sake of the following discussion, we will consider the situation where envelope ejection occurs.)

After emerging from their gaseous cocoon, the system is a detached binary; Star 2 (the secondary) lies inside its Roche lobe as Star 1 (the primary) cools to become a white dwarf. Eventually, the originally less-massive secondary star evolves and fills its Roche lobe, and mass begins to flow in the opposite direction, with $\dot{M}_1 > 0$. In this case a negative feedback mitigates the mass transfer process, because, as Eq. (18.28) implies, the stars will now spiral farther apart (assuming that $M_1$ is still greater than $M_2$) as the Roche lobe around the secondary star expands according to Eq. (18.10). If the mass flow is to persist, either the secondary must expand faster than the Roche lobe grows or the stars must move closer together as angular momentum is removed from the system, either by torques due to stellar winds confined by magnetic fields or by gravitational radiation.[14] Whatever the mechanism, a steady rate of mass transfer from the secondary to the white dwarf is maintained, and the stage is set for the outbursts of a cataclysmic variable, as will be described in Section 18.4.

As the secondary star continues to evolve, another common envelope stage may occur. Figure 18.11 shows an example of the life history of a close binary system that begins with two intermediate-mass stars (between 5 and 9 $M_\odot$) and culminates with two carbon–oxygen white dwarfs in a very tight orbit, circling each other every 15 s to 30 s. The larger, less massive white dwarf (recall Eq. 16.14) overflows its Roche lobe and dissolves into a heavy disk that is accreted by the more massive dwarf. The accumulation of mass pushes the primary white dwarf toward the Chandrasekhar limit, and it explodes as a Type Ia supernova.[15]

---

[14]The reduction of the Sun's angular momentum by the solar wind was described in Section 11.2.

[15]In Section 18.5, we will find that nuclear reactions begin in the core of the white dwarf before the Chandrasekhar limit is reached.

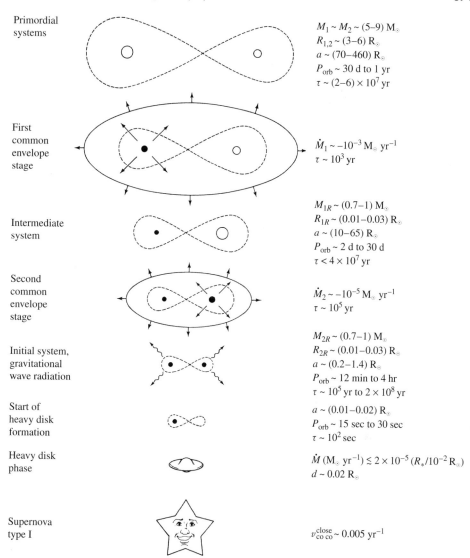

Primordial
systems

$M_1 \sim M_2 \sim (5\text{–}9)\,\text{M}_\odot$
$R_{1,2} \sim (3\text{–}6)\,\text{R}_\odot$
$a \sim (70\text{–}460)\,\text{R}_\odot$
$P_{orb} \sim 30\,\text{d to 1 yr}$
$\tau \sim (2\text{–}6) \times 10^7\,\text{yr}$

First
common
envelope
stage

$\dot{M}_1 \sim -10^{-3}\,\text{M}_\odot\,\text{yr}^{-1}$
$\tau \sim 10^3\,\text{yr}$

Intermediate
system

$M_{1R} \sim (0.7\text{–}1)\,\text{M}_\odot$
$R_{1R} \sim (0.01\text{–}0.03)\,\text{R}_\odot$
$a \sim (10\text{–}65)\,\text{R}_\odot$
$P_{orb} \sim 2\,\text{d to 30 d}$
$\tau < 4 \times 10^7\,\text{yr}$

Second
common
envelope
stage

$\dot{M}_2 \sim -10^{-5}\,\text{M}_\odot\,\text{yr}^{-1}$
$\tau \sim 10^5\,\text{yr}$

Initial system,
gravitational
wave radiation

$M_{2R} \sim (0.7\text{–}1)\,\text{M}_\odot$
$R_{2R} \sim (0.01\text{–}0.03)\,\text{R}_\odot$
$a \sim (0.2\text{–}1.4)\,\text{R}_\odot$
$P_{orb} \sim 12\,\text{min to 4 hr}$
$\tau \sim 10^5\,\text{yr to } 2 \times 10^8\,\text{yr}$

Start of
heavy disk
formation

$a \sim (0.01\text{–}0.02)\,\text{R}_\odot$
$P_{orb} \sim 15\,\text{sec to 30 sec}$
$\tau \sim 10^2\,\text{sec}$

Heavy disk
phase

$\dot{M}\,(\text{M}_\odot\,\text{yr}^{-1}) \lesssim 2 \times 10^{-5}\,(R_*/10^{-2}\,\text{R}_\odot)$
$d \sim 0.02\,\text{R}_\odot$

Supernova
type I

$\nu_{\text{co co}}^{\text{close}} \sim 0.005\,\text{yr}^{-1}$

**FIGURE 18.11**   One possibility for the evolution of a close binary system, ending in a Type Ia
supernova. The masses and radii of the stars, and their orbital separation ($A$), orbital period, and mass
transfer rate, are given for some stages, along with the duration ($\tau$). (Figure adapted from Iben and
Tutukov, *Ap. J. Suppl.*, *54*, 335, 1984.)

## Types of Interacting Binary Systems

There are many types of close binary systems, too many to discuss in any detail. The following list[16] describes the main classes of interacting binaries, together with some of the features that make these systems important for astronomers. Many of the classes are named after the prototype object for that class.

- **Algols**. These are two normal stars (main-sequence stars or subgiants) in a semidetached binary system. They provide checks on stellar properties and evolution, and they yield information on mass loss and mass exchange. Active Algols (the W Serpens stars) provide laboratories for studying rapid (short-lived) stages of stellar and binary star evolution. These systems are important for studying accretion processes and accretion disks. Mass loss from Algols may contribute to the chemical enrichment of the interstellar medium.

- **RS Canum Venaticorum** and **BY Draconis Stars**. These stars are chromospherically active binaries that are important systems for investigating dynamo-driven magnetic activity in cool stars (spectral type F and later). Manifestations of enhanced magnetic activity include starspots, chromospheres, coronae, and flares. These systems also contribute to our understanding of the magnetic activity of the Sun—the so-called *solar–stellar connection*.

- **W Ursae Majoris Contact Systems**. These short-period (0.2–0.8 day) contact binaries display very high levels of magnetic activity and are important stars for studying the stellar dynamo mechanism at extreme levels. The drag of magnetic braking may cause these binaries to coalesce into single stars.

- **Cataclysmic Variables** and **Nova-like Binaries**. These systems have short periods and contain white dwarf components together with cool M-type secondaries that fill their Roche lobes. They provide valuable information on the final stages of stellar evolution. These binaries are also important for studying accretion phenomena and accretion disk properties.

- **X-ray Binaries with Neutron Star and Black Hole Components**. These systems are powerful ($L_x > 10^{28}$ W) X-ray sources that have neutron star or (more rarely) black hole components. The X-rays are due to the accretion of gas onto the degenerate component of the system from a nondegenerate companion. Observations of neutron star systems supplement the information on their structure and evolution that comes from pulsars (such as masses, radii, rotation, and magnetic fields). Systems such as V404 Cygni, A0620−00, and Cygnus X-1 provide evidence for the existence of black holes; see Section 18.6.

- **ζ Aurigae and VV Cephei Systems**. These long-period interacting binaries contain a late-type supergiant component and a hot (usually spectral type B) companion. ζ-Aur

---

[16]Quoted with permission from E. F. Guinan, *Evolutionary Processes in Interacting Binary Stars*, Kondo, Sisteró, and Polidan (eds.), Kluwer Academic Publishers, Dordrecht, 1992. Reprinted by permission of Kluwer Academic Publishers.

systems contain G or K supergiant stars, and VV Cep binaries contain M supergiants. Although not originally interacting binaries, they became so when the more massive star evolved to become a supergiant. When eclipses occur, the atmosphere and wind of the cooler supergiant can be probed as the hotter star passes behind.

- **Symbiotic Binaries**. Symbiotic stars are long-period interacting binaries consisting of an M giant (sometimes a pulsating Mira-type variable) and an accreting component that can be a white dwarf, subdwarf, or low-mass main-sequence star. The common feature of these systems is the accretion of the cool component's wind onto its hot companion. Orbital periods of symbiotic stars typically range between 200 and 1500 days. Several of the symbiotic binaries have the cool component filling its Roche lobe, making them symbiotic Algol systems.

- **Barium and S-Star Binaries**. These stars are thought to be long-period binaries in which the originally more massive component evolved and transferred some of its nuclear-processed gas to the present K or M giant companion (recall the discussion of spectral class S on page 466). The giant stars are thought to have white dwarf companions that are often too cool to be seen in the ultraviolet. These systems are important for studying nucleosynthesis and mass loss in evolved stars.

- **Post-Common-Envelope Binaries**. These binary systems usually contain hot white dwarf or subdwarf components and cooler secondary stars that have presumably passed through the common envelope phase of binary star evolution. The binary nuclei of planetary nebulae are examples of post-common-envelope binaries. These systems are important for studying short-lived stages of stellar evolution.

## 18.4 ■ WHITE DWARFS IN SEMIDETACHED BINARIES

When a white dwarf is the primary component of a semidetached binary system, the result may be a **dwarf nova**, a **classical nova**, or a **supernova**, in order of increasing brilliance. It is somewhat unfortunate that the term *nova* (Latin for "new") appears in each name, because the three types of outbursts employ three very different mechanisms.

### Cataclysmic Variables

Dwarf novae and classical novae belong to the general class of **cataclysmic variables**, of which more than one thousand systems are known to exist. They survive their release of energy (unlike supernovae), and the outburst process can reoccur. Cataclysmic variables are characterized by long quiescent intervals punctuated by outbursts in which the brightness of the system increases by a factor between 10 (for dwarf novae) and $10^6$ (for classical novae). The mean mass of the primary star is $0.86 \ M_\odot$, which is larger than the average of about $0.58 \ M_\odot$ for isolated white dwarfs. The secondary star is usually a main-sequence star of spectral type G or later and is less massive than the primary star.

The two stars orbit each other with periods ranging from 23 minutes to more than five days, although the vast majority have orbital periods of between 78 minutes and 12 hours. Interestingly, a "period gap" exists in the orbital periods of cataclysmic variables between

1.5 hours and 3.25 hours; it is probably due to an abrupt change in angular momentum transfer in the system, associated with a complex interplay of disrupted magnetic braking, gravitational radiation, the changing size of Roche lobes, and the evolution of the stars.

The outbursts are believed to be due to a sudden increase in the rate at which mass flows down through the disk. As the eclipsed disk emerges from behind the secondary star, the radial variation in the disk's temperature can be determined. During an outburst, the disk does indeed appear to be optically thick, with $T \propto r^{-3/4}$, in agreement with Eq. (18.22). But during quiescence the observations are not consistent with the disk model described above, probably because the disk is not completely optically thick when it is cooler and contains less mass.

Additional evidence supporting this view comes from the strong, wide emission lines of hydrogen and helium that are seen in cataclysmic variables during quiescence. These lines are usually doubly peaked, as shown in Fig. 18.12. However, during an eclipse a single emission line is observed, either redshifted or blueshifted. This is what would be expected from a rotating disk of optically thin gas; the Doppler-shifted emission lines produced on the opposite sides of the disk disappear when one side or the other is hidden behind the secondary star.

The source of the emission lines that appear during a cataclysmic variable's quiescent phase is not yet clear. During an outburst, these lines appear in absorption, as would be expected from an optically thick disk that produces absorption lines in the same manner as an optically thick stellar atmosphere. But during quiescence, the rate at which mass flows down through the disk has presumably decreased, making the disk less dense and cooler. At larger radii the disk may then be optically thin and so produce emission lines. Alternatively, there may be a thin layer of hot gas above the disk that produces the emission lines.

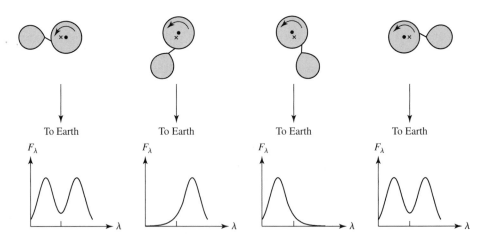

**FIGURE 18.12** A Doppler-shifted emission line at different stages of the eclipse of an accretion disk. The binary system orbits about its center of mass (the "×"), and is observed nearly edge-on. The disk rotates in the direction indicated by the arrow.

**Dwarf Novae**

The first observation of a dwarf nova (U Geminorum) was made in 1855. However, the basic nature of these objects remained elusive until 1974, when Brian Warner at the University of Cape Town showed that the outburst of an eclipsing dwarf nova, Z Chamaeleontis, was due to a brightening of the accretion disk surrounding the white dwarf. Since most of the light from a dwarf nova comes from the accretion disk around the white dwarf, these systems provide astronomers with their best opportunity to study the dynamic structure of accretion disks.[17] Observations of the dwarf nova VW Hydri showed that the outburst at visible wavelengths preceded the ultraviolet brightening by about a day. This indicates that the outburst started in the cooler, outer part of the disk and then spread down to the hotter central regions. For these reasons, astronomers have concluded that the outbursts of dwarf novae are caused by a sudden increase in the rate at which mass flows down through the accretion disk.

---

**Example 18.4.1.** Z Chamaeleontis is a dwarf nova. It consists of an $M_1 = 0.85 \, M_\odot$ white dwarf primary with a radius of $R = 0.0095 \, R_\odot$ and a late M-type main-sequence secondary star of mass $M_2 = 0.17 \, M_\odot$. The orbital period of the system is $P = 0.0745$ day. What does this system look like?

From Kepler's third law, Eq. (2.37), the separation of the two stars is

$$a = \left[ \frac{P^2 G(M_1 + M_2)}{4\pi^2} \right]^{1/3} = 5.22 \times 10^8 \text{ m,}$$

about 75% of the radius of the Sun. The distance between the white dwarf primary and the inner Lagrangian point $L_1$ is given by Eq. (18.9),

$$\ell_1 = a \left[ 0.500 - 0.227 \log_{10} \left( \frac{M_2}{M_1} \right) \right] = 3.44 \times 10^8 \text{ m.}$$

Because the secondary star fills its Roche lobe in a semidetached binary system, the distance between the secondary star and the inner Lagrangian point is a measure of the size of the secondary. For Z Cha,

$$R_2 \approx \ell_2 = a - \ell_1 = 1.78 \times 10^8 \text{ m,}$$

which agrees quite well with the size of an M6 main-sequence star (see Appendix G).

The value of $r_{\text{circ}}$ for this system is, from Eq. (18.25),

$$r_{\text{circ}} = a \left( \frac{\ell_1}{a} \right)^4 \left( 1 + \frac{M_2}{M_1} \right) = 1.18 \times 10^8 \text{ m,}$$

*continued*

---

[17]In some systems, the primary white dwarf has a magnetic field that is sufficiently strong (a few thousand teslas) to prevent the formation of an accretion disk. Instead, the accretion takes place through a magnetically controlled column that funnels mass onto one (or both) of the white dwarf's magnetic poles. These *AM Herculis stars* (or *polars*) will be considered in Section 18.6.

and so a crude estimate of the outer radius of the disk is

$$R_{\text{disk}} \approx 2r_{\text{circ}} = 2.4 \times 10^8 \text{ m},$$

(Eq. 18.26), which is about two-thirds of the way to the inner Lagrangian point. This is in good agreement with observations that indicate that the Z Cha's disk emits very little light from beyond this radius.

The mass transfer rate inferred for Z Cha during an outburst is roughly

$$\dot{M} = 1.3 \times 10^{-9} \text{ M}_\odot \text{ yr}^{-1},$$

or $7.9 \times 10^{13}$ kg s$^{-1}$, which implies a maximum disk temperature of

$$T_{\text{max}} = 0.488 \left( \frac{3GM\dot{M}}{8\pi \sigma R^3} \right)^{1/4} = 4.4 \times 10^4 \text{ K},$$

using Eq. (18.21). Figure 18.13 shows the variation in the disk temperature with radius for Z Cha [calculated from Eq. (18.19)]. Moving from the inner to the outer regions of the disk, the temperature falls from 44,000 K to 8000 K. According to Wien's law, Eq. (3.15), this corresponds to an increase in the peak wavelength of the emitted radiation from 66 nm to 363 nm (from the far to the near portions of the ultraviolet spectrum).

The monochromatic luminosity, $L_\lambda$, for the entire disk can be calculated by integrating Eq. (3.22) for the Planck function, $B_\lambda$, over the disk area and over all directions (recall

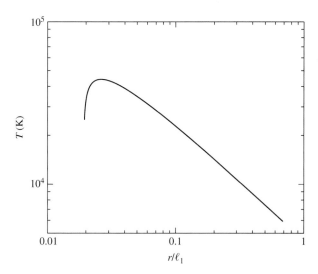

**FIGURE 18.13**    The temperature of the accretion disk calculated for the dwarf nova Z Chamaeleontis. The radius $r$ is given in units of $\ell_1$, the distance from the white dwarf to the inner Lagrangian point. The sudden drop in temperature near the surface of the white dwarf primary is an unrealistic artifact of the assumptions.

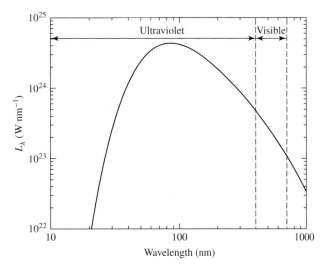

**FIGURE 18.14** The monochromatic luminosity calculated for the accretion disk of the dwarf nova Z Chamaeleontis.

Section 3.5). The resulting graph of the energy emitted per second within wavelength intervals of 1 nm is shown in Fig. 18.14. According to Eq. (18.23), the total luminosity of the accretion disk (integrated over all wavelengths) is

$$L_{\text{disk}} = G\,\frac{M\dot{M}}{2R} = 6.8 \times 10^{26} \text{ W},$$

which exceeds the luminosity of the Sun by about 75%.

An artist's conception of the appearance of Z Cha is shown in Fig. 18.15; see also Fig. 18.6.

## Changes in the Mass Transfer Rate

To date, more than 250 dwarf novae have been discovered. Characteristically, they brighten by between 2 and 6 magnitudes (factors of between 6 and 250 in luminosity) during outbursts that usually last from about 5 to 20 days. These eruptions are separated by quiet intervals of 30–300 days; see Fig. 18.16. Estimates of the rate of mass transfer through the disks of dwarf novae have been obtained by comparing theoretical models with observations of the amount of energy released at different wavelengths. Apparently, during the long quiescent intervals,

$$\dot{M} \approx 10^{12} - 10^{13} \text{ kg s}^{-1} \approx 10^{-11} - 10^{-10} \text{ M}_\odot \text{ yr}^{-1},$$

which increases to

$$\dot{M} \approx 10^{14} - 10^{15} \text{ kg s}^{-1} \approx 10^{-9} - 10^{-8} \text{ M}_\odot \text{ yr}^{-1}$$

**FIGURE 18.15**    An artist's conception of Z Chamaeleontis. [Courtesy of Dale W. Bryner (1935–1999), Weber State University.]

during an outburst. Since the disk luminosity is proportional to $\dot{M}$ (Eq. 18.23), this increase in the mass transfer rate by a factor of 10–100 is consistent with the observed brightening of the system.[18]

The mystery remaining to be solved by astronomers is the origin of the increased rate of mass transfer through the disk of a dwarf nova during an outburst. Possible explanations focus on either an instability in the mass transfer rate from the secondary to the primary star or an instability in the accretion disk itself that periodically dams up and releases the gases flowing through it.

A modulation of the mass transfer rate must depend on the details of the mass flow through the inner Lagrangian point, $L_1$. One possibility is an instability in the outer layers of the secondary star, causing it to periodically overflow its Roche lobe. Such an instability could be powered by the hydrogen partial ionization zone (at $T \approx 10,000$ K) damming up and releasing energy.[19] When one kilogram of H II ions recombines with free electrons, as much as $1.3 \times 10^9$ J is released. If the ionization zone occurs close enough to the surface of the secondary, this could be sufficient to propel some of the overlying stellar material through the $L_1$ point and initiate a dwarf nova outburst. Recall, however, that the secondary star is usually a main-sequence star of spectral type G or later, so the ionization zone may well lie too deep to produce the instability.

The alternative explanation involving an instability in the outer part of the accretion disk also utilizes the hydrogen partial ionization zone. The viscosity of the disk material governs the rate at which mass spirals down through the disk. The lower the viscosity, the lower the resistance to the orbital motion of the disk gases; the inward drift of material decreases, and more matter accumulates in the disk. If the viscosity periodically switches

---

[18]Recall that a difference of 5 magnitudes corresponds to a factor of 100 in brightness; see Eq. (3.4).

[19]This is somewhat reminiscent of the $\kappa$-mechanism that is involved in stellar pulsation; recall Section 14.2.

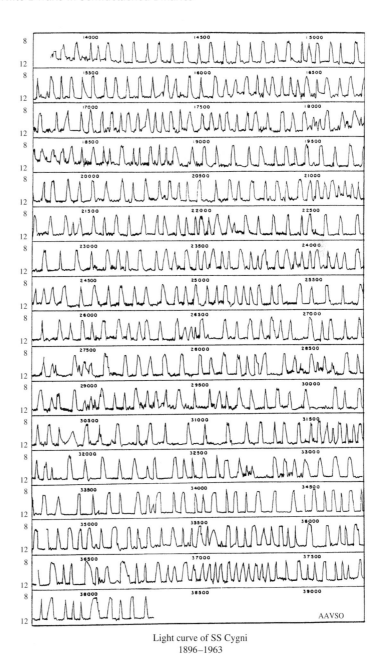

Light curve of SS Cygni
1896–1963

**FIGURE 18.16**   Outbursts of the dwarf nova SS Cygni, about 95 pc away. This light curve, labeled by Julian day at 500-day intervals, covers the years 1896–1963 and was compiled by the American Association of Variable Star Observers. (We acknowledge with thanks the variable star observations from the AAVSO International Database contributed by observers worldwide.)

from a low to a high value, the resulting wave of stored material plunging inward could produce the brightening of the disk observed for dwarf novae. Although the source of the viscosity in accretion disks is poorly understood, it has been suggested that the switch between low and high viscosity may be produced by an instability involving the periodic ionization and recombination of hydrogen in the outer part of the disk where $T \approx 10,000$ K. In such a scenario, the viscosity is roughly proportional to the disk temperature, which in turn depends on the opacity of the disk material. Below $10^4$ K, a plausible chain of reasoning then suggests

neutral hydrogen $\rightarrow$ low opacity

$\rightarrow$ efficient cooling

$\rightarrow$ low temperature

$\rightarrow$ low viscosity

$\rightarrow$ mass retained in the outer disk.

On the other hand, above $10^4$ K,

ionized hydrogen $\rightarrow$ high opacity

$\rightarrow$ inefficient cooling

$\rightarrow$ high temperature

$\rightarrow$ high viscosity

$\rightarrow$ mass released to fall through the disk.

The instability occurs because the accumulation of matter tends to slowly heat the outer disk, while its release results in a rapid cooling. This mechanism should operate only for low accretion rates ($< 10^{12}$ kg s$^{-1} \approx 10^{-11}$ M$_\odot$ yr$^{-1}$), so dwarf novae outbursts should not occur for systems with larger values of $\dot{M}$. This limit is in fact observed and is one reason why most astronomers favor the disk instability explanation of dwarf novae outbursts.

### Classical Novae

Higher accretion rates are associated with classical novae. The earliest record of a nova was that of CK Vulpeculae, which occurred in 1670. Since then hundreds of others have been observed. About 30 novae are detected in the Andromeda galaxy (M31) each year, but only two or three per year can be seen in those regions of our own Milky Way Galaxy that are unobscured by dust. Novae are characterized by a sudden increase in brightness of between 7 and 20 magnitudes, with an average brightening of about 10–12 magnitudes. The rise in luminosity is very rapid, taking only a few days, with a brief pause or *standstill* when the star is about two magnitudes from its maximum brilliance. At its peak, a nova may shine with about $10^5$ L$_\odot$ and release roughly $10^{38}$ J (integrated across all wavelengths) over $\sim 100$ days.

The subsequent decline occurs more slowly over several months, and its rate of decline defines the **speed class** of a nova. A **fast nova** takes a few weeks to dim by two magnitudes, whereas a **slow nova** may take nearly 100 days to decline by the same amount from maximum; see Figs. 18.17 and 18.18. The declines are sometimes punctuated by large fluctuations in brightness, which in extreme cases may take the form of the complete absence of visible light from the nova for a month or so before it reappears. Fast novae are typically three magnitudes brighter than slow novae, but in either case a nova falls to nearly its pre-eruption appearance after a few decades.

During the first few months, the decline in brightness occurs only at visual wavelengths. When observations at infrared and ultraviolet wavelengths are included, the bolometric

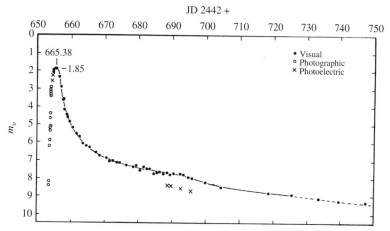

**FIGURE 18.17**   The light curve of V1500 Cyg, a fast nova. (Figure adapted from Young, Corwin, Bryan, and De Vaucouleurs, *Ap. J.*, *209*, 882, 1976.)

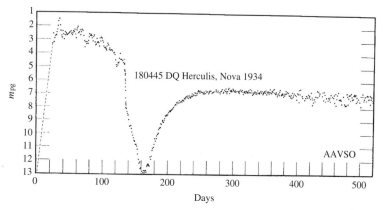

**FIGURE 18.18**   The light curve of DQ Her, a slow nova. The *photographic magnitude*, $m_{pg}$, is measured from the nova's image on photographic plates. (We acknowledge with thanks the variable star observations from the AAVSO International Database contributed by observers worldwide.)

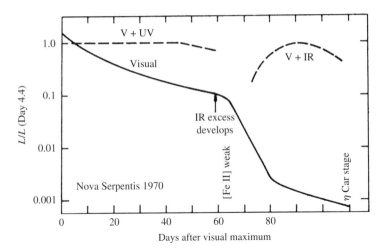

**FIGURE 18.19**   The bolometric luminosity of nova FH Serpentis, in terms of its luminosity at day 4.4. Note that during the first 60 days, the decline in visible energy was almost exactly offset by an increase at ultraviolet wavelengths. Thereafter, the infrared rose as the visible light output was redistributed to infrared wavelengths. (Figure adapted from Gallagher and Starrfield, *Annu. Rev. Astron. Astrophys.*, *16*, 171, 1978. Reproduced with permission from the *Annual Review of Astronomy and Astrophysics*, Volume 16, ©1978 by Annual Reviews Inc.)

luminosity of a nova is found to remain approximately constant for several months following its outburst; see Fig. 18.19. In addition, spectra of novae show that they are accompanied by the ejection of $10^{-5}$ to $10^{-4}$ $M_\odot$ of hot gases at velocities between several hundred and several thousand km s$^{-1}$. The speed of the gases is roughly three times greater for fast novae, but the total mass ejected is about the same for both speed classes. We will see that the changing characteristics of this expanding shell of gas are responsible for the features seen in Fig. 18.19.

The average value of the absolute visual magnitude of a nova in its quiescent state is $M_V = 4.5$. Assuming that the light from such a system comes primarily from the accretion disk around the white dwarf, an estimate of the mass transfer rate for a typical nova can be obtained. (For the purposes of this estimation, visual magnitudes will be used instead of bolometric magnitudes. This means that the mass transfer rate will be slightly underestimated.) From Eq. (3.7), the luminosity of the system is

$$L = 100^{(M_{\text{Sun}} - M_V)/5} \, L_\odot = 1.3 \, L_\odot = 4.9 \times 10^{26} \text{ W}.$$

With this result, the luminosity of an accretion disk (Eq. 18.23) can be solved for the mass transfer rate, giving

$$\dot{M} = \frac{2RL}{GM} = 5.7 \times 10^{13} \text{ kg s}^{-1},$$

or about $9.0 \times 10^{-10}$ $M_\odot$ yr$^{-1}$.

This is in good agreement with the accepted theoretical model of a nova, which incorporates a white dwarf in a semidetached binary system that accretes matter at a rate of about $10^{-8}$ to $10^{-9}$ $M_\odot$ yr$^{-1}$. The hydrogen-rich gases accumulate on the surface of the white dwarf, where they are compressed and heated. At the base of this layer, turbulent mixing enriches the gases with the carbon, nitrogen, and oxygen of the white dwarf. (Without this mixing, the ensuing explosion would be too feeble to eject the mass observed for the expanding shell of hot gases.) Spectroscopic analysis of the shell shows an enrichment of carbon, nitrogen, and oxygen by a factor of 10 to 100 times the solar abundance of these elements.

At the base of this enriched layer of hydrogen, the material is supported by electron degeneracy pressure. When about $10^{-4}$ to $10^{-5}$ $M_\odot$ of hydrogen has accumulated and the temperature at the base reaches a few million kelvins, a shell of CNO-cycle hydrogen burning develops. For highly degenerate matter the pressure is independent of the temperature, so the shell source cannot dampen the reaction rate by expanding and cooling. The result is a runaway thermonuclear reaction, with temperatures reaching $10^8$ K before the electrons lose their degeneracy.[20] When the luminosity exceeds the Eddington limit of about $10^{31}$ W (recall Eq. 10.114), radiation pressure can lift the accreted material and expel it into space. The fast and slow speed classes of novae are likely due to variations in the mass of the white dwarf and in the degree of CNO enrichment of the hydrogen surface layer. The brief standstill that occurs before maximum luminosity is probably an effect of the changing opacity of the ejecta.

The energy that would be released in the complete fusion of a hydrogen layer of $m = 10^{-4}$ $M_\odot$ is $0.007mc^2 \approx 10^{41}$ J, roughly $10^3$ times larger than the energies actually observed. If all of the hydrogen were in fact consumed, the nova would shine for several hundred years. Most of the accumulated material must therefore be propelled into space by the explosion. However, the kinetic energy of the ejecta (far from the nova) is much smaller than the gravitational binding energy of the surface layer, indicating that the total energy given to the ejecta is just barely enough to allow it to escape from the system.

Only about 10% of the hydrogen layer is ejected by the nova explosion. Following this initial **hydrodynamic ejection phase** which dominates for fast novae, hydrostatic equilibrium is established and the **hydrostatic burning phase** begins. During this prolonged stage of hydrogen burning, which is most important for slow novae, energy is produced at a constant rate approximately equal to the Eddington luminosity. The layer above the shell of CNO burning becomes fully convective and expands by a factor of 10 to 100, extending to some $10^9$ m.[21] At the surface of the convective envelope the effective temperature is about $10^5$ K, much less than the $4 \times 10^7$ K in the active CNO shell source below.

Finally, the last of the accreted surface layer is ejected, from between a few months to about a year after the hydrostatic burning phase began. Deprived of fuel, the hydrostatic burning phase ends, and the white dwarf begins to cool. Eventually the binary system reverts to its quiescent configuration and the accretion process begins anew. For accretion rates of

---

[20]This mechanism is similar to the helium core flash that was described in Section 13.2.

[21]The white dwarf remnant may overflow its Roche lobe. The consequences of the resulting disruption of the close binary system are not yet clear.

$10^{-8}$ to $10^{-9}$ $M_\odot$ yr$^{-1}$, it will take some $10^4$ to $10^5$ years to build up another surface layer of $10^{-4}$ $M_\odot$.

The physical character of the *ejected gases* passes through three distinct phases as a consequence of the nova explosion. During the initial **fireball expansion phase**, the material blown off the star in the hydrodynamic ejection phase forms an optically thick "fireball" that radiates as a hot blackbody of 6000–10,000 K. The observed light originates in the "photosphere" of the expanding fireball; at this point, the spectrum of the nova resembles that of an A or F supergiant.

The fireball expansion phase will be examined in Problem 18.13, where you will consider a simple model of a nova for which mass is ejected at a constant rate of $\dot{M}_{\text{eject}}$ at a constant speed $v$. The expanding model photosphere has a radius that initially increases linearly with time and then approaches a limiting value of

$$R_\infty = \frac{3\overline{\kappa}\dot{M}_{\text{eject}}}{8\pi v}. \tag{18.30}$$

If the luminosity, $L$, of the nova is also assumed to be constant, then from Eq. (3.17) the effective temperature of the model photosphere approaches

$$T_\infty = \left(\frac{L}{4\pi\sigma}\right)^{1/4}\left(\frac{8\pi v}{3\overline{\kappa}\dot{M}_{\text{eject}}}\right)^{1/2}. \tag{18.31}$$

For an opacity of $\overline{\kappa} = 0.04$ m$^2$ kg$^{-1}$ [Eq. (9.27) for electron scattering, with pure hydrogen assumed for convenience], a mass ejection rate of $\dot{M}_{\text{eject}} \approx 10^{19}$ kg s$^{-1}$ (about $10^{-4}$ $M_\odot$ yr$^{-1}$), and an ejection speed of $v \approx 1000$ km s$^{-1}$, the fireball's photosphere approaches a limiting radius of about $5 \times 10^{10}$ m, or 1/3 AU. Taking $L$ to be the Eddington limit of about $10^{31}$ W, the effective temperature of the model photosphere approaches a value of nearly 9000 K.

The optically thick fireball phase ends in a few days, at the point of maximum visual brightness. Then, as the shell of gas thrown off by the nova continues to expand, it becomes less and less dense. The rate of mass ejection, $\dot{M}_{\text{eject}}$, has also declined in the hydrostatic burning phase. The result, according to Eqs. (18.30) and (18.31), is that the location of the photosphere moves inward and its temperature increases slightly. Although these general trends are correct, the opacity is in fact very sensitive to the temperature for $T < 10^4$ K (recall Fig. 9.10), and our model is too simplistic to describe the evolution of the nova. More advanced arguments show that as the visual brightness declines, more light is received from the nova at ultraviolet wavelengths. Finally, the shell becomes transparent and the **optically thin phase** begins. The central white dwarf, swollen by its hydrostatic burning phase, now has the appearance of a blue horizontal-branch object located just blueward of the RR Lyrae stars on the H–R diagram. The white dwarf envelope may burn irregularly, resulting in the substantial fluctuations in brightness observed for some novae.

After a few months, when the temperature of the expanding envelope of gases has fallen to about 1000 K, carbon in the ejecta can condense to form dust consisting of graphite grains.[22]

---

[22]The identification of the grain composition comes in part from an infrared emission "bump" at a wavelength of 5 $\mu$m; recall Section 12.1. Novae are natural laboratories for testing theories of grain formation.

**FIGURE 18.20** A 1949 photo of Nova Persei, which exploded in 1901. (Courtesy of Palomar/Caltech.)

This initiates the **dust formation phase**. The resulting dust shell becomes optically thick in roughly 50% of all novae. The visible light from a nova is undiminished by an optically thin shell, but the formation of an optically thick cocoon of dust obscures or completely hides the central white dwarf. In the latter case, the output of visible light suddenly plunges, as seen in Fig. 18.19. The light from the white dwarf is absorbed and re-emitted by the graphite grains, so the optically thick dust shell radiates as a $\sim 900$ K blackbody at infrared wavelengths. In this way, the nova's bolometric luminosity remains constant as long as the white dwarf continues to produce energy at roughly the Eddington rate while in its hydrostatic burning phase. Figure 18.20 shows that the expanding shell may remain visible for years after the hydrostatic burning phase has ended, its gases and dust enriching the interstellar medium.

### Polars: X-Rays from White Dwarf Systems

**AM Herculis stars** (also called **polars**), are semidetached binaries containing white dwarfs with magnetic fields of about 2000 T. The torque produced by the white dwarf's field interacting with the secondary star's envelope results in a nearly synchronous rotation; the two stars perpetually face each other, connected by a stream of hot gas.[23] As this gas approaches the white dwarf, it moves almost straight down toward the surface and forms an accretion *column* a few tens of kilometers across. A shock front occurs above the white dwarf's photosphere, where the gas is decelerated and heated to a temperature of several $10^8$ K. The hot gas emits hard X-ray photons; some escape, and some are absorbed by the photosphere and re-emitted at soft X-ray and ultraviolet wavelengths.

---

[23]If the white dwarf has a somewhat weaker field ($B_s < 1000$ T), or if the stars are farther apart, an accretion disk may form, only to be disrupted near the star (as shown in Fig. 18.22). These systems, called **DQ Herculis stars**, or **intermediate polars**, do not exhibit synchronous rotation.

The visible light observed from these systems is in the form of *cyclotron radiation* emitted by nonrelativistic electrons spiraling along the magnetic field lines of the accretion column. This is the nonrelativistic analog of the synchrotron radiation emitted by relativistic electrons; see Fig. 16.17. In contrast to the continuous spectrum of synchrotron radiation, most of the energy of cyclotron radiation is emitted at the *cyclotron frequency*,

$$\boxed{\nu_c = \frac{eB}{2\pi m_e}.}$$

(18.32)

For $B_s = 1000$ T, $\nu_c = 2.8 \times 10^{13}$ Hz, which is in the infrared. However, a small fraction of the energy is emitted at higher harmonics (multiples) of $\nu_c$ and may be detected at visible wavelengths by astronomers on Earth. The cyclotron radiation is circularly polarized when observed parallel to the direction of the magnetic field lines, and linearly polarized when viewed perpendicular to the field lines.[24] Thus, as the two stars orbit each other (typically every 1 to 2 hours), the measured polarization changes smoothly between being circularly and linearly polarized. In fact, it is this strong variable polarization (up to 30%) that gives polars their name.

## 18.5 ■ TYPE IA SUPERNOVAE

We have seen that there are many differences among the characteristics of individual novae. The peak luminosity, the rate of decline, the presence of rapid fluctuations, and/or the complete disappearance of the nova at visible wavelengths—all of these vary greatly from system to system. On the other hand, another type of cataclysmic variable, the **Type Ia supernova**, varies relatively little and in a systematic way. This means that it is possible to use these exploding stars as calibrated luminosity sources ("standardizable candles"), allowing astronomers to establish the distances to the systems in which they are found.[25]

### Observations

Type Ia supernovae are remarkably consistent in their energy output; at maximum light most Type Ia's reach an average maximum in the blue and visual wavelength bands of

$$\langle M_B \rangle \simeq \langle M_V \rangle \simeq -19.3 \pm 0.03,$$

with a typical spread of less than about 0.3 magnitudes. As can be seen in Fig. 18.21, a clear relationship exists between the peak brightness and the rate of decline in the light curve (the brightest Type Ia's decline the slowest), making it possible to accurately determine the maximum luminosity of an individual Type Ia by measuring the rate of decline. Knowing the luminosity (or absolute magnitude), we can compute the distance to the supernova. Given their tremendous brightness, Type Ia supernovae serve as critically important tools

---

[24]The electric field vector of linearly polarized light oscillates in a single plane, whereas for circularly polarized light this plane of polarization rotates about the direction of travel.

[25]Core-collapse supernovae (Types Ib, Ic, and II) were discussed in detail in Chapter 15.

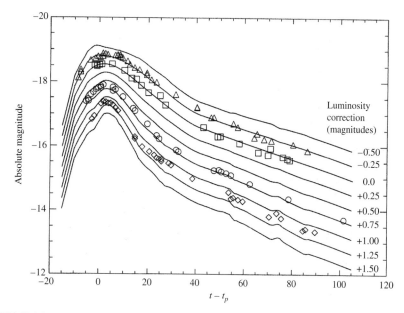

**FIGURE 18.21**   The rate of decline in the light curve of a Type Ia supernova is inversely correlated with the maximum brightness of the light curve. (Figure adapted from Riess, Press, and Kirshner, *Ap. J.*, *438*, L17, 1995.)

for measuring the distances to the galaxies in which they reside. This in turn means that astronomers can probe the structure of the universe to great distances. In fact, Type Ia supernovae played a crucial role in demonstrating that the expansion of the universe is actually accelerating today, 13.7 billion years after the Big Bang, and that nearly three-fourths of the universe consists of **dark energy**.

As was mentioned in Section 15.2, Type Ia supernovae do not exhibit hydrogen lines in their spectra and, instead, show the strong presence of Si II lines, along with neutral and ionized lines of O, Mg, S, Ca and Fe (Fig. 15.6). Given that hydrogen is the most abundant element in the universe, the absence of hydrogen indicates that Type Ia supernovae are evolved objects that have either lost their hydrogen or had it converted to heavier elements, or both. The spectral lines also show P-Cygni profiles, representative of mass loss. In addition, the blueshifted absorption features indicate expansion velocities of the ejecta of $\gtrsim 10^4$ km s$^{-1}$ ($\sim 0.1c$).

### Models of Type Ia Supernovae

Given the remarkable consistency of Type Ia light curves and spectra, it appears that a fairly uniform mechanism must be responsible for these extremely energetic events.

The standard model for Type Ia supernovae assumed by astronomers today is that these events are due to the destruction of a white dwarf star in a binary system. If sufficient mass falls onto the white dwarf, its mass can be driven to near the Chandrasekhar limit, producing a catastrophic explosion. Still unclear at the time this text was written is the exact mechanism (or mechanisms) that trigger the explosion.

Two general scenarios have been proposed. In one scenario, known as **double-degenerate** models, two white dwarf stars exist in a binary orbit. One of the most dramatic predictions of Einstein's general theory of relativity is the existence of **gravitational waves** (or **gravitational radiation**). According to general relativity, mass acts on spacetime, telling it how to curve. If the distribution of a system's mass varies, the resulting changes in the surrounding spacetime curvature may propagate outward as a gravitational wave, carrying energy and angular momentum away from the system. (If the collapse of a star is spherically symmetric, it will not produce gravitational waves; there must be a departure from spherical symmetry.) When applied to a close binary system, general relativity shows that the emission of gravitational radiation will cause the stars to spiral together. If the orbital period is under about 14 hours, the loss of energy via gravitational waves governs the subsequent evolution of a system with solar-mass components. For example, as a white dwarf and a neutron star spiral closer together, the white dwarf may break up and donate some of its mass and angular momentum to its companion. The result could be an isolated millisecond pulsar. A system of two neutron stars, known as the Hulse–Taylor pulsar, has confirmed this prediction of general relativity to incredibly high precision; see the extended discussion beginning on page 703.

In the case where two white dwarf stars are spiraling together, the less massive star (which has the larger radius) will eventually spill over its Roche lobe and be completely torn apart in just a few orbits. The resulting thick disk dumps its C–O-rich material onto the more massive primary. As the mass of the primary grows and nears the Chandrasekhar limit, nuclear reactions begin in the deep interior, eventually destroying the primary white dwarf (this scenario was illustrated in Fig. 18.11).

Double-degenerate models appear to predict about the right number of mergers, consistent with the observed Type Ia supernova rate in galaxies, and they naturally account for the lack of hydrogen in the spectra of Type Ia's. However, computer simulations of nuclear burning suggest that the ignition may be off-center, resulting in ultimate collapse to a neutron star, rather than complete disruption of the white dwarf as a supernova. In addition, it appears that the production of heavy elements may be inconsistent with the relative abundances observed in supernova spectra.

The other general scenario, known as **single-degenerate** models, involves an evolving star in orbit about a white dwarf, much like the models of dwarf novae and novae. However, in this case, the mass falling onto the white dwarf results in complete destruction of the white dwarf in a Type Ia supernova. To date, this set of models is generally favored, but the details of the eruption are still unclear.

One version of single-degenerate models suggests that as the material from the secondary falls onto the primary, the helium in the gas will settle on top of the C–O white dwarf, becoming degenerate. When enough helium has accumulated, a helium flash will occur. Not only will this cause the helium to burn to carbon and oxygen, but it will also send a shock wave downward into the degenerate C–O white dwarf, causing ignition of the degenerate carbon and oxygen.

A second version of the single-degenerate models doesn't invoke degenerate helium burning on the surface but simply has carbon and oxygen igniting in the interior of the white dwarf as the star nears the Chandrasekhar limit, at which point the degenerate gas is no longer able to support the mass of the star. As the star approaches the fatal limit, two-

and three-dimensional simulations suggest that multiple, independent ignition points may occur deep within the core, resulting in nonspherical events.

What happens next is also a matter of significant debate and ongoing research. It is as yet unclear if the resulting burning front of carbon and oxygen occurs at subsonic speeds (known as a **deflagration** event) or if the front accelerates and steepens to become a supersonic burning front (known as a **detonation**, or a true explosion). Precisely how the burning front advances affects the details of the resulting light curve (the maximum luminosity and rate of decay following maximum) as well as the relative abundances of the elements produced as observed in the spectrum. Of course, the question of deflagration versus detonation applies to successful double-degenerate models as well.

In all versions of the single-degenerate scenario, one of the general challenges has been to have just the right rate of accretion from the secondary. If the accretion rate isn't appropriately fine-tuned, the result could be a dwarf nova or a classical nova.

It is possible that both double- and single-degenerate mechanisms may be at work in nature. It is also possible that some single-degenerate events (if they occur) may invoke helium flashes while others may simply ignite carbon and oxygen in the interior without the helium trigger. Perhaps even deflagration and detonation events occur. In any case, the consistency of the light curves ultimately arises from the eruption of a C–O white dwarf near 1.4 $M_\odot$. The variations may arise from slight variations in mass and/or variations in mechanisms.

Much work remains to be done in understanding Type Ia supernovae, which are so critically important to so many aspects of modern astrophysics.

## 18.6 ■ NEUTRON STARS AND BLACK HOLES IN BINARIES

If one of the stars in a close binary system is sufficiently massive that it explodes as a core-collapse supernova, the result may be either a neutron star or a black hole orbiting the companion star. In a semidetached system, hot gas can then spill through the inner Lagrangian point from the distended atmosphere of the companion star onto the compact object. A variety of intriguing phenomena are powered by the energy released when the gas falls down the deep gravitational potential well onto the compact object. As will be seen shortly, many of these systems emit copious quantities of X-rays. In fact, these **binary X-ray systems** shine most strongly in the X-ray region of the electromagnetic spectrum. Other systems may consist of *two* compact objects, such as the binary pulsars.

### Formation of Binaries with Neutron Stars or Black Holes

Whether or not a binary system survives the supernova explosion of one of its component stars depends on the amount of mass ejected from the system.[26] Consider a system initially containing two stars of mass $M_1$ and $M_2$, separated by a distance $a$, that are in circular orbits about their common center of mass. Using Eq. (2.35), we find that the total energy of the

---

[26]We have also seen that asymmetric jets during the formation of a neutron star may give the neutron star a violent kick, which would disrupt the system; recall Section 16.7.

system is

$$E_i = \frac{1}{2}M_1v_1^2 + \frac{1}{2}M_2v_2^2 - G\frac{M_1M_2}{a} = -G\frac{M_1M_2}{2a}. \qquad (18.33)$$

The speeds of the two stars are related by Eq. (7.4), $M_1v_1 = M_2v_2$. Now suppose that Star 1 explodes as a core-collapse supernova, leaving a remnant of mass $M_R$. For a spherically symmetric explosion, there is no change in the velocity of Star 1. Before the spherical shell of ejecta reaches Star 2, its mass acts gravitationally as though it were still on Star 1 (recall Example 2.2.1). So far, the supernova has had no effect on the binary. However, as soon as the shell has swept beyond Star 2, the gravitational influence of the ejecta is no longer detectable.

Thus the main consequence of the supernova on the orbital dynamics of the binary system arises from the ejection of mass, the removal of some of the gravitational glue that was binding the stars together.[27] Since the velocity of Star 2 is initially unchanged and the separation of the two stars remains the same, the total energy of the system after the explosion is now

$$E_f = \frac{1}{2}M_Rv_1^2 + \frac{1}{2}M_2v_2^2 - G\frac{M_RM_2}{a}. \qquad (18.34)$$

If the explosion results in an *unbound* system, then $E_f \geq 0$. It is left as an exercise to show that the mass of the remnant must satisfy

$$\frac{M_R}{M_1 + M_2} \leq \frac{1}{(2 + M_2/M_1)(1 + M_2/M_1)} < \frac{1}{2} \qquad (18.35)$$

for an unbound system. That is, *at least* one-half of the total mass of the binary system must be ejected if the supernova explosion of Star 1 is to disrupt the system. If one-half or more of the system's mass is retained, the result will be a neutron star or a black hole gravitationally bound to a companion star. For a massive companion star ($M_2 \gg M_1$), this is a likely result.

### Capturing Isolated Neutron Stars

It is possible that isolated neutron stars, formed by core-collapse supernovae, may be gravitationally captured during a chance encounter with another star. Because the total energy of two unbound stars is initially greater than zero, some of the excess kinetic energy must be removed for a capture to occur.

If the proximity of the two objects raises a tidal bulge on the nondegenerate star, energy may be dissipated by the damping mechanisms discussed in Section 14.2 for pulsating stars. The outcome of such a **tidal capture** depends on the nearness of the passage and the type of star involved. If the neutron star passes between about 1 to 3 times the radius of the other star, the resulting binary system will have a period ranging from several hours (with a main-sequence star) to several days (with a giant).

---

[27]The direct impact of the supernova blast on the companion star has been neglected, although this too will contribute to disrupting the system.

This tidal capture process is most effective in regions that are extremely densely populated with stars, such as the centers of globular clusters (Section 13.3). It is estimated that in a compact globular cluster, tidal capture could produce up to about ten close binary systems containing a neutron star over a period of some $10^{10}$ years. This is consistent with the number of X-ray sources observed in globular clusters. (The estimated lifetime of a binary X-ray system is on the order of $10^9$ years, so only the most compact globular clusters would be expected to harbor even one X-ray source at a given time; see Problem 18.20.)

An alternative capture mechanism involves three (or more) stars. One of the stars would be gravitationally flung from the system, removing energy and so allowing the capture to take place.

Yet another possibility was envisioned by by Kip Thorne and Anna Żytkow of Caltech in 1977. Although a direct hit would destroy a main-sequence star, the penetration of a neutron star into a giant star would bring it close to the star's degenerate core. The result could be a neutron star orbiting inside the giant star: a system that is known as a **Thorne–Żytkow object**. It is thought that the envelope of the giant star would be quickly expelled, producing a neutron star–white dwarf binary with an orbital period of about 10 minutes. (To date, these objects remain hypothetical.)

## Binary X-Ray Pulsars

Close binary systems containing neutron stars were first identified by their energetic emission of X-rays. The first source of X-rays beyond the Solar System was discovered in 1962 in the constellation Scorpius by a Geiger counter arcing above Earth's atmosphere in a sounding rocket. (X-rays cannot penetrate the atmosphere, so detectors and telescopes designed for X-ray wavelengths must make their observations from space.[28]) This object, called Sco X-1, is now known to be a **binary X-ray pulsar** (also called simply an X-ray pulsar). The periodic eclipse of another X-ray pulsar, Cen X-3 in the constellation Centaurus, revealed its binary nature. [It is important to note that the Crab pulsar also emits X-rays (along with a small number of other isolated pulsars), but the Crab is primarily a radio pulsar that radiates in every region of the electromagnetic spectrum.]

X-ray pulsars are powered by the gravitational potential energy released by accreting matter. Recall from Example 18.1.2 that when mass falls from a great distance to the surface of a neutron star, about 20% of its rest energy is released, an amount that far exceeds the fraction of a percent that would be produced by fusion. The observed X-ray luminosities range up to $10^{31}$ W [the Eddington limit; see Eq. (10.6)]. For a neutron star with a radius of 10 km, the Stefan–Boltzmann equation in the form of Eq. (3.17) shows that the temperature associated with this luminosity is about $2 \times 10^7$ K. According to Wien's law, Eq. (3.19), the spectrum of a blackbody with this temperature would peak at an X-ray wavelength of about 0.15 nm.

X-ray pulsars also emit radio wavelength energy, just like isolated pulsars. However, radio wavelength emissions are easily quelched by the accretion disk in the binary system, and so the radio emissions are not as prominant as they are for isolated pulsars.

---

[28]The first X-ray detector was designed to look for X-rays from the lunar surface, produced when solar wind particles cause the lunar soil to fluoresce. The presence of enormously stronger cosmic X-ray sources came as a surprise to astronomers at the time.

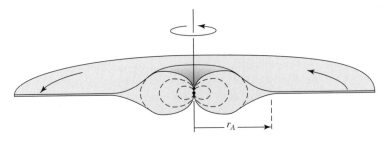

**FIGURE 18.22**   Accreting gas channeled onto a neutron star's magnetic poles, where $r \approx r_A$.

You should recall from Section 16.6 that neutron stars are often accompanied by powerful magnetic fields. In fact, these fields may be sufficiently strong to prevent the accreting matter from even reaching the star's surface. The strength of the neutron star's magnetic dipole field is proportional to $1/r^3$, so the plunging gases encounter a rapidly increasing field. When the magnetic energy density $u_m = B^2/2\mu_0$ (Eq. 11.9) becomes comparable to the kinetic energy density $u_K = \frac{1}{2}\rho v^2$, the magnetic field will channel the infalling ionized gases toward the poles of the neutron star; see Fig. 18.22. This occurs at a distance from the star known as the **Alfvén radius**, $r_A$, where

$$\frac{1}{2}\rho v^2 = \frac{B^2}{2\mu_0}. \tag{18.36}$$

For the special case of spherically symmetric accretion, with the gases starting at rest at a great distance, energy conservation implies that the free-fall velocity is $v = \sqrt{2GM/r}$ for a star of mass $M$. Furthermore, the density and velocity are related to the mass accretion rate, $\dot{M}$, by Eq. (11.3),

$$\dot{M} = 4\pi r^2 \rho v, \tag{18.37}$$

and the radial dependence of the magnetic dipole field strength may be expressed as

$$B(r) = B_s \left( \frac{R}{r} \right)^3, \tag{18.38}$$

where $B_s$ is the surface value of the magnetic field. Inserting these expressions into Eq. (18.36) and solving for the Alfvén radius, we obtain

$$r_A = \left( \frac{8\pi^2 B_s^4 R^{12}}{\mu_0^2 GM\dot{M}^2} \right)^{1/7} \tag{18.39}$$

(the proof is left as an exercise). Of course, the accretion will not actually be spherically symmetric. However, the magnetic field increases so rapidly as the falling matter approaches the star that a more realistic calculation yields nearly the same result: The flow will be disrupted at a *disruption radius* $r_d$,

$$r_d = \alpha r_A, \tag{18.40}$$

with $\alpha \sim 0.5$.

**Example 18.6.1.** Before considering the details of channeled accretion onto a neutron star, let's look at the case of accretion onto the white dwarf considered in Example 18.2.1, for which $M = 0.85$ $M_\odot$, $R = 0.0095$ $R_\odot = 6.6 \times 10^6$ m, and $\dot{M} = 10^{13}$ kg s$^{-1}$ ($1.6 \times 10^{-10}$ $M_\odot$ yr$^{-1}$). Assume that its magnetic field has a surface strength of $B_s = 1000$ T, about 100 times stronger than the typical value for a white dwarf. Then, from Eq. (18.39), the Alfvén radius is

$$r_A = 6.07 \times 10^8 \text{ m.}$$

This is comparable to the separation of the stars in a cataclysmic variable (see Example 18.4.1), so an accretion disk cannot form around a white dwarf with an extremely strong magnetic field. Instead, the mass spilling through the inner Lagrangian point is confined to a stream that narrows as it is magnetically directed toward one (or both) of the poles of the white dwarf. In the absence of an accretion disk, all of the accretion energy will be delivered to the pole(s) of the star, with an accretion luminosity of (Eq. 18.24)

$$L_{\text{acc}} = G\frac{M\dot{M}}{R} = 1.71 \times 10^{26} \text{ W.}$$

### A Polar Analog in a Neutron Star System

**Example 18.6.2.** Consider the case of accretion onto the neutron star described in Example 18.2.1. For this star, $M = 1.4$ $M_\odot$, $R = 10$ km, and $\dot{M} = 10^{14}$ kg s$^{-1}$ ($1.6 \times 10^{-9}$ $M_\odot$ yr$^{-1}$). Furthermore, take the value of the magnetic field at the neutron star's surface to be $B_s = 10^8$ T. The value of the Alfvén radius is then given by Eq. (18.39),

$$r_A = 3.09 \times 10^6 \text{ m.}$$

Although 300 times the radius of the neutron star itself, this is much less than the value of $r_{\text{circ}}$ (Eq. 18.25) that describes the extent of an accretion disk. Thus an accretion disk will form around the neutron star but will be disrupted near the neutron star's surface as shown in Fig. 18.22 (unless the magnetic field is quite weak, roughly $< 10^4$ T). As the accreting gas is funneled onto one of the magnetic poles of the neutron star, it forms an accretion column similar to the one described for polars. In this case, however, the accretion luminosity (Eq. 18.24) is four orders of magnitude greater,

$$L_{\text{acc}} = G\frac{M\dot{M}}{R} = 1.86 \times 10^{30} \text{ W,}$$

close to the Eddington limit of $\sim 10^{31}$ W; see Eq. (10.6). As $L_{\text{acc}}$ approaches $L_{\text{Ed}}$, radiation pressure elevates the shock front to heights reaching $r \sim 2R$. As a result, X-rays are emitted over a large solid angle.

### Eclipsing, Binary X-Ray Pulsar Systems

If the neutron star's magnetic and rotation axes are not aligned, as in Fig. 16.25, the X-ray-emitting region may be eclipsed periodically, and the result is an **eclipsing, binary X-ray pulsar**. Figure 18.23 shows the signal received from Hercules X-1, which exhibits a pulse of X-rays every 1.245 s (the rotation period of the neutron star). Note that the broad pulse (due to the large solid angle of the emission) may occupy ∼ 50% of the pulse period, compared to the sharper radio pulses shown in Fig. 16.19, which take up only 1% to 5% of the pulse period. To date, about 20 binary X-ray pulsars have been found, with periods ranging from 0.15 s to 853 s. As noted on page 590, white dwarfs cannot rotate as rapidly as the lower end of this period range without breaking up. This is one indication that X-ray pulsars are indeed accreting neutron stars.

Further confirmation that most X-ray pulsars are accreting neutron stars comes from the observation that the periods of these objects are slowly decreasing. As time passes, they spin *faster*.[29] The time derivative of the star's rotation period, $\dot{P} \equiv dP/dt$, is related to the rate of change of its angular momentum, $L = I\omega$, by

$$\frac{dL}{dt} = I\frac{d\omega}{dt} = I\frac{d}{dt}\left(\frac{2\pi}{P}\right) = -2\pi I\frac{\dot{P}}{P^2},$$

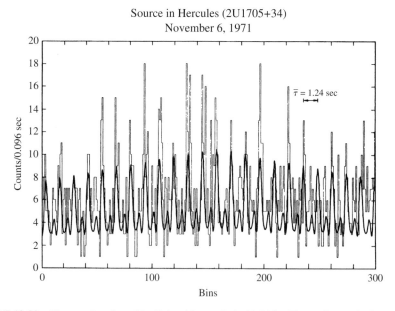

**FIGURE 18.23**   X-ray pulses from Her X-1, with a period of 1.245 s. The peaks are the X-ray counts received from Her X-1 grouped in bins that are 0.096 s wide, and the heavier curve is a fit to the data using sine functions. (Figure adapted from Tananbaum et al., *Ap. J. Lett.*, *174*, L143, 1972.)

---

[29]Recall from Section 16.7 that the periods of radio pulsars *increase* with time as they lose energy due to magnetic dipole radiation.

where $I$ is the moment of inertia of the neutron star. Near the disruption radius, the angular momentum of the gas parcels orbiting in the accretion disk ($L = mvr$) is transferred to the neutron star via magnetic torques. The time derivative of the neutron star's angular momentum is just the rate at which angular momentum arrives at the disruption radius, so at $r = r_d$ we set

$$\frac{dL}{dt} = \dot{M}vr_d,$$

where the orbital velocity at $r = r_d$ is $v = \sqrt{GM/r_d}$ [Eq. (2.33) or (2.34) with $e = 0$ and $a = r_d$ for a circular orbit]. Equating these expressions for $dL/dt$ and using the definitions of the Alfvèn and disruption radii, Eqs. (18.39) and (18.40), results in

$$\frac{\dot{P}}{P} = -\frac{P\sqrt{\alpha}}{2\pi I} \left( \frac{2\sqrt{2}\pi B_s^2 R^6 G^3 M^3 \dot{M}^6}{\mu_0} \right)^{1/7}. \tag{18.41}$$

---

**Example 18.6.3.**   The X-ray pulsar Centaurus X-3 has a period of 4.84 s and an X-ray luminosity of about $L_x = 5 \times 10^{30}$ W. Assuming that it is a 1.4 $M_\odot$ neutron star with a radius of 10 km, its moment of inertia (assuming for simplicity that it is a uniform sphere) is

$$I = \frac{2}{5}MR^2 = 1.11 \times 10^{38} \text{ kg m}^2.$$

Using Eq. (18.24) for the accretion luminosity, we find the mass transfer rate to be

$$\dot{M} = \frac{RL_x}{GM} = 2.69 \times 10^{14} \text{ kg s}^{-1},$$

or $4.27 \times 10^{-9}$ $M_\odot$ yr$^{-1}$. Then, for an assumed magnetic field of $B_s = 10^8$ T and $\alpha = 0.5$, Eq. (18.41) gives the fractional change in the period per second and per year:

$$\frac{\dot{P}}{P} = -2.74 \times 10^{-11} \text{ s}^{-1} = -8.64 \times 10^{-4} \text{ yr}^{-1}.$$

That is, the characteristic time for the period to change is $P/\dot{P} = 1160$ years.

The measured value for Cen X-3 is $\dot{P}/P = -2.8 \times 10^{-4}$ yr$^{-1}$, smaller than our estimate by a factor of 3 but in good agreement with this simple argument. You may verify that if a 0.85 $M_\odot$ white dwarf with a radius of $6.6 \times 10^6$ m and $B_s = 1000$ T is used for the accreting star, rather than a neutron star, then $\dot{P}/P = -1.03 \times 10^{-5}$ yr$^{-1}$. The measured value is larger by a factor of 27. A white dwarf is hundreds of times larger than a neutron star, so it has a much larger moment of inertia and is more difficult to spin up. The substantially better agreement between the neutron star model and the observations obtained for these systems is compelling evidence that neutron stars are the accreting objects in binary X-ray pulsars.

---

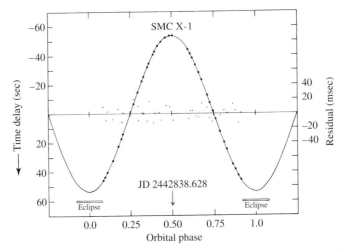

**FIGURE 18.24** Measured pulse arrival times (dots) for the binary X-ray pulsar SMC X-1 as a function of its orbital phase. The curve is for the best-fit circular orbit, and the dots about the straight line show the residuals from the best-fit orbit. (Figure adapted from Primini et al., *Ap. J.*, *217*, 543, 1977.)

As an X-ray pulsar orbits its binary companion, the distance from the pulsar to Earth constantly changes. This results in a cyclic variation in the measured pulse period that is analogous to the Doppler shift of a spectral line observed for a spectroscopic binary (see Section 7.3). Figure 18.24 shows the shift in pulse arrival times as a function of the orbital phase for the X-ray pulsar SMC X-1 in the Small Magellanic Cloud. The orbit for this system is almost perfectly circular, with a radius of 53.5 light-seconds $= 0.107$ AU, less than one-third the size of Mercury's orbit around the Sun.

A complete description of the binary system has been obtained for a small number of *eclipsing* X-ray pulsars with visible companions. Such systems are analogous to double-line, eclipsing, spectroscopic binaries. For example, in the SMC X-1 system the mass of the secondary star is 17.0 $M_\odot$ (with an uncertainty of about 4 $M_\odot$), and its radius is 16.5 $R_\odot$ ($\pm$ 4 $R_\odot$). The masses of the neutron stars have also been determined for these systems. The results are consistent with a neutron star mass of 1.4 $M_\odot$ ($\pm$ 0.2 $M_\odot$), in good agreement with the Chandrasekhar limit.

### X-Ray Bursters

If the magnetic field of the neutron star is too weak ($\ll 10^8$ T) to completely disrupt the accretion disk and funnel the accreting matter onto its magnetic poles, these gases will settle over the surface of the star. Without an accretion column to produce a hot spot, X-ray pulses cannot be produced by the rotation of the neutron star. Instead, calculations indicate that when a layer of hydrogen a few meters thick accumulates on the surface, a shell of hydrogen slowly begins burning about a meter below the surface, with a shell of helium burning ignited another meter below that; see Fig. 18.25.[30] This fusion of helium is explosive and releases

---

[30]This mechanism is reminiscent of the helium shell flashes discussed in Section 13.2.

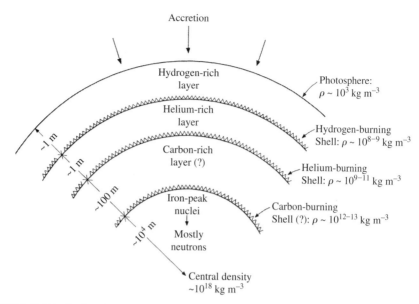

**FIGURE 18.25** Surface layers on an accreting neutron star. (Figure adapted from Joss, *Comments Astrophys.*, 8, 109, 1979.)

a total of $\sim 10^{32}$ J in just a few seconds, with the surface reaching a temperature of about $3 \times 10^7$ K (twice the Sun's central temperature). The resulting blackbody spectrum peaks at X-ray wavelengths, and a flood of X-rays is liberated by this **X-ray burster**. Some of the X-rays may be absorbed by the accretion disk and re-emitted as visible light, so an optical flash is sometimes seen a few seconds after the X-ray burst. As the burst luminosity declines in a matter of seconds, the spectrum matches that of a cooling blackbody with a radius of $\sim 10$ km, consistent with the presence of a neutron star. After a time that can vary from a few hours to a day or more, another layer of hydrogen accumulates and another X-ray burst is triggered.[31] More than 50 X-ray bursters have been found so far. Most are concentrated near the Galactic plane, toward the center of our Galaxy, with some 20% located in old globular clusters.

**Low-Mass and Massive X-Ray Binaries**

From these and other results, astronomers have identified two classes of binary X-ray systems. The more common type consists of those with low-mass secondary stars (late spectral-type stars with $M_2 \leq 2\,M_\odot$). These systems belong to the **low-mass X-ray binaries** (LMXBs). LMXBs produce X-ray bursts rather than pulses, indicating that the neutron star's magnetic field is relatively weak. Because low-mass stars are small, the two stars must orbit more closely if mass is to be transferred from one star to the other. For this reason, the LMXBs have short orbital periods, from 33.5 days down to 11.4 minutes. About

---

[31]It is thought that the gases accreting on X-ray pulsars are constantly undergoing fusion. However, recall from Example 18.1.2 that the energy released in the accretion column will be about 30 times larger, so the energy from fusion will be lost in the glare of the accretion energy.

one-quarter of these systems are found within globular clusters, where the high number density of stars makes the gravitational capture of a neutron star more likely. The neutron stars in LMXBs may also have been formed by the accretion-induced collapse of a white dwarf.

Systems with higher-mass secondaries are referred to as **massive X-ray binaries** (MXRBs). About half of the approximately 130 known MXRBs are X-ray pulsars. With giant or supergiant O and B stars available to fill their Roche lobes, the separation of the stars can be larger and the orbital periods correspondingly longer, from 0.2 days up to 580 days. Even if the secondary star's envelope does not overflow its Roche lobe, the vigorous stellar winds of these stars may still provide the mass transfer rate needed to sustain the production of X-rays. The MXRBs are found near the plane of our Galaxy, where there are young massive stars and ongoing star formation. This is consistent with the idea that an MXRB is the product of the normal evolution of a binary system containing a massive star that survived the supernova explosion of its companion.

So far, only neutron stars have been considered as the accreting object in binary X-ray systems. However, the gravitational potential well is even deeper for matter falling toward a black hole. In this case, up to about 30% of the rest energy of the falling disk material may be emitted as X-rays. In fact, as was discussed in Section 17.3, these systems provide the best evidence for the existence of stellar-mass black holes. The gas spilling through the inner Lagrangian point is heated to millions of kelvins as it spirals down through the black hole's accretion disk (see Fig. 17.24) and so emits X-rays. The identification of a black hole rests on determining that the mass of a compact, X-ray-emitting object exceeds the approximately 3 $M_\odot$ upper limit for the mass of a rapidly rotating neutron star. Thus the procedure for detecting a black hole in a binary X-ray system is similar to that used to measure the masses of neutron stars in these systems.

At present, there are only a handful of X-ray binaries that allow such a dynamical determination of the masses involved. The best cases at the time of this writing are A0620−00, V404 Cygni, Cygnus X-1, and LMC X-3. Since none of these systems exhibit eclipses, the resulting uncertainty about their orbital inclinations means that the masses calculated are lower limits (see Section 7.3). A0620−00 is an **X-ray nova**, powered by the sporadic accretion of material from its companion, a K5 main-sequence star. The relative faintness of the secondary star allows the measurement of the radial velocity of *both* the accretion disk and the companion star. The identification of A0620−00 as a 3.82 ± 0.24 $M_\odot$ black hole seems secure, as was described in Section 17.3 and Problem 17.24. V404 Cyg is also an X-ray nova, where recent measurements persuasively document the presence of a 12 $M_\odot$ black hole (Section 17.3).

The arguments for the other two systems, although strong, are not as conclusive. Neither has a fully developed accretion disk, and so the velocities of both members cannot be determined. Cygnus X-1, perhaps the best-known black hole candidate, is a bright MXRB. Because almost all of the light comes from the secondary, Cyg X-1 is essentially a single-line spectroscopic binary. The identification of Cyg X-1 as a black hole therefore depends on the identification of the secondary star (HDE 226868) as a O9.7 Iab supergiant with a mass of 17.8 $M_\odot$. The most likely result, making reasonable assumptions about this binary system, is that the mass of the compact object in Cyg X-1 is 10.1 $M_\odot$. Even the worst-case argument results in a secure lower limit of 3.4 $M_\odot$, providing the evidence that Cyg X-1 is a black hole.

The secondary star in the LMC X-3 system is a B3 main-sequence star that is orbiting an unseen, more massive companion. Although the lower limit on the mass of the compact companion is 3 $M_\odot$, a more probable mass range is 4–9 $M_\odot$—again, solid evidence for a black hole. Other X-ray binary systems may contain black holes, such as Nova Mus 1991 in the southern constellation Musca (the Fly), LMC X-3 in the Large Magellanic Cloud, and CAL 87 (in the direction of the Large Magellanic Cloud), but the evidence in these cases is not yet as strong.

## SS 433

One more X-ray binary and possible black hole candidate should be mentioned: SS 433, one of the most bizarre objects known to astronomers.[32] In 1978, it was discovered that this object displays *three* sets of emission lines. One set of spectral lines was greatly blueshifted, another set was greatly redshifted, and a third set lacked a significant Doppler shift. Here was an object with three components: Two were approaching and receding, respectively, at one-quarter the speed of light while the third stayed nearly still! The wavelengths of the shifted lines vary with a period of 164 days, while the wavelengths of the nearly stationary lines show a smaller shift with a 13.1-day period. Furthermore, the position of SS 433 lies at the center of a diffuse, elongated shell of gas known as W50, which is probably a supernova remnant.

The 13.1-day period of SS 433 describes the orbit of a compact object (most probably a neutron star, but perhaps a black hole) around the primary. The primary is thought to be a 10–20 $M_\odot$ early-type star with a stellar wind that produces the broad stationary emission lines.[33] Surrounding the compact object is an accretion disk that contributes to the visible light from the system equally with the secondary. A tidal interaction between the disk and the two stars could be responsible for a precessional wobble of the disk that has a period of 164 days, analogous to Earth's 25,770-year precessional wobble discussed in Section 1.3. There is broad agreement that the varying Doppler-shifted emission lines, shown in Fig. 18.26, come from two *relativistic jets* that expel particles at 0.26$c$ in opposite directions along the axis of the disk. The jets are probably powered by the accretion of matter at a rate exceeding the Eddington limit, generating X-rays at a prodigious rate. This could produce a radiation pressure sufficient to expel a portion of the accreting gases at relativistic speeds in the direction of least resistance—perpendicular to the disk. As the disk precesses, two oppositely directed jets sweep out a cone in space every 164 days, resulting in cyclic variations in both the radial velocity of the jets and the observed Doppler shift. The collimation of the jets could be the result of the ionized gases moving along magnetic field lines. The axis of the precessional cone makes an angle of 79° with the line of sight; the cone's axis is also closely aligned with the long axis of the probable supernova remnant, W50. In fact, there are two regions that have been observed to emit X-rays, presumably

---

[32]"SS" stands for the catalog of peculiar emission-line stars compiled by Bruce Stephenson and Nicholas Sanduleak.

[33]For example, one recent measurement of SS 433 favors a 0.8 $M_\odot$ neutron star orbiting a 3.2 $M_\odot$ companion. Some astronomers have suggested that the primary may be a Wolf–Rayet star (Section 15.3) to account for the broad stationary emission lines. Although a substantial percentage of Wolf–Rayet stars are found in binaries, these stars' own energetic winds, rather than the transfer of mass in a close binary system, seem to be responsible for removing most of their hydrogen envelopes.

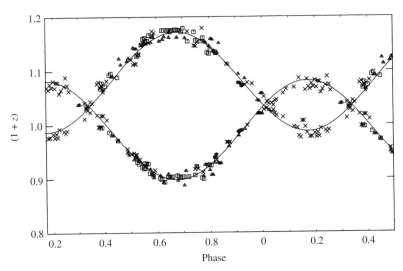

**FIGURE 18.26**    Doppler shifts measured for the emission lines in SS 433. From Eq. (4.38), $z = 0.1$ and $z = 0.2$ correspond to speeds of 28,500 km s$^{-1}$ and 54,100 km s$^{-1}$, respectively. (Figure adapted from Margon, Grandi, and Downes, *Ap. J.*, *241*, 306, 1980.)

where the jets collide with the remnant's gases and heat them to about $10^7$ K. Figure 18.27 shows the general features of this incredible system.

### The Fate of Binary X-Ray Systems

What is the fate of a binary X-ray system? As it reaches the endpoint of its evolution, the secondary star will end up as a white dwarf, neutron star, or black hole. The effect on the system depends on the mass of the secondary star. In low-mass systems (LMXBs), the companion star will become a white dwarf without disturbing the circular orbit of the system. On the other hand, the higher-mass secondary in a MXRB may explode as a supernova. If more than half of the system's mass is retained (Eq. 18.35), a pair of neutron stars will circle each other in orbits that probably have been elongated by the blast. Otherwise, the supernova may disrupt the system and hurl the solitary neutron stars into space. This is consistent with observations that pulsars (like MXRBs) are concentrated near the plane of our Galaxy and may have high space velocities that can exceed 1000 km s$^{-1}$.

### Millisecond Radio Pulsars

The principal way in which a binary system containing two neutron stars can be detected is if at least one of them is a pulsar. Astronomers therefore search for cyclic variations in the measured periods of radio pulsars, analogous to the effect described here for the X-ray pulsars. Although half of all stars in the sky are actually multiple systems, *none* of the first one hundred pulsars discovered belonged to a binary. The first binary pulsar, PSR 1913+16, was discovered in 1974 by American astronomers Russell Hulse and Joseph Taylor, using the Arecibo radio telescope. The search strategy for binary pulsars changed with the 1982

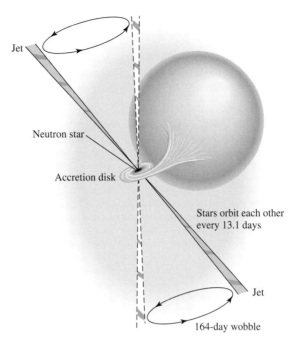

**FIGURE 18.27**   SS 433. The axis of the cone swept out by the precessing jets makes a 79° angle with the line of sight.

discovery by Donald Backer of UC Berkeley and his colleagues of the then-fastest known pulsar, PSR 1937+214. With a period of 1.558 milliseconds, this pulsar spins 642 times each second.[34] Although this astounding rotation rate seemed to indicate a young pulsar, the very small value of the period derivative ($\dot{P} = 1.051054 \times 10^{-19}$) implies a weak magnetic field ($\approx 8.6 \times 10^4$ T; see Example 16.7.2) and a very old pulsar. From Problem 16.22, the age of the pulsar may be estimated as $P/2\dot{P} = 235$ million years, an order of magnitude older than previously discovered pulsars.[35] Although PSR 1937+214 is an isolated pulsar, the paradox of the oldest pulsar also being among the fastest quickly brought astronomers to a surprising conclusion: PSR 1937+214 must once have been a member of a low-mass X-ray binary system. (Recall that, like PSR 1937+214, LMXBs have weak magnetic fields.) Accretion from the secondary star could have spun up the neutron star to its present rapid rate. The neutron star's magnetic field may also have been rejuvenated by this process, although the details of how this might occur are not yet clear.

A likely evolutionary picture has emerged that brings the observations of binary X-ray sources and binary pulsars together. In this scenario, there are two classes of binary pulsars. Those with high-mass companions (neutron stars) have shorter periods and eccentric orbits and probably result from the evolution of a massive X-ray binary system. (Recall that an

---

[34]Middle C on a piano has an audible frequency of 262 Hz. The pulsar's rotation frequency is more than an octave higher, between D[#] and E!

[35]$P/2\dot{P}$ is an estimate of a pulsar's age only if the pulsar's spin has not been affected by accretion.

MXRB that managed to retain more than half of its mass following the supernova of the companion star would produce such a pair of neutron stars with elongated orbits.) The other class of binary pulsars are characterized by low-mass companions (white dwarfs), longer orbital periods, and circular orbits. These are probably the descendants of low-mass X-ray binary systems.

Because LMXBs are common in globular clusters, radio astronomers slued their telescopes toward these targets and discovered more binary and millisecond pulsars (those with periods less than approximately 10 ms). Numerous surveys, including one conducted by the Chandra X-Ray Observatory, suggest that 47 Tuc may have more than 300 neutron stars, approximately 25 of which are millisecond pulsars. The mounting statistics make it clear that most of the globular cluster pulsars are members of binaries and that most (but not all) are millisecond pulsars. (Conversely, most of the known millisecond pulsars have been found in globular clusters.) If these pulsars are the evolutionary product of LMXBs, then how can the absence of a white dwarf companion be explained for a significant minority of them?

## Black Widow Pulsars

An answer may be found from observations of PSR 1957+20. It is a rarity: a binary millisecond pulsar that eclipses its companion, a meager 0.025 $M_\odot$ white dwarf. However, the eclipses last for some 10% of the orbit, implying that the light is blocked by an object larger than the Sun. Significantly, the dispersion of the pulsar signal (see page 595) increases just before and after the eclipse, indicating that the white dwarf is surrounded by ionized gas. The pulsar seems to be *evaporating* its white dwarf companion with its energetic beam of photons and charged particles. Within a few million years, the white dwarf may disappear, devoured by this **black widow pulsar**; see Fig. 18.28.

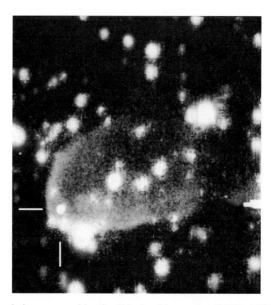

**FIGURE 18.28**    Gas being removed by the "black widow pulsar," PSR 1957+20. The pulsar is at the intersection of the white lines. (Photo courtesy of S. Kulkarni and J. Hester, Caltech.)

Another example of the ablation of an eclipsing millisecond pulsar's companion has been found for PSR 1744−24A in the globular cluster Terzan 5, where the eclipses last for half of the orbital period. It is possible that some of the evaporated material may form a disk of gas and dust around the pulsar that could eventually (after a million years or so) condense and form planets around the pulsar. Or, if the evaporation of the companion star is incomplete, a planet-size remnant could be left orbiting the pulsar.

Mechanisms such as these may be responsible for the three planets thought to be traveling in circular orbits around PSR 1257+12, some 500 pc away in the constellation Virgo. As determined from a careful analysis of pulse arrival times, the innermost planet has a mass of $0.015\ M_\oplus$ that is 0.19 AU from the pulsar, followed by a $3.4\ M_\oplus$ object that is at a distance of 0.36 AU. The outermost planet's mass is $2.8\ M_\oplus$, and it is at a distance of 0.47 AU.

As more millisecond pulsars are discovered, it should become clear whether the foregoing evolutionary picture is correct.

### Double Neutron Star Binaries

A small number of detached binary systems are known to exist in which both both members are neutron stars. As highly relativistic systems with no current mass exchange between the system members, these **double neutron star binaries** are exquisite natural laboratories for the testing of predictions of the General Theory of Relativity.

The first such system discovered is the **Hulse–Taylor pulsar**, PSR 1913+16, with an orbital separation just a little larger than the Sun's diameter. A 30-year study of this system has confirmed the existence of gravitational waves.[36]

Nearly everything is known about the Hulse–Taylor system with incredible precision, as can be seen by inspecting the observational data in Table 18.1. Because of this level of precision, this binary system provides an ideal natural laboratory for testing Einstein's theory of gravity. For example, recall from Section 17.1 that as Mercury passes through the

**TABLE 18.1**   Data for the Hulse–Taylor Pulsar, PSR 1913+16.

| Parameter | Value | Uncertainty |
|---|---|---|
| Pulse Frequency ($\omega$) | 16.94053918425292 Hz | $\pm\ 15 \times 10^{-14}$ Hz |
| Pulse Frequency Derivative ($\dot{\omega}$) | $-2.47583 \times 10^{-15}$ Hz s$^{-1}$ | $\pm\ 3 \times 10^{-20}$ Hz s$^{-1}$ |
| Mass (pulsar) | $1.4414\ M_\odot$ | $\pm\ 0.0002\ M_\odot$ |
| Mass (companion) | $1.3867\ M_\odot$ | $\pm\ 0.0002\ M_\odot$ |
| Eccentricity ($e$) | 0.6171338 | $\pm\ 0.0000004$ |
| Period of Orbit ($P_{orb}$) | 0.322997448930 d | $\pm\ 4 \times 10^{-13}$ d |
| Period of Oribt Derivative ($\dot{P}_{orb}$) | $-2.4056 \times 10^{-12}$ | $\pm\ 0.0051 \times 10^{-12}$ |
| Periastron shift ($\dot{\omega}_{orb}$) | $4.226595°$ yr$^{-1}$ | $\pm\ 0.000005°$ yr$^{-1}$ |

Notes:

1. Data for $\omega$ and $\dot{\omega}$ from epoch January 14, 1986;
see http://www.atnf.csiro.au/research/pulsar/psrcat/

2. Remaining data from J. M. Weisberg and J. H. Taylor (2005).

---

[36]Russell Hulse and Joseph Taylor shared the 1993 Nobel Prize for their discovery of PSR 1913+16.

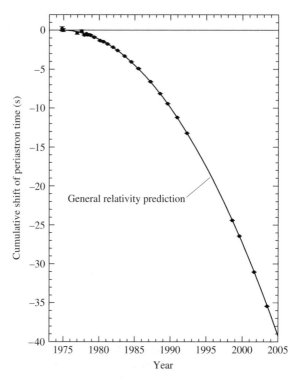

**FIGURE 18.29**   Observations (dots) of the delay in the time of periastron for PSR 1913+16, compared with the prediction of the theory of general relativity (solid line). [Adapted from a figure courtesy of J. M. Weisberg and J. H. Taylor (2005).]

curved spacetime near the Sun, the position of perihelion in its orbit is shifted by $43''$ per century; see Fig. 17.1. For PSR 1913+16, general relativity predicts a similar shift in the point of periastron, where the two neutron stars are nearest each other. The theoretical value is in excellent agreement with the measurement of $4.226595 \pm 0.000005°$ $yr^{-1}$ (35,000 times Mercury's rate of shift). This effect on the orbit is cumulative; with every orbit, the pulsar arrives later and later at the point of periastron. Figure 18.29 shows the incredible agreement between theoretical and observed values of the accumulating time delay.

The most spectacular aspect of the studies of PSR 1913+16 is the confirmation of the existence of gravitational radiation. As the two neutron stars move in their orbits, gravitational waves carry energy away from the system and the orbital period decreases. According to general relativity, the rate at which the orbital period changes as a consequence of the emission of gravitational quadrupole radiation[37] is

$$\dot{P}_{orb} = \frac{d P_{orb}}{dt} = -\frac{96}{5} \frac{G^3 M^2 \mu}{c^5} \left( \frac{4\pi^2}{GM} \right)^{4/3} \frac{f(e)}{P_{orb}^{5/3}}, \qquad (18.42)$$

---

[37]The term *quadrupole* describes the geometry of the emitted gravitational radiation, just as electric dipole radiation describes the electromagnetic radiation emitted by two electric charges moving around each other.

where

$$M = M_1 + M_2$$

$$\mu = \frac{M_1 M_2}{M_1 + M_2}$$

and $f(e)$ describes the effect of the eccentricity of the orbit,

$$f(e) = \left(1 + \frac{73}{24}e^2 + \frac{37}{96}e^4\right)\left(1 - e^2\right)^{-7/2}.$$

(There are also higher-order correction terms that have been neglected here.) Inserting the preceding values for the masses and eccentricity, the theoretical rate of orbital period decay is calculated to be $\dot{P}_{\text{orb,predict}} = -(2.40242 \pm 0.00002) \times 10^{-12}$, which agrees with the measured value of $\dot{P}_{\text{orb,meas}} = -(2.4056 \pm 0.0051) \times 10^{-12}$ to within 0.13%. In presenting the results of an earlier calculation of the orbital period decay in 1984, Joel Weisberg and Joseph Taylor wrote, "It now seems an inescapable conclusion that gravitational radiation exists as predicted by the general relativistic quadrupole formula." Astronomers are fortunate to have caught this superb natural laboratory before it disappears. As the separation of the neutron stars shrinks by about 3 mm per orbit, the system will coalesce some 300 million years in the future.

Another tremendous natural laboratory for testing general relativity was discovered in 2003. This double neutron star system is actually a binary pulsar system. J0737–3039A has a pulse period of $P_A = 0.02269937855615 \pm 6 \times 10^{-14}$ s, and J0737–3039B has a pulse period of $P_B = 2.7734607474 \pm 4 \times 10^{-10}$ s. As with the Hulse–Taylor pulsar, this (thus far) unique system provides a valuable test of orbital precession and gravitational radiation. However, J0737–3039A/B can also test the prediction of delayed arrival times for signals from one pulsar passing through the gravitational well of the other pulsar (recall Fig. 17.4). As their signals interact with each other's magnetic field and with the plasma environment, they provide an opportunity to test theories about plasma physics as well. It may also be possible to measure the moments of inertia of the pulsars, providing important tests of the interior structure models of neutron stars, including their exotic equations of state.

### Short–Hard Gamma Ray Bursts

What will be the consequence of the merger of the two neutron stars? In Section 15.4 two classes of gamma ray bursts were discussed. Extensive observations have confirmed that long–soft gamma ray bursts ($> 2$ s) are extreme examples of core-collapse supernovae (collapsars or supranovas). On the other hand, it is now believed that **short–hard gamma ray bursts** ($< 2$ s) are the result of the mergers of compact objects, either two neutron stars or a neutron star and a black hole.

The first clear detections of mergers of compact objects in binaries were obtained by the Swift and HETE-2 spacecraft in 2005. The July 9, 2005, event in particular also produced a visible-light afterglow that allowed astronomers to unambiguously identify the host galaxy. Short–hard gamma ray bursts emit about 1000 times less energy than the long–soft events do.

The Hulse–Taylor system is destined to produce a short–hard gamma ray burst, although it may or may not be observable from Earth, depending on the orientation of the jet.

## SUGGESTED READING

### General

Backer, Donald C., and Kulkarni, Shrinivas R., "A New Class of Pulsars," *Physics Today*, March 1990.

Cannizzo, John K., and Kaitchuck, Ronald H., "Accretion Disks in Interacting Binary Stars," *Scientific American*, January 1992.

Clark, David H., *The Quest for SS 433*, Viking Penguin Inc., New York, 1985.

Hellier, Coel, *Cataclysmic Variable Stars: How and Why They Vary*, Springer-Verlag, Berlin, 2001.

Kirshner, Robert P., *The Extravagant Universe: Exploding Stars, Dark Energy, and the Accelerating Universe*, Princeton University Press, Princeton, 2002.

Kleppner, Daniel, "The Gem of General Relativity," *Physics Today*, April 1993.

Piran, Tsvi, "Binary Neutron Stars," *Scientific American*, May 1995.

van den Heuvel, Edward P. J., and van Paradijs, Jan, "X-ray Binaries," *Scientific American*, November 1993.

Wheeler, J. Craig, *Cosmic Catastrophies: Supernovae, Gamma-Ray Bursts, and Adventures in Hyperspace*, Cambridge University Press, Cambridge, 2000.

### Technical

Backer, D. C., et al., "A Millisecond Pulsar," *Nature*, *300*, 615, 1982.

Branch, David, and Tammann, G. A., "Type Ia Supernovae as Standard Candles," *Annual Review of Astronomy and Astrophysics*, *30*, 359, 1992.

Cowley, Anne P., "Evidence for Black Holes in Stellar Binary Systems," *Annual Review of Astronomy and Astrophysics*, *30*, 287, 1992.

Damour, Thibault, and Taylor, J. H., "On the Orbital Period Change of the Binary Pulsar PSR 1913+16," *The Astrophysical Journal*, *366*, 501, 1991.

Frank, Juhan, King, Andrew, and Raine, Derek, *Accretion Power in Astrophysics*, Third Edition, Cambridge University Press, Cambridge, 2002.

Hilditch, R. W., *An Introduction to Close Binary Stars*, Cambridge University Press, Cambridge, 2001.

Hillebrandt, Wolfgang, and Niemeyer, Jens C., "Type Ia Supernova Explosion Models," *Annual Review of Astronomy and Astrophysics*, *38*, 191, 2000.

Horne, Keith, and Cook, M. C., "*UBV* Images of the Z Cha Accretion Disc in Outburst," *Monthly Notices of the Royal Astronomical Society*, *214*, 307, 1985.

Iben, Icko, Jr., "The Life and Times of an Intermediate Mass Star—in Isolation/in a Close Binary," *Quarterly Journal of the Royal Astronomical Society*, *26*, 1, 1985.

Iben, Icko, Jr., "Single and Binary Star Evolution," *The Astrophysical Journal Supplement*, *76*, 55, 1991.

Lorimer, D. R., and Kramer, M., *Handbook of Pulsar Astronomy*, Cambridge University Press, Cambridge, 2005.

Lyne, A. G., et al., "A Double-Pulsar System: A Rare Laboratory for Relativistic Gravity and Plasma Physics," *Science*, *303*, 1153, 2004.

Lyne, A. G., and Graham-Smith, F., *Pulsar Astronomy*, Third Edition, Cambridge University Press, Cambridge, 2006.

Margon, Bruce, "Observations of SS 433," *Annual Review of Astronomy and Astrophysics*, *22*, 507, 1984.

Niemeyer, N. C., and Truran, J. W. (eds.), *Type Ia Supernovae: Theory and Cosmology*, Cambridge University Press, Cambridge, 2000.

Petschek, Albert G., *Supernovae*, Springer-Verlag, New York, 1990.

Pringle, J. E., and Wade, R. A. (eds.), *Interacting Binary Stars*, Cambridge University Press, Cambridge, 1985.

Riess, Adam G., Press, William H., and Kirshner, Robert P., "A Precise Distance Indicator: Type Ia Supernova Multicolor Light-Curve Shapes," *The Astrophysical Journal*, *473*, 88, 1996.

Sion, E. M., "White Dwarfs in Cataclysmic Variables," *Publications of the Astronomical Society of the Pacific*, *111*, 532, 1999.

Verbunt, Frank, "Origin and Evolution of X-ray Binaries and Binary Radio Pulsars," *Annual Review of Astronomy and Astrophysics*, *31*, 93, 1993.

Weisberg, J. M., and Taylor, J. H., "Observations of Post-Newtonian Timing Effects in the Binary Pulsar PSR 1913+16," *Physical Review Letters*, *52*, 1348, 1984.

Weisberg, J. M., and Taylor, J. H., "Relativistic Binary Pulsar B1913+16: Thirty Years of Observations and Analysis," in *Binary Radio Pulsars*, Astronomical Society of the Pacific Conference Series, F. A. Rasio and I. H. Stairs (eds.), *328*, 25, 2005.

## PROBLEMS

**18.1** Use the ideal gas law to argue that in a close binary system, the temperature of a star's photosphere is approximately constant along an equipotential surface. What effect could the proximity of the other star have on your argument?

**18.2** Each of the Lagrange points $L_4$ and $L_5$ forms an equilateral triangle with masses $M_1$ and $M_2$ in Fig. 18.3. Use this to confirm the value of the effective gravitational potential at $L_4$ and $L_5$ given in the figure caption.

**18.3** (a) Consider a gas of density $\rho$ moving with velocity $v$ across an area $A$ perpendicular to the flow of the gas. Show that the rate at which mass crosses the area is given by Eq. (18.11).

    (b) Derive Eq. (18.12) for the radius of the intersection of two identical overlapping spheres, when $d \ll R$.

**18.4** Use Eq. (18.19) to show that the maximum disk temperature is found at $r = (49/36)R$ and is equal to $T_{\max} = 0.488 T_{\text{disk}}$.

**18.5** Integrate Eq. (18.15) for the ring luminosity from $r = R$ to $r = \infty$ [with Eq. (18.19) for the disk temperature]. Does your answer agree with Eq. (18.23) for the disk luminosity?

**18.6** Consider an "average" dwarf nova that has a mass transfer rate of

$$\dot{M} = 10^{13.5} \text{ kg s}^{-1} = 5 \times 10^{-10} \text{ M}_\odot \text{ yr}^{-1}$$

during an outburst that lasts for 10 days. Estimate the total energy released and the absolute magnitude of the dwarf nova during the outburst. Use values for Z Cha's white dwarf from Example 18.4.1. Neglect the small amount of light contributed by the primary and secondary stars.

**18.7** Assume that the absolute bolometric magnitude of a dwarf nova during quiescence is 7.5 and that it brightens by three magnitudes during outburst. Using values for Z Cha, estimate the rate of mass transfer through the accretion disk.

**18.8** When the accretion disk in a cataclysmic variable is eclipsed by the secondary star, the blue-shifted emission line is the first to disappear at the beginning of the eclipse, and the redshifted emission line is the last to reappear when the eclipse ends. What does this have to say about the directions of rotation of the binary system and the accretion disk?

**18.9** (a) Show that in a close binary system where angular momentum is conserved, the change in orbital period produced by mass transfer is given by

$$\frac{1}{P} \frac{dP}{dt} = 3\dot{M}_1 \frac{M_1 - M_2}{M_1 M_2}.$$

(b) U Cephei (an Algol system) has an orbital period of 2.49 days that has increased by about 20 s in the past 100 years. The masses of the two stars are $M_1 = 2.9 \text{ M}_\odot$ and $M_2 = 1.4 \text{ M}_\odot$. Assuming that this change is due to the transfer of mass between the two stars in this Algol system, estimate the mass transfer rate. Which of these stars is gaining mass?

**18.10** Algol (the "demon" star, in Arabic) is a semidetached binary. Every 2.87 days, its brilliance is reduced by more than half as it undergoes a deep eclipse, its apparent magnitude dimming from 2.1 to 3.4. The system consists of a B8 main-sequence star and a late-type (G or K) subgiant; the deep eclipses occur when the larger, cooler star (the subgiant) moves in front of its smaller, brighter companion. The "Algol paradox," which troubled astronomers in the first half of the twentieth century, is that according to the ideas of stellar evolution discussed in Section 10.6, the more massive B8 star should have been the first to evolve off the main sequence. What is your solution to this paradox? (The Algol system actually contains a third star that orbits the other two every 1.86 years, but this has nothing to do with the solution to the Algol paradox.)

Algol may be easily found in the constellation Perseus (the Hero, who rescued Andromeda in Greek mythology).

**18.11** Consider a $10^{-4}$ M$_\odot$ layer of hydrogen on the surface of a white dwarf. If this layer were completely fused into helium, how long would the resulting nova last (assuming a luminosity equal to the Eddington luminosity)? What does this say about the amount of hydrogen that actually undergoes fusion during a nova outburst?

**18.12** Consider a layer of $10^{-4}$ M$_\odot$ of hydrogen on the surface of a white dwarf. Compare the gravitational binding energy before the nova outburst to the kinetic energy of the ejected layer when it has traveled far from the white dwarf and has a speed of 1000 km s$^{-1}$.

**18.13** In this problem, you will examine the fireball expansion phase of a nova shell. Suppose that mass is ejected by a nova at a constant rate of $\dot{M}_{\text{eject}}$ and at a constant speed $v$.

(a) Show that the density of the expanding shell at a distance $r$ is

$$\rho = \dot{M}_{\text{eject}}/4\pi r^2 v.$$

(b) Let the mean opacity, $\overline{\kappa}$, of the expanding gases be a constant. Suppose that at some time $t = 0$, the outer radius of the shell was $R$, and the radius of the photosphere, where $\tau = 2/3$, was $R_0$. Show that

$$\frac{1}{R} = \frac{1}{R_0} - \frac{1}{R_\infty},$$

where

$$R_\infty \equiv \frac{3\overline{\kappa}\dot{M}_{\text{eject}}}{8\pi v}.$$

(The reason for the "$\infty$" subscript will soon become clear.)

(c) At some later time $t$, the radius of the shell will be $R + vt$ and the radius of the photosphere will be $R(t)$. Show that

$$\frac{1}{R + vt} = \frac{1}{R(t)} - \frac{1}{R_\infty}.$$

(d) Combine the results from parts (b) and (c) to write

$$R(t) = R_0 + \frac{vt(1 - R_0/R_\infty)^2}{1 + (vt/R_\infty)(1 - R_0/R_\infty)},$$

(e) Argue that terms containing $R_0/R_\infty$ are very small and can be ignored, and so obtain

$$R(t) \simeq \frac{vt}{1 + vt/R_\infty}.$$

(f) Show that the fireball's photosphere initially expands linearly with time and then approaches the limiting value of $R_\infty$, in agreement with Eq. (18.30).

(g) Using the data given in the text following Eq. (18.31), make a graph of $R(t)$ vs. $t$ for the five days after the nova explodes. The "knee" in the graphs marks the end of the linear expansion period; estimate when this occurs. How does this compare with the duration of the optically thick fireball phase of the nova?

**18.14** Use Eq. (18.31) to estimate the photospheric temperature of a nova fireball, adopting the Eddington luminosity for the luminosity of the fireball.

**18.15** Assuming that the hydrostatic-burning phase of a nova lasts for 100 days, find the (constant) rate at which mass is ejected, $\dot{M}_{\text{eject}}$, for a surface layer of $10^{-4}$ $M_\odot$.

**18.16** For each kilogram of a carbon–oxygen composition (30% $^{12}_{6}C$) that is burned to produce iron, $7.3 \times 10^{13}$ J of energy is released. Assuming an initial 1.38 $M_\odot$ white dwarf with a radius of 1600 km, how much iron would have to be produced to cause the star to be gravitationally unbound? How much additional iron would have to be manufactured to produce a Type Ia supernova with an average ejecta speed of 5000 km s$^{-1}$? Take the gravitational potential energy to be $-5.1 \times 10^{43}$ J for a realistic white dwarf model, and express your answers in units of $M_\odot$.

**18.17** Use Eqs. (7.4), (18.33), and (18.34) to derive Eq. (18.35), the condition for a supernova to disrupt a binary system.

**18.18** (a) Show that the Alfvèn radius is given by Eq. (18.39).

(b) Show that $\dot{P}/P$ for the spin-up of an X-ray pulsar is given by Eq. (18.41).

**18.19** Find the value of the magnetic field for which the Alfvèn radius is equal to the radius of the white dwarf found in Example 18.2.1. Do the same thing for the neutron star used in that example.

**18.20** Estimate the lifetime of a binary X-ray system using the information in Example 18.2.1. Take the lifetime to be the time required to transfer a mass of 1 $M_\odot$.

**18.21** The X-ray pulsar 4U0115+63 has a period of 3.61 s and an X-ray luminosity of about $L_x = 3.8 \times 10^{29}$ W. Assuming that it is a 1.4 $M_\odot$ neutron star with a radius of 10 km and a surface magnetic field of $10^8$ T, find its mass transfer rate, $\dot{M}$, and the value of $\dot{P}/P$. Repeat these calculations assuming that this object is a 0.85 $M_\odot$ white dwarf with a radius of $6.6 \times 10^6$ m and a surface magnetic field of 1000 T. For which of these models do you obtain better agreement with the measured value of $\dot{P}/P = -3.2 \times 10^{-5}$ yr$^{-1}$?

**18.22** (a) Use Eq. (18.24) to show that the spin-up rate can be written as

$$\log_{10}\left(-\frac{\dot{P}}{P}\right) = \log_{10}\left(PL_{\mathrm{acc}}^{6/7}\right) + \log_{10}\left[\frac{\sqrt{\alpha}}{2\pi I}\left(\frac{2\sqrt{2\pi}\,B_s^2 R^{12}}{\mu_0 \, G^3 M^3}\right)^{1/7}\right].$$

The term on the left and the first term on the right consist of quantities that can be measured observationally. The second term on the right depends on the specific model (neutron star or white dwarf) of the X-ray pulsar.

(b) Make a graph of $\log_{10}(-\dot{P}/P)$ (vertical axis) vs. $\log_{10}\left(PL_{\mathrm{acc}}^{6/7}\right)$ (horizontal axis). Use the values from Example 18.6.3 to plot two lines, one for a neutron star and one for a white dwarf. Let $\log_{10}\left(PL_{\mathrm{acc}}^{6/7}\right)$ run from 25 to 29.

(c) Use the data in Table 18.2 to plot the positions of six binary X-ray pulsars on your graph. (You will have to convert $-\dot{P}/P$ into units of s$^{-1}$.)

(d) Which model of a binary X-ray pulsar is in better agreement with the data? Comment on the position of Her X-1 on your graph.

**TABLE 18.2**   X-ray Pulsar Data for Problem 18.22. (Data from Rappaport and Joss, *Nature*, 266, 683, 1977, and Joss and Rappaport, *Annu. Rev. Astron. Astrophys.*, 22, 537, 1984.)

| System | $P$ (s) | $L_{\mathrm{acc}}$ ($10^{30}$ W) | $-\dot{P}/P$ (yr$^{-1}$) |
|---|---|---|---|
| SMC X-1 | 0.714 | 50 | $7.1 \times 10^{-4}$ |
| Her X-1 | 1.24 | 1 | $2.9 \times 10^{-6}$ |
| Cen X-3 | 4.84 | 5 | $2.8 \times 10^{-4}$ |
| A0535+26 | 104 | 6 | $3.5 \times 10^{-2}$ |
| GX301−2 | 696 | 0.3 | $7.0 \times 10^{-3}$ |
| 4U0352+30 | 835 | 0.0004 | $1.8 \times 10^{-4}$ |

**18.23** **(a)** Consider an X-ray burster that releases $10^{32}$ J in 5 seconds. If the shape of its peak spectrum is that of a $2 \times 10^7$ K blackbody, estimate the radius of the underlying neutron star.

**(b)** In Problem 17.9 you showed that using the Stefan–Boltzmann formula to find the radius of a compact blackbody can lead to an overestimate of its radius. Use Eq. (17.33) to find a more accurate value for the radius of the neutron star.

**18.24** Make a scale drawing of the SMC X-1 binary pulsar system, including the size of the secondary star. Assuming that the primary is a 1.4 $M_\odot$ neutron star, locate the system's center of mass and its inner Lagrangian point, $L_1$. (You can omit the accretion disk.)

**18.25** The relativistic ($v/c = 0.26$) jets coming from the accretion disk in SS 433 sweep out cones in space as the disk precesses. The central axis of these cones makes an angle of 79° with the line of sight, and the half-angle of each cone is 20°. This means that at some point in the precession cycle, the jets are moving perpendicular to the line of sight. Yet, from Fig. 18.26, the radial velocities obtained from the Doppler-shifted spectral lines do *not* cross at zero radial velocity, but at $\sim 10{,}000$ km s$^{-1}$. Use Eq. (4.32) to explain this discrepancy in terms of a transverse Doppler shift. (You can ignore the speed of the SS 433 binary system itself, which is only about 70 km s$^{-1}$.)

**18.26** The distance to SS 433 is about 5.5 kpc, and the angular separation of SS 433 and the X-ray emitting regions (where the jets interact with the gases of W50) extends as far as 44′. Estimate a lower limit for the amount of time the jets have been active.

**18.27** PSR 1953+29 is a millisecond pulsar with a period of 6.133 ms. The measured period derivative for PSR 1953+29 is $\dot{P} = 3 \times 10^{-20}$. Use Problem 16.22 to estimate the age of this millisecond pulsar, assuming that no accretion has occurred to alter the pulsar's spin. Also, use Eq. (16.33) to estimate the value of this pulsar's magnetic field.

**18.28** Integrate Eq. (18.41) for the spin-up of an X-ray pulsar to estimate the time for a millisecond pulsar to be spun up to a final period of 1 ms from an initial period of 100 s (longer than the longest known pulsar period of 11.7 s, and within the range of X-ray pulsar periods). Assume a 1.4 $M_\odot$ neutron star with a radius of 10 km. Use $B_s = 10^4$ T for the magnetic field and $\dot{M} = 10^{14}$ kg s$^{-1}$ for the mass transfer rate. How much mass is transferred in that time (in kilograms and in solar masses)?

**18.29** The three planets orbiting PSR 1257+12 have orbital periods of 25.34 d, 66.54 d, and 98.22 d. Verify that these objects obey Kepler's third law.

**18.30** **(a)** Use Kepler's third law to find the semimajor axis of the orbit of the binary pulsar PSR 1913+16.

**(b)** What is the change in the semimajor axis after one orbital period of the pulsar?

## COMPUTER PROBLEMS

**18.31** **(a)** Use the StatStar model data on page 283 and Eq. (18.14) to make a graph of $\log_{10} \dot{M}$ (vertical axis) vs. $\log_{10} d$ (horizontal axis). Use the slope of your graph to find how the mass transfer rate, $\dot{M}$, depends on $d$.

**(b)** Use Eqs. (L.1) and (L.2) to show that $\dot{M} \propto d^{4.75}$ near the surface, and so verify that the mass transfer rate increases rapidly with the overlap distance $d$ of two stars. Note that your answer to part (a) will be slightly different from this because of the density-dependence of `tog_bf` (the ratio of the guillotine factor to the gaunt factor) calculated in the `Opacity` routine.

**18.32** Use Eq. (18.19) and Wien's law to make two log-log graphs: (1) the disk temperature, $T(r)$, and (2) the peak wavelength $[\lambda_{max}(r)]$ of the blackbody spectrum, for the accretion disk around the black hole A0620−00 as a function of the radial position $r$. For this system, the mass of the black hole is 3.82 $M_\odot$, the mass of the secondary star 0.36 $M_\odot$, and the period of the orbit is 0.3226 day. Assume $\dot{M} = 10^{14}$ kg s$^{-1}$ (about $10^{-9}$ $M_\odot$ yr$^{-1}$), and use the Schwarzschild radius, $R_S$ (Eq. 17.27), for the radius of the black hole. (On your graph, plot $r/R_S$ rather than $r$.) For a nonrotating black hole, the last stable orbit for a massive particle is at $3R_S$, so use this as the inner edge of the disk. Let the outer edge of the disk be determined by Kepler's third law along with Eqs. (18.25) and (18.26). On your log-log graph of $\lambda_{max}$ vs. $r/R_S$, identify the regions of the disk that emit X-ray, ultraviolet, visible, and infrared radiation.

# PART
# III

# The Solar System

# Physical Processes
# in the Solar System

## 19.1 ■ A BRIEF SURVEY

In Chapter 11 we studied the Sun, by far the largest member of the Solar System, in some detail. The question of its formation, along with the formation of other stars, was discussed in Chapter 12. As we have seen, based on the observations of protostars and very young stars (e.g., HH 30, Vega, $\beta$ Pictoris, and the proplyds in the Orion Nebula; see page 437, along with Figs. 12.19–12.23), it is evident that a natural extension of the formation of many stars includes the accompanying formation of planetary systems that develop within equatorial disks of material that orbit the newborn stars. In fact the first confirmation of an extrasolar planet around a main-sequence star (51 Pegasi) was announced in 1995. After that initial announcement, a total of 155 extrasolar planets were discovered in just the next ten years alone. In Part III we will study one well-known example of a planetary system in some detail, namely our own. We will also consider the growing body of information regarding extrasolar planets. However, it is beyond the scope of this text to describe all of the fascinating details of each of the planets in our Solar System and their moons, not to mention the meteorites, asteroids, comets, Kuiper Belt objects, and interplanetary dust; that is left to the many excellent books dedicated to the subject. Rather, we will consider the basic features of these objects and extrasolar planets in the context of stellar evolution, together with some of the underlying physical processes that have helped to shape them.

### General Characteristics of the Planets

The planets have long been studied from Earth, first with the naked eye and later with telescopes. Since the advent of space flight, we have sent manned and unmanned spacecraft to our Moon, and, with the exception of Pluto, we have visited (with unmanned probes) each of the other planets in the Solar System.

Each of the planets (excluding Pluto, 2003 UB313,[1] and other members of the Kuiper belt) can be thought of as belonging to one of two major groups. The rocky **terrestrial**

---

[1] 2003 UB313 was discovered in January 2005, based on images obtained in 2003 (see page 827). As of May 2006, an official classification of 2003 UB313 as a major planet or a minor planet had not yet been made, nor has a formal name been given to the object. These official designations are made by the International Astronomical Union.

**TABLE 19.1** General Characteristics of the Planets. The range of values for some features of the terrestrial and giant planets ($M_\oplus$ and $R_\oplus$ represent the mass and radius of Earth, respectively).

| Characteristic | Terrestrial | Giant |
|---|---|---|
| Basic form | Rock | Gas/Ice/Rock |
| Mean orbital distance (AU) | 0.39–1.52 | 5.2–30.0 |
| Mean "surface" temperature (K) | 215–733 | 70–165 |
| Mass ($M_\oplus$) | 0.055–1.0 | 14.5–318 |
| Equatorial radius ($R_\oplus$) | 0.38–1.0 | 3.88–11.2 |
| Mean density (kg m$^{-3}$) | 3933–5515 | 687–1638 |
| Sidereal rotation period (equator) | 23.9 h–243 d | 9.9 h–17.2 h |
| Number of known moons | 0–2 | 13–63 |
| Ring systems | no | yes |

**FIGURE 19.1** The relative sizes of the Sun and the planets. From left to right are the Sun, Mercury, Venus, Earth, Mars, Jupiter, Saturn, Uranus, Neptune, and Pluto (with Charon, one of its moons). A tenth planet, 2003 UB313, which is believed to be slightly larger than Pluto, is not shown in this montage. The distances between objects are not to scale.

(or Earth-like) planets include Mercury, Venus, Earth, and Mars, and the **giant** planets (sometimes called **Jovian**, or Jupiter-like) include Jupiter, Saturn, Uranus, and Neptune. The giant planets are also further separated into the gas giants (Jupiter and Saturn) and the ice giants (Uranus and Neptune). The two major groups have a number of very striking differences, as can be seen by looking through Table 19.1. The relative sizes of the planets and the Sun are illustrated in Fig. 19.1, and specific physical and orbital characteristics of each of the planets can be found in Appendix C.

Many of the differences noted in Table 19.1 are directly related to the distances of the planets from the Sun and their corresponding temperatures. In fact, as we shall see, this temperature effect profoundly influenced the evolution of the terrestrial and giant planets by determining the extent of ice formation in the early solar nebula.

## Moons of the Planets

The number of moons orbiting each planet also varies significantly between the terrestrials and the giants. Neither Mercury nor Venus has any moons, Earth has one relatively large moon, and Mars has two tiny satellites. On the other hand, Jupiter, Saturn, Uranus, and

Neptune are known to have at least 63, 47, 27, and 13 moons, respectively. Combined with their ring systems, each of the giant planets possesses a complex orbital system.

With the exception of Pluto[2] and its largest moon, Charon, by far the largest moon in the Solar System *relative to* its parent planet is our own Moon. However, three of the four Galilean moons of Jupiter (Io, Ganymede, and Callisto)[3] and the giant satellite of Saturn (Titan) are physically larger and more massive. In addition, both Ganymede and Titan have radii that are slightly larger than the planet Mercury's even though their masses are somewhat lower.

In some respects, many of the characteristics of the giant moons of the Solar System are similar to those of the terrestrial planets, including active volcanoes on Io and the existence of an atmosphere on Titan. Some of the moons have features unlike anything seen on the planets, however, including the bizarre topography on the surface of Miranda (one of the many moons of Uranus).

## The Asteroid Belt

In 1766, before the discoveries of Uranus, Neptune, and Pluto, Johann Titius (1729–1796) uncovered a simple mathematical sequence representing the orbital distances of the planets from the Sun. The sequence was popularized several years later by Johann Elert Bode (1747–1826) and is now known as the **Titius–Bode rule**, or simply *Bode's rule* (see Table 19.2).

When Bode's rule was proposed, it was realized that the rule "predicted" the existence of an object at a distance of 2.8 AU, between the orbits of Mars and Jupiter. It was after a deliberate search that an Italian monk, Giuseppe Piazzi (1746–1826), discovered the first

**TABLE 19.2**  Predictions of the Titius–Bode Rule. A comparison of the Titius–Bode rule with actual mean orbital distances.

| Planet | Titius–Bode Distance (AU) | | Actual Mean Distance (AU) |
|---|---|---|---|
| Mercury | $(4 + 3 \times 0\ )/10 =$ | 0.4 | 0.39 |
| Venus | $(4 + 3 \times 2^0)/10 =$ | 0.7 | 0.72 |
| Earth | $(4 + 3 \times 2^1)/10 =$ | 1.0 | 1.00 |
| Mars | $(4 + 3 \times 2^2)/10 =$ | 1.6 | 1.52 |
| *Ceres* | $(4 + 3 \times 2^3)/10 =$ | 2.8 | 2.77 |
| Jupiter | $(4 + 3 \times 2^4)/10 =$ | 5.2 | 5.20 |
| Saturn | $(4 + 3 \times 2^5)/10 =$ | 10.0 | 9.58 |
| Uranus | $(4 + 3 \times 2^6)/10 =$ | 19.6 | 19.20 |
| Neptune | $(4 + 3 \times 2^7)/10 =$ | 38.8 | 30.05 |
| Pluto | $(4 + 3 \times 2^8)/10 =$ | 77.2 | 39.48 |
| 2003 UB313 | $(4 + 3 \times 2^9)/10 =$ | 154.0 | 67 |

[2]If it were orbiting one of the other planets rather than the Sun, Pluto would be only the eighth largest moon in the Solar System.

[3]Io, Europa, Ganymede, and Callisto were the four moons discovered by Galileo to be orbiting Jupiter; see Section 2.2.

**asteroid** at approximately that location on January 1, 1801, and named it Ceres, for the patron goddess of Sicily. Today many thousands of asteroids are known, although Ceres is the largest, containing some 30% of the entire mass of the group and having a diameter of roughly 1000 km. Even though there are important exceptions (see Section 22.3), most asteroids orbit the Sun near the ecliptic plane at distances between 2 and 3.5 AU, a region referred to as the **asteroid belt**.

Although Bode's rule agrees reasonably well with the orbits of most of the planets, and it did lead to the "prediction" of Ceres, it fails miserably for objects beyond Uranus. It is widely believed today that Bode's rule is not based on any fundamental physical process and is only a mathematical coincidence. Historically, Bode's rule has often been referred to as Bode's *law*, even though astronomers generally believe today that it is not associated with any basic "law of nature." It is interesting to note, however, that a variation of Bode's rule also works for some of the moons of Jupiter, Saturn, and Uranus. This rephrasing of the rule is explored in Problems 19.2 and 19.3.

While many of the larger moons were certainly formed with their parent planet, others are little more than large rocks that may have been caught in the planet's gravitational field as they wandered by. Many of these rocks are probably captured asteroids.

### The Comets and Kuiper Belt Objects

Another important class of objects that orbit our Sun are the **comets**. Once thought to be atmospheric phenomena, and even harbingers of doom,[4] comets are now known to be dirty snowballs of ices and dust. Their spectacular, long tails are simply the escaped dust and gas of the evaporating ball of ice, being driven away from the Sun by radiation pressure and the solar wind. Some comets, like the famous Halley's comet, have relatively short orbital periods of less than 200 years, whereas the long-period comets can take over one million years to orbit the Sun.

From their orbital characteristics, it seems very likely that the present-day source of the short-period comets is the **Kuiper belt**, a collection of icy objects located predominantly near the plane of the ecliptic and beyond the orbit of Neptune, typically ranging from 30 AU to perhaps 1000 AU or more from the Sun. It is now realized that Pluto and its moon Charon, 2003 UB313, Sedna, and Quaoar are among the largest known members of the family of **Kuiper belt objects** (KBOs), also referred to as **Trans-Neptunian Objects** (TNOs). The long-period comets apparently originate in the **Oort cloud**, an approximately spherically symmetric cloud of cometary nuclei with orbital radii of between 3000 and 100,000 AU. Having spent most of their existence in "deep freeze" at the outer reaches of the Solar System, comets and Kuiper belt objects appear to be ancient remnants of its formation, although perhaps not entirely unaffected by nearly 4.6 billion years of exposure to the environment of space.

### Meteorites

When asteroids collide with one another, they can produce small fragments known as **meteoroids**. If a meteoroid should happen to enter Earth's atmosphere, the heat generated

---

[4]Recall the observations of Tycho Brahe, discussed in Section 2.1.

by friction results in a glowing streak across the sky, referred to as a **meteor**. If the rock survives the trip through the atmosphere and strikes the surface, the remnant is known as a **meteorite**. By analyzing the composition of meteorites, we can learn a great deal about the environment in which they originated.

Another source of meteoritic material is the slow disintegration of comets exposed to the heat of the inner Solar System. When Earth encounters the debris left in a comet's orbit, the result is a **meteor shower** of micrometeorites raining down through the planet's atmosphere.

Finally, the dust remaining in orbit about the Sun due to the disintegration of asteroids and comets produces a faint glow from reflected sunlight. Unfortunately, even the lights of a small town are sufficient to obscure this **zodiacal light**.

## Solar System Formation: A Brief Overview

All of these features of the Solar System can be understood in terms of its initial formation and subsequent evolution. Our present understanding of Solar System evolution is based on the hypothesis that as the Sun was forming from the gravitational collapse of the original solar nebula, the decreasing radius of the cloud resulted in an increasing rate of spin and the accompanying formation of a disk of material. Within this **accretion disk** the temperature varied with distance from the protosun in such a way that rocks were able to consolidate throughout the disk while ices (primarily water) were able to develop only at distances beyond the outer part of the present-day asteroid belt. As a result, the terrestrial planets accreted from collisions of small preplanetary chunks of material, known as **planetesimals**, that were composed exclusively of rock, while the much larger giants benefited from the additional presence of ices in the planetesimals. The higher temperatures in the inner portions of the disk and the lower masses of the terrestrials also inhibited the capture of lighter gases around those planets, while the cooler, more massive giants were able to accumulate significant and, in the cases of Jupiter and Saturn, very massive primordial atmospheres.

Around the newly formed giant planets, smaller local accretion disks were forming some of the moons seen today. Other moons appeared when planetesimals and fragmented asteroids were captured while wandering through the Solar System. In a different mechanism, it appears that our own Moon was produced when a relatively large planetesimal approximately the size of the present-day Mars collided with the young Earth. After the formation of the planets and their moons, the "rain" of remnant material led to heavy cratering. Although the rate of crater formation has decreased significantly since the time of the early Solar System, the process remains ongoing. Evidence of that violent beginning is still readily apparent on many worlds today.

Most of the icy objects that were drifting among the giant worlds without directly colliding with or being captured by one of the giant planets had their orbits dramatically altered through gravitational interactions. Some of the cometary nuclei passing near Uranus or Neptune were catapulted into much larger orbits, characteristic of the present-day Oort cloud, while those that ventured near Jupiter or Saturn were ejected from the Solar System entirely. Other planetesimals that passed near the giant planets were sent inward to collide with the terrestrial planets or the Sun. The icy bodies that formed beyond the orbit of Neptune remain in that region today, constituting the Kuiper belt. Closer to the Sun, just inside the region of ice formation, rocky remnants of the Solar System's formation still reside in the asteroid belt.

As a direct consequence of tidal and viscous interactions with the accretion disk, along with the scattering of planetesimals, Jupiter migrated closer to the Sun than where it initially began forming, whereas Saturn, Uranus, and particularly Neptune migrated farther from the Sun.

As the discussion of this section indicates, significant progress has been made in understanding the makeup and evolution of our local part of the universe. Throughout the remainder of Part III we will be investigating in more detail the objects that populate our Solar System, together with many of the physical processes that have shaped them. Finally, in Chapter 23 we will reconsider the evolutionary scenario described in the last several paragraphs in light of what we have learned about our Solar System and extrasolar planetary systems.

## 19.2 ■ TIDAL FORCES

As we saw in Chapter 2, the force of gravity governs the orbits of the planets and their moons in the form of Kepler's laws. In that study we treated those objects as point masses, under the assumption that they are spherically symmetric (see Example 2.2.1). Important consequences arise from relaxing the constraint of spherical symmetry, however. Since one side of a moon is closer to its parent planet than is the opposite side, the planet's gravitational force on a small test mass must be greatest on the moon's near side. This has the effect of elongating the instantaneous shape of the moon. According to Newton's third law, the same situation must also apply to the near and far sides of the planet because of the gravitational influence of its moon. This *differential* force on an object due to its non-zero size is known as a **tidal force**. The resulting nonspherical shapes of the planet and its moon can actually influence their rotation rates by creating torques. If this tidal force is sufficiently great, it is even possible that the smaller world could be disrupted.

The existence of tides on Earth's surface is well known, particularly for those who live near an ocean. There are two high tides approximately every 24 hours, 53 minutes, depending on local coastal features. Less well known are the tidal bulges of the solid Earth, which measure only about 10 cm in height. Since Earth is significantly more massive than the Moon (approximately 81 times), the bulges on the Moon are much larger, resulting in nearly 20 m of deformation at its surface.

### The Physics of Tides

To better understand how tides arise on Earth, consider the force on a test mass $m_1$ located within the planet at a distance $r$ from the Moon's center of mass,

$$F_m = G\frac{Mm_1}{r^2},$$

where $M$ is the mass of the Moon (see Fig. 19.2). Now consider a second mass $m_2 = m_1 = m$, located at a distance $dr$ from $m_1$ along a line connecting Earth and the Moon. The

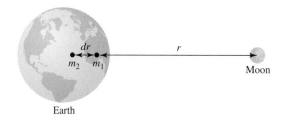

**FIGURE 19.2**   The tidal force on Earth due to the Moon arises because of the varying values for the Moon's gravitational attraction at different locations inside the planet.

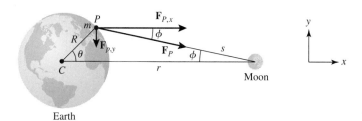

**FIGURE 19.3**   The geometry of the tidal force acting on Earth due to the Moon.

difference in forces (the differential force) between the two test masses is then

$$dF_m = \left(\frac{dF_m}{dr}\right) dr = -2G\frac{Mm}{r^3}\, dr, \tag{19.1}$$

where $dr$ is taken to be the separation between their centers. Note that the differential force decreases more rapidly with distance than does the force of gravity itself, meaning that the closer the test masses are to the Moon, the more pronounced the effect.

The shape of the tidal bulges on Earth can be understood by analyzing the differences in the gravitational force vectors acting at the center of the planet and at some point on its surface (see Fig. 19.3). For simplicity, we will consider only forces in the $x$–$y$ plane. Neglecting rotation, the effects are symmetric about the $x$-axis (the line between the centers of Earth and the Moon). At the center of the planet, the $x$- and $y$-components of the gravitational force on a test mass $m$ due to the Moon are given by

$$F_{C,x} = \frac{GMm}{r^2}, \qquad F_{C,y} = 0,$$

while at point $P$ the components are

$$F_{P,x} = \frac{GMm}{s^2}\cos\phi, \qquad F_{P,y} = -\frac{GMm}{s^2}\sin\phi.$$

The differential force between Earth's center and its surface is

$$\Delta\mathbf{F} = \mathbf{F}_P - \mathbf{F}_C = GMm\left(\frac{\cos\phi}{s^2} - \frac{1}{r^2}\right)\hat{\mathbf{i}} - \frac{GMm}{s^2}\sin\phi\,\hat{\mathbf{j}}.$$

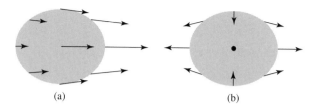

**FIGURE 19.4**   (a) The gravitational force of the Moon on Earth. (b) The differential gravitational force on Earth, relative to its center.

Next, to simplify the solution somewhat, we write $s$ in terms of $r$, $R$, and $\theta$,

$$s^2 = (r - R\cos\theta)^2 + (R\sin\theta)^2 \simeq r^2\left(1 - \frac{2R}{r}\cos\theta\right)$$

where terms of order $R^2/r^2 \ll 1$ have been neglected. Substituting, and recalling that for $x \ll 1$, $(1+x)^{-1} \simeq 1-x$, we find that the differential force becomes

$$\Delta\mathbf{F} \simeq \frac{GMm}{r^2}\left[\cos\phi\left(1 + \frac{2R}{r}\cos\theta\right) - 1\right]\hat{\mathbf{i}}$$

$$-\frac{GMm}{r^2}\left[1 + \frac{2R}{r}\cos\theta\right]\sin\phi\,\hat{\mathbf{j}}. \tag{19.2}$$

Finally, using the first-order relations $\cos\phi \simeq 1$ and $\sin\phi \simeq (R\sin\theta)/r$, we have

$$\Delta\mathbf{F} \simeq \frac{GMmR}{r^3}\left(2\cos\theta\,\hat{\mathbf{i}} - \sin\theta\,\hat{\mathbf{j}}\right). \tag{19.3}$$

Notice the extra factor of 2 in the $x$-component when compared with the $y$-component. You should also compare this result with the expression for the differential force given in Eq. (19.1), noting that here $R$ (the distance between the center and the surface) has replaced $dr$.

The situation described by Eq. (19.3) is illustrated in Fig. 19.4. The actual gravitational force vectors due to the Moon are directed toward the center of mass of the Moon, but the *differential* force vectors act to compress Earth in the $y$-direction and elongate it along the line between their centers of mass, producing tidal bulges. It is the symmetry of the bulges that produces two high tides in a 25-hour period as Earth rotates under an orbiting Moon.

### The Effects of Tides

In reality, Earth's tidal bulges are not directly aligned with the Moon. This is because the rotation period of Earth is shorter than the Moon's orbital period and frictional forces on the surface of the planet drag the bulge axis ahead of the Earth–Moon line. Because friction is a dissipative force, rotational kinetic energy is constantly being lost and Earth's spin rate is continually decreasing. At the present time, Earth's rotation period is lengthening at the rate of $0.0016$ s century$^{-1}$, which, although slow, is measurable.

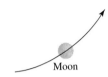

**FIGURE 19.5**   Earth's bulge A is closer to the Moon than is bulge B, resulting in a net torque on the planet. Note that the diagram is not to scale.

The Moon is also known to be drifting away from Earth by 3 to 4 cm yr$^{-1}$. The increasing Earth–Moon distance is determined by bouncing laser beams off the mirrors left on the Moon by the Apollo astronauts in the early 1970s and measuring the round-trip light-travel time.

To see how the decrease in Earth's rotation rate and the increase in the Earth–Moon distance are related, we need only consider the torque exerted on Earth by the Moon's interaction with Earth's tidal bulges. In Fig. 19.5, bulge A leads the Moon and is closer to it than is bulge B. As a result, the force exerted on bulge A by the Moon is greater, resulting in a net torque that is slowing Earth's rotation. At the same time, bulge A is pulling the Moon forward, causing the satellite to move farther out. This complementary behavior is just a consequence of the conservation of angular momentum. Neglecting the dynamical influence of the Sun and the other planets on the Earth–Moon system, no external torques exist to alter its total angular momentum. If Earth's rotational angular momentum is decreasing, the orbital angular momentum of the Moon must necessarily increase.

Because of tidal effects, given sufficient time Earth will slow its rotation enough that the same side of the planet will always face the Moon, just as the Moon now keeps the same "face" toward Earth. In the distant future, if inhabitants on Earth's far side want to take romantic moonlight walks, they will need to take vacations halfway around the world. Calculations indicate that this will happen when the length of the day is about 47 current days long.

### Synchronous Rotation

In the past the Moon was much closer to Earth than it is today, perhaps taking as little as one week to orbit the planet. It is also probable that the Moon's rotation period was once shorter than its orbital period. The Moon's present **1-to-1 synchronous rotation** is due to the same tidal dissipation that is occurring on Earth today. Its rotational period became synchronized with its orbital period more rapidly than Earth's has, simply because it is much smaller and because Earth produces much larger tidal deformations on the Moon than does the Moon on Earth.

Synchronous rotation is common throughout the Solar System.[5] The two moons of Mars, the four Galilean moons of Jupiter (along with Amalthea, a small moon inside Io's orbit),

[5]Some binary star systems are also known to be in synchronous rotation; see Section 18.1.

and most of the moons of Saturn are in synchronous rotation, as are many of the other moons associated with the outer planets. In addition, Pluto and its largest moon, Charon, have reached the final stage of tidal evolution; they are in mutual synchronous rotation, with the same side of Pluto constantly turned toward the same face of Charon.

An interesting and unusual case of tidal evolution is that of Triton, the giant moon of Neptune. Triton is in synchronous rotation and orbits the planet in a retrograde fashion. In this instance, the tidal bulges in Neptune actually work to cause that moon to spiral *toward* the planet rather than away from it. Apparently it will take billions of years before any catastrophic interaction occurs. On the other hand, Phobos (one of the Martian moons) is in a prograde orbit, but with an orbital period of $7^h39^m$ that is *shorter* than the rotation period of Mars ($24^h37^m$). This means that Phobos is inside the planet's **synchronous orbit**, defined to be where the planet's rotation period and the satellite's orbital period are equal.[6] Consequently, it is "outrunning" the tidal bulge axis, and the resulting forces are causing the moon to spiral inward. Phobos's orbit is decaying rapidly enough that if it were to stay intact, it would hit the planet in about 50 million years. The other Martian moon, Deimos, is outside the synchronous orbital radius and is spiraling outward, just as our Moon is.

### Additional Tidal Effects from the Sun

Of course, the Earth–Moon system is not in strict isolation. For instance, the Sun also produces tidal forces that act on Earth. When the Sun, Earth, and the Moon are all aligned (at full Moon or new Moon), the differential forces due to the Sun and the Moon add to create unusually large tidal bulges on Earth, called **spring tides**. At first quarter or third quarter, the Sun, Earth, and Moon form a right angle. In this configuration the tides produced by the Sun and the Moon tend to cancel, and unusually low **neap tides** result.

### The Roche Limit

It is unlikely that Phobos will actually remain intact long enough to strike the planet. Recall that the differential tidal force is proportional to $r^{-3}$; as a moon gets closer to its parent planet, tidal effects become more severe. This means that the shape of a moon in synchronous rotation becomes increasingly elongated. Neglecting any internal cohesion forces (i.e., assuming an idealized fluid object), when the orbital distance has decreased sufficiently, it becomes no longer possible to define a shape for the moon such that the force of gravity is perpendicular to the surface at every point. As a result, the surface will continually flow in the direction of the net gravitational force vector. Oscillations will then develop in the extended structure, and the moon will come apart. The maximum orbital radius for which tidal disruption occurs is known as the **Roche limit**, named for Edouard Roche (1820–1883), who first carried out the analysis in 1850. In his study Roche took into consideration orbital and rotational motion, and he assumed a fluid, prolate spheroid (i.e., a football-shaped moon).

To make an order-of-magnitude estimate of the orbital radius at which a moon will break apart, assume (incorrectly) that this happens when the differential force exceeds the

---

[6]Earth's synchronous orbit is sometimes referred to as geosynchronous orbit. Artificial satellites placed in geosynchronous equatorial orbits remain fixed over the same geographic point on the surface. Communications satellites are generally placed in such orbits.

self-gravitational force holding the moon together. Furthermore, assume for simplicity that the moon and the planet are spherical, and neglect any centrifugal effects. In this case, if the moon is to be tidally disrupted, the inward gravitational acceleration produced by the moon at a point located on its surface closest to the planet must be smaller than the outward differential gravitational acceleration produced by the planet, or

$$\frac{GM_m}{R_m^2} < \frac{2GM_pR_m}{r^3},$$

where $M_p$ and $M_m$ are the masses of the planet and moon, respectively, $R_m$ is the radius of the moon, and $r$ is the distance between the centers of the two worlds. Substituting $M_p = 4\pi R_p^3 \overline{\rho}_p / 3$ and $M_m = 4\pi R_m^3 \overline{\rho}_m / 3$, where $\overline{\rho}_p$ and $\overline{\rho}_m$ are the average densities of the planet and moon, respectively, and solving for $r$, we find that a moon will be tidally disrupted if its orbit is less than

$$r < f_R \left( \frac{\overline{\rho}_p}{\overline{\rho}_m} \right)^{1/3} R_p, \tag{19.4}$$

where, in our case, $f_R = 2^{1/3} = 1.3$. In his more careful analysis, Roche found a larger value for the leading constant of $f_R = 2.456$. The fact that our result gave too small a value for the radius reflects the incorrect assumption that it is the differential force exceeding self-gravity that is ultimately responsible for the disintegration of the satellite. Since oscillations in the body will develop at greater radii, self-gravity is still significantly greater than the differential term at the true Roche limit. (Although not considered in this analysis, self-cohesion of an object that is provided by the electromagnetic force, such as molecular bonds or the formation of a crystal lattice, can also decrease the point at which an object will be disrupted.)

---

**Example 19.2.1.**    The average density of Saturn is 687 kg m$^{-3}$ and its planetary radius is $6.03 \times 10^7$ m. Using a value of $f_R = 2.456$, the Roche limit for a moon having an average density of 1200 kg m$^{-3}$ is $1.23 \times 10^8$ m. Much of the ring system of Saturn lies within this orbital radius given by the Roche limit, and all of Saturn's large moons are farther out. The material within ring systems may be the result of disintegrating or tidally disrupted moons that wandered within the Roche limit.

---

### 19.3 ■ THE PHYSICS OF ATMOSPHERES

Our Solar System today is the result of billions of years of ongoing evolution caused by a host of physical processes. Subtle differences in initial conditions of neighboring planets have led to the very different worlds we see today. We will discuss some of the more frequently encountered atmospheric processes in this section and then, in later chapters, describe the unique characteristics of each planet.

### The Temperatures of the Planets

As has already been mentioned, the temperatures of the planets played a key role in their formation and evolution. During the formation stage, the temperature structure of the solar nebula influenced whether a planet would become a terrestrial or a giant; this point will be discussed more fully in Section 23.2. Temperature also helped to determine the current composition of each planet's atmosphere.

The Stefan–Boltzmann equation (Eq. 3.17) is the most significant factor in determining the present-day temperatures of the planets in the Solar System. Under equilibrium conditions, a planet's total energy content must remain constant. Therefore, all of the energy absorbed by the planet must be re-emitted; if this were not so, the planet's temperature would change with time.

To estimate a planet's equilibrium temperature, assume that the planet is a spherical blackbody of radius $R_p$ and temperature $T_p$ in a circular orbit a distance $D$ away from the Sun. For simplicity, we will assume that the planet's temperature is uniform over its surface[7] and that the planet reflects a fraction $a$ of the incoming sunlight ($a$ is known as the planet's **albedo**). From the condition of thermal equilibrium, the sunlight that is not reflected must be absorbed by the planet and subsequently re-emitted as blackbody radiation. Of course, we will also treat the Sun as a spherical blackbody having an effective temperature $T_\odot = T_e$ and radius $R_\odot$. It is left as an exercise to show that the temperature of the planet is given by

$$T_p = T_\odot (1-a)^{1/4} \sqrt{\frac{R_\odot}{2D}}. \tag{19.5}$$

Note that the temperature of the planet is proportional to the effective temperature of the Sun and does not depend on the size of the planet.

---

**Example 19.3.1.** Using Earth's average value of $a = 0.3$ in Eq. (19.5), the temperature of a blackbody Earth is

$$T_\oplus = 255 \text{ K} = -19°C = -1°F.$$

This value is substantially below the freezing point of water and (fortunately!) is not the correct temperature at the surface of the planet. This analysis neglected the **greenhouse effect**, a significant warming due largely to the water vapor in Earth's atmosphere.[8] According to Wien's law (Eq. 3.15), Earth's blackbody radiation is emitted primarily at infrared wavelengths. This infrared radiation is absorbed and then re-emitted by the atmospheric greenhouse gases, which act as a thermal blanket to warm Earth's surface by about 34°C. A simple model of the greenhouse effect is explored in Problem 19.13. Greenhouse warming on Venus has been much more dramatic; the reasons for this will be discussed in Section 20.2.

---

[7]This assumption is a reasonable approximation if the planet is rapidly rotating or has a circulating atmosphere.
[8]Carbon dioxide, methane, and chlorofluorocarbons also contribute to the greenhouse effect.

## The Chemical Evolution of Planetary Atmospheres

The evolution of a planetary atmosphere is a complex process that depends on the local temperature of the solar nebula during the time of the planet's formation, together with the planet's temperature, gravity, and local chemistry following the formation process. In the case of the terrestrial planets, outgassing from rocks and volcanos also played a role after the development of the initial, primordial atmosphere. On Earth, the development of life has also contributed significantly to the evolution of its atmosphere. Impacting comets and meteorites affect planetary atmospheres as well.

A critical component in the development of an atmosphere is the ability of the planet to retain specific atoms or molecules. Recall from the discussion in Section 8.1 that for a gas in thermal equilibrium, the number of particles having velocities between $v$ and $v + dv$ is given by the Maxwell–Boltzmann velocity distribution (see Eq. 8.1 and Fig. 8.6). At some critical height in an atmosphere, when the number density is low enough that collisions among gas particles become negligible, particles moving upward will travel only under the influence of gravity, following trajectories described by simple projectile motion. Those atoms or molecules that are not moving rapidly enough to escape, or that do not have the correct trajectories, will fall back down into the denser layers and undergo collisions with the gas. On the other hand, particles that are moving upward and have velocities that are sufficiently great will be able to escape the gravitational pull of the planet altogether and move out into interplanetary space. It is this process that can allow the atmospheres of some planets (or at least specific chemical components of those atmospheres) to "leak off." The region in an atmosphere where the mean free path of the particles becomes long enough for them to travel without appreciable collisions is referred to as the **exosphere**.

Because of the high-velocity tail of the Maxwell–Boltzmann distribution, and because of the amount of time that has elapsed since the Solar System formed, if a particular component of the atmosphere is going to escape, it is not necessary that the root-mean-square average velocity of those particles be greater than the escape speed. It is only necessary that a sufficiently large number of particles have speeds greater than $v_{esc}$. As a rough estimate, a planet will have lost a particular component of its atmosphere by now if, for that component (either molecular or atomic),

$$v_{rms} > \frac{1}{6} v_{esc}.$$

Using Eqs. (2.17) and (8.3) for the escape and root-mean-square velocities, respectively, the temperature required for a gas of particles of mass $m$ to escape a planet of mass $M_p$ and radius $R_p$ is approximately

$$T_{esc} > \frac{1}{54} \frac{GM_p m}{k R_p}. \tag{19.6}$$

---

**Example 19.3.2.**    Earth has an atmosphere composed of approximately 78% $N_2$ and 21% $O_2$ by number, while the Moon, which is on average the same distance from the Sun, has no significant atmosphere. From Example 19.3.1, the blackbody equilibrium temperature of an *airless* Earth should be 255 K. Since the Moon's albedo is only 0.07, its blackbody temperature is somewhat higher (274 K). In reality, the vertical temperature structure of Earth's atmosphere is very complex and depends on its hydrodynamic motions together with the ability of various atoms and molecules to absorb radiation. Near the top of the atmosphere, the temperature is also strongly dependent on the amount of solar activity. Within the exosphere the characteristic temperature is about 1000 K.

Consider Earth's ability and the Moon's inability to retain molecular nitrogen. The mass of an $N_2$ molecule is approximately 28 u = $4.7 \times 10^{-26}$ kg, the mass and radius of Earth are $5.9736 \times 10^{24}$ kg and $6.378136 \times 10^6$ m, respectively, and the mass and radius of the Moon are $7.349 \times 10^{22}$ kg and $1.7371 \times 10^6$ m, respectively (see Appendix C). The temperatures required for the nitrogen to escape from each world can now be estimated from Eq. (19.6), giving $T_{esc,\oplus} > 3900$ K and $T_{esc,Moon} > 180$ K. Since Earth's exospheric temperature is cooler and the Moon is warmer than these values, Earth has been able to retain its molecular nitrogen, whereas the Moon could not.

Since $O_2$ is more massive (32 u), even higher temperatures are required for that molecule to escape.

---

### The Loss of Atmospheric Constituents

The loss of specific components of an atmosphere can be understood in more detail by appealing directly to Eq. 8.1, the Maxwell–Boltzmann distribution, $n_v \, dv$. As particles move about randomly in the gas, some of them are traveling approximately vertically upward and therefore have the best chance of escaping. The number of particles with velocities between $v$ and $v + dv$ passing through a horizontal slab of cross-sectional area $A$ and vertical thickness $dz$ during a time interval $dt$ is given by

$$dN_v \, dv = (n_v \, dV) \, dv = A \, dz \, n_v \, dv = A v_z \, dt \, n_v \, dv = C_g A v \, dt \, n_v \, dv,$$

where $C_g$ is a geometrical factor that takes into consideration the requirement that, of all the velocity components of the randomly moving particles, only positive vertical components will be considered. Dividing through by the time interval, we obtain the *rate* at which particles with velocities between $v$ and $v + dv$ are crossing the surface. Furthermore, if we assume that the atmosphere is spherical at the location of the exosphere, so that $A = 4\pi R^2$, then the number of particles per second with speeds between $v$ and $v + dv$ moving vertically upward through the entire exosphere is given by

$$\dot{N}_v \, dv \equiv \frac{dN_v}{dt} \, dv = 4\pi R^2 C_g v n_v \, dv.$$

Finally, to determine the number of particles per second leaving the atmosphere, it is necessary only to consider those particles with sufficiently high velocities, namely $v > v_{esc}$. Substituting Eq. (8.1) and integrating, we have

$$\dot{N} = \frac{n\pi R^2}{4} \left( \frac{m}{2\pi kT} \right)^{3/2} \int_{v_{esc}}^{\infty} 4\pi v^3 e^{-mv^2/2kT} \, dv, \tag{19.7}$$

where $C_g$ has been set equal to 1/16, based on a careful analysis of the geometry of the problem. In Problem 19.16 you will be asked to show that at some height $z$ in the atmosphere, where the particle number density is $n(z)$, Eq. (19.7) reduces to

$$\dot{N}(z) = 4\pi R^2 \nu n(z), \tag{19.8}$$

where

$$\nu \equiv \frac{1}{8}\left(\frac{m}{2\pi kT}\right)^{1/2}\left(v_{esc}^2 + \frac{2kT}{m}\right)e^{-mv_{esc}^2/2kT} \tag{19.9}$$

is an **atmospheric escape parameter** that has units of velocity.

$\nu$ describes the rate at which gas particles of mass $m$ escape across a unit area for a specified number density $n(z)$ in the exosphere. The atmospheric escape parameter can also be thought of as the *effective thickness* of the atmosphere of a certain species that evaporates away (or "leaks off") per second. In Fig. 19.6, $\log_{10}\nu$ is plotted as a function of the mass of specific components in Earth's atmosphere, where a temperature of 1000 K has been used, characteristic of a mean value in the exosphere. For comparison, $\log_{10}\nu$ has also been plotted for the same species using the Moon's escape velocity and a typical temperature of 274 K. Note that of the components listed, only molecular hydrogen and helium have essentially completely escaped Earth's atmosphere, whereas the Moon has lost all of its atmosphere, including the heavier molecules listed.

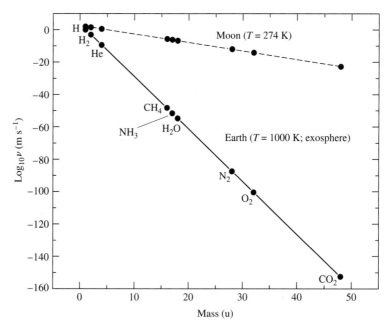

**FIGURE 19.6**    The logarithm of the atmospheric escape parameter, $\nu$, as a function of atomic weight for various chemical species in Earth's atmosphere and on the surface of the Moon. Note that Earth has lost most of its atomic and molecular hydrogen and its helium, while retaining the other molecules listed. The Moon has lost all of its atmosphere.

**Example 19.3.3.**   Equations (19.8) and (19.9) can be used to estimate the amount of time required for molecular nitrogen to escape from Earth's atmosphere. A rough calculation of the total number of $N_2$ molecules in the atmosphere can be made by assuming that the number density decreases approximately exponentially with height, just as the pressure does if the atmosphere is nearly isothermal (see Section 10.4).[9] Then

$$n(z) = n_0 e^{-z/H_P}$$

where $n_0$ is the number density at the surface and $H_P$ is the pressure scale height, given by Eq. (10.70). Using the ideal gas law (Eq. 10.11) with the mass of the nitrogen molecule ($M_{N_2}$) being used for $\mu m_H$, the pressure scale height may be written as

$$H_P = \frac{P}{\rho g} = \frac{kT}{g M_{N_2}}. \tag{19.10}$$

Notice that the pressure scale height is different for particles of different masses. Using characteristic values at Earth's surface ($T = 288$ K and $g = 9.80$ m s$^{-2}$), $H_P = 8.7$ km for molecular nitrogen.

Making the rough assumption that the pressure scale height remains constant with altitude, and neglecting the slight change in $r$ relative to Earth's surface, the number density can now be integrated over the volume of the atmosphere, giving

$$N = 4\pi R_\oplus^2 n_0 H_P.$$

Taking the number density of nitrogen molecules near the surface to be $n_0 = 2 \times 10^{25}$ m$^{-3}$, the total number of nitrogen molecules in the atmosphere is $N = 9 \times 10^{43}$.

From Eq. (19.9) and Fig. 19.6, the atmospheric escape parameter for nitrogen molecules from Earth's exosphere is $v = 4 \times 10^{-88}$ m s$^{-1}$. Also, at the height of the exosphere (approximately 500 km), the mean number density of $N_2$ is $2 \times 10^{11}$ m$^{-3}$. Using Eq. (19.8), we find that the rate at which nitrogen molecules are escaping Earth's atmosphere is approximately $\dot{N} = 4 \times 10^{-62}$ s$^{-1}$. Dividing the total number of available molecules by the rate of loss, the time required to dissipate the nitrogen in Earth's atmosphere is estimated to be

$$t_{N_2} = \frac{N}{\dot{N}} = 2 \times 10^{105} \text{ s} = 6 \times 10^{97} \text{ yr.}$$

It is safe to say that Earth's atmospheric nitrogen is not going to escape any time soon!

The situation is very different for atomic hydrogen in Earth's atmosphere, however. This case will be considered in detail in Problems 19.14 and 19.19.

---

[9]Earth's atmosphere actually differs appreciably from an isothermal approximation. As a result, the estimate of $n(z)$ used here would need to be modified significantly in a more careful analysis. Nevertheless, this "back-of-the-envelope" calculation illustrates many of the basic physical principles involved and yields the correct general conclusion.

Besides the loss of high-velocity particles from the exponential tail of the Maxwell–Boltzmann distribution, other factors also contribute to the dissipation of an atmosphere. Molecular **photodissociation**, caused by the absorption of UV photons in the upper atmosphere, breaks down some molecules into atoms or lighter molecules, with the result that the individual particles have greater speeds (recall the expression for the root-mean-square speed, Eq. 8.3). For instance, $H_2 + \gamma \rightarrow H + H$. The solar wind can also contribute to the loss of particles through collisions in the upper atmosphere, causing direct ejection, or molecular dissociation and subsequent escape. Even heating caused by impacting meteorites and comets can accelerate the loss of atmospheric constituents.

### Gravitational Separation of Atmospheric Constituents

Another feature of atmospheric physics that can affect the loss of certain components is the **gravitational separation** (also known as **chemical differentiation**) of constituents of an atmosphere by weight. In the absence of the continual mixing caused by convection at lower altitudes, composition differences develop with height in the upper atmosphere. This effect can be understood by referring to the expression for the pressure scale height in the form of Eq. (19.10). For a given temperature, the pressure scale height increases as the mass decreases, meaning that the number densities of lighter particles do not diminish as rapidly with $z$. As a result, lighter particles become relatively more abundant in the upper atmosphere, enhancing the likelihood of their escape.

### Circulation Patterns

As is the case with stars (see Section 10.4), convection in planetary atmospheres is driven largely by steep temperature gradients. Near the equator, where the intensity of the sunlight is greatest, the atmosphere heats up and the warm gas rises. The gas then migrates to cooler regions at high latitudes, where it sinks back down again. The cycle closes when the gas returns to the warmer regions near the equator. If the warm air were able to migrate all the way from the equator to the poles before sinking, the global pattern illustrated in Fig. 19.7(a) would occur. This hypothetical circulation pattern is known as **Hadley circulation**.

In reality, the warmer air at higher altitudes is undergoing radiative cooling as it migrates toward the poles. At about 30° N and S latitude, the air has given up enough heat that it sinks and returns to the equator, where it is reheated again. Similarly, the colder air that is migrating from the poles toward the equator at lower altitudes heats and rises at about 55° N and S latitude, returning to the poles where it sinks again. This breaks up the global Hadley circulation pattern into three zonal components, as shown in Fig. 19.7(b).

These general zonal weather patterns are further complicated by the planet's rotation. Since a rotating body does not constitute an inertial reference frame, pseudo-forces, such as the **Coriolis force**, are present.

Assume for simplicity that Earth is perfectly spherical. At a latitude $L$, a point on the surface is located a distance $r_L$ from the rotation axis, given by

$$r_L = R_\oplus \cos L.$$

Letting the angular rotation speed of Earth be $\omega$, the eastward speed of the surface at the

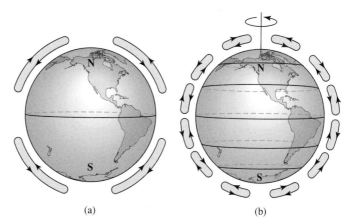

**FIGURE 19.7** (a) The pattern of hypothetical Hadley circulation, caused by warm air rising near the equator and cool air sinking near the poles. (b) General weather circulation on Earth caused by the Hadley cells being broken up by radiative cooling of the warm air migrating toward the poles.

latitude $L$ is just

$$v_L = \omega r = \omega R_{\oplus} \cos L.$$

At the equator ($L = 0°$), this speed is approximately 465 m s$^{-1}$ = 1670 km h$^{-1}$. However, at a latitude of $L = 40°$, the speed is reduced to 1300 km s$^{-1}$. As a result, according to an observer in an inertial reference frame, a person standing still on the surface of Earth at the equator is moving approximately 370 km h$^{-1}$ faster than someone standing still on the surface at a latitude of 40°. This velocity difference with latitude affects weather circulation patterns.

As an illustration of the effect of the Coriolis force, consider the apparent motion of a projectile fired nearly horizontally from the equator northward, as shown in Fig. 19.8(a). Also assume that the elevation of the projectile above the surface of Earth is essentially constant during its flight. From the point of view of an observer on Earth's surface located at the origin of the projectile's motion, the projectile *initially appears* to be traveling straight north since the observer has the same easterly speed as the easterly component of the projectile's velocity vector at the time of launch. However, from the point of view of an observer in an inertial reference frame far above the planet, the direction of the projectile's motion will be northeast precisely because it has an eastward component to its velocity vector.

As the projectile travels north, it will appear to an observer on the ground that the path will deflect toward the east due to some undetected "force" [Fig. 19.8(b)]. However, the observer in an inertial frame understands this observation as being due to the velocity difference of Earth's surface with latitude; the easterly component of the projectile's velocity vector will cause it to "outrun" observers at progressively more northerly latitudes on Earth.

It can be shown that the value of the Coriolis force as measured in a noninertial reference frame fixed to the surface of Earth is given by

$$\boxed{\mathbf{F}_C = -2m\boldsymbol{\omega} \times \mathbf{v},} \tag{19.11}$$

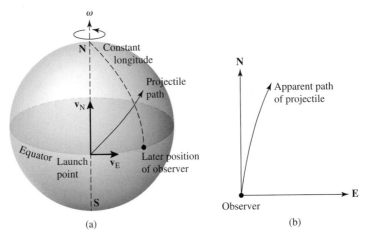

**FIGURE 19.8**   A projectile launched from the equator northward will curve toward the east as seen by an observer on Earth. (a) The view from an observer in an inertial frame above Earth. (b) The apparent trajectory as seen from an observer at rest on the surface of Earth.

where $\boldsymbol{\omega}$ is the angular velocity vector of the planet and $\mathbf{v}$ is the velocity of the projectile with respect to the noninertial reference frame. Clearly, as the velocity of the particle or the angular velocity of the planet increases, the effect of the "force" increases. On Earth, the presence of the Coriolis force causes the large-scale north–south circulation patterns to develop global east–west zonal flows [recall Fig. 19.7(b)]. Nearest the equator, the circulation patterns are generally easterly and are known as the trade winds. In the zonal regions between about 30° and 55°, the prevailing winds are the westerlies, and near the poles, the flow is again generally easterly. The Coriolis force is also responsible for the motions of clouds around high- and low-pressure systems (see Problem 19.21).

### The Complexities of Weather Systems

As anyone who has watched weather forecasts knows, weather circulation patterns on Earth are much more complex than has been described above. Effects such as moisture in the atmosphere, the diversity of land forms, the transport of heat by ocean currents, temperature differences between oceans and land, and even frictional effects between the atmosphere and the surface of the planet all contribute to the complexity of Earth's weather systems.

### SUGGESTED READING

### General

Beatty, J. Kelly, Petersen, Carolyn Collins, and Chaikin, Andrew (eds.), *The New Solar System*, Fourth Edition, Cambridge University Press and Sky Publishing Corporation, Cambridge, MA, 1999.

Booth, Nicholas, *Exploring the Solar System*, Cambridge University Press, Cambridge, 1999.

Consolmagno, Guy J., and Schaefer, Martha W., *Worlds Apart: A Textbook in Planetary Sciences*, Prentice-Hall, Englewood Cliffs, NJ, 1994.

Morrison, David, and Owen, Tobias, *The Planetary System*, Third Edition, Addison-Wesley, San Francisco, 2003.

Trefil, James, *Other Worlds: Images of the Cosmos from Earth and Space*, National Geographic Society, Washington, D.C., 1999.

### Technical

Atreya, S. K., Pollack, J. B., and Matthews, M. S. (eds.), *Origin and Evolution of Planetary and Satellite Atmospheres*, The University of Arizona Press, Tucson, 1989.

de Pater, Imke, and Lissauer, Jack J., *Planetary Sciences*, Cambridge University Press, Cambridge, 2001.

Fowles, Grant R., and Cassiday, George L., *Analytical Mechanics*, Seventh Edition, Thomson Brooks/Cole, Belmont, CA, 2005.

Holton, James R., *An Introduction to Dynamic Meteorology*, Fourth Edition, Elsevier Academic Press, Burlington, MA, 2004.

Houghton, John T., *The Physics of Atmospheres*, Third Edition, Cambridge University Press, Cambridge, 2002.

Lewis, John S., *Physics and Chemistry of the Solar System*, Academic Press, San Diego, 1995.

Lodders, Katharina, and Fegley, Jr., Bruce, *The Planetary Scientist's Companion*, Oxford University Press, New York, 1998.

Manning, Vincent, Boss, Alan P., and Russell, Sara S. (eds.), *Protostars and Planets, IV*, The University of Arizona Press, Tucson, 2000.

Seinfeld, John H., and Pandis, Spyros N., *Atmospheric Chemistry and Physics: From Air Pollution to Climate Change*, John Wiley & Sons, Inc., New York, 1998.

Taylor, Stuart Ross, *Solar System Evolution*, Second Edition, Cambridge University Press, Cambridge, 2001.

### PROBLEMS

**19.1 (a)** Based on the data given in Appendix C, express the masses of the Moon, Io, Europa, Ganymede, Callisto, Titan, Triton, and Pluto in units of the mass of Mercury.

**(b)** Express the radii of these moons and Pluto in units of the radius of Mercury.

**19.2** A second version of Bode's rule (the Blagg–Richardson formulation) is given by

$$r_n = r_0 A^n,$$

where $n$ is the number of the planet in order from the Sun outward (e.g., $n = 1$ for Mercury) and $r_0$ and $A$ are constants.

(a) Plot the position of each planet and that of Ceres on a semilog graph of $\log_{10} r_n$ vs. $n$.

(b) Draw the best-fit straight line through the data on your graph and determine the constants $r_0$ and $A$.

(c) Compare the "predictions" of your fit with the actual values for each planet by calculating the relative error,

$$\frac{r_n - r_{\text{actual}}}{r_{\text{actual}}}.$$

**19.3** Repeat Problem 19.2 for the Galilean moons of Jupiter (Io, Europa, Ganymede, and Callisto). Express their orbital distances in units of the radius of Jupiter. The data for Jupiter and its moons are found in Appendix C.

**19.4** Starting from Eq. (19.2) and using the geometry in Fig. 19.3, derive Eq. (19.3).

**19.5** (a) Assuming for simplicity that Earth is a sphere of constant density, compute the rate of change in rotational angular momentum of Earth due to the tidal influence of the Moon. Is the change positive or negative?

(b) Treating the Moon as a point mass, estimate the rate of change in orbital angular momentum of the Moon. Is this change positive or negative?

(c) Comparing your crude answers to parts (a) and (b), what can you say about the total angular momentum of the Earth–Moon system over time?

**19.6** (a) Make a rough estimate of how long it will take for Earth's rotation period to reach 47 days, at which time it will be synchronized with the Moon's orbital period.

(b) Based on what you know about the evolution of our Sun, will future inhabitants of Earth ever get the opportunity to see the Earth–Moon system completely synchronized (Earth always keeping the same "face" toward the Moon)? Why or why not?

**19.7** (a) Using Kepler's laws, estimate the distance of the Moon from Earth at some time in the distant future when the Earth–Moon system is completely synchronized at 47 days.

(b) As seen from Earth, what will the angular diameter of the Moon be at that time?

(c) Assuming that the Sun's diameter is the same as the present-day value, would a total eclipse of the Sun be possible? Why or why not?

**19.8** (a) Calculate the ratio of the tidal forces on Earth due to the Moon and the Sun.

(b) With the aid of vector diagrams, explain the cause of the strong spring tides and the relatively weak neap tides.

**19.9** Explain the almost complete lack of any tides in the Arctic Ocean at the latitude of Barrow, Alaska (71.3° N).

**19.10** Using the data in Appendix C, estimate the Roche limit for the Mars–Phobos system. Phobos's mean density is 2000 kg m$^{-3}$ and it orbits at a distance of $9.4 \times 10^6$ m. Explain the suggestion that Mars may develop a small ring system in the future.

**19.11** Why aren't spacecraft tidally disrupted when they pass near the giant planets?

**19.12** Including rotation, rederive Eq. (19.4) for the case of a spherical moon in synchronous rotation about a planet. What is your new value for $f_R$? *Hint:* You may find Kepler's third law helpful.

**19.13** (a) Use Eq. (3.17) and simple geometry to derive Eq. (19.5) for the temperature $T_p$ of a planet at a distance $D$ from the Sun.

**(b)** Imagine the greenhouse gases in Earth's atmosphere to be a single layer that is completely transparent to the visible wavelengths of light received from the Sun, but completely opaque to the infrared radiation emitted by the surface of Earth. Assume that the top and bottom surface areas of the layer are each equal to the surface area of the planet and that the temperature at the top of this atmospheric layer, $T_\oplus$, is just the blackbody temperature found in Example 19.3.1. Show that the blackbody radiation emitted by this atmospheric layer results in a warming of Earth's surface to a temperature of $T_{\text{surf}} = 2^{1/4} T_\oplus$. Compare this result with Earth's average surface temperature of $15°C = 59°F$.

**19.14** Using Eq. (19.6), estimate the temperature that would be required for all of the *atomic* hydrogen to escape Earth's atmosphere. Is this consistent with the lack of significant amounts of atomic or molecular hydrogen in the atmosphere? Why or why not?

**19.15** **(a)** Estimate the equilibrium blackbody temperature of Jupiter. Use the data found in Appendix C.

**(b)** Using Eq. (19.6), estimate the temperature that would be required for all of the hydrogen molecules to escape Jupiter's atmosphere since the planet's formation.

**(c)** Based on your answer in part (b), what would you expect the dominant component of the atmosphere to be? Why?

**19.16** Using integration by parts, show that Eqs. (19.8) and (19.9) follow directly from Eq. (19.7).

**19.17** Taking the density of air to be $1.3 \text{ kg m}^{-3}$ near the surface of Earth, show that the number density of nitrogen molecules is approximately $2 \times 10^{25} \text{ m}^{-3}$, as given in Example 19.3.3.

**19.18** Assuming that the mean free path of molecules in Earth's exosphere is sufficiently long ($\sim$ 500 km) to allow them to escape into interplanetary space, use Eq. (9.12) to estimate the number density of molecules in the exosphere. Note that you will need to make an order-of-magnitude estimate of their collision cross sections. Compare your result with the number density of nitrogen molecules quoted in Example 19.3.3. Explain any significant differences between the two values.

**19.19** **(a)** Suppose that Earth once had an atmosphere composed entirely of hydrogen atoms, rather than the molecular nitrogen and oxygen of today. Using Eq. (19.9), calculate the atmospheric escape parameter $\nu$ in this case if the temperature of the exosphere was 1000 K.

**(b)** Using a procedure identical to that of Problem 19.18, estimate the number density of hydrogen atoms in the primordial exosphere.

**(c)** What would have been the rate of loss of hydrogen atoms from the exosphere?

**(d)** Assume that the number of atomic hydrogen atoms in the atmosphere was essentially the same as the number of nitrogen molecules today. Approximately how long would it take for the hydrogen to escape from the planet's atmosphere? Express your answer in years and compare it to the age of Earth. (*Note:* In Section 23.2 we will learn that it appears unlikely that Earth ever had a substantial hydrogen atmosphere.)

**19.20** Calculate the atmospheric escape parameter $\nu$ for atomic hydrogen in Jupiter's exosphere (use $T \sim 1200$ K). Compare your result with the value obtained for Earth (see Fig. 19.6 or the result of Problem 19.19a). *Hint:* Because of numerical limitations on most calculators, you may find it necessary to first determine $\log_{10} \nu$ rather than determining $\nu$ directly.

**19.21** **(a)** Consider the case of a projectile launched from the North Pole toward the equator. With the aid of a diagram, show that the projectile is deflected westward (to the right as viewed from the launch point).

(b) Recalling that a projectile launched from the equator toward the North Pole is also deflected toward the right (eastward), show that the circulation around low-pressure systems is counterclockwise in the Northern Hemisphere.

(c) Which way do low-pressure systems circulate in the Southern Hemisphere?

**19.22** Suppose that a ball of mass $m$ is thrown with a velocity

$$\mathbf{v} = v_x\hat{\mathbf{i}} + v_y\hat{\mathbf{j}} + v_z\hat{\mathbf{k}},$$

where $\hat{\mathbf{i}}$, $\hat{\mathbf{j}}$, and $\hat{\mathbf{k}}$ are unit vectors pointing directly east, north, and upward, respectively, at the point where the ball is thrown. The latitude of the ball is $L$ when it is thrown.

(a) Show that the components of the Coriolis force on the ball are given by

$$\mathbf{F}_C = -2m\omega[(v_z \cos L - v_y \sin L)\hat{\mathbf{i}} + v_x \sin L\,\hat{\mathbf{j}} - v_x \cos L\,\hat{\mathbf{k}}].$$

*Hint:* Be sure to represent the components of the vector $\omega$ in terms of the coordinate system on the surface of Earth defined by $\hat{\mathbf{i}}$, $\hat{\mathbf{j}}$, and $\hat{\mathbf{k}}$, with the origin of the system at the position where the ball was thrown.

(b) What is the value of $\omega$ for Earth?

(c) If the ball is thrown eastward with an initial velocity vector $\mathbf{v} = 30$ m s$^{-1}\hat{\mathbf{i}}$ on the surface of Earth at a latitude of $40°$, what are the components of the acceleration vector that are due to the Coriolis force?

(d) If the ball is thrown northward with $\mathbf{v} = 30$ m s$^{-1}\hat{\mathbf{j}}$, what are the components of the acceleration vector?

(e) If the ball is thrown straight up with $\mathbf{v} = 30$ m s$^{-1}\hat{\mathbf{k}}$, what are the components of the acceleration vector? Give a simple physical explanation for the result.

# CHAPTER

# 20

# The Terrestrial Planets

## 20.1 ■ MERCURY

The four terrestrial planets have a number of characteristics in common, such as being small, rocky, and slowly rotating (see Table 19.1). Our own Moon and several of the moons of the giant planets also share many of those same characteristics. In this chapter we shall focus our attention on the terrestrial planets and their moons, saving our discussion of the giant planets and their systems for Chapter 21.

### The 3-to-2 Spin–Orbit Coupling of Mercury

As we learned in Section 17.1, the innermost planet, Mercury (Fig. 20.1), orbits so close to the Sun (0.39 AU) that Kepler's laws begin to break down. The reason is that spacetime in the vicinity of massive objects is affected in such a way that Newton's familiar inverse-square law (Eq. 2.11) is no longer a completely adequate description of gravity. It was the slow advance of the perihelion point of Mercury's rather eccentric orbit ($e = 0.2056$) that presented one of the first tests of Einstein's general theory of relativity.

The first hint that Mercury's orbit also exhibits another curious feature came in 1965 when Rolf B. Dyce and Gordon H. Pettengill successfully bounced radar signals off the planet using the Arecibo radio telescope. The reflected signals had a spread of wavelengths that revealed Mercury's rotation speed; because of the Doppler effect, radio waves that hit the approaching limb were blueshifted and those that struck the receding limb were redshifted. These observations indicated that Mercury's rotation period was approximately 59 days. More precise measurements made by the **Mariner 10** spacecraft during its repeated flybys of the planet in 1974 and 1975 showed that the rotation period was actually 58.6462 days, exactly two-thirds the length of its sidereal orbital period of 87.95 days.

How this peculiar 3-to-2 relationship between rotation and orbital periods developed can be understood in light of the process of tidal evolution discussed in Section 19.2. At perihelion, Mercury experiences the strongest tidal force, causing the planet to try to align its bulge axis along the line connecting the planet's center of mass to the center of mass of the

(a)          (b)

**FIGURE 20.1**   (a) Mercury, as seen by Mariner 10 when it was 200,000 km from the planet on March 29, 1974. (b) A portion of Caloris Basin can be seen near the terminator (the line separating day and night). Notice the semicircular rings of mountains centered on the impact point at the left-hand edge. (Courtesy NASA/JPL.)

Sun. As a result, due to the tremendous energy dissipation from friction that accompanies tidal distortions, Mercury's spin slowed to the point where the alignment ultimately did occur at perihelion during each orbit; see Fig. 20.2.

### The Surface of the Planet

Pictures returned by Mariner 10 revealed a planet that bears a strong superficial resemblance to the Moon (compare Figs. 20.1 and 20.16). Mercury is a world that is heavily cratered, indicating that it underwent extensive bombardment during its nearly 4.6-billion-year history. Such evidence of violent collisions is commonplace on many worlds, giving us a hint of the Solar System's history. One impact (at what is now known as Caloris Basin) was so large that it created ripples that traveled across the planet and converged on the opposite side to produce a jumbled collection of hills.

A careful comparison of images of the Moon and Mercury shows that Mercury's craters are often separated by regions that are largely devoid of significant cratering. Assuming that the rate of impact was roughly the same on both worlds throughout their histories, and that they formed at approximately the same time, Mercury's surface must have been refreshed

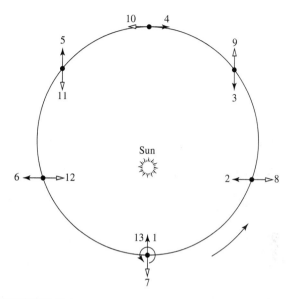

**FIGURE 20.2** The 3-to-2 spin–orbit coupling of Mercury.

more recently (meaning it must be somewhat younger) than most of the Moon's surface. This is consistent with the conclusion that since Mercury is larger and closer to the Sun, it would have cooled off more slowly after formation, and hot, molten material would have been more likely to reach the surface to cover older impact sites.

Given the planet's size and its proximity to the Sun, it is not surprising that Mercury has only a very tenuous atmosphere (recall the analysis of the retention of planetary atmospheres in Section 19.3). Because of Mercury's high temperature on the subsolar side (reaching 825 K) and its relatively low escape velocity (4.3 km s$^{-1}$), atmospheric gases quickly evaporate into space. In fact, its exosphere reaches down to the surface of the planet. What atmosphere it does possess (number densities are less than $10^{11}$ m$^{-3}$) is due to charged nuclei of hydrogen and helium from the strong solar wind that become trapped in its weak magnetic field, together with atoms of oxygen, sodium, potassium, and calcium that have escaped the surface **regolith** (or soil) of the planet. The atoms that leave the regolith may have been liberated by impacting solar wind particles or the vaporization of regolith material by micrometeorites.

Ironically, radar data suggest that this closest planet to the Sun possesses highly reflective volatile material, probably water ice, in permanently shadowed craters near the polar caps.[1] Because tidal interactions have forced the planet's rotation axis to be almost exactly perpendicular to its orbital plane, the polar regions never get more than a very small amount of sunlight. Moreover, with virtually no atmosphere to speak of, Mercury cannot efficiently transport heat away from the equatorial regions. As a result, temperatures near the poles probably never exceed 167 K, and in shadows within craters near the poles, temperatures

---

[1]The NASA 70-m tracking station at Goldstone, California, was used to send the nearly 500-kW signal at a wavelength of 3.5 cm, and the VLA received the reflected beam.

may be as low as 60 K. These temperatures are low enough that any water ice that may have been deposited there by a process such as cometary collisions would sublimate only over very long time scales.

## The Interior

Mercury's relatively high average density (5427 kg m$^{-3}$), when compared to that of the Moon (3350 kg m$^{-3}$), indicates that it must have lost most of its lighter elements and undergone enough gravitational separation to create a fairly dense core. Based on computer simulations first performed in 1987 by Willy Benz, Wayne Slattery, and Alastair G. W. Cameron (1925–2005), it appears that Mercury may have experienced a major collision with a large planetesimal early in its history. The collision was sufficiently energetic that much of the outer, lighter silicate material was removed, leaving behind the iron and nickel that had previously settled to the center of the planet. As a result, after the collision the planet's average density was substantially increased. Estimates place the mass of the impactor at about one-fifth Mercury's current mass, and the speed of the impact at perhaps 20 km s$^{-1}$. Prior to the collision, Mercury's mass may have been twice its present value. Although this may seem to be an ad hoc explanation for Mercury's unusual density, we will soon learn that the early Solar System was a violent place and that massive collisions were simply a part of its evolution.

## Mercury's Weak Magnetic Field

Mercury's rotation, together with its large conducting metallic core, may be responsible for its magnetic field. The maximum strength measured by Mariner 10 was about $4 \times 10^{-7}$ T at an altitude of 330 km, about 100 times weaker than the magnetic field measured near Earth's surface. The mechanism for generating this and other planetary magnetic fields is believed to be the magnetic dynamo, essentially the same process responsible for the Sun's magnetic field (Section 11.3). The difference between the planetary and stellar mechanisms is that a *liquid metallic* conducting core replaces the ionized gas in stars as the source of the field. To date, the details of planetary dynamos are not well understood. In Mercury's case, the fact that the rotation is so slow seems to contradict the idea that a magnetic dynamo is currently in operation. Furthermore, the relatively small size of the planet suggests that its core should have cooled to the point that any molten core would be too insignificant to generate a measurable field. As a result, opponents of a present-day dynamo mechanism suggest that Mercury's magnetic field may be a "frozen-in" remnant of its past, when the planet may have been rotating faster and when it was warmer.

## 20.2 ■ VENUS

Venus, the second planet from the Sun, is sometimes referred to as Earth's sister planet because its mass ($0.815\ M_{\oplus}$) and radius ($0.9488\ R_{\oplus}$) are comparable to Earth's. Despite these basic similarities, the two planets are markedly different in many of their fundamental features.

### Retrograde Rotation

In the 1960s, one of the many unusual features of Venus was discovered. Astronomers learned that the atmospheric circulation was retrograde (the direction opposite its orbital motion), with speeds near 100 m s$^{-1}$ at the cloud tops close to the equator; see Fig. 20.3(a). This inference was based initially on observations of the clouds in its atmosphere; it was confirmed later by measurements of the Doppler shift of spectral lines from sunlight reflected off the clouds. Later, Earth-based radar Doppler measurements of the surface (like those made of Mercury) revealed that the planet itself also rotates retrograde, but 60 times more slowly than its upper atmosphere. The sidereal rotation period of Venus is a very sluggish 243 days; this compares to its orbital period of 224.7 days.

The retrograde rotation of the planet is an interesting puzzle. All of the planets in the Solar System orbit prograde, as do most of their moons. This means that these worlds orbit in a counterclockwise direction as seen from a vantage point above Earth's north pole. Furthermore, with the exceptions of Venus, Uranus, and Pluto, all of the other planets and most of their moons also rotate prograde, just as the Sun does. This agrees with what one would expect from the development of the Solar System out of a spinning disk of material that formed when the Sun did.

Based on detailed analytical and numerical studies of the interactions among Venus, the Sun, and the other planets in the Solar System, Alexandre Correia and Jacques Laskar were able to show that Venus's retrograde rotation can be explained in terms of gravitational perturbations. The many perturbations acting on Venus from the other bodies in the Solar

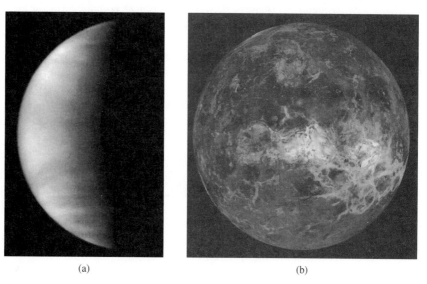

(a)                              (b)

**FIGURE 20.3**    (a) An ultraviolet image of Venus obtained in 1995 by WF/PC 2 onboard the Hubble Space Telescope. Note the "Y"-shaped cloud features at the top of the planet's thick atmosphere. No surface features can be seen in visible or ultraviolet light. (Courtesy of L. Esposito, University of Colorado, Boulder, and NASA.) (b) A composite radar image of the surface of Venus obtained by the Magellan spacecraft that orbited the planet from 1990 to 1994. (Courtesy of NASA/JPL.)

System lead to a chaotic zone in the tilt of the axis between 0° and 90°. As gravitational perturbations cause Venus to pass through that zone, the tilt of its rotation axis can vary dramatically. Within the first several million years following the planet's formation, its very thick atmosphere develops and also starts to influence the outcome of the planet's rotation. This occurs because the thick atmosphere can be significantly affected by tidal forces, and the thick atmosphere can also produce a damping effect. Numerical simulations based on various initial conditions for the rotation period and tilt of Venus's axis most often led to the very slow, retrograde rotation of Venus that is observed today. However, the path to this final state could be through either the flipping of the rotation axis to near 180° or the slowing of the spin rate to zero at an axis tilt of 0°, and then tides producing a slow retrograde rotation.

The dynamical behavior of Venus's atmosphere is also a puzzle. Probes that entered the atmosphere measured the presence of two large Hadley cells (Section 19.3), one in each hemisphere, consistent with the planet's slow rotation rate and the corresponding lack of any significant Coriolis force [$\omega$ is very small in Eq. (19.11)]. However, near the equator the cloud cover circles the planet in just four days, producing the "Y"-shaped cloud patterns evident in Fig. 20.3(a). Such high-speed motions are common in high-altitude jet streams (narrow rivers of air) but are unusual for the bulk of the atmosphere, particularly with such slow underlying rotation.

## The Lack of a Magnetic Field

One consequence of the planet's slow rotation that does agree with expectation is the lack of any measurable magnetic field. The currents within a molten, conducting core are generated by planetary rotation; therefore, one crucial component of the magnetic dynamo mechanism is absent in Venus. Because there is no magnetic field to protect the planet via the Lorentz force (Eq. 11.2), supersonic ions in the solar wind directly strike the upper atmosphere, causing collisional ionization and a standing shock wave at the location where the solar wind particles are abruptly slowed to subsonic speeds.

## The Hot, Thick Atmosphere of Venus

Analysis of the composition of the dense atmosphere, made first by ground-based telescopes and later by Soviet and American probes, revealed that its chief constituent is carbon dioxide ($CO_2$), which makes up about 96.5% of the total number of atoms or molecules, with molecular nitrogen ($N_2$) making up most of the remainder (3.5%). Traces of other molecules are also present, most notably argon (70 ppm),[2] sulfur dioxide ($SO_2$, 60 ppm) carbon monoxide (CO, 50 ppm), and water ($H_2O$, 50 ppm). The probes even detected thick clouds of concentrated sulfuric acid. At the base of the atmosphere the temperature is 740 K, sufficient to melt lead, and the pressure is 90 atm, equal to the pressure at a depth of over 800 m below the surface of Earth's oceans.[3]

The very high surface temperature far exceeds what is expected from a simple blackbody analysis, such as the one performed in Example 19.3.1. It is the large amount of carbon

---

[2] ppm represents parts per million.
[3] 1 atm = $1.013 \times 10^5$ N m$^{-2}$.

dioxide (a greenhouse gas) in the atmosphere that is responsible for the extreme conditions at the surface. The atmosphere is so thick that the optical depth at infrared wavelengths is approximately $\tau = 70$, meaning that the temperature is increased over the blackbody temperature that would be predicted for an *airless* planet at the location of Venus by a factor of nearly $(1 + \tau)^{(1/4)} = 2.9$; see Problem 20.7.

How could Earth's sister planet have developed an atmosphere so different from our own? The formation of terrestrial atmospheres is still not well understood and is an area of active research. However, based on the direct evidence we have of outgassing from Earth's volcanoes and the discovery of volcanoes on both Venus and Mars, it seems likely that at least a portion of a terrestrial planet's atmosphere may arise from volcanic activity. It has also been suggested that significant fractions of the atmospheres of these planets may have been delivered by comets and meteorites. If the later suggestion is true, then understanding the atmospheric evolution of the terrestrial worlds requires a greater understanding of the composition of comets and meteorites, as well as the frequency with which they collide with the worlds of the inner Solar System. Comets and meteorites are discussed in more detail in Sections 22.2 and 22.4, respectively.

Whatever the source of Venus's primordial atmosphere, carbon dioxide is the dominant constituent today, and very little water is present. Conversely, water is abundant in Earth's oceans, but there is very little atmospheric carbon dioxide. What happened to change the relative abundances of those molecules on the two planets? If the two worlds began with similar compositions, as seems likely, given that they formed near one another in the solar nebula and have comparable sizes, then water was probably much more abundant on Venus in the past. In fact, since the luminosity of the zero-age main sequence Sun was only about $0.677\,L_\odot$, much less than it is today (recall Fig. 11.1), Venus may have even had hot water oceans on its surface early in its history. As the Sun's luminosity increased and the planet was bombarded by planetesimals, the surface temperature began to rise and the oceans started to evaporate. The addition of more infrared-absorbing water vapor in the atmosphere triggered a *runaway greenhouse effect*, causing the surface temperature to climb to near 1800 K, hot enough to vaporize the remainder of the water and even melt rock. At the same time, the atmospheric pressure at the surface reached 300 atm. Since $H_2O$ is lighter than $CO_2$, the water migrated to the top of the atmosphere where it was dissociated by solar ultraviolet radiation through the reaction $H_2O + \gamma \rightarrow H + OH$. This UV photodissociation process liberated the lighter hydrogen atoms, allowing most of them to escape from the planet. Since the carbon dioxide remained, it became the dominant species in the atmosphere of Venus.

As is true of any viable scientific theory, it is important that the theory make testable predictions. In the evolutionary scenario for the atmosphere of Venus just described, the photodissociation of water should have left behind altered isotope ratios of hydrogen. Hydrogen has two stable isotopes, $^1_1H$ (or simply hydrogen, H) and $^2_1H$ (deuterium, D), which are chemically identical but differ in mass by a factor of two. On Earth the ratio of the numbers of deuterium atoms to hydrogen atoms is $D/H = 1.57 \times 10^{-4}$. However, within the atmosphere of Venus the ratio is closer to $D/H = 0.016$. The factor-of-100 increase in the atmospheric D/H ratio for Venus relative to Earth is due to the more massive isotope's slower rate of escape (recall Eqs. 19.8 and 19.9). Apparently our understanding of the runaway greenhouse effect on Venus is essentially correct.

### Studying the Surface

Because of the thick cloud cover and inhospitable climate, it has been a difficult job gathering information about the surface of Venus. The Soviet **Venera** missions in the late 1960s through early 1980s were able to descend into the atmosphere of the planet and in some cases land on the surface, operating for short periods of time before succumbing to the environment found there. While on the surface, they returned pictures of their immediate vicinity. The landers also sampled the composition of the atmosphere and surrounding rock, confirming the presence of sulfur in the atmosphere and finding rock of volcanic origin on the surface. Variations in sulfur dioxide content over timescales of decades and the detection of radio bursts characteristic of lightning in the atmosphere support the suggestion of recent volcanic activity. In particular, observations by various spacecraft and ground-based telescopes indicate that the sulfur dioxide content in the atmosphere of Venus has decreased by more than an order of magnitude, with some intermittent fluctuations since the late 1970s. Since ultraviolet radiation converts $SO_2$ to sulfuric acid in the upper atmosphere, the observed decrease in the sulfur dioxide concentration has led some scientists to suggest that a major eruption may have occurred sometime during the 1970s, with a smaller event occuring around 1992.

By far the greatest amount of information about the surface of Venus has come from radar imaging, because radio signals can easily penetrate the atmosphere even though visible and ultraviolet light cannot. Radar studies have been carried out using Earth-based telescopes such as Arecibo, and from orbiters, including the Venera and **Pioneer** series, and most recently from the **Magellan** spacecraft. Launched from the Space Shuttle *Atlantis* in 1989, Magellan's very successful mission lasted until 1994, when it was intentionally sent diving into the atmosphere to gather information about the density structure of the planet's atmosphere. During Magellan's operational lifetime, it mapped 98% of the surface at resolutions of between 75 m and 120 m. A Magellan mosaic of one hemisphere of Venus is shown in Fig. 20.3(b).

During roughly one-half of its mission, Magellan sent back a continuous radio signal to Earth so that scientists could monitor the variation in the signal's wavelength caused by the Doppler effect. As Magellan passed over regions of higher average density, the local gravitational pull would speed up the spacecraft slightly and the wavelength of the signal received at Earth would change. In this way, Magellan was used to generate a detailed gravity map of the planet, covering approximately 95% of its surface.[4]

By combining images of a given region made from two different locations, together with gravity information, scientists have been able to produce detailed three-dimensional images of much of the planet's surface. Figure 20.4 shows Maat Mons, a volcano 8 km high located 0.9 degrees north of the equator. In this image the vertical relief has been exaggerated by a factor of 22.5 to bring out important features. Based on changes in surface reflectivity, variations in rock characteristics become evident. Apparent in the image of Maat Mons are lava flows that extend for hundreds of kilometers from the volcano. Estimates place the age of the surface rock around Maat Mons at less than 10 million years, and it may be much younger.

---

[4]To increase the resolution of the gravity data, flight controllers lowered Magellan's orbit using the previously untried technique of aerobraking; Magellan dipped down into the atmosphere slightly, causing the spacecraft to lose orbital energy via atmospheric drag.

**FIGURE 20.4** Maat Mons is believed to be the tallest volcano on Venus, measuring 8 km in height. The vertical scale has been increased over the horizontal scale by a factor of 22.5. (Courtesy of NASA/JPL.)

It appears that the entire surface of the planet may have been refreshed relatively recently compared to the age of the Solar System. This estimate comes from the number of impact craters found on the surface (e.g., see Fig. 20.5). If we assume that Venus has been struck by impacts at about the same rate as other worlds in the inner Solar System (such as Mercury or our Moon), then from the relatively low number of craters found on the surface of Venus we can conclude that large-scale lava flows must have occurred about 500 million years ago.[5] In support of this conclusion, nearly one thousand volcanic features have been identified on the surface of the planet.

## 20.3 ■ EARTH

The planet for which we have the greatest amount of information by far is of course our Earth (see Fig. 20.6). We have studied its atmosphere, its oceans, and its active geology with a great deal of specificity. We have been able to carefully investigate its extensive biology, from the smallest microbes to the largest plants and animals, and to study the evolutionary processes that have led to our planet's great biodiversity. We have also been able to extend our knowledge by developing follow-up experiments based on information gained from previous studies. This makes investigating our own planet significantly more robust and interactive than investigating other bodies in our Solar System to date.[6]

[5]Absolute age estimates of the Moon will be discussed in detail in Section 20.4.
[6]However, as we will see when we discuss Mars in Section 20.5, humans have begun to conduct extensive robotic studies of that planet based directly on information returned from previous and ongoing missions.

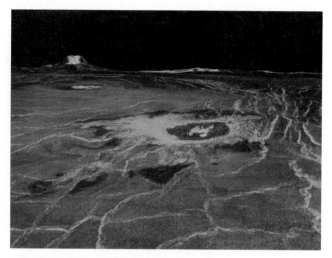

**FIGURE 20.5**    The impact crater Cunitz is visible in this image, with the volcano Gula Mons in the background. The vertical scale has been increased over the horizontal scale by a factor of 22.5. (Courtesy of NASA/JPL.)

(a)                                                           (b)

**FIGURE 20.6**    (a) Earthrise over the limb of the Moon on December 22, 1968. This picture was taken by the astronauts of *Apollo 8*. (b) Earth as seen by the *Apollo 17* astronauts while traveling to the Moon on December 7, 1972. Visible are most of Africa, Saudi Arabia, and the south polar ice cap. (Courtesy of NASA.)

### Our Atmosphere

Beginning early in its history, the bulk of Earth's water condensed to form its oceans. Unlike Venus, however, given Earth's slightly greater distance from the Sun, our planet never got hot enough to turn much of the liquid to vapor (see Eq. 19.5). Therefore, the ensuing runaway greenhouse effect described in Section 20.2 never developed. Instead, the carbon dioxide in the atmosphere was dissolved into water, where it became chemically bound up in carbonate rocks such as limestone. If all of the carbon dioxide trapped within rock today were released into Earth's atmosphere, the amount would be comparable to that currently contained in the atmosphere of Venus.

However, it is also important to recall from Fig. 11.1 that the Sun was significantly less luminous in the early Solar System than it is today. This implies that Earth's surface would have been cooler in the past and its water should have been in the form of ice, even as recently as 2 billion years ago. However, geologic evidence, including fossil records, suggests that Earth's oceans were liquid as early as 3.8 billion years ago. This puzzle has become known as the *faint ancient Sun paradox*. The resolution of this paradox probably lies in details of the greenhouse effect and a different atmospheric composition than exists at the present time.

The present-day atmosphere of Earth is made up of (by number) 78% $N_2$, 21% $O_2$, 1% $H_2O$, and traces of Ar, $CO_2$, and other constituents. The atmosphere owes its current composition in part to the development of life on the planet. For instance, plants process carbon dioxide into oxygen as a by-product of photosynthesis.

### The Greenhouse Effect and Global Warming

Serious concern now exists over the effects of artificially introducing carbon dioxide and other greenhouse gases into Earth's atmosphere by industrial means. To complicate matters, we are simultaneously destroying vast regions of vegetation, such as the Amazon rain forests, that could recycle $CO_2$. The commonly used technique of slash-and-burn clearing of the rain forests also releases tremendous quantities of carbon dioxide into the atmosphere.

To illustrate the problem, recent changes in the abundance of carbon dioxide over Mauna Loa, Hawaii, are shown in Fig. 20.7; the oscillations are due to the annual growing season.

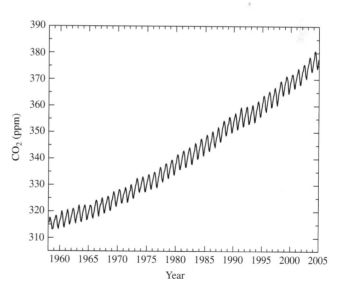

**FIGURE 20.7** The amount of carbon dioxide in parts per million by volume (ppm) over Mauna Loa, Hawaii, as a function of time. (Data from C. D. Keeling, T. P. Whorf, and the Carbon Dioxide Research Group, Scripps Institution of Oceanography, University of California.)

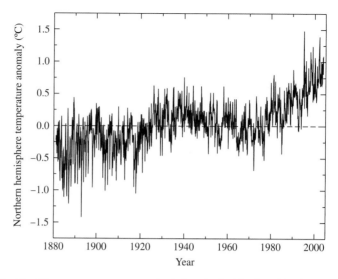

**FIGURE 20.8**    Monthly average temperature deviations in the Northern Hemisphere of Earth from 1881 to 2003. The deviations are measured from a 25-year average between 1951 and 1975. (Data from K.M. Lugina, P.Ya. Groisman, K.Ya. Vinnikov, V.V. Koknaeva, and N.A. Speranskaya, 2004. *In Trends Online: A Compendium of Data on Global Change.* Carbon Dioxide Information Analysis Center, Oak Ridge National Laboratory, U.S. Department of Energy, Oak Ridge, Tennessee, U.S.A.)

Because of the nonlinear behavior of the greenhouse effect and the very complicated physics, chemistry, and meteorology that are involved, accurate computer models are only slowly becoming available. However, despite current limitations in the predictive power of these models, the basic effects of greenhouse gases are understood. As we have learned from Venus, increasing the content of greenhouse gases in an atmosphere will raise its average temperature. The questions are by how much the temperature will increase, and how rapidly it will occur.

Figure 20.8 shows the average temperature deviations in the Northern Hemisphere of Earth between 1881 and 2003 with respect to a 25-yr average of temperatures computed from 1951 through 1975. Evident is a consistently upward trend in the average temperature since about 1970. Whether this upward trend is the start of a long-term steady increase or a fairly short-term fluctuation has been a matter of some debate. However, it is clear that a significant upward trend is currently under way; in fact, seven of the ten hottest years in the twentieth century occurred in the 1990s.

Associated with the effects of **global warming** is evidence that Earth's glaciers are receding world wide. In addition, the Arctic ice cap has thinned significantly since 1970 and Earth's ocean levels have risen. Also supporting the conclusion that human-driven global warming is occurring is the increase in the average ocean surface temperature by approximately 0.5°C since the late 1960s, with warming extending down to depths of several hundred meters. Since the oceans eventually absorb some 84% of the excess heat in the atmosphere, the observation of this temperature increase is significant. The ocean temperature increase also agrees with computer modeling of global climate changes that include the influence of increased greenhouse gas emissions.

Another environmental concern about human activity is the release of chlorofluorocarbons into the atmosphere. These molecules migrate into the upper atmosphere above the North and South Poles, where they are destroying the ozone ($O_3$). Ozone is known to be a major absorber of ultraviolet radiation, and as such, it plays an important role in protecting life on Earth's surface.

Much more research is necessary before we can hope to understand the magnitude of the environmental consequences of human behavior. Unfortunately, by the time more detailed predictions are available, it may not be possible to reverse the trend.

In recognition of the importance of global warming to the inhabitants of our planet, the first-ever "Earth Summit" was held in 1992, involving most of the nations of the world. Officially known as the United Nations Conference on Environment and Development, its purpose was to discuss global environmental concerns. The treaty that came from that summit is the Framework Convention on Climate Change. Then, in December 1997, more than 160 nations met in Kyoto, Japan, to negotiate binding limitations on greenhouse gases for the developed nations. After much debate and compromise, resolutions regarding such things as the emissions of greenhouse gases were finally agreed on. The Kyoto Protocol came into force on February 16, 2005, after being ratified by 157 countries. However, the United States, the world's largest producer of greenhouse gases at the time the Kyoto Protocol went into effect, did not ratify the agreement, citing concerns over its impact on the nation's economy.

### Seismology and Earth's Interior

The structure of Earth's interior can be derived by analyzing the seismic waves generated by earthquakes. Two principal types of waves are produced by earthquakes: **P waves** (pressure or *primary*) are longitudinal waves capable of traveling through both liquids and solids, and **S waves** (shear or *secondary*) are transverse waves that are restricted to traveling through solids only (see Fig. 20.9). Since the velocities and paths of both P and S waves depend on the medium through which they are moving, their detection around the world enables geologists to deduce the structure of our planet.[7] For instance, in regions where only P waves are measured, the absence of S waves implies that there must have been intervening liquid in the path of the wave (see Fig. 20.10). Furthermore, because of the refraction that occurs at boundary interfaces (much like the refraction of light rays at the boundaries between media of differing indices of refraction), **shadow zones** exist where neither type of wave can be detected. Thus, using the data from P and S waves, geologists can map the interior of the planet. Such maps yield information about the depth of the surface **crust** and reveal the existence of a solid **inner core**, a molten **outer core**, and a thick **mantle**.

The behavior of P waves in the outer core implies that its composition is predominantly iron and nickel. This also agrees with the fact that the average density of Earth is $5515 \text{ kg m}^{-3}$, greater than the density of surface rocks (typically $3000 \text{ kg m}^{-3}$) and water ($1000 \text{ kg m}^{-3}$).[8] It is the combination of high temperature ($> 4000$ K) and composition that results in a *liquid outer core*. The transition back to a *solid inner core* occurs because of the extreme pressures found there.

---

[7]An analogous procedure is used to study the interiors of many stars; recall Chapter 14.

[8]Gravitational compression also contributes to the higher value for the average density, relative to surface material.

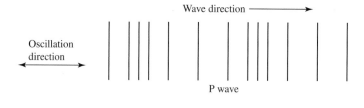

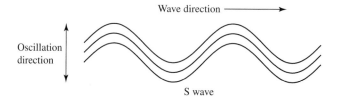

**FIGURE 20.9**   P waves are longitudinal pressure waves capable of traveling through both liquids and solids. S waves are transverse shear waves that can travel only through solids.

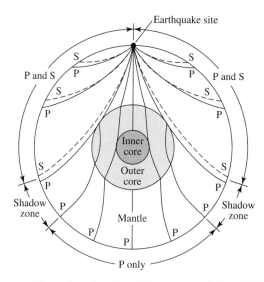

**FIGURE 20.10**   Generated by earthquakes, P and S waves travel through Earth's interior. S waves are unable to traverse the molten outer core. Furthermore, the refraction of P waves at the interface between the outer core and the mantle produces shadow zones.

## Plate Tectonics

Although the presence of volcanos is a feature that Earth shares with Venus and Mars, Earth's present-day **tectonic activity** appears to be unique among the terrestrial planets. This activity has its origin in the dynamic interior of Earth, depicted in Fig. 20.11. Earth's surface layer, known as the **lithosphere**, encompasses both the oceanic and continental crust as well as the outer portion of the mantle. The lithosphere is fractured into **crustal plates**

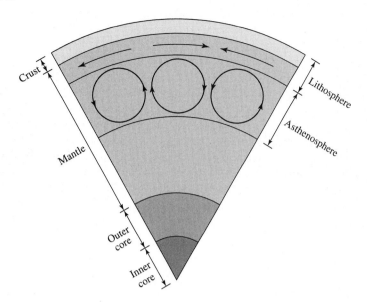

**FIGURE 20.11**   The interior structure of Earth is composed of an inner solid core, an outer liquid core, a mantle, and surface crust. The crust and the outer portion of the mantle make up the lithosphere (containing the surface plates) and the underlying convective asthenosphere. The diagram is not to scale.

(see Fig. 20.12) and rides on the convective, somewhat plastic **asthenosphere**, which is also part of the mantle. As the plates move across the surface of the planet, they crash into or grind against one another, carrying the continents with them.[9] Because of these motions, the Atlantic Ocean is widening at the rate of approximately 3 cm yr$^{-1}$, spreading away from an underwater mountain range that runs the length of the ocean floor. This mid-Atlantic ridge is the location where material from the interior rises to the surface, generating new sea floor as the continents separate (see Fig. 20.13).

Extrapolating the motions of the plates backward in time, geologists believe that there was once one giant supercontinent, known as Pangaea, that broke apart some 200 million years ago into two smaller supercontinents, Laurasia and Gondwanaland. Gondwanaland in turn separated into South America and Africa, and Laurasia divided into Eurasia and North America.

Earth's plate boundaries are generally the sites of active volcanism, mountain building, and frequent earthquakes. For example, when two plates collide, the lighter continental crust overrides the heavier oceanic crust and a **subduction zone** develops, as illustrated in Fig. 20.13. One such location is along the coast of Japan, where its volcanic islands were created as a result of the heat generated by friction as the oceanic crust descended into the interior of Earth. It is at the location of these subduction zones that deep oceanic trenches also develop. If two plates collide that contain continental crust, neither plate will overrun the other; instead, buckling occurs and a mountain range such as the Himalayas is generated.

---

[9]For example, the Pacific and North American plates are currently sliding past one another. The famous San Andreas fault is located on the boundary between these two plates.

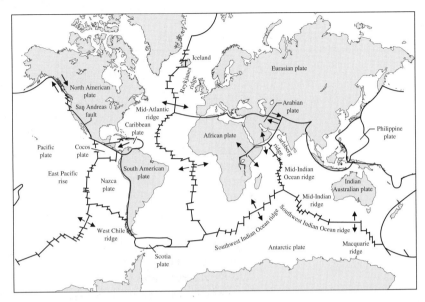

**FIGURE 20.12** The lithosphere is divided into crustal plates that travel across Earth's surface.

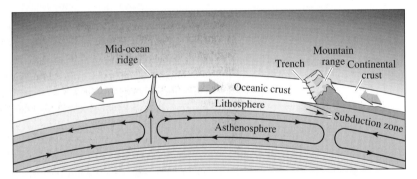

**FIGURE 20.13** The motions of the plates are driven by convection zones in the asthenosphere. A mid-ocean ridge (a rift) occurs where material from below is pushed up to the surface. A subduction zone develops when the lighter continental crust overrides the heavier oceanic crust of two colliding plates.

### Sources of Internal Heat

All of this activity requires one or more sources of energy to sustain itself. Heat is known to be escaping into space through the surface of Earth at a rate of $4 \times 10^{13}$ W, implying an average flux of 0.078 W m$^{-2}$. If the only source of energy in the interior were the heat left over from the formation of the planet almost 4.6 billion years ago, plate tectonic activity would have long since ceased. Other sources of heat augment the *energy budget* of Earth, including the tidal dissipation of its rotational kinetic energy (see Problem 20.10), possible ongoing gravitational separation (releasing gravitational potential energy as heavier constituents sink toward the center of the planet), and the continual radioactive decay of unstable isotopes (believed to be the primary source of the heat).[10] This allows much of the interior to remain somewhat plastic, supporting the large, sluggish convection cells that drive the motions of the crustal plates.

### Earth's Variable Magnetic Field

The presence of a molten iron–nickel outer core together with Earth's relatively rapid rotation rate is consistent with the observation that the planet possesses a global magnetic field, assuming that a dynamo is in operation in the planet's interior. The existence of Earth's magnetic field serves to protect the planet from incoming charged particles in the solar wind, as well as other ionized cosmic rays. Instead of striking the surface, these particles become trapped in the dipole field and bounce back and forth between the North and South poles (Fig. 20.14). Three regions of trapped particles have been identified and are known as the **Van Allen radiation belts**. The innermost belt is composed of protons and is at a height of roughly 4000 km above Earth's surface. Overlapping a portion of the inner belt is a second belt composed of atomic nuclei that were once part of the interstellar medium. The outermost belt is composed of electrons at an altitude of approximately 16,000 km. Particles in the belts that are energetic enough to enter Earth's upper atmosphere near the poles strike atoms and molecules there, causing collisional excitation, ionization, and dissociation. When the atoms or molecules recombine, or when the electrons drop back down to lower energy levels, the subsequent emission of light is observed as the **aurora borealis** (northern lights) and **aurora australis** (southern lights); see Fig. 20.15.

Interestingly, geologic evidence indicates that Earth's magnetic field weakens, reverses polarity, and reestablishes itself on an irregular time scale of some $10^5$ years. This can be seen in the orientation of magnetic minerals trapped in molten rock that later solidified,

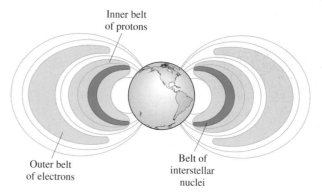

**FIGURE 20.14** The Van Allen radiation belts arise from charged particles becoming trapped in the magnetic field of Earth.

---

[10]In the mid-1800s, Lord Kelvin argued that Earth could not be more than about 80 million years old. His argument was based on the amount of gravitational potential energy the planet could release and the rate at which heat escaped over time. However, his calculation was made before the discovery of radioactivity.

**FIGURE 20.15**    The aurora are due to the collision of high-speed particles with atoms and molecules in Earth's upper atmosphere. (Courtesy of the Geophysical Institute, University of Alaska, Fairbanks.)

such as those sampled on either side of the spreading mid-Atlantic ridge (Figs. 20.12 and 20.13). In this way, a "fossil record" of the direction of the local magnetic field is created. The behavior of Earth's field is not unlike the solar cycle, the flipping of the Sun's magnetic field roughly every 11 years (Section 11.3). Earth's magnetic field is known to be weakening today.

## 20.4 ■ THE MOON

Despite the proximity of the Moon to Earth, the two worlds are very different (see Fig. 20.16). Because of its low surface gravity, the Moon has been unable to retain a significant atmosphere. Without a protective atmosphere, the Moon has suffered impacts by meteorites throughout its history. Along with a large number of smaller impacts, a significant number of very large collisions occurred approximately 700 million years after the Moon formed. These impacts were powerful enough to penetrate its thin crust, allowing molten rock in the interior to flow across the surface. The result was the formation of the many smooth, roughly circular **maria** (or "seas") that can be seen on the surface of the Moon facing Earth. It is the distribution of these maria that has led humans to imagine seeing the face of the "man in the Moon."

### The Moon's Internal Structure

Major advances in our understanding of the Moon's internal structure and evolutionary history have occurred as a result of intense exploration from 1959 through the early 1970s. When the **Apollo** astronauts landed on the Moon, they left seismic detectors designed to measure any *moonquakes* that may be occurring. Many of the very weak quakes that were detected (about magnitude 1 on the Richter scale) were triggered by the tidal strain generated by Earth's gravitational pull. Another class of vibrations has been attributed to the Moon's

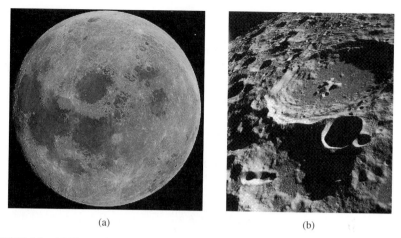

(a)                                        (b)

**FIGURE 20.16**   (a) The surface of the Moon contains heavily cratered highland regions and nearly circular maria that are much less heavily cratered. The portion of the Moon facing Earth is on the left side of the image. (b) A portion of the far side of the Moon showing the extensive cratering found there. The diameter of the large crater is approximately 80 km. This view was obtained by the *Apollo 11* astronauts in 1969. (Courtesy of NASA.)

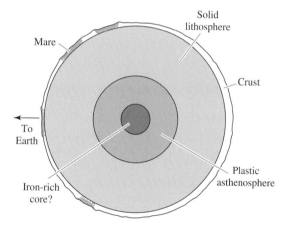

**FIGURE 20.17**   The interior structure of the Moon.

"ringing" after being struck by meteorites. Just as with the analysis of seismic activity on Earth, moonquakes have allowed scientists to develop an understanding of the Moon's interior.

Many of the moonquakes appear to have originated, not at tectonic plate boundaries, but at the interface between the solid, brittle lithosphere and the plastic asthenosphere (see Fig. 20.17). It also appears likely that below the asthenosphere, a small iron-rich core exists as well. This structure is consistent with measurements of the small amount of heat still flowing outward from the Moon's interior, which is responsible for maintaining the plastic nature of the asthenosphere. However, based on the data provided by this seismic activity, Earth-like tectonic activity appears to be absent on the Moon today.

Interestingly, only one mare was seen on the side of the Moon farthest from Earth.[11] This is not because collisions were preferentially occurring on the side of the Moon facing Earth; rather, the crust is actually thinner on the near side. Consequently, impacts on the thin-crust side were more likely to penetrate the crust, allowing interior molten rock to flow over the surface. Because the crust is less dense than material in the Moon's interior, tidal forces have caused the heavier near side to permanently "hang down" toward Earth.

### The Absence of a Global Magnetic Field

Unlike Earth, our Moon has no measurable global magnetic field, apparently because the Moon is small enough to have cooled off much more rapidly than Earth. This evolution has left the Moon as a geologically inactive world today. Furthermore, the Moon's rotation period is more than 27 times longer than Earth's. As a result, there is no evidence of any significant magnetic dynamo in operation, suggesting that if a molten core is present, it is likely to be quite small.[12]

The lack of a global magnetic field on the Moon makes the detection of a weak field around Mercury even more puzzling (recall the discussion on page 740). The two worlds are comparable in mass and radius (the Moon's mass and radius are 23% and 71% of the values for Mercury, respectively), while Mercury's rotation rate is slower by a factor of two. Clearly, much work remains to be done in understanding the details of magnetic field production.

### Moon Rocks

During the 1960s and 1970s, six manned United States Apollo missions returned 382 kg of surface rocks and regolith from the Moon's surface. In addition, three unmanned Soviet Union **Luna** missions returned an addition 0.3 kg of material. The samples were collected from both the maria and the **highland** (or mountainous) regions between the maria. These samples represent the most detailed information we have about the nature of our closest neighbor.

Composition analysis of samples returned from the maria confirm that they are in fact volcanic in origin. The rocks are **basalts**, similar to the kind of volcanic rock found on Earth. The lunar basalts are rich in iron and magnesium, and they also contain glassy structures that are characteristic of rapid cooling. However, unlike Earth basalts, the lunar samples contain no water and a lower percentage of **volatiles** (elements or compounds with low melting and boiling temperatures) relative to **refractories** (higher melting and boiling temperatures).

### Radioactive Dating

Perhaps the most eagerly awaited results of the analysis of the lunar samples were the determinations of their ages. The process is based on measuring the abundances of certain

---

[11]The first observations of the far side of the Moon were made by the Soviet Luna 3 mission in 1959.

[12]Based on the natural remnant magnetization of returned lunar samples and the patchy magnetization detected by satellites, it appears that the Moon once had a global magnetic field. However, there is no evidence of a global field today.

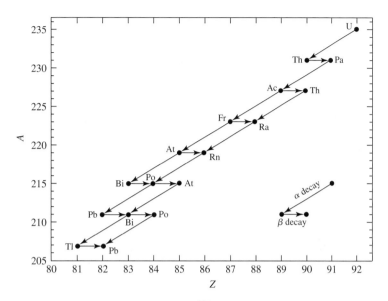

**FIGURE 20.18** The $^{235}_{92}$U decay sequence.

radioactive isotopes and comparing them with the abundances of stable end products of the decay sequence. In this technique of **radioactive dating**, we assume that the "clock" started ticking when the rock solidified, trapping the isotopes inside.

If the half-life of one step in the decay sequence is significantly longer than any of the others, it can be assumed that the original isotope decays directly into the final product with a half-life approximately equal to that of the longest one. For instance, in the decay sequence depicted in Fig. 20.18, which begins with $^{235}_{92}$U and ends with $^{207}_{82}$Pb, the first step, the alpha particle[13] decay $^{235}_{92}$U $\rightarrow$ $^{231}_{90}$Th $+$ $^{4}_{2}$He, has a half-life of $7.04 \times 10^8$ years, while the next slowest step, $^{231}_{91}$Pa $\rightarrow$ $^{227}_{89}$Ac $+$ $^{4}_{2}$He, has a half-life of only $3.276 \times 10^4$ years. As a result, to a good approximation, the half-life of the entire sequence can be taken to be $7.04 \times 10^8$ years. This means that by measuring the relative abundances of the uranium and lead isotopes, we can determine the time required for the transformation.

Some radioactive isotopes that are useful for dating Moon rocks, as well as Earth rocks and meteorites, are given in Table 20.1. Note that the stable products are not necessarily the direct result of a single decay but may be produced after a succession of decays, the longest of which has the quoted half-life.

To understand the method of radioactive dating more fully, suppose that isotope $A$ decays into isotope $B$ (which is stable), either directly or indirectly through a series of steps. From Eq. (15.10) we know that if the number of atoms of $A$ in the sample was initially $N_{A,i}$, then after some time $t$, the number remaining is

$$N_{A,f} = N_{A,i}e^{-\lambda t},$$

---

[13]Recall that helium nuclei ($^{4}_{2}$He) are often referred to as alpha particles ($\alpha$).

**TABLE 20.1**    Radioactive Isotopes with a Half-Life Useful for Determining Geologic Ages.

| Radioactive Parent | Stable Product | Half-Life $(10^9 \text{ yr})$ |
|---|---|---|
| $^{129}_{53}\text{I}$ | $^{129}_{54}\text{Xe}$ | 0.016 |
| $^{235}_{92}\text{U}$ | $^{207}_{82}\text{Pb}$ | 0.704 |
| $^{40}_{19}\text{K}$ | $^{40}_{18}\text{Ar}$ | 1.280 |
| $^{238}_{92}\text{U}$ | $^{206}_{82}\text{Pb}$ | 4.468 |
| $^{232}_{90}\text{Th}$ | $^{208}_{82}\text{Pb}$ | 14.01 |
| $^{176}_{71}\text{Lu}$ | $^{176}_{72}\text{Hf}$ | 37.8 |
| $^{87}_{37}\text{Rb}$ | $^{87}_{38}\text{Sr}$ | 47.5 |
| $^{147}_{62}\text{Sm}$ | $^{143}_{60}\text{Nd}$ | 106.0 |

where

$$\lambda = \frac{\ln 2}{\tau_{1/2}}$$

is the decay constant and $\tau_{1/2}$ is the half-life. Because the total number of atoms of $A$ and $B$ must remain constant over time (even though $A$ is ultimately being converted into $B$), it is necessary that

$$N_{A,f} + N_{B,f} = N_{A,i} + N_{B,i}.$$

Solving for $N_{A,i}$, substituting into the decay equation, and rearranging, we have an expression for the *change* in the number of atoms of $B$ within the sample since it formed:

$$N_B - N_{B,i} = \left(e^{\lambda t} - 1\right) N_A,$$

where $N_A \equiv N_{A,f}$ and $N_B \equiv N_{B,f}$ are the numbers of atoms of species $A$ and $B$ respectively, remaining today. When comparing one sample with another, it is more accurate to evaluate the compositions by using ratios of isotopes: the isotopes of interest relative to a stable third isotope. Representing this third (constant) abundance as $N_C$, we arrive at the relation

$$\frac{N_B}{N_C} = \left(e^{\lambda t} - 1\right) \frac{N_A}{N_C} + \frac{N_{B,i}}{N_C}. \tag{20.1}$$

Equation (20.1) is used to determine the age of a sample by plotting relative abundances of the stable product versus the relative abundances of the radioactive isotope in the sequence at various locations in the rock. The slope $m = e^{\lambda t} - 1$ of the best-fit line is directly related to the age of the sample.

**Example 20.4.1.**   Data for one sample obtained in the lunar highlands, based on the beta decay[14] of rubidium-87 to strontium-87, $^{87}_{37}Rb \rightarrow {}^{87}_{38}Sr + e^- + \bar{\nu}$, are shown in Fig. 20.19. From Eq. (20.1) and Fig. 20.19,

$$m = e^{\lambda t} - 1 = 0.0662,$$

where $\lambda = 0.0146 \times 10^{-9}$ yr$^{-1}$ for $^{87}_{37}Rb$. Solving for $t$, we find that the age of the sample is $4.39 \times 10^9$ yr.

It is important to point out that this procedure assumes that the initial *ratio* $^{87}_{38}Sr/^{86}_{38}Sr$ is a constant throughout the sample, whereas the initial ratio $^{87}_{37}Rb/^{86}_{38}Sr$ may vary somewhat (i.e., the sample is not perfectly homogeneous). This is because $^{86}_{38}Sr$ and $^{87}_{38}Sr$ are chemically identical, allowing them to be bound up in minerals in the same proportions, whereas the proportion of $^{87}_{37}Rb/^{86}_{38}Sr$ need not be constant throughout.

---

The results of radioactive dating are consistent with the view that the lunar maria have relatively young surfaces. As you will demonstrate in Problem 20.14 for one sample returned from the Sea of Tranquility by the *Apollo 11* astronauts in 1969, the ages of the maria (typically 3.1 to $3.8 \times 10^9$ years) are significantly less than those of the highlands. This is consistent with the observation, noted earlier, that relatively few craters can be found in the maria compared to the highland regions.

In sharp contrast, the oldest rocks ever found on Earth date to 3.8 billion years, whereas 90% of the planet's crust is younger than 600 million years. Plate tectonic activity is constantly recycling the surface, carrying old crust down into the mantle and forming new crust to replace it.

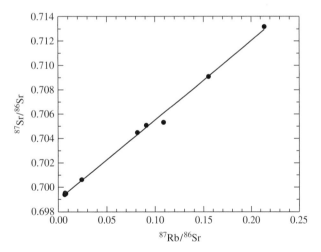

**FIGURE 20.19**   Relative abundance determinations for a sample obtained in the lunar highlands. (Data from D. A. Papanastassiou and G. J. Wasserburg, *Proc. Seventh Lunar Sci. Conf.*, Pergamon Press, New York, 1976.)

---

[14]Recall that an electron is also referred to as a beta particle ($\beta$).

## Late Heavy Bombardment

It is the dating of lunar samples that implies that a spike of **late heavy bombardment** (LHB) occurred roughly 700 million years after the Moon formed. It was during that time that the majority of the cratering occurred in the lunar highlands. During the LHB phase, a small number of very large collisions produced the maria. Over the last 3.8 billion years meteorite impacts have continued, but at a significantly reduced rate. In this way the fairly smooth, relatively uncratered surfaces of the maria have been maintained.

The "time stamp" provided by Moon rocks not only plays an important role in our understanding of the evolution of the Moon but is also crucial in developing a picture of evolution for other planets, as well as an overall formation theory for the Solar System (see Section 23.2). For instance, the recognition of an LHB episode followed by meteorite impacts at roughly a constant rate has helped scientists to conclude that the surface of Venus was refreshed within approximately the last 500 million years. This scenario also suggests that the surface of Mercury is, in general, quite ancient.

## The Formation of the Moon

The question of the Moon's formation has been widely debated. Prior to the Apollo and Luna missions, several models had been proposed. The **fission model** (also sometimes called the **daughter model**), first suggested in 1880 by George Darwin[15] (1845–1912), contended that the Moon was "torn off" from Earth at a time when Earth was spinning more rapidly than it is today. However, the orientation of the Moon's orbital plane is close to the ecliptic (tilted 5.1°), rather than along the plane of Earth's equator as would be expected if the Moon broke away. Furthermore, the lack of any water in the lunar samples, together with the underabundance of other volatiles relative to surface rock on Earth, also contradicts this proposal.

The **co-creation model** (also known as the **sister model**) suggested that the Moon and Earth formed simultaneously, with the Moon coalescing from a small disk of material that developed around the proto-Earth. This idea also fails to explain the composition differences found in the lunar samples.

A third model, the **capture model**, proposed that the Moon was actually formed elsewhere in the solar nebula and was caught in Earth's gravitational field as it drifted by. However, in this scenario the composition differences are not great enough; the Moon and Earth are too similar. For instance, the ratios of stable isotopes of oxygen are nearly identical within lunar and terrestrial samples despite significant differences found in meteorites. Also, the dynamics of such a capture seem unlikely. Since the Moon is fairly large compared to Earth, a third, similarly sized body would need to have been present to take up much of the system's surplus energy, as required for a capture. Having three large objects in close proximity at just the right time seems highly improbable. On the other hand, capture seems to be a likely mechanism for some of the many small moons found throughout the Solar System. In these cases, energy may have been lost through a many-body interaction with other moons already present. Alternatively, orbital energy may have been lost by aerobraking if the captured moon passed through a portion of the planet's atmosphere, much like the

---

[15]George Darwin was the son of Charles Darwin (1809–1882), the author of the theory of Darwinian biological evolution.

maneuver performed with the Magellan spacecraft around Venus. However, our Moon is much too large for any of these mechanisms to work.

In 1975 a fourth model, now known as the **collision model**, was proposed by William K. Hartmann and Don R. Davis. Since that time numerous computer simulations have verified its plausibility (e.g., Fig. 20.20). This model seems to explain many of the problems encountered by the three previous scenarios. The model suggests that a giant object, perhaps

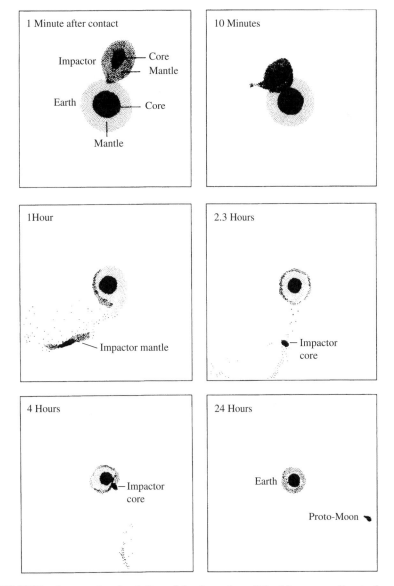

**FIGURE 20.20** A computer simulation of the formation of the Moon according to the collision model. The Earth–Moon system is shown at different times during the simulation. (Figure courtesy of A. G. W. Cameron and W. Benz, Smithsonian Astrophysical Observatory.)

twice the mass of the present-day Mars, collided with Earth almost 4.6 billion years ago, vaporizing much of the impactor and causing a portion of Earth's surface to be ripped away. The pulverized material then formed a disk around Earth that coalesced over a relatively short period of time (estimates range from months to about 100 years). Due to the high temperatures generated during the collision, many of the volatiles present in Earth's crust would have been absent in the condensing debris. Assuming that there had been sufficient time prior to the collision for some gravitational separation to occur within Earth and the impactor, the crusts of the two objects would also have been somewhat deficient in iron, leaving less iron available to form the Moon. The simulations suggest that most of the present-day Moon was produced by silicate-rich material from the mantle of the impactor, and that the impactor's iron-rich core became a part of Earth. This model effectively explains why the Moon's average density is comparable to the density of Earth's uncompressed mantle (i.e., the density that would be measured if the compressional effect of gravity were removed). In this model the collision would have also preserved the similar oxygen-isotope ratios seen on the two worlds.

The collision model is considered by most researchers as the preferred model for the formation of our Moon. Although it appears on first inspection to be a highly unique and perhaps ad hoc way to explain the characteristics of our Moon, recall that a similar scenario also appears to explain the highly dense structure of Mercury (page 740). The existence of Pluto's moon Charon may require a large-scale collision as well (see Section 22.1).

From our investigation of the Moon, it appears that its formation was a violent process. However, numerous questions about the Moon's structure and evolution remain unanswered. It is also apparent that careful studies of our nearest neighbor can shed light on important questions regarding the formation and evolution of Earth and the rest of the Solar System. Perhaps future missions to the Moon would further clarify our understanding of the Solar System.

## 20.5 ■ MARS

Only one-tenth the mass of Earth, the planet Mars has touched our imagination. In 1877 the astronomer Giovanni Virginio Schiaparelli (1835–1910) reported seeing a series of dark lines on the surface of the planet and referred to them as *canali* (naturally occurring channels of water). The term was later misinterpreted to imply that the markings were actually an immense network of artificial *canals* built by an intelligent civilization to irrigate a dying world. In support of this argument is the existence of the seasonally varying polar ice caps, visible in the Hubble Space Telescope image shown in Fig. 20.21(a). It is not difficult to imagine that using smaller telescopes, which were peering at the red planet through Earth's obscuring atmosphere, would have led to the conclusion that canali were present. In an effort to verify these features, Percival Lowell (1855–1916) built an observatory near Flagstaff, Arizona, to carry out a series of careful observations of this nearby world. Other astronomers were somewhat more skeptical of the existence of intelligent life on Mars, and even of the canali. However, the general public seized on the possibility that Martians do (or at least did) live there, leading to a wealth of science fiction literature and films.

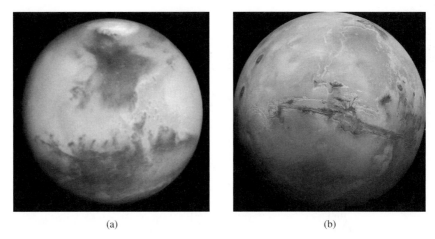

(a)                                                    (b)

**FIGURE 20.21**   (a) An image of Mars obtained using WF/PC 2 onboard the Hubble Space Telescope. The north polar ice cap is clearly visible. (Courtesy of Philip James, University of Toledo; Steven Lee, University of Colorado; and NASA.) (b) A mosaic of 102 Viking Orbiter images obtained in 1976. This perspective places the viewer 2500 km above the surface of the red planet. Valles Marineris (a 3000-km-long canyon system) can be seen near the equator. On the left side of the image, three giant shield volcanos are evident as dark, circular regions. Each volcano is approximately 25 km tall. (Courtesy of U.S. Geological Survey and NASA/JPL.)

## Exploration of the Red Planet

There have been many attempts to study Mars by robotic missions. Early efforts included the **Mariner** flyby missions in the 1960s. In 1975 the **Viking** missions included two orbiters and partnered landers that contained cameras and internal laboratories for the study of Mars surface chemistry. Since the landers did not have any ability to move across the surface, their studies were restricted to the locations where they set down on the planet. The **Mars Global Surveyor**, with its very high-resolution camera entered Mars orbit in 1997 and continues to operate successfully at the time of this writing, as does the **Mars Odyssey**, which arrived in 2001, and the **Mars Express Orbiter**, an ESA mission that reached the red planet in 2003. Another spacecraft, the **Mars Reconnaissance Orbiter**, also began its work around the planet in 2006.

The **Sojourner Rover** of the **Mars Pathfinder** mission (1997) was the first truly mobile lander, able to move short distances across the surface in the vicinity of its lander, the **Carl Sagan Memorial Station**.[16] Then in January 2004, two golf-cart-sized rovers successfully landed on the surface and began extensive exploration of the regions around their landing areas. The **Mars Exploration Rovers**, **Spirit** and **Opportunity**, were originally expected to operate for several months, but they continued to move across the surface of the planet as late as May 2006. The Mars Orbiter has been able to image both rovers from its vantage point in orbit around the planet. Other missions are also planned, including additional orbiters and landers and possible human-crewed missions to Mars.

---

[16]The stationary lander base was renamed the Carl Sagan Memorial Station after landing in honor of Carl Sagan (1934–1996), Solar System researcher, Pulitzer-prize-winning author, and popularizer of astronomy.

## Evidence of Water on Mars

Despite the many studies of Mars from Earth, from Mars orbit, and from its surface, no sign of life has been found on the planet. At first inspection, the images returned by Spirit and Opportunity (see Fig. 20.22), along with images obtained from the Viking landers, give the impression of a dry, dusty world. However, on closer inspection of data returned by Spirit and Opportunity, along with information from the orbiters, research has revealed a fascinating world that, although dry today, once clearly had water flowing across its surface. Apparent in images of the surface from the Mars Orbiter (see Fig. 20.23) are channels that are characteristic of water erosion found on Earth. There is also evidence that huge flash floods may have occurred on the surface of the planet. It appears that lakes of water may have been present on Mars in the distant past as well (Fig. 20.24).

With present-day surface temperatures varying between $-140°C$ ($-220°F$) and $20°C$ ($70°F$), combined with the very low atmospheric pressure found near the surface (typically 0.006 atm), it appears that the liquid water that was present on Mars is now either trapped in a layer of permafrost or frozen in its polar ice caps [Fig. 20.21(a)]. In fact, it is the low atmospheric pressure that makes the existence of persistent water in liquid form impossible on the surface today.

## ALH84001, A Martian Meteorite

Ironically, even though the intense investigations of Mars by robotic spacecraft and landers have thus far failed to identify any evidence that life exists on Mars today or existed there in the past, a meteorite discovered in Allan Hills, Antarctica, in 1984 led to speculation that

(a)

(b)

**FIGURE 20.22**   (a) A panoramic view of the Bonneville crater obtained by Mars Exploration Rover Spirit. (Courtesy of NASA/JPL.) (b) A panoramic view of interesting rock features at Meridiani Planum, near the landing site of Mars Exploration Rover Opportunity. (Courtesy of NASA/JPL.)

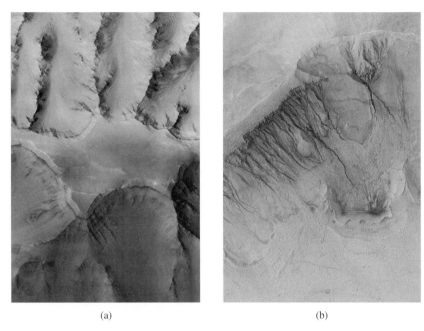

(a)                                                 (b)

**FIGURE 20.23** (a) A portion of Valles Marineris [recall Fig. 20.21(b)] showing evidence of water-caused erosion. (Courtesy of NASA/JPL/Malin Space Science Systems.) (b) Erosion channels seen in an impact crater in Newton Basin in Sirenum Terra, located in the southern hemisphere of Mars. (Courtesy of NASA/JPL/Malin Space Science Systems.)

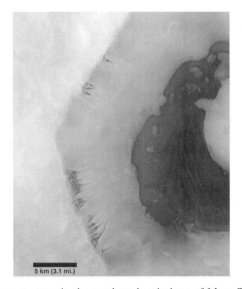

**FIGURE 20.24** An impact crater in the southern hemisphere of Mars. The dark material at the bottom is believed to be sediment deposits from an ancient Martian lake. Seepage into the crater is also evident near the rim of the crater. Dunes are visible in the dark regions as well. (Courtesy of NASA/JPL/Malin Space Science Systems.)

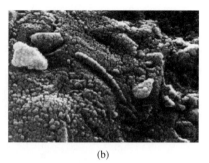

(a)                                                                    (b)

**FIGURE 20.25**    (a) ALH84001 is a Martian meteorite found in Allan Hills, Antarctica, in 1984. (Courtesy of NASA.) (b) An electron microscopy view of a portion of ALH84001 showing tube-like structures less than 1/100 the size of a human hair. Some scientists have argued that these structures represent nanofossils of ancient microbial life on Mars. (Courtesy of NASA.)

evidence existed in that rock from space [Fig. 20.25(a)]. **ALH84001** is the oldest meteorite ever found that originated from the surface of Mars. After forming on Mars 4.5 billion years ago, it was ejected from the surface of the planet 16 million years ago by an energetic collision. After traveling through the inner Solar System, it struck Earth 13,000 years ago and became trapped in the Antarctic ice sheet.[17] Confirmation that the meteorite did indeed originate on Mars comes from comparing its chemical composition with the results of compositional studies conducted by robotic landers.

It was the examination of small amounts of carbonate grains in the meteorite that has led some researchers to suggest that ancient, fossilized Martian microbes may be contained in the rock [Fig. 20.25(b)]. The grains themselves measure less than 200 $\mu$m in size, and what appear to be fossilized microbes are smaller than 1/100 the size of a human hair. In support of the hypothesis that the "nanofossils" are due to ancient microbial life is the presence of organic PAHs in the carbonate, along with oxide and sulfide biominerals. The carbonate grains also appear to have formed in fractures in the rock, possibly in the presence of liquid water.

Most researchers now believe that, although ALH84001 is a fascinating rock that clearly originated on Mars, the evidence is weak that it contains samples of primitive fossilized life. Rather, it could be that the features formed by some inorganic mechanism(s) or that the rock became contaminated as a result of the 13,000 yr it spent on Earth before it was discovered.

### The Polar Caps

Although water ice is certainly present today in the polar caps, the caps are composed primarily of dry ice (frozen carbon dioxide). Mars's axis tilt of 25° and its orbital period of 1.88 yr means that the planet's seasonal variations are similar to Earth's but are roughly twice as long. Consequently, Mars experiences winter and summer seasons corresponding to observed variations in the sizes of the ice caps. It is the dry ice that sublimates during the

---

[17]The ejection and landing ages were determined through cosmic ray exposure that the meteoroid received prior to hitting Earth.

Martian summer and freezes back out again during the winter. The small residual cap that remains during the summer is composed of water ice.

## Chaotic Fluctuations in Mars's Rotation Axis

From numerical simulations designed to investigate the long-term stability of planetary motions, it appears that the orientation of Mars's spin axis fluctuates wildly (chaotically) between about $0°$ and $60°$ over time scales as short as a few million years; the variations are due to gravitational interactions with the Sun and the other planets. If Mars has experienced such large fluctuations in its axis tilt in the past, this would imply that at various times the polar ice caps could completely melt (high tilt angle), whereas at other times the planet's atmosphere might actually freeze out (low tilt angle). The time-variability of the tilt of Mars's spin axis also implies that its current tilt, which is similar to Earth's, is only coincidental.

Interestingly, these simulations of the fluctuations imply that the chaotic behavior does not develop if the effects of general relativity are neglected. It seems that the effects of spacetime curvature discussed in Chapter 17 play an important role in the long-term behavior of planetary orbits and their rotations, even at the distance of Mars's orbit.

Even though it is closer to the Sun, Earth has not experienced the same dramatic oscillations in its axis tilt that Mars seems to have gone through. Apparently Earth's rotation axis is stabilized by our planet's strong tidal interaction with its relatively large moon. Consequently, our planet's climatic variations have been much less pronounced than those on Mars. Amazingly, this seems to imply that the presence of Earth's moon (apparently the result of an accidental collision) is in part responsible for the stable environment that led to the evolution of life on the third planet from the Sun.

## The Thin Atmosphere of Mars

Mars's very thin atmosphere is composed of 95% carbon dioxide and 2.7% molecular nitrogen, by number—percentages very similar to those in the atmosphere of Venus. Unlike the case of Venus, however, the greenhouse effect has very little influence on the current equilibrium temperature of Mars; there simply are not enough molecules present to absorb a significant amount of infrared radiation (the atmospheric pressure at the surface of Venus is 90 atm, 13,000 times greater than the atmospheric pressure at the surface of Mars). In the past, the atmosphere of Mars may have been much more dense, causing the greenhouse effect to be more efficient than it is today. The water that is currently trapped in the ice caps and permafrost would then have been flowing freely, maybe even resulting in rainfall. The water that was present in the atmosphere and on the surface would have absorbed much of the atmospheric carbon dioxide, subsequently locking the $CO_2$ in carbonate rocks. As a result, the greenhouse effect diminished, the global temperature dropped, and the water froze, leaving the dry world we find today.

Shortly after the two Viking landers arrived at Mars in 1975, they began to measure an appreciable drop in atmospheric pressure. This was because winter was coming to the Southern Hemisphere, and carbon dioxide was freezing out of the atmosphere. When spring returned to the south, the atmospheric pressure went back up again. The same behavior was repeated when winter arrived in the Northern Hemisphere.

### Dust Storms

Even though the atmospheric density is quite low near the surface, it is sufficient to produce huge dust storms that sometimes cover the entire surface of Mars. The seasonal storms are driven by high winds and are responsible for the variations in surface hues that can be seen from Earth.[18]

It was during the Viking missions in 1976 that two such major dust storms occurred. Since that time, much of the dust has settled out of the planet's atmosphere, resulting in noticeable changes in its climate. (The absorption of light by dust is the primary source of atmospheric heating.) In fact, the Hubble Space Telescope recorded a decline in the average global temperature of the planet. With the decrease in average temperature, ice-crystal clouds have become more prominent in the planet's lower atmosphere than they were at the time of the Viking missions.

### The Abundance of Iron

The dust on the surface (recall Fig. 20.22) appears reddish in color and contains a relatively high abundance of iron, which oxidizes (rusts) when exposed to the atmosphere. Apparently Mars did not undergo the same degree of gravitational separation that Earth did, possibly because the smaller, more distant planet cooled more rapidly following its formation. However, averaging over the volume of the entire planet reveals that iron is actually underabundant on Mars relative to the other terrestrial planets, as evidenced by its lower average density of 3933 kg m$^{-3}$. The reason for this is not yet understood.

The lack of significant gravitational separation is also consistent with the absence of an appreciable global magnetic field. If an iron core is present, presumably it is quite small and probably not molten.

### Evidence of Past Geological Activity

Even if Mars may not be geologically active today, it certainly has been in the past. Figure 20.21(b) shows Valles Marineris, a 3000-km-long network of canyons near the planet's equator. It appears that Valles Marineris, which is up to 600 km wide in some places and can reach a depth of 8 km, was formed from *faulting* (or fracturing of the crust) in order to relieve stresses that built up in the interior.

Olympus Mons, shown in Fig. 20.26, is a shield volcano that covers an area roughly the size of Utah. The volcano rises 24 km above the surrounding surface and has a huge caldera (a volcanic crater). Geologists believe that Olympus Mons owes its enormous size to a process known as **hot-spot volcanism**, where a weak spot in the crust has allowed molten material to rise to the surface. It is hot-spot volcanism on Earth that is responsible for the creation of the Hawaiian islands.[19] However, in the case of the Hawaiian island chain, motion of the tectonic plate on which the chain rides carries each newly formed volcano away from the hot spot, allowing another one to be created. Today the chain of

---

[18]These seasonal variations were once thought by some astronomers to be evidence of vegetation growing cycles.
[19]The tallest mountain on Earth, measured from its base to its summit, is the Hawaiian Island of Mauna Loa, with a vertical rise above the sea floor of 9.1 km.

**FIGURE 20.26** Olympus Mons is a shield volcano rising 24 km above the surrounding surface. Measured at its base, the diameter of the volcano is more than 500 km. The cliff that rings the volcano in this perspective image is 6 km high. (Courtesy of NASA/JPL.)

mountains that contains the Hawaiian islands actually stretches nearly all the way to Japan, although over time the oldest mountains have undergone significant erosion.[20]

The situation was somewhat different for Olympus Mons. Since Mars has apparently not developed a system of moving tectonic plates, the volcano was not carried off the hot spot where it formed. As a result, it has grown larger and larger as more molten material has made its way to the surface.[21]

### Two Tiny Moons

The two moons of Mars, Phobos and Deimos (Fig. 20.27), were discussed briefly in Section 19.2. Although they were discovered by Asaph Hall (1829–1907) in 1877, Kepler had postulated their existence centuries earlier. His "prediction" was based solely on numerology. Knowing that there were no moons in orbit about Venus, that Earth had one satellite, and that Galileo had recently discovered four moons orbiting Jupiter, Kepler decided it seemed reasonable that Mars ought to have two!

In 1726, 150 years before Hall's actual discovery, Jonathan Swift (1667–1745) wrote in his book *Gulliver's Travels* that astronomers had discovered two satellites orbiting the red planet. His fictitious scientists found that the orbital periods of these moons were 10 hours and $21\frac{1}{2}$ hours, "so that the squares of their periodical times are very near in the same proportion with the cubes of their distance from the centre of Mars, which evidently shows them to be governed by the same law of gravitation that influences the other heavenly bodies." Apparently Swift, who was not a scientist, was aware of scientific discoveries

---

[20]The Yellowstone region, with its geysers, hot springs, and mud volcanoes, is another example of hot-spot volcanism on Earth.

[21]The large volcanos discovered on Venus may have been formed in much the same way as Olympus Mons.

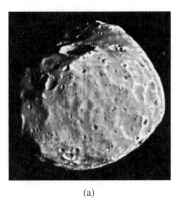
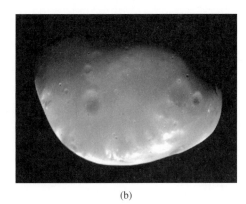

(a)                                                    (b)

**FIGURE 20.27**   The two moons of Mars, (a) Phobos and (b) Deimos, are very similar to asteroids and were probably captured by the planet. (Courtesy of NASA/JPL.)

such as Kepler's third law. The actual orbital periods of Phobos and Deimos are $7^h39^m$ and $30^h17^m$, respectively, remarkably close to the values determined by Swift's astronomers.

Both Phobos and Deimos are small, heavily cratered, elongated rocks. Phobos's longest dimension is a mere 28 km, and Deimos's is even smaller (16 km). It appears likely that the moons are captured asteroids.

## SUGGESTED READING

### General

Beatty, J. Kelly, Petersen, Carolyn Collins, and Chaikin, Andrew (eds.), *The New Solar System*, Fourth Edition, Cambridge University Press and Sky Publishing Corporation, Cambridge, MA, 1999.

Cooper, Henry S. F. Jr., *The Evening Star: Venus Observed*, Farrar, Staus, and Giroux, New York, 1993.

Goldsmith, Donald, and Owen, Tobias, *The Search for Life in the Universe*, Third Edition, University Science Books, Sausalito, CA, 2002.

Jeanloz, Raymond, and Lay, Thorne, "The Core-Mantle Boundary," *Scientific American*, May 1993.

Kargel, Jeffrey S., *Mars—A Warmer, Wetter Planet*, Praxis Publishing Ltd., Chichester, UK, 2004.

Morrison, David, and Owen, Tobias, *The Planetary System*, Third Edition, Addison-Wesley, San Francisco, 2003.

Stofan, Ellen R., "The New Face of Venus," *Sky and Telescope*, August 1993.

### Technical

Atreya, S. K., Pollack, James B., and Matthews, Mildred Shapley (eds.), *Origin and Evolution of Planetary and Satellite Atmospheres*, The University of Arizona Press, Tucson, 1989.

Canup, R. M., and Righter, K. (eds.), *Origin of the Earth and Moon*, The University of Arizona Press, Tucson, 2000.

Correia, Alexandre C. M., and Laskar, Jacques, "The Four Final Rotation States of Venus," *Nature, 411,* 767, 2001.

de Pater, Imke, and Lissauer, Jack J., *Planetary Sciences*, Cambridge University Press, Cambridge, 2001.

Hartmann, W. K., and Davis, D. R., "Satellite-Sized Planetesimals and Lunar Origin," *Icarus, 24,* 504, 1975.

Houghton, John T., *The Physics of Atmospheres*, Third Edition, Cambridge University Press, Cambridge, 2002.

Taylor, Stuart Ross, *Solar System Evolution*, Second Edition, Cambridge University Press, Cambridge, 2001.

## PROBLEMS

**20.1** Assume that radar signals of 10 GHz are used to measure the rotation rates of Mercury and Venus. Using the Doppler effect, determine the relative shifts in frequency for signals returning from the approaching and receding limbs of each planet.

**20.2** What is the ratio of the Sun's tidal force per unit mass on Mercury at perihelion to the Sun's tidal force per unit mass on Earth? How has this difference in tidal effects contributed to differences in the orbital and/or rotational characteristics of the two planets?

**20.3** For Mercury, a slowly rotating planet with no appreciable atmosphere, Eq. (19.5) for a planet's surface temperature must be modified. In particular, the assumption that the temperature is approximately constant over the entire surface of the planet is no longer valid.

   **(a)** Assuming (incorrectly) that Mercury is in synchronous rotation about the Sun, show that the temperature at a latitude $\theta$ north or south of the *subsolar point* (the point on the equator closest to the Sun) is given by

$$T = (\cos\theta)^{1/4}(1-a)^{1/4}T_\odot\sqrt{\frac{R_\odot}{D}}.$$

   Since the planet is actually in a 3-to-2 resonance, this expression is only an approximate description for the temperature at Mercury's surface.

   **(b)** Make a graph of $T$ vs. $\theta$. Mercury's albedo is 0.06.

   **(c)** What is the approximate temperature of the planet at the subsolar point?

   **(d)** At what latitude does the temperature drop to 273 K? This is the freezing point of water at the surface of Earth.

   **(e)** Would you expect to find ice on Mercury at a temperature of 273 K? Why or why not?

**20.4** **(a)** Estimate the angular resolution of the 70-m radio dish of the NASA Goldstone tracking station mentioned in footnote 1 on page 739. Assume that it is operating at a wavelength of 3.5 cm.

   **(b)** What is the angular size of Mercury at inferior conjunction? Assume (incorrectly) for this problem that the planet's orbit is circular.

   **(c)** If the power in the radar signal was approximately uniformly distributed across the cone-shaped beam, how much power actually arrived at the surface of Mercury?

(d) Suppose that all of the radar energy striking the surface of Mercury were reflected isotropically back into a hemisphere. What would be the signal flux received at the VLA?

**20.5** (a) From the data presented in the text, estimate the kinetic energy of the impact that may have been responsible for stripping off the outer layers of Mercury early in the history of the Solar System.

(b) If, prior to the collision, Mercury had twice as much mass as it does today, how much energy would have been required to lift that additional mass off the present planet? Assume that the extra mass had the density of Earth's present-day Moon and that the material was uniformly distributed in a spherical shell around the present-day Mercury. Don't forget to include the energy required to eject the mass of the impactor as well.

(c) Solely on the basis of energy considerations comment on the feasibility of this scenario for the origin of Mercury as we observe it today.

**20.6** Assuming that the atmosphere of Venus is composed of pure carbon dioxide, estimate the number density of molecules at the planet's surface. How many times larger is this value than the number density of nitrogen molecules at the surface of Earth, as quoted in Example 19.3.3?

**20.7** (a) Modeling the greenhouse effect using one atmospheric layer, as was done in Problem 19.13, is equivalent to assuming that the optical depth is about one. If the optical depth is $\tau$, and if we can neglect circulation in the atmosphere, show that the surface temperature should be approximately

$$T_{\text{surf}} = (1 + \tau)^{1/4}\, T_{\text{bb}},$$

where $T_{\text{bb}}$ is the blackbody temperature of an *airless* planet.

(b) The optical depth of Venus's atmosphere is approximately $\tau = 70$. Make an estimate of its surface temperature using this crude greenhouse model. Take the average albedo to be 0.77.

**20.8** Based on the observed rate at which North America and Eurasia are separating from each other, when were the two continents joined together as Laurasia? Assume that the Atlantic Ocean is roughly 4800 km (3000 miles) wide.

**20.9** Using the equation of hydrostatic equilibrium (Eq. 10.6), estimate the pressure at the center of Earth. Detailed computer simulations suggest that the central pressure is $3.7 \times 10^6$ atm.

**20.10** (a) From the data given in Section 19.2, estimate the *rate* at which rotational energy is being dissipated by tidal friction for the case of Earth. *Hint:* This terrestrial problem is similar to the loss of rotational kinetic energy in pulsars; see Eq. (16.30).

(b) What fraction of the total energy being lost from Earth's interior can be accounted for by tidal dissipation of its rotational kinetic energy?

**20.11** Referring to Eq. (11.2) for the Lorentz force and Fig. 20.14, explain why most charged particles bounce back and forth between the North and South Poles of Earth, rather than striking the surface. Use a diagram if necessary. *Hint:* The converging magnetic field lines form magnetic *mirrors* near the North and South Poles. (Magnetic "bottles," which are based on the same principle, are used to confine high-temperature plasmas in laboratories.)

**20.12** The moment of inertia of a planet is used to evaluate its interior structure. In this problem you will construct a simple "two-zone" model of the interior of Earth, assuming spherical symmetry. Take the average densities of the core and mantle to be 10,900 kg m$^{-3}$ and 4500 kg m$^{-3}$, respectively (neglect the thin surface crust).

(a) Using the average density of the entire Earth, determine the radius of the core. Express your answer in units of Earth's radius.

(b) Calculate the *moment-of-inertia ratio* $(I/MR^2)$ for the "two-zone" Earth (the actual value is 0.3315). The moment of inertia for a spherically symmetric mass shell of constant density $\rho$, having inner and outer radii $R_1$ and $R_2$, respectively, is given by

$$I \equiv \int_{\text{vol}} a^2 \, dm = \frac{8\pi\rho}{15} \left( R_2^5 - R_1^5 \right).$$

$a$ is the distance of the mass element $dm$ from the axis of rotation.

(c) Compare your answer in part (b) with the value expected for a solid sphere of constant density. Why are the two values different? Explain.

**20.13** The moment-of-inertia ratio of the Moon is 0.390 (see Problem 20.12).

(a) What does this say about the interior of the Moon?

(b) Is this consistent with the lack of any detectable magnetic field? Why or why not?

**20.14** (a) The *Apollo 11* astronauts, after landing on the Moon on July 20, 1969, returned rocks from the Sea of Tranquility, one of the maria on the near side. Upon their return, the analysis of one rock (basalt 10072) yielded the relative abundances at various locations in the sample; see Table 20.2. Graph the abundance data as $^{143}_{60}\text{Nd}/^{144}_{60}\text{Nd}$ vs. $^{147}_{62}\text{Sm}/^{144}_{60}\text{Nd}$. (Note that the uncertainties listed correspond to the last two significant figures.)

(b) Determine the slope of the best-fit straight line drawn through the data and estimate the age of the lunar sample. Compare your answer with the age of the lunar highland sample, determined in Example 20.4.1 from the data in Fig. 20.19.

**20.15** Estimate the initial rotation period of Earth if the Moon were torn from it, as suggested by the fission model.

**20.16** Estimate the Roche limit for the Earth–Moon system. Express your answer in units of the radius of Earth. Is the Moon in any danger of becoming tidally disrupted?

**20.17** Mars is at its closest approach to the Sun during the summer months in its southern hemisphere.

(a) Using Eq. (19.5), estimate the ratio of the average temperatures on Mars when it is at perihelion and aphelion.

**TABLE 20.2** Results from the Analysis of Basalt 10072, Returned from the Sea of Tranquility by the *Apollo 11* Astronauts in 1969. (Data from D. A. Papanastassiou, D. J. DePaolo, and G. J. Wasserburg, "Rb-Sr and Sm-Nd Chronology and Genealogy of Mare Basalts from the Sea of Tranquility," *Proceedings of the Eighth Lunar Science Conference*, Pergamon Press, New York, 1977.)

| $^{147}_{62}\text{Sm}/^{144}_{60}\text{Nd}$ | $^{143}_{60}\text{Nd}/^{144}_{60}\text{Nd}$ |
|---|---|
| 0.1847 | $0.511721 \pm 18$ |
| 0.1963 | $0.511998 \pm 16$ |
| 0.1980 | $0.512035 \pm 21$ |
| 0.2061 | $0.512238 \pm 17$ |
| 0.2715 | $0.513788 \pm 15$ |
| 0.2879 | $0.514154 \pm 17$ |

(b) Considering the tilt of the planet's rotation axis, describe the seasonal behavior of the two polar ice caps.

**20.18** Assuming that the two moons are in circular orbits, determine the orbital radii of Phobos and Deimos. Express your answers in units of the radius of Mars.

**20.19** Suppose you lived on Mars and watched its moons. If Phobos and Deimos were next to each other one night, what would you see the next night (one Martian day later)? Describe the apparent motions of the two moons. (Both Phobos and Deimos orbit prograde, approximately above the planet's equator.)

# CHAPTER

# 21

# The Realms of the Giant Planets

## 21.1 ■ THE GIANT WORLDS

Excluding the Sun, by far the largest member of the Solar System is **Jupiter**, 317.83 times more massive than Earth. Jupiter and the other three giants, **Saturn**, **Uranus**, and **Neptune**, together contain 99.5% of the entire mass of the planetary system (see Fig. 21.1). Consequently, if we hope to understand the development and evolution of our Solar System, it is vital that we understand these distant worlds.

### The Discovery of the Galilean Moons

Naked-eye observations of Jupiter and Saturn began when human beings first started gazing up at the heavens. But it was in 1610 that Galileo became the first person to look at these planets through a telescope. In so doing, he detected the four large moons of Jupiter, now collectively known as the **Galilean moons**.[1] Galileo also saw Saturn's rings, but because of his telescope's low resolution, he thought that the rings were two large satellites situated on either side of the planet.

### The Discoveries of Uranus and Neptune

It wasn't until 1781 that William Herschel (1738–1822), a German-born musician living in England, made the chance discovery of Uranus. By considering gravitational perturbations affecting the orbit of Uranus, John Couch Adams (1819–1892), a graduate student at Cambridge University, proposed in October 1845 that another planet must exist even farther from the Sun. Using Bode's rule to guess at the distance of this unknown planet from the Sun, Adams predicted its position in the heavens. Unfortunately, when he submitted his work to Sir George Airy, the Astronomer Royal of England, Airy did not believe the conclusions. In June 1846, Urbain Leverrier (1811–1877), a very well-respected French scientist, independently made the same prediction, agreeing with Adams's position to within 1°. Learning of the agreement between the two predictions, Airy began to search for the object. However,

---

[1] The four Galilean moons were also discovered independently by Simon Marius (1570–1624) in 1610.

(a)  (b)

(c)  (d)

**FIGURE 21.1**    The four giant planets. (a) Jupiter and its largest moon, Ganymede. (b) Saturn, with two of its moons, Rhea and Dione, seen near the bottom and right-hand side of the image, respectively. (c) Uranus. (d) Neptune. The images were taken by the Voyager 1 and 2 spacecraft. Notice the oblateness caused by rapid rotation. The image sizes do not correspond to the actual relative sizes of the planets. (Courtesy of NASA/JPL.)

Johann Gottfried Galle (1812–1910) of the Berlin Observatory found Neptune on September 23, 1846, the night after receiving a letter from Leverrier suggesting that he too should look for this new planet. In a very real sense, Neptune was discovered in the mathematical calculations of Adams and Leverrier; Galle merely confirmed their work.

### Missions to the Giant Planets

Since the first observations of these worlds, the efforts of Earth-based astronomers have provided important information about the giant planets and their many satellites. However, many of the data now available have come from spacecraft missions. The first such missions were the **Pioneer 10** and **Pioneer 11** flybys of Jupiter (1973, 1974) and the Pioneer 11 flyby of Saturn (1979). Later, **Voyager 1** and **Voyager 2** embarked on their spectacularly successful "Grand Tour" missions. Both Voyagers, launched from Earth in 1977, visited Jupiter (1979) and Saturn (1980, 1981), and Voyager 2 continued on to Uranus (1986) and Neptune (1989). In each case the encounters with the planets were brief flybys. Today the Pioneer[2] and Voyager spacecraft are on their way out of the Solar System. The Voyager spacecraft (renamed the *Voyager Interstellar Missions*) continue to send back information over immense distances with ever-weakening signals, providing data about the outer reaches of the Solar System, including the interaction between the solar wind and the winds from other stars. In early 2006, Voyager 1 was 8.7 billion miles (14 billion kilometers) from Earth, traveling at a speed of 3.6 AU per year, and Voyager 2 was 6.5 billion miles (10.4 billion kilometers) away, traveling at a speed of 3.3 AU per year. It is believed that Voyager 1 passed the solar wind's termination shock in December 2004, as evidenced by an increase in the strength of the magnetic field in the vicinity of the spacecraft by a factor of 2.5 (recall Example 11.2.1).

The Hubble Space Telescope has also been used to observe the outer planets from Earth orbit. HST has documented significant changes in the planets since the flyby missions of the 1970s and 1980s.

An extended and detailed investigation of the Jovian system began in 1995 when the **Galileo** spacecraft (launched in 1989) entered into orbit around Jupiter. In addition to observing the planet carefully, Galileo completed numerous flybys of the Galilean moons during the eight years it spent in the Jovian system. As a part of the mission, a probe descended into the planet's atmosphere by parachute, sampling the atmosphere's composition and physical conditions.

The **Cassini–Huygens** mission, launched in 1997, entered the Saturnian system July 1, 2004. This dual mission is composed of the Cassini orbiter, which was built by NASA with the high-gain antenna system provided by the Italian Space Agency (ASI), and the Huygens probe, which was built by the European Space Agency (ESA). At the time of this writing in 2006, Cassini is exploring the Saturnian system at length during its four-year mission, including the planet, its moons, and its rings. Huygens descended into the thick atmosphere of Titan, the largest of Saturn's moons on January 14, 2005. Like the Galileo probe, Huygens used a parachute during part of its descent while it made measurements of composition, wind speed, atmospheric structure, and surface features. At an altitude of 40 km, the parachute was be released and the probe fell to the surface. The descent took two hours and 27 minutes, and the probe remained operational on the surface for an additional one hour and 10 minutes, while further observations were made.

## Composition and Structure

Referring to Table 19.1, note that as a class, the giant planets differ markedly from the terrestrials. However, the group can be further subdivided. The gas giants of Jupiter (317.83 $M_\oplus$) and Saturn (95.159 $M_\oplus$) have average compositions that are quite similar to the Sun's,

---

[2]The last signal received from Pioneer 10 occurred on January 23, 2003, almost 31 years after the spacecraft's launch. Pioneer 10 is now more than 8 billion miles from Earth, headed in the general direction of Aldebaran, in the constellation of Taurus, and will be in that star's vicinity in about 2 million years. Pioneer 11 was last heard from in 1995 and is headed in the direction of Aquila.

while the much smaller and more distant ice giants, Uranus (14.536 $M_\oplus$) and Neptune (17.147 $M_\oplus$), have higher proportions of heavier elements. Because each of the giant planets is capable of retaining all of the lighter elements in its atmosphere, this composition difference suggests important differences in their formation.

This conclusion is supported by direct observations of the composition of each giant planet near the cloud tops. Table 21.1 gives the relative number densities of constituents in the giant planet atmospheres; the Sun's photospheric composition is given for comparison. (Note that it is the percentage of atoms or molecules by *number* that is being quoted, not the *mass fraction* that was discussed in Chapter 10.) Jupiter's hydrogen content is somewhat greater than the Sun's, while its helium content is slightly less than solar. Saturn's upper atmosphere is noticeably deficient in helium (96% $H_2$, 3% He), while the other percentages are similar to those of Jupiter. Observations also indicate that whereas their hydrogen and helium contents are intermediate between the Sun and Jupiter, the atmospheres of both Uranus and Neptune are overabundant in methane relative to solar by a factor of 10 or more. Although these studies hint that differences may exist in the interiors of these planets, other observational data and theoretical investigations give us even more information about what is going on inside.

Figure 21.2 shows each planet's radius as a function of its mass. Also plotted are a series of theoretical curves for various mixtures: "H" for pure hydrogen; "H–He" for a hydrogen–helium mixture appropriate for Jupiter and Saturn; "Ice" for a composition of $H_2O$ (water), $CH_4$ (methane), and $NH_3$ (ammonia) ice; and "Rock" for a composition of magnesium, silicon, and iron. The dashed lines correspond to models that follow adiabatic temperature gradients. In particular, the gas models (H, H–He) incorporate the polytropic relationship, $P \propto \rho^2$, appropriate for Coulomb-force pair interactions (recall the discussion of polytropes beginning on page 334). $P \propto \rho^2$ is a reasonable approximation when electron–ion pair interactions are important, because $F \propto q^2$ and the number of charges is proportional to the density of the gas. The solid lines in Fig. 21.2 represent zero-temperature models, corresponding to complete degeneracy (see the discussion of building completely degenerate white dwarf models in Section 16.3). It seems that hydrogen and helium dominate in Jupiter and Saturn, while ices are likely to play key roles in determining the interior structures of Uranus and Neptune.

Notice in Fig. 21.2 that even though Jupiter is more than three times as massive as Saturn, it is only slightly larger than its more distant neighbor. This is because the increased mass

**TABLE 21.1**   Composition of the Atmospheres of the Giant Planets. All values are given as a fractional number density of particles. Jupiter data are from the Galileo probe. Solar photospheric data are provided for comparison. (Data from Table 4.5 of de Pater and Lissauer, *Planetary Sciences*, Cambridge University Press, Cambridge, 2001.)

| Gas | Sun | Jupiter | Saturn | Uranus | Neptune |
|---|---|---|---|---|---|
| $H_2$ | H: 0.835 | $0.864 \pm 0.006$ | $0.963 \pm 0.03$ | $0.85 \pm 0.05$ | $0.85 \pm 0.05$ |
| He | He: 0.195 | $0.157 \pm 0.004$ | $0.034 \pm 0.03$ | $0.18 \pm 0.05$ | $0.18 \pm 0.05$ |
| $H_2O$ | O: $1.70 \times 10^{-3}$ | $2.6 \times 10^{-3}$ | $> 1.70 \times 10^{-3}$? | $> 1.70 \times 10^{-3}$? | $> 1.70 \times 10^{-3}$? |
| $CH_4$ | C: $7.94 \times 10^{-4}$ | $(2.1 \pm 0.2) \times 10^{-3}$ | $(4.5 \pm 2.2) \times 10^{-3}$ | $0.024 \pm 0.01$ | $0.035 \pm 0.010$ |
| $NH_3$ | N: $2.24 \times 10^{-4}$ | $(2.60 \pm 0.3) \times 10^{-4}$ | $(5 \pm 1) \times 10^{-4}$ | $< 2.2 \times 10^{-4}$ | $< 2.2 \times 10^{-4}$ |
| $H_2S$ | S: $3.70 \times 10^{-5}$ | $(2.22 \pm 0.4) \times 10^{-4}$? | $(4 \pm 1) \times 10^{-4}$? | $3.7 \times 10^{-4}$? | $1 \times 10^{-3}$ |

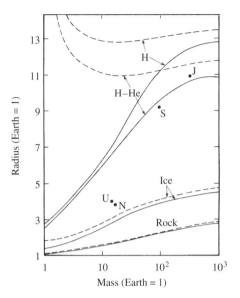

**FIGURE 21.2** Composition and mass are principal elements in determining the radius of a planet. Plotted are the radii of Jupiter (J), Saturn (S), Uranus (U), and Neptune (N) as functions of their masses. Also shown are theoretical curves for various mixtures. The solid lines represent zero-temperature models, and the dashed lines are models that follow adiabatic temperature gradients. (Figure adapted from Stevenson, *Annu. Rev. Earth Planet. Sci.*, *10*, 257, 1982. Reproduced with permission from the *Annual Review of Earth and Planetary Sciences*, Volume 10, ©1982 by Annual Reviews Inc.)

results in increased interior pressure, which in turn leads to changes in the state of the atoms and molecules. [Recall the equation of hydrostatic equilibrium, Eq. (10.6), developed for stellar interiors; it applies to spherically symmetric planets as well.] For models of objects just over three times more massive than Jupiter and having similar compositions, increasing mass actually results in *decreasing* radius, an effect that begins to appear in the solid H–He curve in Fig. 21.2. This is due to the growing contribution of degenerate electron pressure in these cold, massive bodies. (This counterintuitive behavior of degenerate matter was discussed in some detail in Section 16.4 in the context of the sizes of white dwarf stars.)

### The Distribution of Mass Inside the Planets

Other information concerning the distribution of mass in the interior is obtained by observing the motions of moons, rings, and spacecraft. For a spherically symmetric planet, all of the mass acts gravitationally as if it is located at a point in the center, but a rapidly rotating planet produces a more complex gravitational interaction with passing objects. By comparing the actual motion of a spacecraft with what would be expected if the planet were spherically symmetric, it becomes possible to map the mass distribution in the interior in terms of mathematical corrections to a spherical shape. This is just what was done using the Magellan spacecraft around Venus, as discussed in Section 20.2.

One such correction is the **oblateness** of the planet, which describes how flattened it is. Such rotational flattening is readily apparent in Fig. 21.1. For instance, Jupiter's equatorial radius ($R_e$) is 71,493 km and its polar radius ($R_p$) is only 66,855 km at an atmospheric pressure of 1 bar,[3] giving an oblateness of

$$b \equiv \frac{R_e - R_p}{R_e} = 0.064874.$$

The amount of oblateness is a function of the speed of rotation and the rigidity of the interior. The rotation period and oblateness of each of the giant planets are given in Appendix C. Note, however, that since the giant planets are fluid throughout much of their interiors, it is not possible to define a single, unique rotation period; their upper atmospheres tend to rotate *differentially*, just as the Sun does (page 364), while their interiors may rotate at different rates than their surfaces.

The oblateness is related to the first-order correction term in the gravitational potential (the potential energy per unit mass), defined to be

$$\Phi \equiv \frac{U}{m}.$$

For a spherically symmetric mass distribution, $\Phi = -GM/r$, where $r$ is the distance from the center of the planet. However, for a planet that is not exactly spherically symmetric, the gravitational potential can be expanded as an infinite series of the form

$$\Phi(\theta) = -\frac{GM}{r}\left[1 - \left(\frac{R_e}{r}\right)^2 J_2\, P_2(\cos\theta) - \left(\frac{R_e}{r}\right)^4 J_4\, P_4(\cos\theta) - \cdots\right], \quad (21.1)$$

where each succeeding correction term represents a progressively higher-order component of the planet's shape and mass distribution, much like higher-order terms in the familiar Taylor series. Notice that as $r$ increases, each successive higher-order term becomes less significant; as $r \to \infty$, $\Phi$ approaches the form of the spherical potential.

The functions $P_2$, $P_4$, ... are known as **Legendre polynomials** and are encountered frequently in many areas of physics. Each polynomial has $\cos\theta$ as its argument, where $\theta$ is the angle between the rotation axis and the position vector of a point in space (the origin of the coordinate system is centered in the middle of the planet); see Fig. 21.3. Examples of some low-order, even-powered Legendre polynomials are

$$P_0(\cos\theta) = 1$$

$$P_2(\cos\theta) = \frac{1}{2}\left(3\cos^2\theta - 1\right)$$

$$P_4(\cos\theta) = \frac{1}{8}\left(35\cos^4\theta - 30\cos^2\theta + 3\right)$$

$$P_6(\cos\theta) = \frac{1}{16}\left(231\cos^6\theta - 315\cos^4\theta + 105\cos^2\theta - 5\right).$$

The Legendre polynomials are multiplied by *weighting factors*, known as **gravitational moments** ($J_2$, $J_4$, $J_6$, ...), that describe the importance of each polynomial to the overall

[3]1 bar $= 10^5$ N m$^{-2}$

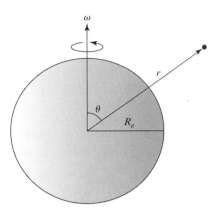

**FIGURE 21.3**   The angle $\theta$ is defined from the rotation axis for the Legendre polynomial expansion of the gravitational potential.

**TABLE 21.2**   Gravitational Moments and Moment-of-Inertia Ratios of the Giant Planets. $R_e$ is the equatorial radius of the giant planet. (Data from Table 1 of Guillot, *Annu. Rev. Earth Planet. Sci.*, *33*, 493, 2005.)

| Moments | Jupiter | Saturn |
|---|---|---|
| $J_2$ | $(1.4697 \pm 0.0001) \times 10^{-2}$ | $(1.6332 \pm 0.0010) \times 10^{-2}$ |
| $J_4$ | $-(5.84 \pm 0.05) \times 10^{-4}$ | $-(9.19 \pm 0.40) \times 10^{-4}$ |
| $J_6$ | $(0.31 \pm 0.20) \times 10^{-4}$ | $(1.04 \pm 0.50) \times 10^{-4}$ |
| $I/MR_e^2$ | 0.258 | 0.220 |

| Moments | Uranus | Neptune |
|---|---|---|
| $J_2$ | $(0.35160 \pm 0.00032) \times 10^{-2}$ | $(0.3539 \pm 0.0010) \times 10^{-2}$ |
| $J_4$ | $-(0.354 \pm 0.041) \times 10^{-4}$ | $-(0.28 \pm 0.22) \times 10^{-4}$ |
| $I/MR_e^2$ | 0.230 | 0.241 |

shape. For example, $J_2$ is related to the planet's oblateness and to its **moment of inertia**.[4] The $J_4$ and $J_6$ terms are more sensitive to the mass distribution in the outer regions of the planet, particularly the equatorial bulge, because the terms have stronger dependence on $R_e$. Because density is more dependent on temperature near the surface of the planet than it is in the deep interior where the gas tends to be degenerate, $J_4$ and $J_6$ also measure the planet's thermal structure. Gravitational moments for the giant planets are given in Table 21.2.

---

**Example 21.1.1.**   The first three higher-order gravitational moments for Jupiter are given in Table 21.2. As a result, the associated expansion terms in Eq. (21.1) have the values shown in Fig. 21.4. The contribution of oblateness to the gravitational potential near the equator

*continued*

---

[4]The moment of inertia has already been discussed for the cases of Earth and the Moon (see Problems 20.12 and 20.13, respectively), and will be explored again for Jupiter (e.g., Problem 21.3).

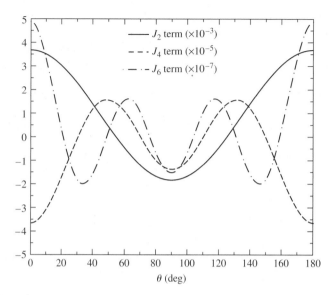

**FIGURE 21.4**  The first three higher-order terms in the gravitational potential expansion for Jupiter when $r = 2R_e$.

($\theta = 90°$) is apparent in the diagram. You should also note that these higher-order correction terms to a spherically symmetric potential are quite small; the first-order correction ($J_2 P_2$) is only on the order of a few tenths of a percent, the second-order term ($J_4 P_4$) is two orders of magnitude smaller than the first-order term, and the third-order term ($J_6 P_6$) is two orders of magnitude smaller than the second-order term.

Related to the gravitational moments is the moment of inertia of the planet. As was described in Problem 20.12, the moment of inertia is given by

$$I \equiv \int_{\text{vol}} a^2 \, dm, \tag{21.2}$$

where $a$ is the distance of the mass element $dm$ from the rotation axis (see Fig. 21.5). For an axially symmetric mass distribution, such as a giant planet rotating about a well-defined axis, it can be shown that $I$ can be expressed in cylindrical coordinates as

$$I = 4\pi \int_{z=0}^{R_p} \int_{a=0}^{a_{\max}(z)} \rho(a, z) \, a^3 \, da \, dz, \tag{21.3}$$

where $z$ is the distance from the center of the planet along the rotation axis to the point where $a$ is measured out to $dm$, and $R_p$ is the polar radius. If we assume that a cross section of the planet along the rotation axis can be approximated by an ellipsoid at the surface, then $a_{\max}$ is related to $z$ by

$$\left(\frac{a_{\max}}{R_e}\right)^2 + \left(\frac{z}{R_p}\right)^2 = 1.$$

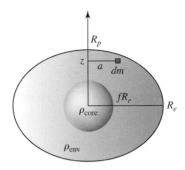

**FIGURE 21.5** A model of an oblate planet with an elliptical cross section that has a spherical core. $\rho_{\mathrm{env}}$ is the density of the envelope and $\rho_{\mathrm{core}}$ is the density of the core. The transition between the two densities occurs at a fraction $f$ of the equatorial radius of the planet.

For a two-component model of an oblate planet having an envelope density of $\rho_{\mathrm{env}}$ and a spherical core with a density of $\rho_{\mathrm{core}}$, and where the transition between the two densities occurs at a fraction of the surface equatorial radius $f$, it can be shown that the moment of inertia is given by

$$I = \frac{8\pi}{15} R_e^4 \left[ R_p \rho_{\mathrm{env}} + f^5 R_e \left( \rho_{\mathrm{core}} - \rho_{\mathrm{env}} \right) \right]. \tag{21.4}$$

Writing $R_p$ in terms of the oblateness $b$ of the planet, we have

$$R_p = R_e(1 - b),$$

Eq. (21.4) becomes

$$I = \frac{8\pi}{15} R_e^5 \left[ (1 - b) \rho_{\mathrm{env}} + f^5 \left( \rho_{\mathrm{core}} - \rho_{\mathrm{env}} \right) \right]. \tag{21.5}$$

Clearly, the moment of inertia depends on the planet's oblateness and the mass distribution throughout the planet. Note that $f$ cannot exceed $f_{\mathrm{max}} = R_p/R_e = 1 - b \leq 1$.

## The Cores of the Planets

All of these data suggest that Jupiter and Saturn have dense cores composed of a thick soup of "rock" (Mg, Si, Fe) and ices. However, although the data suggest dense cores, the masses of the cores are relatively poorly constrained. For example, note that Eq. (21.5) is strongly dependent on $f \leq 1$, and recall that the higher gravitational moments selectively sample the outer envelope of the planet. Based on the available data and numerical models, it appears that Jupiter probably has a rock/ice core of less than about 10 $M_\oplus$, while Saturn's core may be about 15 $M_\oplus$ with an uncertainty of perhaps 50%. (It is possible that the smaller core in Jupiter could be due to some portion of the core having eroded over the age of the planet.) Despite the core masses of Jupiter and Saturn being much greater than the mass of Earth, they constitute only a small fraction of the total mass of each planet. If we assume core masses of 10 $M_\oplus$ and 15 $M_\oplus$ for Jupiter and Saturn, respectively, their cores represent just

3% and 16% of the masses of the two gas giants. Hydrogen and helium make up most of the rest of the mass in each case.

Similar studies of Uranus and Neptune result in core masses comparable to those of Jupiter and Saturn: roughly 13 $M_\oplus$ or so. However, in the cases of Uranus and Neptune, these cores constitute most of the mass of the planets. In particular, both Uranus and Neptune probably have 25% of their mass in the form of rock, 60% to 70% as "ices," and only 5% to 15% in the form of hydrogen or helium gas. Clearly, Uranus and Neptune are not simply smallish versions of their larger siblings, and they are appropriately considered as ice giants rather than gas giants.[5]

## Internal Heat and the Cooling Timescale

Another group of observations that provides hints about the formation and structure of the planets and their subsequent evolution is the detection of heat leaking out from the interior. In the terrestrial planets, the heat generated in the interior is due in large part to the slow decay of radioactive isotopes (page 753). However, this is not sufficient to account for the large quantities of heat coming from the interiors of the giant worlds. For example, as can be seen in Table 21.3, Jupiter absorbs (and re-emits) $5.014 \times 10^{17}$ W of solar radiation, while $3.35 \times 10^{17}$ W of additional power is produced in the interior of the planet. This significantly alters the energy balance and the thermal equilibrium temperature that would result from solar blackbody radiation alone. In Neptune's case, more than one-half of the heat being radiated originates from the interior, explaining why it's effective temperature is very close to that of Uranus, even though Neptune is much farther from the Sun.

One source of internal heat for the giant planets is the gravitational potential energy that was released by gases collapsing onto them during formation. This is just a consequence of the virial theorem (Section 2.4) and is the same Kelvin–Helmholtz mechanism that was discussed in Example 10.3.1 and Section 12.2.

Neglecting any slight differences due to composition and density, for a given specific heat capacity the total thermal energy content of a planet is proportional to its volume (i.e., $\propto R^3$). However, the rate at which heat leaves a planet by blackbody radiation is proportional to surface area ($\propto R^2$). Thus, without an additional source of energy, the timescale for cooling depends on radius as

$$\tau_{\text{cool}} = \frac{\text{total energy content}}{\text{energy loss/time}} \propto R^3/R^2 \propto R.$$

**TABLE 21.3**    Energy Budgets and Effective Temperatures of the Giant Planets. (Data from Table 2 of Guillot, *Annu. Rev. Earth Planet. Sci.*, *33*, 493, 2005.)

| Power or Temperature | Jupiter | Saturn | Uranus | Neptune |
|---|---|---|---|---|
| Absorbed power ($10^{16}$ W) | $50.14 \pm 2.48$ | $11.14 \pm 0.50$ | $0.526 \pm 0.037$ | $0.204 \pm 0.019$ |
| Total emitted power ($10^{16}$ W) | $83.65 \pm 0.84$ | $19.77 \pm 0.32$ | $0.560 \pm 0.011$ | $0.534 \pm 0.029$ |
| Intrinsic power emitted ($10^{16}$ W) | $33.5 \pm 2.6$ | $8.63 \pm 0.60$ | $0.034 \pm 0.038$ | $0.330 \pm 0.035$ |
| Effective temperature (K) | $124.4 \pm 0.3$ | $95.0 \pm 0.4$ | $59.1 \pm 0.3$ | $59.3 \pm 0.8$ |

[5]The term *ices* in this context is somewhat misleading since the $H_2O$, $CH_4$, $NH_3$, and other constituents are actually in a somewhat fluid state under the high pressures found in the interiors of the giant planets.

The characteristic time required for a planet to cool is roughly proportional to the planet's radius. Extrapolating back in time, the giant planets must have been much more luminous when the Solar System was in its infancy; Jupiter may have even glowed visibly.

Since Jupiter is larger than Saturn (as well as being closer to the Sun), it should have remained hotter for a longer period of time and should still be radiating energy into space at a greater rate. In Saturn's case, however, the energy available from the primordial collapse is not sufficient to account for all of the heat now observed to be coming from the planet. The solution to the puzzle of Saturn's additional heat source lies in the observation that its helium is significantly depleted in the upper atmosphere. Referring to Table 21.1, note that helium only accounts for about 3% of the particles in Saturn's upper atmosphere, while the value is closer to 16% for Jupiter and nearly 20% for the Sun. The slow sinking of the heavier helium atoms relative to hydrogen through the atmosphere causes a change in the gravitational potential energy of the planet and the accompanying generation of heat via the virial theorem. This effect has been more pronounced in Saturn because the planet is somewhat cooler.

## Modeling the Interiors of the Giant Planets

Modeling the interiors of the giant planets is done in much the same way it is done for stars; the major difference is the kind of material used in their construction. For example, at the relatively cool temperatures and high pressures of the giant planet interiors, hydrogen takes on a very strange form by terrestrial standards. As we move deeper into the planet, the familiar form of molecular hydrogen becomes so compressed that the molecular bonds are broken and the orbital electrons become shared among the atoms. This is very similar to the behavior of a metal; the hydrogen inside the planet takes on the characteristics of a molten metal, much like mercury at room temperature. This exotic equation of state of hydrogen has been verified in terrestrial laboratories by creating shock waves in the gas that produce temperatures of several thousand kelvins and pressures of millions of atmospheres. It appears that **liquid metallic hydrogen** actually dominates the interiors of Jupiter and Saturn. For Uranus and Neptune, the pressures probably do not get large enough to convert hydrogen into its liquid metallic form, but the ices present in their atmospheres (such as methane and ammonia) become ionized by the pressure.

The interior structures of the giant planets are depicted in Fig. 21.6. The regions labeled "inhomogeneous" for the gas giants are where helium becomes insoluble in hydrogen, and helium-rich droplets form. These droplets then sink deeper into the planet, releasing gravitational potential energy. In the case of Saturn the helium may have settled into the core or formed a shell around the core. Uranus and Neptune have very little hydrogen and helium and are dominated by ices and rock.

## The Upper Atmospheres

In their upper atmospheres, the very colorful and dynamic cloud tops of Jupiter, the more muted hues of Saturn, and the deep blue-greens of Uranus and Neptune owe their beauty to the temperature, composition, rotation, and internal structures of the planets. Observational data, combined with theoretical modeling, suggest that Jupiter's clouds exist in three layers. Clouds in the top layer are composed of ammonia, the next layer is probably composed of ammonium hydrosulfide, and the clouds in the deepest layer are made of water.

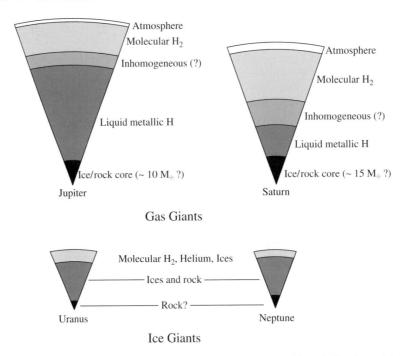

**FIGURE 21.6**    Computer models of the interiors of the giant planets. The relative sizes of the planets are depicted correctly. (Adapted from Guillot, *Annu. Rev. Earth Planet. Sci.*, *33*, 493, 2005.)

The coloration in the clouds of both Jupiter and Saturn is caused by their atmospheres' compositions, although which colors are associated with which molecules remains unclear; suggestions include sulfur, phosphorus, or various organic (carbon-rich) compounds. In Jupiter and Saturn, the bluish regions apparently have higher temperatures, indicating that they lie deeper in the atmosphere. At progressively higher altitudes are brown, white, and red clouds.

Overall, the clouds are located deeper in the atmosphere of Saturn when compared with Jupiter, and hence are not as dramatic. In Uranus and Neptune, reflective clouds of ammonia and sulfur are located deep in the atmosphere. As sunlight passes through the atmosphere, the blue wavelengths are scattered most efficiently by the molecules. In addition, the presence of methane in the atmosphere tends to absorb the red light.

### The Comet P/Shoemaker–Levy 9 Impacts on Jupiter

During July 16–22, 1994, Jupiter took center stage as it got pummeled by the fragments of **Comet P/Shoemaker–Levy 9** (SL9).[6] The comet was discovered in March 1993, although it had apparently been orbiting Jupiter for decades. Extrapolating the comet's orbit back in time, it appears that SL9 broke apart on July 8, 1992, as it passed within 1.6 $R_J$ of Jupiter, well within the planet's Roche limit. [A Hubble Space Telescope view of 21 fragments is

---

[6]The nature of comets will be discussed in some detail in Section 22.2.

**FIGURE 21.7** (a) 21 fragments of SL9 seen on May 17, 1994. The line of cometary nuclei stretches for $1.1 \times 10^6$ km. [Courtesy of H. A. Weaver, T. E. Smith (Space Telescope Science Institute), and NASA.] (b) Hubble Space Telescope images, taken several minutes apart, showing the plume from fragment G on July 18, 1994. (c) Close-up of the fragment G impact site. (Courtesy of Dr. Heidi Hammel, Massachusetts Institute of Technology, and NASA HST.) (d) From left to right, the impact sites in the southern hemisphere of fragments C, A, and E. One of Jupiter's moons (Io) can be seen crossing the planet's disk. (Courtesy of the Hubble Space Telescope Jupiter Imaging Team.)

shown in Fig. 21.7(a).] Astronomers soon realized that the comet fragments would crash into Jupiter in July 1994, possibly providing important clues to the nature of comets and the structure of Jupiter's atmosphere.

Over the week when the collisions occurred, virtually all of the telescopes on Earth (including amateur telescopes) that were in position to view the event, as well as space-based observatories such as the Hubble Space Telescope, Galileo, and Voyager 2, were focused on

Jupiter. Various predictions had indicated that some direct evidence of the impacts might be observable from Earth, but the spectacular display that ensued far exceeded expectations. Figure 21.7(b) shows several images of the enormous plume that rose 3500 km above the cloud tops when fragment G (believed to be the largest) entered the atmosphere of the planet. Even though each of the collisions occurred just beyond our view, on the side of Jupiter away from Earth, the plumes were high enough to make them visible above the limb.[7] The fireballs reached temperatures of 7500 K, greater than the effective temperature of the Sun. Data for the G impact indicate that the temperature cooled to 4000 K after five seconds. Analysis of the data indicated that the largest fragments were no more than 700 m across.

Immediately after each of the larger collisions, scars appeared in the atmosphere greater in diameter than Earth [see Figs. 21.7(c) and (d)]. The dark nature of the marks was probably due to organic molecules rich in sulfur and nitrogen that were present in the atmosphere before the collision. It is also possible that some of the coloration was due to carbon-based compounds, like graphite, that contained silicates delivered by the comet fragments. By December 1994, the marks had been torn apart by the motions in Jupiter's atmosphere, forming a ring around the planet that eventually dissipated completely.

## Atmospheric Dynamics

The most famous atmospheric feature on Jupiter is its **Great Red Spot**, apparent in Figs. 21.1(a) and 21.8(a). This huge anticyclonic storm, which measures roughly one Earth diameter wide by two Earth diameters long, has been observed for more than three centuries. Smaller but similar features can be seen in the atmospheres of each of the giant worlds. Another characteristic shared by these planets is the banded cloud structure following lines of constant latitude. In the case of Uranus the banded cloud features are very difficult to

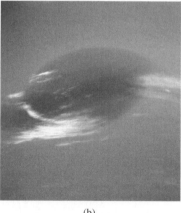

(a)                                              (b)

**FIGURE 21.8**   (a) The Great Red Spot of Jupiter. (b) The Great Dark Spot of Neptune. (Courtesy of NASA/JPL.)

[7]Only the Galileo and Voyager 2 spacecraft had direct views of the impacts.

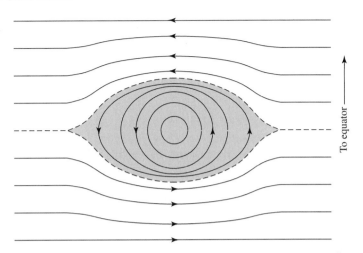

**FIGURE 21.9** Circulation around Jupiter's Great Red Spot is counterclockwise. The anticyclonic storm (which is in the southern hemisphere) is located between two atmospheric bands that are moving in opposite directions. Wind speeds within the Great Red Spot reach 100 m s$^{-1}$, and eddies at the edge of the system circle it in about 7 days.

detect, but they do exist. The circulation within Jupiter's Great Red Spot is attributable to its location between two rivers of atmosphere moving in opposite directions (see Fig. 21.9).

Despite these apparently long-lived features, the atmospheres are very dynamic, with rapid changes occurring on small scales, including rotation around the more stable cyclonic structures. It is worth noting, however, that large features are not necessarily permanent either. For instance, when Voyager 2 visited Neptune in 1989, it discovered the **Great Dark Spot** in the southern hemisphere, shown in Figs. 21.1(d) and 21.8(b). Later, when the planet was observed again by the Hubble Space Telescope in 1994, the Great Dark Spot was gone. Then, in 1995, another dark spot appeared in the northern hemisphere.

Just as the Coriolis force (page 730) redirects the large-scale circulation in Earth's atmosphere from north–south to predominantly east–west flow patterns in each hemisphere, Hadley circulation in the more rapidly rotating giant planets (particularly Jupiter and Saturn) is similarly redirected. However, the atmospheric circulation of Uranus has an interesting aspect not shared with the other giant planets. Unlike any other planet in the Solar System except Pluto, Uranus is almost lying on its side; its rotation axis is tilted 97.9° to the ecliptic. This implies that each pole has the Sun overhead for a portion of its 84-year orbit. During those periods one would expect heat to be transported from the subsolar pole to the one that is in darkness. Yet, when Voyager 2 passed Uranus in 1986, at about the time when one of the poles was pointed toward the Sun, the visible flow patterns were still largely parallel to the planet's equator, due to the planet's rapid rotation and the effects of the Coriolis force. How Uranus was able to transport heat away from the subsolar pole without detectable pole-to-pole flow patterns remains an open question.

Another noticeable difference between Uranus and the rest of the giant worlds is its lack of prominent vortexes. This may correspond to the lack of any detectable heat flow outward from the deep interior. Although it must certainly exist, the rate of heat flow is clearly much less pronounced than in the other three giants.

### Magnetic Fields

The molten iron–nickel core of Earth is the source of its magnetic field. In the giant planets, it is liquid metallic hydrogen that appears to fill that role, at least in Jupiter and Saturn. Rapid rotation generates electric currents in the conducting interiors of the planets. Because the magnetic fields are almost certainly anchored deep in their interiors, measuring the rotation periods of the fields provides a method of determining the rotation periods of their interiors.

In the 1950s, measurements of the radio-wavelength radiation being emitted from Jupiter revealed both thermal and nonthermal components. The thermal radiation is just part of the energy being given off by the planet itself (blackbody radiation). However, the strong nonthermal component was determined to be synchrotron radiation (see Section 4.3) with wavelengths in the decameter (tens of meters) and the decimeter (tenths of meters) ranges. This implies that Jupiter must have a significant magnetic field with relativistic electrons trapped in it. The measured strength of the field is some 19,000 times greater than Earth's field.

Another interesting consequence of the SL9 collisions in Jupiter's southern hemisphere [which all occurred at nearly the same latitude; see Fig. 21.7(d)] was the appearance of an auroral display in the northern hemisphere, not unlike the aurorae seen on Earth; recall Fig. 20.15. Apparently, charged particles near the collision sites acquired sufficient kinetic energy that they traveled along Jupiter's magnetic field lines, colliding with the atmosphere in the north within 45 minutes following the impacts.

The physical extent of Jupiter's magnetic field is enormous. The planet's **magnetosphere**, defined to be the space enveloped by its magnetic field, has a diameter of $3 \times 10^{10}$ m, 210 times the size of the planet and 22 times larger than the Sun. Because of Jupiter's rapid rotation, the charged particles trapped in its field are spread out into a **current sheet** that is situated along *the field's* equator (the field axis is inclined 9.5° to the rotation axis of the planet). Given the large numbers of particles present in Jupiter's current sheet, another source of charged particles beyond those supplied by the solar wind must exist. The solution to this mystery came when the Voyager spacecraft first observed Jupiter's moon Io.

## 21.2 ■ THE MOONS OF THE GIANTS

Many of the most spectacular and fascinating images returned by the Voyager, Galileo, and Cassini–Huygens missions were of the moons of the giant planets, beginning with the Galilean moons of Jupiter (Fig. 21.10). The relative sizes of the Galilean moons are depicted in Fig. 21.11. **Io** (shown in more detail in Fig. 21.12) is the closest of the four large Galilean moons to Jupiter. It is a bizarre-looking yellowish-orange world with as many as nine active volcanoes observed to be erupting simultaneously. **Europa** (Fig. 21.13) is covered with a thin layer of water-ice that is criss-crossed by cracks and nearly devoid of any cratering. **Ganymede** (Fig. 21.14) has a thick ice surface that shows evidence of significant cratering. And finally, **Callisto** (Fig. 21.15) appears to be covered with a layer of dust and has an old and

**FIGURE 21.10**  A "family portrait" of Jupiter and its four largest moons. From nearest Jupiter to farthest are Io, Europa, Ganymede, and Callisto. The portrait is actually a mosaic of a number of Voyager images. (Courtesy of NASA/JPL.)

**FIGURE 21.11**  A mosaic of images obtained by the Galileo spacecraft showing the four Galilean moons of Jupiter. From left to right, and from nearest to Jupiter to farthest away: Io, Europa, Ganymede, Callisto. Here, the moons are depicted in such a way as to show their relative sizes. (Courtesy of NASA/JPL.)

very thick ice crust that has been subjected to extensive bombardment.[8] The characteristics of these worlds are consistent with a decreasing average density with increasing distance from Jupiter, implying that the relative amount of water-ice crust increases with respect to the rock core.

### The Evolution of the Galilean Moons

The increasing percentage of volatiles (principally water-ice) in these worlds at increasing distances from Jupiter suggests that their formation was closely linked to the formation and

---

[8]Recall the discussion in Section 20.4 describing the amount of cratering as a function of surface age.

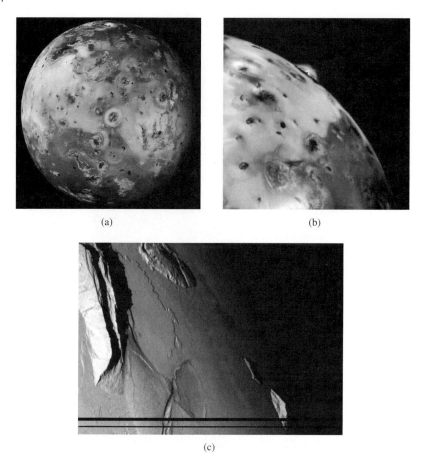

**FIGURE 21.12** (a) The disk of Io shows a large number of volcanic features. (b) A volcano (Prometheus) erupting on the limb of Io. Prometheus was observed to be erupting in every image obtained by the Voyagers (1979) and Galileo (1995–2003). Other volcanic eruptions are not as long-lived. (c) Mountains on Io seen at sunset. The low scarp in the upper left is approximately 250 m high. It is believed that these mountains are produced by uplifted thrust faults. The black lines along the bottom of the image are due to missing data. (Courtesy of NASA/JPL/University of Arizona/Arizona State University.)

subsequent evolution of the planet itself. Given the regular nature of the Galilean satellites, it has been proposed the they may have formed out of the Jupiter *subnebula* while Jupiter was accreting its massive atmosphere. Within this context, recalling that Jupiter must have been hotter in the past than it is today, Io would have been close enough to have had most of its volatiles evaporate away. Moving progressively farther out, Europa would have been able to hold on to some water, Ganymede even more, and Callisto (being the coldest of the Galilean moons at the time of its formation) would have retained the largest percentage of volatiles.

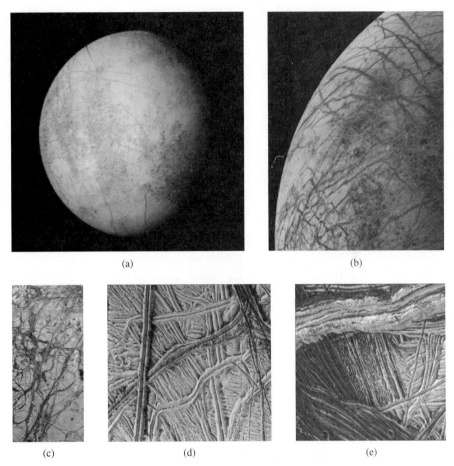

**FIGURE 21.13**  (a) The full disk of Europa. (b) Europa has numerous cracks running across the surface. (c) A close-up of broken ice. (d) Ridged plains. (e) Wedge terrain. (Courtesy of NASA/JPL.)

## The Effects of Tidal Forces on Io

The consequences of this evolution can be seen in each of the Galilean moons. Consider them in sequence beginning with the one closest to Jupiter. Because of its proximity to Jupiter, Io experiences the most severe tidal forces. Even though the moon's rotation period is the same as its orbital period, small deviations from a perfectly circular orbit mean that its orbital velocity is not constant. Consequently, the moon tends to wobble, not quite keeping one side "locked in place" toward Jupiter. This effect is due to the curious resonance that exists among the orbits of Io, Europa, and Ganymede. Their orbital periods form ratios that are approximately 1:2:4, meaning that both Europa and Ganymede perturb Io's orbit at about the same location each time Io orbits the planet. This forces Io's orbit to remain slightly elliptical.

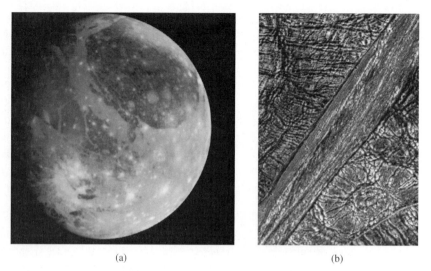

**FIGURE 21.14**    (a) The surface of Ganymede shows significant cratering, indicating that it has not been refreshed as recently as Europa. (b) A close-up view of ridges and grooves prevalent on the surface, indicative of past tectonic activity. The diagonal band is 15 km wide. The circular feature in the lower right portion of the image is probably an impact crater. (Courtesy of NASA/JPL.)

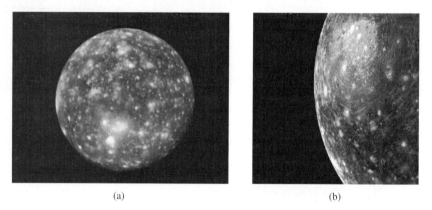

**FIGURE 21.15**    (a) Callisto's surface exhibits extensive cratering. (b) A close-up of a large impact crater known as Valhalla. (Courtesy of NASA/JPL.)

Based on gravitational data from Galileo's close flybys of Io, it appears that Io has an iron-rich core, a molten silicate mantle, and a thin silicate crust (the moon's average density is $3530 \, \mathrm{kg \, m^{-3}}$). This structure suggests that Io was entirely molten at least once, and perhaps numerous times, allowing the moon to become chemically differentiated. Lava flows and lakes of lava, such as Loki Patera (which is larger than the island of Hawaii) are clearly evident on the surface of Io. It is important to note, however, that Io's volcanoes do not operate in quite the same way that Earth-based volcanoes do. Instead, the eruptions may be somewhat more analogous to the geysers seen in such places as Yellowstone National Park. In terrestrial geysers, rapid phase transitions from water to steam force the steam upward at high velocity through cracks in the surface. On Io, sulfur and sulfur dioxide ($SO_2$) probably play the same role. In fact, $SO_2$ has been detected over volcanic vents and in Io's very thin atmosphere. The yellowish-orange surface is due to sulfur raining back down on the moon from its continually erupting volcanoes. The constant eruptions are always regenerating the moon's surface; Io is literally turning itself inside out.

### Io's Interaction with Jupiter's Magnetic Field

All of the Galilean moons are located deep inside Jupiter's magnetosphere, but Io interacts with the magnetic field most strongly. Since Jupiter rotates in just under 10 hours, whereas Io orbits the planet in 1.77 days, Jupiter's magnetic field sweeps past Io at a speed of about 57 km s$^{-1}$. This motion through the magnetic field sets up an electric potential difference across the moon, estimated to be 600 kV.[9] The potential difference acts much like a battery, causing a current of nearly $10^6$ amps to flow back and forth along magnetic field lines between Io and Jupiter. This current flow of charged particles in the magnetic field also generates Joule heating within the moon, analogous to a resistor in a circuit. Roughly $P = IV \sim 6 \times 10^{11}$ W is generated in this way. However, this contribution to the total internal heating of the moon is only a small fraction of the total energy liberated from the surface per second, which is approximately $10^{14}$ W.

That Io must have some interaction with Jupiter's magnetic field has been known for some time. When Jupiter, Io, and Earth are in certain alignments, bursts of decameter-wavelength radiation are detected. Not all the details of the process are yet understood, but the bursts appear to be associated with the electrical current flowing between Jupiter and its volcanic moon.

Io must also be responsible for the excessive number of charged particles trapped in Jupiter's magnetic field, although it is unlikely that they escaped directly from the moon's volcanoes since the ejection speeds are much less than Io's escape velocity. Instead, a process referred to as *sputtering* has been proposed; oxygen and sulfur ions from Jupiter's magnetosphere impacting on the moon's surface or in its atmosphere may provide sufficient energy for other sulfur, oxygen, sodium, and potassium atoms to escape. In fact, clouds of sulfur and sodium (known as the **Io torus**) have been detected around Jupiter at the location of Io's orbit. On the order of $10^{27}$ to $10^{29}$ ions leave Io and enter Jupiter's magnetospheric plasma every second.

### Europa

Europa's surface seems to be continually refreshed. Based on the near absence of cratering, it appears that most of the surface is less than 100 million years old, supporting the idea that a layer of liquid water may exist below the surface. In fact, observations from the Galileo mission indicate an iron-rich core, a silicate mantle, a possible subsurface ocean, and a thin ice crust; the average density of Europa is 3010 kg m$^{-1}$, less than the density of Io. The water ocean/ice crust is collectively about 150 km thick. The source of the heat required to

---

[9]This is just Faraday's law of induction.

keep the subsurface water at least partially melted is probably the weak tidal interactions with Jupiter and the other Galilean moons. The cracks running across the moon's surface appear to be stress fractures induced by those tidal forces combined with tectonic activity.

In 1994 the Hubble Space Telescope detected a thin molecular oxygen atmosphere around Europa. That observation was confirmed by both Galileo and the flyby of the Cassini spacecraft on its way to a rendezvous with Saturn. Cassini also found the presence of atomic hydrogen in Europa's atmosphere. It has been suggested that the atmosphere is due to sputtering of the surface water-ice resulting from the interaction of Europa with Jupiter's magnetosphere.

Given the presence of subsurface heating, a likely source of liquid water, and the probability of organic materials being either intrinsic to the moon or delivered by comets and meteorites, it has been widely speculated that Europa may be a site for the evolution of life. Even though no evidence exists for the presence of life today or in the past on Europa, at the end of its operational lifetime scientists intentionally sent the Galileo spacecraft into the crushing atmosphere of Jupiter on September 21, 2003, in order to avoid any future inadvertent collision with Europa and its possible subsurface ocean.

## Ganymede

Ganymede's surface also shows a complex series of ridges and grooves that strongly suggest some history of tectonic activity on this ice world. This is supported by gravitational data from Galileo indicating a likely partially molten iron core, a silicate lower mantle, an icy upper mantle, an ice crust, and an average density of only $1940 \, \text{kg m}^{-1}$. It has been proposed that before the ice crust became too rigid, convection in the interior was responsible for carrying heat to the surface. This convective motion also caused movement of the surface crust, much like the current action of tectonic plates on Earth. As a result, although the surface is certainly much older and more heavily cratered than Europa's, the surface has been at least partially refreshed during its history.

## Callisto

Callisto apparently cooled and solidified quite rapidly after material accreted out of the local subnebula around Jupiter. As a result, its surface continued to collect dust as the nebula thinned, blanketing the moon with the dark material. Evidence that Callisto solidified quickly is also apparent in the structure of its interior. Models suggest that the interior of the moon is relatively simple, with a partially differentiated interior of ice and rock, an ice-rich crust, and the lowest density of the Galilean moons ($1830 \, \text{kg m}^{-3}$). Having solidified in the early stages of the formation of the Solar System, Callisto was also subject to frequent impacts of the still-abundant objects that traveled among the newly formed planets and moons. Evidence of the nebular dust accretion and the impacts remains today. The whitish-appearing impact craters are the result of ice being exposed during the collisions.

## A Unified Formation of the Galilean Moons

As we have seen, the four Galilean satellites of Jupiter exhibit a trend of decreasing density with distance from their parent planet. Given the evident trends in their properties, including their internal structure (diminishing iron-rich cores and increasing water-ice content with distance from Jupiter), it is apparent that they likely formed systematically with Jupiter, perhaps out of a subnebula around the planet. This is also supported by the fact that each of the Galilean moons orbit prograde and in the equatorial plane of the planet.

### The Smaller Moons of Jupiter

Other, smaller moons also orbit in a prograde direction in the equatorial plane of Jupiter. These additional **regular satellites** may have formed out of the subnebula as well. However, there are a large number of moons around Jupiter that orbit well out of the equatorial plane and, in many cases, in retrograde orbits. It seems that these **irregular satellites** of Jupiter may be captured objects that happened to wander by at some point in time. Still other very small satellites may be **collisional shards** produced by meteoritic collisions with larger satellites.

Space does not allow us to discuss each of these many moons in detail. Similarly, we are unable to discuss most of the smaller moons orbiting the other giant planets; rather, we will focus our attention on the larger moons and a few of the more unusual smaller satellites of Saturn, Uranus, and Neptune.

### Saturn's Titan with Its Thick Atmosphere

When the two Voyager spacecraft reached Saturn in 1980 and 1981, they were directed to examine **Titan**, the second-largest moon in the Solar System (Ganymede being the largest). Ever since Gerard P. Kuiper (1905–1973) detected methane gas around Titan in the 1940s, astronomers have wondered about the nature of this distant, atmospheric world. When the images began arriving, scientists saw a moon with an atmosphere so filled with suspended particles (**aerosols**) that no pictures of its obscured surface were possible.

The joint Cassini–Huygens mission arrived in the Saturnian system in July 2004. After arrival, the Huygens probe detached from the Cassini orbiter and descended to the surface of Titan on January 14, 2005. During its descent, the Huygens probe was able to measure wind speeds of up to $210 \ \mathrm{m \ s^{-1}}$, sample the composition of the atmosphere, and, after passing through the high-altitude smog layer of hydrocarbons, obtain images of the surface (see Fig. 21.16).

The dominant constituent in the atmosphere is nitrogen ($N_2$), which constitutes somewhere between 87% and 99% of the gases. Methane ($CH_4$) makes up between 1% and 6% of the atmosphere, and argon (Ar) constitutes between 0% and 6% of the total. Numerous other species are present in smaller amounts, including molecular hydrogen ($H_2$), carbon monoxide (CO), carbon dioxide ($CO_2$), hydrogen cyanide (HCN), and a host of additional hydrocarbons, such as acetylene ($C_2H_2$), ethylene ($C_2H_4$), ethane ($C_2H_6$), methylacetylene ($C_3H_4$), propane ($C_3H_8$), and diacetylene ($C_4H_2$). The aerosols in the high-altitude smog layer are probably just condensed forms of these compounds.

At the base of the atmosphere, the pressure is approximately 1.5 atm and the temperature is 93 K. With those conditions, methane is able to condense as a liquid and then evaporate again, and thus it plays a role much like that of water on Earth. At the Huygens landing site, the ground was moist, with liquid methane occurring a few centimeters below the surface. It is possible that it had rained methane at that location shortly before the arrival of Huygens. In fact, Huygens sank 10 to 15 cm into the soft ground at the landing site. The

**FIGURE 21.16**   Counterclockwise from upper left: (a) Titan with its thick atmosphere as seen by the Cassini orbiter (Courtesy of NASA/JPL). (b) A mosaic image of the surface from an altitude of 8 km imaged by the Huygens probe (Courtesy of ESA/NASA/JPL/University of Arizona). (c) The surface of Titan with "pebbles" of what is probably water-ice in the foreground. The flat pebble near the middle of the image has a width of 15 cm, and the one to its right has a width of 4 cm. Both pebbles are 85 cm from the Huygens probe camera. (Courtesy of ESA/NASA/JPL/University of Arizona.)

surface water-ice pebbles also show evidence of having a liquid flow across them, much like terrestrial rock pebbles in a dry creek bed. In addition, images obtained during the descent revealed topography that looks like drainage canals leading to low, dark, flat regions that may be lakes (or dried lake beds).

### Mimas and the Herschel Crater

Another member of the Saturnian system, **Mimas**, is a small but fascinating moon (shown in Fig. 21.17). It exhibits a very large impact crater (referred to as Herschel) that is testimony to a collision almost energetic enough to fracture it.[10] Of course Saturn also has numerous

---

[10]More than one researcher has noticed that Mimas bears a strong resemblance to the "Death Star" in the George Lucas film *Star Wars*, Lucasfilm Ltd. Production (1977).

**FIGURE 21.17**   The impact that produced the Herschel crater on Mimas was nearly energetic enough to completely fracture the moon. Mimas is one of the many small moons orbiting Saturn. (Courtesy of NASA/JPL.)

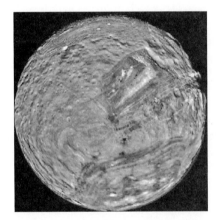

**FIGURE 21.18**   Miranda is one of the moons of Uranus. Its dramatic surface features may be the result of one or more very energetic collisions that fractured the moon. (Courtesy of NASA/JPL.)

other regular and irregular satellites in its system, a few of which will be mentioned in the context of its extensive ring system in Section 21.3.

### The Chaotic Surface of Miranda, a Moon of Uranus

When Voyager 2 reached Uranus in 1986, it encountered another moon that may have suffered a very energetic collision. **Miranda**, which measures only 470 km across, looks like a moon "put together by a committee" (Fig. 21.18). One explanation for its amazing topography is that one or more collisions actually succeeded in breaking the moon apart. When gravity pulled all of the pieces back together, they didn't quite fit. Portions of the rock core tried to settle back to the center of the moon while ice tried to float back to the surface. This proposed rearrangement of the structure of Miranda produced a strange surface with cliffs as tall as 20 km (twice the height of Mount Everest) and features such as the "chevron" that can be seen in the figure.

An alternative explanation for Miranda's topography proposes that tidal forces exerted by Uranus on the small moon caused parts of the surface to be pulled apart. This allowed warmer material in the interior (which was heated by tidal effects) to rise to the surface, producing the ridges and troughs that are observed.

Interestingly, all of the regular moons of Uranus, and, as we will see in the next section, its ring system as well, orbit near the equatorial plane of the planet rather than its orbital plane. Recall that the rotation axis of Uranus is highly inclined to the ecliptic (97.9°), making the orientation of the Uranusian system a puzzle for Solar System dynamicists.

### Neptune's Triton

The last and one of the most unusual moons visited by Voyager 2 was Neptune's largest moon, **Triton** (Fig. 21.19). With a surface temperature of 37 K, it is also the coldest world yet visited. The moon's southern pole is covered with a pinkish frost that is composed almost entirely of nitrogen. Along with the nitrogen frost, other surface ices include $CH_4$, CO, and $CO_2$. Also present are very large "frozen lakes" of water-ice that show very little cratering, indicative of a relatively young age. The water-ice may have erupted from ice volcanos.

During the Voyager 2 flyby, geyser-like jets were detected forcing plumes of gas 8 km up into Triton's tenuous atmosphere, where the plumes were blown down-wind. These plumes may simply be gas rising from a warm source inside the planet, but how they are initiated remains unclear.

The atmosphere of Triton is composed predominantly of nitrogen, like the atmospheres of Earth, Titan, and Pluto. However, unlike Earth and Titan, Triton's atmosphere is extremely thin, with a pressure of only $1.6 \times 10^{-5}$ atm. Much of the atmosphere may be a consequence of the jets of nitrogen gas erupting from the interior of the moon.

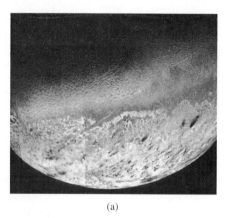

(a)                                    (b)

**FIGURE 21.19**   (a) The southern polar ice cap of Triton, Neptune's largest moon. The dark streaks appear to be a mixture of nitrogen frost and hydrocarbons that was ejected from small volcanos. (b) A water-ice lake that may have been produced by an ice volcano. Given the lack of significant cratering, the surface appears to have been refreshed relatively recently. (Courtesy of NASA/JPL.)

Triton's slowly decaying retrograde orbit was discussed on page 723. Given its highly unusual orbit, both retrograde and inclined to Neptune's equator by 20°, its proximity to the Kuiper Belt, and physical properties similar to other Kuiper Belt objects (such as Pluto), it is widely believed that Triton was captured by Neptune. It is also likely that when the relatively massive Triton was captured, it significantly disrupted Neptune's already-existing system of much smaller satellites. Perhaps the tidal effects that resulted in the moon's present-day circularized and synchronous orbit could have produced sufficient internal heating to cause faulting and its cantaloupe-like terrain.

## 21.3 ■ PLANETARY RING SYSTEMS

Each of the giant planets contains a ring system. Although Saturn's prominent rings were first seen several hundred years ago, the rings of the other planets weren't discovered until the 1970s and 1980s. As we will learn, there are certainly some similarities among the ring systems, but there are significant differences as well.

### The Structure of Saturn's Rings

Arguably, the most well-known feature of the Saturnian system is its spectacular set of rings, seen in Fig. 21.1(b). Based on observations made from Earth, several fairly distinct rings have long been known to exist, labeled (from the outside in) A, B, and C. Between the prominent A and B rings, the **Cassini division** was thought to be virtually devoid of ring material. Another empty region, called the **Encke gap**, was observed within the A ring.

After the planet was visited by the Voyager spacecraft, other rings were discovered. The two spacecraft also revealed previously unexpected complexity in the system. As can be seen in Fig. 21.20(b), instead of large, almost continuous rings, thousands of *ringlets* were discovered; even the Cassini division has a number of rings lying in it, although the number density of the particles is much lower than in the neighboring regions. The F ring proved to be particularly perplexing because it is very narrow and appears to be braided [Fig. 21.21(a)].

The positions of the various rings and the Cassini division are given in Table 21.4; also included, from Example 19.2.1, is the estimate of Saturn's Roche limit for a satellite of density 1200 kg m$^{-3}$. The rings extend out as far as 8 $R_S$ from Saturn, while the disk of the rings is very thin, perhaps only a few tens of meters thick. The presence of vertical ripples in the disk gives the rings the appearance of being about 1 km thick. Because of the thinness of the disk, when it is viewed perpendicular to the plane, the optical depth of the ring system ranges from about 0.1 to 2. It is actually possible to see through the rings in many locations.

When Galileo observed Saturn in 1612 (some two years after his initial observations), he was surprised to discover that the protrusions that he had seen earlier had apparently vanished! We now know that during his later set of observations, Galileo was viewing the rings edge-on, making them undetectable from Earth.

The reason the rings are so thin is easily understood by considering what happens to particles that undergo partially inelastic collisions, as shown in Fig. 21.22. Imagine two particles circling Saturn in the same direction but in orbits that are slightly tilted with respect to each other. If the two particles should collide, the $x$-components of their velocities would

(a)

(b)

(c)

(d)

**FIGURE 21.20** (a) Jupiter's very thin ring. (b) A close-up of Saturn's rings. Even the dark Cassini division is not entirely empty. (c) The rings of Uranus. Because the moving spacecraft was focused on the ring system, the background stars appear as streaks in the image. (d) Neptune was masked to bring out its faint ring system. (Courtesy of NASA/JPL.)

be largely unaffected, but the collision would decrease their $y$-components. The process diminishes the thickness of the disk until other effects start to become important, such as random collisions with incoming particles and perturbations from moons.

### The Composition of Saturn's Rings

Most of the particles that make up the rings are quite small, with the majority having diameters that range from a few centimeters to several meters, although it seems likely that at least some particles with diameters as small as a few micrometers or as large as one kilometer may exist in the system. Size estimates are derived from several pieces of evidence, including the rate at which particles cool off in Saturn's shadow and how efficiently they reflect radar signals of various wavelengths.

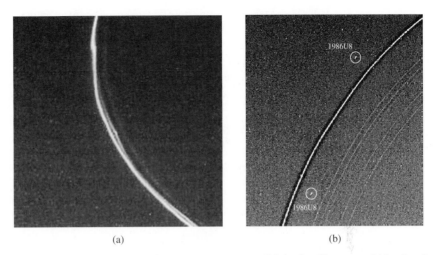

(a)         (b)

**FIGURE 21.21** (a) Saturn's braided F ring. (b) Two small "shepherd" moons orbiting just inside and outside of the $\epsilon$ ring of Uranus. (Courtesy of NASA/JPL.)

**TABLE 21.4** The Positions of Saturn's Ring Features.

| Feature | Position ($R_S$) |
|---|---|
| D Ring | 1.00–1.21 |
| C Ring | 1.21–1.53 |
| B Ring | 1.53–1.95 |
| Cassini Division | 1.95–2.03 |
| A Ring | 2.03–2.26 |
| *Roche Limit* | 2.04 |
| F Ring | 2.33 |
| G Ring | 2.8 |
| E Ring | 3–8 |

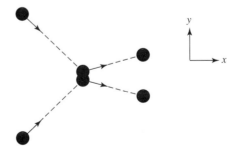

**FIGURE 21.22** The collisions between particles help to keep the rings of Saturn very thin.

It has been known for some time that the material in the Saturnian rings is highly reflective (the rings have albedos in the range 0.2–0.6). Albedo measurements, combined with infrared spectroscopy, provide information about the composition of ring material. It appears that the particles of most of the rings are primarily water-ice, with some dust embedded in them or covering their surfaces. However, the very extended and thin E ring may be composed entirely of dust that is originating from the nearby moon, Enceladus.

### Jupiter's Tenuous Ring System

Jupiter's very tenuous ring system [see Fig. 21.20(a)] has a characteristic optical depth of roughly $10^{-6}$. Three components have been seen: the innermost *toroidal halo*, the *main* ring, and the outermost *gossamer* ring. Taken together they stretch from near Jupiter to about 3 $R_J$. It is believed that the ring material is primarily dust that is constantly resupplied by collisions of micrometeorites with larger objects (tiny moons) in the rings.

### The Rings of Uranus

The rings of Uranus and Neptune [Figs. 21.20(c) and (d)] were first detected indirectly from Earth and later photographed by Voyager 2. On March 10, 1977, astronomers were observing the **occultation** of a background star (Uranus passing in front of the star). They were trying to measure the diameter of Uranus, as well as to gather some information about its atmosphere. Knowing the speed of the planet and the amount of time the star was hidden, they could determine the planet's diameter. Quite unexpectedly, the starlight faded and brightened several times before actually being occulted by the planet. When the star reappeared, the pattern of alternating fading and brightening was repeated, but in reverse order. The astronomers realized that rings were blocking out the star's light. The same procedure was used for Neptune but led to confusing results. In some cases the starlight was blocked on only one side of the planet, leading to the suggestion that only incomplete rings (or arcs) exist around Neptune.

A total of thirteen rings have been detected around Uranus, nine from the ground, two more by Voyager 2 in 1986, and another two by the Hubble Space Telescope in 2004. All of the rings are remarkably narrow, ranging in width from 10 km to 100 km (not unlike the F ring of Saturn), with some of the rings also showing signs of being braided. The two detected using the HST have diameters that are much greater than the other eleven rings, leading some researchers to refer to them as a second Uranian ring system.

The composition of the Uranian rings appears to be very different from that of either Jupiter or Saturn. Reflecting only about 1% of the incident sunlight, the ring material is extremely dark. This is because the rings are composed largely of dust, rather than ice.

Curiously, as mentioned in the previous section, the rings and moons of Uranus lie in the planet's equatorial plane and not along the ecliptic. Recalling that the rotation axis of Uranus is tilted 97.9° with respect to the ecliptic, this implies that the orientation of the orbits of the rings and moons changed after one or more catastrophic impacts dramatically shifted Uranus's axis (if indeed impacts were responsible). Apparently Uranus's rotationally produced equatorial bulge gravitationally affected its satellites and ring material, ultimately reorienting their orbits until the moons and rings were once again aligned with the planet's equator. Similarly, Saturn's rings are also aligned with its equator, despite the planet's equatorial plane being tilted almost 27° to its orbital plane.

### Neptune's Rings

When it reached Neptune, Voyager 2 also found rings orbiting that planet. Like the rings of Uranus, several of the six identified rings are quite narrow, while the others appear to be diffuse sheets of dust. Oddly, the outermost ring, known as Adams,[11] has five discrete regions of concentrated material, like sausages on a string. It was these concentrations that were responsible for the arcs that were deduced from occultations.

### Physical Processes Affecting Ring Systems

The spectacular observations of the Voyagers, Galileo, and Cassini have shown us that the dynamics of ring systems are quite complex. Not all of the features are as yet understood, but many important components have been identified:

- **Collisions** have already been mentioned as the process that maintains the thinness of the rings.

- **Keplerian shear** (or diffusion) spreads the rings out in the system's plane. As more rapidly moving particles in slightly lower orbits overtake more slowly moving particles farther out, collisions between them cause the inner particles to slow somewhat, and they drift closer to the planet. At the same time, the outer particles are accelerated, moving them outward. The process stops when the density of ring particles becomes so low that collisions effectively cease.

- **Shepherd moons** are small moons that reside in or near the edge of the rings, controlling the location of ring boundaries via their gravitational interactions. The narrowness of the F ring of Saturn [recall Fig. 21.21(a)] was understood when the two moons Pandora and Prometheus were discovered to be orbiting just outside and inside the ring, respectively. As the more rapidly moving ring particles pass Pandora, the moon's gravitational pull slows them down, causing the particles to drift inward. When Prometheus overtakes the ring particles, it pulls them forward, speeding them up and causing them to move outward. As a result, the F ring is confined to a narrow region just 100 km wide. Another shepherd moon (Atlas) defines the sharp outer edge of the A ring. Shepherd moons have also been discovered guiding one of the rings of Uranus; see Fig. 21.21(b).

- **Orbital resonances** between moons and ring particles in specific orbits can act to deplete or enhance particle concentrations. (It is also necessary that the orbital plane of the moon align with the ring plane.) For instance, there exists a 2:1 orbital resonance between Mimas and particles at the inner edge of the Cassini division. In other words, a particle in that location orbits twice for every orbit of Mimas. Since an inferior conjunction of such a particle with Mimas always occurs at the same position, gravitational perturbations of the particle's orbit produced by Mimas become cumulative, implying that the moon tends to force the particle into an elliptical orbit.

---

[11]The Adams ring, along with the Leverrier and Galle rings, were named for the mathematical and observational discoverers of Neptune.

As the particle begins to cross the more circular orbits of particles at other radii, collisions become more likely. The outcome is that the particle has been removed from its original orbit and relocated in another part of the system (see Problem 21.13).

- **Spiral density waves**, first proposed by Peter Goldreich and Scott Tremaine in the late 1970s, are set up by moons as a consequence of orbital resonance. Gravitational perturbations can cause particles at different orbital radii to bunch up, effectively increasing their gravitational influence on other nearby particles in the disk. Those neighboring particles in turn are drawn toward the increase in density, extending the enhancement. If the moon responsible for the resonance is beyond the edge of the disk, the wave of density enhancement spirals outward. Since the density is larger in the wave, the probability of collision increases. Keplerian shear then causes the number density of particles near resonance orbits to decrease. This process helps explain the width of the Cassini division.[12]

- **The Poynting–Robertson effect** (a consequence of the headlight effect discussed in Example 4.3.3) can cause ring particles to spiral in toward the planet. When particles in the rings absorb sunlight, they must re-radiate that energy again if they are to remain in thermal equilibrium. The original light was emitted from the Sun isotropically, but in the Sun's rest frame the re-radiated light is concentrated in the direction of motion of the particle. Since the re-radiated light carries away momentum as well as energy, the particle slows down and its orbit decays. This process is explored in more detail in Problems 21.15 and 21.16.

- **Plasma drag** is a consequence of the collisions of ring particles with charged particles trapped in the planet's magnetic field. Since the magnetic field is anchored inside the planet, it must revolve with the rotation period of the planet. If the ring particles are inside the planet's synchronous orbit (as most rings are), the particles will overtake the magnetic field plasma, and collisions will slow the particles down. The particles will then spiral in toward the planet, just as with the Poynting–Robertson effect. If the ring particles are outside the synchronous orbit, they will spiral outward.

- **Atmospheric drag** occurs as particles approach the outer reaches of the planet's atmosphere. This effect quickly causes the particles to spiral down into the planet.

- **Radial spokes** have been observed in the rings of Saturn and are attributed to the interaction of charged dust particles with the planet's magnetic field. These spokes move through the ring system with the rotation period of the planet, rather than with the orbital period of the rings. It appears that some dust particles acquire a net electrostatic charge as a result of their frequent collisions with other dust particles. This causes the dust to become trapped in magnetic field lines tens of meters above the plane of the rings. Sunlight that is scattered from the suspended particles produces the observed spokes.

[12]Spiral density waves also play an important role in the structure of spiral galaxies. They will be discussed in more detail in that context in Section 25.3.

- **Warping** of the disk is caused by the gravitational influences of the Sun and the planet's moons. If the Sun or the moons are not in exactly the same plane as the ring, particles in the ring are pulled out of the ring plane.

## Ring Formation

The formation of planetary rings is still not fully understood. A major problem lies in the timescales involved in maintaining rings against processes that tend to disperse or destroy them. Are rings long-lived or transient phenomena? One idea, first suggested by Pierre-Simon Laplace (1749–1827) and Immanuel Kant (1724–1804) in the late 1700s, argues that rings are nebular in origin; they were formed at the same time that the planets accreted. Since most of Saturn's rings are composed largely of water-ice, while the rings of Jupiter, Uranus, and Neptune contain primarily nonvolatile substances (silicates and carbon), Saturn must have cooled more rapidly, before the water could escape. Although this idea could account for the spectacular Saturnian system, while also explaining the composition and greater sparsity of the rings of the other giant planets, it is difficult to understand how the systems could be maintained for more than 4.5 billion years.

It is also possible that ring systems arise due to tidal forces; if moons were drawn inside the planet's Roche limit, or if a comet or meteoroid ventured too close, tidal forces would fracture the objects, producing a new ring system. However, tidal disruption should leave intact rocky fragments as large as tens of kilometers in diameter. Grinding and meteoritic impacts would eventually break down the remnants, but such processes are extremely slow. On the other hand, loosely packed icy objects, such as comets, may be broken into smaller pieces by tidal disruptions (recall Shoemaker–Levy 9).

The discovery of the giant, outermost ring of Uranus by the HST was accompanied by the discovery of another moon, Mab, in the same orbit as the ring. It seems that when Mab is hit by meteorites, material ejected from the moon replenishes the giant ring, suggesting that the source for this ring, at least, has been identified.

Clearly, much work remains to be done before we can claim to understand the complexities of planetary rings.

## SUGGESTED READING

### General

Beatty, J. Kelly, Petersen, Carolyn Collins, and Chaikin, Andrew (eds.), *The New Solar System*, Fourth Edition, Cambridge University Press and Sky Publishing Corporation, Cambridge, MA, 1999.

Booth, Nicholas, *Exploring the Solar System*, Cambridge University Press, Cambridge, 1996.

Goldsmith, Donald, and Owen, Tobias, *The Search for Life in the Universe*, Third Edition, University Science Books, Sausalito, CA, 2002.

Morrison, David, and Owen, Tobias, *The Planetary System*, Third Edition, Addison-Wesley, San Francisco, 2003.

Trefil, James, *Other Worlds: Images of the Cosmos from Earth and Space*, National Geographic, Washington, D.C., 1999.

## Technical

Asplund, M., Grevesse, N., and Sauval, A. J., "The Solar Chemical Composition," *Cosmic Abundances as Records of Stellar Evolution and Nucleosynthesis in Honor of David L. Lambert*, Barnes, Thomas G. III, and Bash, Frank N. (eds), Astronomical Society of the Pacific Conference Series, *336*, 25, 2005.

Atreya, S. K., Pollack, J. B., and Matthews, M. S. (eds.), *Origin and Evolution of Planetary and Satellite Atmospheres*, University of Arizona Press, Tucson, 1989.

de Pater, Imke, and Lissauer, Jack J., *Planetary Sciences*, Cambridge University Press, Cambridge, 2001.

Greenberg, Richard, and Brahic, André (eds.), *Planetary Rings*, University of Arizona Press, Tucson, 1984.

Guillot, Tristan, "The Interiors of Giant Planets: Models and Outstanding Questions," *Annual Review of Earth and Planetary Sciences*, *33*, 493, 2005.

Houghton, John T., *The Physics of Atmospheres*, Third Edition, Cambridge University Press, Cambridge, 2002.

Hubbard, W. B., Burrows, A., and Lunine, J. I., "Theory of Giant Planets," *Annual Review of Astronomy and Astrophysics*, *40*, 103, 2002.

Kivelson, Margaret G. (ed.), *The Solar System: Observations and Interpretations*, Prentice-Hall, Englewood Cliffs, NJ, 1986.

Lewis, John S., *Physics and Chemistry of the Solar System*, Academic Press, San Diego, 1995.

Mannings, Vincent, Boss, Alan P., and Russell, Sara S. (eds.), *Protostars and Planets, IV*, University of Arizona Press, Tucson, 2000.

Saumon, D., and Guillot, T., "Shock Compression of Deuterium and the Interiors of Jupiter and Saturn," *The Astrophysical Journal*, *609*, 1170, 2004.

Taylor, Stuart Ross, *Solar System Evolution*, Second Edition, Cambridge University Press, Cambridge, 2001.

## PROBLEMS

**21.1** Estimate the pressures at the centers of Jupiter and Saturn. Compare your answers to the Sun's central gas pressure.

**21.2** Analytic functions can be derived for the pressure and density structure in the interior of Jupiter if an approximate relationship between pressure and density is assumed. A reasonable choice for a composition of pure molecular hydrogen is

$$P(r) = K\rho^2(r),$$

where $K$ is a constant. This type of analytic model is known as a polytrope (see page 334ff for a discussion of polytropic stellar models).

(a) By substituting the expression for the pressure into the hydrostatic equilibrium equation (Eq. 10.6) and differentiating, show that a second-order differential equation for the density can be obtained, namely

$$\frac{d^2\rho}{dr^2} + \frac{2}{r}\frac{d\rho}{dr} + \left(\frac{2\pi G}{K}\right)\rho = 0.$$

(b) Show that the equation is satisfied by

$$\rho(r) = \rho_c\left(\frac{\sin kr}{kr}\right),$$

where $\rho_c$ is the density at the center of the planet and

$$k \equiv \left(\frac{2\pi G}{K}\right)^{1/2}.$$

(c) Taking the average radius of Jupiter to be $R_J = 6.99 \times 10^7$ m and assuming that the density goes to zero at the surface (i.e., $kR_J = \pi$), determine the values of $k$ and $K$.

(d) Integrate Eq. (10.7) using the analytical solution for Jupiter's density as a function of radius to find an expression for the planet's interior mass, $M_r$, written in terms of $r$ and $\rho_c$. *Hint:*

$$\int r(\sin kr)\,dr = \frac{1}{k^2}\sin kr - \frac{r}{k}\cos kr.$$

(e) Using the boundary condition that $M_r = M_J$ at the surface, estimate the planet's central density. (The value obtained in this problem is lower than the result found from detailed numerical calculations, 1500 kg m$^{-3}$. One major reason for the difference is that Jupiter's composition is not purely molecular hydrogen.)

(f) Make separate plots of the density and interior mass as functions of radius.

(g) What is the central pressure of your model of Jupiter? (One detailed model gives a value of $8 \times 10^{12}$ N m$^{-2}$.)

**21.3** (a) Assuming spherical symmetry and the density distribution for the polytropic model of Jupiter given in Problem 21.2b, show that the moment of inertia is given by

$$I = \frac{8\rho_c}{3\pi}\left(1 - \frac{6}{\pi^2}\right)R_J^5.$$

*Hint:* Since this analytical model does not assume constant density, you will need to integrate over concentric rings to find the moment of inertia about Jupiter's rotation axis. Recall that $I \equiv \int_{\text{vol}} a^2\,dm$, where $a = r\sin\theta$ is the distance from the rotation axis to the ring of mass $dm$, and

$$dm = \rho(r)\,dV = \rho(r)\,2\pi ar\,d\theta\,dr.$$

(b) Using the estimated value for the central density of Jupiter obtained in Problem 21.2e, calculate the planet's moment-of-inertia ratio. *Hint:* The moment-of-inertia ratio was first introduced in Problem 20.12.

(c) Compare your answer to part (b) with the measured value given in Table 21.2. What does this result say about the true density distribution within the planet relative to the analytical model?

**21.4** **(a)** Show that Eq. (21.5) follows directly from Eq. (21.3) in cylindrical coordinates for the case of a two-component model planet with two constant densities (see Fig. 21.5). Assume that the outer component is oblate and the inner component is spherical.

**(b)** Verify that Eq. (21.5) reduces to the familiar case of

$$I_{\text{sphere}} = \frac{2}{5} M R^2$$

for a spherically symmetric planet of constant density.

**21.5** **(a)** Derive an equation for the mass of the core of the two-component planetary model shown in Fig. 21.5. You should express your answer in terms of the fractional equatorial radius, $f R_e$, and the constant density of the core.

**(b)** Assume that Jupiter has a 10 $M_\oplus$ core and that the average density of the core is 15,000 kg m$^{-3}$. Determine $f$, the ratio of the equatorial radius of the planet's core to the equatorial radius of its surface.

**(c)** What is the average envelope density in this two-component model?

**(d)** Determine the moment-of-inertia ratio $(I/M R_e^2)$ for this two-component model.

**(e)** Compare your answer in part (d) to the measured value of the moment-of-inertia ratio for Jupiter given in Table 21.2. What can you say about the mass distribution of Jupiter compared to the analytical model?

**21.6** Estimate the angular diameter of Jupiter's magnetosphere as viewed from Earth at opposition. Compare your answer with the angular diameter of the full Moon.

**21.7** Suppose that fragment G of comet Shoemaker–Levy 9 measured 700 m in diameter. If this fragment had an average density of 200 kg m$^{-3}$, estimate its kinetic energy just before it entered the planet's atmosphere. You may assume that it struck the atmosphere with a speed equal to the planet's escape speed. Express your answer in joules and megatons of TNT (1 MTon $= 4.2 \times 10^{15}$ J).

**21.8** **(a)** On the *same scale*, plot (1) the first-order correction term to the gravitational potential of Saturn as a function of $\theta$ [i.e., the $J_2$ term in Eq. (21.1)], (2) the second-order correction term, and (3) the sum of the two terms. (Similar plots for Jupiter are shown in Fig. 21.4; note that the plots for Jupiter use different scales.) Assume that the observer is a distance $r = 2R_e$ from the planet.

**(b)** For which angle(s) is the gravitational potential largest? smallest? By what percent do these values for the gravitational potential deviate from the case of spherical symmetry (the zeroth-order term)?

**21.9** **(a)** Estimate the amount of energy radiated by Jupiter over the last 4.55 billion years (see Eq. 10.23).

**(b)** Estimate the *rate* of energy output from Jupiter due to gravitational collapse alone, assuming that the rate has been constant over its lifetime.

**(c)** Compare your answer for part (b) with the value for the intrinsic power output that was given in the text. What does this say about the rate of energy output in the past? Discuss the implications for the evolution of the Galilean moons.

**21.10** Estimate the blackbody temperature of Neptune, taking into consideration that one-half of all the energy radiated by the planet is due to internal energy sources. Compare your answer with the measured value of 59.3 $\pm$ 1.0 K.

**21.11** Assume that all of the ions escaping Io are sulfur ions. Assuming also that this rate has been constant over the last 4.55 billion years, estimate the amount of mass lost from the moon since its formation. Compare your answer with Io's present mass ($8.932 \times 10^{22}$ kg).

**21.12** **(a)** Make a rough estimate of the mass contained in Saturn's rings. Assume that the rings have a constant mass density and that the disk is 30 m thick with an inner radius of 1.5 $R_S$ and an outer radius of 3 $R_S$ (neglect the E ring). Assume also that all of the ring particles are water-ice spheres of radius 1 cm and that the optical depth of the disk is unity. The density of the particles is approximately 1000 kg m$^{-3}$. *Hint:* Refer to Eq. (9.12) to estimate the number density of water-ice spheres.

   **(b)** If all of the material in Saturn's rings were contained in a sphere having an average density of 1000 kg m$^{-3}$, what would the radius of the sphere be? For comparison, the radius and mass of Mimas are 196 km and $4.55 \times 10^{19}$ kg, respectively, and it has an average density of 1440 kg m$^{-3}$.

**21.13** Carefully sketch the orbits of Mimas and a characteristic Saturnian ring particle that is locked in a 2:1 orbital resonance with the moon. Show qualitatively that the resonance produces an elliptical orbit.

**21.14** Calculate the position of Saturn's synchronous orbit. Are any of its rings located outside of that radius? If so, which ones?

**21.15** A dust grain orbiting the Sun (or in a planetary ring system) absorbs and then re-emits solar radiation. Since the light is radiated from the Sun isotropically and re-emitted by the grain preferentially in the direction of motion, the particle is decelerated (it loses angular momentum) and spirals in toward the object it is orbiting. This process (known as the Poynting–Robertson effect) is just a consequence of the headlight effect discussed in Example 4.3.3.

   **(a)** If a dust grain orbiting the Sun absorbs 100% of the energy that strikes it and all of the energy is then re-radiated so that thermal equilibrium is maintained, what is the luminosity of the grain? Assume that the particle's cross-sectional area is $\sigma_g$ and its distance from the Sun is $r$.

   **(b)** Show that the *rate* at which angular momentum is lost from a grain is given by

$$\frac{d\mathcal{L}}{dt} = -\frac{\sigma_g}{4\pi r^2}\frac{L_\odot}{mc^2}\mathcal{L}, \tag{21.6}$$

   where $m$ and $\mathcal{L} = mvr$ are the mass and angular momentum of the grain, respectively, and $L_\odot$ is the luminosity of the Sun. *Hint:* Think of radiated photons as carrying an *effective mass* away from the grain; the effective mass of a photon is just $m_\gamma = E_\gamma/c^2$.

**21.16** **(a)** Beginning with Eq. (21.6), show that the time required for a spherical particle of radius $R$ and density $\rho$ to spiral into Saturn from an initial orbital radius $R_0$ is given by

$$t_{\text{Saturn}} = \frac{8\pi\rho c^2}{3L_\odot} R r_S^2 \ln\left(\frac{R_0}{R_S}\right),$$

   where $R_S$ is the radius of the planet and $r_S$ is its distance from the Sun. Assume that the orbit of the particle is approximately circular at all times and that it is always a constant distance from the Sun.

   **(b)** The E ring is known to contain dust particles having average radii of 1 $\mu$m. If the density of the particles is 3000 kg m$^{-3}$, how long would it take for a typical particle to spiral into the planet from an initial distance of 5 $R_S$?

(c) Compare your answer in part (b) to the estimated age of the Solar System. Could the E ring be a permanent feature of the Saturnian system without a source to replenish the ring? Note that the small moon Enceladus orbits Saturn in the E ring.

**21.17** The mass and radius of Miranda are $8 \times 10^{19}$ kg and 236 km, respectively.

(a) What is the escape velocity from the surface of Miranda?

(b) What would be the speed of a small object freely falling toward Uranus when it crossed the orbit of Miranda? Assume that the object started falling toward Uranus from rest, infinitely far from the planet. Neglect any effects due to the orbital motion of the planet around the Sun. Miranda's orbital radius is $1.299 \times 10^8$ m.

(c) Using Eq. (10.22), estimate the amount of energy needed to pulverize that moon.

(d) Suppose that a spherical object with a density of 2000 kg m$^{-3}$ were to collide with Miranda, completely destroying it. If the object hit Miranda with the speed found in part (b), what would the object's radius need to be? *Note:* For the purposes of this "back-of-the-envelope" calculation, you need not be concerned with the energy that would be expended in pulverizing the impacting object.

# CHAPTER

# 22 Minor Bodies of the Solar System

## 22.1 ■ PLUTO AND CHARON

The success of the mathematical prediction of Neptune's position (page 775) led astronomers to consider the possibility that a ninth planet existed even farther from the Sun. Based on *perceived* anomalies in the orbits of Uranus and Neptune, the search began in the late nineteenth century. Finally, on February 18, 1930, after a systematic and tedious search, Clyde W. Tombaugh (1906–1997) discovered a small 15th-magnitude object orbiting the Sun. The new object was classified as a planet and named Pluto for the Roman god of the underworld.[1] It turns out that even though Pluto was discovered near its predicted position, the prediction was invalid because it was founded on statistically insignificant apparent deviations in the orbits of the other planets.

Pluto is unlike any of the terrestrial or giant planets of the Solar System; in fact, it bears much more resemblance to Neptune's moon Triton than to any of the planets discussed in Chapter 20 or 21. Its 248.5-year orbit is very eccentric ($e = 0.25$). At perihelion it is only 29.7 AU from the Sun (actually closer than Neptune), while it is 49.3 AU away at aphelion. Its orbit is also inclined significantly from the ecliptic ($17°$).

Despite the fact that Pluto is a Neptune-crossing object, it is not in any danger of colliding with the giant world. Pluto is protected from that fate by a 3:2 orbital resonance with Neptune. Consequently, Pluto is never near perihelion when it is in conjunction with Neptune, and the two planets never get any closer than about 17 AU. Pluto actually approaches Uranus more closely, coming within 11 AU.

### The Discovery of Charon

Many of the most basic characteristics of Pluto, such as its mass and radius, were poorly determined until its largest moon, **Charon**, was discovered in 1978.[2] Figure 22.1 shows a

---

[1]The name Pluto was suggested by Venetia Burney, who was then an 11-year-old English schoolgirl.
[2]Prior to the discovery of Charon, Pluto's radius was uncertain to within a factor of 4, and its mass wasn't known to better than a factor of 100.

**FIGURE 22.1**    Pluto and its three moons. Charon was discovered in 1978, and the other two moons were detected by the Hubble Space Telescope's Advanced Camera for Surveys in 2005. [Courtesy of NASA, ESA, H. Weaver (JHU/APL), A. Stern (SwRI), and the Hubble Space Telescope Pluto Companion Search Team.]

Hubble Space Telescope image of the Pluto system. Pluto and Charon orbit the system's center of mass in 6.39 d with a separation of only $1.964 \times 10^7$ m (just slightly more than 1/20 of the distance between Earth and the Moon). Using Kepler's third law, the combined mass of the system is only $0.00247 \, M_\oplus$. Of course, to determine each mass individually requires knowledge of the ratio of their separations from the system's center of mass, which yields a mass ratio of $M_{\text{Charon}}/M_{\text{Pluto}} = 0.124$. From these data, the mass of Pluto has been estimated to be $1.3 \times 10^{22}$ kg, and Charon's mass is roughly $1.6 \times 10^{21}$ kg; for comparison, Triton's mass is $2.14 \times 10^{22}$ kg.

### The Densities and Compositions of Pluto and Charon

Shortly after the discovery of Charon, astronomers realized that a rare **eclipse season** would occur between 1985 and 1990. Since the orbital plane of the Pluto–Charon system is inclined $122.5°$ to their orbit around the Sun, observers on Earth see the system edge-on only for brief intervals once every 124 years. Fortuitously, Pluto was also at perihelion in 1989. The next eclipse season will not occur until the twenty-second century.

   The duration of the occultations provided the information necessary to determine the radius of Charon. Pluto's radius has been determined to be 1137 km, making it only about two-thirds the size of our Moon, but the occultation data indicate that Charon's radius is about 600 km. This means that Pluto's average density is about 2110 kg m$^{-3}$ and Charon's density is roughly 1770 kg m$^{-3}$. These data seem to indicate that Pluto and Charon are probably made of frozen ices and rock, with Pluto having a somewhat higher proportion of rock than the majority of the moons of the giant planets. The best map of the surface of Pluto obtained to date is shown in Fig. 22.2.

   It is worth noting that Triton's density is 2050 kg m$^{-3}$, very similar to that of Pluto.

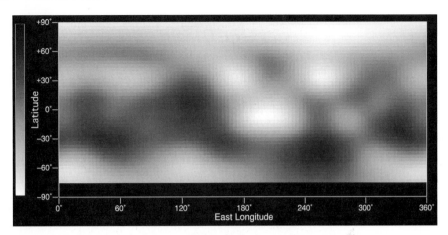

**FIGURE 22.2** The surface of Pluto as seen by the Hubble Space Telescope in 1994. 85% of the surface is represented; the south polar region was pointed away from Earth when the map was constructed. The observed features may be basins and craters, or perhaps frost from various ices. [Courtesy of Alan Stern (Southwest Research Institute), Marc Buie (Lowell Observatory), NASA, and ESA.]

### The Possible Formation of Charon by a Large Impact

With a mass that is almost 1/8 that of Pluto, Charon is proportionately the most massive moon in the Solar System relative to its parent planet.[3] It appears that Charon must have formed as a result of a large impact on Pluto, similar to the way scientists believe our Moon formed around Earth (recall Fig. 20.20). It is also possible that the two moons discovered in 2005 are additional products of that collision. The impactor probably had a mass of between 0.2 and 1 $M_{Pluto}$.

### Complete Spin–Orbit Coupling

Pluto and Charon also have another interesting dynamical characteristic, briefly mentioned in Section 19.2: Both objects have rotation periods that are exactly the same as their orbital period about their mutual center of mass. Since they spin in the same direction as their orbital motion, Pluto and Charon keep the same faces toward each other at all times; they are completely locked in a synchronous orbit. The tidal forces between these two small worlds have resulted in the final state of lowest energy. Because they are fully locked, the tidal forces do not produce the constantly changing bulges seen in other systems, such as Earth's tidal bulges produced by the Moon. Therefore, the frictional heat losses and angular momentum transfer in operation elsewhere have now ceased for the Pluto–Charon interaction.

A necessary consequence of the locked, synchronous orbit is that Charon is located directly over Pluto's equator. If this weren't the case, the orbital motion would carry Charon alternately north and south of Pluto's equator, and any deviation from spherical symmetry between the two worlds would result in constantly changing tidal forces. Since the orbital

---

[3]Our Moon, which is the second-largest satellite in the Solar System relative to its parent planet, contains only about 1/81 the mass of Earth.

plane of the system is inclined 122.5° to its orbit around the Sun, this implies that both Pluto and Charon rotate retrograde. Recall that Uranus also rotates retrograde and that its ring system and regular satellites are located directly over its equator; that orientation has also been attributed to tidal forces.

### A Frozen Surface and a Changing Atmosphere

In 1992 Tobias C. Owen and collaborators used the United Kingdom Infrared Telescope on Mauna Kea, Hawaii, to carry out a spectroscopic study of Pluto's surface. Their work revealed that the surface is covered with frozen nitrogen ($N_2$), which constitutes some 97% of the total area, with carbon monoxide ice (CO) and methane ice ($CH_4$) each accounting for 1% to 2%, similar to the surface of Triton. Oddly, Charon's surface appears to be composed primarily of water-ice; no molecular nitrogen, carbon monoxide, or methane ices or gases have been detected on Charon.

When Pluto occulted a faint star in 1988, a very tenuous atmosphere was detected, with a surface pressure of about $10^{-5}$ atm. The atmosphere is dominated by $N_2$, with $CH_4$ and CO probably making up roughly 0.2% of the total by number, consistent with the composition of the surface ice and the rate of sublimation of the various species. Curiously, when Pluto occulted another star in 2002, measurements of the pressure and scale height of the atmosphere had doubled, implying that Pluto's atmosphere had become significantly thicker over the 14-year period.

It has been suggested that the atmosphere of this tiny, distant world is not permanent. The 1988 observation of its atmosphere was made near perihelion when the planet's temperature was near its maximum value of approximately 40 K. At this temperature, ices on the surface are able to undergo partial sublimation. The atmosphere apparently thickened between 1988 and 2002 because ices continued to sublimate, releasing additional gases into the atmosphere. However, as the planet moves back toward aphelion, the atmosphere will probably "freeze out" again.

### A Rendezvous with Pluto

In January 2006, NASA launched the **New Horizons** flyby mission. If all goes according to plan, New Horizons will pass near Pluto in 2015, giving us the first close-up look at this tiny, distant planet and its moons. Whether Pluto will still have an atmosphere at the time of the flyby is one of the many interesting questions waiting to be answered about this Trans-Neptunian quadruple system.

### 22.2 ■ COMETS AND KUIPER BELT OBJECTS

Comets such as Comet Mrkos (Fig. 11.25) and Comet Halley [Fig. 22.3(a)], have been observed frequently throughout history. In fact, the periodic visits of Comet Halley have been recorded during each passage through the inner Solar System since at least 240 B.C. (its orbital period is 76 years).[4] Because of their unusual appearance when near perihelion, bright

---

[4]This famous comet has probably been making its periodic tours of the inner Solar System for 23,000 years, but it was Edmond Halley who first realized that these were repeated observations of the same object. The proof of his hypothesis was accomplished with the aid of Newton and his newly developed mechanics; see Section 2.2.

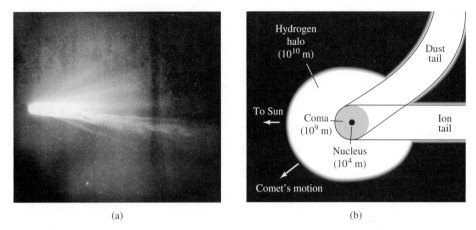

**FIGURE 22.3**   (a) Comet Halley clearly exhibited both a dust tail (curved) and an ion tail (straight) during its most recent trip through the inner Solar System. This image was obtained by the Michigan Schmidt telescope at Cerro Tololo Interamerican Observatory on April 12, 1986. Visible is a detachment event occurring in the ion tail. (Courtesy of NASA/JPL.) (b) The anatomy of a comet.

comets have long been associated with mystery and forces beyond human comprehension. While many people believed that comets foretold coming evils, others considered them messengers of good news. The artist Giotto di Bondone (1266–1337) depicted the "Star of Bethlehem" as a comet in his work *Adoration of the Magi*, which adorns the interior of the Scrovegni Chapel in Padua, Italy (Fig. 22.4). The painting dates from 1303, just two years after Comet Halley had last appeared.

## A Model of a Comet

In 1950 Fred L. Whipple (1906–2004) proposed a model of comets that successfully explained most of their physical characteristics, including the development of **tails** when they enter the inner Solar System; see Fig. 22.3(b). He suggested that a "dirty snowball," roughly 10 km across, lies at the center of the comet. This dirty snowball constitutes the comet's **nucleus**. The nucleus is composed of ices and embedded dust grains. As the nucleus moves from the cold environment of the outer Solar System into the warmer regions near the Sun, the ices begin to sublimate. The released dust and gas then expand outward to produce a $10^9$-m cloud of gas and dust known as the **coma**. Subsequently, the material in the coma interacts with the sunlight and solar wind, producing the very long, familiar tails (up to 1 AU in length) that are always associated with our images of comets. We now know that a hydrogen gas **halo** (or envelope) also surrounds the coma and can have a diameter of $10^{10}$ m. When the comet leaves the inner Solar System, the temperature decreases sufficiently that the rate of sublimation diminishes significantly and the halo, coma, and tails disappear. Cometary activity does not stop entirely, however; small outbursts can occur from time to time as heat from the Sun travels inward through the nucleus, releasing highly volatile gases.

**FIGURE 22.4**   *Adoration of the Magi*, by Giotto di Bondone (1266–1337). The painting adorns the interior of the Scrovegni Chapel in Padua, Italy. Work on the painting began two years after the 1301 appearance of Comet Halley.

## The Dynamics of Comet Tails

Comet tails are always directed away from the Sun, as depicted in Fig. 22.5. Two independent mechanisms are responsible for the structure of the tails: radiation pressure on the liberated dust grains, and the interaction of ions with the solar wind and the Sun's magnetic field.

First consider the effect of radiation pressure on grains. For an *idealized* spherical dust grain of radius $R$ that is located a distance $r$ from the Sun and that absorbs all of the incident light that strikes it, Eq. (3.13) can be used to calculate the outward force of radiation pressure on the grain. The factor of $\cos \theta$ means that the grain's cross-sectional area $\sigma = \pi R^2$ should be used in calculating the force. In other words, for the case of complete absorption the same force would be exerted if the grain were replaced by a circular disk with an identical radius, oriented perpendicular to the light. Using $\langle S \rangle = L_\odot / 4\pi r^2$ for the magnitude of the time-averaged Poynting vector, the force on the grain due to radiation pressure is

$$F_{\text{rad}} = \frac{\langle S \rangle \sigma}{c} = \frac{L_\odot}{4\pi r^2} \frac{\pi R^2}{c}. \tag{22.1}$$

Of course the Sun's gravity is also acting on the grain. If the density of the grain is $\rho$, its mass is

$$m_{\text{grain}} = \frac{4}{3} \pi R^3 \rho$$

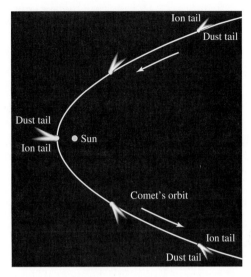

**FIGURE 22.5** The curved dust tail and the straight ion tail are always directed away from the Sun.

and the magnitude of the gravitational force acting on it is given by

$$F_g = \frac{GM_\odot m_{grain}}{r^2} = \frac{4\pi GM_\odot \rho R^3}{3r^2},$$

where $F_g$ is directed inward toward the star. Now, the ratio of the magnitudes of these forces is

$$\frac{F_g}{F_{rad}} = \frac{16\pi GM_\odot R\rho c}{3L_\odot}. \tag{22.2}$$

Because both gravity and light obey an inverse square law, this ratio does not depend on the distance $r$ from the Sun.

The magnitudes of the gravitational and radiation pressure forces on a cometary dust grain will be equal if the radius $R$ of the grain has the critical value

$$R_{crit} = \frac{3L_\odot}{16\pi GM_\odot \rho c}. \tag{22.3}$$

Grains *smaller* than $R_{crit}$ will experience a net outward force and will spiral away from the Sun. The curvature of the dust tail arises because of the decrease in the orbital speeds of the grains with increasing distance from the Sun. For a typical density of $\rho = 3000$ kg m$^{-3}$, the critical radius of the dust grain is found to be $R_{crit} = 1.91 \times 10^{-7}$ m $= 191$ nm.

Grains larger than $R_{crit}$ will continue to orbit the Sun. However, a competing process, the Poynting–Robertson effect, causes larger grains to spiral slowly in toward the Sun. As will be shown in Problem 22.2, the time required for a spherical particle of radius $R$ and density $\rho$ to spiral into the Sun from an initial orbital radius $r$ is given by[5]

$$t_{\text{Sun}} = \frac{4\pi\rho c^2}{3L_\odot} R r^2.$$ (22.4)

Because $R_{\text{crit}}$ is comparable to the wavelengths $\lambda$ of light emitted by the star, the actual situation is more complicated than this simple analysis suggests. The smallest grains ($R \ll \lambda$) are inefficient absorbers of light and have absorption cross sections that are much smaller than $\pi R^2$. Furthermore, the dust grains will scatter some of the incident light rather than absorb it. The effect of scattering on the radiation pressure force depends on the composition and geometry of the dust grains and on the wavelength of the light.

### The Compositions of Comets

It is the scattering of light that gives dust tails their white or yellowish appearance. On the other hand, the color of an ion tail is blue because $CO^+$ ions absorb and reradiate solar photons at wavelengths near 420 nm. However, $CO^+$ is certainly not the only species identified in comets. To date, a rich variety of atoms, molecules, and ions have been discovered spectroscopically, including some molecules that are quite complex. During its 1986 perihelion passage, the inner coma of Comet Halley was found to contain roughly (by number) 80% $H_2O$, 10% CO, 3.5% $CO_2$, a few percent $(H_2CO)_n$ (polymerized formaldehyde), 1% methanol ($CH_3OH$), and traces of other compounds. A partial list of chemical species that have been found in comets is given in Table 22.1.

### Disconnection Events

The straight ion (or plasma) tail owes its structure to a complex interaction among the comet, the solar wind, and the Sun's magnetic field.[6] Since the comet is an obstruction in the path of the solar wind and because the relative speed of the solar wind and the comet exceeds both the local sound speed (Eq. 10.84) and the Alfvén speed (Eq. 11.11), a shock front develops in the direction of motion.[7] As matter piles up at this *bow shock*, ions in the coma get trapped in the Sun's magnetic field, loading the field down. This causes the magnetic field to wrap around the nucleus. The cometary ions circle the field lines and trail behind the nucleus in the antisolar direction. A **disconnection event** occurs when the comet encounters a reversal in the solar magnetic field (the boundary between magnetic field directions is known as a *sector boundary*). During a disconnection event, the ion tail breaks away and a new one forms in its place. Disconnection events are quite common and are evident in the image of Halley's comet in Fig. 22.3(a), as well as in the time sequence of Comet Hyakutake in March 1996 (Fig. 22.6).

### Robotic Investigations of Comets

Whipple's dirty-snowball hypothesis was dramatically verified in March 1986, when an international armada of spacecraft rendezvoused with Comet Halley. The fleet was made up of

---

[5]Recall that the Poynting–Robertson effect is also important in understanding the dynamics of Saturn's rings.
[6]The model of a cometary *magnetotail* was first proposed by Hannes Alfvén in 1957.
[7]Shocks were first discussed in Section 11.2.

**TABLE 22.1**   A Partial List of Chemical Species Found in Comets. Various isotopomers have also been detected, such as HDO (deuterium replacing one of the hydrogen atoms in $H_2O$).

| Atoms | Molecules | Ions |
|-------|-----------|------|
| H | CH | $H^+$ |
| C | $C_2$ | $C^+$ |
| O | CN | $Ca^+$ |
| Na | CO | $CH^+$ |
| Mg | CS | $CN^+$ |
| Al | NH | $CO^+$ |
| Si | $N_2$ | $N_2^+$ |
| S | OH | $OH^+$ |
| K | $S_2$ | $H_2O^+$ |
| Ca | $H_2O$ | $H_2S^+$ |
| Ti | HCH | $CO_2^+$ |
| V | HCN | $H_3O^+$ |
| Cr | HCO | $H_3S^+$ |
| Mn | $NH_2$ | $CH_3OH_2^+$ |
| Fe | $C_3$ | |
| Co | OCS | |
| Ni | $H_2CO$ | |
| Cu | $H_2CS$ | |
| | $NH_3$ | |
| | $NH_4$ | |
| | $CH_3OH$ | |
| | $CH_3CN$ | |
| | $(H_2CO)_n$ | |

two spacecraft from Japan (**Suisei** and **Sakigake**), two from the former Soviet Union (**Vega 1** and **Vega 2**),[8] one from the European Space Agency (**Giotto**, named for the twelfth-century Italian painter; recall page 817), and one from the United States (the **International Cometary Explorer**, which flew past Comet Giacobini–Zinner six months earlier).[9] A number of spacecraft that were already flying other missions temporarily trained their instruments on the comet as well, even though they did not actually go out and meet the famous visitor.

On March 6 and 9, from distances of 8900 km and 8000 km, respectively, Vega 1 and Vega 2 were able to obtain low-resolution pictures of the nucleus of the comet. Relaying the Vega telemetry data to the European Space Agency, the Giotto scientists were then able to guide their spacecraft to an even closer encounter. As a direct result of extraordinary

---

[8]The two Soviet missions traveled to Venus first, releasing probes into its atmosphere.
[9]ICE was originally named the **International Sun–Earth Explorer 3** when it was launched in 1978. After many years on another mission, the spacecraft was reassigned to investigate the two comets.

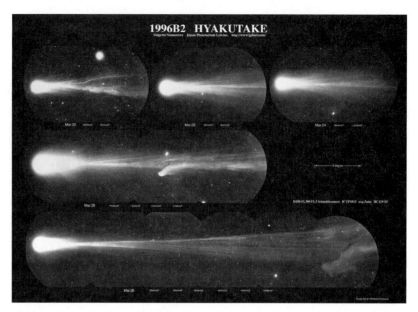

**FIGURE 22.6**    A spectacular example of a disconnection event in the March 25 image of this sequence of images of Comet Hyakutake in 1996. (©Shigemi Numazawa/Atlas Photo Bank/Photo Researchers, Inc.)

international cooperation, Giotto came within 596 km of the nucleus on March 14 (the mission goal was 540 km).[10]

Because of the high relative velocity of Giotto and Halley (68.4 km s$^{-1}$), even a collision with a small dust grain from the nucleus could cause serious damage to the spacecraft. Therefore, to protect the instruments on board, Giotto was equipped with a 50-kg shield made of aluminum and plastic Kevlar. Despite these precautions, just seven seconds before its closest approach to the nucleus, a dust particle struck the spacecraft off-axis, causing it to wobble severely. Stability was restored again one-half hour later.

Near its closest approach, the Giotto camera recorded the image of the nucleus seen in Fig. 22.7. The size of Whipple's dirty snowball is approximately 15 km × 7.2 km × 7.2 km, and it is shaped roughly like a potato. The surface is extremely dark, with an albedo of between 0.02 and 0.04. Repeated trips near the Sun have apparently left behind a layer of dust and possibly organic material as the ices evaporated away.

Visible in Fig. 22.7 are dust jets (ejected streams of material) located on the Sunward side of the nucleus. The positions of the jets probably correspond to thin regions in the dark covering on the surface through which trapped, heated gases in the interior are able to escape. As the nucleus rotates, other surfaces are exposed to the Sun, and new jets develop.[11]

---

[10]"The Halley Encounters," by Rüdeger Reinhard, in *The New Solar System*, Third Edition, Beatty and Chaikin (eds.), Cambridge University Press and Sky Publishing Corporation, Cambridge, MA, 1990, pp. 207–216, offers a fascinating recounting of the flybys.

[11]The nucleus of Comet Halley appears to have two rotation periods (2.2 days and 7.4 days) corresponding to motions about different axes of this irregularly shaped object.

**FIGURE 22.7**  The nucleus of Comet Halley as seen by the Giotto spacecraft. Evident are jets of gases on the sunward side of the nucleus. (Image from Reitsema et al., *20th ESLAB Symp., ESA SP-250, Vol. II*, 351, 1986.)

It has been estimated that during the flybys, jets were present on 15% of the surface, with gas and dust discharge rates of approximately $2 \times 10^4$ kg s$^{-1}$ and $5 \times 10^3$ kg s$^{-1}$, respectively. Because of the reaction forces produced by the jets, comets tend to have slightly erratic orbits.

By considering these nongravitational perturbations to its orbit, the mass of Comet Halley's nucleus has been estimated at between $5 \times 10^{13}$ kg and $10^{14}$ kg. If these crude estimates turn out to be correct, the average density of the nucleus is less than 1000 kg m$^{-3}$ and may be as low as 100 kg m$^{-3}$. It is likely that gases and dust escaping from the icy body have left a porous, honeycomb structure inside the nucleus with an average density close to that of new-fallen snow. (Recall that the Shoemaker–Levy 9 fragments that hit Jupiter also appear to have been very loosely packed, with densities of about 600 kg m$^{-3}$; see Section 21.1.)

The passage of the International Cometary Explorer through the tail of Comet Giacobini–Zinner and the extensive international collaboration to rendezvous with Comet Halley were the first two of several missions to comets that have occurred since the mid-1980s. In 2001 the experimental ion-propulsion spacecraft **Deep Space 1** obtained images of the 10-km-long nucleus of Comet Borrelly from a vantage point of 2200 km. In addition, on January 2, 2004, **Stardust** passed within 250 km of Comet Wild 2, obtaining high-resolution images of the surface. Stardust also captured dust from the comet and, on January 15, 2006, returned the dust to Earth for analysis.

A very dramatic encounter occurred on July 4, 2005, when **Deep Impact** sent a 370-kg impactor into Comet Tempel 1, a 7.6 km by 4.9 km object, at a speed of 10.2 km s$^{-1}$. Images of Comet Tempel 1 obtained by Deep Impact are shown in Fig. 22.8. The impactor created

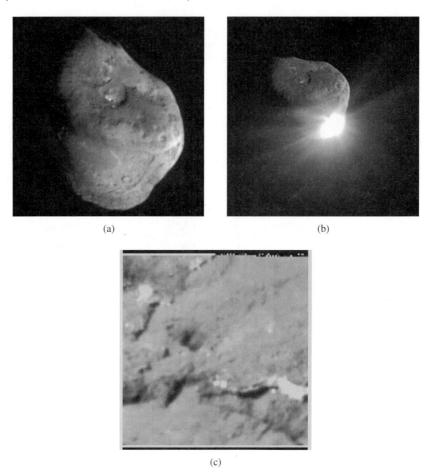

(a)                    (b)

(c)

**FIGURE 22.8** (a) Comet Tempel 1 seen from Deep Impact five minutes before the impactor struck the comet. The nucleus measures 5 km by 11 km. (b) 67 seconds after impact. (c) The surface of Tempel 1, 20 seconds before impact. The image was obtained by the impactor's targeting sensor. Features as small as 4 m can be identified. (NASA/JPL-Caltech/UMD.)

a crater in the comet's nucleus that allowed scientists to study the interior. Analysis of the data identified a variety of compounds, including water, carbon dioxide, hydrogen cyanide, methyl cyanide, polycyclic aromatic hydrocarbons (PAHs), and other organic molecules, as well as minerals such as olivine, calcite, iron sulfite, and aluminum oxide. The $10^7$ kg of material excavated from the near surface of the comet has the consistency of a very fine sand, or perhaps even the consistency of talcum powder. It also seems that the comet's density is roughly that of powder snow, loosely held together by its weak gravity. With a mass of $7.2 \times 10^{13}$ kg, Tempel 1 has a density of only 600 kg m$^{-3}$. The low density suggests that Tempel 1 may be a very porous rubble pile.

In the future, ESA's **Rosetta** spacecraft, which was launched in 2004, is scheduled to rendezvous with Comet Churyumov–Gerasimenko in 2014 and then spend almost two years carefully studying the comet from a low orbit around the nucleus. The mission will also land a probe on the surface of the comet. During its extended tour, Rosetta will fly alongside the comet as it moves into the inner Solar System, is heated by the Sun, and begins releasing trapped volatiles from its interior. Among other goals, this exhaustive study may help us to determine whether or not the "seeding" of Earth with cometary organic material may have helped life develop on our planet.

### Sun-Grazing Comets

While many comets that venture into the inner Solar System suffer relatively small amounts of sublimation near perihelion compared with their masses, other comets may experience more severe consequences. For instance, when Comet West moved through the inner Solar System in 1976, its nucleus broke into four separate pieces. Comet Kohoutek also split apart in 1974. Perhaps more impressive are the **Sun-grazing comets**. While it has been intensely studying the Sun, the LASCO instrument onboard the SOHO spacecraft has discovered more than 1000 comets that make close approaches to the Sun. In some instances the comets' orbits cause them to plunge into the Sun, as occurred with Comet SOHO-6, shown in Fig. 22.9.

### The Oort Cloud

Halley is one example of a class of comets known as **short-period comets** with orbital periods of less than 200 years. Short-period comets are found near the ecliptic and return to the inner Solar System repeatedly. The **long-period comets** have orbital periods of greater than 200 years, and some may take from 100,000 to 1 million years or more to return. In 1950, based on a very careful statistical study of their apparently random orbits, Jan Oort (1900–1992) concluded that long-period comets originate in a distant distribution of cometary nuclei now known as the **Oort cloud**. Although the Oort cloud has never been

**FIGURE 22.9**   An image of Comet SOHO-6 plunging into the Sun (lower left-hand side). The disk of the Sun is covered so that the Sun's corona and coronal mass ejections can be studied. The image was obtained by the LASCO instrument onboard SOHO on December 23, 1996. [SOHO (ESA & NASA).]

observed, its existence seems certain. As we noted in Section 19.1, the reservoir of nuclei appears to be located between 3000 AU and 100,000 AU from the Sun and probably contains $10^{12}$–$10^{13}$ members with a total mass on the order of 100 $M_\oplus$. For comparison, the nearest stars are approximately 275,000 AU away. The inner Oort cloud (3000 AU to 20,000 AU) may be slightly concentrated along the ecliptic, while the outer Oort cloud (20,000 AU to 100,000 AU) has a nearly spherical distribution of cometary nuclei.

The comets in the Oort cloud probably did not form at their present locations. Instead, they may be ancient planetesimals that coalesced near the ecliptic in the vicinity of Uranus and Neptune. After repeated gravitational interactions with the ice giants, the nuclei were catapulted out to their current distances. Because these nuclei were so far from the Sun, passing stars and gas clouds ultimately randomized their orbits, resulting in the nearly spherical distribution that exists in the outer cloud today. With the inner cloud being deeper in the Sun's gravitational well, these cometary nuclei did not become quite as randomly distributed. Therefore, comets in the inner Oort cloud were able to retain some history of their original locations when the Solar System was young. It is probably the gravitational perturbations of other stars and gas clouds that cause some of the cometary nuclei to start their long falls into the inner Solar System.

### The Kuiper Belt

Since the orbits of short-period comets lie preferentially near the ecliptic, it seems unlikely that these objects originated in the Oort cloud. Kenneth E. Edgeworth (1880–1972) in 1949 and Kuiper in 1951 independently proposed that a second collection of cometary nuclei might be located close to the plane of the ecliptic.

In August 1992, 1992 $QB_1$, a 23rd-magnitude object, was discovered by Jane Luu and David Jewitt 44 AU from the Sun and having an orbital period of 289 years. Seven months later a second 23rd-magnitude object (1993 FW) was discovered at nearly the same distance from the Sun. Assuming these objects have albedos characteristic of typical cometary nuclei (3% to 4%), then they must have diameters of approximately 200 km in order to appear as bright as they do. That would make them about one-tenth the size of Pluto. By early 2006, telescopic surveys employing sensitive CCD cameras had resulted in the discovery of more than 900 similar objects beyond Neptune's orbit.

Now known as the **Kuiper belt**,[12] this disk of cometary nuclei extends from 30 to 50 AU from the Sun; the semimajor axis of Neptune's orbit is 30 AU. Some members appear to have particularly eccentric orbits that may reach out to 1000 AU at aphelion, however. Noting their location beyond the outermost ice giant, these **Kuiper Belt Objects** (KBOs) are sometimes alternatively referred to as **Trans-Neptunian Objects** (TNOs).

### A Kuiper Belt Object Larger Than Pluto

As more and more KBOs have been discovered, a number of them have been found to have diameters somewhat smaller than, but comparable to, that of Pluto (see Table 22.2). However, in 2005, astronomers Mike Brown (Caltech), Chad Trujillo (Gemini Observatory), and David Rabinowitz (Yale University) announced the discovery of the first object in the

---

[12]In recognition of the independent suggestion of Edgeworth, this collection of objects is sometimes referred to as the **Edgeworth–Kuiper belt**.

**TABLE 22.2**   A List of the Largest Known Kuiper Belt Objects, as of May 2006. Many of the listed diameters are quite uncertain.

| Name | Diameter (km) | Period (yr) | $a$ (AU) | $e$ | $i$ (deg) |
|------|--------------|-------------|----------|-----|-----------|
| 2003 UB313 | 2400 | 559 | 67.89 | 0.4378 | 43.99 |
| Pluto | 2274 | 248 | 39.48 | 0.2488 | 17.16 |
| Sedna* | 1600 | 12,300 | 531.7 | 0.857 | 11.93 |
| Orcus | 1500 | 247 | 39.39 | 0.220 | 20.6 |
| Charon | 1270 | 248 | 39.48 | 0.2488 | 17.16 |
| 2005 FY9 | 1250 | 309 | 45.71 | 0.155 | 29.0 |
| 2003 EL61 | 1200 | 285 | 43.34 | 0.189 | 28.2 |
| Quaoar | 1200 | 287 | 43.55 | 0.035 | 8.0 |
| Ixion | 1070 | 249 | 39.62 | 0.241 | 19.6 |
| Varuna | 900 | 282 | 42.95 | 0.052 | 17.2 |
| 2002 AW197 | 890 | 326 | 47.37 | 0.131 | 24.4 |

* Sedna has an orbit that is much larger than the classical Kuiper belt.

Kuiper belt known to be larger than Pluto. 2003 UB313 was discovered on January 5, 2005, from data that were collected in a sky survey in 2003. With an orbital period of $P = 559$ yr, a semimajor axis of $a = 68$ AU, an orbital eccentricity of $e = 0.44$, and an inclination with respect to the ecliptic of $i = 44°$, the orbital characteristics of 2003 UB313 are why it took so long to find this large object in the Kuiper belt. Since most surveys looking for KBOs have been focused near the plane of the ecliptic, it was surprising to find 2003 UB313 with such a large inclination. HST observations indicate that the diameter of 2003 UB313 is 2400 km, making it about 6% larger than Pluto. The spectrum of 2003 UB313, shown in Fig. 22.10, is also strikingly similar to Pluto's, suggesting a surface composition dominated by frozen methane. A moon has been detected orbiting 2003 UB313 as well.

The discovery of 2003 UB313 has rekindled the debate over what officially constitutes a planet. Should 2003 UB313 be designated as a planet since it is larger than Pluto? Should Pluto be removed from the planet classification? Interestingly, prior to the discovery of large KBOs, no formal scientific definition of a planet had ever been established; it had more or less been assumed that we will recognize a planet when we see one! At the time of writing, this is an issue that the International Astronomical Union is struggling to resolve. Given that society has long come to see Pluto as a planet, the decision may very well be beyond the formal definitions of scientists. Whatever the resolution of these questions, it is clear that in light of the apparent size and composition of 2003 UB313 and many of the other KBOs, Pluto and Charon are certainly Kuiper belt objects; they just happen to be among the largest known members. After all, Pluto and Charon have more in common with very large cometary nuclei than with a "typical" planet–moon system. Neptune's unusual moon, Triton, is probably a captured KBO as well.

### Classes of Kuiper Belt Objects

As the number of known KBOs has continued to grow, it has become evident that they fall into three different groups based on their orbital characteristics. **Classical KBOs** are

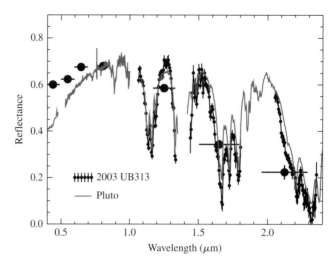

**FIGURE 22.10**    The reflection spectrum of 2003 UB313 (individual points) compared to the spectrum of Pluto (gray line). Absorption features of methane dominate the spectrum. The large points are data from *BVRIJHK* photometry. [Courtesy of Mike Brown (Caltech), Chad Trujillo (Gemini Observatory), and David Rabinowitz (Yale University).]

those that orbit between 30 and 50 AU from the Sun, most have semimajor axes between 42 and 48 AU. The orbital inclinations of classical KBOs tend to be less than 30°. It has been suggested that the cutoff of classical KBOs at 50 AU may be due to a passing star early in the formation of the Solar System. **Scattered KBOs** have much higher orbital eccentricities than the classical KBOs and were probably pumped up to those orbits by gravitational interactions with the ice giants, most notably Neptune. 2003 UB313 is one example of a scattered KBO. The perihelion distances of scattered KBOs are characteristically about 35 AU, and they tend to have greater orbital inclinations than the classical KBOs. In addition, it is likely that the scattered KBOs are at least one source of the short-period comets. Finally, a class of **resonant KBOs** exists that have orbital resonances with Neptune. As we noted in Section 22.1, Pluto is locked in a 3:2 orbital resonance with Neptune that protects it from ever colliding with the ice giant. As a result, Pluto (and Charon) are resonant KBOs. In fact, KBOs that have 3:2 orbital resonances with Neptune are referred to as **Plutinos**. Orbital resonances of 4:3, 5:3, and 2:1 have been observed for KBOs as well.

### Centaurs

Other fairly large icy bodies have been discovered orbiting the Sun as well. In 1977, an object known as 2060 Chiron was detected in an orbit that carries it from inside Saturn's orbit out to the orbit of Uranus. Chiron's diameter has been estimated to be between 200 km and 370 km. A slightly smaller object, nicknamed "Son of Chiron" but officially called 5145 Pholus, has also been seen orbiting among the outer planets, between 8.7 and 32 AU. Originally classified as an asteroid, Chiron brightened unexpectedly in 1988 and developed a measurable coma. Such behavior defines this object as a comet rather than a rocky asteroid. Chiron and 5145 Pholus are two examples of a class of objects known as **Centaurs** that

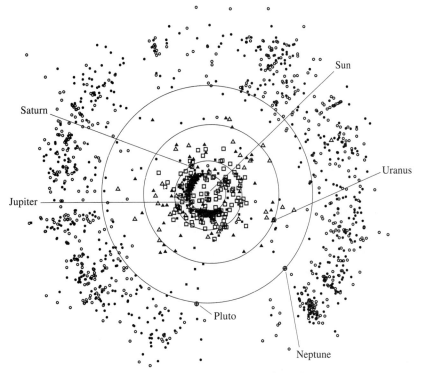

Plot prepared by the Minor Planet Center (2005 Sept 8).

**FIGURE 22.11** The positions of known classical and resonant Kuiper belt objects (circles), Centaurs and scattered KBOs (triangles), comets (squares), and Trojan asteroids from the orbit of Jupiter outward on September 8, 2005. The position of each object has been projected onto the plane of the ecliptic. Open symbols are objects observed at only one opposition; filled symbols are objects that have been observed at multiple oppositions. For the comets, filled squares represent numbered periodic comets; other comets are indicated by open squares. The classical Kuiper belt is clearly evident beyond the orbit of Neptune. In this orientation, the planets and most of the other objects orbit counterclockwise, and the vernal equinox is to the right. (Adapted from a figure courtesy of Gareth Williams, Minor Planet Center.)

appear to be KBOs which were scattered into the region of the planetary orbits. Centaurs may eventually become short-period comets.

Figure 22.11 shows the locations of KBOs, Centaurs, comets, and Jupiter's Trojan asteroids (see Section 22.3) on September 8, 2005, projected onto the plane of the ecliptic. Because of the projection effect and the very large sizes of the symbols relative to the actual sizes of the objects, the outer Solar System appears to be more crowded than it really is.

## The Implications for Water in the Inner Solar System

As we shall see in Section 23.2, it seems unlikely that the terrestrial planets could have condensed out of the warm inner solar nebula with large abundances of volatiles such as water. It has been suggested that much of the water that is found in the oceans of Earth, that is trapped in the permafrost and ice caps of Mars, and that probably existed on Venus in the past could have been delivered to those worlds by impacting comets after the planets formed. However, "the devil is in the details." After careful examination of the composition of several comets that have been explored by spacecraft, it has been noted that the deuterium-to-hydrogen (D/H) ratio in the comets exceeds that in Earth's oceans by at least a factor of two. In fact, the D/H ratio is more characteristic of the interstellar medium than of the terrestrial oceans. From the small sample of comets that have been closely investigated to date, it appears that another source for the water in Earth's oceans must be determined. On the other hand, it may be that the sample is biased, consisting only of objects that probably derived from the Oort cloud, rather than the Kuiper belt. Of course, it is also possible that the delivery of water to Earth was a prolonged process, involving a variety of mechanisms, including comets, asteroids with relatively high water content (Section 22.3), water-rich meteorites (Section 22.4), and planetesimals (Section 23.2).

## 22.3 ■ ASTEROIDS

As was already mentioned in Section 19.1, **asteroids** (sometimes referred to as **minor planets**) usually occupy orbits that are closer to the Sun than most comets. The vast majority of asteroids can be found in a belt situated between the orbits of Mars and Jupiter. Since the discovery of Ceres in 1801, several hundred thousand asteroids have been cataloged, and it may be that the total number is in excess of $10^7$. However, despite their large numbers, the combined mass of all the asteroids may be as low as $5 \times 10^{-4}$ $M_\oplus$. A close-up view of asteroid 243 Ida and its moon, Dactyl, is shown in Fig. 22.12. (The number designates the order in which the asteroid was discovered. Ceres is designated 1 Ceres.)

**FIGURE 22.12**   243 Ida and its moon, Dactyl, as seen by the Galileo spacecraft on August 28, 1993, during its journey to Jupiter. 243 Ida is 55 km long, and Dactyl (100 km from Ida at the time of the flyby) is somewhat egg-shaped, measuring 1.6 km by 1.2 km. Surface features as small as 30 m are visible on Ida. Galileo was about 10,500 km from Ida when this image was taken. (Courtesy of NASA/JPL.)

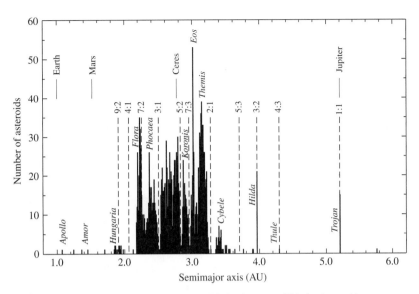

**FIGURE 22.13**   The distribution of 1796 asteroids in the asteroid belt. Asteroid group names and orbital resonances with Jupiter are also shown. Kirkwood gaps are evident at numerous resonance locations, and enhancements in the number of asteroids are apparent at other resonance locations. (Data from Williams, *Asteroids II*, Binzel, Gehrels, and Matthews (eds.), University of Arizona Press, Tucson, 1989.)

### The Kirkwood Gaps in the Asteroid Belt

The distribution of asteroids in the belt is not completely uniform or even smoothly varying with distance from the Sun. Instead, for various values of the orbital semimajor axis, asteroids are either conspicuously absent or overabundant (see Fig. 22.13). These positions correspond to orbital resonances with Jupiter, analogous to the resonances in Saturn's rings that are produced by its moons, most notably Mimas. Regions where asteroids are underabundant are known as the **Kirkwood gaps**, the most prominent being at 3.3 AU (a 2:1 resonance of orbital periods) and at 2.5 AU (a 3:1 resonance). In reality, physical gaps in the belt, equivalent to gaps in Saturn's rings such as the Cassini division, do not actually exist. Instead, the varying eccentricities and orbital inclinations of the asteroids tend to smear out the gaps somewhat, populating them with objects that are transients at those radii. The locations of the asteroids (and of some comets) on September 8, 2005, projected onto the plane of the ecliptic, are shown in Fig. 22.14.

### The Trojan Asteroids

In some cases, resonances with Jupiter correspond to local increases in the number of asteroids. A particularly interesting resonance group is the **Trojan asteroids** (1:1), which occupy the same orbit as Jupiter but either lead or trail the planet by 60°, as illustrated in Fig. 22.15 and evident in Figs. 22.11 and 22.14. In addition, at least one asteroid is orbiting the Sun at the trailing 60° position in Mars's orbit, and two are in the lead 60° position

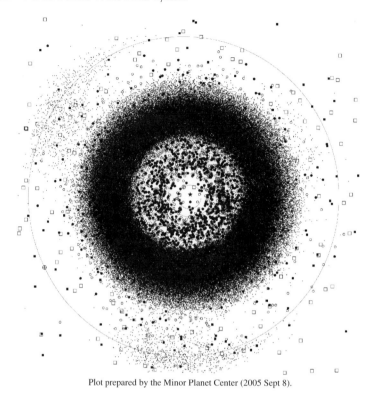

Plot prepared by the Minor Planet Center (2005 Sept 8).

**FIGURE 22.14**    The distribution of minor bodies in the inner Solar System on September 8, 2005, projected onto the ecliptic. Approximately 237,000 objects are shown in this plot. The outer orbit is that of Jupiter, and the asteroid belt is clearly visible. The belt is not actually saturated with asteroids; rather, the symbols representing them are vastly larger than the objects themselves. The location of each planet is marked by a ⊕ sign; Jupiter is seen in the lower left of the diagram. Jupiter's Trojan asteroids are evident in the large "clouds" that lead and trail Jupiter by 60°. The orbits of the terrestrial planets are visible among the clutter of the Amors, Apollos, and Atens. Comets are indicated by squares, as in Fig. 22.11. In this orientation the planets and most of the other objects orbit counterclockwise, and the vernal equinox is to the right. (Courtesy of Gareth Williams, Minor Planet Center.)

of Neptune's orbit. These asteroids are found in regions of unusual gravitational stability (gravitational "wells") that are established by the combined influence of the Sun and Jupiter. The positions are the $L_4$ and $L_5$ Lagrangian points, which are locations of equilibrium that exist in a three-body system when one of the bodies (in this case an asteroid) is much smaller than the other two. Recall that Lagrangian points were discussed in detail in Section 18.1; they play an important role in the evolution of some binary star systems.

### The Amors, Apollos, and Atens

Other special groups of asteroids are those that have orbits among the terrestrial planets. The **Amors** are located between the orbits of Mars and Earth, the **Apollos** cross Earth's orbit as they approach perihelion, and the **Atens** have semimajor axes that are less than 1 AU,

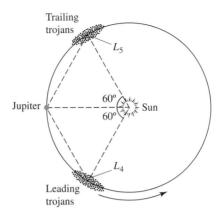

**FIGURE 22.15**  Trojan asteroids are located in Jupiter's orbit, either leading or trailing the planet by 60°. The occupied positions are two of the five Lagrangian points in the Sun–Jupiter system.

although they can cross Earth's orbit near aphelion. It appears that many of these objects were probably main-belt asteroids at one time, but perturbations with Jupiter reoriented their orbits. Some of the Earth-crossing objects could also be extinct cometary nuclei that have lost most of their volatiles after repeated trips near the Sun. Since the Apollo and Aten asteroids intersect Earth's orbit, there is always the possibility that a collision could occur.

### Hirayama Families

In 1918, the Japanese astronomer Kiyotsugu Hirayama (1874–1943) pointed out associations of asteroids that occupy nearly identical orbits. Today, more than 100 **Hirayama families** (also known as **asteroid families**) have been identified. It is believed that each family was once a single larger asteroid that suffered a catastrophic collision. With collision speeds that can reach 5 km s$^{-1}$, the available energy is more than enough to crush rock and cause pieces of the original asteroid to escape. If the collisional energy is not sufficient, only a portion of the surface may escape, or, after fracture, the self-gravity of the debris could cause the asteroid to reform again as a rubble pile. The Infrared Astronomical Satellite observed dust bands that seem to be associated with some of the major Hirayama families.

### Rendezvousing with Asteroids

The first asteroids to be visited by a spacecraft were 951 Gaspra and 243 Ida in 1991 and 1993, respectively (Ida and its moon, Dactyl, are shown in Fig. 22.12). The flybys occurred as the Galileo spacecraft passed through the asteroid belt while on its trip to Jupiter. Just as astronomers had expected, the irregularly shaped asteroids show evidence of having sustained numerous meteoritic impacts throughout their existence. In fact, the number of impacts suggests that Gaspra (a member of the Flora family) was probably broken off from a larger asteroid 200 million years ago. On the other hand, age estimates for Ida (a member of the Koronis family) vary. Given the small size of Dactyl, it is unlikely that the moon could have existed for more than 100 million years without getting destroyed by a major collision. However, based on the high crater density on Ida's surface, it appears that Ida may be as old as 1 billion years. Assuming that the two objects were created together from the breakup of a larger body, the resolution to the puzzle may rest with an increased rate of cratering from debris created when the larger object was destroyed. Even though the number density of asteroids in the belt is very low, the expected frequency of collisions is such that very few would have been lucky enough to avoid a major impact sometime during the Solar System's history.

With the discovery of Dactyl orbiting Ida, it is possible to estimate the mass of Ida from Kepler's third law. Unfortunately, because of the high relative speed of the flyby ($12.4 \, \text{km s}^{-1}$) and the spacecraft's trajectory relative to the orbit of Dactyl (the angle between the trajectory and the orbit was about $8°$), only an approximate range of orbits were derived from the data. The results suggest that the mass of Ida is approximately 3 to $4 \times 10^{16}$ kg, giving an average density of between 2200 and 2900 kg m$^{-3}$.

The **Near Earth Asteroid Rendezvous mission** (NEAR–Shoemaker)[13] was launched in 1996. On its way to its ultimate destination of asteroid 433 Eros, NEAR–Shoemaker also made a 10-km s$^{-1}$ flyby of 253 Mathilde. Based on the gravitational perturbations of Mathilde on the spacecraft, it was determined that the average density of the asteroid is only 1300 kg m$^{-3}$, indicating that this asteroid is likely a very heavily fractured rubble pile that has been broken apart by multiple collisions and only loosely reassembled by its own gravity.

When NEAR–Shoemaker arrived at 433 Eros on February 14, 2000, it entered into orbit around the asteroid and began a year-long intensive study of the object (see Fig. 22.16).

**FIGURE 22.16**   A composite image of the two hemispheres of 433 Eros as observed from orbit around the asteroid. Eros is heavily covered with regolith and shows significant evidence of cratering. Eros is one of the largest near-Earth asteroids, measuring 33 km long by 8 km wide by 8 km thick. (Courtesy of NASA/Johns Hopkins University Applied Physics Laboratory.)

---

[13]The mission was renamed in flight in honor of the late Eugene M. Shoemaker, planetary scientist and co-discoverer of the Shoemaker–Levy 9 comet. Shoemaker had always said that he wanted to hit 433 Eros with a rock hammer to see what was inside.

**FIGURE 22.17**   The surface of Eros from an altitude of 250 m. The image is 12 m across. The image was taken during the February 12, 2001, descent of the NEAR–Shoemaker spacecraft. (Courtesy of NASA/Johns Hopkins University Applied Physics Laboratory.)

During the orbital mission, the spacecraft studied Eros's gravitational field and obtained information about its surface composition. Although not designed as a lander, after one year in orbit, NEAR–Shoemaker survived an intentional landing on the surface of the asteroid at a speed of about 1.6 m s$^{-1}$, and was able to transmit information back to Earth for another week. During the descent phase of the mission, NEAR–Shoemaker returned many close-up images, including the one shown in Fig. 22.17.

Measurements obtained during the mission indicate that the density of Eros is 2670 kg m$^{-3}$ and that it has probably been fractured, but not to the point of being a rubble pile like Mathilde. The interior of Eros appears to have a porosity of 25% or so. From measurements of radioactivity and from gamma-ray spectroscopy measurements (see Fig. 22.18), it appears that Eros contains K, Th, U, Fe, O, Si, and Mg, as expected for this primitive object.

### Classes of Asteroids

It was in the 1930s that astronomers first realized that asteroids vary in color. By observing the spectrum of reflected sunlight, it is possible to identify absorption bands that provide important information about the surface compositions of these objects. Information can also be obtained by studying their albedos. The composition of asteroids is now known to vary significantly, but a general trend exists with increasing distance from the Sun (see Fig. 22.19). Some of the major classes of asteroids are

- **S-type** asteroids reside in the inner part of the belt (2–3.5 AU) and make up roughly one-sixth of all the known asteroids. Their surfaces are dominated by a mixture of iron- or magnesium-rich silicates, together with pure metallic iron–nickel. They tend to have a low abundance of volatiles, appear somewhat reddish, and have moderate albedos (0.1–0.2). Gaspra, Ida, and Eros are S-type asteroids.

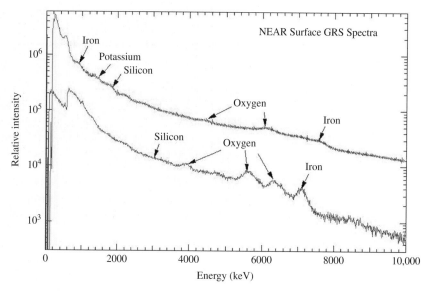

**FIGURE 22.18**    The gamma-ray spectrum of 433 Eros obtained by the NEAR–Shoemaker space-craft after it landed on the surface of the asteroid. Two different gamma-ray detectors obtained the two spectra shown. (Adopted from a figure courtesy of NASA/Johns Hopkins University Applied Physics Laboratory.)

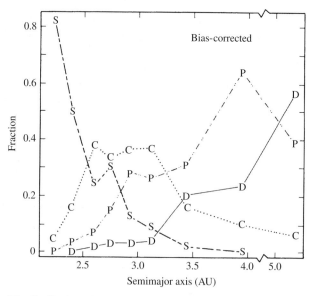

**FIGURE 22.19**    The distribution of major asteroid types with distance from the Sun. [Figure adapted from Gradie, Chapman, and Tedesco, *Asteroids II*, Binzel, Gehrels, and Matthews (eds.), University of Arizona Press, Tucson, 1989.]

- **M-type** asteroids are very metal-rich with absorption spectra dominated by iron and nickel. They appear slightly reddish, and they have moderate albedos (0.10–0.18). M-types are preferentially located in the inner portion of the belt, among the S-types (2–3.5 AU).

- **C-type** asteroids constitute perhaps three-fourths of all the minor planets. These objects are located predominantly near 3 AU but can be found throughout the main belt (2–4 AU). They are very dark, with albedos in the range 0.03–0.07, and they appear to be rich in carbonaceous material. Two-thirds of the C-types also contain significant quantities of volatiles, particularly water. Mathilde is a C-type asteroid.

- **P-type** asteroids are located near the outer edge of the main belt and beyond (3–5 AU), peaking in population near 4 AU. They have a slightly reddish appearance and low albedos (0.02–0.06). Their surfaces may contain a significant abundance of ancient organic compounds, which are also present in comets.

- **D-type** asteroids are much like P-types, except that they have a redder appearance and are located farther from the Sun. The Trojan asteroids are dominated by D-types. Some of Jupiter's smaller moons also exhibit spectra similar to D-type asteroids.

It seems likely that the differences in asteroid types with distance from the Sun are largely a result of the process of condensation out of the solar nebula (to be discussed in more detail in Section 23.2). Closer to the Sun, near the inner edge of the asteroid belt where the temperature was higher, more refractory compounds (like silicon) condensed out while volatiles such as water and organic compounds could not. Farther from the Sun, temperatures had decreased sufficiently to allow more volatile compounds to condense and become part of the asteroids in that region. Many of the C-type asteroids in the middle of the belt appear to be *hydrated* (meaning that water is present in these objects), whereas the more distant P- and D-types may contain water-ice, like most of the moons of the outer Solar System. Interestingly, 1 Ceres, at a distance from the Sun of 2.77 AU, and the largest asteroid in the belt, appears to be nearly spherical and may have a water-ice mantle.

Evidently the majority of the minor planets in the inner part of the belt have been subjected to significant gravitational separation during their lifetimes, including most or all of the S-type asteroids. The unusual and very metal-rich M-types have also been profoundly altered by evolution since formation. It is generally believed that M's represent the cores of much larger parent asteroids that became chemically differentiated and were later shattered by cataclysmic collisions, exposing the core.

At least one asteroid, 4 Vesta, appears to have a surface that is covered with basalt (rock formed from lava flows). Vesta has a radius of 250 km and is the third-largest asteroid known, behind 1 Ceres and 2 Pallas. It seems that magma developed in the interior and eventually found its way through cracks to the surface, where it solidified. Vesta also has an impact crater large enough to have exposed the subsurface mantle.

### Internal Heating

As Vesta and the S- and M-type asteroids suggest, the interiors of at least some asteroids must have become molten for a period of time during their lives, raising the question of the source of the heat. Being small objects, the asteroids readily radiate their interior heat into space, so they should have cooled off rather quickly after formation, too quickly to allow for significant gravitational separation (recall that $\tau_{cool} \propto R$; see Section 21.1). Furthermore, the very long-half-life radioactive isotopes that are, in large part, responsible for maintaining the hot interior of Earth could not generate heat rapidly enough to melt the interior of an asteroid. It has been suggested that a relatively short, intense burst of heat could be produced if a shorter-half-life isotope were available in sufficient abundance. A likely candidate is $^{26}_{13}$Al, with a half-life of 716,000 years:

$$^{26}_{13}\text{Al} \rightarrow ^{26}_{12}\text{Mg} + e^+ + \nu_e. \tag{22.5}$$

One difficulty with this suggestion is that in order to be effective in melting the interior of an asteroid, the aluminum must be incorporated relatively rapidly into a forming asteroid after the aluminum is produced (in just a small number of half-lives). This places a severe constraint on the formation timescale for the Solar System.

A second problem with the radioactive isotope solution lies in the apparent trend from chemically differentiated, volatile-poor asteroids in the inner belt to hydrated asteroids around 3.2 AU and icy bodies near Jupiter. This distribution seems to imply that $^{26}_{13}$Al was preferentially included in asteroids in the inner belt if it is the source of heat that led to chemical differentiation.

## 22.4 ■ METEORITES

In the early morning hours of February 8, 1969, residents in the region around Chihuahua City, Mexico, saw a bright blue-white light that streaked across the sky. As they watched, the light broke into two parts, each in turn exploding into a spectacular display of glowing fragments. Sonic booms were also heard accompanying the light show. It was reported that some observers even believed that the world was coming to an end. Rocks rained down on the countryside over an area that measured 50 km by 10 km (known as a **strewnfield**). The next day the first meteorite was discovered in the small village of Pueblito de Allende. All of the more than two tons of specimens collected from this meteor shower are now collectively referred to as the **Allende meteorite**. Many of the Allende stones were taken to the NASA Lunar Receiving Laboratory in Houston, Texas, for study.[14] One sample of the Allende meteorite is shown in Fig. 22.20.[15]

The observed streaks of light were produced by the frictional heating of the meteorite surfaces by Earth's atmosphere, causing the meteorites to glow. Although the outsides of the samples were covered by **fusion crusts** produced by the frictional heating, the interiors of the samples were unaffected. When a meteorite passes through the atmosphere, its damaged surface flakes off almost as quickly as it forms.

---

[14]The Lunar Receiving Laboratory was preparing to analyze the Moon rocks that were to be collected later that year by the *Apollo* astronauts.

[15]After attending a lecture about a meteoritic fall given by two Yale professors in Connecticut in 1807, President Thomas Jefferson (1743–1826) reportedly commented, "I could more easily believe that two Yankee professors could lie than that stones could fall from Heaven." Jefferson was, himself, a well-respected amateur scientist.

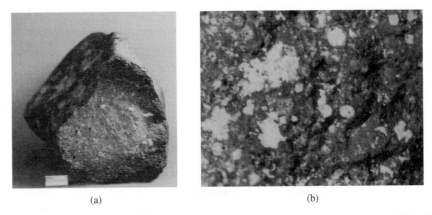

(a)                                                    (b)

**FIGURE 22.20**    (a) A sample of the Allende meteorite. The surface has a fusion crust. (b) A close-up of a portion of the interior of the sample showing CAIs and chondrules embedded in a matrix. (Courtesy of Smithsonian Astrophysical Observatory.)

### The Age and Composition of the Allende Meteorite

A very precise chronometer for determining ages of events in the formation of the Solar System is available by comparing the relative abundances of two stable isotopes of lead that can be identified in meteorites, $^{207}_{82}Pb$ and $^{206}_{82}Pb$. These isotopes are ultimately produced by independent sequences of decays that begin with $^{235}_{92}U$ (half-life of 0.704 Gyr) and $^{238}_{92}U$ (half-life of 4.47 Gyr), respectively. By using this **Pb–Pb system**, scientists have deduced an age for the Allende meteorite of $4.566 \pm 0.002$ Gyr, which is very close to the solar model age of the Sun (4.57 Gyr) given in Section 11.1.[16] It seems that the Allende meteorite is a nearly primordial remnant of the early solar nebula (as are other meteorites).

A chemical analysis of the samples revealed that the meteorite's composition is close to solar (similar to the Sun's photosphere), with some exceptions; the most volatile elements (H, He, C, N, O, Ne, and Ar) are underabundant, and lithium (Li) was found to be overabundant. The relative underabundance of volatiles can be understood by assuming that the Allende meteorite condensed out of the inner portion of the solar nebula where the temperature was too high for those elements to be included in solar concentrations.[17] Allende's lithium content is probably overabundant relative to the Sun because the Sun has actually destroyed much of its own complement of that element during the star's lifetime.

### CAIs and Chondrules

Contained in the Allende samples are two types of nodules embedded in a **matrix** of dark silicate material. The **calcium- and aluminum-rich inclusions** (CAIs, also known as **refractory inclusions**) are small pockets of material ranging in size from microscopic to 10 cm in diameter that are relatively overabundant in calcium, aluminum, and titanium when compared with the remainder of the meteorite. This is significant because they are the

---

[16]Refer to Section 20.4 for a discussion of radioactive dating.

[17]Of course, light gases such as hydrogen and helium easily escape low-mass objects such as meteorites.

most refractory (least volatile) of the primary elements in meteoritic material. It seems that the CAIs have undergone repeated episodes of evaporation and condensation. **Chondrules** are spherical objects (1–5 mm across) made predominantly of $SiO_2$, $MgO$, and $FeO$, which seem to have cooled very rapidly from a molten state. Apparently no more than one melting and cooling event occurred for a given chondrule, and some chondrules may have been only partially molten.

A particularly intriguing discovery in the Allende CAIs is the overabundance of $^{26}_{12}Mg$. Because this particular nuclide is produced by the radioactive decay of $^{26}_{13}Al$ (recall Eq. 22.5), which is known to be produced by supernovae, the meteorite may have formed out of material significantly enriched with supernova ejecta. Moreover, because the half-life of $^{26}_{13}Al$ is relatively short by astronomical timescales, the meteorite must have formed within a few million years or so following the production of the $^{26}_{13}Al$. This suggests that a supernova shock wave may have triggered the collapse of the solar nebula. Because the material from the supernova should not be expected to mix thoroughly with the original nebula, regions of enhanced abundance would probably exist out of which objects such as the Allende meteorite could form. An alternative mechanism for the production of the required $^{26}_{13}Al$ has also been proposed: Intense flares during pre-main-sequence T-Tauri and FU Orionis phases appear capable of synthesizing $^{26}_{13}Al$. This mechanism seems to eliminate the need for a possibly ad hoc supernova trigger. More will be said about the formation and evolution of the Solar System in Section 23.2.

### Carbonaceous and Ordinary Chondrites

The Allende meteorite is one example of a class of primitive specimens known as **carbonaceous chondrites**, so named because they are rich in organic compounds and contain chondrules. They may also include appreciable amounts of water in their silicate matrix. The matrix even records the existence of a fairly strong primordial magnetic field (about equal in strength to the value of Earth's present-day field). **Ordinary chondrites** contain fewer volatile materials than the carbonaceous chondrites, implying that they formed in a somewhat warmer environment. Both general types of chondrites are **chemically undifferentiated stony meteorites**.

### Chemically Differentiated Meteorites

Several forms of **chemically differentiated meteorites** have also been discovered. Igneous stones, known as **achondrites**, do not contain any inclusions or chondrules; instead, they were formed entirely out of molten rock. **Iron** meteorites do not contain any stony (silicate) material, but they may be composed of up to 20% nickel. About three-quarters of all iron meteorites have long iron–nickel crystalline structures, up to several centimeters long and known as **Widmanstätten patterns**, that could have developed only if the crystal cooled very slowly over millions of years.[18] **Stony–iron** meteorites contain stony inclusions in a matrix of iron–nickel. **Stones** (chondrites and achondrites) make up about 96% of all the meteorites that hit Earth, irons account for about 3% of the total, and stony–irons make up the remainder (1%).

---

[18]The patterns were named for Count Alois von Widmanstätten, director of the Imperial Porcelain Works in Vienna, who discovered them in 1808.

### Sources of Meteorites

The vast majority of all meteorites probably originate from asteroids, either chipped off their parents or liberated from the deep interior during a catastrophic collision. For a sufficiently large asteroid, significant gravitational separation may have occurred, as suggested by the M-type minor planets. The exposed metallic cores are the source of the irons, and the core–rock interface is the source of the stony–irons. Other asteroids underwent very little chemical alteration during their lives and may account for the chondrites.

The reflection spectra of asteroids can be compared with meteorite samples to test whether the asteroids could be the source of objects striking Earth. Figure 22.21 shows the strong correlations between the spectra of some asteroids and meteorites. Note, for instance, that the asteroid 176 Iduna has a spectrum very similar to that of the carbonaceous chondrite, Mighel, while the basaltic surface of 4 Vesta agrees well with that of the achondrite meteorite, Kapoeta.

An unusual achondrite was discovered on the ice cap of Antarctica in 1982.[19] It has the chemical makeup of rocks collected from the lunar highlands by the *Apollo* astronauts. Clearly this achondrite was ejected from the Moon instead of from an asteroid. Because the escape velocity of the Moon is much larger than the escape velocities of asteroids, the discovery was certainly unexpected. Even more surprising, a small handful of meteorites have been discovered whose ages date back only 1.3 Gyr. Because these stones are much younger than the surface of the Moon, they must have originated on a body that has been

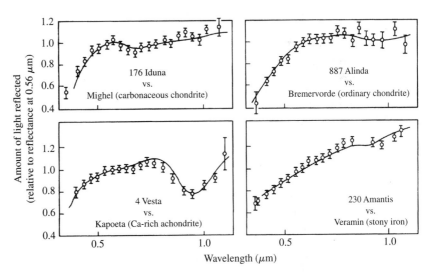

**FIGURE 22.21**   Comparisons between the infrared spectra of asteroids and meteorites. The reflection data for the asteroids are depicted by open circles with attached error bars. The laboratory spectra of the meteorites are given as solid curves. (Adapted from a figure courtesy of C. R. Chapman, in *The New Solar System*, Third Edition, Beatty and Chaikin (eds.), Cambridge University Press and Sky Publishing, Cambridge, MA, 1990.)

---

[19]Antarctica is an excellent site for finding meteorites. Any rock lying on the surface of a glacier is almost certainly extraterrestrial in origin.

**TABLE 22.3**    The Dates and Parent Bodies of Principal Meteor Showers.

| Shower | Approximate Date | Parent Body |
|--------|------------------|-------------|
| Quadrantid | January 3 | (unknown) |
| Lyrid | April 21 | Comet 1861 I |
| Eta Aquarid | May 4 | Comet Halley |
| Delta Aquarid | July 30 | (unknown) |
| Perseid | August 11 | Comet Swift–Tuttle |
| Draconid | October 9 | Comet Giacobini–Zinner |
| Orionid | October 20 | Comet Halley |
| Taurid | October 31 | Comet Encke |
| Andromedid | November 14 | Comet Biela |
| Leonid | November 16 | Comet 1866 I |
| Geminid | December 13 | Asteroid 3200 Phaeton |

geologically active more recently. The only real candidate is Mars, with an escape velocity of 5 km s$^{-1}$. At least one of the meteorites has inclusions of shock-melted glass that contain noble gases and nitrogen in the same proportions as the Martian atmosphere. However, also recall that Mars has produced at least one very old meteorite, ALH84001, discussed in Section 20.5 on page 764.

A number of **meteor showers** occur near the same dates every year, during which time meteors seem to emanate from a fixed position on the celestial sphere, known as a **radiant**. The source of the meteorites is debris left in the orbits of comets or asteroids that happen to intersect Earth's orbit. As Earth passes through the body's orbit, material rains down as if that material were coming from a position in the sky that Earth happens to be moving toward at the time; hence the radiant. Most parent bodies of meteor showers are comets, although at least one object, 3200 Phaeton, is classified as an asteroid. Meteor showers are named for the constellation in which their radiants lie. A list of the principal meteor showers, their approximate dates of maximum, and the parent object (if known) is given in Table 22.3.

### A History of Collisions with Earth

By now it should be apparent that objects throughout the Solar System have been subjected to numerous and sometimes violent collisions; Earth is no exception. Even as recently as 50,000 years ago, an iron meteorite, estimated to be 50 m in diameter, hit the ground in Arizona, producing a crater 1.2 km wide and 200 m deep (Fig. 22.22). There is also strong evidence to support the hypothesis that a stony asteroid exploded in the atmosphere above Siberia in 1908 (an episode known as the Tunguska event). The detonation leveled trees in a radial pattern for 15 km in every direction. It is even reported that the blast wave knocked a man off his porch 60 km from the epicenter and that the explosion was audible at distances of up to 1000 km. Estimates place the energy released during the Tunguska event at $5 \times 10^{17}$ J, equivalent to a nuclear explosion of 12 MTons.

Could other, even more energetic collisions have produced catastrophic consequences for life on Earth at the time? In about 1950, Ralph Baldwin suggested that meteoritic impacts

**FIGURE 22.22**   The 50,000-year-old Meteor Crater (also known as Barringer's Crater) in Arizona is 1.2 km in diameter and 200 m deep. It was produced by an iron meteorite estimated to be 50 m in diameter. (Courtesy of D. J. Roddy and K. Zeller, USGS.)

could have been responsible for the mass extinctions of many species seen in the paleontological record. Support for this hypothesis came in 1979 when geologist Walter Alvarez and his father, Luis Alvarez (1911–1988), a Nobel Prize winner in physics, announced the discovery of high abundances of iridium in a dark-colored clay that was located in the geologic strata at the Cretaceous–Tertiary boundary (commonly called the K–T boundary). The K–T boundary corresponds in time to the extinction, 65 million years ago, of 70% of the species then in existence, including the last of the dinosaurs. Since their original discovery in the Appenine Mountains of Italy, the anomalously high iridium concentrations have been seen throughout the world at the K–T boundary.

The significance of iridium is that it is rare in rock found near Earth's surface. This is because iridium is readily soluble in molten iron (it is a **siderophile**) and as such participated in the chemical differentiation of heavy elements sinking toward the core of Earth. However, iridium is fairly common in iron-rich meteorites. The amount of iridium present in the K–T clay strata, where it is thousands of times more abundant than is typical of ordinary rock, is consistent with an impact by a stony asteroid that measured 6 to 10 km across (or perhaps the object was a slightly larger comet).[20] An impactor of this size should have produced a crater some 100 to 200 km in diameter.

Shocked mineral grains have also been found worldwide at the K–T boundary but are most abundant in North America, suggesting that the impact (or impacts) could have occurred there. Attention has focused on an ancient impact site along the northern coast of the Yucatan peninsula, near the town of Chicxulub. A nearly semicircular structure at least 180 km across is located there and, based on radioactive dating, appears to be of the correct age.[21] Based on the size of the crater, the energy of the impact is estimated to have been $4 \times 10^{22}$ J, the equivalent of $10^{13}$ tons of TNT. An impact of that magnitude at the Chicxulub site could also account for evidence of an enormous tidal wave (a **tsunami**) that apparently traveled as far north as central Texas.

How could such an ocean impact have led to mass extinctions? If a meteorite of the size suggested hit in the ocean, it would vaporize a large amount of water. Some of this

[20]For comparison, the vertical rise of a typical mountain in the Rockies is about 1.5 km above the valley floor, and ocean depths are approximately 6 km.
[21]Some scientists have suggested that the diameter of the crater may be more like 300 km.

water would wash out airborne dust while the rest of the moisture would increase the greenhouse effect. As the temperature rose, even more water would evaporate into the atmosphere. Global atmospheric and ocean surface temperatures could rise by as much as 10 K through this enhanced greenhouse effect. (Recall the discussion on page 743 of the runaway greenhouse effect on Venus.)

Alternatively, if a major impact were to occur on land, a tremendous amount of dust would be injected into the atmosphere. As a consequence, the albedo would increase and more solar radiation would be reflected back into space, cooling the surface.[22]

In either case, as the meteorite passed through the atmosphere, the enormous amount of kinetic energy available in the impactor would have produced searing heat and generated devastating fires. It would also have reacted with appreciable amounts of nitrogen, producing nitrogen oxides and nitric acid. The ensuing acid rain would have damaged delicate land-based and aquatic ecosystems, killing vegetation and destroying much of the remaining food source. Carbon soot is found in the K–T clay layer, and there is also geologic evidence suggesting that flowering plants were destroyed in some regions for periods of at least several thousand years. Regardless of whether the impact occurred on land or in an ocean, the global environment would have been dramatically affected.

Even if asteroids or comets did not kill the dinosaurs and other creatures, there is clear evidence that major impacts have occurred in the past. By some estimates, the probability of the occurrence, during our lifetimes, of a cataclysmic impact that would be capable of destroying civilizations is perhaps as high as one in a few thousand. In light of this rather surprising statistic, some scientists have suggested that we should build a global asteroid–comet defense system. Although no definite plans have yet been formulated, conferences have been held to discuss the possibility.

### The Basic Building Blocks of Life

Ironically, even though impactors have been proposed as the mass murderers of some life forms on Earth, a number of carbonaceous chondrites have been found to contain many of the basic building blocks of life. Seventy-four amino acids have been found in one meteorite alone (the Murchison meteorite, which fell in Australia in 1972). Of those, seventeen are important in terrestrial biology. In addition to the amino acids, all four of the bases that cross-link the double helix of the DNA molecule (guanine, adenine, cytosine, and thymine), and the fifth base that is important in cross-linking in RNA (uracil), have been discovered in the Murchison meteorite. Other molecules important to life on Earth (such as fatty acids) have also been found in carbonaceous chondrites. Of course, it is a long way from producing relatively simple amino acids and cross-linking bases to the generation of the extremely complex DNA and RNA molecules, but these discoveries indicate that the fundamental chemistry necessary to start the process can occur in an extraterrestrial environment.

---

[22]It has been suggested that such a situation could also arise following a large-scale nuclear war. This scenario has been referred to as "nuclear winter."

## SUGGESTED READING

### General

Beatty, J. Kelly, Petersen, Carolyn Collins, and Chaikin, Andrew (eds.), *The New Solar System*, Fourth Edition, Cambridge University Press and Sky Publishing Corporation, Cambridge, MA, 1999.

Canavan, Gregory H., and Solem, Johndale, "Interception of Near-Earth Objects," *Mercury*, May/June 1992.

Goldsmith, Donald, and Owen, Tobias, *The Search for Life in the Universe*, Third Edition, University Science Books, Sausalito, CA, 2002.

Morrison, David, "The Spaceguard Survey: Protecting the Earth from Cosmic Impacts," *Mercury*, May/June 1992.

Morrison, David, and Owen, Tobias, *The Planetary System*, Third Edition, Addison-Wesley, San Francisco, 2003.

Sagan, Carl, and Druyan, Ann, *Comet*, Pocket Books, New York, 1985.

Smith, Fran, "A Collision over Collisions: A Tale of Astronomy and Politics," *Mercury*, May/June 1992.

### Technical

Bottke, William F., Cellino, Alberto, Paolicchi, Paolo, and Binzel, Richard P. (eds.), *Asteroids III*, University of Arizona Press, Tucson, 2002.

Brown, M. E., Trujillo, C. A., and Rabinowitz, D. L., "Discovery of a Planet-Sized Object in the Scattered Kuiper Belt," *The Astrophysical Journal*, *635*, L97, 2005.

de Pater, Imke, and Lissauer, Jack J., *Planetary Sciences*, Cambridge University Press, Cambridge, 2001.

Festou, Michel C., Keller, H. Uwe, and Weaver, Harold A. (eds.), *Comets II*, University of Arizona Press, Tucson, 2005.

Gilmour, Jamie, "The Solar System's First Clocks," *Science*, *297*, 1658, 2002.

Luu, Jane X., and Jewitt, David C., "Kuiper Belt Objects: Relics from the Accretion Disk of the Sun," *Annual Review of Astronomy and Astrophysics*, *40*, 63, 2002.

Mendis, D. A., "A Postencounter View of Comets," *Annual Review of Astronomy and Astrophysics*, *26*, 11, 1988.

*Minor Planet Center*, http://cfa-www.harvard.edu/cfa/ps/mpc.html.

Praderie F., Grewing, M., and Pottasch, S. R. (eds.), "Halley's Comet," *Astronomy and Astrophysics*, *187*, 1987.

Ryan, "Asteroid Fragmentation and Evolution of Asteroids," *Annual Review of Earth and Planetary Sciences*, *28*, 367, 2000.

Stern, S. A., "The Pluto–Charon System," *Annual Review of Astronomy and Astrophysics*, *30*, 185, 1992.

Taylor, Stuart Ross, *Solar System Evolution*, Second Edition, Cambridge University Press, Cambridge, 2001.

## PROBLEMS

**22.1** (a) Assume that a spherical dust grain located 1 AU from the Sun has a radius of 100 nm and a density of $3000 \text{ kg m}^{-3}$. In the absence of gravity, estimate the acceleration of that grain due to radiation pressure. Assume that the solar radiation is completely absorbed.

(b) What is the gravitational acceleration on the grain?

**22.2** In Problem 21.16, the Poynting–Robertson effect was shown to be important in understanding the dynamics of ring systems. The Poynting–Robertson effect, together with radiation pressure, is also important in clearing the Solar System of dust left behind by comets and colliding asteroids (the dust that is responsible for the zodiacal light).

(a) Beginning with Eq. (21.6), found in Problem 21.15, show that the time required for a spherical particle of radius $R$ and density $\rho$ to spiral into the Sun from an initial orbital radius of $r \gg R_{\odot}$ is given by Eq. (22.4). Assume that the orbit of the dust grain is approximately circular at all times.

(b) Find the radius of the largest spherical particle that could have spiraled into the Sun from the orbit of Mars during the Solar System's 4.57-billion-year history. Take the density of the dust grain to be $3000 \text{ kg m}^{-3}$.

**22.3** Estimate the amount of mass lost by Comet Halley during its most recent trip through the inner Solar System. Take into consideration the fact that the comet exhibits significant activity only during a short period of time near perihelion (an interval of approximately one year). Compare your answer with the total amount of mass present in the nucleus. Assuming that the mass loss rates are the same for each trip, how many more trips might the comet be able to make before it becomes extinct?

**22.4** In the text it was mentioned that *nongravitational perturbations* were used to estimate the mass of Comet Halley. How might this be done?

**22.5** Comet 1943 I, which last passed through perihelion on February 27, 1991, has an orbital period of 512 years and an orbital eccentricity of 0.999914. This is one member of the class of Sun-grazing comets.

(a) What is the comet's semimajor axis?

(b) Determine its perihelion and aphelion distances from the Sun.

(c) What is the most likely source of this object, the Oort cloud or the Kuiper belt?

**22.6** Using Kepler's laws, verify that the 2:1 and 3:1 orbital resonances of Jupiter correspond to the two prominent Kirkwood gaps indicated in Fig. 22.13.

**22.7** Vesta orbits the Sun at a distance 2.362 AU and has an albedo of 0.38 (unusually reflective for an asteroid).

(a) Estimate Vesta's blackbody temperature, assuming that the temperature is uniform across the asteroid's surface.

(b) If Vesta's radius is 250 km, how much energy does it radiate from its surface every second?

**22.8** Figure 22.14 makes it appear that the asteroid belt is saturated with objects. In this problem we will consider the fraction of the volume actually occupied by asteroids.

(a) If there are 300,000 large asteroids between 2 AU and 3 AU from the Sun, and each asteroid is assumed to be spherical with a radius of 100 km, determine the total volume occupied by the asteroids considered here.

(b) Model the region in which these asteroids orbit as an annulus with an inner radius of 2 AU, an outer radius of 3 AU, and a thickness of 2 $R_\odot$. Determine the volume of the region.

(c) What is the ratio of the volume occupied by asteroids to the volume of the region in which they orbit?

(d) Comment on the validity of a spaceship needing to maneuver quickly through a dense population of asteroids as frequently depicted in popular science fiction movies.

**22.9** In this problem you will estimate the amount of energy released per second by the radioactive decay of $^{26}_{13}$Al inside Vesta during its lifetime.

(a) Vesta has a radius of 250 km and a density of 2900 kg m$^{-3}$. Assuming spherical symmetry, estimate the asteroid's mass.

(b) Assume for the moment that the asteroid is composed entirely of silicon atoms. Estimate the total number of atoms inside Vesta. The mass of one silicon atom is approximately 28 u.

(c) The mass of $^{26}_{13}$Al is 25.986892 u and the mass of $^{26}_{12}$Mg is 25.982594 u. How much energy is released in the decay of one aluminum atom? Express your answer in joules.

(d) The ratio of $^{26}_{13}$Al to all aluminum atoms formed in a supernova is about $5 \times 10^{-5}$, and aluminum constitutes approximately 8680 ppm (parts per million) of the atoms in a chondritic meteorite. Assuming that these values apply to Vesta, estimate the number of $^{26}_{13}$Al atoms originally present in the asteroid.

(e) Find an expression for the amount of energy released per second in the decay of $^{26}_{13}$Al within Vesta as a function of time, and plot your results over the first $5 \times 10^7$ years on semilog graph paper. You may find Eq. (15.9) useful.

(f) How much time was required after the formation of Vesta before energy production due to the radioactive decay of $^{26}_{13}$Al dropped to $1 \times 10^{13}$ W, comparable to the current rate of energy output from the asteroid? See Problem 22.7.

**22.10** With the aid of a diagram, explain why it is best to observe a meteor shower between 2 A.M. and dawn instead of in the early evening. *Hint:* Consider the velocities of the infalling meteors and the orbital and rotational motions of Earth.

**22.11** Suppose that the Tunguska event was caused by an asteroid colliding with Earth. Assume that the density of the object was 2000 kg m$^{-3}$ and that it exploded above the surface of the planet traveling at a rate equal to Earth's escape velocity. If all of the energy of the explosion was derived from the asteroid's kinetic energy, estimate the mass and radius of the impacting body (assume spherical symmetry).

# CHAPTER

# 23

# Formation of Planetary Systems

## 23.1 ■ CHARACTERISTICS OF EXTRASOLAR PLANETARY SYSTEMS

You will recall that in Section 7.4 we discussed several methods that have been used to detect extrasolar planets (sometimes referred to as *exoplanets*). With the rapid increase in the number of known planets beyond our own Solar System, important new information is being gathered concerning how planetary systems form and evolve. In addition to these systems being interesting to study in their own right, this increase in knowledge about extrasolar planets helps to inform us about our own Solar System.

### Detections through the Reflex Radial Velocity Technique

As mentioned in Section 7.4, the most effective method of discovering extrasolar planets to date has been through the measurement of the reflex radial velocity of the parent star. With the exception of the pulsar, PSR 1257+12, 51 Pegasi was the first star (other than our Sun) found to have a planet in orbit around it. Michel Mayor and Didier Queloz of the Geneva Observatory made the announcement in October 1995 of a planet with a period of $P = 4.23077$ d in a nearly circular orbit ($e < 0.01$) around 51 Peg (a more recent radial velocity curve of 51 Peg obtained by Geoffrey Marcy and his collaborators is shown in Fig. 23.1). Since the system is not eclipsing, and the planet is too faint to be visually identified, the inclination of the orbit of the planet ($i$) is unknown. As a result, only the quantity $m \sin i$ can be determined for the planet from radial velocity measurements (see, for example, Eq. 7.5). Given that the parent star is a near twin of our Sun, with a spectral classification of G2V–G3V, implying a stellar mass of approximately 1 $M_\odot$, the lower mass limit of the orbiting planet is obtained from the maximum radial velocity wobble of the star.

---

**Example 23.1.1.**   To determine the minimum mass of the planet orbiting 51 Peg, we must first determine its orbital velocity. From Kepler's third law (Eq. 2.37), and assuming that the mass of the star is $m_{51} = 1$ $M_\odot$ and that the planet's mass, $m$, is insignificant ($m \ll m_{51}$), we find

$$a = \left[ \frac{G P^2 (m_{51} + m)}{4\pi^2} \right]^{1/3} = 7.65 \times 10^9 \text{ m} = 0.051 \text{ AU}.$$

Since the orbit of the planet is nearly circular, the orbital speed of the planet is

$$v = 2\pi a/P = 131 \text{ km s}^{-1}.$$

Employing Eq. (7.5), and noting from Fig. 23.1 that the amplitude of the star's observed radial velocity is $v_{r,\text{max}} = v_{51} \sin i = 56.04 \text{ m s}^{-1}$, we find that

$$m \sin i = m_{51} \frac{v_{51} \sin i}{v} = 8.48 \times 10^{26} \text{ kg} = 0.45 \text{ M}_J,$$

where $\text{M}_J$ is the mass of Jupiter. Since $\sin i \leq 1$, the mass of the planet, 51 Peg b, must be greater than 0.45 $\text{M}_J$.

51 Peg b is one example of a "hot Jupiter," one of a number of extrasolar planets that have been discovered having Jupiter-class masses but orbiting very close to their parent star.

---

## Multi-Planet Systems

A number of extrasolar planetary systems have been found through the radial velocity technique to have multiple planets in orbit about the central stars. An example of one such system is $\upsilon$ Andromedae; see Fig. 23.2. After the orbital perturbations due to the 4.6-d orbit of one planet were removed from the radial velocity curve of the star, evidence remained of additional perturbations. The $\upsilon$ And system contains at least three planets with orbital periods of 4.6 d, 241 d, and 1284 d, with $m \sin i$'s of 0.69 $\text{M}_J$, 1.89 $\text{M}_J$, and 3.75 $\text{M}_J$, respectively. The mass of the F8V parent star is estimated to be 1.3 $\text{M}_\odot$ (Appendix G).

As of May 2006, 193 extrasolar planets have been detected in 165 planetary systems. While most of the planetary systems have had just one planet detected in them so far, 20 systems are known to be multi-planet systems.

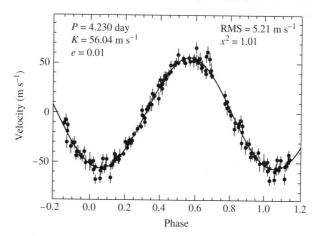

**FIGURE 23.1** The radial velocity measurements of 51 Pegasi, revealing the presence of a planet orbiting only 0.051 AU from the star. The sinusoidal shape of the velocity curve is evidence of a very low orbital eccentricity; recall the discussion in Section 7.3. (Figure adapted from Marcy, et al., *Ap. J.*, *481*, 926, 1997.)

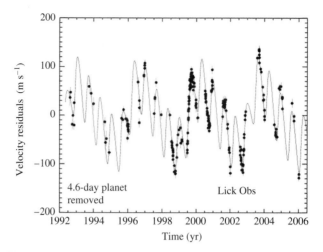

**FIGURE 23.2** The residuals in the radial velocity measurements of $\upsilon$ Andromedae after the gravitational perturbations of the 4.6-day planet have been removed. The data reveal the presence of at least three planets orbiting $\upsilon$ And. (Adapted from a figure provided by Debra A. Fischer, private communication.)

## The Mass Distribution of Extrasolar Planets

Initially, the radial velocity technique was able to discover only very massive (Jupiter-class) planets in close-in orbits around their parent stars. One of the reasons for this selection effect is that these objects exert the greatest gravitational influence on their parent star and generate the largest reflex radial velocities. The other reason is that a star must be observed over a time interval greater than the orbital period of the planet before the existence of the planet can be confirmed. As the amount of time increases for the systems being surveyed, the longer time-line data have allowed researchers to find lower-mass planets and planets orbiting farther from the star. The lowest-mass planet discovered to date is in a multiple system orbiting Gliese 876 and has an $m \sin i = 0.023$ $M_{\rm J}$, which is just 7.3 $M_{\oplus}$. The largest orbit detected thus far using the reflex motion technique is in the multiple system 55 Cancri, with a semimajor axis of 5.257 AU and an orbital period of 4517 d = 12.37 yr.

Over time, this selection effect is systematically diminishing. As is evident from statistical studies of the systems investigated so far, nature seems able to produce planets with a range of masses, with the lowest-mass planets being the most common. When binned by mass interval (see Fig. 23.3), the number of planets in each mass bin varies as

$$\frac{dN}{dM} \propto M^{-1}. \tag{23.1}$$

## The Distribution of Orbital Eccentricities

It is also interesting to note the relationship between orbital eccentricity ($e$) and semimajor axis for extrasolar planets (Fig. 23.4). Those planets that are orbiting close to their parent star

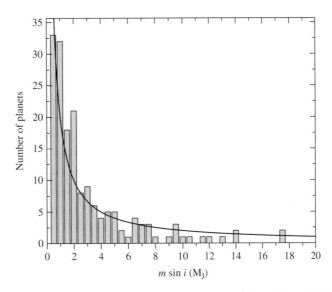

**FIGURE 23.3** The number of planets in mass bins of interval 0.5 $M_J$. The solid line is given by Eq. (23.1). (Data from *The Extrasolar Planets Encyclopedia*, http://exoplanet.eu, maintained by Jean Schneider.)

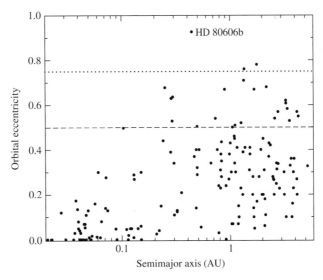

**FIGURE 23.4** The orbital eccentricities of the known extrasolar planets as a function of semi-major axis. Only 3 planets ($< 2\%$) are known to have eccentricities greater than 0.75, and less than 15% have eccentricities greater than 0.5. (Data from *The Extrasolar Planets Encyclopedia*, http://exoplanet.eu, maintained by Jean Schneider.)

tend to have circularized orbits (or at least orbits with smaller eccentricities). Planets orbiting farther from their parent star may have high orbital eccentricities, with the maximum value determined to date being a planet orbiting HD 80606 with $e = 0.927$ and a semimajor axis of 0.439 AU. However, from the data obtained thus far, only 15% of planets are known to have eccentricities of greater than 0.5, and less than 2% have eccentricities in excess of 0.75.

With the small number of high-eccentricity planets, it is important to ask whether or not there is something unique about the systems in which they are found. HD 80606 turns out to be one member of a wide stellar binary system, the other member being HD 80607. These two G5V stars are nearly identical and slightly smaller than the Sun. The two stars are also separated by a projected distance of 2000 AU. It has been suggested that the gravitational perturbations exerted on the planet, HD 80606b, by HD 80607 may have pumped its orbit up to its current very high eccentricity. In support of this suggestion, another planet with a high eccentricity ($e = 0.67$), 16 Cyg Bb, is also a member of a binary star system. However, the timescale for the gravitational perturbations provided by HD 80607 that would cause the orbital eccentricity of HD 80606b to significantly increase is estimated to be 1 Gyr. This long period of time comes from the necessary resonant alignment of the second star with the planet. The 1-Gyr timescale must be compared against the 1-Myr timescale provided by the general relativistic effect of the advance of the periastron of HD 80606b's orbit due to its parent star (recall the discussion of the advance of perihelion of Mercury's orbit in Section 17.1). It is argued that the general relativistic effects would completely overwhelm the perturbations from HD 80607 unless there were a third body in the HD 80606 system with an orbital period of roughly 100 yr that could also gravitationally influence HD 80606b. So far, a third object has not been discovered.

Two conclusions may be drawn from these data: (1) Planets with orbital periods of less than 5 days tend to have the smallest eccentricities ($e < 0.17$, with 80% of those having $e < 0.1$), probably due to strong tidal interactions with the parent star, and (2) planets sufficiently far from the parent star may have fairly large orbital eccentricities, but typically less than about 0.5. It seems that our own Solar System is somewhat unique, at least compared to the systems studied to date, in that our planets tend to have orbital eccentricities that are very small (excluding the Kuiper belt objects).

## The Trend toward High Metallicity

An additional important trend has also been emerging from the extrasolar planetary system data obtained to date. It appears that there is a strong tendency for planetary systems to preferentially form around metal-rich (Population I) stars. One way to quantify the **metallicity** is by comparing the ratios of iron to hydrogen in stars relative to our Sun, defining the metallicity to be

$$\text{[Fe/H]} \equiv \log_{10}\left[\frac{(N_{\text{Fe}}/N_{\text{H}})_{\text{star}}}{(N_{\text{Fe}}/N_{\text{H}})_{\odot}}\right], \tag{23.2}$$

where $N_{\text{Fe}}$ and $N_{\text{H}}$ represent the *number* of iron and hydrogen atoms, respectively. Stars with [Fe/H] < 0 are metal-poor relative to the Sun, and stars with [Fe/H] > 0 are relatively

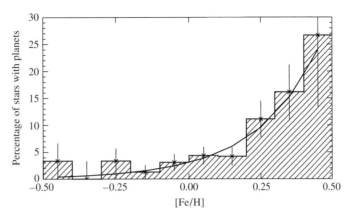

**FIGURE 23.5** The percentage of stars found to have planetary systems, relative to the number of stars investigated in each metallicity bin. The solid curve is given by Eq. (23.3). (Figure adapted from Fischer and Valenti, *Ap. J.*, *622*, 1102, 2005.)

metal-rich. For comparison, extremely metal-poor (Population II) stars in the Milky Way Galaxy have been measured with values of [Fe/H] as low as $-5.4$, while the highest values for metal-rich stars are about 0.6.

As can be seen in Fig. 23.5, most stars with planetary systems detected so far tend to be metal-rich compared to the Sun. Those stars that do have a metallicity lower than the solar value are only moderately lower. The data in Fig. 23.5 are plotted as the percentage of stars in a given metallicity bin that were well studied and found to have planetary systems. According to the sample of 1040 F, G, and K stars used in the study, the data seem to be well-fit by the relationship

$$\mathcal{P} = 0.03 \times 10^{2.0[Fe/H]}, \tag{23.3}$$

where $\mathcal{P}$ is the probability of a star having a detectable planetary system.

### Measuring Radii and Densities Using Transits

The transit of a planet across the disk of the parent star provides further information about the planet (recall Fig. 7.12). From the timing of the eclipse, and using atmospheric models of the star that include limb darkening, it is possible to determine the planet's radius. Of course, once the radius is determined, the planet's average density may also be computed. From the small number of systems where this has been possible, it appears that the Jupiter-class planets have densities that are similar to those of the gas giants in our Solar System (see Fig. 23.6). However, some of the so-called "hot Jupiters" that orbit close to the parent star appear to be somewhat inflated (e.g., HD 209458b and OGLE-TR-10b). The simple answer to explain the effect, namely the higher surface temperature due to the planet's proximity to the parent star, doesn't seem to explain all of the close-in systems, so apparently another source (or sources) of heat is required to puff up the planets. Some of the suggestions for solving the puzzle include tidal dissipation due to ongoing circularization of the orbit (perhaps involving another undetected object trying to simultaneously pump up the orbital

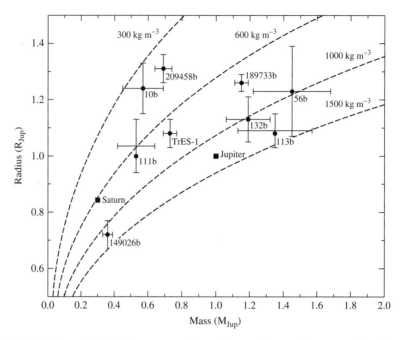

**FIGURE 23.6** The relationship between radius and mass for transiting extrasolar planets. The dashed lines correspond to specific average mass densities. (Adapted from a figure provided by Debra A. Fischer, private communication.)

eccentricity), a misalignment between the planet's orbital plane and the equator of the star, and dissipation of atmospheric currents in the planet as its gas migrates from the hot, substellar point to the cooler region on the back of the planet (Hadley circulation).

At least one planet appears to have a massive rocky core. The G0IV star HD 149026 has a transiting "hot Saturn," with an $m \sin i = 0.36$ $M_J$. From the transits, the orbital inclination has been determined to be $85.3° \pm 1.0°$, allowing a determination of the mass of the planet (not just a lower limit) of $0.36$ $M_J = 1.2$ $M_S$, where $M_S$ designates the mass of Saturn. The timing of the transits also yields a radius of $0.725 \pm 0.05$ $R_J$ for the planet, implying an average density of 1253 kg m$^{-3}$, which is 94% the density of Jupiter but 1.8 times the density of Saturn. The star itself has a mass and radius of $1.3 \pm 0.1$ $M_\odot$ and 1.45 $R_\odot$, respectively. In addition, the star's metallicity is [Fe/H] $= 0.36$, making it a significantly metal-rich star. Based on computer models of the planet's interior, it appears that the planet possesses a 67 $M_\oplus$ core composed of elements heavier than hydrogen and helium, assuming that the core density is 10,500 kg m$^{-3}$, which is believed to be similar to Saturn's core. If the core density is only 5500 kg m$^{-3}$, then the calculated core mass would be even larger (78 $M_\oplus$).

### The Detection of an Extrasolar Planet Atmosphere

Transits of extrasolar planets across the disks of their parent stars also provide for the possibility of detecting extrasolar planetary atmospheres. The first planet for which this was accomplished was HD 209458b. David Charbonneau and his collaborators were able to detect the spectroscopic signature of sodium at its resonance doublet wavelength of 589.3 nm by noting differences in the spectrum of the star as the planet passed in front of it. The starlight passing through the planet's atmosphere produced an enhanced absorption feature at that wavelength. The effect was very subtle, with deepening of the absorption feature by only an additional factor of $2.32 \pm 10^{-4}$ relative to adjacent wavelength bands during the transit. It has been proposed that the spectral signatures of water, methane, and carbon monoxide may be able to be detected in this way as well.

### Distinguishing Extrasolar Planets from Brown Dwarfs

With the detection of a few extrasolar planets having masses more than a factor of ten larger than the mass of Jupiter, the question is again raised concerning the definition of a planet. At the low-mass end, large Kuiper belt objects such as Pluto have been classified as planets. At the upper end, what distinguishes a planet from a brown dwarf?

Two different criteria have been proposed to answer this question. One suggestion is tied to the formation process of planets and stars. As we discussed in Chapter 12, stars form from the gravitational collapse of a gas cloud. As we will explore further in the next section, planets are generally believed to form from a bottom-up accretion process, although there has been speculation that gravitational collapse in the star's accretion disk may also produce planets. One proposed definition of planet is that it is an object that forms through a process beginning with the bottom-up accretion of planetesimals, whereas a brown dwarf forms directly from gravitational collapse. The challenge with such a definition is determining after the fact how a particular object may have formed.

A second criterion that has been proposed is based on whether or not the object that forms is massive enough ever to have had nuclear fusion occur in its core. Computer models of very low-mass objects indicate that if the mass of the object is greater than 13 $M_J$, deuterium can burn while the object is forming. The rate of energy production would not be sufficient to stabilize the object during gravitational collapse, but deuterium burning can be sufficient to affect the luminosity of the object during collapse. At the other end, as mentioned in Section 10.6, stars with mass of at least 0.072 $M_\odot$ (75 $M_J$) for solar composition undergo nuclear fusion at a sufficient rate to stabilize them at the low-mass end of the main sequence. Thus, it is proposed that brown dwarfs should be considered as being those objects having masses between these two limits (13 $M_J < M_{bd} < 75$ $M_J$); in other words, brown dwarfs are "stars" that burn some deuterium but never reach a stable nuclear-burning phase during contraction. Given the difficulty with the formation-mechanism criterion, the nuclear-reaction/mass-based criterion is generally favored.

### An Image of an Extrasolar Planet

In 2004 the first image of an extrasolar planet was obtained by Gael Chauvin and collaborators, using the European Southern Observatory's Very Large Telescope with an infrared detector; see Fig. 23.7. The parent star is a 25 $M_J$ brown dwarf of spectral type M8.5, known as 2MASSWJ1207334−393254, or 2M1207 for short! The system was also resolved later by the Hubble Space Telescope's NICMOS instrument. The planet resides 55 AU from the brown dwarf and has an estimated mass of $5 \pm 2$ $M_J$. From the infrared observations, the spectral type of the planet is between L5 and L9.5.

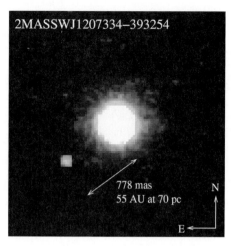

**FIGURE 23.7**    The first image obtained of an extrasolar planet. The planet is orbiting the brown dwarf 2MASSWJ1207334−393254. (Image courtesy of the European Southern Observatory.)

### Future Space-Based Planet Searches

Given the dramatic success since the mid 1990s in detecting planetary companions of main-sequence stars, a number of projects are planned to further the search using space-based observatories:

- **COROT** (COnvection, ROtation, and planetary Transits) is a joint mission of France, ESA, Germany, Spain, Belgium, and Brazil that is designed to study stellar seismology and search for planetary transits. COROT is scheduled for launch in 2006.

- NASA's **Kepler** mission is slated for launch in 2008 and will search for transits of Earth-sized planets across their parent stars' disks. Specifically, the Kepler mission hopes to identify Earth-like planets in the habitable zone around solar-type stars out to a distance of about 1 kpc.

- The **SIM PlanetQuest** mission, scheduled for launch in 2011, is designed to obtain high-precision astrometric data (see Section 6.5). One of SIM's primary missions is to search for nearby extrasolar Earth-sized planets.

- The data obtained from Kepler and SIM will provide input data for another NASA mission, known as the **Terrestrial Planet Finder** (TPF). TPF, as it is currently envisioned, will be made of two complementary component missions: a *visible-light coronagraph*, scheduled for launch around 2014, and an *infrared nulling interferometer* that will be composed of five individual spacecraft flying in precise formation (to be launched before 2020). Together, the two components of the TPF should be able to identify Earth-like planets and measure their atmospheric chemistries. One goal of TPF is to try to detect the signatures of life in the atmospheres of other Earth-like planets.

- Sometime in 2015 or later, ESA plans to launch **Darwin**, a free-flying array of six infrared telescopes that will also act as an infrared nulling interferometer.

With the great focus on planetary searches currently under way from the ground and from space, and with additional space-based missions planned for the future, the tremendous advances in this field of modern astrophysics can be expected to continue.

## 23.2 ■ PLANETARY SYSTEM FORMATION AND EVOLUTION

The question of how Earth and the Solar System formed has intrigued humans in all cultures for thousands of years. In 1778 Georges-Louis Leclerc, Comte de Buffon (1707–1788) proposed that a giant comet collided with the Sun, causing the ejection of a disk of material that ultimately condensed to form the planets. Competing tidal theories argued that a close encounter with a passing star ripped material from the Sun. Unfortunately, each of these theories suffers from a number of difficulties, including inadequate energy, composition differences between the planets and the Sun, and the sheer improbability of such an event. Another class of theories suggested that the Sun accreted planetary material from interstellar space, taking care of the difficulty of composition differences between the Sun and the planets, but not those among the planets themselves. Yet another class of theories, the basis of today's models, argue for the simultaneous formation of the Sun and the planets out of the same nebula. Among the early proponents of these so-called nebular theories were René Descartes (1596–1650), Immanuel Kant (1724–1804), and Pierre-Simon, Marquis de Laplace (1749–1827).

Although a significant number of problems remain to be solved, there is now some sense of convergence on the basic components of planetary system formation. Throughout Part III (as well as in the rest of the book to this point), we have presented clues related to critical features of a comprehensive model, some obvious and others more subtle. Before discussing our present understanding of the formation of planetary systems, we will review some of these clues and the questions they raise.

### Accretion Disks and Debris Disks

In Chapter 12 we introduced the wide range of observational data related to the formation and pre-main-sequence evolution of stars. It is clear from both observational and theoretical studies that stars form from the gravitational collapse of clouds of gas and dust. If a collapsing cloud contains any angular momentum at all (which it surely will), the collapse leads to the formation of an accretion disk around the growing protostar, as explored in Problem 12.18.

As a direct observational consequence of the conservation of angular momentum, numerous examples of accretion disk formation have been discovered and studied in detail, including the many proplyds observed in the Orion Nebula and elsewhere (Fig. 12.23) and the jets and Herbig–Haro objects associated with young protostars (Figs. 12.18 and 12.19). In addition, there is growing evidence that clumps of material exist in these disks.

There is also substantial evidence of **debris disks** around older stars, such as $\beta$ Pictoris (Fig. 12.21). The implication is that material is left over in the disk after the star has finished forming. Debris disks may be the extrasolar analogs to the asteroid belt and the Kuiper belt.

### Angular Momentum Distribution in the Solar System

However, one problem that has frustrated most attempts to put together an adequate picture of how our own Solar System developed concerns the present-day *distribution* of its angular momentum. In Problem 2.6 a simple calculation of the angular momentum in the Sun and Jupiter revealed that the orbital angular momentum of that planet exceeds the rotational angular momentum of the Sun by roughly a factor of twenty. A more detailed analysis shows that even though the Sun contains 99.9% of the mass, it contains only about 1% of the angular momentum of the entire Solar System, and most of the remainder is associated with Jupiter.[1] To complicate matters further, the Sun's spin axis is tilted 7° with respect to the average angular momentum vector of the planets, making it hard to envision how such a distribution of angular momentum could develop.

An additional interesting component of the angular momentum question concerns the amount of angular momentum possessed by other stars. It turns out that, on average, main-sequence stars that are more massive rotate much more rapidly and contain more angular momentum per unit mass than do less massive ones. Moreover, as can be seen in Fig. 23.8, a very discernible break occurs in the amount of angular momentum per unit mass as a function of mass near spectral class A5. If the total angular momentum of the Solar System were included, rather than just the angular momentum of the Sun, the trend along the upper

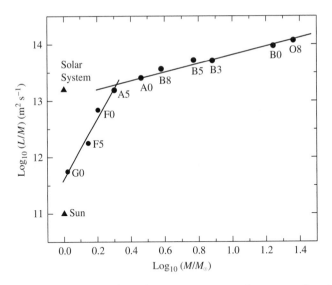

**FIGURE 23.8**    The average amount of angular momentum per unit mass as a function of mass for stars on the main sequence. The Sun's value and the total for the entire Solar System are indicated by triangles. Best-fit straight lines have been indicated for stars A5 and earlier, as well as for stars A5 and later (not including the Sun).

---

[1]The difference between the result obtained in Problem 2.6 and the one quoted here lies in the assumption made in Problem 2.6 that the Sun is a solid, constant-density sphere having a moment of inertia of $0.4 M_\odot R_\odot^2$. In fact, the Sun does not rotate as a rigid body, and because it is centrally condensed, its moment of inertia is closer to $0.073 M_\odot R_\odot^2$.

end of the main sequence would extend to include our Solar System as well (recall that the Sun is a G2 star).

A portion of the angular momentum problem may be solved by the transport of angular momentum outward via plasma drag in a corotating magnetic field. Charged particles trapped in the protosun's field would have been dragged along as the field swept through space. In response, the protosun's rotation speed slowed because of the torque exerted on it by the magnetic field lines. In addition, much of the rotational angular momentum of the newly formed Sun was probably also carried away by the particles in the solar wind (you may recall a similar discussion in Chapter 11; see Fig. 11.27). In support of these mechanisms is Fig. 23.8. The change in slope of the angular-momentum-per-unit-mass curve corresponds well with the onset of surface convection in low-mass stars, which in turn is linked to the development of coronae and mass loss. Other mechanisms for angular momentum transport will be discussed later.

### Composition Trends throughout the Solar System

We have already seen that lower-mass stars with metallicities similar to or greater than the solar value seem able to form planetary systems routinely. Therefore, the process of planetary system formation must be robust. The process must also be capable of producing systems with planets that are far from the parent star and systems where the planets are very close in.

A crucial piece of any successful theory must be the ability to explain the clear composition trends that exist among the planets in our Solar System (see, for example, Table 19.1). The inner terrestrial planets are small, generally volatile-poor, and dominated by rocky material, while the gas and ice giants contain an abundance of volatile material. Moreover, even though the ice giants Uranus and Neptune contain substantial volatiles, the gas giants Jupiter and Saturn contain the overwhelming majority of volatile material in the Solar System.

The moons of the giant planets also exhibit composition trends. In going from Jupiter out to Neptune, the progression is from rocky moons to increasingly icy bodies, first containing water-ice and then methane- and nitrogen-ice. The pattern even includes such objects as the asteroids, the Centaurs, the Kuiper belt objects, and other cometary nuclei. It is particularly important to note that a composition trend also exists across the asteroid belt itself. Even on the smaller scale of Jupiter's system of satellites, the Galilean moons change from volcanic Io to the thick-ice surface of Callisto.

### The Temperature Gradient in the Solar Nebula

Apparently, either a composition gradient or a temperature gradient (or both) must have existed in the early solar nebula while these objects were forming. For instance, the observations just described could be accounted for if the temperature of the nebular disk had decreased sufficiently across the asteroid belt. In that case, water would not have condensed in the region of the terrestrial planets but could have condensed in the form of ice in the vicinity of the giant planets. Another temperature gradient associated with the formation of Jupiter could help to explain the formation of the Galilean moons from the Jupiter subnebula.

Recall from Section 18.2 that an accretion disk that forms in a binary star system has a well-determined temperature gradient [$T \propto r^{-3/4}$; see Eq. (18.17)]. An analogous sort

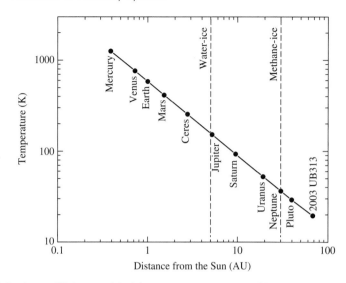

**FIGURE 23.9**    An equilibrium model of the temperature structure of the early solar nebula. Water-ice was able to condense out of the nebula in those regions beyond approximately 5 AU, and methane-ice could condense out of the nebula beyond 30 AU. The positions of the planets and Ceres represent their present-day locations.

of temperature structure should have existed in the solar nebula as well. The temperature structure for an equilibrium solar nebula model is shown in Fig. 23.9. Even though specific features of the distribution may change with more sophisticated modeling (by including time dependence, turbulence, and magnetic fields), it seems apparent that the condensation temperature of water-ice must be reached at some point near the current position of Jupiter, perhaps in the outer portions of the main asteroid belt (roughly 5 AU). The position in the solar nebula where water-ice could form has been variously referred to as the "snow line," the "ice line," or, more dramatically, the "blizzard line."

We have also learned that the environments around newly forming stars can be very dynamic places, with mass accretion and mass loss happening at virtually the same time in T-Tauri systems. During FU Orionis events, the environment around the star can become particularly active, with significant outbursts of energy occurring because of greatly increased mass accretion rates. It also seems certain that these environments will have complex magnetic fields that would lead to frequent and intense flares, analogous to the solar flares on our Sun that are produced by magnetic field reconnection events (Section 11.3).

### Consequences of Heavy Bombardment

We have also seen throughout Part III that, at least within our own Solar System, the formation of the Sun was accompanied by the formation of a wide range of objects, including small rocky planets, gas giants, ice giants, moons, rings, asteroids, comets, Kuiper belt objects, meteoroids, and dust.

Of course, it is readily apparent that our Solar System is riddled with evidence of collisions in the past, leaving cratered surfaces on objects of all sizes, from planets and moons to asteroids and comets. As a consequence, any formation theory must also be able to account for the obvious, heavy bombardment endured by bodies in the early Solar System. As discussed in Chapter 20, the high mass density of Mercury and the extremely volatile-poor composition of the Moon strongly suggest that both of these worlds were directly influenced by cataclysmic collisions involving very large planetesimals (the Moon's formation is tied to just such a collision with Earth). Heavy surface cratering shows that collisions continued even after their surfaces formed, with a brief episode of late heavy bombardment about 700 Myr after the formation of the Moon. Features such as the enormous Herschel crater on Mimas and the bizarre surface of Miranda testify to the fact that the other bodies in the Solar System underwent the same intense barrage from planetesimals.

Another consequence of the heavy bombardment by planetesimals is the variety of present-day orientations for the spin axes of the planets. The extreme examples of the retrograde rotations of Uranus and Pluto have already been discussed, but the other planets must have had their rotation axes shifted as well. Assuming that the planets did form out of a flattened nebular disk, the inherent angular momentum of the system would have resulted in rotation axes being initially aligned nearly perpendicular to the plane of the disk. Because this is not the case today, some event (or events) must have occurred to alter the directions of the planets' rotational angular momentum vectors. With the exception of Venus's and Mars's complex tidal interactions with the Sun and the other planets, the only likely mechanism suggested to date that can naturally account for the range of orientations observed requires collisions of planets or protoplanets with large planetesimals.

### The Distribution of Mass within Planetary Systems

Other features of the present-day Solar System that should be explained in a model of Solar System formation include the relatively small mass of Mars compared with its neighbors, the very small amount of mass present in the asteroid belt, and the existence of the Oort cloud and the Kuiper belt.

Furthermore, if we are to seek a general, unifying model of planetary system formation that includes our own Solar System as one example, it is necessary to understand the distributions of planets in other systems. Particularly perplexing when first discovered was the existence of "hot Jupiters" such as 51 Peg b. How could a gas giant form and survive so close to its parent star? In our own Solar System, none of the giant planets resides closer to the Sun than 5.2 AU.

### Formation Timescales

One aspect of all formation theories that cannot be neglected are constraints imposed by timescales:

- The collapse of a molecular cloud was discussed in Section 12.2. In Fig. 12.9 we saw that once a collapse is initiated, on the order of $10^5$ years is required for the formation of a protosun and nebular disk.

- The onset of violent T-Tauri and FU Orionis activity and extensive mass loss follows the initial collapse in some $10^5$ to $10^7$ years (see Section 12.3). This means that any nebular gas and dust that has not been accreted into a planetesimal or a protoplanet will be swept away within about 10 Myr, terminating further formation of large planets.

- The presence of $^{26}_{13}$Al in carbonaceous chondrites indicates that these meteorites must have been formed within a few million years after the creation of the aluminum, whether it was through a supernova detonation or through flares during FU Orionis activity. Otherwise, all of the radioactive nuclides that were created would have decayed into $^{26}_{12}$Mg. This observation puts severe constraints on condensation rates in the early solar nebula.

- The oldest meteorites, including Allende, date back to near 4.566 Gyr, while the age of the Sun itself is 4.57 Gyr. Clearly these oldest meteorites must have formed rapidly within the solar nebula.

- The ages of rocks returned from the Moon show that the surface of that body must have solidified some 100 Myr after the collapse of the solar nebula. Similar constraints exist on the formation of the surface of Mars judging on the basis of the age of the Martian meteorite, ALH84001.

- The lunar surface underwent a spike of late heavy bombardment about 700 Myr after the Moon formed.

- As we will learn later, as planets grow in accretion nebulae, they tend to migrate inward due to tidal interactions with the nebula and viscosity effects. It is estimated that a planetesimal could drift all the way into its parent star from a distance of 5 AU within roughly 1 to 10 Myr.

- A rather loose constraint on any model requires that all of the planets, moons, asteroids, Kuiper belt objects, and comets must be fully formed today, 4.57 Gyr years after the process started. Although this may seem trivial, not all models of Solar System formation have been successful in creating planets this rapidly!

### The Gravitational Instability Formation Mechanism

Two general, competing mechanisms have been proposed for the formation of planets within the accretion disks of proto- and pre-main-sequence stars. One mechanism is based on the idea that planets (or perhaps brown dwarfs) could form in accretion disks in a manner analogous to star formation. In regions where there may be a greater density of material in the disk, self-collapse could result. As the mass accumulates in that region, its gravitational influence on the surrounding disk increases, causing additional material to accrete onto the newly forming planet. This mechanism could even result in a local subnebula accretion disk forming around the protoplanet that could lead to the creation of moons and/or ring systems.

While this "top-down" **gravitational instability** mechanism has several attractive features, including simplicity and being strongly analogous to the formation of protostars, its general applicability suffers from numerous difficulties. By observations of other accretion disks, along with T-Tauri accretion and mass-loss rates, and combined with detailed numerical simulations, it appears that the solar nebula's lifetime would not have been sufficient to allow objects like Uranus and Neptune to grow quickly enough to attain the masses we observe before the nebula was depleted. This mechanism also does not explain the large number of other, smaller objects that are present in our Solar System and are likely to exist in other planetary systems as well (recall the $\beta$ Pic debris disk). In addition, the gravitational instability mechanism doesn't appear to readily account for the mass distribution of extrasolar planets, the correlation between planetary system formation and metallicity, or the wide range in the densities and core sizes of planets, both within our Solar System and among the extrasolar planets.

### The Accretion Formation Mechanism

An alternative model, and the one general favored by most astronomers, is that planets grow from the "bottom up" through a process of **accretion** of smaller building blocks. Based on all of the observational and theoretical information presently available, it appears that a reasonable description of the formation of planetary systems can now be given. What follows is a possible scenario for the formation of our own Solar System, although references to general aspects of planetary system formation will also be made. It is important to note, however, that because of the complexity of the problem, revisions in the model (both minor and major) are likely to occur in the future.

### The Formation of the Solar System: An Example

Within an interstellar gas and dust cloud (perhaps a giant molecular cloud), the Jeans condition was satisfied locally, and a portion of the cloud began to collapse and fragment (see Eq. 12.14 for the Jeans mass). The most massive segments evolved rapidly into stars on the upper end of the main sequence, while less massive pieces either were still in the process of collapsing or had not yet started to collapse. Within a period of a few million years or less, the most massive stars would have lived out their entire lives and died in spectacular supernovae explosions.[2]

As the expanding nebulae from one or more of the supernovae traveled out through space at a velocity of roughly $0.1c$, the gases cooled and became less dense. It may have been during this time that the most refractory elements began to condense out of the supernova remnants, including calcium, aluminum, and titanium, the ingredients of the CAIs that would eventually be discovered in carbonaceous chondrites that would fall to Earth billions of years later. When a supernova remnant encountered one of the cooler, denser components of the cloud that had not yet collapsed, the remnant began to break up into "fingers" of gas and dust that penetrated the nebula unevenly. The small cloud fragment would have also been compressed by the shock wave of the high-speed supernova remnant when the expanding nebula collided with the cooler gas. It is possible that this compression may have even helped trigger the collapse of the small cloud. In any case, the material in the solar nebula was now enriched with elements synthesized in the exploded star.

Assuming that the solar nebula possessed some initial angular momentum, conservation of angular momentum demands that the cloud "spun up" as it collapsed, producing a protosun surrounded by a disk of gas and dust. In fact, the disk itself probably formed more rapidly than the star did, causing much of the mass of the growing protosun to be funneled through the disk first. Although this important point is not entirely resolved, it has

---

[2]Recall the extremely disparate timescales for high-mass and low-mass stellar evolution; see Tables 12.1 and 13.1.

been estimated that the solar nebular disk may have contained a few hundredths of a solar mass of material, with the remaining 1 $M_\odot$ of the nebula ending up in the protosun. At the very least, a minimum amount of mass must have ended up in the nebular disk to form the planets and other objects that exist today. Such a disk is referred to as the **Minimum Mass Solar Nebula** (see Problem 23.4).

### The Hill Radius

Within the nebular disk, small grains with icy mantles were able to collide and stick together randomly. When objects of appreciable size were able to develop in the disk, they began to gravitationally influence other material in their areas.

To quantify the influence that these growing planetesimals had, we can define the **Hill radius**, $R_H$, to be that distance from the planetesimal where the orbital period of a test particle around the planetesimal is equal to the orbital period of the planetesimal around the Sun.

Assuming a circular orbit, the orbital period of a test particle ($m_t$) around an object of mass $M$ ($M \gg m_t$) at a distance $R$ is given by Kepler's third law (Eq. 2.37) as

$$P \simeq 2\pi \sqrt{\frac{R^3}{GM}}.$$

At a distance $a$ from the Sun, the orbital period of the growing planetesimal around the Sun equals the orbital period of a massless test particle around the planetesimal at the Hill radius when

$$\sqrt{\frac{a^3}{M_\odot}} = \sqrt{\frac{R_H^3}{M}}.$$

Thus, the Hill radius is given by

$$\boxed{R_H = \left(\frac{M}{M_\odot}\right)^{1/3} a.} \qquad (23.4)$$

Rewriting in terms of the density of the Sun and the density of the planetesimal (assumed to be spherical), the Hill radius becomes

$$R_H = R/\alpha \qquad (23.5)$$

where $R$ is the radius of planetesimal and

$$\alpha \equiv \left(\frac{\rho_\odot}{\rho}\right)^{1/3} \frac{R_\odot}{a}.$$

The physical significance of the Hill radius is that if a particle comes within about one Hill radius of a planetesimal with a relative velocity that is sufficiently low, the particle can become gravitationally bound to the planetesimal. In this way, the planetesimal acquires the mass of the particle and continues to grow. Of course, as the planetesimal's radius grows, so does its Hill radius.

**Example 23.2.1.** For a planetesimal of density $\rho = 800 \, \text{kg m}^{-3}$ and radius 10 km, located 5 AU from the Sun ($\rho_\odot = 1410 \, \text{kg m}^{-3}$), the planetesimal's Hill radius would be

$$R_H = R/\alpha = R \left( \frac{\rho}{\rho_\odot} \right)^{1/3} \left( \frac{a}{R_\odot} \right) = 8.9 \times 10^6 \, \text{m} = 1.4 \, \text{R}_\oplus.$$

This planetesimal is similar to present-day cometary nuclei.

### The Formation of the Gas and Ice Giants

As the low-energy collisions continued, progressively larger planetesimals were able to form. In the innermost regions of the disk the accreting particles were composed of CAIs, silicates (some in the form of chondrules), iron, and nickel; relatively volatile materials were unable to condense out of the nebula because of the high temperatures in that region. At distances greater than 5 AU from the growing protosun, just inside the present-day orbit of Jupiter, the nebula became sufficiently cool that water-ice could form as well. The result was that water-ice could also be included in the growing planetesimals beyond that distance. Even farther out (perhaps near 30 AU, the present-day orbit of Neptune), methane-ice also participated in the development of planetesimals. The location of the "snow line" where water-ice could form is shown in Fig. 23.10 (recall also Fig. 23.9).

The object that grew most rapidly was Jupiter. Thanks to the presence of water-ice along with rocky materials, and with a nebula that was sufficiently dense in its region, Jupiter's core reached a mass of between 10 and 15 $M_\oplus$. At that point the planet's gravitational influence became great enough that it started to collect the gases in its vicinity (principally hydrogen and helium). In effect, this created a localized subnebula, complete with its own accretion disk. The outcome was the formation of the massive planet we see today, together with the Galilean satellites. Heat generated in the gravitational collapse of Jupiter, combined with tidal effects, led to the eventual evolution of its moons. Astronomers believe that the entire process of forming Jupiter required on the order of $10^6$ years, halting when the gas was depleted.

As we will see shortly, the formation of the massive Jupiter had a significant impact on the other three planetesimals that had also grown to significant size beyond the snow

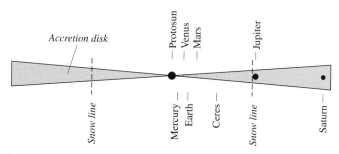

**FIGURE 23.10** A schematic drawing of the solar nebular disk, indicating the position of the water-ice "snow line" 5 AU from the protosun. Methane-ice began forming at roughly 30 AU from the protosun as well. The protosun, the protoplanets, and Ceres are located at their relative present-day distances, but their relative sizes are not correct.

line. Although Saturn, Uranus, and Neptune all developed cores of 10 to 15 $M_\oplus$, they were somewhat farther out in the nebula where the density was lower. As a result, they were unable to acquire the amount of gas that Jupiter captured in the same period of time.

## The Formation of the Terrestrial Planets and the Asteroids

In the inner portion of the solar nebula the temperatures were too warm to allow the volatiles to condense out and participate in the formation of planetesimals. But as the nebula cooled, the most refractory elements were able to condense out to form the CAIs.[3] Next to condense were the silicates and other equally refractory materials.

The slow relative velocities of silicate grains in nearly identical orbits resulted in low-energy collisions that promoted grain growth. Eventually, a hierarchy of planetesimal sizes developed. Computer simulations suggest that in the region of the terrestrial planets, along with a large number of smaller objects, there may have been as many as 100 planetesimals roughly the size of the Moon, 10 with masses comparable to Mercury's, and several as large as Mars. However, during the accretion process, most of these large planetesimals became incorporated into Venus and Earth. When the forming planets became massive enough, internal heat that was generated by decaying radioactive isotopes, together with energy released during collisions, started the process of gravitational separation. The results were the chemically differentiated worlds we see today.

With the formation of the massive Jupiter just beyond 5 AU from the Sun, gravitational perturbations began to influence the orbits of planetesimals in the region. In particular, most of the objects in the present-day asteroid belt had their orbits "pumped up" into progressively more and more eccentric orbits until some of them were absorbed by Jupiter or the other developing planets or were sent crashing into the Sun, while most were ejected from the Solar System entirely. This process stole material from the "feeding zones" near Mars and in the asteroid belt, resulting in a smallish fourth planet and very little mass in the belt. Perhaps only 3% of the original mass near Mars's orbit remained and only 0.02% of the mass in the region of the belt. Continued perturbations from Jupiter meant that the remaining belt of planetesimals had rather high relative velocities and were never able to consolidate into a single object. In fact, the high relative velocities imply that collisions cause fracturing, rather than growth.

As planetesimals continued to move throughout the forming Solar System, other collisions occurred. Some of the largest planetesimals in the inner Solar System collided with Mercury, removing its low-density mantle, and some struck Earth, forming the Earth–Moon system. Still other planetesimals of significant mass crashed into Mars and the outer planets, changing the orientations of their axes. Apparently, some of the planetesimals were also captured as moons or were torn apart by the giant planets when they wandered inside the planets' Roche limits.

Long before the terrestrial planets finished "feeding" on planetesimals in their regions of the disk, however, the evolving Sun reached the stage of thermonuclear ignition in its core, initiating the T-Tauri phase. At this point the infall of material from the disk was reversed by the strong stellar wind that ensued, and any gases and dust that had not yet collected into planetesimals were driven out of the inner Solar System.

---

[3]It is only in the innermost part of the Solar System that the nebula was warm enough to form CAIs in the first place.

### The Process of Migration

The accretion scenario described above is not without its own challenges. For instance, a long-standing problem has to do with the formation of the ice giants. At their current positions in the Solar System, it appears that the solar nebula would not have been dense enough to allow them to reach their present-day masses before the remaining gas was swept away by the T-Tauri wind. In addition, how is the episode of late heavy bombardment to be explained as a spike in collision rates roughly 3.8 Gyr ago? The apparent solution to both of these problems seems to lie in understanding a perplexing problem with many extrasolar planets.

With the discovery of "hot Jupiters" in extrasolar planetary systems, scientists realized that planets must be able to migrate inward while they are forming, and Jupiter is no different. Computer simulations of Solar System evolution suggest that Jupiter formed about 0.5 AU farther out in the nebula than its current position.

One mechanism by which inward migration of Jupiter (and extrasolar planets) could occur involves gravitational torques between the planet and the disk.[4] In this mechanism, initial deviations from axial symmetry produce density waves in the disk (density waves were mentioned in Section 21.3 in connection with the dynamics of Saturn's rings and will be discussed again extensively in connection with galaxy dynamics; see Section 25.3). The gravitational interaction between a growing planet and density waves results in the simultaneous transfer of angular momentum outward and mass inward. This so-called **Type I migration** mechanism can be shown to be proportional to mass, implying that as the planet accretes more material, it moves more rapidly toward its parent star. It may be that this can actually cause some planets to collide with the star on a timescale of one to ten million years.

However, it initially appeared that the timescale for Type I migration was too short compared with the runaway accretion of gases onto the growing Jupiter; in other words, Jupiter would crash into the Sun before it could fully form. It also appeared that Jupiter couldn't grow rapidly enough to reach its present size before the nebula was dissipated by the T-Tauri wind.

The solution to these problems may rest with the migration process itself. As the growing planet moves through the solar nebula, it continually encounters fresh material to "feed on." If the planet remained in a fixed orbit, it would quickly consume all of the available gas within several Hill radii and would grow only slowly after that. Migration allows it to move through the disk without creating a significant gap in the nebula.

It has also been shown that viscosity within the disk can cause objects to migrate inward. This **Type II migration** mechanism causes slowly orbiting particles farther out to speed up because of collisions with higher-velocity particles occupying slightly smaller orbits.[5] The loss of kinetic energy by the inner particles causes them to spiral inward. Type II migration can become the more significant, if slower, migration process when a gap is opened up in the disk.

---

[4] Peter Goldreich and Scott Tremaine suggested in 1980 that this mechanism would be important in the dynamical evolution of accretion disks. Their paper was published some fifteen years before the first confirmed detection of an extrasolar planet.

[5] Recall that the problem of angular momentum transport in a disk was also discussed in Chapter 18.

Outward migration is also possible. In this case, the scattering of planetesimals inward results in migration outward. Whether inward or outward migration occurs depends on the density of the nebula and the abundance of planetesimals.

Applying the mechanisms of migration to the evolution of our own Solar System, it appears that Jupiter not only influenced objects interior to its present-day orbit but also was influential in causing Saturn, Uranus, and Neptune to migrate outward. It seems that Uranus and Neptune initially formed their cores in a region of the nebula with a greater density, just as Jupiter and Saturn did. However, because of outward migration, they were able to put on only a small amount of extra gas and remain today as ice giants, rather than gas giants.

### Resonance Effects in the Early Solar System

Assuming that Jupiter originally formed at about 5.7 AU from the Sun as some simulations suggest, and that Saturn formed perhaps 1 AU closer to the Sun than its current position, the two gas giants would have moved through a critical resonance as Jupiter migrated inward and Saturn migrated outward. When the orbital periods of the two planets reached a 2:1 resonance (i.e., the orbital period of Saturn was exactly twice the orbital period of Jupiter), their gravitational influences on other objects in the Solar System would have periodically combined at the same points in their orbits, causing significant perturbations to orbits of objects in the asteroid belt and in the Kuiper belt.[6] Computer simulations suggest that this resonance effect may have occurred about 700 Myr after the formation of the inner planets and our Moon. It seems plausible that the passage of Jupiter and Saturn through this 2:1 resonance may have caused the episode of late heavy bombardment that is now recorded on the surface of the Moon.

As a consequence of Neptune's outward migration, Neptune swept up some of the remaining planetesimals, trapping them in 3-to-2 orbital resonances with the planet as it moved outward. It may be that Pluto and the other Plutinos were caught up in this outward migration. The orbits of the scattered Kuiper belt objects were also likely to have been perturbed by the migration of Neptune. The classical KBOs were probably far enough from Neptune not to be as drastically affected by its migration. In fact, the Kuiper belt may be the Solar System's analog to debris disks seen around other stars.

Similarly, the Oort cloud cometary nuclei are likely to be planetesimals that were scattered more severely by Uranus and Neptune. Once sufficiently far from the Sun, scattered cometary nuclei had their orbits randomized by passing stars and interstellar clouds.

### The Formation of CAIs and Chondrules

A particularly challenging problem with the model of Solar System formation described above is the presence of chondrules mixed in with CAIs in a matrix of hydrated and carbon-bearing minerals in chondritic meteorites. Both the chondrules and the CAIs have certainly been exposed to intense heat, but the matrix has clearly never been heated to temperatures greater than a few hundred kelvins. Because silicates require lower temperatures to condense out of the solar nebula, the chondrules probably formed after the CAIs. Silicate dust grains likely formed out of the nebula, coalescing into small clumps through repeated collisions.

---

[6]Recall the effect of Mimas on the ring system of Saturn (the Cassini division) and the effect of Jupiter on the asteroid belt (the Kirkwood gaps).

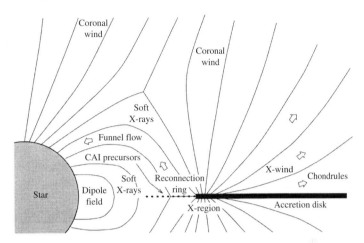

**FIGURE 23.11**  A schematic diagram of the X-wind model. (Figure adapted from Shu et al., *Ap. J.*, *548*, 1029, 2001.)

However, they could not have formed initially as molten droplets but, instead, were melted after formation.

Currently, the most plausible scenario suggests that powerful flares during FU Orionis events may be responsible for the melting or partial melting of chondrules and CAIs. As the inner edge of the accretion disk moves in and out on timescales of 30 years or so (perhaps associated with magnetic field activity), the silicate grains are exposed to flash heating by flares resulting from reconnection events. In the rarefied environment at the interior edge of the nebula, the droplets are able to cool rather quickly, perhaps between 100 and 2000 K per hour. Frank Shu and his colleagues have suggested that the metamorphosed chondrules may be launched back into the planet-forming region of the nebula by an X-wind, similar to the wind responsible for the ejection of Herbig–Haro objects in jets (see Fig. 23.11). In the planet-forming region of the nebula, the chondrules and CAIs are incorporated into the matrix. This model of chondrule formation implies that the solar nebula was a very dynamic system indeed.

Although much work remains to be done in fully developing our understanding of Solar System formation and the formation extrasolar planetary systems, tremendous progress has been made in this very complex area of research.

## SUGGESTED READING

### General

Basri, Gibor, "A Decade of Brown Dwarfs," *Sky and Telescope*, *109*, 34, May, 2005.

Naeye, Robert, "Planetary Harmony," *Sky and Telescope*, *109*, 45, January, 2005.

Marcy, Geoffrey, et al., "California and Carnegie Planet Search,"
    `http://exoplanets.org`.

Schneider, Jean, "The Extrasolar Planets Encyclopedia," `http://exoplanet.eu`.

**Technical**

Alibert, Y., Mordasini, C., Benz, W., and Winisdoerffer, C., "Models of Giant Planet Formation with Migration and Disc Evolution," *Astronomy and Astrophysics*, *434*, 343, 2005.

Beaulieu, J. P., Lecavelier des Etangs, A., and Terquem, C. (eds.), *Extrasolar Planets: Today and Tomorrow*, Astronomical Society of the Pacific Conference Proceedings, *321*, San Francisco, 2004.

Bodenheimer, Peter, and Lin, D. N. C., "Implications of Extrasolar Planets for Understanding Planet Formation," *Annual Review of Earth and Planetary Sciences*, *30*, 113, 2002.

Butler, R. Paul, et al., "Evidence for Multiple Companions to υ Andromedae," *The Astrophysical Journal*, *526*, 916, 1999.

Canup, R. M., and Righter, K. (eds.), *Origin of the Earth and Moon*, University of Arizona Press, Tucson, 2000.

Charbonneau, David, et al., "Detection of an Extrasolar Planet Atmosphere," *The Astrophysical Journal*, *568*, 277, 2002.

de Pater, Imke, and Lissauer, Jack J., *Planetary Sciences*, Cambridge University Press, Cambridge, 2001.

Fischer, Debra A., and Valenti, Jeff, "The Planet–Metallicity Correlation," *The Astrophysical Journal*, *622*, 1102, 2005.

Goldreich, Peter, and Tremaine, Scott, "Disk-Satellite Interactions," *The Astrophysical Journal*, *241*, 425, 1980.

Goldreich, Peter, Lithwick, Yoram, and Sari, Re'em, "Planet Formation by Coagulation: A Focus on Uranus and Neptune," *Annual Review of Astronomy and Astrophysics*, *42*, 549, 2004.

Gomes, R., Levison, H. F., Tsiganis, K., and Morbidelli, A., "Origin of the Cataclysmic Late Heavy Bombardment of the Terrestrial Planets," *Nature*, *435*, 466, 2005.

Lecar, Myron, Franklin, Fred A., Holman, Matthew J., and Murray, Norman W., "Chaos in the Solar System," *Annual Review of Astronomy and Astrophysics*, *39*, 581, 2001.

Mannings, Vincent, Boss, Alan P., and Russell, Sara S. (eds.), *Protostars and Planets, IV*, University of Arizona Press, Tucson, 2000.

Marcy, Geoffrey, et al., "Observed Properties of Exoplanets: Masses, Orbits, and Metallicities," *Progress of Theoretical Physics Supplement*, *158*, 1, 2005.

Mayor, M., and Queloz, D., "A Jupiter-Mass Companion to a Solar-Type Star," *Nature*, *378*, 355, 1995.

Shu, Frank H., Shang, Hsien, Gounelle, Matthieu, Glassgold, Alfred E., and Lee, Typhoon, "The Origin of Chondrules and Refractory Inclusions in Chondritic Meteorites," *The Astrophysical Journal*, *548*, 1029, 2001.

Taylor, Stuart Ross, *Solar System Evolution*, Second Edition, Cambridge University Press, Cambridge, 2001.

## PROBLEMS

**23.1** **(a)** If the actual separation between HD 80606 and HD 80607 is 2000 AU, determine the orbital period of the binary star system. *Hint:* You may want to refer to the data in Appendix G to estimate the masses of the two stars.

**(b)** How many orbits will the planet HD 80806b make around its parent star in the time that the two stars complete one orbit about their common center of mass? The semimajor axis of the planet's orbit is 0.44 AU.

**(c)** What is the ratio of the force of gravity exerted on HD 80606b by its parent star to that exerted by HD 80607 when it is aligned between the two stars?

**23.2** Compare Pluto with asteroids and cometary nuclei. Comment on the significance of any differences in view of the evolutionary model discussed in the chapter.

**23.3** **(a)** If all of the angular momentum that is tied up in the rest of the Solar System could be returned to the Sun, what would its rotation period be (assume rigid-body rotation)? Refer to the data in Fig. 23.8. The moment-of-inertia ratio of the Sun is 0.073.

**(b)** What would the equatorial velocity of the photosphere be?

**(c)** How short could the rotation period be before material would be thrown off from the Sun's equator?

**23.4** The Minimum Mass Solar Nebula is the smallest nebula that could be formed and still have sufficient mass to create all of the objects in the Solar System. Make a rough estimate of the mass of the Minimum Mass Solar Nebula.

**23.5** Estimate the present-day Hill radius of Jupiter. Express your answer in terms of the radius of Jupiter, as well as in astronomical units.

**23.6** HD 63454 is a K4V star known to have an extrasolar planet orbiting it in a circular orbit with an orbital period of 2.81782 d, producing a maximum radial reflex velocity of 64.3 m s$^{-1}$ relative to the center of mass of the star. The distance to HD 63454 is 35.80 pc. Consulting Appendix G, determine

**(a)** the semimajor axis of the planet's orbit.

**(b)** the minimum mass of the planet.

**(c)** the maximum astrometric wobble of the star due to the planet's pull, expressed in arcseconds.

**23.7** 14 Her is a K0V star located 18.1 pc from Earth. The extrasolar planet orbiting the star has an orbital period of 1796.4 d with an orbital eccentricity of 0.338. Consulting Appendix G, determine

**(a)** the semimajor axis of the planet's orbit.

**(b)** the maximum separation of the planet from the center of mass of its parent star.

**(c)** the velocity of the planet in its orbit at closest approach to the star.

**23.8** Explain why the high metallicities of systems with known extrasolar planets support the hypothesis that planets form from the "bottom up" by mass accretion of planetesimals.

**23.9** Assume that Jupiter and Saturn formed 5.7 AU and 8.6 AU from the Sun, respectively. Show that if the planets simultaneously migrated to their present orbital distances, they passed through a 2:1 orbital period resonance.

# PART
# IV

# Galaxies and the Universe

# CHAPTER

# 24

# The Milky Way Galaxy

**24.1**  *Counting the Stars in the Sky*
**24.2**  *The Morphology of the Galaxy*
**24.3**  *The Kinematics of the Milky Way*
**24.4**  *The Galactic Center*

## 24.1 ■ COUNTING THE STARS IN THE SKY

As we learned in the first two chapters of this text, human beings have long looked up at the heavens and contemplated its vastness, proposing various models to explain its form. In some civilizations the stars were believed to be located on a celestial sphere that rotated majestically above a fixed, central Earth. When Galileo made his first telescopic observations of the night sky in 1610, we started down a long road that has dramatically expanded our view of the universe.

In this chapter we will explore the complex system of stars, dust, gas, and dark matter known as the **Milky Way Galaxy**.[1] Although it is possible to get at least a general idea about the nature of other galaxies from our external viewpoint, studying our own Galaxy has proved to be very challenging. As we will learn, we live in a disk of stars, dust, and gas that severely impacts our ability to "see" beyond our relative stellar neighborhood when we look along the plane of the disk. The problem is most severe when looking toward the center of the Galaxy in the constellation Sagittarius. In Section 24.1 we will discover that studying the distribution of stars while considering the effects of extinction provides us with our first hint of what the Milky Way looks like from an outside perspective. In Section 24.2, a detailed description of the many varying components of the Galaxy will be presented.

Much of what we know today about the formation and evolution of the Milky Way is encoded in the motions of the Galaxy's constituents, especially when combined with information about variations in composition. Unfortunately, measuring the motions of the stars and gas in the Galaxy is done from an observing platform (Earth) that is itself undergoing a complex motion that involves the orbit of Earth around the Sun and the Sun's elaborate path around the Galaxy. In Section 24.3 we will investigate these motions, allowing us to move from a description of motions relative to the Sun to motions relative to the center of

---

[1]Throughout the remainder of this text, we may refer to the Milky Way Galaxy alternately in the shortened forms "the Galaxy" and "our Galaxy."

the Galaxy. We will also be led to the remarkable conclusion that the luminous, baryonic matter in the Galaxy is only a small fraction of what our Galaxy is composed of.

Finally, in Section 24.4, we will probe the center of the Milky Way and study the exotic environment found there, including indisputable evidence for a supermassive black hole.

Throughout the remainder of Part IV we will investigate other, more distant galaxies; we will study their morphologies and evolution, as well as the evolution of our own Milky Way. We will also study the large-scale structure of the universe and trace our developing understanding of the earliest moments of the universe and its ultimate fate.

**Historical Models of the Milky Way Galaxy**

As can be seen by even a casual observation of the dark night sky, an almost continuous band of light appears to circle Earth, inclined about 60° with respect to the celestial equator (see Fig. 24.1). It was Galileo who first realized that this Milky Way is a vast collection of individual stars. In the mid-1700s, in order to explain its circular distribution across the heavens, Immanuel Kant (1724–1804) and Thomas Wright (1711–1786) proposed that the Galaxy must be a stellar disk and that our Solar System is merely one component within that disk. Then, in the 1780s, William Herschel (1738–1822) produced a map of the Milky Way based crudely on counting the numbers of stars that he could observe in 683 regions of the sky (see Fig. 24.2). In his analysis of the data, Herschel assumed that (a) all stars

**FIGURE 24.1** A mosaic of the Milky Way showing the presence of dust lanes. (Courtesy of The Observatories of the Carnegie Institution of Washington.)

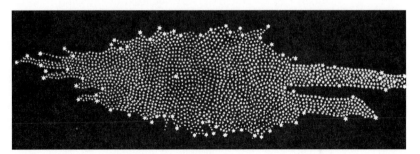

**FIGURE 24.2** William Herschel's map of the Milky Way Galaxy, based on a qualitative analysis of star counts. He believed that the Sun (indicated by a larger star) resided near the center of the stellar system. (Courtesy of Yerkes Observatory.)

have approximately the same absolute magnitude, (b) the number density of stars in space is roughly constant, (c) there is nothing between the stars to obscure them, and (d) he could see to the edges of the stellar distribution. From his data, Herschel concluded that the Sun had to be very near the center of the distribution and that the dimensions measured along the plane of the disk were some five times greater than the disk's vertical thickness.

Jacobus C. Kapteyn (1851–1922) essentially confirmed Herschel's model of the Galaxy, again using the technique of star counting. However, through the use of more quantitative methods, Kapteyn was able to specify a distance scale for his model of the Galaxy. The **Kapteyn universe**, as it is now called, was a flattened spheroidal system with a steadily decreasing stellar density with increasing distance from the center. A depiction of the Kapteyn universe is shown in Fig. 24.3. In the plane of the Galaxy and at a distance of some 800 pc from the center, the number density of stars had decreased from its central value by a factor of two. On an axis passing through the center and perpendicular to the central plane, the number density decreased by 50% over a distance of only 150 pc. The number density diminished to 1% of its central value at distances of 8500 pc and 1700 pc in the plane and perpendicular to the plane, respectively. Kapteyn concluded that the Sun was located 38 pc north of the Galactic midplane and 650 pc from the center, measured along the Galactic midplane.

To get some idea of how Kapteyn arrived at his nearly heliocentric (or Sun-centered) model of the universe, recall the equation for the distance to a star whose absolute magnitude is known (Eq. 3.5):

$$d = 10^{(m-M+5)/5}.$$

Assuming a value for $M$ (for instance, if the spectral class and luminosity class are known), and measuring $m$ at a telescope, the distance modulus $m - M$ and the distance $d$ are readily obtained. And given the known coordinates of the star on the celestial sphere, its three-dimensional position relative to Earth is determined.

In actuality, since the number of stars in any given region is so great, it is impractical to estimate the distance to each individual star in the way just described. Instead, a statistical approach is used that is based on counting the number of stars in a specified region down to a predetermined limiting apparent magnitude. From this counting procedure, the number density of stars at a given distance from the Sun can be estimated. We will discuss some

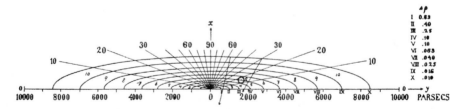

**FIGURE 24.3** The Kapteyn universe. Surfaces of constant stellar number density are indicated around the Galactic center. Note that the open circle does not represent the position of the Sun derived by Kapteyn. Rather, the open circle was used as an estimate from which Kapteyn began his analysis of the available data. (Figure from Kapteyn, *Ap. J.*, 55, 302, 1922.)

of the details of this method shortly. In his original study, Kapteyn used over 200 selected regions of the sky.

During the years between 1915 and 1919, shortly before Kapteyn's model was published, Harlow Shapley (1885–1972) estimated the distances to 93 globular clusters using RR Lyrae and W Virginis variable stars (Section 14.1). Since these stars are easily identified in the clusters through their periodic variations in luminosity, it is a relatively simple matter to use their absolute magnitudes (obtained from a period–luminosity relation such as Eq. 14.1) to estimate their distances from the Sun. The distances to the variable stars correspond to the distances to the clusters in which they reside.

In analyzing his data, Shapley recognized that the globular clusters are not distributed uniformly throughout space but are found preferentially in a region of the sky that is centered in the constellation of Sagittarius, at a distance that he determined to be 15 kpc from the Sun. Furthermore, he estimated that the most distant clusters are more than 70 kpc from the Sun, over 55 kpc beyond the center. As a result, by assuming that the extent of the globular clusters was the same as the rest of the Galaxy, Shapley believed that the diameter of the Galaxy was on the order of 100 kpc, close to ten times the value proposed by Kapteyn. Shapley's picture of the Galaxy also differed from Kapteyn's in another important way: Kapteyn's model located the Sun relatively near the center of the distribution of stars, whereas Shapley's Galactic center was much farther away.

We know today that both Kapteyn and Shapley were in error; Kapteyn's universe was too small and the Sun was too near the center, and Shapley's Galactic model was too large. Surprisingly, both models erred in part for the same reason: the failure to include in their distance estimates the effects of interstellar extinction due to dust and gas. Kapteyn's selected regions were largely within the Galactic disk where extinction effects are most severe; as a result, he was unable to see the most distant portions of the Milky Way, causing him to underestimate its size. The problem is analogous to someone on Earth trying to see the surrounding land while standing in a dense fog with limited visibility. Shapley, on the other hand, chose to study objects that are generally found well above and below the plane of the Milky Way and that are inherently bright, making them visible from great distances. It is in directions perpendicular to the disk that interstellar extinction is least important, although it cannot be neglected entirely. Unfortunately, errors in the calibration of the period–luminosity relation used by Shapley led to overestimates of the distances to the clusters. The calibration errors were traced to the effects of interstellar extinction (see Section 27.1).

Interestingly, Kapteyn was aware of the errors that interstellar extinction could introduce, but he was unable to find any quantitative evidence for the effect, even though researchers suspected that dust might be responsible for the dark bands seen running across the Milky Way (see Fig. 24.1). Further evidence for strong extinction could also be found in Shapley's own data; no globular clusters were visible within a region between approximately $\pm 10°$ of the Galactic plane, called the **zone of avoidance**. Shapley suggested that globular clusters were apparently absent in the zone of avoidance because strong gravitational tidal forces disrupted the objects in that region. In reality, interstellar extinction is so severe within the zone of avoidance that the very bright clusters are simply undetectable. Clearly the problems encountered by Kapteyn and Shapley in deducing the structure of our Galaxy point out the difficulty of determining its general morphology from a nearly fixed location within the

Galaxy's disk. Unfortunately, given the immense distances involved, we are not likely to be able to reach another, more favorable vantage point any time soon!

### The Effects of Interstellar Extinction

To see how interstellar extinction directly affects estimates of stellar distances, we must modify Eq. (3.5), as was done in Section 12.1. Starting from Eq. (12.1) and solving for $d$, we find

$$d = 10^{(m_\lambda - M_\lambda - A_\lambda + 5)/5} = d'10^{-A_\lambda/5}, \tag{24.1}$$

where $d' = 10^{(m_\lambda - M_\lambda + 5)/5}$ is the erroneous estimate of distance made when extinction is neglected, and $A_\lambda$ is the amount of extinction in magnitudes, and as a function of wavelength, that has occurred between the star and Earth. Since $A_\lambda \geq 0$ in all cases (after all, extinction cannot make a star appear brighter), $d \leq d'$; the true distance is always less than the apparent distance.

In the disk of the Milky Way the typical rate of extinction in visible wavelengths is 1 magnitude kpc$^{-1}$, although that value can vary dramatically if the line of sight includes distinct nebulae such as giant molecular clouds. Fortunately, it is often possible to estimate the amount of extinction by considering how dust affects the color of a star (interstellar reddening); see Section 12.1.

---

**Example 24.1.1.**   Suppose that a B0 main-sequence star with an absolute visual magnitude of $M_V = -4.0$ is observed to have an apparent visual magnitude of $V = +8.2$. Neglecting interstellar extinction (i.e., assuming naively that $A_V = 0$), the distance to the star would be estimated to be

$$d' = 10^{(V - M_V + 5)/5} = 2800 \text{ pc}.$$

However, if it is known by some independent means (such as reddening) that the amount of extinction along the line of sight is 1 mag kpc$^{-1}$, then $A_V = kd$ mag, where $k = 10^{-3}$ mag pc$^{-1}$ and $d$ is measured in pc. This gives

$$d = 10^{(V - M_V - kd + 5)/5} = 2800 \times 10^{-kd/5} \text{ pc},$$

which may be solved iteratively or graphically, giving a true distance to the star of $d = 1400$ pc.

In this case the distance to the star would have been overestimated by almost a factor of two if the effects of interstellar extinction were not properly accounted for.

---

### Differential and Integrated Star Counts

As we have already mentioned, Kapteyn's method of star counting was not based on directly determining $d$ for individual stars. Rather, the numbers of stars visible in selected regions of the sky are counted over a specified apparent magnitude range. Alternatively, all stars in the regions brighter than a chosen limit of apparent magnitude can be counted. These approaches are known as **differential** and **integrated star counts,** respectively. The technique of star counting is still used today to determine the number density of stars in the sky. The distribution depends on a variety of parameters, including direction, distance, chemical composition, and spectral classification. Such information is very helpful to astronomers in their efforts to understand the structure and evolution of the Milky Way Galaxy.

Let $n_M(M, S, \Omega, r)\, dM$ be the number density of stars with absolute magnitudes between $M$ and $M + dM$ and attribute $S$ that lie within a solid angle $\Omega$ in a specific direction at a distance $r$ from the observer ($S$ could be composition or the Morgan–Keenan spectral class discussed in Section 8.2, for example). According to the notation used here, $n_M$ has units of $\mathrm{pc}^{-3}\mathrm{mag}^{-1}$, and the actual number density of stars having attribute $S$ that lie within a solid angle $\Omega$ and located a distance $r$ from the observer is given by

$$n(S, \Omega, r) = \int_{-\infty}^{\infty} n_M(M, S, \Omega, r)\, dM. \tag{24.2}$$

If other specific information is also desired, those attributes could be included in $S$ as well. However, the amount of data required to carry out star count analyses is formidable and often prohibitive; as the number of variables increases, so does the amount of data required. In his original study, Kapteyn considered *general* star counts that tracked absolute magnitude only, regardless of spectral class.

If the number density $n_M(M, S, \Omega, r)\, dM$ is integrated over the volume of a cone defined by the solid angle $\Omega$ and extending from the observer at $r = 0$ to some distance $r = d$, the result is $N_M(M, S, \Omega, d)\, dM$, the total number of stars with absolute magnitudes in the range $M$ to $M + dM$ that are found within that conical volume of space (see Fig. 24.4). Using $dV = \Omega r^2\, dr$ in spherical coordinates, this is

$$N_M(M, S, \Omega, d)\, dM = \left[ \int_0^d n_M(M, S, \Omega, r)\, \Omega r^2\, dr \right] dM. \tag{24.3}$$

Equation (24.3) is the general expression for the *integrated star count,* written in terms of

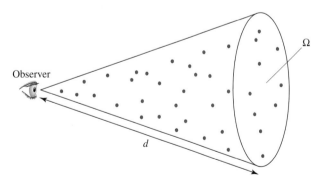

**FIGURE 24.4** An observer on Earth counting the number of stars within a specified range of Morgan–Keenan spectral and luminosity types out to a distance $d$ located within a cone of solid angle $\Omega$. When the effects of interstellar extinction are considered, this is equivalent to counting the same group of stars down to a corresponding apparent magnitude, $m$.

the limiting distance, $d$. Note that this means that $n_M\,dM$ can be obtained from $N_M\,dM$ (with limiting distance $r$) by differentiating:

$$n_M(M, S, \Omega, r)\,dM = \frac{1}{\Omega r^2}\frac{dN_M\,dM}{dr}.$$

Of course, those stars sharing the same absolute magnitude will have different apparent magnitudes because they lie at different distances from us. We can use Eq. (24.1) to replace the limiting distance, $d$, with the apparent magnitude $m$. This results in $\overline{N}_M(M, S, \Omega, m)\,dM$, the integrated star count written in terms of the limiting magnitude, $m$. (Note that the "bar" designation now indicates that the integrated star count is a function of $m$ rather than $d$.) Thus $\overline{N}_M(M, S, \Omega, m)\,dM$ is the total number of stars with absolute magnitudes in the range $M$ to $M + dM$ that appear brighter than the limiting magnitude, $m$.

If the limiting magnitude is increased slightly, then the limiting distance becomes correspondingly greater and the conical volume of space is extended to include more stars. The increase in the number of included stars is

$$\left[\frac{d\overline{N}_M(M, S, \Omega, m)}{dm}\,dm\right]dM.$$

This defines the *differential star count*,

$$A_M(M, S, \Omega, m)\,dM\,dm \equiv \frac{d\overline{N}_M(M, S, \Omega, m)}{dm}\,dM\,dm, \tag{24.4}$$

the number of stars with an absolute magnitude between $M$ and $M + dM$ that are found within a solid angle $\Omega$ and have apparent magnitudes in the range between $m$ and $m + dm$.

As a simple (and unrealistic) illustration of the use of integrated and differential star counts, consider the case of an infinite universe of uniform stellar density [i.e., $n_M(M, S, \Omega, r) = n_M(M, S) = $ constant] and no interstellar extinction ($A = 0$). Then Eq. (24.3) becomes, after canceling $dM$,

$$N_M(M, S, \Omega, d) = n_M(M, S)\,\Omega \int_0^d r^2\,dr = \frac{\Omega d^3}{3} n_M(M, S).$$

(Note that when the last expression is considered over all directions, $\Omega = 4\pi$ and $\Omega d^3/3$ is just the volume of a sphere of radius $d$.) Expressing $d$ in units of parsecs and writing it in terms of the apparent magnitude $m$ (Eq. 24.1), we have

$$\overline{N}_M(M, S, \Omega, m) = \frac{\Omega}{3} n_M(M, S)\,10^{3(m-M+5)/5}$$

$$= \frac{\Omega}{3} n_M(M, S)\,e^{\ln 10^{3(m-M+5)/5}}$$

$$= \frac{\Omega}{3} n_M(M, S)\,e^{[3(m-M+5)/5]\ln 10}.$$

Now, the differential star count formula is, from Eq. (24.4),

$$
\begin{aligned}
A_M(M, S, \Omega, m) &= \frac{d\overline{N}_M(M, S, \Omega, m)}{dm} \\
&= \frac{\ln 10}{5} \Omega\, n_M(M, S)\, 10^{3(m-M+5)/5} \\
&= \frac{3\ln 10}{5} \overline{N}_M(M, S, \Omega, m).
\end{aligned}
\tag{24.5}
$$

If either $\overline{N}_M(M, S, \Omega, m)$ or $A_M(M, S, \Omega, m)$ is known from observations, the equations can be used to determine the spatial number density $n_M(M, S)\, dM$.

The constant-density model just described suffers from a dramatic flaw. When it is used to calculate the amount of light received at Earth due to the stars contained in the solid angle $\Omega$, the result diverges exponentially as $m$ increases (Problem 24.4). This implies an infinite amount of light arriving from infinitely far away! This dilemma is one expression of **Olbers's paradox**, a problem known since the time of Kepler but brought to the attention of the general public by Heinrich Olbers (1758–1840). If we restrict our attention to the Milky Way Galaxy only, the solution rests in its finite size and nonconstant stellar number density. However, the resolution of Olbers's paradox is not as simple when applied to the universe as a whole, as we will learn in Chapter 29.

The modern process of gathering star count data involves the automated use of CCD detectors to determine $\overline{N}_M$ or $A_M$. Traditionally, these data are then combined with stellar number densities in the neighborhood of the Sun to estimate the number density of stars of a given spectral type in other regions of the Galaxy. More recently, an iterative computer modeling approach has been used with some success. A general model, based in part on observations of other galaxies believed to be similar to the Milky Way (see Section 25.2), is superimposed on the data; through successive iterations, the density function, the amount of interstellar extinction, and variations in composition with position are fine-tuned until a satisfactory match to the original data is obtained. However, given the complexity of the data and analysis, significant uncertainties remain in our picture of stellar distributions within the Milky Way.

## 24.2 ■ THE MORPHOLOGY OF THE GALAXY

Armed with numerous examples of other galaxies beyond the Milky Way, together with the data obtained from star counts and information gathered from various distance indicators, abundance analyses, and so on, astronomers have been able to piece together a model of the structure of our own Galaxy. However, it is important to point out that many of the details of the model remain uncertain and may even change in significant ways as more information becomes available. With that caveat, we will attempt to describe our current understanding of the general morphology of the Galaxy in this section, leaving details of its complex motions and fascinating nucleus until later in the chapter.

## Distance to the Galactic Center

Just as Herschel and Kapteyn had believed, the Milky Way is a galaxy that possesses a disk of stars, of which the Sun is one member. However, as Shapley had suspected, the Sun does not reside near the center of the disk but is actually located roughly one-third of the way out from the middle. As viewed from Earth, the center of the disk is in the direction of the constellation of Sagittarius corresponding to a very compact emission source known as Sgr A$^\star$ (see Section 24.4) at the J2000.0 equatorial coordinates

$$\alpha_{\text{Sgr A}^\star} = \quad 17^\text{h}45^\text{m}40.0409^\text{s} \tag{24.6}$$

$$\delta_{\text{Sgr A}^\star} = -29°00'28.118''. \tag{24.7}$$

Face-on and edge-on diagrams of the Galaxy are given in Figs. 24.5 and 24.6, respectively. Some of the details of the Galaxy's components (to be discussed later) are presented in Table 24.1.

The Sun's distance from the center of the Galaxy, known as the **solar Galactocentric distance**, $R_0$, has been revised downward many times since Shapley's first estimate of 15 kpc. In 1985 the International Astronomical Union (IAU) recommended a standardized value of $R_0 = 8.5$ kpc for the purpose of allowing direct comparisons of Galactic structure among various researchers. ($R_0$ is often used to normalize other distances in the Galaxy.) However, a number of studies have found that the value of $R_0$ is about 8 kpc. For example, in 2003 Frank Eisenhauer and colleagues determined a value of $7.94 \pm 0.42$ kpc based on astrometric and spectroscopic measurements of S2, the closest star to the Galactic center. Given the remaining uncertainties in the distance to the Galactic center, and given the

**FIGURE 24.5**    An artist's depiction of the Milky Way Galaxy seen face-on. The shapes of the spiral arms and the length of the bar associated with the central bulge are based on currently available data. The position of the Sun is shown. [Courtesy of NASA/JPL-Caltech/R. Hurt (SSC).]

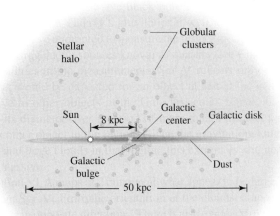

**FIGURE 24.6** An edge-on diagram of the Galaxy, not strictly to scale; see Table 24.1.

common use of 8 kpc in the literature, in this text we will adopt a value of

$$R_0 = 8 \text{ kpc} \tag{24.8}$$

for the solar Galactocentric distance.

The full diameter of the disk, including the dust, gas, and stars, is believed to be roughly 50 kpc, with estimates ranging from 40 to 50 kpc. Furthermore, it appears that the disk may not be completely cylindrically symmetric. Rather, the disk may be somewhat elliptical, with a ratio of the lengths of the minor and major axes of about 0.9. The Sun is probably located near the major axis of the disk. The **solar circle** is defined to be a perfect circle of radius $R_0$.

### The Structure of the Thin and Thick Disks

The disk is actually composed of two major components. The **thin disk**, which is composed of relatively young stars, dust, and gas, has a vertical scale height of $z_{\text{thin}} \simeq 350$ pc and is the region of current star formation (recall that one scale height is the distance over which the number density decreases by $e^{-1}$). A portion of the thin disk (sometimes referred to as the young thin disk) also corresponds to the central plane of the Galactic dust and gas distribution; it has a scale height of perhaps 90 pc, although some researchers have found a scale height as small as 35 pc. The **thick disk**, which is probably an older population of stars, has a scale height of approximately $z_{\text{thick}} \simeq 1000$ pc. The number of stars per unit volume in the thick disk is only about 8.5% of that in the thin disk at the Galactic midplane. When the thin and thick disks are combined, empirical fits to the stellar number density derived from star count data give

$$n(z, R) = n_0 \left( e^{-z/z_{\text{thin}}} + 0.085 e^{-z/z_{\text{thick}}} \right) e^{-R/h_R}, \tag{24.9}$$

**TABLE 24.1**   Approximate Values for Various Parameters Associated with the Components of the Milky Way Galaxy. Definitions and details are discussed in the text.

| | Disks | | |
|---|---|---|---|
| | Neutral Gas | Thin Disk | Thick Disk |
| $M$ ($10^{10}$ M$_\odot$) | $0.5^a$ | 6 | 0.2 to 0.4 |
| $L_B$ ($10^{10}$ L$_\odot$)$^b$ | — | 1.8 | 0.02 |
| $M/L_B$ (M$_\odot$/L$_\odot$) | — | 3 | — |
| Radius (kpc) | 25 | 25 | 25 |
| Form | $e^{-z/h_z}$ | $e^{-z/h_z}$ | $e^{-z/h_z}$ |
| Scale height (kpc) | $< 0.1$ | 0.35 | 1 |
| $\sigma_w$ (km s$^{-1}$) | 5 | 16 | 35 |
| [Fe/H] | $> +0.1$ | $-0.5$ to $+0.3$ | $-2.2$ to $-0.5$ |
| Age (Gyr) | $\lesssim 10$ | $8^c$ | $10^d$ |

| | Spheroids | | |
|---|---|---|---|
| | Central Bulge$^e$ | Stellar Halo | Dark-Matter Halo |
| $M$ ($10^{10}$ M$_\odot$) | 1 | 0.3 | $190^{+360}_{-170}{}^f$ |
| $L_B$ ($10^{10}$ L$_\odot$)$^b$ | 0.3 | 0.1 | 0 |
| $M/L_B$ (M$_\odot$/L$_\odot$) | 3 | $\sim 1$ | — |
| Radius (kpc) | 4 | $> 100$ | $> 230$ |
| Form | boxy with bar | $r^{-3.5}$ | $(r/a)^{-1} (1 + r/a)^{-2}$ |
| Scale height (kpc) | 0.1 to $0.5^g$ | 3 | 170 |
| $\sigma_w$ (km s$^{-1}$) | 55 to $130^h$ | 95 | — |
| [Fe/H] | $-2$ to 0.5 | $< -5.4$ to $-0.5$ | — |
| Age (Gyr) | $< 0.2$ to 10 | 11 to 13 | $\sim 13.5$ |

$^a$ $M_{\text{dust}}/M_{\text{gas}} \simeq 0.007$.

$^b$ The total luminosity of the Galaxy is $L_{B,\text{tot}} = 2.3 \pm 0.6 \times 10^{10}$ L$_\odot$, $L_{\text{bol,tot}} = 3.6 \times 10^{10}$ L$_\odot$ ($\sim 30\%$ in IR).

$^c$ Some open clusters associated with the thin disk may exceed 10 Gyr.

$^d$ Major star formation in the thick disk may have occurred 7–8 Gyr ago.

$^e$ The mass of the black hole in Sgr A$^\star$ is $M_{\text{bh}} = 3.7 \pm 0.2 \times 10^6$ M$_\odot$.

$^f$ $M = 5.4^{+0.2}_{-3.6} \times 10^{11}$ M$_\odot$ within 50 kpc of the center.

$^g$ Bulge scale heights depend on age of stars: 100 pc for young stars, 500 pc for old stars.

$^h$ Dispersions increase from 55 km s$^{-1}$ at 5 pc to 130 km s$^{-1}$ at 200 pc.

where $z$ is the vertical height above the midplane of the Galaxy, $R$ is the radial distance[2] from the Galactic center, $h_R > 2.25$ kpc is the disk scale length, and $n_0 \sim 0.02$ stars pc$^{-3}$ for the absolute magnitude range $4.5 \leq M_V \leq 9.5$. It should be pointed out that the relative density coefficient, the scale heights, and the disk scale length are all somewhat uncertain;

---

[2] In general, we reserve the use of $R$ to denote the cylindrical coordinate radius within the disk and the use of $r$ to represent the spherical coordinate radius, both measured from the Galactic center.

different researchers give somewhat different values for these parameters. The Sun is a member of the thin disk and is currently located about 30 pc above the midplane.

The **luminosity density** (the luminosity per unit volume) of the thin disk is often modeled with the functional form

$$L(R, z) = L_0 e^{-R/h_R} \operatorname{sech}^2(z/z_0), \tag{24.10}$$

where

$$\operatorname{sech}(z/z_0) = \frac{2}{e^{z/z_0} + e^{-z/z_0}}$$

is the hyperbolic secant function. For the thin disk $z_0 = 2z_{\text{thin}}$ and $L_0 \simeq 0.05 \, L_\odot \, \text{pc}^{-3}$.

### The Age–Metallicity Relation

The thin and thick disks not only are identifiable by separate scale heights and stellar number densities but are further distinguished by the chemical compositions and kinematic properties of their members. We will discuss composition effects now but will delay our discussion of kinematics until the next section.

As we noted in Section 13.3, stars are generally classified according to the relative abundance of heavier elements; Population I stars are metal-rich, with $Z \sim 0.02$, Population II stars are metal-poor, with $Z \sim 0.001$, and Population III stars are essentially devoid of metals, with $Z \sim 0$. In reality, a wide range of metallicities exists in stars. At one end are the extreme Population I stars, and on the other, the hypothetical Population III stars (if they still exist). Between Population I and Population II stars are the **intermediate** (or, alternatively and suggestively, **disk**) population stars.

To more carefully quantify the important parameter of composition, the *ratio* of iron to hydrogen has become almost universally adopted by researchers because iron lines are generally readily identifiable in stellar spectra. During a supernova detonation (particularly of Type Ia), iron is ejected, enriching the interstellar medium. New stars can then be created with a greater abundance of iron in their atmospheres than in their predecessors. As a result, iron content should correlate with stellar age, the youngest, most recently formed stars having the highest relative abundance of iron. The iron-to-hydrogen ratio in the atmosphere of a star is compared with the Sun's value through the expression

$$[\text{Fe/H}] \equiv \log_{10} \left[ \frac{(N_{\text{Fe}}/N_{\text{H}})_{\text{star}}}{(N_{\text{Fe}}/N_{\text{H}})_\odot} \right], \tag{24.11}$$

a quantity often referred to as the **metallicity**, which was first introduced in Section 23.1. Stars with abundances identical to the Sun's have $[\text{Fe/H}] = 0.0$, less-metal-rich stars have negative values, and more-metal-rich stars have positive values. Values ranging from $-5.4$ for old, extremely metal-poor stars to about 0.6 for young, extremely metal-rich stars have been measured in our Galaxy. According to studies of the main-sequence turn-off points in clusters (both globular and galactic), metal-rich stars tend to be younger than metal-poor ones of similar spectral type. The apparent correlation between age and composition is referred to as the **age–metallicity relation**.

However, in many situations the correlation between age and [Fe/H] may not be as reliable as first believed. For example, significant numbers of Type Ia supernovae do not appear until roughly $10^9$ years after star formation begins, and since Type Ia supernovae are responsible for most of the iron production, iron is not available in large quantities to enrich the interstellar medium. Furthermore, mixing of the interstellar medium after a SN Ia event may not be complete. In other words, a local region of the ISM may become enriched in iron after $10^9$ years, while another region may not experience the same level of enrichment. Therefore, according to the age–metallicity relation, the iron-rich region would subsequently produce stars that appear younger, when in fact both regions are the same age.

A second measure of ISM enrichment (and age) is based on [O/H], defined in a way completely analogous to Eq. (24.11). Since core-collapse supernovae appear after only $10^7$ years following the onset of star formation (recall that they are the result of massive-star evolution) and they produce a higher abundance of oxygen relative to iron, [O/H] may also be used to determine the ages of Galactic components; some astronomers use [O/Fe] for the same purpose.

To complicate matters further, accurate age estimates depend critically on precise values for distance moduli, which are needed to determine cluster main-sequence turn-off points (Section 13.3); errors of 0.1 in distance moduli lead to 10% errors in age. Needless to say, the important task of Galactic age determinations remains a major challenge in astronomy.

### Age Estimates of the Thin and Thick Disks

In the thin disk, typical values for the iron–hydrogen metallicity ratio are in the range $-0.5 < [\text{Fe/H}] < 0.3$, while for the majority of stars in the thick disk, $-0.6 < [\text{Fe/H}] < -0.4$ is more characteristic, although some thick-disk members may have metallicities at least as low as $[\text{Fe/H}] \sim -1.6$.

According to various age determinations, the stellar members of the thin disk are probably significantly younger than their thick-disk counterparts. It appears that star formation began in the thin disk about 8 Gyr ago and is ongoing today. This conclusion is supported by the observations of white dwarf stars in the thin disk and theoretical estimates of their cooling times (Section 16.5). There is also some evidence that star formation in the thin disk may not have been continuous over time but may have come in bursts with intervening gaps of several billion years. On the other hand, star formation in the thick disk appears to have predated the onset of star formation in the thin disk by two to three billion years. It is generally believed that the episode of thick-disk star formation spanned the time interval between 10 and 11 Gyr ago.

### Mass-to-Light Ratios

Based on data from star counts and orbital motions, the estimated stellar mass of the thin disk is roughly $6 \times 10^{10}$ M$_\odot$ with another $0.5 \times 10^{10}$ M$_\odot$ of dust and gas. Furthermore, the luminosity in the blue-wavelength band is $L_B = 1.8 \times 10^{10}$ L$_\odot$. When the first of these parameters is divided by the second, the resulting **mass-to-light ratio** is $M/L_B \approx 3$ M$_\odot$/L$_\odot$. This quantity gives us information about the kinds of stars responsible for the generation of the light.

We learned from Fig. 7.7 and Problem 10.27 that along the main sequence, a star's luminosity depends rather strongly on its mass, with

$$\frac{L}{L_\odot} = \left( \frac{M}{M_\odot} \right)^\alpha, \tag{24.12}$$

where $\alpha \simeq 4$ above about 0.5 $M_\odot$ and $\alpha \simeq 2.3$ for less massive stars. Assuming that most of the stars in the disk are main-sequence stars, an "average" stellar mass can be estimated. Substituting the observed mass-to-light ratio and solving for the mass, we have

$$\langle M \rangle = 3^{1/(1-\alpha)} \, M_\odot.$$

Assuming that $\alpha \simeq 4$, we find that $\langle M \rangle \simeq 0.7$ $M_\odot$. Apparently the total luminosity of the disk is dominated by stars somewhat less massive than the Sun. This should not be surprising since, as we saw in Fig. 12.12, the initial mass function indicates that many more low-mass than high-mass stars are created out of the interstellar medium, which is consistent with the fact that dwarf M stars are the most common class of stars in the vicinity of the Sun.

The $B$-band luminosity of the thick disk is about $2 \times 10^8$ $L_\odot$, or 1% of the value of the thin disk (which explains why the thick disk has been so difficult to detect). The mass of the thick disk is probably about 2 to $4 \times 10^9$ $M_\odot$, or approximately 3% of the thin-disk mass.

### Spiral Structure

Significant structure exists within the disk itself. When neutral hydrogen clouds or relatively young objects such as O and B stars, H II regions, and galactic (open) clusters are used as tracers of Galactic structure, a **spiral structure** emerges, giving the disk the appearance of a pinwheel. When other galaxies that possess distinct disks are observed in blue light (the dominant visible-wavelength band of hot, bright, young, high-mass main-sequence stars), these galaxies often exhibit similarly beautiful spiral structure. One such example is the great spiral galaxy of Andromeda, shown in Fig. 24.7. However, when the galaxies are viewed in the red light characteristic of older, low-mass stars, the spiral structure is less pronounced. It appears that spirals are associated with ongoing star formation and that older stars have had ample time to drift out of the spiral pattern (see Section 25.3). As we can see in Fig. 24.8, the Sun seems to be located close to, but not actually in, one of the spiral arm features known as the **Orion–Cygnus arm**, or simply the **Orion arm** (this feature is also known as the **Orion spur** since it is probably not a full spiral arm structure). Spiral arms get their names from the constellations in which they are observed.

The interstellar gas and dust clouds that plagued Kapteyn's attempts at determining the overall structure of the Galaxy and that are clearly evident in Fig. 24.1 are primarily located near the midplane and found preferentially in the spiral arms. If it were possible to view our Galaxy from a vantage point outside the disk but along the plane, it would probably appear similar to NGC 891,[3] the galaxy shown in Fig. 24.9.

---

[3] NGC is a designation indicating that the galaxy is a member of the *New General Catalog*, discussed on page 940.

**FIGURE 24.7**    The Andromeda galaxy (also known as M31 or NGC 224) is a spiral galaxy believed to be much like our own. M31 is located 770 kpc from the Milky Way Galaxy. (From Sandage and Bedke, *The Carnegie Atlas of Galaxies*, The Carnegie Institution of Washington, Washington, D.C., 1994.)

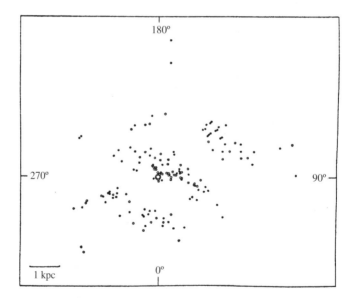

**FIGURE 24.8**    The spatial distribution of young Galactic clusters and H II regions reveals the presence of spiral arms within the disk of the Galaxy. The Sun is shown at the origin of the diagram (the intersection of the degree tick marks). The Sun is located near the Orion–Cygnus arm. [Figure adapted from Becker and Fenkart, *The Spiral Structure of Our Galaxy*, Becker and Contopoulos (eds.), D. Reidel Publishing Company, Dordrecht, 1970.]

**FIGURE 24.9** NGC 891, seen edge-on, clearly shows a thin dust band in the plane of the disk. The Milky Way Galaxy probably appears much like NGC 891 when viewed from a distant vantage point. (From Sandage and Bedke, *The Carnegie Atlas of Galaxies*, The Carnegie Institution of Washington, Washington, D.C., 1994.)

### Interstellar Gas and Dust

As we discussed in Section 12.1, gas and dust clouds exist in the Milky Way with a range of masses, temperatures, and densities. It is from these clouds that new stars are ultimately formed. Astronomers have been able to map out the overall distribution of dust and gas within the Milky Way by measuring the effects of obscuration by and emissivity of dust, as well as the location of 21-cm H I emission, and by using the CO molecule as a tracer of $H_2$. Molecular hydrogen and cool dust are found predominantly in the regions 3 to 8 kpc and 3 to 7 kpc from the Galactic center, respectively (i.e., inside the solar circle), while atomic hydrogen can be found in the region from 3 kpc out to the edge of the Galactic disk (25 kpc). It appears that $H_2$ and the dust are most tightly confined to the plane of the Galaxy, with vertical scale heights above or below the midplane of perhaps 90 pc or less. This is only about 25% of the value for stars in the thin disk and on the order of 9% of the scale height of thick-disk stars. In the region near the Sun, the scale height for atomic hydrogen is approximately 160 pc. The total mass of H I is estimated to be $4 \times 10^9$ $M_\odot$ and the mass of $H_2$ is approximately $10^9$ $M_\odot$. In the solar neighborhood the total mass density of gas is 0.04 $M_\odot$ $pc^{-3}$, of which atomic hydrogen accounts for approximately 77%, molecules contribute 17%, and ions add an additional 6%.

At distances beyond 12 kpc from the center of the Galaxy, the scale height of H I increases dramatically, reaching a value of more than 900 pc. In addition, the distribution of H I in the outer reaches of the Galaxy is no longer strictly confined to the plane but, rather, exhibits a well-defined **warp** that reaches a maximum angle of deviation from the plane of 15°. A map of the H I distribution at a Galactic radius of 13.6 kpc is shown in Fig. 24.10. The

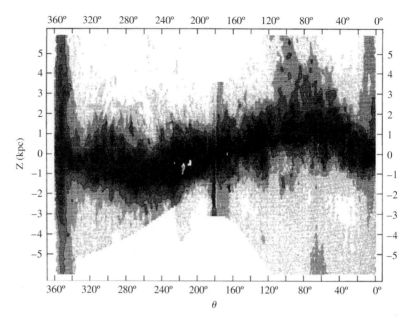

**FIGURE 24.10**   The H I warp in the Milky Way at a distance of 13.6 kpc from the Galactic center. The Galaxy's midplane is located at $z = 0$, and the direction toward the center of the Milky Way is at $0°$. (Figure adapted from Burton and te Lintel Hekkert, *Astron. Astrophys. Suppl. Ser.*, 65, 427, 1986.)

line of sight toward the center of the Milky Way is labeled as $0°$ in the figure. Warped H I distributions such as the one observed in our own Galaxy appear to be common features in other spiral galaxies as well, including Andromeda (in some spiral galaxies the warp angle can reach $90°$). Although we do not yet fully understand the dynamics that lead to the creation of warps (they do not seem to be the result of simple gravitational perturbations from one or more external galaxies, for example), they do seem to be associated with the distribution of mass in the outer regions of the Galaxy, beyond where most of the luminosity is produced.

Hydrogen clouds can also be found at high latitudes. Although some of these clouds have positive radial velocities, implying that they are moving away from the disk, the majority possess large negative radial velocities (up to $400$ km s$^{-1}$ or more), as measured by their 21-cm emission. There appear to be two types of sources responsible for these **high-velocity clouds**. In one process, clouds of gas ejected from supernovae are driven to large values of $z$, where they eventually cool and rain back down onto the Galactic plane. This suggestion is known as the **Galactic fountain model**. It also appears that the Galaxy is accreting gas from intergalactic space, as well as from a number of its small satellite galaxies; hence the predominance of negative-radial-velocity clouds.

In addition, a very hot, tenuous gas exists at distances up to or exceeding 70 kpc from the Galactic center. The **Far Ultraviolet Spectroscopic Explorer** (FUSE) has been able to detect O VI absorption lines in the spectra of distant extragalactic sources and halo stars produced when the light passes through the gas in the Galactic halo. Using O VI as a tracer of the hydrogen gas, the strengths of the O VI absorption lines imply a number density of hydrogen of $n_H \sim 10^{-11}$ m$^{-3}$. Assuming that the distribution is approximately spherical with a radius of $R \sim 70$ kpc leads to an estimate of the mass of the gas of $M \simeq 4 \times 10^8$ M$_\odot$. If the gas distribution is somewhat flattened, the total mass could exceed $10^9$ M$_\odot$. In order to support the gas against gravitational collapse, the gas temperature must be every hot. It is estimated that the temperature of this hot, tenuous gas is in excess of $10^6$ K, with the high gas temperature perhaps being due to collisions between infalling gas and existing gas in the Galaxy. The gas is sometimes referred to as a **coronal gas**, suggestive of its high temperature.

### The Disruption of Satellite Galaxies

Another unusual high-latitude feature is the **Magellanic Stream**, a narrow band of H I emission stretching more than 180° across the sky and trailing the Southern Hemisphere's Magellanic Clouds (recall that the Large and Small Magellanic Clouds are small satellite galaxies of the Milky Way; the LMC and SMC are located 52 and 61 kpc from Earth, respectively).[4] The Magellanic Stream appears to be the result of a tidal encounter of the Magellanic Clouds with the Milky Way some 200 Myr ago. It has also been suggested that the Magellanic Stream may owe some of its structure to the interaction with our Galaxy's hot coronal gas.

Other satellite galaxies have also tidally interacted with the Milky Way, and still others are currently doing so. For example, in 1995 Rodrigo Ibata and his colleagues announced the discovery of a previously unknown dwarf spheriodal galaxy in Sagittarius. At a distance of only 24 kpc from Earth and 16 kpc from the center of the Milky Way, the Sagittarius dwarf spheroidal is the closest galaxy to Earth. It is clearly elongated with the long axis directed toward the center of the Galaxy and, with a radial velocity of 140 km s$^{-1}$, has had only a few orbital encounters with the Milky Way. Evidently the Sagittarius dwarf spheroidal galaxy, along with its globular clusters, is being incorporated into the Milky Way Galaxy.

Using the **2-Micron All Sky Survey** (2MASS) catalog, researchers have also identified an overdensity of stars in the constellation Canis Major near the plane of the Milky Way. A group of globular clusters and open clusters are associated with this overdensity in both position and radial velocity. This feature strongly suggests that another dwarf satellite galaxy was integrated into the Milky Way in the past and may now be a part of the thick disk.

The unusual globular cluster $\omega$ Centauri also seems to be the remnant of a dwarf galaxy that has been subsumed by the Milky Way. $\omega$ Cen is the largest and brightest globular cluster visible from Earth and has an unusually high surface brightness. It appears that this globular cluster is the stripped core of another former satellite galaxy. (It has been suggested that the globular clusters M54 and NGC 2419 were also once dwarf galaxies that suffered the same fate as $\omega$ Cen.)

### The Galactic Bulge

Although the vertical scale height of the thin disk is near 350 pc in the vicinity of the Sun, that value increases somewhat toward the inner regions of the Galaxy, where the disk meets the **Galactic bulge**. The bulge is not simply an extension of the disk but is an independent

---

[4]The LMC contains 30 Dorodus and SN 1987A, both discussed in Chapter 15.

**FIGURE 24.11**   An infrared view of the Galaxy, as seen by COBE. The image was produced from observations at 1.2 $\mu$m, 2.2 $\mu$m, and 3.4 $\mu$m, and extends 96° either side of the Galactic center. (Courtesy of the COBE Science Working Group and NASA's Goddard Space Flight Center.)

component of the Galaxy. The mass of the bulge is believed to be roughly $10^{10}$ $M_\odot$ and its $B$-band luminosity is near $3 \times 10^9$ $L_\odot$. This gives a mass-to-light ratio of 3 $M_\odot/L_\odot$, a value comparable to that found for the thin disk.

The boxy (or elongated) bulge is evident in the **COsmic Background Explorer satellite** (COBE) image shown in Fig. 24.11. The image, which was produced by combining observations at 1.2 $\mu$m, 2.2 $\mu$m, and 3.4 $\mu$m, should be compared to the optical picture of NGC 891, shown in Fig. 24.9. Using the COBE data, together with observations of RR Lyraes and K and M giants, we find that the variation in the number density of stars in the bulge corresponds to a vertical scale height that ranges from 100 pc to 500 pc, depending on the ages of the stars used to make the determination; younger stars yield smaller scale heights.

The surface brightness, $I$, of the bulge (measured in units of $L_\odot$ pc$^{-2}$) exhibits an approximate radial dependence of the form

$$\log_{10}\left[\frac{I(r)}{I_e}\right] = -3.3307 \left[\left(\frac{r}{r_e}\right)^{1/4} - 1\right], \tag{24.13}$$

which is often referred to as a $r^{1/4}$ **law** [the law was first formulated by Gerard de Vaucouleurs (1918–1995) in 1948 and is also referred to as a **de Vaucouleurs profile**]. $r_e$ is a reference radius (called the **effective radius**) and $I_e$ is the surface brightness at $r_e$. Formally, $r_e$ is defined to be that radius within which one-half of the bulge's light is emitted. At the infrared wavelength of 12 $\mu$m, star count data from the **InfraRed Astronomical Satellite** (IRAS) suggests an effective radius of about 0.7 kpc. Similar results were found by COBE. (We will find that the de Vaucouleurs profile is a general distribution for relaxed, spheroidal systems.)

A serious difficulty in observing the properties of the bulge rests in the large amount of extinction at visible wavelengths due to the dust between the Sun and the Galactic center. The total amount of extinction within several degrees of the center can be more than 30 magnitudes. However, a number of lines of sight exist for which the amount of extinction is minimal. The most well known of these is **Baade's window**, which Walter Baade (1893–

1960) discovered in 1944 while observing the globular cluster NGC 6522. Baade realized that by observing in that region of the sky, he was able to see RR Lyraes that were actually beyond the Galactic center. Baade's window is 3.9° below the Galactic center, and the line of sight passes within 550 pc of the center. It is believed that NGC 6522 is located within the bulge.

From the observational evidence, the chemical abundances of stars in the bulge vary significantly, ranging from quite metal-poor to very metal-rich; $-2 < $ [Fe/H] $< 0.5$. In fact, based on the chemical abundances, it appears that three somewhat distinct age groupings exist in the central bulge. One set of stars appear to be very young, with ages less than 200 Myr, a second set has ages between 200 Myr and 7 Gyr, and the third set tends to be older than 7 Gyr (perhaps up to 10 Gyr or older).

In a trend that on initial reflection appears to be counterintuitive, the oldest stars in the bulge tend to have the highest metallicities, while the youngest stars have a fairly uniform distribution of metallicities across the range from $-2$ to $0.5$. This pattern is probably due to a burst of massive star formation when the Galaxy was young. Apparently core-collapse supernovae enriched the interstellar medium early in the life of the bulge, implying that subsequent generations of stars contained an enhanced abundance of heavier elements. The more uniform distribution of metallicity in recent generations of stars could be the result of fresh, infalling material. It is this type of complication that must be considered carefully when employing the age–metallicity relation discussed on page 885.

### The Milky Way's Central Bar

Although it was originally thought to be essentially spheroidal in nature, a number of observing campaigns and database studies have determined that the budge contains a distinct **bar**.[5] The Milky Way's central bar is clearly depicted in the artist's drawing of the Galaxy shown in Fig. 24.5. The bar has a radius (one-half length) from the galactic center of $4.4 \pm 0.5$ kpc and is oriented at an angle of $\phi = 44° \pm 10°$ with respect to the line-of-sight angle from Earth to the Galactic center. It also seems that the bar is somewhat thicker in the plane of the Galaxy than in the $z$ direction; the dimension ratios are approximately 1:0.5:0.4. (As we will learn in Chapter 25, bars are dynamically stable structures and common features of many spiral galaxies.)

### The 3-kpc Expanding Arm

A unique feature in the inner regions of the Galaxy that is most easily observed at the 21-cm wavelength of H I is the **3-kpc expanding arm**, a gas cloud that is moving toward us at roughly 50 km s$^{-1}$. Once believed to be the product of a gigantic explosion in the center of the Galaxy, the rapidly moving structure is now thought to be a consequence of the presence of the stellar bar. Rather than being driven away from the center in an explosive event that would require an unrealistic $10^{52}$ J of energy, the gas cloud is merely in a very elliptical orbit about the Galactic center resulting from gravitational perturbations from the bar.

---

[5]One such database study is the analysis of the GLIMPSE (Galactic Legacy Mid-Plane Survey Extraordinaire) point source catalog of some 30 million infrared sources in the direction of the inner Galaxy, produced by the Spitzer Space Telescope. The Galactic bulge is much more transparent to infrared wavelengths than to visible wavelengths.

### The Stellar Halo and Globular Cluster System

The last luminous component of the Galaxy is the **stellar halo** (or simply the halo), composed of the globular clusters and those **field stars** (stars that are not members of clusters) that have large velocity components perpendicular to the Galactic plane. These field stars are often referred to as **high-velocity stars** since, as we will learn in the next section, their velocity components differ significantly from those of the Sun. Most of the globular clusters and the high-velocity stars can reach positions that are far above or below the plane of the Galaxy.

Although it appeared to Shapley that all of the known globular clusters were distributed nearly spherically about the Galactic center, it has become apparent that two distinct spatial distributions exist, delineated by metallicity. Older, metal-poor clusters whose members have [Fe/H] < −0.8 belong to an extended spherical halo of stars, while younger clusters with [Fe/H] > −0.8 form a much flatter distribution and may even be associated with the thick disk. The spatial distribution of the two metallicity groups is shown in Fig. 24.12. The notable exception is the well-studied globular cluster 47 Tucana (47 Tuc; also known as NGC 104), which is located 3.2 kpc below the Galactic plane and has an unusually high metallicity of [Fe/H] = −0.67. Some astronomers have argued that 47 Tuc is a member of the halo population, while others consider 47 Tuc to be a member of the thick disk. A color–magnitude diagram of 47 Tuc was presented in Fig. 13.20(a).

Our Galaxy is known to contain at least 150 globular clusters with distances from the center of the Milky Way ranging from 500 pc to 120 kpc. The youngest globular clusters appear to be about 11 Gyr old, and the oldest are probably a little over 13 Gyr old.[6] It now appears that a significant age spread of two billion years or so exists between the youngest and oldest members of the halo.

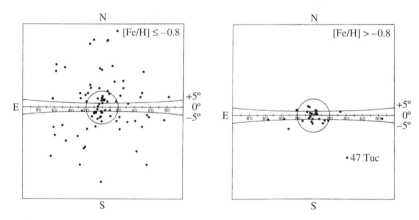

**FIGURE 24.12**    Metal-poor globular clusters form a nearly spherical distribution about the Galactic center, while more metal-rich clusters are found preferentially near the plane of the Galaxy, possibly associated with the thick disk. (Figure adapted from Zinn, *Ap. J.*, *293*, 424, 1985.)

[6]Krauss and Chaboyer, *Science*, *299*, 65, 2003, find a best-fit age for the globular clusters of 12.6 Gyr, with lower and upper limits of 10.4 Gyr and 16 Gyr, respectively, at 95% confidence levels. Gratton et al., *Aston. Astrophys.*, *408*, 529, 2003, find ages for NGC 6397, NGC 6752, and 47 Tuc of 13.9 ± 1.1 Gyr, 13.9 ± 1.1 Gyr, and 11.2 ± 1.1 Gyr, respectively. For comparison, the age of the universe is 13.7 ± 0.2 Gyr.

Although 144 of the globular clusters are found within 42 kpc of the Galactic center, 6 globular clusters have been found between 69 and 123 kpc of the Galactic center. Some astronomers have suggested that those six most distant clusters may have been captured by the Milky Way or may be dwarf spheroidal galaxies, much as $\omega$ Cen and the Sagittarius dwarf galaxy may have been. If we do not include these very remote objects, the metal-poor clusters seem to be confined to a halo with a radius of approximately 42 kpc. However, the detection of extremely distant and luminous field stars suggests that a stellar halo radius of 50 kpc may be more appropriate.

The number density profile of the metal-poor globular clusters and the field stars in the halo has the form

$$n_{\text{halo}}(r) = n_{0,\text{halo}}(r/a)^{-3.5}, \tag{24.14}$$

where $n_{0,\text{halo}} \simeq 4 \times 10^{-5}$ pc$^{-3}$ is roughly 0.2% of the thin disk's midplane value (see Eq. 24.9). The scale length ($a$) of the number density distribution is several thousand parsecs. At visible wavelengths, the effective radius of the halo $r^{1/4}$ law (Eq. 24.13) is $r_e = 2.7$ kpc.

Metal-rich globular clusters seem to have many characteristics in common with thick-disk field stars, including spatial distribution. Just like the thick-disk field stars, the vertical scale height of these clusters is on the order of 1 kpc, unlike the much larger scale length of the metal-poor clusters.

However, based on studies that use RR Lyraes as tracers of other field stars in the halo, there is some question about whether field stars actually share the same spatial distribution pattern as the corresponding metal-poor clusters. Instead, the field stars may occupy a volume that is appreciably flattened, perhaps as much as $c/a \sim 0.6$, where $c$ is the minor axis of a spheroid in the direction perpendicular to the Galactic plane, and $a$ is the major axis (see Fig. 24.13). Such a discrepancy in distributions, if real, could have a significant influence on an ultimate model of Galaxy formation and evolution. On the other hand, other

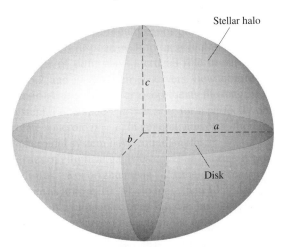

**FIGURE 24.13**   The general shape of a triaxial spheroidal system, where the three axes of the spheroid are assumed to have $a \geq b \geq c$. The Galactic disk (shown foreshortened by perspective) is roughly circular.

astronomers believe that a single value of $0.8 \leq c/a \leq 0.9$ probably applies to the field stars and clusters alike.

Based largely on star counts, the total *stellar* mass density in the solar neighborhood is approximately $0.05 \ M_\odot \ pc^{-3}$. Of that total, the contribution due to high-velocity halo stars is on the order of 0.2%, or $1 \times 10^{-4} \ M_\odot \ pc^{-3}$. Combining this with the spatial density power law yields a total estimated mass of the stellar halo on the order of $1 \times 10^9 \ M_\odot$, of which roughly 1% is the combined mass of the globular clusters; the remainder is locked up in field stars. The $B$-band luminosity of the halo is estimated to be $1 \times 10^9 \ L_\odot$, giving a mass-to-light ratio on the order of unity (the exact value is quite uncertain).

When all of the components of the Galaxy are considered together, its total $B$-band luminosity is $L_{B,\text{tot}} = 2.3 \pm 0.6 \times 10^{10} \ L_\odot$. However, approximately one-third of the Galaxy's emission is in the infrared part of the electromagnetic spectrum and is primarily associated with interstellar dust. Including IR emissions, the bolometric luminosity of the Galaxy is $L_{\text{bol}} = 3.6 \times 10^{10} \ L_\odot$.

### The Dark Matter Halo

When the masses of all of the luminous components of the Galaxy are combined (the thin and thick disks, the interstellar dust and gas, the Galactic bulge, the stellar halo, and the bar), the total mass of luminous matter in the Galaxy is estimated to be $9 \times 10^{10} \ M_\odot$. As we shall see in the next section, this value is in good agreement with the orbital motion of the Sun about the Galactic center, but it does not explain the orbits of stars and gas at Galactocentric distances much greater than $R_0$. Apparently there is still another crucial component to the overall structure of our Galaxy. Along with influencing orbital motions, this unseen element of Galactic structure may also be responsible for the generation of the warps seen in the H I distribution near the outer edges of the luminous disk.

This **dark matter halo** seems to be roughly spherically distributed, enveloping the stellar halo and extending out to at least 230 kpc. Based on its gravitational influence over luminous matter, the dark matter halo has an apparent mass distribution of the form

$$\rho(r) = \frac{\rho_0}{(r/a)\,(1 + r/a)^2}, \tag{24.15}$$

where $\rho_0$ and $a$ are chosen by fitting the mass distribution in the dark matter halo (estimates of $\rho_0$ and $a$ will be made in Problem 24.24). This functional dependence behaves as $1/r$ when $r \ll a$ and as $1/r^3$ when $r \gg a$. The mass of the dark matter halo may be as great as $5.4 \times 10^{11} \ M_\odot$ within 50 kpc of the Galactic center and $1.9 \times 10^{12} \ M_\odot$ within 230 kpc of the Galactic center. It appears that the dark matter halo accounts for about 95% of the entire mass of the Galaxy. [The empirical justification for a dark matter density distribution such as Eq. (24.15) will be discussed beginning on page 914.]

The composition of the dark matter halo is still a mystery. It cannot be in the form of interstellar dust, because dust betrays its presence through the extinction of starlight. Furthermore, the dark matter halo cannot be composed of gas, because absorption lines would be apparent when observing halo stars. One possible class of candidates are referred to as **weakly interacting massive particles** (WIMPs). WIMPs would not contribute to the overall luminosity of the Galaxy, but they would affect it through their gravitational

**FIGURE 24.14** Gravitational lensing (focusing) of starlight produced by an intervening MACHO.

interactions. In support of a WIMP-dominated Galaxy, theoretical considerations related to the formation and subsequent evolution of the universe suggest that nonbaryonic matter (constituents of matter other than protons, neutrons, and related massive particles) may constitute the majority of the dark matter halo's mass, an issue that will be discussed in more detail in Chapters 29 and 30. It had been suggested that neutrinos (which are leptons) fill the bill, although it now appears that they aren't massive enough to explain the amount of mass that must be present in the dark matter halo. Perhaps currently hypothetical (and undetected) particles are responsible, such as the suggested supersymmetric particles known as **neutralinos** that are implied in some grand unified theories of particle physics.

A competing hypothesis for the possible makeup of dark matter suggests that **massive compact halo objects** (MACHOs) may be responsible. MACHOs that could supply the unseen mass may be in the form of white dwarfs, neutron stars, black holes, or less exotic red dwarfs or brown dwarfs. Some searches for MACHOs have been based on the general-relativistic prediction that starlight is deflected as it passes near a massive object (Section 17.1). If a MACHO is located between a distant star and Earth, the light from the star can be focused with the MACHO acting as a **gravitational lens**, as shown in Fig. 24.14. In 1993, while observing stars in the Large Magellanic Cloud, astronomers detected the telltale signature of temporary brightening that is predicted to occur when a MACHO passes through the line of sight. Figure 24.15 shows the light curves of one event that lasted for 33 days. From a statistical analysis of the small number of such events, it appears that there are too few MACHOs to account for a significant fraction of the mass contained in the dark matter halo.

That conclusion is also supported by searches for white dwarfs and small red dwarf stars, carried out using the Hubble Space Telescope. In those studies, the number of stars detected is too small to constitute a major component of the dark matter in the Galaxy. Based on the faintest (deepest) searches yet conducted at the time of writing, white dwarfs can contribute no more than 10% of the dark matter halo's mass, and red dwarfs no more than 6% of it.

### The Galactic Magnetic Field

Like many other astrophysical environments, the Galaxy possesses a magnetic field. The orientation and strength of the field can be measured in a variety of ways, including the Zeeman effect and the polarization of visible- and radio-wavelength electromagnetic radiation by reflection from interstellar grains aligned with the field. It appears that within the disk, the field tends to follow the Galaxy's spiral arms and has a typical strength of 0.4 nT. The field strength in the stellar halo may be an order of magnitude weaker than in the disk, and the field strength near the Galactic center may reach 1 $\mu$T (see Section 24.4).

Although the global Galactic magnetic field is quite weak relative to the magnetic field near the surface of Earth (about 50 $\mu$T), it probably plays a significant role in the structure

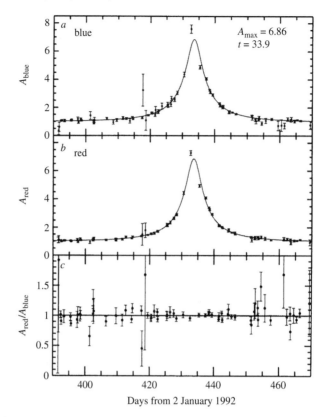

**FIGURE 24.15**    The light curve of a star in the LMC brightened over a period of 33 days, apparently because a MACHO passed through the line of sight. The data are shown for (a) blue light, (b) red light, and (c) the ratio of blue light to red light. (Figure adapted from Alcock et al., *Nature*, *365*, 621, 1993. Reprinted with permission, © 1993, Macmillan Magazines Limited.)

and evolution of the Milky Way. This can be seen by considering its energy density, which appears to be comparable (perhaps equal) to the thermal energy density of gas within the disk (see Problem 24.9).

## 24.3 ■ THE KINEMATICS OF THE MILKY WAY

One of the great challenges in understanding the nature of our Galaxy lies in determining its internal kinematics. In this section, we will discuss how astronomers have been able to uncover the complexities of motions in the Milky Way from our vantage point in orbit around a moving Sun.

### The Galactic Coordinate System

The Galactic midplane is not aligned with the plane of the celestial equator but is inclined at an angle of $62.87°$ to it. As a result, rather than using the Earth-based equatorial coordinate system, it is more convenient to introduce a new coordinate system when discussing the structure and kinematics of the Milky Way Galaxy. The **Galactic coordinate system** exploits the natural symmetry introduced by the existence of the Galactic disk.

The intersection of the midplane of the Galaxy with the celestial sphere forms what is very nearly a great circle, known as the **Galactic equator**.[7] This orientation is depicted in Fig. 24.16. **Galactic latitude** ($b$) and **Galactic longitude** ($\ell$) are defined from a vantage point taken to be the Sun, as shown in Fig. 24.17. Galactic latitude is measured in degrees

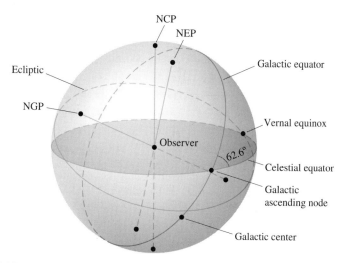

**FIGURE 24.16** The relative orientations of the celestial equator, the ecliptic, and the Galactic equator as they appear on the celestial sphere. Note that from the vantage point depicted in the figure, the north celestial pole (NCP), the north ecliptic pole (NEP), and north Galactic pole (NGP) are all on the front of the celestial sphere.

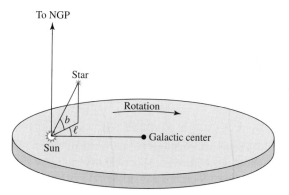

**FIGURE 24.17** The definition of the Galactic coordinates $\ell$ and $b$. The direction of rotation of the Galaxy is also labeled.

---

[7]Technically, the Galactic midplane does not exactly trace out a great circle, because the Sun is not precisely in the midplane; however, the deviation is very small. Recall from Section 24.2 that the Sun is located just 30 pc above the plane.

north or south of the Galactic equator along a great circle that passes through the **north Galactic pole**. Galactic longitude (also in degrees) is measured east along the Galactic equator, beginning *near* the Galactic center, to the point of intersection with the great circle used to measure Galactic latitude.

By international convention, the J2000.0 equatorial coordinates of the north Galactic pole ($b = 90°$) are

$$\alpha_{NGP} = 12^h51^m26.28^s$$

$$\delta_{NGP} = 27°7'41.7'',$$

and the origin of the Galactic coordinate system ($\ell_0 = 0°$, $b_0 = 0°$) corresponds to

$$\alpha_0 = \quad 17^h45^m37.20^s$$

$$\delta_0 = -28°56'9.6''.$$

Note that the center of the Galaxy ($\alpha_{Sgr\ A^*}$, $\delta_{Sgr\ A^*}$) given by Eqs. (24.6) and (24.7) is very close to, but not exactly aligned with, ($\ell_0 = 0°$, $b_0 = 0°$).

Two other useful positions on the sky are also worth specifying in both coordinate systems. The location of the north celestial pole ($\delta_{NCP} = 90°$), given in J2000.0 Galactic coordinates, is

$$\ell_{NCP} = 123°55'55.2''$$

$$b_{NCP} = \quad 27°7'41.7''.$$

And the intersection of the celestial equator with the Galactic equator moving eastward from negative to positive declination (the *ascending node*) is given in equatorial coordinates by

$$\alpha_{ascending} = 18^h51^m24^s$$

$$\delta_{ascending} = 0°$$

and in Galactic coordinates by

$$\ell_{ascending} = 33.0°$$

$$b_{ascending} = 0°.$$

The transformations between equatorial and Galactic coordinates involves the application of the methods of spherical trigonometry. To make the transformation from equatorial to Galactic coordinates (assuming epoch J2000.0):

$$\sin b = \sin \delta_{NGP} \sin \delta + \cos \delta_{NGP} \cos \delta \cos(\alpha - \alpha_{NGP}) \quad (24.16)$$

$$\cos b \sin(\ell_{NCP} - \ell) = \cos \delta \sin(\alpha - \alpha_{NGP}) \quad (24.17)$$

$$\cos b \cos(\ell_{NCP} - \ell) = \cos \delta_{NGP} \sin \delta - \sin \delta_{NGP} \cos \delta \cos(\alpha - \alpha_{NGP}). \quad (24.18)$$

For transformations from Galactic to equatorial coordinates (again assuming epoch J2000.0):

$$\sin \delta = \sin \delta_{\mathrm{NGP}} \sin b + \cos \delta_{\mathrm{NGP}} \cos b \cos(\ell_{\mathrm{NCP}} - \ell) \tag{24.19}$$

$$\cos \delta \sin(\alpha - \alpha_{\mathrm{NGP}}) = \cos b \sin(\ell_{\mathrm{NCP}} - \ell) \tag{24.20}$$

$$\cos \delta \cos(\alpha - \alpha_{\mathrm{NGP}}) = \cos \delta_{\mathrm{NGP}} \sin b - \sin \delta_{\mathrm{NGP}} \cos b \cos(\ell_{\mathrm{NGP}} - \ell). \tag{24.21}$$

Note that care must be taken in inverting these expressions to solve for $\alpha$ or $\ell$, because computers and electronic calculators may return the wrong (default) quadrant when the calculation of an inverse trigonometric function is performed. A graphical form of the transformations that does not suffer from this multiplicity is presented in Fig. 24.18.

### A Cylindrical Coordinate System for Galactic Motions

The motions of stars in the solar neighborhood allow us to glean important clues regarding the large-scale structure of the Galaxy. Although the Galactic coordinate system is useful for representing the locations of objects within the Galaxy as seen from Earth, it is not the most convenient choice for studying kinematics and dynamics. One reason is that the Sun, which is the origin of the Galactic system, is itself moving about the center of the Galaxy. In addition, a coordinate system centered on the Sun constitutes a noninertial reference frame with respect to Galactic motions.

Therefore, to complement the Galactic coordinate system, a cylindrical coordinate system is used that places the center of the Galaxy at the origin. In this system the radial coordinate $R$ increases outward, the angular coordinate $\theta$ is pointed in the direction of rotation of the Galaxy, and the vertical coordinate $z$ increases to the north (see Fig. 24.19). Furthermore, the corresponding velocity components are traditionally labeled

$$\Pi \equiv \frac{dR}{dt}, \qquad \Theta \equiv R\frac{d\theta}{dt}, \qquad Z \equiv \frac{dz}{dt}. \tag{24.22}$$

It is worth noting that this set of directional choices results in a left-handed coordinate system instead of a more conventional right-handed one. This occurs because, when viewed from the north Galactic pole, the Galaxy rotates clockwise, rather than counterclockwise. (Fortunately, we are not going to be concerned with cross products in this analysis. Recall that cross products are right-handed by definition.)

### Peculiar Motions and the Local Standard of Rest

Since all of our observations are made from Earth (or at least relatively near it, in the case of satellites), and because we can transform those observations to Sun-centered ones by removing any effects resulting from the rotational and orbital motions of Earth or motions of the spacecraft, we will consider the Sun as the site of all observations of the Galaxy. In particular, since the Earth–Sun distance is very small when compared to any distances on a Galactic scale, we need not concern ourselves with the change in position, only with changes in velocity.

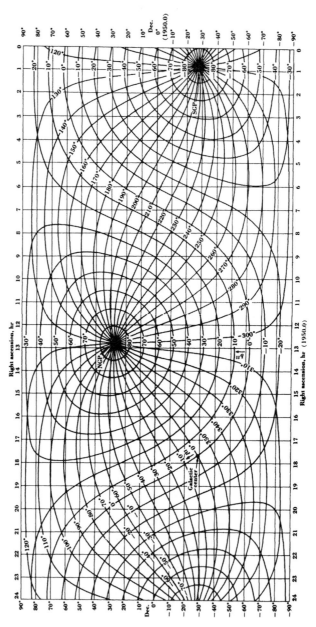

**FIGURE 24.18**   A chart for converting between equatorial and Galactic coordinates. (From Kraus, *Radio Astronomy*, Second Edition, Cygnus-Quasar Books, 1986, used by permission.)

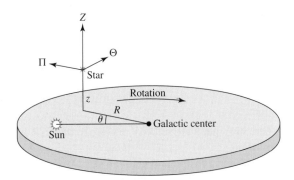

**FIGURE 24.19** The cylindrical coordinate system used to analyze Galactic kinematics. (Note that some authors define the radial coordinate to be positive toward the Galactic center. It should also be noted that the coordinate system defined here is *not* a right-handed coordinate system.)

Before we can hope to make a final transformation from the Sun to the center of the Galaxy, however, we must also understand the Sun's motion. In fact, the Sun does not follow a simple planar orbit or even a closed nonplanar orbit; instead, as it moves about the Galactic center, it is also currently moving slowly inward (in the negative-$R$ direction) and farther north, away from the midplane (positive-$z$ direction).

To investigate the motion of the Sun and the other stars in the solar neighborhood, we will first define the **dynamical local standard of rest** (dynamical LSR) to be a point that is *instantaneously* centered on the Sun and moving in a *perfectly circular orbit* along the solar circle about the Galactic center. An alternative definition for the LSR known as the **kinematic local standard of rest** is based on the average motions of stars in the solar neighborhood. Although with the right choice of reference stars, dynamical and kinematic LSRs agree quite well, it can be shown that the kinematic LSR systematically lags behind the dynamical LSR. Throughout the remainder of this text, "the LSR" will always refer to the dynamical LSR.

The velocity components of the LSR must be

$$\Pi_{LSR} \equiv 0, \qquad \Theta_{LSR} \equiv \Theta_0, \qquad Z_{LSR} \equiv 0,$$

where $\Theta_0 \equiv \Theta(R_0)$ and $R_0$ is the solar Galactocentric distance. Note that once the LSR is chosen, the Sun immediately begins to drift away from it, implying that we would effectively need to redefine the reference point constantly. In reality this is not a significant problem because (fortunately) the 230-Myr orbital period of the LSR is very long compared to the time since modern telescopic observations began (and even longer compared to the lifetime of a typical research grant); consequently, there has not been sufficient time for the effect to become noticeable.

The velocity of a star *relative to the LSR* is known as the star's **peculiar velocity** and is given by

$$\mathbf{V} = (V_R, V_\theta, V_z) \equiv (u, v, w), \tag{24.23}$$

where

$$u = \Pi - \Pi_{\text{LSR}} = \Pi, \tag{24.24}$$

$$v = \Theta - \Theta_{\text{LSR}} = \Theta - \Theta_0, \tag{24.25}$$

$$w = Z - Z_{\text{LSR}} = Z. \tag{24.26}$$

The Sun's peculiar velocity relative to the LSR is generally referred to simply as the **solar motion**.

The average of $u$ and $w$ for all stars in the solar neighborhood, excluding the Sun, should be nearly zero if we assume an axisymmetric Galaxy. The reason is that with symmetries about both the rotation axis and the midplane, there should be as many stars moving inward as outward, and there should be as many stars moving toward the north Galactic pole as toward the south Galactic pole. In reality this is not quite true because the Galaxy is not precisely axisymmetric, but the error is not significant for our purposes here. As a result, we shall assume that, summing over a sample of $N$ nearby stars,

$$\langle u \rangle = \frac{1}{N} \sum_{i=1}^{N} u_i \simeq 0, \tag{24.27}$$

$$\langle w \rangle = \frac{1}{N} \sum_{i=1}^{N} w_i \simeq 0. \tag{24.28}$$

The same assumption cannot be made for the $v$ component, however. To see why, consider the situations depicted in Fig. 24.20. Stars with different average orbital radii must follow paths that bring them very close to the LSR if they are to be considered members of the solar neighborhood and eligible for inclusion in the calculation of $\langle v \rangle$. If we consider the special case that $u = 0$ and $w = 0$ for all stars in our sample, then the stars must be at either their most distant point from the Galactic center (**apogalacticon**) or their closest approach to the Galactic center (**perigalacticon**) when they coincide with the LSR; this is true of $A$ and $B$, respectively. Then, for the two stars to follow their specified orbits, it is necessary

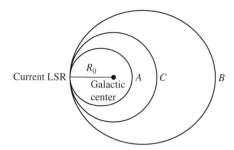

**FIGURE 24.20**   The orbits of three hypothetical stars intersecting at the LSR. $A$ and $B$ represent stars in elliptical orbits with semimajor axes $a_A < R_0$ and $a_B > R_0$, respectively. $C$ represents a star following the perfectly circular path of the LSR.

that $\Theta_A (R_0) < \Theta_0$ and $\Theta_B (R_0) > \Theta_0$. This implies that $v_A < 0$ and $v_B > 0$. Finally, since more stars reside inside the solar Galactocentric distance than beyond it (see, for example, Eq. 24.9),

$$\langle v \rangle < 0. \tag{24.29}$$

(This explains why the kinematic LSR systematically lags behind the dynamical LSR, as mentioned previously.)

The velocity that is measured for a particular star relative to the Sun is just the difference between the star's peculiar velocity and the solar motion with respect to the LSR, or

$$\Delta u \equiv u - u_\odot, \qquad \Delta v \equiv v - v_\odot, \qquad \Delta w \equiv w - w_\odot.$$

Using the average values of the stellar peculiar velocity components given by Eqs. (24.27), (24.28), and (24.29), and solving for the solar motion, we have

$$u_\odot = -\langle \Delta u \rangle, \tag{24.30}$$

$$v_\odot = \langle v \rangle - \langle \Delta v \rangle, \tag{24.31}$$

$$w_\odot = -\langle \Delta w \rangle. \tag{24.32}$$

The $u$ and $w$ components of the solar motion simply reflect the averaged *relative* velocities of the other stars with respect to the Sun in the $R$ and $z$ directions. Qualitatively, these stars appear to be "streaming" past the Sun as it moves through space (see Problem 24.13).

To find the $v$ component of the solar motion, we must first determine the average value of $v$ for stars in the solar neighborhood. Unfortunately, this requires a theory of Galactic motion that is beyond the scope of the present argument. However, qualitatively the procedure involves deriving an analytical expression for $\langle v \rangle$ in terms of the radial variation in the number density of stars in the solar neighborhood. The justification for this relationship lies in the argument made concerning the lag of stellar motions behind the LSR due to the increase in number density with decreasing Galactocentric distance. The result is an equation of the form

$$\langle v \rangle = C \sigma_u^2,$$

where $C$ is a constant and

$$\sigma_u \equiv \langle u^2 \rangle^{1/2}$$

measures the spread in the $R$ components of the peculiar velocities of stars in the solar neighborhood with respect to the LSR; of course, $\sigma_u^2 = \langle \Pi^2 \rangle$.

Note that $\sigma_u$ is related to the standard deviation of the velocity distribution, defined as

$$\text{standard deviation} \equiv \frac{1}{\sqrt{N}} \left[ \sum_{i=1}^{N} (u - \langle u \rangle)^2 \right]^{1/2}.$$

In the special case that $\langle u \rangle = 0$, then $\sigma_u = \langle u^2 \rangle^{1/2}$ is identical to the standard deviation. $\sigma_u$ is known as the **velocity dispersion** in $u$.

Turning again to Fig. 24.20, we can see why $\sigma_u$ should correlate with $\langle v \rangle$. A stellar sample that produces a larger dispersion in $u$ means that a wider range of elliptical orbits are included. This results in a more negative average value of $v$ for the sample because of the larger fraction of the stellar population with $R < R_0$; there are more stars in the sample with $v < 0$ than with $v > 0$. Alternatively, as $\sigma_u^2$ decreases, fewer stars will have orbits that are appreciably noncircular and $\langle v \rangle$ will approach zero. From Eq. (24.31),

$$\langle \Delta v \rangle = C\sigma_u^2 - v_\odot$$

and $-v_\odot$ is simply the ordinate intercept on a graph of $\langle \Delta v \rangle$ versus $\sigma_u^2$; see Problem 24.14. $C$ is the slope of the linear relationship.

The components of the Sun's peculiar velocity are

$$u_\odot = -10.0 \pm 0.4 \text{ km s}^{-1}, \tag{24.33}$$

$$v_\odot = \phantom{-}5.2 \pm 0.6 \text{ km s}^{-1}, \tag{24.34}$$

$$w_\odot = \phantom{-}7.2 \pm 0.4 \text{ km s}^{-1}, \tag{24.35}$$

so that relative to the LSR, the Sun is moving (a) toward the Galactic center, (b) more rapidly in the direction of Galactic rotation, and (c) north out of the Galactic plane. Overall, the solar motion is approximately 13.4 km s$^{-1}$ toward a point in the constellation of Hercules. The point toward which the Sun is moving is called the **solar apex**; the point away from which the Sun is retreating is the **solar antapex** (located in Columba). It is important to note, however, that the exact value for the solar motion and the location of the solar apex depend on the choice of reference stars.

Now that the solar motion is known, the velocities of stars relative to the Sun can be transformed into peculiar motions relative to the LSR. It is then possible to plot one component of peculiar motion against another for a specified sample of stars in the solar neighborhood in order to obtain important information about their kinematics. Such plots result in patterns known as **velocity ellipsoids**. As depicted in a diagram of $u$ versus $v$ (Fig. 24.21), when young, metal-rich main-sequence A stars are used, the range in velocities about the LSR is fairly limited (a small dispersion), for older K giants a wider variation in both $u$ and $v$ is observed, and when old, metal-poor red dwarfs are plotted, the spread is even larger (a large dispersion). The same general behavior is seen in plots of $w$ versus $v$, whereas a much more symmetric diagram results when $w$ is plotted against $u$.

Specific features of Fig. 24.21 are worth discussing in more detail. First is the very noticeable relationship between metallicity and velocity dispersion, called a **velocity–metallicity relation**. When the velocity–metallicity and age–metallicity relations are combined, the velocity ellipsoids suggest that the oldest stars in the Galaxy have the widest range of peculiar velocities, a trend that is evident in all three coordinates. Because stars with the smallest peculiar velocities do not drift away from the LSR as quickly, they must occupy orbits that are similar to that of the LSR, implying that these young stars are members of the thin disk. On the other hand, the stars with the largest peculiar velocities follow very different paths about the center of the Galaxy. In particular, stars with large $|w|$ must be passing through

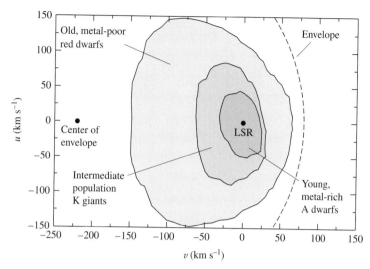

**FIGURE 24.21** A schematic diagram of the peculiar velocity components $v$ and $u$ for stars in the solar neighborhood. The innermost contour represents metal-rich main-sequence A stars, the middle contour depicts older K giants, and the outer contour indicates very metal-poor red dwarfs. The LSR is located at $(v, u) = (0, 0)$. An enveloping circle with an approximate radius of 300 km s$^{-1}$ and centered at $v = -220$ km s$^{-1}$ reveals the orbital velocity of the LSR.

the solar neighborhood on trajectories that will carry them to great distances above and below the disk. These old, metal-poor stars are the high-velocity stars that were mentioned in the last section as being members of the stellar halo.

A second common feature of peculiar velocity diagrams is the clear asymmetries in the velocity ellipsoids along the $v$-axis as a function of metallicity or age (this effect is known as **asymmetric drift**). Few stars are observed with $v > +65$ km s$^{-1}$, but there are metal-poor RR Lyraes and subdwarfs with $v < -250$ km s$^{-1}$. In fact, as suggested by the dashed line in Fig. 24.21, a nearly circular "envelope" with a radius of roughly 300 km s$^{-1}$ can be drawn around the high-velocity stars. The center of the velocity envelope appears to be near $v = -220$ km s$^{-1}$ for both $u$–$v$ and $w$–$v$ diagrams. If on average the stellar halo is rotating very slowly (if at all), then the orbital velocity of the LSR should reveal itself as a point of symmetry along the $v$-axis. This is because halo stars with $\Theta \simeq 0$ (no velocity component in the direction of Galactic rotation) should exhibit peculiar $v$ velocities that simply reflect the motion of the LSR (i.e., $v \simeq -\Theta_0$). Stars that have orbital components that are in the opposite sense from the overall Galactic rotation direction have $v < -\Theta_0$. By this argument it appears that the orbital speed of the LSR is

$$\Theta_0 (R_0) = 220 \text{ km s}^{-1}, \tag{24.36}$$

the presently accepted IAU standard. This value has also been measured using groups of external galaxies for reference.[8]

---

[8]Kuijken and Tremaine (1994) have suggested that the IAU value of $\Theta_0$ may be too large. Based on a set of self-consistent solutions to various Galactic parameters, they argue for $\Theta_0 = 180$ km s$^{-1}$.

**Example 24.3.1.**   An estimate of the mass of the Milky Way Galaxy interior to the solar Galactocentric distance can be made using Kepler's third law, together with $R_0$ and $\Theta_0$. Using $R_0 = 8$ kpc and $\Theta_0 = 220$ km s$^{-1}$, the orbital period of the LSR is

$$P_{\text{LSR}} = \frac{2\pi R_0}{\Theta_0} = 230 \text{ Myr.}$$

Assuming that the mass of the Galaxy within the solar circle is much greater than the mass of a test particle orbiting along with the LSR, and that the bulk of the Galaxy's mass is distributed spherically symmetrically, Kepler's third law (Eq. 2.37) gives

$$M_{\text{LSR}} = \frac{4\pi^2 R_0^3}{G P_{\text{LSR}}^2} = 8.8 \times 10^{10} \text{ M}_\odot.$$

This value compares well with the estimate of the mass of luminous matter quoted on page 896, but it is much less than the total mass estimate of the Galaxy when the dark matter halo is included.

In 1927, Jan Oort (1900–1992) proposed that since no stars had been observed with $v > +65$ km s$^{-1}$, the escape velocity of the Galaxy must be $\Theta_0 + 65$ km s$^{-1} \sim 300$ km s$^{-1}$ relative to the Galactic center. In fact, we know today that a small number of extremely high-velocity stars do exist in the solar neighborhood with speeds of $\sim 500$ km s$^{-1}$ relative to the center of the Galaxy. Since these stars have not escaped from the Galaxy, it seems that the strong asymmetry near $v \sim +65$ km s$^{-1}$ simply points to a deficiency in very high-velocity stars; see Problem 24.15.

### Differential Galactic Rotation and Oort's Constants

In 1927 Jan Oort also derived a series of relations that have become the framework with which astronomers have attempted to determine the differential rotation curve of the Galactic disk. To simplify the discussion, we will assume that all motions are circular about the center of the Galaxy.

Consider the situation depicted in Fig. 24.22. Assume that the Sun (at $O$) and a star or some other object (at $S$) are orbiting the Galactic center (at $C$) in the Galactic midplane. The velocity vector that is measured between the Sun and the star at point $S$ is the *relative velocity* between the two objects. Therefore, in order to compare the observed-velocity vector to the object's true velocity with respect to the Galactic center, it is necessary to consider the difference between the star's motion and the Sun's motion. Of course, in practice it is not the relative space motion that is measured directly but the radial velocity and the proper motion, with the proper motion being converted into the transverse velocity if the distance $d$ to the star is known.

If the line of sight is in the direction of the Galactic longitude $\ell$, and if $\Theta(R)$ is the orbital velocity curve as a function of distance from the Galactic center, then the relative radial and transverse velocities of the star are, respectively,

$$v_r = \Theta \cos \alpha - \Theta_0 \sin \ell,$$

$$v_t = \Theta \sin \alpha - \Theta_0 \cos \ell,$$

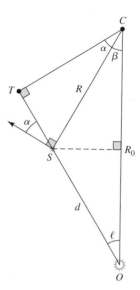

**FIGURE 24.22** The geometry of analyzing differential rotation in the Galactic plane. The Sun is at point $O$, the center of the Galaxy is located at $C$, and the star is at $S$, located a distance $d$ from the Sun. $\ell$ is the Galactic longitude of the star at $S$, and $\alpha$ and $\beta$ are auxiliary angles. The directions of motion reflect the clockwise rotation of the Galaxy as viewed from the NGP.

where $\Theta_0$ is the orbital velocity of the Sun in the idealized case of perfectly circular motion (actually the orbital velocity of the LSR) and $\alpha$ is defined in the figure. Defining the *angular-velocity curve* to be

$$\Omega(R) \equiv \frac{\Theta(R)}{R},$$

the relative radial and transverse velocities become

$$v_r = \Omega R \cos\alpha - \Omega_0 R_0 \sin\ell,$$

$$v_t = \Omega R \sin\alpha - \Omega_0 R_0 \cos\ell.$$

Now, by referring to the geometry of Fig. 24.22 and considering the right triangle $\triangle OTC$, we find

$$R \cos\alpha = R_0 \sin\ell,$$

$$R \sin\alpha = R_0 \cos\ell - d.$$

Substituting these relations into the previous expressions, we have

$$v_r = (\Omega - \Omega_0) R_0 \sin\ell, \tag{24.37}$$

$$v_t = (\Omega - \Omega_0) R_0 \cos\ell - \Omega d. \tag{24.38}$$

Equations (24.37) and (24.38) are valid as long as the assumption of circular motion is justified.

Although the Sun's motion around the Galactic center is not perfectly circular, its peculiar velocity relative to the LSR is small compared to $\Theta_0$ (only about 6%). As a result, to a first approximation, Eqs. (24.37) and (24.38) provide a reasonable estimate of $\Omega = \Omega(R)$ if the other parameters are known. However, $d$ is in general very difficult to measure unless the object is close enough to yield to trigonometric parallax, or perhaps some other reasonably reliable distance estimate may be applicable (for instance, the star may be a Cepheid variable).

Another complication arises because of the effects of interstellar extinction. Our ability to observe Galactic structure to great distances is severely limited at visual wavelengths. Unless we are making observations in relatively unobscured directions such as Baade's window, we are restricted to seeing stars out to a few thousand parsecs from the Sun. One important exception to this constraint (to be discussed in more detail below) is the 21-cm-wavelength band of H I; virtually the entire Galaxy is optically thin to 21-cm radiation, making that wavelength band a valuable tool for studying Galactic structure.

Because of the distance limitation at optical wavelengths, Oort derived a set of approximate equations for $v_r$ and $v_t$ that are valid only in the region near the Sun. Despite this restriction, these alternative formulae are still able to provide a surprising amount of information about the large-scale structure of the Galaxy.

We make the assumption here that $\Omega(R)$ is a smoothly varying function of $R$ so that the Taylor expansion of $\Omega(R)$ about $\Omega_0(R_0)$ is given by

$$\Omega(R) = \Omega_0(R_0) + \left.\frac{d\Omega}{dR}\right|_{R_0} (R - R_0) + \cdots .$$

Thus, to first order, the *difference* between $\Omega$ and $\Omega_0$ is

$$\Omega - \Omega_0 \simeq \left.\frac{d\Omega}{dR}\right|_{R_0} (R - R_0),$$

and the *approximate value* of $\Omega$ is

$$\Omega \simeq \Omega_0.$$

If we also make use of the identity $\Omega = \Theta/R$, Eqs. (24.37) and (24.38) become (after some rearrangement)

$$v_r \simeq \left[ \left.\frac{d\Theta}{dR}\right|_{R_0} - \frac{\Theta_0}{R_0} \right] (R - R_0) \sin \ell,$$

$$v_t \simeq \left[ \left.\frac{d\Theta}{dR}\right|_{R_0} - \frac{\Theta_0}{R_0} \right] (R - R_0) \cos \ell - \Omega_0 d.$$

From Fig. 24.22 it is clear that

$$R_0 = d \cos \ell + R \cos \beta \simeq d \cos \ell + R,$$

the latter result being due to the small-angle approximation $\cos \beta \simeq 1$, since $d \ll R_0$ implies that $\beta \ll 1$ radian. Finally, using the appropriate trigonometric identities and defining the **Oort constants**

$$A \equiv -\frac{1}{2}\left[\left.\frac{d\Theta}{dR}\right|_{R_0} - \frac{\Theta_0}{R_0}\right],\tag{24.39}$$

$$B \equiv -\frac{1}{2}\left[\left.\frac{d\Theta}{dR}\right|_{R_0} + \frac{\Theta_0}{R_0}\right],\tag{24.40}$$

we have

$$v_r \simeq Ad \sin 2\ell,\tag{24.41}$$

$$v_t \simeq Ad \cos 2\ell + Bd.\tag{24.42}$$

To understand the functional dependence of the Oort formulae on Galactic longitude, consider the orbits of nearby stars shown in Fig. 24.23(a). For stars in the directions $\ell = 0°$ and $\ell = 180°$, the lines of sight are perpendicular to their motions relative to the LSR. As a result, the radial velocity must be zero. For $\ell = 90°$ or $270°$, the stars being observed are in essentially the same circular orbit as the Sun and are moving with the same speed, so again $v_r = 0$ km s$^{-1}$. At intermediate angles the situation is somewhat more complicated, however. For instance, if we assume that in the neighborhood of the Sun, $\Omega(R)$ is monotonically decreasing outward, then at $\ell = 45°$, the star being observed is closer to the Galactic center and is "outrunning" the Sun; hence a positive radial velocity would be measured. For $\ell = 135°$, the Sun is "overtaking" the star and a negative radial velocity results. At $\ell = 225°$, the Sun is moving away from the star, producing a positive radial velocity, and at $\ell = 315°$, the star is gaining on the Sun, causing the observed radial velocity to be negative. The overall result is the double sine function shown in Fig. 24.23(b). A similar analysis shows that the transverse-velocity curve is a double cosine function plus an additive constant; see Problem 24.17. For a sample of stars, all with similar distances $d$, the amplitudes of the $v_r$ and $v_t$ curves (Eqs. 24.41 and 24.42, respectively) give $A$, and the vertical offset in $v_t$ gives $B$.

It is now possible to derive several important relationships between the Oort constants $A$ and $B$, and the *local* parameters of Galactic rotation, $R_0, \Theta_0, \Omega_0 = \Theta_0/R_0$, and $(d\Theta/dR)_{R_0}$. For instance, from Eqs. (24.39) and (24.40), we immediately find that

$$\Omega_0 = A - B,\tag{24.43}$$

$$\left.\frac{d\Theta}{dR}\right|_{R_0} = -(A + B).\tag{24.44}$$

Yet another useful relation can be found by considering the largest radial velocity seen along the line of sight at a constant Galactic longitude $\ell$. Turning again to Fig. 24.22, the star with the maximum observable radial velocity will be located at point $T$, the position

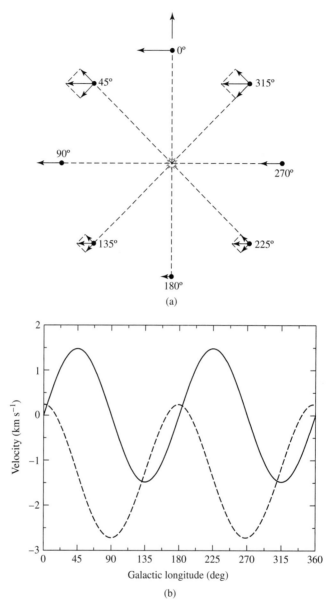

(a)

(b)

**FIGURE 24.23** (a) The differential rotation of stars near the Sun is revealed through the dependence of radial and transverse velocities on Galactic longitude. (b) Radial velocity is proportional to $\sin 2\ell$ (solid line), and transverse velocity is a function of $\cos 2\ell$ (dashed line). The curves depict stars located 100 pc from the Sun with $A = 14.8$ km s$^{-1}$ kpc$^{-1}$ and $B = -12.4$ km s$^{-1}$ kpc$^{-1}$.

where $\alpha = 0°$. It is at this *tangent point* that the distance to the center of the Galaxy will be a minimum and $\Theta(R)$ will be a maximum (if we can assume that $\Theta(R)$ is monotonically decreasing from the center outward). Moreover, the orbital-velocity vector is directed along the line of sight at that position. This minimum distance from the center of the Galaxy is given by

$$R_{\min} = R_0 \sin \ell,$$

and the maximum radial velocity is

$$v_{r,\max} = \Theta(R_{\min}) - \Theta_0(R_0) \sin \ell.$$

If we now restrict our observations to Galactic longitudes near but less than 90°, or near but greater than 270° (i.e., inside the solar circle), then $d \ll R_0$, $R \sim R_0$, and $\Theta(R)$ can be expressed in terms of a Taylor expansion about $\Theta_0$:

$$\Theta(R_{\min}) = \Theta_0(R_0) + \left.\frac{d\Theta}{dR}\right|_{R_0} (R_{\min} - R_0) + \cdots.$$

Substituting into the expression for $v_{r,\max}$, retaining first-order terms, and making use of Eq. (24.39), we find

$$v_{r,\max} \simeq 2AR_0 \left(1 - \sin \ell\right). \tag{24.45}$$

One last relation, which we will not attempt to derive here but merely include for completeness, associates $A$ and $B$ with the dispersions of peculiar velocities in the $R$ and $\theta$ directions:

$$\frac{-B}{A-B} = \frac{\sigma_v^2}{\sigma_u^2}. \tag{24.46}$$

Equations (24.43–24.46) place additional constraints on the values of $R_0$ and $\Theta_0$ beyond the direct observations discussed above. In fact, because $A$ and $B$ provide critical information about Galactic differential rotation in the solar neighborhood, considerable effort has gone into determining these constants. Based on results from the Hipparcos astrometry mission, values of

$$A = 14.8 \pm 0.8 \text{ km s}^{-1} \text{ kpc}^{-1} \tag{24.47}$$

$$B = -12.4 \pm 0.6 \text{ km s}^{-1} \text{ kpc}^{-1} \tag{24.48}$$

appear to be consistent with the available data, although there remains debate over the best choices for $A$ and $B$.

### Hydrogen 21-cm Line as a Probe of Galactic Structure

To determine the large-scale velocity structure of the Galactic disk, we must return to the more general expressions for $v_r$ and $v_t$ that do not rely on first-order Taylor series expansions.

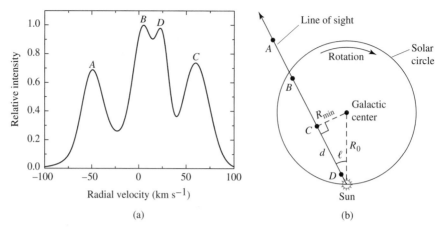

**FIGURE 24.24** (a) A typical 21-cm H I line profile. (b) The line profile is produced by observing several gas clouds along a particular line of sight. Because of differential Galactic rotation, each cloud has a different radial velocity relative to the Sun.

As has already been mentioned, 21-cm emission from H I is able to penetrate virtually the entire Galaxy, making it an indispensable tool in probing the structure of the Milky Way. By measuring $v_r$ as a function of $\ell$, the Galactic rotation curve can be determined from Eq. (24.37), provided that the distance of the emitting region from the Sun can be found.

Figure 24.24(a) shows a typical intensity distribution for the 21-cm emission line of H I along a particular line of sight. When a specific cloud is encountered along the line of sight, the wavelength of the radiation from that cloud is Doppler shifted because of the effects of differential Galactic rotation. Furthermore, the intensity of the radiation at a given wavelength (or velocity) is proportional to the number of hydrogen atoms along the line of sight in the cloud. The peaks of the line profile shown in Fig. 24.24(a) correspond to the clouds shown in Fig. 24.24(b).

The principal difficulty in using 21-cm radiation to determine $\Omega(R)$, and hence $\Theta(R)$, lies in measuring $d$. This problem can be overcome by selecting the largest radial velocity measured along each line of sight, which must originate in the region $R_{min}$ from the Galactic center, implying that $d = R_0 \cos \ell$. By measuring $v_{r,\text{max}}$ for $0° < \ell < 90°$ and $270° < \ell < 360°$, we can determine the rotation curve within the solar Galactocentric radius. Unfortunately, this technique does not work for Galactic longitudes $90° < \ell < 270°$ because there is no unique orbit for which a maximum radial velocity can be observed. The method also tends to break down near $\ell = 90°$ and $\ell = 270°$ because $v_r$ becomes rather insensitive to changes in distance from the Sun. For longitudes within approximately 20° of the Galactic center, further problems develop; clouds that have markedly noncircular motions exist in that region, perhaps because of the gravitational perturbations of the central bar, so the assumptions underlying the preceding analysis are not valid.

### The Flat Rotation Curve and Evidence of Dark Matter

To measure $\Theta(R)$ for $R > R_0$, we must rely on objects available in the Galactic plane, such as Cepheids, for which we can directly obtain distances. These data suggest that the

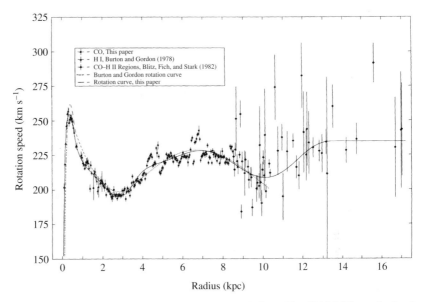

**FIGURE 24.25** The rotation curve of the Milky Way Galaxy. The 1985 IAU standard values of $R_0 = 8.5$ kpc and $\Theta_0 = 220$ km s$^{-1}$ have been assumed. (Figure adapted from Clemens, *Ap. J.*, *295*, 422, 1985.)

rotation curve of the Galaxy does not decrease significantly with distance beyond $R_0$ and may actually increase somewhat (meaning that for the Oort constants, $A < -B$ near $R_0$). Combining all of the available data, a possible form for the rotation curve of the Galaxy is shown in Fig. 24.25.

It came as a great surprise to astronomers to discover that the Galactic rotation curve is essentially constant beyond $R_0$. According to Newtonian mechanics, if most of the mass were interior to the solar circle, the rotation curve should drop off as $\Theta \propto R^{-1/2}$, a behavior referred to as Keplerian motion (see Problem 24.18). The fact that it does not implies that a significant amount of mass exists in the Galaxy beyond $R_0$. This result was particularly unexpected since most of the luminosity in the Galaxy is produced by matter residing inside the solar Galactocentric radius.

The data for the Milky Way are supported by observations of other spiral galaxies, such as those obtained by Vera Rubin and her collaborators in the late 1970s. Figure 24.26 shows a spectrograph slit superimposed on NGC 2998, a galaxy in the constellation of Ursa Major, 96 Mpc from Earth. Below that image is a portion of the spectrum in a wavelength region near H$\alpha$. The left side of the slit recorded blueshifted light, and the light on the right side was redshifted. The Doppler shifts were then translated into radial velocities, and a corresponding rotation curve was determined.

Similar rotation curves have also been measured for a number of other spiral galaxies (see Fig. 24.27). With the exception of the innermost regions (to be discussed in Section 24.4 for the Milky Way), there is a rapid rise in rotation speed with distance out to a few kiloparsecs from the center. This type of rotation is referred to as **rigid-body rotation** because when $\Theta \propto R$, $\Omega = \Theta/R$ is a constant and all stars have the same orbital period about the Galactic

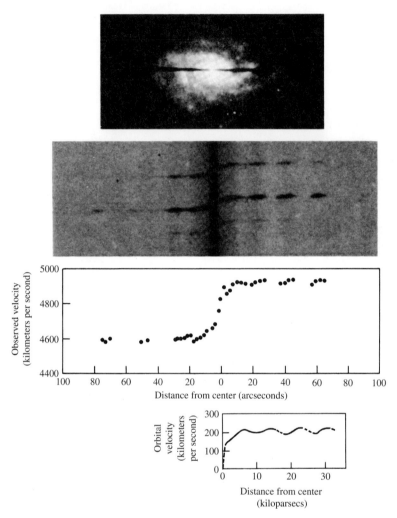

**FIGURE 24.26**   The rotation curve of NGC 2998 was measured using a slit spectrograph. The Hα wavelength region is shown. Note that overall, the entire galaxy is receding from us at a speed of 4800 km s$^{-1}$. (Adapted from a figure courtesy of Vera Rubin, 1983.)

center, just as a rigid object would. Beyond a few kiloparsecs, nearly **flat rotation curves** continue out to the edge of the measurements.

Since Galactic rotation depends on the distribution of mass, a great deal can be learned about the matter in galaxies by studying these curves. For instance, rigid-body rotation near the Galactic center implies that the mass must be roughly spherically distributed and the density nearly constant (see Problem 24.20). On the other hand, flat rotation curves suggest that the bulk of the mass in the outer portions of the Galaxy are spherically distributed with a density law that is proportional to $r^{-2}$.

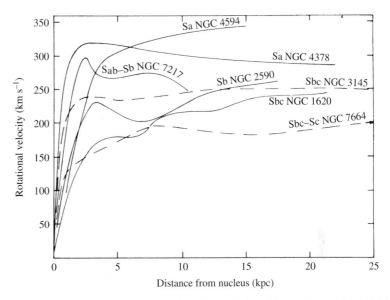

**FIGURE 24.27** A series of rotation curves for spiral galaxies. (Figure adapted from Rubin, Ford, and Thonnard, *Ap. J. Lett.*, 225, L107, 1978.)

To see this, assume that $\Theta(r) = V$, where $V$ is a constant.[9] Then, from the equation for centripetal force and Newton's law of gravity, the force acting on a star of mass $m$ due to the mass $M_r$ of the Galaxy interior to the star's position at $r$ is

$$\frac{mV^2}{r} = \frac{GM_r m}{r^2},$$

if spherical symmetry is assumed. Solving for $M_r$,

$$M_r = \frac{V^2 r}{G}, \tag{24.49}$$

and differentiating with respect to the radius of the distribution,

$$\frac{dM_r}{dr} = \frac{V^2}{G}.$$

If we now borrow the equation for mass conservation in a spherically symmetric system from stellar structure theory, Eq. (10.7),

$$\frac{dM_r}{dr} = 4\pi r^2 \rho,$$

---

[9]We are using $r$ for a spherically symmetric mass distribution here, rather than $R$ for cylindrical rotation in the Galactic plane. However, to obtain a rotation curve within the Galactic plane, we need only consider the special case of $r = R$.

we see that the mass density in the outer regions of the Galaxy must vary as

$$\rho(r) = \frac{V^2}{4\pi G r^2}. \tag{24.50}$$

This $r^{-2}$ density dependence is very different from the form determined by star counts in the portion of the Galaxy beyond the solar Galactocentric radius. Recall from Eq. (24.14) that the number density of stars in the luminous stellar halo is believed to vary as $r^{-3.5}$, a much more rapid drop-off than is evident from the flat rotation curve. It was this discrepancy that so surprised astronomers. As we mentioned at the end of Section 24.2, it appears that the majority of the mass in the Galaxy is in the form of nonluminous (dark) matter. Only through its gravitational influence on the luminous component of our Galaxy and satellite galaxies like the LMC and the SMC, and through gravitational lensing of light from background sources, does the dark matter make its presence known.

One modification to Eq. (24.50) that has been made by many researchers is to force the density function to approach a constant value near the center, rather than diverge. Such a model is also consistent with the observational evidence of rigid-body rotation. As a result, one commonly used density profile for the Milky Way's dark matter halo is assumed to be of the form

$$\rho(r) = \frac{\rho_0}{1 + (r/a)^2}, \tag{24.51}$$

where $\rho_0$ and $a$ are chosen as parametric fits to the overall rotation curve. Note that for $r \gg a$, the $r^{-2}$ dependence is obtained, and $\rho \sim$ constant when $r \ll a$. A similar profile is often used for modeling other galaxies as well, with different choices for $\rho_0$ and $a$.

It is important to point out that Eq. (24.51) cannot be correct to arbitrarily large values of $r$. The reason for this is that the total amount of mass in the Galaxy would increase without bound since $M_r \propto r$. As a result, the density function for the dark matter halo must eventually terminate or at least decrease sufficiently rapidly that the mass integral $\int_0^\infty \rho(r) 4\pi r^2 \, dr$ remains finite.

An alternative form of the dark matter halo density distribution was proposed by Julio Navarro, Carlos Frenk, and Simon White in 1996. Using a commonly assumed form of dark matter dynamics known as **cold dark matter** (CDM), Navarro, Frenk, and White ran numerical simulations of the formation of dark matter halos over a wide range of size and mass scales, ranging from dwarf galaxies to rich clusters of galaxies. Their simulations revealed that a "universal" profile of the form

$$\rho_{\mathrm{NFW}}(r) = \frac{\rho_0}{(r/a)(1 + r/a)^2} \tag{24.52}$$

was applicable over an enormous range with appropriate choices of $\rho_0$ and $a$ (this is the profile first given by Eq. 24.15). The NFW density profile behaves approximately like a $1/r^2$ profile over much of the halo but is shallower ($\sim 1/r$) near the center and steeper ($\sim 1/r^3$) near the edge of the halo. Even though the NFW profile decreases more rapidly than Eq. (24.51) with increasing $r$, it can be shown that the total mass contained within the

NFW profile is still not bound (see Problem 24.23). In reality, other galaxies exist in our universe besides the Milky Way, and their mass density functions may overlap our own. As a result, although galaxies appear to be separate luminous objects, their dark matter halos may actually merge in intergalactic space.

## A Component Model of the Milky Way Galaxy

Based on the mass density functions from star counts and kinematics, astronomers have been able to construct approximate models of the overall rotation curve of the Galactic disk. One such model is shown in Fig. 24.28. Note that the observational data show that the rotation curve is decreasing slightly in the solar neighborhood ($R_0$); hence the negative value for $d\Theta/dR = -(A + B)$ near the Sun. Although the model does not reproduce all of the fine structure that is present in the velocity data (which may be due to local density variations in the thin disk, such as spiral arms), the overall correspondence between the model and observations is quite good. Notice in particular the rigid-body rotation near the center, the local maximum due to the combined effects of the central bulge, the stellar halo and the dark matter halo, and the eventual flat rotation at large values of $R$.

## Some Methods for Determining Distances

Before leaving the topic of Galactic kinematics, it is appropriate to discuss motion-based methods of determining distances within the Milky Way, the most important of which is the **moving cluster method**. Since in a stellar cluster the stars are gravitationally bound to one another, they move through space collectively. By recording the changing positions of members of the group over time due to the cluster's bulk motion, it is possible to determine their directions of motion. After removing the effects of the Sun's peculiar motion, tracing

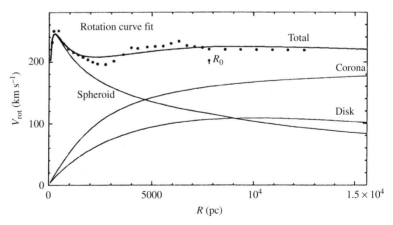

**FIGURE 24.28** One model of the rotation curve of the Galaxy. The mass distribution of each Galactic component contributes to the overall velocity structure of the disk. The dots represent observational data. Note that the "spheroid" represents the bulge and stellar halo combined and the "corona" represents the dark matter halo. (Figure adapted from Gilmore, King, and van der Kruit, *The Milky Way as a Galaxy*, University Science Books, Mill Valley, CA, 1990.)

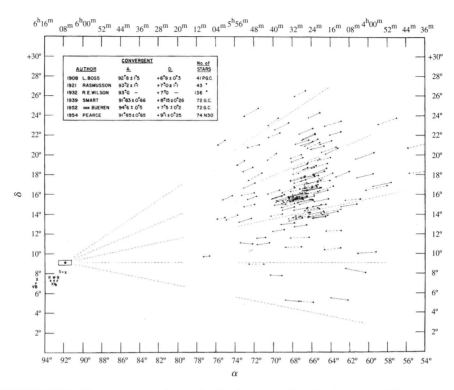

**FIGURE 24.29**   The apparent motion of the Hyades across the celestial sphere. (Figure adapted from *Elementary Astronomy* by Otto Struve, Beverly Lynds, and Helen Pillans. Copyright © 1959 by Oxford University Press, Inc. Renewed 1987 by Beverly T. Lynds. Reprinted by permission of the publisher.)

these directional vectors through space gives the appearance that each star is moving toward (or coming from) a common point known as the **convergent point**. This is simply the illusion that parallel lines appear to converge at infinity, a phenomenon familiar to anyone who has looked down the length of a long, straight set of railroad tracks. A diagram of the motion of the Hyades galactic cluster, found in the constellation of Taurus, is shown in Fig. 24.29.

From the geometry illustrated in Fig. 24.30, the angle between the cluster and the convergent point, as seen from the Sun, must be the same as the angle between the line of sight to the cluster and its space velocity vector, **v** (this statement is valid only because the convergent point is taken to be at infinity). Now, decomposing the space velocity into its perpendicular components, the radial velocity is given by $v_r = v \cos \phi$, and the transverse velocity is $v_t = v \sin \phi$. Combining yields

$$v_t = v_r \tan \phi.$$

Since the transverse velocity is observed as the proper motion, $\mu = v_t/d$, the distance to the cluster can be determined from knowledge of the direction to the convergent point $\phi$,

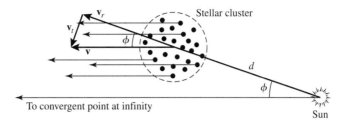

**FIGURE 24.30**  The space motion of the cluster is directed toward the convergent point. This velocity vector may be decoupled into its radial and transverse components.

the average radial velocity of the cluster members, and the average of their proper motions. Solving for $d$, we get

$$d = \frac{\langle v_r \rangle \tan \phi}{\langle \mu \rangle},$$

(24.53)

where standard SI units have been used for $d$, $v_r$, and $\mu$. Expressing $d$, $v_r$, and $\mu$ in the more commonly used units of pc, km s$^{-1}$, and arcsec yr$^{-1}$, respectively, Eq. (24.53) can be written as

$$d \text{ (pc)} = \frac{\langle v_r \rangle \tan \phi}{4.74 \langle \mu'' \rangle}.$$

(24.54)

This technique has been used to determine the distances to several clusters, including the Hyades with its approximately 200 members, the Ursa Major group (60 stars), and the Scorpio–Centaurus group (100 stars); the most important of these is the Hyades. The distance to the Hyades is found to be $46 \pm 2$ pc, in excellent agreement with other determinations such as those made by trigonometric parallax (the Hipparcos space astrometry mission data yielded a value of 47 pc to the center of the Hyades).

Once the distance to the Hyades was determined, the absolute magnitudes of its members could be found, providing an important calibration of its main sequence. By comparing the apparent magnitudes of other cluster H–R diagram main sequences to the Hyades, it is possible to find the distance moduli of those clusters, as illustrated schematically in Fig. 24.31. Assuming that the amount of interstellar extinction is known (from reddening data, for instance), the distances to those clusters can be determined. This distance technique, first discussed in Section 13.3, is known as **main-sequence fitting** and is similar to the method of spectroscopic parallax discussed at the end of Section 8.2. However, main-sequence fitting is a more precise procedure because it relies on a large number of stars along the main sequence rather than on a single object, significantly reducing statistical errors. Identifying RR Lyraes in clusters of known distances then provides a means of determining the intrinsic luminosities of these stars more accurately. Once the luminosities of the RR Lyraes have been calibrated, they can be used to determine other distances, such as to globular clusters. Historically, the Hyades provided the foundation for virtually all distance estimates, both Galactic and extragalactic, beyond about 100 pc from Earth, although this is changing with

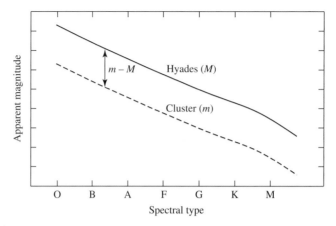

**FIGURE 24.31**   The distance modulus of a cluster can be determined by shifting the cluster's main sequence vertically in the H–R diagram until it coincides with the known absolute magnitude of the Hyades' main sequence.

space astrometric missions such as the European Space Agency's Hipparcos (1989–1993) and Gaia (launch 2011) missions, and NASA's SIM PlanetQuest (launch 2011).

One last distance determination technique worth mentioning is **secular parallax**. Recall from the discussion of trigonometric parallax in Section 3.1 that the ability to measure distances directly from the motion of Earth around the Sun relies on the 2-AU diameter of Earth's orbit. If the length of the baseline could be increased, it would be possible to extend the method to objects farther away. This is accomplished by using the peculiar motion of the Sun with respect to a group of stars having similar properties, such as similar spectral types, distances, and space motions. The overall solar motion of $13.4 \, \mathrm{km \, s^{-1}}$ is equivalent to $2.8 \, \mathrm{AU \, yr^{-1}}$; allowing the effect to build up for several years provides a significantly larger baseline than is available from Earth's annual orbit about the Sun. This larger baseline can then be used to determine an average distance to the group of stars. However, it is important to remember that the measured solar motion does depend on the group of stars used as a reference. We will not discuss the details of the method in this text.

## 24.4 ■ THE GALACTIC CENTER

Observations of the center of our Galaxy pose a particular challenge. This is because the abundance of gas and dust in the Galactic plane results in more than 30 magnitudes of extinction at visible wavelengths. Located only 30 pc above the midplane and 8 kpc from the center, the line of sight from the Sun to the center traverses nearly the maximum possible amount of interstellar material. It is interesting to note, however, that because the solar peculiar velocity has an appreciable component perpendicular to the Galactic plane ($w_\odot = 7.2 \, \mathrm{km \, s^{-1}}$), the Sun will reach a height of approximately 85 pc above the plane in 15 million years (see Problem 24.27); this will place it above most of the obscuring material.

If humanity is still inhabiting Earth at that time, our descendants will enjoy a spectacular view of a dense stellar cluster near the center of our Galaxy.

## The Mass Distribution Near the Galactic Center

To peer into the central regions of the Galaxy today, we are forced to make most of our observations in wavelengths longer than about 1 $\mu$m (i.e., infrared, microwave, and radio), or in X-rays and gamma rays. Despite our difficulty of clearly seeing the center of the Galaxy in visible wavelengths, we have been able to construct an image of the **Galactic nucleus** that depicts a history of violent events and exotic phenomena.

One IR wavelength band commonly used for investigations of the core of our Galaxy is centered at 2.2 $\mu$m, the so-called *K band*. This wavelength band is employed because the large number of old Population I K and M giant stars ($T_e \sim 4000$ K) that exist in the central region of the Galaxy are readily observable at 2.2 $\mu$m. When we use the *K* band to study the brightness distribution of the central cluster and use an appropriate mass-to-light ratio ($\sim 1$ $M_\odot/L_\odot$), it appears that the mass density of stars rises toward the center as $r^{-1.8}$ down to a radius of between 0.1 pc and 1 pc. This is roughly the type of distribution that would be expected on dynamical grounds (the central region of the Galaxy is interior to the region of "rigid-body" rotation discussed on page 915).

Since these stars are very close together, particularly when compared with the distance between stars in the solar neighborhood, close encounters are fairly frequent, occurring on average once every $10^6$ years or so. Because of the gravitational perturbations produced by these close encounters, the constant exchange of mechanical energy between stars has generated a nearly *isothermal* velocity distribution; in other words, the stars in the sample have a velocity distribution just like the particles in an isothermal gas, meaning that the velocity distribution is approximately Maxwellian. In a truly isothermal *stellar gas*, the mass density distribution is $r^{-2}$, close to the $r^{-1.8}$ variation that is observed. Recall that this is also the spherical density distribution required for flat rotation curves [when all stars have the same orbital velocity and $M_r \propto r$; Eq. (24.49)].

The observed density distribution from the isothermal stellar gas is inconsistent with measurements of the velocities of stars within 2 pc of the center, however. Kristen Sellgren, Martina T. McGinn, and their colleagues made one such set of observations in the late 1980s using the 2.3-$\mu$m molecular absorption band of CO found in the spectra of the cool K and M giants. They found that although the velocity distribution is fairly isothermal from several hundred parsecs down to a couple of parsecs from the center, velocities begin to increase significantly as the distance to the center continues to decrease. This suggests that either the stellar density must rise substantially faster than $r^{-2}$ toward the center (at least as steeply as $r^{-2.7}$) or there must be a great deal of mass occupying a very small volume near the middle of the cluster.

More recently, Rainer Schödel, Reinhard Genzel, and their research group were able to follow the orbits of stars very close to the Galactic center. In particular, the star known as S2 has an orbital period of 15.2 yr, an orbital eccentricity of $e = 0.87$, and a perigalacticon distance of $1.8 \times 10^{13}$ m $= 120$ AU (17 light-hours). This size is only a few times the semimajor axis of Pluto's orbit! Figure 24.32 shows the orbit of S2 against the central stellar cluster of the Galaxy.

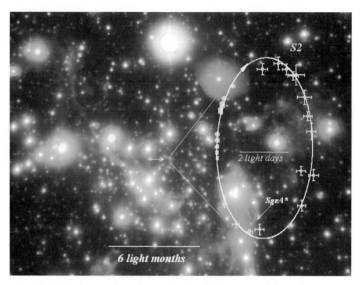

**FIGURE 24.32**   The orbit of S2 about the center of the Milky Way Galaxy. The center is designated as Sgr A*. (Courtesy of Reinhard Genzel and Rainer Schödel.)

---

**Example 24.4.1.**   The semimajor axis of S2's orbit is (from Eq. 2.5)

$$a_{S2} = \frac{r_p}{1 - e} = 1.4 \times 10^{14} \text{ m}.$$

From Kepler's third law, Eq. (2.37), the mass interior to S2's orbit must be about

$$M = \frac{4\pi^2 a_{S2}^3}{G P^2} \simeq 7 \times 10^{36} \text{ kg} \simeq 3.5 \times 10^6 \text{ M}_\odot.$$

A more precise calculation gives

$$M = 3.7 \pm 0.2 \times 10^6 \text{ M}_\odot.$$

---

Figure 24.33 shows estimates of the amount of mass interior to $r$ based on measurements of objects at varying distances from the center.

The luminosity distribution of the stars near the center of the Galaxy peaks within a few arcseconds ($\sim 0.1$ pc) of an infrared object known as **IRS 16** (for infrared source). It was during a lunar occultation that IRS 16 was resolved into at least 15 very luminous point-like sources that are most likely individual stars. These sources appear to be hot stars with brightnesses in excess of $10^6$ $L_\odot$. They may be O and B stars, but with luminosities that are much greater than expected for normal stars. Furthermore, their ultraviolet fluxes are absorbed by the surrounding gas and dust and reradiated in the infrared. It has been suggested that these objects may be Wolf–Rayet stars, which are extremely rare in other parts of the Galaxy (recall Section 15.1). Since Wolf–Rayets are massive stars, they must have evolved to their present states very rapidly. If these stars are in fact Wolf–Rayets, then

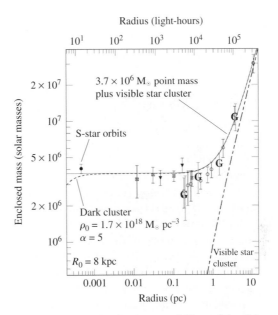

**FIGURE 24.33** The interior mass function for the central 10 pc of the Galaxy. Note that the curve is consistent with a mass distribution $M_r \propto r$ beyond about 5 pc but that interior to 2 pc the distribution levels off, approaching a constant nonzero value of $3.7 \times 10^6$ M$_\odot$. "Dark cluster" refers to a hypothetical object. Note that the predictions of a dark cluster model at the center of the Galaxy do not agree with the observational data. (Adapted from a figure courtesy of Reinhard Genzel and Rainer Schödel. For a discussion of an earlier version of this diagram, see Schödel, et al., *Nature, 419*, 694, 2002.)

a burst of star formation must have occurred within the past 10 million years. However, an episode of recent star formation seems to contradict the facts that the density of gas and dust is very low and that there is no evidence of ongoing star formation in that region today. On the other hand, in further support of the Wolf–Rayet idea is the presence of high-velocity gas near IRS 16 ($\sim 700$ km s$^{-1}$), which some researchers suggest may have been ejected as a stellar wind by one of the stars. As an alternative explanation, other astronomers point out that it is also possible that the high-velocity gas could simply be falling in toward the center of the Galaxy rather than being ejected from it.[10] The velocity structure of the gas does not appear to be consistent with the acceleration expected of Wolf–Rayet winds with increasing distance from the star; instead, the gas velocity appears to decrease with distance.

IRS 16 does not appear to have sufficient mass to account for the rise in orbital velocities near the center. It seems that whatever the origin of this extremely localized mass distribution, it must have an overall luminosity that remains below our threshold of detectability. One possibility is that the mass is composed of a very dense cluster of brown dwarfs and/or more massive neutron stars. Even in the case of neutron stars, however,

[10]Determining whether the gas is falling in toward the center of the Galaxy or being ejected from it requires information about the orientation of the gas trajectories relative to the line of sight, something that is often difficult to determine.

interactions with high-mass stars would eject the lower-mass neutron stars from the center in less than $10^8$ years. This would have the additional effect that the most massive members of the cluster would "sink" to the center of the Galaxy's gravitational well, which is not observed.

### Radio Sources in Sagittarius

It was Karl Jansky (1905–1950) who, in the 1930s, first discovered that a radio source was located in the direction of Sagittarius, but extensive observations of the radio Galaxy were delayed until after World War II.[11] Since that time the Galactic center has been the site of numerous studies at radio wavelengths.

Radio observations of H I clouds reveal a **nuclear disk** of neutral gas occupying a region between a few hundred parsecs and about 1 kpc from the center. The disk appears to be tilted slightly with respect to the Galactic plane and contains clouds with significant noncircular motions. Observations of the nuclear disk are responsible for the mass estimates of the Galactic center beyond 100 pc that are shown in Fig. 24.33.

**Galactic lobes** of ionized gas have also been observed within the central few hundred parsecs. Elongated, with dimensions of 10 pc by 200 pc, the lobes are oriented nearly perpendicular to the Galactic plane and may represent material being ejected from the center. However, the data may also be consistent with infalling material.

One of the more unusual features of the radio emission originating in the central region of the Galaxy is the set of filaments that stretch for 20 pc from the center in a direction perpendicular to the Galactic plane and then make an almost right-angle turn; see Fig. 24.34. Even a casual inspection of this 20-cm radio structure, located near $\ell \sim 0.18°$, suggests that magnetic fields may be responsible for the unusual pattern. In fact, the radiation is linearly

**FIGURE 24.34**    A view of the central 60 pc $\times$ 60 pc of the Galaxy. This image was made using 20-cm radiation produced by synchrotron radiation. Sgr A is the central, radio-bright region. (Figure from Yusef-Zadeh, Morris, and Chance, *Nature*, *310*, 557, 1984, and NRAO.)

---

[11] Major advances in radio and microwave electronics that occurred during World War II helped to advance radio astronomy in the years that followed.

polarized and appears to be synchrotron emission. From the intensity of the radiation and the amount of polarization, the strength of the Galactic magnetic field in that region is probably between $10^{-8}$ and $10^{-6}$ T, roughly two to four orders of magnitude weaker than Earth's magnetic field.

A similar feature has also been discovered south of the Galactic plane. Together, these structures seem to be part of a "halo" of streamers and filaments that may correspond to mass outflow from the center of the Galaxy. A satisfactory model of the source of the relativistic electrons that are spiraling around the magnetic field lines, or of the source of the field itself, has not yet been developed.

The inner 8 pc of the Galaxy contains the radio source designated as **Sagittarius A** (Sgr A). With the increase in resolving power provided by the technique of very long baseline interferometry and the use of telescopes such as the Very Large Array, it has become possible to study the structure of the Sgr A complex down to angular resolutions of 0.2 milliarcsecond, corresponding to linear dimensions of less than 2 AU.

The largest feature of the complex is the **molecular circumnuclear ring**, a doughnut-shaped structure with an inner radius of 2 pc and an outer radius of 8 pc that is inclined some 20° with respect to the Galactic plane. The inner edge of the ring exhibits some warping, and the ring thickness increases from 0.5 pc at its inner edge to 2 pc at a Galactocentric distance of 7 pc. This molecular ring has been observed at various wavelengths associated with several atoms and molecules, including H I, $H_2$, C II, O I, OH, CO, HCN, and CS. The ring is rotating about the Galactic center at a speed of approximately 110 km s$^{-1}$, a value that is nearly independent of radius. From the collisional excitation of the molecules and the intensity of the emission, the estimated mass of the portion of the ring located between 2 and 5 pc is $1 \times 10^4 \ M_\odot$ to $3 \times 10^4 \ M_\odot$.

It is apparent that the ring is unlike any other molecular region known to exist in our Galaxy. For instance, the temperatures of individual molecular clouds that exist throughout much of the ring increase from less than 300 K near 4.5 pc to more than 400 K near the inner edge. At the same time, the number density of hydrogen molecules increases from $1.5 \times 10^{10}$ m$^{-3}$ to $5 \times 10^{10}$ m$^{-3}$ over the same distance. These values should be compared with those of more typical molecular clouds, such as the giant molecular clouds discussed in Section 12.1. In GMCs that are located several kiloparsecs from the center of the Galaxy, $T \sim 15$ K and $n_{H_2} \sim 10^8$ m$^{-3}$.

The circumnuclear ring also shows evidence that some violent event occurred near the Galactic center in the relatively recent past. The inner edge of the ring is very sharp; the number density of particles inside the central cavity is 10 to 100 times less than in the ring itself. Such a strong density discontinuity cannot be an equilibrium feature of the ring because the ring's internal turbulence would destroy the discontinuity in less than $10^5$ years. Furthermore, the gas within the cavity is largely ionized while the gas in the ring is in the form of neutral atoms and molecules. It is estimated that the amount of energy required to clear out the cavity is on the order of $10^{44}$ J, a value characteristic of a supernova explosion.

The ring shows other evidence of some past violent episode as well. For instance, material in the ring is very clumpy, a situation that cannot be maintained indefinitely because of the relatively rapid smoothing effects of cloud–cloud collisions. Also, a study of the hydroxyl molecule (OH) in portions of the ring indicates that the temperature of nearly 2000 K is consistent with strong shocks that rapidly heated the molecules to temperatures well above the values seen elsewhere in the ring. Furthermore, OH radicals can be produced when water molecules ($H_2O$) are torn apart by energetic events such as shocks.

Besides the circumnuclear ring, several other components have also been discovered within the Sgr A complex. **Sgr A East** is a nonthermal source with a shell-like appearance (a "nonthermal" source is one that *does not* emit its electromagnetic radiation in the form of a blackbody spectrum; one type of nonthermal emission is synchrotron radiation). It is generally accepted that Sgr A East is a young supernova remnant that may be between 100 and 5000 years old. **Sgr A West**, which is located $1.5'$ from the center of Sgr A East, is an unusual H II region (a thermal source) that looks very much like a "mini-spiral." Finally, **Sgr A\*** (pronounced "Sagittarius A star") is a strong, unresolved radio point source that is located near the center of Sgr A West. A radio image of Sgr A West and Sgr A\* is shown in Fig. 24.35(a).

The location of the Sgr A complex corresponds very closely with the brightness peak of the central stellar cluster, suggesting strongly that this region marks the center of the Galaxy. In fact, one component of IRS 16 (named IRS 16 Center) is only about $1''$ west of Sgr A\*. If the two objects are the same distance from Earth and not simply aligned along the line of sight, then their angular separation translates into a linear separation of only 0.04 pc. For reasons we will soon discuss more fully, it now appears that Sgr A\* is the actual center of the Milky Way.

Although Sgr A West appears superficially to be much like the large-scale spiral pattern seen in the disk of the Milky Way, it is fundamentally very different. The western arc is just the ionized inner edge of the circumnebular ring, which is rotating about the center at

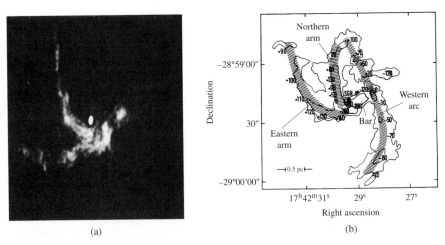

(a)                                        (b)

**FIGURE 24.35**    (a) A VLA 6-cm image of the inner $40''$ of the Galaxy, showing the Sgr A West mini-spiral. Sgr A\* (a radio point source) is the bright oval at the center of the image. (Figure from Lo, *Science*, *233*, 1394, 1986.) (b) An intensity contour map of the Sgr A West region. Radial-velocity measurements of the gas are also indicated on the map (in units of km s$^{-1}$). (Figure from Genzel and Townes, *Annu. Rev. Astron. Astrophys.*, *25*, 377, 1987. Reproduced with permission from the *Annual Review of Astronomy and Astrophysics*, Volume 25, ©1987 by Annual Reviews Inc.)

$110 \text{ km s}^{-1}$. The velocities recorded on the map in Fig. 24.35(b) are the measured radial velocities. Portions of the arc that are moving nearly along the line of sight record the largest radial velocities (positive or negative), and the parts of the arc that are moving perpendicular to the line of sight have no radial-velocity components.

The other components of Sgr A West appear to be ionized filaments of matter that may be both rotating about and falling in toward the central "bar" and the vicinity near Sgr A*. These features may also be associated with a central magnetic field that has an estimated strength of 30 nT.

As was seen for the stellar velocities, the velocities of the ionized gas within the cavity increase dramatically toward the center, from $110 \text{ km s}^{-1}$ at the inner edge of the circum-nebular ring to $700 \text{ km s}^{-1}$ at 0.1 pc.

---

**Example 24.4.2.** The motion of gas in the central regions of the Galaxy can be used to estimate the amount of mass interior to the location of the gas. A gas cloud 0.3 pc from the center has a measured velocity of $260 \text{ km s}^{-1}$. If the cloud is in orbit about the center, Eq. (24.49) gives

$$M_r = \frac{v^2 r}{G} = 4.7 \times 10^6 \ M_\odot.$$

---

### An X-Ray Source in Sgr A

To within the positional error bars, the radio Sgr A West region (including Sgr A*) also coincides with a small continuous X-ray source. Although the estimates are highly uncertain due to extensive absorption along the line of sight, the X-ray source appears to have a characteristic temperature of $T \sim 10^8$ K and luminosities of $10^{28}$ W in the 2–6 keV energy band ("soft" X-rays) and $2 \times 10^{31}$ W between 10 keV and 10 MeV ("hard" X-rays). Because the X-ray source is also highly variable, it must be composed of one or a few objects whose diameters are less than about 0.1 pc. The upper limit on the linear size is derived from the minimum amount of time required for information to travel across the object, assuming that the information travels at the speed of light. If one side of the source begins changing brightness, the opposite side can learn about the change and begin contributing to it only after a time $\Delta t \geq d/c$, where $d$ is the diameter; measuring $\Delta t$ places the upper limit of 0.1 pc on $d$.

### The Supermassive Black Hole in Sgr A*

As we saw in Example 24.4.1, it has been possible to follow the orbits of individual stars to within about 120 AU of Sgr A*, providing us with critical data for accurately calculating the mass in the innermost region of the Galactic center. It has also been possible to place an upper limit of less than 2 AU on the size of Sgr A*. From these data, it appears that Sgr A* can only be a **supermassive black hole** with a mass of

$$M_{\text{Sgr A}^*} = 3.7 \pm 0.2 \times 10^6 \ M_\odot.$$

This implies that the Schwarzschild radius of the black hole is (Eq. 17.27)

$$R_{\text{Sgr A}^\star} = \frac{2GM_{\text{bh}}}{c^2} = 0.08 \text{ AU} = 16 \text{ R}_\odot,$$

which remains below the roughly 2-AU resolution limit of current observations.

Producing the infrared radiation detected from the ring and maintaining the degree of ionization seen inside the cavity require an ultraviolet luminosity of $10^7 \text{ L}_\odot$ with an effective blackbody temperature of 35,000 K; the temperature is deduced from the strengths of the IR emission lines of S IV, O III, and Ar III. The gas absorbs some of the UV radiation, ionizing it and producing the unusual H II region, Sgr A West. The dust absorbs the remainder of the UV photons and reradiates the light at IR wavelengths.

Could a supermassive black hole be responsible for the luminosity? Observations of the number density of particles and the velocity structure inside the cavity of the circumnuclear ring suggest that matter is accreting onto the Galactic center at a rate of $\dot{M} = 10^{-3}$ to $10^{-2} \text{ M}_\odot \text{ yr}^{-1}$. This matter must release gravitational potential energy as its distance from the center decreases.

It is possible to make a rough estimate of the luminosity that could be produced by an accreting supermassive black hole in Sgr A$^\star$. If we consider a Newtonian view of energy released when a particle of mass $M$ spirals in through an accretion disk from an initial radius of $r_i$ to a final radius of $r_f$, then according to the virial theorem, the amount of energy radiated should be one-half the change in potential energy, or

$$E = \frac{1}{2} \left( \frac{GM_{\text{bh}}M}{r_f} - \frac{GM_{\text{bh}}M}{r_i} \right),$$

where $M_{\text{bh}}$ is the mass of the black hole. Assuming that $r_i \gg r_f$ and $r_f = R_S$ (the Schwarzschild radius), then

$$E = \frac{1}{2} \frac{GM_{\text{bh}}M}{R_S}.$$

Taking the luminosity as $L = dE/dt$ and the mass accretion rate as $\dot{M} = dM/dt$, and substituting the expression for the Schwarzschild radius, we have

$$L = \frac{1}{4}\dot{M}c^2, \tag{24.55}$$

a result that is independent of both the mass and the radius of the black hole. Now, the minimum mass-accretion rate required to generate $10^7 \text{ L}_\odot$ is

$$\dot{M} = \frac{4L}{c^2} = 1.7 \times 10^{17} \text{ kg s}^{-1} = 2.7 \times 10^{-6} \text{ M}_\odot \text{ yr}^{-1}.$$

The observed accretion rate of $10^{-3}$ to $10^{-2} \text{ M}_\odot \text{ yr}^{-1}$ is more than sufficient to produce the luminosity seen in Sgr A West and Sgr A$^\star$. Curiously, high-resolution observations of the Sgr A$^\star$ region alone suggests an upper limit of luminosity from Sgr A$^\star$ of less than $3 \times 10^4 \text{ L}_\odot$.

Assuming that there is a supermassive black hole at the center of Sgr A* and that it is actually at the center of our Galaxy's gravitational well, then it should remain nearly motionless. Studies that have attempted to measure the proper motion of Sgr A* with respect to extremely distant extragalactic sources known as quasars (see Chapter 28) indicate that to within the uncertainties in the measurements, Sgr A* reflects only the peculiar motion of the Sun, just what would be expected for an object at rest at the center of the Galaxy. Furthermore, relative to other objects in the Galactic center, Sgr A* has moved at most a small fraction of its diameter over a period of more than a decade, a rate that is much slower than for any other object in the region. Given the gravitational tugs that it certainly receives from stars and large clouds in that region, this extremely sluggish motion implies that Sgr A* is very massive.

Very high-resolution VLA maps at 2 cm reveal what appear to be large clouds of ionized gas only 0.06 pc from Sgr A*. From the orientation of the clouds, it appears that they were ejected from the center in opposite directions. There is also evidence of a trail of hot, ionized gas coming from IRS 7, a red supergiant star less than 0.3 pc from the center. The direction of the stream points away from Sgr A*, suggesting that a strong wind or perhaps intense UV radiation from the center is blowing the material away from the star.

As has already been mentioned, although the Galactic center appears relatively quiet today, it must have experienced rather violent episodes in the recent past. One possible mechanism that could produce these periodic events would be the tidal disruption of a passing star and the subsequent infall of matter. As the matter falls toward the supermassive black hole in the center, it collects onto an accretion disk, releasing a tremendous amount of gravitational potential energy accompanied by a dramatic increase in luminosity (see Section 18.2). The entire episode may last for only a few years, but if such events occurred on the order of every $10^4$ to $10^5$ years, enough energy would be released to keep the central cavity ionized and the circumnuclear disk turbulent.

Sgr A* is not entirely inactive, however. From studies of the Galactic center using the Chandra X-Ray Observatory and the XMM-Newton Observatory, it appears that flares occur on average about once per day. Lasting for up to an hour or so, luminosities can reach peak values of $3.6 \times 10^{28}$ W, more than 160 times the quiescent X-ray level of Sgr A*.

Many other galaxies (Andromeda, for example) appear to have supermassive black holes in their centers and yet are relatively quiet (see Section 25.2). On the other hand, other galaxies have extremely active nuclei. These active galactic nuclei probably derive their energies from central "engines" that are powered by supermassive black holes. Active galactic nuclei will be discussed in more detail in Chapter 28.

### High-Energy Emission Lines from Near the Galactic Center

Before leaving the discussion of the Galactic center, it is worth mentioning the detection of two significant high-energy emission lines that originate in the region of the Sgr A complex. The first of these results from an electron colliding with a positron, causing their mutual annihilation and the corresponding production of two photons, each of energy 511 keV.[12] Since it is believed that black holes can help produce positrons in the space surrounding

---

[12]511 keV is the rest mass energy of an electron or a positron.

them, the presence of the 511-keV line seemed to support the existence of a black hole in Sgr A*. However, the efficient production of positrons in the numbers needed to explain the enormous flux of $10^{44}$ photons per second ($L_{511} \sim 5 \times 10^4 \, L_\odot$) requires a black hole smaller than the one envisioned for the Galactic center (perhaps only a few hundred solar masses). This is because high accretion-disk temperatures are required for the production of the particles, and the temperature of the disk increases with decreasing radius (see Eq. 18.19).

Due to the poor angular resolution of the original detections, it was unclear whether Sgr A* could properly be considered the source of the 511-keV photons. On October 13–14, 1990, the Soviet spacecraft GRANAT, with its higher-resolution imaging capability, discovered that the source of the 511-keV photons was not Sgr A* but a previously known X-ray emitter, 1E1740.7−2942, first detected by the Einstein satellite in 1979.[13] Nicknamed "the Einstein source," 1E1740.7−2942 is located some 45′ from Sgr A*, more than 300 pc from the center. With its accretion-disk plasma temperature of $10^9$ K and variable luminosity, the Einstein source appears to be a very strong candidate for a stellar black hole.

The second high-energy emission line detected in the region of the Galactic center is the 1.8-MeV line produced by the decay of $^{26}_{13}$Al to $^{26}_{12}$Mg. Since $^{26}_{13}$Al has a half-life of 716,000 years and is produced only in relatively small amounts in supernovae, novae, and possibly Wolf–Rayet stars, the presence of an estimated $\sim 5 \, M_\odot$ of the radioactive isotope seems to suggest that a large number of supernovae have occurred in that region over the past $10^5$ to $10^6$ years. The Galactic center is clearly an extremely dynamic environment.

## SUGGESTED READING

### General

Henbest, Nigel, and Couper, Heather, *The Guide to the Galaxy*, Cambridge University Press, Cambridge, 1994.

Mateo, Mario, "Searching for Dark Matter," *Sky and Telescope*, January, 1994.

Oort, Jan, "Exploring the Nuclei of Galaxies, Including Our Own," *Mercury*, March/April, 1992.

Trimble, Virginia, and Parker, Samantha, "Meet the Milky Way," *Sky and Telescope*, January, 1995.

Waller, William H., "Redesigning the Milky Way," *Sky and Telescope*, September, 2004.

### Technical

Benjamin, R. A., et al., "First GLIMPSE Results on the Stellar Structure of the Galaxy," *The Astrophysical Journal*, *630*, L149, 2005.

Bertin, Giuseppe, *Dynamics of Galaxies*, Cambridge University Press, Cambridge, 2000.

Binney, James, and Merrifield, Michael, *Galactic Astronomy*, Princeton University Press, Princeton, 1998.

---

[13]The source gets its name from its right ascension and declination.

Binney, James, and Tremaine, Scott, *Galactic Dynamics*, Princeton University Press, Princeton, 1987.

Eisenhauer, F., et al., "A Geometric Determination of the Distance to the Galactic Center," *The Astrophysical Journal*, *597*, L121, 2003.

Freeman, Ken, and Bland-Hawthorn, Joss, "The New Galaxy: Signatures of Its Formation," *Annual Review of Astronomy and Astrophysics*, *40*, 487, 2002.

Gilmore, Gerard, King, Ivan R., and van der Kruit, Pieter C., *The Milky Way as a Galaxy*, University Science Books, Mill Valley, CA, 1990.

Ibata, Rodrigo A., Gilmore, Gerard, and Irwin, Michael J., "Sagittarius: The Nearest Dwarf Galaxy," *Monthly Notices of the Royal Astronomical Society*, *277*, 781, 1995.

Kuijken, Konrad, and Tremaine, Scott, "On the Ellipticity of the Galactic Disk," *The Astrophysical Journal*, *421*, 178, 1994.

Majewski, S. R., "Galactic Structure Surveys and the Evolution of the Milky Way," *Annual Review of Astronomy and Astrophysics*, *31*, 1993.

Navarro, Julio F., Frenk, Carlos F., and White, Simon, D. M., "The Structure of Cold Dark Matter Halos," *The Astrophysical Journal*, *462*, 563, 1996.

Reid, Mark, "The Distance to the Center of the Galaxy," *Annual Review of Astronomy and Astrophysics*, *31*, 345, 1993.

Reid, Mark, "High-Velocity White Dwarfs and Galactic Structure," *Annual Review of Astronomy and Astrophysics*, *43*, 247, 2005.

Schödel, R., et al., "A Star in a 15.2-year Orbit around the Supermassive Black Hole at the Centre of the Milky Way," *Nature*, *419*, 694, 2002.

Sparke, Linda S., and Gallagher, John S., *Galaxies in the Universe: An Introduction*, Cambridge University Press, Cambridge, 2000.

## PROBLEMS

**24.1** Approximately how many times has the Sun circled the center of the Galaxy since the star's formation?

**24.2** **(a)** What fraction of the total $B$-band luminosity of the Galaxy is produced by each of the stellar components? Refer to Table 24.1.

**(b)** What is the total bolometric magnitude of the Galaxy?

**24.3** The globular cluster IAU C0923−545 has an integrated apparent visual magnitude of $V = +13.0$ and an integrated absolute visual magnitude of $M_V = −4.15$. It is located 9.0 kpc from Earth and is 11.9 kpc from the Galactic center, just 0.5 kpc south of the Galactic midplane.

**(a)** Estimate the amount of interstellar extinction between IAU C0923−545 and Earth.

**(b)** What is the amount of interstellar extinction per kiloparsec?

**24.4** Using the differential star count formula for an infinite universe of constant stellar number density and no interstellar extinction (Eq. 24.5), show that the amount of light arriving at Earth from a cone of solid angle $\Omega$ diverges exponentially as the length of the cone increases without bound (or, equivalently, as $m$ approaches infinity). Assume that all stars in the field have the same absolute magnitude $M$. *Hint:* You may find the discussion leading up to Eq. (3.5) helpful.

**TABLE 24.2** Hypothetical Differential Star Count Data.

| $V$ | $\log_{10} A_M$ | $V$ | $\log_{10} A_M$ |
|---|---|---|---|
| 4 | −2.31 | 12 | 2.24 |
| 5 | −1.71 | 13 | 2.59 |
| 6 | −1.11 | 14 | 2.94 |
| 7 | −0.51 | 15 | 3.29 |
| 8 | 0.09 | 16 | 3.89 |
| 9 | 0.69 | 17 | 4.49 |
| 10 | 1.29 | 18 | 5.09 |
| 11 | 1.89 | 19 | 5.69 |

**24.5** **(a)** From Eq. (24.5), derive an expression for $\log_{10} A_M(m)$ as a function of $m$ for stars of the same absolute magnitude and M–K spectral classification, assuming a constant stellar number density.

**(b)** If observations are made in apparent magnitude bins separated by one (i.e., $\delta m = 1$), calculate

$$\Delta \log_{10} A_M(m) \equiv \log_{10} A_M(m+1) - \log_{10} A_M(m).$$

**(c)** If the results of observations show that $\Delta \log_{10} A_M(m)$ is always less than the result found in part (b), what can you conclude about the distribution of stars in the region under investigation? [Recall that Eq. (24.5) applies to the case of an infinite universe of constant stellar number density and no interstellar extinction.]

**24.6** **(a)** Plot $\log_{10} A_M$ as a function of $V$ for the hypothetical data given in Table 24.2. Assume that all stars included in the differential star counts are main-sequence A stars of absolute visual magnitude $M_V = 2$.

**(b)** Assuming a constant density of stars out to at least $V = 11$, how much interstellar extinction is present to that limit (express your answer in magnitudes)? *Hint:* Consider the slope of the curve. You may also find the results of Problem 24.5 helpful.

**(c)** What is the distance to the stars corresponding to $V = 11$?

**(d)** If the solid angle over which the data were collected is 0.75 square degrees, or $2.3 \times 10^{-4}$ sr, estimate the number density $n_M(M, S)$ of A stars out to $V = 11$.

**(e)** Give two possible explanations for the change in slope between $V = 11$ and $V = 15$.

**24.7** **(a)** Assume that a cloud of gas and dust is encountered along the line of sight for the data given in Table 24.2 and plotted in Problem 24.6. Assume also that the stellar number density found in Problem 24.6 is constant along the entire line of sight. Estimate the amount of extinction (in magnitudes) that is due to the cloud. *Hint:* How would the graph change if the cloud were not present? The cloud's presence may be revealed through reddening.

**(b)** If the density of gas and dust in the cloud leads to an extinction rate of 10 mag kpc$^{-1}$, what is the length of the cloud along the line of sight?

Graphs of $\log_{10} A_M$ vs. $m$ that demonstrate changes in slope and then resume the original slope at larger values of $m$ are referred to as Wolf diagrams, after Maximilian Wolf (1863–1932), who first used them to explore the properties of interstellar clouds.

**24.8** **(a)** Plot the thin disk's luminosity density (Eq. 24.10) as a function of $z$ for $R = 8$ kpc.

**(b)** Prove that for $z \gg z_0$,

$$L(R, z) \simeq 4L_0 e^{-R/h_R} e^{-2z/z_0}$$

and so $z_0 = 2z_{\text{thin}}$ is the effective scale height of the luminosity density function.

**24.9** **(a)** From the data given in Table 24.1, and, using a typical value for the temperature of hydrogen in the interstellar medium of 15 K, estimate the average thermal energy density of hydrogen gas in the disk of the Galaxy. For this problem, assume that the disk has a radius of 8 kpc and a height of 160 pc.

   **(b)** Using Eq. (11.9), estimate the energy density of the magnetic field in the spiral arms. Compare your answer with the thermal energy density of the gas. Would you expect the magnetic field to play a significant role in the structure of the Galaxy? Why or why not?

**24.10** What are the J2000.0 Galactic coordinates of Sgr A*?

**24.11** Use Eqs. (24.16–24.18) to determine the Galactic coordinates of the following objects. (You may wish to refer to Fig. 24.18 to verify your answers.)

   **(a)** The north celestial pole

   **(b)** The vernal equinox

   **(c)** Deneb (see Appendix E)

**24.12** **(a)** Estimate the height ($z$) above or below the Galactic plane for both M13 ($\ell = 59.0°$, $b = 40.9°$) and the Orion nebula ($\ell = 209.0°$, $b = -19.4°$). M13 and the Orion Nebula are 7 kpc and 450 pc from Earth, respectively.

   **(b)** To which components of the Galaxy do these objects probably belong? Explain your answers.

**24.13** **(a)** Consider a sample of stars that lie in the Galactic plane and are distributed in a circle about the LSR, as shown in Fig. 24.36. For the purpose of this problem, assume also that these stars are at rest with respect to the LSR (of course, this could not actually occur in such a

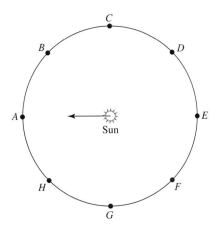

**FIGURE 24.36** A set of stars distributed in a circle about the LSR. The circle is assumed to be in the Galactic plane, and the stars are at rest with respect to the LSR. The solar motion is in the direction of Star $A$.

dynamic system). With the Sun located at the position of the LSR and the solar motion in the direction of Star $A$ as indicated, sketch the velocity vectors associated with the apparent motion of each star, as seen from the Sun. Label the apex and antapex on your diagram.

**(b)** Sketch the radial-velocity and transverse-velocity components of each star's apparent motion on the diagram used in part (a).

**(c)** Describe how you might locate the apex of the solar motion given the radial-velocity data of a large sample of stars in the solar neighborhood.

**(d)** How would you identify the solar apex from proper motion data of stars in the solar neighborhood?

**24.14** Figure 24.37 illustrates older data derived from a kinematic study of the Milky Way. From the data presented, what was the estimate of $v_\odot$ in the early 1960s?

**24.15** **(a)** Assuming (incorrectly) that the high-velocity stars known to Oort in 1927 are near the escape speed from the Galaxy, estimate the mass of the Milky Way. For simplicity, take the directions of the velocity vectors to be radially away from the Galactic center and assume that all of the mass is spherically distributed and is interior to $R_0$. (This calculation is meant only to be an order-of-magnitude estimate.) Compare your answer with the mass estimate given in Example 24.3.1.

**(b)** Repeat your calculation using the extremely high-velocity stars discussed on page 908. What could account for the extra mass compared to your answer in part (a)?

**(c)** Comment on the difficulty of determining the true mass of the Galaxy on the basis of observations of stars in the solar neighborhood.

**24.16** Starting with Eqs. (24.37) and (24.38), derive Eqs. (24.41) and (24.42), showing each step explicitly.

**24.17** Referring to Eq. (24.42) and Fig. 24.23, explain the functional dependence of transverse velocity on Galactic longitude for stars near the Sun.

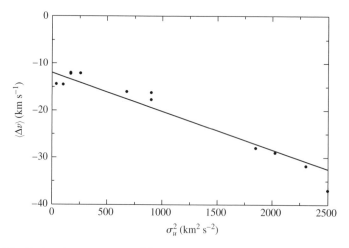

**FIGURE 24.37**    Each point represents a different sample of objects, including for instance supergiants, carbon stars, white dwarfs, Cepheids, and planetary nebulae. [Data from Delhaye, *Galactic Structure*, Blaauw and Schmidt (eds.), University of Chicago Press, Chicago, 1965.]

**24.18** **(a)** Beginning with Kepler's third law (Eq. 2.37), derive an expression for $\Theta(R)$, assuming that the Sun travels in a Keplerian orbit about the center of the Galaxy.

**(b)** From your result in part (a), derive analytic expressions for the Oort constants $A$ and $B$.

**(c)** Determine numerical values for $A$ and $B$ in the solar neighborhood, assuming $R_0 = 8$ kpc and $\Theta_0 = 220$ km s$^{-1}$. Express your answers in units of km s$^{-1}$ kpc$^{-1}$.

**(d)** Do your answers in part (c) agree with the measured values for the Milky Way Galaxy? Why or why not?

**24.19** **(a)** Estimate $d\Theta/dR$ in the solar neighborhood, assuming that the Oort constants $A$ and $B$ are $+14.8$ and $-12.4$ km s$^{-1}$ kpc, respectively. What does this say about the variation of $\Theta$ with $R$ in the region near the Sun?

**(b)** If $A$ and $B$ were $+13$ and $-13$ km s$^{-1}$ kpc, respectively, what would the value of $d\Theta/dR$ be? What would this say about the shape of the rotation curve in the solar neighborhood?

**24.20** **(a)** Show that rigid-body rotation near the Galactic center is consistent with a spherically symmetric mass distribution of constant density.

**(b)** Is the distribution of mass in the dark matter halo (Eq. 24.51) consistent with rigid-body rotation near the Galactic center? Why or why not?

**24.21** Using the result of the "back-of-the-envelope" calculation for the density of dark matter (Eq. 24.50), estimate the mass density of dark matter in the solar neighborhood. Express your answer in units of kg m$^{-3}$, M$_\odot$ pc$^{-3}$, and M$_\odot$ AU$^{-3}$. How does your answer compare with the stellar mass density in the solar neighborhood?

**24.22** **(a)** Assuming that Eq. (24.51) is valid for any arbitrary distance from the center of the Galaxy, show that the amount of dark matter interior to a radius $r$ is given by the expression

$$M_r = 4\pi \rho_0 a^2 \left[ r - a \tan^{-1}\left(\frac{r}{a}\right) \right].$$

**(b)** If $5.4 \times 10^{11}$ M$_\odot$ of dark matter is located within 50 kpc of the Galactic center, determine $\rho_0$ in units of M$_\odot$ pc$^{-1}$. Repeat your calculation if $1.9 \times 10^{12}$ M$_\odot$ is located within 230 kpc of the Galactic center. Assume that $a = 2.8$ kpc.

**24.23** Using Eq. (24.52) for the density profile of the dark matter halo, show that

**(a)** $\rho_{NFW} \propto r^{-1}$ for $r \ll a$ and $\rho_{NFW} \propto r^{-3}$ for $r \gg a$.

**(b)** the integral of the mass from $r = 0$ to $r \to \infty$ is infinite.

**24.24** Using data provided in the text for the mass of the dark matter halo interior to 50 kpc and interior to 230 kpc, estimate the values for the constants $\rho_0$ and $a$ in the NFW version of the dark matter halo density profile (Eq. 24.52).

**24.25** **(a)** From the information given in Table 24.1 and in the text, determine the approximate mass-to-light ratio of the Galaxy interior to a radius of 25 kpc from the center.

**(b)** Repeat your calculation for a radius of 100 kpc. What can you conclude about the effect that dark matter might have on the average mass-to-light ratio of the universe?

**24.26** The $r^{-2}$ dependence of Coulomb's electrostatic force law allows the construction of Gauss's law for electric fields, which has the form

$$\oint \mathbf{E} \cdot d\mathbf{A} = \frac{Q_{in}}{\epsilon_0},$$

where the integral is taken over a closed surface that bounds the enclosed charge, $Q_{in}$. Because Newton's gravitational force law also varies as $r^{-2}$, it is possible to derive a gravitational "Gauss's law." The form of this gravitational version is

$$\oint \mathbf{g} \cdot d\mathbf{A} = -4\pi G M_{in}, \tag{24.56}$$

where the integral is over a closed surface that bounds the mass $M_{in}$, and $\mathbf{g}$ is the local acceleration of gravity at the position of $d\mathbf{A}$. The differential area vector ($d\mathbf{A}$) is assumed to be normal to the surface everywhere and is directed outward, away from the enclosed volume.

Show that if a spherical gravitational Gaussian surface is employed that is centered on and surrounds a spherically symmetric mass distribution, Eq. (24.56) can be used to solve for $\mathbf{g}$. The result is the usual gravitational acceleration vector around a spherically symmetric mass.

**24.27** We learned in Sections 24.2 and 24.3 that the Sun is currently located 30 pc north of the Galactic midplane and moving away from it with a velocity $w_\odot = 7.2$ km s$^{-1}$. The $z$ component of the gravitational acceleration vector is directed toward the midplane, so the Sun's peculiar velocity in the $z$ direction must be decreasing. Eventually the direction of motion will reverse and the Sun will pass through the midplane heading in the opposite direction. At that time the direction of the $z$ component of the gravitational acceleration vector will also reverse, ultimately causing the Sun to move northward again. This oscillatory behavior above and below the midplane has a well-defined period and amplitude that we will estimate in this problem.

Assume that the disk of the Milky Way has a radius that is much larger than its thickness. In this case, as long as we confine ourselves to regions near the midplane, the disk appears to be infinite in the $z = 0$ plane. Consequently, the gravitational acceleration vector is always oriented in the $\pm z$ direction. We will neglect the radial acceleration component in this problem.

(a) By constructing an appropriate Gaussian surface and using Eq. (24.56), derive an expression for the gravitational acceleration vector at a height $z$ above the midplane, assuming that the Sun always remains inside the disk of constant density $\rho$.

(b) Using Newton's second law, show that the motion of the Sun in the $z$ direction can be described by a differential equation of the form

$$\frac{d^2 z}{dt^2} + kz = 0.$$

Express $k$ in terms of $\rho$ and $G$. This is just the familiar equation for simple harmonic motion.

(c) Find general expressions for $z$ and $w$ as functions of time.

(d) If the total mass density in the solar neighborhood (including stars, gas, dust, and dark matter) is $0.15$ M$_\odot$ pc$^{-3}$, estimate the oscillation period.

(e) By combining the current determinations of $z_\odot$ and $w_\odot$, estimate the amplitude of the solar oscillation and compare your answer with the vertical scale height of the thin disk.

(f) Approximately how many vertical oscillations does the Sun execute during one orbital period around the Galactic center?

**24.28** Show that

$$d = \frac{\langle v_r \rangle \tan \phi}{\langle \mu \rangle},$$

leads to Eq. (24.54) with the appropriate change in units.

**24.29** Refer to the data in Appendix E for this problem.

    **(a)** What angle does Altair's space motion vector make with its radial-velocity vector?

    **(b)** Find Altair's transverse velocity and space motion relative to the Sun.

**24.30** Using Newtonian gravity, estimate the amount of energy required to move $10^7$ $M_\odot$ from a position just above the event horizon of the supermassive black hole at the center of the Galaxy to 3 kpc, the present location of the expanding arm. Compare your answer to the amount of energy liberated in a typical Type II supernova.

**24.31** If the accretion rate at the Galactic center is $10^{-3}$ $M_\odot$ $yr^{-1}$ and if it has remained constant over the past 5 billion years, how much mass has fallen into the center over that period of time? Compare your answer with the estimated mass of a possible supermassive black hole residing in the center of our Galaxy.

**24.32** **(a)** Compute the lowest possible density of Sgr A$^\star$ based on the data obtained from the orbit of S2. Assume a spherically symmetric mass distribution.

    **(b)** Assuming a mass of $3.7 \times 10^6$ $M_\odot$ and a radius of 1 AU (roughly the current limit of resolution of the center of the Milky Way), estimate the density of Sgr A$^\star$. Express your answer in kg $m^{-3}$, $M_\odot$ $AU^{-3}$, and $M_\odot$ $pc^{-3}$.

**24.33** Using the data found in the text, calculate the speed of S2 when it is closest to Sgr A$^\star$.

**24.34** Using Newtonian gravity, estimate the Roche limit of a supermassive black hole of mass $3.7 \times 10^6$ $M_\odot$ (assume that a 1 $M_\odot$ main-sequence star is tidally disrupted). How does your answer compare with the black hole's Schwarzschild radius? *Hint:* Begin with Eq. (19.4), substituting the appropriate average densities and radii.

**24.35** Estimate the Eddington luminosity of a black hole with the mass of Sgr A$^\star$. What is the ratio of the upper limit of the bolometric luminosity of Sgr A$^\star$ to its Eddington luminosity?

**24.36** In this problem you will construct a crude model for the mass distribution and velocity curve in the inner 1 kpc of the Galaxy. Assume that a point (a black hole) of mass $M_0 = 3.7 \times 10^6$ $M_\odot$ is located at the center of the Galaxy and that the remainder of the mass has an isothermal density distribution that varies as $r^{-2}$.

    **(a)** Show that if the mass distribution is spherically symmetric, the mass interior to a radius $r$ can be expressed as a function of the form

$$M_r = kr + M_0,$$

    where $k$ is a constant to be determined.

    **(b)** Assuming perfectly circular motion and Newtonian gravity, show that the orbital velocity curve is given by

$$v = \left[ G \left( k + \frac{M_0}{r} \right) \right]^{1/2}.$$

    **(c)** If the orbital velocity is 110 km $s^{-1}$ at 2 pc, determine a value for $k$.

    **(d)** Plot $\log_{10} M_r$ as a function of $\log_{10} r$ over the range 0.01 pc $< r <$ 1 kpc. Express $M_r$ in solar units and $r$ in parsecs. Your graph should be qualitatively similar to the observational data depicted in Fig. 24.33.

    **(e)** Plot $v$ as a function of $\log_{10} r$ over the range 0.01 pc $< r <$ 1 kpc. Express $v$ in km $s^{-1}$ and $r$ in parsecs. At what radius does the contribution of the central point mass begin to become significant?

# The Nature of Galaxies

## 25.1 ■ THE HUBBLE SEQUENCE

As was mentioned at the beginning of Chapter 24, it was in the middle of the eighteenth century that Kant and Wright first suggested that the Milky Way represents a finite-sized disk-like system of stars. In the two centuries of scientific investigation since their proposal, we have indeed come to learn that a major component of our Galaxy is well represented by a disk of stars that also contains a significant amount of gas and dust. As an extension of their philosophical argument about the nature of the Galaxy, Kant went on to suggest that if the Milky Way is limited in extent, perhaps the diffuse and very faint "elliptical nebulae" seen in the night sky might actually be extremely distant disk-like systems, similar to our own but well beyond its boundary.[1] He called these objects *island universes*.

### Cataloging the Island Universes

The true nature of the island universes became a matter of much investigation, and extensive catalogs of these objects were collected. One such catalog we owe to Charles Messier (1730–1817), who, while hunting for comets, recorded 103 fuzzy objects that could otherwise be confused with the intended targets of his search.[2] Although many of the members of the **Messier catalog** are truly gaseous nebulae contained within the Milky Way (such as the Crab supernova remnant and the Orion Nebula, M1 and M42, respectively), and others are stellar clusters (for instance, the Pleiades open cluster is M45 and the great globular cluster in Hercules is M13), the nature of other nebulae, such as M31 in Andromeda (Fig. 24.7), was unknown.

Another catalog of nebulae was produced by William Herschel and subsequently expanded by his son, Sir John Herschel (1792–1871), to include the southern hemisphere.

---

[1] Although today we generally use the word *nebula* to refer to gas and dust clouds, it was originally used to describe any fuzzy patch of light in the sky that could not be resolved into a clear ensemble of stars.
[2] Later, other astronomers added seven additional members to the original list, bringing the total number of Messier objects to 110. The Messier catalog is found in Appendix H.

Later, J. L. E. Dreyer (1852–1926) published the **New General Catalog** (NGC), which was based on the work of the Herschels and contained almost 8000 objects. Like Messier's catalog, the NGC includes many entries that are either gaseous nebulae or stellar clusters located within the Milky Way. However, the true nature of other objects in the catalog remained in question.[3]

It was in 1845 that William Parsons, the third Earl of Rosse (1800–1867), built what was then the largest telescope in the world. Located in Ireland and nicknamed the "Leviathan," the 72-in (1.8-m) instrument was able to resolve, for the first time, the spiral structure in some nebulae. Their pinwheel appearance strongly suggested that these *spiral nebulae* may be rotating. This suspicion was eventually verified by Vesto M. Slipher (1875–1969) in 1912 when he detected Doppler-shifted spectral lines in a number of these objects.

### The Great Shapley–Curtis Debate

The argument over the nature of the nebulae centered on their distances from us and the relative size of the Galaxy. Many astronomers believed that the spiral nebulae resided within the confines of the Milky Way, and others favored the view that they were really Kant's island universes. On April 26, 1920, at the National Academy of Sciences in Washington, D.C., Harlow Shapley of the Mount Wilson Observatory and Heber D. Curtis (1872–1932) of the Lick Observatory met to argue the merits of each point of view. In what has become known as the *Great Debate* in astronomy, Shapley supported the idea that the nebulae are members of our Galaxy. Curtis, on the other hand, was a proponent of the extragalactic interpretation of the data, believing that the nebulae were physically much like the Milky Way, but separated from it.

One of Shapley's strongest points was based on the apparent magnitudes of novae observed in M31. He argued that if the disk of Andromeda were as large as the Milky Way (approximately 100 kpc in diameter by his own recent estimates), then its angular size in the sky ($\sim 3° \times 1°$) would imply a distance to the nebula that was so large as to make the luminosities of the novae in M31 much greater than those found in the Milky Way.

His second major point was based on data of Adrian van Maanen (1884–1946), a well-respected observer, whose proper-motion measurements of M101 seemed to suggest an angular rotation rate of $0.02'' \text{ yr}^{-1}$. If M101 had a diameter similar to Shapley's estimate for the Milky Way, then points near its outer edge would have rotational speeds far in excess of those observed within the Milky Way.

In defense of the extragalactic hypothesis, Curtis argued that the novae observed in spiral nebulae must be at least 150 kpc away from us in order to have intrinsic brightnesses comparable to those in the Milky Way. At this distance, M31 would be similar in size to Kapteyn's much smaller estimate of the diameter of the Galaxy, rather than to Shapley's estimate (recall the discussion in Section 24.1). He also argued that the large radial velocities measured for many spiral nebulae seemed to indicate that they could not remain gravitationally bound within a Kapteyn-model Milky Way. Furthermore, assuming that the transverse velocities of the nebulae are similar in value to their radial velocities, then if the nebulae were close enough to be located within the Milky Way it should be possible to

[3]It is worth pointing out that most of the members in the Messier catalog are contained within the NGC; for instance, M31 is also designated as NGC 224.

measure their proper motions across the sky; however, no such motions had been detected. Finally, for those spiral nebulae that are oriented edge-on, dark absorption regions can be seen. Curtis suggested that if the Milky Way had a similar dark layer, the zone of avoidance would be easily explained.

In the end, neither set of arguments proved to be definitive and the Great Debate served more to highlight the issues than to solve the dilemma. Although, as we now know, errors existed on both sides of the controversy, Shapley's arguments were perhaps the more flawed. Part of the problem rested in his overestimate of the size of the Milky Way's disk. The other difficulty was in his reliance on van Maanen's data, which van Maanen himself later showed to be incorrect. In fact, proper-motion studies were unable to measure any rotation of M101.

The debate was finally settled in 1923 when Edwin Hubble (1889–1953) detected Cepheid variable stars in M31 using the 100-inch telescope at Mount Wilson. By measuring their apparent magnitudes and determining their absolute magnitudes via the period–luminosity relation (Eq. 14.1), he was able to use the distance modulus $m - M$ (Eq. 3.5) to calculate the distance to Andromeda. Hubble's original value of 285 kpc is approximately 2.7 times smaller than the modern estimate of 770 kpc, but it was still good enough to show definitively that the spiral nebulae are indeed island universes.

The process of scientific exploration that led to the realization that our Solar System is not located near the center of the Milky Way, and that the Galaxy is only one of a countless number of galaxies in the universe, represented an advance in our understanding of the universe comparable to that of the Copernican Revolution. Although many people contributed to that new understanding, it was in large part Hubble's pivotal work in determining extragalactic distances that set the stage for many subsequent investigations of galaxies, the largest stellar systems in the universe.[4]

### The Classification of Galaxies

Now that the extragalactic nature of the galaxies had been established, work began on determining their physical properties. As a first step in understanding any new collection of objects, it is necessary to classify them according to their intrinsic characteristics, akin to the zoological classification of various species of animals. Once again Hubble played a key role. In his 1926 paper "Extra-Galactic Nebulae," and later in his book *The Realm of the Nebulae*, Hubble proposed that galaxies be grouped into three primary categories based on their overall appearance. This morphological classification scheme, known as the **Hubble sequence**, divides galaxies into **ellipticals** (E's), **spirals**, and **irregulars** (Irr's).[5] The spirals are further subdivided into two parallel sequences, the **normal spirals** (S's), and the **barred spirals** (SB's). A transitional class of galaxies between ellipticals and spirals, known as **lenticulars**, can be either normal (S0's) or barred (SB0's). Hubble then arranged his morphological sequence in the form of a **tuning-fork diagram**, shown in Fig. 25.1, which explicitly shows the two types of spirals. A galaxy's **Hubble type** is its designation along the Hubble sequence.

---

[4]The challenging and important process of determining the extragalactic distance scale, to which Hubble made another critical contribution, is the subject of Sections 27.1 and 27.2.

[5]A more modern classification designates true irregulars as Ir, as shown in Fig. 25.1.

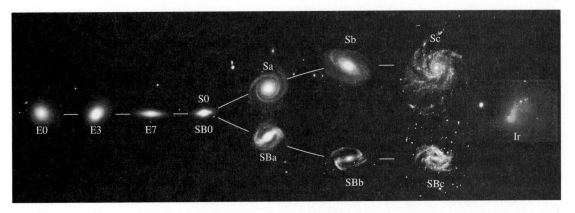

**FIGURE 25.1** Hubble's tuning-fork diagram of galaxy types.

Hubble originally thought (incorrectly) that the tuning-fork diagram could be interpreted as an evolutionary sequence for galaxies. As a result, he referred to galaxies toward the left of the diagram as *early types* and to those toward the right as *late types*, terminology that is still in widespread use today.

Within the category of ellipticals, Hubble made divisions based on the observed **ellipticity** of the galaxy, defined by

$$\epsilon \equiv 1 - \beta/\alpha, \tag{25.1}$$

where $\alpha$ and $\beta$ are the *apparent* major and minor axes of the ellipse, respectively, projected onto the plane of the sky. The Hubble type is then quoted in terms of $10\epsilon$. Ellipticals range from a spherical distribution of stars, E0, to a highly flattened distribution, E7. Galaxies with ellipticities greater than $\epsilon = 0.7$ have never been observed, implying that no E galaxies with intrinsic ellipticities greater than 0.7 appear to exist.

It is important to realize that a difficulty immediately arises with this scheme: The apparent ellipticity may not correspond well to an actual ellipticity since the orientation of the spheroid[6] to our line of sight plays a crucial role in our observations. This effect can be seen in Figs. 25.2 and 25.3, when observers view **oblate** and **prolate** galaxies, respectively. Note that, in general, $a \geq b \geq c$ represents the lengths of the three axes of a **triaxial** spheroidal system. For a sphere $a = b = c$, for a perfectly oblate spheroid $a = b$, and for a perfectly prolate spheroid $b = c$. In general, however, there is no requirement that any of the axes of a spheroidal system have equal lengths.

The physical properties of elliptical galaxies cover an enormous range. Their absolute $B$ magnitudes may be as dim as $-8$ or brighter than $-23$, their masses (including both luminous and dark matter) vary from as little as $10^7 \, M_\odot$ to more than $10^{13} \, M_\odot$, and their diameters can be as small as a few tenths of a kiloparsec or as large as hundreds of kiloparsecs. The giant elliptical galaxies are among the largest objects in the universe, while the smallest dwarfs are comparable in size to a typical globular cluster. The lenticular galaxies have masses and luminosities comparable to the larger ellipticals. Although the giant ellipticals and the

---

[6]The term *spheroid* is commonly used to mean *axisymmetric ellipsoid*.

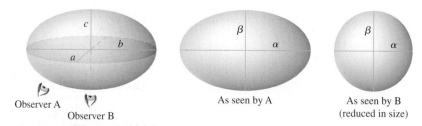

**FIGURE 25.2**  An oblate spheroidal galaxy has axis lengths $a = b$ and $c < a$. If $c/a = 0.6$, the apparent shape resembles an E4 galaxy ($\beta/\alpha = 0.6$) when seen by observer A. The same galaxy appears as an E0 when seen by observer B ($\beta/\alpha = 1$).

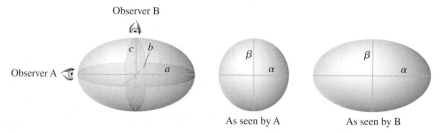

**FIGURE 25.3**  A prolate spheroidal galaxy has axis lengths $b = c$ and $a > b$. If $b/a = 0.6$, the apparent shape resembles an E0 galaxy when seen by observer A. The same galaxy appears as an E4 when seen by observer B.

lenticulars are the easiest to observe, the dwarfs are by far the most numerous. Typical examples of elliptical galaxies are shown in Fig. 25.4, along with an S0 and an SB0. The physical characteristics of these galaxies will be discussed in more detail in Section 25.4.

Hubble subdivided the spiral sequences into Sa, Sab, Sb, Sbc, Sc, and SBa, SBab, SBb, SBbc, SBc. The galaxies with the most prominent bulges (the largest bulge-to-disk luminosity ratios, $L_{\text{bulge}}/L_{\text{disk}} \sim 0.3$), the most tightly wound spiral arms (with pitch angles[7] of approximately $6°$), and the smoothest distribution of stars in the arms are classified as Sa's (or SBa's), while Sc's (or SBc's) have smaller bulge-to-disk ratios ($L_{\text{bulge}}/L_{\text{disk}} \sim 0.05$), more loosely wound spiral arms ($\sim 18°$), and spiral arms that resolve into clumps of stars and H II regions. Examples of normal and barred spiral galaxies are shown in Figs. 25.5 and 25.6, respectively. M31 (see Fig. 24.7) and NGC 891 (see Fig. 24.9) are Sb's, whereas the Milky Way is probably an SBbc (Fig. 24.5 reflects the evidence for a central bar in our Galaxy).

Sa–Sc (SBa–SBc) galaxies tend to have much smaller variations in their physical parameters than do ellipticals. On average, spirals also tend to be among the largest galaxies in the universe, with absolute $B$ magnitudes from $-16$ to less than $-23$, masses (including both luminous and dark matter) between $10^9$ $M_\odot$ and $10^{12}$ $M_\odot$, and disk diameters of 5 to 100 kpc.

---

[7]The *pitch angle* is defined as the angle between the tangent to the spiral arm and the tangent to a perfect circle, measured at the point where the arm and the circle intersect.

(a)     (b)     (c)

(d)     (e)     (f)

**FIGURE 25.4** Typical early-type galaxies. (a) IC 4296 (E0), (b) NGC 4365 (E3), (c) NGC 4564 (E6), (d) NGC 4623 (E7), (e) NGC 4251 (S0), (f) NGC 4340 (RSB0). (Images from Sandage and Bedke, *The Carnegie Atlas of Galaxies*, Carnegie Institution of Washington, Washington, D.C., 1994.)

Hubble split the remaining category of irregulars into Irr I if there was at least some hint of an organized structure, such as spiral arms, and Irr II for the most extremely disorganized structures. Both the Large Magellanic Cloud and the Small Magellanic Cloud are examples of Irr I galaxies, while M82 (NGC 3034) is an example of an Irr II; see Fig. 25.7.

Irregular galaxies have a wide range of characteristics, although they tend not to be particularly large. Typically their absolute $B$ magnitudes vary from $-13$ to $-20$, they have masses of between $10^8$ $M_\odot$ and $10^{10}$ $M_\odot$, and their diameters range from 1 to 10 kpc. Most irregulars also tend to have noticeable bars that are often off-center.

Since the publication of Hubble's tuning-fork diagram, astronomers have made numerous modifications to his original classification scheme. For instance, Gerard de Vaucouleurs suggested the elimination of the irregular classifications, Irr I and Irr II, in favor of the addition of other morphological classes later than Sc (or SBc). Those galaxies that were binned into Irr I have been designated Sd (SBd), Sm (SBm), or Im (where m stands for Magellanic type); for instance, the LMC is classified as an SBm, and the SMC is an Im. The truly irregular galaxies are simply designated Ir, such as M82 (which is why Ir is used as the designation in Fig. 25.1). Sandage and Brucato further suggested that the Ir class should more appropriately be termed **amorphous** to indicate the lack of any organized structure. As we shall see, spirals of Hubble-type Sd and later tend to be significantly smaller than earlier-type spirals; consequently, they are sometimes referred to as **dwarf spirals**.

**FIGURE 25.5** Typical normal spirals. (a) NGC 7096 (Sa(r)I), (b) M81/NGC 3031 (Sb(r)I–II), (c) M101/NGC 5457/Pinwheel (Sc(s)I), (d) M104/NGC 4594/Sombrero (Sa/Sb) seen nearly edge on. (Images from Sandage and Bedke, *The Carnegie Atlas of Galaxies*, Carnegie Institution of Washington, Washington, D.C., 1994.)

In order to make finer distinctions between normal and barred spirals, de Vaucouleurs had also suggested referring to normal spirals as SA rather than simply S. Intermediate types with weak bars are then characterized as SAB, and strongly barred galaxies are SB.

As a further refinement to the system, the lenticular galaxies are also sometimes subdivided according to the amount of dust absorption in their disks. $S0_1$ galaxies have no discernable dust their disks, while $S0_3$ galaxies have significant amounts of dust, and similarly for $SB0_1$ through $SB0_3$.

Thus, the modern sequence from early ellipticals through normal late-type galaxies is (see also Fig. 25.1)

$$E0, E1, \ldots, E7, S0_1, S0_2, S0_3, Sa, Sab, Sb, Sbc, Sc, Scd, Sd, Sm, Im, Ir.$$

A similar sequence exists for barred spirals.

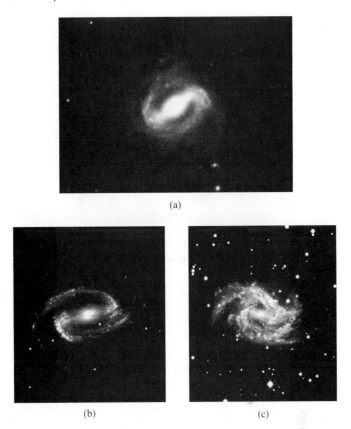

**FIGURE 25.6** Typical barred spirals. (a) NGC 175 (SBab(s) I–II), (b) NGC 1300 (SBb(s) I), (c) NGC 2525 (SBc(s)II). (Images from Sandage and Bedke, *The Carnegie Atlas of Galaxies*, Carnegie Institution of Washington, Washington, D.C., 1994.)

Sidney van den Bergh introduced the **luminosity class** for spirals. The class ranges from I through V, with I representing those spirals with well-defined arms; galaxies with the least distinct arms are classified as V. M31 is classified as SbI–II (intermediate between I and II), the Milky Way Galaxy is an SBbcI–II, M101 is an ScI, the LMC is an SBmIII, and the SMC is an ImIV–V. Except for the largest elliptical galaxies (to be discussed further in Section 25.4), the Milky Way and M31 are among some of the largest and brightest galaxies in the universe. It is important to note, however, that despite its name, luminosity class does not necessarily correlate well with absolute magnitude.

Besides their striking arms, spirals also show an amazing array of more complex and subtle features. While some galaxies have spiral arms that can be followed nearly all the way into the center, others have arms that appear to terminate at the location of an inner ring. Special designations further help to classify these systems. M101 [Fig. 25.5(c)] is a galaxy of the former type and is labeled as an Sc(s)I, where (s) designates that the spiral can be traced to the center of the galaxy. On the other hand, NGC 7096 [Fig. 25.5(a)] and M81 [Fig. 25.5(b)] are galaxies of the later type and are classified as Sa(r)I and Sb(r)I–II,

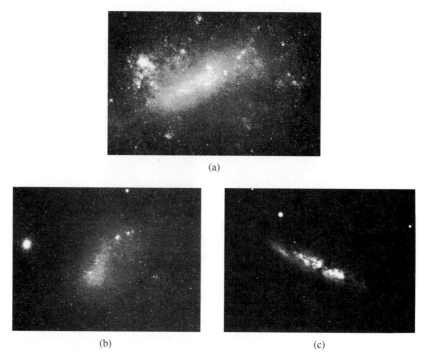

(a)

(b)                                    (c)

**FIGURE 25.7**    Examples of irregular galaxies. (a) The Large Magellanic Cloud (Irr I/SBmIII), (b) the Small Magellanic Cloud (Irr I/ImIV–V), (c) M82/NGC 3034 (Irr II/Ir/Amorphous). (Images from Sandage and Bedke, *The Carnegie Atlas of Galaxies*, Carnegie Institution of Washington, Washington, D.C., 1994.)

respectively, where (r) indicates an inner ring. Galaxies may also have outer rings that can be identified. One such example is NGC 4340 [Fig. 25.4(f)], which is designated as an RSB0, where the R prefix stands for outer ring. In some cases, galaxies may have both an inner ring and an outer ring.

Given the enormous range in galaxy morphological types, it is not surprising that any classification scheme will necessarily be complex. Fortunately, the Hubble sequence, with its various permutations and enhancements, has greatly facilitated the process of trying to understand the nature of galaxies.

## 25.2 ■ SPIRAL AND IRREGULAR GALAXIES

Hubble's classification scheme for late-type galaxies has proved to be very successful in organizing our study of these objects. Not only do bulge-to-disk ratios, the tightness of the spiral arms, and the ability to resolve the arms into stars and H II regions all correlate well with Hubble type, but so do a host of other physical parameters.

Tables 25.1 and 25.2 summarize the characteristics of late-type galaxies, the details of which will be discussed later. Although the spread in parameters can be quite large, trends in Hubble type are clearly evident. For instance, if we compare an Sa galaxy with an Sc

**TABLE 25.1**   Characteristics of Early Spiral Galaxies.

|  | Sa | Sb | Sc |
|---|---|---|---|
| $M_B$ | $-17$ to $-23$ | $-17$ to $-23$ | $-16$ to $-22$ |
| $M$ ($M_\odot$) | $10^9$–$10^{12}$ | $10^9$–$10^{12}$ | $10^9$–$10^{12}$ |
| $\langle L_{\text{bulge}}/L_{\text{total}} \rangle_B$ | 0.3 | 0.13 | 0.05 |
| Diameter ($D_{25}$, kpc) | 5–100 | 5–100 | 5–100 |
| $\langle M/L_B \rangle$ ($M_\odot/L_\odot$) | $6.2 \pm 0.6$ | $4.5 \pm 0.4$ | $2.6 \pm 0.2$ |
| $\langle V_{\text{max}} \rangle$ (km s$^{-1}$) | 299 | 222 | 175 |
| $V_{\text{max}}$ range (km s$^{-1}$) | 163–367 | 144–330 | 99–304 |
| pitch angle | $\sim 6°$ | $\sim 12°$ | $\sim 18°$ |
| $\langle B - V \rangle$ | 0.75 | 0.64 | 0.52 |
| $\langle M_{\text{gas}}/M_{\text{total}} \rangle$ | 0.04 | 0.08 | 0.16 |
| $\langle M_{H_2}/M_{H\,I} \rangle$ | $2.2 \pm 0.6$ (Sab) | $1.8 \pm 0.3$ | $0.73 \pm 0.13$ |
| $\langle S_N \rangle$ | $1.2 \pm 0.2$ | $1.2 \pm 0.2$ | $0.5 \pm 0.2$ |

**TABLE 25.2**   Characteristics of Late Spiral and Irregular Galaxies.

|  | Sd/Sm | Im/Ir |
|---|---|---|
| $M_B$ | $-15$ to $-20$ | $-13$ to $-18$ |
| $M$ ($M_\odot$) | $10^8$–$10^{10}$ | $10^8$–$10^{10}$ |
| Diameter ($D_{25}$, kpc) | 0.5–50 | 0.5–50 |
| $\langle M/L_B \rangle$ ($M_\odot/L_\odot$) | $\sim 1$ | $\sim 1$ |
| $V_{\text{max}}$ range (km s$^{-1}$) | 80–120 | 50–70 |
| $\langle B - V \rangle$ | 0.47 | 0.37 |
| $\langle M_{\text{gas}}/M_{\text{total}} \rangle$ | 0.25 (Scd) | 0.5–0.9 |
| $\langle M_{H_2}/M_{H\,I} \rangle$ | 0.03–0.3 | $\sim 0$ |
| $\langle S_N \rangle$ | $0.5 \pm 0.2$ | $0.5 \pm 0.2$ |

galaxy of comparable luminosity, the Sa will be more massive (larger $M/L_B$), have a higher peak in its rotation curve ($V_{\text{max}}$), have a smaller mass fraction of gas and dust, and contain a higher proportion of older, red stars.

**The *K*-Correction**

Before we can consider these correlations in more detail, it is necessary to discuss the problems associated determining the brightnesses of galaxies. As is the case with determining absolute magnitudes for stars, calculating the absolute magnitudes of galaxies requires making corrections to their observed apparent magnitudes if we are to properly account for the effects of extinction, both within the Milky Way and within the target galaxy (see Eq. 12.1). (Extinction is generally negligible in the nearly empty space between galaxies.) Furthermore, for extragalactic objects another important correction must be considered as well.

Since most galaxies are observed to have measurable redshifts (the topic of Section 27.2), some or even most of the light that would normally fall within the wavelength band of interest, say the $B$ band, would be redshifted to longer-wavelength regions. Accounting for this effect is known as the **$K$-correction**. As we will learn, the $K$-correction is most severe for very distant galaxies. If this effect were not considered, possible conclusions about the evolution of galaxies would be likely to be in error (see Sections 26.2 and 29.4).

### The Brightness of the Background Sky

Another problem that arises when making observations of faint galaxies, or when measuring their outermost regions, is the competition with the brightness of the background sky. The dimly glowing night sky has an average surface brightness of about $\mu_{sky} = 22$ $B$-mag arcsec$^{-2}$ (measured in the $B$ wavelength band). Sources of this background light include light pollution from nearby cities, photochemical reactions in Earth's upper atmosphere, the zodiacal light, unresolved stars in the Milky Way, and unresolved galaxies. However, in modern photometric studies using CCDs, the surface brightnesses of galaxies can be measured down to levels of $29$ $B$-mag arcsec$^{-2}$ or fainter. Consequently, to accurately determine the light distribution of a galaxy at these extremely faint levels, it is necessary to subtract the contribution from the background sky.

### Isophotes and the de Vaucouleurs Profile

Once the sky subtraction is performed, it becomes possible to map contours of constant surface brightness. Such contours are known as **isophotes** (lines of constant photon number). In specifying the "radius" of a galaxy, it is necessary to define the surface brightness of the isophote being used to determine that radius. We encountered this problem in trying to identify a radius for our own Galaxy in Chapter 24; no definite cutoff is known to exist in either the exponential distribution of the disk or the $r^{1/4}$ distribution of the spheroid. One commonly used radius, introduced by Erik Holmberg, a Swedish astronomer, is the **Holmberg radius**, $r_H$, defined to be the projected length of the semimajor axis of an ellipsoid having an isophotal surface brightness of $\mu_H = 26.5$ $B$-mag arcsec$^{-2}$. A second standard radius in frequent use is the **effective radius**, $r_e$, the projected radius within which one-half of the galaxy's light is emitted.[8] The surface brightness level at $r_e$, designated $\mu_e$, depends on the distribution of the surface brightness with radius.

For the bulges of spiral galaxies, and for large ellipticals, the surface brightness distribution typically follows an $r^{1/4}$ law given by

$$\mu(r) = \mu_e + 8.3268 \left[ \left( \frac{r}{r_e} \right)^{1/4} - 1 \right]. \qquad (25.2)$$

This is just the $r^{1/4}$ de Vaucouleurs profile discussed in Section 24.2 (Eq. 24.13), written in units of mag arcsec$^{-2}$, rather than $L_\odot$ pc$^{-2}$. It is left as an exercise (Problem 25.5) to show that these two forms are equivalent.

---

[8]The effective radius was first introduced on page 892.

Disks are frequently modeled with an exponential decay, similar to the one given in Eq. (24.10) for our Galaxy. However, as with the $r^{1/4}$ law, the disk luminosity per unit area can be written in units of mag arcsec$^{-2}$ as

$$\mu(r) = \mu_0 + 1.09 \left(\frac{r}{h_r}\right), \tag{25.3}$$

where $h_r$ is the characteristic scale length of the disk along its midplane (see Problem 25.6).

A generalized version of the $r^{1/4}$ law is frequently used in which $1/4$ is replaced by $1/n$. The resulting **generalized de Vaucouleurs profile**, also known as the **Sérsic profile** (named for José Luis Sérsic), has the form

$$\mu(r) = \mu_e + 8.3268 \left[ \left(\frac{r}{r_e}\right)^{1/n} - 1 \right], \tag{25.4}$$

where $\mu_e$, $r_e$, and $n$ are all free parameters used to obtain the best possible fit to the actual surface brightness profile. Note that when $n = 1$, Eq. (25.3) is a special case of Eq. (25.4), where $\mu_0$ and $h_r$ are written in terms of $\mu_e$ and $r_e$, respectively.

### The Rotation Curves of Galaxies

While surface brightness profiles, such as the generalized de Vaucouleurs profile, sample the distribution of luminous matter in a galaxy, they do not reveal the distribution of the galaxy's dark matter. A direct means to determine the distribution of all matter, luminous and dark, is to measure the galaxy's rotation curve, as was discussed for the case of the Milky Way in Chapter 24. Examples of rotation curves, averaged over groups of galaxies of a given $M_B$ range and Hubble type, are shown in Fig. 25.8. You might also refer again to Fig. 24.27, which shows actual rotation curves for galaxies of various Hubble types.

When rotation curves are compared with either luminosity or Hubble type, a number of correlations are found. For instance, with increasing luminosity in the $B$ band, $L_B$, the rotation curves tend to rise more rapidly with distance from the center and peak at higher maximum velocities ($V_{max}$). For galaxies of equal $B$-band luminosities, spirals of earlier type have larger values of $V_{max}$. Within a given Hubble type, galaxies that are more luminous have larger values of $V_{max}$. Also, for a given value of $V_{max}$, the rotation curves tend to rise slightly more rapidly with radius for galaxies of progressively earlier type. The fact that galaxies of different Hubble types, and therefore very different bulge-to-disk luminosity ratios, exhibit rotation curves that are very similar in form if not in amplitude suggests that the shapes of their gravitational potentials do not necessarily follow the distribution of luminous matter. This behavior is believed to be a signature of the existence of dark matter in these galaxies.

Although the maximum rotational velocity within the disk increases for earlier-type galaxies, a wide range in $V_{max}$ exists for each type. For typical samples of spirals of type Sc and earlier, the mean maximum rotation velocities are $V_{max} = 299$ km s$^{-1}$ for Sa, 222 km s$^{-1}$ for Sb, and 175 km s$^{-1}$ for Sc, while the ranges in values are 163 km s$^{-1}$ to

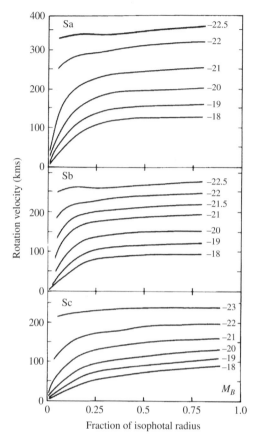

**FIGURE 25.8**    Variations in the average rotation curves of Sa, Sb, and Sc galaxies for various values of absolute magnitude in the $B$ band. (Figure adapted from Rubin et al., *Ap. J.*, *289*, 81, 1985.)

367 km s$^{-1}$, 144 km s$^{-1}$ to 330 km s$^{-1}$, and 99 km s$^{-1}$ to 304 km s$^{-1}$, respectively. Notice that the value of $V_{max} \simeq 250$ km s$^{-1}$ for the Milky Way Galaxy (believed to be an SBbc; see Fig. 24.25) is only slightly greater than the mean value of $V_{max}$ for Sb's.

The corresponding maximum rotation velocities for irregular galaxies is significantly lower than it is for earlier-type spirals, typically ranging from 50 to 70 km s$^{-1}$. This seems to suggest that a minimum rotation speed of roughly 50 to 100 km s$^{-1}$ may be required for the development of a well-organized spiral pattern. The slower rotation velocities of Im's imply that their values of the rotational angular momentum per unit mass are only about 10% of the value found for our Galaxy in the solar neighborhood.

### The Tully–Fisher Relation

As the preceding discussion suggests, a relationship exists between the luminosity of a spiral galaxy and its maximum rotation velocity. This correlation, now known as the **Tully–Fisher relation**, was first determined in 1977 by R. Brent Tully and J. Richard Fisher when

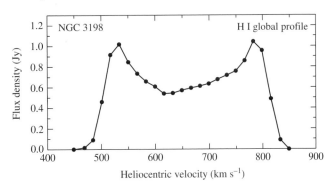

**FIGURE 25.9** The global H I profile for NGC 3198. (Data from Begeman, *Astron. Astrophys.*, *223*, 81, 1991.)

they measured the Doppler-broadened 21-cm radio emission line of neutral hydrogen in a sample of spiral galaxies. When the 21-cm line is sampled across the entire galaxy at one time, this line typically displays a double peak, as shown in Fig. 25.9. The double peak arises because of the flat rotation curve of the galaxy, which generally has the highest rotational velocity in the flat part of the curve. Since so much of the H I participates in the rotation at this maximum velocity, the flux density is greatest at this value. The double peak occurs because a portion of the disk is rotating toward the observer, causing the line to be blueshifted, and a portion of the disk is rotating away from the observer, resulting in a redshifted line. The average radial velocity of the galaxy relative to the observer is the midpoint value between the two peaks.

The shift $\Delta\lambda$ of a peak from its rest wavelength is given by Eq. (4.39),

$$\frac{\Delta\lambda}{\lambda_{\text{rest}}} \simeq \frac{v_r}{c} = \frac{V\sin i}{c}.$$

Here, $v_r$ is the radial velocity, and $i$ is the angle of inclination between the observer's line of sight and the direction perpendicular to the galactic plane (so $i = 90°$ when viewing the galaxy edge-on).

Figure 25.10 shows the Tully–Fisher relation as $M_B$ vs. $V_{\text{max}}$ for a sample of Sa, Sb, and Sc galaxies. Notice the shift to lower values of $V_{\text{max}}$ for galaxies of later Hubble type but with similar $M_B$'s. When the data are fitted with linear relations that depend on Hubble type, we find that

$$M_B = -9.95\log_{10} V_{\text{max}} + 3.15 \quad \text{(Sa)}, \tag{25.5}$$

$$M_B = -10.2\log_{10} V_{\text{max}} + 2.71 \quad \text{(Sb)}, \tag{25.6}$$

$$M_B = -11.0\log_{10} V_{\text{max}} + 3.31 \quad \text{(Sc)}. \tag{25.7}$$

The Tully–Fisher relation can be further refined and tightened if observations are made at infrared wavelengths. This offers two advantages. Observing at dust-penetrating infrared

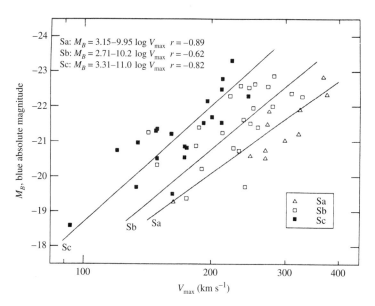

**FIGURE 25.10** The Tully–Fisher relation for early spiral galaxies. (Figure adapted from Rubin et al., *Ap. J.*, *289*, 81, 1985.)

wavelengths reduces extinction by a factor of 10. In addition, the infrared light comes primarily from late-type giant stars that are good tracers of the galaxy's overall luminous mass distribution; the $B$ band tends to emphasize young, hot stars in regions of recent star formation. One expression of the Tully–Fisher relation in the infrared $H$ wavelength band (1.66 $\mu$m), devised by Pierce and Tully (1992), is

$$M_H^i = -9.50(\log_{10} W_R^i - 2.50) - 21.67 \pm 0.08. \qquad (25.8)$$

$W_R^i$ is a measure of the rotation of the galaxy defined as

$$W_R^i \equiv (W_{20} - W_{\rm rand})/\sin i, \qquad (25.9)$$

where $W_{20}$ is the velocity difference between the blueshifted and redshifted emission in the $H$ band when the intensity of the emission is 20% of its blue and red peak values. $W_{\rm rand}$ is a measure of the random velocities superimposed on observed velocities due to noncircular orbital motions in the galaxy.[9] Finally, $i$ is the inclination angle of the plane of the galaxy. An example of the $H$-band Tully–Fisher relation (Eq. 25.8) for galaxies in three clusters is shown in Fig. 25.11.

Although the exact form of the Tully–Fisher relation depends on the distribution of mass within galaxies, as well as on variations in their mass-to-light ratios, we can still gain some insight into its origin. As is evident in Figs. 24.27 and 25.8, spiral galaxies have nearly

---

[9]For a Gaussian random velocity distribution, it can be shown that $W_{\rm rand}$ is related to the velocity dispersion $\sigma$ by $W_{\rm rand} = 3.6\sigma$.

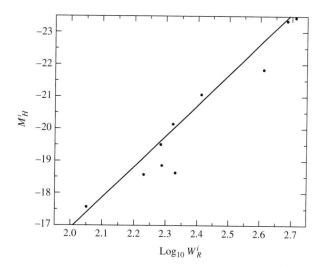

**FIGURE 25.11**   The Tully–Fisher relation in the infrared $H$-band (1.66 $\mu$m) as given by Eq. (25.8). The data are for galaxies in the Local, Sculptor, and M81 groups of galaxies. (Data from Pierce and Tully, *Ap. J., 387*, 47, 1992.)

flat rotation curves beyond a few kiloparsecs. In Section 24.3 we saw that this implies that the mass contained within a radius $r$ is given by Eq. (24.49). Evaluating this for the entire galaxy ($r \rightarrow R$ and $M_r \rightarrow M$) results in

$$M = \frac{V_{\max}^2 R}{G}, \tag{25.10}$$

where the maximum rotation speed, $V_{\max}$, is equated with the flat portion of the rotation curve. Now, *if* the mass-to-light ratio has the same value for all spirals ($M/L \equiv 1/C_{\text{ML}}$, where $C_{\text{ML}}$ is a constant), then

$$L = C_{\text{ML}} \frac{V_{\max}^2 R}{G}.$$

Finally, if we make the crude assumption that all spirals have the same surface brightness at their centers, then $L/R^2 \equiv C_{\text{SB}}$, where $C_{\text{SB}}$ is another constant. Eliminating $R$ from the expression for the luminosity, we obtain

$$L = \frac{C_{\text{ML}}^2}{C_{\text{SB}}} \frac{V_{\max}^4}{G^2} = C V_{\max}^4,$$

where $C$ incorporates the other constants. The absolute magnitude comes from Eq. (3.8),

$$M = M_{\text{Sun}} - 2.5 \log_{10} \left( \frac{L}{L_{\odot}} \right)$$

$$= M_{\text{Sun}} - 2.5 \log_{10} V_{\max}^4 - 2.5 \log_{10} C + 2.5 \log_{10} L_{\odot}$$

$$= -10 \log_{10} V_{\max} + \text{constant}.$$

Although the additive (zero-point) constant remains unevaluated, this simple argument nearly reproduces the leading coefficients of Eqs. (25.5–25.8), the slopes of the $B$-band Tully–Fisher relations for Sa's, Sb's, and Sc's, and the slope of the collective relationship in the $H$ band.

### Radius–Luminosity Relation

Another important pattern also emerges in the data of early-type spiral galaxies (e.g., Sa–Sc): Radius increases with increasing luminosity, independent of Hubble type. At the disk radius ($R_{25}$) corresponding to a surface-brightness level of 25 $B$-mag arcsec$^{-2}$, the data are well represented by the linear relationship

$$\log_{10} R_{25} = -0.249 M_B - 4.00, \qquad (25.11)$$

where $R_{25}$ is measured in units of kpc.

### Masses and Mass-to-Light Ratios

Through the use of the Tully–Fisher relation (Eqs. 25.5–25.7), combined with Eqs. (25.11) and (24.49), it is now possible to estimate both the masses of early-type spiral galaxies and their mass-to-light ratios interior to $R_{25}$. While masses range from greater than $10^9$ $M_\odot$ to at most about $10^{12}$ $M_\odot$, there is only a very weak relationship between a galaxy's mass and its Hubble type (on average, Sc's are *slightly* less massive than Sa's), as might be expected, given the lack of any dependence on Hubble type in the correlation between $R_{25}$ and $M_B$. However, a correlation does exist for the mass-to-light ratios, with the average value of $M/L_B$ decreasing with progressively later Hubble type; $\langle M/L_B \rangle = 6.2 \pm 0.6$ for Sa's, $4.5 \pm 0.4$ for Sb's, and $2.6 \pm 0.2$ for Sc's.

### Colors and the Abundance of Gas and Dust

The trend in $M/L_B$ suggests that Sc's tend to have a greater fraction of massive main-sequence stars relative to earlier spirals (recall from the mass–luminosity relation shown in Fig. 7.7 that upper main-sequence stars have low mass-to-light ratios). If this is the case, we should also expect Sc's to be bluer than Sa's and Sb's, which is just what is observed. The mean values of the color index, $B - V$, decrease with later Hubble types: 0.75 for Sa's, 0.64 for Sb's, and 0.52 for Sc's. For successively later-type galaxies, progressively greater portions of the overall light from spirals is emitted in bluer wavelength regions, implying an increasingly greater fraction of younger, more massive, main-sequence stars.[10]

Irregulars tend to be the bluest of all galaxies represented by the Hubble sequence, with characteristic values of $B - V \sim 0.4$. Furthermore, Ir's often get bluer toward their centers, rather than redder as is the case for early-type spirals. This suggests that irregulars are still actively manufacturing stars in their central regions. For instance, the LMC and the SMC still appear to be making blue globular clusters in their disks.

---

[10]It is an unfortunate consequence of the historical development of the terminology that galaxies of later Hubble type are actually dominated by stars lying on the early (upper) part of the main sequence.

Since blue main-sequence stars are short-lived, they must have formed relatively recently. Presumably an abundant supply of gas and dust exists in Sc's from which these stars can be produced. Based on the analysis of 21-cm radiation, $H\alpha$ emission, and CO emission (which is a good tracer of $H_2$), we find that the mass fraction of gas relative to the total mass interior to $R_{25}$ increases steadily from Sa's to Sc's and later. For instance, $\langle M_{gas}/M_{total}\rangle = 0.04$ for Sa's, 0.08 for Sb's, 0.16 for Sc's, and 0.25 for Scd's. This agrees with the observation that the brightness of H II emission also increases with later Hubble types.

The relative amounts of atomic and molecular hydrogen also change with Hubble type. For Sab's, $\langle M_{H_2}/M_{H\,I}\rangle = 2.2 \pm 0.6$, decreasing to $1.8 \pm 0.3$ for Sb's, $0.73 \pm 0.13$ for Sc's, and $0.29 \pm 0.07$ for Scd's. This observation has been interpreted as implying that Sa's are somewhat more centrally condensed, containing correspondingly deeper gravitational wells in which gas can collect and combine to form molecules. This means that the interstellar medium of Sa's is dominated by molecular gas while the interstellar medium of Scd's is primarily composed of atomic hydrogen. Overall, the amount of molecular hydrogen in spirals can range from $5 \times 10^{10}$ $M_\odot$ for the most massive galaxies to as little as $10^6$ $M_\odot$ for dwarf spirals. The mass of dust is characteristically 150 to 600 times lower than the mass of gas in the ISM.

It is dust that is primarily responsible for the far-infrared (FIR) luminosities of galaxies, although synchrotron radiation and the emissions of stars can also contribute to the total. From observations made by IRAS, astronomers have discovered that $L_{FIR}/L_B = 0.07$ for M31 (an SbI–II), 0.2 for M33 (an Sc(s)II–II), 0.4 for M101 (an Sc(s)I), 0.18 for the LMC (an SBmIII), and 0.09 for the SMC (an ImIV–V). In general, Sc's have a larger fraction of infrared emission than Sa's or Sb's, consistent with the other observations of correspondingly greater fractions of their masses being in the form of gas and dust. Interestingly, SB's tend to have greater infrared luminosities than do normal S's.

### Metallicity Gradients and Color Gradients of Spirals

Not only is there a dependence of color on Hubble type, but individual spiral galaxies also exhibit *color gradients*, with their bulges generally being redder than their disks. This arises for two reasons: metallicity gradients and star formation activity. The average number of electrons per atom is larger for metal-rich stars than for metal-poor stars. Since ionization and the orbital transitions of electrons contribute to the opacity in stellar photospheres, the opacity in metal-rich stars is greater. Because it is more difficult for light generated in the interior to escape from a star with a higher-opacity photosphere, the star will tend to "puff up"; its radius will increase, with a corresponding decrease in effective temperature. Hence, a higher-opacity star will be redder than a lower-opacity star, all else being equal (this effect was first discussed in Example 9.5.4 of Section 9.5). The redness of bulges argues for those regions being more metal-rich than are the portions of disks farther from the center, as is the case in our own Galaxy. In fact, within the Milky Way, metallicity gradients have been measured at Galactocentric radii of between 4 and 14 kpc with values of $d\,[\text{He/H}]\,/dr = -0.01 \pm 0.008$ dex kpc$^{-1}$, $d\,[\text{O/H}]\,/dr = -0.07 \pm 0.015$ dex kpc$^{-1}$, and $d\,[\text{Fe/H}]\,/dr = -0.01$ to $-0.05$ dex kpc$^{-1}$.[11]

---

[11] "dex" refers to the logarithmic nature of the metallicity term in the gradient; recall Eq. (24.11).

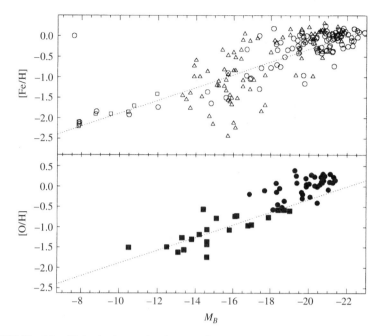

**FIGURE 25.12** Metallicity is observed to increase linearly with absolute magnitude. The symbols in the upper panel correspond to dE's (dwarf ellipticals, open squares) and ellipticals (open circles and open triangles). In the lower panel, spirals are represented by filled circles and irregulars are indicated by filled squares. (Figure adapted from Zaritsky, Kennicutt, and Huchra, *Ap. J.*, *420*, 87, 1994.)

Star formation, the second major cause of color gradients, implies that the disks of spiral galaxies are more actively involved in star-making than are their bulges. This is consistent with the distribution of gas and dust in the galaxies. The spheroidal components of spirals usually contain much less gas and have correspondingly lower star formation rates than their disks. Since disks are able to produce young, hot, blue stars at a greater rate, the spheroids appear relatively redder and a color gradient is established.

Color gradients have been observed within the spheroids of spiral galaxies as well, such as NGC 7814. The galaxies' spheroidal components become bluer with increasing radius. This is also the case for the Milky Way, with the more metal-rich, redder globular clusters orbiting closer to the Galactic center.

Observations indicate that metallicity correlates with the absolute magnitudes of galaxies; [Fe/H] and [O/H] both increase with $M_B$ (see Fig. 25.12). Apparently chemical enrichment was somehow more efficient in luminous, massive galaxies. Composition enrichment histories and gradients have significant implications for galaxy formation theories, as will be discussed in Section 26.2.

### X-Ray Luminosity

The X-ray luminosities of galaxies also provide some information concerning their evolution. In spirals, luminosities in the wavelength region sampled by the Einstein satellite (with photon energies of 0.2 to 3.5 keV) typically range from $L_X = 10^{31}$ W to $10^{34}$ W. A surprisingly tight correlation exists between X-ray and $B$-band luminosities ($L_X/L_B \simeq 10^{-7}$), which has been interpreted as implying that the X-rays are due to a class of objects that constitutes an approximately constant fraction of the population of all objects in spirals. The suspected sources are X-ray binaries. It is probable that supernova remnants also contribute to the X-ray emission.

**Supermassive Black Holes**

Observations of stellar and gas motions near the centers of some spirals strongly suggest the presence of supermassive black holes. For instance, near the center of M31, $M/L$ exceeds 35 $M_\odot/L_\odot$, indicating a large amount of nonluminous matter confined to a small region. Rotational-velocity measurements can be used to estimate the *dynamical mass* of the central black hole of M31 in the same way it was done for Sgr A$^\star$ in the Galactic center (see Section 24.4). A precise determination based on kinematic studies of the triple nucleus of M31 (Fig. 25.13) gives a mass of $1.4^{+0.9}_{-0.3} \times 10^8$ $M_\odot$ for the central supermassive black hole.

Another (although less precise) method of determining the mass of a central supermassive black hole uses the velocity dispersion to obtain a mass estimate via the virial theorem. According to Eq. (2.44), the time-averaged kinetic and potential energies of stars in the galaxy's central region are related by

$$\frac{1}{2}\left\langle \frac{d^2 I}{dt^2} \right\rangle - 2\langle K \rangle = \langle U \rangle,$$

**FIGURE 25.13** An HST image of the triple nucleus of the Andromeda galaxy (M31). Two groupings of red stars are evident in this image. The brighter grouping of red stars on the left is P1, and the dimmer grouping is P2. P1 and P2 are believed to be portions of an elongated disk of stars orbiting the galaxy's supermassive black hole. P1 is farthest from the black hole (the apocenter), and P2 is nearest the black hole (the pericenter). A blue cluster of stars, known as P3, is superimposed on P2. P3 is believed to be orbiting the galaxy's supermassive black hole. [Courtesy of NASA, ESA, and T. Lauer (NOAO/AURA/NSF).]

where $I$ is the region's moment of inertia. If the galaxy is in equilibrium, then $\langle d^2 I/dt^2 \rangle = 0$, resulting in the usual statement of the virial theorem,

$$-2\langle K \rangle = \langle U \rangle .$$

Furthermore, for a large number of stars, the central bulge will look the same (in a statistical sense) at any time, and the time-averaging can be dropped. So for $N$ stars,

$$-2\sum_{i=1}^{N} \frac{1}{2} m_i v_i^2 = U,$$

where $U$ is given by Eq. (2.43).

For simplicity, we restrict our attention to a spherical cluster of radius $R$ with $N$ stars, each of mass $m$, so the total mass of the bulge is $M = Nm$. Dividing the above expression by $N$ produces

$$-\frac{m}{N}\sum_{i=1}^{N} v_i^2 = \frac{U}{N}. \tag{25.12}$$

Of course, astronomers actually measure the radial component of the velocity vector (a galaxy is too far away to allow for detection of proper motions). Moreover, an astronomer is just as likely to see a star moving in the radial direction as in either of the other two perpendicular directions. With the brackets denoting an average value,[12]

$$\langle v^2 \rangle = \langle v_r^2 \rangle + \langle v_\theta^2 \rangle + \langle v_\phi^2 \rangle = 3\langle v_r^2 \rangle ,$$

so

$$\frac{1}{N}\sum_{i=1}^{N} v_i^2 = \langle v^2 \rangle = 3\langle v_r^2 \rangle = 3\sigma_r^2,$$

where $\sigma_r$ is the dispersion in the radial velocity (see Section 24.3). Inserting this result into Eq. (25.12), and using Eq. (10.22) for the (approximate) potential energy of a spherical distribution of total mass $M$ and radius $R$, leads to

$$-3m\sigma_r^2 \approx -\frac{3}{5}\frac{GM^2}{NR}.$$

Using $M = Nm$ and solving for the mass give

$$\boxed{M_{\text{virial}} \approx \frac{5R\sigma_r^2}{G},} \tag{25.13}$$

where the mass obtained in this way is called the **virial mass**.

---

[12]This procedure is identical to the one that was used in Section 10.2 to derive the pressure integral.

It is important to note that in order to determine an accurate value for the mass, an appropriate choice of $R$ must be made. As the observations move farther from the black hole, contributions to the total mass increase from surrounding stars and gas. $R$ must be chosen to be within the black hole's "sphere of influence."

---

**Example 25.2.1.**   Equation (25.13) can be used to estimate a virial mass for the central black hole of M32, a companion galaxy of M31. From Fig. 25.14, the central radial-velocity dispersion is approximately $162 \text{ km s}^{-1}$. This means that within $0.1''$ (approximately 0.4 pc) of the center, there is a total mass of roughly

$$M_{\text{virial}} \sim 1 \times 10^7 \ M_\odot.$$

A more accurate estimate, based on the rotation curve, gives a value of between $1.5 \times 10^6 \ M_\odot$ and $5 \times 10^6 \ M_\odot$.

---

It is important to note that central supermassive black holes are not restricted to late-type galaxies. For example, based on observations made using the Hubble Space Telescope, it appears that the giant elliptical galaxy M87 (NGC 4476) also contains a $3.2 \pm 0.9 \times 10^9 \ M_\odot$ black hole. HST was able to resolve a disk of material within M87 that has rotational speeds reaching $550 \text{ km s}^{-1}$; the disk itself is orbiting a central region no larger than our Solar System. Given that the distance to M87 is nearly 20 Mpc, this is a remarkable observation. (M87 is also known to have a relativistic jet that is believed to be powered by the supermassive black hole at its center; see Fig. 28.10.)

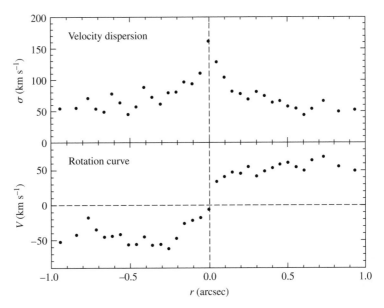

**FIGURE 25.14**   The stellar velocity dispersion and rotational velocities of stars near the center of M32. Given the distance to M32 of 770 kpc, $1''$ corresponds to a linear distance from the center of 3.7 pc. (Data from Joseph, et al., *Ap. J.*, *550*, 668, 2001.)

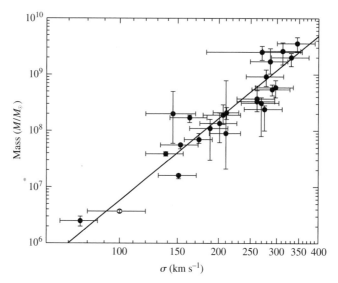

**FIGURE 25.15** The relationship between the mass of a galaxy's supermassive black hole and the velocity dispersion, $\sigma$, of the galaxy's spheroid. The data set includes elliptical, lenticular, and spiral galaxies. (Data from Ferrarese and Ford, *Space Science Reviews*, *116*, 523, 2005. The open symbol represents the Milky Way with mass data from Ghez, et. al, *Ap. J.*, *620*, 744, 2005.)

As the number of known central supermassive black holes has increased, it has become possible to look for relationships between the black holes and their host galaxies. A very intriguing and useful correlation has been discovered between the mass of the supermassive black hole in a galaxy's center and the velocity dispersion of the stars within the galaxy. The relationship is illustrated in Fig. 25.15 and is given by the best-fit power law,

$$M_{bh} = \alpha(\sigma/\sigma_0)^{\beta}, \tag{25.14}$$

where $\sigma$ is the velocity dispersion in units of km s$^{-1}$, $\alpha = (1.66 \pm 0.24) \times 10^8$ M$_{\odot}$, $\beta = 4.86 \pm 0.43$, and $\sigma_0 = 200$ km s$^{-1}$. The velocity dispersion is measured for the stellar population near the black hole. Correlations between the mass of supermassive black holes and other bulk galaxy parameters, such as the luminosity of the bulge, have also been uncovered. Apparently a fundamental link exists between the formation of the central supermassive black hole in a galaxy and the overall formation of the galaxy itself. Just what that link is remains to be determined. (Galaxy formation will be discussed in more detail in Chapter 26.)

Evidence continues to mount that supermassive black holes may exist in the centers of most galaxies. The dramatic activity associated with the central supermassive black holes of M87 and other galaxies is the subject of Chapter 28.

### Specific Frequency of Globular Clusters

Finally, it is worth discussing the abundance of globular clusters in late-type galaxies, which, when compared with the numbers of globular clusters in ellipticals, has important implications for theories of galactic formation and evolution. It appears that galaxies that are more spheroidally dominant (i.e., earlier Hubble types) were more efficient at forming globular clusters during their early histories. Although virtually all galaxies appear to contain some globular clusters, the number of clusters within a galaxy seems to increase with total galaxy luminosity and with progressively earlier Hubble type.

To compare cluster systems among Hubble types more directly, the globular-cluster counts are usually normalized to a standard absolute magnitude for the parent galaxy of $M_V = -15$ mag (in the visual wavelength band). If $N_t$ is the total number of globular clusters in a galaxy, then the **specific frequency** of globular clusters is defined as

$$S_N = N_t \frac{L_{15}}{L_V}$$

$$= N_t \, 10^{0.4(M_V+15)}, \tag{25.15}$$

where $L_V$ is the galaxy's luminosity and $L_{15}$ is the reference luminosity corresponding to an absolute visual magnitude of $M_V = -15$ mag [Eq. (3.7) was used to arrive at the second expression]. As can be seen in Fig. 25.16, for Sc galaxies and later, the average value of $S_N$ is in the range of $0.5 \pm 0.2$, while for Sa's and Sb's, $\langle S_N \rangle$ increases to $1.2 \pm 0.2$. $\langle S_N \rangle$ is even larger for elliptical galaxies, meaning that they have more clusters per unit luminosity than do spirals. In fact, by far the largest number of globular clusters per unit luminosity belongs to the gigantic elliptical galaxies (cD's) that are frequently found near the centers of large clusters of galaxies (see Sections 25.4 and 27.3). It seems that the number of globular clusters offers important clues to galaxy formation theories.

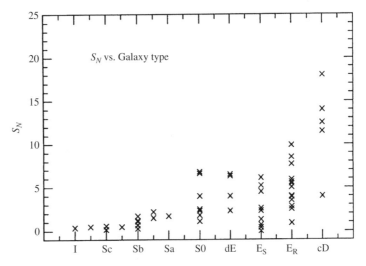

**FIGURE 25.16** The association of the specific frequency of globular clusters with Hubble type. cD galaxies are the largest elliptical galaxies in the universe; dE's are dwarf ellipticals. (Figure adapted from Harris, *Annu. Rev. Astron. Astrophys.*, 29, 543, 1991. Reproduced by permission from the *Annual Review of Astronomy and Astrophysics*, Volume 29, ©1991 by Annual Reviews Inc.)

## 25.3 ■ SPIRAL STRUCTURE

Galaxies exhibit a rich variety of spiral structure, which may vary in number of arms and how tightly wound they are, degree of smoothness in the distribution of stars and gas, surface brightness, and the existence or lack of bars. The most majestic spiral galaxies, known as **grand-design spirals**, usually have two very symmetric and well-defined arms. One of the best-known examples of a grand-design spiral is M51 (NGC 5194), shown in Fig. 25.17. Also called the Whirlpool galaxy, M51 has a companion galaxy (NGC 5195, visible in the image) near the end of one of the spiral arms.

However, not all spirals are grand designs with two distinct arms. For instance, M101 [Fig. 25.5(c)] has four arms, and NGC 2841 (Fig. 25.18) has a series of partial arm fragments. Galaxies like NGC 2841, which do not possess well-defined spiral arms that are traceable over a significant angular distance, are called **flocculent spirals**. Only about 10% of all spirals are considered grand-design galaxies, another 60% are multiple-arm galaxies, and the remaining 30% are flocculent galaxies.

The optical images of spiral galaxies are dominated by their arms. This is because very luminous O and B main-sequence stars and H II regions are found preferentially in the arms. Since massive OB stars are short-lived objects relative to the characteristic rotation period of a galaxy, spiral structure must correspond to regions of active star formation. (For instance, the ages of OB stars are on the order of 10 Myr, compared to 230 Myr for the orbital period of the local standard of rest in the Milky Way; see Table 13.1 and Example 24.3.1, respectively.)

**FIGURE 25.17**   The Whirlpool galaxy, M51 (NGC 5194), is an Sbc(s)I–II grand-design spiral located in the constellation Canes Venatici (the Hunting Dogs of Boötes, just below Ursa Major). Also visible is its companion NGC 5195, situated near the end of one of the spiral arms. (Image from Sandage and Bedke, *The Carnegie Atlas of Galaxies*, Carnegie Institution of Washington, Washington, D.C., 1994.)

**FIGURE 25.18** The Sb galaxy NGC 2841 is an example of a flocculent spiral. (Image from Sandage and Bedke, *The Carnegie Atlas of Galaxies*, Carnegie Institution of Washington, Washington, D.C., 1994.)

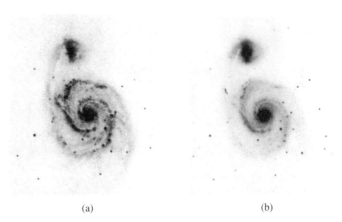

(a)                                          (b)

**FIGURE 25.19** M51 as seen in (a) blue light and (b) red light. (Figures from Elmegreen, *Ap. J. Suppl.*, 47, 229, 1981.)

Careful inspection of the various images found in this chapter reveals that dust bands are also evident in the spiral arms; M51 (Fig. 25.17) is a particularly good example. Notice that the dust bands tend to reside on the inner (concave) edges of the arms. Observation of 21-cm H I emission indicates that gas clouds are also more prevalent near the inner edges of the arms.

As Fig. 25.19 illustrates, when spiral galaxies are observed in red light, the arms become much broader and less pronounced, although they still remain detectable. Since observations

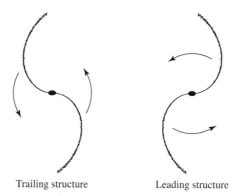

Trailing structure                    Leading structure

**FIGURE 25.20**    Trailing and leading spiral arm structures.

at red wavelengths emphasize the emission of long-lived, lower-mass main-sequence stars and red giants, this implies that the bulk of the disk is dominated by older stars. However, despite the dominance of lower-mass stars between the spiral arms, observations indicate that there is still an increase in the number density of older stars (and an accompanying increase in mass density) within the spiral arms.

### Trailing and Leading Spiral Arms

Although the general appearance of spiral galaxies suggests that their arms are *trailing*, meaning that the tips of the arms point in the opposite direction from the direction of rotation (see Fig. 25.20), verifying this is not always a simple matter. Distinguishing between trailing and *leading* spiral arms requires a determination of the orientation of the plane of the galaxy relative to our line of sight so that radial-velocity measurements can be unambiguously interpreted in terms of the direction of galaxy rotation. In almost all cases where such a clear determination can be made, it does appear that spiral arms are trailing. However, in one case, NGC 4622, two arms are going one way and another arm is winding in the opposite direction; at least one of these arms must be leading. It has also been suggested that M31 (Andromeda) has one tightly wound leading arm. In each case it is likely that the cause of the leading spiral is a tidal encounter with a retrograde-moving object (M32 in the case of Andromeda).

### The Winding Problem

Given that spiral galaxies are commonplace within the universe, it is natural to ask what causes spiral structure, and whether spiral arms are long-lived (with lifetimes comparable to the age of a galaxy) or transient.

One problem immediately arises when the nature of spiral structure is considered; *material arms* composed of a fixed set of identifiable stars and gas clouds would necessarily "wind up" on a timescale that is short compared to the age of the galaxy. This so-called **winding problem** can be understood by considering a set of stars that are originally along a single line, but at varying distances from the center of the galaxy, as shown in Fig. 25.21(a). Since the disk of a spiral galaxy rotates differentially (except very near the center), the outer

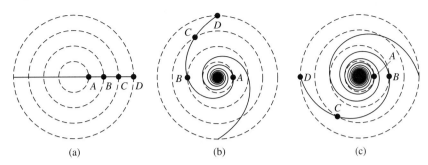

**FIGURE 25.21** The winding problem for material arms. The arms become progressively more tightly wound as time goes on. A flat rotation curve is assumed with $R_B = 2R_A$, $R_C = 3R_A$, and $R_D = 4R_A$. (a) The stars start out in a line at time $t = 0$. (b) After star $A$ has completed one orbit. (c) After star $A$ has completed two orbits.

stars will require more time to complete one orbit than will stars with smaller orbital radii.[13] This effect will of course lead to a natural generation of trailing spiral arms. However, after only a few orbits, the spiral arms will become too tightly wound to be observed; this situation is depicted in Figs. 25.21(b) and 25.21(c). Another mechanism is needed to explain persistent spiral structure.

### The Lin–Shu Density Wave Theory

In the mid-1960s, the American astronomers C. C. Lin and Frank Shu proposed that spiral structure arises because of the presence of long-lived **quasistatic density waves**. Density waves consist of regions in the galactic disk where the mass density is greater than average, perhaps by 10% to 20%. Stars, dust, and gas clouds move through the density waves during their orbits around the galactic center, much like cars slowly working their way through a traffic jam on a highway. Lin and Shu suggested that when the galaxy is viewed in a noninertial reference frame that is rotating with a specific angular speed $\Omega_{gp}$, known as the **global pattern speed**, the spiral wave pattern appears to be stationary, as shown in Fig. 25.22. This does not imply that the motions of the stars are also stationary in that frame, however. Stars near the center of the galaxy can have orbital periods that are shorter than the density wave pattern (or $\Omega > \Omega_{gp}$) and so they will overtake a spiral arm, move through it, and continue on until they encounter the next arm. Stars sufficiently far from the center of the galaxy will be moving more slowly than the density wave pattern and will be overtaken by it ($\Omega < \Omega_{gp}$). At a specific distance from the center, called the **corotation radius** ($R_c$), the stars and density waves will move together. In this noninertial reference frame in which the density wave pattern is static, the stars with $R < R_c$ will appear to pass through the arms moving in one direction, while stars with $R > R_c$ will appear to be moving in the opposite sense.

The Lin–Shu hypothesis helps to explain many of the observations concerning spiral structure that have been discussed—for instance, the ordering of H I clouds and dust bands

---

[13]For flat rotation curves, the orbital velocities of stars are nearly independent of distance from the center, but the *angular velocity*, $\Omega = v/R$, still decreases with increasing distance.

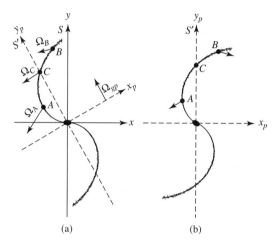

(a)                    (b)

**FIGURE 25.22**   (a) A galaxy with trailing spiral arms as seen in an inertial reference frame $S$ in which a quasistatic density wave is moving with a global angular pattern speed $\Omega_{gp}$. Star $A$ has an angular speed $\Omega_A > \Omega_{gp}$, Star $B$ has an angular speed $\Omega_B < \Omega_{gp}$, and Star $C$ is corotating with the density wave, meaning that $\Omega_C = \Omega_{gp}$. (b) The motions of the stars as seen in a noninertial reference frame $S'$, corotating with the density wave.

on the inner (trailing) edges of spiral arms, the existence of young, massive stars and H II regions throughout the arms, and an abundance of old, red stars in the remainder of the disk. Apparently, as dust and gas clouds within the corotation radius overtake a density wave, they are compressed by the effects of the increase in local mass density. This causes some of the clouds to satisfy the Jeans criterion (Eqs. 12.14 and 12.16) and begin to collapse, resulting in the formation of new stars. Since this process takes some time (approximately $10^5$ yr for a 15 $M_\odot$ star; see Table 12.1), the appearance of new stars will occur within the arm slightly "downstream" from dust and gas clouds at the edge of the wave. The birth of the brightest and bluest new stars, the massive O and B stars, will result in the creation of H II regions as the UV ionizing radiation moves through the interstellar medium. Because massive stars have relatively short lifetimes, they will die before they can move entirely out of the density wave in which they were born. Less massive, redder stars will be able to live much longer (some longer than the current age of the galaxy) and so will continue through the density wave and become distributed throughout the disk. Local maxima in the number density of red dwarfs within spiral arms are due to the presence of the density wave during a subsequent passage, which causes the stars to collect at the bottom of the wave's gravitational potential well. Of course, the same scenario could also occur on the outer (leading) edges of spiral arms outside the corotation radius. However, it is likely that less dust and gas will be found in these outer regions of the galaxy.

In principle, the density wave theory also suggests a solution to the winding problem. The problem arose because we considered material arms (arms composed of a fixed set of stars). If, instead, the stars are allowed to pass through a quasistatic density wave, then the problem has been changed to one of establishing and maintaining the wave of enhanced density. This has been the focus of considerable research since the Lin–Shu hypothesis was proposed.

### Small-Amplitude Orbital Perturbations

We now turn our attention to developing the basic ideas of how the orbital motions of individual stars about the galactic center can result in spiral-shaped regions of enhanced density according to the Lin–Shu hypothesis. We begin by considering the general motion of a star (or gas cloud) in an axially symmetric gravitational field that is also symmetric about the galactic midplane. This implies that we are assuming here that density waves make an insignificant contribution to the gravitational field, an assumption that may not be valid in all spiral galaxies. However, the assumption does simplify the analysis considerably.

From Fig. 25.23, the position of a star at some general point above the galactic midplane can be written in the form

$$\mathbf{r} = R\hat{\mathbf{e}}_R + z\hat{\mathbf{e}}_z, \tag{25.16}$$

where $\hat{\mathbf{e}}_R$, $\hat{\mathbf{e}}_\phi$, and $\hat{\mathbf{e}}_z$ are the unit vectors in cylindrical coordinates.[14] To convert between rectangular and cylindrical coordinates, we see that

$$x = R\cos\phi, \qquad y = R\sin\phi, \qquad z = z,$$

and

$$\hat{\mathbf{e}}_R = \hat{\mathbf{i}}\cos\phi + \hat{\mathbf{j}}\sin\phi, \qquad \hat{\mathbf{e}}_\phi = -\hat{\mathbf{i}}\sin\phi + \hat{\mathbf{j}}\cos\phi, \qquad \hat{\mathbf{e}}_z = \hat{\mathbf{k}}.$$

For a star of mass $M$, Newton's second law of motion in cylindrical coordinates is

$$M\frac{d^2\mathbf{r}}{dt^2} = \mathbf{F}_g(R, z),$$

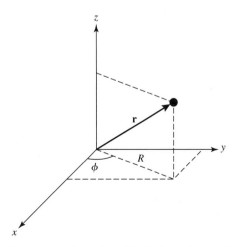

**FIGURE 25.23** The position vector of a star in cylindrical coordinates. The origin of the coordinate system is at the center of the galaxy, and the $x$–$y$ plane is in the galaxy's midplane.

[14]Note that we have chosen to use a traditional right-handed coordinate system here, rather than the left-handed one that we used to analyze the kinematics of the Milky Way Galaxy in Section 24.3. To avoid confusion, we will use $\phi$ rather than $\theta$ to designate the azimuthal angle.

where $\mathbf{F}_g$ is the gravitational force on the star, and the $\phi$-dependence is neglected due to the assumption of axial symmetry. Replacing the force with the negative of the gradient of the gravitational potential energy (recall the discussion in Section 2.2), yields

$$M\frac{d^2\mathbf{r}}{dt^2} = -\nabla U(R, z) = -\frac{\partial U}{\partial R}\hat{\mathbf{e}}_R - \frac{1}{R}\frac{\partial U}{\partial \phi}\hat{\mathbf{e}}_\phi - \frac{\partial U}{\partial z}\hat{\mathbf{e}}_z.$$

Dividing both sides of the last expression by the star's mass, writing the result in terms of the gravitational potential[15] $\Phi \equiv U/M$, and noting that $U$ is independent of $\phi$ give

$$\frac{d^2\mathbf{r}}{dt^2} = -\frac{\partial \Phi}{\partial R}\hat{\mathbf{e}}_R - \frac{\partial \Phi}{\partial z}\hat{\mathbf{e}}_z. \tag{25.17}$$

It is left as an exercise (Problem 25.13) to show that the star's acceleration vector (the left-hand side of Eq. 25.17) can be written in the form

$$\frac{d^2\mathbf{r}}{dt^2} = \left(\ddot{R} - R\dot{\phi}^2\right)\hat{\mathbf{e}}_R + \frac{1}{R}\frac{\partial\left(R^2\dot{\phi}\right)}{\partial t}\hat{\mathbf{e}}_\phi + \ddot{z}\hat{\mathbf{e}}_z, \tag{25.18}$$

where the dots represent time derivatives. Comparing Eqs. (25.17) and (25.18), we see that the motion in each coordinate is given by

$$\ddot{R} - R\dot{\phi}^2 = -\frac{\partial \Phi}{\partial R}, \tag{25.19}$$

$$\frac{1}{R}\frac{\partial\left(R^2\dot{\phi}\right)}{\partial t} = 0, \tag{25.20}$$

$$\ddot{z} = -\frac{\partial \Phi}{\partial z}. \tag{25.21}$$

Because we have assumed axial symmetry, there is no component of the force vector in the $\hat{\mathbf{e}}_\phi$ direction and consequently no component of torque along the $z$-axis (recall that $\boldsymbol{\tau} = \mathbf{r} \times \mathbf{F}$). Therefore, the $z$ component of the star's orbital angular momentum ($L_z$) must be constant since $\tau_z = dL_z/dt = 0$ (i.e., angular momentum is conserved throughout the star's motion). This is exactly what Eq. (25.20) implies, since $R^2\dot{\phi} = Rv_\phi = L_z/M = $ constant. Defining

$$J_z \equiv L_z/M = R^2\dot{\phi}$$

to be the (constant) $z$ component of the orbital angular momentum *per unit mass* of the star, we have that

$$\dot{\phi} = \frac{J_z}{R^2}$$

and

$$R\dot{\phi}^2 = \frac{J_z^2}{R^3}.$$

[15]Recall previous discussions of $\Phi$ in Sections 18.1 and 21.1.

Substituting into Eq. (25.19) gives

$$\ddot{R} = -\frac{\partial \Phi}{\partial R} + \frac{J_z^2}{R^3}.$$

To simplify the expression, we can define an *effective gravitational potential* that incorporates a term representing the portion of the star's kinetic energy per unit mass that is associated with its azimuthal motion ($v_\phi = R\dot{\phi} = J_z/R$):

$$\Phi_{\text{eff}}(R, z) \equiv \Phi(R, z) + \frac{J_z^2}{2R^2}. \tag{25.22}$$

Now the radial and vertical equations of motion can be written in the forms

$$\ddot{R} = -\frac{\partial \Phi_{\text{eff}}}{\partial R}, \tag{25.23}$$

$$\ddot{z} = -\frac{\partial \Phi_{\text{eff}}}{\partial z}, \tag{25.24}$$

the later expression coming from Eq. (25.21) with the azimuthal term in $\Phi_{\text{eff}}$ being independent of $z$.

To solve Eqs. (25.23) and (25.24) for the motion of the star through the galaxy, we must first determine the behavior of $\Phi_{\text{eff}}$. Specifically, it would be helpful to know where the minima of the effective gravitational potential are located, since the star should attempt to settle into an orbit of minimum possible energy. In other words, we want to find the values of $R$ and $z$ for which

$$\frac{\partial \Phi_{\text{eff}}}{\partial R} = 0, \tag{25.25}$$

$$\frac{\partial \Phi_{\text{eff}}}{\partial z} = 0. \tag{25.26}$$

Given our assumption that the gravitational potential is symmetric about the midplane, it is clear that the second condition is satisfied at $z = 0$, because $\Phi$, and therefore $\Phi_{\text{eff}}$, must be either a local maximum or a local minimum there. Since $\Phi$ is always less than zero (cf., Eq. 21.1) and is assumed to be identically zero at infinity, it is necessary that $\Phi_{\text{eff}}$ be a minimum at $z = 0$.

The physical significance of the minimum in $\Phi_{\text{eff}}$ with respect to $R$ can be uncovered by realizing that since $J_z$ is a constant for the star's motion,

$$\frac{\partial \Phi_{\text{eff}}}{\partial R} = \frac{\partial \Phi}{\partial R} - \frac{J_z^2}{R^3} = 0$$

for some radius $R_m$ in the galaxy's midplane ($z = 0$). This gives

$$\left.\frac{\partial \Phi}{\partial R}\right|_{(R_m, 0)} = \frac{J_z^2(R_m, 0)}{R_m^3}. \tag{25.27}$$

But since $J_z = R v_\phi$, the right-hand side of the last expression is just

$$
\frac{J_z^2}{R_m^3} = \left. \frac{v_\phi^2}{R} \right|_{R_m},
$$

the centripetal acceleration for perfectly circular motion. Furthermore, the left-hand side of Eq. (25.27) is simply the radial component of the gradient of the true gravitational potential, which is nothing more than the negative of the $R$ component of the force per unit mass exerted on the star. Therefore, the last expression is the familiar equation of perfectly circular motion,

$$
F_R (R_m) = - \left. \frac{M v_\phi^2}{R} \right|_{R_m},
$$

and the minimum value for $\Phi_{\text{eff}}$ occurs when the star is executing perfectly circular orbital motion in the midplane of the spiral galaxy.

This is precisely the assumption we made in the last chapter regarding the motion of the local standard of rest at the position of the Sun. However, in Section 24.3 we also discovered that the Sun itself does not actually execute perfectly circular motion, but it does exhibit a peculiar motion with respect to the LSR, known as its solar motion. As a result, to learn more about the motion of our Sun (or, more generally, about the motions of stars in the plane of any spiral galaxy), it is necessary to explore deviations from the minimum value of $\Phi_{\text{eff}}$.

To understand these first-order effects, we will expand $\Phi_{\text{eff}}$ about its minimum position $(R_m, 0)$ by means of a two-dimensional Taylor series. (This is similar to the procedure outlined in Problem 14.9.) Letting

$$
\rho \equiv R - R_m
$$

and using the subscript $m$ to indicate that the leading constant and partial derivatives are being evaluated at the minimum position, we obtain

$$
\Phi_{\text{eff}} (R, z) = \Phi_{\text{eff},m} + \left. \frac{\partial \Phi_{\text{eff}}}{\partial R} \right|_m \rho + \left. \frac{\partial \Phi_{\text{eff}}}{\partial z} \right|_m z + \frac{1}{2} \left. \frac{\partial^2 \Phi_{\text{eff}}}{\partial R \partial z} \right|_m \rho z
$$

$$
+ \frac{1}{2} \left. \frac{\partial^2 \Phi_{\text{eff}}}{\partial R^2} \right|_m \rho^2 + \frac{1}{2} \left. \frac{\partial^2 \Phi_{\text{eff}}}{\partial z^2} \right|_m z^2 + \cdots. \tag{25.28}
$$

The first term on the right-hand side is the constant minimum value in $\Phi_{\text{eff}}$, and the two first-derivative terms are identically zero, since they were used to identify the point about which the expansion is being carried out (see Eqs. 25.25 and 25.26). The mixed-partial-derivative term is also identically zero because of the symmetry of $\Phi_{\text{eff}}$ about the $z = 0$

plane. Defining the constants

$$\kappa^2 \equiv \left. \frac{\partial^2 \Phi_{\text{eff}}}{\partial R^2} \right|_m , \tag{25.29}$$

$$\nu^2 \equiv \left. \frac{\partial^2 \Phi_{\text{eff}}}{\partial z^2} \right|_m , \tag{25.30}$$

and neglecting the remaining higher-order terms, the effective gravitational potential becomes

$$\Phi_{\text{eff}}(R, z) \simeq \Phi_{\text{eff},m} + \frac{1}{2}\kappa^2 \rho^2 + \frac{1}{2}\nu^2 z^2. \tag{25.31}$$

Finally, noting that $\ddot{\rho} = \ddot{R}$ and recalling Eqs. (25.23) and (25.24), we arrive at two of the three first-order expansions for the equations of motion about a perfectly circular orbit in the galaxy's midplane:

$$\ddot{\rho} \simeq -\kappa^2 \rho, \tag{25.32}$$

$$\ddot{z} \simeq -\nu^2 z, \tag{25.33}$$

which are the familiar differential equations of simple harmonic motion. Physically, Eqs. (25.32) and (25.33) represent the components of a star's acceleration relative to a point that is executing perfectly circular motion.

Equation (25.32) can be solved to give

$$\rho(t) = R(t) - R_m = A_R \sin \kappa t, \tag{25.34}$$

where $\kappa$ is called the **epicycle frequency**.[16] $R_m$ is the radius of the energy-minimum circular orbit, and $A_R$ is the amplitude of the radial oscillation. We have arbitrarily assumed that the star is passing through equilibrium and moving outward at some time $t = 0$.

Equation (25.33) is identical to the result obtained in Problem 24.27 for the sinusoidal oscillations of the Sun above and below the Galactic midplane, with $\nu$ being the **vertical oscillation frequency**. The star's position along the $z$-axis is given by

$$z(t) = A_z \sin(\nu t + \zeta), \tag{25.35}$$

where $A_z$ is the amplitude of the oscillation in the $z$ direction and $\zeta$ is a general phase shift between $\rho(t)$ and $z(t)$.

To help visualize the relationship between $\Phi_{\text{eff}}$, as given in Eq. (25.31) and the equations of motion (Eqs. 25.34 and 25.35), imagine that the star is located inside a gravitational well described by a surface in the three-dimensional "space" of $(R, z, \Phi_{\text{eff}})$, depicted in Fig. 25.24. (The shape is similar to that of a bed sheet held at its corners, with $(R = R_m, z = 0)$ located at the lowest point.) Now consider the special case where the star is constrained to remain in the midplane ($A_z = 0$) but can oscillate about the path

---

[16]The choice of the term *epicycle frequency* will be described shortly.

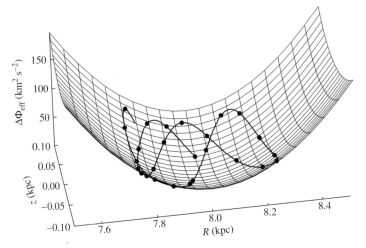

**FIGURE 25.24**   The effective gravitational potential well for a star that is executing a general first-order simple harmonic oscillation about a perfectly circular orbit near the midplane of a disk galaxy. In this case, the star is assumed to be oscillating about the equilibrium position ($R_m = 8$ kpc, $z = 0$). $\Delta\Phi_{\rm eff} \equiv \Phi_{\rm eff} - \Phi_{{\rm eff},m}$.

of perfectly circular motion. This implies that the star is able to move only along the $R$-axis in Fig. 25.24. When the star is displaced from the minimum value of $\Phi_{\rm eff}$, a restoring force (given per unit mass by Eq. 25.32) will try to bring the star back toward equilibrium. Since the force is always directed toward the position of minimum $\Phi_{\rm eff}$, the star will accelerate toward the bottom of the potential well, overshoot the equilibrium position, and climb back up the opposite side. The star will then turn around and fall back down toward the bottom of the well again, continually executing simple harmonic motion. The same type of behavior would occur if the star were orbiting in a perfect circle projected onto the midplane ($A_R = 0$), but with superimposed vertical oscillations. However, since in general $\kappa \neq \nu$ and $\zeta \neq 0$, the general motion within the well can be quite complicated, as Fig. 25.24 illustrates.

We now have two expressions (Eqs. 25.34 and 25.35) that describe the motion of a star about an equilibrium position ($R = R_m$, $z = 0$) that is moving in a circular orbit. To complete our description of the approximate motion of the star, consider its azimuthal orbital angular speed, given by

$$\dot{\phi} = \frac{v_\phi}{R(t)} = \frac{J_z}{[R(t)]^2}.$$

But $R(t) = R_m + \rho(t) = R_m(1 + \rho(t)/R_m)$. Assuming that $\rho(t) \ll R_m$, as is required of our approximations, and using the binomial expansion theorem to first order,[17]

$$\dot{\phi} \approx \frac{J_z}{R_m^2}\left(1 - 2\frac{\rho(t)}{R_m}\right).$$

---

[17] For $\delta \ll 1$, $(1 + \delta)^n \approx 1 + n\delta$.

Substituting the expression for $\rho(t)$ from Eq. (25.34) and integrating with respect to time, we find that

$$\phi(t) = \phi_0 + \frac{J_z}{R_m^2}t + \frac{2J_z}{\kappa R_m^3}A_R\cos\kappa t = \phi_0 + \Omega t + \frac{2\Omega}{\kappa R_m}A_R\cos\kappa t,$$

where $\Omega \equiv J_z/R_m^2$. The first two terms in this expression correspond to the perfectly circular orbit traced out by the equilibrium point, moving at a constant angular speed, $\Omega$. The last term represents the oscillation of the star about the equilibrium point in the $\phi$ direction.

Finally, defining

$$\chi(t) \equiv [\phi(t) - (\phi_0 + \Omega t)]\,R_m$$

to be the difference in azimuthal position between the star and the equilibrium point, we have

$$\chi(t) = \frac{2\Omega}{\kappa}A_R\cos\kappa t. \qquad (25.36)$$

The three equations—(25.34), (25.35), and (25.36)—represent the motions of the star in the $R$, $z$, and $\phi$ coordinates, respectively, about an equilibrium position that is moving in a perfect circle around the center of the galaxy and in the galactic midplane.

$\kappa$ takes its name from an epicycle model for the first-order perturbation, as shown in Fig. 25.25. In general, in an inertial reference frame the star's orbit is not closed, but produces

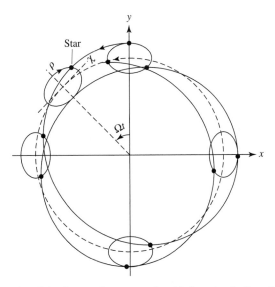

**FIGURE 25.25**   In an inertial reference frame a star's orbital motion in the galactic midplane (solid line) forms a nonclosing rosette pattern. In the first-order approximation, the motion can be imagined as being the combination of a retrograde orbit about an epicycle and the prograde orbit of the center of the epicycle about a perfect circle (dashed line). The dimensions of the epicycle have been exaggerated by a factor of five to illustrate the effect.

a rosette pattern. However, the star can be imagined as being located on an epicycle, with the center of the epicycle corresponding to the equilibrium position. As the star moves in a retrograde direction about the epicycle, it is carried alternately closer to and then farther from the galactic center. The epicycle is also oval in shape and has an axial ratio that is given by the ratios of the amplitudes of the oscillations in $\chi$ and $\rho$, or $2\Omega/\kappa$. The $(\chi, \rho)$ coordinate system of the epicycle rotates about the galaxy's center with the angular speed $\Omega$ (the angular speed of the equilibrium point) and is indicated in Fig. 25.25.[18]

---

**Example 25.3.1.**    As we saw in Section 24.3, our Sun is moving relative to the perfectly circular motion of the LSR, as reflected in the Sun's peculiar solar motion. Since $u_\odot \neq 0$ (the $R$ component of its peculiar motion), the Sun must have a nonzero epicycle frequency. Information about the radial component of the Sun's peculiar motion is also contained in the Oort constants, Eqs. (24.39) and (24.40), since they involve derivatives of the Sun's orbital speed, $\Theta_0$, with respect to Galactocentric radius, $R_0$.

From the condition for perfectly circular motion, Eq. (25.27) gives

$$\frac{\partial \Phi}{\partial R} = \frac{J_z^2}{R^3} = \frac{\Theta_0^2}{R_0},$$

where the last expression arises because $J_z = R_0 \Theta_0$ for the orbital angular momentum per unit mass at the solar Galactocentric distance. Substituting into the expression for the square of the epicycle frequency (Eq. 25.29) and making use of Eq. (25.22), we find that the solar epicycle frequency is

$$\kappa_0^2 = 2 \frac{\Theta_0}{R_0} \left[ \frac{\Theta_0}{R_0} + \left. \frac{\partial \Theta_0}{\partial R} \right|_{R_0} \right]. \tag{25.37}$$

Rewriting this in terms of the Oort constants produces

$$\kappa_0^2 = -4B(A - B). \tag{25.38}$$

Using the values for the Oort constants obtained in Eqs. (24.47) and (24.48), $A = 14.8$ km s$^{-1}$ kpc$^{-1}$ and $B = -12.4$ km s$^{-1}$ kpc$^{-1}$, respectively, the epicycle frequency for our Sun is $\kappa_0 = 36.7$ km s$^{-1}$ kpc$^{-1}$ = $1.2 \times 10^{-15}$ rad s$^{-1}$. (Note that when kiloparsecs are converted to kilometers, $\kappa_0$ has units of rad s$^{-1}$.) This value for the epicycle frequency corresponds to an oscillation period of $P = 2\pi/\kappa_0 = 170$ Myr.

The ratio of the Sun's epicycle frequency to its orbital angular speed (or orbital frequency, $\Omega_0 = \Theta_0/R_0 = A - B$; Eq. 24.43) is

$$\frac{\kappa_0}{\Omega_0} = 2 \left( \frac{-B}{A - B} \right)^{1/2} = 1.35.$$

Therefore, the Sun executes 1.35 epicycle oscillations for every orbit around the Galactic center.

---

[18]This model bears a strong resemblance to the epicycle–deferent models of planetary motion devised by Hipparchus and Ptolemy, illustrated in Fig. 1.3. However, in those ancient planetary models, the epicycles were assumed to be perfectly circular.

The axial ratio of the Sun's epicycle is given by

$$\frac{\chi_{max}}{\rho_{max}} = \frac{2\Omega_0}{\kappa_0} \simeq 1.5.$$

## Closed Orbits in Noninertial Frames

The number of oscillations per orbit about the galaxy's center is equal to the ratio of the star's epicycle frequency to its orbital angular speed. If the ratio $\kappa/\Omega$ is a ratio of integers, the orbit is closed. Like the Sun's, however, most stellar orbits are not closed and a rosette pattern results. But in a noninertial reference frame that is rotating with a *local* angular pattern speed, $\Omega_{\ell p} = \Omega$, relative to the inertial frame, the star's path would appear to be very simple, tracing out a *closed orbit* that is retrograde and centered at a distance $R_m$ from the galaxy's center (see Fig. 25.25). Such a reference frame corresponds to the epicycle's own coordinate system in which the equilibrium point is stationary and the closed path simply traces the epicycle itself.

To obtain a closed orbit in a noninertial reference frame, we need not necessarily choose an angular pattern speed that equals the unperturbed orbital angular speed $\Omega$. Instead, we could choose to have the star complete $n$ orbits as seen in the rotating frame while executing $m$ epicycle oscillations (where $n$ and $m$ are positive or negative integers), after which time the star would be back at its starting point. That is, we could choose $m(\Omega - \Omega_{\ell p}) = n\kappa$, or

$$\Omega_{\ell p}(R) = \Omega(R) - \frac{n}{m}\kappa(R). \tag{25.39}$$

Note that this is a local pattern speed, so $\Omega_{\ell p}$ is a function of $R$. Although in principle, there are an infinite number of local pattern speeds at each $R$, only a small number of values for $n$ and $m$ produce substantial enhancements in mass density.[19]

Figure 25.26 shows the rotation of a coordinate system for which the local pattern speed corresponds to $(n = 1, m = 2)$, together with the position of the star at one point. Figure 25.27 shows the motion of the star as seen in four rotating reference frames with $(n, m) = (0, 1)$, $(1, 2)$, $(2, 3)$, and $(1, 4)$. The position of the star that was indicated in Fig. 25.26 corresponds to the same position in Fig. 25.27(a).

Now imagine a large number of stars at various distances $R$ from the center of a spiral galaxy, all observed in a reference frame rotating with the *global* angular pattern speed $\Omega_{gp}$. If we consider the case $(n, m) = (1, 2)$ and *if* the local pattern speed $\Omega_{\ell p} = \Omega(R) - \kappa(R)/2$ is a constant for all values of $R$, then we can set $\Omega_{gp} = \Omega_{\ell p}$. Seen from the noninertial frame, the resulting orbital patterns could be nested with their major axes aligned, as illustrated in Fig. 25.28(a). The structure that results bears a significant resemblance to the bars present in roughly two-thirds of all spiral galaxies. Of course, we could also orient each successive oval-shaped orbit so that its major axis is rotated slightly relative to the one immediately interior to it. In this example the result is a trailing two-armed grand-design spiral wave

---

[19]This problem is analogous to situations encountered within the Solar System. For example, the orbital resonances of Saturn's moons (primarily Mimas) with particles in the planet's ring system produce gaps in the rings. In addition, small integer ratios between Jupiter's orbital period and the orbital periods of asteroids result in either increases or decreases in the number of asteroids having certain orbital radii.

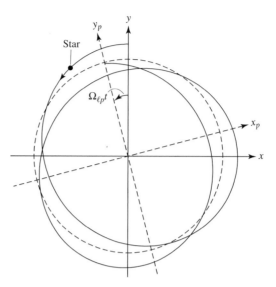

**FIGURE 25.26**   A noninertial coordinate system, $(x_p, y_p)$, rotating within a fixed inertial frame $(x, y)$ with the local angular pattern speed $\Omega_{\ell p} = \Omega - \kappa/2$, for which $(n = 1, m = 2)$. The dashed line corresponds to the perfectly circular motion of the equilibrium point, and the solid line represents the orbital motion of the star in the galaxy's inertial reference frame. The position of the star on the diagram, along with the position of the noninertial coordinate system, corresponds to one-eighth of the orbital period of the equilibrium point.

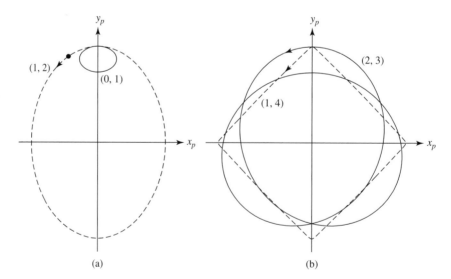

**FIGURE 25.27**   Stars appear to trace out closed orbits in noninertial coordinate systems that rotate with appropriately chosen angular pattern speeds. The motion of a star as it would appear in coordinate systems having local angular pattern speeds corresponding to $(n, m)$ values of (a) $(0, 1)$ = solid line, $(1, 2)$ = dashed line, (b) $(2, 3)$ = solid line, and $(1, 4)$ = dashed line.

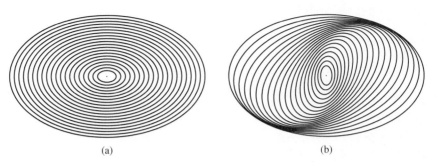

(a)                                                    (b)

**FIGURE 25.28** (a) Nested oval orbits with aligned major axes, as seen in a reference frame rotating with the global angular pattern speed ($n = 1, m = 2$), or $\Omega_{gp} = \Omega - \kappa/2$. The result is a bar-like structure. (b) Each oval is rotated relative to the orbit immediately interior to it. The result is a two-armed grand-design spiral density wave.

pattern, shown in Fig. 25.28(b). Twisting the ovals in the opposite sense would result in a leading two-armed spiral. The two-armed spiral M51 (Fig. 25.17) is an example of a trailing-arm, ($n = 1, m = 2$) pattern structure, while the four-armed spiral M101 [Fig. 25.5(c)] is an ($n = 1, m = 4$) system. Patterns with $m = 2$ are the most common type of density wave structure.

It is important to remember that the individual stars are following their own orbits in an inertial reference frame and that these orbits appear only as simplified oval shapes in a reference frame that is rotating with the local angular pattern speed. Furthermore, even in the rotating frame the stars themselves are still moving along the oval orbits. Only the spiral pattern appears to be static in that frame (if we still assume that $\Omega_{gp} = \Omega_{\ell p}$, independent of $R$); in the nonrotating inertial frame the spiral pattern will appear to move with an angular speed of $\Omega_{gp}$. It is the "traffic jam" of stars becoming packed together where their oval orbits approach one another that leads to the density waves.

The stability of the structures shown in Fig. 25.28 depends crucially on whether $\Omega_{\ell p} = \Omega(R) - \kappa(R)/2$ is actually independent of $R$—that is, on whether there is an appropriate global value of $\Omega_{gp}$. Figure 25.29 shows curves of $\Omega(R) - n\kappa(R)/m$ with several ratios of $n/m$ for one model of our Galaxy. Notice that $\Omega - \kappa/2$ is nearly flat over a wide range of values for $R$, a fact first realized by the Swedish astronomer Bertil Lindblad (1895–1965). The same general behavior is exhibited by a large number of spiral galaxies and probably accounts for the prevalence of two-armed spirals. Of course, $\Omega - \kappa/2$ is not exactly constant with respect to galactocentric radius, so some drifting of epicycle orbits relative to one another does occur, leading once again to a winding problem. However, in this case the winding occurs for density waves rather than material waves, and because the relative drift is slower, the winding takes about five times longer to develop. If a means can be found to stabilize the galaxy against this remnant winding effect, the original Lin–Shu hypothesis of a quasistatic density wave will be realized.

### Lindblad, Corotation, and Ultraharmonic Resonances

Since $\Omega - \kappa/2$ is not exactly constant across the entire galactic disk, stars at varying radii do not encounter the spiral arms at precisely the same point in their epicyclic paths on

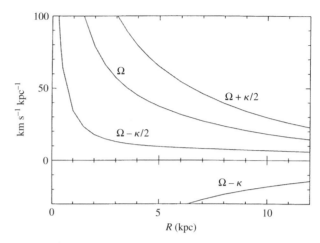

**FIGURE 25.29**    The Bahcall–Soneira model of our Galaxy has been used to construct the functions $\Omega_{\ell p} = \Omega - n\kappa/m$ for various ratios of $n/m$. Note the nearly constant curve for $n/m = 1/2$ over much of the Galactic disk. (Figure adapted from Binney and Tremaine, *Galactic Astronomy*, Princeton University Press, Princeton, NJ, 1987.)

successive passages through the density waves. However, there are certain radii for which this can happen and a resonance develops, analogous to the resonance that can occur when a spring is forced to oscillate at its natural frequency. For instance, if a star is at its maximum value of $\chi$ (Eq. 25.36) each time it encounters a density wave, the perturbation it experiences due to the local increase in density and gravitational potential will always be in the same sense, and the effect will be cumulative. A similar argument applies to each of the local pattern speeds, such as $\Omega_{\ell p} = \Omega(R)$ and $\Omega_{\ell p} = \Omega(R) + \kappa/2$.[20]

This amplification can occur at several radii in a galaxy, depending on its mass distribution and its resulting rotation curve. An **inner Lindblad resonance** exists when the local angular pattern speed of the star equals the global angular pattern speed of the density wave for the case when $\Omega_{gp} = \Omega - \kappa/2$ for $n/m = 1/2$. It is possible that either zero, one, or two inner Lindblad radii may exist for a given galaxy, depending on the shape of the rotation curve. A **corotation resonance** can occur if $\Omega_{gp} = \Omega$ for some value of $R$, and an **outer Lindblad resonance** might exist if $\Omega_{gp} = \Omega + \kappa/2$. The **ultraharmonic resonance** may also develop for $\Omega_{gp} = \Omega - \kappa/4$. The inner rings of galaxies such as NGC 7096 and M81 [Figs. 25.5(a) and (b)] are apparently due to either inner Lindblad resonances or ultraharmonic resonances, while the outer rings of galaxies like NGC 4340 [Fig. 25.4(f)] are found at the outer Lindblad resonance. The positions of the Lindblad and corotation resonances are depicted in Fig. 25.30 for two different rotation curves.

Resonances have the effect that epicycle oscillation amplitudes increase dramatically. As a result, collisions of gas clouds should also increase significantly at resonance positions, and energy will be dissipated. Consequently, resonance locations (if they exist in a specific galaxy) can actually lead to damping of spiral waves unless other processes are capable of

---

[20]Note that we are now going beyond the assumption made on page 969 that the density waves make an insignificant contribution to the gravitational field.

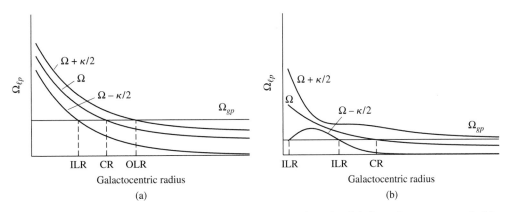

**FIGURE 25.30** The existence of resonance radii depends on the global angular pattern speed of the arms and the shape of the galaxy's rotation curve. (a) A galaxy with a single inner Lindblad resonance (ILR), a corotation resonance (CR), and an outer Lindblad resonance (OLR). (b) A galaxy with two ILRs, a CR, and no OLR. Note that for sufficiently large values of $\Omega_{gp}$, there may not be any ILRs.

continually building the waves back up again. This situation is analogous to that of stellar pulsation (Chapter 14); regions of partial ionization (primarily hydrogen and helium) drive pulsations, while other regions of the star damp the oscillations.

### Nonlinear Effects in Density Wave Theory

Despite a great deal of work, a complete understanding of density waves has not yet been fully realized. For instance, various effects not considered in the simple model presented here may play important roles, such as nonlinear (higher-order) terms in $\Phi_{eff}$ (recall Eq. 25.28). Also, the waves themselves alter the gravitational potential in which they originate so that azimuthal symmetry breaks down.

One important driving mechanism in a number of grand-design spirals, including M51 (Fig. 25.17), is the presence of a companion galaxy that triggers spiral structure through tidal interactions (see Section 26.1 and Fig. 26.10). A more detailed discussion of the physical driving mechanisms that help to establish and maintain density waves is beyond the level of the current discussion, but it is worth pointing out that the techniques for studying spiral density waves have much in common with the theoretical procedures used to investigate stellar pulsation; both linear and nonlinear models have been employed in the investigation of spiral structure.

### Numerical N-Body Simulations

An important example of a nonlinear $N$-body simulation of a rotating disk is one that was calculated in an early work by F. Hohl in 1971 (see Fig. 25.31). In that calculation, $10^5$ stars were initially placed in axisymmetric orbits ($t = 0$), but as the simulation progressed, Hohl's model proved to be very unstable against the development of the $m = 2$ pattern (also known as the $m = 2$ mode), and a two-armed spiral density wave developed ($t = 2.0$). As the simulation continued, the disk became "hot," meaning that the velocity dispersion of

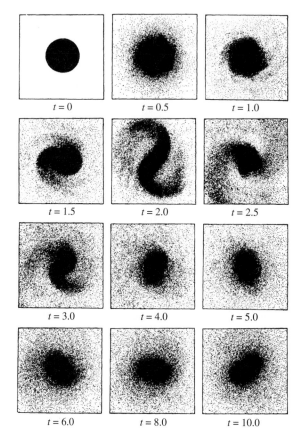

$t = 0$     $t = 0.5$     $t = 1.0$

$t = 1.5$     $t = 2.0$     $t = 2.5$

$t = 3.0$     $t = 4.0$     $t = 5.0$

$t = 6.0$     $t = 8.0$     $t = 10.0$

**FIGURE 25.31**   An early study of a rotating disk with an $N$-body simulation using $10^5$ stars. The disk began with complete axial symmetry and quickly developed an $m = 2$ mode instability. Eventually the disk "heated up," destroying the spiral arms but leaving a long-lived bar. (Figure from Hohl, *Ap. J.*, *168*, 343, 1971.)

its stars became large relative to their orbital velocities, and the spiral structure dissipated. Interestingly, a bar instability persisted throughout the rest of the simulation, and the final structure bears a strong resemblance to an SB0 galaxy.

Such bar instabilities have proved themselves to be common features in $N$-body calculations. Apparently a rotating disk is highly susceptible to a bar-mode instability, at least on computers. Of course, there is strong support for the idea that bar modes are favored in real galaxies as well, since roughly two-thirds of all disk galaxies do exhibit bar-like structures in their centers.

Hohl's work does suggest one possible means of stabilizing disks against various mode instabilities: the presence of high-velocity dispersions of its stars. In his original simulations, Hohl's bar heated the disk, destroying the $m = 2$ trailing-arm mode. As we will see in Section 26.1, tidal interactions and/or mergers may also play important roles in heating the disks of spiral galaxies.

In modern $N$-body simulations, models have also included the effects of highly centralized masses, such as the gravitational influence of supermassive black holes. Various researchers have shown that the inclusion of sufficiently condensed and massive central mass concentrations can weaken, and even destroy, the bar instability. Most studies find that a mass on the order of a few percent of the mass of the galaxy's disk is required in order to affect the bar.

### Stochastic, Self-Propagating Star Formation

Although most of the discussion found in this section has focused on the Lin–Shu hypothesis of quasistatic density waves and the existence of grand-design spirals, many of the galaxies observed in the universe are flocculent spirals. It may be that these objects are composed of a linear combination of several stable density perturbations (somewhat reminiscent of modes in stellar oscillations), along with sufficient patchiness in the interstellar medium to give the appearance of less well-defined spirals. Alternatively, it may be that an entirely different mechanism is responsible for the spiral structure seen in these galaxies.

In 1976 M. W. Mueller and W. David Arnett proposed a theory of spiral structure for flocculent spirals known as **stochastic, self-propagating star formation (SSPSF)**. In their theory they imagine spiral structure arising from outbursts of star formation that propagate across the galaxy. When one region of the galaxy undergoes a star formation episode, its most massive stars will age rapidly, producing core-collapse supernovae. The supernovae shock waves will travel through the interstellar medium, triggering the collapse of other gas clouds in nearby regions, where further star formation will occur, allowing the process to continue. This scheme has been compared to a forest fire, with the flames jumping from tree to tree. Spiral structure arises when the differential rotation of the galaxy draws these newly "lit" regions into trailing arms.

Although SSPSF has been successful in producing flocculent spiral structure in computer simulations, it is unable to account for the transitions from dust lanes to OB stars to red stars across the spiral arms of grand-design spirals, so it probably cannot explain those galaxies. Of course, it may well be that we will need both of these theories (and perhaps others) to understand the abundance of spiral structure found throughout the universe.

## 25.4 ■ ELLIPTICAL GALAXIES

Although Hubble type correlates well with a wide variety of physical parameters for late-type galaxies, thereby ensuring its continued usefulness for those systems, the Hubble-type designation for early galaxies (which is based solely on apparent ellipticity) has shown itself to be virtually irrelevant in terms of trying to categorize other characteristics. As a result, subtype distinctions are made for ellipticals that are independent of the ellipticity, focusing instead on other morphological features, such as the size, absolute magnitude, and surface brightness of the image. Once thought to be the simplest of the major galaxy types, since the 1980s ellipticals have come to be seen as remarkably diverse and complex. As we shall see in Section 26.1, some of this complexity may arise, at least in part, from strong environmental evolution, possibly involving tidal interactions or mergers with neighboring galaxies.

## Morphological Classes of Elliptical Galaxies

Today a number of separate morphological classes are commonly used to distinguish among the elliptical galaxies:

- **cD galaxies** are immense, but rare, bright objects that sometimes measure nearly 1 Mpc across and are usually found only near the centers of large, dense clusters of galaxies. Their absolute $B$ magnitudes range from less than $-22$ mag to $-25$ mag, and they have masses of between $10^{13}$ and $10^{14}$ $M_\odot$. cD galaxies are characterized by having central regions with high surface brightnesses ($\mu = 18$ $B$-mag arcsec$^{-2}$) and very extended, diffuse envelopes ($\mu = 26$ to $27$ $B$-mag arcsec$^{-2}$). They may also possess tens of thousands of globular clusters, with typical specific frequencies, $S_N$, of 15 (see Eq. 25.15). Furthermore, these galaxies are known to have very high mass-to-light ratios, sometimes exceeding 750 $M_\odot/L_\odot$, implying large quantities of dark matter.

- **Normal elliptical galaxies** are centrally condensed objects with relatively high central surface brightnesses. They include the giant ellipticals (gE's), intermediate-luminosity ellipticals (E's), and compact ellipticals (cE's). The absolute $B$ magnitudes of normal E's range from $-15$ to $-23$, masses of between $10^8$ and $10^{13}$ $M_\odot$, diameters from less than 1 kpc to nearly 200 kpc, mass-to-light ratios from 7 to more than 100 $M_\odot/L_\odot$, and specific frequencies of globular clusters in the range of 1 to 10. Lenticular galaxies (S0's and SB0's) are often grouped with normal E's.

- **Dwarf elliptical galaxies** (dE's) have surface brightnesses that tend to be much lower than those of cE's of the same absolute magnitude. The absolute $B$ magnitudes of dE's fall between $-13$ and $-19$, they have typical masses of $10^7$ to $10^9$ $M_\odot$, and their diameters are on the order of 1 to 10 kpc. Their metallicities also tend to be lower than for normal E's. The average value of the specific frequency of globular clusters is $\langle S_N \rangle = 4.8 \pm 1.0$, which is still higher than for spirals.

- **Dwarf spheroidal galaxies** (dSph's) are extremely low-luminosity, low-surface-brightness objects that have been detected only in the vicinity of the Milky Way. Their absolute $B$ magnitudes are only $-8$ to $-15$ mag, their masses are roughly $10^7$ to $10^8$ $M_\odot$, and their diameters are between 0.1 and 0.5 kpc.

- **Blue compact dwarf galaxies** (BCD's) are small galaxies that are unusually blue, with color indices ranging from $\langle B - V \rangle = 0.0$ to $0.3$. This corresponds to main-sequence stars of spectral class A, indicating that these galaxies are undergoing particularly vigorous star formation. They have absolute $B$ magnitudes of $-14$ to $-17$, masses on the order of $10^9$ $M_\odot$, and diameters of less than 3 kpc. As is expected for very active star formation, BCD's also have a large abundance of gas, with $M_{\text{H I}} = 10^8$ $M_\odot$ and $M_{\text{H II}} = 10^6$ $M_\odot$ constituting roughly 15% to 20% of the entire mass of the galaxy. They also have correspondingly low mass-to-light ratios; in an extreme case, ESO400−G43 has $M/L_B = 0.1$ despite the dominance of dark matter at large radii.

**TABLE 25.3** Characteristic Data for cD, Elliptical, and Lenticular Galaxies.

|  | cD | E | S0/SB0 |
|---|---|---|---|
| $M_B$ | $-22$ to $-25$ | $-15$ to $-23$ | $-17$ to $-22$ |
| $M$ $(M_\odot)$ | $10^{13}$–$10^{14}$ | $10^8$–$10^{13}$ | $10^{10}$–$10^{12}$ |
| Diameter $(D_{25}$, kpc) | 300–1000 | 1–200 | 10–100 |
| $\langle M/L_B \rangle$ $(M_\odot/L_\odot)$ | $> 100$ | 10–100 | $\sim 10$ |
| $\langle S_N \rangle$ | $\sim 15$ | $\sim 5$ | $\sim 5$ |

**TABLE 25.4** Characteristic Data for Dwarf Elliptical, Dwarf Spheroidal, and Blue Compact Dwarf Galaxies.

|  | dE | dSph | BCD |
|---|---|---|---|
| $M_B$ | $-13$ to $-19$ | $-8$ to $-15$ | $-14$ to $-17$ |
| $M$ $(M_\odot)$ | $10^7$–$10^9$ | $10^7$–$10^8$ | $\sim 10^9$ |
| Diameter $(D_{25}$, kpc) | 1–10 | 0.1–0.5 | $< 3$ |
| $\langle M/L_B \rangle$ $(M_\odot/L_\odot)$ | $\sim 10$ | 5–100 | 0.1–10 |
| $\langle S_N \rangle$ | $4.8 \pm 1.0$ | — | — |

Tables 25.3 and 25.4 summarize some of the characteristics of early-type galaxies. Lenticular galaxies have also been included in the list, since they appear to have much in common with many ellipticals.

### Surface Brightness Profiles

cD's and normal ellipticals have surface brightness profiles that closely follow the $r^{1/4}$ law (Eq. 25.2). As an example, the brightness profile of NGC 3379, a gE galaxy, is shown in Fig. 25.32. However, as the mass of the galaxy decreases, there is a fairly smooth transition over to an exponential profile (Eq. 25.3). This is particularly true of dE's and dSph's.

### Dust and Gas in Elliptical Galaxies

The low gravitational binding energy of dE's and dSph's means that it would be very difficult for these galaxies to retain a significant amount of gas, which is indeed what is observed. In fact, dSph's are virtually devoid of gas. Consequently, dE's and dSph's are not actively forming stars today. Moreover, they have very low metallicities, with values similar to those found in globular clusters. It has been suggested that these systems lost most of their gas via supernova-driven mass loss or by **ram-pressure stripping** as the galaxies passed through the gas found in clusters of galaxies (see Section 27.3).

For many years it was believed that all elliptical galaxies had been largely stripped of any dust or gas that had not yet formed into stars or, alternatively, that star formation had proceeded very efficiently during the earliest history of these galaxies, depleting all available gas. However, we have now come to realize that gas and dust are present within most normal elliptical galaxies, albeit at somewhat diminished levels relative to spirals.

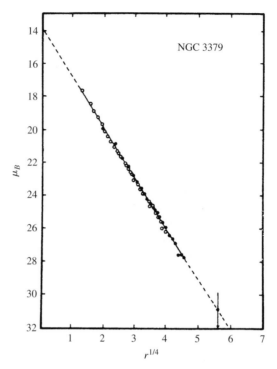

**FIGURE 25.32**   The surface brightness profile of the giant elliptical galaxy NGC 3379 is well represented by an $r^{1/4}$ law. (Figure adapted from de Vaucouleurs and Capaccioli, *Ap. J. Supp.*, *40*, 699, 1979.)

The hydrogen gas that is present in E/S0's is usually evident in several forms. A $10^7$ K X-ray-emitting hot-gas component of $10^8$ to $10^{10}$ $M_\odot$ makes up the majority of all the gas that is present in a typical E/S0 galaxy. (Possible heating mechanisms include supernovae and stellar winds.) A warm-gas component ($10^4$ to $10^5$ $M_\odot$ at $10^4$ K), observable in H$\alpha$, is also present in the form of H II regions. Furthermore, $10^7$ to $10^9$ $M_\odot$ of cold H I ($\sim 10^2$ K) is detectable at 21 cm. CO emission studies indicate that $10^7$ to $10^9$ $M_\odot$ of molecular hydrogen can be found in early-type galaxies as well.

Astronomers have also discovered that perhaps 50% of all elliptical galaxies contain appreciable amounts of dust, with typical masses in the range of $10^5$ to $10^6$ $M_\odot$. One intriguing observation is that there exist dust lanes that are essentially randomly oriented with respect to their parent galaxy's optical-light axes. Frequently, the dust is even found to be counter-rotating relative to the rotation direction of other (stellar) components. Apparently, much of this gas was acquired after the initial formation of the galaxy. More will be said about the dynamics of dust rings in ellipticals in Section 26.1.

## Metallicity Gradients and Color Gradients

As we saw in Fig. 25.12 during our discussion of spirals, the metallicity of elliptical galaxies is well correlated with luminosity; brighter galaxies have higher overall metal content.

Furthermore, as with spiral galaxies, metallicity gradients and color gradients also exist in ellipticals. Generally, the central regions of E's are redder and more metal-rich than are regions at larger radii. This trend can be expressed in terms of the gradient of the mass fraction of metals found in the distribution of stars; typically $d \log_{10} Z / d \log_{10} r \approx -0.13$. The same is also true for lenticular galaxies. In addition, the disks of S0's are bluer than their bulges, as is the case for galaxies of later Hubble type [for S0's, $(U - V)_{\text{disk}} - (U - V)_{\text{bulge}} \sim -0.1$ to $-0.5$]. Any successful theory of galaxy formation must incorporate the available observations concerning chemical enrichment.

### The Faber–Jackson Relation

One relationship that dE's, dSph's, normal E's, and the bulges of spirals all have in common is the correlation between their central radial-velocity dispersion and $M_B$. This relation is shown in Fig. 25.33. To see why the relation arises, we begin with the result of the virial theorem for the central radial-velocity dispersion of a spherically symmetric mass distribution (Eq. 25.13). In that case we made the simple approximation that a mass $M$ was uniformly distributed throughout a volume of radius $R$. Now, if we further assume that the mass-to-light ratio is essentially constant for all galaxies and that their average surface brightnesses are all equal (assumptions that were also made in "deriving" the Tully–Fisher relation), we arrive at the result

$$L \propto \sigma_0^4, \tag{25.40}$$

where $\sigma_r$ has been replaced by $\sigma_0$ to signify that the central value for the radial-velocity dispersion is being assumed. This relationship was first identified by Sandra Faber and Robert Jackson and is now referred to as the **Faber–Jackson relation**. Expressing the luminosity in solar units, taking the logarithm of both sides of the equation, and writing

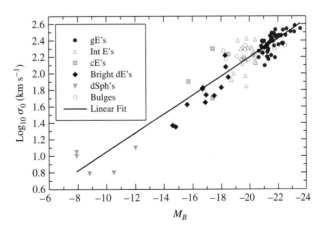

**FIGURE 25.33** The Faber–Jackson relation represents a correlation between central velocity dispersion and luminosity. The relation is a consequence of the virial theorem. (Data from Bender et al., *Ap. J.*, *399*, 462, 1992.)

$\log_{10}(L/L_\odot)$ as a difference in $M_B$'s via Eq. (3.8), we find the linear relationship between $\log_{10} \sigma_0$ and $M_B$ illustrated in Fig. 25.33, namely

$$\log_{10} \sigma_0 = -0.1M_B + \text{constant}. \tag{25.41}$$

[Note the resemblance to the Tully–Fisher relations given by Eqs. (25.5–25.7) if those expressions are solved for $\log_{10} V_{\text{max}}$.]

## The Fundamental Plane

As is readily apparent by inspecting Fig. 25.33, there is considerable scatter in the data. This is further reflected in the fact that the slope of the best-fit line through the data differs slightly from the one found in Eq. (25.41). Apparently Eq. (25.40) is not strictly true, which should not be surprising given the assumptions made in its development. Depending on the sample set used, $L \propto \sigma_0^\alpha$, with $3 < \alpha < 5$.

In an effort to find a tighter fit to the data, astronomers have introduced a second parameter into the expression, the effective radius. One representation of this fit is

$$L \propto \sigma_0^{2.65} r_e^{0.65}. \tag{25.42}$$

Here, galaxies are visualized as residing on a two-dimensional "surface" in the three-dimensional "space" represented by the coordinates $L$, $\sigma_0$, and $r_e$. Known as the **fundamental plane**, Eq. (25.42) combines the contributions of a galaxy's gravitational well ($\sigma_0$) with its radius and luminosity. Alternatively, the fundamental plane can be written in terms of the effective surface brightness of a galaxy at its effective radius:

$$r_e \propto \sigma_0^{1.24} I_e^{-0.82}. \tag{25.43}$$

The fundamental plane appears to represent the whole family of elliptical galaxies and seems to imply something about the formation of these systems.

## The Effects of Rotation

Parameterizations such as the fundamental plane do a much better job of classifying ellipticals than Hubble's original scheme based on ellipticity. But then, what is the source of a galaxy's shape? Although the answer is not yet entirely clear, it is evident that most ellipticals are not purely oblate or prolate rotators with two axes, but are triaxial, meaning that there is no single preferred axis of rotation. We have already encountered this situation in the randomness of the dust lanes found in at least one-half of all ellipticals. Further evidence exists in the frequent observations of counter-rotating stellar cores in as many as 25% of larger ellipticals. It appears that in at least some of these cases, material in the form of gas, dust, globular clusters, or dwarf galaxies has been captured sometime since the galaxy's formation. This situation will be discussed in more detail in Section 26.1.

When the rotational velocities of elliptical galaxies are measured, we find that the more luminous ones have mean rotational velocities that are much less than their velocity dispersions, implying that the shapes of these galaxies are not due to rotation. Instead, their shapes are caused by the anisotropic velocity dispersion of stars in the galaxies.[21] In one particularly extreme case, the luminous E4 galaxy, NGC 1600 ($M_B = -22.87$), has $V_{rot} = 1.9 \pm 2.3$ km s$^{-1}$ and $V_{rot}/\sigma < 0.013$; it has no unique axis of rotation that is statistically significant.

If the ellipticity, $\epsilon$ (Eq. 25.1), of a galaxy were due to its being an ideal, oblate rotator with an isotropic stellar velocity distribution, then it can be shown that

$$\left(\frac{V_{rot}}{\sigma}\right)_{isotropic} \approx \left(\frac{\epsilon}{1-\epsilon}\right)^{1/2}.$$

For NGC 1600 this means that if its ellipticity of $\epsilon = 0.4$ were due to pure rotation, then $V_{rot}/\sigma$ should be approximately 0.8. Defining the rotation parameter $(V/\sigma)^*$ to be

$$(V/\sigma)^* \equiv \frac{(V_{rot}/\sigma)_{observed}}{(V_{rot}/\sigma)_{isotropic}}$$

gives $(V/\sigma)^* < 0.016$ for NGC 1600. Although the distinction is somewhat arbitrary, a galaxy is considered to be primarily **rotationally supported** if $(V/\sigma)^* > 0.7$.

Bright E's and gE's have typical values of $\langle(V/\sigma)^*\rangle \approx 0.4$ and are **pressure-supported**, meaning that their shapes are due primarily to random stellar motions. In addition, the diffuse, faint dwarf E's have predominantly anisotropic velocity dispersions and are also pressure-supported. On the other hand, galaxies for which $-18 > M_B > -20.5$ mag have $\langle(V/\sigma)^*\rangle \approx 0.9$, implying that they are largely rotationally supported (this includes the cE's). Interestingly, the bulges of spiral galaxies tend to be rotationally supported as well.

### Correlations with Boxiness or Diskiness

In 1988 Ralf Bender, Jean-Luc Nieto (1950–1992), and their collaborators proposed that many of the characteristics of elliptical galaxies can be understood in terms of the degree of **boxiness** or **diskiness** that their isophotal surfaces exhibit. As a means of quantifying the deviation from an elliptical shape, the shape of an isophotal contour (defined for some specific value of $\mu$) is written in polar coordinates as a Fourier series of the form

$$a(\theta) = a_0 + a_2 \cos(2\theta) + a_4 \cos(4\theta) + \cdots, \tag{25.44}$$

where $a$ is the contour's radius and the angle $\theta$ is measured counterclockwise from the major axis of the ellipsoid. The first term in the expansion represents the shape of a perfect circle, the second term corresponds to the amount of ellipticity, and the third term is associated with the degree of boxiness. If $a_4 < 0$, the isophotal surface tends toward a "boxy" appearance, and if $a_4 > 0$, the surface tends toward being "disky" (see Fig. 25.34). Typically, $|a_4/a_0| \sim 0.01$, with deviations from perfect ellipses routinely measured to an accuracy of approximately 0.5%.

Figure 25.35 shows the correlations of various quantities with the isophotal Fourier parameter $a_4$, measured in units of $100a_4/a_0$. The upper left-hand panel illustrates that disky galaxies tend to be rotationally supported [with larger values of $(V/\sigma)^*$], while boxy

---

[21]An anisotropic velocity distribution is one for which one or two directions of motion are preferred, rather than being completely random in all three dimensions, as in an isotropic distribution.

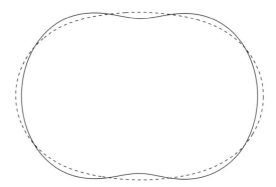

**FIGURE 25.34** Two examples of galaxies with $\epsilon = 0.4$. $|a_4/a_0|$ has been set to the rather large value of 0.03 to illustrate the effect. The solid line represents a "boxy" galaxy ($a_4/a_0 < 0$), and the dashed line represents a "disky" galaxy ($a_4/a_0 > 0$).

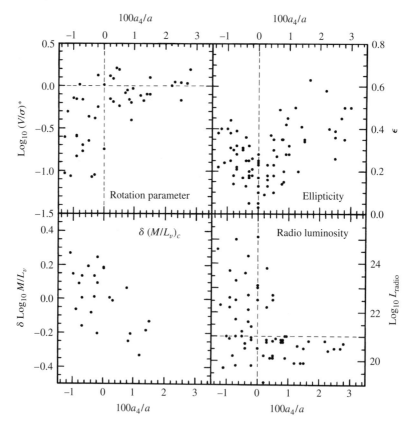

**FIGURE 25.35** The correlations of various parameters with $a_4$ (isophotal shape). "Disky" ellipticals have $a_4 > 0$, while "boxy" ellipticals have $a_4 < 0$. (Figure adapted from Kormendy and Djorgovski, *Annu. Rev. Astron. Astrophys.*, 27, 235, 1989. Reproduced with permission from the *Annual Review of Astronomy and Astrophysics*, Volume 27, ©1989 by Annual Reviews Inc.)

galaxies are largely pressure-supported. The lower left-hand panel indicates the deviation of the core mass-to-light ratio (measured in the $V$ band) from the average value for galaxies of the same luminosity; boxy galaxies tend to have higher-than-average ratios, while disky galaxies have lower-than-average values. The panel in the lower right-hand corner shows that boxy ellipticals tend to be much brighter at radio wavelengths than are disky galaxies. However, the spread in the radio luminosities of boxy galaxies is quite large, while the variation in radio luminosity for disky galaxies is much smaller. The same sort of correlation is found in other wavelength regions as well (such as X-ray, UV, and H$\alpha$). For disky E's, the sources of X-rays tend to be consistent with compact sources only, as is the case for S0's, while boxy E's show evidence of diffuse sources (i.e., hot gas). Finally, the upper right-hand panel indicates that ellipticity (and therefore Hubble type) tends to increase with the absolute value of $a_4$ ($|a_4|$). However, this suggests that two galaxies of identical Hubble type could have very different characteristics when other physical parameters are considered. For instance, for two E4 galaxies ($\epsilon = 0.4$), one could be rotationally supported ($a_4 > 0$) while the other is pressure-supported ($a_4 < 0$), and the former galaxy would probably have a core mass-to-light ratio lower than average while the latter galaxy would have a higher-than-average core mass-to-light ratio. Also, the former galaxy would have a low radio luminosity, and the latter might have a much higher radio luminosity.

Other parameters also seem to be associated with the boxiness or diskiness of a galaxy. Whereas rotationally supported disky galaxies have relatively large ratios of rotational angular momentum to mass (specific angular momentum), pressure-supported boxy galaxies have rather low ratios. Boxy galaxies are frequently observed to contain counter-rotating cores, while disky galaxies rarely possess counter-rotating cores. Finally, disky galaxies generally have rotational symmetries that are oblate, and boxy galaxies tend to be much more triaxial.

Although it is likely that the degree of boxiness among galaxies represents a continuum rather than two distinctly different classes of objects, apparently some 90% of all ellipticals are generally disky in nature. It has been suggested that disky galaxies may simply represent an extension of the sequence of S0's to progressively smaller and smaller disk-to-bulge ratios, and that some disky galaxies are actually misclassified S0's. However, boxy galaxies may well represent a tell-tale signature of some level of environmental evolution, such as tidal interactions or mergers (Section 26.1).

### The Relative Numbers of Galaxies of Various Hubble Types

Before leaving the discussion of morphology, it is important to address the question of the relative numbers of galaxies of various Hubble types. This is usually represented by the **luminosity function**, $\phi(M)\,dM$, defined to be the number of galaxies in a particular sample that have absolute magnitudes between $M$ and $M + dM$. Figure 25.36 shows luminosity functions for two samples; the top graph is for galaxies located in the vicinity of the Milky Way and the bottom graph is for a sample of galaxies in the Virgo cluster of galaxies (see Section 27.3). The total luminosity function in either environment is the sum of the individual luminosity functions of each Hubble type.

In an attempt to find a general analytic fit to galactic luminosity functions, Paul Schechter proposed the functional form

$$\phi(L)\,dL \sim L^{\alpha} e^{-L/L^*}\,dL, \tag{25.45}$$

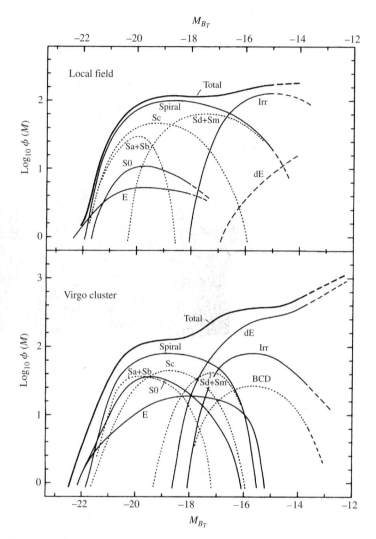

**FIGURE 25.36** Luminosity functions for two samples of galaxies. The top panel is based on a sample of galaxies near the Milky Way Galaxy, and the bottom panel is for galaxies located in the Virgo cluster of galaxies. The zero point of the $\log_{10} \phi(M)$ axis is arbitrary. (Figure adapted from Binggeli, Sandage, and Tammann, *Annu. Rev. Astron. Astrophys.*, **26**, 509, 1988. Reproduced with permission from the *Annual Review of Astronomy and Astrophysics*, Volume 26, ©1988 by Annual Reviews Inc.)

which can also be written in the equivalent form

$$\phi(M)\,dM \sim 10^{-0.4(\alpha+1)M} e^{-10^{0.4(M^*-M)}}\,dM. \qquad (25.46)$$

In both forms, $\alpha$ and $L^*$ (or $M^*$) are free parameters that are used to obtain the best possible fit to the available data. For the data in Fig. 25.36, $\alpha = -1.0$ and $M_B^* = -21$ for the local field of galaxies near the Milky Way, while $\alpha = -1.24 \pm 0.02$ and $M_B^* = -21 \pm 0.7$ for the Virgo cluster. Although the values are similar, it is clear that there is not a universal luminosity function. Instead, each luminosity function depends on the environment of the particular sample of galaxies.

As can be seen for either environment depicted in Fig. 25.36, dE's and dwarf irregulars represent the largest fraction of all galaxies, even though S's and E's are the most prominent in terms of both cluster luminosity and mass. Careful inspection of the figures also shows another important detail: Although spirals represent the largest fraction of bright galaxies in each case, there is a somewhat higher proportion of ellipticals in the Virgo cluster. When the Virgo cluster is compared with the much larger and more densely populated Coma cluster, the relative numbers of spirals and ellipticals change dramatically. In the Virgo cluster we find 12% E's, 26% S0's, and 62% S+Ir, while in the Coma cluster, the proportions are 44% E's, 49% S0's, and only 7% S+Ir. Once again there is evidence that environment plays a role in galaxy formation and/or evolution, an issue we will explore in more detail in Chapter 26.

## SUGGESTED READING

### General

Ferris, Timothy, *Galaxies*, Stewart, Tabori, and Chang, New York, 1982.

Silk, Joseph, *The Big Bang*, Third Edition, W. H. Freeman and Company, New York, 2001.

Waller, William H., and Hodge, Paul W., *Galaxies and the Cosmic Frontier*, Harvard University Press, Cambridge, MA, 2003.

### Technical

Athanassoula, E., Lambert, J. C., and Dehnen, W., "Can Bars Be Destroyed by a Central Mass Concentration? – I. Simulations," *Monthly Notices of the Royal Astronomical Society*, *363*, 496, 2005.

Bender, Ralf, Burstein, David, and Faber, S. M., "Dynamically Hot Galaxies. I. Structural Properties," *The Astrophysical Journal*, *399*, 462, 1992.

Bender et al., "HST STIS Spectroscopy of the Triple Nucleus of M31: Two Nested Disks in Keplerian Rotation around a Supermassive Black Hole," *The Astrophysical Journal*, *631*, 280, 2005.

Bertin, Giuseppe, *Dynamics of Galaxies*, Cambridge University Press, Cambridge, 2000.

Binggeli, Bruno, Sandage, Allan, and Tammann, G. A., "The Luminosity Function of Galaxies," *Annual Review of Astronomy and Astrophysics*, *26*, 1988.

Binney, James, and Merrifield, Michael, *Galactic Astronomy*, Princeton University Press, Princeton, NJ, 1998.

Binney, James, and Tremaine, Scott, *Galactic Dynamics*, Princeton University Press, Princeton, NJ, 1987.

Combes, F., Boissé, P., Mazure, A., and Blanchard, A., *Galaxies and Cosmology*, Second Edition, Springer, 2002.

de Vaucouleurs, Gerard, *Third Reference Catalogue of Bright Galaxies*, Springer-Verlag, New York, 1991.

de Zeeuw, Tim, and Franx, Marijn, "Structure and Dynamics of Elliptical Galaxies," *Annual Review of Astronomy and Astrophysics*, *29*, 1991.

Elmegreen, Debra Meloy, *Galaxies and Galactic Structure*, Prentice-Hall, Upper Saddle River, NJ, 1998.

Gallagher, John S. III, and Wyse, Rosemary F. G., "Dwarf Spheroidal Galaxies: Keystones of Galaxy Evolution," *Publications of the Astronomical Society of the Pacific*, *106*, 1225, 1994.

Gilmore, Gerard, King, Ivan R., and van der Kruit, Pieter C., *The Milky Way as a Galaxy*, University Science Books, Mill Valley, CA, 1990.

Hubble, Edwin, "Extra-Galactic Nebulae," *The Astrophysical Journal*, *64*, 321, 1926.

Hubble, Edwin, *The Realm of the Nebulae*, Yale University Press, New Haven, CT, 1936.

Kormendy, John, and Djorgovski, S., "Surface Photometry and the Structure of Elliptical Galaxies," *Annual Review of Astronomy and Astrophysics*, *27*, 1989.

Kormendy, John, and Richstone, Douglas, "Inward Bound: The Search for Supermassive Black Holes in Galaxy Nuclei," *Annual Review of Astronomy and Astrophysics*, *33*, 1995.

Mateo, Mario, "Dwarf Galaxies in the Local Group," *Annual Review of Astronomy and Astrophysics*, *36*, 435, 1998.

Pierce, Michael J., and Tully, R. Brent, "Luminosity–Line Width Relations and the Extragalactic Distance Scale. I. Absolute Calibration," *The Astrophysical Journal*, *387*, 47, 1992.

Rubin, Vera C., Burstein, David, Ford, W. Kent Jr., and Thonnard, Norbert, "Rotation Velocities of 16 Sa Galaxies and a Comparison of Sa, Sb, and Sc Rotation Properties," *The Astrophysical Journal*, *289*, 81, 1985.

Sandage, Allan, "The Classification of Galaxies: Early History and Ongoing Developments," *Annual Review of Astronomy and Astrophysics*, *43*, 581, 2005.

Sandage, Allan, and Bedke, John, *The Carnegie Atlas of Galaxies*, Carnegie Institution of Washington, Washington, D.C., 1994.

Sofue, Yoshiaki, and Rubin, Vera, "Rotation Curves of Spiral Galaxies," *Annual Review of Astronomy and Astrophysics*, *39*, 137, 2001.

Sparke, Linda S., and Gallagher, John S., *Galaxies in the Universe: An Introduction*, Cambridge University Press, Cambridge, 2000.

van den Bergh, Sidney, *Galaxy Morphology and Classification*, Cambridge University Press, 1998.

Wyse, Rosemary F. G., Gilmore, Gerard, and Franx, Marijn, "Galactic Bulges," *Annual Review of Astronomy and Astrophysics*, *35*, 637, 1997.

## PROBLEMS

**25.1** Using Shapley's assumption that M101 has a diameter of 100 kpc, and adopting van Maanen's flawed observation of a measurable rotational proper motion, estimate the speed of a point at the edge of the galaxy and compare it to the characteristic rotation speed of the Milky Way.

**25.2** (a) The absolute magnitude of M101, an Sc galaxy, is $-21.51$ in the $B$ band. Using Eq. (25.11), estimate its isophotal radius ($R_{25}$) at 25 $B$-mag arcsec$^{-2}$.

(b) Use the Tully–Fisher relation (Eq. 25.7) to estimate the rotational velocity of M101.

(c) Estimate the angular rotation speed of a star at $R_{25}$, in units of arcsec yr$^{-1}$.

(d) Could van Maanen have detected the rate of rotation of M101? How long would it take for the galaxy to rotate through $1''$?

**25.3** (a) From the data given in Table 24.1, estimate $M_B$ for our Galaxy. *Hint:* Be sure to include $M_B$ for the Sun in your calculations, not $M_{\text{bol}}$; see Appendix G.

(b) Using the Tully–Fisher relation, calculate the maximum rotation speed of the Galactic disk and compare your answer with the observational value given in Section 24.3.

**25.4** Neglecting the effects of extinction and the $K$-correction, show that the surface brightness of a galaxy is independent of its distance from the observer.

**25.5** Prove that Eq. (25.2) follows directly from Eq. (24.13).

**25.6** Prove that Eq. (25.3) follows directly from Eq. (24.10).

**25.7** (a) Show that Eq. (24.13) can be written as

$$I(r) = I_e e^{-b[(r/r_e)^{1/4} - 1]},$$

where $b = 7.67$.

(b) Integrating over the entire surface brightness profile, show that the luminosity of the galaxy can be expressed in terms of $r_e$ and $I_e$ as

$$L_{\text{tot}} = \int_0^\infty 2\pi r\, I(r)\, dr = 8! \, \frac{e^{7.67}}{(7.67)^8} \pi r_e^2 I_e \simeq 7.22\pi r_e^2 I_e. \qquad (25.47)$$

*Hint:* $\int_0^\infty e^{-x} x^7 \, dx = \Gamma(8) = 7!$.

(c) Recalling that $I_e \equiv I(r_e)$, show that if the integration of Eq. (25.47) is carried over $0 \le r \le r_e$ rather than over $0 \le r < \infty$, then the resulting luminosity is $\frac{1}{2} L_{\text{tot}}$, consistent with the definition of $r_e$.

**25.8** NGC 2639 is an Sa galaxy with a measured maximum rotational velocity of 324 km s$^{-1}$ and an apparent magnitude of $B = 12.22$ mag (after making corrections for extinction).

(a) Estimate its absolute magnitude in the $B$ band from the Tully–Fisher relation.

(b) Determine the distance to NGC 2639 using its distance modulus.

(c) What is the galaxy's radius ($R_{25}$) at a surface brightness level of 25 $B$-mag arcsec$^{-2}$?

(d) Find the mass of NGC 2639 that is interior to $R_{25}$.

(e) What is the luminosity of the galaxy in the $B$ band? (Refer to the hint in Problem 25.3.)

(f) Calculate the mass-to-light ratio for NGC 2639 in the $B$ band, interior to $R_{25}$.

**25.9** Referring to the color indices ($\langle B - V \rangle$) given in Table 25.1 and Appendix G, estimate the *average* (or integrated) spectral classification of main-sequence stars in spiral galaxies of types Sa, Sb, and Sc.

**25.10** Use the rotation curve data in Fig. 25.14 to estimate the mass within the central $1''$ of the center of M32. Compare your answer with the value obtained using the velocity dispersion data in Example 25.2.1 and with the estimated range quoted in that example.

**25.11** **(a)** From the data shown in Fig. 25.37 for the stellar rotational velocities near the center of M31, estimate the amount of mass within $1''$ of the center of the galaxy. Compare your answer with the value quoted in the text.

**(b)** Estimate the amount of mass within the central $1''$ based on the velocity dispersion data.

**(c)** Comment on the source of the asymmetries evident in Fig. 25.37. Recall Fig. 25.13.

**25.12** Beginning with the general expression for the position vector in rectangular coordinates

$$\mathbf{r} = x\hat{\mathbf{i}} + y\hat{\mathbf{j}} + z\hat{\mathbf{k}},$$

show that the vector can be represented in cylindrical coordinates by Eq. (25.16).

**25.13** Show that the acceleration vector is given by Eq. (25.18) in cylindrical coordinates. *Hint:* Note that the unit vectors $\hat{\mathbf{e}}_R$ and $\hat{\mathbf{e}}_\phi$ are position-dependent and therefore time-dependent. You may find their relationships with rectangular-coordinate unit vectors helpful.

**25.14** Using the solar motion data given in Chapter 24, estimate the amplitude of the Sun's excursion in the radial direction relative to a perfectly circular orbit. Assume that the Sun is currently at the midpoint of its oscillation. Does your result represent a minimum or a maximum estimate of the actual deviation?

**25.15** **(a)** From the information given in the text, derive Eq. (25.37) for the square of the solar epicycle frequency.

**(b)** Show that Eq. (25.38) follows directly from Eq. (25.37).

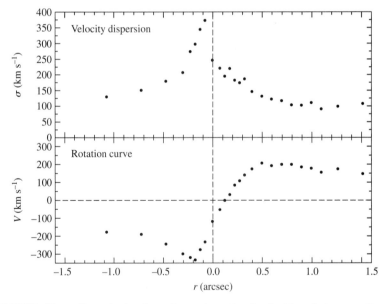

**FIGURE 25.37**    The stellar velocity dispersion and rotational velocities of stars near the center of M31, measured along the major axis of the bulge. Given the distance to Andromeda of 770 kpc, $1''$ corresponds to a linear distance from the center of 3.7 pc. (Data from Bender et al., *Ap. J.*, *631*, 280, 2005.)

**25.16** Determine the epicycle axis ratio, $\chi_{max}/\rho_{max}$, for a Keplerian orbit (a point mass orbiting a central, spherically symmetric mass). *Hint:* Begin with Eq. (25.37).

**25.17** Show that if the surface brightness of an elliptical galaxy follows the $r^{1/4}$ law given by Eq. (24.13), then the average surface brightness over the area of a circular disk of radius $r_e$ is given by

$$\langle I \rangle = 3.607 I_e.$$

*Hint:* Begin by rewriting the $r^{1/4}$ law in the form

$$I(r) = I_e e^{-\alpha\left[(r/r_e)^{1/4} - 1\right]}$$

and recall the definition of an integral average given in Problem 2.9. You may also find it helpful to write your integral in such a way that the limits of integration extend from zero to infinity. This can be done by considering the definition of $r_e$.

**25.18** Show that the Holmberg radius of a galaxy obeying the $r^{1/4}$ law is related to the galaxy's effective radius, $r_e$, and the corresponding isophotal surface brightness, $\mu_e$, by

$$r_H = r_e (4.18 - 0.12\mu_e)^4.$$

**25.19** **(a)** Use the result of Problem 25.17 to show that $\langle \mu \rangle = \mu_e - 1.393$ for an elliptical galaxy that follows the $r^{1/4}$ law.

**(b)** NGC 3091 has an effective radius of 10.07 kpc in the $B$ band and an average surface brightness within the effective radius of 21.52 $B$-mag arcsec$^{-2}$. From this information, determine $\mu_e$ in the $B$ band.

**(c)** What is the Holmberg radius of NGC 3091? You may find the result of Problem 25.18 useful.

**25.20** According to the virial theorem, the central radial-velocity dispersion is related to the mass and size of the galaxy by $\sigma_r^2 \propto M/R$ (see Eq. 25.13). Use arguments similar to those for the Tully–Fisher relation to show that $L \propto \sigma_r^4$, which is the Faber–Jackson relation, Eq. (25.40).

**25.21** **(a)** From the data given in Fig. 25.33, estimate the slope of the curve that represents the best-fit linear relationship.

**(b)** How does the slope in Fig. 25.33 compare with Eq. (25.41)? Why wouldn't you expect them to be exactly the same?

**25.22** **(a)** It is estimated that M31 has approximately 350 globular clusters. If its absolute visual magnitude is $-21.7$, estimate the specific frequency for its clusters.

**(b)** NGC 3311 is a cD galaxy with an estimated 17,000 globular clusters and an absolute visual magnitude of $-22.4$. Estimate the specific frequency of clusters in this galaxy.

**(c)** Discuss the problem of globular cluster statistics in the suggestion that cD galaxies are due to mergers of already formed spiral galaxies.

**25.23** **(a)** Find a general expression for $a_2$, the coefficient of the first-order Fourier term in Eq. (25.44), written in terms of $a_0$ and $\epsilon$ (see Eq. 25.1). Assume that all higher-order terms are identically zero for this part of the problem.

**(b)** Make a polar-coordinate plot of $a$ as a function of $\theta$ for an E4 galaxy with $a_0 = 30$ kpc and $a_2$ determined from the relationship found in part (a). Again assume that all higher-order terms are identically zero.

    **(c)** Make a polar-coordinate plot for the same E4 galaxy, but with $a_4 = 0.1a_0$.

    **(d)** Make a polar-coordinate plot for the same E4 galaxy, but with $a_4 = -0.1a_0$.

    **(e)** Comment on the general appearance of your last two plots. Which one looks more like a lenticular galaxy?

**25.24** Plot $\log_{10} \phi(M)$, the logarithm of the Schechter luminosity function, for both the local field of galaxies near the Milky Way and the Virgo cluster over the range $-23 < M_B < -12$ (see Eq. 25.46). Use the values of $\alpha$ and $M^*$ given in the text. To compare your results with those given in Fig. 25.36, shift your data so that $\log_{10} \phi(-23) = 0$ for both groups of galaxies.

# CHAPTER
# 26

# Galactic Evolution

## 26.1 ■ INTERACTIONS OF GALAXIES

With the exquisite vision of the heavens provided by modern ground-based and space-based observatories, it has become increasingly apparent that galaxies are not "island universes"; they do not evolve in isolation.

### Evidence of Interactions

Nearly all galaxies belong to clusters, and the galaxies take up a larger fraction of the cluster's volume than do the stars in a stellar cluster. We also know that the spacing between galaxies is typically only 100 times larger than the size of the galaxies themselves (see Section 27.3). Densely populated clusters, such as the Coma cluster (Figs. 26.1 and 26.2), have a higher proportion of early-type galaxies (ellipticals) in their centers than they do in their outer, less dense regions. As was already mentioned in Section 25.4, the central regions of these rich, regularly shaped clusters also have a higher proportion of early-type galaxies than the centers of less populated, amorphous-shaped irregular clusters, such as the Hercules cluster (Fig. 26.3). These observations seem to correlate with the increased probability of interactions and/or mergers between galaxies in regions of higher galaxy number density. Interactions tend to increase the velocity dispersions of stars in the galaxies involved, possibly destroying disk structures in late-type galaxies and causing the galaxies to relax to early-type $r^{1/4}$ profiles (Eq. 25.2).

A VLA radio survey of the H I layer of galactic disks found that at least 50% of all disk galaxies display warped disks. Also, more than half of all elliptical galaxies harbor discrete shells of stars.[1] Some disk warping may be due to tidal interactions with smaller satellite galaxies, and, as we shall see, shells in ellipticals are signatures of mergers.

Observations also suggest that hot, X-ray-emitting gas occupies much of the space between the galaxies in rich clusters and has a mass equal to or exceeding the mass of all of the cluster's stars (Section 27.3). It seems that the gravitational influence of interacting galaxies is largely responsible for removing the gas from the individual galaxies that make up the cluster, while still leaving the gas trapped in the cluster's overall gravitational well.

---

[1] See Binney (1992) and Barnes and Hernquist (1992) for further details on these statistics.

**FIGURE 26.1**    The center of the Coma cluster. The width of this view is about 18 arcmin. (Courtesy of National Optical Astronomy Observatories.)

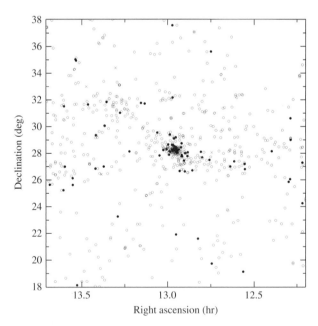

**FIGURE 26.2**    The Coma cluster of galaxies, showing the ellipticals (filled circles) and spirals (open circles). Note that the scale is much larger than the width of the image in Fig. 26.1.

**FIGURE 26.3** The Hercules cluster of galaxies. (Courtesy of National Optical Astronomy Observatories.)

Although only a small fraction of the galaxies' gas is removed in a direct collision, mergers may initiate a burst of star formation that produces stellar mass loss and supernovae, leading to a galactic superwind capable of liberating a large amount of gas.

This evidence and that cited in Chapter 25 suggest that interactions between galaxies play an important role in their evolution. In the hierarchical "bottom-up" scenario of galaxy formation to be described in the next section, large galaxies are thought to be formed by mergers and the gravitational capture of smaller entities. In this view, the interactions seen today between galaxies are simply a natural extension of their formative years. Figure 26.4 shows one of many examples of galactic interactions. In this section, we will examine such obvious examples, as well as more subtle phenomena.

### Dynamical Friction

What happens when galaxies collide? Given that stars are generally spread very far apart in galaxies, the chance of even a single stellar collision is quite small (see Problem 26.1). Instead, interactions between stars will be gravitational in nature. To see this, imagine that an object (a globular cluster or small galaxy) of mass $M$ is moving through an infinite collection of stars, gas clouds, and dark matter with a constant mass density, $\rho$. We will assume that the mass of each object in the background "sea" of material is much less than $M$, so $M$ continues moving in a straight line instead of being deflected. In the absence of collisions, it might be thought that $M$ would move unimpeded. However, as $M$ moves forward, the other objects are gravitationally pulled toward its path, with the closest ones feeling the largest force. As shown in Fig. 26.5, this produces a region of enhanced density along the path, with a high-density "wake" trailing $M$. The result, known as **dynamical friction**, is a net gravitational force on $M$ that opposes its motion. Kinetic energy is transferred from $M$ to the surrounding material as $M$'s speed is reduced.

**FIGURE 26.4**   The plumes on the opposite sides of NGC 520 are evidence of a tidal interaction, possibly ending in the merger of the two colliding disk galaxies. Note the diagonally oriented dust lane. (Courtesy of Gemini Observatory/AURA.)

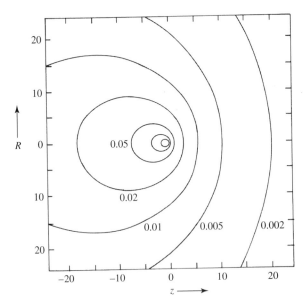

**FIGURE 26.5**   The fractional enhancement in the density of stars caused by the motion of a mass $M$ in the positive $z$ direction. (Figure adapted from Mulder, *Astron. Astrophys.*, *117*, 9, 1983.)

The derivation of an expression for dynamical friction is beyond the scope of this text, but it is easy to see that the physical quantities involved are the mass, $M$, and speed, $v_M$, of the massive object, and the mass density of the surrounding material, $\rho$. As a result, the expression for the force must contain $GM$, $v_M$, and $\rho$ only.[2] It is left as a problem to show that there is only one combination of these variables that has the units of force, so the expression for the force of dynamical friction, $f_d$, must look like

$$f_d \simeq C \, \frac{G^2 M^2 \rho}{v_M^2},$$

(26.1)

where $C$ is dimensionless. Here, $C$ is not a constant, but a function that depends on how $v_M$ compares with the velocity dispersion, $\sigma$, of the surrounding medium. For $v_M \sim 3\sigma$, typical values of $C$ are 23 for the LMC, 76 for globular clusters, and 160 for ellipticals.[3]

Through a careful examination of Eq. (26.1), it is possible to see why the various terms enter as they do. Clearly the dynamical friction must be proportional to the mass density of stars. Assuming that the relative numbers of objects of various masses do not change, doubling $\rho$ means doubling the total number of objects, which would in turn double the gravitational force on $M$. The mass $M$ itself is squared; one power comes from its role in producing the high-density wake that trails behind it, and the other from the gravitational force on $M$ produced by the enhanced density. Finally, consider the velocity-squared term in the denominator. If $M$ moves twice as fast, it will spend only half as much time near a given object, and so the impulse $\int \mathbf{F}\, dt = \Delta \mathbf{p}$ given to that object is only half as great. Consequently, the density enhancement develops only half as rapidly, and $M$ will be twice as far away by the time the enhancement arises. Thus the $v_M^2$ in the denominator comes from the inverse-square law of gravity. This last point means that slow encounters are much more effective at decreasing the speed of an intruding mass.

To get an estimate of the timescale associated with the effects of dynamical friction acting on a galaxy's globular clusters, recall that flat rotation curves imply that the density of the dark matter halo may be approximated most simply by Eq. (24.50),

$$\rho(r) = \frac{v_M^2}{4\pi G r^2}.$$

Inserting this expression into Eq. (26.1), we find that the dynamical friction acting on a cluster is

$$f_d = C \, \frac{G^2 M^2 \rho(r)}{v_M^2} = C \, \frac{G M^2}{4\pi r^2}.$$

If the cluster's orbit is circular and of radius $r$, its orbital angular momentum is just $L = M v_M r$. Since dynamical friction acts tangentially to the orbit and opposes the cluster's

---

[2]Because the only time that $M$ arises is in connection with Newton's law of gravity, the combination $GM$ will be inseparable in any derivation.
[3]You may refer to Binney and Tremaine (1987) for a detailed discussion of dynamical friction.

motion, a torque of magnitude $\tau = r f_d$ is exerted on the cluster. The torque in turn reduces the cluster's angular momentum according to

$$\frac{dL}{dt} = \tau.$$

Recalling that a flat rotation curve implies that the orbital speed $v_M$ is essentially constant, differentiating the angular momentum and substituting for the torque give

$$M v_M \frac{dr}{dt} = -rC \frac{GM^2}{4\pi r^2}.$$

Integrating the last equation leads to an expression describing the time required for the globular cluster to spiral into the center of the host galaxy from an initial radius $r_i$, or

$$\int_{r_i}^{0} r \, dr = -\frac{CGM}{4\pi v_M} \int_{0}^{t_c} dt.$$

Solving for the cluster lifetime, $t_c$, yields

$$t_c = \frac{2\pi v_M r_i^2}{CGM}. \tag{26.2}$$

Equation (26.2) can be inverted to find the most distant cluster that could have been captured within the estimated age of the galaxy:

$$r_{max} = \sqrt{\frac{t_{max} CGM}{2\pi v_M}} \tag{26.3}$$

---

**Example 26.1.1.** Consider a globular cluster that orbits the Andromeda galaxy (M31). Assume that the cluster's mass is $M = 5 \times 10^6$ M$_\odot$ and its velocity is $v_M = 250$ km s$^{-1}$, typical of the rotation curve in the outer part of the galaxy. If the age of M31 is approximately 13 Gyr, then Eq. (26.3) implies that $r_{max} = 3.7$ kpc. This means that globular clusters of $5 \times 10^6$ M$_\odot$ that were originally within approximately 4 kpc of the center of Andromeda would have spiraled into its nucleus by now.

According to Eq. (26.3), $r_{max} \propto M^{1/2}$, implying that clusters with masses greater than $5 \times 10^6$ M$_\odot$ could have been gathered from greater distances. This may help to explain why there are no very massive globular clusters remaining around M31 today.

---

Not only are globular clusters affected by dynamical friction, but satellite galaxies are as well. Recall that on page 891 we learned that a stream of material has been tidally stripped from the Magellanic Clouds. In fact, it appears that dynamical friction will ultimately cause the Magellanic Clouds to merge with the Milky Way some 14 billion years in the future. This is a fate that has already befallen the Sagittarius dwarf spheroidal galaxy, the remnant of the dwarf galaxy in Canis Major, and possibly a progenitor dwarf galaxy of $\omega$ Centauri. In fact, any giant galaxy will probably devour numerous satellite galaxies during its lifetime.

The process of satellite accretion has a variety of possible consequences. For instance, the gravitational torques involved in the merger of a satellite galaxy in a retrograde orbit may produce the counter-rotating cores that are observed in some elliptical galaxies (recall Section 25.4). In a disk galaxy, mergers may excite global modes of oscillation (similar to those in pulsating stars; see Chapter 14). Mergers are also capable of producing the sort of featureless disks that are characteristic of Ir (amorphous) galaxies. The importance of mergers in the evolution of galaxies will be explored in Section 26.2.

## Rapid Encounters

We now turn to another type of encounter, one that occurs so rapidly between two galaxies that their stars do not have time to respond. Even in the special case where the two systems pass through one another, there is no significant dynamical friction because there is no appreciable density enhancement. In this *impulse approximation*, the stars barely have time to alter their positions. As a result, the internal potential energy, $U$, of each galaxy is unchanged by the collision. However, the gravitational work that each galaxy performs on the other has increased the internal kinetic energies of both galaxies in a random way. This internal kinetic energy comes at the expense of the overall kinetic energies of the galaxies' motions with respect to one another. The amount of internal energy gained by the galaxies depends on the nearness of the approach of the galaxies involved, with the energy declining rapidly for more distant encounters. (Roughly, the energy gained is inversely proportional to the fourth power of the distance of closest approach of the galaxies.)

Suppose that one of the galaxies gains some internal kinetic energy during the interaction. To determine how it will respond, assume that the galaxy was initially in equilibrium, meaning that it satisfied the virial theorem prior to the encounter (recall Eqs. 2.46 and 2.47), and its initial kinetic, potential, and total energies were related by $2K_i = -U_i = -2E_i$. Imagine that during the encounter, the galaxy's internal kinetic energy increased from $K_i$ to $K_i + \Delta K$. Because its potential energy has remained essentially constant, the galaxy's total energy has increased to $E_f = E_i + \Delta K$. As a result, the galaxy has been thrown out of virial equilibrium. When equilibrium is finally reestablished (after a timescale of a few orbital periods), the final kinetic energy must, according to the virial theorem, be

$$K_f = -E_f = -(E_i + \Delta K) = K_i - \Delta K.$$

Thus, as equilibrium is regained, the internal kinetic energy of the galaxy actually *decreases* by $2\Delta K$ from the value it had just after the collision.

How does the galaxy accomplish this reduction? One way to regain equilibrium is to convert the excess kinetic energy into an increased (less negative) gravitational potential energy. The galaxy expands slightly as the separation between its masses increases. Another way of reducing the kinetic energy is for the most energetic components to carry it out away from the galaxy in the form of a stream of stars and gas. This evaporation cools the galaxy and moves it toward a new equilibrium. In fact, both of these processes may occur, and which one dominates is determined by the specific circumstances of the collision. For instance, a high-speed, nearly head-on collision may produce a **ring galaxy**, such as the Cartwheel shown in Fig. 26.6. A numerical computer simulation and interpretation of this interaction are left as an exercise (Problem 26.20).

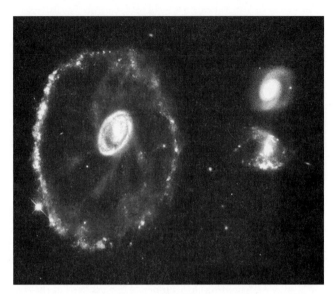

**FIGURE 26.6**    An HST image of the ring galaxy known as the Cartwheel (A0035−335). The diameter of the ring is about 46 kpc. As the ring expands at 89 km s$^{-1}$, it triggers bursts of star formation. It is not clear which of the two galaxies at right may have been the intruder. [Courtesy of Kirk Borne (STScI), and NASA.]

If two galaxies are gravitationally bound to one another, they will eventually merge, given enough time. Because of their extended mass distributions, galaxies do not follow the same trajectories they would if they were point masses; some orbital energy is diverted into increasing the galaxies' internal energy, and the orbit shrinks a bit. Tidal forces may also dissipate the orbital energy by pulling stars and gas out of one or both galaxies, a process known as **tidal stripping**. This is probably the cause of the Magellanic Stream.

An instructive way of thinking about this loss of galactic material involves using the idea of gravitational equipotential surfaces, discussed in Section 18.1 for a close binary system. Figure 18.3 shows the equipotential lines for a rotating system of two masses. Although the figure is for point masses in a circular orbit, it contains valuable insights for the present case of extended galaxies in noncircular orbits. In particular, it is useful to talk about the *tidal radii*, $\ell_1$ and $\ell_2$, of the galaxies. They are the distances from the center of each galaxy to the inner Lagrangian point, $L_1$, given by (approximately, in this noncircular case) Eqs. (18.9) and (18.10). As the two galaxies move about one another, the shape of the equipotential surfaces and the values of the tidal radii are constantly changing. If stars or clouds of gas extend beyond a galaxy's tidal radius, they have a tendency to escape from the gravitational potential well. (This is analogous to a star overflowing its Roche lobe, as discussed in Section 18.1.) As a result, the density of galactic material undergoes a sharp decline beyond the tidal radius.

**Polar-ring galaxies** and **dust-lane ellipticals** are normal galaxies that are orbited by rings of gas, dust, and stars that were stripped from other galaxies as they passed by or merged. Polar rings typically contain $10^9$ M$_\odot$ or more of gas and are found only around elliptical or S0 galaxies. It is estimated that some 5% of all S0 galaxies have or have had a polar ring. Figure 26.7 shows the polar-ring galaxy NGC 4650A, an S0 galaxy with a ring that is nearly perpendicular to its long axis (inclined just 7°).

Because the rings respond to the gravitational field of the galaxies, astronomers use them as probes of the three-dimensional distribution of matter, both luminous and dark, in their host galaxies. The results of studying several polar-ring galaxies show that they are surrounded by spherical dark halos. An example of a well-known dust-lane elliptical is Centaurus A (NGC 5128), shown in Fig. 26.8. Cen A is also a powerful radio source and will be discussed in that context in Section 28.1.

An extreme case of a gaseous ring (no stars) is the one that encircles the central two galaxies of the M96 group in the constellation Leo, as shown in Fig. 26.9. The gas is in an

**FIGURE 26.7**   The polar-ring galaxy NGC 4650A. [Courtesy of J. Gallagher (University of Wisconsin–Madison) and the Hubble Heritage Team (AURA/STScI/NASA).]

**FIGURE 26.8**   Centaurus A (NGC 5128) is a dust-lane elliptical galaxy. (Courtesy of NOAO Cerro Tololo Interamerican Observatory.)

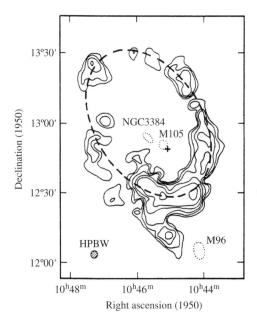

**FIGURE 26.9**  21-cm radio contours showing the ring of neutral hydrogen around the galaxies M105 (a giant elliptical) and NGC 3384 (an S0) in the M96 group. Note that the ring is more than 1° wide on the sky. (Figure adapted from Schneider, *Warped Disks and Inclined Rings around Galaxies*, Cambridge University Press, Cambridge, 1991.)

elliptical orbit around the two galaxies; the eccentricity of the orbit is $e = 0.402$, and the cross ("+") in the figure marks the location of the orbit's focus. Recall from Section 2.3 that an elliptical orbit is the result of an inverse-square force law, which means that the gas ring in the M96 group must lie *beyond* any dark-matter halos surrounding the two galaxies. The semimajor axis of the orbit is $a = 101$ kpc, so the maximum extent of any dark halo must be given by Eq. (2.5),

$$r_p = a(1 - e) = 60 \text{ kpc},$$

the perigalacticon distance. Note, however, that this calculation must be viewed cautiously because the ring is merely a locus of gas and may not in fact coincide with the orbit followed by a single particle.

### Modeling Interactions with N-Body Simulations

The merger of two galaxies is a complicated affair. If the relative speed of the galaxies is substantially greater than the velocity dispersions of their stars, the collision will not result in a merger. Conversely, if the relative speed of the collision is less than the velocity dispersion of the stars in one of the galaxies, a merger is inevitable. The situation is particularly complicated when the relative velocity of the galaxies is comparable to the velocity dispersion of stars in one of the galaxies. This situation is best studied today by numerical experiments involving *N*-body simulations, like the ones that were used to investigate the bar instability in a disk of stars (Section 25.3). In these computations, Newton's second law is used to follow the motions of the stars through a sequence of time intervals.[4]

One of the first such efforts was made in 1972 by Alar and Juri Toomre. Each galaxy consisted of a massive nucleus surrounded by concentric rings of disk stars. To simplify the calculations, the stars in the disk were allowed to feel the gravitational pull of the galactic nuclei, but not of each other. With their program, the Toomre brothers succeeded in demonstrating how interactions could explain the appearance of galaxies such as M51 (the Whirlpool), which has a bridge of stars and gas that appears to extend to its neighbor galaxy NGC 5195. Figure 25.17 shows this system, and Fig. 26.10 suggests how it could have been produced. Also note the development of the grand-design spiral arms that are so prominent in M51. In addition, the Toomres were successful in reproducing the appearance of **tidal-tail galaxies**, such as the pair NGC 4038 and NGC 4039 (the "Antennae") shown in Fig. 26.11. It is remarkable that the rapid advances in computing power since the early 1970s now allow a program similar to the Toomres' to be run on a personal computer. Such a program, adapted from Schroeder and Comins (1988), is described in Appendix M.

In general, only close, slow encounters produce bridges and tails. The effect is most pronounced when the orbital angular speed of one of the galaxies matches the angular speed of some of the stars in the other galaxy's disk. The resulting orbital resonance, which acts on both the near and far sides of the disk, allows the tidal forces to be especially effective. Two bulges tend to develop on opposite sides of one (or both) of the galaxies, similar to Earth's tidal bulges. Tidal stripping then pulls out streams of stars and gas as the two galaxies pirouette around one another. When conditions are right, the stars and gas torn from the near side will form an apparent bridge, while, because of angular momentum conservation, the material stripped from the far side moves off to form a curving tail.

Modern computer codes include the effects of dark matter and the gravitational inter-actions (*self-gravity*) of individual stars and gas clouds. Figures 26.12 and 26.13 show the results of one such calculation that has reproduced the appearance of the tidal-tail galaxies NGC 4038 and NGC 4039. These codes have also been used to simulate the merger of a disk galaxy (with a central bulge and halo) and a satellite galaxy with 10% of the mass of the primary. The presence of a dark-matter halo significantly decreases the timescale of a merger because it allows the galaxies to interact over much larger distances. The satellite is devoured and absorbed in only two revolutions of the disk galaxy.

It is worth recalling that roughly one-half of the stars in the Milky Way's outer stellar halo move in retrograde orbits, which is probably a consequence of tidal stripping and/or mergers of other dwarf galaxies. Recall also (Section 24.2) that there are two different spatial distributions of globular clusters around the Milky Way. The inner clusters are old and vary little in age, while the outer population has a wider age spread. The inner clusters also tend to have a preferential direction for their orbits around the Galaxy, while the outer clusters have randomly oriented orbits. Again, it appears likely that these outer clusters formed elsewhere and were captured later by the Milky Way. In fact, some researchers

---

[4]In 1941, Holmberg used an *analog computer* to investigate the tidal interactions of galaxies. He made "galaxies" out of light bulbs and used the inverse-square law of light to mimic the same behavior for gravity. With the aid of a light meter, he was able to analyze the gravitational influence of one model galaxy on another.

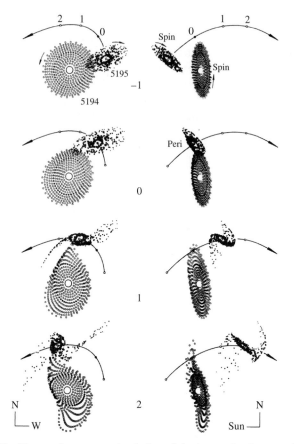

**FIGURE 26.10**   The Toomres' computer simulation of the interaction between M51 (NGC 5194) and NGC 5195. Note from the side view that the "bridge" does not actually connect the two galaxies. You may also note the resulting warped disk. [Figure adapted from Toomre, *The Large Scale Structure of the Universe*, Longair and Einasto (eds.), Reidel, Dordrecht, 1978.]

have suggested that the young globular clusters Palomar 12 and Ruprecht 106 were tidally stripped from the Large Magellanic Cloud.

Finally, if a satellite galaxy is moving at an angle with respect to the plane of the disk, the merger can result in a warped disk that persists for 3 to 5 billion years. More than 50% of galactic disks (including the Milky Way's) display a warp in their gas distributions (although some stars may participate in the warp as well), and most of these galaxies do not appear to have nearby companions. The absence of a companion might argue in favor of a merger, under the assumption that the companion was enveloped by the primary galaxy. However, most disks show no explicit evidence that a merger has occurred. It has also been shown that warped disks can arise simply as a consequence of the triaxial nature of the surrounding dark halos. The importance of the role of mergers or close encounters in producing warped disks has yet to be determined.

**FIGURE 26.11** The Antennae galaxies, NGC 4038 and NGC 4039, and their tidal tails. [Courtesy of Brad Whitmore (STScI) and NASA.]

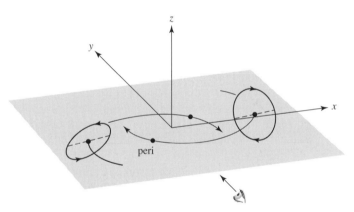

**FIGURE 26.12** The orbital geometry used to calculate the formation of the tidal-tail galaxies shown in Fig. 26.11. The disks are shown in their initial positions. Each disk is initially inclined by 60° to the orbital plane, and a dashed line indicates the intersection of each disk with the orbital plane. The positions of closest approach are identified ("peri"). The viewing direction for Fig. 26.13 is indicated as along the positive $y$-axis. (Figure adapted from Barnes, *Ap. J.*, *331*, 699, 1988.)

## Starburst Galaxies

In 1972 Richard Larson and Beatrice Tinsley found that strongly interacting galaxies tend to be bluer than isolated ones of the same type. They attributed the excess blue light to hot newborn stars and argued that tidal interactions have induced vigorous bursts of star formation in these galaxies. The increased luminosity is difficult to detect because, like all star formation, it is shrouded in thick clouds of gas and dust (recall Section 12.2). The light emitted by the young stars at visible and ultraviolet wavelengths is absorbed and then reradiated in the infrared. In 1983 the Infrared Astronomy Satellite (IRAS) found that these

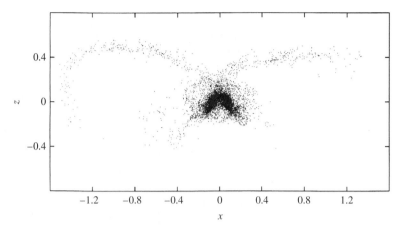

**FIGURE 26.13**    The result of Joshua Barnes's simulation of the formation of the tidal-tail galaxies NGC 4038 and NGC 4039. A distance of 1 on the $x$- and $z$-axes corresponds to 40 kpc. (Figure adapted from Barnes, *Ap. J.*, *331*, 699, 1988.)

**starburst galaxies** are extremely bright at infrared wavelengths, pouring out up to 98% of their energy in this portion of the spectrum. (In comparison, the Milky Way emits 30% of its luminosity in the infrared, and M31 only a few percent.) Most of the star formation is confined to within about one kiloparsec or so of the galaxy's center, where radio telescopes have detected vast clouds containing some $10^9$ to $10^{12}$ $M_\odot$ of hydrogen gas that serves as fuel. Between 10 and 300 $M_\odot$ of gas is converted into stars each year in a starburst galaxy, while only two or three stars are formed annually in the Milky Way. The clouds typically contain sufficient hydrogen to support this rate of star formation for about $10^8$ to $10^9$ years, although a given burst may last for only 20 million years or so.

The observational evidence also shows that starbursts do not occur exclusively in the nuclei of interacting galaxies. Although this is often the case, many disk-wide starbursts are also known. In this case, the problem is how star formation can be triggered nearly simultaneously over such a wide area.

Since hydrogen clouds were initially distributed throughout the galactic disk, one puzzle facing astronomers is how a violent interaction with another galaxy could have removed more than 90% of the clouds' angular momentum, allowing the gas to become concentrated in the galactic center. An answer may be provided by numerical simulations of colliding galaxies. These studies indicate that the gas and stars react differently to the impact of an intruding galaxy, with the gas tending to move out in front of the stars as they orbit the galactic center. The stars' gravity then pulls back on the gas, and the resulting torque on the gas reduces its angular momentum, causing it to plunge toward the galactic center. As the two galaxies begin to merge, more angular momentum is lost. Shock fronts then compress the gas, and a burst of star formation begins.

Figure 26.14 shows the starburst galaxy M82/NGC 3034, about 3.2 Mpc away [see also Fig. 25.7(c)]. From its unusual appearance, astronomers once thought that this galaxy was exploding. The current interpretation is that between $10^7$ and $10^8$ years ago, the inner 400 pc of M82 began a tremendous episode of star formation that is still continuing. The center of M82 contains a wealth of OB stars and supernova remnants that have expelled more

**FIGURE 26.14**   The starburst galaxy M82 (NGC 3034). This composite image was obtained by the Subaru Telescope's FOCAS instrument in $B$, $V$, and H$\alpha$ wavelengths. [Courtesy of National Astronomical Observatory of Japan (NAOJ).]

than $10^7$ M$_\odot$ of gas from the plane of the galaxy. All of this violent activity may have been triggered by a tidal interaction with M81, a spiral galaxy that is now about 36 kpc away from M82 (as projected on the plane of the sky) and linked to it by a gaseous bridge of neutral hydrogen. In fact, the center of M81 is known to be a significant source of X-rays, with an X-ray luminosity of $L_X = 1.6 \times 10^{33}$ W. This has been interpreted as implying that M81 has a central black hole that is being "fed" by infalling gas at the rate of $10^{-5}$ to $10^{-4}$ M$_\odot$ yr$^{-1}$. This gas infall may also be a consequence of the tidal interaction with M82.

### Mergers in Elliptical and cD Galaxies

The importance of interactions in galactic evolution is most obvious for elliptical galaxies. The case is especially compelling for the cD galaxies typically found at the centers of rich, regular clusters. More than half of the known cD's have multiple nuclei that move differently than the galaxy as a whole. Although the typical orbital speed of a star in a cD galaxy is $\sim 300$ km s$^{-1}$, the multiple nuclei move with relative velocities of $\sim 1000$ km s$^{-1}$. All of this is taken as evidence that cD galaxies are the products of galactic mergers that supplied the supernumerary nuclei and the expansive halo of stars. Their frequent occurrence at the bottom of a cluster's gravitational potential well makes collisions and other close encounters all the more probable. Figure 26.15 is a remarkable photograph of three smaller galaxies passing through a cD at the center of the cluster Abell 2199. (In this case their velocities are so large that these galaxies will not be captured during this passage through the cD.) Recall from Fig. 25.16 and Table 25.3 that cD's also have unusually high specific frequencies of globular clusters, suggesting that a large number of clusters have been captured throughout the evolution of a typical cD galaxy.

Normal elliptical galaxies show evidence of interactions and mergers as well, particularly the boxy E's. As was mentioned in Section 25.4, about 25% of all ellipticals have very different velocity fields in their cores when compared with their outer regions. Observations

**FIGURE 26.15**   The passage of three small galaxies through a giant elliptical in the cluster Abell 2199. (Courtesy of Whipple Observatory and Harvard CfA.)

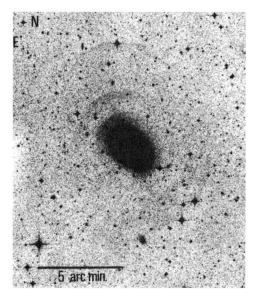

**FIGURE 26.16**   Faint concentric shells around the elliptical galaxy NGC 3923. (Figure from Malin and Carter, *Nature*, *285*, 643, 1980. Anglo-Australian Observatory photo by David Malin.)

also reveal that 56% of all ellipticals have faint shells (as do 32% of all S0's). As shown in Fig. 26.16, otherwise normal ellipticals may have as many as twenty large concentric arcs of very low surface brightness, both outside and inside the galaxy. Numerical experiments find that these shells may be the result of head-on (or nearly head-on) collisions with smaller galaxies that have only a few percent of the mass of the larger elliptical. In essence, the captured stars slosh back and forth in the elliptical's gravitational well, like water rocking back and forth in a large bowl. Arcs are seen where the stars slow and reverse their courses. The natural spread in the stellar kinetic and potential energies creates the concentric arcs.

From the evidence discussed here and in Section 25.4, it seems likely that many, and perhaps all, large elliptical galaxies are strongly influenced by mergers. Galaxy mergers seem to produce results that are strikingly similar to elliptical galaxies in many ways. Photometry of the inner regions of cD galaxies and other obvious merger remnants confirms that their surface brightness profiles follow the $r^{1/4}$ law (Eq. 25.2) that is obeyed by most large ellipticals. Computer simulations bolster this view by showing that $r^{1/4}$ brightness profiles are normal outcomes of mergers. Calculations of mergers can also reproduce the observations of distinctly different velocity fields inside and outside the galactic core.

It is important to note, however, that dwarf ellipticals and dwarf spheroidals, the most common types of elliptical galaxies, do not appear to result from mergers.

### Binary Supermassive Black Holes

An intriguing consequence of the merging of two large galaxies, whether they are ellipticals or spirals, is the apparently inevitable formation of a binary system of supermassive black holes. If each galaxy involved in the merger originally contained a supermassive black hole, the two black holes would migrate toward the center of the potential well of the combined system as a direct result of dynamical friction. Depending on their trajectories, the two black holes would probably enter into a well-separated binary system. One such system appears to be NGC 6420, which has two very strong X-ray sources (believed to be supermassive black holes) that are currently separated by a distance of about 1 kpc (see Fig. 26.17).

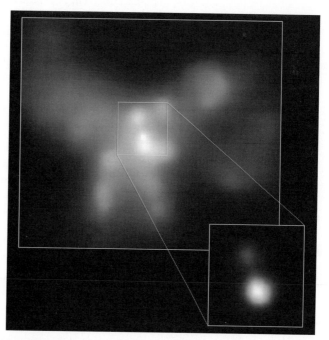

**FIGURE 26.17** A Chandra X-Ray Observatory image of two supermassive black holes near the center of the "butterfly-shaped" galaxy, NGC 6240. (Courtesy of NASA/CXC/MPE/S. Komossa, et al.)

As the two supermassive black holes orbit around one another, they will have a profound and disturbing time-dependent gravitational influence on stars near the center of the merged galaxy. As with other three-body systems where one member is much less massive than the other two, it is probable that individual stars will be ejected from the central region of the galaxy. If the total mass of stars ejected in this way becomes comparable to the mass of the black holes, enough angular momentum and energy will be carried away by the stars that the two black holes will spiral in toward one another. When the black holes become sufficiently close, they will produce large amounts of gravitational radiation that will further rob them of the energy and angular momentum needed to stay separated (recall the discussion of gravitational radiation from binary pulsar systems in Section 18.6). The final outcome of the process will be a merger of the two black holes, producing an even larger beast in the center of the merged galaxy.

It has been suggested that such a mechanism is a probable route to growing the super-massive black holes that seem to be so prevalent in the centers of large galaxies.

## 26.2 ■ THE FORMATION OF GALAXIES

As we have seen so far in our discussions of the Milky Way and other galaxies, galactic structures are complex and varied. The luminous components of elliptical galaxies are dominated by spheroidal mass distributions that are primarily composed of old stars, while spiral galaxies contain both relatively old spheroids and appreciable disks of younger generations of stars, dust, and gas. Important differences even exist within the designations of early- and late-type galaxies, such as the degree of diskiness or boxiness of normal ellipticals, the existence of dwarf spheroidals, and the relative dominance of the bulges and disks of spirals, to name a few. Furthermore, the presence of large quantities of dark matter, accounting for 90% or more of the mass in many galaxies, also plays a critical role in determining their overall structures.

### The Eggen, Lynden–Bell, Sandage Collapse Model

It was in 1962 that Olin J. Eggen (1919–1998), Donald Lynden-Bell, and Allan R. Sandage presented an important early attempt at modeling the evolution of our Galaxy, often referred to as the **ELS collapse model**. Their work was based on observed correlations between the metallicity of stars in the solar neighborhood, and their orbital eccentricities and orbital angular momenta. Eggen, Lynden-Bell, and Sandage noted that the most metal-poor stars tend to have the highest eccentricities, the largest $w$ components of their peculiar motions (Section 24.3), and the lowest angular momenta about the rotation axis of the Galaxy. On the other hand, metal-rich stars tend to exist in nearly circular orbits and are confined to regions near the plane of the Galaxy.

To explain the kinematic and chemical properties of stars in the solar neighborhood, ELS suggested that the Milky Way Galaxy formed from the rapid collapse of a large proto-Galactic nebula. The oldest halo stars formed early in the collapse process while still on nearly radial trajectories, resulting in their highly elliptical orbits above and below the Galactic plane. As a further consequence of their rapid formation, the model predicts that the halo stars are naturally very metal-poor (Population II) since the interstellar medium had not yet had time to become enriched by the by-products of stellar nucleosynthesis. However, as the first generations of massive stars generated heavier elements in their interiors, underwent supernova explosions, and ejected metal-rich material back into the ISM, the ISM evolved chemically over time.

As the proto-Galactic cloud continued to fall inward, the model predicts that the rapid collapse slowed when collisions between gas and dust particles became more frequent and the kinetic energy of infall was dissipated (converted into the thermal energy of random particle motions). Furthermore, the presence of angular momentum in the original proto-Galactic nebula meant that the cloud began to rotate more quickly as the radius decreased. The combination of the increased dissipation and the increased angular speed led to the development of a disk of chemically enriched gas from which Population I stars continue to form today.

---

**Example 26.2.1.**   The time required for the free-fall collapse of the proto-Galactic cloud as envisioned by Eggen, Lynden-Bell, and Sandage can be estimated from Eq. (12.26), which we first used to study the collapse of a protostellar cloud. Assume that the proto-Galactic cloud contained some $5 \times 10^{11}$ $M_\odot$, the estimated mass of the Milky Way Galaxy within a nearly spherical volume of radius 50 kpc (which includes the dark matter halo; see Table 24.1). If we further assume that the mass was uniformly distributed over the sphere, then the initial density of the cloud was

$$\rho_0 = \frac{3M}{4\pi r^3} = 8 \times 10^{-23} \text{ kg m}^{-3}.$$

Substituting into Eq. (12.26) gives

$$t_{\text{ff}} = \left( \frac{3\pi}{32} \frac{1}{G\rho_0} \right)^{1/2} = 200 \text{ Myr}.$$

Of course, if the nebula were initially somewhat centrally condensed, the inner portions of the Galaxy would collapse more rapidly than the outer, rarefied regions (see Problem 26.8). This may explain the existence of the very old stellar population within the bulge. The high metal abundance of bulge stars would arise if the first, massive, short-lived stars could quickly enrich the relatively dense ISM in that part of the Galaxy. Recall that the lifetimes of the most massive stars are on the order of one million years, much shorter than the estimated free-fall timescale.

---

## Problems with the ELS Model

Although the model does account for many of the basic features found in the structure of the Milky Way, this *top-down* approach involving the differentiation of a single, immense proto-Galactic cloud does not explain several important aspects of our current understanding of the Galaxy's morphology. For instance, given an initial rotation of the proto-Galactic cloud, essentially all halo stars and globular clusters should be moving in the same general direction, albeit with highly eccentric orbits about the Galactic center. However, astronomers have come to realize that approximately one-half of all outer-halo stars are in retrograde orbits (see Section 24.3) and the net rotational velocity of the outer halo is roughly $0 \ \text{km s}^{-1}$. On the other hand, stars in the inner halo, along with the inner globular clusters, appear to have a small net rotational velocity. (Our current understanding of the kinematics of halo stars and clusters seems to suggest that the early environment of the Galaxy was fairly turbulent and clumpy.)

A second problem with the standard ELS collapse model is the apparent age spread among the globular clusters and halo stars. If the approximately 2-billion-year variation in ages discussed in Chapter 24 is real (the age range being perhaps 11 to 13 Gyr), then the collapse must have taken roughly an order of magnitude longer to complete than proposed by Eggen, Lynden-Bell, and Sandage. The model also does not readily explain the existence of a multicomponent disk having differing ages. Recall that the young disk is probably about 8 Gyr old, whereas the age of the thick disk may be 10 Gyr (both being significantly younger than the halo).

Yet another difficulty lies in the compositional variation found between globular clusters. The clusters located nearest the Galactic center are generally the most metal-rich and oldest, while the clusters in the outer halo exhibit a wider variation in metallicity and tend to be younger. The clusters also seem to form two spatial distributions; one set is associated with the spheroid, and the other may more properly be affiliated with the thick disk.

The problems that have developed with the early ELS view of the formation of the Milky Way suggest that our understanding of its formation and subsequent evolution must be revised or is otherwise incomplete. Furthermore, the rich variety of galaxies, along with their ongoing dynamical evolution via mutual interactions and mergers, poses interesting challenges to the development of an overall, coherent theory of galactic evolution. Although at the time this text was written, such a theory had not yet reached the same level of maturity that exists in our understanding of stellar evolution, important features have begun to emerge.

### The Stellar Birthrate Function

As we have seen in the ELS model, any theory of galaxy formation must be able to explain the rate of formation of stars of various masses, as well as the chemical evolution of the interstellar medium. Since the ISM is enriched by mass loss via stellar winds and supernovae of various types, the theory must also incorporate the rates of stellar evolution and the chemical yields of stars.

One problem immediately arises in this regard: Although astronomers have been able to develop a reasonable description of the evolution of individual stars, our understanding of stellar birthrates is not yet complete. It is customary to express the **stellar birthrate function**, $B(M, t)$, in terms of the **star formation rate** (SFR), $\psi(t)$, and the **initial mass function** (IMF), $\xi(M)$, in the form

$$B(M, t) \, dM \, dt = \psi(t) \, \xi(M) \, dM \, dt, \tag{26.4}$$

where $M$ is the stellar mass and $t$ is time. $B(M, t)$ represents the number of stars per unit volume (or per unit surface area in the case of the Galactic disk) with masses between $M$ and $M + dM$ that are formed out of the ISM during the time interval between $t$ and $t + dt$. The SFR describes the rate per unit volume at which mass in the ISM is being

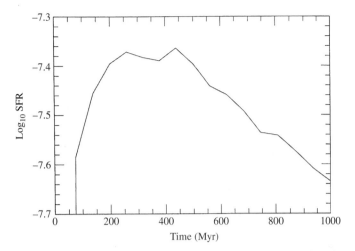

**FIGURE 26.18** A model of the total star formation rate (in units of $M_\odot \ pc^{-2} \ yr^{-1}$) for the disk of the Milky Way Galaxy as a function of time. (Figure adapted from Burkert, Truran, and Hensler, *Ap. J., 391,* 651, 1992.)

converted into stars; the present value for the SFR within the Galactic disk is believed to be $5.0 \pm 0.5 \ M_\odot \ pc^{-2} \ Gyr^{-1}$, integrated over the $z$ direction (this corresponds to the two to three stars formed per year that was mentioned on page 1012). Finally, the IMF, which was first discussed in Section 12.3 (see Fig. 12.12), represents the relative numbers of stars that form in each mass interval.

To understand the birthrates of stars and their ensuing contribution to the chemical evolution of the ISM, different researchers have made various assumptions about the SFR. For instance, some astronomers have assumed that the SFR is time-independent, while others have argued for an exponentially decreasing function with time, or perhaps one that is proportional to some power of the surface mass density of the Galactic disk. One computer simulation of the evolution of the disk of our Galaxy, performed by Andreas Burkert, James W. Truran, and G. Hensler (1992), produced an SFR that reached a maximum value and then decreased with time as the available gas and dust in the ISM was consumed (see Fig. 26.18). Other studies have considered the possibility that the SFR may be highly variable in both space and time. Such a situation could occur because of short-timescale starburst activity, for instance.

The IMF is often modeled as a power-law fit of the form

$$\xi(M) = \frac{dN}{dM} = CM^{-(1+x)}, \tag{26.5}$$

where $x$ may take on different values for various mass ranges and $C$ is a normalization constant. The first attempt to derive an IMF for the solar neighborhood was carried out by Edwin E. Salpeter in 1955, where he argued for a value of $x = 1.35$. According to more recent determinations, it appears that $x = 1.8$ may be a better fit for stars in the approximate mass range 7 $M_\odot$ to 35 $M_\odot$ (and perhaps down as low as 2 $M_\odot$). For stars more massive than about 40 $M_\odot$, $x = 4$ may be required, implying that the production of massive stars drops off very rapidly with increasing mass. For lower-mass stars, $x$ is very difficult to determine through observational studies. The complications arise because the IMF must be decoupled from the **present-day mass function** and the SFR, which itself can be quite complicated. It has been suggested that the IMF may flatten considerably at low masses. Unfortunately, an exact form of the IMF is not yet known, and it is not even clear whether the IMF varies with time or location.

## The G-Dwarf Problem

Attempts to use the SFR and IMF to model the chemical evolution of the Galactic disk in the solar neighborhood have resulted in predictions that do not always agree with observations. If we assume that the first generation of stars was born without any metals ($Z_0 = 0$; Population III) and the chemical evolution of the ISM occurs within a *closed box* (meaning that no gas or dust is allowed to enter or leave the system being modeled), then the calculations predict too many stars of low metallicity when compared with observations. For instance, this simple model suggests that roughly one-half of the stars in the solar neighborhood should have values for $Z$ that are less than one-quarter of the solar value ($Z_\odot \simeq 0.02$). However, only about 2% of the F and G main-sequence stars have such low $Z$ values. This is known as the **G-dwarf problem**.

One possible solution to the G-dwarf problem, referred to as **prompt initial enhancement**, is to assume that the disk of our Galaxy formed with $Z_0 \neq 0$, which could occur if heavy-element enrichment of the ISM resulted from rapidly evolving massive stars before the gas and dust settled into the disk. A second suggestion is that the disk accumulated mass over a significant period of time, perhaps even continuing to the present (in other words, the closed-box assumption is invalid). In this scenario a substantial infall of metal-poor material onto the Galactic disk has occurred since its initial formation; as the gas entered the system, it mixed with the metal-enriched ISM. Since a lower initial mass density would imply fewer stars formed during the early history of the disk, fewer metal-poor stars would be observed today. Yet another proposal argues that the IMF was different in the early history of the Galaxy, and a larger fraction of more massive stars were formed with correspondingly fewer low-mass stars. Since massive stars are short-lived, this hypothesis would result in fewer metal-poor stars today.

## A Dissipative Collapse Model

Another issue that must be addressed by a comprehensive theory of galactic evolution is the question of a free-fall collapse versus a slow, dissipative one. A free-fall collapse is governed by the free-fall timescale, sometimes alternatively referred to as the **dynamical timescale**, given by Eq. (12.26) (see Example 26.2.1). On the other hand, a dissipative collapse can be described in terms of the time necessary for the nebula to cool significantly. If the **cooling timescale**, $t_{cool}$, is much less than the free-fall timescale, then the cloud will not be pressure-supported and the collapse will be rapid (i.e., essentially in free-fall). However, if the cooling time exceeds the free-fall timescale, the gas cannot radiate its energy away fast enough to allow for a rapid collapse, and the gravitational potential energy that is released during the collapse will heat the nebula adiabatically. This situation is yet another example of the virial theorem.

To estimate the cooling timescale, we must first determine the characteristic amount of thermal kinetic energy contained within each particle in the gas. According to the virial theorem, if we assume that the gas is in quasistatic equilibrium, the average thermal kinetic energy of the gas must be related to its potential energy by Eq. (2.46),

$$-2 \langle K \rangle = \langle U \rangle.$$

If we further assume that the gas has a mean molecular weight of $\mu$ and contains $N$ particles, then the virial theorem gives

$$-2N \frac{1}{2} \mu m_H \langle v^2 \rangle = -\frac{3}{5} \frac{GM^2}{R},$$

where $m = \mu m_H$ is the average mass of a single particle, $R$ is the radius of the nebula, and $M = N \mu m_H$ is the nebula's mass. In the last expression, we used a gravitationally bound, spherical mass distribution of constant density to estimate the potential energy of the system; see Eq. (10.22). Solving for the velocity dispersion $\sigma = \langle v^2 \rangle^{1/2}$ gives

$$\sigma = \left( \frac{3}{5} \frac{GM}{R} \right)^{1/2}. \tag{26.6}$$

Note the similarity of this expression to Eq. (25.13) for the virial mass of a spherical cluster of stars.

Now we may determine a characteristic temperature of the gas, known as the **virial temperature**, by equating the typical kinetic energy of a gas particle to its thermal energy, or

$$\frac{1}{2} \mu m_H \sigma^2 = \frac{3}{2} k T_{\text{virial}},$$

which gives

$$T_{\text{virial}} = \frac{\mu m_H \sigma^2}{3k}. \tag{26.7}$$

Finally, to estimate the cooling time we must determine the rate at which energy can be radiated away from the gas. This is done by expressing the cooling rate per unit volume as

$$r_{\text{cool}} = n^2 \Lambda(T),$$

where $r_{\text{cool}}$ has units of energy per unit time per unit volume, $n$ is the number density of particles in the gas, and $\Lambda(T)$ is a quantum mechanical **cooling function**. $\Lambda(T)$, shown in Fig. 26.19, includes the same physical processes that were discussed in the context of stellar opacity in Section 9.2 (bound–bound, bound–free, free–free, and electron scattering). The two "bumps" in the curve just above $10^4$ K and near $10^5$ K correspond to the ionization/recombination temperatures of hydrogen and helium, respectively. Above about $10^6$ K, the cooling is due to thermal bremsstrahlung and Compton scattering. The $n^2$ dependence in the expression for $r_{\text{cool}}$ can be understood in terms of the interactions between

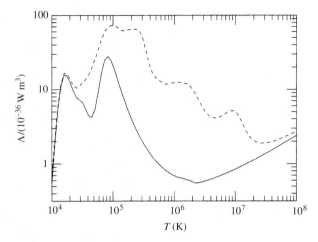

**FIGURE 26.19**    The cooling function $\Lambda(T)$. The solid line corresponds to a gas mixture of 90% hydrogen and 10% helium, by number. The dashed line is for solar abundances. (Figure adapted from Binney and Tremaine, *Galactic Dynamics*, Princeton University Press, Princeton, NJ, 1987.)

pairs of particles in the gas; collisions excite ions, atoms, or molecules, which then radiate the energy away in the form of photons, cooling the gas.

If all of the energy in the cloud is radiated away in a time $t_{\rm cool}$, then

$$r_{\rm cool} V t_{\rm cool} = \frac{3}{2} N k T_{\rm virial},$$

where $V$ is the volume of the cloud. Solving for the cooling time, we have

$$t_{\rm cool} = \frac{3}{2} \frac{k T_{\rm virial}}{n \Lambda}. \tag{26.8}$$

---

**Example 26.2.2.**    Assume for simplicity that the proto-Galactic nebula discussed in Example 26.2.1 was initially composed of 90% hydrogen and 10% helium, by number (this corresponds to $X \simeq 0.7$ and $Y \simeq 0.3$).[5] If we assume complete ionization, then the mean molecular weight is given by Eq. (10.16), or $\mu \simeq 0.6$. Also, according to Eq. (26.6), the initial velocity dispersion of the particles in the gas was approximately $\sigma = 160$ km s$^{-1}$. Then, substituting into Eq. (26.7), the virial temperature of the gas at the time of collapse was roughly

$$T_{\rm virial} \simeq 6 \times 10^5 \text{ K}.$$

(At this temperature our assumption that the gas was completely ionized is certainly valid.)

---

[5]Note that we are assuming here that *all* of the mass is in the form of baryonic matter. This means that we are neglecting the influence of dark matter in this example.

The number density of particles in the gas is given by

$$n = \frac{\rho}{\mu m_H} = \frac{3M}{4\pi R^3 \mu m_H} \sim 5 \times 10^4 \text{ m}^{-3}.$$

This value should be compared to the typical number densities found in Galactic molecular clouds today, which are on the order of $n \sim 10^8$ to $10^9$ m$^{-3}$ (see Section 12.1).

Now, from Fig. 26.19, $\Lambda \sim 10^{-36}$ W m$^3$ and the cooling time for the cloud (Eq. 26.8) is found to be

$$t_{\text{cool}} = 8 \text{ Myr}.$$

Clearly, in this case $t_{\text{cool}} \ll t_{\text{ff}}$. Apparently the proto-Galactic nebula was capable of radiating energy away at a rate sufficient to allow for a free-fall collapse.

---

It is instructive to consider the situation where $t_{\text{cool}} > t_{\text{ff}}$. In this case the nebula is unable to efficiently radiate away the gravitational potential energy that is released by the collapse. As a result, the cloud's temperature would rise adiabatically as the cloud shrinks, resulting in an increasing internal pressure and a halt to the collapse. After the collapse has halted, the virial theorem governs the equilibrium conditions of the cloud. For the approximate values of $T \sim 10^6$ K and $n \sim 5 \times 10^4$ m$^{-3}$ characteristic of a protogalactic cloud at the time of formation of the first galaxies, the upper limit on the mass that can cool and collapse is on the order of $10^{12}$ M$_\odot$, with a corresponding radius of about 60 kpc. In regions of the cloud where the gas temperature had decreased to the level of hydrogen recombination, $T \sim 10^4$ K, the mass limit becomes $\sim 10^8$ M$_\odot$. Thus the galaxies that are observed today would be expected to have masses in the range from $10^8$ M$_\odot$ to $10^{12}$ M$_\odot$. The lower limit corresponds fairly well with the values of the smallest dwarf elliptical galaxies, and the upper limit agrees with the values measured for the most massive giant spiral galaxies (Sa–Sc). Although some giant ellipticals and cD galaxies exceed $10^{12}$ M$_\odot$, they have certainly been affected significantly by mergers throughout their histories and are not in virial equilibrium near their outer edges.

Although the proto-Galactic cloud was able to radiate away the initial release of gravitational energy content from the system, shortly after the collapse began a new source of energy became available. The deaths of the first generations of very massive stars meant that supernova shock waves moved through the ISM at speeds on the order of 0.1$c$. As the expanding shells struck the gas, the gas was reheated to temperatures of a few million kelvins, slowing the rate of collapse somewhat. However, calculations of the supernova production rate that are based on estimates of the IMF seem to indicate that even this new source of energy was unable to slow the collapse appreciably.

## The Hierarchical Merger Model

How then can we explain the apparent age and metallicity differences among the globular clusters, as well as the existence of distinct components of varying ages within the disk of our Galaxy (i.e., the thick disk and the younger thin disk)? The answer appears to be that not only did galaxy-building involve the top-down process originally envisioned by Eggen, Lynden-Bell, and Sandage, but it also incorporated a *bottom-up* **hierarchical process of mergers**.

With the realization in the 1970s and 1980s that mergers play an important role in galactic evolution, and because of observational and theoretical developments regarding the nature of the early universe, the hierarchical merger scenario has received a great deal of attention. Scientists now believe that shortly after the birth of the universe via the Big Bang, density fluctuations existed in the overall distribution of matter. (The details of the evolution of the universe as a whole will be discussed in Chapter 30.) Our current understanding of those fluctuations suggests that the most common density perturbations occurred on the smallest mass scales. Consequently, density fluctuations involving $10^6$ to $10^8$ $M_\odot$ were much more common than those for $10^{12}$ $M_\odot$ or more.

First consider the formation of the Milky Way as an example of the hierarchical merger process. As these $10^6$ to $10^8$ $M_\odot$ proto-Galactic fragments were gravitationally attracted to one another, they began to merge into a growing spheroidal mass distribution. Initially, many of the fragments evolved in virtual isolation, forming stars and, in some cases, globular clusters in their centers. As a result, they developed their own chemical histories and unique abundance signatures. In the inner regions of the growing spheroid, where the density of matter was greater, its rate of collapse and subsequent evolution would have been more rapid. This resulted in the production of the oldest stars that are observed today, together with a greater degree of chemical enrichment (hence the old, metal-rich central bulge). In the rarefied outer regions of the Galaxy, chemical evolution and star formation would have been much slower.

According to the hierarchical model, collisions and tidal interactions between merging fragments disrupted the majority of the fragments and left exposed the globular cluster cores of others. Furthermore, in this model the disrupted systems would have led to the present distribution of the field halo stars, while leaving the remaining globular clusters scattered throughout the spheroid. Those proto-Galactic fragments that were initially moving in a retrograde direction relative to the eventual orbital motion of the Galactic disk and inner halo produced the net zero rotation of the outer halo that is observed today.

Certainly the rate of collisions would have been greater near the center of the Galaxy, disrupting those proto-Galactic fragments first and building the bulge more rapidly than the halo. Consequently, according to this picture, the spheroidal component of the Galaxy can be considered as forming from the inside out.

The globular clusters still present in the Galaxy today probably total only some 10% of the number that originally formed from proto-Galactic fragments. The other 90% were disrupted by collisions and tidal interactions during the early merger process and by the subsequent ongoing effects of dynamical friction. This may help to explain the relative uniformity in the masses of globular clusters observed today (approximately $10^5$ to $10^6$ $M_\odot$). Low-mass globular clusters would have had small gravitational binding energies, allowing them to be disrupted comparatively easily and rapidly when the Galaxy was young. On the other hand, recall from Eq. (26.1) that dynamical friction is strongly dependent on the mass of the cluster ($f_d \propto M^2$) so that massive clusters would have spiraled rapidly into the inner regions of the Galaxy where stronger and more frequent interactions ultimately disrupted them as well.

It is important to note that because of their isolation in the outer reaches of the Galaxy and the slower rate of evolution there, the proto-Galactic fragments in that region would have evolved almost like individual dwarf galaxies for a time. In fact, the significant number of dSph galaxies still present in the Local Group are assumed to be surviving proto-Galactic fragments. In addition, there is clear evidence of ongoing mergers today, such as the dwarf spheroidal Sagittarius galaxy and the Magellanic Stream, indicating that the construction of the Milky Way's halo is still a work in progress.

**Formation of the Thick Disk**

As the gas clouds of disrupted proto-Galactic fragments collided, the collapse became largely dissipative. This means that the gas began to settle slowly toward the central regions of the Galaxy. However, because of the presence of some initial angular momentum in the system, introduced through torques from other neighboring protogalactic clouds, the collapsing material eventually became rotationally supported and settled into a disk about the Galactic center. Of course, the already-formed halo stars did not participate in the collapse to the disk, because their collisional cross sections were now too small to allow them to interact appreciably, except through gravitational forces.

One model of thick-disk formation suggests that the thick disk may have formed around the Galactic midplane with a characteristic temperature of $T \sim 10^6$ K. By equating the kinetic energy of a typical particle in the gas to its gravitational potential energy above the midplane of the disk, the approximate scale height, $h$, of the disk of gas can be estimated.

To determine the local acceleration of gravity, $g$, at a height, $h$, above the midplane, imagine that the disk has a mass density, $\rho$, given by

$$\rho(h) = \rho_0 e^{-z/h},$$

where $\rho_0$ is the mass density in the Galactic midplane (see Eq. 24.9). Now, according to the gravitational version of "Gauss's law" [see Eq. (24.56) in Problem 24.26],

$$\oint \mathbf{g} \cdot d\mathbf{A} = -4\pi G M_{\text{in}},$$

where the integral is over a closed surface that bounds the mass $M_{\text{in}}$, and $\mathbf{g}$ is the local acceleration of gravity at the position of $d\mathbf{A}$. If $h$ is much smaller than the diameter of the disk, then for a Gaussian cylinder of height $2h$ and cross-sectional area $A$, centered on the midplane (see Fig. 26.20),

$$2Ag = 4\pi G M_{\text{in}},$$

where $M_{\text{in}}$ is the amount of disk mass contained within the cylinder. $M_{\text{in}}$ can be estimated by integrating the mass density throughout the volume of the cylinder, or

$$M_{\text{in}} = 2 \int_0^h \rho_0 e^{-z/h} A \, dz = 1.26 \rho_0 A h.$$

Substituting into our previous expression, we have that the local acceleration of gravity at a height, $h$, is given by

$$g(h) = 2.53\pi G \rho_0 h.$$

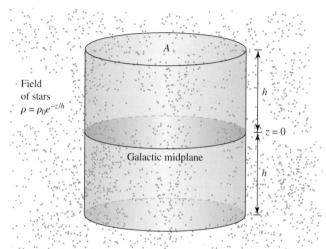

**FIGURE 26.20**    A Gaussian cylinder located entirely within the field of stars of the Galactic disk.

Next, the gravitational potential energy of a particle of mass, $m$, at a height, $h$, above the midplane is given by

$$U(h) = \int_0^h mg(z)\,dz = 1.26\pi\,Gm\rho_0 h^2.$$

Equating the potential energy to the average thermal kinetic energy of a particle, $K = 3kT/2$, we find that

$$h(T) = \left(\frac{3kT}{2.53\pi\,Gm\rho_0}\right)^{1/2}. \tag{26.9}$$

---

**Example 26.2.3.**    If the gas in the proto-Galactic thick disk had a characteristic temperature of $T \sim 10^6$ K, and if we assume that the central mass density was comparable to the value that is estimated today for the solar neighborhood,

$$\rho_0 \simeq 0.15\ \text{M}_\odot\ \text{pc}^{-3} = 1.0 \times 10^{-20}\ \text{kg m}^{-3},$$

then

$$h\left(10^6\ \text{K}\right) \simeq 2.2\ \text{kpc},$$

where we have used the mass of a hydrogen atom for $m$. From Table 24.1, the measured value for the scale height of the thick disk is approximately 1 kpc.

---

In regions where the gas was locally more dense, it cooled more rapidly, since $t_{\text{cool}} \propto n^{-1}$ (see Eq. 26.8). This was accomplished first through thermal bremsstrahlung and Compton scattering, and then, when the temperature reached $\sim 10^4$ K, via the radiation emitted by hydrogen atoms. This meant that once the hydrogen recombination temperature was reached, H I clouds could form and begin producing stars. Within a few million years, the most massive stars underwent core-collapse supernova detonations and their shocks began to reheat the gas between the molecular clouds, maintaining the temperature of the intercloud gas at roughly $10^6$ K. At the same time, the production of iron in the supernovae raised the metallicity from an initial value of [Fe/H] $< -5.4$ (perhaps originally enriched somewhat by the stars that formed in the proto-Galactic fragments) to [Fe/H] $= -0.5$. About 400 million years after the first stars were created in the thick disk, star formation nearly ceased. In total, a few percent of the mass of the gas was converted into stars during this thick-disk-producing period of the Galaxy's evolution.

A modified version of the thick-disk formation model described above suggests that the infalling gas was initially much cooler. This meant that the gas (and dust) was able to settle onto the midplane with a much smaller scale height, similar to today's thin disk. Star formation was then able to proceed due to the greater local density of gas and dust. However, as a direct result of a significant merger event with a proto-Galactic fragment some 10 Gyr ago, the disk was reheated by the energy of the interaction, causing it to puff up to its present 1-kpc scale height.

### Formation of the Thin Disk

After the formation of the thick disk, cool molecular gas continued to settle onto the midplane with a scale height of approximately 600 pc. During the next several billion years, star formation occurred in the thin disk.

The process of maintaining the scale height was essentially a self-regulating one. If the disk became thinner, its mass density would increase. This in turn would cause the SFR to increase, producing more supernovae and reheating the disk's intercloud gas component. The ensuing expansion of the disk would again decrease the SFR, yielding fewer supernovae, and the disk would cool and shrink. However, despite the self-regulating process, as the gas was depleted in the ISM the SFR decreased from about 0.04 $M_\odot$ pc$^{-3}$ Myr$^{-1}$ to 0.004 $M_\odot$ pc$^{-3}$ Myr$^{-1}$. At the same time, the metallicity continued to rise, reaching a value of approximately [Fe/H] $= 0.3$. Because of the decrease in the SFR, the thickness of the disk decreased to about 350 pc, the scale height of today's thin disk. During the development of the thin disk, some 80% of the available gas was consumed in the form of stars.

Finally, as the remaining gas continued to cool, it settled into an inner, metal-rich and gas-rich component of the thin disk with a scale height of less than 100 pc. Today most ongoing star formation occurs in this young, inner portion of the thin disk, the component in which the Sun resides.

### The Existence of Young Stars in the Central Bulge

The existence of young stars in the central bulge of our Galaxy can be understood in the context of the evolution just described by arguing for recent mergers with gas-rich satellite galaxies. When those galaxies were disrupted by tidal interactions with the Milky Way, their gas settled into the disk and the center of the Galaxy, ultimately forming new stars. It also appears that the Milky Way's central bar plays a role in the migration of dust and gas into the inner portion of the Galaxy by generating dynamical instabilities as it rotates.

## Metallicity Gradients

The hierarchical merger scenario just outlined predicts that metallicity gradients ought to exist in galaxies that have undergone a dissipative collapse. If a galaxy is more metal-rich in its center than it is near the outskirts of the system, then a color gradient should also exist. Because of the enhancement of opacity with metallicity (see Section 25.2), the galaxy would be redder in its center than it is farther out.

Of course, the strength of the metallicity and color gradients can be diminished or even destroyed by sufficiently frequent and energetic mergers with other galaxies. For instance, many starburst galaxies actually have inverted color gradients and appear bluer in their centers. This is because of the large SFR that resulted from the sudden influx of gas-rich material in the galactic center when another galaxy was disrupted, or from the effects of tidal torques that acted on the starburst galaxy itself, causing its own gas to spiral into the center. It is worth noting that weaker metallicity gradients are observed in boxy ellipticals than in disky ones, another possible indication that boxy E's have undergone significant merger activity during their lifetimes.

## The Formation of Elliptical Galaxies

Although we do not yet understand all of the complex details of galactic evolution, the nature of the Hubble sequence may ultimately correspond to the mass of the individual galaxies, the efficiency with which the galaxies made stars, and the relative importance of free-fall collapse, dissipative collapse, and mergers. As an example, it appears that many ellipticals may have formed the majority of their stars early in the galaxy-building process, before the gas had a chance to settle into a disk, whereas late-type galaxies took a more leisurely pace. You may also recall that current observations indicate that later Hubble types have a higher relative abundance of gas and dust in their disks than galaxies of earlier Hubble type (see Tables 25.1 and 25.2).

However, as first mentioned at the end of Section 26.1, other E's probably formed from the collisions of already-existing spirals. The energy involved in the collision would destroy the disks of both galaxies and cause the merged system to relax to the characteristic $r^{1/4}$ distribution of an elliptical.

Although $N$-body simulations have been able to produce such a result (see Fig. 26.21), some questions remain. For instance, the large number of globular clusters in E's relative to spirals present a serious difficulty in arguing that cataclysmic collisions are the cause of all large elliptical galaxies [Fig. 25.16 and Eq. (25.15)]. On the other hand, perhaps mergers can actually produce globular clusters by triggering star formation in clouds, in which case the larger specific frequency of globular clusters may not be a problem. It is also possible that many of the observed globular clusters are captured dwarf spheroidal galaxies, just as $\omega$ Cen appears to be in the Milky Way.

From the discussion at the beginning of Section 26.1, recall that elliptical galaxies are much more abundant relative to spirals in the centers of dense, rich clusters of galaxies, whereas spirals dominate in less dense clusters and near the periphery of rich clusters. (This **morphology–density relation** was first reported by Alan Dressler in 1980.) This effect may be partly explained by the increased likelihood of interactions in regions where galaxies are more tightly packed, destroying spirals and forming ellipticals.

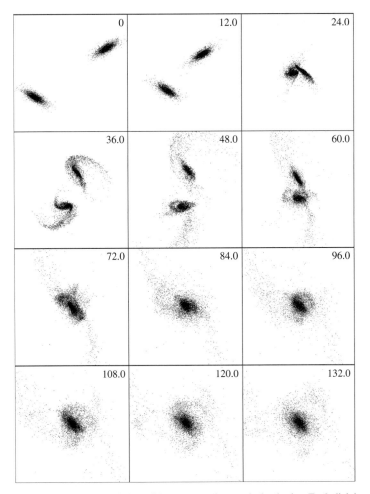

**FIGURE 26.21** An *N*-body simulation of the merger of two spiral galaxies. Each disk is represented by 16,384 particles, and each bulge contains 4096 particles. The result is an elliptical galaxy with an $r^{1/4}$ profile. (Figure adapted from Hernquist, *Ap. J.*, *409*, 548, 1993.)

However, a competing hypothesis has also been suggested—namely, that ellipticals tend to develop preferentially near the bottoms of deep gravitational potential wells, even in the absence of interactions. Lower mass-density fluctuations in the early universe may have resulted in spiral galaxies, and the smallest fluctuations led to the formation of dwarf systems. If this is the case, then the large number of dSph's and dE's that exist has a natural explanation in the much larger number of smaller fluctuations that formed in the early universe. This mechanism could help explain galactic morphology if the initial density fluctuations in the early universe were largest in what later became the centers of rich clusters. Because the gravitational potential well in those regions would have been deeper, the probability of collisions between protogalactic clouds should have been correspondingly greater as well.

Perhaps the frequency of collisions between protogalactic clouds is correlated with galaxy morphology.

## Galaxy Formation in the Early Universe

Given the large number of fundamental problems that still remain in our development of a coherent theory of galaxy formation, it is fortunate that a means exists (at least in principle) for testing our ideas by observing galactic evolution through time. Because of the finite speed of light, when astronomers look farther and farther out into space, we are literally looking farther and farther back in time. For instance, galaxies that are 1 Mpc away emitted their light more than 3 million years ago. This means that the light we observe today coming from the galaxy is itself more than 3 million years old; we are "seeing" the galaxy 3 million years in the past.

In 1978, Harvey Butcher and Augustus Oemler, Jr., noted that there appeared to be an overabundance of blue galaxies in two distant clusters. They speculated that there may have been a significant evolution in galaxies over time, leading to the types of objects we see closer to us today. The **Butcher–Oemler effect**, as it is now called, suggests that galaxies in the early universe were bluer on average than they are today, indicating an increased level of star formation. The effect has been confirmed in numerous recent studies.

The morphology–density relation has also been shown to be time-dependent. As observations probe earlier times (more distant galaxy clusters), elliptical, and lenticular galaxies become less abundant relative to spiral galaxies, suggesting an evolution from later Hubble types to earlier types over time. This is just what would be expected if some earlier Hubble-type galaxies form from the mergers of spirals. This is also consistent with the overall picture of hierarchical galaxy building.

In 2004, the Space Telescope Science Institute released the Hubble Ultra Deep Field (HUDF) image shown in Fig. 26.22. The HUDF image is actually a composite of two images, one taken by the Advanced Camera for Surveys (ACS) and the other from the Near-Infrared Camera and Multi-Object Spectrometer (NICMOS). The region of the sky contained in the HUDF image is only 3 arcmin on a side (about 1/10 the size of the full moon), centered on ($\alpha = 3^h32^m40.0^s$, $\delta = -27°48'00''$) in the constellation of Fornax. The image required a total exposure time of 11.3 days for ACS (obtained between September 24, 2003, and January 16, 2004) and 4.5 days for NICMOS (over the period between September 3, 2003, and November 27, 2003).

The HUDF reveals some very distant galaxies as they existed just 400 Myr to 800 Myr after the Big Bang. As is evident in the close-up of a portion of the HUDF shown in Fig. 26.23, the very distant galaxies appear quite different from the relatively nearby majestic spirals and ellipticals that are seen in the present-day universe.

Observations like these suggest that the abundance of strange-looking, remote, blue galaxies seen in the early universe may be the building blocks of today's Hubble sequence of galaxies. They probably represent the proto-galactic fragments responsible for the hierarchical mergers that are still occuring today at a much diminished rate.

In Section 30.2 we will turn our attention to the earliest formation of structure in the universe, including the clumping of dark matter that likely formed the potential wells in which the clusters of galaxies and their diverse membership evolved. Certainly a great deal

**FIGURE 26.22**   The Hubble Ultra Deep Field (HUDF), obtained by combining images from ACS and NICMOS onboard the HST. This HUDF image contains an estimated 10,000 galaxies covering a region 3 arcmin square (about 1/10 the size of the full moon) in the constellation Fornax. Some of the galaxies in the HUDF are so far away that we are seeing them less than 1 Gyr after the Big Bang. [Courtesy of NASA, ESA, S. Beckwith (STScI) and the HUDF Team.]

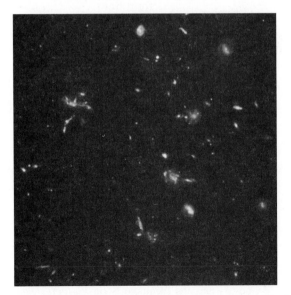

**FIGURE 26.23**   A close-up of a portion of the HUDF image shown in Fig. 26.22. Many of the galaxies in this image are extremely distant—and therefore extremely young. [Courtesy of NASA, ESA, S. Beckwith (STScI) and the HUDF Team.]

of work remains to be done before we can hope to fully understand the complex histories of galaxies, but it appears that many of the basic ingredients of an evolution theory may be in place.

## SUGGESTED READING

### General

Barnes, Joshua, Hernquist, Lars, and Schweizer, Francois, "Colliding Galaxies," *Scientific American*, August 1991.

Bothun, Gregory D., "Beyond the Hubble Sequence," *Sky and Telescope*, May, 2000.

Chiappini, Cristina, "Tracing the Milky Way's History," *Sky and Telescope*, October, 2004.

Dressler, Alan, "Observing Galaxies through Time," *Sky and Telescope*, August 1991.

Keel, William C., "Crashing Galaxies, Cosmic Fireworks," *Sky and Telescope*, January, 1989.

Miller, M. Coleman, Reynolds, Christopher S., and Krishnamurthi, Anita, "Supermassive Black Holes: Shaping Their Surroundings," *Sky and Telescope*, April, 2005.

Schroeder, Michael C., and Comins, Neil F., "Galactic Collisions on Your Computer," *Astronomy*, December, 1988.

Silk, Joseph, *The Big Bang*, Third Edition, W. H. Freeman and Company, New York, 2001.

*Space Telescope Science Institute*, http://www.stsci.edu.

Toomre, Alar, and Toomre, Juri, "Violent Tides between Galaxies," *Scientific American*, December, 1973.

van den Bergh, Sidney, and Hesser, James E., "How the Milky Way Formed," *Scientific American*, January, 1993.

### Technical

Arp, H. C., "Atlas of Peculiar Galaxies," *The Astrophysical Journal Supplement*, *14*, 1, 1966.

Barnes, Joshua E., and Hernquist, Lars, "Computer Models of Colliding Galaxies," *Physics Today*, March, 1993.

Bender, Ralf, et al., "HST STIS Spectroscopy of the Triple Nucleus of M31: Two Nested Disks in Keplerian Rotation around a Supermassive Black Hole," *The Astrophysical Journal*, *631*, 280, 2005.

Bertin, Giuseppe, *Dynamics of Galaxies*, Cambridge University Press, Cambridge, 2000.

Binney, James, "Warps," *Annual Review of Astronomy and Astrophysics*, *30*, 51, 1992.

Binney, James, and Merrifield, Michael, *Galactic Astronomy*, Princeton University Press, Princeton, NJ, 1998.

Binney, James, and Tremaine, Scott, *Galactic Dynamics*, Princeton University Press, Princeton, NJ, 1987.

Burkert, A., Truran, J. W., and Hensler, G., "The Collapse of Our Galaxy and the Formation of the Galactic Disk," *The Astrophysical Journal*, *391*, 651, 1992.

Butcher, Harvey, and Oemler, Augustus Jr., "The Evolution of Galaxies in Clusters. I. ISIT Photometry of Cl 0024+1654 and 3C 295," *The Astrophysical Journal, 219*, 18, 1978.

Chabrier, Gilles, "Galactic Stellar and Substellar Initial Mass Function," *Publications of the Astronomical Society of the Pacific, 115*, 763, 2003.

Dressler, Alan, "Galaxy Morphology in Rich Clusters: Implications for the Formation and Evolution of Galaxies," *The Astrophysical Journal, 236*, 351, 1980.

Eggen, O. J., Lynden-Bell, D., and Sandage, A. R., "Evidence from the Motions of Old Stars That the Galaxy Collapsed," *The Astrophysical Journal, 136*, 748, 1962.

Elmegreen, Bruce G., and Scalo, John, "The Effects of Star Formation History on the Inferred Stellar Initial Mass Function," *The Astrophysical Journal, 636*, 149, 2006.

Elmegreen, Debra Meloy, *Galaxies and Galactic Structure*, Prentice-Hall, Upper Saddle River, NJ, 1998.

Kormendy, John, and Kennicutt, Robert C., Jr., "Secular Evolution and the Formation of Pseudobulges in Disk Galaxies," *Annual Review of Astronomy and Astrophysics, 42*, 603, 2004.

Larson, Richard B., "Galaxy Building," *Publications of the Astronomical Society of the Pacific, 102*, 709, 1990.

Mo, H. J., Mao, Shude, and White, Simon D. M., "The Formation of Galactic Discs," *The Monthly Notices of the Royal Astronomical Society, 295*, 319, 1998.

Silk, Joseph, "Formation and Evolution of Disk Galaxies," *Astrophysics and Space Science, 284*, 663, 2003.

Spinrad, Hyron, *Galaxy Formation and Evolution*, Praxis Publishing Ltd., Chichester, 2005.

Toomre, Alar, and Toomre, Juri, "Galactic Bridges and Tails," *The Astrophysical Journal, 178*, 623, 1972.

## PROBLEMS

**26.1** (a) The mass density of stars in the neighborhood of the Sun is approximately $0.05 \, M_\odot \, pc^{-3}$ (see Section 24.2). Assuming that the mass density is constant and that all of the stars are main-sequence M stars, estimate the fraction of the Galactic disk's volume that is occupied by stars.

(b) Suppose that an intruder star (a main-sequence M star) travels perpendicularly through the Galactic disk. What are the odds of the intruder colliding with another star during its passage through the disk? Take the thickness of the disk to be approximately 1 kpc.

**26.2** Write a general expression for the force of dynamical friction as

$$f_d \simeq C \, (GM)^a (v_M)^b \rho^c,$$

where $C$ is dimensionless, and $a$, $b$, and $c$ are constants. Set up a system of three linear equations for $a$, $b$, and $c$ such that $f_d$ has units of force. Solve the system and show that Eq. (26.1) is obtained.

**26.3** Use the age of the Milky Way's oldest globular clusters from Section 24.2 and the orbital velocity of the local standard of rest to estimate the greatest distance from which globular clusters could have spiraled into the nucleus due to dynamical friction. Use $5 \times 10^6$ $M_\odot$ for the cluster's mass. Compare your answer with the size of the Milky Way's central bulge.

**26.4** **(a)** The Large Magellanic Cloud, which has a mass of about $2 \times 10^{10}$ $M_\odot$, orbits the Milky Way at a distance of 51 kpc. Assuming that the Milky Way's dark-matter halo and flat rotation curve extend out to the LMC, estimate how much time it will take for the LMC to spiral into the Milky Way (take $C = 23$). (Note that the Small Magellanic Cloud may be gravitationally bound to the LMC, but its mass is an order of magnitude smaller; therefore, the SMC will be neglected in this problem.)

**(b)** Compare this rough estimate with the value quoted in the text for the merger time of the LMC (which was based on a detailed computer simulation of the LMC's dynamical interaction with our Galaxy).

**26.5** Use Eq. (18.10) and the information in Problem 26.4 to estimate the tidal radius of the Large Magellanic Cloud as it orbits the Milky Way. Take the mass of the Milky Way's dark-matter halo to be $5.4 \times 10^{11}$ $M_\odot$ from Table 24.1. How does your answer compare with the size of the LMC, which has an angular diameter of $460'$?

**26.6** The Small Magellanic Cloud is 60 kpc away and has one-tenth the mass of the LMC (see Problem 26.4). The SMC has an angular diameter of $150'$, and its angular separation from the LMC is $21°$.

**(a)** What is the distance between the LMC and the SMC?

**(b)** Ignoring the influence of the Milky Way, estimate the present tidal radii of both the LMC and the SMC due to the other.

**(c)** The Magellanic Clouds are in eccentric orbits around one another, and during the last few hundred million years they have been moving apart with an average speed of about 110 km s$^{-1}$. How recently did the LMC extend beyond its tidal radius? How recently did the SMC extend beyond its tidal radius? Assuming that the Magellanic Stream was formed by tidal stripping when the LMC and SMC were close to each other, when would you estimate that the Magellanic Stream was formed? (*Note:* Following this mutual tidal stripping, the gas was pulled from the LMC/SMC system by our Galaxy.)

**26.7** The gaseous ring orbiting the central galaxies of the M96 group has an orbital period of $4.1 \times 10^9$ years.

**(a)** What is the combined mass (in $M_\odot$) of the two galaxies?

**(b)** The combined visual luminosity of the central galaxies M105 and NGC 3384 is $L_V = 2.37 \times 10^{10}$ $L_\odot$ (from Schneider, *Ap. J.*, *343*, 94, 1989). What is the mass-to-light ratio for the pair?

**26.8** Suppose that the original density distribution of the proto-Galactic nebula envisioned by Eggen, Lynden-Bell, and Sandage had a radial functional dependence similar to the dark-matter halo [assume the form of Eq. (24.51) with $r \gg a$]. Beginning directly with the radial equation of motion,

$$\frac{d^2r}{dt^2} = -\frac{GM_r}{r^2},$$

show that the free-fall time would be proportional to radius.

*Hints:* Multiply both sides of the equation by $v = dr/dt$ and use the relation

$$v\left(\frac{dv}{dt}\right) = \frac{1}{2}\frac{dv^2}{dt}.$$

You may also find the following definite integral helpful:

$$\int_0^1 \frac{du}{\sqrt{\ln\left(\frac{1}{u}\right)}} = \sqrt{\pi}.$$

**26.9** (a) Assume an initial mass function of the form given in Eq. (26.5). If $x = 1.8$, calculate the ratio of the number of stars that are formed in the mass range between 2 $M_\odot$ and 3 $M_\odot$ to those formed with masses between 10 $M_\odot$ and 11 $M_\odot$.

(b) Beginning with the initial mass function and using the mass–luminosity relation for main-sequence stars, Eq. (24.12) (see also Fig. 7.7 and Problem 10.27), derive an expression for the number of main-sequence stars formed per unit luminosity interval, $dN/dL$.

(c) If $x = 1.8$ and $\alpha = 4$, calculate the ratio of the number of stars that are formed with main-sequence luminosities between 2 $L_\odot$ and 3 $L_\odot$ to the number of those formed with luminosities between 10 $L_\odot$ and 11 $L_\odot$.

(d) Compare your answers in parts (a) and (c), and explain the results in terms of the physical characteristics of stars along the main sequence.

**26.10** Use the value for the star formation rate given on page 1019 to estimate the number of stars currently being formed per year in the Milky Way Galaxy. Compare your answer to the value quoted in the text. Clearly describe the assumptions (if any) that you made in arriving at your estimate.

**26.11** (a) Using the Jeans mass (Eq. 12.14), estimate the upper mass limit of galaxies that are in virial equilibrium. Use the virial temperature and the number density determined in Example 26.2.2.

(b) Assuming that the least massive galaxies could have formed out of a gas with an initial temperature near the ionization temperature of hydrogen, estimate the lower mass cutoff.

(c) By using the Jeans length (Eq. 12.16), estimate the radius of the largest galaxies that are in virial equilibrium and compare your answer to the approximate size of our Galaxy.

**26.12** (a) By equating the cooling timescale to the free-fall timescale, show that the maximum mass of a protogalactic nebula is given by

$$M = \frac{25}{32} \frac{\Lambda^2}{G^3 \mu^4 m_H^4 R}.$$

(b) Estimate the maximum mass of a protogalactic nebula that can undergo a free-fall collapse if $R = 60$ kpc. Assume that $\Lambda \simeq 10^{-37}$ W m$^3$.

**26.13** (a) M87 is a giant elliptical galaxy near the center of the Virgo cluster of galaxies. It is estimated that its mass is approximately $3 \times 10^{13}$ $M_\odot$ within a radius of 300 kpc. How long would it take for a star near the outer edge of the galaxy to orbit the center once? Compare your answer to the approximate age of our Galaxy.

(b) Based on your result from part (a) and assuming that M87 has been capturing smaller satellite galaxies up until the present time, would you expect the outer portions of M87 to be in virial equilibrium? Why or why not?

**26.14** (a) From the data given in Problem 26.13, calculate the velocity of a star near the outer edge of the galaxy that may be following a circular orbit.

(b) Assuming that the velocity you found in part (a) is characteristic of the velocity dispersion of material near the outer edge of the galaxy, estimate the virial temperature of M87.

(c) What is the characteristic cooling time for M87? Use $\Lambda \simeq 10^{-37}$ W m$^3$.

(d) If the mass in the outer portions of M87 were to undergo a radial free-fall collapse, estimate the free-fall timescale for such a collapse and compare your answer with the cooling timescale.

**26.15** Estimate the scale height associated with a Galactic disk having a temperature of $10^4$ K. With which component of our Galaxy might such a disk correspond today?

**26.16** It has been pointed out many times that given the current rate of star formation, the disk of our Galaxy will be depleted of interstellar gas in the relatively near future. From the data given in Table 24.1 and Section 26.2, estimate how much longer the Milky Way can effectively create new stars without an appreciable infall of new material. How might the period of star formation be prolonged?

**26.17** (a) From the data in Table 25.4, estimate the escape speed from a dwarf elliptical galaxy.

(b) Compare your answer in part (a) with the speed of the ejecta from a typical Type II supernova.

(c) Compare your answer in part (a) with the thermal velocities of hydrogen ions in a gas having a temperature of $10^6$ K.

(d) How might you explain the evolution of ancient, gas-poor systems such as dE's and dSph's?

## COMPUTER PROBLEMS

**26.18** Figure 25.17 shows the Whirlpool galaxy, M51, and its companion, NGC 5195. Use the program `Galaxy`, described in Appendix M, to reproduce the appearance of this system. Use 10 rings of 50 stars each, and let the intruder galaxy have one-fourth the mass of the target galaxy. Use $(x, y, z) = (30, -30, 0)$ for the initial position of the intruder galaxy, and use $(v_x, v_y, v_z) = (0, 0.34, 0.34)$ for its velocity (in the units used by the program). Follow the motion for 648 million years (540 time steps).

(a) When does the calculated appearance of the target and host galaxies resemble Fig. 25.17? Indicate the simulation times.

(b) What happens to the system near the end of this time period?

(c) Rerun the program, changing the initial $x$ coordinate from 30 to $-30$ (and leaving everything else the same). Describe the difference in the outcomes.

**26.19** `Galaxy` was used in Problem 26.18 to reproduce some the features of the Whirlpool galaxy, shown in Fig. 25.17. Run the program again, using the same values of the mass, initial position, and initial velocity of the intruder galaxy, but use only *two* rings with 24 stars each.

(a) Compare the initial speed of the intruder galaxy with the orbital speeds of the nearby stars. Describe (in words) how the spiral pattern forms. Follow the motion for 360 million years (300 time steps).

(b) Rerun the program, changing the initial $x$ coordinate from 30 to $-30$ (and leaving everything else the same). Describe and physically explain the differences between this and the behavior of the stars in part (a).

**26.20** Figure 26.6 shows the Cartwheel galaxy, the result of a high-speed, head-on collision. Use the program `Galaxy`, described in Appendix M, to investigate how a nearly head-on collision may have produced this ring galaxy. Use 10 rings of 24 stars each, and assume that the galaxies involved have equal masses.

**(a)** Try an exact head-on collision first. Let $(x, y, z) = (0, 0, 35)$ for the initial position of the intruder galaxy, and let $(v_x, v_y, v_z) = (0, 0, -1)$ for its velocity (in the units used by the program). Follow the motion for 78 million years (65 time steps), and describe what happens. Could such a process be responsible for Fig. 26.6?

**(b)** Explain how the force of gravity, always acting inward, could have caused the rings of stars to *expand*.

**(c)** Investigate what happens if the initial value of the $x$ coordinate varies between 0 and 10 (with everything else unchanged). Which values (roughly) do the best job of reproducing the appearance of a ring galaxy with the least disruption of the target galaxy?

# CHAPTER
# 27

# The Structure of the Universe

## 27.1 ■ THE EXTRAGALACTIC DISTANCE SCALE

The drama of the heavens unfolds on a two-dimensional stage. Observations readily supply the two coordinates, right ascension and declination, that specify an object's position on the celestial sphere. However, the object's distance is not so easily obtained. Astronomers today use many methods to estimate the distances of remote galaxies. In this section we will introduce a variety of techniques; some depend on the properties of individual stars, some use other objects (globular clusters, planetary nebulae, supernovae, HII regions), and some rely on the statistical properties of whole galaxies and galaxy clusters. The third dimension of distance is doubly significant because, as we saw at the end of Section 26.2, when astronomers peer deeper and deeper into space, they are looking back in time at increasingly ancient light. It is vital to understand not only a galaxy's position in space, but its depth in time as well.

### Unveiling the Third Dimension

The techniques used to determine distances within the solar neighborhood were described in Section 3.1. In 1761, the method of trigonometric parallax was used to measure the distance to Venus, thereby calibrating the size of Kepler's Solar System. When Friedrich Wilhelm Bessel measured the subtle annual shift in the position of the star 61 Cygni in 1862, he combined the parallax method with his knowledge of the true size of Earth's orbit to discover that 61 Cygni is 650,000 times farther away than the Sun. Today, the surveyor's method of trigonometric parallax can reach out to a kiloparsec or so.[1] We have also seen how the moving cluster method (Section 24.3) made it possible to determine the distance of the Hyades star cluster. From there, the technique of main-sequence fitting (Sections 13.3 and 24.3) can be used to find the distances to open clusters out to about 7 kpc by comparing their main sequences on an H–R diagram with that of the Hyades cluster. The repeated application

---

[1] The distances to the nearer planets can be determined within a few kilometers by measuring the round-trip time taken by radar waves sent from Earth and reflected back from the planet's surface. However, because the strength of the reflected wave diminishes rapidly (as $1/r^4$), this method is limited to within about a light-hour, or $10^9$ km.

of a variety of methods using this pattern of calibration and measurement constitutes the steps of the **extragalactic distance scale**, also referred to as the **cosmological distance ladder**.

### The Wilson–Bappu Effect

Spectroscopic parallax (discussed in Section 8.2) can in principle provide reliable distances to remote stars out to about 7 Mpc away, although in practice it is employed within a few hundred kiloparsecs, far enough to reach the Magellanic Clouds. Some stars have specific features in their spectra that allow their absolute magnitudes, and hence their distances, to be calculated. For example, the K absorption line of calcium can be quite broad, reaching maximum strength at spectral type K0 (Section 8.1). In late-type stars with chromospheres (types G, K, and M), a narrow emission line is seen, centered on the wide K absorption line. The width of this emission line is strongly correlated with a star's absolute visual magnitude. This **Wilson–Bappu effect**, discovered in 1956 by Olin C. Wilson, Jr. (1909–1994) and M. K. Vainu Bappu (1927–1982), is valid for stars over a range of 15 magnitudes.

### The Cepheid Distance Scale

For more distant objects, astronomers turn to the period–luminosity relation for Cepheids, discovered by Henrietta Leavitt (Section 14.1). Before this relation could be used, it had to be calibrated by finding the distance to a classical Cepheid. Unfortunately, the nearest one, Polaris, was too far away ($\approx 200$ pc) for trigonometric parallax to be useful. In 1913, however, Ejnar Hertzsprung used the Sun's motion of 16.5 km s$^{-1}$ with respect to the local standard of rest (Section 24.3) to provide a longer baseline for parallax measurements. It was this technique of **secular parallax** that enabled him to determine the average distance to a classical Cepheid with a period of 6.6 days. Hertzsprung then used this information to calibrate the period–luminosity relation; Harlow Shapley soon carried out a similar procedure. Since then, the distances to more Cepheids have been measured by parallax and other methods, and the period–luminosity relation has become well established. As mentioned in Section 14.1, astronomers today use a period–luminosity–color relation such as

$$M_{\langle V \rangle} = -3.53 \log_{10} P_d - 2.13 + 2.13(B - V), \qquad (27.1)$$

in the $V$ band to account for the finite width of the instability strip on the H–R diagram [you should compare this with the period–luminosity relation, Eq. (14.1) in the $V$ band]. Here, $P_d$ is the pulsation period in units of days, and $B - V$ is the color index; for classical Cepheids, $B - V \approx 0.4$ to 1.1. After the star's absolute magnitude has been calculated, it can be combined with the star's apparent magnitude to give its distance modulus; Eq. (24.1) then provides the star's distance *if the extinction (in magnitudes) is known*.

Cepheid variable stars immediately proved their value as stellar yardsticks. It was in 1917 that Shapley measured the distances to Population II Cepheids in globular clusters, thereby determining his estimates of 100 kpc for the diameter of the Galaxy and 15 kpc for the distance of the Sun from its center.[2] Then, in 1923, Edwin Hubble discovered several Cepheids in the Andromeda galaxy and announced that it was 285 kpc away. (The modern value is 770 kpc; see Section 25.1.) It was Hubble's series of observations that established M31 as an external galaxy, not a smaller nebula within the borders of the Milky Way.

It should be noted that the existence of a correlation between the period of a Cepheid's oscillations and its absolute magnitude meant that these stars could be used as standard candles to determine distances many years *before* the physical processes that cause the pulsations were understood. This involves some risk, and in this case there is a cautionary tale worth telling. At the time, it was not known that *three* types of pulsating stars were being used to determine the size of the Galaxy and the distance to M31. Furthermore, although the existence of interstellar dust clouds had been established by Edward Barnard (1857–1923) in the early years of the twentieth century, the existence of diffuse interstellar dust and gas capable of extinguishing starlight had yet to be demonstrated. Leavitt's variables in the Magellanic Clouds were classical Cepheids (Population I stars), while those observed in the globular clusters by Shapley were W Virginis and RR Lyrae stars (both Population II).[3] The W Virginis stars are about 1.5 magnitudes fainter than classical Cepheids of the same period, which corresponds to their being less luminous by a factor of 4. When Hertzsprung and later Shapley calibrated the period–luminosity relation, they used nearby classical Cepheids but neglected the effect of extinction. Neither realized that dust in the Galactic disk dimmed these stars by (coincidentally) about 1.5 magnitudes.

Thus by chance, the resulting period–luminosity relation was just about right when used with W Virginis stars. However, it would give an underluminous result with classical Cepheids. When Edwin Hubble used this calibration to find the distances to M31 and other galaxies (which were observed out of the obscuring plane of the Galaxy), the apparent magnitudes of his Cepheids were correct but his estimates of the absolute magnitudes were too large. Consequently, their distance moduli ($m - M$) and the distances themselves were underestimated. The stars were thought to be dimmer and closer, rather than brighter and farther away, and so their parent galaxies were determined to be at roughly *half* of their actual distances and *half* their actual sizes.

Harlow Shapley fared no better with his observations of the (unknown to him) W Virginis stars in the globular clusters. Although he unwittingly used the right variety of star for the calibration of the period–luminosity relation, Shapley's observations of W Virginis stars were made in systems near the Galactic plane. Extinction due to dust within the disk dimmed the starlight and increased the stars' apparent magnitudes. Because the importance of extinction was not yet recognized, Shapley incorrectly attributed his stars' faintness to their remoteness. As a result, his distances to the globular clusters were too great, as was the size he calculated for the Galaxy.

Something was obviously wrong. It appeared that the Milky Way was much larger than any of the other galaxies surveyed. This is just the sort of anthropomorphic preeminence that makes astronomers uncomfortable. Robert Trumpler's (1886–1956) proof of interstellar

---

[2] Recall that Shapley's values for the dimensions of the Milky Way have since been revised substantially, an ongoing process described in Section 24.2.

[3] Shapley used clusters containing both W Virginis and RR Lyrae stars to calibrate a period–luminosity relation for the RR Lyraes. He then used observations of RR Lyrae stars in those clusters lacking the brighter W Virginis variables, and so the error was propagated.

extinction in 1930 remedied part of the problem. The rest of the puzzle was solved in 1952 when Walter Baade announced that there are two types of Cepheids: the classical Cepheids and the intrinsically fainter W Virginis stars.[4] With these corrections, the measured distances and sizes of the other galaxies doubled and the Milky Way was reduced to being a bit smaller than M31.

In the early 1990s, the calibration of the period–luminosity relation was carried out using data obtained by the Hipparcos mission, described in Section 3.1. Hipparcos astronomers measured the parallax angles of 273 Cepheids and used the resulting distances to derive the period–luminosity relation quoted in Eq. (14.1). This required the cooperation of amateur astronomers, whose careful observations of light curves made it possible to measure the brightness of these variables at the appropriate phases. This was the first period–luminosity relation to be obtained from a direct distance measurement, and it is interesting that the result was nearly identical to an expression obtained in 1968 by American astronomer Allan Sandage and his Swiss colleague Gustav Tammann. Although further adjustments are sure to follow, Cepheids certainly provide a firm foundation for measuring the distances to other galaxies.

Interstellar extinction is still the largest source of error when Cepheids are used as standard candles, and there may also be a weak dependence on metallicity. The extinction problem can be reduced by observing these stars at infrared wavelengths, since infrared light more readily penetrates dusty regions (recall Fig. 14.6). However, because Cepheids are about three magnitudes dimmer in the infrared, astronomers have not stopped searching for them at visible wavelengths.

The Cepheid-based distances to galaxies within the Virgo cluster of galaxies ranges from 15 Mpc to 25 Mpc. The variation in these values is primarily due to the large spatial extent of the Virgo cluster along the line of sight, and to the fact that the spiral galaxies in which the Cepheids are observed are found primarily on the outskirts of the cluster and not in its core. To date, the most remote classical Cepheids known are 29 Mpc distant and reside in NGC 3370, a spiral galaxy in the direction of the constellation Leo.[5] Overall, the uncertainty in the distances obtained using classical Cepheids as standard candles ranges from 7% for the LMC to perhaps 15% for more distant galaxies.

## Supernovae as Distance Indicators

Supernovae can be used in several ways to measure extragalactic distances. Suppose the angular extent, $\theta(t)$, of a supernova's photosphere is observed. The angular velocity of the expanding gases, $\omega = \Delta\theta/\Delta t$, can be found by comparing two observations separated by a time $\Delta t$. If $d$ is the distance to the supernova, then the transverse velocity of the expanding photosphere is $v_\theta = \omega d$. Assuming that the expansion is spherically symmetric, this transverse velocity should be equal to the ejecta's radial velocity, $v_{ej}$, obtained from the supernova's Doppler-shifted spectral lines. Then the distance to the supernova is

$$d = \frac{v_{ej}}{\omega}.$$

---

[4]Baade had described Population I and II stars in 1944 and had wondered at the time whether there might be two corresponding varieties of Cepheids.

[5]See Riess et al. (2005) for details of the observations of these remote Cepheids.

Most supernovae are too distant for this method to be employed, so another strategy is adopted. It assumes that the expanding shell of hot gases radiates as a blackbody (to a first approximation). Then the supernova's luminosity is given by the Stefan–Boltzmann law, Eq. (3.17),

$$L = 4\pi R^2(t)\sigma T_e^4,$$

where $R(t)$ is the radius of the expanding photosphere and $t$ is the age of the supernova. If we assume that the ejecta's radial velocity has remained nearly constant, then $R(t) = v_{ej}t$. The effective temperature of the photosphere comes from the characteristics of its blackbody spectrum. Once the luminosity is found, it can be converted to an absolute magnitude (via Eq. 3.8) and then used to find the distance to the supernova by comparing it with the observed apparent magnitude. Of course, the photosphere of the expanding shell of a supernova is neither perfectly spherical nor a perfect blackbody. Difficulties with accurate values for interstellar extinction plague both methods, but the problem is more acute for core-collapse supernovae (Types Ib, Ic, and II), which are found near sites of recent star formation. Typical uncertainties in the distances obtained range from 15% (for M101) to 25% (for the Virgo cluster of galaxies).

---

**Example 27.1.1.**    Twenty-five days after maximum light, the spectrum of an "average" Type Ia supernova is that of a blackbody with an effective temperature of $6000 \pm 1000$ K. The speed of the shell's photosphere, obtained from its Doppler-shifted absorption lines, is $9500 \pm 500$ km s$^{-1}$. If the rise time to maximum light is $17 \pm 3$ days, then when the average Type Ia supernova is $t = 42$ days old, its luminosity is

$$L \approx 4\pi \left(v_{ej}t\right)^2 \sigma T_e^4 = 1.10 \times 10^{36} \text{ W}. \tag{27.2}$$

From Eq. (3.8) its absolute bolometric magnitude at this time is

$$M_{bol} = M_{Sun} - 2.5\log_{10}(L/L_\odot) \approx -18.9.$$

---

### Type Ia Light Curves

The most important way of using supernovae to measure distance takes advantage of the similarity of Type Ia light curves. As discussed in Section 18.5, these supernovae have blue and visual absolute magnitudes of $\langle M_B \rangle \simeq \langle M_V \rangle \simeq -19.3 \pm 0.03$ at maximum light. If the peak magnitude of a Type Ia can be determined, its distance may then be determined.

Much effort has been devoted to improving our understanding of Type Ia light curves; recall Fig. 18.21. Fortunately there is a well-defined inverse correlation between a Type Ia's maximum brightness and the rate of decline of its light curve, and astronomers have learned how to use this information to more precisely determine the supernova's intrinsic peak luminosity.

In practice, a supernova is observed over time at several wavelengths. The **multicolor light curve shapes (MCLS) method** then compares the shape of the light curve to a family of parameterized template curves that allows the absolute magnitude of the supernova at maximum brightness to be determined, even if the supernova was not caught at its peak

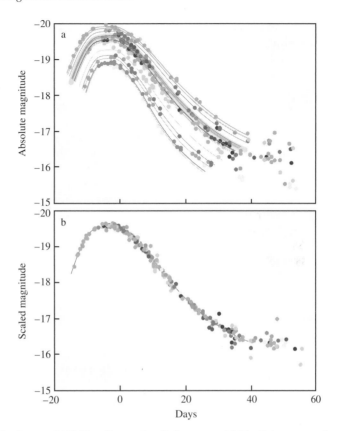

**FIGURE 27.1** Low-redshift Type Ia template light curves. (a) The light curves of several Type Ia supernovae, as measured. (b) The light curves after applying the time scale stretch factor. The blue absolute magnitude is displayed on the vertical axis. (Figure adapted from Perlmutter, *Physics Today*, *56*, No. 4, 53, 2003.)

brilliance. The MLCS method also allows the reddening and dimming effect of interstellar dust to be detected and removed.

Another approach is the **stretch method**, which fits the $B$ and $V$ magnitude light curves with a single template light curve that has been stretched (or compressed) in time (see Fig. 27.1). The peak magnitude is then determined by the stretch factor. These techniques allow astronomers to use Type Ia supernovae to determine distances with an uncertainty approaching just 5%, corresponding to an uncertainty in the distance modulus, $m - M$, of 0.1 magnitude.

---

**Example 27.1.2.** The Type Ia supernova SN 1963p in the galaxy NGC 1084 had an apparent blue magnitude of $B = 14.0$ at peak brilliance. Then, with an extinction of 0.49 mag to that galaxy, the distance to the supernova is approximately

$$d = 10^{(m-M-A+5)/5} = 41.9 \text{ Mpc.}$$

---

Because Type Ia supernovae are about 13.3 magnitudes brighter than the brightest Cepheid variables ($-19.3$ compared to $-6$), this method is capable of reaching out more than 500 times farther than with Cepheids, providing estimates to truly cosmological distances exceeding 1000 Mpc.[6] Sophisticated supernova search programs are carried out that more than compensate for the unlikely probability of detecting an explosion in a specific galaxy by scanning large numbers of galaxies in a wide field of view. Near the end of the twentieth century, two teams of astronomers made careful observations of high-redshift Type Ia supernovae and discovered that the expansion of the universe is accelerating, a story that will be recounted in detail in Section 29.4. The Supernova/Acceleration Probe (SNAP) spacecraft is currently being designed and may be launched in 2010. Its mission will be to observe thousands of distant Type Ia supernovae and to map several percent of the sky for gravitational lenses every year.

## Using Novae in Distance Determinations

From the sizes of their expanding photospheres, novae can be used in the same manner as supernovae to derive distances. In addition, although there is a wide variation in the absolute magnitudes of novae at peak light, there is a relation between a nova's maximum visual magnitude, $M_V^{\text{max}}$, and the time it takes for its visible light to decline by two magnitudes.[7] Consequently, novae can also serve as standard candles.

The physical reason why this relationship holds for novae is that more massive, smaller white dwarfs produce greater compression and heating of the accreted gases on their surfaces, and so a runaway thermonuclear reaction may be initiated with a smaller accumulated mass; see Section 18.5. The less massive surface layers are more readily ejected, and so the nova declines more rapidly. Writing the average rate of decline over the first 2 mag as $\dot{m}$ (in mag d$^{-1}$), this relation can be expressed as

$$M_V^{\text{max}} = -9.96 - 2.31 \log_{10} \dot{m} \tag{27.3}$$

for Galactic novae, with an uncertainty of about $\pm 0.4$ mag. After fading by two magnitudes, the brightest novae are about as luminous as the brightest Cepheids, so these two methods span the same region of space (to about 20 Mpc, just past the Virgo cluster of galaxies).

## Secondary Distance Indicators

With the exception of techniques associated with unpredictable supernovae, in order for astronomers to reach farther into space and to more remote galaxies, *secondary* methods must be used to measure distance. Unlike the *primary* methods already described, these secondary distance indicators require a galaxy with an established distance for their calibration. One way of seeing farther involves using the brightest objects in a galaxy. For example, the three brightest giant H II regions in a galaxy have been used to provide a standard candle. These regions, which may contain up to $10^9$ $M_\odot$ of ionized hydrogen, are visible at great distances. Measurements of the angular sizes of the H II regions and the apparent magnitude of the galaxy can be compared with similar measurements for other

---

[6]Type II supernovae are dimmer than Type Ia's by about 2 mag and, therefore, at best can be seen only about 40% as far away.

[7]There are several variants of this relation that use the time for the nova to decline by 2 mag, 3 mag, etc.

galaxies with known distances. This makes it possible to calculate the H II regions' linear sizes (in parsecs), along with the galaxy's absolute magnitude. However, since the diameter of an H II region is difficult to define unambiguously, this procedure is relatively insensitive to distance and must be used carefully.

The brightest red supergiants seem to have about the same absolute visual magnitude in all galaxies, as shown by careful studies of the brightest galactic stars carried out by Roberta Humphreys. Apparently, mass loss reduces the brightest red supergiants to about the same maximum mass, which results in their having about the same luminosity. Sampling a number of galaxies revealed the average visual magnitude of the three brightest red stars to be $M_V = -8.0$. Because individual stars must be resolved in order for this method to be used, its range is the same as for spectroscopic parallax, about 7 Mpc.

### The Globular Cluster Luminosity Function

The limited sampling inherent in these methods ("the three brightest …") can lead to errors. It is statistically more secure to take as complete an inventory as possible of some class of objects associated with a galaxy, and then describe how the objects vary with magnitude. For example, Fig. 27.2 shows the **globular cluster luminosity function**, $\varphi(M_B)$, for the globular clusters around four giant elliptical galaxies in the Virgo cluster.[8] In the figure, $\varphi(M_B)\, dM_B$ is the number of globular clusters having absolute blue magnitudes between $M_B$ and $M_B + dM_B$. The distribution is well described by a Gaussian function (the solid

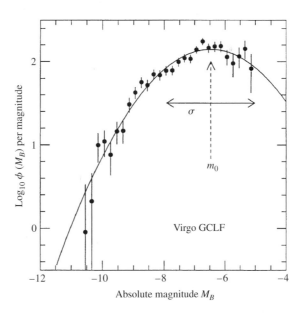

**FIGURE 27.2** The luminosity function for the globular clusters around four giant elliptical galaxies in the Virgo cluster. About 2000 clusters brighter than $B = 26.2$ were used. A distance of 17 Mpc was adopted for the Virgo cluster. [Adapted from a figure courtesy of Jacoby (KPNO/NOAO), as provided by Harris (McMaster Univ.) for *Publ. Astron. Soc. Pac.*, *104*, 599, 1992.]

[8]Luminosity functions were introduced in Section 25.4.

line in the figure) that has a well-defined peak at a *turnover magnitude* of $M_0 \simeq -6.5$. (Notice that the value of $M_0$ depends on the distance to the Virgo cluster used in calculating the absolute magnitudes.) The value of the turnover magnitude provides a standard candle that can be used to find the distance to the globular clusters surrounding another galaxy.

The procedure is to measure the luminosity function for the galaxy under investigation and to compare its apparent turnover magnitude, $m_0$, with $M_0$ for the Virgo cluster. The best results are achieved with galaxies having large numbers of globular clusters, such as giant ellipticals; recall Fig. 25.16. It is desirable that the calculated $\varphi(m_B)$ extend well past the turnover point, although there are (less desirable) ways of fitting just the brighter end of the luminosity function. Overall, this method should yield a value for a galaxy's distance modulus that is uncertain by only about 0.4 mag, corresponding to an uncertainty in distance of about 20%. Globular clusters are visible from vast distances, and in the future this technique may reach out beyond the Virgo cluster to 50 Mpc.

Unfortunately, it is not yet certain whether there is a *universal* globular cluster luminosity function that applies to all types of galaxies, although for nine galaxies (including M31 and the Milky Way), the average value is $M_0 = -6.6 \pm 0.26$. The physical basis for this apparent agreement is unclear. However, globular clusters are ancient collections of stars, and it is reasonable to suppose that the subsequent evolution of the galaxies to which they are attached may not have significantly affected the individual clusters.

### The Planetary Nebula Luminosity Function

A similar statistical analysis can be performed for a galaxy's planetary nebulae. Figure 27.3 shows the **planetary nebula luminosity function (PNLF)** for planetary nebulae in the Leo I group of galaxies as a function of their absolute magnitudes at a wavelength of

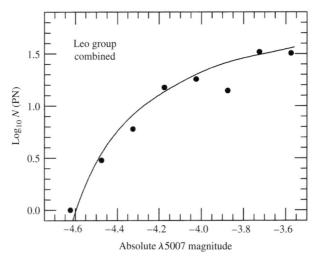

**FIGURE 27.3**   The planetary nebula luminosity function for the Leo I group of galaxies. (Figure adapted from Ciardullo, Jacoby, and Ford, *Ap. J.*, *344*, 715, 1989.)

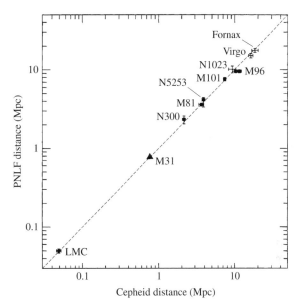

**FIGURE 27.4** A comparison between distances obtained using the planetary nebula luminosity function and Cepheids. (Figure adapted from Jacoby, "Future Directions for the Planetary Nebula Luminosity Function" in *Harmonizing Cosmic Distance Scales in a Post-Hipparcos Era, ASP Conference Series, 167,* 175, 1999.)

500.7 nm.[9] Using extragalactic planetary nebulae in this way is a reliable method of finding the distances to elliptical galaxies within about 20 Mpc. As a larger sample of galaxies was studied, the value determined for the cutoff decreased from the one depicted in Fig 27.3 to $M_{5007} = -4.53$. This can be adopted as a standard candle for the brightest planetary nebula. If the promise of this method is fulfilled, it should reach out to a distance of some 50 Mpc. Figure 27.4 shows the good agreement obtained between distances from the planetary nebula luminosity function and from Cepheid variables.

### The Surface Brightness Fluctuation Method

Astronomers turn to the global properties of galaxies to probe even farther, up to 100 Mpc or more. One promising approach is to take advantage of how a detector such as a CCD camera records the appearance of a galaxy. Some pixels will record more stars than others due to spatial fluctuations in the galaxy's surface brightness, but the overall appearance should become smoother with increasing distance. The results of a statistical analysis describe the magnitude of the pixel-to-pixel variation, and this is correlated with the galaxy's distance. With the Hubble Space Telescope, the **surface brightness fluctuation method** could reach out to 125 Mpc, but it is usually applied more locally.

---

[9]This is a forbidden line of oxygen, [O III]. The observations are made through a filter with a bandwidth of about 3 nm; see Section 3.6.

## The Tully–Fisher Relation

The Tully–Fisher relation for spiral galaxies (see Section 25.2) also provides a valuable tool for determining extragalactic distances. As discussed in Section 25.2, this is a relation between the luminosity of a spiral galaxy and its maximum rotation velocity. For the sake of completeness, we include an example of this important technique.

---

**Example 27.1.3.**   For the galaxy M81, the quantity $W_R^i = 484$ km s$^{-1}$. From Eq. (25.8), the infrared absolute magnitude is

$$M_H = -9.50(\log_{10} W_R^i - 2.50) - 21.67 = -23.43.$$

The apparent $H$ magnitude of M81 is $H = 4.29$ (already corrected for extinction), and so according to the Tully–Fisher relation, the distance to M81 is

$$d = 10^{(H - M_H + 5)/5} = 3.50 \text{ Mpc}.$$

---

The appeal of the Tully–Fisher method of distance determination lies in its accuracy (typically $\pm 0.4$ mag in the infrared, although for carefully selected targets the uncertainty may be reduced to $\pm 0.1$ mag) and its great range (as far as 100 Mpc). Furthermore, nearby spirals whose distances have been accurately measured by Cepheids can be used for its calibration. The Tully–Fisher method has been applied to 161 spiral galaxies belonging to the Virgo Cluster. The results have produced a three-dimensional map of the cluster, and they show that the Virgo Cluster is extended along the line of sight by about four times its diameter on the sky. The Tully–Fisher method is also being used to study the Coma cluster of galaxies, about 90 Mpc away, and has been applied to distances well beyond that. It is one of the most widely used methods of measuring extragalactic distances.

## The $D$–$\sigma$ Relation

The analogous correlation for elliptical galaxies, the Faber–Jackson relation ($L \propto \sigma_r^4$), shows considerable scatter; recall Fig. 25.33. More recently, an improvement has appeared, the **$D$–$\sigma$ relation**, that relates the velocity dispersion to the diameter, $D$, of an elliptical galaxy. More precisely, $D$ is the galaxy's angular diameter out to a surface brightness level of 20.75 $B$-mag arcsec$^{-2}$.[10] Because the surface brightness is independent of the galaxy's distance[11] (recall Sections 6.1 and 9.1), $D$ is inversely proportional to the galaxy's distance, $d$. If the galaxy were twice as far away, its angular diameter would be half as large. In this way, $D$ provides a *standard ruler* rather than a standard candle. There is a tighter relationship between $\sigma$ and $D$ than there is between $\sigma$ and $L$; see Fig. 27.5. An empirical relation between $\log_{10} D$ and $\log_{10} \sigma$ for the galaxies in a cluster (all at about the same distance) is

$$\log_{10} D = 1.333 \log \sigma + C, \tag{27.4}$$

---

[10]Actually, this is just another representation of the *fundamental plane* discussed in Section 25.4, since it relates radius, surface brightness, and velocity dispersion.

[11]This statement is technically true only if the galaxy is closer than a cosmological distance of roughly 500 Mpc, at which point spacetime curvature effects become significant (see Chapter 29).

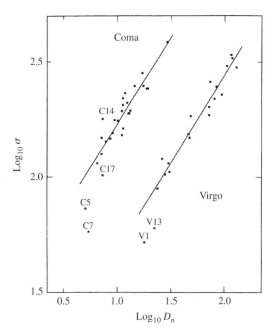

**FIGURE 27.5** A logarithmic plot of diameters $D$ (in arcseconds) and velocity dispersions $\sigma$ (in km s$^{-1}$) for galaxies in the Virgo and Coma clusters. (Figure adapted from Dressler et al., *Ap. J., 313*, 42, 1987.)

where the value of the constant $C$ depends on the distance to the cluster of galaxies. Unfortunately, there are no bright elliptical galaxies available for the accurate calibration of this $D$–$\sigma$ method by primary distance indicators such as Cepheids. However, because the slopes of the lines in the figure are very nearly the same, the vertical distance between the lines for two different clusters is

$$\log_{10} D_1 - \log_{10} D_2 = C_1 - C_2.$$

Then, because $D$ is inversely proportional to a galaxy's distance, $d$, the *relative* distances between the two clusters can be found by using

$$\frac{d_2}{d_1} = \frac{D_1}{D_2} = 10^{C_1 - C_2}. \tag{27.5}$$

---

**Example 27.1.4.** In Eq. (27.4), the value of $C$ for the Virgo cluster is $-1.237$, and for the Coma cluster it is $-1.967$. Thus the ratio of the distance of the Coma cluster of galaxies to that of the Virgo cluster is

$$\frac{d_{\text{Coma}}}{d_{\text{Virgo}}} = 10^{C_{\text{Virgo}} - C_{\text{Coma}}} = 5.37.$$

The Coma cluster is more than 5 times farther away than the Virgo cluster.

---

The $D$–$\sigma$ relation is a powerful tool for studying the distribution of groups of galaxies in space. Because of the inherent brightness of giant elliptical galaxies, this technique has the potential of exceeding the range of the Tully–Fisher relation.

### The Brightest Galaxies in Clusters

Just as the brightest giant H II regions or the brightest red supergiants have been used to determine the distance to individual galaxies, the brightest galaxy in a cluster of galaxies can be used to obtain the cluster's distance. Figure 27.6 shows a composite galaxy luminosity function, similar to the ones given in Fig. 25.36. The sharp decline in the function at the bright end (small $M$) shows that the absolute magnitude of the brightest galaxy can be determined with some accuracy. The average value of the absolute visual magnitude for the brightest galaxy in ten nearby clusters is $M_V = -22.83 \pm 0.61$. This is 3.2 magnitudes brighter than a Type Ia supernova at peak brilliance, so in principle this method should reach four times farther than with supernovae, to over 4000 Mpc. The light received from such a distance is truly ancient, having traveled through space for 13 billion years.

The danger in using this method to look out to such great distances is that galaxies and clusters of galaxies both evolve, as discussed in Section 26.2. For instance, the brightest

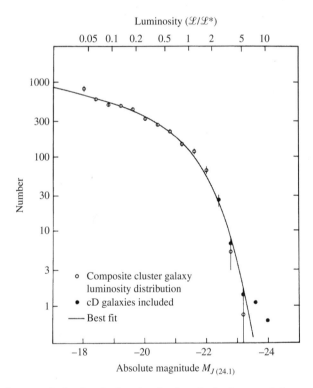

**FIGURE 27.6**    A composite luminosity function for the galaxies in several clusters. (Figure adapted from Schechter, *Ap. J.*, *203*, 297, 1976.)

galaxy in a contemporary cluster may not have existed in its present form billions of years ago. As was mentioned in Section 26.1, the giant cD galaxies often found at the centers of rich clusters are probably the result of galactic mergers. Therefore, using a general galaxy luminosity function calibrated with nearby clusters may not be appropriate for very distant, and consequently younger-appearing, clusters.

## A Summary of Distance Indicators

By now it should be clear that the extragalactic distance scale is not a simple ladder, with just one sequence of steps possible for measuring the greatest distances. Different astronomers use different techniques and choose different calibrations. There are variations with every method, and the relations described above are chosen only to be representative. As a summary, Table 27.1 lists the distance to the Virgo cluster of galaxies as determined by many of these methods. It also shows the uncertainty (in magnitudes) associated with each of these techniques. Overall, the distances listed are in good agreement, and we will adopt 16 Mpc as the distance to the Virgo cluster (see Problem 27.19). These methods are sufficiently accurate to determine the features of our corner of the universe, out to a distance of a few hundred Mpc.

**TABLE 27.1**    Distance Indicators. (Adapted from Jacoby et al., *Publ. Astron. Soc. Pac.*, *104*, 599, 1992.)

| Method | Uncertainty for Single Galaxy (mag) | Distance to Virgo Cluster (Mpc) | Range (Mpc) |
|---|---|---|---|
| Cepheids | 0.16 | $15 - 25$ | 29 |
| Novae | 0.4 | $21.1 \pm 3.9$ | 20 |
| Planetary nebula luminosity function | 0.3 | $15.4 \pm 1.1$ | 50 |
| Globular cluster luminosity function | 0.4 | $18.8 \pm 3.8$ | 50 |
| Surface brightness fluctuations | 0.3 | $15.9 \pm 0.9$ | 50 |
| Tully–Fisher relation | 0.4 | $15.8 \pm 1.5$ | $> 100$ |
| $D$–$\sigma$ relation | 0.5 | $16.8 \pm 2.4$ | $> 100$ |
| Type Ia supernovae | 0.10 | $19.4 \pm 5.0$ | $> 1000$ |

## 27.2 ■ THE EXPANSION OF THE UNIVERSE

In the first decade of the twentieth century, even before the extragalactic nature of spiral nebulae was understood, astronomers began making systematic observations of the radial velocities of galaxies by measuring their Doppler-shifted spectral lines. It was hoped that if the motions of these objects were found to be random, then the Sun's motion in the Milky Way should be related to the vector sum of the radial velocities of the nebulae.[12] It was V. M. Slipher, working at Lowell Observatory, who first discovered that this plan was doomed to failure. The velocities of the nebulae were *not* random. Instead, most of the spectra showed redshifted spectral lines. Slipher announced in 1914 that most of the 12 galaxies he had surveyed were rapidly moving *away* from Earth, although Andromeda's blueshifted spectrum showed it was approaching Earth at nearly 300 km s$^{-1}$. It was quickly realized that these galaxies not only were moving away from Earth but were moving away from each other as well. Astronomers began to talk about these galactic motions in terms of an expansion. Meanwhile, Slipher continued his measurements of radial velocities. By 1925 he had examined 40 galaxies and confirmed that spectra showing redshifted lines were much more common than those exhibiting blueshifts. Slipher concluded that nearly every galaxy he examined was rapidly receding from Earth.

### Hubble's Law of Universal Expansion

1925 was also the year that Hubble discovered Cepheid variable stars in M31, thereby establishing that the Andromeda "nebula" was in fact an external galaxy. Hubble continued his search for Cepheids, determining the distances to 18 galaxies. He combined his results with Slipher's velocities and discovered that a galaxy's recessional velocity, $v$, was proportional to its distance, $d$. In 1929 Hubble presented his results in a paper, "A Relation between Distance and Radial Velocity among Extra-Galactic Nebulae," at a meeting of the National Academy of Sciences. This relation,

$$v = H_0 d, \tag{27.6}$$

is today called **Hubble's law**, and $H_0$ is the **Hubble constant**. Usually $v$ is given in km s$^{-1}$ and $d$ in Mpc, so $H_0$ has units of km s$^{-1}$ Mpc$^{-1}$.

Hubble realized that he had discovered an exceptionally powerful way of finding distances to remote galaxies simply by measuring their redshifts. His early results show a straight line drawn through a scattering of points on a graph of velocity versus distance; see Fig. 27.7.

Hubble continued to compile distances and redshifts to strengthen this relationship, with much of the work done by his assistant, Milton Humason (1891–1972). Humason literally worked his way up to the top at the Mt. Wilson Observatory. He started out as a mule packer when the observatory was being constructed and served as a restaurant busboy, janitor, and night assistant at Mt. Wilson. After Humason got permission to do some observing on the

---

[12]This is basically the same procedure that was described in Section 24.3 for using nearby stars to find the Sun's peculiar motion relative to the local standard of rest; see Eqs. (24.30–24.32).

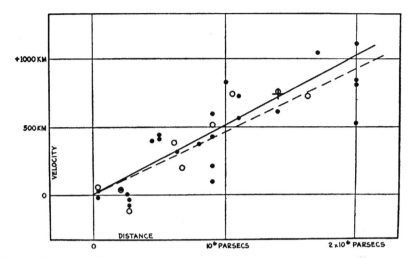

**FIGURE 27.7** Hubble's 1936 velocity–distance relation. The two lines use different corrections for the Sun's motion. (Note: The vertical units should be km s$^{-1}$.) (Figure from Hubble, *Realm of the Nebulae*, Yale University Press, New Haven, CT, ©1936.)

smaller telescopes, Hubble was so impressed with the results that Humason ended up as his assistant. Humason exposed and measured most of the photographic plates himself, and by 1934 the distances and velocities of 32 galaxies had been obtained. The expansion of the universe became an observational fact of life. Figure 27.8 shows the redshift of the H and K lines of Ca II for five galaxies.

Interestingly, the idea of a universal expansion was simultaneously being developed by theorists. In 1917, Willem de Sitter (1872–1935), a Dutch astronomer, used Einstein's theory of general relativity to describe a universe that was expanding.[13] Although de Sitter's solution of Einstein's equations described an empty universe devoid of matter, it did predict a redshift that increased with the distance from the light's origin. Hubble was aware of de Sitter's work and, in his 1929 paper, stated that its "outstanding feature . . . is the possibility that the velocity–distance relation may represent the de Sitter effect." Other theorists later found additional solutions that also indicated a universal expansion, but astronomers were unaware of their results until 1930. Einstein himself initially favored a static universe that was neither expanding nor contracting. However, the observations of Hubble and Humason forced Einstein to abandon this view in 1930. Figure 27.9 shows Einstein and Hubble at the Mt. Wilson Observatory.[14]

### The Expansion of Space and the Hubble Flow

To understand what "the expansion of the universe" really means, suppose that Earth were to expand, doubling in size during an hour's time. Figure 27.10 shows the effect on the

[13]This and other models of the universe will be described in more detail in Chapter 29.

[14]During a tour of the observatory, Elsa Einstein was told how the telescope was used to explore the structure of the universe. She replied, "Well, well, my husband does that on the back of an old envelope."

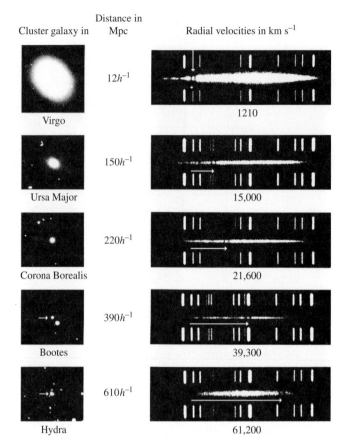

**FIGURE 27.8**   The appearance and redshifts of the H and K lines of calcium for five galaxies. (Courtesy of Palomar/Caltech.)

United States with two maps, aligned from the point of view of Salt Lake City. Initially, Yellowstone National Park is 500 km from Salt Lake City, Albuquerque is 1000 km away, and Minneapolis is at 2000 km. An hour later, Yellowstone is 1000 km away, Albuquerque is at 2000 km, and Minneapolis is at 4000 km, so a Salt Lake City resident would find that Yellowstone drifted away from her at 500 km hr$^{-1}$, Albuquerque moved away at 1000 km hr$^{-1}$, and Minneapolis receded at 2000 km hr$^{-1}$. Thus a recessional velocity that is proportional to distance is the natural result of an expansion that is both isotropic and homogeneous (same magnitude in every direction and at every location). Of course, observers in Yellowstone, Albuquerque, and Minneapolis would come to the same conclusion. Everyone involved in the expansion sees everyone else moving away with a velocity that obeys Hubble's law.

There is a vital distinction between the velocity of a galaxy through space (called its **peculiar velocity**) and its **recessional velocity** due to the expansion of the universe. As will be described in more detail in Section 29.1, a galaxy's recessional velocity is *not* due to its motion through space; instead, the galaxy is being carried along *with* the surrounding space

**FIGURE 27.9** Einstein and Hubble at the Mt. Wilson Observatory. (Reproduced by permission of The Huntington Library, San Marino, CA.)

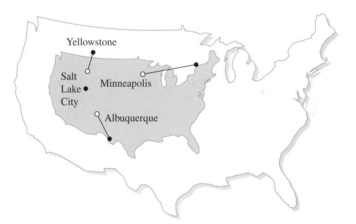

**FIGURE 27.10** The effect of a doubling of the size of Earth, from the perspective of an astronomer in Salt Lake City.

as the universe expands. The motion of galaxies as they participate in the expansion of the universe is referred to as the **Hubble flow**. In the same manner, a galaxy's **cosmological redshift** is produced by the expansion as the wavelength of the light emitted by the galaxy is stretched along with the space through which the light travels.[15] For this reason, the cosmological redshift is not related to the galaxy's recessional velocity by the Doppler shift equations derived in Section 4.3. Those equations were derived for a static, Euclidean spacetime and do not include the effects of the expanding, curved spacetime of our universe.

[15]The cosmological redshift will be derived in Section 29.4.

Nevertheless, astronomers frequently use Eq. (4.38) to translate a measured redshift, $z$, into the radial velocity a galaxy would have, *as if* it had a peculiar velocity (moving through space) instead of its actual recessional velocity (moving along with expanding space). For instance, the statement in Example 4.3.2 that "quasar SDSS 1030+0524 appears to be moving away from us at more than 96% of the speed of light" must be interpreted in this way. Furthermore, for $z \leq 2$, the distance estimate obtained by using Eq. (4.38) with Hubble's law,

$$d \simeq \frac{c}{H_0} \frac{(z+1)^2 - 1}{(z+1)^2 + 1},\tag{27.7}$$

differs from the actual proper distance by less than 5%.[16] When $z \ll 1$, the expression for the distance assumes the nonrelativistic form

$$d = \frac{cz}{H_0},\tag{27.8}$$

as would be found by using Eq. (4.39) with Hubble's law. [As we will see in Section 29.4, Eq. (27.8) can lead to significant errors in the distance for roughly $z > 0.13$.] In Chapter 29, we will revisit the question of how distances are measured in an expanding universe.

It is important to realize that although the universe is expanding, this does not mean, for example, that the orbits of the planets around the Sun have been expanding. Gravitationally bound systems do not participate in the universal expansion. There is also no compelling evidence that the constants that govern the fundamental laws of physics (such as Newton's gravitational constant, $G$) were once different from their present values. Thus the sizes of atoms, planetary systems, and galaxies have *not* changed because of the expansion of space (although the latter two may have certainly gone through evolutionary changes).

### The Value of the Hubble Constant

In principle, Hubble's law can be used to find the distance to any galaxy whose redshift can be measured. The stumbling block to carrying out this procedure has been the uncertainty in the Hubble constant. Through the end of the twentieth century, $H_0$ was known only to within a factor of two, between 50 km s$^{-1}$ Mpc$^{-1}$ and 100 km s$^{-1}$ Mpc$^{-1}$.

Historically, the difficulty in determining the value of $H_0$ arose from having to use remote galaxies for its calibration. A major source of the diverse values of $H_0$ obtained by different researchers lay in their choice and use of secondary distance indicators when measuring remote galaxies. There are also large-scale motions of galaxies relative to the Hubble flow that have yet to be sorted out, as will be described in Section 27.3.

In addition, there is a selection effect called a **Malmquist bias** that must be guarded against. This can occur when an astronomer uses a magnitude-limited sample of objects, looking only at those objects brighter than a certain *apparent* magnitude. At larger distances, only the intrinsically brightest objects will be included in the sample, which (if not properly corrected for) will skew the statistics.

---

[16]This statement assumes that the universe is "flat"; see Section 29.1 and Problem 29.57. Observations by missions such as Boomerang and WMAP indicate that space is very nearly flat.

As has been traditionally done in order to incorporate the uncertainty in $H_0$, in this text we will define a dimensionless parameter, $h$, through the expression

$$H_0 = 100h \text{ km s}^{-1} \text{ Mpc}^{-1}. \tag{27.9}$$

Due to early uncertainty with $H_0$, the value of $h$ was known only to fall somewhere between about 0.5 and 1. However, as the end of the twentieth century approached, astronomers began to converge on the middle ground of values of $H_0$.

## The Big Bang

Since the universe is expanding, it must have been smaller in the past than it is now. Imagine viewing a filmed history of the universe, watching the galaxies fly farther and farther apart. Now run the film *backward*, in effect reversing the direction of time. Seen in reverse, all of the galaxies are approaching one another. According to the Hubble law, a galaxy twice as far away is approaching twice as fast. The inescapable conclusion is that all of the galaxies (*and all of space!*) will simultaneously converge to a single point. As everything in the universe rapidly converged (still running the film backward), it heated to extremely high temperatures. The expansion of the universe from a single point is known as the **Big Bang**.

The hot early universe was filled with blackbody radiation, and as the universe expanded that radiation cooled to become the **cosmic microwave background (CMB)**, microwaves that we detect arriving from all directions in the sky. Observations of the CMB provide convincing evidence for the hot Big Bang. The **Wilkinson Microwave Anisotropy Probe (WMAP)**,[17] launched in 2001, studied the minute fluctuations (anisotropies) of the CMB. In 2003 the first WMAP results were released, the details of which will be discussed in Chapters 29 and 30. This ushered in a new era of precision cosmology, and the uncertainty in the Hubble constant (and $h$) fell from a factor of two to just 10%. The value of $h$ determined by the WMAP data,[18] and the standard value adopted in this text, is

$$[h]_{\text{WMAP}} = 0.71 \, ^{+0.04}_{-0.03}. \tag{27.10}$$

Throughout the remainder of this chapter, when a quantity involving $h$ is evaluated, $[h]_{\text{WMAP}}$ will be assumed. In conventional units, the Hubble constant is

$$H_0 = 3.24 \times 10^{-18} h \text{ s}^{-1} \tag{27.11}$$

so

$$[H_0]_{\text{WMAP}} = 2.30 \times 10^{-18} \text{ s}^{-1}. \tag{27.12}$$

[17]David Wilkinson (1935–2002) made pioneering contributions to many CMB experiments, including COBE and WMAP. The MAP satellite was renamed "WMAP" after it was launched in 2001.

[18]Parameters published in Table 3 ("Best" Cosmological Parameters) of Bennett et al., 2003, such as $H_0$, are placed inside a square bracket with a "WMAP" subscript, while quantities calculated using these parameters are not. A list of these parameters may be found in Appendix N. Although identified in the text as "WMAP parameters," they are a combination of WMAP results with measurements from projects such as COBE, the Cosmic Background Imager (CBI), the Arcminute Cosmology Bolometer Array Receiver (ACBAR), and the 2dF Galaxy Redshift Survey (2dFGRS).

Thus a galaxy with a measured recessional velocity of 1000 km s$^{-1}$ is at a distance of $d = v/H_0 = 10h^{-1}$ Mpc, or $d = 14.1$ Mpc.

To estimate how long ago the Big Bang occurred, let $t_H$ be the time that has elapsed since the Big Bang. This is the time required for a galaxy to travel to its present distance $d$ while moving at its recessional velocity, $v$, as given by Hubble's law. Assuming (incorrectly) that $v$ has remained constant,

$$d = v\, t_H = H_0 d\, t_H, \tag{27.13}$$

and so the **Hubble time** is

$$t_H \equiv \frac{1}{H_0} = 3.09 \times 10^{17} h^{-1} \text{ s} = 9.78 \times 10^9 h^{-1} \text{ yr.} \tag{27.14}$$

Using WMAP values,

$$t_H = 4.35 \times 10^{17} \text{ s} = 1.38 \times 10^{10} \text{ yr.} \tag{27.15}$$

Thus, as a rough estimate, the age of the universe is about 13.8 Gyr.


## 27.3 ■ CLUSTERS OF GALAXIES

Astronomers hold as a fundamental tenet that, on the largest scales, the universe is both homogeneous and isotropic, appearing the same at all locations and in all directions (an assumption called the **cosmological principle**). However, this is certainly not the case at smaller scales.

### The Classification of Clusters

As has been mentioned a number of times already in this text, galaxies are not strewn randomly throughout the universe; instead, nearly all galaxies are found in associations, either in **groups** or in **clusters**.[19] In both types, the galaxies are gravitationally bound to one another and orbit the system's barycenter (center of mass).

Groups generally have less than 50 members and are about $1.4h^{-1}$ Mpc across. The galaxies of a group have a velocity dispersion of about 150 km s$^{-1}$, and the mass of an average group is on the order of $2 \times 10^{13} h^{-1}$ M$_\odot$, obtained from the virial theorem (Eq. 25.13). Furthermore, the mass-to-light ratio of a typical group is about $260h$ M$_\odot$/L$_\odot$, which is indicative of large amounts of dark matter.

Clusters, on the other hand, may contain from approximately 50 galaxies (a **poor** cluster) to thousands of galaxies (a **rich** cluster), within a region of space about $6h^{-1}$ Mpc in diameter. The individual galaxies in a cluster move more rapidly with respect to other members than do galaxies residing in a group. The characteristic velocity dispersion of a cluster is 800 km s$^{-1}$, possibly exceeding 1000 km s$^{-1}$ for very rich clusters. A typical cluster's virial mass is around $1 \times 10^{15} h^{-1}$ M$_\odot$, and its mass-to-light ratio is roughly $400h$ M$_\odot$/L$_\odot$, again

---

[19]Some galaxies are not affiliated with a known group or cluster, but they represent rare exceptions.

indicative of large amounts of dark matter. Clusters of galaxies are further classified as **regular** (spherical and centrally condensed) and **irregular**.

In this section, we will describe how groups of galaxies, clusters of galaxies, and clusters of clusters (called **superclusters**) make up the basic structure of the universe.

### The Local Group

As shown in Fig. 27.11, about 35 galaxies are known to lie within roughly 1 Mpc of the Milky Way. This collection of galaxies is known as the **Local Group**. Its **zero velocity surface** (where the galaxies would eventually turn around if their motions were suddenly directed outward) is about 1.2 Mpc from the barycenter. Its most prominent members are its three spiral galaxies: our Galaxy, M31 (the Andromeda galaxy), and M33 (in the constellation Triangulum). The Large and Small Magellanic Clouds are the next most luminous; they are two of the 13 irregular galaxies in the Local Group. [The LMC and SMC are the two prominent features below the Galactic equator in Fig. 6.28(c).] The remaining galaxies are dwarf ellipticals or dwarf spheroidals, which are very small and quite faint. It is obvious from Fig. 27.11 that many of these galaxies have accumulated around the Milky Way and Andromeda galaxies, which are on opposite sides of the Local Group, about $r = 770$ kpc apart.[20] (Even at this distance, M31 is about 2.5° wide on the sky, five times the Moon's diameter.) In addition to the Magellanic Clouds, nine of the dwarf ellipticals and dwarf

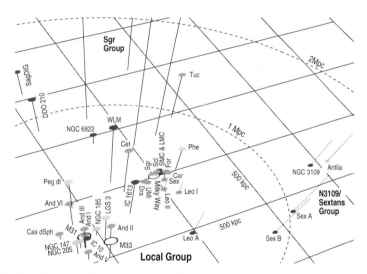

**FIGURE 27.11** The Local Group of galaxies. The dashed circles are centered on the barycenter of the Local Group and have radii of 1 and 2 Mpc. (The barycenter is 462 kpc from us in the direction of galactic coordinate $\ell = 121.7°$ and $b = -23.1°$.) Solid lines indicate distances above the (arbitrarily chosen) plane, and dashed lines show distances below the plane. (Figure from Grebel, "The Local Group," in *Microlensing 2000: A New Era of Microlensing Astrophysics, A. S. P. Conference Series, 239,* 280, 2001.)

---

[20]This accumulation can also be seen in Hubble's own velocity–distance diagram, Fig. 27.7; the Local Group is represented by the eight data points near the origin.

spheroidals are situated near the Milky Way. Interestingly, several of them are found in the Magellanic Stream, the long ribbon of neutral hydrogen gas that was tidally stripped from the Magellanic Clouds some 200 million years ago. It should also be noted that all three spiral galaxies have warped disks that resemble an integral sign ($\int$) when viewed edge on (recall the map of the H I warp in our Galaxy shown in Fig. 24.10). In fact, the line of sight from Earth passes twice through some parts of the disk of M33.

The Andromeda and Milky Way galaxies are approaching each other with a relative velocity of $v = 119$ km s$^{-1}$.[21] Apparently, the gravitational attraction between them has overcome their tendency to expand along with the Hubble flow. As a result, they will collide in approximately $t_c = r/v = 6.3$ billion years.[22] Indeed, astronomers have found that the entire Local Group is still in a state of collapse.

Because the Milky Way and Andromeda dominate the Local Group, producing some 90% of its luminosity, their motion toward each other provides an opportunity to estimate their combined mass. We can assume that following the Big Bang, the two newly formed galaxies were initially moving apart.[23] At some time in the past, their gravitational attraction halted and reversed their original recessional motion. The Milky Way and M31 are therefore in orbit about one another with an eccentricity $e \simeq 1$ (i.e., on a collision course). From conservation of energy, the orbital speed and separation are related by

$$v^2 = GM \left( \frac{2}{r} - \frac{1}{a} \right),$$

(Eq. 2.36), where $M$ is the total mass of the two galaxies. The orbit's semimajor axis, $a$, is related to its period, $P$, by Kepler's third law, Eq. (2.37):

$$P^2 = \frac{4\pi^2}{GM} a^3.$$

Combining these relations to eliminate $a$ leads to

$$v^2 - \frac{2GM}{r} + \left( \frac{2\pi GM}{P} \right)^{2/3} = 0. \qquad (27.16)$$

In this equation,[24] $r = 770$ kpc and $v = 119$ km s$^{-1}$. To obtain the total mass $M$, we need an estimate of the orbital period, $P$. When the galaxies do meet in the future, they will have returned to their initial configuration, and one orbital period will have elapsed since the Big Bang. Let us therefore use

$$P = t_H + t_c,$$

[21]M31 is approaching the Sun with a speed of 297 km s$^{-1}$, but a large fraction of this is due to the Sun's orbital motion around the Milky Way.

[22]This suggests the possibility that they might then merge to form a giant elliptical galaxy.

[23]Remember, in the absence of peculiar velocities, an *expansion* means that every point in space moves directly away from every other point.

[24]Equation (27.16) is a cubic equation for $M^{1/3}$ that has one real root and two complex roots. The real root is the one with physical significance.

where $t_c = r/v$ is roughly the time until they collide. This is an overestimate of the period, because the Milky Way and M31 will accelerate as they approach their rendezvous. For the WMAP value $h = 0.71$ (so $H_0 = 71$ km s$^{-1}$ Mpc$^{-1}$), the total mass is found to be $M = 7.9 \times 10^{42}$ kg $= 4.0 \times 10^{12}$ M$_\odot$.

This estimate of the mass of the Local Group is much more than the luminous mass observed in these galaxies. The $B$-band luminosity of the Milky Way is about $2.3 \times 10^{10}$ L$_\odot$ (see Table 24.1), and M31 is roughly twice as bright. Thus, the mass-to-light ratio is estimated to be $M/L = 57$ M$_\odot$/L$_\odot$. Note that a smaller period gives a larger mass. This means that, because the value of $P$ was overestimated, $M/L$ has been *underestimated*. These numbers are much larger than the value of $M/L \simeq 3$ M$_\odot$/L$_\odot$ for the luminous matter in the Milky Way's thin disk and central bulge (Section 24.2). Our WMAP result of $M/L \simeq 57$ M$_\odot$/L$_\odot$ is consistent with the estimates quoted in Section 24.2 when the Milky Way's dark halo is included. Apparently, astronomers have seen less than approximately 10% of the matter that makes up the Milky Way and Andromeda galaxies.

A careful study of the Magellanic Stream supports this conclusion. The tip of this flow of neutral hydrogen gas is plunging toward the Galaxy with a speed of 220 km s$^{-1}$. Researchers attempting to reproduce the dynamics of the stream obtain the best results with a model that has a halo of dark matter extending at least 100 kpc from the Galaxy. The Galactic mass-to-light ratio for these models exceeds 80 M$_\odot$/L$_\odot$.

### Other Groups within 10 Mpc of the Local Group

The Local Group is indeed a spatial concentration of galaxies. Although there are some 35 galaxies in the Local Group within a volume of space about a megaparsec across, the next nearest galaxies are found in the Sculptor group (six members, 1.8 Mpc away) and in the M81 group (eight members, 3.1 Mpc away). The six spiral galaxies of the Sculptor group span about 20° on the night sky. You could just cover them with your outstretched hand held at arm's length.[25] Other associations within 10 Mpc include the Centaurus group (17 members, 3.5 Mpc away) and the M101 group (5 members, 7.7 Mpc away). The M66 and M96 groups are separated by only 7° in the sky. Together they contain about 10 galaxies, and both are about 9.4 Mpc from Earth. And finally, the last group within 10 Mpc of Earth is the NGC 1023 group (6 members, 9.5 Mpc away). Altogether, there are about twenty small groups of galaxies that are closer to us than the Virgo cluster. Most galaxies are found in such small groups and poor clusters. It is estimated that, at most, 20% of all galaxies inhabit rich clusters like the Virgo cluster.

Figure 27.12 shows the positions of the nearest galaxies. Centered on the Milky Way, the diagram reaches out to just past the M101 group. There are many more dwarfs than there are giant galaxies, and both types are found generally in the same regions. There are also some extended volumes of space, termed **voids**, where galaxies are apparently absent. Another striking feature is that the galaxies show a tendency to lie along a two-dimensional plane ($SGZ = 0$) that forms part of the Local Supercluster of galaxies centered on the Virgo cluster (to be discussed later).[26]

---

[25]It is fascinating to realize that many astronomical objects occupy a significant area of the sky. If only our eyes had the sensitivity to see them!

[26]"SG" denotes a supergalactic coordinate system created by de Vaucouleurs.

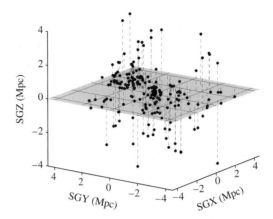

**FIGURE 27.12**    Galaxies near the Milky Way. Most of the positions (except for M31 and galaxies in its subgroup) were found using redshifts and the Hubble law. The *SGY*-axis points toward the Virgo cluster.

### The Virgo Cluster: A Rich, Irregular Cluster

The **Virgo cluster** of galaxies was first recognized in the eighteenth century by William Herschel. Located where the constellations of Virgo and Coma Berenices (Berenice's Hair) meet, the Virgo cluster covers a $10° \times 10°$ region of the sky.[27] The center of the cluster is located about 16 Mpc from Earth.[28] This rich, irregular cluster is a sprawling collection of approximately 250 large galaxies and more than 2000 smaller ones, contained within a region about 3 Mpc across. Interestingly, at least seven of these galaxies display blueshifted spectral lines, as their peculiar velocities of approach overpower the receding Hubble flow.

Like most irregular clusters, the Virgo cluster is made up of all types of galaxies. Although the four brightest are giant elliptical galaxies, ellipticals make up only 19% of the cluster's 205 brightest galaxies. Spirals dominate overall, accounting for 68% of the top 205. However, roughly equal numbers of spirals and dwarf ellipticals are found within 6° of the cluster's center, with the ellipticals becoming increasingly common near the center. In fact, the center of the Virgo cluster is dominated by three of the cluster's four giant elliptical galaxies (M84, M86, and M87; see Figs. 27.13 and 28.10). The diameters of these galaxies are comparable to the distance between our Galaxy and Andromeda, so each of these giant ellipticals is nearly the size of the entire Local Group.

M87, a giant E1 elliptical, is the largest and brightest galaxy in the Virgo cluster. Like many other luminous elliptical galaxies, it contains roughly $10^{10}$ M$_\odot$ of hot ($\sim 10^7$ K) gas that has accumulated through normal stellar mass loss mechanisms. The gas loses energy by the free–free emission of X-ray photons. This thermal bremsstrahlung process, first discussed in Section 9.2, produces a characteristic spectrum that can be readily identified

---

[27]Berenice, wife of the Egyptian king Ptolemy III, offered her golden tresses as a sacrifice to the goddess Aphrodite in gratitude for her husband's victory over the Assyrians.

[28]This value of the distance to the Virgo cluster is in the best agreement with the values in Table 27.1; see Problem 27.19.

**FIGURE 27.13** The center of the Virgo cluster, showing the giant ellipticals M84 (right) and M86 (center). (Courtesy of National Optical Astronomy Observatories.)

and used to estimate the mass of the galaxy [see Eq. (27.18) and Fig. 27.17]. Although the analysis is complicated, observations of the spectrum of the gas can provide an idea of how its temperature, $T(r)$, and mass density, $\rho(r)$, vary with $r$, the distance from the center of the galaxy. Fortunately, the nearly spherical shape of M87 simplifies the geometry. The gas is in hydrostatic equilibrium (to a good approximation), so from Eq. (10.6),

$$\frac{dP}{dr} = -G\frac{M_r\rho}{r^2}.$$

Substituting the ideal gas law, Eq. (10.11), for the pressure, $P$, assuming constant $\mu$, and solving for the interior mass, $M_r$, produce

$$M_r = -\frac{kTr}{\mu m_H G}\left(\frac{\partial \ln \rho}{\partial \ln r} + \frac{\partial \ln T}{\partial \ln r}\right). \qquad (27.17)$$

All of the quantities on the right-hand side can be roughly evaluated from observations of the X-rays emitted by the hot gas, and used to calculate the interior mass on the left-hand side. Note that $M_r$ is the *total* interior mass, both luminous and dark. (The derivatives are themselves negative, so $M_r$ is positive.)

One result for M87 shows that $M_r$ increases *linearly* with radius out to about 300 kpc. This is the same signature seen for the distribution of dark matter in spiral galaxies. The total mass contained within $r = 300$ kpc is $M_r \simeq 3 \times 10^{13}$ M$_\odot$, with a central density of $1.5 \times 10^{-2}$ M$_\odot$ pc$^{-3}$. The corresponding mass-to-light ratio is $M/L \simeq 750$ M$_\odot$/L$_\odot$. This is 250 times the mass-to-light ratio for the stars in the Milky Way's Galactic bulge and thin disk (Section 24.2) and indicates that over 99% of M87's mass is dark matter. As we will see in Section 28.1, M87 is *not* a typical elliptical galaxy. However, studies of other elliptical galaxies have found that many of them also contain large amounts (90% or more) of dark matter.

### The Coma Cluster: A Rich, Regular Cluster

The nearest rich, regular cluster of galaxies is the **Coma cluster**. It is about 15° north (in declination) of the Virgo cluster, in the constellation Coma Berenices. Recall from Example 27.1.4 that the Coma cluster is about 5.4 times farther away than is the Virgo cluster, or about 90 Mpc from Earth. The cluster's angular diameter is about 4°, which corresponds to a linear diameter of 6 Mpc. The Coma cluster consists of perhaps 10,000 galaxies, most of them dwarf ellipticals that are too faint to be seen. Generally, in a rich, regular cluster the vast majority of the galaxies are ellipticals and S0's. This is the case for the Coma cluster. It contains over 1000 bright galaxies, but only 15% of them are spirals and irregulars. At the cluster's center are two large, luminous cD ellipticals; see Fig. 26.1.

### Evidence for the Evolution of Galaxies

Recall that the predominance of early-type galaxies in a cluster may be due to the increased likelihood of interactions (the morphology–density relation). Perhaps in the past, more spirals existed in the Coma cluster and other rich, regular clusters, but tidal interactions and mergers destroyed the spiral morphology, leaving ellipticals and S0's.[29] Alternatively, early-type galaxies may simply form as a result of hierarchical mergers of protogalaxies near the bottom of the cluster's gravitational well, in regions where dark matter would naturally tend to collect.

In support of the interaction view, Fig. 27.14 shows the center of the very rich cluster CL 0939+4713, with a redshift of $z = 0.41$, implying a distance of $1230h^{-1} = 1730$ Mpc. We are seeing this cluster as it was about $4.0h^{-1} = 5.6$ billion years ago, when many late-type spirals were found in this central region. This is unlike the center of any high-density cluster belonging to the present epoch. Most of the spirals in Fig. 27.14 show at least some evidence of an interaction. The frequency of interactions and mergers is perhaps an order of magnitude greater than would be found in a contemporary rich cluster.

Looking even more deeply into the universe, astronomers have discovered a remote cluster in the constellation Serpens at $z \approx 1.2$, or $1970h^{-1} = 2770$ Mpc. This cluster, shown in Fig. 27.15, is seen as it was about $6.4h^{-1} = 9.0$ billion years ago. It contains apparently mature ellipticals but very few normal spirals. Instead, there is an abundance of odd galactic fragments having a bluish color (a signature of active star formation; recall the Butcher–Oemler effect). It is not clear whether these are the debris of former spirals, or spirals still in the process of coalescing. Further observations should reveal more details about the formation and disruption of spiral galaxies in the dynamic environment of such crowded young clusters.

### A Preponderance of Matter between the Galaxies

In 1933 Fritz Zwicky measured the radial velocities of galaxies in the Coma cluster from their Doppler-shifted spectra. From those observations he calculated the dispersion in their

---

[29]The greater frequency of spirals in young clusters was discussed at the end of Section 26.2.

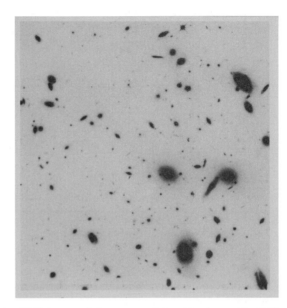

**FIGURE 27.14** An HST image of the center of the rich cluster CL 0939+4713. (Figure from Dressler et al., *Ap. J. Lett.*, *435*, L23, 1994.)

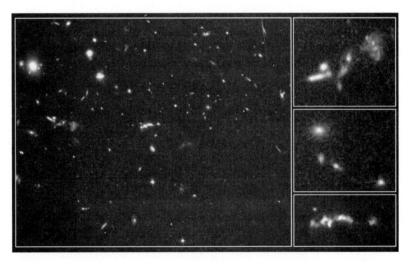

**FIGURE 27.15** An HST view of a young cluster of galaxies, centered on the peculiar radio galaxy 3C 324 (also seen at bottom right) at $z = 1.2$. At center right is a pair of normal-appearing elliptical galaxies with a few faint companions, while at top right are some galactic fragments that may become, or may have once been, spirals. [Courtesy of Mark Dickinson (STScI) and NASA.]

radial velocities, now known to be $\sigma_r = 977$ km s$^{-1}$. Zwicky then used the virial theorem to estimate the mass of the Coma cluster. Overall, the Coma cluster's intensity profile, $I(r)$, follows a characteristic $r^{1/4}$ law (Eq. 24.13) like those that describe the intensity profiles of the Milky Way's bulge and halo, and elliptical galaxies. Presumably, like the stars that make up elliptical galaxies and the spheroidal components of spirals, the galaxies in the Coma cluster have become dynamically relaxed and are in an equilibrium configuration. This makes the virial theorem an appropriate method to use.

---

**Example 27.3.1.**   In the case of the Coma cluster, the dispersion in the radial velocity is 977 km s$^{-1}$. With the cluster radius $R = 3$ Mpc, Eq. (25.13) leads to a mass of

$$M \approx \frac{5\sigma_r^2 R}{G} = 3.3 \times 10^{15} \, M_\odot$$

for the Coma cluster. Since the visual luminosity of the Coma cluster is about $5 \times 10^{12} \, L_\odot$, the mass-to-light ratio of the cluster is $M/L \approx 660 \, M_\odot/L_\odot$.[30]

---

Zwicky understood the significance of this result and stated that in the Coma cluster, "the total mass ... considerably exceeds the sum of the masses of individual galaxies." He realized that there is not enough visible mass to bind the cluster together. If not for the presence of a large amount of unseen matter, the galaxies in the Coma cluster would have dispersed long ago. Forty years later, when the flat rotation curve of the Andromeda galaxy was measured, other astronomers also realized the significance of Zwicky's result.

### The Hot, Intracluster Gas

A portion of Zwicky's "missing mass" was discovered with the High Energy Astronomical Observatory (HEAO) series of satellites that were first launched in 1977. They revealed that many clusters of galaxies emit X-rays from much of the cluster's volume. These satellites, together with optical observations, indicated that clusters of galaxies contain an **intracluster medium**. The intracluster medium has two components. One is a diffuse, irregular distribution of stars. The other component is a hot **intracluster gas** that is distributed more or less homogeneously, occupying the space between the galaxies and filling the cluster's gravitational potential well. The X-ray luminosities lie in the range $10^{36}$ to $10^{38}$ W, with richer clusters shining more brightly in X-rays. The core of the Virgo cluster contains about $5 \times 10^{13} \, M_\odot$ of X-ray emitting gas, and the core of the Coma cluster holds $3 \times 10^{13} \, M_\odot$. Figure 27.16 shows the X-ray emission from the hot gas that fills the Coma cluster. Typically the mass of the gas is several times greater than the combined mass of all stars in the cluster's galaxies.

Thermal bremsstrahlung, the same mechanism that produces the X-rays coming from the hot gas within individual galaxies (described earlier for M87 in the Virgo cluster), is at work here as well. For fully ionized hydrogen gas, the energy emitted per unit volume per unit time between frequencies $\nu$ and $\nu + d\nu$ is given by

$$\ell_\nu \, d\nu = 5.44 \times 10^{-52} \left(4\pi n_e^2\right) T^{-1/2} e^{-h\nu/kT} \, d\nu \text{ W m}^{-3}, \tag{27.18}$$

---

[30]A more detailed calculation results in a mass-to-light ratio about half as large as our estimate.

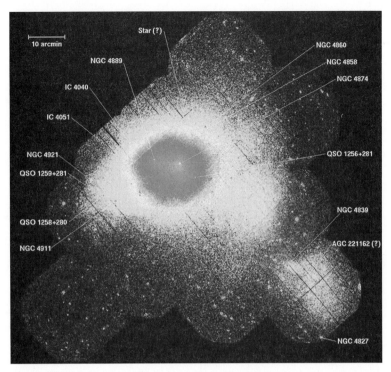

**FIGURE 27.16** An X-ray Multi-Mirror (XMM) telescope image of the Coma cluster taken by the European Space Agency's XMM-Newton space observatory. The mosaic image shows X-rays in the 0.3–2.0-keV range from the Coma cluster's intracluster gas and occupies a field of view of $1.5° \times 1.33°$. The intensity of the X-rays increases continuously toward the center of the cluster. The part of the Coma cluster shown in Fig. 26.1 occupies the most intense central area of this image. The irregular shape of the emission indicates that the Coma cluster may have formed from the merger of several subclusters. (Image courtesy of U. Briel, Max Planck Institut für extraterrestrische Physik, Garching, Germany, and the European Space Agency.)

where $T$ is the gas temperature and $n_e$ is the number density of free electrons. One version of this spectrum for the Coma cluster is shown in Fig. 27.17. The total amount of energy emitted per second per unit volume at all frequencies (the *luminosity density*, $\mathcal{L}_{\text{vol}}$) is obtained by integrating $\ell_\nu$ over frequency. This results in

$$\mathcal{L}_{\text{vol}} = 1.42 \times 10^{-40} n_e^2 T^{1/2} \text{ W m}^{-3}. \tag{27.19}$$

**Example 27.3.2.** Equation (27.19) can be used to estimate the mass of the Coma cluster's intracluster gas. For simplicity, we will assume that the cluster can be modeled as an isothermal sphere of hot ionized hydrogen gas. Figure 27.16 indicates that the central core radiates most strongly, so we will take the radius to be $R = 1.5$ Mpc (one-half of the cluster's actual radius). The gas is optically thin, meaning that any photon emitted along the line of sight can be observed. The gas temperature comes from the X-ray spectrum of the gas,

*continued*

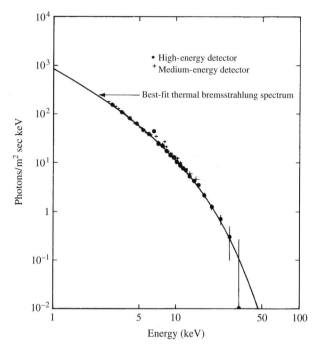

**FIGURE 27.17**    Thermal bremsstrahlung spectrum (line) for 88 million K. The points are observations of X-rays from the Coma cluster's intracluster gas. Photon energy is plotted on the horizontal axis. (Figure adapted from Henriksen and Mushotzky, *Ap. J.*, *302*, 287, 1986.)

shown in Fig. 27.17. Using Eq. (27.18), the best fit to the data (the dots on the figure) is obtained with a temperature of $8.8 \times 10^7$ K.

Using Eq. (27.19), $L_x$ can be written as

$$L_x = \frac{4}{3}\pi R^3 \mathcal{L}_{\text{vol}}. \tag{27.20}$$

Since the X-ray luminosity of the gas is $L_x = 5 \times 10^{37}$ W, the value of $n_e$, the number of free electrons per $m^{-3}$, is

$$n_e = \left[ \frac{3L_x}{4\pi R^3 T^{1/2}(1.42 \times 10^{-40}\ \text{W m}^{-3})} \right]^{1/2} = 300\ \text{m}^{-3}.$$

The intracluster gas is several million times less dense than the giant molecular clouds described in Section 12.1, for which $n_{H_2} \sim 10^8$ to $10^9\ \text{m}^{-3}$.

For ionized hydrogen, there is one proton for every electron, so the total mass of the gas is

$$M_{\text{gas}} = \frac{4}{3}\pi R^3 n_e m_H = 1.05 \times 10^{14}\ M_\odot,$$

a slight overestimate of the value of $3 \times 10^{13}\ M_\odot$ quoted previously.

Let's compare the foregoing mass of the intracluster gas with the luminous mass of the Coma cluster. Using the mass-to-light ratio for the stars in the Milky Way's Galactic bulge and thin disk, $M/L \simeq 3 \ M_\odot/L_\odot$, and the visual luminosity of the Coma cluster, $L_V = 5 \times 10^{12} \ L_\odot$, the visible mass of the Coma cluster is approximately $1.5 \times 10^{13} \ M_\odot$. According to this estimate, there is roughly 7 times more intracluster gas in the Coma cluster than there is in its galaxies' stars. However, the $10^{14} \ M_\odot$ of intracluster gas is still only a few percent of the total mass of the Coma cluster.

---

The X-ray spectrum of the intracluster gas also displays emission lines of highly ionized iron (e.g., Fe XXV and Fe XXVI), silicon, and neon. This indicates that the gas has been processed through the stars in the cluster's galaxies, becoming enriched in heavy elements via stellar nucleosynthesis. How, then, did so much mass escape from the cluster's galaxies?

Mergers must have occurred more frequently in the early history of the cluster. When the cluster was still forming during its first few billion years, it was more dispersed. According to the virial theorem, expressed in the form of Eq. (25.13), the cluster's galaxies would have been moving more slowly, thereby increasing the dynamical friction and the chance of a merger. The large amounts of intracluster gas found in many rich clusters were probably ejected during these early galactic interactions, or by bursts of star formation. Once this process was under way, it would have been enhanced by ram-pressure stripping. When a galaxy moves at several thousand kilometers per second through the intracluster gas, it encounters a furious wind that is capable of stripping its gas away.

### The Existence of Superclusters

Next in the hierarchy of galactic clustering is the **supercluster**. As the name suggests, this is a clustering of clusters on a grand scale (up to about 100 Mpc). It may be that virtually every galaxy belongs to a supercluster; certainly every known rich cluster is found in one.

The Virgo cluster is near the center of the **Local Supercluster**. This disk of clusters has the shape of a flattened ellipsoid, a pancake that contains most of the galaxies lying within about 20 Mpc of the Virgo cluster (including the Local Group, which is located near the edge of the supercluster). Figure 27.18 shows the Local Supercluster as the linear grouping of galaxies extending from the center of the figure. The view that is depicted is parallel to the plane of the supercluster, so you are looking at the pancake edge-on. Compare this view with Fig. 27.12 and note that the concentration of galaxies along a plane continues in Fig. 27.18. Of the other known superclusters, two of the most prominent are the Perseus–Pisces supercluster, named for the northern constellations in which it is found, and the Hydra–Centaurus supercluster in the southern sky. The Perseus–Pisces supercluster is about $50h^{-1}$ Mpc away and has a threadlike appearance, a planar, filamentary structure that is about $40h^{-1}$ Mpc long; see Fig. 27.25. The Hydra–Centaurus supercluster is at $30h^{-1}$ Mpc, in roughly the opposite direction.

Because the Local Group resides near the edge of the Local Supercluster, it might be expected that the supercluster's gravitational pull would be detectable. In 1958 de Vaucouleurs found that the Local Group is moving away less rapidly from the Virgo cluster than would be expected solely from the expansion of the universe. At a distance of 16 Mpc, a pure Hubble flow would produce a recessional velocity of $v = H_0 d = 1600h$ km s$^{-1}$. The difference between the actual velocity of the Local Group relative to the Virgo cluster and

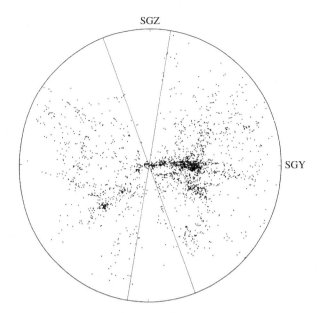

**FIGURE 27.18**    The distribution of 2175 bright galaxies out to roughly 50 Mpc. The Local Super-cluster extends to the right, with the Milky Way (at center) located near the edge of the supercluster. The plane of the Milky Way bisects the "empty" slices; galaxies within these slices are hidden from view by Galactic dust and gas (the zone of avoidance). (Figure from Tully, *Ap. J.*, *257*, 389, 1982.)

the Hubble flow, called the **Virgocentric peculiar velocity** of the Local Group, is estimated to be $168 \pm 50$ km s$^{-1}$ by Sandage and Tammann (1990). An argument similar in spirit to that used for the approaching paths of the Milky Way and M31 can now be invoked to esti-mate the mass and the mass-to-light ratio of the Local Supercluster. The results show that the supercluster's mass is about $8 \times 10^{14} h^{-1}$ M$_\odot$. The corresponding mass-to-light ratio is $M/L \simeq 400h$ M$_\odot$/L$_\odot$, providing more evidence for a preponderance of dark matter.

## Large-Scale Motions Relative to the Hubble Flow

The Virgocentric peculiar velocity is a minor perturbation in a much larger scale inhomo-geneity in the Hubble flow. As we will see in Section 29.2, there is a large-scale streaming motion (relative to the Hubble flow) that carries the Milky Way, the Local Group, the Virgo cluster, and thousands of other galaxies through space in the direction of the constella-tion Centaurus. The peculiar velocity of the Local Group relative to the Hubble flow is 627 km s$^{-1}$. This riverlike motion of thousands of galaxies extends at least $40h^{-1}$ Mpc both upstream and downstream. Astronomers would like to use this streaming to deduce the lo-cation(s) of the mass, visible or dark, capable of exerting such an immense gravitational tug. The Hydra–Centaurus supercluster, which is in the direction of the flow, is *not* responsible, since it too is moving along with the flow of the rest of the galaxies. This implies that the source of the motion lies beyond that supercluster.

In the 1980s, American astronomers Alan Dressler and Sandra Faber calculated the presence of a **Great Attractor** (GA), a diffuse collection of clusters spread over a wide

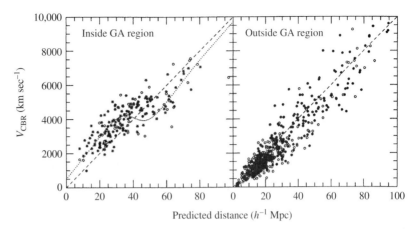

**FIGURE 27.19** Left: velocities of galaxies in the Centaurus region, compared with the Hubble flow (dashed line). The dotted line shows the theoretical variation in velocity produced by a model of the Great Attractor. The Hydra–Centaurus supercluster is centered at about $30h^{-1}$ Mpc. Right: comparison figure for galaxies observed in another direction. (Figure adapted from Dressler and Faber, *Ap. J. Lett.*, *354*, L45, 1990.)

(60°) region of the sky. According to their calculations, the GA lies in the same plane as the Local Supercluster, about $42h^{-1}$ Mpc away in the direction of $\ell = 309°$, $b = 18°$ (Galactic coordinates) in the constellation Centaurus. The mass of the Great Attractor is estimated to be about $2 \times 10^{16}h^{-1}$ $M_{\odot}$, but there are just 7,500 galaxies known in that region—too few to account for so much mass. This suggests that approximately 90% of the mass of the GA may be in the form of dark matter. Alternatively, because the Great Attractor lies behind the plane of our galaxy and is obscured by dust, there could be a large hidden supercluster of galaxies in that direction. If so, it would be centered on a cluster of galaxies called Abell 3627.

Figure 27.19 shows the velocities of E and S0 galaxies in the region of the Great Attractor; the figure's dotted line is the galactic velocity field calculated for a specific model of the GA. On the near side of the GA, the velocities clearly exceed that of the unperturbed Hubble flow (the diagonal dashed line in the figure). There are hints of a "back flow" of the farside of the GA (velocities less than the Hubble flow), but this has been disputed by other investigators.[31] These astronomers believe that the velocity excess actually continues *beyond* the proposed location of the GA and suggest other (or additional) concentrations of matter. Whether or not the Great Attractor exists as described, the existence of large-scale streaming motions relative to the Hubble flow seemed undeniable.

Another possibility is that the Great Attractor is not solely responsible for the large-scale streaming motion. The **Shapley concentration** of galaxies is probably the most massive collection of galaxies in our neighborhood of the universe; see Fig. 27.20. With a core consisting of a gravitationally bound concentration of some 20 rich clusters of galaxies, the Shapley concentration has a mass of a few $\times 10^{16}h^{-1}$ $M_{\odot}$ and lies within 10° of the

---

[31] See, for example, a similar diagram in Mathewson, Ford, and Buchhorn (1992).

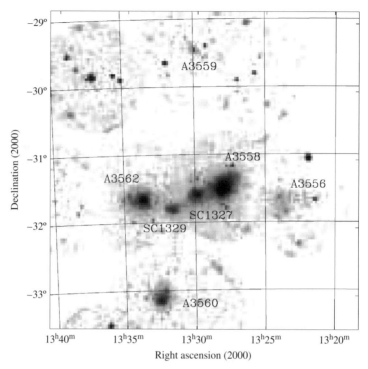

**FIGURE 27.20**   A large-scale view of the center of the Shapley concentration, centered on the galaxy A3558 and its companions. The entire collection of galaxies is about 10° across in the sky. (Figure from Bardelli et al., *Astron. Astrophys.*, *396*, 2002.)

direction of the Local Group's motion and very close to the direction of the Great Attractor. However, its distance of $140h^{-1}$ Mpc is three times the distance calculated for the Great Attractor, so it cannot be identified with the GA. Still, it is probably responsible for 10–15% of the net acceleration of the Local Group.

Astronomers believe that the large-scale bulk motions cannot continue on increasingly larger scales because that would violate the cosmological principle that the universe is isotropic. The direction such a large-scale bulk motion would be a preferred direction in the universe. Recent observations based on Type Ia supernova distances (which have a smaller uncertainly than the Tulley–Fisher distances often used in these studies) indicate that these vast bulk flows eventually converge with the Hubble flow (with no peculiar velocity) at a distance of about $50$–$60h^{-1}$ Mpc. In fact, some observations using Type Ia supernovae fail to find any evidence at all of large-scale coherent motion! We will have to wait for more complete and precise observations before the issue can be settled.

### Bubbles and Voids: Structure on the Largest Scales

To find how the universe looks when examined at even larger scales, extensive surveys of the sky have been conducted. During the 1950s, the Palomar Observatory Sky Survey recorded the northern portion of the heavens on glass photographic plates, each covering a 6° × 6°

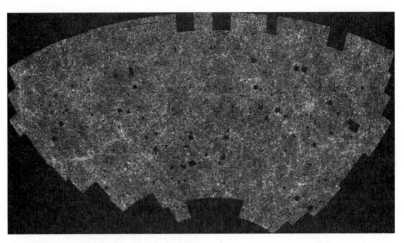

**FIGURE 27.21** Two million galaxies, compiled by the United Kingdom Schmidt Telescope sky survey over a region of 4300 square degrees, centered on the south Galactic pole. (Figure from Maddox et al., *MNRAS*, *242*, 43p, 1990.)

area. In this study, two plates were made for each area of sky, recorded on emulsions that were more sensitive to either red or blue light, thus giving an indication of the temperatures of the objects photographed in addition to their relative brightnesses. Figure 27.21 shows the result of a more recent survey of the southern skies, made with the United Kingdom Schmidt Telescope. This map displays some two million galaxies between apparent magnitudes 17 and 20.5 that are found within 4300 square degrees of the sky, centered on the south Galactic pole. It is not a photograph; rather, each dot has an intensity (black to white) representing between zero (black) and 20 (white) galaxies. It is obvious that the galaxies are not randomly distributed across the sky. The small, bright regions are clusters, and the longer, filamentary strands are superclusters.

The human eye and brain have a talent for finding patterns, even when none exist. In the case of Fig. 27.21, the difficulty is compounded because it is a two-dimensional depiction of the sky. All of the galaxies are projected onto the plane of the sky, so true associations cannot be distinguished from line-of-sight coincidences. Three-dimensional structures may be hidden or distorted when flattened into two dimensions. In the 1980s, several groups of astronomers began making three-dimensional maps of the universe by using the redshifts of galaxies and the Hubble law to provide the distance of the galaxies. **Redshift surveys** such as these revealed the existence of huge voids, immense versions of the vacant regions seen in Fig. 27.12. These voids measure up to 100 Mpc across and are vaguely spherical, rather than flattened like the sheets occupied by superclusters. The space inside a void is empty of bright spirals and ellipticals, although a few faint dwarf ellipticals may be present.

Most of the early surveys sampled just one area of the sky, so it was not possible to get an overall picture of the distribution of galaxies. To remedy this, Margaret Geller and John Huchra of the Harvard–Smithsonian Center for Astrophysics (CfA), along with their co-workers, carried out surveys of wedge-shaped slices of the heavens; see Fig. 27.22. Like making parallel cuts through a block of Swiss cheese, this technique shows where the empty

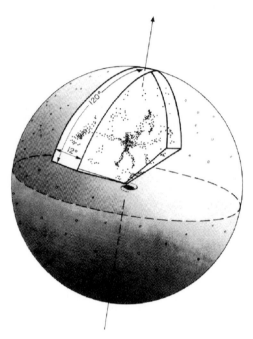

**FIGURE 27.22**    Two adjacent 6°-wide wedges used in the CfA redshift survey. (Figure from Geller, *Mercury*, May/June 1990.)

regions are and how they relate to one another. During a night's observation, the rotation of Earth carries the telescope through about 9 hours of right ascension, and redshifts are measured for the galaxies that cross the telescope's field of view. This procedure is repeated until all of the galaxies in a 6° declination slice have been measured, out to a radial velocity of some 15,000 km s$^{-1}$ (a distance of about $150h^{-1}$ Mpc, according to Hubble's law).[32]

Figures 27.23(a) and (b) show two of the first slices, completed in 1985. Because of the uncertainty in the value of $H_0$ at the time, the distance is expressed in terms of a radial velocity $cz$, where $z$ is the measured redshift (cf., Eq. 27.8). Each dot on the plots is a galaxy. Study the figures and try to imagine a three-dimensional distribution of galaxies that would have these cross sections. Clearly, the galaxies are not randomly distributed. The slices also show that galaxies do not form a network of threadlike filaments that meander through space. If they did, then a slice would be much more likely to cut across a filament than along it (think of cutting through a bowl of cooked spaghetti). There are simply too many connected features, and too few isolated clumps, for this to be true.

Another terrestrial analogy does supply a structure that matches these observations. A slice through a froth of soap bubbles would look very much like what is seen here. Figures 27.23(a) and (b) show that galaxies lie on two-dimensional sheets that form the walls of bubble-shaped regions of space and that inside the bubbles are huge voids.[33] One

---

[32]For distances over roughly $100h^{-1}$ Mpc, only the brighter galaxies could be measured, producing a sparser diagram near the curved edge.

[33]Another possibility is that the voids are connected like the interior of a sponge, with thin sheets of galaxies separating the passageways.

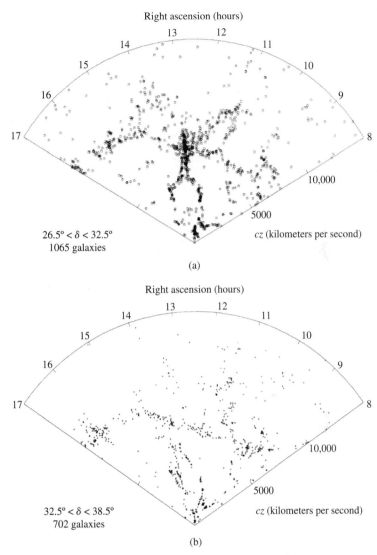

FIGURE 27.23   Two slices from the CfA redshift survey. Right ascension is indicated along the top of each figure. (a) $26.5° < \delta < 32.5°$. (b) $32.5° < \delta < 38.5°$. (Figures adapted from Geller, *Mercury*, May/June 1990.)

such large void is at $\alpha = 15^h$, $v = 7500$ km s$^{-1}$ ($75h^{-1}$ Mpc away) and has a diameter of 5000 km s$^{-1}$ ($50h^{-1}$ Mpc). In Fig. 27.23(a), the Coma cluster is the torso of the humanlike figure at the center of the slice.[34] Apparently, rich clusters and superclusters are found at

---

[34]The large peculiar velocities of the cluster's galaxies cause them to be spread out in the radial velocity direction. This somewhat elongates the Coma cluster's appearance.

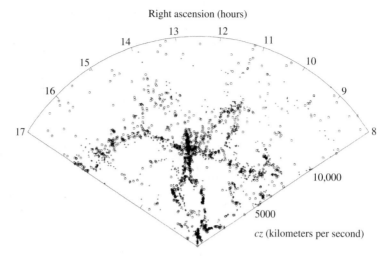

**FIGURE 27.24**   A combination of the first two slices of the CfA redshift survey, for $26.5° < \delta < 38.5°$. (Figure adapted from Geller, *Mercury*, May/June 1990.)

the intersections of bubbles, and long structures like the Perseus–Pisces supercluster may lie along the surface of adjoining bubbles. The filamentary features show where the survey slices through the sheets of galaxies that form the walls of the voids. This interpretation is strengthened by Fig. 27.24, which shows the combination of the first two slices on the same diagram. Not only do the features show a general correspondence, but they are slightly offset; some features of the combined slice are somewhat thicker than those of the individual slices. This is because the slices are not perpendicular to the walls of the voids. Instead, the curvature of a wall causes adjacent slices to intersect the wall at slightly different locations.

Figure 27.25 shows a combination of results from the CfA redshift survey of the northern skies with a southern survey carried out by L. Nicolaci da Costa and co-workers. The long arc at the top is the Great Wall, with its counterpart, the Southern Wall, running diagonally across the bottom. At the time of this writing, these are the largest known structures in the universe, and they may consist of the walls of several adjacent bubbles. The Great Wall passes through the arms of the homunculus in Fig. 27.23. It extends for *at least* $150h^{-1}$ Mpc in right ascension and *at least* $60h^{-1}$ Mpc in declination (perpendicular to the diagram) but is less than $5h^{-1}$ Mpc thick. The Great Wall disappears into regions of the sky that are obscured by dust (on one end) or not yet mapped (on the other end). It may actually continue on to join up with the Southern Wall, which contains the Perseus–Pisces supercluster.

The deepest survey to date is the 2dF Galaxy Redshift Survey (2dFGRS) carried out by the Anglo-Australian Observatory. It produced a three-dimensional map of 220,929 galaxies brighter than $m = -19.5$ and covering 5% of the sky. The survey penetrated out to approximately 1 Gpc, to a redshift of $z \approx 0.3$. Figure 27.26 shows a 3° slice that cuts through the north (left) and south (right) galactic poles. It includes 63,000 galaxies and clearly shows the galaxies collecting on the walls of immense voids. The spongelike structure of the universe is apparent in Fig. 27.27, which illustrates how the clusters and superclusters of galaxies clump together.

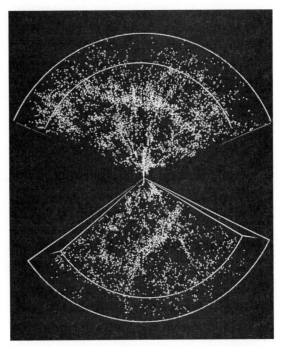

**FIGURE 27.25** A cross section of the universe with $cz \leq 12{,}000$ km s$^{-1}$ (within $120h^{-1}$ Mpc of Earth), showing 9325 galaxies. The coordinates of the top slice are $8^h < \alpha < 17^h$ and $8.5° < \delta < 44.5°$, those of the bottom slice $20.8^h < \alpha < 4^h$ and $-40° < \delta < 2.5°$. (Figure from da Costa et al., *Ap. J. Lett.*, *424*, L2, 1994.)

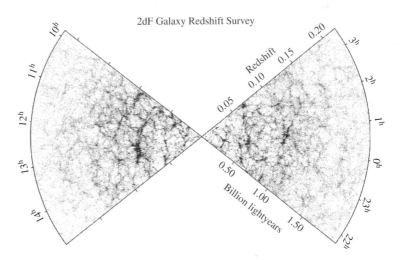

**FIGURE 27.26** A 3° slice through the 2dF Galaxy Redshift Survey. It displays 62,559 galaxies. The north galactic pole is on the left, and the south galactic pole is on the right. (Figure adapted from Colless et al., arXiv:astro-ph/0306581, 2003.)

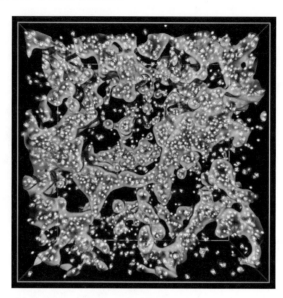

**FIGURE 27.27**    A cubical volume of space with sides 100 Mpc long. The clusters and superclusters of galaxies within are joined together, illustrating the spongelike structure of space. (Figure courtesy of the 2dF Galaxy Redshift Survey Team.)

The ongoing Sloan Digital Sky Survey (SDSS) is the most recent advance in deep surveys of the universe. The first phase of the project, completed in June 2005, imaged more than 20% of the sky at five wavelengths and recorded some 200 million objects. The SDSS obtained spectra of over 675,000 galaxies, 90,000 quasars, and 185,000 stars. The SDSS utilizes a 2.5-meter telescope on Apache Point, New Mexico. Its 120-megapixel camera can photograph 1.5 square degrees of sky at a time (eight times the area of the full moon). When completed, the SDSS will have produced a three-dimensional map of more than 25% of the sky. Such an extensive map should provide a more coherent and detailed view of the large-scale structures of the universe and supply clues concerning their origin. Figure 27.28, based on the 2dFGRS and SDSS catalogs, shows the clusters of galaxies belonging to the Pisces–Cetus supercluster of galaxies. The lines connecting the 19 clusters of galaxies (at an average redshift of $z = 0.0591$) outline the cosmic web of interconnected filaments that make up this supercluster.

### Quantifying Large-Scale Structure

Another method of inferring the presence of large features is to use the available data to mathematically describe the clustering of galaxies. If galaxies were uniformly distributed through space with a number density $n$, then the probability $dP$ of finding a galaxy within a volume $dV$ would be the same everywhere, $dP = n \, dV$.[35] However, galaxies are not scattered about randomly. The probability of finding a galaxy within a volume $dV$ at distance

---

[35]We assume that $dV$ is sufficiently small enough that $dP \leq 1$.

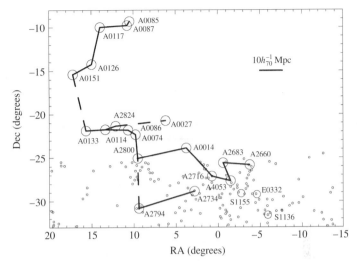

**FIGURE 27.28** Clusters of galaxies belonging to the Pisces-Cetus supercluster of galaxies. The lines connecting the clusters are between $20h^{-1}$ Mpc and $25h^{-1}$ Mpc long. (Figure adapted from Porter and Raychaudhury, *Mon. Not. R. Astron. Soc.*, *364*, 1387, 2005.)

$r$ from a specified galaxy then becomes

$$dP = n[1 + \xi(r)]\,dV, \tag{27.21}$$

where $n$ is the average number density of galaxies and $\xi(r)$ is a *two-point correlation function* that describes whether the galaxies are more concentrated ($\xi > 0$) or more dispersed ($\xi < 0$) than average. For 118,149 galaxies from the SDSS and separations of $0.1h^{-1}$ Mpc $< r <$ $16h^{-1}$ Mpc, the **correlation function** is observed to be of the form

$$\xi(r) = \left(\frac{r}{r_0}\right)^{-\gamma}, \tag{27.22}$$

where $r_0$, the correlation length, is $r_0 = 5.77h^{-1}$ Mpc, and $\gamma = 1.8$. The correlation length varies with the luminosities of the galaxies and ranges from $4.7h^{-1}$ Mpc to $7.4h^{-1}$ Mpc, brighter galaxies having the longer correlation lengths.

As surveys such as the SDSS sample more remote regions of space (say, within a sphere of radius $R_s$), the behavior of $r_0$ with sample depth $R_s$ is of some interest. If, as some have suggested, the universe has a fractal structure and is scale-free, then $r_0$ should increase proportionally with $R_s$. (If the universe looks essentially the same at all scales, then deeper observations must reveal ever larger versions of smaller structures.) However, the cosmological principle of homogeneity requires that $r_0$ become constant for sufficiently deep samples, because beyond some $R_s$ there are no larger structures to be found.

The search is under way to find the value of $R_s$ where the universe becomes homogeneous. Although some studies, such as the one shown in Fig. 27.29, have apparently found the transition to homogeneity, others have not. There is general agreement that the universe

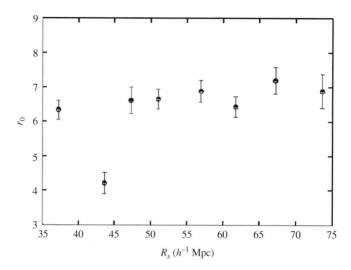

**FIGURE 27.29**   The correlation length $r_0$ as a function of the sample depth $R_s$ for the CfA-II catalog (a galactic redshift catalog compiled by the Harvard Center for Astrophysics). The flat plateau on the right indicates a transition to homogeneity for this study at about $R_s \approx 60$–$70h^{-1}$ Mpc. (Figure adapted from Martinez et al., *Ap. J.*, 554, L5, 2001.)

has a fractal structure on scales out to $\approx 30h^{-1}$ Mpc. (This does *not* mean that the universe is a fractal, because the self-similarity holds only for smaller scales and not for larger ones.) The prevailing view is that the transition to homogeneity occurs somewhere around or past $R_s \approx 30h^{-1}$ Mpc, perhaps as far as $R_s \approx 200h^{-1}$ Mpc. This scale is consistent with the extent of the largest coherent structures in the universe, the cosmic network of interconnected filaments of galaxies that fill the universe (see Fig. 27.28).

## What Were the Seeds of Structure?

It is not easy to imagine how the pattern of galaxies revealed by the redshift surveys could have come about from an initially random distribution of galaxies. By chance, of course, some areas would have had more galaxies than others. The denser regions would have gravitationally attracted still more galaxies, leaving the less dense volumes empty. Although gravity does act to increase the difference between high- and low-density regions, it is not enough to explain the voids. The large voids are some 100 Mpc across, but the peculiar velocities of galaxies are only on the order of hundreds of kilometers per second. Using a characteristic peculiar velocity of $\sim 600$ km s$^{-1}$, the time for a galaxy to cross a void is at least 160 Gyr. This is much longer than the age of the universe, $t_H$, and means that the galaxies could *not* have formed inside the voids and then moved out. Instead, the galaxies must have formed nearly in their present pattern. Apparently, the seeds of the structure we see today were present from the earliest moments of the universe. The question of the origin of this amazing arrangement of galaxies will be taken up again in Chapter 30.

## SUGGESTED READING

### General

Berendzen, Richard, Hart, Richard, and Seeley, Daniel, *Man Discovers the Galaxies*, Science History Publications, New York, 1976.

Ferris, Timothy, *The Red Limit*, Second Edition, Quill, New York, 1983.

Lake, George, "Cosmology of the Local Group," *Sky and Telescope*, December 1992.

Landy, Stephen D., "Mapping the Universe," *Scientific American*, June 1999.

Osterbrock, Donald E., Gwinn, Joel A., and Brashear, Ronald S., "Edwin Hubble and the Expanding Universe," *Scientific American*, July 1993.

Silk, Joseph, *The Big Bang*, Third Edition, W. H. Freeman and Company, New York, 2001.

Waller, William H., and Hodge, Paul W., *Galaxies and the Cosmic Frontier*, Harvard University Press, Cambridge, 2003.

### Technical

Binney, James, and Tremaine, Scott, *Galactic Dynamics*, Princeton University Press, Princeton, NJ, 1987.

Courteau, S., and Dekel, A., "Cosmic Flows: A Status Report," *Astrophysical Ages and Time Scales*, ASP Conference Series, von Hippel, T., Simpson, C., and Manset, N. (eds.), *245*, 584, 2001.

Dressler, Alan, et al., "Spectroscopy and Photometry of Elliptical Galaxies. I. A New Distance Estimator," *The Astrophysical Journal*, *313*, 42, 1987.

Dressler, Alan, and Faber, S. M., "New Measurements of Distances to Spirals in the Great Attractor: Further Confirmation of the Large-Scale Flow," *The Astrophysical Journal Letters*, *354*, L45, 1990.

Heck, A., and Caputo, F. (eds.), *Post-Hipparcos Cosmic Candles*, Kluwer Academic Publishers, Dordrecht, 1999.

Hubble, Edwin, "A Relation between Distance and Radial Velocity among Extra-Galactic Nebulae," *The Proceedings of the National Academy of Sciences*, *15*, 168, 1929.

Jacoby, George H., et al., "A Critical Review of Selected Techniques for Measuring Extragalactic Distances," *Publications of the Astronomical Society of the Pacific*, *104*, 599, 1992.

Jones, Bernard J. T., et al., "Scaling Laws in the Distribution of Galaxies," *Reviews of Modern Physics*, *76*, 1211, 2005.

Leibundgut, B., and Tammann, G. A., "Supernova Studies III. The Calibration of the Absolute Magnitude of Supernovae of Type Ia," *Astronomy and Astrophysics*, *230*, 81, 1990.

Mathewson, D. S., Ford, V. L., and Buchhorn, M., "No Back-Side Infall into the Great Attractor," *The Astrophysical Journal Letters*, *389*, L5, 1992.

Peebles, P. J. E., *Principles of Physical Cosmology*, Princeton University Press, Princeton, NJ, 1993.

Riess, Adam G., et al., "Cepheid Calibrations from the Hubble Space Telescope of the Luminosity of Two Recent Type Ia Supernovae and a Redetermination of the Hubble Constant," *The Astrophysical Journal*, *627*, 579, 2005.

Rowan-Robinson, Michael, *The Cosmological Distance Ladder*, W. H. Freeman, New York, 1985.

Solanes, José M., et al., "The Three Dimensional Structure of the Virgo Cluster from Tully–Fisher and HI Data," *The Astronomical Journal*, *124*, 2440, 2002.

Sparke, Linda S., and Gallagher, John S., *Galaxies in the Universe: An Introduction*, Cambridge University Press, Cambridge, 2000.

Tremaine, Scott, "The Dynamical Evidence for Dark Matter," *Physics Today*, February 1992.

Willick, J. A., "Cosmic Velocities 2000: A Review," arXiv:astro-ph/0003232 v1, 2000.

Zehavi, Idit, et al., "On Departures from a Power Law in the Galaxy Correlation Function," *The Astrophysical Journal*, *608*, 16, 2004.

## PROBLEMS

**27.1** By equating the period–luminosity relation, Eq. (14.1), and the period–luminosity–color relation, Eq. (27.1), estimate the range of the $B - V$ color index for Cepheid variable stars.

**27.2** The central light ring produced by SN 1987A, shown in Fig. 15.14, gave astronomers a unique chance to determine its distance (and therefore the distance to the Large Magellanic Cloud). The supernova heated the ring of gas, causing it to glow. The ring, which has an angular diameter of $1.66''$ (long axis), is presumed to be circular but tilted from the perspective of Earth. Light was received from the near side of the ring 340 days before light arrived from its far side.

  **(a)** Carefully measure the photograph, and determine the angle between the plane of the ring and the plane of the sky.

  **(b)** Use the time delay to determine the diameter of the ring in parsecs, and compare your result with the caption to Fig. 15.14.

  **(c)** Use trigonometry to find the distance to SN 1987A.

**27.3** The three brightest red stars in the galaxy M101 (the "Pinwheel" galaxy; see Fig. 25.5) have visual magnitudes of $V = 20.9$. Assuming that there is 0.3 mag of extinction, what is the distance to M101? How does this compare to the distance of 7.5 Mpc found using classical Cepheids?

**27.4** Show that an uncertainty in the distance modulus $m - M$ of 0.4 mag corresponds to an uncertainty in the distance of about 20%. What uncertainty in the distance modulus would produce a 5% uncertainty in the distance? What uncertainty in the distance modulus would produce a 50% uncertainty in the distance?

**27.5** In Eq. (27.4), the value of $C$ for the Fornax cluster of galaxies is $C = -1.264$. What is the ratio of the distances to the Virgo and Fornax clusters? to the Coma and Fornax clusters?

**27.6** Use the solid line in Hubble's velocity–distance diagram, Fig. 27.7, to determine his value of $H_0$. Why was his result so different from today's value?

**27.7** Use the relative motion of the Andromeda and Milky Way galaxies to estimate their total mass for the case $h = 0.5$. What is the corresponding mass-to-light ratio of these galaxies?

**27.8** The Magellanic Stream orbits the Milky Way and extends from 50 kpc to 100 kpc from the Galactic center.

(a) Consider a clump of gas in the stream in a circular orbit about the Galactic center. Take the radius of the orbit to be 75 kpc and the orbital speed to be 244 km s$^{-1}$. Treating the Galaxy and the mass clump as point masses, estimate the mass of the Milky Way. What is the corresponding mass-to-light ratio?

(b) Suppose a clump of gas at the tip of the stream starts with zero radial velocity at a distance of 100 kpc and reaches a radial velocity of $-220$ km s$^{-1}$ after falling down to 50 kpc from the Galactic center. Assuming that the transverse (orbital) velocity of the clump of gas has not changed, use conservation of energy to estimate the mass of the Milky Way. Find the corresponding mass-to-light ratio. As before, treat the Galaxy and the clump of gas as point masses.

**27.9** Assuming that the Sculptor group of galaxies occupies a spherical volume of space, find the difference in magnitude between two identical objects located at the very front and back of the group.

**27.10** Derive Eq. (27.17).

**27.11** *Suppose* that the densities of the dark matter and interstellar gas in M87 have the same $r$-dependence (i.e., they are proportional). Use Eq. (27.17) to show that the gas is isothermal. *Hint:* Assume $T \propto r^\alpha$ and show $\alpha = 0$. Use flat rotation curves and make use of Eqs. (24.49) and (24.50). [The controversial assumption that $T(r) =$ constant has been used in some studies of dark matter in elliptical galaxies.]

**27.12** Like the Coma cluster, the Virgo cluster contains a large amount of hot (70 million K) intra-cluster gas that emits X-rays.

(a) If the X-ray luminosity of the Virgo cluster's intracluster gas is about $1.5 \times 10^{36}$ W, use Eq. (27.20) to find the electron number density and the mass of the gas. Assume that the Virgo cluster is a sphere of radius 1.5 Mpc that is filled with completely ionized hydrogen.

(b) Use $L_V = 1.2 \times 10^{12}\ L_\odot$ for the visual luminosity of the Virgo cluster to estimate the cluster's luminous mass. How does this compare with your answer to part (a) for the mass of the intracluster gas?

(c) Assuming that the gas has no energy source and that it is simply losing energy via thermal bremsstrahlung, use Eq. (10.17) for the average kinetic energy per gas particle (protons and electrons) to estimate how long it will take for the gas to lose all of its energy. (Assume that the X-ray luminosity remains constant throughout your calculation.) How does your answer compare with the Hubble time, $t_H$?

**27.13** Estimate how long a galaxy in the Coma cluster would take to travel from one side of the cluster to the other. Assume that the galaxy moves with a constant speed equal to the cluster's radial velocity dispersion. How does this compare with the Hubble time, $t_H$? What can you conclude about whether the galaxies in the Coma cluster are gravitationally bound?

**27.14** For the galaxies in the Virgo cluster, the dispersion in the radial velocity is $\sigma_r = 666$ km s$^{-1}$. Use the virial theorem to estimate the mass of the Virgo cluster.

**27.15** Suppose that the galaxies in the Coma cluster are all moving *away* from the cluster's center but that the measured value of the radial velocity dispersion is unchanged. (For example, directing each galaxy's present velocity radially outward would accomplish this.) In this case, the cluster would not be in equilibrium. What is the sign of the term $\langle d^2 I/dt^2 \rangle$ in the virial theorem, Eq. (2.44)? Explain how neglecting this term would affect your estimate of the mass of the Coma cluster. How would your answer be different if all of the galaxies were moving *toward* the cluster's center?

**27.16** For the galaxy NGC 5585, the quantity $2v_r / \sin i = 218$ km s$^{-1}$, and its apparent $H$ magnitude is $H = 9.55$ (already corrected for extinction). Use the Tully–Fisher method to determine the distance to this galaxy.

**27.17** The brightest galaxy in the cluster A1060 has an apparent visual magnitude of $V = 10.99$. Estimate the distance to the cluster. Use the uncertainty in the average absolute magnitude of the brightest galaxy to determine how far off your answer could be.

**27.18** The Galactic coordinates of the direction of the large-scale streaming motion of galaxies in the direction of the Great Attractor are $\ell = 309°$, $b = 18°$. Convert these to equatorial coordinates, and use a star chart to confirm that your answer is in the constellation of Centaurus.

## COMPUTER PROBLEM

**27.19** Given the results in Table 27.1 for the distance to the Virgo cluster of galaxies, the problem arises of how $d_{ave}$, the average distance, should be chosen.

(a) $d_{ave}$ might be chosen to minimize

$$\Delta \equiv \sum_i \left( \frac{d_i - d_{ave}}{\delta_i} \right)^2 ,$$

where $\delta_i$ is the uncertainty in the distance $d_i$ determined by the $i$th method in the table (e.g., for the estimate using novae, $d = 21.1$ Mpc and $\delta = 3.9$ Mpc). (There are echoes of the least-squares fit to a straight line in this approach.) Make a graph of $\Delta$ vs. $d_{ave}$ for values of $d_{ave}$ between 15 and 22 Mpc, and determine (to the nearest 0.1 Mpc) the value of $d_{ave}$ that minimizes $\Delta$.

(b) The *weighted mean* value is

$$d_w = \frac{\sum_i (d_i/\delta_i^2)}{\sum_i (1/\delta_i^2)}.$$

Calculate the weighted mean for the values in Table 27.1, and compare your answer with what you found in part (a).

(c) Prove that your answers to parts (a) and (b) will always agree.

# CHAPTER

# 28

# Active Galaxies

## 28.1 ■ OBSERVATIONS OF ACTIVE GALAXIES

The story of modern astrophysics is one of a dynamically evolving universe. On every scale, from planets to stars to galaxies, the objects that are present in this era differ from what they were during previous epochs. As we study the ancient light that arrives from distant corners of the universe, we are able to examine how galaxies looked and behaved in their youth. These observations reveal a level of activity in the centers of young, remote galaxies that is rarely found in nearer galactic nuclei.

### Seyfert Galaxies

The first hint of the violent heritage of today's galaxies was found by Edward A. Fath (1880–1959), who in 1908 was observing the spectra of "spiral nebulae." Although most showed an absorption-line spectrum produced by the combined light of the galaxy's stars, NGC 1068 displayed six bright emission lines. In 1926 Edwin Hubble recorded the emission lines of this and two other galaxies. Seventeen years later Carl K. Seyfert (1911–1960) reported that a small percentage of galaxies have very bright nuclei that are the source of broad emission lines produced by atoms in a wide range of ionization states. These nuclei are nearly stellar in appearance.

Today these objects are known as **Seyfert galaxies**, with spectra that are categorized into one of two classes. **Seyfert 1** galaxies have very broad emission lines that include both allowed lines (H I, He I, He II) and narrower forbidden lines (such as [O III]).[1] Seyfert 1 galaxies generally have "narrow" allowed lines as well, although even the narrow lines are broad compared to the spectral lines exhibited by normal galaxies. The width of the lines is attributed to Doppler broadening, indicating that the allowed lines originate from sources with speeds typically between 1000 and 5000 km s$^{-1}$, while the forbidden lines correspond to speeds of around 500 km s$^{-1}$. **Seyfert 2** galaxies have only narrow lines (both permitted

---

[1]Recall from Section 11.2 that forbidden lines involve low-probability transitions in atoms and are an indication of low gas densities.

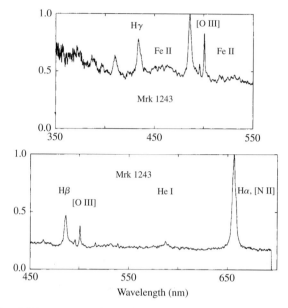

**FIGURE 28.1**    The visible spectrum of Mrk 1243, a Seyfert 1 galaxy. (Figure adapted from Oster-brock, *QJRAS*, *25*, 1, 1984.)

and forbidden), with characteristic speeds of about 500 km s$^{-1}$. Every spectrum also shows a featureless continuum that is devoid of lines, originating from a small central source. The great luminosity of a Seyfert 1 galaxy arises from this continuum, which often overwhelms the combined light of all of the galaxy's stars. The continuum observed for a Seyfert 2 is significantly less luminous.

Figures 28.1 and 28.2, respectively, show the visible spectra of Mrk 1243 (a Seyfert 1) and Mrk 1157 (a Seyfert 2), where "Mrk" indicates an entry in the galaxy catalog of E. B. Markarian (1913–1985), produced in 1968. Some spectra display both broad and narrow permitted lines, and so they are classified as an intermediate type such as Seyfert 1.5. However, it is important to emphasize that this is a spectral classification. The spectra of a few Seyfert galaxies have changed nearly from type 1.5 to type 2 in a matter of years, although the broad H$\alpha$ emission line has rarely if ever completely disappeared.

The galaxies known to emit the most X-ray energy are Seyferts of types 1 and 1.5. The X-ray emission is quite variable, and can change appreciably on timescales ranging from days to hours. In contrast, X-rays are less frequently measured for Seyfert 2 galaxies. An analysis of the hard X-rays that are observed for Seyfert 2s indicates that the "missing" X-rays have been absorbed by intervening material with huge hydrogen column densities[2] of between $10^{26}$ and $10^{28}$ m$^{-2}$.

Seyferts make up only a few tenths of a percent of all field galaxies. It is interesting that at least 90% of the Seyferts close enough to be resolved by telescopes are spiral galaxies, typically of types Sb or SBb. They are frequently accompanied by other galaxies with which

---

[2]Recall from Section 9.5 that line profiles can be used to calculate the column density of the absorbing material.

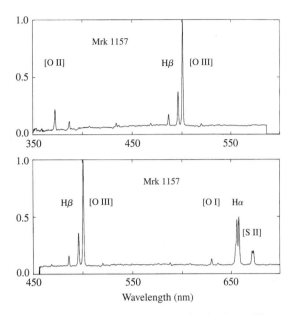

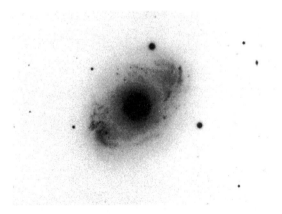

**FIGURE 28.2**  The visible spectrum of Mrk 1157, a Seyfert 2 galaxy. (Figure adapted from Oster-brock, *QJRAS*, *25*, 1, 1984.)

**FIGURE 28.3**  A long exposure of the Seyfert 1 (or 1.5) galaxy NGC 4151 showing the galactic disk around its bright nucleus. (Image from Sandage and Bedke, *The Carnegie Atlas of Galaxies*, Carnegie Institution of Washington, Washington, D.C., 1994.)

they may be gravitationally interacting. Figure 28.3 is a long-exposure view of the Seyfert galaxy NGC 4151 (type Sab) that shows the galactic disk around its bright nucleus.

## The Spectra of Active Galactic Nuclei

Seyferts belong to the general class of galaxies with **active galactic nuclei**, or AGN for short. Other members of this class, such as radio galaxies, quasars, and blazars, will be introduced in the discussion that follows.

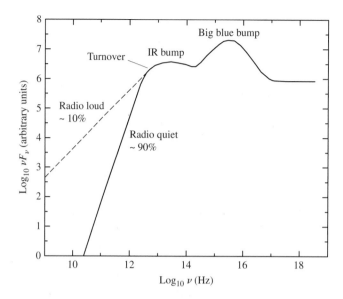

**FIGURE 28.4**    A sketch of the continuum observed for many types of AGNs.

Figure 28.4 is a rough schematic of the continuum observed for many types of AGNs (note that the logarithm of the product $\nu F_\nu$ is plotted on the figure's vertical axis). The most notable feature of this **spectral energy distribution** (SED) is its persistence over some 10 orders of magnitude in frequency. This wide spectrum is markedly different from the thermal (blackbody) spectrum of a star or the combined spectra of a galaxy of stars.

When AGNs were first studied, it was thought that their spectra were quite flat. Accordingly, a power law of the form

$$\boxed{F_\nu \propto \nu^{-\alpha}} \tag{28.1}$$

was used to describe the monochromatic energy flux, $F_\nu$.[3] The **spectral index**,[4] $\alpha$, was believed to have a value of $\alpha \simeq 1$.

The power received within any frequency interval between $\nu_1$ and $\nu_2$ is

$$L_{\text{interval}} \propto \int_{\nu_1}^{\nu_2} F_\nu \, d\nu = \int_{\nu_1}^{\nu_2} \nu F_\nu \, \frac{d\nu}{\nu} = \ln 10 \int_{\nu_1}^{\nu_2} \nu F_\nu \, d\log_{10} \nu, \tag{28.2}$$

so that equal areas under a graph of $\nu F_\nu$ vs. $\log_{10} \nu$ correspond to equal amounts of energy; hence the reason for plotting $\log_{10} \nu F_\nu$ on the ordinate in Fig. 28.4. A value of $\alpha \simeq 1$ reflects the horizontal trend seen to the right of the turnover in Fig. 28.4.

The continuous spectra of AGNs are now known to be more complicated, involving a mix of thermal and nonthermal emission. However, Eq. (28.1) is still used to *parameterize*

---

[3]Recall from Section 3.5 that $F_\nu \, d\nu$ is the amount of energy with a frequency between $\nu$ and $\nu + d\nu$ that arrives per unit area per second on a detector aimed at the source.

[4]*Warning:* Some authors define the spectral index with the opposite sign.

the continuum. The spectral index typically has a value between 0.5 and 2 that usually increases with increasing frequency, so the curve of $\log_{10} \nu F_\nu$ vs. $\log_{10} \nu$ in Fig. 28.4 is generally concave downward. In fact, the value of $\alpha$ is constant over only a limited range of frequencies, such as in the infrared and visible regions of the spectrum. The shape and polarization of the visible-UV spectrum indicates that it can sometimes be decomposed into contributions from thermal sources (blackbody spectrum, low polarization) and nonthermal sources (power-law spectrum, significant polarization). The thermal component appears as the **big blue bump** in Fig. 28.4, which can contain an appreciable amount of the bolometric luminosity of the source. It is generally believed that the emission from the big blue bump is due to an optically thick accretion disk, although some researchers have suggested that free–free emission may be responsible. Also evident is a thermal **infrared bump** to the left of the big blue bump; it is probably due to emission from warm ($T \lesssim 2000$ K) dust grains.

A pure power-law spectrum (with constant $\alpha$) is the signature of synchrotron radiation, which is frequently encountered in astronomical situations involving relativistic electrons and magnetic fields (cf., Examples 4.3.3 and 16.7.1). As shown in Fig. 28.5, a synchrotron spectrum is produced by the combined radiation emitted by individual electrons as they spiral around magnetic field lines. If the distribution of the individual electron energies obeys a power law, then the resulting synchrotron spectrum is described by Eq. (28.1). However, the synchrotron spectrum does not continue to rise without limit as the frequency decreases. At a transition frequency, the spectrum turns over and varies as $\nu^{5/2}$ (spectral

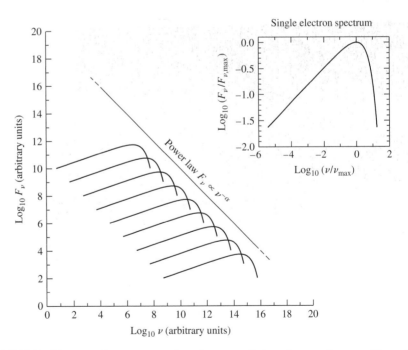

**FIGURE 28.5**   The power-law spectrum of synchrotron radiation, shown as the sum of the radiation produced by individual electrons as they spiral around magnetic field lines. The spectrum of a single electron is at the upper right. The turnover at low frequencies is not shown.

index $\alpha = -2.5$). This occurs because the plasma of spiraling electrons becomes opaque to its own synchrotron radiation, an effect known as **synchrotron self-absorption**.

In some SEDs, the "turnover" evident in the schematic continuum spectrum in Fig. 28.4 may be due to synchrotron self-absorption. However, the thermal contributions to the continuum spectrum evident in the infrared bump suggest that in other cases, the turnover may be due to the long-wavelength Rayleigh–Jeans portion of the blackbody spectrum produced by the warm dust grains. It is possible that the steeper, low-frequency spectra of radio-quiet AGNs are due to the thermal spectrum of dust grains, while the shallower, low-frequency spectra of radio-loud AGNs may be due to a combination of thermal and nonthermal emission.

## Radio Galaxies

After World War II the science of radio astronomy that was started by Karl Jansky made rapid progress, led by astronomers in Australia and England. The first discrete source of strong radio waves (other than the Sun) was discovered in the constellation Cygnus and was named Cygnus A (a modern VLA radio image of Cyg A is shown in Fig. 28.6). Using the accurate position provided by English radio astronomer F. Graham Smith, the team of Walter Baade and Rudolph Minkowski (1895–1976) was able to find the optical counterpart of Cyg A. It is a peculiar-looking cD galaxy whose center is apparently encircled by a ring of dust (Fig. 28.7 shows an optical image of Cyg A obtained using the Hubble Space Telescope). Cyg A's spectrum shows a redshift of $z = \Delta\lambda/\lambda_{\rm rest} = 0.057$, corresponding to a recessional velocity of 16,600 km s$^{-1}$ (Eq. 4.38). From Hubble's law in the form of Eq. (27.7), the distance to Cyg A is about $170h^{-1}$ Mpc (implying a distance of 240 Mpc if $h = [h]_{\rm WMAP}$). Considering that Cyg A is the brightest radio source beyond the Milky Way, this distance is surprisingly large. In fact, the only discrete radio sources brighter than

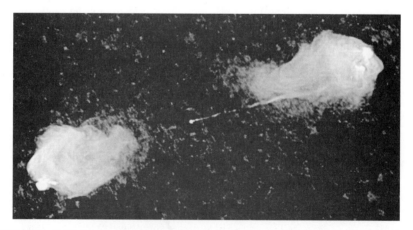

**FIGURE 28.6**    A VLA radio image of Cygnus A, showing the two radio lobes separated by about $100h^{-1}$ kpc and the jet extending from the galaxy to the right-hand lobe. Cyg A is a narrow-line radio galaxy. The central cD galaxy does not show up on this radio picture (see Fig. 28.7). The width of the image is about 2 arcminutes. [Courtesy of R. A. Perley, J. W. Dreher, and J. J. Cowan (NRAO/AUI).]

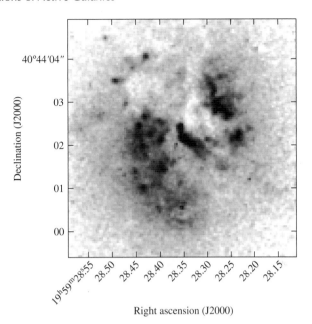

$40°44'04''$ —

03 —

02 —

01 —

00 —

Declination (J2000)

$19^h59^m28^s55$   28.50   28.45   28.40   28.35   28.30   28.25   28.20   28.15

Right ascension (J2000)

**FIGURE 28.7**    A continuum HST image of Cygnus A (3C 405) taken at a wavelength of 622 nm. (Figure adapted from Jackson, et al., *MNRAS, 301*, 131, 1998.)

Cyg A are the Sun and Cassiopeia A, the nearby (3 kpc) remnant of a Type II supernova. To be detected so far away, Cyg A must pour out enormous amounts of radio energy.

Cyg A is one example of a class of galaxies, called **radio galaxies**, that are extremely bright at radio wavelengths.

---

**Example 28.1.1.**    The radio energy emitted by Cygnus A can be estimated by using its distance of $d = 170h^{-1}$ Mpc together with the observed value of the monochromatic flux at a radio frequency of 1400 MHz,

$$F_{1400} = 1.255 \times 10^{-23} \text{ W m}^{-2} \text{ Hz}^{-1} = 1255 \text{ Jy.}$$

The radio spectrum follows the power law of Eq. (28.1) with $\alpha \simeq 0.8$, so $F_\nu \propto \nu^{-0.8}$. That is, we can write

$$F_\nu = F_{1400} \left( \frac{\nu}{1400 \text{ MHz}} \right)^{-0.8}.$$

The radio luminosity can be found by integrating the monochromatic flux given by Eq. (3.29) over the range of radio frequencies found in Table 3.1. The upper frequency limit is taken to be $\nu_2 = 3 \times 10^9$ Hz, corresponding to a radio wavelength of 0.1 m. As is shown in Problem 28.20, the power-law behavior of the radio spectrum does not continue to $\nu = 0$. Instead, the flux received from Cygnus A declines when the frequency falls below about

*continued*

$\nu_1 = 10^7$ Hz. With these limits, the radio luminosity is approximately

$$L_{\text{radio}} = 4\pi d^2 \int_{\nu_1}^{\nu_2} F_\nu \, d\nu = 4\pi d^2 F_{1400} \int_{\nu_1}^{\nu_2} \left(\frac{\nu}{1400 \text{ MHz}}\right)^{-0.8} d\nu = 2.4 \times 10^{37} h^{-2} \text{ W}.$$

Using the WMAP value of $[h]_{\text{WMAP}} = 0.71$, the radio luminosity of Cygnus A is estimated to be $L_{\text{radio}} = 4.8 \times 10^{37}$ W. This is several million times more radio energy than is
produced by a normal galaxy such as M31 and is roughly three times the energy produced
at *all* wavelengths by the Milky Way.

---

Like Seyfert galaxies, radio galaxies may also be divided into two classes: **broad-line
radio galaxies** (BLRGs, corresponding to Seyfert 1s) and **narrow-line radio galaxies**
(NLRGs, corresponding to Seyfert 2s). BLRGs have bright, starlike nuclei surrounded by
very faint, hazy envelopes. NLRGs, on the other hand, are giant or supergiant elliptical
galaxies (types cD, D, and E); Cyg A is a NLRG.

Despite their similarities, there are obvious differences between Seyferts and radio galaxies. Although Seyfert nuclei emit some radio energy, they are relatively quiet at radio wavelengths compared with radio galaxies. Furthermore, while nearly all Seyferts are spiral
galaxies, none of the strong radio galaxies are spirals.

### Radio Lobes and Jets

A radio galaxy may display extended **radio lobes**, as in Fig. 28.6, or it may radiate its energy
both from a compact **core** in its nucleus and from a **halo** that is about the size of the visible
galaxy or larger. The optical cD galaxy in Fig. 28.7 is the central dot in the Cyg A radio
image shown in Fig. 28.6. The optical galaxy is flanked by two huge radio lobes that are
the sources of the tremendous radio luminosity estimated in Example 28.1.1. Each of the
lobes has a diameter of about $17h^{-1}$ kpc.

Observations reveal that one of the lobes is connected to the central galaxy of Cyg A
by a collimated **jet** that spans the roughly $50h^{-1}$ kpc of space separating the galaxy from
the lobe. (Since the orientations of jets and radio lobes are not well determined, the values
for their sizes quoted here are projected distances on the plane of the sky.) At least half
of the stronger radio galaxies have detectable jets, as do more than three-quarters of the
weaker sources. The jets associated with the powerful sources tend to be one-sided (like
Cyg A's), while those found in less luminous radio galaxies are typically two-sided. One
reason for this is that the stronger radio galaxies can be seen at greater distances, and so a
dim counterjet may go undetected. (Reasons why counterjets may not appear as luminous
to the observer will be discussed in Section 28.3.)

Figure 28.8 shows the strong jet and weak counterjet of the elliptical galaxy NGC 6251
at several radio frequencies. Note that the Moon's angular diameter would just fit into either
of the two rectangular boxes at the top of the figure, while the scale of the bottom box is
just a few milliarcseconds across. It is remarkable that the jet can be traced essentially all
the way to the core of the galaxy.

Other radio jets are not as straight as those of Cyg A or NGC 6251. Figure 28.9 shows the
windblown appearance of the jets emanating from NGC 1265, produced by that galaxy's
motion through the intracluster gas of the Perseus cluster.

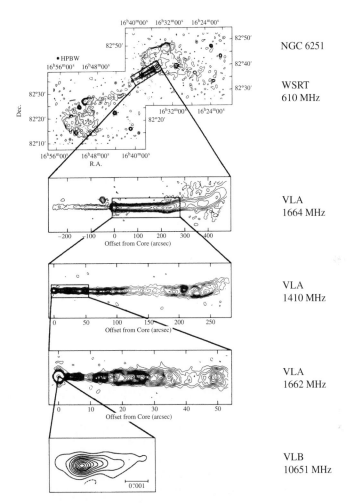

**FIGURE 28.8**   The jet and counterjet (second panel) of the radio galaxy NGC 6251. (Figure adapted from Bridle and Perley, *Annu. Rev. Astron. Astrophys.*, *22*, 319, 1984. Reproduced by permission from the Annual Review of Astronomy and Astrophysics, Volume 22, ©1984 by Annual Reviews Inc.)

Following Cyg A, many more radio galaxies were discovered. One of these is M87, the giant elliptical (E1) galaxy described in Section 27.3 that lies at the center of the Virgo cluster. With an apparent visual magnitude of $V = 8.7$, M87 is one of the brighter-appearing galaxies in the sky. Figure 28.10 shows two HST views of M87, also known as Virgo A to radio astronomers. Its prominent jet, shown at the right, was discovered optically in 1917. The jet extends from the galaxy some 1.5 kpc into one of its radio lobes. The jet also displays evenly spaced knots that are bright at radio, visible, and X-ray wavelengths. The X-ray luminosity of M87, including the jet, is roughly $10^{36}$ W. This is about 50 times greater than M87's radio luminosity. The inset shows the spiral-shaped disk of hot gas that is at the core of M87. There is also evidence for a faint counterjet extending away from M87 in the direction opposite that of the dominant jet.

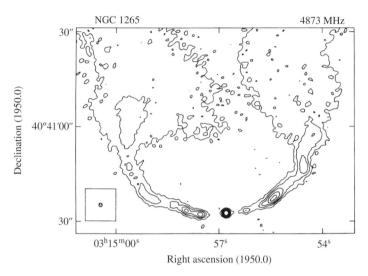

**FIGURE 28.9**    The radio appearance of NGC 1265, with its jets swept back by that galaxy's motion through the surrounding intracluster gas. (Figure adapted from O'Dea and Owen, *Ap. J.*, *301*, 841, 1986.)

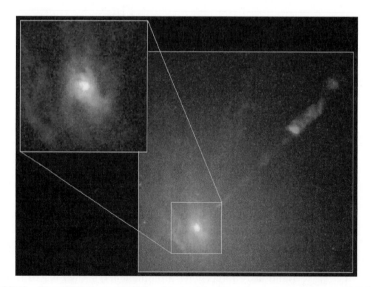

**FIGURE 28.10**    Two HST views of M87 and its jet. The inset shows the spiral-shaped disk of hot gas at the center of M87. [Courtesy of H. Ford (STScI/Johns Hopkins U.); R. Harms (Applied Research Corp.); Z. Tsvetanov, A. Davidsen, and G. Kriss (Johns Hopkins U.); R. Bohlin and G. Hartig (STScI); L. Dressel and A. K. Kochhar (Applied Research Corp.); and Bruce Margon (U. Washington).]

**FIGURE 28.11** The visual and radio appearance (superimposed contour lines) of the radio galaxy Centaurus A. (Courtesy of NRAO.)

One of the largest radio galaxies known is 3C 236 (the "3C" designates a listing in the **Third Cambridge Catalog** of radio sources). With a redshift of $z = 0.0988$, its distance is about $280h^{-1}$ Mpc, according to Hubble's law. The radio lobes of 3C 236 are separated by more than $1.5h^{-1}$ Mpc, projected onto the plane of the sky, while its radio jet is only $400h^{-1}$ pc long.

The closest example of an AGN is Centaurus A (NGC 5128), at a distance of $4.7h^{-1}$ Mpc. Figure 28.11 shows an optical image of Cen A, an E2 galaxy girded by a thick dust lane. Superimposed on the photograph is a radio map showing the radio lobes. (This figure corresponds roughly to the central region of Fig. 6.19.) Like M87, Cen A has a jet extending from its nucleus containing several knots of radio and X-ray emission. Although Cen A is in our astronomical backyard, radio galaxies on average are roughly 100 times less abundant than Seyferts in regions that are nearby in cosmological terms.

### The Discovery of Quasars

As radio telescopes discovered increasing numbers of radio sources in the late 1950s, the task of identifying these sources with known objects became more important. In 1960 Thomas Matthews and Allan Sandage were searching for an optical counterpart to another radio source, 3C 48. They found a 16th-magnitude starlike object whose unique spectrum displayed broad emission lines that could not be identified with any known element or molecule. In Sandage's words, "The thing was exceedingly weird." In 1963 a similarly weird spectrum was found for another radio source with a stellar appearance, 3C 273. Figure 28.12 shows 3C 273 and its jet, which extends a projected distance of $39h^{-1}$ kpc from the nucleus.

**FIGURE 28.12**   The quasar 3C 273 and its jet. [Courtesy of NASA and J. Bahcall (IAS).]

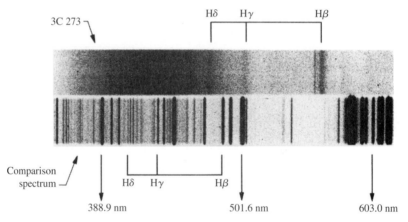

**FIGURE 28.13**   The $z = 0.158$ redshift of the quasar 3C 273. (Adapted from a figure courtesy of Palomar/Caltech.)

3C 48, 3C 273, and other, similar sources were classified as **quasi-stellar radio sources** (QSRs), which became known as **quasars**. But names are not a substitute for understanding, and an understanding of quasars proved to be elusive. Later that year, the mystery lifted somewhat when Dutch astronomer Maarten Schmidt recognized that the pattern of the broad emission lines of 3C 273 was the same as the pattern of the Balmer lines of hydrogen. These familiar lines had been severely redshifted ($z = 0.158$) to unfamiliar wavelengths, making their identification difficult; see Fig. 28.13. This Doppler shift means that 3C 273 is receding from Earth at 14.6% of the speed of light. According to Hubble's law, this places 3C 273 at a distance of about $440h^{-1}$ Mpc. At Caltech, Schmidt's associates Jesse

Greenstein (1909–2002) and Thomas Matthews calculated that 3C 48 has an even greater redshift, $z = 0.367$, corresponding to a radial velocity of $0.303c$ and a Hubble distance of just over $900h^{-1}$ Mpc. Astronomers realized that 3C 48 was one of the most distant objects yet discovered in the universe.

### Quasar Luminosities

A quasar's radio emission may come either from radio lobes or from a central source in its core. Quasars are so far away that in optical images most appear as overwhelmingly bright, starlike nuclei surrounded by faint fuzzy halos. In some cases, a fuzzy halo can be resolved into a faint parent galaxy. To be visible from such great distances, quasars must be exceptionally powerful.

---

**Example 28.1.2.**   Equation (3.6) can be used to obtain the absolute visual magnitude of the quasar 3C 273, which has an apparent visual magnitude of $V = 12.8$. Adopting $[h]_{\text{WMAP}} = 0.71$ yields a distance of $d \simeq 620$ Mpc, implying that

$$M_V = V - 5 \log_{10}\left(\frac{d}{10 \text{ pc}}\right) = -26.2.$$

This value can be used to obtain an estimate of the luminosity of the quasar at visual wavelengths. Using $M_{\text{Sun}} = 4.82$ for the Sun's absolute visual magnitude, Eq. (3.7) gives an estimate of the quasar's visual luminosity:

$$L_V \approx 100^{(M_{\text{Sun}} - M_V)/5} L_\odot = 2.6 \times 10^{12} L_\odot = 1 \times 10^{39} \text{ W}.$$

The radio energy emitted by 3C 273 can be estimated from its distance and the value of the monochromatic flux at a radio frequency of 1400 MHz, $F_{1400} = 4.64 \times 10^{-25}$ W m$^{-2}$ Hz$^{-1}$ = 46.4 Jy. The radio spectrum follows the power law of Eq. (28.1) with a spectral index of $\alpha \simeq 0.24$. Integrating the monochromatic flux from $\nu_1 \simeq 0$ to $\nu_2 = 3$ GHz gives

$$L_{\text{radio}} = 4\pi d^2 \int_{\nu_1}^{\nu_2} F_\nu \, d\nu = 7 \times 10^{36} \text{ W}.$$

---

The bolometric luminosities inferred for quasars range from about $10^{38}$ W to more than $10^{41}$ W, with $5 \times 10^{39}$ W being a typical value; see Fig. 28.16. This implies that the most luminous quasars are on the order of $10^5$ times more energetic than a normal galaxy like our own Milky Way.

### Quasar Spectra

The monochromatic flux of 3C 273 is shown in Fig. 28.14. This continuous spectrum spans nearly 15 orders of magnitude in frequency, very broad compared with the sharply peaked blackbody spectrum of a star. The gentle decline at the low-frequency end of the spectrum reflects the larger-than-average spectral index ($\alpha = 0.24$) for 3C 273 in this regime. (At low frequencies, the spectrum of 3C 273 is dominated by radiation from its jet rather than

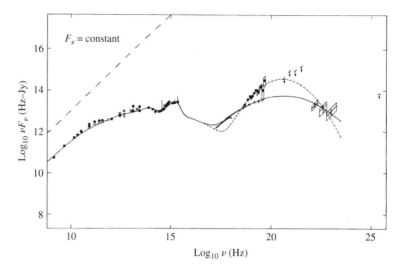

**FIGURE 28.14**    The spectrum of 3C 273, after the Doppler shift of the frequencies due to the Hubble flow has been removed. A horizontal line would correspond to a spectral index of $\alpha = 1$; for reference, the diagonal dashed line shows the slope for $F_\nu = $ constant. The two lines on the right correspond to 3C 273 during quiescence and during an outburst. (Figure adapted from Perry, Ward, and Jones, *MNRAS*, *228*, 623, 1987.)

from its core.) For most other quasars, the spectrum at the low-frequency end falls off more abruptly (smaller $\alpha$). A typical spectrum turns over in the far infrared at a frequency of about $5 \times 10^{12}$ Hz, possibly due to dust and/or synchrotron self-absorption. Also, although some quasars are most luminous at infrared wavelengths and others peak in X-rays, the peak power output of 3C 273 is in the form of low-energy gamma rays.

Quasars emit an excess of ultraviolet light relative to stars and so are quite blue in appearance. For example, the color indices of 3C 48 are $U - B = -0.61$ and $B - V = 0.38$ (you should note that this lies well above the positions of main-sequence stars on the color–color diagram in Fig. 3.11). In Fig. 28.14, this ultraviolet excess is indicated by the big blue bump between roughly $10^{14}$ Hz and $10^{16}$ Hz. A big blue bump is a feature of most (but not all) quasar spectra.

Absorption lines may also be present in some quasar spectra. In particular, Doppler-broadened absorption lines, found in up to 10% of the spectra of quasars, originate from sources with speeds exceeding $10^4$ km s$^{-1}$. These lines are believed to be associated with the quasar itself. Many additional narrow absorption lines are typically seen in the spectra of quasars with high redshifts ($z > 2.2$) due to the Lyman series of hydrogen (see Section 5.3) and metals such as C IV and Mg II. These lines would normally appear at ultraviolet wavelengths but have been redshifted into the visible spectrum by the recessional velocity of the absorbing material. The absorption lines of a given quasar can be placed into different groups that share common redshifts. Furthermore, the redshifts of these narrow absorption lines are nearly always less than the redshift of the quasar's emission lines. The various groupings of lines are thought to arise from clouds of intervening material that lie between the quasar and Earth, as will be discussed in Section 28.4.

## Quasi-Stellar Objects

The distinctive appearance of quasars, starlike with an excess of ultraviolet light, led astronomers to search for more objects fitting this description. In fact, choosing those objects with $U - B < -0.4$ results in a nearly complete list of possible quasars (those at very high $z$ are redder), which must then be confirmed by a spectroscopic analysis. Researchers discovered that about 90% of the confirmed quasar candidates, and AGNs in general, are relatively radio-quiet. For this reason, most of these objects are technically referred to as **quasi-stellar objects** (QSOs), rather than quasars (QSRs).

## Quasar Terminology

Today, the term *quasar* has come to be used almost universally for both radio-loud QSRs and radio-quiet QSOs. As a result, it is common to encounter the descriptions *radio-loud quasars* and *radio-quiet quasars*. However, it is also sometimes the case that QSO is used as an abbreviation for *quasar*. The terminology can be confusing in the literature, so it is important for you to understand the context in which the term is being used. In this text, we will generally use *quasar* to represent both classes of objects, making the distinction between radio-loud and radio-quiet when necessary.[5]

## Ultraluminous Infrared Galaxies

Nearly all quasars have spectra similar to those of broad-line radio galaxies and Seyfert 1s, with bright power-law continua and broad emission lines (both allowed lines and narrower forbidden lines). Seyfert 2 spectra, with their narrow emission lines (both permitted and forbidden), appear to have no counterparts among the quasars. However, some astronomers argue that a subset of the galaxies that were cataloged by the IRAS satellite as being ultraluminous at infrared wavelengths, known as **ultraluminous infrared galaxies** (ULIRGs), should be considered quasars of type 2 rather than starburst galaxies; recall Section 26.1. It is suggested that the infrared light results from dust that absorbs and reradiates the light from the quasar nucleus.

## The High Cosmological Redshifts of Quasars

The **Sloan Digital Sky Survey** (SDSS) has cataloged 46,420 quasars. The brightest entry in the catalog in the $i$ band (centered on a wavelength of 748.1 nm) is the object SDSS 17100.62+641209.0 at a redshift of $z = 2.7356$, having $M_i = -30.242$. The most distant quasar in the catalog is SDSS 023137.65−072854.4 at a redshift of $z = 5.4135$, implying a recessional velocity of more than $0.95c$. In fact, there are 520 quasars in the SDSS catalog with redshifts greater than $z = 4$.

For such large cosmological redshifts, we must abandon using the Hubble law in the forms of Eqs. (27.7) and (27.8) to determine distances. Cosmological redshifts are caused by the expansion of the space through which the light travels, so for extremely large distances the total elongation of the wavelength depends on how the expansion of the universe has

---

[5]It has been pointed out by more than one astronomer that this confusing terminology is also a bit contradictory; to say that a particular quasar is radio-quiet is equivalent to saying that we are discussing a *radio-quiet* quasi-stellar *radio source* (based on the original definition of *quasar*)!

changed with time. The rate of expansion is changing in response to all of the matter and energy in the universe. For this reason, it is customary to quote the redshift, $z$, rather than an actual distance determination. You should keep in mind, however, that the fractional change in wavelength for a cosmological redshift is the same as the fractional change in the size of the universe, $R$, since the time when the light was emitted.[6] That is,

$$z = \frac{\lambda_{obs} - \lambda_{emitted}}{\lambda_{emitted}} = \frac{R_{obs} - R_{emitted}}{R_{emitted}},$$

which gives

$$\boxed{\frac{R_{obs}}{R_{emitted}} = 1 + z.}$$
(28.3)

Thus a redshift of $z = 3$ means that the universe is now four times larger than when the light was emitted.

### Evidence for Quasar Evolution

As was pointed out in Section 26.2, the regions that are cosmologically close to us make up "today's universe." In observing these regions, astronomers can study galaxies as they appear here and now in the present epoch. When looking deeper into the universe, however, we see ancient photons that have been traveling for a long time from a source that may have changed significantly since the light began its journey. This essentially means that looking farther into space implies peering farther into the past. Thus telescopes serve as time machines, providing a window to the early universe.

Bright quasars were certainly more common at earlier epochs than they are now, as evidenced by observations at large and small $z$, respectively. Several factors could contribute to the greater space density of luminous quasars in the past. Both the total number of quasars and their luminosities may have been different then, and it is obviously a difficult task to disentangle these influences. A further complication is introduced by the expansion of the universe. The universe is larger today than it was at a redshift $z$ by a factor of $1 + z$, so the space density of quasars would be greater in the past even if their numbers and luminosities have remained constant. To avoid unnecessary confusion caused by the expansion of the universe, astronomers have defined a **comoving space density** that mathematically removes the effect of the expanding universe. The number of objects per $Mpc^3$ at a redshift $z$ is divided by $(1 + z)^3$, scaling the space density down to the value it would have today (at $z = 0$). The comoving space density of a constant number of nonevolving objects does not change as the universe expands, and so a change in this density implies that the number of objects is varying or that the objects are evolving (or both).

Statistical studies indicate that there are more than 1000 times as many quasars per $Mpc^3$ (comoving space density) brighter than $M_B = -25.9$ at $z = 2$ than there are today ($z = 0$). However, there is strong evidence that the total number of quasars has *not* changed significantly from the present ($z = 0$) back to roughly $z = 2$. Figure 28.15 shows several

---

[6]This fractional change can be measured by the average separation of its constituents, for instance. The meaning of "the size of the universe" will be made more precise in Section 29.1.

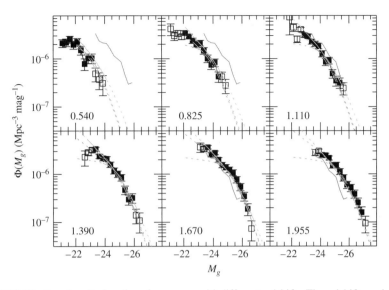

**FIGURE 28.15**   Luminosity functions for quasars with different redshifts. The redshifts are indicated in each frame; for instance, in the upper-left-hand frame, $z = 0.540$. The jagged line that is present in every frame (upper right in $z = 0.540$) represents the data for $z = 1.390$. Note that the population is brighter at greater redshifts. Data are from the Sloan Digital Sky Survey (SDSS) and the 2-degree Field survey (2dF) for 5645 quasars. The $g$ band is centered on 480 nm. (Figure adapted from Richards et al., *MNRAS*, *360*, 839, 2005.)

luminosity functions, $\Phi$, for quasars in different redshift intervals, where $\Phi(M_g)$ is the number of quasars per $\text{Mpc}^{-3}$ (comoving) that have an absolute magnitude between $M_g$ and $M_g + dM_g$. Note that for $z < 2$, the curves would overlap if they were shifted horizontally along the $M_g$-axis. This indicates that for $z < 2$, the populations of quasars with different redshifts differ only in their luminosities, not in their comoving space densities. If this is so, then the scarcity of bright quasars today is an evolutionary effect, caused by a decrease in their luminosities with time. This luminosity evolution of quasars is shown in Fig. 28.16.[7] Apparently, a picture in which a constant number of quasars grows dimmer as the universe expands is consistent with the observations for $z < 2$.

The situation becomes more complicated between $z = 2$ and $z = 3$. Astronomers can study the birth and evolution of quasars out to $z \sim 6$. Statistical surveys at both optical and X-ray wavelengths show that the comoving space density of AGNs peaks at a redshift of approximately $z \approx 2.5$ and then drops off for $z > 3$; see Fig. 28.17. These studies indicate that the comoving space density declines by roughly a factor of 10 from its peak value by $z \approx 4$.

This high-$z$ deficit in the number of quasars could reflect a growth phase of supermassive black holes that power the nascent AGN. You may recall from Fig. 25.15 that a well-defined relationship exists between the mass of a supermassive black hole and the velocity dispersion of the spheroid of a galaxy, suggesting that as the mass of the galaxy grows and the velocity

---

[7]Note that Figure 28.16 assumes a pre-WMAP value of $h = 0.5$ and a specific model for the expansion of the universe that corresponds to a "flat universe" (a deceleration parameter of $q_0 = 0.5$; see Eq. 29.56).

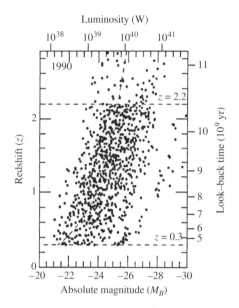

**FIGURE 28.16**    The dimming of quasars with time. For $z < 0.3$, there are too few nearby objects to provide an adequate sample for this figure. The empty region at the upper left has not been sampled by observations in this study. [Figure adapted from Boyle, *The Environment and Evolution of Galaxies*, Shull and Thronson (eds.), Kluwer Academic Publishers, Dordrecht, 1993.]

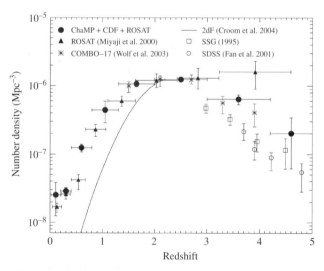

**FIGURE 28.17**    The comoving space density of active galactic nuclei (number per cubic megaparsec) as a function of redshift. (Figure adapted from Silverman et al., *Ap. J.*, *624*, 630, 2005.)

dispersion of its spheroid increases, so does the mass of the central supermassive black hole. Evidence for this evolutionary relationship also appears in Fig. 26.17, which shows a Chandra X-ray image of two merging supermassive black holes in NGC 6240.

There is evidence, including interactions in observed quasars, suggesting that an individual quasar "event" lasts only for a galactic dynamical timescale (the dynamical timescale, essentially the characteristic free-fall or orbital time, was introduced in Section 26.2). Figure 28.18 shows the quasar PKS 2349−014 and a companion galaxy about the size of the LMC (the small bright spot just above the quasar). The thin curved wisps that almost surround the quasar are probably the result of a tidal interaction between the quasar and the companion galaxy. The companion is so close that it will probably merge with PKS 2349−014 in the near future. A portion of the diffuse nebulosity centered on the quasar may (or may not) represent a host galaxy in which it resides.

In 2003, James S. Dunlop, Ross J. McLure, and their colleagues reported on an extensive study of the morphology of the host galaxies of 33 radio-loud quasars, radio-quiet quasars, and radio galaxies in the redshift band $0.1 < z < 0.25$. The study was conducted by using the Hubble Space Telescope in combination with VLA radio imaging. The research team was able to conclude that all of the galaxies in their sample associated with radio-loud quasars or radio galaxies are massive ellipticals. Of the 13 radio-quiet quasars in the sample, 9 are hosted by massive ellipticals while the remaining 4 are in disk/bulge systems. Furthermore, of the 4 disk/bulge systems, the luminosities of 2 of them are dominated by their bulge components, implying that 11 of the 13 radio-quiet quasars (or $\sim 85\%$) are associated with galaxies that are predominantly spheroidal. In addition, the 2 disk-dominated galaxies are the sites of the lowest-luminosity AGNs in the sample by far, and they may be more appropriately considered Seyfert 2 galaxies. From the systems investigated in this study, it appears that all of the true quasars and radio galaxies in the sample are hosted by massive ellipticals that are nearly indistinguishable from lower-$z$ quiescent galaxies typically found near the centers of rich clusters.

**FIGURE 28.18**   The quasar PKS 2349−014 in a gravitational interaction with a companion galaxy. (Figure from Bahcall, Kirhakos, and Schneider, *Ap. J. Lett.*, *447*, L1, 1995. Courtesy of J. Bahcall, Institute for Advanced Study, NASA.)

The study also revealed that all of the radio-loud quasars contained central supermassive black holes of at least $10^9$ $M_\odot$ and that the radio-quiet quasars contained black holes with masses in excess of $5 \times 10^8$ $M_\odot$. It appears from this work that radio-loud systems are much less abundant than radio-quiet systems (10% vs. 90%) simply because the radio-loud systems require more massive central black holes to power the strong radio energy emission. However, although the investigation did find a broad correlation between increasing black hole mass and increasing radio luminosity, the most luminous radio sources cannot be attributed to black hole mass alone. Rather, there is speculation that black hole rotation also may be required to power the strongest radio sources.

In a second, statistical study of 12,698 quasars with redshifts in the range $0.1 < z < 2.1$ found in the Sloan Digital Sky Survey quasar catalog, McLure and Dunlop considered the evolution of black hole masses with increasing redshift. They found that black hole masses sufficient to power quasars were in place by $z \sim 2$. They further determined that all of the central black holes had masses in the range $10^7$ $M_\odot < M_{bh} < 3 \times 10^9$ $M_\odot$, where the upper limit corresponds to the most massive black holes yet found in the local universe (specifically, in M87 and Cygnus A).

The SDSS quasar study was also able to point out that quasar bolometric luminosities increase steadily with redshift from roughly $0.15 L_{Ed}$ at $z \sim 0.2$ to $0.5 L_{Ed}$ at $z \sim 2.0$, where $L_{Ed}$ is the Eddington luminosity (Eq. 10.114). It is also evident from the data that the Eddington luminosity limit remains valid at the high-$z$ end of the study.

### Timescales of AGN Variability

The energy produced by many of the AGNs discussed above (excluding NLRGs and Seyfert 2s) can vary on short timescales. The luminosity of the broad emission lines and continuum of some Seyfert 1 galaxies and quasars can change by a factor of 2 within a few months, weeks, or even days, although there is little or no corresponding variation in the narrow lines. The variation in broad emission lines typically lags behind the continuum variation over similar timescales. There are also variations of a few percent in the visible and X-ray output of Seyfert 1s and quasars on timescales as short as a few minutes, with X-ray fluctuations typically the most rapid. At the other end of the scale, there may be changes of a longer duration. For example, Fig. 28.19 shows that around the year 1937, the quasar 3C 279 brightened by a factor of 250 at visible wavelengths during an outburst that lasted for several years.

### Polarization of the Emission

Quasars typically show low degrees of polarization. At visible wavelengths, the degree of linear polarization is usually less than 3% for both radio-quiet and radio-loud objects, although it may reach up to 35% polarization for a few objects. The radio emission from high-polarization quasars comes mostly from a compact core; such quasars are called **core-dominant radio sources** (and also **compact sources**). These quasars are less polarized at radio wavelengths than the **lobe-dominant sources**, which may reach up to 60% linear polarization. The degree of linear polarization of AGN radio jets is typically 40% but may exceed 50% within a small region. (As discussed in Section 16.7, synchrotron radiation is highly linearly polarized. The lower polarization of the core-dominant sources is probably

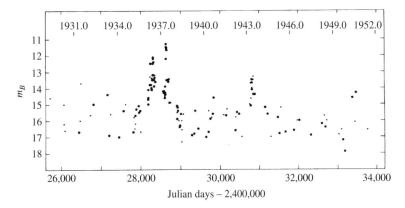

**FIGURE 28.19** The variation in the apparent magnitude of the quasar 3C 279, based on an examination of archival astronomical photographs. (Figure adapted from Eachus and Liller, *Ap. J. Lett.*, *200*, L61, 1975.)

due to synchrotron self-absorption.) Figure 28.20 shows the orientation of the magnetic field of the quasar 3C 47, as obtained from polarization measurements.

### Fanaroff–Riley Luminosity Classes

In 1974, B. L. Fanaroff and J. M. Riley suggested that the radio-loud AGNs in the third Cambridge catalog could be categorized into two general luminosity classes. Fanaroff and Riley defined Class I objects as those for which the ratio of the distance between the brightest spots of radio emission on either side of the center (excluding the central source) to the full extent of the radio source is less than 0.5; Class II objects have a ratio greater than 0.5. An example of an FR I galaxy is NGC 1265, shown in Fig. 28.9, and Cyg A is a classic example of an FR II galaxy (Fig. 28.6). Quasars are also FR II objects.

From the classification scheme, it becomes apparent that FR I sources have diminishing radio luminosity with increasing distance from the center of the jets, while FR IIs tend to be most radio-bright at the ends of the lobes. It is also common that FR I galaxies have two recognizable radio jets, while FR II galaxies often exhibit only a single identifiable jet (the counterjet is either very weak or undetectable). Furthermore, FR I galaxies may have curved jets, while FR II jets tend to be straight.

Also intriguing, and the reason why this morphological classification scheme is referred to as a luminosity classification, is that there is also a rather clear demarcation between FR I and FR II classes in terms of the specific luminosity. Sources having a specific luminosity at 1.4 GHz of less than $10^{25}$ W Hz$^{-1}$ are identified as FR Is, and those with greater specific luminosities are inevitably classified as FR IIs.

### Blazars

The properties of rapid variability and a high degree of linear polarization at visible wavelengths define the class of AGNs known as **blazars**. The most well-known object in this class is BL Lacertae, found in the northern constellation of Lacerta (the Lizard). BL Lac

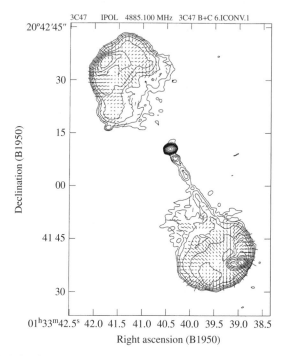

**FIGURE 28.20**    Polarization mapping of the magnetic field of the quasar 3C 47. Both lobes are highly polarized. (Figure adapted from Fernini et al., *Ap. J.*, *381*, 63, 1991.)

was originally classified as a variable star because of its irregular variations in brightness; hence the variable star type of designation. In a week's time BL Lac would double its luminosity, and it would change by a factor of 15 as the months passed. But although BL Lac has a stellar appearance, its spectrum shows only a featureless continuum with very weak emission and absorption lines. Careful observations reveal that the bright, starlike nucleus of BL Lac is surrounded by a fuzzy halo that has a spectrum similar to that of an elliptical galaxy.

**BL Lac objects** are a subclass of blazars that are characterized by their rapid time-variability. Remarkably, their luminosities may change by up to 30% in just 24 hours and by a factor of 100 over a longer time period. BL Lacs are also distinguished by their strongly polarized power-law continua (30–40% linear polarization) that are nearly devoid of emission lines. However, observations of a few faint spectral lines have revealed high redshifts, so that, like quasars, BL Lacs are at cosmological distances. Of those BL Lacs that have been resolved, about 90% appear to reside in elliptical galaxies. Joining the BL Lac objects in the blazar classification are the **optically violently variable quasars** (OVVs). They are similar to the BL Lacs except that they are typically much more luminous, and their spectra may display broad emission lines.

### LINERs

A final class of objects worth mentioning consists of the so-called **Low Ionization Nuclear Emission-line Regions** (LINERs). These galaxies have very low luminosities in their nuclei, but with fairly strong emission lines of low-ionization species, such as the forbidden lines of [O I] and [N II]. The spectra of LINERs seem similar to the low-luminosity end of the Seyfert 2 class, and LINER signatures are detected in many (perhaps a majority of) spiral galaxies in high-sensitivity studies. These low-ionization lines are also detectable in starburst galaxies and in H II regions, and so it is unclear whether LINERs truly represent a low-luminosity limit of the AGN phenomena.

### A Summary of AGN Classifications

This section has introduced a large number of objects with some commonalities and some apparent differences. Before proceeding to a discussion of how we might unify the description of AGN phenomena, we briefly summarize the objects in Table 28.1.

**TABLE 28.1**   A Summary of AGN Classes.

| Class | Sub-class | Description |
|---|---|---|
| Seyferts | Type 1 | broad and narrow emission lines, weak radio emission, X-ray emission, spiral galaxies, variable |
| | Type 2 | narrow emission lines only, weak radio emission, weak X-ray emission, spiral galaxies, not variable |
| Quasars | Radio-loud (QSR) | broad and narrow emission lines, strong radio emission, some polarization, FR II, variable |
| | Radio-quiet (QSO) | broad and narrow emission lines, weak radio emission, weak polarization, variable |
| Radio Galaxies | BLRG | broad and narrow emission lines, strong radio emission, FR II, weak polarization, elliptical galaxies, variable |
| | NLRG | narrow emission lines only, strong radio emission, FR I and FR II, no polarization, elliptical galaxies, not variable |
| Blazars | BL Lacs | almost devoid of emission lines, strong radio emission, strong polarization, rapid variability, 90% in ellipticals |
| | OVV quasars | broad and narrow emission lines, strong radio emission, strong polarization, rapid variability, much more luminous than BL Lacs |
| ULIRGs | | possibly dust-enshrouded quasars, alternatively may be starburst phenomena |
| LINERs | | similar to low-luminosity Seyfert 2, low-ionization emission lines, in many (perhaps majority of) spiral galaxies, alternatively may be starburst phenomena or H II region emission |

## 28.2 ■ A UNIFIED MODEL OF ACTIVE GALACTIC NUCLEI

The last section presented a bewildering array of observations of AGNs. Although there were many similarities, such as a bright compact nucleus, a wide continuum, and time variability, there were also many differences, including the presence or absence of broad emission lines, and the strength of radio and X-ray emission. The question is: Are the types of AGNs fundamentally different or fundamentally the same?

It now seems likely that active galactic nuclei are all powered by the same general engine, accretion onto central supermassive black holes (see page 959ff). Accordingly, the observational differences are due to the different orientations of the objects as viewed from Earth and to the different rates of accretion and masses of the central black holes. The presence of radio lobes is then something in addition to, and consistent with, the basic model.

Although there is not yet general agreement on some of the specific details of a unified model of AGNs, the model does serve to provide a framework for organizing the observations of AGNs and their interpretations. Any model should be self-consistent, meaning that its constitutive elements are all in harmony. Also, as with all viable theories and models in science, a successful AGN model should have the power to predict the results of new observational tests, in addition to explaining all of the observations made previously. Successful predictions have been made on the basis of the idea of unification, and it appears that the basic features of a unified model of AGNs are in hand. The purpose of this section is to demonstrate how the features of such a model may be deduced and to describe the best idea of what an AGN probably looks like.

### Toward a Unified Model of AGNs

First, we will examine two pieces of evidence suggesting that the pursuit of a unified model is indeed justified. Figure 28.21 shows $L_{H\alpha}$, the luminosity in the H$\alpha$ emission line, and $L_{FC}$, the luminosity of the featureless continuum at a wavelength near 480 nm, plotted for a variety of AGNs (excluding blazars). If the hydrogen emission lines are produced via the photoionization of hydrogen atoms by the continuum radiation and the atoms' subsequent recombination, then the two luminosities should be proportional, and a straight line with a slope of 1 should be found on a log–log graph. The slope of the dashed line is 1.05, confirming that $L_{H\alpha} \propto L_{FC}$. This result implies a common origin for the hydrogen emission lines, both broad and narrow, that are observed in AGNs for Seyfert 1 and 2 galaxies, broad- and narrow-line radio galaxies, and radio-loud and radio-quiet quasars.

Another piece of evidence for a unified model comes from an observation reported by Robert Antonucci and Joseph Miller in 1985. When they observed NGC 1068 (a Seyfert 2) in polarized light, they found a Seyfert 1 spectrum with broad emission lines. This and similar cases discovered since then imply that within these Seyfert 2s are Seyfert 1 nuclei that are hidden from the direct view of Earth by some optically thick material. The diminished Seyfert 1 spectrum (normally overwhelmed by the direct Seyfert 2 spectrum) comes from light that reaches us indirectly by reflection from the interstellar medium outside the nucleus. This reflection would also contribute to the observed linear polarization, when the electric field vector is perpendicular to the radio axis. The orientation of the AGN relative to the line of sight from Earth will be an important factor in the unified model to be described.

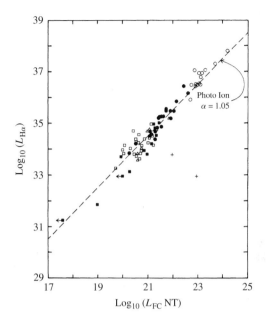

**FIGURE 28.21**   The luminosity in the Hα emission line versus the luminosity of the featureless continuum at a wavelength near 480 nm (the "NT" stands for "nonthermal"). The symbols are quasars (open circles), Seyfert 1s (filled circles), Seyfert 2s (open squares), NLRGs (triangles), and more Seyfert 2s and NLRGs (filled squares). (Figure adapted from Shuder, *Ap. J.*, *244*, 12, 1981.)

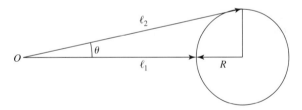

**FIGURE 28.22**   The brightening of a sphere as seen by a distant observer at point $O$.

### The Nature of the Central Engine

The most important clue to the nature of the central engine that powers AGNs is their rapid time variability. Consider an optically thick sphere of radius $R$ that simultaneously (in its own rest frame) brightens everywhere; see Fig. 28.22. The news of the change reaches a distant observer first from the nearest part of the sphere after traveling a distance $\ell_1$, and last from the edge or limb after traveling a distance $\ell_2$. (The back of the sphere isn't seen.) Using

$$\ell_2 = \frac{\ell_1 + R}{\cos \theta} \simeq \ell_1 + R$$

for $R \ll \ell_1$ and $\cos\theta \simeq 1$, the light from the limb of the sphere must travel an additional distance of $\ell_2 - \ell_1 \simeq R$. The brightening is thus smeared out over a time interval $\Delta t = R/c$. In this way, the rapidity of a luminosity change can be used to set an upper limit on the size of the object involved. The high recessional speeds of quasars mean that relativity also plays a role. It is left as an exercise to show that if the sphere described above were moving away from Earth with a velocity $v$, then its radius as determined on Earth would be

$$R = c\Delta t \sqrt{1 - \frac{v^2}{c^2}} = \frac{c\Delta t}{\gamma}, \tag{28.4}$$

where $\gamma$ is the Lorentz factor defined by Eq. (4.20). Using $\Delta t = 1$ hr for a typical value, and taking $\gamma = 1$ for convenience, the radius of the emitting region is no more than

$$R \simeq \frac{c\Delta t}{\gamma} = 1.1 \times 10^{12} \text{ m} = 7.2 \text{ AU}.$$

Considering that AGNs are the most luminous objects known, this is an incredibly small size. Whatever powers an active galactic nucleus would fit comfortably within our Solar System!

The typical quasar luminosity of $5 \times 10^{39}$ W is equivalent to more than 360 Milky Way galaxies. However, there is an upper limit to the luminosity, $L$, of any spherically symmetric object that is in equilibrium. It must be less than the Eddington limit, $L < L_{Ed}$, where, from Eq. (10.114),

$$L_{Ed} \simeq 1.5 \times 10^{31} \text{ W} \left(\frac{M}{M_\odot}\right).$$

For a luminosity of $L = 5 \times 10^{39}$ W, this provides a lower limit for the mass:

$$M > \frac{L}{1.5 \times 10^{31} \text{ W}} \, M_\odot = 3.3 \times 10^8 \, M_\odot. \tag{28.5}$$

Finding such a large amount of mass in such a small space is clear evidence for a supermassive black hole. Using Eq. (17.27) for the Schwarzschild radius, the mass of a black hole with the radius $R$ found in Eq. (28.4) is

$$M = \frac{Rc^2}{2G} = 3.7 \times 10^8 \, M_\odot.$$

The fact that these two mass estimates are of the same order of magnitude is enough to support the idea that supermassive black holes are involved in powering AGNs. For the rest of this section, we will assume a value of $10^8 \, M_\odot$ for a typical mass, which corresponds to a Schwarzschild radius of $R_S \simeq 3 \times 10^{11}$ m $\simeq 2$ AU $\simeq 10^{-5}$ pc.

### Generating Luminosity through Accretion

The most efficient way of generating energy is by the release of gravitational potential energy through mass accretion. As was shown in Example 18.1.2 for matter falling onto the surface of a 1.4 $M_\odot$ neutron star, about 21% of the rest energy is released. However, dropping matter straight down onto a black hole is very *inefficient* because there is no surface for the mass to strike. Instead, according to an observer at a great distance, a freely falling mass would slow to a halt and then disappear as it approached the Schwarzschild radius, $R_S$. On the other hand, as matter spirals in toward a black hole through an accretion disk, a substantial fraction of the rest energy can be released as viscosity converts kinetic energy into heat and radiation.

For a nonrotating black hole, the smallest stable circular orbit for a massive particle (and therefore the inner edge of an accretion disk) is at $r = 3R_S$. At this location, theoretical calculations show that the gravitational binding energy is 5.72% of the particle's rest-mass energy, so mass spiraling down through an accretion disk would release this much energy. The situation is even more favorable for a rotating black hole because the event horizon is located at a smaller $r$. For the most rapidly rotating black hole possible, both the event horizon and the smallest stable prograde orbit are at $r = 0.5R_S$ (the smallest stable retrograde orbit is at $4.5R_S$). The gravitational binding energy in this case of maximal rotation is calculated to be 42.3% of a particle's rest mass.

The **accretion luminosity** generated by a mass accretion rate, $\dot{M}$, through the disk may be written as

$$L_{\text{disk}} = \eta \dot{M} c^2, \tag{28.6}$$

where $\eta$ is the efficiency of the process, $0.0572 \le \eta \le 0.423$. (For comparison, we found in Example 18.1.2 that the efficiencies for accretion onto a 0.85 $M_\odot$ white dwarf and a 1.4 $M_\odot$ neutron star are $\eta = 1.9 \times 10^{-4}$ and 0.21, respectively.)

The accretion of matter through a disk around a rapidly rotating black hole is an extremely efficient way of producing large amounts of energy. Furthermore, the smallest stable prograde orbit lies inside the ergosphere of a rapidly rotating black hole, and frame dragging guarantees that the accreting matter will rotate along with the black hole (recall Section 17.3). For these reasons, most astronomers believe that an accretion disk around a supermassive black hole is an essential ingredient of a unified model of AGNs. Figure 28.10 shows the spiral-shaped disk of gas that lies at the center of M87. The inner edge of the disk is rotating with a speed of about 550 km s$^{-1}$, causing the light from the lower right-hand edge of the disk to be blueshifted (approaching), while the light from the upper-right is redshifted (receding). The central supermassive black hole is calculated to have a mass of about $3 \times 10^9$ $M_\odot$.

Recall from Example 18.2.1 that the inner regions of accretion disks around white dwarfs and neutron stars are bright at ultraviolet and X-ray wavelengths, respectively. It might be expected that an accretion disk around a supermassive black hole would be a source of photons of even higher energies, but this is not the case. Because they are supported by degeneracy pressure, white dwarfs and neutron stars obey the mass–volume relation, Eq. (16.14), which states that these stars become smaller with increasing mass. Therefore, the accretion disks around more massive white dwarfs and neutron stars penetrate deeper into their gravitational potential wells. The Schwarzschild radius, however, increases with increasing mass, and so the characteristic disk temperature, $T_{\text{disk}}$ (Eq. 18.20), decreases as the mass of the black hole increases. (Recall that the equation for $T_{\text{disk}}$ was derived using Newtonian physics, but in this case a full relativistic treatment is clearly warranted.)

To see this, we will assume a rapidly rotating black hole and adopt $R = 0.5R_S = GM/c^2$ for the location of the inner edge of the accretion disk. Substituting $R$ into Eq. (18.20), the characteristic disk temperature becomes

$$T_{\text{disk}} = \left( \frac{3c^6 \dot{M}}{8\pi \sigma G^2 M^2} \right)^{1/4}. \tag{28.7}$$

For a disk radiating at a fraction $f_{\text{Ed}}$ of the Eddington limit,

$$f_{\text{Ed}} \equiv L_{\text{disk}}/L_{\text{Ed}}. \tag{28.8}$$

Equations (10.114) and (28.6) give

$$\eta \dot{M} c^2 = f_{\text{Ed}} \frac{4\pi G c}{\overline{\kappa}} M,$$

or

$$\dot{M} = \frac{f_{\text{Ed}}}{\eta} \frac{4\pi G}{\overline{\kappa} c} M. \tag{28.9}$$

Substituting this expression into Eq. (28.7) shows that

$$T_{\text{disk}} = \left( \frac{3c^5 f_{\text{Ed}}}{2\overline{\kappa}\sigma G M \eta} \right)^{1/4}, \tag{28.10}$$

and so for the disk temperature, $T_{\text{disk}} \propto M^{-1/4}$.

---

**Example 28.2.1.**   Consider an accretion disk around a rapidly rotating supermassive black hole of $10^8 \ M_\odot$. The value of $f_{\text{Ed}}$ is probably close to 1 for luminous quasars and roughly between 0.01 and 0.1 for Seyfert galaxies. In this example, let the disk luminosity be equal to the Eddington limit ($f_{\text{Ed}} = 1$), so from Eq. (10.114), $L = 1.5 \times 10^{39}$ W. Also, we will adopt $\eta = 0.1$ as a representative accretion efficiency. The mass accretion rate required to maintain the disk luminosity is

$$\dot{M} = \frac{f_{\text{Ed}}}{\eta} \frac{4\pi G}{\overline{\kappa} c} M = 1.64 \times 10^{23} \text{ kg s}^{-1} = 2.60 \ M_\odot \text{ yr}^{-1}.$$

Luminous quasars must be fed at a rate of around 1 to 10 $M_\odot$ yr$^{-1}$. Less luminous AGNs may have correspondingly smaller appetites. The characteristic disk temperature is

$$T_{\text{disk}} = \left( \frac{3c^5 f_{\text{Ed}}}{2\overline{\kappa}\sigma G M \eta} \right)^{1/4} = 7.30 \times 10^5 \text{ K},$$

where Eq. (9.27) with $X = 0.7$ has been used for the opacity due to electron scattering. According to Wien's displacement law (Eq. 3.19), the spectrum of a blackbody with this temperature peaks at a wavelength of 39.7 nm, in the extreme ultraviolet region of the electromagnetic spectrum.

Although this expression for $T_{\text{disk}}$ is at best a rough estimate of the characteristic disk temperature, temperatures of several hundred thousand kelvins agree with the results of more realistic disk calculations.

---

It is thought that the big blue bump observed in the spectra of quasars is the thermal signature of an underlying accretion disk. However, although accretion is believed to provide the power for AGNs, the theoretical spectrum of an accretion disk (such as that shown in Fig. 18.14) cannot account for the wide continuum that is actually observed.

### Structure of the Accretion Disk

A detailed model of the accretion disk around a supermassive black hole is difficult to derive because the high luminosities involved must have a significant effect on the disk's structure. Theoretical calculations indicate that the structure of the accretion disk depends on $f_{Ed}$ (Eq. 28.8). Several possible structures have been identified. If $f_{Ed} < 0.01$, then the density of the disk is too small for efficient cooling. The energy generated by the disk's viscosity (internal friction) cannot be radiated away efficiently, and the disk puffs up into an **ion torus** that is supported by the pressure of the hot ions. Part or all of the disk would then resemble a doughnut around the central black hole. Values of $0.01 < f_{Ed} < 0.1$ or so imply a geometrically thin disk, as described in Section 18.2 for close binary systems (recall that, by definition, at any radial distance $r$ in a thin disk, the vertical height $h \ll r$). As the value of $L_{disk}$ becomes super-Eddington ($f_{Ed} > 1$), the radiation pressure that is generated balances the force of gravity and the photons are capable of supporting the matter in an inflated **radiation torus**.

One scenario involves a composite disk that has three regions, as shown in Fig. 28.23. Within about $1000R_S$ of the center, radiation pressure exceeds the gas pressure, resulting in a thick, hot disk. This is the probable origin of the big blue bump in the continuous spectrum. Exterior to this, reaching out to some $10^5 R_S$ ($\simeq 1$ pc for $M = 10^8$ M$_\odot$), is a thin disk that is supported by gas pressure. This part of the disk flares outward, becoming thicker with increasing radius. The concave surface of the outer disk means that it can be irradiated by the central source or the thick, hot portion of the inner disk, resulting in a wind flowing

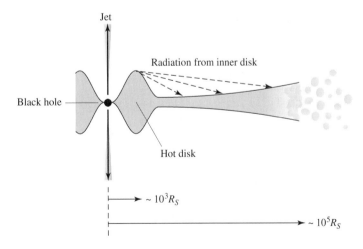

**FIGURE 28.23**   A schematic structure of the accretion disk in an AGN. The radial direction is not drawn to scale.

outward from the disk. Finally, beyond about $10^5 R_S$, the thick disk breaks up into numerous small clouds.

Unfortunately, some problems with this picture remain. For example, the values of $f_{Ed}$ quoted above for Seyferts appear to be incompatible with models of thick disks.

### The Implications of AGN Spectra

For the blazars, at least, there is widespread agreement that the continuous spectrum, with its power-law form and significant polarization, is produced by synchrotron radiation. As we saw in Section 16.7 for pulsars, synchrotron radiation is produced when relativistic charged particles, such as electrons, spiral around magnetic field lines. (For objects other than blazars, the situation is more complicated. Recall that the big blue bump observed in the continua of other types of AGNs is believed to be thermal radiation. In addition, dust emission plays an important role in the infrared.)

With a wide range of photon energies available for ionizing atoms, synchrotron radiation can account for the variety of ionization states observed in the emission line spectra of AGNs. For example, a number of ionization states have been seen for forbidden lines, including [O I] and [Fe X]. Furthermore, synchrotron radiation can be up to 70% linearly polarized, in agreement with the high degrees of polarization observed for some AGNs.

### Producing a Relativistic Outflow of Charged Particles

It might seem surprising to find a magnetic field involved in this situation. Although it is theoretically possible for an isolated black hole to have a magnetic field, it is unlikely to occur naturally. This is because although the three attributes of a black hole (its mass, angular momentum, and electric charge) can be combined to produce a magnetic field, black holes should be essentially electrically neutral, since any net charge acquired by a black hole would be rapidly canceled as it attracted charge of the opposite sign. However, the ionized disk material is highly conducting, so there can be a magnetic field that is generated by the accretion disk as the disk orbits the black hole.

It may be that, in a manner reminiscent of the mechanism described for pulsars in Section 16.7, the varying magnetic field near the surface of the disk induces a large electric field that is capable of accelerating charged particles away from the disk. As the particles move outward, they are accelerated to relativistic speeds while they spiral around the magnetic field lines that rotate with the disk. Because the field lines are anchored to the conducting disk, the particle energy ultimately comes at the expense of the accretion energy.

There is another source of energy that taps the rotational energy of the black hole itself, first described by Roger Blandford and Roman Znajek. Detailed calculations show that the rotating black hole can be thought of as a spinning conductor in a magnetic field; see Fig. 28.24. Just as the motion of a conducting wire through a magnetic field will produce an electromotive force (emf) between its ends, the rotation of a black hole in a magnetic field will produce a potential difference between its poles and its equator. The effective resistance of the rotating black hole between its poles and equator is about 30 $\Omega$. In this picture, the black hole acts like an immense battery connected to a 30-$\Omega$ resistor. Power is extracted from the black hole's rotational energy, just as if a current of charged particles were flowing through the resistor in response to the voltage difference. As a result, the black

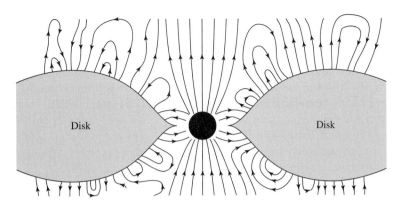

**FIGURE 28.24**   An accretion disk and its magnetic field orbiting a rotating black hole.

hole's rate of spin is reduced. The power generated by the **Blandford–Znajek mechanism** is approximately

$$P \simeq \frac{4\pi}{\mu_0} B^2 R_S^2 c \tag{28.11}$$

$$= 2.7 \times 10^{38} \text{ W} = 7.1 \times 10^{11} \text{ L}_\odot$$

for a $10^8$ $M_\odot$ black hole with $R_S = 3 \times 10^{11}$ m and a magnetic field of 1 T. The energy is in the form of electromagnetic radiation and a flow of relativistic pairs of electrons and positrons. Up to 9.2% of the rest energy of a maximally rotating black hole may be extracted in this manner. This is another important source of energy, comparable to that obtained by disk accretion.

The two processes just described appear capable of producing a relativistic outflow of charged particles, although the mechanism is still uncertain. As the electrons spiral around the magnetic field lines, they emit the synchrotron radiation that contributes to the continuous spectrum of AGNs. (You will recall from Section 28.1 that the observed power-law synchrotron spectra of AGNs imply that the distribution of electron energies is described by a power law. Just how this distribution comes about is as yet unknown, but power-law synchrotron spectra are also observed for supernova remnants such as the Crab Nebula.)

### The Generation of X-Rays

AGNs can be very bright in X-rays,[8] and several mechanisms are usually invoked to explain the excess over that produced directly by synchrotron radiation. The high-frequency end of the accretion disk spectrum may be sufficient to account for the soft (low-energy) X-rays. Lower-energy photons from other sources may also be scattered to much higher

---

[8]X-rays from AGNs are probably responsible for the *X-ray background*, a uniform glow of X-ray photons that pervades the sky. Discovered in 1962, the X-ray background extends into the gamma-ray region of the electromagnetic spectrum.

energies by collisions with relativistic electrons. As the name suggests, this **inverse Compton scattering** is the reverse of the Compton scattering process described in Section 5.2. In addition, inverse Compton scattering may produce the gamma rays coming from the quasar 3C 273. Thermal bremsstrahlung, the mechanism that produces the X-ray emission observed in clusters of galaxies, has a characteristic spectrum (Fig. 27.17) that could also be consistent with observations of X-rays from AGNs.

### Broad-Line and Narrow-Line Emission

The characteristic broad emission lines (when present) and narrow emission lines of AGNs are the result of photoionization by the continuum radiation. A careful examination of these lines reveals much about the conditions under which they were formed. All of the broad lines arise from permitted atomic transitions, but none of them involve the forbidden transitions seen in some narrow lines. The broad H$\alpha$ and H$\beta$ lines vary on timescales of a month or less, while the narrow lines seem to vary little, if at all. This evidence, along with the discovery that Seyfert 2 galaxies may harbor Seyfert 1 nuclei that are hidden from direct view by some obscuring material, indicates that the broad and narrow lines in the spectra of AGNs originate in different regions under different conditions.

### The Broad-Line Region

The broad emission lines observed in the spectrum of many AGNs are formed in a **broad-line region** that is relatively close to the center. A study of the Seyfert galaxy NGC 4151 revealed that when the intensity of the continuum radiation varied, most of the broad emission lines responded very quickly, within a month or less and perhaps as quickly as one week. Light can travel a distance of nearly $10^{15}$ m in 30 days, so this provides a rough estimate of the radius of the broad-line regions for Seyferts and BLRGs. The variation of the lines in quasars takes place more slowly, so their broad-line regions may be larger by a factor of four or so. A study of the broad Fe II emission lines that are usually present indicates that the temperature in the broad-line region is $\sim 10^4$ K. Other lines indicate that the number density of electrons probably lies between $10^{15}$ m$^{-3}$ and $10^{16}$ m$^{-3}$. Forbidden lines will not be seen with large number densities such as these because of the frequent collisions between the atoms. Atoms and ions with electrons in the long-lived metastable states that give rise to forbidden lines are deexcited by collisions before downward radiative transitions can occur. As a result, the forbidden lines are much weaker than the permitted lines.

There is widespread agreement that the broad-line region must be clumpy, containing partially ionized clouds of gas, rather than being homogenous. The optically thick clouds that actually produce the emission lines fill only about 1% of the available volume and probably have a flattened distribution. These regions of high density may be surrounded by a rarefied, high-temperature medium that prevents the clouds from dispersing.

According to the unified model, the various types of observed AGN phenomena (see Table 28.1) derive from different viewing angles of the central engine and surrounding environment. The unified model postulates that a large, optically thick torus of gas and dust surrounds the clouds of the broad-line region. This is presumably what conceals the broad-line region and the central source from direct view when observing a Seyfert 2 galaxy; see Fig. 28.25. In this case, the continuum and emission lines must reach the observer indirectly

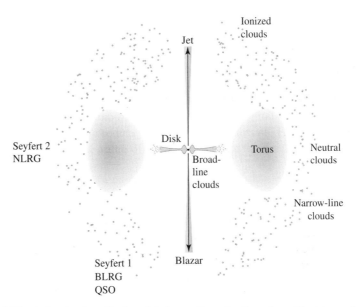

**FIGURE 28.25**   A sketch of a unified model of an active galactic nucleus. The jets would be present in a radio-loud AGN. A typical observer's point of view is indicated for AGNs of various types.

by reflected light, which explains why the continuum of Seyfert 2s is much fainter than the continuum of Seyfert 1s. Overall, the light received directly from the central nucleus makes Seyfert 1 galaxies generally brighter than Seyfert 2s. The torus is also thought to be opaque to soft X-rays because the X-rays observed for Seyfert 2s are hard (with energies above approximately 5 keV).

### Determining Black Hole Masses in Broad-Line Regions

The broad emission lines also indicate that the clouds orbit the central supermassive black hole. In fact, taking the 5000 km s$^{-1}$ width of the lines to be an orbital velocity and using $r = 10^{15}$ m as an orbital radius provides an estimate of the central mass. From Eq. (24.49),

$$M_{\text{bh}} = \frac{rv^2}{G} = 1.9 \times 10^8 \ M_\odot,$$

which is consistent with previous mass estimates.

A second technique for determining the masses of the central black holes of AGNs in broad-line regions is based on measuring the lag time between changes in brightness of the continuum and emission lines. This **reverberation mapping** technique combines the measured time delay, $\tau$, with the root-mean-square width of the emission line, $\sigma_{\text{line}}$, giving a mass estimate of

$$M_{\text{bh}} = \frac{f c \tau \sigma_{\text{line}}^2}{G}, \tag{28.12}$$

where $c$ is the speed of light and $f$ is a factor that depends on the structure, kinematics, and

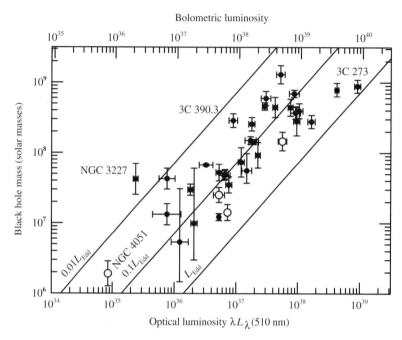

**FIGURE 28.26** The masses of the supermassive black holes in AGNs as a function of their luminosities. The solid diagonal lines represent lines of constant fractions of the Eddington luminosity. The masses were determined by the reverberation mapping technique. The bolometric luminosities are given on the top axis, and the optical luminosities (centered at 510 nm) are given on the bottom axis. (Figure courtesy of Bradley Peterson.)

orientation of the broad-line region. Clearly $c\tau$ is a measure of the size of the broad-line region. Of course it is important to be consistent in terms of which ionized species are used in making the measurement (in other words, which emission line is used in the study) because highly ionized species tend to have the shortest lag times, suggesting that the level of ionization in the broad-line region is dependent on distance from the central source.

By comparison with other black hole–spheroid relations, such as the mass–velocity dispersion relation displayed in Fig. 25.15 for a sample of resolved galactic centers, the scaling factor in Eq. (28.12) is found to have the value $f \approx 5.5$. The reverberation mapping method requires long-term observations to determine the lag times, and high spectral resolution, but it does not require spatially resolving the central region of the host galaxy. Thus, the reverberation technique holds great promise for measuring black hole masses in AGNs that are at high $z$. Figure 28.26 shows the results of a study of AGN broad-line emission regions.

### The Narrow-Line Region

Outside the opaque torus is the **narrow-line region** where the narrow emission lines originate. The number density of electrons in the narrow-line region is only about $10^{10}$ m$^{-3}$, comparable to the values found in planetary nebulae and dense H II regions. The narrow-line region contains more mass than the broad-line region, and both permitted and forbidden lines can be formed in such an environment. They reveal a temperature of approximately

$10^4$ K. Like the broad-line region, the region that generates the narrow-line spectrum is clumpy. It is probably composed of a more or less spherical distribution of clouds. The clouds that are far enough above or below the plane of the obscuring torus can be illuminated and photoionized by the continuum radiation from the center. Other clouds have their lines of sight to the central source blocked by the opaque torus, and so they remain neutral.

In fact, if the narrow-line region can be treated as a clumpy H II region, then Eq. (12.32) for the Strömgren radius can be used to estimate the fraction of the narrow-line region that is occupied by clouds. If the clouds occupy a fraction $\epsilon$ (referred to as the **filling factor**) of the volume of the narrow-line region, then Eq. (12.32) can be modified to produce an estimate of the radius of that region,

$$r_{\text{NLR}} \approx \left( \frac{3N}{4\pi \alpha_{\text{qm}} \epsilon} \right)^{1/3} \frac{1}{n_e^{2/3}} . \tag{28.13}$$

In this case $N$ is the number of photons per second produced by the central source of the AGN that have enough energy to ionize hydrogen from the ground state, and $\alpha_{\text{qm}}$ is a quantum-mechanical recombination coefficient (*not* the spectral index), as given on page 432.

---

**Example 28.2.2.**   To estimate the filling factor of the narrow-line region, we will assume an AGN luminosity of $L = 5 \times 10^{39}$ W. The continuum includes photons with a wide range of energies. We will assume that the monochromatic energy flux (Eq. 28.1) obeys a power law with a spectral index of $\alpha = 1$. Recalling Example 28.1.1, the flux is related to the luminosity by

$$L = 4\pi d^2 \int_{\nu_1}^{\nu_2} F_\nu \, d\nu = \int_{\nu_1}^{\nu_2} C\nu^{-1} \, d\nu$$

with $\nu_1 = 10^{10}$ Hz and $\nu_2 = 10^{25}$ Hz for the range of frequencies of the continuous spectrum (cf., Fig. 28.14), and $C$ is a constant to be determined.[9] Evaluating the integral and solving for $C$ gives

$$C = \frac{L}{\ln(\nu_2/\nu_1)} = \frac{L}{\ln 10^{15}} = 0.029L.$$

We are now ready to find $N$, the number of photons emitted per second with an energy $E_H > 13.6$ eV, or a frequency $\nu_H > 3.29 \times 10^{15}$ Hz, required to ionize hydrogen from the ground state. Dividing the monochromatic energy flux by the energy per photon, $E_{\text{photon}} = h\nu$, results in

$$N = \int_{\nu_H}^{\nu_2} \frac{C\nu^{-1}}{h\nu} \, d\nu = \int_{\nu_H}^{\nu_2} \frac{0.029L}{h\nu^2} \, d\nu \simeq \frac{0.029L}{h\nu_H} = 6.64 \times 10^{55} \ \text{s}^{-1},$$

where $\nu_2 \gg \nu_H$.

*continued*

---

[9]The constant $C$ includes the leading factor of $4\pi d^2$.

Observations of the nearest Seyfert 2 galaxies show narrow-line regions with diameters between roughly 100 and 1000 pc. If we let $r_{\mathrm{NLR}} = 200$ pc, $n_e = 10^{10}$ m$^{-3}$, and $\alpha_{\mathrm{qm}} = 3.1 \times 10^{-19}$ m$^3$ s$^{-1}$ (as on page 432), the filling factor of the narrow-line region is approximately

$$\epsilon \approx \frac{3N}{4\pi\,\alpha_{\mathrm{qm}}} \frac{1}{n_e^2 r_{\mathrm{NLR}}^3} = 2.2 \times 10^{-2}.$$

Thus clouds occupy roughly 2% of the volume of the narrow-line region.

---

The profiles of the narrow emission lines seen in Seyfert 2s often have extended blue wings, indicating that the clouds are moving toward us relative to the galactic nucleus. This is usually interpreted as a radial flow of the clouds away from the center. The light from the clouds moving away from us on the far side of the AGN is presumably diminished by extinction.

An outward flow of clouds in the region that produces narrow emission lines could be driven by a combination of radiation pressure and a wind coming from the accretion disk, as mentioned previously, or the outflow could be associated with the material in radio jets. Figure 28.27 shows an image obtained by the Hubble Space Telescope of the narrow-line

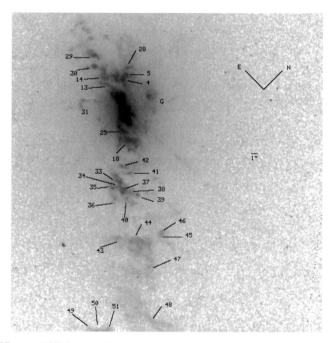

**FIGURE 28.27**    An HST image of the narrow-line region of the Seyfert 1 galaxy, NGC 4151. Numerous clouds are evident in a biconical distribution. The clouds to the southwest are approaching the observer relative to the nucleus, and the clouds to the northeast have recessional velocities. There is some evidence that the clouds may be associated with the galaxy's radio jets. An angular scale of $1''$ is indicated on the image; at the distance to NGC 4151, $1''$ corresponds to a projected linear distance of 63 pc. The labels correspond to clouds identified in the paper. (Figure from Kaiser et al., *Ap. J.*, *528*, 260, 2000.)

region for the Seyfert 1 galaxy, NGC 4151. Distinct emission clouds are clearly evident in this high-resolution image. It is also evident that the optical emission falls within two conical distributions extending to the northeast and the southwest of the center of the galaxy. When a radio map is overlaid on the image, the radio emission also falls along the same axis as the biconical optical emission.

Similar observations have been made of Mrk 3, a Seyfert 2 galaxy. In this case there is additional evidence that the narrow-line region is composed of expanding shells around the radio jets. It has been suggested that the material forming the radio jets is expanding at near $0.1c$ from the center of the galaxy. As the jets move through the interstellar material, the gas is ionized at a temperature of $\sim 10^7$ K. The overheated gas expands outward away from the jets, energizing the gas near the surface of the expanding shell, which then produces the narrow-line emission region.

A further consequence of the expanding jet model of producing the narrow-line region is that the region may be relatively short-lived, at least in the case of the relatively weak Seyfert systems. Since the radio jets of Seyfert galaxies are typically only a few kiloparsecs in length, this implies that ages of roughly $10^4$ to $10^5$ years can be deduced based on the expansion velocity of the jets. If that is the case, Seyfert phenomena may be relatively transient events, perhaps caused by the temporary feeding of the galaxy's supermassive black hole.

**A Summary of the Unified Model of AGNs**

The preceding details comprise what must be considered as a rough sketch of a unified model of an AGN. Its central engine is an accretion disk orbiting a rotating, supermassive black hole. The AGN is powered by the conversion of gravitational potential energy into synchrotron radiation, although the rotational kinetic energy of the black hole may also serve as an important energy source. The structure of the accretion disk depends on the ratio of the accretion luminosity to the Eddington limit. To supply the observed luminosities, the most energetic AGNs must accrete between about 1 and 10 $M_\odot$ yr$^{-1}$. The perspective of the observer, together with the mass accretion rate and mass of the black hole, largely determines whether the AGN is called a Seyfert 1, a Seyfert 2, a BLRG, a NLRG, or a radio-loud or radio-quiet quasar.

Although many of the details of the unified model have not yet been fully confirmed, the unified model does appear to provide an important framework for describing many of the general characteristics of active galaxies. For example, Fig. 28.28 shows an amazing HST image of NGC 4261, an elliptical radio galaxy in the Virgo cluster that is classified optically as a LINER. The core of this radio-loud object shows a bright nucleus surrounded by a large, obscuring torus that is perpendicular to the radio jets. The central object is probably a $10^7$ $M_\odot$ black hole, although the HST image does not have the resolution to confirm this. The torus has a radius of about 70 pc ($2 \times 10^{18}$ m), and the jets reach out some 15 kpc from the nucleus.[10]

---

[10]The Schwarzschild radius of a $10^7$ $M_\odot$ black hole is $2.95 \times 10^{10}$ m, smaller than the torus by nearly eight orders of magnitude. The black hole would be less than a nanometer wide on the scale of Fig. 28.28.

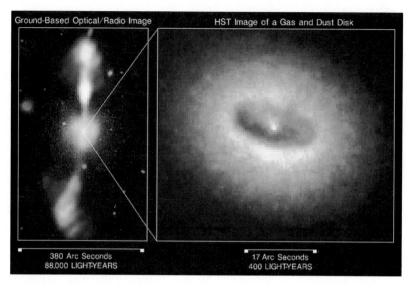

**FIGURE 28.28**    Two views of NGC 4261. Left: a composite optical and radio image from the ground, showing the radio jets. Right: an optical image from the HST, showing the dusty torus around the nucleus. (Courtesy of NASA.)

## 28.3 ■ RADIO LOBES AND JETS

There is a basic division of active galaxies into objects that are radio-loud and those that are radio-quiet. Radio-loud sources usually consist of a radio core, one or two detectable jets, and two dominant radio lobes. The radio-quiet sources are less luminous at radio wavelengths by a factor of $10^3$ to $10^4$, consisting of a weak radio core and perhaps a feeble jet. The increased level of activity in radio-loud AGNs is not confined to radio wavelengths, however; they also tend to be about three times brighter in X-rays than their radio-quiet cousins.

### The Generation of Jets

The radio lobes are produced by jets of charged particles ejected from the central nucleus of the AGN at relativistic speeds; recall Fig. 28.8. These particles are accelerated away from the nucleus in two opposite directions, powered by the energy of accretion and/or by the extraction of rotational kinetic energy from the black hole via the Blandford–Znajek mechanism. The jet must be electrically neutral overall, but it is not clear whether the ejected material consists of electrons and ions or an electron–positron plasma. The latter, being less massive, would be more easily accelerated. The disk's magnetic field is coupled ("frozen in") to this flow of charged particles. The resulting magnetic torques may remove angular momentum from the disk, which would allow the accreting material to move inward through the disk.

The incredible narrowness and straightness of some jets means that a collimating process must be at work very near the central engine powering the jet. A thick, hot accretion

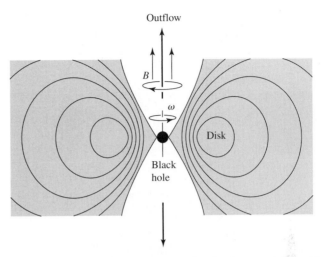

**FIGURE 28.29** A schematic showing the collimation of outflowing material by a thick, hot accretion disk. The loops represent contours of constant disk density.

disk around the black hole could provide natural collimation by funneling the outflowing particles, as shown in Fig. 28.29. Because the accreting material retains some angular momentum as it spirals inward through the disk, it will tend to pile up at the smallest orbit that is compatible with its angular momentum. Inside this "centrifugal barrier" there may be a relatively empty cavity that can act as a nozzle, directing the accreting gases outward along the walls of the cavity. However, producing highly relativistic jets, as frequently observed, appear to be difficult to accomplish with this nozzle mechanism.

Alternatively, magnetohydrodynamic (MHD) effects could play an important role in accelerating and collimating the relativistic flows. Unfortunately, details of MHD mechanisms have not yet been fully developed either. Whatever the specific details of the collimation of jets, their straightness is likely to be linked to the rotating supermassive black holes at the hearts of AGNs.

**The Formation of Radio Lobes**

As a jet of material travels outward, its energy primarily resides in the kinetic energy of the particles. However, the jet encounters resistance as it penetrates the interstellar medium within the host galaxy and the intergalactic medium beyond. As a result, the material at the head of the jet is slowed, and a shock front forms there. The accumulation and deceleration of particles at the shock front cause the directed energy of the jet to become disordered as the particles "splash back" to form a large lobe in which the energy may be shared equally by the kinetic and magnetic energy. The problem of calculating the motion of a jet through the intergalactic medium is so complicated that extensive numerical simulations are required to model the process. Figure 28.30 shows a series of computer simulations of jets with various initial energies working their way through the intergalactic medium.

The motion of the charged particles and the magnetic fields within the lobes of radio-loud objects contain an enormous amount of energy. For Cygnus A, the energy of each

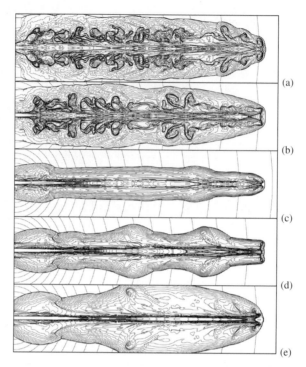

**FIGURE 28.30**    Numerical simulation of electron–positron plasma jets moving through the intergalactic medium, which is assumed to be decreasing in density with increasing distance from the source of the jets (left-hand side of each frame). The frames correspond to initial Lorentz factors ($\gamma$) at the source of the jets of (a) 2.0, (b) 2.5, (c) 5.0, (d) 7.0, and (e) 10.0. Somewhat different behaviors are seen in the simulations when the jet material is assumed to be composed of electrons and protons. (Figure from Carvalho and O'Dea, *Ap. J. Suppl.*, *141*, 371, 2002.)

lobe is estimated to be approximately $10^{53}$ to $10^{54}$ J, equivalent to the energy liberated by $10^7$ supernovae.

---

**Example 28.3.1.**    Assuming that each radio lobe of Cyg A contains an energy of $E_{\text{lobe}} = 10^{53}$ J, and adopting $h = [h]_{\text{WMAP}} = 0.71$ for the values given in Example 28.1.1 for Cyg A, the lifetime of the radio lobes can be estimated. With Cyg A's radio luminosity of $L_{\text{radio}} = 4.8 \times 10^{37}$ W, the time to radiate away the energy stored in its radio lobes is

$$t_{\text{lobe}} = \frac{E_{\text{lobe}}}{L_{\text{radio}}} = 66 \text{ Myr}.$$

Generally, the lifetime of the radio emission from radio lobes ranges from $10^7$ to more than $10^8$ years.

---

The average strength of the magnetic field in the lobes can be estimated by making the common assumption that the energy is shared equally between the kinetic and magnetic energy. From Eq. (11.9), the magnetic energy stored per unit volume is $u_m = B^2/2\mu_0$. If the volume of the lobe is $V_{\text{lobe}}$, then

$$\frac{1}{2} E_{\text{lobe}} = u_m V_{\text{lobe}} = \frac{B^2 V_{\text{lobe}}}{2\mu_0}$$

or

$$B = \sqrt{\frac{\mu_0 E_{\text{lobe}}}{V_{\text{lobe}}}}. \tag{28.14}$$

---

**Example 28.3.2.**    Assume that each of Cyg A's radio lobes can be modeled as a sphere of radius $R = 8.5$ kpc $= 2.6 \times 10^{20}$ m, characteristic of the size of the lobes. With $E_{\text{lobe}} = 10^{53}$ J, the average value of the magnetic field in the lobes is estimated to be

$$B = \sqrt{\frac{\mu_0 E_{\text{lobe}}}{\frac{4}{3}\pi R_{\text{lobe}}^3}} \approx 41 \text{ nT}.$$

A value of order 10 nT is typical of the bright emission regions ("hot spots" that are a few kpc across) found in radio lobes. In diffuse radio lobes, the value may be more than an order of magnitude smaller, while the field strength in the radio core is probably around 100 nT.

---

### Accelerating the Charged Particles in the Jets

The observations of jets are made possible by inefficiencies in the transport of particles and energy out to the radio lobes. The spectra of the radio lobes and jets follow a power law, with a typical spectral index of $\alpha \simeq 0.65$. The presence of power-law spectra and a high degree of linear polarization strongly suggest that the energy emitted by the lobes and jets comes from synchrotron radiation.

The loss of energy by synchrotron radiation is unavoidable, and in fact the relativistic electrons in jets will radiate away their energy after just 10,000 years or so. This implies that there is not nearly enough time for particles to travel out to the larger radio lobes; for example, for the large radio galaxy 3C 236, the journey would take several million years, even at the speed of light. This long travel time and the long lifetime of radio lobes imply that there must be some mechanism for accelerating particles in the jets and radio lobes. As one possibility, shock waves may accelerate charged particles by magnetically squeezing them, reflecting them back and forth inside the shock. Radiation pressure may also play a role, but it alone is not enough to generate the necessary acceleration.

### Superluminal Velocities

Although the standard model of jets and radio lobes requires a steady supply of charged particles moving at relativistic speeds, evidence for such high velocities is difficult to obtain. The absence of spectral lines in a power-law spectrum means that the relativistic velocity of the jet material cannot be measured directly but must be inferred from indirect evidence. The most compelling argument for relativistic speeds involves radio observations of material ejected from the cores of several AGNs with so-called **superluminal velocities**. This effect is observed within about 100 pc of the AGN's center and probably continues farther out.

---

**Example 28.3.3.**    Figure 28.31 is a radio view of the core of the quasar 3C 273 that shows a blob of radio emission moving away from the nucleus with an angular velocity of $\mu = 0.0008''$ yr$^{-1}$. Assuming that the radio knot is traveling in the plane of the sky, perpendicular to the line of sight, and using a distance of $d = 440h^{-1}$ Mpc for 3C 273, the *apparent* transverse velocity of the blob away from the nucleus is, from Eq. (1.5),

$$v_{\text{app}} = d\mu = 1.67 \times 10^9 h^{-1} \text{ m s}^{-1} = 5.57h^{-1}c.$$

If $h = [h]_{\text{WMAP}}$, we find that $v_{\text{app}} = 7.85c$. This is clearly unphysical, and so the assumption of motion perpendicular to the line of sight must be wrong.

---

Figure 28.32 shows how the motion of the knot toward the observer can resolve this dilemma. Suppose a source is traveling with a velocity $v$ (the actual speed of the source, not its apparent speed) at an angle $\phi$ measured from the line of sight. A photon is emitted along the line of sight at time $t = 0$ when the source is a distance $d$ from Earth. At a later time ($t_e$), another photon is emitted when the distance to Earth is $d - vt_e \cos \phi$. The first

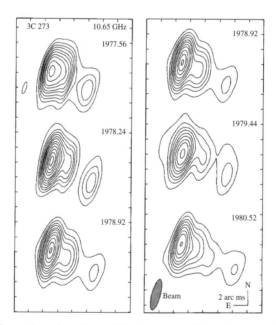

**FIGURE 28.31**    The motion of a radio-emitting knot ejected from the core of the quasar 3C 273. The dates of the observations are recorded as fractions of a year, and the third image has been repeated for clarity. (Figure adapted from Pearson et al., *Nature*, *290*, 365, 1981. Reprinted by permission from *Nature*, Vol. 290, pp. 365–368. Copyright 1981 Macmillan Magazines Limited.)

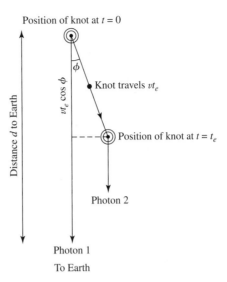

**FIGURE 28.32**   Two photons emitted at $t = 0$ and $t = t_e$ by a source moving with speed $v$.

photon reaches Earth at time $t_1$, where

$$t_1 = \frac{d}{c}.$$

The second photon arrives at Earth at time

$$t_2 = t_e + \frac{d - vt_e \cos\phi}{c}.$$

The time on Earth between the reception of the two photons is thus

$$\Delta t = t_2 - t_1 = t_e \left(1 - \frac{v}{c}\cos\phi\right),$$

a time that is *shorter* than $t_e$. The apparent transverse velocity measured on Earth is then

$$v_{\mathrm{app}} = \frac{vt_e \sin\phi}{\Delta t} = \frac{v \sin\phi}{1 - (v/c)\cos\phi}.$$

Solving this for $v/c$ results in

$$\boxed{\frac{v}{c} = \frac{v_{\mathrm{app}}/c}{\sin\phi + (v_{\mathrm{app}}/c)\cos\phi}.} \tag{28.15}$$

It is left as an exercise to show that $v/c < 1$ for angles satisfying

$$\frac{v_{\mathrm{app}}^2/c^2 - 1}{v_{\mathrm{app}}^2/c^2 + 1} < \cos\phi < 1, \tag{28.16}$$

and that the smallest possible value of $v/c$ for the source is

$$\frac{v_{\min}}{c} = \sqrt{\frac{v_{\text{app}}^2/c^2}{1 + v_{\text{app}}^2/c^2}}, \tag{28.17}$$

which occurs at an angle $\phi_{\min}$ given by

$$\cot \phi_{\min} = \frac{v_{\text{app}}}{c}. \tag{28.18}$$

This minimum value of $v/c$ corresponds to a minimum Lorentz factor (Eq. 4.20) of the source of

$$\gamma_{\min} = \frac{1}{\sqrt{1 - v_{\min}^2/c^2}} \simeq \sqrt{1 + v_{\text{app}}^2/c^2} = \frac{1}{\sin \phi_{\min}}. \tag{28.19}$$

---

**Example 28.3.4.** Referring to Example 28.3.3, since the actual speed of the radio knot ejected by 3C 273 must be less than $c$, as required by special relativity, $\phi$ must be less than

$$\phi_{\min} = \cot^{-1}\left(\frac{v_{\text{app}}}{c}\right) = 7.26°.$$

That is, the knot must be approaching Earth within $7.26°$ of the line of sight. From Eq. (28.17), the lower limit of the knot's speed is $v_{\min} = 0.992c$. Therefore, from Eq. (28.19), $\gamma_{\min} = 7.92$.

The minimum value of the Lorentz factor inferred for other superluminal sources ranges between $\gamma_{\min} = 4$ and $12$ for $h = [h]_{\text{WMAP}}$. 3C 273 and similar examples provide compelling evidence that the central cores of AGNs can accelerate material to relativistic speeds.

---

### Relativistic Beaming and One-Sided Jets

The headlight effect described in Section 4.3 will be involved whenever a source of light moves with a relativistic speed ($\gamma \gg 1$). All of the light emitted into the forward hemisphere in the rest frame of the source is concentrated into a narrow cone in the observer's rest frame. The cone's half-angle, $\theta$, is given by Eq. (4.43), $\sin \theta = 1/\gamma$. Comparing this with Eq. (28.19) above shows that if the source is approaching Earth with a relativistic velocity within the angle $\phi_{\min}$ of the line of sight, this relativistic beaming effect will cause it to appear much brighter than expected and it will appear to be moving with a superluminal speed across the plane of the sky. Interestingly, nearly all AGNs showing superluminal motions are surrounded by large, dim halos that may be radio lobes seen end-on. Blazars may be quasars or radio galaxies viewed with the jet coming directly (or nearly so) toward the observer. Their very rapid time variability could then be exaggerated by the relativistic Doppler shift; see Eq. (4.31). Any luminosity variations due to a source within the relativistic jet would be observed to occur approximately $2\gamma$ times more rapidly by astronomers on Earth (see Problem 28.12).

Conversely, a relativistic source moving away from us will appear unusually dim (recall Fig. 28.6). All of the jets showing superluminal motion are one-sided, even when the

AGNs exhibit two radio lobes. It is expected that the central engines of AGNs produce two oppositely directed jets; however, relativistic beaming seems to explain why the jets appear to be only one-sided.

## The Roles of Galactic Companions

The galactic companions of AGNs may play an important role in supplying them with the fuel. Most Seyferts (at least 90%) are spiral galaxies, and many have close neighbors with whom they may be interacting. Gravitational perturbations produce the distorted appearance frequently seen in those Seyferts close enough to be studied, as evidenced by the appearance of bars and/or outer rings. As described in Section 26.1, these interactions could produce gravitational torques on the gas in a Seyfert galaxy, drastically reducing its angular momentum and sending the gas plunging into the galactic center. The result would be the delivery of a fresh supply of fuel to the Seyfert nucleus to be accreted by the black hole. The concentration of gas could also result in a burst of star formation around the nucleus. Furthermore, if a merger with a galactic companion occurs, the subsequent disruption could produce an elliptical galaxy with an active nucleus, resulting in a young radio galaxy.

Mergers are certainly important for quasars as well. Some low-redshift quasars show evidence of past interactions (see Fig. 28.18), and mergers were undoubtedly more common in the early universe than they are today. Since galaxies are believed to have contained more gas when they were young, mergers may have resulted in the infall of large amounts of gas that could have contributed to the growth of a central supermassive black hole as the gas simultaneously fueled its activity. In addition, mergers probably resulted in the coalescing of supermassive black holes, producing even larger central engines. As the masses of the black holes grew, so did the number of quasars and their energy output, until the fuel powering the engines was largely consumed.

## AGN Evolution

What happens when a quasar runs out of fuel? In broad terms, the diminishing fuel supply of an energetic object could lead to its transformation into a less luminous form. For example, Cen A has huge radio lobes (see Fig. 28.11) but is a weak radio source. It was probably much more luminous in the past but is now fading away. On the other hand, a lesser luminosity could be explained by a less massive black hole rather than a smaller accretion rate. Our Milky Way does not have a $10^8$ $M_\odot$ black hole at its center, although there is a more modest one of $3.7 \times 10^6$ $M_\odot$ (Section 24.4). If, as has been conjectured, every large galaxy comparable to the Milky Way has a supermassive black hole of at least $10^6$ $M_\odot$, then low-level galactic activity may be a common occurrence.

One large impediment to understanding the evolution of active galaxies is our current lack of knowledge about their lifetimes. Some researchers find that around $z = 2$, the number of luminous AGNs decreases toward the present-day epoch with a characteristic decay time of $\tau \simeq 2h^{-1}$ Gyr. However, this is only an *upper limit* to the lifetime of an AGN. A single AGN may remain active this long, or the individual lifetimes may be much shorter, say between $10^7$ and $10^8$ years, the typical timescale needed to radiate away the energy stored in a radio lobe. In this latter case, $\tau$ would describe the statistical changes in a population of active galaxies, rather than the behavior of a single individual. A galaxy may then experience just one, or several, brief episodes of activity during its history as mergers refuel the central engine. It may be that Seyfert galaxies experience recurring episodes of activity, for example (recall the discussion on page 1121).

## 28.4 ■ USING QUASARS TO PROBE THE UNIVERSE

Quasars are among the most distant visible objects in the universe, and so they offer a unique opportunity to probe the intervening space. Clouds of gas, galaxies, and dark matter can all affect the light from a quasar during its journey to Earth. By decoding the clues supplied by observations of quasars, astronomers can learn a great deal about the perturbing objects along the line of sight.

### Gravitational Lensing and Multiple Images of Quasars

In 1919, the same year that Eddington measured the bending of starlight as it passed near the Sun and verified Einstein's general theory of relativity (see Section 17.1), Sir Oliver Lodge (1851–1940), an English physicist, put forth the possibility of using a gravitational lens to focus starlight (recall the discussion of gravitational lensing; see Fig. 24.14). During the 1920s, astronomers began to consider how light passing through the curved spacetime surrounding a massive object could produce multiple images of the source. Then, in 1937, Fritz Zwicky (1898–1974) proposed that gravitational lensing by a galaxy would be much more likely than gravitational lensing by individual stars. By the 1970s the search was on for a multiply imaged quasar, and in 1979 the quasar Q0957+561 was discovered to appear twice in the sky. As shown in Fig. 28.33, the two images are separated by 6.15″, and each

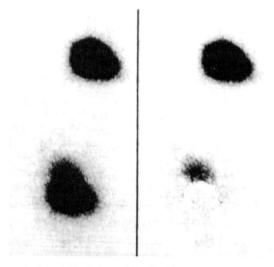

**FIGURE 28.33**    An optical (negative) view of the double quasar Q0957+561. The photo on the left shows the two gravitationally lensed images. The fuzz extending upward from the bottom image is the lensing galaxy. On the right, the upper image has been subtracted from the lower image to reveal the lensing galaxy more clearly. (Figure from Stockton, *Ap. J. Lett.*, *242*, L141, 1980.)

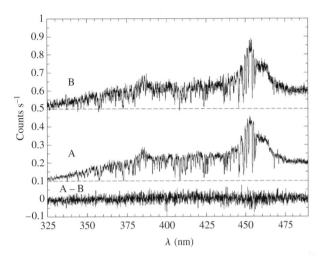

**FIGURE 28.34**   Spectra of the images of the quasar Q0142−100 formed by a gravitational lens. The bottom panel shows the difference between the two spectra. (Figure adapted from Smette et al., *Ap. J., 389*, 39, 1992.)

shows a quasar with a redshift of $z = 1.41$. The gravitational lens is due to the gravity of an intervening giant cD galaxy with $z = 0.36$ that is between the two images and $0.8''$ away from one of them. In addition to having the same redshift, both images have the same two bright emission lines and many absorption features in common. Both images also show the same radio core and jet structure.

Figure 28.34 shows spectra of the two images of another quasar, Q0142−100, that are also formed by a gravitational lens. Like optical lenses, gravitational lenses can magnify and increase an object's brightness (recall Fig. 24.15). The brighter of the two images of Q0142−100 has the appearance of being one of the most luminous quasars known because of the gravitational amplification of the image. (The difference in the apparent magnitudes of the images of Q0142−100 is about $\Delta m_V = 2.12$.)

## The Geometry of Gravitational Lensing

Gravitational lensing results when light follows the straightest possible worldline (a geodesic, see Section 17.2) as it travels through the curved spacetime around a massive object. It is analogous to the normal refraction of light by a glass lens that occurs as the light crosses the lens surface, passing from one index of refraction, $n$, to another, where $n \equiv c/v$ is just the ratio of the speed of light in a vacuum to its speed, $v$, in the medium. Outside of a spherical object of mass $M$ (which is equivalent to a point mass), the coordinate speed of light in the radial direction is given by Eq. (17.28),

$$\frac{dr}{dt} = c \left( 1 - \frac{2GM}{rc^2} \right)$$

so the effective "index of refraction" is

$$n = \frac{c}{dr/dt} = \left( 1 - \frac{2GM}{rc^2} \right)^{-1} \simeq 1 + \frac{2GM}{rc^2}$$

for radially traveling light, assuming that $2GM/rc^2 \ll 1$. At a distance of $10^4$ pc from a galaxy with a mass of $10^{11}$ $M_\odot$, the effective index of refraction is $n = 1 + 9.6 \times 10^{-7}$. (Of course, the light passing by the point mass will never be traveling exactly radially. This was merely used to estimate the magnitude of the effect of gravity in a gravitational lens.) Obviously, the deviation of the light from a straight line will be extremely small.

Figure 28.35 shows the path taken by light from a source at point $S$, as it is deflected through an angle, $\phi$, by the gravitational lens due to a point mass, $M$, at point $L$. The light arrives at the position of the observer at point $O$.

In Problem 17.6, you found that the angular deviation of a photon passing a distance $r_0$ (very nearly the distance of closest approach) from a mass $M$ was

$$\phi = \frac{4GM}{r_0 c^2} \text{ rad} \tag{28.20}$$

[this includes the factor of 2 mentioned in part (c) of that problem]. The distance to the source is $d_S/\cos\beta \simeq d_S$, where $\beta \ll 1$, and $d_L$ is the distance to the lensing mass. It is then a matter of simple trigonometry (left as an exercise) to show that the angle $\theta$ between the lensing mass and the image of the source must satisfy the equation

$$\theta^2 - \beta\theta - \frac{4GM}{c^2}\left(\frac{d_S - d_L}{d_S d_L}\right) = 0, \tag{28.21}$$

where $\theta$ and $\beta$ are measured in radians.

The quadratic equation (Eq. 28.21) indicates that for the geometry shown in the figure, there will be two solutions for $\theta$, and so two images will be formed by the gravitational lens. Designating these solutions as $\theta_1$ and $\theta_2$, these angles can be measured observationally and then used to find the values of $\beta$ and $M$. The results are

$$\beta = \theta_1 + \theta_2 \tag{28.22}$$

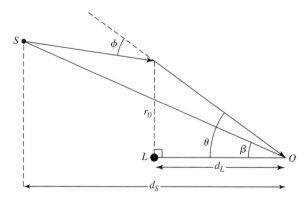

**FIGURE 28.35**   The geometry for a gravitational lens. Light from the source, $S$, passes within a distance of approximately $r_0$ of a lensing point mass at $L$ on its way to an observer at $O$. The angles involved are actually just a fraction of a degree, and so $r_0$ is very nearly the distance of closest approach.

and

$$M = -\frac{\theta_1 \theta_2 c^2}{4G} \left( \frac{d_S d_L}{d_S - d_L} \right). \tag{28.23}$$

Referring back to Fig. 28.35, note that Eq. (28.22) implies that $\theta_1$ and $\theta_2$ have opposite signs. As a result, the two images are formed on opposite sides of the gravitational lens, so $M$ will be positive.

---

**Example 28.4.1.** For the quasar Q0957+561 shown in Fig. 28.33, $\theta_1 = 5.35'' = 2.59 \times 10^{-5}$ rad, and $\theta_2 = -0.8'' = -3.88 \times 10^{-6}$ rad. (Which angle assumes the minus sign is arbitrary.) From the quasar's redshift of $z_S = 1.41$ and the gravitational lens redshift of $z_L = 0.36$, the Hubble law gives the corresponding distances of $d_S = 2120h^{-1}$ Mpc and $d_L = 890h^{-1}$ Mpc. Then Eq. (28.23) gives

$$M = -\frac{\theta_1 \theta_2 c^2}{4G} \left( \frac{d_S d_L}{d_S - d_L} \right) = 8.1 \times 10^{11} h^{-1} \, M_\odot.$$

This is in good agreement with a value of $M = 8.7 \times 10^{11} h^{-1} \, M_\odot$ obtained with a more accurate treatment of the mass distribution of the lensing galaxy.

---

**Einstein Rings and Crosses**

If a quasar or other bright source lies exactly along the line of sight to the lensing mass, then it will be imaged as an **Einstein ring** encircling the lens (this phenomenon was described by Einstein in 1936). In this case, $\beta = 0$ in Fig. 28.35, and so Eq. (28.21) can be solved immediately for the angular radius of the Einstein ring,

$$\theta_E = \sqrt{\frac{4GM}{c^2} \left( \frac{d_S - d_L}{d_S d_L} \right)} \text{ rad.} \tag{28.24}$$

Of course, for a point source, the chance of an exact alignment with the lensing mass is essentially zero. For an extended source, the requirements for an Einstein ring are that $\beta < \theta_E$ and that the line of sight through the lensing mass must pierce the extended source. Figure 28.36 shows a calculation of a partial ring—the image of a slightly off-center source. The first Einstein ring to be discovered, MG1131+0456, was found at radio wavelengths by the VLA. Figure 28.37 shows the radio appearance of the ring, which is thought to be the image of a radio galaxy lensed by an elliptical galaxy.

The value of $\theta_E$ can be calculated for any gravitational lens, regardless of the alignment of the lens and the source. Although the image may not be a ring, $\theta_E$ does provide a useful parameter for describing the properties of any gravitational lens. If $\beta < \theta_E$, as shown in Fig. 28.35, there will be two images formed by the point mass. If $\beta \gg \theta_E$, the position and brightness of the source are only slightly altered, but a secondary image appears close to the lensing mass that is reduced in angular size by a factor of $(\theta_E / \beta)^4$.

A point mass is clearly a crude representation of an actual galaxy. A better model of the lensing galaxy is provided by an isothermal sphere around a central core, similar to the

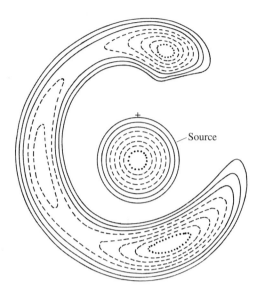

**FIGURE 28.36**    A calculation of the image of a slightly off-center spherical galaxy formed by a lensing mass located at the cross ("+"). (Figure adapted from Chitre and Narasimha, *Gravitational Lenses*, Springer-Verlag, Berlin, 1989.)

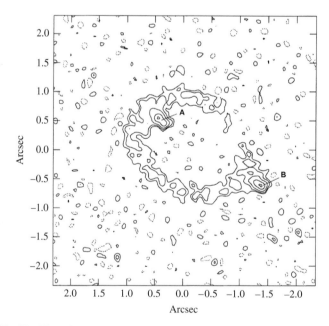

**FIGURE 28.37**    The Einstein ring MG1131+0456. The knot labeled A is produced by the core of the imaged radio galaxy, and the knot labeled B represents one of its lobes. (Figure adapted from Hewitt et al., *Nature, 333*, 537, 1988. Courtesy of J. Hewitt.)

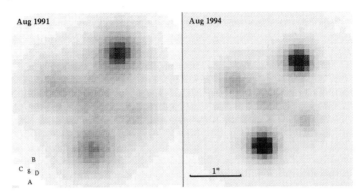

**FIGURE 28.38** The Einstein cross Q2237+031, as observed in August 1991 (left) and August 1994 (right). The cross consists of 4 images of the quasar (labeled A–D), with the lensing galaxy (labeled g) at the center. (Courtesy of Geraint Lewis and Mike Irwin.)

model used for the central bulge of the Milky Way; see Section 24.4. Another improvement is to depart from spherical symmetry and use an isothermal ellipsoid, which can produce either three or five images (an extended distribution of mass will produce an odd number of images). The **Einstein cross** shown in Fig. 28.38 includes four images of a distant quasar (Q2237+031, at $z = 1.69$) that is lensed by a nearby ($z = 0.04$) spiral galaxy. There is probably also a fifth faint central image that is overwhelmed by the lensing galaxy at the center of the cross. Note that image A has brightened by 0.5 mag in the 3-year interval between the photos.

**Luminous Arcs in Galaxy Clusters**

Another striking example of gravitational lensing is the formation of arcs by light passing through a cluster of galaxies. One such arc in the cluster Abell 370 is shown in Fig. 28.39. Up to 60 additional "arclets" and several distorted distant background galaxies have also been observed in that cluster. The source of the large arc must be a resolved object such as a galaxy rather than the starlike nucleus of a quasar. According to one model of Abell 370, the lensing mass (visible galaxies and dark matter) needed to produce the images in Abell 370 is at most about $5 \times 10^{14} \ M_\odot$. Taken with the combined luminosity of a few $\times \ 10^{11} \ L_\odot$ for the lensing galaxies, this implies a mass-to-light ratio of at least $1000 \ M_\odot/L_\odot$, indicating the presence of large amounts of dark matter.

Abell 370 is an unusual cluster in that it is sufficiently centrally condensed to produce such arcs. The dark matter in most clusters is probably more widely distributed, producing weak lensing effects just strong enough to distort the appearance of distant galaxies seen beyond the cluster. Figure 28.40 shows a spectacular example of multiple arclets that are lensed images of background galaxies produced by the cluster Abell 2218. Such weak lensing can also cause an apparent bunching of quasars, so statistical studies of the clustering of objects in the very early universe must take this effect into account.

**Time Variability of Multiple Images**

An interesting effect occurs when the source for a pair of images increases its luminosity. Because the light from the source takes different paths on its way to the observer, there

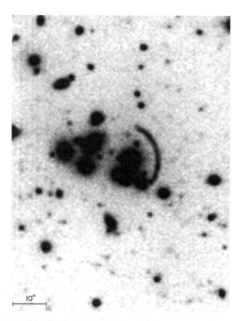

**FIGURE 28.39**    A gravitationally lensed arc, about 20″ long, produced by the cluster Abell 370. (Figure from Lynds and Petrosian, *Ap. J.*, *336*, 1, 1989. Courtesy of National Optical Astronomy Observatories/R. Lynds.)

**FIGURE 28.40**    An HST view of the gravitationally lensed images of background galaxies produced by the cluster Abell 2218. [Courtesy of W. Couch (U. of South Wales), R. Ellis (Cambridge U.), and NASA.]

will be a time delay between the brightening of the lensed images. A time delay of about 1.4–1.5 yr has been measured for the original double quasar, Q0957+561. Nonperiodic celestial events usually catch astronomers by surprise, but this time delay puts astronomers in the unique situation of knowing in advance how a lensed quasar will behave.

It turns out that the time delay is also inversely proportional to the Hubble constant. This offers a way of determining the value of $H_0$ that is independent of any other distance measurement. At the cosmological distances of quasars, their recessional velocities should completely dominate their peculiar velocities through space. One study using Q0957+561 concluded that $H_0 = 69 \pm 21$ km s$^{-1}$ Mpc$^{-1}$, assuming that the lensing galaxy contains a substantial amount of dark matter. The result is in excellent agreement with the WMAP value of $H_0 = 71^{+4}_{-3}$ km s$^{-1}$ Mpc$^{-1}$.

### The Lyman-Alpha Forest

The spectra of high-redshift quasars always display a large number of narrow absorption lines superimposed on the quasar's continuous spectrum (these lines are in addition to any broad absorption lines that are associated with the quasar itself). These narrow lines are formed when the light from a quasar passes through material (an interstellar cloud, a galactic halo) that happens to lie along the line of sight. If the absorbing material is far from Earth, its recessional motion will cause these absorption lines to be strongly redshifted. Furthermore, if the light passes through more than one cloud or galactic halo during its trip to Earth, different sets of absorption lines will be seen. Each set of lines corresponds to the redshift of a particular cloud or halo.

There are two classes of narrow absorption lines in quasar spectra:

- The **Lyman-$\alpha$ forest** is a dense thicket of hydrogen absorption lines. These lines are believed to be formed in intergalactic clouds and display a variety of redshifts. Absorption by primordial ionized helium (He II) has also been detected.

- Lines are also formed by **ionized metals**, primarily carbon (C IV) and magnesium (Mg II), together with silicon, iron, aluminum, nitrogen, and oxygen. The mix of elements is similar to that found in the interstellar medium of the Milky Way, indicating that the material has been processed through stars and enriched in heavy elements. These lines are thought to be formed in the extended halos or disks of galaxies found along the line of sight to the quasar.

Most of these lines are normally found at ultraviolet wavelengths, when the absorbing material is moving at a small fraction of the speed of light relative to Earth (i.e., has a small redshift). They are rarely seen from the ground because Earth's atmosphere absorbs most ultraviolet wavelengths. However, if the absorbing material is receding fast enough, the Doppler effect can shift ultraviolet lines to visible wavelengths, where the atmosphere is transparent. For this reason, these absorption lines are seen from the ground only in the spectra of highly redshifted quasars.

---

**Example 28.4.2.** The rest wavelength of the ultraviolet Lyman-$\alpha$ line of hydrogen is $\lambda_{\text{Ly}\alpha} = 121.6$ nm (Section 5.3). To determine the redshift required to bring this line into the visible region of the electromagnetic spectrum, we can use the definition of $z$,

$$z = \frac{\lambda_{\text{obs}} - \lambda_{\text{rest}}}{\lambda_{\text{rest}}}.$$

*continued*

Using $\lambda_{\text{rest}} = \lambda_{\text{Ly}\alpha}$ and $\lambda_{\text{obs}} = 400$ nm for the blue end of the visible spectrum, we require a redshift of

$$z > \frac{400 \text{ nm} - 121.6 \text{ nm}}{121.6 \text{ nm}} \simeq 2.3,$$

just to bring the Ly$\alpha$ line to the edge of the visible spectrum. Actually, some near-ultraviolet light can penetrate Earth's atmosphere, so the Ly$\alpha$ line can be observed when $z > 1.7$ for the absorbing material.

---

Typically, the spectrum of a high-redshift quasar contains a strong Lyman-$\alpha$ emission line produced by the quasar itself, and perhaps some 50 Ly$\alpha$ absorption lines at shorter wavelengths (smaller redshifts); see Fig. 28.41. Each one of these lines is from a different intergalactic cloud of hydrogen (and presumably helium) encountered by the quasar's continuum radiation on its journey to Earth. As explained in Section 9.5, the Ly$\alpha$ line profile can be used to calculate the column density of the neutral hydrogen atoms in the cloud that produces each line. A typical result is $10^{18}$ m$^{-2}$. In other words, a hollow tube having a cross-sectional area of 1 m$^2$ that crossed completely through the cloud would contain $10^{18}$ neutral hydrogen atoms. Such a cloud would be extremely transparent to the ultraviolet radiation that is normally present throughout space. As a result, this ultraviolet background can penetrate the cloud and keep it almost completely ionized. Calculations indicate that only one hydrogen atom in $10^5$ remains neutral in the cloud and is capable of absorbing an ultraviolet photon.

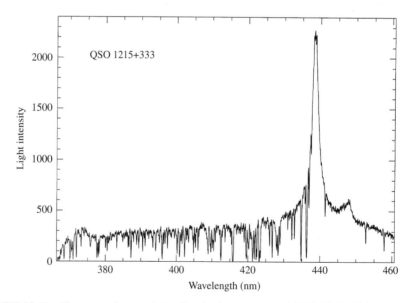

**FIGURE 28.41**   The strong Ly$\alpha$ emission line in the spectrum of QSO 1215+333, with the Ly$\alpha$ forest of absorption lines at shorter wavelengths. (Adapted from a figure courtesy of J. Bechtold, Steward Observatory, University of Arizona.)

We deduce the size of the intergalactic clouds by comparing the Ly$\alpha$ forest in the spectra of pairs of lensed quasars. Many of the absorption lines are seen in both spectra, but some are not. This indicates that the clouds are, on average, about the size of the lensing galaxy. From the total calculated column density of hydrogen (ionized as well as neutral), the mass of a typical cloud probably lies between $10^7$ and $10^8$ $M_\odot$. At the temperature estimated for a typical cloud by some astronomers (approximately $3 \times 10^4$ K), its self-gravity would be too weak to keep it from dispersing. It may be held together by the pressure of a less dense (but hotter) external intergalactic medium or by the presence of dark matter within the cloud.

### Ionized Metal Absorption Lines in Quasars

The narrow absorption lines produced by ionized metals in quasar spectra have a different origin. They can be divided into two groups as observed from Earth's surface, corresponding to two different redshift ranges. Below roughly $z = 1.5$, the Mg II lines dominate, accompanied by Si II, C II, Fe II, and Al II, because they fall within the wavelength window that can be seen from the ground (the Mg II lines are probably produced in the halos of normal galaxies or in regions of star formation.). The C IV lines, together with Si IV, N V, and O IV, are common between about $z = 1.2$ and $z = 3.5$, however.

The distribution of redshifts of these lines is in general agreement with the expected distribution of galaxies at that earlier time when the universe was smaller by a factor of $1 + z$, assuming that the galactic halos are typically some 30–50 kpc across. In fact, some Mg II systems with $z < 1$ have been clearly identified with foreground galaxies seen in direct images. The C IV lines probably come from clouds in young galaxies that are strongly ionized by young, hot OB stars. These narrow metal lines indicate lower-than-solar abundances of heavy elements, consistent with their origin in young galaxies that may still be in the process of forming.

### The Density Distribution of Intergalactic Clouds

The comoving space density of intergalactic clouds appears to have been greater in the past than it is today, so the number of clouds has been decreasing as the universe ages. A statistical analysis of the clouds' redshifts reveals little evidence that the clouds tend to be grouped in clusters. Instead, they appear to be distributed randomly throughout space. In particular, there do not appear to be large voids in the distribution of these intergalactic clouds similar to those described in Section 27.3 for galaxies (the significance of this is not yet clear). The distribution of the He II is similarly uncertain.

## SUGGESTED READING

### General

Courvoisier, Thierry J.-L., and Robson, E. Ian, "The Quasar 3C 273," *Scientific American*, June 1991.

Finkbeiner, Ann, "Active Galactic Nuclei: Sorting Out the Mess," *Sky and Telescope*, August 1992.

Levi, Barbara Goss, "Space-based Telescopes See Primordial Helium in Spectra of Distant Quasars," *Physics Today*, October 1995.

Miley, George K., and Chambers, Kenneth C., "The Most Distant Radio Galaxies," *Scientific American*, June 1993.

Preston, Richard, *First Light*, New American Library, New York, 1988.

Schild, Rudolph, E., "Gravity Is My Telescope," *Sky and Telescope*, April 1991.

Voit, G. Mark, "The Rise and Fall of Quasars," *Sky and Telescope*, May 1999.

## Technical

Antonucci, Robert, "Unified Models for Active Galactic Nuclei and Quasars," in *Annual Review of Astronomy and Astrophysics*, *31*, 473, 1993.

Balsara, Dinshaw S., and Norman, Michael L., "Three-Dimensional Hydrodynamic Simulations of Narrow-Angle-Tail Radio Sources. I. The Begelman, Rees, and Blandford Model," *The Astrophysical Journal*, *393*, 631, 1992.

Binney James, and Merrifield, Michael, *Galactic Astronomy*, Princeton University Press, Princeton, 1998.

Blandford, R. D., and Narayan, R., "Cosmological Applications of Gravitational Lensing," *Annual Review of Astronomy and Astrophysics*, *30*, 311, 1992.

Carvalho, Joel C., and O'Dea, Christopher P., "Evolution of Global Properties of Powerful Radio Sources. II. Hydrodynamical Simulations in a Declining Density Atmosphere and Source Energetics," *The Astrophysical Journal Supplement Series*, *141*, 371, 2002.

Collin-Souffrin, Suzy, "Observations and Their Implications for the Inner Parsec of AGN," *Central Activity in Galaxies*, Aa. Sandqvist and T. P. Ray (eds.), Springer-Verlag, Berlin, 1993.

Dunlop, J. S., et al., "Quasars, Their Host Galaxies, and Their Central Black Holes," *Monthly Notices of the Royal Astronomical Society*, *340*, 1095, 2003.

Hartwick. F. D. A., and Schade, David, "The Space Distribution of Quasars," *Annual Review of Astronomy and Astrophysics*, *28*, 437, 1990.

Kembhavi, Ajit K., and Narlikar, Jayant V., *Quasars and Active Galactic Nuclei: An Introduction*, Cambridge University Press, Cambridge, 1999.

King, Andrew R., Frank, Juhan, and Raine, Derek Jeffery, *Accretion Power in Astrophysics*, Third Edition, Cambridge University Press, Cambridge, 2002.

Krolik, Julian H., *Active Galactic Nuclei: From the Central Black Hole to the Galactic Environment*, Princeton University Press, Princeton, NJ, 1999.

Osterbrock, Donald E., and Ferland, Gary J., *Astrophysics of Gaseous Nebulae and Active Galactic Nuclei*, Second Edition, University Science Books, Sausalito, CA, 2005.

Perry, Judith J., "Activity in Galactic Nuclei," *Central Activity in Galaxies*, Aa. Sandqvist and T. P. Ray (eds.), Springer-Verlag, Berlin, 1993.

Peterson, Bradley M., *An Introduction to Active Galactic Nuclei*, Cambridge University Press, Cambridge, 1997.

Silverman, J. D., et al., "Comoving Space Density of X-Ray-Selected Active Galactic Nuclei," *The Astrophysical Journal*, *624*, 630, 2005.

*Sloan Digital Sky Survey,* http://www.sdss.org.

Sparke, Linda S., and Gallagher, John S., *Galaxies in the Universe: An Introduction,* Cambridge University Press, Cambridge, 2000.

Tyson, Anthony, "Mapping Dark Matter with Gravitational Lenses," *Physics Today,* June 1992.

## PROBLEMS

**28.1** The radio galaxy Centaurus A has a redshift of $z = 0.00157$. The monochromatic flux of Cen A is $F_\nu = 912$ Jy at a frequency of 1400 MHz. Using $\alpha = 0.6$ for its spectral index, estimate the radio luminosity of Cen A.

**28.2** Use Fig. 28.14 to calculate the spectral index, $\alpha$, of the quasar 3C 273 at a radio frequency of 1400 MHz. Compare your answer with the value given in Example 28.1.2.

**28.3** For a temperature of $7.3 \times 10^5$ K, make a graph of the Planck function (Eq. 3.24), plotting $\log_{10} \nu B_\nu(T)$ vs. $\log_{10} \nu$ for $\log_{10} \nu$ between 15.5 and 17.5. How does the behavior of your graph of a blackbody compare with that of Fig. 28.14 for the continuous spectrum of the quasar 3C 273?

**28.4** A rough idea of how the population of quasars may have been different in the past can be gained by mathematically modeling the dimming of quasars as they age.

(a) Consider the case where the total number of quasars has remained constant back to $z = 2.2$, and suppose that the average luminosity, $L$, of a quasar with a redshift $z$ has the form

$$L = L_0(1 + z)^a,$$

where $L_0$ is the luminosity at $z = 0$ (today). Use Fig. 28.16 to estimate the value of the constant $a$.

(b) From your answer to part (a), how much more luminous is an average quasar at $z = 2$ than at $z = 0$?

**28.5** Use Newtonian physics to calculate the values of the average density and "surface gravity" for a $10^8$ $M_\odot$ black hole. Compare these values with those for the Sun.

**28.6** Use the Newtonian expression for the disk luminosity, Eq. (18.23), to estimate the efficiency of the accretion luminosity of a disk around a nonrotating black hole ($R = 3R_S$). Repeat this for a maximally rotating black hole ($R = 0.5R_S$).

**28.7** The maximum possible angular momentum for an electrically neutral rotating black hole is

$$L_{\max} = \frac{GM^2}{c}$$

(see Problem 17.19). Use Newtonian physics to make estimates for this problem.

(a) What is the maximum angular velocity, $\omega_{\max}$, for a $M = 10^8$ $M_\odot$ black hole? Use $MR_S^2$ as an estimate of the black hole's moment of inertia, where $R_S$ is the Schwarzschild radius.

(b) Consider a straight wire with a length $\ell = R_S$ that rotates about one end with angular velocity $\omega_{\max}$ perpendicular to a uniform magnetic field of $B = 1$ T. What is the induced voltage between the ends of the wire?

(c) If a battery with the voltage found in part (b) were connected to a wire with a resistance of 30 $\Omega$, how much power would be dissipated by the wire?

**28.8** Repeat the derivation of Eq. (12.32) for the Strömgren radius, including the effect of the ionized gas filling only a fraction, $\epsilon$, of the total volume, and so arrive at Eq. (28.13) for the radius of the narrow-line region.

**28.9** Use the values in Example 28.2.2 to find the fraction of all photons in the continuum (frequencies between $10^{10}$ Hz and $10^{25}$ Hz) that are capable of ionizing hydrogen in the ground state.

**28.10** Starting with Eq. (28.15) for superluminal motion, derive Eqs. (28.16), (28.17), (28.18), and (28.19).

**28.11** Consider material that is ejected from a quasar directly toward Earth.

   **(a)** If the redshift of the quasar is $z_Q$ and the redshift of the ejecta is $z_{ej}$, show that the speed of the ejecta relative to the quasar is given by

$$\frac{v}{c} = \frac{(1+z_Q)^2 - (1+z_{ej})^2}{(1+z_Q)^2 + (1+z_{ej})^2}.$$

   **(b)** Consider a radio-emitting knot ejected from the quasar 3C 273 directly toward Earth. If astronomers measure the speed of the approaching knot at $v = 0.9842c$, what is the speed of the knot relative to the quasar? From the frame of reference of the quasar, what is the value of the knot's Lorentz factor?

**28.12** Consider a relativistic ($\gamma \gg 1$) blazar jet that is coming directly toward the observer. If there is a time variation $\Delta t_{rest}$ in the rest frame of the jet, use Eq. (4.31) to show that the variation observed at Earth is approximately

$$\Delta t_{obs} \simeq \frac{\Delta t_{rest}}{2\gamma}.$$

**28.13** Estimate the effective index of refraction for light passing within $10^4$ pc of a spherical cluster of galaxies of a total mass of $10^{14}$ M$_\odot$.

**28.14** Verify that Eq. (28.20) gives the correct numerical value for the angular deflection of a light ray that grazes the Sun's surface.

**28.15** Use the geometry of the gravitational lens shown in Fig. 28.35 and the value of $\phi$ given by Eq. (28.20) to derive Eqs. (28.21), (28.22), and (28.23). *Hint:* Start by showing that

$$\frac{\sin(\theta - \beta)}{d_S - d_L} = \frac{\sin \phi}{d_S}$$

for the small angles involved in this problem.

**28.16** Two images of the quasar Q0142−100 (also known as UM 673) are formed by a gravitational lens. The redshift of the quasar is $z = 2.727$, and the redshift of the imaging galaxy is $z = 0.493$. The images are separated by 2.22″, and the lensing galaxy is along the line between the two images, 0.8″ from one of them. Estimate the mass of the lensing galaxy.

**28.17** The Einstein ring MG1654+1346 has a diameter of 2.1″. The redshift of the source is $z = 1.74$, and the redshift of the lensing galaxy is $z = 0.25$. Estimate the mass of the lensing galaxy.

**28.18** In Section 24.2, the detection of a MACHO by its gravitational microlensing of a star in the Large Magellanic Cloud was described. Assume that a MACHO with a mass ten times that of Jupiter orbits halfway between Earth and the LMC and that it moves perpendicular to our line of sight to the LMC. How much time will it take for the MACHO to move through an angle of $2\theta_E$ across a lensed star in the LMC? Take the distance to the LMC to be 52 kpc and the orbital velocity of the MACHO to be 220 km s$^{-1}$. Neglect the motion of Earth and the LMC in this problem. Comment on a comparison of your answer with the time shown in Fig. 24.15.

**28.19** When a small object approaches a much more massive object, the smaller object can be tidally disrupted. The distance of closest approach before being tidally disrupted is the Roche limit (see Eq. 19.4). If the small object is a star and the large object is a supermassive black hole, the Roche limit is given by

$$r_R = 2.4 \left( \frac{\overline{\rho}_{\text{BH}}}{\overline{\rho}_\star} \right)^{1/3} R_S,$$

where $R_S$ is the Schwarzschild radius, $\overline{\rho}_{\text{BH}}$ is the density of the black hole, and $\overline{\rho}_\star$ is the average density of the star.

(a) Setting the average density of the supermassive black hole equal to its mass divided by the volume contained within the Schwarzschild radius, derive an expression for the mass of a black hole that would have $r_R = R_S$.

(b) If the Sun were to fall into a supermassive black hole, what maximum mass could the black hole have if the Sun would be tidally disrupted before crossing the event horizon? Compare your answer to the mass estimates of typical supermassive black holes in galactic nuclei.

(c) If the supermassive black hole exceeded the mass found in part (b), what would be the implications in terms of liberating the gravitational potential energy of the infalling star? Could infalling stars effectively power AGNs in this case?

## COMPUTER PROBLEM

**28.20** Table 28.2 gives values of the monochromatic flux for Cygnus A at several radio wavelengths.

(a) Make a graph of $\log_{10} F_\nu$ vs. $\log_{10} \nu$, and determine the value of the spectral index at $\log_{10} \nu = 8$ for the power law given by Eq. (28.1).

(b) Use a simple trapezoid rule to integrate the area under the curve of $F_\nu$ vs. $\nu$ (*not* $\log_{10} F_\nu$ vs. $\log_{10} \nu$!) for the data given, and use your answer to estimate the radio luminosity of Cygnus A.

**TABLE 28.2** Cygnus A Data for Problem 28.20.

| $\log_{10} \nu$ (Hz) | $\log_{10} F_\nu$ (W m$^{-2}$ Hz$^{-1}$) |
|---|---|
| 7.0 | −21.88 |
| 7.3 | −21.55 |
| 7.7 | −21.67 |
| 8.0 | −21.86 |
| 8.3 | −22.09 |
| 8.7 | −22.38 |
| 9.0 | −22.63 |
| 9.3 | −22.96 |
| 9.7 | −23.43 |
| 10.0 | −23.79 |

# CHAPTER
# 29

# Cosmology

## 29.1 ■ NEWTONIAN COSMOLOGY

On December 27, 1831, a ship sailed out of Plymouth, England, on a voyage around the world that would last nearly five years. Only 90 feet long, the *Beagle* was crowded with 74 people, one of whom was Charles Darwin. During stops in South America, the Galapagos Islands, Tahiti, New Zealand, and Australia, he exercised his formidable powers of observation. In 1859, after two decades of careful study and reflection, Darwin published *On the Origin of Species by Means of Natural Selection, or the Preservation of Favored Races in the Struggle for Life*. For the first time, people began to comprehend their own origins.

Other discoveries followed during the next one hundred years, with careful observations and brilliant deductions uncovering more about our beginnings. The elucidation of DNA and plate tectonics revealed the mechanisms by which we and our planet evolved. The ideas of stellar nucleosynthesis explained the manufacture of the chemical elements by stars, implying the origin of our corporeal bodies and the ground on which we walk. Even the universe itself was found to be expanding. Then, in 1964, two researchers at Bell Laboratories measured the afterglow of the Big Bang, confirming the explosive origin of everything in existence. It is difficult to imagine a more breathtaking leap from ignorance to self-knowledge than that which occurred during this century of discovery.

Cosmology, taken as a whole, is the study of the origin and evolution of the universe. In this chapter, cosmology will be considered from several different perspectives. To help develop our intuition, this section will discuss the expansion of the universe from the point of view of Newtonian mechanics, without the complications (or insights) provided by general relativity or the modern ideas of particle physics. The discovery and implications of the cosmic microwave background radiation are described in Section 29.2, followed, in Section 29.3, by an introduction to the geometry of the universe as explained by general relativity. Section 29.4 describes how some of the key parameters of cosmology may be measured observationally. The intriguing theories and speculations provided by particle physics will be reserved for Chapter 30.

### Olbers's Paradox

Newton believed in an infinite static universe filled with a uniform scattering of stars. If the distribution of matter did not extend forever, he realized, then it would collapse inward due to its own self-gravity. However, Newton's contemporary, Edmund Halley, worried about a sky filled with an infinite number of stars. Why then, asked Halley, is the sky dark at night?

This question was put in its strongest form by a German physician, Heinrich Olbers (1758–1840). Olbers argued in 1823 that if we live in an infinite, transparent universe filled with stars, then in any direction one looks in the night sky, one's line of sight will fall on the surface of a star. (Similarly, if standing in an infinite forest, in every direction you look you will see a tree.) This conclusion is valid regardless of whether the stars are uniformly distributed, as Newton believed, or grouped in galaxies. Olbers's argument was so strong that its disagreement with the obvious fact that the night sky is indeed dark became known as **Olbers's paradox**.[1]

Olbers believed that the answer to this paradox was that space is not transparent. The ideas of thermodynamics were still being developed at that time, and Olbers could not appreciate that his explanation was incorrect. The flaw was that any obscuring matter hiding the stars beyond would be heated up by the starlight until it glowed as brightly as a stellar surface. Surprisingly, the first essentially correct answer came from American poet and author Edgar Allan Poe (1809–1849). Poe proposed that because light has a finite speed and the universe is not infinitely old, the light from the most distant sources has not yet arrived. This solution was independently put on a firm scientific foundation by William Thomson (Lord Kelvin, 1824–1907). In more modern terms, the solution to Olbers's paradox is that our universe is simply too young for it to be filled with light.[2]

### Developing Our Intuition: A Newtonian Approach

You will soon discover that this chapter contains more mathematics than the preceding material.[3] The reward for the necessary effort is a *quantitative* description of how the universe unfolded that is much more satisfying than qualitative storytelling. Although general relativity is required for a complete understanding of the structure and evolution of the universe, it is useful to develop some intuition by first considering the expansion of the universe from a Newtonian point of view.

### The Cosmological Principle

In Section 27.2 it was argued that Hubble's law, Eq. (27.6), is a natural outcome of an expanding universe that is both *isotropic* and *homogeneous*, appearing the same in all directions and at all locations. This crucial assumption of an isotropic and homogeneous universe is called the **cosmological principle**. To show that the expansion of the universe

---

[1]You should recall the earlier discussion of Olbers's paradox in Section 24.1.

[2]It is sometimes argued that the cosmological redshift caused by the expansion of the universe is responsible for the darkness of the night sky because it shifts starlight out of the visible spectrum. In fact, this effect is much too small to contribute significantly to a dark night sky.

[3]In some derivations, many of the intermediate steps have been omitted. Filling in these mathematical gaps has been relegated to several of the end-of-chapter problems.

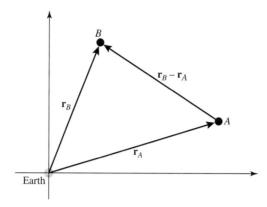

**FIGURE 29.1**    The expansion of the universe, with Earth at the origin.

appears the same to all observers at all locations, let Earth be at the origin of the coordinate system shown in Fig. 29.1, and consider two galaxies, $A$ and $B$, located at positions $\mathbf{r}_A$ and $\mathbf{r}_B$, respectively. According to the Hubble law, the recessional velocities of the two galaxies are described by the vectors

$$\mathbf{v}_A = H_0 \mathbf{r}_A$$

and

$$\mathbf{v}_B = H_0 \mathbf{r}_B.$$

The recessional velocity of Galaxy $B$ as seen by an observer in Galaxy $A$ is therefore

$$\mathbf{v}_B - \mathbf{v}_A = H_0 \mathbf{r}_B - H_0 \mathbf{r}_A = H_0(\mathbf{r}_B - \mathbf{r}_A),$$

so the observer in Galaxy $A$ sees all of the other galaxies in the universe moving away with recessional velocities described by the *same* Hubble law as on Earth.

Although the value of the Hubble "constant," $H_0$, is assumed to be the same for all observers, it is actually a function of time, $H(t)$. If the present time is $t_0$, then $H_0 \equiv H(t_0)$.

### A Simple Pressureless "Dust" Model of the Universe

To develop an understanding of how the expansion of the universe varies with time, imagine a universe filled with a pressureless "dust" of uniform density, $\rho(t)$, and choose an arbitrary point for the origin. Unlike the actual universe, this model universe is both perfectly isotropic and homogeneous at all scales. The pressureless dust represents all of the matter in the universe after being homogenized and uniformly dispersed. It should not be confused with the physical dust grains found throughout the interstellar medium. There are no photons or neutrinos in this single-component model of the universe.

As the universe expands, the dust is carried radially outward from the origin. Let $r(t)$ be the radius of a thin spherical shell of mass $m$ at time $t$; see Fig. 29.2. This shell of mass

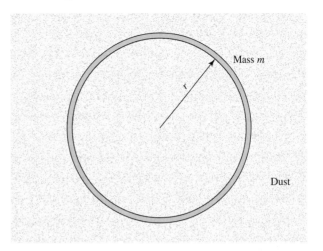

**FIGURE 29.2** Spherical mass shell in a dust-filled universe.

expands along with the universe with recessional velocity $v(t) = dr(t)/dt$, so it always contains the same dust particles. Then the mechanical energy $E$ of the shell is

$$K(t) + U(t) = E.$$

As the shell expands, the gravitational pull from the mass inside causes the kinetic energy, $K$, to decrease while the gravitational potential energy, $U$, increases.[4] However, by conservation of energy, the total energy, $E$, of the shell does not change as the shell moves outward. For future convenience, the total energy of the shell is written in terms of two constants, $k$ and $\varpi$, such that $E = -\frac{1}{2}mkc^2\varpi^2$. The constant $k$ has units of $(\text{length})^{-2}$; its meaning will be explored in Section 29.3. The other constant, $\varpi$ ("varpi"), labels this particular mass shell and may be thought of as the *present* radius of the shell, so $r(t_0) = \varpi$. The statement of the conservation of the mass shell's energy is then

$$\frac{1}{2}mv^2(t) - G\frac{M_r m}{r(t)} = -\frac{1}{2}mkc^2\varpi^2. \tag{29.1}$$

In the left-hand side of Eq. (29.1), $M_r$ is the mass interior to the shell,

$$M_r = \frac{4}{3}\pi r^3(t)\rho(t).$$

Although the radius of the shell and the density of the dust are continually changing, the combination $r^3(t)\rho(t)$ does not vary because the mass interior to a specific shell remains constant as the universe expands. Canceling $m$ and substituting for $M_r$ in Eq. (29.1) gives

$$v^2 - \frac{8}{3}\pi G\rho r^2 = -kc^2\varpi^2. \tag{29.2}$$

---

[4]Recall from Section 10.1 that the mass outside the shell does not contribute to the gravitational force on the shell.

The constant $k$ determines the ultimate fate of the universe:

- If $k > 0$, the total energy of the shell is negative, and the universe is *bounded*, or **closed**. In this case, the expansion will someday halt and reverse itself.

- If $k < 0$, the total energy of the shell is positive, and the universe is *unbounded*, or **open**. In this case, the expansion will continue forever.

- If $k = 0$, the total energy of the shell is zero, and the universe is **flat**, neither open nor closed. In this case, the expansion will continue to slow down, coming to a halt only as $t \rightarrow \infty$ and the universe is infinitely dispersed.

The Newtonian cosmology of this section always takes place in a flat spacetime, with a metric given by Eq. (17.19). The terms *closed*, *open*, and *flat* above should be understood as describing the *dynamics* of the universal expansion. In Section 29.3, these terms will be reinterpreted to describe the geometry of spacetime.

The cosmological principle requires that the expansion proceed in the same way for all shells; for example, the time required for every shell to double its distance from the origin is assumed to be the same. This means that the radius of a particular shell (identified by $\varpi$) at any time can be written as

$$r(t) = R(t)\varpi. \tag{29.3}$$

In this expression, $r(t)$ is called the **coordinate distance**. Because $\varpi$ labels a shell and follows it as it expands, $\varpi$ is referred to as a **comoving coordinate**; see Fig. 29.3. $R(t)$ is a dimensionless **scale factor** (the same for all shells) that describes the expansion; $R(t_0) = 1$ corresponds to $r(t_0) = \varpi$. The scale factor $R$ is equal to $R_{\text{emitted}}/R_{\text{obs}}$ in Eq. (28.3). Thus

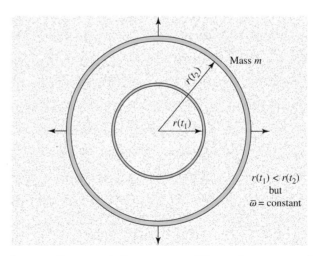

**FIGURE 29.3**  An expanding mass shell seen at two different times, $t_1 < t_2$. As the mass shell expands, its comoving coordinate, $\varpi$, is the same at times $t_1$ and $t_2$, while $r(t_1) < r(t_2)$.

$R$ and the redshift $z$ are related by

$$R = \frac{1}{1+z},$$  (29.4)

For example, looking back to a redshift of $z = 3$ implies a universe for which the scale factor was $R = 1/4$.

The previous statement that $r^3 \rho$ does not vary for a specific shell means that $R^3 \rho$ remains constant for *all* shells. That is, since $R(t_0) = 1$,

$$R^3(t)\rho(t) = R^3(t_0)\rho(t_0) = \rho_0,$$  (29.5)

where $\rho_0$ is the density of the dust-filled universe at the present time. Using Eq. (29.4), this can also be written as

$$\boxed{\rho(z) = \rho_0(1+z)^3,}$$  (29.6)

which gives the average density of the universe as observed at redshift $z$. You are cautioned that Eqs. (29.5) and (29.6) are valid only for a universe consisting of pressureless dust. The more general counterpart of Eq. (29.5) will be derived at the end of this section.

### The Evolution of the Pressureless "Dust" Universe

The evolution of our Newtonian universe, which can be described by the time behavior of the scale factor $R(t)$, will be investigated in the next few pages. A brief flurry of mathematics will produce the needed tools. The first step is to write the Hubble parameter, $H(t)$, in terms of the scale factor. The Hubble law is

$$v(t) = H(t)r(t) = H(t)R(t)\varpi.$$  (29.7)

Because $v(t)$ is the time derivative of $r(t)$, Eq. (29.3) gives

$$v(t) = \frac{dR(t)}{dt}\varpi.$$

Comparing this with Eq. (29.7) shows that

$$H(t) = \frac{1}{R(t)}\frac{dR(t)}{dt}.$$  (29.8)

Inserting Eqs. (29.3) and (29.7) into Eq. (29.2) and canceling the $\varpi^2$ results in

$$\left(H^2 - \frac{8}{3}\pi G\rho\right)R^2 = -kc^2,$$  (29.9)

or, employing Eq. (29.8),

$$\left[\left(\frac{1}{R}\frac{dR}{dt}\right)^2 - \frac{8}{3}\pi G\rho\right]R^2 = -kc^2.$$  (29.10)

The left-hand sides of Eqs. (29.9) and (29.10) apply to *all shells* and involve the functions of time $H(t)$, $\rho(t)$, and $R(t)$, while the right-hand sides are constant (the same for all positions and times). Using Eq. (29.5), Eq. (29.10) can be written in terms of $R$ and $t$ only:

$$\left(\frac{dR}{dt}\right)^2 - \frac{8\pi G\rho_0}{3R} = -kc^2. \tag{29.11}$$

This result, along with Eqs. (29.9) and (29.10), will be used to describe the expansion of the universe.

Now we are ready to examine the motion of mass shells in the three cases of a flat, closed, or open universe. First, consider the case of a flat universe ($k = 0$), corresponding to each shell expanding at exactly its escape velocity. The value of the density that will result in a value of $k = 0$ is known as the **critical density**, $\rho_c(t)$. From Eq. (29.9),

$$\boxed{\rho_c(t) = \frac{3H^2(t)}{8\pi G}.} \tag{29.12}$$

To evaluate this at the present time, it is useful to recall (Eq. 27.11) that the Hubble constant in conventional units is

$$H_0 = 100h \text{ km s}^{-1} \text{ Mpc}^{-1} = 3.24 \times 10^{-18}h \text{ s}^{-1} \tag{29.13}$$

which, using the WMAP[5] value of

$$[h]_{\text{WMAP}} = 0.71,$$

is

$$[H_0]_{\text{WMAP}} = 71 \text{ km s}^{-1} \text{ Mpc}^{-1} = 2.30 \times 10^{-18} \text{ s}^{-1}. \tag{29.14}$$

The present value of the critical density, $\rho_{c,0}$, is then

$$\boxed{\rho_{c,0} = \frac{3H_0^2}{8\pi G}} \tag{29.15}$$

$$= 1.88 \times 10^{-26}h^2 \text{ kg m}^{-3},$$

with a WMAP value of

$$\rho_{c,0} = 9.47 \times 10^{-27} \text{ kg m}^{-3}. \tag{29.16}$$

This is equivalent to about six hydrogen atoms per cubic meter. However, the WMAP value of the average density of **baryonic matter** in the universe is about 4% of the critical density,

$$\rho_{b,0} = 4.17 \times 10^{-28} \text{ kg m}^{-3} \qquad \text{(for } h = 0.71\text{)}, \tag{29.17}$$

---

[5]The WMAP (Wilkinson Microwave Anisotropy Probe) measurements of cosmological parameters were introduced in Section 27.2 and will be more fully described in Section 30.2.

or 1 hydrogen atom per 4 m$^3$ of space. By "baryonic matter," we mean matter made of *baryons* (e.g., protons and neutrons); hence the "*b*" subscript designating baryonic matter in Eq. (29.17).[6] As will be explained in Section 29.2, this value is consistent with that obtained from comparing the theoretical and observed abundances of light elements, such as helium-3 and lithium-7, that were formed in the early universe. The density of nonbaryonic dark matter, which is of unknown composition, is not included in the value of $\rho_{b,0}$. Nonbaryonic dark matter is revealed only by its gravitational influence on baryonic matter. Presumably it interacts very weakly (if at all) with photons and charged particles via the electromagnetic force, so it does not absorb, emit, or scatter appreciable amounts of light. Our model universe of pressureless dust includes both types of matter, baryonic and nonbaryonic, luminous and dark.

The ratio of a measured density to the critical density is an important parameter in cosmology. Accordingly, it is useful to define the **density parameter**,

$$\Omega(t) \equiv \frac{\rho(t)}{\rho_c(t)} = \frac{8\pi G\rho(t)}{3H^2(t)}, \tag{29.18}$$

which has a present value of

$$\Omega_0 = \frac{\rho_0}{\rho_{c,0}} = \frac{8\pi G\rho_0}{3H_0^2}. \tag{29.19}$$

Table 29.1 shows the mass-to-light ratios of a variety of astronomical systems and the density parameters derived for them. With the exception of Big Bang nucleosynthesis, these values were obtained by studying gravitational effects and thus include both baryonic and dark matter. There is a significant trend that more extensive systems have larger mass-to-light ratios and density parameters, but, as shown in Fig. 29.4, for the largest systems the density parameters seem to reach a "ceiling" at a maximum value of $\Omega_0 \simeq 0.3$. This is consistent with the WMAP result for the value of the average density of all types of matter, baryonic and dark:

$$[\Omega_{m,0}]_{\text{WMAP}} = \left(0.135\,^{+0.008}_{-0.009}\right) h^{-2} = 0.27 \pm 0.04 \qquad \text{(for } h = 0.71\text{).} \tag{29.20}$$

This corresponds to a mass density of

$$\rho_{m,0} = 2.56 \times 10^{-27} \text{ kg m}^{-3} \qquad \text{(for } h = 0.71\text{)} \tag{29.21}$$

The "m" subscript, which stands for "mass," anticipates models of the universe with more than one component. This subscript will be suppressed for the present one-component model.

---

[6]Actually, anything made of known particles whose density obeys Eq. (29.5) counts as baryonic matter. This excludes photons and neutrinos because, as we shall see in Section 29.2, a gas of either of these particles does not obey Eq. (29.5). Such a gas exerts a pressure and therefore is inconsistent with our model universe of pressureless dust.

**TABLE 29.1**    Mass-to-Light Ratios and Density Parameters, Measured for a Variety of Systems. The complicated dependence on $h$ for the values from the X-ray halo of M87 and Local Group timing is not shown. (Adapted from Binney and Tremaine, *Galactic Dynamics*, Princeton University Press, Princeton, NJ, 1987, and Schramm, *Physica Scripta, T36*, 22, 1991.)

| Method | $M/L$ $(M_\odot/L_\odot)$ | $\Omega_0$ |
|---|---|---|
| Solar neighborhood | 3 | $0.002h^{-1}$ |
| Elliptical galaxy cores | $12h$ | 0.007 |
| Local escape speed | 30 | $0.018h^{-1}$ |
| Satellite galaxies | 30 | $0.018h^{-1}$ |
| Magellanic Stream | > 80 | $> 0.05h^{-1}$ |
| X-ray halo of M87 | > 750 | $> 0.46h^{-1}$ |
| Local Group timing | 100 | $0.06h^{-1}$ |
| Groups of galaxies | $260h$ | 0.16 |
| Clusters of galaxies | $400h$ | 0.25 |
| Gravitational lenses | — | $0.1 - 0.3$ |
| Big Bang nucleosynthesis | — | $0.065 \pm 0.045$ |

The WMAP value of the density parameter for baryonic matter is

$$[\Omega_{b,0}]_{\text{WMAP}} = (0.0224 \pm 0.0009)h^{-2} = 0.044 \pm 0.004 \qquad (\text{for } h = 0.71). \qquad (29.22)$$

Thus, according to the WMAP results, baryonic matter accounts for only about 16% of the matter in the universe; the other 84% is some sort of nonbaryonic dark matter.

The general characteristics of the expansion of our model universe composed of pressureless dust can now be determined. First note that, from Eqs. (29.6) and (29.19),

$$\frac{\Omega}{\Omega_0} = \frac{\rho}{\rho_0}\frac{H_0^2}{H^2} = (1+z)^3\frac{H_0^2}{H^2},$$

so

$$\Omega H^2 = (1+z)^3\Omega_0 H_0^2. \qquad (29.23)$$

Another relation between $\Omega$ and $H$ comes from combining the density parameter, Eq. (29.18), with Eq. (29.9):

$$H^2(1 - \Omega)R^2 = -kc^2 \qquad (29.24)$$

As a special case at $t = t_0$,

$$H_0^2(1 - \Omega_0) = -kc^2. \qquad (29.25)$$

This confirms that:

- If $\Omega_0 > 1$, then $k > 0$ and the universe is closed.

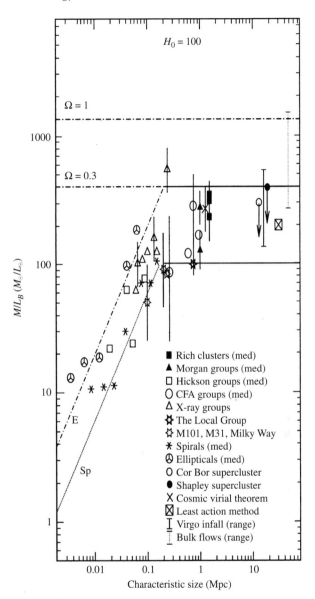

**FIGURE 29.4** The mass-to-light ratio as a function of the characteristic size of a variety of systems. $H_0$ was taken to be $100 \text{ km s}^{-1} \text{ Mpc}^{-1}$ for this figure prior to publication of the WMAP results. (Figure adapted from Dodelson, *Modern Cosmology*, Academic Press, New York, 2003, with permission from Elsevier. Data from Bahcall et al., *Ap. J.*, 541, 1, 2000.)

- If $\Omega_0 < 1$, then $k < 0$ and the universe is open.

- If $\Omega_0 = 1$, then $k = 0$ and the universe is flat.

Remember that we are now dealing with a simple model of a one-component universe of pressureless dust. Later we will study more realistic multicomponent models, which will show that a measurement of the mass density parameter alone is not enough for us to draw any conclusions about the ultimate fate of our physical universe.

Equating Eqs. (29.24) and (29.25), and using (29.4), we find

$$H^2(1 - \Omega) = H_0^2(1 - \Omega_0)(1 + z)^2. \tag{29.26}$$

Thus we have two equations, Eqs. (29.23) and (29.26), with the two unknowns $\Omega$ and $H$. These may be easily solved to find

$$H = H_0(1 + z)\,(1 + \Omega_0 z)^{1/2} \tag{29.27}$$

and

$$\Omega = \left(\frac{1+z}{1+\Omega_0 z}\right)\Omega_0 = 1 + \frac{\Omega_0 - 1}{1 + \Omega_0 z}. \tag{29.28}$$

Equation (29.27) implies that at very early times, as $R \to 0$ and $z \to \infty$, the Hubble parameter $H \to \infty$. Equation (29.28) shows that the sign of $\Omega - 1$ does not change, and in particular that if $\Omega = 1$ at any time, then $\Omega = 1$ at all times. The character of the universe does not change as the universe evolves; it is either always closed, always open, or always flat. Equation (29.28) also shows that at very early times, as $z \to \infty$, the density parameter $\Omega \to 1$ regardless of today's value of $\Omega_0$. *Therefore, the early universe was essentially flat;* see Fig. 29.5. The assumption of a flat early universe will greatly simplify the description of the first few minutes of the universe.

---

**Example 29.1.1.**   Later we will find that when the universe was about 3 minutes old, protons and neutrons combined to form helium nuclei. This occurred at a redshift of $z = 3.68 \times 10^8$. Using the WMAP value of $[\Omega_{m,0}]_{\text{WMAP}} = 0.27$ for $\Omega_0$, we find that at the time of helium formation, the value of $\Omega$ was

$$\Omega = 1 + \frac{\Omega_0 - 1}{1 + \Omega_0 z} = 1 + \frac{0.27 - 1}{1 + (0.27)(3.68 \times 10^8)} = 0.99999999265. \tag{29.29}$$

At even earlier times the value of $\Omega$ contains a much longer string of nines. During the late twentieth century, it appeared absurd to theoreticians that a mechanism would exist to fine-tune $\Omega$ to a value so very close to unity without having an exactly flat universe with $\Omega = 1$. And yet, observational measurements of the value of the density parameter continued to hover around $\Omega_0 \simeq 0.3$. The solution to this puzzle will be described in Section 29.3, where we will find that the theoreticians and the observers were *both* correct.

---

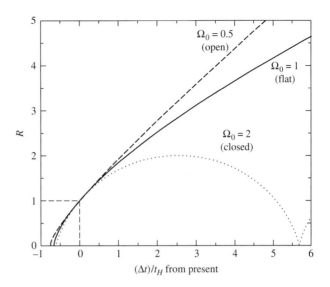

**FIGURE 29.5** The evolution of the scale factor, $R$, for three model universes—open ($\Omega_0 = 0.5$), flat ($\Omega_0 = 1$), and closed ($\Omega_0 = 2$)—as a function of time, measured from the present. The dotted lines locate the position of today's universe on the three curves. At the present time ($R = 1$) all three universes have the same value of $H_0$, as exhibited by the curves having the same slope. For the early universe ($R < 1$) there is little difference among the kinematic behaviors of a flat, a closed, and an open universe because the early universe was essentially flat. The elapsed time $\Delta t$ is in units of the Hubble time, $t_H$.

The expansion of a flat, one-component universe of pressureless dust as a function of time may be found by solving Eq. (29.11) with $k = 0$ (so $\rho_0 = \rho_{c,0}$ and $\Omega_0 = 1$):

$$\left(\frac{dR}{dt}\right)^2 = \frac{8\pi G\rho_{c,0}}{3R}.$$

Taking the square root of each side and integrating (with $R = 0$ at $t = 0$) gives

$$\int_0^R \sqrt{R'}\, dR' = \sqrt{\frac{8\pi G\rho_{c,0}}{3}} \int_0^t dt'$$

or

$$R_{\text{flat}} = (6\pi G\rho_{c,0})^{1/3} t^{2/3} \tag{29.30}$$

$$= \left(\frac{3}{2}\right)^{2/3} \left(\frac{t}{t_H}\right)^{2/3} \qquad \text{(for } \Omega_0 = 1\text{)}, \tag{29.31}$$

where the last expression was obtained by using Eq. (29.15), and $t_H \equiv 1/H_0$ for the Hubble time (Eq. 27.14). The increase in $R$ for $\Omega_0 = 1$ is shown in Fig. 29.5, with time in units of the Hubble time.

If $\Omega_0 \neq 1$, the density is not equal to the critical density and Eq. (29.11) is more difficult to solve. If $\Omega_0 > 1$, the universe is closed and the solution can be expressed in parametric form as

$$R_{closed} = \frac{4\pi G\rho_0}{3kc^2} [1 - \cos(x)] \tag{29.32}$$

$$= \frac{1}{2} \frac{\Omega_0}{\Omega_0 - 1} [1 - \cos(x)] \tag{29.33}$$

and

$$t_{closed} = \frac{4\pi G\rho_0}{3k^{3/2}c^3} [x - \sin(x)] \tag{29.34}$$

$$= \frac{1}{2H_0} \frac{\Omega_0}{(\Omega_0 - 1)^{3/2}} [x - \sin(x)], \tag{29.35}$$

where the variable $x \geq 0$ merely parameterizes the solution. [This may be easily verified by direct substitution into Eq. (29.11); see Problem 29.3.] The behavior of this solution with $\Omega_0 = 2$ is shown in Fig. 29.5. The "bounce" that occurs after the contraction of the universe is a mathematical artifact and does not imply an endless sequence of oscillating universes.

On the other hand, if $\Omega_0 < 1$, the universe is open and the parametric form of the solution of Eq. (29.11) is

$$R_{open} = \frac{4\pi G\rho_0}{3|k|c^2} [\cosh(x) - 1] \tag{29.36}$$

$$= \frac{1}{2} \frac{\Omega_0}{1 - \Omega_0} [\cosh(x) - 1] \tag{29.37}$$

and

$$t_{open} = \frac{4\pi G\rho_0}{3|k|^{3/2}c^3} [\sinh(x) - x] \tag{29.38}$$

$$= \frac{1}{2H_0} \frac{\Omega_0}{(1 - \Omega_0)^{3/2}} [\sinh(x) - x]. \tag{29.39}$$

Recall that the hyperbolic cosine is defined as $\cosh(x) \equiv (e^x + e^{-x})/2 \geq 1$. Similarly, the hyperbolic sine is given by $\sinh(x) \equiv (e^x - e^{-x})/2 \geq x$, so $R_{open}$ increases monotonically with $t$. See Fig. 29.5 for the appearance of the solution with $\Omega_0 = 0.5$. If $\Omega_0 \leq 1$, then the universe will continue to expand forever.

## The Age of the Pressureless "Dust" Universe

We are now ready to calculate the age of the universe as a function of the redshift $z$. Before continuing, a note of caution should be sounded about referring to any time $t$ as the "age of the universe." The laws of physics, as we presently understand them, cannot remain valid under the extreme conditions that must prevail as $t \to 0$. In using $t$ as a measure of the time since the Big Bang, we must always keep in mind that this is an *extrapolated time* and cannot be taken literally at the earliest instants ($t < 10^{-43}$ s; see Section 30.1).

Keeping this admonition in mind, we now proceed by using Eq. (29.4) to replace $R$ by $1/(1+z)$ in Eq. (29.31) for a flat universe. The age of a flat universe (in units of the Hubble time) that is observed at redshift $z$ is then found to be

$$\frac{t_{\text{flat}}(z)}{t_H} = \frac{2}{3}\frac{1}{(1+z)^{3/2}} \qquad (\text{for } \Omega_0 = 1). \qquad (29.40)$$

Replacing $R$ by $1/(1+z)$ in Eq. (29.33) for a closed universe and using Eq. (29.35) to eliminate $x$ leads to

$$\frac{t_{\text{closed}}(z)}{t_H} = \frac{\Omega_0}{2(\Omega_0-1)^{3/2}}\left[\cos^{-1}\left(\frac{\Omega_0 z - \Omega_0 + 2}{\Omega_0 z + \Omega_0}\right) - \frac{2\sqrt{(\Omega_0-1)(\Omega_0 z+1)}}{\Omega_0(1+z)}\right]$$

$$(\text{for } \Omega_0 > 1). \qquad (29.41)$$

Following a similar procedure using Eq. (29.37) for an open universe and using Eq. (29.39) to eliminate $x$ results in

$$\frac{t_{\text{open}}(z)}{t_H} = \frac{\Omega_0}{2(1-\Omega_0)^{3/2}}\left[-\cosh^{-1}\left(\frac{\Omega_0 z - \Omega_0 + 2}{\Omega_0 z + \Omega_0}\right) + \frac{2\sqrt{(1-\Omega_0)(\Omega_0 z+1)}}{\Omega_0(1+z)}\right]$$

$$(\text{for } \Omega_0 < 1). \qquad (29.42)$$

In the limit of large redshift, Eqs. (29.40) through (29.42) all reduce to

$$\frac{t(z)}{t_H} = \frac{2}{3}\frac{1}{(1+z)^{3/2}\Omega_0^{1/2}}, \qquad (29.43)$$

where the higher-order terms may be neglected for $\Omega_0 \neq 1$ (see Problem 29.11). Because the early universe was flat to a very good approximation, precise observations are required to determine whether the universe is flat, closed, or open. In Section 29.4, the observational aspects of cosmology will be considered in more detail.

The current age of the universe, $t_0$, may be easily found by setting $z = 0$ in Eqs. (29.40–29.42) to find

$$\frac{t_{\text{flat},0}}{t_H} = \frac{2}{3} \qquad (\text{for } \Omega_0 = 1) \qquad (29.44)$$

for a flat universe,

$$\frac{t_{\text{closed},0}}{t_H} = \frac{\Omega_0}{2(\Omega_0-1)^{3/2}}\left[\cos^{-1}\left(\frac{2}{\Omega_0}-1\right) - \frac{2\sqrt{\Omega_0-1}}{\Omega_0}\right] \qquad (\text{for } \Omega_0 > 1) \qquad (29.45)$$

for a closed universe, and

$$\frac{t_{\text{open},0}}{t_H} = \frac{\Omega_0}{2(1-\Omega_0)^{3/2}}\left[-\cosh^{-1}\left(\frac{2}{\Omega_0}-1\right) + \frac{2\sqrt{1-\Omega_0}}{\Omega_0}\right] \qquad (\text{for } \Omega_0 < 1) \qquad (29.46)$$

for an open universe.

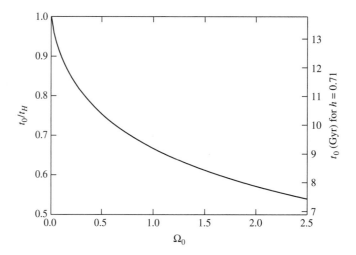

**FIGURE 29.6**   The age of the universe as a function of the density parameter, $\Omega_0$. The age is expressed as a fraction of the Hubble time, $t_H \simeq 10^{10}h^{-1}$ yr. The right axis shows the age in billions of years for $h = 0.71$.

The age of the universe for these models, expressed as a fraction of the Hubble time, is shown in Fig. 29.6. As we will see in Section 30.1, according to the inflation scenario the universe should be essentially flat, a scenario supported by recent observations. If the average density of the universe is equal to the critical density, then the age of the universe is two-thirds of the Hubble time. Using the WMAP value of $[h]_{\text{WMAP}} = 0.71$ gives an age of about 9.2 Gyr. Although this result is less than the currently accepted value of 13.7 Gyr, it is remarkable that this simple model of an expanding universe of pressureless dust produces ages that are in rough accordance with the mean age of the oldest globular clusters, 11.5 billion years.[7]

Piecing together the history of the universe is a bit like working a jigsaw puzzle. As progress is made and the pattern begins to emerge, the shapes of the missing pieces are defined by those already in place. For our simple model of a pressureless dust universe, the pieces overlap a bit; the age of the oldest stars is greater than the age of the universe. However, there is more to the universe than pressureless dust. For example, the universe is filled with photons—about two billion photons for every baryon. Photons, baryons, and other constituents will play a role in resolving the discrepancy between the age of the oldest globular clusters and the age of the universe.

**The Lookback Time**

The **lookback time**, $t_L$, is defined as how far back in time we are looking when we view an object with redshift $z$. This is just the difference between the present age of the universe and its age at time $t(z)$,

$$t_L = t_0 - t(z). \tag{29.47}$$

---

[7]A more accurate comparison between the age of the universe and the ages of globular clusters will be made in Section 29.3.

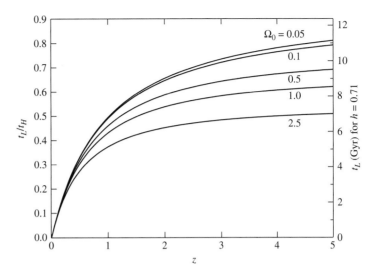

**FIGURE 29.7** The lookback time as a function of the redshift, $z$, for a range of values of the density parameter, $\Omega_0$. The lookback time is expressed as a fraction of the Hubble time, $t_H \simeq 10^{10}h^{-1}$ yr. The right axis shows the lookback time in billions of years if $h = 0.71$.

For example, for a flat universe, Eqs. (29.40) and (29.44) show that the lookback time is, in units of the Hubble time,

$$\frac{t_L}{t_H} = \frac{2}{3}\left[1 - \frac{1}{(1+z)^{3/2}}\right] \qquad \text{(for } \Omega_0 = 1\text{)}. \qquad (29.48)$$

Figure 29.7 shows the lookback times for flat, closed, and open models of the universe.

---

**Example 29.1.2.** In Example 4.3.2, the redshift of the quasar SDSS 1030+0524 was found to be $z = 6.28$. Assuming a flat universe of pressureless dust, Eq. (29.48) shows that the lookback time to this quasar is

$$\frac{t_L}{t_H} = \frac{2}{3}(1 - 0.0509) = 0.633.$$

Since the age of a flat universe is $t_0 = 2t_H/3$,

$$\frac{t_L}{t_0} = 0.949.$$

This means that only 5% of the history of the universe had unfolded when the light left this quasar. At that time the universe was smaller by about a factor of 7 when, according to Eq. (29.4), the scale factor was

$$R = \frac{1}{1+z} = 0.137.$$

---

**Extending Our Simple Model to Include Pressure**

Let's take stock of the basic equations we have derived so far and generalize them a bit, anticipating some features of the equations of general relativity we will encounter in Section 29.3. We start with Eq. (29.10),

$$\left[ \left( \frac{1}{R}\frac{dR}{dt} \right)^2 - \frac{8}{3}\pi G\rho \right] R^2 = -kc^2.$$

With a nod toward Einstein's $E_{rest} = mc^2$, we broaden the meaning of the density $\rho$ to include matter in all of its forms. For nonrelativistic particles, $\rho$ is the usual mass density. For relativistic particles, such as photons and neutrinos, $\rho$ is the **equivalent mass density**—that is, the energy density divided by $c^2$.

Equation (29.5), $R^3\rho = \rho_0$, describes the conservation of mass within the expanding shell. Again acknowledging the equivalence of mass and energy, this equation is also a statement of the conservation of energy for a pressureless dust universe.

A thermodynamic argument supplies the generalization of Eq. (29.5) for models of the universe that incorporate pressure-producing components. Imagine a universe filled with a fluid (dust, photons, etc.) of uniform density $\rho$, pressure $P$, and temperature $T$, and choose an arbitrary point for the origin. Let $r$ be the radius of a comoving spherical surface, centered on the origin.[8] We will employ the first law of thermodynamics, which applies the law of conservation of internal energy, $U$, to the fluid within the expanding sphere:

$$dU = dQ - dW. \tag{29.49}$$

First note that the entire universe has the same temperature, so there can be no heat flow: $dQ = 0$. That is, the expansion of the universe is adiabatic. Any change in internal energy must be produced by work done by the fluid. Writing the result as a time derivative,

$$\frac{dU}{dt} = -\frac{dW}{dt} = -P\frac{dV}{dt}.$$

and substituting $V = \frac{4}{3}\pi r^3$, we obtain

$$\frac{dU}{dt} = -\frac{4}{3}\pi P\frac{d(r^3)}{dt}.$$

If we define the internal energy per unit volume $u$ as

$$u = \frac{U}{\frac{4}{3}\pi r^3},$$

then we find

$$\frac{d(r^3 u)}{dt} = -P\frac{d(r^3)}{dt}.$$

---

[8]We need not worry about photons leaving the sphere, because the cosmological principle assures us that the same number of equivalent photons will enter the sphere.

Writing $u$ in terms of the equivalent mass density $\rho$,

$$\rho = \frac{u}{c^2},$$

gives

$$\frac{d(r^3\rho)}{dt} = -\frac{P}{c^2}\frac{d(r^3)}{dt}.$$

Finally, using $r = R\varpi$ (Eq. 29.3), we obtain the **fluid equation**,

$$\frac{d(R^3\rho)}{dt} = -\frac{P}{c^2}\frac{d(R^3)}{dt}. \tag{29.50}$$

For a universe of pressureless dust, $P = 0$ so $R^3\rho = $ constant, in agreement with Eq. (29.5).

An equation describing the acceleration of the universal expansion can be obtained by multiplying Eq. (29.10) by $R$ and then taking a time derivative. Using Eq. (29.50) to replace $d(\rho R^3)/dt$ and using Eq. (29.10) to eliminate the $-kc^2$, we arrive at the **acceleration equation**

$$\frac{d^2 R}{dt^2} = -\frac{4}{3}\pi G\left(\rho + \frac{3P}{c^2}\right)R. \tag{29.51}$$

Note that the effect of the pressure $P$ is to slow down the expansion (assuming $P > 0$). If this seems counterintuitive, recall that because the pressure is the same everywhere in the universe, both inside and outside the shell, there is no pressure gradient to exert a net force on the expanding sphere. The answer lies in the motion of the particles that creates the fluid's pressure. The equivalent mass of the particles' kinetic energy creates a gravitational attraction that slows down the expansion just as their actual mass does. In fact, the assumption that $P = 0$ is valid for much of the history of the universe. For instance, you will find in Problem 29.14 that $\rho \gg P/c^2$ in today's universe.

Equation (29.51) is an illustration of **Birkhoff's theorem**. In 1923 the American mathematician G. D. Birkhoff (1884–1944) proved quite generally that for a spherically symmetric distribution of matter, Einstein's field equations have a unique solution. As a corollary, the acceleration of an expanding shell in our fluid universe is determined solely by the fluid lying within the shell.[9] Equation (29.51) shows that the acceleration does not depend on any factors other than $\rho$, $P$, and $R$. Because Birkhoff's theorem holds even when general relativity is included, it is quite important in the study of cosmology.

Equations (29.10), (29.50), and (29.51) have three unknowns: $R$, $\rho$, and $P$. However, the equations are not independent; any two may be used to derive the third. To solve these two equations for $R$, $\rho$, and $P$, we need a third relation, an *equation of state*, that links the variables. Such an equation of state can be written generally as

$$P = wu = w\rho c^2, \tag{29.52}$$

[9]You may recall that a similar result was obtained in Problem 10.2.

where $w$ is a constant. In words, the pressure is proportional to the energy density of the fluid. For example, for mass in the form of pressureless dust, $w_m = 0$, and for blackbody radiation, with the equation of state $P_{rad} = u_{rad}/3$ (Eq. 9.11), we have $w_{rad} = 1/3$. Inserting the general equation of state, Eq. (29.52), into the fluid equation, Eq. (29.50), quickly produces the relation (Problem 29.15)

$$R^{3(1+w)}\rho = \text{constant} = \rho_0, \tag{29.53}$$

where $\rho_0$ is the present value of the mass density (or equivalent mass density). For pressureless dust ($w_m = 0$), we recover Eq. (29.5), $R^3\rho_m = \rho_{m,0}$.

### The Deceleration Parameter

Finally, we introduce a useful dimensionless quantity that describes the acceleration of the universal expansion: the **deceleration parameter**, $q(t)$, which is defined as

$$q(t) \equiv -\frac{R(t)\,[d^2R(t)/dt^2]}{[dR(t)/dt]^2}. \tag{29.54}$$

Both the name and the minus sign (to ensure that $q > 0$ for a deceleration) betray the certainty of twentieth-century astronomers that the expansion of the universe must be slowing down with time.[10] It is left to Problem 29.16 to show that for a pressureless dust universe,

$$q(t) = \frac{1}{2}\Omega(t), \tag{29.55}$$

and so at the present time,

$$q_0 = \frac{1}{2}\Omega_0. \tag{29.56}$$

Thus, for a pressureless dust universe, $q_0 = 0.5$ for a flat universe, while $q_0 > 0.5$ and $q_0 < 0.5$ correspond to a closed and an open universe, respectively.

## 29.2 ■ THE COSMIC MICROWAVE BACKGROUND

In 1946 George Gamow was pondering the cosmic abundances of the elements. Realizing that the newborn, dense universe must have been hot enough for a burst of nuclear reactions to occur, he proposed that a sequence of reactions in the very early universe could explain the measured cosmic abundance curve (shown in Fig. 15.16). Gamow, together with Ralph Alpher, published this idea two years later.[11] Still later, however, detailed calculations by

---

[10]In Section 29.3 this sense of certainty will be toppled.

[11]Gamow, a Russian émigré, was famous for his Puckish sense of humor. When this paper first appeared, Gamow added Hans Bethe as a co-author (without Bethe's knowledge). Gamow thought it would be appropriate that a paper on cosmic beginnings be authored by Alpher ($\alpha$), Bethe ($\beta$), and Gamow ($\gamma$), a play on the first three letters of the Greek alphabet.

Alpher and Robert Herman (1914–1997) showed that Gamow's idea was flawed because there were roadblocks to assembling succeedingly heavier nuclei simply by adding protons or neutrons. There are no stable nuclei with five or eight nucleons, leaving $_2^4$He as the heaviest element that can be formed as Gamow proposed. (A small amount of an isotope of lithium, $_3^7$Li, is also formed in the early universe by the nuclear fusion of $_2^4$He with $_1^3$H and the fusion of $_2^4$He with $_2^3$He. The latter produces $_4^7$Be, which radioactively decays to $_3^7$Li.)

At that time, there was also a large problem with the idea of a hot, dense universe coming into existence approximately one Hubble time ago. Edwin Hubble's original value of his constant was $H_0 = 500$ km s$^{-1}$ Mpc$^{-1}$ (see Fig. 27.7), which corresponds to $t_H = 1/H_0 = 10^9$ yr for the age of the universe. This is only a fraction of Earth's age, which in 1928 had been radioactively dated as several billion years. By the late 1940s further measurements indicated that $1/H_0 = 1.8 \times 10^9$ yr, which was still embarrassingly short. It was certainly hard to understand how Earth could be older than the universe.

### The Steady-State Model of the Universe

In 1946 at Cambridge University, Hermann Bondi (1919–2005), Thomas Gold (1920–2004), and Fred Hoyle (1915–2001) attempted to find an alternative to Gamow's unpalatable Big Bang universe.[12] In papers published in 1948 and 1949, they proposed their model of a **steady-state universe**. It extended the cosmological principle to include time, stating that, in addition to the universe being isotropic and homogeneous, it also appears the same *at all times*. A steady-state universe has no beginning and no end. It is infinitely old, and as it expands, a continuous creation of matter is required to maintain the average density of the universe at its present level. This changes the interpretation of the Hubble time; rather than the characteristic age of the universe, $t_H$ becomes a characteristic time for the creation of matter. If the universe roughly doubles in size in time $t_H$, then its volume becomes eight times greater, and so the rate of matter creation required to maintain the universe as it is today is approximately $8\rho_0/t_H = 8H_0\rho_0$. Just a few hydrogen atoms per cubic meter of space would need to be created every ten billion years, a rate far too small to be measured experimentally. In the original steady-state models, the "when," "where," and "how" of the spontaneous appearance of new matter (in violation of the law of conservation of mass–energy) were questions left unanswered. The appeal of the steady-state universe was its resolution of the timescale problem.[13]

Just as Gamow and his collaborators tried to explain the cosmic abundance curve (Fig. 15.16) by using the nuclear reactions that would accompany a Big Bang, Hoyle sought an explanation in the nuclear reactions that took place inside stars. He joined forces with two English colleagues, Geoffrey and Margaret Burbidge (a theoretical physicist and an astronomer, respectively), and William Fowler (1911–1995), an American physicist. In

---

[12]Ironically, it was Fred Hoyle who came up with the term *Big Bang*. He used it derisively in a 1950 BBC radio broadcast when he said, "This big bang idea seemed to me to be unsatisfactory even before examination showed that it leads to serious difficulties. For when we look at our own Galaxy there is not the smallest sign that such an explosion ever occurred."

[13]The solution to the short Hubble time appeared in 1952, when Walter Baade discovered that there were two types of Cepheid variable stars. This revised the period–luminosity relation that forms the foundation of the cosmic distance scale; see Section 27.1.

1957 they published their seminal paper, referred to as B²FH, that laid out the theory of stellar nucleosynthesis as described in Chapters 10 and 13.

The B²FH analysis was a success, and its results were compatible with both the Big Bang and steady-state cosmologies. During the 1950s both theories had their supporters and detractors. However, in addition to the unanswered questions about the continuous creation of matter, the steady-state theory had a serious problem explaining the large amount of helium observed in the universe. Astronomers had established that about one-quarter $(0.274 \pm 0.016)$ of the baryonic mass of the universe is in the form of helium. When compared to the cosmic abundances of the heavier elements, it was clear that stellar nucleosynthesis could not account for the amount of helium observed, especially considering that carbon, nitrogen, and oxygen are the results of exhaustively burning the star's helium core (recall Section 13.2). Gamow, Alpher, and Herman had shown that the Big Bang could at least explain the abundance of helium, but where was the proof that such a violent event had ever occurred? To invoke an event that was apparently beyond the reach of investigation seemed unscientific to many astronomers.

### The Cooling of the Universe after the Big Bang

A key idea of the $\alpha$–$\beta$–$\gamma$ paper was that the dense, early universe must have been very hot. In this hot, dense universe, the mean free path of photons would have been short enough to maintain thermodynamic equilibrium. Although an expanding universe cannot be precisely in equilibrium, this assumption of thermodynamic equilibrium is extremely good. As was described in Section 9.2, under these conditions the radiation field has a blackbody spectrum. In 1948 Alpher and Herman published their description of how this blackbody radiation would have cooled as the universe expanded, and they predicted that the universe should now be filled with blackbody radiation at a temperature of 5 K.

The cooling of the blackbody radiation can be derived by considering its energy density $u = aT^4$ (Eq. 9.7). According to the fluid equation (Eq. 29.53) with $w_{rad} = 1/3$ for blackbody radiation and $R(t_0) = 1$,

$$R^{3(1+w_{rad})} u_{rad} = R^4 u_{rad} = u_{rad,0}. \tag{29.57}$$

The energy density today, $u_{rad,0}$, is smaller than the earlier value $u_{rad}$ by a factor of $R^4$; a factor of $R^3$ is due to the fact that the volume of the universe has increased since then, and the other factor of $R$ comes from the lesser energy of today's longer-wavelength photons ($E_{photon} = hc/\lambda$), a result of the cosmological redshift. Thus

$$R^4 a T^4 = a T_0^4,$$

and we find that the present blackbody temperature must be related to the temperature at an earlier time by

$$RT = T_0. \tag{29.58}$$

That is, the product of the scale factor and the blackbody temperature remains constant as the universe expands. When the universe was half as large, it was twice as hot.

An order-of-magnitude estimate of the present blackbody temperature of the universe may be calculated by considering the temperature and baryonic mass density that must have prevailed in the early universe when helium was being formed. The fusion of hydrogen nuclei requires roughly that $T \simeq 10^9$ K and $\rho_b \simeq 10^{-2}$ kg m$^{-3}$. If the temperature were any higher, the deuterium nuclei involved in the fusion chain would have undergone photodissociation due to the presence of energetic blackbody radiation, whereas a lower temperature would have made the Coulomb barrier between the nuclei too difficult to overcome. The quoted density is needed to produce the observed amounts of $^3_2$He and other nuclei. From Eqs. (29.5) and (29.17), the value of the scale factor at the time of the helium formation was roughly

$$R \simeq \left( \frac{\rho_{b,0}}{\rho_b} \right)^{1/3} = 3.47 \times 10^{-9}.$$

At that time, the universe was only a few billionths of its present size. Combining the scale factor with $T(R) = 10^9$ K, the present temperature of the blackbody radiation can be estimated from Eq. (29.58) as

$$T_0 = RT(R) \simeq 3.47 \text{ K},$$

similar to Alpher and Hermann's original estimate of 5 K in 1948. Wien's law then gives the peak wavelength of the blackbody spectrum as

$$\lambda_{\max} = \frac{0.00290 \text{ m K}}{T_0} \simeq 8.36 \times 10^{-4} \text{ m}.$$

### The Discovery of the Cosmic Microwave Background

Sixteen years after Alpher and Herman predicted that the universe had cooled to 5 K and was filled with blackbody radiation, Robert Dicke (1916–1997) of Princeton University and his postdoctoral student P. J. E. Peebles unknowingly followed in their footsteps. In 1964 Peebles calculated that the blackbody radiation left over from the Big Bang should have a temperature of about 10 K. Unlike Alpher and Herman, however, Dicke was interested in searching for the relic radiation. He didn't realize that this cosmic background radiation had just been found by two radio astronomers working a few miles away at Bell Laboratories in Holmdel, New Jersey.[14] Arno Penzias and Robert Wilson were working with a huge horn reflector antenna (shown in Fig. 29.8) that had been used to communicate with the new Telstar satellite. Despite a year of effort, the two men had been unable to get rid of a persistent hiss in the signal. The hiss came continually from all directions in the sky and remained even after Penzias and Wilson had scrubbed their antenna clean, taped over seams and rivets, and removed two pigeons that had nested inside the horn.[15] They knew that a 3 K blackbody would produce their interference but were unaware of any possible source until Penzias learned of Peeble's calculation of a 10-K background. Penzias called

---

[14]Interestingly, it was in Holmdel that Karl Jansky built his first radio telescope.
[15]The pigeons were freed 60 miles away. Unfortunately, they proved to be homing pigeons and had to be removed again.

**FIGURE 29.8**    Robert Wilson and Arno Penzias standing in front of the antenna used to first identify the cosmic microwave background. (Courtesy of AT&T Archives.)

Dicke and invited him to Holmdel, and in 1965 the pieces of the puzzle finally fell together. Penzias and Wilson had detected the blackbody radiation that fills the universe, with a peak wavelength of $\lambda_{max} = 1.06$ mm in the microwave region of the electromagnetic spectrum. This afterglow of the Big Bang is now known as the **cosmic microwave background**, often abbreviated as the CMB. Dicke, Peebles, and their co-workers at Princeton immediately wrote a note for the *Astrophysical Journal Letters* detailing the theory of the cosmic microwave background that strongly supported the Big Bang cosmology, while Penzias and Wilson wrote an accompanying letter, discreetly titled "A Measurement of Excess Antenna Temperature at 4080 Megacycles per Second," that described their discovery.

The discovery of the CMB was a death knell for steady-state cosmology. As further measurements at other wavelengths confirmed that the shape of the CMB spectrum was that of a blackbody, the number of astronomers supporting the Big Bang theory swelled while those favoring a steady-state universe dwindled. In 1991 a striking measurement of the cosmic microwave background was obtained by the COBE satellite (see Section 6.4). The COBE measurement of the spectrum of the CMB is shown in Fig. 29.9. The data points (whose errors are smaller than the points themselves) fall almost perfectly on the theoretical spectrum of a 2.725-K blackbody. The Planck function $B_\nu(T)$ in Fig. 29.9 peaks at a frequency of 160 GHz, corresponding to the frequency version of Wien's law,[16]

$$\frac{\nu_{max}}{T} = 5.88 \times 10^{10} \text{ Hz K}^{-1}. \tag{29.59}$$

The WMAP value for the CMB is

$$\boxed{[T_0]_{\text{WMAP}} = 2.725 \pm 0.002 \text{ K,}} \tag{29.60}$$

remarkably close to our estimate of 3.47 K obtained from the simple estimate on page 1165.

---

[16]This version of Wien's law may be found from Eq. (3.24) by setting $dB_\nu(T)/d\nu = 0$.

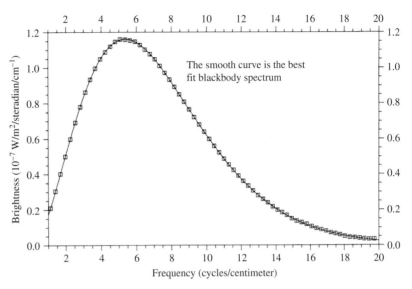

**FIGURE 29.9**  The COBE measurement of the spectrum of the cosmic microwave background, which is that of a blackbody with a temperature of 2.725 K. The horizontal axis (frequency) is actually $1/\lambda$ (cm$^{-1}$); the spectrum peaks at a frequency of 160 GHz (5.35 cycles per centimeter). (Figure adapted from Mather et al., *Ap. J. Lett.*, *354*, L37, 1990. Courtesy of NASA/GSFC and the COBE Science Working Group.)

### The Dipole Anisotropy of the CMB

The cosmic microwave background suffuses the entire universe. It does not emanate from any object but, rather, originated in the Big Bang, when the entire universe was essentially located at a single point (or event) of spacetime. For this reason, all observers at rest with respect to the Hubble flow (no peculiar velocity) see the same spectrum for the CMB, with the same intensity in all directions (the CMB is isotropic). In particular, two observers in different galaxies that are being carried apart by the Hubble flow see the same blackbody spectrum.[17]

However, there is a Doppler shift of the CMB caused by an observer's peculiar velocity through space, relative to the Hubble flow. Using Wien's law, a shift in wavelength can be expressed as a change in the temperature of the blackbody radiation. For example, a slight blueshift (smaller $\lambda_{max}$) would correspond to a slightly higher temperature. Suppose an observer at rest relative to the Hubble flow determines that the cosmic microwave background has a temperature $T_{rest}$. Then, as shown in Problem 29.21, the temperature measured by an observer with a peculiar velocity $v$ relative to the Hubble flow is

$$T_{moving} = \frac{T_{rest}\sqrt{1 - v^2/c^2}}{1 - (v/c)\cos\theta},\tag{29.61}$$

[17]It is assumed that the two observers agree on the age of the universe when their measurements are made. Of course, each galaxy views the other as it appeared at an earlier time.

where $\theta$ is the angle between the direction of observation and the direction of motion. Both observers see a blackbody spectrum, but the moving observer measures a slightly hotter temperature in the forward direction ($\theta = 0$) and a slightly cooler temperature in the opposite direction. If the peculiar velocity is $v \ll c$, then

$$T_{\text{moving}} \simeq T_{\text{rest}} \left( 1 + \frac{v}{c} \cos \theta \right) \tag{29.62}$$

(the proof is left as an exercise). The second term on the right-hand side, called the **dipole anisotropy** of the CMB, has been detected and measured; see Fig. 29.10. The temperature variation indicates that the peculiar velocity of the Sun relative to the Hubble flow is $370.6 \pm 0.4$ km s$^{-1}$ in the direction $(\alpha, \delta) = (11.2^{\text{h}}, -7°)$, between the constellations of Leo and Crater. Of course, the Sun is orbiting the Galaxy, and the Milky Way is moving within the Local Group of galaxies. When these motions are accounted for, the peculiar motion of the Local Group relative to the Hubble flow is about 627 km s$^{-1}$ toward $(\alpha, \delta) = (11.1^{\text{h}}, -27°)$, in the middle of the constellation Hydra. From this observation and measurements of the velocities of other galaxies and clusters of galaxies, astronomers have discovered a large-scale streaming motion of thousands of galaxies at $\sim 600$ km s$^{-1}$ in the direction of $(\alpha, \delta) = (13.3^{\text{h}}, -44°)$ in the constellation Centaurus. The Hydra–Centaurus supercluster is also being carried along in this riverlike perturbation of the Hubble flow; see Section 27.3.

After the dipole anisotropy has been subtracted from the CMB, the remaining radiation is incredibly isotropic, having nearly equal intensity in all directions. Sensitive instruments, however, have revealed that the CMB does have hotter and cooler areas. The CMB appears as a patchwork of small regions, about 1° degree or less in diameter, where the temperature departs from the average value $(T_0)$ by about one part in $10^5$. Careful observations and analyses of these regions by WMAP and various ground-based and balloon-borne experiments have produced the first precision measurement of cosmological parameters, as will

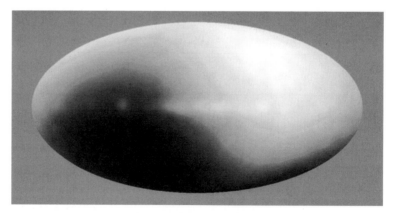

**FIGURE 29.10**   The dipole anisotropy in the CMB caused by the Sun's peculiar velocity, shown in Galactic coordinates. The figure is a combination of observations made at 53 GHz and 90 GHz. Lighter shades are somewhat hotter, and darker shades are somewhat cooler, than the 2.725-K CMB. The horizontal bright feature is due to Galactic sources. (Figure from Smoot et al., *Ap. J. Lett., 371,* L1, 1991.)

be described in Section 30.2. We will continue to use the WMAP values of these parameters throughout this and the next chapter.

### The Sunyaev–Zel'dovich Effect

It should be emphasized that an observer in a galaxy being carried along with the Hubble flow (no peculiar velocity) does not measure a Doppler shift of the CMB. An observer in a distant galaxy receding from us at an appreciable fraction of the speed of light sees the same CMB spectrum that we do. Evidence of this is produced when low-energy photons of the CMB pass through the hot ($\simeq 10^8$ K) ionized intracluster gas in a rich cluster of galaxies. A small fraction (typically $10^{-3}$ to $10^{-2}$) of the photons are scattered to higher energies by the high-energy electrons in the gas. This **inverse Compton scattering** increases the frequency of a scattered photon by an average amount $\Delta \nu$ of

$$\frac{\overline{\Delta \nu}}{\nu} = 4 \frac{kT_e}{m_e c^2}, \tag{29.63}$$

where $T_e$ is the temperature of the electron gas (see Problem 29.23). The resulting distortion of the CMB spectrum, shown in Fig. 29.11, is called the thermal **Sunyaev–Zel'dovich effect** [named for two Russians, astrophysicist Rashid Sunyaev and physicist Yakov Zel'dovich (1914–1987)].[18] Although the spectrum no longer has the precise shape of a blackbody, its translation to higher frequencies may be used to define an effective decrease $\Delta T$ in the temperature $T_0$ of the CMB of approximately[19]

$$\frac{\Delta T}{T_0} \simeq -2 \frac{kT_e}{m_e c^2} \tau \tag{29.64}$$

where $\tau$ is the optical depth of the intracluster gas along the line of sight. Typical values of $\Delta T / T_0$ are a few times $10^{-4}$. Observations of the Sunyaev–Zel'dovich effect for many clusters of galaxies confirm that it is independent of the cluster's redshift, as expected if the CMB spectrum observed at a cluster is not affected by the cluster's recessional velocity. Figure 29.12 shows the Sunyaev–Zel'dovich effect surrounding two clusters of galaxies. In addition to confirming the cosmological nature of the CMB, the Sunyaev–Zel'dovich effect is a promising probe of the properties and evolution of rich clusters of galaxies in the early universe.

### Does the CMB Constitute a Preferred Frame of Reference?

It may seem to you that the frame of reference in which the CMB appears isotropic is a preferred frame of reference that is *truly motionless*, in violation of the postulates of special relativity as described in Section 4.2. We should remember, however, that general relativity is concerned with *local* inertial reference frames (cf. Section 17.1). There is no single frame

---

[18]If the cluster has a peculiar velocity, the bulk motion of the intracluster gas produces an additional Doppler shift of the scattered photons. This smaller perturbation of the CMB spectrum is called the kinetic Sunyaev–Zel'dovich effect.

[19]The leading coefficient in Eq. (29.64) is equal to $-2$ only on the Rayleigh–Jeans (long-wavelength) side of the spectrum; cf. Eq. (3.20).

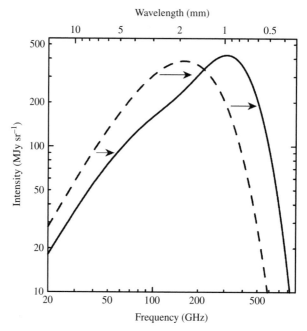

**FIGURE 29.11**   The undistorted CMB spectrum (dashed line) and the spectrum distorted by the Sunyaev–Zel'dovich effect (solid line). In a rich cluster of galaxies, CMB photons may be scattered to higher frequencies by colliding with the electrons in the hot intracluster gas. For frequencies less than the peak frequency, more photons are scattered out of a frequency interval than into it, so the intensity at that frequency decreases. Similarly, for frequencies greater than the peak frequency, fewer photons are scattered out of a frequency interval than into it, so the intensity at that frequency increases. The net result is a shift of the CMB spectrum to higher frequencies. The calculated distortion has been exaggerated by employing a fictional cluster 1000 times more massive than a typical rich cluster of galaxies. (Figure adapted from Carlstrom, Holder, and Reese, *Annu. Rev. Astron. Astrophys.*, *40*, 646, 2002. Reproduced with permission from the *Annual Review of Astronomy and Astrophysics*, Volume 40, ©2002 by Annual Reviews Inc.)

of reference that is capable of covering the entire universe. Although we can define a local reference frame at rest relative to the glow of the CMB, we can also define a local reference frame at rest relative to the motions of nearby stars (the local standard of rest described in Section 24.3). Both of these local inertial reference frames measure velocities relative to an arbitrary standard of rest and do not violate the tenets of relativity.

## A Two-Component Model of the Universe

Recall that in Section 29.1 we considered the expansion of a universe with a single component, pressureless dust, that was slowing down due to its own self-gravity. According to the relativistic equivalence of mass and energy, however, the effect of the cosmic microwave background on the expansion must also be included. In fact, we now know that the gravitational effect of the CMB photons dominated the dynamics of the early universe, although their effect is completely negligible in the present universe.

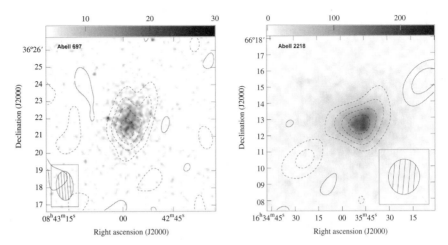

**FIGURE 29.12** Radio contours showing the Sunyaev–Zel'dovich effect superimposed on ROSAT images of the clusters Abell 697 ($\Delta T = 1047$ $\mu$K, $z = 0.282$) and Abell 2218 ($\Delta T = 797$ $\mu$K, $z = 0.171$). The contour interval is 60 $\mu$Jy (left) and 80 $\mu$Jy (right). The dashed contours indicate a decrease in the received radio flux density. (Figure adapted from Jones et al., *MNRAS*, *357*, 518, 2005.)

To incorporate this new feature, we introduce a two-component model of the universe, one that includes both the total density of matter (baryonic and dark), $\rho_m$, and the equivalent mass density of relativistic particles (such as neutrinos and CMB photons), $\rho_{\text{rel}}$. It is the equation of state $P = wu$ (Eq. 29.52) that determines whether we count a particle as matter ($w_m = 0$), a relativistic particle (a photon or neutrino, for which $w_{\text{rel}} = 1/3$), or (in Section 29.3) dark energy ($w_\Lambda = -1$). (The gravitational effect of the neutrinos' mass clearly persists; however, we will neglect the neutrinos' contribution to the value of $\Omega_{m,0}$ of roughly 0.003.)

Equation (29.53) shows that particles belonging to different categories are diluted differently by the expansion of the universe. Of course, at earlier epochs when the universe was much hotter, even massive particles were relativistic. For example, an electron gas at $T > 6 \times 10^9$ K has $kT > m_e c^2$, implying that the electron gas is relativistic and its equation of state is described by $w_{\text{rel}} = 1/3$ (see Problem 29.24). However, we will ignore such complications for the remainder of this chapter and will consider photons and neutrinos as our only relativistic particles.

With both matter and relativistic particles included, Eq. (29.10) becomes

$$\left[ \left( \frac{1}{R} \frac{dR}{dt} \right)^2 - \frac{8}{3} \pi G \left( \rho_m + \rho_{\text{rel}} \right) \right] R^2 = -kc^2. \tag{29.65}$$

The equivalent mass density of the CMB photons comes from the energy density of blackbody radiation, Eq. (9.7),

$$u_{\text{rad}} = aT^4, \tag{29.66}$$

where $a$ is the radiation constant. We will rewrite this in the form

$$u_{\mathrm{rad}} = \frac{1}{2} g_{\mathrm{rad}} a T^4, \tag{29.67}$$

where $g_{\mathrm{rad}}$ is the number of **degrees of freedom** of a photon. The value of $g$ reflects the number of spin states $n_{\mathrm{spin}}$ and the possible existence of an antiparticle ($n_{\mathrm{anti}} = 1$ or 2). A photon is its own antiparticle ($n_{\mathrm{anti}} = 1$) and can exist in $n_{\mathrm{spin}} = 2$ spin states, corresponding to its two possible polarizations with its spin parallel or antiparallel to its motion. Thus

$$g_{\mathrm{rad}} = 2 \tag{29.68}$$

for photons, as expected.

**Neutrino Decoupling**

We will neglect the small mass of the other relativistic particle we are considering, the neutrino, and treat it as a massless particle. The very early universe was sufficiently dense that neutrinos attained thermal equilibrium, with a spectrum very similar to that of blackbody radiation, Eq. (3.22), except the "−1" in the denominator of that equation becomes a "+1" for neutrinos. This occurs because photons are bosons, described by Bose–Einstein statistics, while neutrinos are fermions, described by Fermi–Dirac statistics; recall Section 10.2. Although the cosmic neutrino background has yet to be observed (not surprising given the challenge of detecting solar neutrinos), we have confidence that it exists.

Recall that there are three types (or flavors) of neutrinos—electron neutrinos, muon neutrinos, and tau neutrinos—and that each neutrino has a corresponding antineutrino. The total energy density of all three flavors is given by

$$u_\nu = 3 \times \frac{7}{8} \times a T_\nu^4 = 2.625 \, a T_\nu^4, \tag{29.69}$$

where the 7/8 derives from the "+1" in the expression for Fermi–Dirac statistics, and $T_\nu$ is the temperature of the neutrinos. As before, we write this as

$$u_\nu = \frac{1}{2} \left( \frac{7}{8} \right) g_\nu a T_\nu^4, \tag{29.70}$$

where

$$g_\nu = 6 \tag{29.71}$$

for neutrinos.

In general,

$$g = (\# \text{ types}) n_{\mathrm{anti}} n_{\mathrm{spin}}. \tag{29.72}$$

There is an antineutrino for each of the three types of neutrino so $n_{\mathrm{anti}} = 2$, and neutrinos have one spin state (all neutrinos are left-handed) so $n_{\mathrm{spin}} = 1$. We therefore recover $g_\nu = 3 \times 2 \times 1 = 6$.

The usual $T$ in cosmology is always taken to be the temperature of the blackbody photons. However, the $T_\nu$ in Eq. (29.70) is the temperature of the neutrinos. For $T > 3.5 \times 10^{10}$ K, these temperatures are the same, and $T = T_\nu$. However, as the temperature dropped below about $3.5 \times 10^{10}$ K, the expansion of the universe diluted the number density of neutrinos, and the neutrinos ceased to interact with other particles. Essentially, the cosmos expanded faster than the neutrino interaction rate, and the neutrinos decoupled from the other constituents of the universe. Since the time of **neutrino decoupling**, the neutrinos have expanded and cooled at their own rate, independently of the CMB.

### The Energy Density of Relativistic Particles

Because the annihilation of electrons with positrons continued to supply energy to the photons (via $e^- + e^+ \rightarrow \gamma + \gamma$) but not to the neutrinos, the neutrino temperature is somewhat less than the temperature of the CMB photons. Although it is beyond the scope of this book, it can be shown that $T_\nu$ is related to the temperature $T$ of the CMB photons by

$$T_\nu = \left(\frac{4}{11}\right)^{1/3} T. \tag{29.73}$$

The total neutrino energy density is therefore[20]

$$u_\nu = \frac{1}{2}\left(\frac{7}{8}\right) g_\nu \left(\frac{4}{11}\right)^{4/3} aT^4 = 0.681 aT^4. \tag{29.74}$$

Thus the energy density for relativistic particles, both photons and neutrinos, is

$$u_{\text{rel}} = \frac{1}{2} g_* aT^4, \tag{29.75}$$

where

$$g_* = g_{\text{rad}} + \left(\frac{7}{8}\right) g_\nu \left(\frac{4}{11}\right)^{4/3} = 3.363 \tag{29.76}$$

is the **effective number of degrees of freedom** of the relativistic particles. We also define the equivalent mass density of relativistic particles as

$$\rho_{\text{rel}} = \frac{u_{\text{rel}}}{c^2} = \frac{g_* aT^4}{2c^2}. \tag{29.77}$$

This value of $g_*$ is valid back to the end of electron–positron annihilation, at about $t = 1.3$ s. For the higher temperatures of the very early universe ($t < 1$ s) discussed in Chapter 30, however, we will encounter a greater number of relativistic particles, and the value of $g_*$ will grow accordingly.

---

[20]The factors of 7/8 and $(4/11)^{4/3}$ do not describe degrees of freedom, and therefore we separate them from the definition of $g_\nu$.

Employing Eq. (29.8), Eq. (29.65) becomes

$$H^2 \left[1 - (\Omega_m + \Omega_{\text{rel}})\right] R^2 = -kc^2, \tag{29.78}$$

where

$$\Omega_m = \frac{\rho_m}{\rho_c} = \frac{8\pi G \rho_m}{3H^2} \tag{29.79}$$

is the density parameter for matter (both baryonic and dark), and

$$\Omega_{\text{rel}} = \frac{\rho_{\text{rel}}}{\rho_c} = \frac{8\pi G \rho_{\text{rel}}}{3H^2} = \frac{4\pi G g_* a T^4}{3H^2 c^2} \tag{29.80}$$

is the density parameter for relativistic particles (both photons and neutrinos). Note that Eq. (29.78) implies that for a flat ($k = 0$) two-component universe, $\Omega_m + \Omega_{\text{rel}} = 1$. Inserting $T_0 = 2.725$ K, we find that

$$\Omega_{\text{rel},0} = 8.24 \times 10^{-5},$$

which is very small compared with $[\Omega_{m,0}]_{\text{WMAP}} = 0.27$.

### Transition from the Radiation Era to the Matter Era

Recalling that $w_{\text{rel}} = 1/3$ for relativistic particles, Eq. (29.53) yields

$$R^4 \rho_{\text{rel}} = \rho_{\text{rel},0}, \tag{29.81}$$

which shows how the equivalent mass density of relativistic particles varies with the scale factor $R$. By comparing this with Eq. (29.5),

$$R^3 \rho_m = \rho_{m,0}, \tag{29.82}$$

for massive particles, we notice that $\rho_{\text{rel}}$ increases more rapidly than the mass density $\rho_m$ as the scale factor becomes smaller. As $R \to 0$ in the early universe, therefore, there must have been an early era when the radiation (i.e., all relativistic particles, not just photons and neutrinos) dominated and governed the expansion of the universe. The transition from this **radiation era** to the present **matter era** occurred when the scale factor satisfied $\rho_{\text{rel}} = \rho_m$, or $\Omega_{\text{rel}} = \Omega_m$. From Eqs. (29.79–29.82), the equality of $\Omega_{\text{rel}}$ and $\Omega_m$ occurred when the scale factor was

$$R_{r,m} = \frac{\Omega_{\text{rel},0}}{\Omega_{m,0}} = 4.16 \times 10^{-5} \Omega_{m,0}^{-1} h^{-2},$$

with a WMAP value of

$$R_{r,m} = 3.05 \times 10^{-4}.$$

This corresponds to a redshift (Eq. 29.4) of

$$z_{r,m} = \frac{1}{R_{r,m}} - 1 = 2.41 \times 10^4 \Omega_{m,0} h^2,$$

which for WMAP values is

$$z_{r,m} = 3270.$$

This is in very good agreement with the WMAP result,

$$[z_{r,m}]_{\text{WMAP}} = 3233 \, {}^{+194}_{-210},$$

for the redshift when the the universe passed from being radiation-dominated to being matter-dominated.

Using $RT = T_0$ (Eq. 29.58), the temperature at this transition was

$$T_{r,m} = \frac{T_0}{R_{r,m}} = 6.56 \times 10^4 \Omega_{m,0} h^2 \text{ K},$$

or

$$T_{r,m} = 8920 \text{ K}$$

using WMAP values. Thus, when the universe had cooled to 8920 K and typical separations were some $4 \times 10^{-5}$ of their present extent, relativistic particles ceased to govern the cosmic expansion, and matter assumed a dominant role.

### Expansion in the Two-Component Model

We are now ready to determine how the early universe expanded with time. To discover how the scale factor, $R$, behaved during the radiation era, we begin by substituting Eqs. (29.81) and (29.82) into Eq. (29.65) to find

$$\left[ \left( \frac{dR}{dt} \right)^2 - \frac{8}{3}\pi G \left( \frac{\rho_{m,0}}{R} + \frac{\rho_{\text{rel},0}}{R^2} \right) \right] = -kc^2. \tag{29.83}$$

Because the early universe was essentially flat, we can set $k = 0$ and use a bit of algebra to obtain

$$\int_0^R \frac{R' \, dR'}{\sqrt{\rho_{m,0} R' + \rho_{\text{rel},0}}} = \sqrt{\frac{8\pi G}{3}} \int_0^t dt'.$$

Integrating this eventually yields an expression for the age of the universe as a function of the scale factor $R$:

$$t(R) = \frac{2}{3} \frac{R_{r,m}^{3/2}}{H_0 \sqrt{\Omega_{m,0}}} \left[ 2 + \left( \frac{R}{R_{r,m}} - 2 \right) \sqrt{\frac{R}{R_{r,m}} + 1} \right], \tag{29.84}$$

where

$$\frac{2}{3} \frac{R_{r,m}^{3/2}}{H_0 \sqrt{\Omega_{m,0}}} = 5.51 \times 10^{10} h^{-4} \Omega_{m,0}^{-2} \text{ s} = 1.75 \times 10^3 h^{-4} \Omega_{m,0}^{-2} \text{ yr}.$$

The time $t_{r,m}$ of the transition from a radiation-dominated to a matter-dominated universe may be found by setting $R/R_{r,m} = 1$ and using WMAP values of $h = 0.71$ and $\Omega_{m,0} = 0.27$ to obtain

$$t_{r,m} = 1.74 \times 10^{12} \text{ s} = 5.52 \times 10^4 \text{ yr.} \tag{29.85}$$

The form of Eq. (29.84) becomes simpler deep in the radiation era, when $R \ll R_{r,m}$. It is left for you (Problem 29.27) to show that in this limit the factors of $\Omega_{m,0}$ cancel, resulting in

$$R(t) = \left( \frac{16\pi G g_* a}{3c^2} \right)^{1/4} T_0\, t^{1/2} \tag{29.86}$$

$$= (1.51 \times 10^{-10} \text{ s}^{-1/2}) g_*^{1/4} t^{1/2}. \tag{29.87}$$

This shows that during the radiation era, $R \propto t^{1/2}$. Using $T = T_0/R$ quickly reveals the temperature deep in the radiation era:

$$T(t) = \left( \frac{3c^2}{16\pi G g_* a} \right)^{1/4} t^{-1/2} \tag{29.88}$$

$$= (1.81 \times 10^{10} \text{ K s}^{1/2}) g_*^{-1/4} t^{-1/2}. \tag{29.89}$$

At the other extreme, for $R \gg R_{r,m}$, Eq. (29.84) becomes

$$t(R) = \frac{2}{3} \frac{R^{3/2}}{H_0 \sqrt{\Omega_{m,0}}} \tag{29.90}$$

so

$$R(t) = \left( \frac{3}{2} H_0 t \sqrt{\Omega_{m,0}} \right)^{2/3} = \left( \frac{3\sqrt{\Omega_{m,0}}}{2} \right)^{2/3} \left( \frac{t}{t_H} \right)^{2/3}, \tag{29.91}$$

using $t_H = 1/H_0$ for the Hubble time. As expected, this displays the $R \propto t^{2/3}$ dependence we found earlier in Eq. (29.30) for a flat universe of pressureless dust. Equation (29.90) can be expressed in terms of $z$ using $R = 1/(1 + z)$ to obtain

$$\frac{t(z)}{t_H} = \frac{2}{3} \frac{1}{(1 + z)^{3/2}} \frac{1}{\sqrt{\Omega_{m,0}}}, \tag{29.92}$$

which may be compared with Eq. (29.40) for a flat universe of pressureless dust. Evaluating these for $R = 1$ ($z = 0$) and using WMAP values gives the age of the universe as 12.5 billion years, a billion years more than the mean age of the oldest globular clusters (see page 1158). However, as we will see later, this estimate of the age of the universe is still about one billion years too short, as determined by a full analysis of the WMAP results.

### Big Bang Nucleosynthesis

The process that manufactured the lightest elements in the early universe is known as **Big Bang nucleosynthesis**. We are now prepared to ask, why is approximately one-quarter of the mass of the universe in the form of helium? The temperature at time $t$ during the radiation era is given by Eq. (29.89). At a temperature just below $10^{12}$ K ($t \sim 10^{-4}$ s), the universe contained a mixture of photons ($\gamma$), electron–positron pairs, and electron and muon neutrinos and their antiparticles ($\nu_e, \nu_\mu, \bar{\nu}_e, \bar{\nu}_\mu$). There were also a smaller number of protons and neutrons, about five for every $10^{10}$ photons, that were constantly being transformed into each other via the reactions

$$n \rightleftharpoons p^+ + e^- + \bar{\nu}_e \tag{29.93}$$

$$n + e^+ \rightleftharpoons p^+ + \bar{\nu}_e \tag{29.94}$$

$$n + \nu_e \rightleftharpoons p^+ + e^-. \tag{29.95}$$

These constant conversions were easily accomplished because the mass difference between a proton and a neutron is only

$$(m_p - m_n)c^2 = 1.293 \text{ MeV},$$

while the characteristic thermal energy of particles at $10^{12}$ K is $kT \simeq 86$ MeV. The Boltzmann equation (Eq. 8.6) gives the equilibrium ratio of the number density of neutrons, $n_n$, to the number density of protons, $n_p$, as[21]

$$\frac{n_n}{n_p} = e^{-(m_p - m_n)c^2/kT}. \tag{29.96}$$

At $10^{12}$ K, this ratio is 0.985. The numbers of neutrons and protons were nearly equal because the mass difference between the protons and neutrons is negligible at such a high temperature.

As the universe expanded and the temperature fell, the ratio of the number densities continued to be given by Eq. (29.96) as long as reactions (29.93–29.95) proceeded fast enough to reach equilibrium. Detailed calculations show, however, that when the temperature had declined to about $10^{10}$ K, the timescale for these reactions exceeded the characteristic timescale of the expansion given by $1/H(t) = 2t$; see Problem 29.26.

At a bit above $10^{10}$ K, the reaction rates decreased significantly, for two reasons. First, the expansion had reduced the energy of the neutrinos until they were unable to participate in reactions (29.93–29.95). Also, shortly thereafter, the characteristic thermal energy of the photons, $kT$, fell below the 1.022-MeV threshold for creating electron–positron pairs via the pair-production process $\gamma \rightarrow e^- + e^+$ (see Problem 29.30). As a result, the electrons and positrons annihilated each other without being replaced, leaving only a small remainder of excess electrons. For these reasons, the neutrons could not be replenished as fast as they were destroyed, and there was not enough time for these reactions to reach equilibrium. In a sense, the creation of new neutrons could not keep up with the rate of expansion

---

[21]The equal statistical weights of the neutron and proton cancel in Eq. (8.6).

of the universe. The ratio of the number densities then became "frozen" at its value of $n_n/n_p = 0.223$ when $T \simeq 10^{10}$ K.

At this point, there were 223 neutrons for every 1000 protons (or 446 neutrons for every 2000 protons, which will be used in what follows), and essentially no more neutrons were being created. The beta decay reaction [the forward reaction in Eq. (29.93)] continued to operate, however, converting neutrons into protons with a half-life of $\tau_{1/2} = 614$ s = 10.2 min. It was not yet possible for the protons and neutrons to combine to form deuterium nuclei ($_1^2$H) via

$$p^+ + n \rightleftharpoons {}_1^2\text{H} + \gamma,$$

because at temperatures exceeding $10^9$ K, the energetic radiation quickly dissociated the nuclei (recall page 1165). As a result, the neutrons and protons remained separated until the temperature had dropped from $10^{10}$ K to $10^9$ K. According to Eq. (29.89) with $g_* = 3.363$, this took approximately

$$t(10^9 \text{ K}) - t(10^{10} \text{ K}) = 178 \text{ s} - 1.78 \text{ s} \approx 176 \text{ s}.$$

From the law of radioactive decay, Eq. (15.10), in this amount of time the 446 neutrons mentioned previously declined to 366, and the number of protons rose to 2080.

Below $10^9$ K the neutrons and protons readily combined to form as many deuterium nuclei as possible. A number of reactions then led to the formation of $_2^4$He, the most tightly bound nucleus involved in Big Bang nucleosynthesis. The most efficient reactions leading to $_2^4$He include

$$_1^2\text{H} + {}_1^2\text{H} \rightleftharpoons {}_1^3\text{H} + {}_1^1\text{H}$$
$$_1^3\text{H} + {}_1^2\text{H} \rightleftharpoons {}_2^4\text{He} + n$$

and

$$_1^2\text{H} + {}_1^2\text{H} \rightleftharpoons {}_2^3\text{He} + n$$
$$_2^3\text{He} + {}_1^2\text{H} \rightleftharpoons {}_2^4\text{He} + {}_1^1\text{H}.$$

[Note that these reactions differ from those of the pp chain (Eqs. 10.37–10.45), which produce helium in the cores of stars.[22]] No other nuclei were formed with abundances approaching that of $_2^4$He, although there were traces of $_1^2$H, $_2^3$He, and $_3^7$Li (from the reaction $_2^4$He + $_1^3$H $\rightarrow$ $_3^7$Li + $\gamma$). Figure 29.13 shows the network of reactions involved in Big Bang nucleosynthesis. Our sample of 366 neutrons and 2080 protons could form 183 $_2^4$He nuclei, with 1714 protons ($_1^1$H) left over. Because a $_2^4$He nucleus is four times more massive than a $_1^1$H nucleus, the preceding analysis shows that the mass fraction of $_2^4$He in the universe should have been about

$$\frac{4(183)}{1714 + 4(183)} = 0.299.$$

---

[22]The most obvious route to helium-4 is $_1^2$H + $_1^2$H $\rightleftharpoons$ $_2^4$He + $\gamma$. However, this is a "forbidden" reaction and has a negligibly small cross section.

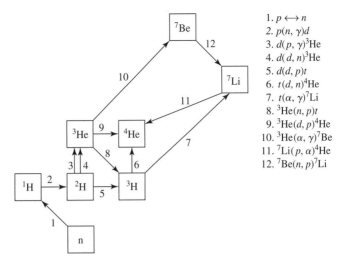

**FIGURE 29.13** The reaction network that is responsible for Big Bang nucleosynthesis. The letter "*d*" stands for deuterium, and "*t*" stands for tritium. (Figure adapted from Nollett and Burles, *Phys. Rev. D, 61*, 123505, 2000.)

This rough estimate is consistent with the primordial percentage of helium inferred from observations, between 23% and 24%.

Because essentially all of the available neutrons were incorporated in the helium-4 ($^4_2$He) nuclei, the abundance of helium-4 was insensitive to the density of the universe at the time. However, the amounts of deuterium, helium-3, and lithium-7 manufactured in this way depend sensitively on the density of ordinary matter at the time of the reactions. Figure 29.14 shows the abundances of these nuclei as a function of the prevailing *present* density of baryonic matter. Comparing the theoretical curves with the observations makes it apparent that the present density of baryonic matter probably lies between 2 and $5 \times 10^{-28}$ kg m$^{-3}$, only a few percent of the critical density of $1.88 \times 10^{-26} h^2$ kg m$^{-3}$. This explanation of the abundances of the light elements that were not manufactured by stars is one of the greatest achievements of the Big Bang theory.

### The Origin of the Cosmic Microwave Background

Now that we have described the nature of the universal expansion, let's return to the question of the origin of the CMB. When we observe the cosmic microwave background, what are we actually viewing? The copious electrons in the hot environment of the very early universe obstructed the photons of the cosmic microwave background, allowing them to travel only relatively short distances before being scattered. The scattering of photons by free electrons kept the electrons and photons in thermal equilibrium, meaning that they had the same temperature.[23] However, as the expansion of the universe diluted the number density of

---

[23]The CMB photons are also scattered by free protons, but the cross section for photon–proton scattering is smaller than the Thomson cross section for electrons, Eq. (9.20), by a factor of $m_e^2/m_p^2$, so it can be neglected. The Coulomb interaction between the electrons and protons kept the protons in thermal equilibrium with the electrons and photons.

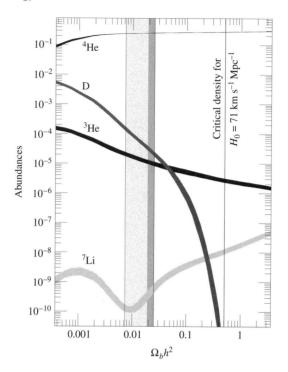

**FIGURE 29.14** The calculated mass abundances of helium-4, deuterium, helium-3, and lithium-7 as a function of the present density of baryonic matter in the universe. The wide bar delineates the consistency interval, the range of values of $\Omega_{b,0}h^2$ that agree with the observed abundances. The narrow dark stripe at the right edge of the consistency interval corresponds to the abundances of primeval deuterium measured using the Lyman-$\alpha$ forest of absorption lines in high-$z$ molecular clouds observed in front of quasars. The WMAP value of $\Omega_{b,0}h^2 = 0.0224$ runs down the center of the dark stripe, and the WMAP value of the critical density ($\Omega_{b,0}h^2 = 1h^2 = 0.504$) is shown at the right. Note that the agreement between the theoretical and observed abundances spans nine orders of magnitude. (Figure adapted from Schramm and Turner, *Rev. Mod. Phys.*, *70*, 303, 1998.)

free electrons, the average time between scatterings of a photon by an electron gradually approached the characteristic timescale of the universal expansion,

$$\tau_{\exp}(t) \equiv \left( \frac{1}{R(t)} \frac{dR(t)}{dt} \right)^{-1} = \frac{1}{H(t)}.$$

This expression is analogous to that of the pressure scale height, Eq. (10.69). As the time of **decoupling** approached, the photons became increasingly disengaged from the electrons.[24]

If the electrons had remained free, decoupling would have occurred when the universe was about 20 million years old (see Problem 29.32). However, when the universe was only some one million years old ($10^{13}$ s), another important event altered the opacity of the

---

[24]You may recall the discussion of neutrino decoupling on page 1173.

universe and rendered it transparent. The independent evolution of radiation and matter began when the temperature had cooled sufficiently to allow the free electrons to combine with nuclei of hydrogen and helium. This formation of neutral atoms is sometimes referred to as **recombination**, an oddly inappropriate term since the electrons and nuclei had never been previously combined into atoms! The loss of free electrons and the resulting drop in opacity completed the decoupling of radiation and matter, freeing the photons to roam unhindered throughout a newly transparent universe. The photons of the cosmic microwave background that we observe today were last scattered during the time of recombination.

### The Surface of Last Scattering

We define the **surface of last scattering** as a spherical surface, centered on the Earth, from which the CMB photons just now arriving at Earth were last scattered before beginning their unimpeded journey to us. (Of course, other observers in the universe are at the centers of their own surfaces of last scattering.) The surface of last scattering is the farthest redshift we can possibly observe at this moment in time. More accurately, because recombination did not happen all at once, the surface of last scattering has a thickness $\Delta z$. Just as the light from the Sun was last scattered from somewhere within its photosphere, the CMB photons originated within a layer, the "surface" of last scattering.[25] The surface of last scattering can therefore be thought of as a curtain that screens everything prior to decoupling from the direct view of astronomers. The earliest moments of the universe are hidden behind this veil and must be investigated indirectly.

### The Conditions at Recombination

The temperature at recombination can be estimated through use of the Saha equation (Eq. 8.8),

$$\frac{N_{II}}{N_I} = \frac{2Z_{II}}{n_e Z_I} \left( \frac{2\pi m_e kT}{h^2} \right)^{3/2} e^{-\chi_I/kT}.$$

Assuming (incorrectly) a composition of pure hydrogen for simplicity, we use $Z_I = 2$ and $Z_{II} = 1$ as in Example 8.1.4. It is useful to define $f$ to be the fraction of hydrogen atoms that are ionized, so

$$f = \frac{N_{II}}{N_I + N_{II}} = \frac{N_{II}/N_I}{1 + N_{II}/N_I}, \tag{29.97}$$

or

$$\frac{N_{II}}{N_I} = \frac{f}{1 - f}. \tag{29.98}$$

---

[25]The difference between the Sun's photosphere and the surface of last scattering is that the photosphere has a spatial thickness, while the surface of last scattering has a thickness in terms of redshift or, equivalently, time.

For ionized hydrogen there is one free electron for every proton, $n_e = n_p$, so the number density of free electrons depends on $f$ as

$$n_e = n_p = f(n_p + n_H) = \frac{f\rho_b}{m_H}, \tag{29.99}$$

where $\rho_b$ is the density of baryonic matter. Note that in obtaining Eq. (29.99) from Eq. (29.97), $N_I$ corresponds to $n_H$, the number density of neutral hydrogen atoms, and $N_{II}$ corresponds to $n_p$, the number density of protons (ionized hydrogen atoms). Using Eq. (29.82), we can write this as

$$n_e(R) = \frac{f\rho_{b,0}}{m_H R^3}, \tag{29.100}$$

Substituting Eqs. (29.98) and (29.100) into the Saha equation, together with Eq. (29.58) for the blackbody temperature, we find

$$\frac{f}{1-f} = \frac{m_H R^3}{f\rho_{b,0}} \left(\frac{2\pi m_e kT_0}{h^2 R}\right)^{3/2} e^{-\chi_I R/kT_0}, \tag{29.101}$$

where $T_0 = 2.725$ K and $\chi_I = 13.6$ eV. This can be solved numerically to find that the universe had cooled sufficiently for one-half of its electrons and protons to combine to form atomic hydrogen ($f = 0.5$) when the value of the scale factor was approximately $R \approx 7.25 \times 10^{-4}$ ($z \approx 1380$), corresponding to a temperature of about 3760 K [again from Eq. (29.58)].

More precisely, the WMAP value for the redshift at the time of decoupling (i.e., the surface of last scattering) is

$$[z_{dec}]_{WMAP} = 1089 \pm 1.$$

We will adopt this as the value of the redshift for both recombination and decoupling. Using Eqs. (29.4) and (29.58) yields a temperature at recombination of

$$T_{dec} = T_0(1 + z_{dec}) = 2970 \text{ K}.$$

This is lower than our estimate of 3760 K because the photons created by the formation of some atoms were then absorbed by other atoms, putting these atoms into excited states from which they were easier to ionize. Thus a slightly cooler temperature was needed to complete the recombination process. It is important to remember that at times earlier than recombination, the radiation and matter shared a common temperature, whereas after recombination, the temperatures of the radiation and matter must be distinguished. For the remainder of the text, it is the radiation temperature (the temperature of the CMB) that will be of interest after recombination.

The WMAP value for the time at which recombination and decoupling occurred is

$$[t_{dec}]_{WMAP} = 379^{+8}_{-7} \text{ kyr.} \tag{29.102}$$

Of course, these events did not occur at a single instant of time; the WMAP value of the decoupling time interval is

$$[\Delta t_{\text{dec}}]_{\text{WMAP}} = 118^{+3}_{-2} \text{ kyr.}$$

This corresponds to the surface of last scattering having a thickness (in redshift) of

$$[\Delta z_{\text{dec}}]_{\text{WMAP}} = 195 \pm 2.$$

### The Dawn of Precision Cosmology

Before ending this section, it is worth taking a few moments to reflect on the power of simple mathematics to illuminate the earliest stages of the universe. Before the discovery of the cosmic microwave background, theorizing about the early universe was not considered to be legitimate science by many physicists. The point is well made by Steven Weinberg in his book *The First Three Minutes*:

> This is often the way it is in physics—our mistake is not that we take our theories too seriously, but that we do not take them seriously enough. It is always hard to realize that these numbers and equations we play with at our desks have something to do with the real world. . . . The most important thing accomplished by the ultimate discovery of the 3° K radiation background in 1965 was to force us all to take seriously the idea that there *was* an early universe.

The results of the Wilkinson Microwave Anisotropy Probe have now taken cosmology to a level that was hardly imaginable when the first edition of *An Introduction to Modern Astrophysics* was published in 1996. In remarks made at the WMAP press conference on February 11, 2003, astrophysicist John Bahcall said,

> Every astronomer will remember where he or she was when they first heard the WMAP results. I certainly will. . . . For cosmology, the formal announcement today represents a "rite of passage" from speculation to precision science.

## 29.3 ■ RELATIVISTIC COSMOLOGY

The appearance of objects at truly cosmological distances is affected by the curvature of the spacetime through which the light travels on its way to Earth. The geometrical properties of the universe as a whole will be best understood by starting with simple analogies before proceeding to the more comprehensive description that only Einstein's general theory of relativity can provide.

### Euclidean, Elliptic, and Hyperbolic Geometries

The foundations of plane geometry (appropriate for a flat universe) were laid by Euclid sometime around 300 B.C. Euclid's *Elements* consists of 13 "books" (chapters) that contain 465 theorems. These theorems, in turn, are derived from just five postulates that embody self-evident truths and so are stated without proof. In Euclid's words,

1. It is possible to draw a straight line from any point to any point.

2. It is possible to produce [extend] a finite straight line continuously in a straight line.

3. It is possible to describe a circle with any center and distance [radius].

4. All right angles are equal to one another.

Euclid's statement of his fifth postulate is awkward and opaque:

5. If a straight line falling on two straight lines makes the interior angles on the same side less than two right angles, the two straight lines, if produced indefinitely, meet on the side on which the angles are less than the two right angles.

A moment's thought will show that Euclid's fifth postulate concerns the behavior of parallel lines.[26] It is equivalent to the following statement, due to an English mathematician, John Playfair (1748–1819):

5a. Given, in a plane, a line $L$ and a point $P$ not on $L$, then through $P$ there exists one and only one line parallel to $L$.

Figure 29.15(a) illustrates Playfair's version of the parallel postulate in Euclidean geometry.

The cumbersome statement of Euclid's fifth postulate led some mathematicians to suspect that it might be derived from the first four. In the nineteenth century, mathematicians mounted an attack on the parallel postulate by denying it, in hopes that they would reach a contradiction. Such a proof by contradiction would make the fifth postulate dependent on the rest of geometry and so abolish its stature as an independent postulate. To their astonishment, instead of arriving at a contradiction as they expected, several renowned mathematicians realized that they had developed *alternative* geometries, each as self-consistent as Euclid's.

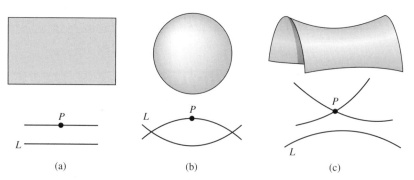

**FIGURE 29.15**   The parallel postulate, illustrated for three alternative geometries: (a) Euclidean, (b) elliptic, and (c) hyperbolic.

---

[26]Euclid defines parallel lines as "straight lines which, being in the same plane and being produced indefinitely in both directions, do not meet one another in either direction."

Foremost among these mathematicians were Carl Frederich Gauss (1777–1855), Nikolai Lobachevski (1793–1856), János Bolyai (1802–1860), and Bernhard Riemann (1826–1866). There are *three* different fifth postulates, each of which leads to a perfectly well-behaved geometry. In 1868 it was proved that the two additional, non-Euclidean geometries are as logically consistent as Euclid's original version. In addition to Playfair's version of the parallel postulate given above for Euclidean geometry, there is the fifth postulate of the elliptic geometry developed by Riemann:

5b. Given, in a plane, a line $L$ and a point $P$ not on $L$, then through $P$ there exists *no line* parallel to $L$.

As Fig. 29.15(b) shows, this describes the geometry of the surface of a sphere, where two lines that both start out perpendicular to the sphere's equator meet at its poles. In elliptic geometry, the angles of a triangle add up to *more* than 180°, and the circumference of a circle is less than $2\pi r$.

On the other hand, the fifth postulate of the hyperbolic geometry, developed independently by Gauss, Bolyai, and Lobachevski, is more generous than Euclid's:

5c. Given, in a plane, a line $L$ and a point $P$ not on $L$, then through $P$ there exist *at least two lines* parallel to $L$.

Figure 29.15(c) shows this geometry applied to a saddle-shaped hyperboloid. Neither of the two lines shown passing through point $P$ intersects the line $L$, and it is possible to draw (in this example) infinitely many more. In hyperbolic geometry, the angles of a triangle add up to *less* than 180°, and the circumference of a circle exceeds $2\pi r$.

The logical independence of the parallel postulate means that it cannot be derived from Euclid's earlier postulates. Which geometry to adopt is an arbitrary choice, since all three are equally valid from a mathematical viewpoint. Which of these three geometries describes the spatial structure of the physical universe is a question that must be answered *empirically*, by observation. Gauss himself conducted such an experiment beginning in 1820, when he directed a survey of the Germanic state of Hanover. After carefully measuring the distances between three mountaintops that formed a huge triangle (the longest side was 107 km), he determined that the sum of the triangle's angles was 180°—within experimental error. Gauss's experiment was not sensitive enough to measure the curvature of space near Earth's surface.

### The Robertson–Walker Metric for Curved Spacetime

The spacetime surrounding any massive object is indeed curved. As described in Section 17.2, the spatial curvature is revealed in the radial term of the Schwarzschild metric, Eq. (17.22). However, the Schwarzschild metric is valid only outside matter. We will have to find another metric to describe the spacetime of the dust-filled universe of Section 29.1. Although a derivation of this metric is beyond the scope of this book, the following argument should help elucidate some of its properties. Our search for the metric is made somewhat easier by the cosmological principle. In a homogeneous and isotropic universe, although the curvature of space may change with time, the curvature must have the same value everywhere at a given time since the Big Bang.

Let's begin by considering the curvature of a two-dimensional surface. "Curvature" has a precise mathematical meaning; for example, a sphere of radius $R$ has a constant curvature, $K$, that is defined to be $K \equiv 1/R^2$. Gauss realized that the curvature of any surface can be determined *locally*, on an arbitrarily small patch of its surface. It is instructive to imagine how the curvature of a sphere of radius $R$ might be measured by a tiny (but talented) ant on the sphere's surface. It is free to roam about the sphere's surface but is unable to gain an outside perspective and view the sphere as a whole. How could the inquisitive ant determine the sphere's curvature?

Starting at the north pole of the sphere (call this point $P$; see Fig. 29.16), the ant could mark a series of other points all of which are a distance $D$ from $P$. When connected, these points form a circle with $P$ at its center. The ant now measures the circumference, $C_{\text{meas}}$, of the circle and compares it with the expected value of $C_{\text{exp}} = 2\pi D$. However, the two values do not agree; since $D = R\theta$,

$$C_{\text{exp}} = 2\pi R\theta$$

while

$$C_{\text{meas}} = 2\pi R \sin\theta = 2\pi R \sin(D/R).$$

When the ant divides the fractional discrepancy between these two values of the circumference by the expected area of the circle, $A_{\text{exp}} = \pi D^2$, and then multiplies by $6\pi$, the result (in the limit $D \to 0$) is the curvature of the sphere. That is, the ant starts with

$$6\pi \cdot \frac{C_{\text{exp}} - C_{\text{meas}}}{C_{\text{exp}} A_{\text{exp}}} = 6\pi \cdot \frac{2\pi D - C_{\text{meas}}}{(2\pi D)(\pi D^2)} = \frac{3}{\pi} \frac{2\pi D - C_{\text{meas}}}{D^3}.$$

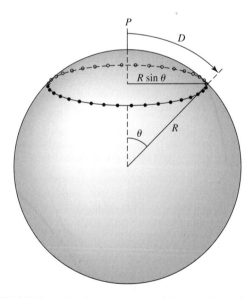

**FIGURE 29.16**    A local measurement of the curvature of a sphere.

Substituting for $C_{\text{meas}}$, the clever ant then uses a Taylor series for $\sin(D/R)$,

$$\frac{3}{\pi} \frac{2\pi D - 2\pi R \sin(D/R)}{D^3} = \frac{6}{D^3} \left\{ D - R \left[ \frac{D}{R} - \frac{1}{3!} \left( \frac{D}{R} \right)^3 + \frac{1}{5!} \left( \frac{D}{R} \right)^5 + \cdots \right] \right\}$$

$$= \frac{1}{R^2} - \frac{1}{20} \frac{D^2}{R^4} + \cdots .$$

In the limit $D \to 0$, this is $1/R^2$, the curvature of the sphere. In fact, the prescription

$$K = \frac{3}{\pi} \lim_{D \to 0} \frac{2\pi D - C_{\text{meas}}}{D^3} \tag{29.103}$$

can be used to calculate the curvature at any point on a two-dimensional surface; see Fig. 29.17. For a flat plane, $K = 0$, while for a saddle-shaped hyperboloid, $K$ is *negative* because the measured circumference exceeds $2\pi D$.[27]

The next step toward the spacetime metric that describes a uniform dust-filled universe comes from considering how a small distance is measured in two dimensions. For a flat plane, polar coordinates are an appropriate choice of variables, and the differential distance $d\ell$ between two nearby points $P_1$ and $P_2$ on the plane [see Fig. 29.18(a)] is given by

$$(d\ell)^2 = (dr)^2 + (r \, d\phi)^2.$$

Polar coordinates can also be used to measure the differential distance between two nearby points on the surface of a sphere. As an example, we will return to the surface of the

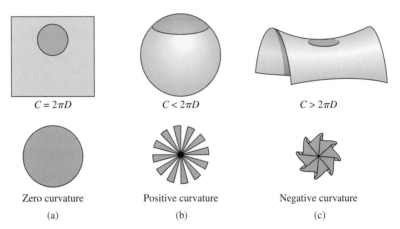

| $C = 2\pi D$ | $C < 2\pi D$ | $C > 2\pi D$ |
| Zero curvature | Positive curvature | Negative curvature |
| (a) | (b) | (c) |

**FIGURE 29.17**   Calculating the curvature of a surface in three geometries: (a) a flat plane, (b) the surface of a sphere, and (c) the surface of a hyperboloid.

---

[27]Unfortunately, it is not possible to visualize the negative curvature of the hyperboloid because its surface (unlike a sphere's) has infinite area. We cannot "step back" to view it as a whole.

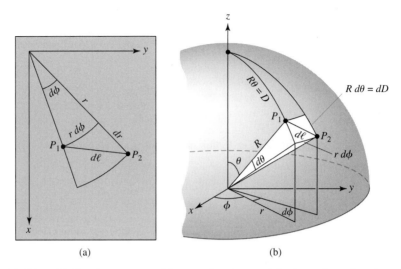

**FIGURE 29.18**    $d\ell$ as measured for (a) a flat plane and (b) the surface of a sphere.

sphere of radius $R$ and curvature $K = 1/R^2$ considered previously. Then, as shown in Fig. 29.18(b), the distance $d\ell$ between two points $P_1$ and $P_2$ on the sphere is now given by

$$(d\ell)^2 = (dD)^2 + (r\,d\phi)^2 = (R\,d\theta)^2 + (r\,d\phi)^2.$$

But $r = R\sin\theta$, so $dr = R\cos\theta\,d\theta$ and

$$R\,d\theta = \frac{dr}{\cos\theta} = \frac{R\,dr}{\sqrt{R^2 - r^2}} = \frac{dr}{\sqrt{1 - r^2/R^2}}.$$

The differential distance on the sphere's surface may therefore be written as

$$(d\ell)^2 = \left(\frac{dr}{\sqrt{1 - r^2/R^2}}\right)^2 + (r\,d\phi)^2,$$

in terms of the plane polar coordinates $r$ and $\phi$. More generally, in terms of the curvature $K$ of a two-dimensional surface,

$$(d\ell)^2 = \left(\frac{dr}{\sqrt{1 - Kr^2}}\right)^2 + (r\,d\phi)^2.$$

The extension to three dimensions is accomplished simply by making a transition from polar to spherical coordinates,

$$(d\ell)^2 = \left(\frac{dr}{\sqrt{1 - Kr^2}}\right)^2 + (r\,d\theta)^2 + (r\sin\theta\,d\phi)^2, \qquad (29.104)$$

where $r$ now measures the radial distance from the origin.

Equation (29.104) shows the effect of the curvature of our three-dimensional universe on spatial distances. Although the curvature of a two-dimensional spherical surface can be seen by stepping back from the sphere and viewing it immersed in its three-dimensional surroundings, there is unfortunately no fourth spatial dimension into which we can step back to view the curvature of our universe.

The final step toward the spacetime metric comes from recalling that by "distance," we mean the *proper distance* between two spacetime events that occur simultaneously according to an observer. In an expanding universe, the positions of two points must be recorded at the same time if their separation is to have any meaning. In an isotropic, homogeneous universe, there is no reason why time should pass at different rates at different locations; consequently, the temporal term should simply be $c\,dt$.[28] If we take

$$(ds)^2 = (c\,dt)^2 - \left(\frac{dr}{\sqrt{1 - Kr^2}}\right)^2 - (r\,d\theta)^2 - (r\sin\theta\,d\phi)^2$$

as the metric for an isotropic, homogeneous universe, then Eq. (17.18) for the proper distance agrees with Eq. (29.104). That is, the differential proper distance is just $d\mathcal{L} = \sqrt{-(ds)^2}$ with $dt = 0$.

All that remains is to express this metric in terms of the dimensionless scale factor, $R(t)$, defined by $r(t) = R(t)\varpi$ (Eq. 29.3). Because the expansion of the universe affects all of its geometric properties, including its curvature, it is useful to define the time-dependent curvature in terms of a time-independent constant, $k$, as

$$K(t) \equiv \frac{k}{R^2(t)}. \tag{29.105}$$

These substitutions for $r$ and $K$ result in

$$(ds)^2 = (c\,dt)^2 - R^2(t)\left[\left(\frac{d\varpi}{\sqrt{1 - k\varpi^2}}\right)^2 + (\varpi\,d\theta)^2 + (\varpi\sin\theta\,d\phi)^2\right], \tag{29.106}$$

which is known as the **Robertson–Walker metric**. The Robertson–Walker metric determines the spacetime interval between two events in an isotropic, homogeneous universe, just as the Schwarzschild metric, Eq. (17.22), is used to measure the interval between two events in the curved spacetime surrounding a massive object. The name honors Howard Percy Robertson (1903–1961) and Arthur Geoffrey Walker (1909–2001), who independently demonstrated, in the mid-1930s, that this is the most general metric possible for describing an isotropic and homogeneous universe.

In fact, we will use the same technique to define $\varpi$ for a curved spacetime that we used in Section 17.2 to specify the radial coordinate $r$. From the Robertson–Walker metric, the area today ($t = t_0$ so $R(t_0) = 1$) of a spherical surface centered on the point $\varpi = 0$ is $4\pi\varpi^2$. By definition, this surface is located at the coordinate $\varpi$. It is important to remember that

---

[28]You should compare this with the Schwarzschild metric, Eq. (17.22), where the presence of a central mass produces a more complicated time dependence.

the $\varpi$ in Eq. (29.106) is a *comoving* coordinate that follows a given object as the universe expands. Furthermore, the time, $t$, is a *universal time* that essentially measures the time that has elapsed since the Big Bang. This is not an absolute time, but merely reflects a choice of how the clocks of distant observers are to be synchronized. For example, observers at various locations throughout the universe could, theoretically, synchronize their clocks using the cosmological principle by noting the precise temperature of the cosmic microwave background or the exact value of the Hubble constant.

## The Friedmann Equation

Solving Einstein's field equations (Eq. 17.15) for an isotropic, homogeneous universe leads to a description of the dynamic evolution of the universe in the form of a differential equation for the scale factor, $R(t)$. This is known as the **Friedmann equation**,

$$\left[ \left( \frac{1}{R} \frac{dR}{dt} \right)^2 - \frac{8}{3} \pi G \rho \right] R^2 = -kc^2, \tag{29.107}$$

named for the Russian meteorologist and mathematician Aleksandr Friedmann (1888–1925).[29] In 1922 Friedmann solved Einstein's field equations for an isotropic, homogeneous universe to obtain this equation for a nonstatic universe. The same equation was derived independently in 1927 by the Belgian cleric Abbé Georges Lemaître (1894–1966).[30]

## The Cosmological Constant

Einstein realized that, as originally conceived, his field equations could not produce a static universe. Hubble's discovery of the expanding universe had not yet been made, so in 1917 Einstein modified his equations by adding an ad hoc term that contained the **cosmological constant**, $\Lambda$.[31] With this addition, the general solution of Einstein's field equations is

$$\left[ \left( \frac{1}{R} \frac{dR}{dt} \right)^2 - \frac{8}{3} \pi G \rho - \frac{1}{3} \Lambda c^2 \right] R^2 = -kc^2. \tag{29.108}$$

Except for the cosmological constant, this is the same as the Friedmann equation. The additional term containing $\Lambda$ would result from the Newtonian cosmology of Section 29.1 if a potential energy term,

$$U_\Lambda \equiv -\frac{1}{6} \Lambda m c^2 r^2,$$

---

[29] According to general relativity, the density $\rho$ includes the equivalent mass density of photons and neutrinos, as in Eq. (29.77).

[30] Lemaître was the first person to propose that the present universe evolved from a highly dense beginning and so is sometimes called "the father of the Big Bang."

[31] Some authors incorporate a factor of $c^2$ into the definition of $\Lambda$; in this text, $\Lambda$ has units of (length)$^{-2}$.

were added to the left-hand side of Eq. (29.1). The conservation of mechanical energy applied to an expanding shell of mass $m$ (described on page 1146) then becomes

$$\frac{1}{2}mv^2 - G\frac{M_r m}{r} - \frac{1}{6}\Lambda mc^2 r^2 = -\frac{1}{2}mkc^2\varpi^2.$$

The force due to this new potential is

$$\mathbf{F}_\Lambda = -\frac{\partial U_\Lambda}{\partial r}\,\hat{\mathbf{r}} = \frac{1}{3}\Lambda mc^2 r\,\hat{\mathbf{r}} \qquad (29.109)$$

(recall Eq. 2.15), which is radially outward for $\Lambda > 0$. In effect, a positive cosmological constant produces a repulsive force on the mass shell. This allowed Einstein to achieve his goal of balancing his static, closed universe against a gravitational collapse in an (unstable) equilibrium. Later, after the expansion of the universe had been discovered, Einstein expressed his regret at including the $\Lambda$-term in his field equations and has been quoted as referring to it as "the biggest blunder of my life."

A nonzero cosmological constant implies that space would be curved even in an empty universe that is devoid of matter, an idea that Einstein disliked because it conflicted with his ideas concerning mass as the cause of spacetime curvature. Ironically, as mentioned in Section 27.2, Willem de Sitter used Einstein's field equations with the $\Lambda$-term to describe an expanding, empty universe, with the expansion powered by the cosmological constant, and Hubble viewed the distance-dependent redshift of de Sitter's universe as theoretical support for an expanding universe in his 1929 paper.

### The Effects of Dark Energy

Nature has the final say, and in the late 1990s, observations forced astronomers to recognize a nonzero cosmological constant. Although we will defer a discussion of the physical origin of the cosmological constant to the next chapter, for the sake of discussion we will give the physical source of the cosmological constant a name: **dark energy**.[32]

We begin by rewriting the Friedmann equation (including the $\Lambda$-term) in a form that makes it explicit that we are dealing with a three-component universe of mass (baryonic and dark), relativistic particles (photons and neutrinos), and dark energy,

$$\left[\left(\frac{1}{R}\frac{dR}{dt}\right)^2 - \frac{8}{3}\pi G(\rho_m + \rho_{\text{rel}}) - \frac{1}{3}\Lambda c^2\right]R^2 = -kc^2. \qquad (29.110)$$

The **fluid equation** (Eq. 29.50) also emerges from solving Einstein's field equations with the cosmological constant included, as

$$\frac{d(R^3\rho)}{dt} = -\frac{P}{c^2}\frac{d(R^3)}{dt}, \qquad (29.111)$$

where $\rho$ and $P$ are the density and pressure due to every component of the universe (including, as we shall see, dark energy). Note that although $\Lambda$ was included in the field equations,

---

[32]The term *dark energy* originated with cosmologist Michael Turner.

it does not appear in the fluid equation. As was done in Section 29.1, the Friedmann and fluid equations can be combined to produce the **acceleration equation**,

$$\frac{d^2 R}{dt^2} = \left\{ -\frac{4}{3}\pi G \left[ \rho_m + \rho_{\text{rel}} + \frac{3(P_m + P_{\text{rel}})}{c^2} \right] + \frac{1}{3}\Lambda c^2 \right\} R. \tag{29.112}$$

(Although $P_m = 0$ for a pressureless dust universe, it is included in the acceleration equation for the sake of completeness.)

If we define the equivalent mass density of the dark energy to be

$$\rho_\Lambda \equiv \frac{\Lambda c^2}{8\pi G} = \text{constant} = \rho_{\Lambda,0}, \tag{29.113}$$

then the Friedmann equation becomes, in parallel with Eq. (29.65),

$$\left[ \left( \frac{1}{R}\frac{dR}{dt} \right)^2 - \frac{8}{3}\pi G(\rho_m + \rho_{\text{rel}} + \rho_\Lambda) \right] R^2 = -kc^2. \tag{29.114}$$

Because $\rho_\Lambda$ remains constant as the universe expands, more and more dark energy must continually appear to fill the increasing volume.[33]

The pressure due to dark energy, as calculated from Eq. (29.111), is

$$P_\Lambda = -\rho_\Lambda c^2. \tag{29.115}$$

Thus $w_\Lambda = -1$ in the general equation of state $P = w\rho c^2$ (Eq. 29.52). This equation of state is unlike any other we have encountered. A positive cosmological constant corresponds to a positive mass density and a negative pressure! Substituting the expressions for $\rho_\Lambda$ and $P_\Lambda$ into Eq. (29.112), the acceleration equation, yields

$$\frac{d^2 R}{dt^2} = \left\{ -\frac{4}{3}\pi G \left[ \rho_m + \rho_{\text{rel}} + \rho_\Lambda + \frac{3(P_m + P_{\text{rel}} + P_\Lambda)}{c^2} \right] \right\} R. \tag{29.116}$$

With the inclusion of $\rho_\Lambda$ and $P_\Lambda$, these equations have the same form as their Newtonian counterparts, Eqs. (29.10), (29.50), and (29.51). However, the interpretation of the constant $k$ has changed. In Section 29.1, $k$ was related to the mechanical energy of an expanding mass shell by Eq. (29.1). Here, it is seen to be the present value of the curvature of the universe [Eq. (29.105) with $R = 1$].[34]

Using Eq. (29.8) and recalling from Eq. (29.12) that $3H^3/8\pi G = \rho_c$, the Friedmann equation can be written as

$$H^2 \left[ 1 - (\Omega_m + \Omega_{\text{rel}} + \Omega_\Lambda) \right] R^2 = -kc^2, \tag{29.117}$$

---

[33]There are more general models in which the cosmological constant, $\Lambda$, is not really constant. In these models, dark energy is replaced by *quintessence* ("fifth element"), a time-dependent energy density. We will not pursue these models.

[34]In some texts, the comoving coordinate, $\varpi$, is scaled so that $k$ takes on the values of 0, $+1$, or $-1$ for a flat, closed, or open universe.

where

$$\Omega_\Lambda = \frac{\rho_\Lambda}{\rho_c} = \frac{\Lambda c^2}{3H^2}. \tag{29.118}$$

We define the **total density parameter** as[35]

$$\Omega \equiv \Omega_m + \Omega_{\text{rel}} + \Omega_\Lambda. \tag{29.119}$$

The Friedmann equation then becomes

$$H^2(1 - \Omega)R^2 = -kc^2, \tag{29.120}$$

so for a flat universe ($k = 0$), we must have $\Omega(t) = 1$. It is useful to note that, as a special case at $t = t_0$,

$$H_0^2(1 - \Omega_0) = -kc^2. \tag{29.121}$$

Using Eqs. (29.120) and (29.121) along with Eqs. (29.79), (29.80), and (29.118), the Hubble parameter as a function of the redshift $z$ is found to be

$$H = H_0(1 + z) \left[ \Omega_{m,0}(1 + z) + \Omega_{\text{rel},0}(1 + z)^2 + \frac{\Omega_{\Lambda,0}}{(1 + z)^2} + 1 - \Omega_0 \right]^{1/2} \tag{29.122}$$

(cf. Eq. 29.27).

The WMAP values for $\Omega_{m,0}$, $\Omega_{\text{rel},0}$, and $\Omega_{\Lambda,0}$ are

$$
\begin{aligned}
[\Omega_{m,0}]_{\text{WMAP}} &= 0.27 \pm 0.04, \\
\Omega_{\text{rel},0} &= 8.24 \times 10^{-5}, \\
[\Omega_{\Lambda,0}]_{\text{WMAP}} &= 0.73 \pm 0.04,
\end{aligned}
$$

where the value of $\Omega_{\text{rel},0}$ was obtained from Eq. (29.80). Adding these results from the Wilkinson Microwave Anisotropy Probe reveals that

$$\Omega_0 = \Omega_{m,0} + \Omega_{\text{rel},0} + \Omega_{\Lambda,0} = 1;$$

that is, the universe is flat ($k = 0$), and dark energy now dominates the expansion of the universe. More precisely, the WMAP result is

$$[\Omega_0]_{\text{WMAP}} = 1.02 \pm 0.02,$$

which is *consistent* with $\Omega_0 = 1$. We can never prove the $\Omega_0$ is *exactly* 1 because any measurement will necessarily be accompanied by a small (we hope) uncertainty.

---

[35]The symbol "$\Omega$" without a subscript "$m$," "rel," or "$\Lambda$" always denotes the total density parameter for all of the components of that particular model.

The deceleration parameter (Eq. 29.54) may be written as (Problem 29.42)

$$q(t) = \frac{1}{2} \sum_i (1 + 3w_i)\Omega_i(t),$$ (29.123)

where $w$ is the coefficient from the equation of state $P_i = w_i \rho_i c^2$ and the "$i$" subscript identifies one of the components of the universe (i.e., pressureless dust, relativistic particles, or dark energy). Using $w_m = 0$, $w_{rel} = 1/3$, and $w_\Lambda = -1$, we obtain

$$q(t) = \frac{1}{2}\Omega_m(t) + \Omega_{rel}(t) - \Omega_\Lambda(t).$$ (29.124)

With WMAP values we find that the current value of the deceleration parameter is

$$q_0 = -0.60.$$

The minus sign indicates that expansion of the universe is now accelerating![36]
    Assuming $\Lambda > 0$, the equivalent mass density of dark energy $\rho_\Lambda$ adds to the effect of the other densities in the Friedmann equation (Eq. 29.114) on the curvature of the universe, $k$. However, in the acceleration equation, the negative pressure $P_\Lambda$ opposes the gravitational effect of a positive $\rho_\Lambda$ and acts to *increase* the acceleration of the universe, as seen by the positive term $\Lambda c^2/3$ added to the acceleration equation, Eq (29.112). The presence of a cosmological constant therefore decouples the geometry of the universe (open, closed, flat), which is described by $k$, from the dynamics of the universe, which are governed by the interplay of $\rho_m$, $\rho_{rel}$, and $\rho_\Lambda$. As we shall see, our universe may be flat ($k = 0$), yet it may have an accelerating expansion—a combination that is not possible within the more restricted model of a one-component universe of pressureless dust.

### The $\Lambda$ Era

The $\Lambda$-terms in Eqs. (29.114) and (29.116) became dominant as the scale factor $R$ increased because $\rho_\Lambda$ is constant while the mass density $\rho_m \propto R^{-3}$ and $\rho_{rel} \propto R^{-4}$. Just as the radiation era yielded to the mass era when the universe was about 55,000 years old (Eq. 29.85), the mass era has segued into the $\Lambda$ era. Dark energy now governs the expansion of the universe. The transition from the matter era to the present $\Lambda$ era occurred when the scale factor satisfied $\rho_m = \rho_\Lambda$. Since $\rho_\Lambda$ is constant, Eq. (29.82) can be used to show that the transition value of $R$ is

$$R_{m,\Lambda} = \left(\frac{\Omega_{m,0}}{\Omega_{\Lambda,0}}\right)^{1/3}.$$ (29.125)

Inserting WMAP values gives

$$R_{m,\Lambda} = 0.72,$$ (29.126)

---

[36]By "accelerating," we mean $d^2 R/dt^2 > 0$.

which corresponds to a redshift of

$$z_{m,\Lambda} = \frac{1}{R_{m,\Lambda}} - 1 = 0.39. \qquad (29.127)$$

The acceleration equation with WMAP values shows (Problem 29.43) that the acceleration of the universe changed sign (from negative to positive) when the scale factor was

$$R_{\text{accel}} = 2^{-1/3} R_{m,\Lambda} = 0.57,$$

corresponding to

$$z_{\text{accel}} = 0.76,$$

meaning that the acceleration became positive *before* the $\Lambda$-term dominated the Friedmann equation. As $R \to 0$ we can deduce that the effects of the cosmological constant are negligible in the early universe because $\rho_m \propto R^{-3}$ and $\rho_{\text{rel}} \propto R^{-4}$ (Eqs. 29.81 and 29.82) while $\rho_\Lambda$ remains constant. The happy outcome is that *all of the results for the early universe obtained in Sections 29.1 and 29.2 are valid for the present relativistic cosmology as well*; see Problem 29.40.

The behavior of the scale factor $R$ for a flat universe can be found by setting $k = 0$ in the Friedmann equation (Eq. 29.114). A little algebra and Eqs. (29.79) and (29.118) lead to

$$t = \sqrt{\frac{3}{8\pi G}} \int_0^R \frac{R' \, dR'}{\sqrt{\rho_{m,0} R' + \rho_{\text{rel},0} + \rho_{\Lambda,0} R'^4}}. \qquad (29.128)$$

Although this can be integrated numerically, it has no simple analytic solution. Figure 29.19 displays a numerical solution of Eq. (29.128) using WMAP values, showing the different behaviors of the scale factor in the radiation, matter, and $\Lambda$ eras.

To make further progress, we will neglect the reign of relativistic particles during the first 55,000 years or so of the universe by setting $\rho_{\text{rel},0} = 0$. Integrating eventually yields, for $k = 0$,

$$t(R) = \frac{2}{3} \frac{1}{H_0\sqrt{\Omega_{\Lambda,0}}} \ln\left[\sqrt{\left(\frac{\Omega_{\Lambda,0}}{\Omega_{m,0}}\right) R^3} + \sqrt{1 + \left(\frac{\Omega_{\Lambda,0}}{\Omega_{m,0}}\right) R^3}\right]. \qquad (29.129)$$

The present age of the universe may be obtained by substituting $R = 1$ into Eq. (29.129). Using the WMAP values for $\Omega_{m,0}$ and $\Omega_{\Lambda,0}$, we obtain

$$t_0 = 4.32 \times 10^{17} \text{ s} = 1.37 \times 10^{10} \text{ yr}.$$

This is in good agreement with the best determination of the age of the universe currently available, the published WMAP value of [37]

$$\boxed{[t_0]_{\text{WMAP}} = 13.7 \pm 0.2 \text{ Gyr.}}$$

[37] When the first edition of this text was written, the authors could not have anticipated that the age of the universe, to this precision, would be included in the second edition!

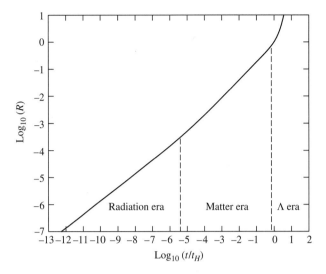

**FIGURE 29.19**   A logarithmic graph of the scale factor $R$ as a function of time. During the radiation era, $R \propto t^{1/2}$; during the matter era, $R \propto t^{2/3}$; and during the $\Lambda$ era, $R$ grows exponentially.

The acceleration of the universe changed sign (from negative to positive) when $R = R_{\text{accel}} = 0.57$. According to Eq. (29.129), the expansion of the universe began to speed up when its age was

$$t_{\text{accel}} = 2.23 \times 10^{17} \text{ s} = 7.08 \text{ Gyr.}$$

Thus the expansion of the universe has been accelerating for approximately the second half of its existence. For this reason, $t_0$ is very nearly equal to the Hubble time:

$$t_0 = 0.993 \, t_H.$$

At the present epoch, the effects of deceleration during the radiation and matter eras and acceleration during the $\Lambda$ era nearly cancel, so the age of the universe is what we would have calculated for a constant rate of expansion.

Equation (29.129) can be inverted to obtain, for $k = 0$,

$$R(t) = \left( \frac{\Omega_{m,0}}{4 \, \Omega_{\Lambda,0}} \right)^{1/3} \left( e^{3H_0 t \sqrt{\Omega_{\Lambda,0}}/2} - e^{-3H_0 t \sqrt{\Omega_{\Lambda,0}}/2} \right)^{2/3} \tag{29.130}$$

$$= \left( \frac{\Omega_{m,0}}{\Omega_{\Lambda,0}} \right)^{1/3} \sinh^{2/3} \left( \frac{3}{2} H_0 t \sqrt{\Omega_{\Lambda,0}} \right) \tag{29.131}$$

Figure 29.20 shows the evolution of the scale factor as a function of time. In the limit of $3H_0 t \sqrt{\Omega_{\Lambda,0}}/2 \ll 1$ (essentially when $t \ll t_H$), Eq. (29.131) reduces to Eq. (29.91),

$$R(t) \simeq \left( \frac{3}{2} H_0 t \sqrt{\Omega_{m,0}} \right)^{2/3} = \left( \frac{3 \sqrt{\Omega_{m,0}}}{2} \right)^{2/3} \left( \frac{t}{t_H} \right)^{2/3}, \tag{29.132}$$

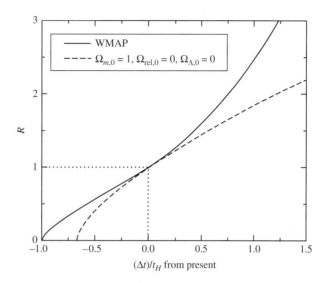

**FIGURE 29.20** The scale factor $R$ as a function of time, measured from the present, for a WMAP universe with $t_0 \simeq t_H$, and a flat, one-component universe of pressureless dust with $t_0 = 2t_H/3$ (Eq. 29.44). The dotted lines locate the position of today's universe on the two curves.

as appropriate for a matter-dominated universe. But when $t \gg t_H$, the second exponential in Eq. (29.131) can be neglected, leaving

$$R(t) \simeq \left( \frac{\Omega_{m,0}}{4\,\Omega_{\Lambda,0}} \right)^{1/3} e^{H_0 t \sqrt{\Omega_{\Lambda,0}}}. \tag{29.133}$$

When the cosmological constant dominates the Friedmann equation, the scale factor $R$ grows exponentially with a characteristic time of $t_H/\sqrt{\Omega_{\Lambda,0}}$. As we will discover in the next section, an exponentially expanding universe has profound implications for the long-term future of observational astronomy.

### Model Universes on the $\Omega_{m,0}$–$\Omega_{\Lambda,0}$ Plane

Every model of a three-component universe (matter, relativistic particles, and dark energy) is specified by the values of the three density parameters $\Omega_{m,0}$, $\Omega_{\text{rel},0}$, and $\Omega_{\Lambda,0}$. At the present epoch $\Omega_{\text{rel},0}$ is negligible, so we can consider a two-dimensional plot of $\Omega_{m,0}$ vs. $\Omega_{\Lambda,0}$, as shown in Fig. 29.21. The $\Omega_{m,0}$–$\Omega_{\Lambda,0}$ plane is divided into several regions. The Friedmann equation (Eq. 29.117) shows that the line $\Omega_{m,0} + \Omega_{\Lambda,0} = 1$ determines the sign of $k$ and so divides the $\Omega_{m,0}$–$\Omega_{\Lambda,0}$ plane into open and closed universes. The sign of the deceleration parameter (Eq. 29.123) is determined by the sign of the quantity $\Omega_{m,0} - 2\Omega_{\Lambda,0}$, so the line $\Omega_{m,0} - 2\Omega_{\Lambda,0} = 0$ divides the $\Omega_{m,0}$–$\Omega_{\Lambda,0}$ plane into accelerating and decelerating universes.

Although it seems that our universe will expand forever, driven by dark energy, it is easy to conceive of other model universes that will eventually recollapse. This includes any universe with $\Omega_{\Lambda,0} < 0$, as well as universes with $\Omega_{\lambda,0} > 0$ but containing enough matter

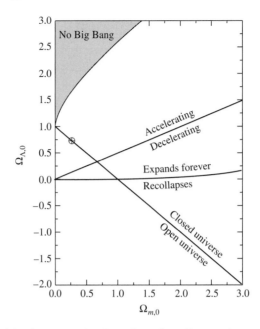

**FIGURE 29.21**    Model universes on the $\Omega_{m,0}$–$\Omega_{\Lambda,0}$ plane. Every point on this plane represents a possible universe. The point ($\Omega_{m,0} = 0.27$, $\Omega_{\Lambda,0} = 0.73$) is indicated by the circle.

to bring the expansion to a halt before dark energy dominates. When the expansion stops, $dR/dt = 0$. Equation (29.114), together with Eqs. (29.81), (29.82), (29.113), and (29.15), can be used to express $dR/dt$ as

$$\left(\frac{dR}{dt}\right)^2 = H_0^2 \left(\frac{\Omega_{m,0}}{R} + \frac{\Omega_{\text{rel},0}}{R^2} + \Omega_{\Lambda,0}R^2 + 1 - \Omega_{m,0} - \Omega_{\text{rel},0} - \Omega_{\Lambda,0}\right). \qquad (29.134)$$

Setting the left-hand side equal to zero, canceling the $H_0^2$, and neglecting the radiation era, we have a cubic equation for the scale factor $R$.

$$\frac{\Omega_{m,0}}{R} + \Omega_{\Lambda,0}R^2 + 1 - \Omega_{m,0} - \Omega_{\Lambda,0} = 0. \qquad (29.135)$$

We want to know when this cubic equation for $R$ has a positive, real root. It turns out that if $\Omega_{m,0} < 1$ and $\Omega_{\lambda,0} > 0$, Eq. (29.135) has no positive, real root, and we conclude that these universes will expand forever.[38] But a universe with $\Omega_{m,0} > 1$ will expand forever only if

$$\Omega_{\Lambda,0} > 4\Omega_{m,0} \left\{\cos\left[\frac{1}{3}\cos^{-1}\left(\frac{1}{\Omega_{m,0}} - 1\right) + \frac{4\pi}{3}\right]\right\}^3.$$

[38]You are referred to Felten and Isaacman (1986) for a complete analysis of this diagram.

On this diagram, lines of constant age $t_0$ are roughly diagonal from the lower left to the upper right, and $t_0$ increases from lower right to upper left. In fact, $t_0$ becomes *infinite* as we approach a line across the upper-left corner of the diagram where there is an (unstable) equilibrium between the inward pull of gravity and the outward push of dark energy. Models on this line are infinitely old, meaning that they never unfolded from a hot, dense Big Bang. This line is given by

$$\Omega_{\Lambda,0} = 4\Omega_{m,0} \left\{ \cos \left[ \frac{1}{3} \cos^{-1} \left( \frac{1}{\Omega_{m,0}} - 1 \right) \right] \right\}^3,$$

if $\Omega_{m,0} > 0.5$; otherwise, "cos" should be replaced by "cosh." Models beyond this line represent "bounce" universes that are now on the rebound from an earlier collapse. We merely state the result that these "bounce" models have a maximum redshift (at the bounce) that satisfies

$$z_{\text{bounce}} \leq 2 \cos \left[ \frac{1}{3} \cos^{-1} \left( \frac{1 - \Omega_{m,0}}{\Omega_{m,0}} \right) \right] - 1 \tag{29.136}$$

if $\Omega_{m,0} > 0.5$; otherwise, "cos" should be replaced by "cosh." Because objects are observed beyond this maximum redshift, these "bounce" models may be rejected.

Astronomers have the task of determining which point on Fig. 29.21 represents *our* universe. This task is possible because the dynamics of the expansion of the universe determines $q_0 = \Omega_{m,0}/2 - \Omega_{\Lambda,0}$, and the geometry of the universe determines $\Omega_0 = \Omega_{m,0} + \Omega_{\Lambda,0}$. In the next section we will learn how $q_0$ is measured, but the evaluation of $\Omega_0$ will have to wait until Chapter 30.

## 29.4 ■ OBSERVATIONAL COSMOLOGY

Most of the key parameters of cosmology encountered in the previous section, such as $H_0$, $q_0$, and the various $\Omega_0$'s, are not quantities that can be directly measured by astronomers. Observers are primarily limited to measuring the spectrum, redshift, radiant flux, and polarization of the starlight from a distant object. We now proceed to link these observables to the theoretical framework we have erected.

### The Origin of the Cosmological Redshift

Let's begin by uncovering the origin of the cosmological redshift. We start with the Robertson–Walker metric, Eq. (29.106), with $ds = 0$ for a light ray (recall Section 17.2), and $d\theta = d\phi = 0$ for a radial path traveled from the point of the light's emission at comoving coordinate $\varpi_e$ to its arrival at Earth at $\varpi = 0$. Taking the negative square root (so $\varpi$ decreases with increasing time) gives

$$\frac{-c\,dt}{R(t)} = \frac{d\varpi}{\sqrt{1 - k\varpi^2}}.$$

Integrating this from a larger $\varpi_{\text{far}}$ at an initial time $t_i$ to a smaller $\varpi_{\text{near}}$ at time $t_f$, we have

$$\int_{t_i}^{t_f} \frac{c\,dt}{R(t)} = -\int_{\varpi_{\text{far}}}^{\varpi_{\text{near}}} \frac{d\varpi}{\sqrt{1-k\varpi^2}} = \int_{\varpi_{\text{near}}}^{\varpi_{\text{far}}} \frac{d\varpi}{\sqrt{1-k\varpi^2}}. \qquad (29.137)$$

A moment's thought reveals that the same result describes an outwardly moving light ray. Suppose that one crest of the light wave was emitted at time $t_e$ and received at $t_0$, and the next wave crest was emitted at $t_e + \Delta t_e$ and received at $t_0 + \Delta t_0$. These times, which describe how long it takes for the successive crests of the light wave to travel to Earth, satisfy the relations

$$\int_{t_e}^{t_0} \frac{c\,dt}{R(t)} = \int_0^{\varpi_e} \frac{d\varpi}{\sqrt{1-k\varpi^2}} \qquad (29.138)$$

for the first crest and

$$\int_{t_e+\Delta t_e}^{t_0+\Delta t_0} \frac{c\,dt}{R(t)} = \int_0^{\varpi_e} \frac{d\varpi}{\sqrt{1-k\varpi^2}} \qquad (29.139)$$

for the next. The right-hand sides are the same, since the comoving coordinate of an object does not change as the universe expands (assuming its peculiar velocity is negligible). Subtracting Eq. (29.138) from Eq. (29.139) produces

$$\int_{t_e+\Delta t_e}^{t_0+\Delta t_0} \frac{dt}{R(t)} - \int_{t_e}^{t_0} \frac{dt}{R(t)} = 0. \qquad (29.140)$$

But

$$\int_{t_e+\Delta t_e}^{t_0+\Delta t_0} \frac{dt}{R(t)} = \int_{t_e+\Delta t_e}^{t_e} \frac{dt}{R(t)} + \int_{t_e}^{t_0} \frac{dt}{R(t)} + \int_{t_0}^{t_0+\Delta t_0} \frac{dt}{R(t)},$$

so

$$\int_{t_0}^{t_0+\Delta t_0} \frac{dt}{R(t)} - \int_{t_e}^{t_e+\Delta t_e} \frac{dt}{R(t)} = 0.$$

Any change in $R(t)$ during the time intervals $\Delta t_e$ and $\Delta t_0$ between the emission of the two successive wave crests can safely be neglected. This allows us to treat $R(t)$ as a constant with respect to the time integration, so that, using $R(t_0) = 1$,

$$\Delta t_0 = \frac{\Delta t_e}{R(t_e)}. \qquad (29.141)$$

The times $\Delta t_e$ and $\Delta t_0$ are just the periods of the emitted and received light waves and are related to their wavelengths by $\lambda = c\Delta t$. Making this substitution into Eq. (29.141) and using the definition of the redshift $z$ (Eq. 4.34) results in the expression for the cosmological redshift,

$$\frac{1}{R(t_e)} = \frac{\lambda_0}{\lambda_e} = 1 + z. \qquad (29.142)$$

This derivation shows that the cosmological redshift is due to the wavelength of a photon expanding along with the space through which the photon moves during its journey to Earth. Equation (29.142) is just Eq. (29.4) discussed previously (this result was also discussed in the context of the distances to QSOs on page 1100). Combining Eqs. (29.141) and (29.142) results in the formula for **cosmological time dilation**,

$$\boxed{\frac{\Delta t_0}{\Delta t_e} = 1 + z.} \tag{29.143}$$

Note that these relations for the cosmological redshift and time dilation hold regardless of the functional form of the scale factor, $R(t)$. Experimental confirmation of cosmological time dilation has been frustrated by the lack of a reliable natural clock located at cosmological distances. However, a measurement of cosmological time dilation has been made using the temporal changes in the spectrum of a moderate-redshift ($z = 0.361$) Type Ia supernova. The results were consistent with Eq. (29.143). [See Foley et al. (2005) for details of this measurement of cosmological time dilation.]

### Distances to the Most Remote Objects in the Universe

To make further progress in our quest to survey the universe and determine its geometry and dynamics, we must next learn how to measure the distance to objects in the most remote regions of the universe. The proper distance of an object from Earth can be found from the Robertson–Walker metric. Recall that the differential proper distance (Eq. 17.18) is just $\sqrt{-(ds)^2}$ with $dt = 0$. Furthermore, if the comoving coordinate of the object is $\varpi$ (with Earth at $\varpi = 0$), then $d\theta = d\phi = 0$ along a radial line from Earth to the object. Inserting these into the Robertson–Walker metric (Eq. 29.106), we can find the proper distance, $d_p(t)$, to the object at time $t$ by integrating

$$d_p(t) = R(t) \int_0^{\varpi} \frac{d\varpi'}{\sqrt{1 - k\varpi'^2}}. \tag{29.144}$$

Using Eq. (29.138), this becomes

$$d_p(t) = R(t) \int_{t_e}^{t_0} \frac{c\, dt'}{R(t')}. \tag{29.145}$$

The physical meaning of this is readily apparent. As the photon moves in from $\varpi_e$, in each interval of time $dt$ it travels a small distance of $c\, dt$. These intervals cannot simply be added up, because the universe expands as the photon travels. Dividing $c\, dt$ by the scale factor at the time, $R(t)$, converts this small distance to what it would be at the present time, $t_0$. Integrating then yields the proper distance from $\varpi_e$ to $\varpi = 0$ today, at time $t_0$. Multiplying by the scale factor $R(t)$ then converts this to the distance at some other time $t$. It is worth emphasizing that the current value of the proper distance, $d_{p,0} \equiv d_p(t_0)$, to an object is how far away it is *today*, and not its distance when its light was emitted. As long as the object has zero peculiar velocity (constant comoving coordinate $\varpi$), it suffices to find $d_{p,0}$ because the proper distance at any other time can be obtained from

$$d_p(t) = R(t)d_{p,0}. \tag{29.146}$$

In particular, if the object's redshift is $z$, then its distance at time $t_e$, when its light was emitted, is

$$d_p(t_e) = d_{p,0}R(t_e) = \frac{d_{p,0}}{1+z}. \tag{29.147}$$

Integrating Eq. (29.144) and using $R(t_0) = 1$ shows that the expression for the present proper distance in a flat universe is

$$d_{p,0} = \varpi \qquad \text{(for } k = 0\text{)}, \tag{29.148}$$

the expression for a closed universe is

$$d_{p,0} = \frac{1}{\sqrt{k}} \sin^{-1}(\varpi\sqrt{k}) \qquad \text{(for } k > 0\text{)}, \tag{29.149}$$

and the expression for an open universe is

$$d_{p,0} = \frac{1}{\sqrt{|k|}} \sinh^{-1}(\varpi\sqrt{|k|}) \qquad \text{(for } k < 0\text{)}. \tag{29.150}$$

In a flat universe, the present proper distance to an object is just its coordinate distance $d_{c,0} = \varpi$ (cf. Eq. 29.3). However, the coordinate distance will not agree with the proper distance if $k \neq 0$. Because $\sin^{-1}(x) \geq x$, in a closed universe ($k > 0$) the proper distance to an object is *greater* than its coordinate distance. Similarly, $\sinh^{-1}(x) \leq x$, so in an open universe ($k < 0$) the proper distance to an object is *less* than its coordinate distance. Later we will find expressions for the proper distance $d_{p,0}$ and the comoving coordinate $\varpi$ as functions of the redshift $z$. First, however, we pause to examine the above equations for $d_{p,0}$ because measuring distances in an expanding universe has some interesting aspects.

Distances in a closed universe ($k > 0$) are especially interesting. Solving Eq. (29.149) for $\varpi$ gives

$$\varpi = \frac{1}{\sqrt{k}} \sin(d_{p,0}\sqrt{k}). \tag{29.151}$$

Note that in a closed universe there is a maximum value of the comoving coordinate of $\varpi_{\max} = 1/\sqrt{k}$. Also, there are an infinite number of distances along a radial line to the same point $X$ in space, located, say, at $\varpi_X$. If $d_{p,0}$ is one value of the present proper distance to $X$ at time $t$, then for any integer $n$, $d_{p,0} + 2\pi n/\sqrt{k}$ will also bring us back to $X$, with the same value of $\varpi_X$. The extra multiples of $2\pi/\sqrt{k}$ correspond to traversing the circumference of the universe $n$ times before stopping at $X$. However, such a journey, running circles around the universe, would not be physically possible. As it happens, a photon emitted at $t = 0$ in a closed, matter-dominated universe with $\Lambda = 0$ would return to its starting point just as the universe ended in a Big Crunch; see Problem 29.47.

In a similar manner, our tiny ant that lives on the surface of a sphere could walk from pole to pole to pole, encircling the sphere $n$ times before proceeding on to its final destination. This illustrates that although there is no boundary to a closed universe, it contains only a finite amount of space, like the unbounded surface of a sphere. Furthermore, a closed universe curves back on itself; moving outward from Earth (or from any other choice of origin), the farthest you can get from home is a point where $\varpi = 1/\sqrt{k}$. From that point, a step in *any* direction brings you closer to where you started.

You should be warned that suggestive phrases like "the circumference of the universe" and "curves back on itself" do not imply a curved path through three-dimensional space, since there is no deviation from a radial line as $\varpi$ increases. Despite these caveats, we can define the circumference of a closed universe (including the time-dependence) as

$$C_{\text{univ}}(t) = \frac{2\pi R(t)}{\sqrt{k}} \tag{29.152}$$

which is the proper distance along a radial line that brings you back to your starting point. This expression for the circumference is consistent with our definition of curvature since, from Eq. (29.105), the radius of curvature of a closed universe at time $t$ is $1/\sqrt{K(t)} = R(t)/\sqrt{k}$. Nevertheless, the radius of curvature must not be thought of as the radius of an actual circular path.

### The Particle Horizon and the Horizon Distance

As the universe expands and ages, photons from increasingly distant objects have more time to complete their journey to Earth. This means that as time increases, we might expect that more of the universe will come into causal contact with the observer. The proper distance to the farthest observable point (called the **particle horizon**) at time $t$ is the **horizon distance**, $d_h(t)$. Note that two points separated by a distance greater than $d_h$ are not in causal contact. Thus $d_h$ may be thought of as the *diameter* of the largest causally connected region.

We will now derive an expression for $d_h(t)$, the size of the observable universe as a function of time. (It is important to note that because the farthest observable point moves outward through increasingly larger values of $\varpi$, $d_h(t)$ is *not* proportional to $R(t)$.) Consider an observer at the origin ($\varpi = 0$), and let the particle horizon for this observer be located at $\varpi_e$ at time $t$. This means that a photon emitted at $\varpi_e$ at $t = 0$ would reach the origin at time $t$. With an appropriate change of limits in Eq. (29.145), the horizon distance at time $t$ is found to be

$$d_h(t) = R(t) \int_0^t \frac{c\,dt'}{R(t')}. \tag{29.153}$$

First we consider distances in the early universe, when the effect of dark energy was negligible (so $\Lambda$ may be set equal to zero). During the radiation era, the universe was essentially flat and the scale factor was of the form $R(t) = Ct^{1/2}$, where $C$ is a constant (see Eq. 29.87). Inserting this into Eq. (29.153) gives

$$d_h(t) = 2ct \qquad \text{(radiation era)}. \tag{29.154}$$

After the radiation era, the expansion of the universe was governed by the effects of matter and, later, dark energy.

For the matter era, assuming a flat universe, the scale factor is given by Eq. (29.91), which is of the form $R(t) = Ct^{2/3}$, where again $C$ is a constant. (Since the radiation era lasted only 55,000 years following the Big Bang, for the purposes of the following calculations we will ignore radiation and set the lower limit to $t = 0$.) Substituting this into Eq. (29.153) results in

$$d_h(t) = 3ct \qquad \text{(for } k = 0\text{)}. \tag{29.155}$$

Using Eqs. (29.4) and (29.90), this expression can be rewritten in terms of the redshift as

$$d_h(z) = \frac{2c}{H_0\sqrt{\Omega_{m,0}}} \frac{1}{(1+z)^{3/2}} \qquad \text{(for } k = 0\text{)}. \tag{29.156}$$

We can obtain a rough estimate for the present horizon distance by setting $z = 0$ to obtain

$$d_{h,0} \approx \frac{2c}{H_0\sqrt{\Omega_{m,0}}} = 5.02 \times 10^{26} \text{ m} = 16{,}300 \text{ Mpc} = 16.3 \text{ Gpc} \tag{29.157}$$

using WMAP values.

Finally, in the $\Lambda$ era, we substitute Eq. (29.131) for the scale factor into Eq. (29.153) for the horizon distance to get, for $k = 0$,

$$d_h(t) = \left(\frac{\Omega_{m,0}}{\Omega_{\Lambda,0}}\right)^{1/3} \sinh^{2/3}\left(\frac{3}{2}H_0 t \sqrt{\Omega_{\Lambda,0}}\right) \int_0^t \frac{c\,dt'}{\left(\frac{\Omega_{m,0}}{\Omega_{\Lambda,0}}\right)^{1/3} \sinh^{2/3}\left(\frac{3}{2}H_0 t' \sqrt{\Omega_{\Lambda,0}}\right)}. \tag{29.158}$$

This has no simple analytic solution and must be integrated numerically. Using $t_0 = 13.7$ billion years, we calculate that at the present time, the distance to the particle horizon in a flat universe is

$$d_{h,0} = 4.50 \times 10^{26} \text{ m} = 14{,}600 \text{ Mpc} = 14.6 \text{ Gpc} \tag{29.159}$$

Figure 29.22 uses WMAP values and shows $d_h$, the size of the observable universe, as a function of time. Of course, when viewing an object near the particle horizon, astronomers see it as it was when the light was emitted, not as it would appear in today's universe.

Note that the distance to the particle horizon in Eqs. (29.154) and (29.155) is proportional to $t$, while the scale factor in the radiation and matter eras is proportional to $t^{1/2}$ and $t^{2/3}$, respectively. This means that during those eras the size of the observable universe increased more rapidly than the universe expanded, so the universe became increasingly causally connected as it aged. However, the integral in Eq. (29.158)—without the term in front—is just the present distance to the point that will be at the particle horizon at time $t$, as we can see by comparing Eq. (29.153) with Eq. (29.145), evaluated at $t_0$ with $t_e = 0$. As $t \to \infty$, this integral converges to 19.3 Gpc. This means that the proper distance *today* to the farthest

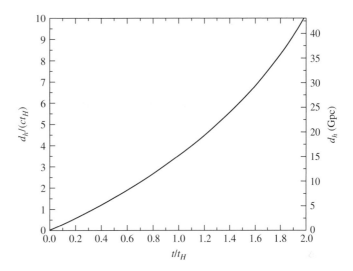

**FIGURE 29.22**   The proper distance from Earth to the particle horizon as a function of time, using WMAP values. The horizon distance is expressed as a fraction of $ct_H$. The right axis shows $d_h$ in billions of parsecs.

object that will *ever* be observable in the future is 19.3 Gpc. Everything within a sphere, centered on Earth, of radius 19.3 Gpc will eventually become visible, while everything beyond will be forever hidden. In the future, the particle horizon and the scale factor will both grow exponentially as $e^{H_0 t \sqrt{\Omega_{\Lambda,0}}}$ (Eq. 29.133).

Ultimately, an object located at the particle horizon will remain at the particle horizon as the universe expands. The particle horizon will never catch up to any object that is presently more than 19.3 Gpc away, and so its light will *never* reach us. What will we observe when looking at an object at the ultimate ($t \to \infty$) particle horizon? Although photons from the object will continue to arrive, they will be increasingly deeply redshifted, and their arrival rate will decline toward zero due to cosmological time dilation (Eq. 29.143). Thus the object will fade from view, apparently frozen in time, as its redshift diverges to infinity. This bears a striking similarity to the description in Section 17.3 of how we would view an astronaut falling into a black hole, although the physics of the two situations is completely different.

---

**Example 29.4.1.**   Helium nuclei were being formed when the temperature was roughly $10^9$ K and $t = 178$ s. From Eq. (29.58), the scale factor at that time was $R = 2.73 \times 10^{-9}$. From Eq. (29.154), the horizon distance was then about

$$d_h(t) = 2ct = 1.07 \times 10^{11} \text{ m} = 0.7 \text{ AU}.$$

This is the diameter of a causally connected region, which we call $\mathcal{C}$, when the universe was 178 s old.

The region $\mathcal{C}$ (which has a comoving boundary, so it always contains the same mass) has been expanding along with the rest of the universe since $t = 178$ s. How large is $\mathcal{C}$

*continued*

today? Assuming a flat universe, Eq. (29.146) shows that $\mathcal{C}$ has expanded by a factor of $1/R = 3.66 \times 10^8$, with a present diameter of

$$\frac{d_h(t)}{R(t)} = 3.92 \times 10^{19} \text{ m},$$

about 1.3 kpc. In other words, the largest region that was causally connected at $t = 178$ s is now just over a kiloparsec across, roughly $8.7 \times 10^{-8}$ of the present horizon distance, $d_{h,0}$. This illustrates that as the early universe aged, the amount of material that is in causal contact increased dramatically. Today's causally connected universe extends far beyond $\mathcal{C}$ because, since $t = 178$ s, light from more distant regions has had time to arrive and causally connect those regions with $\mathcal{C}$. The comoving boundary of $\mathcal{C}$ cannot keep up with the more rapid recession of the particle horizon.

---

### The Arrival of Photons

You may be wondering, if the scale factor $R$ was zero at the Big Bang and everything was right next to everything else, then why has it taken the age of the universe for a Big Bang photon to reach Earth? What is the path followed by the photon?

In the following discussion, the actual complications of the Big Bang will be neglected. Instead, we will consider a perfectly transparent, expanding, flat universe where a single photon is emitted at comoving coordinate $\varpi_e$ at time $t = 0$. What, then, is the proper distance of that photon from our position ($\varpi = 0$) at a later time $t$? The coordinate, $\varpi$, of the photon at time $t$ may be found from Eq. (29.137) with $k = 0$,

$$\int_0^t \frac{c\,dt'}{R(t')} = \int_\varpi^{\varpi_e} d\varpi'. \tag{29.160}$$

To simplify the calculation, we will ignore the $\Lambda$ era and adopt a flat, matter-dominated universe with a scale factor given by Eq. (29.132),

$$R(t) = \left( \frac{3}{2} H_0 t \sqrt{\Omega_{m,0}} \right)^{2/3}. \tag{29.161}$$

Setting $R = 1$, the age of this model universe is found to be

$$t_0 = \frac{2}{3 H_0 \sqrt{\Omega_{m,0}}}, \tag{29.162}$$

so the scale factor is simply

$$R(t) = \left( \frac{t}{t_0} \right)^{2/3}.$$

Integrating Eq. (29.160) yields

$$\varpi = \varpi_e - 3ct_0 \left( \frac{t}{t_0} \right)^{1/3}. \tag{29.163}$$

We can evaluate $\varpi_e$ by noting that at $t = t_0$, $\varpi = 0$; thus

$$\varpi_e = 3ct_0, \tag{29.164}$$

the present horizon distance for this model (Eq. 29.155). Inserting this value for the photon's starting point into Eq. (29.163) and multiplying both sides by the scale factor $R(t)$ shows that the proper distance of the photon from Earth is, as a function of time,

$$d_p(t) = 3ct_0 \left[ \left(\frac{t}{t_0}\right)^{2/3} - \left(\frac{t}{t_0}\right) \right] \tag{29.165}$$

for our model flat universe.

Since the Big Bang, the entire system of comoving coordinates has been stretching out from its compact beginning. Indeed, the "Big Stretch" would be a more appropriate term than the "Big Bang" for this event! As shown in Fig. 29.23, the initial expansion of the universe actually carried the photon *away* from Earth. Although the photon's comoving coordinate was always decreasing from an initial value of $\varpi_e$ toward Earth's position at $\varpi = 0$, the scale factor $R(t)$ increased so rapidly that at first the proper distance between the photon and Earth increased with time. This means that a photon emitted from the present particle horizon at $t = 0$ is only now reaching Earth. Photons emitted from a greater $\varpi$, beyond the present particle horizon, have yet to arrive—and in fact may *never* arrive if $\varpi$ is sufficiently large that the exponential expansion of the universe ultimately carries the photon away from Earth.

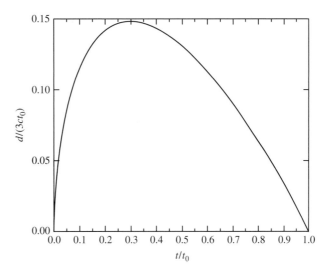

**FIGURE 29.23** The proper distance from Earth of a photon emitted from the present particle horizon at the time of the Big Bang. The photon's proper distance is expressed as a fraction of $3ct_0$.

## The Maximum Visible Age of a Source

The previous calculation assumes that the photon was emitted at $t = 0$. Is it possible that the exponential expansion of space could carry a presently visible object away from Earth so fast that the object would never again be seen in the sky from some future time forward? To answer this, consider an object (say, a galaxy) that is now visible, meaning that its light was emitted at an earlier time $t_e$ and it arrives today at $t_0$. Assuming that the galaxy is still visible at some time in the future, we will let the time of the emission of its future photons be $t_i$ and their arrival time here be $t_f$, where $t_e < t_i$ and $t_0 < t_f$. Applying these conditions to Eq. (29.140) gives

$$\int_{t_e}^{t_0} \frac{dt}{R(t)} = \int_{t_i}^{t_f} \frac{dt}{R(t)}, \tag{29.166}$$

where the scale factor $R$ is given by Eq. (29.131) for a flat universe ($k = 0$). Because the scale factor increases monotonically, it may be, for sufficiently large $t_i$, that *no* value of $t_f$ can satisfy this equality. In that case, a photon emitted at time $t_i$ will never reach Earth. The latest time of emission, $t_{mva}$, for photons to eventually reach us (the **maximum visible age** of the source) may be found by setting $t_f = \infty$.[39]

Just like an object located at the ultimate particle horizon, as photons from this galaxy continue to arrive, they will be increasingly deeply redshifted and their flux will drop toward zero. The galaxy will fade from view, apparently frozen in time, as its redshift diverges to infinity. The farther a source is, the sooner it will fade away. This places a fundamental limit on extragalactic astronomy. We will never be able to watch galaxies continue to age and evolve through the millenia. Figure 29.24 shows that if the redshift of an object is roughly larger than 1.8, then $t_{mva} < t_H$, and we will never see it even as it appears *today*. That is, the light emitted by the object today will never reach Earth because those photons will eventually be carried away from us by the accelerating Hubble flow. Objects with a redshift in the range 5–10 can be observed only as they appeared when the universe

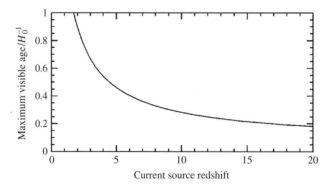

**FIGURE 29.24** The maximum visible age of a source, in units of $t_H = 1/H_0$, as a function of its current redshift. (Figure adapted from Loeb, *Phys. Rev. D*, *65*, 047301, 2002.)

[39]Actually, $t_{mva}$ is the age of the universe when the source emitted the last photons to reach Earth.

was approximately 4 to 6 billion years old. As more light is carried away from Earth by the expansion of the universe, the observable sky will become increasingly empty. Similarly, any signal we send toward a galaxy with $z \approx 1.8$ or more will never arrive. Because there can be no contact between that galaxy and Earth anytime in the future, we are no longer in causal contact with it. As the universe ages, it is becoming causally fragmented, with one region no longer capable of influencing another.[40]

### The Comoving Coordinate $\varpi(z)$

Returning to Eqs. (29.148–29.150), we want to express the comoving coordinate $\varpi$ as a function of the redshift $z$. We begin by using Eq. (29.145) to find another expression for the present proper distance $d_{p,0}$. Writing $dt = dR/(dR/dt)$, Eq. (29.145) may be written as

$$d_{p,0} = \int_{R(t_e)}^{R(t_0)} \frac{c\,dR}{R(dR/dt)}.$$

It is useful to use $R(t_0) = 1$, $R(t_e) = 1/(1+z)$, $dR = -R^2\,dz$ [from differentiating Eq. (29.4), $R = 1/(1+z)$] and Eq. (29.8) to define the dimensionless integral

$$I(z) = H_0 \int_{\frac{1}{1+z}}^{1} \frac{dR}{R(dR/dt)} = H_0 \int_0^z \frac{dz'}{H(z')}. \qquad (29.167)$$

Using Eq. (29.122), we obtain

$$I(z) \equiv \int_0^z \frac{dz'}{\sqrt{\Omega_{m,0}(1+z')^3 + \Omega_{\text{rel},0}(1+z')^4 + \Omega_{\Lambda,0} + (1 - \Omega_0)(1+z')^2}}. \qquad (29.168)$$

With this definition of the integral $I(z)$, the present proper distance is

$$d_{p,0}(z) = \frac{c}{H_0} I(z). \qquad (29.169)$$

Comparing this with Eqs. (29.148–29.150) and using Eq. (29.121) for $k$, we find our expressions for the comoving coordinate $\varpi(z)$:

$$\varpi(z) = \frac{c}{H_0} I(z) \qquad (\Omega_0 = 1) \qquad (29.170)$$

$$= \frac{c}{H_0\sqrt{\Omega_0 - 1}} \sin\left[ I(z)\sqrt{\Omega_0 - 1} \right] \qquad (\Omega_0 > 1) \qquad (29.171)$$

$$= \frac{c}{H_0\sqrt{1 - \Omega_0}} \sinh\left[ I(z)\sqrt{1 - \Omega_0} \right] \qquad (\Omega_0 < 1). \qquad (29.172)$$

---

[40]Recall that gravitationally bound systems do not participate in the expansion of the universe, so our Solar System and Galaxy will not become causally fragmented.

These exact expressions must be evaluated numerically. For later reference, we define

$$S(z) \equiv I(z) \qquad (\Omega_0 = 1) \tag{29.173}$$

$$\equiv \frac{1}{\sqrt{\Omega_0 - 1}} \sin\left[I(z)\sqrt{\Omega_0 - 1}\right] \qquad (\Omega_0 > 1) \tag{29.174}$$

$$\equiv \frac{1}{\sqrt{1 - \Omega_0}} \sinh\left[I(z)\sqrt{1 - \Omega_0}\right] \qquad (\Omega_0 < 1), \tag{29.175}$$

so we may simply write

$$\varpi(z) = \frac{c}{H_0} S(z). \tag{29.176}$$

Note that because $\sin(x) = x - x^3/3! + x^5/5! + \cdots$ and $\sinh(x) = x + x^3/3! + x^5/5! + \cdots$ we have the useful approximation (to second order in $z$) that $S(z) \simeq I(z)$. Thus

$$\varpi(z) \simeq \frac{c}{H_0} I(z). \qquad \text{(for } z \ll 1) \tag{29.177}$$

Because the comoving coordinate $\varpi$ is so important in observational cosmology, it will be helpful to find an approximate expression for the integral $I(z)$. (Again we will ignore the brief radiation era, so $\Omega_{\mathrm{rel},0} = 0$ and $\Omega_0 = \Omega_{m,0} + \Omega_{\Lambda,0}$.) The integrand may be expressed as a Taylor series about $z = 0$ as

$$I(z) = \int_0^z \left\{1 - (1 + q_0)z' + \left[\frac{1}{2} + 2q_0 + \frac{3}{2}q_0^2 + \frac{1}{2}(1 - \Omega_0)\right]z'^2 + \cdots\right\} dz' \tag{29.178}$$

where we have used (Eq. 29.124) for the deceleration parameter, $q_0 = \frac{1}{2}\Omega_{m,0} - \Omega_{\Lambda,0}$. Integrating gives our result,

$$I(z) = z - \frac{1}{2}(1 + q_0)z^2 + \left[\frac{1}{6} + \frac{2}{3}q_0 + \frac{1}{2}q_0^2 + \frac{1}{6}(1 - \Omega_0)\right]z^3 + \cdots . \tag{29.179}$$

Eqs. (29.170–29.172) then provide series expressions for the comoving coordinate $\varpi$ of an object observed at a redshift $z$.

Note that the squared term in Eq. (29.179) involves only $q_0$ and thus depends only on the dynamics of the expanding universe, while the cubed term involves both $q_0$ and $k$ (through Eq. 29.121) and so depends on both the dynamics and the geometry of the cosmos. A further simplification comes from using only the first two terms of the series expression for $I(z)$ along with Eq (29.177) to obtain, to second order in $z$,

$$\varpi \simeq \frac{cz}{H_0}\left[1 - \frac{1}{2}(1 + q_0)z\right] \qquad \text{(for } z \ll 1). \tag{29.180}$$

Equation (29.180) is valid regardless of whether or not the universe is flat and whether or not the cosmological constant, $\Lambda$, is nonzero. In fact, Eq. (29.180) can be derived very generally, without reference to the Friedmann equation or any specific model of the universe. The procedure, which is outlined in Problem 29.52, uses the fact that the deceleration parameter is defined as a second time derivative of the scale factor; recall Eq. (29.54).

## The Proper Distance

We have arrived at our goal, an approximate expression for the proper distance of an object *at the present time*. According to Eq. (29.169), this is

$$d_{p,0} \simeq \frac{cz}{H_0} \left[ 1 - \frac{1}{2}(1 + q_0)z \right] \qquad \text{(for } z \ll 1\text{)}. \qquad (29.181)$$

The first term is just the Hubble law, after applying Eq. (4.39) for the redshift. Since $q_0 = \Omega_{m,0}/2 - \Omega_{\Lambda,0}$, we see from the second term that larger values of $\Omega_{m,0}$ imply smaller distances (more mass to slow down the expansion of the universe), as do smaller values of $\Omega_{\Lambda,0}$ (less dark energy pressure to speed up the expansion). As we shall see, the second term involves a departure from the linearity of the Hubble law that can be used to determine the deceleration parameter, $q_0$. For $q_0 = -0.6$, the second term is 10% of the first when $z = 0.13$. This is the reason for the statement on page 1056 that Eq. (27.8) should be abandoned for values of $z$ greater than 0.13.

## The Luminosity Distance

Now we are ready for the concept of the luminosity distance (as measured by the inverse square law, Eq. 3.2). Then we will be prepared to describe some of the classic observational tests of cosmological models. First, we associate the source of the emitted photons with the rate at which energy arrives at a telescope's detectors. Suppose that a radiant flux $F$ is measured for a source of light with a known luminosity $L$. (For now, we will assume that $F$ is a bolometric flux, measured over all wavelengths.) Then the inverse square law, Eq. (3.2), can be used to define the **luminosity distance**, $d_L$, of the star by

$$\boxed{d_L^2 \equiv \frac{L}{4\pi F}.} \qquad (29.182)$$

Let a source of light be located at the origin ($\varpi = 0$) of a comoving coordinate system. The source emits photons that arrive at a spherical surface around the origin for which $\varpi = \text{constant} > 0$. From the Robertson–Walker metric, Eq. (29.106), the surface area of the sphere at the present time ($R = 1$) is $4\pi\varpi^2$; see Fig. 29.25.[41] After traveling out to $\varpi$ from the source, the photons will be spread over this surface area, and so the radiant flux will diminish as $1/\varpi^2$. Two effects, in addition to the inverse square law, act to further reduce the value of the radiant flux measured at this sphere. The cosmological redshift, Eq. (29.142), shows that the energy of each photon, $E_{\text{photon}} = hc/\lambda$, is reduced by a factor of $1 + z$. Also, cosmological time dilation, Eq. (29.143), affects the average time interval *between* photons emitted by the source. This means that the rate at which the photons arrive at the sphere is less than the rate at which they leave the source by another factor of $1 + z$. Combining these effects, the radiant flux at the sphere's surface is

$$F = \frac{L}{4\pi\varpi^2(1+z)^2}.$$

[41]We emphasize that although the surface area of a sphere centered at the origin is $4\pi[R(t)\varpi]^2$, the proper distance of the surface from the origin is *not* $R(t)\varpi$ when $k \neq 0$; see Eqs. (29.149) and (29.150).

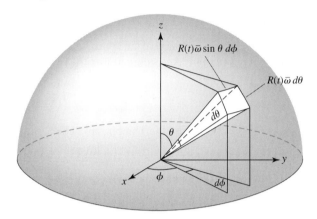

**FIGURE 29.25**   An element of area on the surface of a sphere centered at $\varpi = 0$. Integrating over the angles $\theta$ and $\phi$ shows that the surface area of the sphere is $4\pi[R(t)\varpi]^2$.

Substituting this into Eq. (29.182), we find that

$$d_L = \varpi(1 + z), \tag{29.183}$$

where $\varpi$ must be evaluated numerically using Eq. (29.168) and Eqs. (29.170–29.172). The luminosity distance, $d_L$, is the distance actually measured by the distance modulus $m - M$. Although the luminosity distance is *not* the same as either the present proper distance (Eqs. 29.148–29.150) or the coordinate distance (Eq. 29.3), the three distances do agree for $z \ll 1$.

Equation (29.176) shows that the luminosity distance is exactly given by

$$d_L(z) = \frac{c}{H_0}(1 + z)S(z). \tag{29.184}$$

Using our approximation to second order in $z$, Eq. (29.180),

$$d_L(z) \simeq \frac{cz}{H_0}\left[1 + \frac{1}{2}(1 - q_0)z\right] \qquad \text{(for } z \ll 1\text{)}. \tag{29.185}$$

Comparing this with Eq. (29.181), we see that the luminosity distance is approximately equal to the proper distance only when $z$ is very small and the first term in each expansion dominates. For larger values of $z$, $d_p(z) < d_L(z)$. Ultimately, the exact expressions for both $d_p(z)$ and $d_L(z)$ are the most useful: Eqs. (29.169) and (29.184), respectively, evaluated by numerical integration.

### The Redshift–Magnitude Relation

We are finally in a position to describe some of the most exciting observational tests of cosmology. The **redshift–magnitude relation** comes from using the luminosity distance with Eq. (3.6) for the distance modulus,

$$m - M = 5\log_{10}(d_L/10 \text{ pc}) \tag{29.186}$$

Equation (29.184), along with $H_0 = 100h$ km s$^{-1}$ Mpc$^{-1}$ (Eq. 29.13), quickly produces

$$m - M = 5\log_{10}\left[\frac{c}{(100 \text{ km s}^{-1} \text{ Mpc}^{-1})(10 \text{ pc})}\right] - 5\log_{10}(h)$$

$$+ 5\log_{10}(1+z) + 5\log_{10}[S(z)]$$

$$= 42.38 - 5\log_{10}(h) + 5\log_{10}(1+z) + 5\log_{10}[S(z)]. \qquad (29.187)$$

In the same manner, using the approximate Eq. (29.185) for the luminosity distance with Eq. (29.186) leads to, for $z \ll 1$,

$$m - M \simeq 5\log_{10}\left[\frac{c}{(100 \text{ km s}^{-1} \text{ Mpc}^{-1})(10 \text{ pc})}\right] - 5\log_{10}(h)$$

$$+ 5\log_{10}(z) + 5\log_{10}\left[1 + \frac{1}{2}(1-q_0)z\right] \qquad \text{(for } z \ll 1\text{)}.$$

Expanding the last term on the right in a Taylor series about $z = 0$ and keeping only the first-order terms in $z$ results in

$$m - M \simeq 42.38 - 5\log_{10}(h) + 5\log_{10}(z) + 1.086(1-q_0)z \qquad \text{(for } z \ll 1\text{)}. \qquad (29.188)$$

Figure 29.26 shows the redshift $z$ plotted on a logarithmic scale as a function of $m - M$. For $z \ll 1$, the redshift–magnitude relation is linear. Observations confirm the linearity of the $\log_{10}(z)$ term (which is just the Hubble law) for small $z$. Then, at larger $z$, the fourth term on the right-hand side of Eq. (29.188), $1.086(1-q_0)z$, will cause the line to curve upward. Accurately measuring this departure from a straight line allows the value of the deceleration

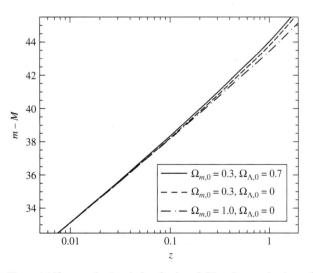

**FIGURE 29.26**   The redshift–magnitude relation for $h = 0.71$ and several values of $\Omega_{m,0}$ and $\Omega_{\Lambda,0}$.

parameter to be determined. At still larger $z$, the curve is sensitive to the individual values of $\Omega_{m,0}$ and $\Omega_{\Lambda,0}$.

The cosmological redshift affects the measurement of an object's spectrum because these observations are usually made within a specific wavelength region. For example, observations made in the $V$-band at 550 nm can be affected as the cosmological redshift brings shorter-wavelength radiation into the $V$ band. This effect can be corrected for by adding a compensating term called the $K$-correction to Eq. (29.188) if the spectrum, $I_\lambda$, of the object is known.[42]

In the mid-1990's two rival teams of astronomers—the Supernova Cosmology Project (SCP, led by Saul Perlmutter) and the High-Z Supernova Search Team (HZSNS Team, headed by Brian Schmidt)—made observations of Type Ia supernovae at cosmological distances. Both teams were astonished to discover that supernovae observed with a redshift of $z \approx 0.5$ were about 0.25 magnitude *dimmer* than expected for a universe with $\Omega_{m,0} \simeq 0.3$ and $\Lambda = 0$. The supernovae were farther away than they would be in this canonical decelerating universe. The possibility of an accelerating universe and a nonzero cosmological constant immediately leapt to their minds, but it took nearly a year of intense work to eliminate several plausible alternative explanations. For example, about 20% of the light from a distant supernova could be absorbed at high $z$ by a hypothetical "gray dust."[43] Or perhaps evolutionary effects were misleading the astronomers, since at high $z$ we observe a younger generation of supernovae, formed in a younger galactic environment where heavy elements were less abundant. One by one, these possibilities were considered and eliminated. The redshift–magnitude diagram in Fig. 29.27 shows a more recent compilation of the results of these two teams. Both groups found that their analyses ruled out a flat universe with $\Omega_{m,0} = 1$ and $\Lambda = 0$ (the scenario championed by most theorists at the time) and were also incompatible with an open universe having $\Omega_{m,0} \simeq 0.3$ and $\Lambda = 0$. Instead, their findings favored a universe with $\Omega_{m,0} \simeq 0.3$ and $\Omega_{\Lambda,0} = 0.7$. Figure 29.28 shows, for each team, the location on the $\Omega_{m,0}$–$\Omega_{\Lambda,0}$ plane of the most likely set of values that are consistent with the high-$z$ supernovae results. Their evidence for a nonzero cosmological constant is persuasive.

If we look at supernovae beyond $z_{accel} = 0.76$, when the universe started accelerating, we should find the signature of a decelerating universe. Figure 29.29 shows further results of observations of high-$z$ supernovae, including six with $z > 1.25$. It includes SN 1997ff, the most distant supernova yet observed at $z = 1.7$. SN 1997ff and others in this high-$z$ sample appear *brighter* than they would if the universe had expanded at a constant rate (i.e., with $\Omega = 0$), as expected for the deceleration phase of the early universe. These observations rule out the alternative explanations of acceleration, "gray dust" and evolutionary effects, as indicated in the figure. Clearly, astronomers will have to grapple with the implications of an accelerating universe whose dynamics are dominated by dark energy.

It is perhaps surprising that the values of the Hubble constant obtained from the redshift–magnitude diagrams ($H_0 \sim 70 \pm 10$ km s$^{-1}$ Mpc$^{-1}$) are not better determined. The spread in the values of $H_0$ obtained by various groups using supernovae is due to their different calibrations of Cepheid distances. This systematic uncertainty does not affect the values

---

[42]The $K$-correction was previously discussed in Section 25.2.

[43]A "gray dust" of large grains would absorb light nearly equally at all wavelengths and so would produce no detectable reddening of a supernova's spectrum.

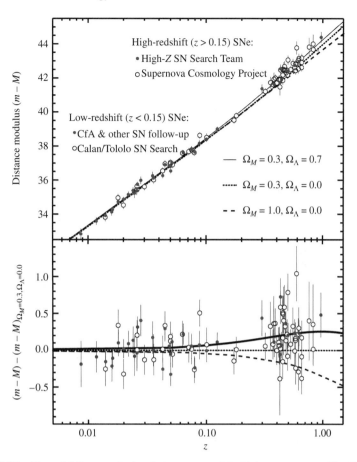

**FIGURE 29.27** The redshift–magnitude relation measured for high-$z$ supernovae. The $K$-correction has been applied to the apparent magnitudes. The lower graph shows the data after subtracting the theoretical curve for $\Omega_{m,0} = 0.3$, $\Omega_{\lambda,0} = 0$. [Figure adapted from Perlmutter and Schmidt, *Supernovae and Gamma-Ray Bursters*, K. Weiler (ed.), *Lecture Notes in Physics*, *598*, 195, 2003. Data from Perlmutter et al, *Ap. J.*, *517*, 565, 1999 (SCP) and Riess et al, *A. J.*, *116*, 1009, 1998 (HZSNS).]

of $\Omega_{m,0}$ and $\Omega_{\Lambda,0}$ because these are determined by the *departures from linearity* in the redshift–magnitude diagram.

### Angular Diameter Distance

Another measure of an object's distance may be found by comparing its linear diameter $D$ (assumed known) with its observed angular diameter $\theta$ (assumed small). The **angular diameter distance**, $d_A$ is then defined to be

$$d_A \equiv \frac{D}{\theta}. \tag{29.189}$$

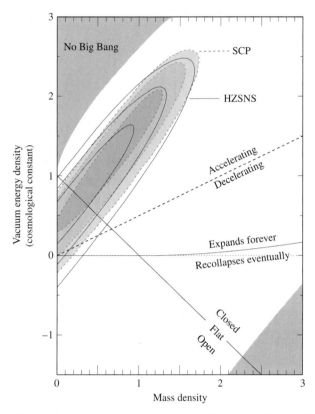

**FIGURE 29.28**    The location of the most probable values of $\Omega_{m,0}$ and $\Omega_{\Lambda,0}$ for high-$z$ supernovae. The results of the SCP and the HZSNS Team are superimposed. [Figure adapted from Perlmutter and Schmidt, *Supernovae and Gamma-Ray Bursters*, K. Weiler (ed.), *Lecture Notes in Physics*, *598*, 195, 2003. Data from Perlmutter et al, *Ap. J.*, *517*, 565, 1999 (SCP) and Riess et al, *A. J.*, *116*, 1009, 1998 (HZSNS).]

To place this into context with our previous results, consider a galaxy of redshift $z$ located at comoving coordinate $\varpi$. We can use Eq. (17.18), $d\mathcal{L} = \sqrt{-(ds)^2}$, to find an expression for $D$, the proper distance from one side of the galaxy to the other. Integrating the Robertson–Walker metric (Eq. 29.106) across the galaxy in the plane of the sky with $dt = d\varpi = d\phi = 0$, we obtain

$$D = R(t_e)\varpi\,\theta = \frac{\varpi\,\theta}{1+z}.$$

[Note from Eq. (29.146) that for a flat universe this is the usual Euclidean geometric relation.] Of course, $D$ is the galaxy's diameter at the time $t_e$, when the light we observe was emitted. Since the light from the galaxy traveled a radial path to Earth, $\theta$ is the angular size of the galaxy as measured by astronomers. Equation (29.176) can be used to express the diameter

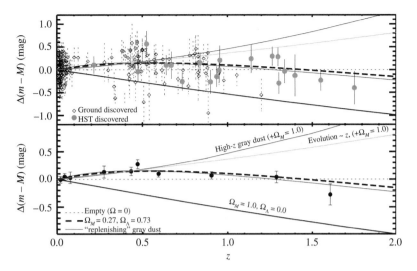

**FIGURE 29.29** The redshift–magnitude relation measured for very high-$z$ supernovae. The $K$-correction has been applied to the apparent magnitudes, and the theoretical curve for $\Omega = 0$ (a "coasting universe") has been subtracted from the data. The lower graph illustrates the averages of binned data (grouped according to redshift) and compares them to curves of alternative models incorporating "gray dust" or evolutionary effects for supernovae. (Figure adapted from Riess et al, *Ap. J., 607*, 665, 2004.)

as

$$D = \frac{c}{H_0} \frac{S(z)\theta}{1 + z}. \tag{29.190}$$

Thus the angular diameter distance is

$$d_A = \frac{c}{H_0} \frac{S(z)}{1 + z}. \tag{29.191}$$

From Eq. (29.183), we find that the angular diameter distance and the luminosity distance are related by

$$d_A = \frac{d_L}{(1 + z)^2}. \tag{29.192}$$

Figure 29.30 shows a graph of $\theta$ in units of $H_0 D/c$,

$$\frac{c\theta}{H_0 D} = \frac{(1 + z)}{S(z)}, \tag{29.193}$$

as a function of the redshift $z$ for several model universes. It is surprising that the angular diameter of a galaxy does *not* continue to decrease with distance. In fact, beyond a certain redshift, the angular size actually *increases* with distance. This is due to the universe acting

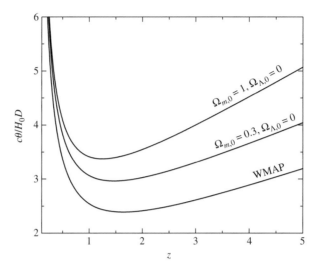

**FIGURE 29.30**    The angular diameter $\theta$ of a galaxy in units of $H_0 D/c$ for several values of $\Omega_{m,0}$ and $\Omega_{\Lambda,0}$.

as a sort of gravitational lens, enlarging the appearance of a galaxy beyond what would be expected in a static Euclidean universe. In principle, observations of galaxies of a known linear diameter $D$ would allow observers to determine which values of cosmological parameters are preferred. In practice, however, galaxies do not have sharp boundaries, and they evolve as the universe ages. As of this writing, the most productive use of the angular diameter distance has been coupled with observations of the Sunyaev–Zel'dovich effect.

The Sunyaev–Zel'dovich effect provides an independent determination of the Hubble constant. Measurements of $\Delta T/T_0$ (Eq. 29.64) along with the X-ray flux $F_X$ and temperature $T_e$ of the intracluster gas in rich clusters of galaxies can be used to model the physical properties of the cluster. Comparing the calculated diameter $D$ of the cluster with its measured angular diameter $\theta$ yields $d_A$, the cluster's angular diameter distance. On the other hand, the measured X-ray flux from the cluster and Eq. (27.20) for the X-ray luminosity of the intracluster gas determine the cluster's luminosity distance. The connection between these two distances, Eq. (29.192), can then be used to calculate the Hubble constant. In Problem 29.55 you will derive

$$H_0 = Cf(z)\frac{F_X T_e^{3/2}}{\theta(\Delta T/T_0)^2},$$

where $f(z)$ is a function of the redshift $z$ of the cluster and $C$ is a constant. One team of astronomers measured[44] the value of the Hubble constant for five clusters and obtained an average of $H_0 = 65$ km s$^{-1}$ Mpc$^{-1}$, consistent with the values obtained from other recent measurements.

[44] See Jones et al. (2005) for details of how $H_0$ was determined.

## SUGGESTED READING

### General

Alpher, R. A., and Herman, R. C., "Evolution of the Universe," *Nature, 162*, 774, 1948.

Goldsmith, Donald, *The Runaway Universe*, Perseus Publishing, Cambridge, MA, 2000.

Harrison, Edward, *Cosmology*, Second Edition, Cambridge University Press, Cambridge, 2000.

Harrison, Edward, *Darkness at Night*, Harvard University Press, Cambridge, MA, 1987.

Kirshner, Robert, *The Extravagant Universe*, Princeton University Press, Princeton, NJ, 2002.

Krauss, Lawrence M., "Cosmological Antigravity," *Scientific American*, January 1999.

Lineweaver, Charles, and Davis, Tamara, "Misconceptions about the Big Bang," *Scientific American*, March 2005.

Silk, Joseph, *The Big Bang*, Third Edition, W. H. Freeman and Company, New York, 2001.

Silk, Joseph, *On the Shores of the Unknown*, Cambridge University Press, Cambridge, 2005.

Weinberg, Steven, *The First Three Minutes*, Second Edition, Basic Books, New York, 1988.

### Technical

Bennett, C. L., "First-Year *Wilkinson Microwave Anisotropy Probe* (*WMAP*) Observations: Preliminary Maps and Basic Results," *The Astrophysical Journal Supplement, 148*, 1, 2003.

Binney, James, and Tremaine, Scott, *Galactic Dynamics*, Princeton University Press, Princeton, NJ, 1987.

Carlstrom, John E., Holder, Gilbert P., and Reese, Erik D., "Cosmology with the Sunyaev–Zel'dovich Effect," *Annual Review of Astronomy and Astrophysics, 40*, 643, 2002.

Davis, Tamara M., Lineweaver, Charles H., and Webb, John K., "Solutions to the Tethered Galaxy Problem in an Expanding Universe and the Observation of Receding Blueshifted Objects," *American Journal of Physics*, 71, 358, 2003.

Dicke, R. H., Peebles, P. J. E., Roll, P. G., and Wilkinson, D. T., "Cosmic Black-body Radiation," *The Astrophysical Journal, 142*, 414, 1965.

Felten, James E., and Isaacman, Richard, "Scale factors $R(t)$ and Critical Values of the Cosmological Constant Lambda in Friedmann Universes," *Reviews of Modern Physics, 58*, 689, 1986.

Foley, Ryan J., et al., "A Definitive Measurement of Time Dilation in the Spectral Evolution of the Moderate-Redshift Type Ia Supernova 1997ex," *The Astrophysical Journal Letters, 626*, L11, 2005.

Garnavich, Peter M., et al., "Supernova Limits on the Cosmic Equation of State," *The Astrophysical Journal, 509*, 74, 1998.

Gott, J. Richard III, et al., "A Map of the Universe," *The Astrophysical Journal, 624*, 463, 2005.

Harrison, Edward R., "Why Is the Sky Dark at Night?" *Physics Today*, February 1974.

Jones, Michael E., et al., "$H_0$ from an Orientation-Unbiased Sample of SZ and X-ray Clusters," *Monthly Notices of the Royal Astronomical Society*, *357*, 518, 2005.

Kolb, Edward W., and Turner, Michael S., *The Early Universe*, Westview Press, 1994.

Loeb, Abraham, "Long-Term Future of Extragalactic Astronomy," *Physical Review D*, *65*, 047301, 2002.

Peacock, John A., *Cosmological Physics*, Cambridge University Press, Cambridge, 1999.

Peebles, P. J. E., *Principles of Physical Cosmology*, Princeton University Press, Princeton, NJ, 1993.

Penzias, A. A., and Wilson, R. W., "A Measurement of Excess Antenna Temperature at 4080 Mc/s," *The Astrophysical Journal*, *142*, 419, 1965.

Perlmutter, S., et al., "Measurements of $\Omega$ and $\Lambda$ from 42 High-Redshift Supernovae," *The Astrophysical Journal*, *517*, 565, 1999.

Perlmutter, S., and Schmidt, B. P., "Measuring Cosmology with Supernovae," *Supernovae and Gamma-Ray Bursters*, K. Weiler (ed.), *Lecture Notes in Physics*, *598*, 195, 2003.

Raine, D. J., and Thomas, E. G., *An Introduction to the Science of Cosmology*, Institute of Physics Publishing, Philadelphia, 2001.

Riess, Adam G., et al., "Type Ia Supernova Discoveries at $z > 1$ from the *Hubble Space Telescope*: Evidence for Past Deceleration and Constraints on Dark Energy Evolution," *The Astrophysical Journal*, *607*, 665, 2004.

Riess, Adam G., et al., "Observational Evidence from Supernovae for an Accelerating Universe and a Cosmological Constant," *The Astronomical Journal*, *116*, 1009, 1998.

Ryden, Barbara, *Introduction to Cosmology*, Addison-Wesley, New York, 2003.

Spinrad, Hyron, Dey, Arjun, and Graham, James R., "Keck Observations of the Most Distant Galaxy: 8C 1435+63 at $z = 4.25$," *The Astrophysical Journal Letters*, *438*, L51, 1995.

## PROBLEMS

**29.1** It might be argued that the inverse square law for light, Eq. (3.2), would provide a solution to Olbers's paradox. To see that this is not so, consider a uniform distribution of stars with $n$ stars per unit volume, each of luminosity $L$. Imagine that two thin, spherical shells of stars with radii $r_1$ and $r_2$ are centered on Earth; let the thickness of each shell be $\Delta r$. Show that the same energy flux reaches Earth from each shell.

**29.2** Suppose that all of the matter in the universe were converted into energy in the form of blackbody radiation. Take the average density of matter to be the WMAP value for the density of baryonic matter, $\rho_{b,0}$. Use Eq. (9.7) for the energy density of blackbody radiation to find the temperature of the universe in this situation. At what wavelength would the blackbody spectrum peak? In what region of the electromagnetic spectrum is this wavelength found? Explain how your result may be applied to Olbers's paradox.

**29.3** Show by substitution that Eqs. (29.32) and (29.34) are solutions to Eq. (29.11) for a closed universe ($k > 0$).

**29.4** Show by substitution that Eqs. (29.36) and (29.38) are solutions to Eq. (29.11) for an open universe ($k < 0$).

**29.5** Derive Eqs. (29.33) and (29.35) from Eqs. (29.32) and (29.34), respectively.

**29.6** Derive Eqs. (29.37) and (29.39) from Eqs. (29.36) and (29.38), respectively.

**29.7** (a) Use Eq. (29.11) to find an expression for the maximum scale factor $R$ in a closed universe. Does your answer agree with Eq. (29.33)?

 (b) Find the lifetime of a closed universe (expressed as a multiple of the Hubble time, $t_H$) as a function of the density parameter, $\Omega_0$.

**29.8** Derive Eqs. (29.40–29.42) for the age of the universe using Eqs. (29.31), (29.33), (29.35), (29.37), (29.39), and (29.4).

**29.9** Consider a one-component universe of pressureless dust.

 (a) Show that

$$\Omega(t) = \frac{\rho(t)}{\rho_c(t)} = 1 + \frac{kc^2}{(dR/dt)^2}, \qquad (29.194)$$

which describes how $\Omega$ varies with time. What does this have to say about the nature of the early universe?

 (b) Show that $dR/dt \to \infty$ as $t \to 0$. What does this say about the difference between a closed, a flat, and an open universe at very early times?

**29.10** For a one-component universe of pressureless dust, show that

$$\frac{1}{\Omega} - 1 = \left(\frac{1}{\Omega_0} - 1\right)(1 + z)^{-1}. \qquad (29.195)$$

What happens as $z$ increases?

**29.11** Show that in the limit $(1 + z) \gg 1/\Omega_0$, Eq. (29.41) reduces to Eq. (29.43). *Hint:* First write Eq. (29.41) in terms of a variable $u \equiv 1/[\Omega_0(1 + z)]$, and then expand the equation in a Taylor series about $u = 0$. You may find

$$\cos^{-1}(1 - x) = \sqrt{2}x^{1/2} + \frac{\sqrt{2}}{12}x^{3/2} + \cdots$$

for $x \ll 1$ to be useful.

**29.12** Derive the acceleration equation, Eq. (29.51).

**29.13** Assuming that $P = 0$, show that Eq. (29.51) for the acceleration of a mass shell can be found from Newton's second law by considering the gravitational force on an expanding shell.

**29.14** Consider a model of the universe consisting of neutral hydrogen atoms for which the average (rms) speed of the atoms is 600 km s$^{-1}$ (approximately the speed of the Local Group relative to the Hubble Flow). Show that $\rho \gg P/c^2$ for the gas. For an adiabatically expanding universe, for what value of $R$ and $z$ will $\rho = P/c^2$?

**29.15** By inserting the equation of state $P = w\rho c^2$ into the fluid equation, Eq. (29.50), show that $R^{3(1+w)}\rho = \text{constant} = \rho_0$, where $\rho_0$ is the present value of $\rho$.

**29.16** Show that for a pressureless dust universe, $q(t) = \frac{1}{2}\Omega(t)$, which is Eq. (29.55).

**29.17** The deuterium ($_1^2$H) nucleus is not very tightly bound.

(a) Calculate the binding energy of the deuterium nucleus, using values of $m_H = 1.007825$ u, $m_n = 1.008665$ u, and $m_D = 2.014102$ u.

(b) What is the wavelength of a photon with this energy?

(c) From Wien's law, at what temperature is this the characteristic energy of a blackbody photon?

**29.18** The carbon absorption lines that are formed when the light from a distant quasar, Q1331+70, passes through an intergalactic cloud have been measured by Antoinette Songaila and her colleagues. The relative strengths of the lines indicate that the temperature of the cloud is $7.4 \pm 0.8$ K, and the lines show a redshift of $z = 1.776$. How does the temperature of the cloud compare with the temperature of the CMB at that redshift? (If there are sources of heating for the cloud in addition to the CMB, then its temperature must be considered as an *upper limit* to the temperature of the CMB.)

**29.19** In 1941, microwave observations detected absorption lines due to cyanogen molecules (CN) in molecular clouds. A cyanogen molecule has three first excited rotational states, each of which is degenerate and has an energy that is $4.8 \times 10^{-4}$ eV above the ground state. An analysis of the absorption lines shows that for every 100 molecules in the ground state, there are 27 others that are in one of the three first excited states. Assuming that the molecular clouds are in thermal equilibrium with the CMB, use the Boltzmann equation (Eq. 8.6) to estimate the temperature of the CMB.

**29.20** Channel 6 on your television consists of radio waves with wavelengths between 3.41 m and 3.66 m. Consider a 25,000-watt television station located 70 km from your home. Use Eq. (9.5) for the energy density of blackbody radiation to estimate the ratio of the number of channel 6 photons to the number of CMB photons that your television antenna picks up in this wavelength band. (*Hint:* For the television broadcast, recall that the energy density of an electromagnetic wave is related to the time-averaged Poynting vector by $u = \langle S \rangle / c$.)

**29.21** Use Eq. (4.32) for the relativistic Doppler shift to derive Eq. (29.61). Show that Eq. (29.61) reduces to Eq. (29.62) when $v \ll c$.

**29.22** Calculate the magnitude of the variation in the temperature of the CMB due to the Sun's peculiar velocity.

**29.23** In this problem you will approximate the physics that produces the Sunyaev–Zel'dovich effect.

(a) First, estimate the shift in the frequency of a low-energy CMB photon as it is scattered by a high-energy electron (inverse Compton scattering) of speed $v_e$ in the hot intracluster gas of a rich cluster of galaxies. Although photons can be scattered from any direction into any direction, we will consider the four equally likely situations shown in Fig. 29.31. Show that the *average* shift in the frequency of these four photons is

$$\frac{\Delta \nu}{\nu} = \frac{v_e^2}{c^2} = 3 \frac{kT_e}{m_e c^2},$$

where $T_e$ is the temperature of the electron gas. Evaluate this expression using $T_e = 10^8$ K, and explain why you could assume that in the rest frame of the electron, the change in the wavelength ($\sim \lambda_C$) of the photon could be neglected.

(b) About what fraction of the CMB photons will be scattered as they pass through the intracluster gas? Assume an electron number density of $n_e = 10^4$ m$^{-3}$ and a cluster radius of 3 Mpc.

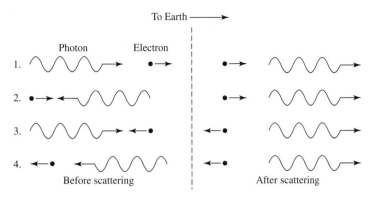

**FIGURE 29.31**  Inverse Compton scattering of a CMB photon by a high-energy electron.

(c) Use the increase $\Delta\nu$ of the peak frequency with Wien's law (Eq. 29.59) to obtain an approximate expression for the effective decrease in the temperature of the CMB, $\Delta T/T_0$ (the Sunyaev–Zel'dovich effect).

**29.24**  Show that in the general equation of state $P = wu$ (Eq. 29.52), $w = 1/3$ for relativistic particles ($E \gg mc^2$). *Hint:* The pressure integral, Eq. (10.8), may prove useful.

**29.25**  Consider a comoving sphere whose surface expands with the universe. Let it be centered at the origin and filled with CMB photons. Show that Eq. (29.81), $R^4\rho_{\text{rel}} = \rho_{\text{rel},0}$, is consistent with the conservation of energy within the sphere.

**29.26**  Some quantities obey an exponential time-behavior of the form $f(t) = f_0 e^{t/\tau}$, where $\tau$ is the *characteristic time* for the system under consideration.

(a) Show that

$$\tau = \left(\frac{1}{f}\frac{df}{dt}\right)^{-1}.$$

This expression can be used to define a characteristic time for any function, regardless of whether its behavior is exponential. (See the analogous discussion of the pressure scale height, $H_P$, in Section 10.4.)

(b) Use the scale factor, $R(t)$, to show that the characteristic time for the expansion of the universe is $\tau_{\text{exp}}(t) = 1/H(t)$.

(c) Assuming a flat universe containing only matter and radiation, find an expression (valid in both the radiation era and the matter era) for the characteristic expansion time $\tau_{\text{exp}}$ as a function of the scale factor $R$.

**29.27**  (a) Show that deep in the radiation era when $R \ll R_{r,m}$, Eq. (29.84) is well approximated by Eq. (29.86).

(b) Solve the Friedmann equation for a flat, one-component universe that contains only relativistic particles, and compare your result with Eq. (29.86).

**29.28**  Use a procedure similar to that used to obtain Eq. (29.28) to show that a one-component universe of relativistic particles is flat in the limit $z \to \infty$.

**29.29** Assuming that the present density of baryonic matter is given by Eq. (29.17), what was the density of matter at the time of Big Bang nucleosynthesis, when $T \sim 10^{10}$ K?

**29.30** One factor that contributed to the cessation of the reactions that formed neutrons at roughly $10^{10}$ K [Eqs. (29.93–29.95)] was the annihilation of electron–positron pairs that occurred at that time. When the temperature became too low, the electron–positron pairs could not be replaced by pair production. (This removed the supply of electrons that could combine with protons to form neutrons.) By setting the characteristic thermal energy of a photon, $kT$, equal to the rest energy of an electron–positron pair, estimate the temperature below which an annihilated pair will not readily be replaced.

**29.31** In this problem, you will show that when the temperature of the universe was about $10^9$ K, all of the neutrons would have combined with protons to form helium nuclei.

(a) Using arguments similar to those leading up to Eq. (9.12), show that the number of collisions between a neutron and a proton that occur within a time $\Delta t$ is $n_p \sigma v \Delta t$, where $n_p$ is the number density of protons, $\sigma$ is the neutron's collision cross section, and $v$ is the speed of the neutron.

(b) Evaluate $n_p \sigma v \Delta t$. If the result is $\gg 1$, then each neutron had ample opportunity to combine with a proton. Let $\Delta t$ be the characteristic timescale of the universe at the time of helium formation, and use $\sigma = \pi (2r)^2$, where $r \simeq 10^{-15}$ m is the radius of a neutron. The number density of protons can be estimated from the baryonic mass density when $T = 10^9$ K.

**29.32** (a) Use the cross section for electron scattering (Eq. 9.20) to find an expression for the average time between scatterings of a photon by free electrons.

(b) Assuming that the electrons remain free, at what value of $R$ and $z$ will the average time between scatterings equal the characteristic expansion time $\tau_{\text{exp}}$ of Problem 29.26? (Use WMAP values.) What is the age of the universe when this occurs? This is when a flat universe of matter and radiation would have become transparent due solely to its expansion (no recombination).

**29.33** Solve Eq. (29.101) for a composition of pure hydrogen to find the temperature when half of the electrons and protons have combined to form neutral atoms.

**29.34** Calculate the time of decoupling, $t_{\text{dec}}$, for a universe of matter and radiation using the WMAP values for $z_{\text{dec}} = 1089$ and other quantities. Compare your answer with the WMAP result of $379 \, ^{+8}_{-7}$ kyr.

**29.35** Using WMAP values for a universe of matter and radiation, estimate the time interval $\Delta t$ between when recombination began (say, when 99% of the hydrogen atoms were ionized) and when recombination ended (say, when 1% of the hydrogen atoms were ionized). What is the difference $\Delta z$ between the values of the redshift $z$ at these two times? This is the thickness (in terms of $z$) of the "surface" of last scattering. Compare your answers with the WMAP results of $\Delta t = 118 \, ^{+3}_{-2}$ kyr and $\Delta z = 195 \pm 2$. Assume a composition of pure hydrogen.

**29.36** Suppose that Earth were a perfectly smooth sphere. If you drew a circle of radius $D = 100$ meters on Earth's surface, what discrepancy would you find between the expected and measured values of the circle's circumference?

**29.37** Follow a procedure similar to that used to obtain Eq. (29.28) to derive an expression for the total density parameter, $\Omega(z)$, as a function of $z$. Verify that your expression reduces to

Eq. (29.28) if $\Omega_{rel,0} = \Omega_{\Lambda,0} = 0$. What does your expression say about the geometry of an early, three-component universe?

**29.38** Use the Robertson–Walker metric, Eq. (29.106), to show that the proper area ($dt = 0$) of a spherical surface, centered at the origin and passing through comoving coordinate $\varpi$, is $4\pi[R(t)\varpi]^2$.

**29.39** Einstein originally introduced the cosmological constant $\Lambda$ to stabilize his model of a pressureless dust universe against expansion or contraction.

    **(a)** Find an expression for $\Lambda$ in terms of the density $\rho_m$ of a static model of a pressureless dust universe with a cosmological constant.

    **(b)** Find an expression for the curvature $k$ for this static model. Is this model universe closed, open, or flat?

    **(c)** Explain why Einstein's static model is in an unstable equilibrium, so any departure from equilibrium (expansion or contraction) will tend to increase.

**29.40** Evaluate $\Omega_m$, $\Omega_{rel}$, and $\Omega_\Lambda$ at the time of decoupling ($z = 1089$) using WMAP values.

**29.41** Show that Eq. (29.122) may be written as

$$H = H_0 \left[ \sum_i \Omega_{i,0}(1 + z)^{3(1+w_i)} + (1 - \Omega_0)(1 + z)^2 \right]^{1/2}, \qquad (29.196)$$

where $w$ is the coefficient from the equation of state $P_i = w_i \rho_i c^2$ and the "$i$" subscripts identify one of the components of the universe (i.e., pressureless dust, relativistic particles, or dark energy).

**29.42** Derive Eq. (29.123) for a general expression of the deceleration parameter,

$$q(t) = \frac{1}{2} \sum_i (1 + 3w_i)\Omega_i(t).$$

**29.43** Use the acceleration equation to show that the acceleration of the universe changed sign (from negative to positive) when the scale factor was

$$R_{accel} = \left( \frac{\Omega_{m,0}}{2\Omega_{\Lambda,0}} \right)^{1/3}.$$

Evaluate the value of $R_{accel}$ and $z_{accel}$ at this time with WMAP values.

**29.44** **(a)** Use Eq. (29.129) to find an expression for the lookback time, $t_L$, as a function of the redshift $z$.

    **(b)** Figure 28.17 shows the comoving space density of active galactic nuclei (AGN) as a function of redshift. Using your expression for the lookback time with WMAP values, replot the "ChoMP + CDF + ROSAT" data (marked with filled circles) with $t_L/t_H$ on the horizontal axis (the lookback time as a fraction of the Hubble time). How would you characterize the decline in the space density of AGN with increasing lookback time?

**29.45** The cosmological constant becomes dominant as the scale factor $R$ becomes increasingly larger in the $\Lambda$ era.

    **(a)** Show that the Hubble parameter is a constant in a flat universe deep in the $\Lambda$ era.

(b) Suppose that, starting today ($t = t_0$, when $R = 1$), only the cosmological constant contributes to the Friedmann equation.[45] Solve the Friedmann equation and show that for $\Lambda > 0$, the scale factor will increase exponentially.

(c) Use WMAP values to evaluate the characteristic time for the exponential expansion (cf. Problem 29.26).

**29.46** In the matter era, the distance to the particle horizon for a closed, one-component universe of pressureless dust is given by

$$d_h(z) = \frac{c}{H_0(1+z)\sqrt{\Omega_0 - 1}} \cos^{-1}\left[1 - \frac{2(\Omega_0 - 1)}{\Omega_0(1+z)}\right]. \qquad (29.197)$$

In this problem, you will derive this expression for $d_h$. First change variables in Eq. (29.153) to obtain

$$d_h(t) = R(t) \int_0^{\frac{1}{1+z}} \frac{c\,dR}{R(dR/dt)},$$

where the limits of integration range from $R = 0$ (at $t = 0$) to $R = 1/(1+z)$ (at time $t$). Then show that

$$\left(\frac{dR}{dt}\right)^2 = H_0^2\left(\frac{\Omega_0}{R} - \Omega_0 + 1\right),$$

and make this substitution into the denominator of the integral. You may find

$$\int \frac{dx}{\sqrt{bx - ax^2}} = \frac{1}{\sqrt{a}}\left[\cos^{-1}\left(1 - \frac{2ax}{b}\right) - \frac{\pi}{2}\right]$$

to be useful.

**29.47** Use the results of Problem 29.46 to find the ratio of the distance to the particle horizon to the circumference of a closed, matter-dominated universe. What happens at very early times, as $z \to \infty$? Show that at the time of maximum expansion (just before the closed universe begins to collapse), this ratio is equal to one-half. This means that at the moment of the Big Crunch ending the collapse, the particle horizon encompasses the entire universe.

**29.48** Using Eq. (29.163) for the comoving coordinate, $\varpi$, of a photon just now arriving from the present particle horizon in a flat universe, find the maximum proper distance of the photon during its journey. Express your answer as a fraction of the model's particle horizon, $d_{h,0} = 3ct_0$. At what time ($t/t_0$) is the photon at this distance? Carefully explain the meaning of the phrase "just now arriving from the present particle horizon."

**29.49** Consider the (unrealistic) model of a flat, one-component universe of pressureless dust, as described in Section 29.1.

(a) Show that for this model,

$$\varpi = \frac{2c}{H_0}\left(1 - \frac{1}{\sqrt{1+z}}\right). \qquad (29.198)$$

(b) Find an expression for the proper distance to an object with redshift $z$ for this model.

[45] Such a universe is called a *de Sitter universe*, named for the Dutch mathematician Willem de Sitter (1872–1934).

(c) Find an expression for the horizon distance in this model. Evaluate this using WMAP values, and compare your result with the more accurate value of Eq. (29.159).

**29.50** Derive Mattig's relation for a one-component universe of pressureless dust,

$$\varpi = \frac{2c}{H_0} \frac{1}{1+z} \frac{1}{\Omega_0^2} \left[ \Omega_0 z - (2 - \Omega_0)(\sqrt{1 + \Omega_0 z} - 1) \right], \tag{29.199}$$

or in terms of the deceleration parameter $q_0 = \Omega_0/2$ (valid for this one-component model),

$$\varpi = \frac{c}{H_0} \frac{1}{1+z} \frac{1}{q_0^2} \left[ q_0 z - (1 - q_0)(\sqrt{1 + 2q_0 z} - 1) \right]. \tag{29.200}$$

Show that this is valid for a flat, an open, and a closed one-component universe of pressureless dust. Note that Eq. (29.198) in the previous problem and Eq. (29.199) are in agreement when $\Omega_0 = 1$.

**29.51** Show that Eq. (29.178) is the Taylor-series expansion of Eq. (29.168) about $z = 0$.

**29.52** In this problem, you will carry out a general derivation of Eq. (29.180) without reference to any specific model of the universe.

(a) Expand the scale factor, $R(t)$, in a general Taylor series about the present time, $t_0$, and obtain

$$R(t) = R(t_0) + \frac{dR}{dt}\bigg|_{t_0} (t - t_0) + \frac{1}{2} \frac{d^2 R}{dt^2}\bigg|_{t_0} (t - t_0)^2 + \cdots$$

$$= 1 - H_0(t_0 - t) - \frac{1}{2} H_0^2 q_0 (t_0 - t)^2 + \cdots . \tag{29.201}$$

(b) Show that $1/R(t)$ is then given by

$$\frac{1}{R(t)} = 1 + H_0(t_0 - t) + H_0^2 \left(1 + \frac{1}{2} q_0\right)(t_0 - t)^2 + \cdots .$$

[*Hint:* Use long division to divide 1 by $R(t)$.]

(c) Use Eq. (29.142), $1/R(t) = 1 + z$, with the result of part (b) to write the expansion for $z$ about the present time. Now solve for $t_0 - t$ and express the result as a series in $z$ to get

$$t_0 - t = \frac{z}{H_0} - \left(1 + \frac{1}{2} q_0\right) \frac{z^2}{H_0} + \cdots .$$

(d) Consider a photon that is emitted at comoving coordinate $\varpi$ at time $t$ and received on Earth at the present time $t_0$. Use an equation similar to Eq. (29.138) to find an approximate expression for $\varpi$. You need only use the first two terms in the expansion in part (b) for $1/R(t)$ in the left-hand integral. For the right-hand integral, use a two-term Taylor series for $1/\sqrt{1 - k\varpi^2}$. You should find that

$$\varpi = c(t_0 - t) \left[ 1 + \frac{1}{2} H_0(t_0 - t) \right] + \cdots .$$

(e) By substituting the expression for $t_0 - t$ from part (c) into this equation for $\varpi$, show that, to second order in $z$,

$$\varpi = \frac{cz}{H_0} \left[ 1 - \frac{1}{2}(1 + q_0)z \right] + \cdots .$$

This result is very important because it does not rely on any particular model of the universe. It is valid even if the cosmological constant, $\Lambda$, is not equal to zero.

**29.53** Use Eq. (29.196) from Problem 29.41 to show that the luminosity distance may be written as

$$d_L = \frac{c(1+z)}{H_0\sqrt{|1-\Omega_0|}}\, \text{sinn}\left\{ \sqrt{|1-\Omega_0|} \int_0^z \frac{dz'}{\left[ \sum_i \Omega_{i,0}(1+z')^{3(1+w_i)} + (1-\Omega_0)\,(1+z')^2 \right]^{1/2}} \right\},$$

(29.202)

where

$$\text{sinn}(x) \equiv \sinh(x) \qquad \text{if } \Omega_0 < 1,$$

$$\equiv x \qquad \text{if } \Omega_0 = 1,$$

$$\equiv \sin(x) \qquad \text{if } \Omega_0 > 1.$$

This is Eq. (1) of Garnavich et al. (1998), which explains how this equation may be used with the distance modulus, Eq. (29.186), to place limits on the value of $w_\Lambda$ for the equation of state of dark energy. The WMAP result is that $w_\Lambda < -0.78$.

**29.54** Assume a flat, one-component universe of pressureless dust for this problem.

(a) Show that the angular diameter observed for an extended object of linear diameter $D$ at redshift $z$ is

$$\theta = \frac{H_0 D}{2c} \frac{(1+z)^{3/2}}{\sqrt{1+z}-1}.$$

(b) Find the value of the redshift for which $\theta$ is a minimum. Compare your result with Fig. 29.30.

(c) What is the smallest value of $\theta$ when observing a cluster of galaxies with a diameter of 1 Mpc? Use $h = 0.71$.

**29.55** Model the hot intragalactic gas in a rich cluster of galaxies as a homogeneous sphere of radius $R$ and temperature $T_e$. Let $F_X$ be the X-ray flux observed for the gas, and let $\theta$ be the angular diameter of the gas, as observed from Earth.

(a) Show that if the Sunyaev–Zel'dovich effect, $\Delta T/T_0$, is measured for the cluster, then the Hubble constant may be calculated as

$$H_0 = Cf(z) \frac{F_X T_e^{3/2}}{\theta (\Delta T/T_0)^2}.$$

where $f(z)$ is a function of the redshift $z$ of the cluster and $C$ is a constant factor you must determine.

(b) Use the data in Fig. 29.12 for the clusters, together with $\theta = 46''$, $F_x = 5.77 \times 10^{-15}\,Wm^{-2}$, and $kT_e = 8.0$ keV for Abell 697, and $\theta = 69''$, $F_x = 7.16 \times 10^{-15}\,Wm^{-2}$, and $kT_e = 7.2$ keV for Abell 2218 (from Jones et al., 2005) to evaluate $h$, the Hubble parameter. How do your answers compare with the WMAP value of $h = 0.71$?

## COMPUTER PROBLEMS

**29.56** In Eq. (29.166), the limits of integration on the left-hand side span a time interval that lies entirely in the past. Use Eq. (29.132) to approximate the scale factor on the left-hand side. On the right-hand side, consider the case of $t_{mva} > t_0$, so the limits of integration span a time interval that lies entirely in the future. Use Eq. (29.133) for the scale factor on the right-hand side. [Note that very distant objects have already emitted the last photon that will ever reach us ($t_{mva} < t_0$), so we must restrict our attention to nearer objects to ensure that $t_{mva} > t_0$.] Show that

$$t_{mva} \simeq \frac{t_H}{\sqrt{\Omega_{\Lambda,0}}} \ln \left[ \left( \frac{\sqrt{\Omega_{m,0}}}{2\sqrt{\Omega_{\Lambda,0}}} \right)^{1/3} \left( \frac{\sqrt{1+z}}{\sqrt{1+z}-1} \right) \right].$$

Using WMAP values, what is the largest redshift for which $t_{mva} > t_H$? Find the maximum visible age, in units of $t_H$, for sources at values of $z$ of 0.1, 0.5, 1, and 1.5.

**29.57** Use WMAP values for this problem.

**(a)** On a single graph, plot the luminosity distance, the present proper distance, the angular diameter distance, the relativistic Hubble law distance estimate (Eq. 27.7), and the non-relativistic Hubble law distance (Eq. 27.8) for values of $z$ between 0 and 4. Express these distances in units of $c/H_0$.

**(b)** From your results, determine the value of $z$ when the relativistic Hubble law distance estimate differs from the proper distance by more than 10%.

**29.58** A distant radio galaxy, 8C 1435+63, has a redshift of $z = 4.25$. Assume WMAP values for this problem.

**(a)** How old was the universe at this redshift? Express your answer both in terms of years and as a fraction of the present age of the universe.

**(b)** What is the present proper distance (in Mpc) to 8C 1435+63?

**(c)** What was the proper distance (in Mpc) to 8C 1435+63 when its light was emitted?

**(d)** What is the luminosity distance to 8C 1435+63?

**(e)** What is the angular diameter distance to 8C 1435+63?

**(f)** The angular diameter of the nucleus of 8C 1435+63 is about 5″. What is the linear diameter of the galaxy (in units of kpc)?

**(g)** Suppose the galaxy's redshift were $z = 1$. What would its linear diameter be [using the same angular diameter as in part (d)]?

Further information about 8C 1435+63, which may be the progenitor of a cD elliptical galaxy, can be found in Spinrad, Dey, and Graham (1995).

# CHAPTER
# 30

# The Early Universe

**30.1**  *The Very Early Universe and Inflation*
**30.2**  *The Origin of Structure*

## 30.1 ■ THE VERY EARLY UNIVERSE AND INFLATION

In Chapter 29 we developed the mathematical machinery that describes the expansion of an isotropic, homogeneous universe consisting of nonrelativistic particles (baryonic and dark), relativistic particles (photons and neutrinos), and dark energy. This chapter deals with the emergence of structure from the featureless cauldron of the Big Bang.

Much happened during the universe's first fraction of a second, and the nature of that earliest environment is still only partially understood. The ideas of modern particle physics that describe these early epochs have also made successful predictions about the types and numbers of elementary particles that exist today. It is to these theories that we turn (with some measure of confidence) in hopes of glimpsing the engines of creation.

### Fundamental Particles

According to the **Standard Model** of particle physics, there are three kinds of fundamental (not composite) particles.

- The **leptons** are the charged leptons $e^\pm$, $\mu^\pm$, and $\tau^\pm$, and the neutrinos $\nu_e$, $\nu_\mu$, and $\nu_\tau$, together with their antineutrinos $\bar{\nu}_e$, $\bar{\nu}_\mu$, and $\bar{\nu}_\tau$. Leptons are fermions.

- There are six **quarks**: up, down, strange, charm, bottom, and top, along with their antiquarks. Each quark comes in three "colors" (three choices of an internal degree of freedom). Particles made of quarks are called **hadrons**. There are two types of hadrons: **baryons** (made of three quarks) and **mesons** (formed by a quark–antiquark pair). Baryons are fermions, while mesons are bosons.

- The **force-carrying particles** consist of the photon, eight different gluons (particles that mediate the strong interaction and bind quarks together), three vector gauge bosons ($W^\pm$ and $Z^0$) that mediate the weak interaction, and the scalar Higgs boson (which has yet to be confirmed by experiment). All of these particles are bosons.

Table 30.1 summarizes some of the properties of these particles. Current technology can reproduce the temperatures, energies, and densities that prevailed back to the **quark–hadron**

1230

**TABLE 30.1** Particles of the Standard Model in the Relativistic (High-Temperature) Limit. See Eq. (29.72) and the accompanying discussion for $n_{anti}$ and $n_{spin}$; $f$ is the fermion factor in Eq. (30.2).

| Particle | # Types | $n_{anti}$ | $n_{spin}$ | $g$ | $f$ | Contribution to $g_*$ |
|---|---|---|---|---|---|---|
| Photon | 1 | 1 | 2 | 2 | 1 | 2.00 |
| Charged leptons | 3 | 2 | 2 | 12 | 7/8 | 10.50 |
| Neutrinos | 3 | 2 | 1 | 6 | 7/8 | 5.25 |
| Quarks | 18 | 2 | 2 | 72 | 7/8 | 63.00 |
| Gluons | 8 | 1 | 2 | 16 | 1 | 16.00 |
| Vector gauge bosons | 3 | 1 | 3 | 9 | 1 | 9.00 |
| Scalar Higgs boson | 1 | 1 | 1 | 1 | 1 | 1.00 |

**TABLE 30.2** Eras and Events in the Early Universe. The values are approximate.

| Era or Event | Time | Temperature ($kT$) |
|---|---|---|
| Planck era | $< 5 \times 10^{-44}$ s | $> 10^{19}$ GeV |
| Planck transition | $5 \times 10^{-44}$ s | $10^{19}$ GeV |
| Grand unification era | $5 \times 10^{-44}$ s to $10^{-36}$ s | $10^{19}$ GeV to $10^{15}$ GeV |
| Inflation | $10^{-36}$ s to $10^{-34}$ s | $10^{15}$ GeV |
| Electroweak era | $10^{-34}$ s to $10^{-11}$ s | $10^{15}$ GeV to 100 GeV |
| Electroweak transition | $10^{-11}$ s | 100 GeV |
| Quark era | $10^{-11}$ s to $10^{-5}$ s | 100 GeV to 200 MeV |
| Quark–hadron transition | $10^{-5}$ s | 200 MeV |
| Neutrino decoupling | 0.1 s | 3 MeV |
| Electron–positron annihilation | 1.3 s | 1 MeV |

**transition** at about $10^{-5}$ s, when a plasma of free quarks and gluons condensed to form hadrons, including the familiar proton and neutron.

Table 30.2 shows the typical times and temperatures ($kT$) of different eras during the first second after the Big Bang. (We remind you that any "age of the universe" that is quoted has been obtained from a backward extrapolation in time, using a Friedmann model of the universe.) The temperature as a function of time is given by Eq. (29.89), which we reproduce here:

$$T(t) = (1.81 \times 10^{10} \text{ K s}^{1/2}) g_*^{-1/4} t^{-1/2}. \tag{30.1}$$

The value of $g_*$, the effective number of degrees of freedom of all of the relativistic particles present at temperature $T$, is given by

$$g_* = \sum_{\text{bosons } i} g_i + \frac{7}{8} \sum_{\text{fermions } i} g_i \left( \frac{T_\nu}{T} \right)^4. \tag{30.2}$$

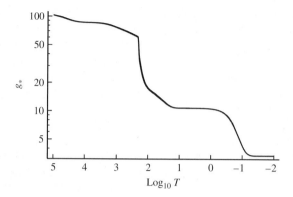

**FIGURE 30.1**    The effective number of degrees of freedom, $g_*$, as a function of temperature. The temperature is actually $kT$ in units of MeV. The quark-hadron transition is chosen to occur at 200 MeV. (Figure adapted from Coleman and Roos, *Phys. Rev. D, 68*, 027702, 2003.)

The factor $(T_\nu/T)^4 = (4/11)^{4/3}$ (Eq. 29.73) is applied only to neutrinos, and only if the temperature is below that of neutrino decoupling ($kT \approx 3$ MeV).

Figure 30.1 shows the variation in the value of $g_*$ as a function of $kT$ in units of MeV. The present value of $g_* = 3.363$ (Eq. 29.76) at the right side of the graph is due to the photon and the three species of neutrinos and their antineutrinos. Between neutrino decoupling at about 3 MeV and $\sim 100$ MeV, the electron and positron are the only additional relativistic particles, yielding $g_* = 10.75$. Above about 200 MeV, the hadrons are dissociated into their component quarks and gluons, which contribute more to the value of $g_*$. Finally, by $kT \sim 300$ GeV (beyond the left side of the graph), all of the particles of the standard model are relativistic and $g* = 106.75$. No new particles masses are predicted within the range of energies of the electroweak era, so we may adopt a value of $g_* \simeq 100$ throughout this time interval. (The electroweak era is sometimes referred to as the *great desert* because of the paucity of new particles.) At even earlier times, it is estimated for the simplest model that $g_* \simeq 160$. Fortunately, $T(t)$ is sensitive only to the fourth root of $g_*$, so precise values of $g_*$ are not required for the purpose of estimation.

## Hot and Cold Dark Matter

Missing from Table 30.2, of course, are the particles that make up dark matter. Determining the composition of dark matter is one of the greatest challenges facing cosmologists today. Some dark matter may consist of ordinary baryonic matter, although the total amount of baryonic matter (both shining and dark) cannot make up more than 4% of the total mass of the universe. The WMAP results and constraints imposed by the Big Bang nucleosynthesis of the light elements are inconsistent with much larger amounts of baryonic matter (see Fig. 29.14). The gravitational microlensing of stars in the Large Magellanic Cloud by MACHOs (massive compact halo objects) described in Section 24.2 may demonstrate that some dark baryonic matter is hiding in galactic halos, perhaps in the form of brown dwarfs or stellar-mass black holes. However, a statistical analysis of the microlensing events indicates that only about 19% of the mass of the Milky Way's dark matter halo can be explained by MACHOs.

Dark matter candidates are usually divided into two categories, **hot dark matter** (HDM) and **cold dark matter** (CDM). Hot dark matter consists of particles moving with relativistic velocities.[1] The leading candidates for hot dark matter are massive neutrinos (which are leptons). Recall from Section 10.3 that the best estimate for the upper limit to the electron neutrino's mass is $2.2 \text{ eV}/c^2$. Cold dark matter candidates are hypothetical particles that move slowly, such as the WIMPs (weakly interacting massive particles) that hardly interact with normal matter except through their gravitational attraction.[2] It has been suggested that the rest energy of WIMPs may range from 10 GeV (about 10 times more massive than a proton) up to several TeV ($10^{12}$ eV). Another hypothetical cold dark matter candidate is the **axion**, a low-mass boson that is extremely lightweight ($mc^2 \approx 10^{-5}$ eV). If axions do exist, and if they constitute most of the dark matter, then they are by far the most numerous type of particle in existence. Although there is as yet no evidence for the existence of WIMPs or axions, the search for them continues.

The distinction between hot and cold dark matter is important because it is difficult for the relativistic hot dark matter to clump together gravitationally and participate in the formation of structure in the early universe. For this reason, as we shall see in Section 30.2, models that incorporate cold dark matter are currently favored. In fact, the standard model in cosmology is referred to as the **ΛCDM model** because it includes both the cosmological constant and cold dark matter.

### The Planck Limits on Time, Mass, and Length

The earliest time that can be addressed by current physical theory is the **Planck time**,

$$t_P \equiv \sqrt{\frac{\hbar G}{c^5}} = 5.39 \times 10^{-44} \text{ s}. \tag{30.3}$$

The Planck time is the only combination of fundamental constants that has units of time and, as such, is a characteristic quantity in fundamental theories. It contains Planck's constant, Newton's gravitational constant, and the speed of light, a melding of quantum mechanics, gravitation, and relativity that has yet to be achieved in a unified theory.

The physicist's toolbox of theories cannot penetrate this barrier in time. Some insight into the limits imposed by the Planck time comes from Heisenberg's uncertainty principle, Eq. (5.19). Consider a black hole of mass $M$, the most compact region within which a mass can be contained. The central singularity is confined within the event horizon, so the uncertainty in its position is just its Schwarzschild radius,

$$\Delta x \approx R_S.$$

Thus the uncertainty in its momentum is

$$\Delta p \approx \frac{\hbar}{\Delta x} \approx \frac{\hbar}{R_S}.$$

---

[1] More precisely, it is the internal velocity dispersion of the particles that is relativistic.
[2] WIMPs were previously discussed as a candidate for a component of the Milky Way's dark matter halo in Section 24.2.

For a very low-mass primordial black hole that formed immediately after the Big Bang, $R_S$ will be extremely small and so $\Delta p$ will be quite large. In the relativistic limit, the energy of the singularity is, from Eq. (4.48), $E = pc$, and so the uncertainty in the energy is

$$\Delta E = (\Delta p)c \approx \frac{\hbar c}{R_S}.$$

The expected value of the energy will be at least as large as $\Delta E$, so the kinetic energy of the singularity is approximately $K \approx \hbar c / R_S$. There will be a conflict between the quantum and classical descriptions of gravity when this energy is comparable to the magnitude of the (Newtonian) gravitational energy of the black hole, which, according to Eq. (10.22), is approximately $U \approx -GM^2/R_S$. Is the black hole bound, or could the singularity emerge from behind the event horizon in violation of the "Law of Cosmic Censorship" discussed in Section 17.3? This confrontation between the application of quantum and classical physics to gravitation may be estimated to occur when the sum of the kinetic energy, $K$, and the potential energy, $U$, is zero:

$$\frac{\hbar c}{R_S} - \frac{GM^2}{R_S} = 0.$$

The resulting mass is called the **Planck mass**,

$$m_P \equiv \sqrt{\frac{\hbar c}{G}} = 2.18 \times 10^{-8} \text{ kg}. \tag{30.4}$$

As was mentioned in Problem 17.21, this is an estimate of the least massive primordial black hole. Inserting the Planck mass into Eq. (17.27) for the Schwarzschild radius, $R_S = 2GM/c^2$, neglecting the factor of two in this order-of-magnitude estimate, and solving for $R_S$ results in a characteristic length called the **Planck length**,

$$\ell_P \equiv \sqrt{\frac{\hbar G}{c^3}} = 1.62 \times 10^{-35} \text{ m}. \tag{30.5}$$

The Planck time, $t_P = \ell_P/c$, is just the time required for light to travel the Planck length. The Planck length is therefore the diameter of a causally connected region (the horizon distance) at that time; see Eq. (29.154). Although the foregoing argument is only a little better than a dimensional analysis, it does reveal the significance of combining these fundamental constants to produce a characteristic time, mass, and length.

Equation (30.1) indicates that the temperature at the Planck time was nearly $10^{32}$ K, so the typical thermal energy of a particle was $kT \sim 10^{19}$ GeV! The description of the universe before the Planck time staggers even the strongest imagination. The universe would have been a collection of primordial black holes that were continually forming, evaporating, and reforming. In the process, different regions of spacetime were rapidly connecting and disconnecting, giving it a foamlike structure. Our current physical theories break down at times earlier than $t_P$, and in fact the very notion of space and time as separate concepts dissolves before the Planck time. A quantum theory of gravity capable of describing this convoluted arena in which space and time have lost their familiar, separate identities has yet to be invented. After the Planck time, spacetime began to take on a more coherent structure as greater portions of it became causally connected. Exactly how time itself emerged from the Big Bang is a question to be pondered by physicists and philosophers alike.[3]

## Unification and Spontaneous Symmetry Breaking

It is an article of faith for physicists that before the Planck time, the four fundamental forces of nature (the gravitational force, the electromagnetic force, and the strong and weak nuclear forces) were merged into one all-encompassing **Theory of Everything** (TOE); see Fig. 30.2. Although the rough outlines of such a theory are still a matter of conjecture, the theory must have certain mathematical symmetries guaranteeing that the four forces were conjoined. When the universe reached the Planck time, the single, all-encompassing TOE force spontaneously separated into the gravitational force (as described by Einstein's theory of general relativity) and a unified version of the three remaining forces. This process is called **spontaneous symmetry breaking**, a term that refers to the changes in the mathematical symmetries of the theory's equations.[4]

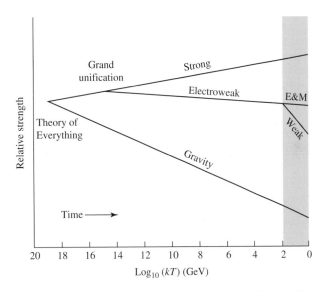

**FIGURE 30.2** The unification of the four fundamental forces, according to a Theory of Everything. The electromagnetic and weak forces were unified at times earlier than $10^{-11}$ s, and the electroweak and strong forces were unified before $10^{-34}$ s. All four forces were joined at times earlier than the Planck time, $t_P = 5 \times 10^{-44}$ s. "E&M" stands for the electromagnetic force. Only the electroweak unification has been tested experimentally (shaded region).

---

[3]For example, is the statement that "time was created by the Big Bang" meaningful? Because any act of creation involves a time sequence ("first it isn't, then it is"), can time itself be created?

[4]A physical example of spontaneous symmetry breaking is a marble rolling around the rim of a roulette wheel. Initially, the marble is equally likely to be found above any of the numbered bins, but as the marble loses energy it falls into one of them and becomes trapped.

The mathematics involved in spontaneous symmetry breaking is similar to that which describes a familiar phase transition in ordinary matter. For example, above a temperature of 273 K, water is in a liquid state. Assuming that there is no preferred direction imposed from outside, the water molecules are randomly oriented in all directions. When the temperature falls below 273 K, however, the water undergoes a phase transition and freezes. The symmetry of the liquid water is broken because the arrangement of $H_2O$ molecules in the lattice of an ice crystal has preferred directions along the axes of the lattice. Just as there is a release of latent heat when water freezes, energy was released in the episodes of spontaneous symmetry breaking in the early universe.

The theories (there are several variants) that describe the joining of the remaining three forces are called **grand unified theories** (GUTs). The simplest GUT, proposed in 1973 by the American physicists Sheldon Glashow and Howard Georgi (both at Harvard), is known as SU(5). The GUTs have had some successes, such as providing a fundamental explanation for the equal magnitudes of the proton and electron charge. They have also had some setbacks; the failure to detect SU(5)'s predicted decay of the proton probably eliminates it as a successful GUT. The moral is that specific predictions based on grand unified theories should be viewed with some caution.

The unification of the strong and weak nuclear forces and the electromagnetic force would have lasted until the temperature of the universe had fallen to about $10^{29}$ K, when the characteristic thermal energy ($kT$) of a particle was about $10^{15}$ GeV and the universe was some $10^{-36}$ s old. At this point, following another episode of spontaneous symmetry breaking, the strong nuclear force parted company with the electroweak (combined electromagnetic and weak) force.

The theory of the electroweak unification was worked out in the 1960s by three physicists: Sheldon Glashow and Steven Weinberg (American) and Abdus Salam (Pakistani). They described how electromagnetic and weak forces were united when the temperature exceeded about $2 \times 10^{15}$ K, at roughly $10^{-11}$ s.[5] At this temperature, the characteristic thermal energy of a particle is about 150 GeV. Their theory predicted the existence of three new particles (the three vector gauge bosons $W^\pm$ and $Z^0$) that mediate the weak force, just as photons convey the electromagnetic force. Above $kT \sim$ a few hundred GeV, the vector gauge bosons become massless, and so become indistinguishable from photons. The electromagnetic and weak forces are then unified. When the temperature falls below about $2 \times 10^{15}$ K, a spontaneous symmetry breaking endows the vector gauge bosons with mass. When these particles were discovered in the 1980s, the agreement between experiment and theory provided a striking confirmation of the electroweak unification. This success provides encouragement to physicists working on GUTs and TOEs.

## Problems with the Standard Theory of the Big Bang

We now return to the Planck time to discover how spontaneous symmetry breaking, with its concomitant release of specific heat, affected the expansion of the early universe. As we move forward from the Planck time, we are confronted by three problems with the simple picture of the Big Bang we have considered so far.

---

[5]This is the earliest time that can be probed experimentally.

- *Why is the cosmic background radiation so smooth?* Consider two CMB photons arriving at the observer from opposite sides of the sky. It would seem that the regions from which photons came could never have communicated with each other, since the photons have just passed each other. In fact, the problem is even more severe. At the time of decoupling, regions of the CMB that are now separated by more than 2° on the sky could never have been in causal contact (see Problem 30.11). The spacetime diagram shown in Fig. 30.3 illustrates the situation. Photons arriving now at Earth, Point *A*, were emitted at the time of recombination from sources at Points *B* and *C*, which lie on the past light cone of Point *A*. However, when extended back to the Big Bang, the past light cones of Points *B* and *C* do not intersect. Points *B* and *C* are not in causal contact, since each is outside the other's particle horizon. The smoothness of the CMB is then very difficult to explain, especially since the variations that have been found are so very small ($\delta T/T \sim 10^{-5}$). The problem of finding the mechanism that allowed the universe's left hand to know what its right hand was doing is called the **horizon problem**.

- *Why is the universe so nearly flat* ($\Omega_0 \simeq 1$)? Recall from Example 29.1.1 that if $\Omega_0$ is roughly of order unity, then at very early times ($z \gg 1$) the density parameter was essentially equal to one. For the simple model of a universe of pressureless dust considered in Section 29.1,

$$\frac{1}{\Omega_0} - 1 = \left(\frac{1}{\Omega} - 1\right)(1+z) \tag{30.6}$$

(see Problem 29.10). If $\Omega$ had differed slightly from one in the very early universe, today's value, $\Omega_0$, would vary significantly from unity. Suppose, at the time of decoupling ($[t_{\text{dec}}]_{\text{WMAP}} = 379$ kyr at $[z]_{\text{WMAP}} = 1089$), that the density parameter had

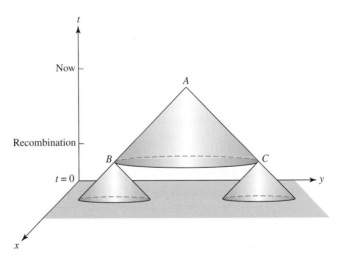

**FIGURE 30.3** The horizon problem. The same CMB spectrum is observed at Point *A* from Points *B* and *C*, although *B* and *C* have never been in causal contact.

differed from unity by just 0.0009. If $\Omega_{dec} = 0.9991$, then according to Eq. (30.6), today's value of the density parameter would be $\Omega_0 \simeq 0.5$; and if $\Omega_{dec} = 1.0009$, then $\Omega_0 \simeq 50$. Such underdense and overdense universes have profound dynamical implications. For example, if $\Omega_{dec} = 0.5$, then after another few hundred thousand years the universe and its contents would be too widely dispersed and expanding too fast to allow the formation of galaxies or stars. On the other hand, if $\Omega_{dec} = 2$, then after a few hundred thousand years the expansion would have reversed itself, and the universe would have ended in a Big Crunch. Our existence on Earth reflects this fine-tuning of $\Omega = 1$ in the early universe. Otherwise, we would not be here today to wonder about it! The problem of the origin of our nearly flat universe is called the **flatness problem**.

- *Why have we found no magnetic monopoles?* A magnetic monopole is a single magnetic charge, tantamount to an isolated magnetic pole. When the universe experienced an episode of spontaneous symmetry breaking at the end of the GUTs era of unification, a choice of "direction" was made at each point in space, associated with the quantum field involved (called the Higgs field). This direction was not one of spatial orientation but, rather, involved a choice of certain discrete values (analogous to quantum numbers) for the theory describing the field. **Defects** arose at locations where a discontinuity in this choice of values existed, and a pointlike discontinuity corresponds to a **magnetic monopole**.[6]

  Although a magnetic monopole might be $\sim 10^{16}$ times more massive than a proton ($mc^2 \sim 10^{16}$ GeV), there was sufficient energy at the end of the GUTs era to create copious numbers of magnetic monopoles. Yet despite many searches for magnetic monopoles, only one experiment has found any evidence for them, and then only one possible monopole candidate was detected (on Valentine's Day in 1982). There have not been any other detections to confirm that result. If our understanding of grand unification is correct in predicting the creation of magnetic monopoles, then where are they? This is called the **monopole problem**.

By itself, the Big Bang explanation for the beginning of the universe is incapable of meaningfully answering these questions. Of course, one possible answer is "That's just the way it is"—that the universe started out smooth and flat because of some set of initial conditions at $t = 0$. Cosmologists would prefer to find a natural sequence of events that inevitably led to the universe as it appears today, instead of specifying, in an ad hoc way, the values of certain variables at the beginning to make everything come out right.

### Inflation

In 1980 an American astronomer working at Stanford University, Alan Guth, proposed a single solution. According to Guth, the Big Bang picture is essentially correct, but during its first fraction of a second, when $t$ was barely greater than zero, the whole universe was much more compact than is described by the standard Big Bang. At that time, every point was close

---

[6]Defects can arise any time symmetry breaking occurs, but those from the end of the GUTs episode have received the most attention.

enough to every other point to be in causal contact, and the entire universe had achieved thermodynamic equilibrium. Then there was a tremendous spurt of *exponential expansion* that smoothed out the universe, rendering it exactly flat with $\Omega_0 = 1$ (or *very* nearly so). Guth called this period of exponential growth **inflation**. This episode of inflation would explain the general isotropy of the universe, and the smoothness of the cosmic background radiation in particular. After inflation, the expansion of the universe proceeded as in the standard Big Bang model.

The details of Guth's inflationary proposal involve ideas at the frontiers of particle physics. In fact, many variants on the original inflationary scenario have been proposed. As observations accumulate, one model may fall out of favor, perhaps to be revived later as more information becomes available. Much research is still ongoing in this area, and the version of inflation described in the following discussion may well be wrong in some of its specifics. However, most cosmologists believe that, because it gives such satisfying answers to so many questions, some form of inflation must have taken place in the early universe.

**Virtual Particles and the Energy of the Vacuum**

The particle energies of $\sim 10^{19}$ GeV at the Planck time were so great that the familiar baryons (such as protons, neutrons, and their antiparticles) present at much lower temperatures were absent. We need not concern ourselves with the highly relativistic particles that were present then. Instead, we will concentrate on another significant component of the very early universe, the energy of the **vacuum**. The term *vacuum* requires some explanation. In everyday language, it means a complete emptiness, devoid of any matter or energy. However, physicists use *vacuum* to describe the ground state of a system. For instance, in Example 5.4.2 an electron confined to a small region was calculated to have a certain minimum energy, its ground-state energy. This energy cannot be lost or extracted from the electron, since there is no lower quantum state to which the electron can make a transition. The existence of the vacuum has been verified by observations of the **Casimir effect** (named for Dutch physicist Hendrick Casimir, 1909–2000). Two uncharged parallel flat conducting plates with a very small separation will alter the properties of the vacuum between the plates. This change in the vacuum creates an attractive force between the plates, which has been measured. Unfortunately, the Casimir effect cannot be used to calculate the value of the energy density of the vacuum.

We can crudely estimate the value of the energy density of the vacuum using the uncertainty principles $\Delta x \, \Delta p \approx \hbar$ (Eq. 5.19) and $\Delta E \, \Delta t \approx \hbar$ (Eq. 5.20). The vacuum can be modeled as a place where matter–antimatter pairs of particles are constantly being created and annihilated. These particles cannot be directly observed and their energy cannot be tapped; they are known as **virtual particles**. They borrow their rest energy $\Delta E$ from the vacuum and are annihilated in such a short time $\Delta t$ that they escape detection.

Let's consider a virtual particle of mass $m \approx (\Delta E)/c^2$ confined to a cubical box of side $L \approx \Delta x$ with a particle lifetime of

$$\Delta t \approx \hbar/\Delta E \approx \hbar/mc^2.$$

In addition, the particle's speed is approximately

$$v \approx \frac{\hbar}{m \Delta x} \approx \frac{\hbar}{mL}$$

(see Example 5.4.2). Since the farthest the particle can travel in time $\Delta t$ is $v\Delta t$, we set $L = v\Delta t$ to be certain the particle's motion does not carry it outside the box. Thus

$$L = v\Delta t \approx \frac{\hbar}{mL} \frac{\hbar}{mc^2}.$$

Solving for $L$, we find

$$L \approx \frac{\hbar}{mc}.$$

The energy density of the vacuum must be capable of creating a pair of particles in the box so that conservation rules can be maintained (such as electric charge). That is, the energy density of the vacuum must be at least

$$u_{\text{vac}} \approx \frac{2mc^2}{L^3} \approx \frac{2m^4c^5}{\hbar^3}.$$

The greatest mass for each particle in the pair may be taken to be the Planck mass, $m_P = \sqrt{\hbar c/G}$ (Eq. 30.4). Our estimate of the energy density of the vacuum is thus

$$u_{\text{vac}} \approx \frac{2m_P^4 c^5}{\hbar^3} \approx \frac{2c^7}{\hbar G^2},$$

or $u_{\text{vac}} \approx 10^{114}$ J m$^{-3}$. Of course this is only a rough estimate. More sophisticated calculations (which also invoke an arbitrary cut-off to avoid an infinite answer) result in a value of $u_{\text{vac}} \approx 10^{111}$ J m$^{-3}$.

Einstein's general relativity is responsive to every form of energy, including the energy of the vacuum. Furthermore, the energy density of the vacuum is a *constant*, just as for dark energy. The fluid equation (Eq. 29.50) shows that the constant vacuum energy density would produce a negative pressure

$$P_{\text{vac}} = -\rho_{\text{vac}} c^2 = -u_{\text{vac}}, \tag{30.7}$$

an equation of state with $w = -1$ (cf. Eq. 29.52). It is natural to identify dark energy as the energy density of the vacuum. However, the value of the vacuum energy density is huge compared to the energy density of dark energy. From Eq. (29.113),

$$u_{\text{dark}} \equiv \rho_\Lambda c^2 = \rho_{c,0}\Omega_{\Lambda,0}c^2 = 6.22 \times 10^{-10} \text{ J m}^{-3}. \tag{30.8}$$

If we identify dark energy as the energy density of the vacuum, we are faced with a daunting discrepancy of some 120 orders of magnitude between the theoretical and observed values. The huge value calculated for the vacuum energy density is in sharp conflict with the observed rate of the Hubble flow. If every cubic meter were accompanied by approximately

$10^{111}$ J of vacuum energy as the universe expanded and its volume grew, the universe would have expanded so rapidly that no galaxies or stars could ever have coalesced under the influence of gravity.

A plausible physical mechanism for reducing the value calculated for today's vacuum energy density to the observed value of $u_{\text{dark}}$ has yet to be found. According to some advanced particle theories, bosons and fermions should make contributions to the vacuum energy of opposite signs, and so cancel to yield zero vacuum energy. If the cancelation were not perfect, a small observed residual vacuum energy density could result. Why should the cancelation be effective for the first 120 decimal places and then break down? So far the combined efforts of cosmologists and particle physicists have shed little light on this mystery. Nevertheless, we will continue to identify dark energy as the energy density of the vacuum and to accept its presently observed value of nearly zero.

### The False Vacuum

At the end of the GUTs epoch, when $t \sim 10^{-36}$ s and the temperature was $T \sim 10^{28}$ K, the universe entered an extremely peculiar state called the **false vacuum**. The false vacuum that existed when the universe was approximately $10^{-36}$ s old was not a true vacuum, so the universe was not in the state with the lowest possible energy density. Instead, the universe had entered a supercooled state in which its temperature had plummeted *far below* $10^{28}$ K, the temperature for spontaneous symmetry breaking. Supercooling happens when a phase transition proceeds much more slowly than the cooling rate. For example, water can be supercooled to more than 20 K below its freezing point without changing phase. Since freezing would release the water's latent heat, there is a state of lower energy density (ice) available to the supercooled water. However, the supercooled water persists in a liquid state of higher energy density. Similarly, the universe persisted in its false vacuum state of unbroken symmetry with a high energy density, even though a spontaneous symmetry breaking to a true vacuum with zero energy density was energetically favorable. According to grand unified theories, the false vacuum had a constant energy density on the order of[7]

$$u_{\text{fv}} \approx 10^{105} \text{ TeV m}^{-3} = 1.6 \times 10^{98} \text{ J m}^{-3}. \tag{30.9}$$

### Quantum Fluctuations and the Onset of Inflation

When $t < 10^{-36}$ s, the universal expansion was governed by relativistic particles. Inflation began when quantum fluctuations governed by the uncertainty principle, Eq. (5.20), allowed a small region of space to enter a true vacuum state in a universe otherwise filled with false vacuum. Although the pressure within the bubble of true vacuum was essentially zero, it was surrounded outside by the negative pressure of the false vacuum. The greater pressure inside the bubble caused the bubble to grow at an astounding rate. Thus at roughly $t_i = 10^{-36}$ s, the constant energy density of the false vacuum became dominant in the acceleration equation,

---

[7]It is important to recall that dark energy plays an insignificant role at the time of inflation. If dark energy is identified as the energy density of the vacuum, then the mechanism of inflation requires an additional source of vacuum energy, called the **inflaton field**, that is depleted (or nearly so) by the time inflation ends.

Eq. (29.116). This equation then became of the form

$$\frac{d^2 R}{dt^2} = \left[ -\frac{4}{3}\pi G \left( \frac{u_{fv}}{c^2} + \frac{3 P_{fv}}{c^2} \right) \right] R. \tag{30.10}$$

Using $P_{fv} = -u_{fv}$ from Eq. (30.7), this becomes

$$\frac{d^2 R}{dt^2} = \frac{8\pi G u_{fv}}{3c^2} R. \tag{30.11}$$

The exponentially growing solution is

$$R(t) = R(t_i) e^{t/\tau_i}, \tag{30.12}$$

where $R(t_i)$ is the initial value of the scale factor when inflation began, and $\tau_i$ is the *inflation time scale* given by

$$\tau_i = \sqrt{\frac{3c^2}{8\pi G u_{fv}}} \approx 10^{-36} \text{ s}. \tag{30.13}$$

Over what time interval $\Delta t$ did inflation occur? Fundamental physics cannot yet supply the answer, but we can estimate the *minimum* time interval by examining what inflation accomplished: allowing our present universe to be causally connected with a uniform cosmic microwave background. At the initial time $t_i = 10^{-36}$ s, the horizon distance was only

$$d_i = 2ct_i = 6 \times 10^{-28} \text{ m}, \tag{30.14}$$

from Eq. (29.154). By the time inflation ended at time $t_f = t_i + \Delta t$, the horizon must have expanded at least far enough to include today's observable universe to account for the uniformity of the CMB. We can therefore write the final horizon distance, at the end of inflation, as the present horizon distance, $d_{h,0} = 14.6$ Gpc (Eq. 29.159), multiplied by the scale factor $R(t_f)$. From Eq. (29.87),

$$d_f = d_{h,0} R(t_f) = (4.5 \times 10^{26} \text{ m})(2.04 \times 10^{-10} \text{ s}^{-1/2})(t_i + \Delta t_{min})^{1/2}. \tag{30.15}$$

The minimum time interval consistent with the uniform CMB may then be found by solving

$$2ct_i e^{\Delta t_{min}/\tau} = (4.5 \times 10^{26} \text{ m})(2.04 \times 10^{-10} \text{ s}^{-1/2})\sqrt{\tau} \left( \frac{t_i}{\tau} + \frac{\Delta t_{min}}{\tau} \right)^{1/2}. \tag{30.16}$$

This may be solved numerically to obtain

$$\frac{\Delta t_{min}}{\tau} = 62.4 \tag{30.17}$$

or

$$\Delta t_{min} = 6.25 \times 10^{-35} \text{ s}. \tag{30.18}$$

During this brief time interval, the size of the observable universe grew from $6 \times 10^{-28}$ m to 0.73 m! There is no reason to believe that the universe conspired to have inflation last exactly long enough to endow only the presently observable universe with a uniform CMB. It is customary to assume that the inflationary phase lasted for $10^{-34}$ s or more, during which the size of the universe increased by a factor of approximately $e^{100} \simeq 3 \times 10^{43}$. This is the roughest of estimates; the actual factor could have easily exceeded $10^{50}$, a value that is also quoted.

Today's observable universe began inside a small bubble of true vacuum. We will take the initial diameter of the bubble to be the particle horizon distance at $10^{-36}$ s, just $d_i = 6 \times 10^{-28}$ m across. At the beginning of inflation, every point in the bubble lay within the particle horizon of every other point, and its contents were in thermodynamic equilibrium. The isotropy and homogeneity evident in our universe today were established at that early time. The bubble's final diameter, assuming inflation ended at $10^{-34}$ s, was

$$d_f = e^{100} d_i = 1.6 \times 10^{16} \text{ m}.$$

What fraction of this inflated bubble was occupied by the material that makes up the presently observable universe? To answer this question, we will work backward. Assuming a flat universe, the horizon distance to the farthest point that can be observed today is $d_{h,0} = 4.5 \times 10^{26}$ m (Eq. 29.159). According to Eq. (29.87), the scale factor at $10^{-34}$ s was $R = 4.9 \times 10^{-27}$. This means that the diameter of the presently observable universe at that time was, from Eq. (29.148),

$$2R d_{h,0} = 4.4 \text{ m} = 2.8 \times 10^{-16} d_f.$$

This implies that the small bubble of true vacuum expanded to such an extent that when inflation ended at $10^{-34}$ s, today's observable universe occupied about 5 parts in $10^{46}$ of its volume. If the size of the presently observable universe immediately after inflation is represented by a hydrogen atom, then the inflated bubble of true vacuum in which it was immersed would be a sphere of radius 200 km!

This episode of inflationary growth came to a halt when spontaneous symmetry breaking finally brought an end to the GUTs epoch, and the strong nuclear force became distinct from the electroweak force. The elevated energy density of the false vacuum was then released, like the release of latent heat that occurs in freezing. This energy reheated the universe to nearly $10^{28}$ K, its preinflation value, and generated a burst of particle–antiparticle creation. From this point onward, the universe developed as previously described in the standard Big Bang picture. As shown in Fig. 30.4, the only difference lies in the first $10^{-34}$ s or so. But in a single stroke, this brief instant of inflation resolves the problems with the standard Big Bang.

### Solutions to the Problems of the Standard Big Bang Theory

The exponential growth of our small bubble of true vacuum carried most of its volume far beyond the boundaries of today's universe. Nevertheless, because the bubble's volume was in thermodynamic equilibrium before inflation, the spectrum of the CMB is extremely smooth, thereby solving the horizon problem.

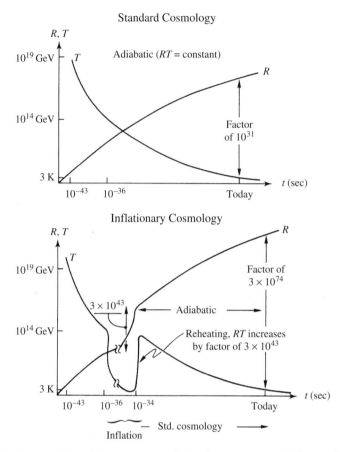

**FIGURE 30.4**    The evolution of the temperature of the universe and the scale factor, without and with inflation. Except for the bottom value, the temperature is given in terms of $kT$. [Figure adapted from Edward W. Kolb and Michael S. Turner, *The Early Universe* (page 274), ©1990 by Addison-Wesley Publishing Company, Inc., Reading, MA. Reprinted by permission of the publisher.]

Inflation also resolves the flatness problem. As the small bubble of true vacuum inflated, the scale factor $R$ increased by a factor of $e^{100}$ or more. Equations (29.194) and (30.12) show that during the episode of inflation, the density parameter would have varied according to

$$\Omega(t) = 1 + \frac{kc^2}{(dR/dt)^2} = 1 + \frac{kc^2\tau_i^2}{R^2(t)}.$$

The dramatic increase in $R$ by at least 43 orders of magnitude forced $\Omega \to 1$. In the inflation model, the density parameter and curvature after inflation may be safely specified as $\Omega = 1$ and $k = 0$; the inflationary universe is *flat*. Note that this solution to the flatness problem arises naturally from the physics, instead of having to be specified as an ad hoc condition at the beginning.

The monopoles problem is also solved. Recall that at the end of the GUTs era, it is expected that defects arose at locations where a discontinuity in the choice of an internal

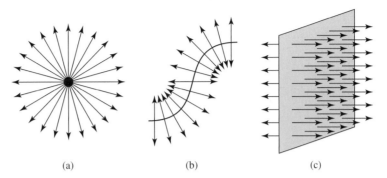

**FIGURE 30.5**   Types of defects: (a) magnetic monopoles, (b) cosmic strings, and (c) domain walls. The arrows illustrate the geometry of the discontinuity involved in each type of defect.

degree of freedom existed. These defects are remnants of the false vacuum, still in a state of unbroken symmetry with an energy density given by Eq. (30.9). Thus it is thought that they should be extremely massive. A pointlike discontinuity corresponds to a magnetic monopole, and it is conjectured that other discontinuities created additional relics. For example, a linear discontinuity gives rise to a **cosmic string**, whereas a planar discontinuity is called a **domain wall**. Figure 30.5 illustrates the geometry of these defects. A fourth type of defect called a **texture** is not shown, since the discontinuity is not localized to a point, line, or plane; it is more like a diffuse three-dimensional clump.

Monopoles and domain walls are so massive that they would easily overwhelm the gravitational effects of ordinary matter if they had not been diluted to insignificance by the inflation of the universe. These defects probably arose at the boundaries between homogeneous regions—that is, near the surfaces of the bubbles of true vacuum. It is thought that each bubble contained only a few monopoles or domain walls. As seen above, when the episode of inflation ended, the presently observable universe occupied an insignificant fraction of its bubble's volume (a few parts in $10^{46}$), so it is expected that magnetic monopoles and domain walls should be exceedingly rare in today's universe.

Cosmic strings and textures could be more common in the present universe, however. A typical cosmic string, a long crack in space filled with false vacuum, might have a linear mass density of $\sim 10^{21}$ kg m$^{-1}$. In the simplest theories, either these thin tubes of false vacuum are infinitely long, or they form closed loops that are conjectured to be hundreds or thousands of parsecs across. A cosmic string might be detected through its influence as a gravitational lens, bending the starlight that passes by on either side to form a double image of the source. Tangled strings would curl and twist as they slowly straighten, forming loops that would sometimes cut the strings up into smaller loops. It is thought that these small loops eventually decay by emitting gravitational waves.

### Matter–Antimatter Asymmetry

Another challenge cosmologists face is explaining why the universe consists of matter rather than antimatter. All but about 0.01% of cosmic rays which sample our Galaxy are matter rather than antimatter. The few antiparticles that are detected can be explained as the result of high-energy interactions, such as the creation of proton–antiproton pairs when two energetic protons collide. Rare ultra-high-energy cosmic rays (UHECR), with energies exceeding $10^{11}$ GeV, may (or may not) originate in M87 in the Virgo cluster. (The source is only suggested, and the trajectory of UHECRs is complicated by the Milky Way's magnetic field.) Observations of UHECRs reveal no evidence of antimatter. Neither is there any sign, in the form of gamma rays of a certain energy, of the violent annihilation reactions that would occur if substantial concentrations of matter and antimatter were to collide. This indicates that within the region of space extending at least as far as the Galactic halo, and perhaps as far as the Virgo cluster, some 16 Mpc away, the universe is overwhelmingly composed of matter.

The explanation of this **matter–antimatter asymmetry** comes from a combination of the details of grand unified theories and the cooling of an expanding universe following inflation. As we have seen, any particles that were present before inflation would have been diluted to insignificance by the exponential expansion. All of the particles in our universe today originated in the burst of particle–antiparticle production that was fueled by the energy (latent heat) released by the false vacuum.[8] The universe was filled with a soup of quarks, leptons, photons, and even more exotic hypothetical particles simply denoted $X$ bosons, and their antiparticles $\overline{X}$. The spontaneous symmetry breaking that ended the GUTs epoch endowed the $X$ particles with mass, just as the vector gauge bosons gained mass when the electroweak unification ended. These extremely massive $X$ particles ($m_X c^2 \sim 10^{15}$ GeV) are not present at the much lower energies that characterize the universe today.

According to the grand unified theories, the $X$ and $\overline{X}$ particles were present in equal numbers and could be transformed into pairs of quarks and antiquarks, respectively, according to

$$X \rightleftharpoons q + q \tag{30.19}$$

$$\overline{X} \rightleftharpoons \overline{q} + \overline{q}. \tag{30.20}$$

It is thought that the first of these reactions occurred slightly more frequently than the second.[9]

A related matter–antimatter asymmetry regarding reaction rates has been seen experimentally. One type of particle, called a *kaon*, can decay into another particle, called a *pion*, by the reactions $K \rightarrow \pi^- + e^+ + \nu_e$ and $K \rightarrow \pi^+ + e^- + \overline{\nu}_e$. The first reaction occurs slightly (but measurably) more frequently than the second.

Initially the temperature was so high that reactions (30.19) and (30.20) occurred equally in both directions. But as the universe cooled and the characteristic particle energy dropped, there was not enough energy available to replace the $X$s and $\overline{X}$s as fast as they decayed. A permanent excess of quarks over antiquarks developed that survived the electroweak symmetry breaking at $t \sim 10^{-11}$ s and lasted until the universe was a few microseconds old. The universe was then cool enough ($T \sim 2 \times 10^{12}$ K) for the quark–hadron transition, when quarks and antiquarks combined to produce huge numbers of baryons and antibaryons, with a slight excess of baryons. Following this, there was a firestorm of particle–antiparticle annihilation that eliminated practically all of the antimatter, leaving only the small excess

---

[8]Alan Guth has referred to this as the "ultimate free lunch."

[9]If you have some acquaintance with modern physics, you will recognize violations of both baryon number and CP (charge and parity) symmetry here. Both violations are integral features of grand unified theories.

of baryons that constitutes the visible matter in the universe today. The barrage of photons that was unleashed has since been cooled by the expansion of the universe to become the cosmic background radiation. Nearly all of the photons in the universe are from the CMB; the number produced by other sources (such as stars) pales in comparison. The ratio of the number of baryons in the universe to the number of photons is about $5 \times 10^{-10}$ (see Problem 30.1). Because the annihilation of a baryon and an antibaryon produces two photons, this ratio implies that there was roughly one unpaired baryon for every one billion baryon–antibaryon pairs. These unpaired baryons were the tiny residue of matter that survived annihilation to make up the material world.

### The CMB and the Decoupling of Matter and Radiation

For several thousand years following the formation of helium nuclei, the universe remained a hot broth of photons, hydrogen and helium nuclei, and electrons. Radiation dominated the dynamics of the expansion, and the scale factor of the universe increased as $R \propto t^{1/2}$. Finally, at $t \simeq 2 \times 10^{12}$ s, the CMB had become sufficiently diluted that massive particles began to govern the expansion. The radiation era came to a close and the matter era began at a temperature of roughly 9000 K.

The prevalence of free electrons continued to obstruct the CMB photons until the temperature fell to about 3000 K, when electrons combined with the nuclei to form neutral atoms. The resulting drop in the opacity decoupled the radiation from the matter in the universe and allowed the photons to roam freely. As we shall see, this decoupling had dramatic implications for the collapse of higher-density regions and the subsequent formation of structure in the universe.

### 30.2 ■ THE ORIGIN OF STRUCTURE

Even a momentary glimpse at the night sky provides convincing evidence that the universe is not perfectly homogeneous. Structure abounds on all scales, from planets to superclusters of galaxies. A typical galaxy is about $10^5$ times more dense than the universe as a whole, so the present universe on a small scale is rather lumpy. This final section of the text deals with the emergence of structure from the featureless cauldron of the Big Bang, with galaxies and clusters of galaxies revealing the presence of early density fluctuations.

### Adiabatic and Isothermal Density Fluctuations

The early universe (before recombination) was filled with a hot plasma of photons and particles. The electrons and photons were tightly coupled by Compton scattering and electromagnetic interactions, and the electrons coupled the protons to the photons. As a result, the photons, electrons, and protons moved together to form a **photon–baryon fluid**. Heisenberg's uncertainty principle guarantees that, because the positions of particles are indeterminate, there must have been density inhomogeneities in the early universe. These disturbances were nearly the same on all scales, meaning that the density fluctuations caused almost the same magnitude perturbation of the gravitational potential. We say that the density fluctuations were nearly **scale-invariant**.[10] Their masses are thought to have ranged anywhere from the mass of stars to the mass of clusters of galaxies, with the less massive fluctuations being more common.

There are two types of density fluctuations. In principle, there can be independent fluctuations of the four components of the early universe: photons, baryons, neutrinos, and dark matter. If the fluctuations are proportional to one another in such a way that there is no energy exchange between them, they are called **adiabatic fluctuations** or **curvature fluctuations** (because the nonzero density fluctuation affects the local curvature of space-time). On the other hand, for **isothermal fluctuations** or **isocurvature fluctuations**, the sum of the four density fluctuation is zero. They involve a *potential* density increase that is suppressed or "frozen in." For example, a spatial variation in the types of particles present may potentially produce a pressure difference, but this could not be converted into an actual density variation while particle motion was inhibited by the close interaction of matter with the uniform radiation field. Any collection of density fluctuations can be expressed as the sum of adiabatic fluctuations and isothermal fluctuations.

The distinction between adiabatic and isothermal fluctuations continued until decoupling, when radiation and matter went their separate ways. Freed from the drag of the surrounding radiation, particles in isothermal fluctuations could respond to differences in pressure and move to produce true density perturbations. The distinction between adiabatic and isothermal fluctuations then vanished. The importance of isothermal fluctuations lies in their ability to act as a deep freeze, preserving potential density fluctuations and protecting them from the dissipation process suffered by the adiabatic fluctuations (described later). Because nothing significant happened to the isothermal fluctuations until decoupling, we will concentrate on the story of their adiabatic counterparts.

### The Development of Adiabatic Density Fluctuations

Quantum mechanical density fluctuations guarantee that, even in its first instant, the Big Bang was not perfectly smooth. Everywhere the density of regions of space vacillated randomly between being underdense and overdense. However, the baryonic matter, coupled to the photons, remained evenly distributed throughout the plasma. At about $10^{-36}$ s, inflation suddenly stretched the universe at an exponential rate so rapid that the size of the overdense and underdense regions vastly exceeded the particle horizon at that time. These density fluctuations were essentially frozen because they were not causally connected. The motion of their particles was inhibited because they could not respond to conditions beyond their horizon. The region therefore passively participated in the expansion of the universe.

To understand how a super-horizon-sized region of enhanced density evolved with time as the universe expanded, we will concentrate on a particularly simple case and imagine that, within an otherwise flat universe of density $\rho(t)$, there was a spherical region with an increased density of $\rho'(t) > \rho(t)$; see Fig. 30.6. Because the dynamics of the spherical density fluctuation depended only on the mass contained within it (recall Birkhoff's theorem of Section 29.1), it evolved as a separate entity, like a miniature closed universe. Assuming

---

[10]We say *nearly* scale-invariant because inflation predicts that the fluctuations were a bit weaker on smaller mass scales. WMAP observations confirming this prediction were announced in March 2006, supporting the inflationary model.

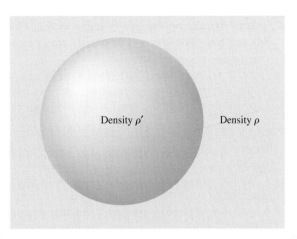

**FIGURE 30.6** A spherical region of enhanced density $\rho'$, in a flat universe of density $\rho$.

that the Hubble flow has always been uniform, the region of higher density expanded at the same rate as its surroundings, so both regions can be described by the same scale factor, $R(t)$, and the same $H(t)$.[11] Equation (29.9) then shows that

$$H^2 R^2 - \frac{8}{3}\pi G\rho' R^2 = -kc^2$$

for the high-density region, and

$$H^2 R^2 - \frac{8}{3}\pi G\rho R^2 = 0$$

for the flat universe outside. Subtracting these results and dividing by $8\pi G\rho/3$ shows that the density fluctuation $\delta\rho/\rho$ is

$$\frac{\delta\rho}{\rho} = \frac{\rho' - \rho}{\rho} = \frac{3kc^2}{8\pi G\rho R^2}. \tag{30.21}$$

During the radiation era, $\rho = \rho_{\text{rad}} \propto R^{-4}$ (Eq. 29.57) and the scale factor varied as $R \propto t^{1/2}$ (Eq. 29.86). Applying these dependencies to Eq. (30.21), we find that the density fluctuation is described by

$$\frac{\delta\rho}{\rho} = \left(\frac{\delta\rho}{\rho}\right)_i \left(\frac{t}{t_i}\right) \qquad \text{(radiation era)}, \tag{30.22}$$

where $(\delta\rho/\rho)_i$ is the value $\delta\rho/\rho$ at some initial time, $t_i$. The amplitude of our super-horizon-sized adiabatic fluctuation therefore increased linearly with time during the radiation era. (An isothermal fluctuation remained frozen with $\delta\rho/\rho = 0$.)

[11]There is actually a slight difference in the scale factors, but for the early universe this can be neglected.

Similarly, during the matter era, $\rho \propto R^{-3}$ (Eq. 29.5) and the scale factor varied as $R \propto t^{2/3}$ for a flat universe (Eq. 29.31), so

$$\frac{\delta\rho}{\rho} = \left(\frac{\delta\rho}{\rho}\right)_i \left(\frac{t}{t_i}\right)^{2/3} \qquad \text{(matter era)}. \qquad (30.23)$$

Note that the increase in $\delta\rho/\rho$ did not involve separation of the high-density region from the Hubble flow. The increase in the density perturbation was due to the difference in the rates at which the density decreased inside and outside the fluctuation as the universe expanded.

We now move ahead to a later time in the radiation era, when the particle horizon had expanded to encompass the entire adiabatic density fluctuation. The region of enhanced density is now causally connected and can respond as a whole to the physical conditions inside. From this point onward, the fate of the fluctuation is determined by the relative values of its mass and the Jeans mass.

The evolution of the Jeans mass as the universe expanded is of primary importance to our story. From Eq. (12.14) for a static (not expanding) medium, the minimum mass required for $\delta\rho/\rho$ to increase with time is[12]

$$M_J \simeq \left(\frac{5kT}{G\mu m_H}\right)^{3/2} \left(\frac{3}{4\pi\rho}\right)^{1/2}. \qquad (30.24)$$

The same expression for the Jeans mass is valid for an expanding universe, although the consequences of falling below this minimum mass are different, as will be explained shortly.

The Jeans mass can be written in terms of the adiabatic sound speed. In general, the speed of sound $v_s$ is given by

$$v_s = \sqrt{\frac{\partial P}{\partial \rho}}. \qquad (30.25)$$

For the familiar adiabatic sound waves in air, $P = C\rho^\gamma$, where $C$ is a constant, and we recover Eq. (10.84),

$$v_s = \sqrt{\frac{\gamma P}{\rho}} = \sqrt{\frac{5kT}{3\mu m_H}}, \qquad (30.26)$$

where the last expression comes from the ideal gas law, Eq. (10.11), with $\gamma = 5/3$ for an ideal monatomic gas. Rearranging and substituting into Eq. (30.24), the Jeans mass becomes

$$M_J = \frac{9\rho}{2\pi^{1/2}} \frac{v_s^3}{(G\rho)^{3/2}}. \qquad (30.27)$$

The temperature dependence of the Jeans mass changed dramatically at the time of recombination. Although it is not obvious from Eq. (30.27), a more careful derivation of the Jeans

---

[12]From the derivation of the Jeans mass in Section 12.2, it should be noted that $M_J$ must be in the form of baryonic matter. For example, the use of $K = 3/2\, NkT$ in the derivation of $M_J$ may not be applied to nonbaryonic particles.

mass applicable to the time before recombination reveals that the $\rho$ in the numerator is $\rho_b$, the baryonic mass density. Because the temperatures of radiation and matter were equal during this epoch, Eqs. (29.5) and (29.58) show that $\rho_b = \rho_{b,0}/R^3 = \rho_{b,0}T^3/T_0^3$. However, the $\rho$ in the denominator of Eq. (30.27) includes the dominant effect of the photons, so $\rho \propto T^4$ in the denominator (Eq. 29.77). Furthermore, before recombination $P_{rel} = u_{rel}/3 = \rho_{rel}c^2/3$ (Eq. 9.11), so the speed of sound was

$$ v_s = \sqrt{\frac{\partial P}{\partial \rho}} = \frac{c}{\sqrt{3}}, \qquad (30.28) $$

about 58 percent of the speed of light. The result, as shown in Fig. 30.7, is that the Jeans mass was proportional to $T^{-3}$ until the time of recombination approached. After recombination, the density in the denominator reverted to a mass density, and the sound speed plunged to assume its ideal gas value. Because radiation and matter are decoupled after recombination, the mass density is independent of the temperature. The only temperature dependence is that of the sound speed, Eq. (30.26), and so after recombination the Jeans mass is proportional to $T^{3/2}$. Figure 30.7 shows the temperature dependence of the Jeans mass. In reality, the presence of dark matter levels off the sharp peak before the time of recombination to produce a plateau at about $M_J \simeq 10^{16}\ M_\odot$.

Before recombination occurred, the Jeans mass was about an order of magnitude *greater* than the amount of baryonic matter contained within a causally connected region (see Problem 30.7), so the mass of a sub-horizon-sized density fluctuation was always less than the Jeans mass. Consequently, during this period the growth of an adiabatic density fluctuation was suppressed; in other words, $\delta\rho/\rho$ remained constant until the passage to the matter era.

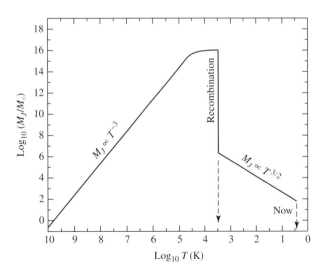

**FIGURE 30.7** The variation of the Jeans mass, $M_J$, with temperature as the universe expands. The sharp peak at the time of recombination is actually leveled off by the presence of dark matter at about $M_J \simeq 10^{16}\ M_\odot$.

## Acoustic Oscillations and Damping

From the time an adiabatic fluctuation became sub-horizon-sized until recombination, it underwent acoustic oscillations reminiscent of stellar pulsation (see Section 14.2). Sound waves traversed its volume at a substantial fraction of the speed of light ($c/\sqrt{3}$) until recombination occurred. These sound waves produced regions of compression (slightly hotter) and rarefaction (slightly cooler) that left their imprint on the CMB when decoupling occurred. We will return to these immensely useful acoustic oscillations after our discussion of the fate of the density fluctuations.

The smaller adiabatic fluctuations did not survive the phase of acoustic oscillations, since photons could diffuse and leak out of compressed regions when the wavelengths of the sound waves were short enough. The close interaction between the escaping radiation and matter then produced a smoothing effect that damped out the short-wavelength compressions.

## The Minimum Mass Required to Survive Damping

To determine how large a fluctuation needed to be to survive this damping, consider a region of enhanced density just prior to the time of decoupling, at $t_{\text{dec}} = 3.79 \times 10^5$ yr (Eq. 29.102). As a rough estimate, the density fluctuation would survive if the size of the region were larger than the distance that a diffusing, random-walking photon could travel during the age of the universe up to decoupling. The mean free path of the photons is given by Eq. (9.12), $\ell = 1/n\sigma$, where $\sigma = 6.65 \times 10^{-29}$ m$^2$ is the electron-scattering cross section (Eq. 9.20) and $n$ is the electron number density. The number of scatterings suffered by a photon in the age of the universe is then

$$N = \frac{t_{\text{dec}}}{\ell/c} = n\sigma c t_{\text{dec}}.$$

The average displacement from the photon's starting point can be obtained from Eq. (9.29),

$$d = \ell\sqrt{N} = \frac{\sqrt{n\sigma c t_{\text{dec}}}}{n\sigma} = \sqrt{\frac{c t_{\text{dec}}}{n\sigma}}.$$

A sound wave with a wavelength longer than $d$ could survive the damping effect of the diffusing photons. The damping of the shorter-wavelength oscillations is known as **Silk damping**, named for American astronomer Joseph Silk.

The minimum mass of a surviving density fluctuation must have been roughly the mass contained within a sphere of radius $d$. Equations (29.5) and (29.4) show that the baryonic mass density at the time of decoupling (when $z = 1089$) was

$$\rho_b = \frac{\rho_{b,0}}{R^3} = \rho_{b,0}(1+z)^3 \simeq 5.4 \times 10^{-19} \text{ kg m}^{-3}, \tag{30.29}$$

where the present baryonic mass density, $\rho_{b,0}$, is given by (Eq. 29.17). The number density of electrons can now be found from $n \simeq \rho_b/m_H$, assuming, for convenience, composition of pure hydrogen (one electron for every proton). Setting $\rho = \rho_b$, the mass contained in this spherical volume was then about

$$\frac{4}{3}\pi d^3 \rho_b = \frac{4\pi}{3\rho_b^{1/2}} \left( \frac{m_H c t_{dec}}{\sigma} \right)^{3/2} = 7.7 \times 10^{13} \ M_\odot.$$

Any adiabatic fluctuations with a mass smaller than this did not survive past recombination. It is interesting that this value of several $\times 10^{13} \ M_\odot$ is reminiscent of the mass of an elliptical cD galaxy or a small cluster of galaxies.

### The Isothermal Density Fluctuations

We now return our attention to the isothermal density fluctuations. Density perturbations less massive than some $10^{13} \ M_\odot$ existed after recombination only because the "frozen" isothermal fluctuations were preserved from the dissipative effects of the acoustic oscillation phase. After recombination, these fluctuations were transformed into actual regions of enhanced density. The pressure perturbations, formerly restrained by the radiation field, were freed to move particles and produce true density differences. Consequently, there is no further need to distinguish between adiabatic and isothermal density fluctuations.

### The Jeans Mass after Recombination

The deviations from the average density were fairly small at the time of recombination ($\delta\rho/\rho \sim 10^{-4}$, as will be seen later), but the fluctuations encompassed various amounts of mass, presumably ranging from stellar to galactic cluster values. We can evaluate the Jeans mass just after recombination, when $T \simeq 2970$ K and the density is given by Eq. (30.29). With a mean molecular weight of $\mu = 0.584$ for an assumed composition of $X = 0.77$ and $Y = 0.23$ (see Eq. 10.16), the value of the Jeans mass was

$$M_J \simeq \left( \frac{5kT}{G\mu m_H} \right)^{3/2} \left( \frac{3}{4\pi\rho} \right)^{1/2} \simeq 1.9 \times 10^6 \ M_\odot,$$

comparable to the mass of a globular cluster. After recombination, fluctuations with masses exceeding about $10^6 \ M_\odot$ were amplified. The rate of growth of the density perturbations is just that given by Eq. (30.23), $\delta\rho/\rho = (\delta\rho/\rho)_i (t/t_i)^{2/3}$.

### The Timing of Structure Formation

It is remarkable that the two characteristic mass values that have emerged from our analysis of density fluctuations, $10^6 \ M_\odot$ and $10^{13} \ M_\odot$, neatly span a range of masses that includes most of the galactic objects of interest, from stellar clusters to clusters of galaxies. We will now follow the evolution of these higher-density regions and investigate whether they could have produced the objects that are actually observed in the universe.

---

**Example 30.2.1.** The highest-redshift quasar is J114816.64+525150.3 (as of this writing). Discovered by the Sloan Digital Sky Survey, its redshift is $z = 6.43$, corresponding to $R_q = 1/(1+z) = 0.135$ (the $q$ subscript denotes "quasar"). Using Eq. (29.129) reveals

*continued*

that the age of the universe when the light left the quasar is, using WMAP values, $2.75 \times 10^{16}$ s $= 872$ Myr. We will assume that the gravitational collapse of a region can begin when $\delta\rho/\rho \approx 1$. At this point, nonlinear effects become important, and the region separates from the Hubble flow and starts to collapse. Setting $(\delta\rho/\rho)_q = 1$ for the collapse that formed this farthest quasar, Eq. (30.23) can be used to find the size of the density fluctuation at the time of recombination, $[t_{\mathrm{dec}}]_{\mathrm{WMAP}} = 3.79 \times 10^5$ yr (see Eq. 29.102). The result is

$$\left(\frac{\delta\rho}{\rho}\right)_{\mathrm{rec}} = \left(\frac{\delta\rho}{\rho}\right)_q \left(\frac{t_{\mathrm{dec}}}{t_q}\right)^{2/3} \simeq 0.005.$$

To manufacture the most distant quasars, there must have been density fluctuations of several tenths of a percent at the time decoupling occurred.

---

The results of this example are contradicted by careful observations of the anisotropies in the cosmic background radiation. At the time of decoupling, photons would have experienced a gravitational redshift as they climbed out of the gravitational potential wells that surrounded the largest fluctuations. (These largest fluctuations are discussed further on page 1268.) The radiation from regions of enhanced density should appear slightly cooler than the background CMB, while radiation from lower-density regions should appear a bit hotter. These fluctuations in the temperature of the CMB have been measured to be in the range of $\delta T/T \simeq 1\text{--}7.5 \times 10^{-5}$. These properties can be translated into density variations by using Eqs. (29.5) and (29.58), which together imply that $\rho \propto T^3$. It is left as an exercise to show that the linearized version of this relationship is

$$\frac{\delta\rho}{\rho} = 3\frac{\delta T}{T}. \tag{30.30}$$

The resulting values of $\delta\rho/\rho \simeq 0.3\text{--}2.3 \times 10^{-4}$ are about two orders of magnitude less than those predicted from arguments like the one made in Example 30.2.1. The accumulations of baryonic matter represented by these temperature fluctuations could not have grown fast enough by themselves to collapse and form the large-scale structures that are common in the present universe. There must have been density fluctuations hundreds of times greater than these at the time of recombination, even though this simple analysis of baryonic matter does not detect them.

Nonbaryonic dark matter may provide the solution to this puzzle. If this dark matter has a negligible interaction with radiation, it could have started to accumulate long before decoupling occurred, at the beginning of the matter era.[13] (Recall that baryonic fluctuations could not grow during this time because the close interaction of radiation and baryonic matter resulted in a sound speed of $c/\sqrt{3}$, which in turn produced a prohibitively high value of the baryonic Jeans mass.) During the matter era, these concentrations of dark matter would have grown at the rate given by Eq. (30.23), $\delta\rho/\rho = (\delta\rho/\rho)_i(t/t_i)^{2/3}$. By the time of decoupling, the relative density enhancement of the dark matter could have reached the level of several tenths of a percent, as required by Example 30.2.1. After decoupling,

---

[13]During the earlier radiation era, the universe was expanding too fast to allow any sub-horizon-sized density fluctuations to grow, regardless of their composition.

**TABLE 30.3** Redshifts for Structure Formation. Approximate redshifts at the time of the formation of various structures. (Adapted from Peebles, *Principles of Physical Cosmology*, Princeton University Press, Princeton, NJ, 1993.)

| Structure | Redshift |
|---|---|
| Density fluctuations in CMB | $[z_{dec}]_{WMAP} = 1089$ |
| Spheroids of galaxies | $z \sim 20$ |
| The first engines of active galactic nuclei | $z \geq 10$ |
| The intergalactic medium | $z \sim 10$ |
| Dark halos of galaxies | $z \sim 5$ |
| The first 10% of heavy elements | $z \geq 3$ |
| Rich clusters of galaxies | $z \sim 2$ |
| Thin disks of spiral galaxies | $z \sim 1$ |
| Superclusters, walls, and voids | $z \sim 1$ |

baryonic matter would have been gravitationally attracted to these clumps of dark matter. It is thought that the sizes of the resulting baryonic fluctuations would quickly catch up with those of the underlying dark matter and that thereafter they would share the same value of $\delta\rho/\rho$ as they grew to form the concentrations of ordinary matter that are visible throughout the universe. When $\delta\rho/\rho \approx 1$, these regions of dark and baryonic matter would detach from the Hubble flow and begin their collapse to form the spheroidal components of galaxies, the central engines of quasars, and other early-forming structures; see Table 30.3. However, the time constraints are severe, and it is still not clear whether the collapse of regions of dark matter can account for the earliest observed structures.

### Determining When the First Stars and Galaxies Formed

A galaxy discovered far beyond the galactic cluster Abell 2218 illustrates just how tight the timing for galaxy formation is. With a redshift of approximately $z \simeq 7$, this is the most remote galaxy found so far. The light we observe left this galaxy when the universe was just $1/(1 + z) = 0.125$ of its present size. Using Eq. (29.129) with WMAP values, this corresponds to a time just 780 million years after the Big Bang, just 5.7% of the present age of the universe. The galaxy is tiny, perhaps just 600 pc in diameter. Nevertheless, it is a site of extremely active star formation. It is fortunate that Abell 2218 served as a gravitational lens that magnified the image of this farthest galaxy and allowed it to be detected. The presence of such a young galaxy serves as a severe constraint on theories of structure formation. Because it formed less than 1 billion years after the Big Bang, it must have formed extremely rapidly (perhaps by the process of dissipative collapse discussed in Section 26.2).

One clue pointing to the time when the first stars and galaxies were formed comes from observations of the Lyman-$\alpha$ forest in high-$z$ quasars (recall Section 28.4). The closely spaced Ly$\alpha$ absorption lines are produced by clouds of neutral hydrogen of smaller redshift that lie between the quasar and Earth. Neutral hydrogen is a very efficient absorber of 122-nm (UV) photons. However, if, after recombination at $z = 1089$, all of the hydrogen in the intergalactic medium remained neutral, then almost all of the Ly$\alpha$ forest should be reduced to zero due to absorption by the intergalactic neutral hydrogen. In effect, the neutral hydrogen in the intergalactic medium would act as an absorbing cloud with a continually declining redshift in the direction of Earth. This indicates that the hydrogen in the intergalactic medium is not neutral but almost completely ionized.

We conclude that after recombination, the universe entered a "Dark Age," before the first stars and galaxies had formed and started to shine. Then UV radiation from the first generation of stars and AGN reionized the universe, and it has remained ionized until the present time. If we can find a flattening of the Ly$\alpha$ forest for a high-$z$ quasar, then we know that we are observing that quasar during the **epoch of reionization**. This flat region in a quasar's spectrum is called the **Gunn–Peterson trough**, named for the astronomers who predicted it, James Gunn (American) and Bruce Peterson (Australian). Figure 30.8 shows the Gunn–Peterson trough for four high-$z$ quasars. The relatively rapid diminishing of the Ly$\alpha$ forest indicates that the reionization of the intergalactic neutral hydrogen was finished by $z = 6$ and that its final stages occurred very quickly. We may very well be penetrating back to the end of the "Dark Age."

According to the WMAP results, the first stars ignited about 200 million years after the Big Bang. The Spitzer Space Telescope may have detected the light (shifted to infrared wavelengths) from the first generation of objects that initiated the reionization of the universe. Following a deep, 10-hour exposure in the direction of the constellation Draco, all known objects were carefully subtracted from the image. An infrared background remained, with blobs that may be the glow from the very first stars (Population III), as shown in Fig. 30.9.

### Top-Down Galaxy Formation and Hot Dark Matter

As noted previously, it is intriguing that the two mass values to emerge from recombination roughly span the mass range of galactic structure. The Jeans mass of $M_J \simeq 10^6$ M$_\odot$ is typical of a globular cluster of stars. In addition, the lower limit for the mass of adiabatic fluctuations that survived the acoustic oscillation phase before recombination, $10^{13}$ M$_\odot$, is characteristic of an elliptical cD galaxy or a small cluster of galaxies. (Adiabatic density fluctuations and hot dark matter are thought to behave similarly, because they both tend to resist clumping through the diffusion of photons and fast-moving particles.)

The resulting top-down process of galaxy formation involves the breakup of larger structures. The problem with this process is that the breakup may occur too late to be consistent with the observed times of formation of the earliest galaxies. Any clumping that occurs involves large amounts of mass ($10^{13}$ to $10^{15}$ M$_\odot$, representative of galactic clusters and superclusters) and so must be involved in a top-down process for forming galaxies from the breakup of larger entities.

### Bottom-Up Galaxy Formation and Cold Dark Matter

On the other hand, cold dark matter is slow moving and should accumulate much more easily. Concentrations of cold dark matter can begin to collect on a variety of mass scales when the matter era begins, resulting in a bottom-up scenario for forming galaxies from the assembly of smaller components. The difficulty with the cold dark matter model is that numerical simulations of the early universe have trouble reproducing the galactic voids found in redshift surveys, such as those shown in Fig. 27.23. It has been found that if

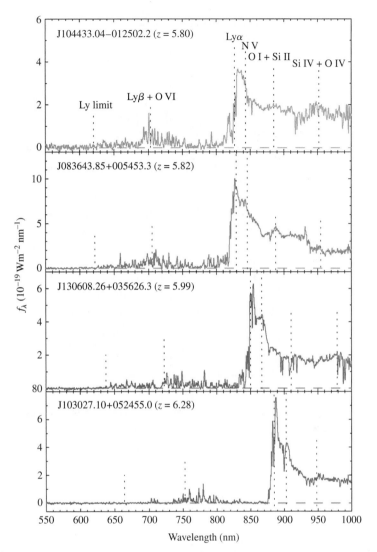

**FIGURE 30.8** The Gunn–Peterson trough observed in four high-$z$ quasars illustrates the rapid suppression of the Ly$\alpha$ forest with increasing $z$. This indicates that the abundance of ionized hydrogen has declined significantly from $z \sim 5$ to 6 and that the universe is approaching the reionization epoch at $z \sim 6$. Note that cgs units and angstroms are used in this figure. (Figure adapted from Becker et al., *A. J.*, *122*, 2850, 2001.)

galaxy formation is *biased*, so galaxies tend to form more readily in overdense regions, this problem with cold dark matter can be overcome. Figure 30.10 shows numerical simulations of the formation of structure from the bottom-up clumping of cold dark matter and from the top-down fragmentation of hot dark matter. As explained in Sections 24.2 and 26.2, current observations and interpretations strongly favor a bottom-up process for the formation of galaxies. The bright galaxies we observe today were assembled from fragments at high redshift.

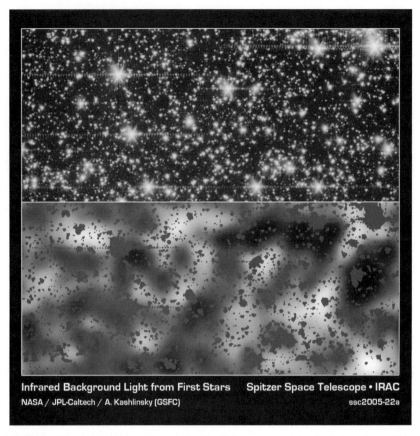

**FIGURE 30.9**   The top panel shows the results of a 10-hour exposure by the Spitzer Space Telescope. The lower panel shows the infrared emission, perhaps from the first generation of stars, that remains when all known objects have been subtracted.

The presence of high-redshift galaxies and quasars presents severe challenges for every theory of structure formation. The Sloan Digital Sky Survey has discovered several luminous quasars with $z > 6$ that indicate the existence of supermassive black holes of $10^9$ to $10^{10}$ $M_\odot$ when the universe was less than 1 Gyr old. It is significant that the most realistic numerical simulation of structure formation can reproduce these very early components of the universe. The Virgo consortium, an international group from the United Kingdom, Germany, Japan, Canada, and the United States, computed the behavior of $\sim 10^{10}$ particles as they evolved from $z = 127$ to form some $2 \times 10^7$ galaxies in a cube with sides of $500h^{-1}$ Mpc, a project the consortium called the Millennium Simulation. The results (see Fig. 30.11) demonstrate that a few massive black holes can form quickly enough (by a few $10^8$ years) to account for the SDSS observations. These black holes evolve to become the most massive galaxies, which reside at the center of the largest clusters of galaxies.

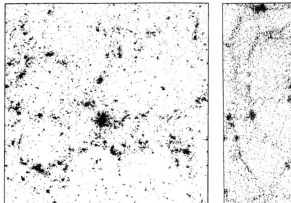

**FIGURE 30.10**  The results of numerical simulations of the formation of structure in the universe from the bottom-up clumping of cold dark matter (left), and the top-down fragmentation of hot dark matter (right). Both models assume $\Omega_0 = 1$ and $h = 0.5$. (Figure from Frenk, *Physica Scripta*, *T36*, 70, 1991.)

### Cosmic Harmonics and Acoustic Oscillations

"The music of earth sings through a thousand holes.
The music of man is made on flutes and instruments.
What makes the music of heaven?"
— Chuang Tzu (4th century B.C.)[14]

We now return to the acoustic oscillations mentioned on page 1252. Recall that before recombination, the photons, electrons, and protons moved together to form a photon–baryon fluid. Quantum mechanical density fluctuations caused the density of regions of space to vary randomly between being underdense and overdense. At about $10^{-36}$ s, inflation suddenly stretched the universe at an exponential rate, so fast that the size of the overdense and underdense regions vastly exceeded the particle horizon at that time. These density fluctuations were essentially frozen; they could not respond as a whole because they were not causally connected.

When the particle horizon finally grew large enough to engulf the density fluctuations, the fluctuations were set free to react to their environment. At this time, when the universe was roughly $10^5$ years old, the universe became filled with sound waves that traveled through the plasma. As the density of a region was enhanced by a compression, it was heated by a small amount ($\delta T/T \approx 10^{-5}$) until radiation pressure reversed the motion. Similarly, as the density was diminished by an expansion, it cooled. The small amplitude of the sound waves means that a passing wave will cause the photon–baryon fluid to execute simple harmonic motion. This motion continued until the time of decoupling, when the electrons combined with the protons and heavier nuclei formed in Big Bang nucleosynthesis. The

[14]Quotation from *The Way of Chuang Tzu*, Thomas Merton (trans.), New Directions, 1965.

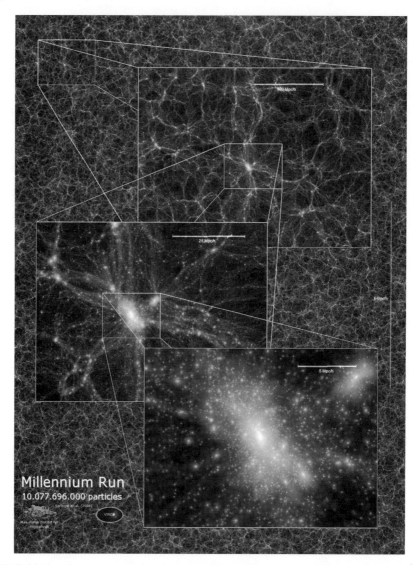

**FIGURE 30.11**   The Millennium Simulation of 200 million galaxies. This is a 15 Mpc $h^{-1}$ thick slice of the results at $z = 0$. The panels zoom in by a factor of four. The maximum magnification shows the substructure of a rich cluster of galaxies. (Figure courtesy of the Virgo consortium.)

photons of the CMB, now released to traverse the universe freely, carried the signature of the acoustic oscillations. Some areas of the CMB appear slightly hotter (higher frequency), others slightly cooler.

The frequency of the photons is also affected by the **Sachs–Wolfe effect** as the photons lose energy when they climb out of the gravitational potential wells of the density fluctuations. Overdense regions cause the photon frequency to be reduced slightly (hence appearing cooler), whereas underdense regions have the opposite effect.

$-200$    $T(\mu K)$    $+200$

**FIGURE 30.12** The temperature fluctuations in the CMB observed by the WMAP orbiting observatory. (Figure adapted from Bennett et al., *Ap. J. Suppl.*, *148*, 1, 2003.)

The overall pattern of CMB temperature variations on the celestial sphere (see Fig. 30.12) can be expressed as the sum of spherical harmonic functions $Y_m^\ell(\theta, \phi)$, encountered previously in Chapter 14. Referring to Fig. 14.16, note that the pattern for a given $Y_m^\ell(\theta, \phi)$ has $\ell$ nodal lines [where $Y_m^\ell(\theta, \phi) = 0$], with $|m|$ of these lines passing through the figure's north pole ($\phi = 0$).

The temperature fluctuation in the direction of the angular position $(\theta, \phi)$ is

$$\frac{\delta T(\theta, \phi)}{T} = \sum_{\ell=1}^{\infty} \sum_{m=-\ell}^{\ell} a_{\ell,m} Y_m^\ell(\theta, \phi).$$

The coefficients $a_{\ell,m}$ are in general complex numbers because the spherical harmonics themselves are complex. The value of the $a_{\ell,m}$ can be determined by observing the CMB and making measurements of $\delta T/T$ in all directions. The $\ell = 1$ is the dipole anisotropy of the CMB (the Doppler shift of the CMB caused by an observer's peculiar velocity through space relative to the Hubble flow; see Eq. 29.62), which is ignored in the following analysis.

To remove the effect of the arbitrary choice of the direction of $\phi = 0$, an angular average over the $2\ell + 1$ values of $m$ can be taken:

$$C_\ell = \frac{1}{2\ell + 1} \sum_{-\ell}^{\ell} |a_{\ell,m}|^2 = \frac{1}{2\ell + 1} \sum_{-\ell}^{\ell} a_{\ell,m} a_{\ell,m}^*,$$

where $a^*$ is the complex conjugate of $a$. The angular power spectrum is then defined to be $\ell(\ell + 1)C_\ell/2\pi$. Note that every term in the summation is $\geq 0$, so both positive and negative temperature fluctuations contribute to the angular power spectrum; they do not cancel each other.

The values of $C_\ell$ contain much information about the physical conditions and constituents of the early universe. Figure 30.13 shows the angular power spectrum of the CMB and the best-fit theoretical $\Lambda$CDM model. This is an amazing result. In musical terms, it is analogous to the waveform generated by displaying the sound of an instrument (a flute, say) on an

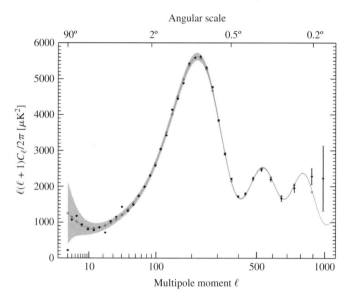

**FIGURE 30.13** The angular power spectrum of the temperature fluctuations in the CMB. The solid line is the best-fit $\Lambda$CDM model. [Figure adapted from Hinshaw et al., *Ap. J.* (submitted), 2006, and courtesy of NASA and the WMAP Science Team.]

oscilloscope. The waveform identifies the fundamental and harmonic frequencies of the flute, and the height of the peaks reveals the relative power of the harmonics. Similarly, Fig. 30.13 shows the fundamental (the highest peak) and two harmonics of the acoustic oscillations of the early universe. Just as the air inside the flute has several modes of vibration, so does the CMB on the celestial sphere.

## Fine-Tuning Our Understanding of Cosmic Harmonics

But why does the cosmic symphony involve these particular harmonics? Why is there a low, flat plain for small values of $\ell$, a fundamental peak around $\ell \simeq 200$ (which corresponds to an angular size of about $1°$ on the sky), and then a few harmonics peaks whose heights decline as $\ell$ spproaches 1000? These are difficult questions because we are actually looking at a cross section of the sound waves, where they intersect the surface of last scattering (a complication we will conveniently ignore). Nevertheless, we can use some simple models to understand the main features of the angular power spectrum.

Let's begin with the first peak, at $\ell \simeq 200$. We want to determine the angular diameter of the largest region that could have responded to a sound wave. For the region to respond as a whole, a sound wave must have been able to travel across the region by the time of decoupling $t_{\text{dec}}$. This is known as the **sonic horizon distance**. Because the speed of sound before decoupling is proportional to the speed of light, the sonic horizon distance $d_s$ is related to the particle horizon distance $d_h$ by $d_s = d_h/\sqrt{3}$. The time of radiation-matter equality is much earlier than the time of decoupling, $t_{r,m} \ll t_{\text{dec}}$ [see Eqs. (29.85) and (29.102)], so the acoustic oscillations occurred primarily during the matter era. Employing Eq. (29.155) for the particle horizon distance during the matter era, $d_h = 3ct$, the sonic horizon distance is thus

$$d_s(t) = ct\sqrt{3}.$$

Using Eq. (29.102) for the time of decoupling, we obtain

$$d_s(t_{\text{dec}}) = ct_{\text{dec}}\sqrt{3} = 6.21 \times 10^{21} \text{ m} = 201 \text{ kpc}.$$

We now estimate the proper distance to the surface of last scattering, the origin of the CMB photons and the site of our region of diameter $d_s(t_{\text{dec}})$. For a flat universe, the proper distance $d_p$ and angular distance $d_A$ are related by [cf. Eqs. (29.169), (29.173), and (29.191)],

$$d_p = d_A(1+z) = \frac{D(1+z)}{\theta}, \qquad (30.31)$$

where $\theta$ is the angular diameter of an object of physical diameter $D$ located at redshift $z$. Using Eq. (29.145) and $R(t) = Ct^{2/3}$ for the matter era [Eq. (29.91), $C$ is a constant], we obtain the proper distance

$$d_p = Ct_0^{2/3} \int_{t_{\text{dec}}}^{t_0} \frac{c\,dt}{Ct^{2/3}} = 3ct_0 \left[1 - \left(\frac{t_{\text{dec}}}{t_0}\right)^{1/3}\right] = \frac{D(1+z)}{\theta}. \qquad (30.32)$$

Setting $D = d_s(t_{\text{dec}})$ and using $[z_{\text{dec}}]_{\text{WMAP}} = 1089$ along with the WMAP values of $t_{\text{dec}}$ and $t_0$, we find that

$$\theta = 1.79 \times 10^{-2} \text{ rad} = 1.03°,$$

which corresponds to $\ell \simeq \pi/\theta = 175$, reasonably close to the observed value of $\ell \simeq 200$ for the first acoustic peak. A more detailed analysis shows that the location of the first peak is sensitive to the value of $\Omega_0$ and is given by $\ell \simeq 200/\sqrt{\Omega_0}$. In effect, we are looking at a region of known size $d_s$ through the gravitational lens of the overall geometry of the universe, as determined by $\Omega_0$, and comparing its angular size with our theoretical expectations. The Boomerang observatory,[15] carried aloft by a stratospheric long-duration balloon, was one of the first to accomplish this measurement. Figure 30.14 illustrates the comparison of Boomerang's observations with theoretical simulations for a closed, a flat, and an open universe. The results convincingly demonstrate that the universe is indeed very nearly flat.

## A Simple Model of Acoustic Oscillations

To investigate further the acoustic oscillations, we will consider a model problem that still retains the essential physics. Consider a cylinder of cross-sectional area $A$ and length $2L$ that is filled with a gas; see Fig. 30.15. In order to model the ebb and flow of the photon–baryon

[15]The acronym stands for Balloon Observations Of Millimetric Extragalactic Radiation ANisotropy and Geophysics.

**25°**

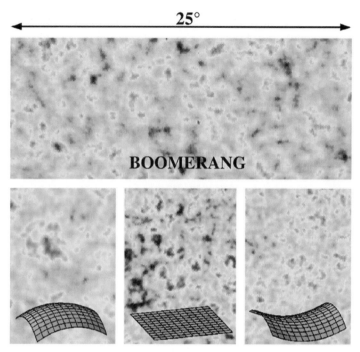

**BOOMERANG**

**FIGURE 30.14** A comparison of Boomerang's observations (top) of temperature fluctuations in the CMB with theoretical simulations for a closed (left), flat (center), and open (right) universe. The results show that the universe is very nearly flat. (Image courtesy of the Boomerang Collaboration.)

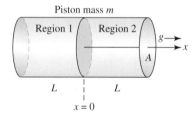

**FIGURE 30.15** A cylinder of length $2L$ and cross-sectional area $A$ filled with a gas. A movable piston is at the middle of the cylinder.

fluid as a sound wave passes through it, we want to study how the gas in the cylinder can experience simple harmonic motion. Of course the gas in the cylinder will form a standing sound wave, but we want to approximate the physics rather than write down the exact answer. To this end, let there be a movable piston in the middle of the cylinder. The piston represents the inertia of the baryons, so we will set the mass of the piston equal to the mass of the gas in the cylinder. If the equilibrium values of the pressure and density are $P_0$ and $\rho_0$, respectively, then the mass of the piston is $m = 2LA\rho_0$.

If the piston is displaced, the density on either side will change by a small amount $\Delta\rho$ from equilibrium. This is accompanied by a pressure difference

$$\Delta P = P_1 - P_2,$$

where, to first order,

$$P_1 = P_0 + \frac{dP}{d\rho}\Delta\rho_1$$

and

$$P_2 = P_0 + \frac{dP}{d\rho}\Delta\rho_2.$$

Using Eq. (30.25) we have

$$\Delta P = \frac{dP}{d\rho}(\Delta\rho_1 - \Delta\rho_2) = v_s^2(\Delta\rho_1 - \Delta\rho_2).$$

For a displacement $x$ of the piston, and noting that the mass of the gas on either side of the piston does not change,

$$\Delta\rho_1 = \rho_1 - \rho_0 = \rho_0\left(\frac{\rho_1}{\rho_0} - 1\right) = \rho_0\left(\frac{L}{L+x} - 1\right)$$

and similarly

$$\Delta\rho_2 = \rho_2 - \rho_0 = \rho_0\left(\frac{\rho_2}{\rho_0} - 1\right) = \rho_0\left(\frac{L}{L-x} - 1\right).$$

Thus

$$\Delta P = v_s^2\rho_0\left(\frac{1}{1+x/L} - \frac{1}{1-x/L}\right)$$

so, to first order in $x$,

$$\Delta P = -2v_s^2\rho_0\left(\frac{x}{L}\right).$$

Newton's second law for the piston of mass $m = 2LA\rho_0$ is then

$$m\frac{d^2x}{dt^2} = A\Delta P$$

or

$$2LA\rho_0\frac{d^2x}{dt^2} = -2v_s^2A\rho_0\left(\frac{x}{L}\right).$$

Solving for the acceleration of the piston, we find the equation of motion to be

$$\frac{d^2x}{dt^2} = -\frac{v_s^2}{L^2}x.$$

The resulting simple harmonic motion of the piston,

$$x = x_0 \sin(\omega t),$$

has angular frequency

$$\omega = \frac{v_s}{L}. \tag{30.33}$$

This is in reasonable agreement with the fundamental frequency for an organ pipe closed at both ends, $\omega = \pi v_s/2L$. This shows that more extensive density fluctuations (larger $L$) will oscillate with a lower frequency.

Now we will add a uniform gravitational field of strength $g$ directed in the positive-$x$ direction. This represents the effect of a concentration of dark matter on the baryons in the photon–baryon fluid. Because dark matter is not subject to the intense radiation pressure of the CMB, it can form clumps capable of gravitationally assisting or resisting the motion of the fluid. Returning to our cylinder, Newton's second law now gives

$$m\frac{d^2x}{dt^2} = A\Delta P + mg,$$

which leads to

$$\frac{d^2x}{dt^2} = -\frac{v_s^2}{L^2}x + g.$$

To solve this differential equation, define

$$y \equiv x - \frac{L^2}{v_s^2}g$$

so Newton's second law becomes

$$\frac{d^2y}{dt^2} = -\frac{v_s^2}{L^2}y.$$

As before, the piston executes simple harmonic motion with angular frequency $\omega = v_s/L$, but now the oscillations are about $y = 0$, which corresponds to an equilibrium position of $x_{eq} = L^2g/v_s^2 > 0$. This means that the compressions in Region 2, in the direction of the gravitational field, are of greater magnitude than are the rarefactions in Region 2. The equilibrium density in Region 2 (when $y = 0$) is

$$\rho_{2,eq} = \rho_0 \left(\frac{L}{L - x_{eq}}\right), \tag{30.34}$$

which is greater than $\rho_0$, the average density of the gas in the cylinder.

We will take Region 2 to be a model of the oscillation of the photon–baryon fluid surrounding a local concentration of dark matter. The model indicates that the starting state of a density fluctuation is more likely to be one that would fall inward toward the dark matter rather than expand outward. The reason for this bias is that the quantum density fluctuations varied randomly between being underdense ($\rho_2 < \rho_0$ in our model) and overdense ($\rho_2 > \rho_0$) relative to the average density of the universe ($\rho_0$ in our model), and not relative to the density of the equilibrium configuration in the presence of dark matter ($\rho_{2,\text{eq}}$). Similarly, if the initial position of our model's piston were chosen randomly about $x = 0$, we would be more likely to get a model for which Region 2 would initially move in the $+x$ direction (in the direction of the gravitational field) because $x_{\text{eq}} > 0$ introduces a bias. Therefore, in the early universe, collapsing initial configurations were favored over expanding ones. Because the compressions in the direction of the gravitational field are of greater magnitude than are the rarefactions, we expect the peaks of the angular power spectrum to be enhanced for odd harmonics (compressions), and diminished for even harmonics (rarefactions).

### The Implications of the Angular Power Spectrum Peaks

As shown in Fig. 30.16, the first peak in the angular power spectrum of the CMB (Fig. 30.13) is due to the compression of a large region (large $L$) that reached its maximum compression at the time of decoupling. The first trough is produced by a smaller region (smaller $L$) that started oscillating earlier, when it became sub-horizon-sized. This smaller region was able to oscillate more quickly (larger $\omega$, from Eq. 30.33), so that it arrived at $\delta T = 0$ at the

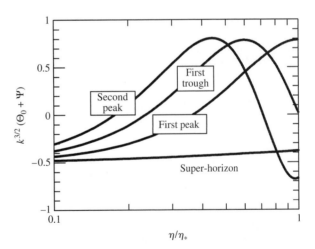

**FIGURE 30.16**   Several modes of the acoustic oscillations in the early universe. The sum of the potentials plotted on the vertical axis can be interpreted as $\delta T$, so positive values indicate a compression of the photon–baryon fluid and negative values indicate a rarefaction. The time $\eta$ is plotted on the horizontal axis as a fraction of the time at recombination ($\eta_*$), so $\eta/\eta_* = 1$ at recombination. The oscillations are not symmetric about $\delta T = 0$ because the baryons are attracted to concentrations of dark matter. [Figure adapted from Dodelson, *Neutrinos, Flavor Physics, and Precision Cosmology: Fourth Tropical Workshop on Particle Physics and Cosmology*, José F. Nieves and Raymond R. Volkas (eds.), AIP Conference Proceedings, *689*, 184, 2003.]

time of decoupling, where $\delta T$ is the variation in temperature from equilibrium. The second peak is due to the oscillation of a still smaller region that passed through its maximum compression and then reached its maximum rarefaction at the time of decoupling. Note that the magnitude of $\delta T$ at decoupling for the first peak is greater than the magnitude of $\delta T$ at decoupling for the second peak. This is due to the biasing effect of the gravitational pull on the baryons by local concentrations of dark matter. As a result, $\delta T$ for the compression is greater (in magnitude) than that for the rarefaction. Thus the height of the second peak is less than that of the first peak, and the relative suppression of the second peak increases as the value of $\Omega_b$ increases. Comparison of the heights of the first and second harmonic peaks reveals the density of baryonic matter in the universe.

The third peak (the third harmonic) is due to an oscillation that reaches its second maximum compression at the time of decoupling. Just as the relative heights of the first two peaks indicate the amount of baryonic matter in the universe, the height of the third peak is sensitive to the density of dark matter. In essence, the location and heights of the three peaks can be used to determine, with some accuracy, the values of the three density parameters $\Omega_0$, $\Omega_{b,0}$, and $\Omega_{dark,0}$. The comparable heights of the second and third peaks indicate that most of the matter in the universe is in the form of dark matter.

The line labeled "Super-Horizon" in Fig. 30.16 describes the behavior of a density fluctuation so vast that it remained outside the particle horizon until recombination, so it never experienced acoustic oscillations. The CMB photons from such a super-horizon region were affected by both the primordial temperature fluctuation and the gravitational redshift of the Sachs-Wolfe effect as they climbed out of the gravitational potential wells. The angular size of these regions corresponds roughly to $\ell < 50$, so the left side of the angular power spectrum shows only a low, flat plain for these values of $\ell$.

Figure 30.17 shows that on the right side of the angular power spectrum, the peaks die off for $\ell > 1000$. This is the Silk damping that occurs when the random-walking photons can diffuse between the compressions and rarefactions of the short-wavelength sound waves and so decrease their amplitudes.

As you admire the harmonic structure of these celestial sound waves, it is striking to realize that the very existence of harmonics strongly supports the idea of inflation. The regularity in this harmonic structure means that the oscillations of all density fluctuations of a given size (and so having the same frequency) must reach their maximum compressions and rarefactions *at the same time*. This requires that they begin their oscillations simultaneously. Inflation provides the natural mechanism of generating density fluctuations and expanding them to super-horizon size. They remain "frozen" until the particle horizon grows large enough to encompass them and allow them to oscillate. Without this inflationary mechanism, a collection of density fluctuations that was generated randomly in time would not be in phase, and a series of evenly spaced peaks would not be observed in the power spectrum.

Figure 30.18 shows the effect on the power spectrum of varying several cosmological parameters. You may find it illuminating to download the CMBFAST program (listed in the references) that will quickly and accurately calculate the angular power spectrum for a choice of model input parameters. Future missions are being planned to better resolve the higher harmonic peaks and so further constrain the models. Some theoretical models of structure formation have already been decisively eliminated by comparing them with the observed angular power spectrum. For example, exotic relics of the false vacuum such as

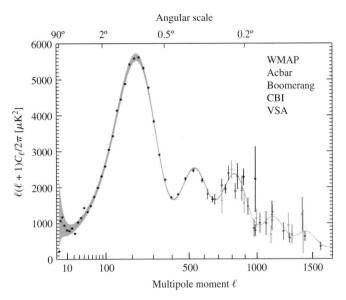

**FIGURE 30.17** The results of recent measurements of the angular power system by a variety of projects. The solid line is the best fit to the WMAP data. [Figure adapted from Hinshaw et al., *Ap. J.* (submitted), 2006, and courtesy of NASA and the WMAP Science Team.]

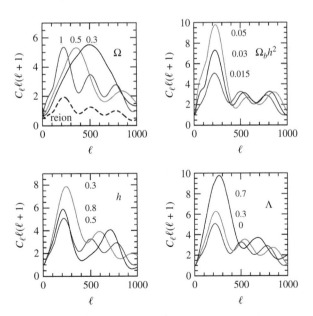

**FIGURE 30.18** Theoretical calculations of the angular power spectrum for a variety of cosmological parameters. For the lower right panel, $\Omega_0 = 1$. (Figure adapted from Kamionkowski and Kosowsky, *Annu. Rev. Nucl. Part. Sci.*, *49*, 77, 1999.)

monopoles and domain walls were once considered candidates for the "seeds" of structure formation. However, their power spectrum is quite different from the observed spectrum. These relics may or may not exist, but they do not play a significant role in the growth of structure.

### Polarization of the CMB Anisotropies

In March 2006, the latest WMAP findings were announced, including an analysis of the polarization of the CMB anisotropies. The magnitude of the polarization signal is less than 1% of the size of the temperature fluctuations, so this was an extremely demanding measurement. Complications such as polarized foreground emission due to Galactic synchrotron and thermal dust emission had to be carefully accounted for. The polarization originates in the Thompson scattering of the CMB photons by the free electrons in the photon–baryon fluid. The top of Fig. 30.19 shows how the electric field of incident unpolarized radiation (in the $+y$ direction) causes an electron to vibrate. This produces scattered light in the $+x$ and $+z$ directions that is linearly polarized. However, for an isotropic radiation field coming at the electron from all directions, there will be no net polarization. The observed partial polarization of the CMB therefore requires a radiation field that is more intense along one axis than along another axis. This is shown at the bottom of Fig. 30.19, where the incident

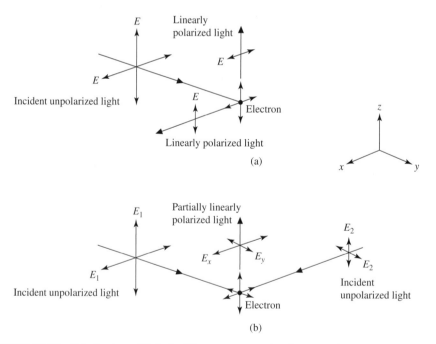

**FIGURE 30.19** Polarization of light by Thompson scattering from an electron. In (a), incident unpolarized light from one direction produces linearly polarized light upon scattering, when observed in the $x$–$z$ plane. In (b), incident unpolarized light of different intensities ($E_1 > E_2$) from two directions produces partially linearly polarized light ($E_x > E_y$) when observed along the $z$-axis.

light in the $+y$ direction is more intense than that in the $+x$ direction ($E_1 > E_2$). We say that such a radiation field has a *quadrupole moment*. As a result, the scattered radiation along the $z$-axis is partially polarized in the $x$ direction.

If Thompson scattering occurs frequently, then the photon directions quickly become randomized, and the quadrupole moment of the radiation field cannot be sustained. This means that the observed polarization must be established at the time of decoupling, when the universe became transparent and Thompson scattering ceased. After decoupling there are no free electrons to scatter the CMB photons until the first generation of stars reionizes the universe. This occurs at a much lower redshift than decoupling ($[z_{\text{dec}}]_{\text{WMAP}} = 1089$), so the signature of reionization can be gleaned from the polarization measurements. The results show that reionization took place at a redshift of $z = 10.9^{+2.7}_{-2.3}$.

There are two ways of producing a local quadrupole moment in the CMB: by the flow of the photon–baryon fluid and by the effect of gravitational radiation. The acoustic oscillation of the photon–baryon fluid implies that there is a velocity gradient within the fluid. For simplicity, consider the situation where the fluid is moving radially inward toward a local concentration of dark matter (see Fig. 30.20). From the frame of reference of an electron in the fluid, the neighboring elements of fluid are all moving toward it, with (in this example) the fluid in the radial direction approaching more rapidly than the fluid in the transverse direction. The relativistic headlight effect (Section 4.3) then causes the radiation from the radial directions to be more intense than the radiation in the transverse directions. This produces the quadrupole moment in the local radiation field that is responsible for partially polarizing the photons as they are scattered by the electron. The polarization pattern (the direction of the observed electric field of the CMB photons) is directed radially from a cooler temperature fluctuation ($\Delta T < 0$) and forms tangential loops around a warmer fluctuation ($\Delta T > 0$). This pattern is called an *E-mode*. [Notice in Fig. 30.19(b) that the electric field of the partially polarized light is along the direction of the less intense source of light and perpendicular to the direction of the more intense source.]

The second means of creating a local quadrupole moment in the CMB is through the effect of the gravitational radiation produced by the primordial density fluctuations. When a gravitational wave passes through the photon–baryon fluid, the space containing the

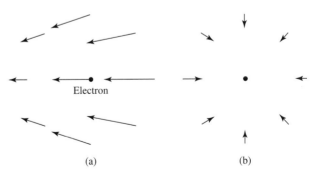

**FIGURE 30.20**   (a) The motion of an electron in the photon–baryon fluid as the fluid moves radially inward toward its maximum compression. (b) The velocity field seen from the rest frame of the electron.

photons will be stretched along one axis while it is compressed along a perpendicular axis. The resulting crowding and thinning of the photons produce a quadrupole moment in the radiation field, and hence the observed partial polarization of the CMB. In addition to the E-mode pattern of polarization, gravitational waves can produce a curling pattern called *B-modes* that spirals outward from the temperature fluctuations (the direction of the spiral depending on the sign of the temperature fluctuation). Any polarization pattern can be decomposed into its E-mode and B-mode contributions, so observations of the polarization of the CMB carry much information about the conditions at the time of decoupling.

Figure 30.21 shows the WMAP measurements of the CMB polarization in five frequency bands. First of all, note that the vertical axis is logarithmic and displays the *square root* of the power spectrum displayed in Fig. 30.13. For $\ell > 50$ (beyond the low, flat plain of the Sachs–Wolfe effect) we expect to see a correlation between the TT and EE curves. This is because the velocity gradients of the acoustic oscillations should be greatest for modes that are near their equilibrium configurations, rather than at their maximum compression or rarefaction, at the time of decoupling. (For a harmonic wave, the largest velocity gradients are found where the wave displacement is a maximum, while the maximum compression or rarefaction of the medium occurs at the points of zero displacement.) Thus the peaks in the angular power spectrum should be out of phase with the E-mode peaks. For example, the

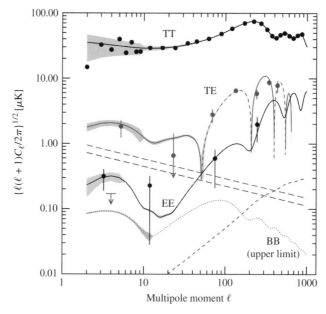

**FIGURE 30.21**   Plots of the WMAP measurements of the angular power spectrum (TT), the temperature/E-mode correlation (TE), the E-mode polarization (EE), and the B-mode polarization (BB). WMAP found no evidence for B-modes, so the upper limit for them is shown as a dotted line. The curves are for the best-fit cosmological model. [Figure adapted from Page et al., *Ap. J.* (submitted), 2006, and courtesy of NASA and the WMAP Science Team.]

first peak in the angular power spectrum should correspond to a minimum in the E-mode curve, and this is indeed what is observed. The curve labeled "TE" shows the expected correlation between the temperature and E-mode observations.

To date, WMAP has found no evidence for the B-modes, but their expected signal is faint and the WMAP observatory and the European Space Agency's Planck spacecraft (planned to launch in 2007) will continue to search for the B-modes. A definitive measurement of the B-modes for $\ell < 100$ would provide insights about the physics of the early universe at energies of $10^{15}$–$10^{16}$ GeV, back to the end of inflation.

The detection of B-modes would also impact an alternative cosmology, the cyclic model. This radical (but testable) alternative to the Big Bang/inflationary scenario proposes a cyclic universe in which the large-scale structure of the universe was set up *before* the Big Bang, during a slow contraction phase that followed an even earlier expansion. The observed energy we attribute to the Big Bang was instead released by the collision of two universes residing on two parallel three-dimensional membranes separated by a fourth dimension. The Big Bang/inflation model and the cyclic model offer the only possible mechanisms for producing nearly scale-invariant, adiabatic fluctuations (as are observed by WMAP). The collision of membranes is incapable of generating B-modes in the polarization of the CMB, so the detection of B-modes would rule out this imaginative model and at the same time confirm the inflationary model.[16]

### Evidence of Harmonics in Large-Scale Structure

If the quantum density fluctuations, amplified by concentrations of dark matter, really are the "seeds" of structure formation in the universe, and if the density fluctuations reveal an acoustic harmonic structure, then it seems reasonable to ask if that harmonic structure was imprinted on the large-scale structure of the universe. Incredibly, the answer is yes. A large sample of 46,748 galaxies from the Sloan Digital Sky Survey was examined to see if there was any correlation in their positions. There was a slight excess of galaxies separated by $100h^{-1}$ Mpc, or about 140 Mpc for $[h]_{\text{WMAP}} = 0.71$. This is in excellent agreement with the separation predicted for the first harmonic sound wave when the growth of structure due to the universal expansion is taken into account. Astronomers have another powerful statistical tool to apply to the secrets of the early universe.

Because the location of the first peak is sensitive to the value of $\Omega_0 = \Omega_{m,0} + \Omega_{\Lambda,0}$, we can plot the WMAP results on the $\Omega_{m,0}$–$\Omega_{\Lambda,0}$ plane together with the results from high-$z$ supernovae (recall Fig. 29.28). We also add the results of studies of X-rays from clusters of galaxies by the Chandra orbiting observatory, which, when combined with gravitational lens data, allow $\Omega_{m,0}$ to be determined. The result, shown in Fig. 30.22, displays a beautiful concordance—a spectacular agreement with the WMAP values obtained by using theoretical models to reproduce the observed angular power spectrum.

The science of precision cosmology has arrived.

---

[16]The cyclic model does not deny the expansion of the universe from a hot, dense state 13.7 billion years ago. If evidence favoring the cyclic model is discovered, only Chapter 30 of this book will have to be rewritten. You are referred to Steinhardt and Turok (2002) for details of their cyclic model of the universe.

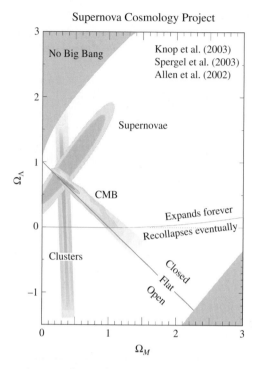

**FIGURE 30.22** The cosmic concordance of the results of three different observing projects on the $\Omega_{m,0}$–$\Omega_{\Lambda,0}$ plane. (Adapted from a figure courtesy of the Supernova Cosmology Project.)

## The Question

In our astrophysics class, a student once asked, "Why are we here?" The answer is as amazing to us as it was to the class:

We are here because, 13.7 billion years ago, the universe borrowed energy from the vacuum to create vast amounts of matter and antimatter in nearly equal numbers. Most of it annihilated and filled the universe with photons. Less than one part per billion survived to form protons and neutrons and then the hydrogen and helium that make up most of everything there is. Some of this hydrogen and helium collapsed to make the first generation of massive stars, which produced the first batch of heavy elements in their central nuclear fires. These stars exploded and enriched the interstellar clouds that would form the next generation of stars. Finally, about 4.6 billion years ago, one particular cloud in one particular galaxy collapsed to form our Sun and its planetary system. Life arose on the third planet, based on the hydrogen, carbon, nitrogen, oxygen, and other elements found in the protostellar cloud. The development of life transformed Earth's atmosphere and allowed life to move onto land. Sixty-five million years ago, a fortunate collision with a large meteoroid hastened the demise of the dinosaurs and allowed small, furry mammals to take center stage. Primitive men and women evolved and moved out of Africa to conquer the world with their new knowledge of tools, language, and agriculture. After raising food on the land, your ancestors, your parents, and then you consumed this food and breathed the air. Your own body is a collection of the atoms that were created billions of years earlier in the interiors of stars, the fraction of a fraction of a percent of normal matter that escaped annihilation in the first microsecond of the universe. Your life and everything in the world around you is intimately tied to countless aspects of modern astrophysics.

## SUGGESTED READING

### General

Abbott, Larry, "The Mystery of the Cosmological Constant," *Scientific American*, May 1988.

Bucher, Martin A., and Spergel, David N., "Inflation in a Low-Density Universe," *Scientific American*, January 1999.

Guth, Alan, *The Inflationary Universe: The Quest for a New Theory of Cosmic Origins*, Addison-Wesley, Reading, MA, 1997.

Hawking, Stephen W., *A Brief History of Time*, Bantam Books, Toronto, 1988.

Hedman, Matthew, "Polarization of the Cosmic Microwave Background," *American Scientist*, *93*, 236, 2005.

Hu, Wayne, and White, Martin, "The Cosmic Symphony," *Scientific American*, February 2004.

Lemonick, Michael D., "Before the Big Bang," *Discover*, February 2004.

Riordan, Michael, and Schramm, David N., *The Shadows of Creation*, W. H. Freeman and Company, New York, 1991.

Seife, Charles, *Alpha & Omega: The Search for the Beginning and End of the Universe*, Viking, New York, 2003.

Silk, Joseph, *The Big Bang*, Third Edition, W. H. Freeman and Company, New York, 2001.

Strauss, Michael A., "Reading the Blueprints of Creation," *Scientific American*, February 2004.

**Technical**

Allen, Steven W., "Cosmological Constraints from Chandra Observations of Galaxy Clusters," *Philosophical Transactions of the Royal Society of London. Series A.*, *360*, 2005, 2002.

Bennett, C. L., et al., "First-Year *Wilkinson Microwave Anisotropy Probe* (*WMAP*) Observations: Preliminary Maps and Basic Results," *The Astrophysical Journal Supplement Series*, *148*, 1, 2003.

Carroll, Sean M., Press, William H., and Turner, Edwin L., "The Cosmological Constant," *Annual Review of Astronomy and Astrophysics*, *30*, 499, 1992.

Dodelson, Scott, "Coherent Phase Argument for Inflation," *Neutrinos, Flavor Physics, and Precision Cosmology: Fourth Tropical Workshop on Particle Physics and Cosmology*, José F. Nieves and Raymond R. Volkas (eds.), AIP Conference Proceedings, *689*, 184, 2003.

Eisenstein, Daniel J., et al., "Detection of the Baryon Acoustic Peak in the Large-Scale Correlation Function of SDSS Luminous Red Galaxies," *The Astrophysical Journal*, *633*, 560, 2005.

Fukugita, Masataka, and Peebles, J. P. E., "The Cosmic Energy Inventory," *The Astrophysical Journal*, *616*, 643, 2004.

Gott, J. Richard III, et al., "A Map of the Universe," *The Astrophysical Journal*, *624*, 463, 2005.

Hinshaw, G., et al., "First-Year *Wilkinson Microwave Anisotropy Probe* (*WMAP*) Observations: The Angular Power Spectrum," *The Astrophysical Journal Supplement Series*, *148*, 135, 2003.

Hinshaw, G., et al., "Third-Year *Wilkinson Microwave Anisotropy Probe* (*WMAP*) Observations: Temperature Analysis," submitted to *The Astrophysical Journal*, 2006.

Hu, Wayne, "The Physics of Microwave Background Anisotropies," http://background.uchicago.edu/.

Hu, Wayne, Sugiyama, Naoshi, and Silk, Joseph, "The Physics of Microwave Background Anisotropies," *Nature*, *386*, 1997.

Kolb, Edward W., and Turner, Michael S., *The Early Universe*, Addison-Wesley, Redwood City, CA, 1990.

Page, L., et al., "Three-Year *Wilkinson Microwave Anisotropy Probe* (*WMAP*) Observations: Polarization Analysis," submitted to *The Astrophysical Journal*, 2006.

Raine, D. J., and Thomas, E. G., *An Introduction to the Science of Cosmology*, Institute of Physics Publishing, Bristol, England, 2001.

Roos, Matts, *Cosmology*, Third Edition, John Wiley & Sons, West Sussex, England, 2003.

Ryden, Barbara, *Introduction to Cosmology*, Addison-Wesley, San Francisco, 2003.

Seljak, Uros, and Zaldarriaga, Matias, CMBFAST software,
http://www.cmbfast.org/.

Steinhardt, Paul J., and Turok, Neil, "A Cyclic Model of the Universe," *Science*, *296*, 1436, 2002.

Springel, Volker, et al., "Simulations of the Formation, Evolution, and Clustering of Galaxies and Quasars," *Nature*, *435*, 629, 2005.

## PROBLEMS

**30.1** In Problem 9.3, you derived Eq. (9.65) for the number density of blackbody photons:

$$n = \frac{u}{2.70kT} = \frac{aT^3}{2.70k}.$$

Use this result along with the baryon density, $\rho_{b,0}$, to estimate the ratio of the number of baryons to the number of photons in today's universe. For convenience, assume that the universe is made solely of hydrogen.

**30.2** **(a)** Suppose observations of the microlensing of stars in the Large Magellanic Cloud showed that most of the dark matter in the Milky Way's dark halo is in the form of ordinary brown dwarfs and Jupiter-sized objects. Explain whether this would favor hot or cold dark matter for the nonbaryonic matter in the universe, and why.

**(b)** In fact, observations indicate that this is probably *not* the case: Less than 20% of the Milky Way's dark halo is in the form of ordinary brown dwarfs and Jupiter-sized objects. How do these observations change the conclusion you reached in part (a)?

**30.3** An example of spontaneous symmetry breaking can be obtained by considering a small ball of mass $m = 1/9.8$ kg that is free to roll on a surface whose height (in meters) is given by

$$h(x) = kx^2 + \varepsilon x^4,$$

where $k = \pm 1$ m$^{-1}$ and $\varepsilon = 0.5$ m$^{-3}$. The gravitational potential energy of the ball is then $V(x) = mgh(x)$.

**(a)** Make two graphs of $V(x)$ from $x = -2$ m to $x = 2$ m, one for each choice of sign for $k$.

**(b)** The case of $k = 1$ m$^{-1}$ corresponds to the symmetric false vacuum. Where is the point of equilibrium? Is this point stable or unstable? (In the case of stable equilibrium, if the ball is displaced slightly, it will return to the equilibrium point.)

**(c)** The case of $k = -1$ m$^{-1}$ corresponds to the broken symmetry of the true vacuum. Where are the three points of equilibrium? Which of these points are stable, and which are unstable?

**(d)** For the case of $k = -1$ m$^{-1}$, consider the ball at rest at the origin. What are the implications of the uncertainty principle, Eq. (5.20), for the ball remaining in that position? In what ways is this situation analogous to that of the supercooled false vacuum just before inflation?

**30.4** Compare the energy density of CMB photons at the end of the GUTs epoch to that of the false vacuum.

**30.5** According to the standard Big Bang cosmology, what was the value of the scale factor, $R$, in the radiation era at the Planck time? When inflation is considered (take $t/\tau_i = 100$), what was the value of the scale factor at the Planck time? For both situations, what was the size of the presently observable universe at the Planck time?

**30.6** Estimate the thickness of a typical cosmic string.

**30.7** (a) Estimate the mass of baryonic matter contained within a causally connected region (diameter = the horizon distance) when $T = 10^9$ K during the radiation era. Express your answer in solar units. How does this compare with the Jeans mass shown in Fig. 30.7?

(b) Show that during the radiation era, the mass of the baryonic matter inside this causally connected region varied as $T^{-3}$. What does this say about the relative magnitudes of the region's mass and the Jeans mass throughout the radiation era?

**30.8** Estimate the mass of baryonic matter contained within a causally connected region at the time of the transition from the radiation era to the matter era. Express your answer in solar units. (Use WMAP values.) From the magnitude of your answer, did most of the baryonic density fluctuations become sub-horizon-sized during the radiation era or during the matter era?

**30.9** Show that if a baryonic density perturbation at the present time is $(\delta\rho/\rho)_0 = 1$, then its value at a redshift $z$ in the matter era was $\delta\rho/\rho = (1 + z)^{-1}$. By what factor does any $\delta\rho/\rho < 1$ increase between a redshift $z_1$ and a redshift $z_2$ $(z_1 > z_2)$ in the matter era? You may assume that the universe is nearly flat and that $z < 1100$ for baryonic fluctuations.

**30.10** Follow the linearization procedure of Example 14.3.1 and show that the variations in the temperature of the CMB are related to the baryonic density fluctuations by Eq. (30.30).

**30.11** Assuming a flat universe, find the angular size of the largest causally connected region of the CMB. *Hint:* We see this region as it existed at the time of decoupling, when the CMB photons were set free.

**30.12** Estimate the value of $\ell$ when Silk damping becomes important for the angular power spectrum of the CMB anisotropies.

**30.13** According to the WMAP CMB polarization results, how old was the universe when the first stars ignited?

# APPENDIX

# C

# Solar System Data

**Planetary Physical Data**

| Planet | Mass[a] $(M_\oplus)$ | Equatorial Radius[b] $(R_\oplus)$ | Average Density $(\text{kg m}^{-3})$ | Sidereal Rotation Period (d) | Oblateness $(R_e - R_p)/R_e$ | Bond Albedo |
|---|---|---|---|---|---|---|
| Mercury | 0.05528 | 0.3825 | 5427 | 58.6462 | 0.00000 | 0.119 |
| Venus | 0.81500 | 0.9488 | 5243 | 243.018 | 0.00000 | 0.750 |
| Earth | 1.00000 | 1.0000 | 5515 | 0.997271 | 0.0033396 | 0.306 |
| Mars | 0.10745 | 0.5326 | 3933 | 1.02596 | 0.006476 | 0.250 |
| Ceres (*dwarf planet*) | 0.00016 | 0.076 | 2100 | 0.378 | 0.0380 | 0.1 |
| Jupiter | 317.83 | 11.209 | 1326 | 0.4135 | 0.064874 | 0.343 |
| Saturn | 95.159 | 9.4492 | 687 | 0.4438 | 0.097962 | 0.342 |
| Uranus | 14.536 | 4.0073 | 1270 | 0.7183 | 0.022927 | 0.300 |
| Neptune | 17.147 | 3.8826 | 1638 | 0.6713 | 0.017081 | 0.290 |
| Pluto (*dwarf planet*) | 0.0021 | 0.178 | 2110 | 6.3872 | 0.0000 | 0.4 – 0.6 |
| Eris[c] (*dwarf planet*) | 0.002? | 0.188 | 2100? | | | 0.6? |

**Planetary Orbital and Satellite Data**

| Planet | Semimajor Axis (AU) | Orbital Eccentricity | Sidereal Orbital Period (yr) | Orbital Inclination to Ecliptic (°) | Equatorial Inclination to Orbit (°) | Number Natural Satellites |
|---|---|---|---|---|---|---|
| Mercury | 0.3871 | 0.2056 | 0.2408 | 7.00 | 0.01 | 0 |
| Venus | 0.7233 | 0.0067 | 0.6152 | 3.39 | 177.36 | 0 |
| Earth | 1.0000 | 0.0167 | 1.0000 | 0.000 | 23.45 | 1 |
| Mars | 1.5236 | 0.0935 | 1.8808 | 1.850 | 25.19 | 2 |
| Ceres (*dwarf planet*) | 2.767 | 0.097 | 4.603 | 9.73 | | 0 |
| Jupiter | 5.2044 | 0.0489 | 11.8618 | 1.304 | 3.13 | 63 |
| Saturn | 9.5826 | 0.0565 | 29.4567 | 2.485 | 26.73 | 47 |
| Uranus | 19.2012 | 0.0457 | 84.0107 | 0.772 | 97.77 | 27 |
| Neptune | 30.0476 | 0.0113 | 164.79 | 1.769 | 28.32 | 13 |
| Pluto (*dwarf planet*) | 39.4817 | 0.2488 | 247.68 | 17.16 | 122.53 | 3 |
| Eris[c] (*dwarf planet*) | 67.89 | 0.4378 | 559 | 43.99 | | 1 |

[a] $M_\oplus = 5.9736 \times 10^{24}$ kg
[b] $R_\oplus = 6.378136 \times 10^6$ m
[c] Eris was formerly known as 2003 UB313

### Data of Selected Major Satellites

| Satellite | Parent Planet | Mass ($10^{22}$ kg) | Radius ($10^3$ km) | Density (kg m$^{-3}$) | Orbital Period (d) | Semimajor Axis ($10^3$ km) |
|-----------|---------------|---------------------|--------------------|-----------------------|--------------------|----------------------------|
| Moon | Earth | 7.349 | 1.7371 | 3350 | 27.322 | 384.4 |
| Io | Jupiter | 8.932 | 1.8216 | 3530 | 1.769 | 421.6 |
| Europa | Jupiter | 4.800 | 1.5608 | 3010 | 3.551 | 670.9 |
| Ganymede | Jupiter | 14.819 | 2.6312 | 1940 | 7.155 | 1070.4 |
| Callisto | Jupiter | 10.759 | 2.4103 | 1830 | 16.689 | 1882.7 |
| Titan | Saturn | 13.455 | 2.575 | 1881 | 15.945 | 1221.8 |
| Triton | Neptune | 2.14 | 1.3534 | 2050 | 5.877 | 354.8 |

# D

# The Constellations

| Latin Name | Genitive | Abbrev. | Translation | R. A. h | Dec. deg |
|---|---|---|---|---|---|
| Andromeda | Andromedae | And | Princess of Ethiopia | 1 | +40 |
| Antlia | Antliae | Ant | Air Pump | 10 | −35 |
| Apus | Apodis | Aps | Bird of Paradise | 16 | −75 |
| Aquarius | Aquarii | Aqr | Water Bearer | 23 | −15 |
| Aquila | Aquilae | Aql | Eagle | 20 | +5 |
| Ara | Arae | Ara | Altar | 17 | −55 |
| Aries | Arietis | Ari | Ram | 3 | +20 |
| Auriga | Aurigae | Aur | Charioteer | 6 | +40 |
| Boötes | Boötis | Boo | Herdsman | 15 | +30 |
| Caelum | Caeli | Cae | Chisel | 5 | −40 |
| Camelopardalis | Camelopardis | Cam | Giraffe | 6 | +70 |
| Cancer | Cancri | Cnc | Crab | 9 | +20 |
| Canes Venatici | Canum Venaticorum | CVn | Hunting Dogs | 13 | +40 |
| Canis Major | Canis Majoris | CMa | Big Dog | 7 | −20 |
| Canis Minor | Canis Minoris | CMi | Little Dog | 8 | +5 |
| Capricornus | Capricorni | Cap | Goat | 21 | −20 |
| Carina | Carinae | Car | Ship's Keel | 9 | −60 |
| Cassiopeia | Cassiopeiae | Cas | Queen of Ethiopia | 1 | +60 |
| Centaurus | Centauri | Cen | Centaur | 13 | −50 |
| Cepheus | Cephei | Cep | King of Ethiopia | 22 | +70 |
| Cetus | Ceti | Cet | Sea Monster (whale) | 2 | −10 |
| Chamaeleon | Chamaeleontis | Cha | Chameleon | 11 | −80 |
| Circinus | Circini | Cir | Compass | 15 | −60 |
| Columba | Columbae | Col | Dove | 6 | −35 |
| Coma Berenices | Comae Berenices | Com | Berenice's Hair | 13 | +20 |
| Corona Australis | Coronae Australis | CrA | Southern Crown | 19 | −40 |
| Corona Borealis | Coronae Borealis | CrB | Northern Crown | 16 | +30 |
| Corvus | Corvi | Crv | Crow | 12 | −20 |
| Crater | Crateris | Crt | Cup | 11 | −15 |
| Crux | Crucis | Cru | Southern Cross | 12 | −60 |
| Cygnus | Cygni | Cyg | Swan | 21 | +40 |
| Delphinus | Delphini | Del | Dolphin, Porpoise | 21 | +10 |
| Dorado | Doradus | Dor | Swordfish | 5 | −65 |
| Draco | Draconis | Dra | Dragon | 17 | +65 |
| Equuleus | Equulei | Equ | Little Horse | 21 | +10 |
| Eridanus | Eridani | Eri | River Eridanus | 3 | −20 |
| Fornax | Fornacis | For | Furnace | 3 | −30 |
| Gemini | Geminorum | Gem | Twins | 7 | +20 |
| Grus | Gruis | Gru | Crane | 22 | −45 |
| Hercules | Herculis | Her | Son of Zeus | 17 | +30 |

| Latin Name | Genitive | Abbrev. | Translation | R. A. h | Dec. deg |
|---|---|---|---|---|---|
| Horologium | Horologii | Hor | Clock | 3 | −60 |
| Hydra | Hydrae | Hya | Water Snake | 10 | −20 |
| Hydrus | Hydri | Hyi | Sea Serpent | 2 | −75 |
| Indus | Indi | Ind | Indian | 21 | −55 |
| Lacerta | Lacertae | Lac | Lizard | 22 | +45 |
| Leo | Leonis | Leo | Lion | 11 | +15 |
| Leo Minor | Leonis Minoris | LMi | Little Lion | 10 | +35 |
| Lepus | Leporis | Lep | Hare | 6 | −20 |
| Libra | Librae | Lib | Balance, Scales | 15 | −15 |
| Lupus | Lupi | Lup | Wolf | 15 | −45 |
| Lynx | Lyncis | Lyn | Lynx | 8 | +45 |
| Lyra | Lyrae | Lyr | Lyre, Harp | 19 | +40 |
| Mensa | Mensae | Men | Table, Mountain | 5 | −80 |
| Microscopium | Microscopii | Mic | Microscope | 21 | −35 |
| Monoceros | Monocerotis | Mon | Unicorn | 7 | −5 |
| Musca | Muscae | Mus | Fly | 12 | −70 |
| Norma | Normae | Nor | Square, Level | 16 | −50 |
| Octans | Octantis | Oct | Octant | 22 | −85 |
| Ophiuchus | Ophiuchi | Oph | Serpent-bearer | 17 | 0 |
| Orion | Orionis | Ori | Hunter | 5 | +5 |
| Pavo | Pavonis | Pav | Peacock | 20 | −65 |
| Pegasus | Pegasi | Peg | Winged Horse | 22 | +20 |
| Perseus | Persei | Per | Rescuer of Andromeda | 3 | +45 |
| Phoenix | Phoenicis | Phe | Phoenix | 1 | −50 |
| Pictor | Pictoris | Pic | Painter, Easel | 6 | −55 |
| Pisces | Piscium | Psc | Fish | 1 | +15 |
| Piscis Austrinus | Piscis Austrini | PsA | Southern Fish | 22 | −30 |
| Puppis | Puppis | Pup | Ship's Stern | 8 | −40 |
| Pyxis | Pyxidis | Pyx | Ship's Compass | 9 | −30 |
| Reticulum | Reticuli | Ret | Net | 4 | −60 |
| Sagitta | Sagittae | Sge | Arrow | 20 | +10 |
| Sagittarius | Sagittarii | Sgr | Archer | 19 | −25 |
| Scorpius | Scorpii | Sco | Scorpion | 17 | −40 |
| Sculptor | Sculptoris | Scl | Sculptor | 0 | −30 |
| Scutum | Scuti | Sct | Shield | 19 | −10 |
| Serpens | Serpentis | Ser | Serpent | 17 | 0 |
| Sextans | Sextantis | Sex | Sextant | 10 | 0 |
| Taurus | Tauri | Tau | Bull | 4 | +15 |
| Telescopium | Telescopii | Tel | Telescope | 19 | −50 |
| Triangulum | Trianguli | Tri | Triangle | 2 | +30 |
| Triangulum Australe | Trianguli Australis | TrA | Southern Triangle | 16 | −65 |
| Tucana | Tucanae | Tuc | Toucan | 0 | −65 |
| Ursa Major | Ursae Majoris | UMa | Big Bear | 11 | +50 |
| Ursa Minor | Ursae Minoris | UMi | Little Bear | 15 | +70 |
| Vela | Velorum | Vel | Ship's Sai | 9 | −50 |
| Virgo | Virginis | Vir | Maiden, Virgin | 13 | 0 |
| Volans | Volantis | Vol | Flying Fish | 8 | −70 |
| Vulpecula | Vulpeculae | Vul | Little Fox | 20 | +25 |

# E

# The Brightest Stars

| Name | Star | Spectral Class A | Spectral Class B | $V^a$ A | $V^a$ B | $M_V$ A | $M_V$ B |
|---|---|---|---|---|---|---|---|
| Sirius | α CMa | A1 V | wd$^b$ | −1.44 | +8.7 | +1.45 | +11.6 |
| Canopus | α Car | F0 Ib | | −0.62 | | −5.53 | |
| Arcturus | α Boo | K2 II Ip | | −0.05 | | −0.31 | |
| Rigel Kentaurus | α Cen | G2 V | K0 V | −0.01 | +1.3 | +4.34 | +5.7 |
| Vega | α Lyr | A0 V | | +0.03 | | +0.58 | |
| Capella$^c$ | α Aur | M1 III | M1 V | +0.08 | +10.2 | −0.48 | +9.5 |
| Rigel | β Ori | B8 Ia | B9 | +0.18 | +6.6 | −6.69 | −0.4 |
| Procyon | α CMi | F5 IV–V | wd$^b$ | +0.40 | +10.7 | +2.68 | +13.0 |
| Betelgeuse | α Ori | M2Ib | | +0.45v | | −5.14 | |
| Achernar | α Eri | B3 Vp | | +0.45 | | −2.77 | |
| Hadar | β Cen | B1 III | ? | +0.61 | +4 | −5.42 | −0.8 |
| Altair | α Aql | A7 IV–V | | +0.76 | | +2.20 | |
| Acrux | α Cru | B0.5 IV | B3 | +0.77 | +1.9 | −4.19 | −3.5 |
| Aldebaran | α Tau | K5 III | M2V | +0.87 | +13 | −0.63 | +12 |
| Spica | α Vir | B1 V | | +0.98v | | −3.55 | |
| Antares | α Sco | M1 Ib | B4eV | +1.06v | +5.1 | −5.58 | −0.3 |
| Pollux | β Gem | K0 III | | +1.16 | | +1.09 | |
| Fomalhaut | α PsA | A3 V | K4V | +1.17 | +6.5 | +1.74 | +7.3 |
| Deneb | α Cyg | A2 Ia | | +1.25 | | −8.73 | |
| Mimosa | β Cru | B0.5 III | B2 V | +1.25v | | −3.92 | |

$^a$ Values labeled v designate variable stars.

$^b$ wd represents a white dwarf star.

$^c$ Capella has a third member of spectral class M5 V, $V = +13.7$, and $M_V = +13$.

| Name | R. A.[a] (h m s) | Dec.[a] (° ′ ″) | Parallax[b] (″) | Distance[c] (pc) | Proper Motion[d] (″ yr⁻¹) | Radial Velocity (km s⁻¹) |
|---|---|---|---|---|---|---|
| Sirius | 06 45 08.92 | −16 41 58.0 | 0.37921(158) | 2.64 | 1.33942 | −7.7 |
| Canopus | 06 23 57.11 | −52 41 44.4 | 0.01043(53) | 95.88 | 0.03098 | +20.5 |
| Arcturus | 14 15 39.67 | +19 10 56.7 | 0.08885(74) | 11.26 | 2.27887 | −5.2 |
| Rigel Kentaurus | 14 39 36.50 | −60 50 02.3 | 0.74212(140) | 1.35 | 3.70962 | −24.6 |
| Vega | 18 36 56.34 | +38 47 01.3 | 0.12893(55) | 7.76 | 0.35077 | −13.9 |
| Capella | 05 16 41.36 | +45 59 52.8 | 0.07729(89) | 12.94 | 0.43375 | +30.2 |
| Rigel | 05 14 32.27 | −08 12 05.9 | 0.00422(81) | 237 | 0.00195 | +20.7 |
| Procyon | 07 39 18.12 | +05 13 30.0 | 0.28593(88) | 3.50 | 1.25850 | −3.2 |
| Betelgeuse | 05 55 10.31 | +07 24 25.4 | 0.00763(164) | 131 | 0.02941 | +21.0 |
| Achernar | 01 37 42.85 | −57 14 12.3 | 0.02268(57) | 44.09 | 0.09672 | +19 |
| Hadar | 14 03 49.40 | −60 22 22.9 | 0.00621(56) | 161 | 0.04221 | −12 |
| Altair | 19 50 47.0 | +08 52 06.0 | 0.19444(94) | 5.14 | 0.66092 | −26.3 |
| Acrux | 12 26 35.90 | −63 05 56.7 | 0.01017(67) | 98.33 | 0.03831 | −11.2 |
| Aldebaran | 04 35 55.24 | +16 30 33.5 | 0.05009(95) | 19.96 | 0.19950 | +54.1 |
| Spica | 13 25 11.58 | −11 09 40.8 | 0.01244(86) | 80.39 | 0.05304 | +1.0 |
| Antares | 16 29 24.46 | −26 25 55.2 | 0.00540(168) | 185 | 0.02534 | −3.2 |
| Pollux | 07 45 18.95 | +28 01 34.3 | 0.09674(87) | 10.34 | 0.62737 | +3.3 |
| Fomalhaut | 22 57 39.05 | −29 37 20.1 | 0.13008(92) | 7.69 | 0.36790 | +6.5 |
| Deneb | 20 41 25.91 | +45 16 49.2 | 0.00101(57) | 990 | 0.00220 | −4.6 |
| Mimosa | 12 47 43.26 | −59 41 19.5 | 0.00925(61) | 108 | 0.04991 | +10.3 |

[a] Right ascension and declination are given in epoch J2000.0.

[b] Parallax data are from the Hipparcos Space Astrometry Mission. Uncertainties are in parentheses; for instance, the parallax of Sirius is $0.37921'' \pm 0.00158''$.

[c] Distance was calculated from the parallax measurement.

[d] Proper motion data are from the Hipparcos Space Astrometry Mission.

# APPENDIX

# F

# The Nearest Stars

| Name | HIP[a] | Spectral Class | $V^b$ | $M_V$ | $B - V$ | Parallax[c] ('') | Distance[d] (pc) |
|------|--------|----------------|-------|-------|---------|------------------|------------------|
| Proxima Centauri ($\alpha$ Cen C) | 70890 | M5 Ve | 11.01 | 15.45 | +1.81 | 0.77233(242) | 1.29 |
| $\alpha$ Cen B | 71681 | K1 V | 1.35 | 5.70 | +0.88 | 0.74212(140) | 1.35 |
| $\alpha$ Cen A | 71683 | G2 V | −0.01 | 4.34 | +0.71 | 0.74212(140) | 1.35 |
| Barnard's Star | 87937 | M5 V | 9.54 | 13.24 | +1.57 | 0.54901(158) | 1.82 |
| Gl 411 | 54035 | M2 Ve | 7.49 | 10.46 | +1.50 | 0.39240(91) | 2.55 |
| Sirius A ($\alpha$ CMa) | 32349 | A1 V | −1.44 | 1.45 | +0.01 | 0.37921(158) | 2.64 |
| Sirius B ($\alpha$ CMa) | | wd (DA) | 8.44 | 11.33 | −0.03 | 0.37921(158)[e] | 2.64[e] |
| Gl 729 | 92403 | M4.5 Ve | 10.37 | 13.00 | +1.51 | 0.33648(182) | 2.97 |
| $\epsilon$ Eri | 16537 | K2 V | 3.72 | 6.18 | +0.88 | 0.31075(85) | 3.22 |
| Gl 887 | 114046 | M2 Ve | 7.35 | 9.76 | +1.48 | 0.30390(87) | 3.29 |
| Ross 128 (Gl 447) | 57548 | M4.5 V | 11.12 | 13.50 | +1.75 | 0.29958(220) | 3.34 |
| 61 Cyg A (Gl 820) | 104214 | K5 Ve | 5.20 | 7.49 | +1.07 | 0.28713(151) | 3.48 |
| Procyon A ($\alpha$ CMi) | 37279 | F5 IV–V | 0.40 | 2.68 | +0.43 | 0.28593(88) | 3.50 |
| Procyon B ($\alpha$ CMi) | | wd | 10.7 | 13.0 | +0.00 | 0.28593(88)[e] | 3.50[e] |
| 61 Cyg B (Gl 820B) | 104217 | K7 Ve | 6.05 | 8.33 | +1.31 | 0.28542(72) | 3.50 |
| Gl 725B | 91772 | M5 V | 9.70 | 11.97 | +1.56 | 0.28448(501) | 3.52 |
| Gl 725A | 91768 | M4 V | 8.94 | 11.18 | +1.50 | 0.28028(257) | 3.57 |
| GX And | 1475 | M2 V | 8.09 | 10.33 | +1.56 | 0.28027(105) | 3.57 |
| $\epsilon$ Ind | 108870 | K5 Ve | 4.69 | 6.89 | +1.06 | 0.27576(69) | 3.63 |
| $\tau$ Cet | 8102 | G8 Vp | 3.49 | 5.68 | +0.73 | 0.27417(80) | 3.65 |
| Gl 54.1 | 5643 | M5.5 Ve | 12.10 | 14.25 | +1.85 | 0.26905(757) | 3.72 |
| Luyten's Star (Gl 237) | 36208 | M3.5 | 9.84 | 11.94 | +1.57 | 0.26326(143) | 3.80 |
| Kapteyn's Star | 24186 | M0 V | 8.86 | 10.89 | +1.55 | 0.25526(86) | 3.92 |
| AX Mic | 105090 | M0 Ve | 6.69 | 8.71 | +1.40 | 0.25337(113) | 3.95 |
| Kruger 60 | 110893 | M2 V | 9.59 | 11.58 | +1.61 | 0.24952(303) | 4.01 |
| Ross 614 (GL 234A) | 30920 | M4.5 Ve | 11.12 | 13.05 | +1.69 | 0.24289(264) | 4.12 |

[a] HIP designates the Hipparcos catalog number.

[b] Values labeled v designate variable stars.

[c] Parallax data are from the Hipparcos Space Astrometry Mission. Uncertainties are in parentheses; for instance, the parallax of Proxima Centauri is $0.77233'' \pm 0.00242''$.

[d] Distances were calculated from the Hipparcos parallax data.

[e] Parallax and distance taken to be that of bright companion.

| Name | R. A.[a] (h m s) | dec.[a] (° ′ ″) | Proper Motion | | Radial. Velocity. (km s⁻¹) | |
|---|---|---|---|---|---|---|
| | | | R. A.[b] (″ yr⁻¹) | dec.[b] (″ yr⁻¹) | | |
| Proxima Centauri | 14 29 42.95 | −62 40 46.1 | −3.77564(152) | 0.76816(182) | −33.4 | |
| α Cen B | 14 39 35.08 | −60 50 13.8 | −3.60035(2610) | 0.95211(1975) | −23.4 | |
| α Cen A | 14 39 36.50 | −60 50 02.3 | −3.67819(151) | 0.48184(124) | −23.4 | |
| Barnard's Star | 17 57 48.50 | +04 41 36.2 | −0.79784(161) | 10.32693(129) | −112.3 | |
| Gl 411 | 11 03 20.19 | +35 58 11.6 | −0.58020(77) | −4.76709(77) | −85. | |
| Sirius A | 06 45 08.92 | −16 41 58.0 | −0.54601(133) | −1.22308(124) | −7.7 | |
| Gl 729 | 18 49 49.36 | −23 50 10.4 | 0.63755(222) | −0.19247(145) | −7.0 | |
| ε Eri | 03 32 55.84 | −09 27 29.7 | −0.97644(98) | 0.01797(91) | +13. | |
| Gl 887 | 23 05 52.04 | −35 51 11.1 | 6.76726(70) | 1.32666(74) | −6.4 | |
| Ross 128 | 11 47 44.40 | +00 48 16.4 | 0.60562(214) | −1.21923(186) | −31. | |
| 61 Cyg A | 21 06 53.94 | +38 44 57.9 | 4.15510(95) | 3.25890(119) | −65. | |
| Procyon A | 07 39 18.12 | +05 13 30.0 | −0.71657(88) | −1.03458(38) | −3.2 | |
| 61 Cyg B | 21 06 55.26 | +38 44 31.4 | 4.10740(43) | 3.14372(59) | −65. | |
| Gl 725B | 18 42 46.90 | +59 37 36.6 | −1.39320(1150) | 1.84573(1202) | +1. | |
| Gl 725A | 18 42 46.69 | +59 37 49.4 | −1.32688(310) | 1.80212(358) | −1. | |
| GX And | 00 18 22.89 | +44 01 22.6 | 2.88892(75) | 0.41058(63) | +13.5 | |
| ε Ind | 22 03 21.66 | −56 47 09.5 | 3.95997(55) | −2.53884(42) | −40.4 | |
| τ Cet | 01 44 04.08 | −15 56 14.9 | −1.72182(83) | 0.85407(80) | −16.4 | |
| Gl 54.1 | 01 12 30.64 | −16 59 56.3 | 1.21009(521) | 0.64695(391) | +37.0 | |
| Luyten's Star | 07 27 24.50 | +05 13 32.8 | 0.57127(141) | −3.69425(90) | +18. | |
| Kapteyn's Star | 05 11 40.58 | −45 01 06.3 | 6.50605(95) | −5.73139(90) | +242.8 | |
| AX Mic | 21 17 15.27 | −38 52 02.5 | −3.25900(128) | −1.14699(56) | +23. | |
| Kruger 60 | 22 27 59.47 | +57 41 45.1 | −0.87023(300) | −0.47110(297) | −34. | |
| Ross 614 | 06 29 23.40 | −02 48 50.3 | 0.69473(300) | −0.61862(248) | +23.2 | |

[a] Right ascension and declination are given in epoch J2000.0.

[b] Proper-motion data are from the Hipparcos Space Astrometry Mission. Uncertainties are in parentheses; for instance, the proper motion of Proxima Centauri in right ascension is −3.77564″ yr⁻¹ ± 0.00152″ yr⁻¹.

# APPENDIX

# G

# Stellar Data

| Sp. Type | $T_e$ (K) | $L/L_\odot$ | $R/R_\odot$ | $M/M_\odot$ | $M_{bol}$ | BC | $M_V$ | $U - B$ | $B - V$ |
|---|---|---|---|---|---|---|---|---|---|
| O5 | 42000 | 499000 | 13.4 | 60 | −9.51 | −4.40 | −5.1 | −1.19 | −0.33 |
| O6 | 39500 | 324000 | 12.2 | 37 | −9.04 | −3.93 | −5.1 | −1.17 | −0.33 |
| O7 | 37500 | 216000 | 11.0 | — | −8.60 | −3.68 | −4.9 | −1.15 | −0.32 |
| O8 | 35800 | 147000 | 10.0 | 23 | −8.18 | −3.54 | −4.6 | −1.14 | −0.32 |
| B0 | 30000 | 32500 | 6.7 | 17.5 | −6.54 | −3.16 | −3.4 | −1.08 | −0.30 |
| B1 | 25400 | 9950 | 5.2 | — | −5.26 | −2.70 | −2.6 | −0.95 | −0.26 |
| B2 | 20900 | 2920 | 4.1 | — | −3.92 | −2.35 | −1.6 | −0.84 | −0.24 |
| B3 | 18800 | 1580 | 3.8 | 7.6 | −3.26 | −1.94 | −1.3 | −0.71 | −0.20 |
| B5 | 15200 | 480 | 3.2 | 5.9 | −1.96 | −1.46 | −0.5 | −0.58 | −0.17 |
| B6 | 13700 | 272 | 2.9 | — | −1.35 | −1.21 | −0.1 | −0.50 | −0.15 |
| B7 | 12500 | 160 | 2.7 | — | −0.77 | −1.02 | +0.3 | −0.43 | −0.13 |
| B8 | 11400 | 96.7 | 2.5 | 3.8 | −0.22 | −0.80 | +0.6 | −0.34 | −0.11 |
| B9 | 10500 | 60.7 | 2.3 | — | +0.28 | −0.51 | +0.8 | −0.20 | −0.07 |
| A0 | 9800 | 39.4 | 2.2 | 2.9 | +0.75 | −0.30 | +1.1 | −0.02 | −0.02 |
| A1 | 9400 | 30.3 | 2.1 | — | +1.04 | −0.23 | +1.3 | +0.02 | +0.01 |
| A2 | 9020 | 23.6 | 2.0 | — | +1.31 | −0.20 | +1.5 | +0.05 | +0.05 |
| A5 | 8190 | 12.3 | 1.8 | 2.0 | +2.02 | −0.15 | +2.2 | +0.10 | +0.15 |
| A8 | 7600 | 7.13 | 1.5 | — | +2.61 | −0.10 | +2.7 | +0.09 | +0.25 |
| F0 | 7300 | 5.21 | 1.4 | 1.6 | +2.95 | −0.09 | +3.0 | +0.03 | +0.30 |
| F2 | 7050 | 3.89 | 1.3 | — | +3.27 | −0.11 | +3.4 | +0.00 | +0.35 |
| F5 | 6650 | 2.56 | 1.2 | 1.4 | +3.72 | −0.14 | +3.9 | −0.02 | +0.44 |
| F8 | 6250 | 1.68 | 1.1 | — | +4.18 | −0.16 | +4.3 | +0.02 | +0.52 |

**Main-Sequence Stars (Luminosity Class V)**

**Main-Sequence Stars (Luminosity Class V)**

| Sp. Type | $T_e$ (K) | $L/L_\odot$ | $R/R_\odot$ | $M/M_\odot$ | $M_{bol}$ | BC | $M_V$ | $U - B$ | $B - V$ |
|---|---|---|---|---|---|---|---|---|---|
| G0 | 5940 | 1.25 | 1.06 | 1.05 | +4.50 | −0.18 | +4.7 | +0.06 | +0.58 |
| G2 | 5790 | 1.07 | 1.03 | — | +4.66 | −0.20 | +4.9 | +0.12 | +0.63 |
| Sun[a] | 5777 | 1.00 | 1.00 | 1.00 | +4.74 | −0.08 | +4.82 | +0.195 | +0.650 |
| G8 | 5310 | 0.656 | 0.96 | — | +5.20 | −0.40 | +5.6 | +0.30 | +0.74 |
| | | | | | | | | | |
| K0 | 5150 | 0.552 | 0.93 | 0.79 | +5.39 | −0.31 | +5.7 | +0.45 | +0.81 |
| K1 | 4990 | 0.461 | 0.91 | — | +5.58 | −0.37 | +6.0 | +0.54 | +0.86 |
| K3 | 4690 | 0.318 | 0.86 | — | +5.98 | −0.50 | +6.5 | +0.80 | +0.96 |
| K4 | 4540 | 0.263 | 0.83 | — | +6.19 | −0.55 | +6.7 | — | +1.05 |
| K5 | 4410 | 0.216 | 0.80 | 0.67 | +6.40 | −0.72 | +7.1 | +0.98 | +1.15 |
| K7 | 4150 | 0.145 | 0.74 | — | +6.84 | −1.01 | +7.8 | +1.21 | +1.33 |
| | | | | | | | | | |
| M0 | 3840 | 0.077 | 0.63 | 0.51 | +7.52 | −1.38 | +8.9 | +1.22 | +1.40 |
| M1 | 3660 | 0.050 | 0.56 | — | +7.99 | −1.62 | +9.6 | +1.21 | +1.46 |
| M2 | 3520 | 0.032 | 0.48 | 0.40 | +8.47 | −1.89 | +10.4 | +1.18 | +1.49 |
| M3 | 3400 | 0.020 | 0.41 | — | +8.97 | −2.15 | +11.1 | +1.16 | +1.51 |
| M4 | 3290 | 0.013 | 0.35 | — | +9.49 | −2.38 | +11.9 | +1.15 | +1.54 |
| M5 | 3170 | 0.0076 | 0.29 | 0.21 | +10.1 | −2.73 | +12.8 | +1.24 | +1.64 |
| M6 | 3030 | 0.0044 | 0.24 | — | +10.6 | −3.21 | +13.8 | +1.32 | +1.73 |
| M7 | 2860 | 0.0025 | 0.20 | — | +11.3 | −3.46 | +14.7 | +1.40 | +1.80 |

[a]Values adopted in this text.

## Giant Stars (Luminosity Class III)

| Sp. Type | $T_e$ (K) | $L/L_\odot$ | $R/R_\odot$ | $M/M_\odot$ | $M_{bol}$ | $BC$ | $M_V$ | $U - B$ | $B - V$ |
|---|---|---|---|---|---|---|---|---|---|
| O5 | 39400 | 741000 | 18.5 | — | −9.94 | −4.05 | −5.9 | −1.18 | −0.32 |
| O6 | 37800 | 519000 | 16.8 | — | −9.55 | −3.80 | −5.7 | −1.17 | −0.32 |
| O7 | 36500 | 375000 | 15.4 | — | −9.20 | −3.58 | −5.6 | −1.14 | −0.32 |
| O8 | 35000 | 277000 | 14.3 | — | −8.87 | −3.39 | −5.5 | −1.13 | −0.31 |
| B0 | 29200 | 84700 | 11.4 | 20 | −7.58 | −2.88 | −4.7 | −1.08 | −0.29 |
| B1 | 24500 | 32200 | 10.0 | — | −6.53 | −2.43 | −4.1 | −0.97 | −0.26 |
| B2 | 20200 | 11100 | 8.6 | — | −5.38 | −2.02 | −3.4 | −0.91 | −0.24 |
| B3 | 18300 | 6400 | 8.0 | — | −4.78 | −1.60 | −3.2 | −0.74 | −0.20 |
| B5 | 15100 | 2080 | 6.7 | 7 | −3.56 | −1.30 | −2.3 | −0.58 | −0.17 |
| B6 | 13800 | 1200 | 6.1 | — | −2.96 | −1.13 | −1.8 | −0.51 | −0.15 |
| B7 | 12700 | 710 | 5.5 | — | −2.38 | −0.97 | −1.4 | −0.44 | −0.13 |
| B8 | 11700 | 425 | 5.0 | — | −1.83 | −0.82 | −1.0 | −0.37 | −0.11 |
| B9 | 10900 | 263 | 4.5 | — | −1.31 | −0.71 | −0.6 | −0.20 | −0.07 |
| A0 | 10200 | 169 | 4.1 | 4 | −0.83 | −0.42 | −0.4 | −0.07 | −0.03 |
| A1 | 9820 | 129 | 3.9 | — | −0.53 | −0.29 | −0.2 | +0.07 | +0.01 |
| A2 | 9460 | 100 | 3.7 | — | −0.26 | −0.20 | −0.1 | +0.06 | +0.05 |
| A5 | 8550 | 52 | 3.3 | — | +0.44 | −0.14 | +0.6 | +0.11 | +0.15 |
| A8 | 7830 | 33 | 3.1 | — | +0.95 | −0.10 | +1.0 | +0.10 | +0.25 |
| F0 | 7400 | 27 | 3.2 | — | +1.17 | −0.11 | +1.3 | +0.08 | +0.30 |
| F2 | 7000 | 24 | 3.3 | — | +1.31 | −0.11 | +1.4 | +0.08 | +0.35 |
| F5 | 6410 | 22 | 3.8 | — | +1.37 | −0.14 | +1.5 | +0.09 | +0.43 |
| G0 | 5470 | 29 | 6.0 | 1.0 | +1.10 | −0.20 | +1.3 | +0.21 | +0.65 |
| G2 | 5300 | 31 | 6.7 | — | +1.00 | −0.27 | +1.3 | +0.39 | +0.77 |
| G8 | 4800 | 44 | 9.6 | — | +0.63 | −0.42 | +1.0 | +0.70 | +0.94 |
| K0 | 4660 | 50 | 10.9 | 1.1 | +0.48 | −0.50 | +1.0 | +0.84 | +1.00 |
| K1 | 4510 | 58 | 12.5 | — | +0.32 | −0.55 | +0.9 | +1.01 | +1.07 |
| K3 | 4260 | 79 | 16.4 | — | −0.01 | −0.76 | +0.8 | +1.39 | +1.27 |
| K4 | 4150 | 93 | 18.7 | — | −0.18 | −0.94 | +0.8 | — | +1.38 |
| K5 | 4050 | 110 | 21.4 | 1.2 | −0.36 | −1.02 | +0.7 | +1.81 | +1.50 |
| K7 | 3870 | 154 | 27.6 | — | −0.73 | −1.17 | +0.4 | +1.83 | +1.53 |
| M0 | 3690 | 256 | 39.3 | 1.2 | −1.28 | −1.25 | +0.0 | +1.87 | +1.56 |
| M1 | 3600 | 355 | 48.6 | — | −1.64 | −1.44 | −0.2 | +1.88 | +1.58 |
| M2 | 3540 | 483 | 58.5 | 1.3 | −1.97 | −1.62 | −0.4 | +1.89 | +1.60 |
| M3 | 3480 | 643 | 69.7 | — | −2.28 | −1.87 | −0.4 | +1.88 | +1.61 |
| M4 | 3440 | 841 | 82.0 | — | −2.57 | −2.22 | −0.4 | +1.73 | +1.62 |
| M5 | 3380 | 1100 | 96.7 | — | −2.86 | −2.48 | −0.4 | +1.58 | +1.63 |
| M6 | 3330 | 1470 | 116 | — | −3.18 | −2.73 | −0.4 | +1.16 | +1.52 |

**Supergiant Stars (Luminosity Class Approximately Iab)**

| Sp. Type | $T_e$ (K) | $L/L_\odot$ | $R/R_\odot$ | $M/M_\odot$ | $M_{bol}$ | BC | $M_V$ | $U-B$ | $B-V$ |
|---|---|---|---|---|---|---|---|---|---|
| O5 | 40900 | 1140000 | 21.2 | 70 | −10.40 | −3.87 | −6.5 | −1.17 | −0.31 |
| O6 | 38500 | 998000 | 22.4 | 40 | −10.26 | −3.74 | −6.5 | −1.16 | −0.31 |
| O7 | 36200 | 877000 | 23.8 | — | −10.12 | −3.48 | −6.6 | −1.14 | −0.31 |
| O8 | 34000 | 769000 | 25.3 | 28 | −9.98 | −3.35 | −6.6 | −1.13 | −0.29 |
| B0 | 26200 | 429000 | 31.7 | 25 | −9.34 | −2.49 | −6.9 | −1.06 | −0.23 |
| B1 | 21400 | 261000 | 37.3 | — | −8.80 | −1.87 | −6.9 | −1.00 | −0.19 |
| B2 | 17600 | 157000 | 42.8 | — | −8.25 | −1.58 | −6.7 | −0.94 | −0.17 |
| B3 | 16000 | 123000 | 45.8 | — | −7.99 | −1.26 | −6.7 | −0.83 | −0.13 |
| B5 | 13600 | 79100 | 51.1 | 20 | −7.51 | −0.95 | −6.6 | −0.72 | −0.10 |
| B6 | 12600 | 65200 | 53.8 | — | −7.30 | −0.88 | −6.4 | −0.69 | −0.08 |
| B7 | 11800 | 54800 | 56.4 | — | −7.11 | −0.78 | −6.3 | −0.64 | −0.05 |
| B8 | 11100 | 47200 | 58.9 | — | −6.95 | −0.66 | −6.3 | −0.56 | −0.03 |
| B9 | 10500 | 41600 | 61.8 | — | −6.81 | −0.52 | −6.3 | −0.50 | −0.02 |
| A0 | 9980 | 37500 | 64.9 | 16 | −6.70 | −0.41 | −6.3 | −0.38 | −0.01 |
| A1 | 9660 | 35400 | 67.3 | — | −6.63 | −0.32 | −6.3 | −0.29 | +0.02 |
| A2 | 9380 | 33700 | 69.7 | — | −6.58 | −0.28 | −6.3 | −0.25 | +0.03 |
| A5 | 8610 | 30500 | 78.6 | 13 | −6.47 | −0.13 | −6.3 | −0.07 | +0.09 |
| A8 | 7910 | 29100 | 91.1 | — | −6.42 | −0.03 | −6.4 | +0.11 | +0.14 |
| F0 | 7460 | 28800 | 102 | 12 | −6.41 | −0.01 | −6.4 | +0.15 | +0.17 |
| F2 | 7030 | 28700 | 114 | — | −6.41 | 0.00 | −6.4 | +0.18 | +0.23 |
| F5 | 6370 | 29100 | 140 | 10 | −6.42 | −0.03 | −6.4 | +0.27 | +0.32 |
| F8 | 5750 | 29700 | 174 | — | −6.44 | −0.09 | −6.4 | +0.41 | +0.56 |
| G0 | 5370 | 30300 | 202 | 10 | −6.47 | −0.15 | −6.3 | +0.52 | +0.76 |
| G2 | 5190 | 30800 | 218 | — | −6.48 | −0.21 | −6.3 | +0.63 | +0.87 |
| G8 | 4700 | 32400 | 272 | — | −6.54 | −0.42 | −6.1 | +1.07 | +1.15 |
| K0 | 4550 | 33100 | 293 | 13 | −6.56 | −0.50 | −6.1 | +1.17 | +1.24 |
| K1 | 4430 | 34000 | 314 | — | −6.59 | −0.56 | −6.0 | +1.28 | +1.30 |
| K3 | 4190 | 36100 | 362 | — | −6.66 | −0.75 | −5.9 | +1.60 | +1.46 |
| K4 | 4090 | 37500 | 386 | — | −6.70 | −0.90 | −5.8 | — | +1.53 |
| K5 | 3990 | 39200 | 415 | 13 | −6.74 | −1.01 | −5.7 | +1.80 | +1.60 |
| K7 | 3830 | 43200 | 473 | — | −6.85 | −1.20 | −5.6 | +1.84 | +1.63 |
| M0 | 3620 | 51900 | 579 | 13 | −7.05 | −1.29 | −5.8 | +1.90 | +1.67 |
| M1 | 3490 | 60300 | 672 | — | −7.21 | −1.38 | −5.8 | +1.90 | +1.69 |
| M2 | 3370 | 72100 | 791 | 19 | −7.41 | −1.62 | −5.8 | +1.95 | +1.71 |
| M3 | 3210 | 89500 | 967 | — | −7.64 | −2.13 | −5.5 | +1.95 | +1.69 |
| M4 | 3060 | 117000 | 1220 | — | −7.93 | −2.75 | −5.2 | +2.00 | +1.76 |
| M5 | 2880 | 165000 | 1640 | 24 | −8.31 | −3.47 | −4.8 | +1.60 | +1.80 |
| M6 | 2710 | 264000 | 2340 | — | −8.82 | −3.90 | −4.9 | — | — |

# The Messier Catalog

| M | NGC | Name | Const. | $m_V{}^a$ | R. A.$^b$ h | m | Dec.$^b$ ° | ′ | Type$^c$ |
|---|---|---|---|---|---|---|---|---|---|
| 1 | 1952 | Crab | Tau | 8.4: | 5 | 34.5 | +22 | 01 | SNR |
| 2 | 7089 | | Aqr | 6.5 | 21 | 33.5 | −0 | 49 | GC |
| 3 | 5272 | | CVn | 6.4 | 13 | 42.2 | +28 | 23 | GC |
| 4 | 6121 | | Sco | 5.9 | 16 | 23.6 | −26 | 32 | GC |
| 5 | 5904 | | Ser | 5.8 | 15 | 18.6 | +2 | 05 | GC |
| 6 | 6405 | | Sco | 4.2 | 17 | 40.1 | −32 | 13 | OC |
| 7 | 6475 | | Sco | 3.3 | 17 | 53.9 | −34 | 49 | OC |
| 8 | 6523 | Lagoon | Sgr | 5.8: | 18 | 03.8 | −24 | 23 | N |
| 9 | 6333 | | Oph | 7.9: | 17 | 19.2 | −18 | 31 | GC |
| 10 | 6254 | | Oph | 6.6 | 16 | 57.1 | −4 | 06 | GC |
| 11 | 6705 | | Sct | 5.8 | 18 | 51.1 | −6 | 16 | OC |
| 12 | 6218 | | Oph | 6.6 | 16 | 47.2 | −1 | 57 | GC |
| 13 | 6205 | | Her | 5.9 | 16 | 41.7 | +36 | 28 | GC |
| 14 | 6402 | | Oph | 7.6 | 17 | 37.6 | −3 | 15 | GC |
| 15 | 7078 | | Peg | 6.4 | 21 | 30.0 | +12 | 10 | GC |
| 16 | 6611 | | Ser | 6.0 | 18 | 18.8 | −13 | 47 | OC |
| 17 | 6618 | Swan$^d$ | Sgr | 7: | 18 | 20.8 | −16 | 11 | N |
| 18 | 6613 | | Sgr | 6.9 | 18 | 19.9 | −17 | 08 | OC |
| 19 | 6273 | | Oph | 7.2 | 17 | 02.6 | −26 | 16 | GC |
| 20 | 6514 | Trifid | Sgr | 8.5: | 18 | 02.6 | −23 | 02 | N |
| 21 | 6531 | | Sgr | 5.9 | 18 | 04.6 | −22 | 30 | OC |
| 22 | 6656 | | Sgr | 5.1 | 18 | 36.4 | −23 | 54 | GC |
| 23 | 6494 | | Sgr | 5.5 | 17 | 56.8 | −19 | 01 | OC |
| 24 | 6603 | | Sgr | 4.5: | 18 | 16.9 | −18 | 29 | OC |
| 25 | | | Sgr | 4.6 | 18 | 31.6 | −19 | 15 | OC |
| 26 | 6694 | | Sct | 8.0 | 18 | 45.2 | −9 | 24 | OC |
| 27 | 6853 | Dumbbell | Vul | 8.1: | 19 | 59.6 | +22 | 43 | PN |
| 28 | 6626 | | Sgr | 6.9: | 18 | 24.5 | −24 | 52 | GC |
| 29 | 6913 | | Cyg | 6.6 | 20 | 23.9 | +38 | 32 | OC |
| 30 | 7099 | | Cap | 7.5 | 21 | 40.4 | −23 | 11 | GC |
| 31 | 224 | Andromeda | And | 3.4 | 0 | 42.7 | +41 | 16 | SbI–II |
| 32 | 221 | | And | 8.2 | 0 | 42.7 | +40 | 52 | cE2 |
| 33 | 598 | Triangulum | Tri | 5.7 | 1 | 33.9 | +30 | 39 | Sc(s)II–III |
| 34 | 1039 | | Per | 5.2 | 2 | 42.0 | +42 | 47 | OC |
| 35 | 2168 | | Gem | 5.1 | 6 | 08.9 | +24 | 20 | OC |
| 36 | 1960 | | Aur | 6.0 | 5 | 36.1 | +34 | 08 | OC |
| 37 | 2099 | | Aur | 5.6 | 5 | 52.4 | +32 | 33 | OC |
| 38 | 1912 | | Aur | 6.4 | 5 | 28.7 | +35 | 50 | OC |
| 39 | 7092 | | Cyg | 4.6 | 21 | 32.2 | +48 | 26 | OC |
| 40 | | | UMa | 8: | 12 | 22.4 | +58 | 05 | DS |

| M | NGC | Name | Const. | $m_V{}^a$ | R. A.$^b$ h | R. A.$^b$ m | Dec.$^b$ ° | Dec.$^b$ ′ | Type$^c$ |
|---|-----|------|--------|-----------|---|---|---|---|------|
| 41 | 2287 | | CMa | 4.5 | 6 | 47.0 | −20 | 44 | OC |
| 42 | 1976 | Orion$^e$ | Ori | 4: | 5 | 35.3 | −5 | 23 | N |
| 43 | 1982 | | Ori | 9: | 5 | 35.6 | −5 | 16 | N |
| 44 | 2632 | Praesepe | Cnc | 3.1 | 8 | 40.1 | +19 | 59 | OC |
| 45 | | Pleiades | Tau | 1.2 | 3 | 47.0 | +24 | 07 | OC |
| 46 | 2437 | | Pup | 6.1 | 7 | 41.8 | −14 | 49 | OC |
| 47 | 2422 | | Pup | 4.4 | 7 | 36.6 | −14 | 30 | OC |
| 48 | 2548 | | Hya | 5.8 | 8 | 13.8 | −5 | 48 | OC |
| 49 | 4472 | | Vir | 8.4 | 12 | 29.8 | +8 | 00 | E2 |
| 50 | 2323 | | Mon | 5.9 | 7 | 03.2 | −8 | 20 | OC |
| 51 | 5194 | Whirlpool$^f$ | CVn | 8.1 | 13 | 29.9 | +47 | 12 | Sbc(s)I–II |
| 52 | 7654 | | Cas | 6.9 | 23 | 24.2 | +61 | 35 | OC |
| 53 | 5024 | | Com | 7.7 | 13 | 12.9 | +18 | 10 | GC |
| 54 | 6715 | | Sgr | 7.7 | 18 | 55.1 | −30 | 29 | GC |
| 55 | 6809 | | Sgr | 7.0 | 19 | 40.0 | −30 | 58 | GC |
| 56 | 6779 | | Lyr | 8.2 | 19 | 16.6 | +30 | 11 | GC |
| 57 | 6720 | Ring | Lyr | 9.0: | 18 | 53.6 | +33 | 02 | PN |
| 58 | 4579 | | Vir | 9.8 | 12 | 37.7 | +11 | 49 | Sab(s)II |
| 59 | 4621 | | Vir | 9.8 | 12 | 42.0 | +11 | 39 | E5 |
| 60 | 4649 | | Vir | 8.8 | 12 | 43.7 | +11 | 33 | E2 |
| 61 | 4303 | | Vir | 9.7 | 12 | 21.9 | +4 | 28 | Sc(s)I |
| 62 | 6266 | | Oph | 6.6 | 17 | 01.2 | −30 | 07 | GC |
| 63 | 5055 | Sunflower | CVn | 8.6 | 13 | 15.8 | +42 | 02 | Sbc(s)II–III |
| 64 | 4826 | Evil Eye | Com | 8.5 | 12 | 56.7 | +21 | 41 | Sab(s)II |
| 65 | 3623 | | Leo | 9.3 | 11 | 18.9 | +13 | 05 | Sa(s)I |
| 66 | 3627 | | Leo | 9.0 | 11 | 20.2 | +12 | 59 | Sb(s)II |
| 67 | 2682 | | Cnc | 6.9 | 8 | 50.4 | +11 | 49 | OC |
| 68 | 4590 | | Hya | 8.2 | 12 | 39.5 | −26 | 45 | GC |
| 69 | 6637 | | Sgr | 7.7 | 18 | 31.4 | −32 | 21 | GC |
| 70 | 6681 | | Sgr | 8.1 | 18 | 43.2 | −32 | 18 | GC |
| 71 | 6838 | | Sge | 8.3 | 19 | 53.8 | +18 | 47 | GC |
| 72 | 6981 | | Aqr | 9.4 | 20 | 53.5 | −12 | 32 | GC |
| 73 | 6994 | | Aqr | 9.1 | 20 | 58.9 | −12 | 38 | OC |
| 74 | 628 | | Psc | 9.2 | 1 | 36.7 | +15 | 47 | Sc(s)I |
| 75 | 6864 | | Sgr | 8.6 | 20 | 06.1 | −21 | 55 | GC |
| 76 | 650/651 | | Per | 11.5: | 1 | 42.3 | +51 | 34 | PN |
| 77 | 1068 | | Cet | 8.8 | 2 | 42.7 | −0 | 01 | Sb(rs)II |
| 78 | 2068 | | Ori | 8: | 5 | 46.7 | +0 | 03 | N |
| 79 | 1904 | | Lep | 8.0 | 5 | 24.5 | −24 | 33 | GC |
| 80 | 6093 | | Sco | 7.2 | 16 | 17.0 | −22 | 59 | GC |
| 81 | 3031 | | UMa | 6.8 | 9 | 55.6 | +69 | 04 | Sb(r)I–II |
| 82 | 3034 | | UMa | 8.4 | 9 | 55.8 | +69 | 41 | Amorph |
| 83 | 5236 | | Hya | 7.6: | 13 | 37.0 | −29 | 52 | SBc(s)II |
| 84 | 4374 | | Vir | 9.3 | 12 | 25.1 | +12 | 53 | E1 |
| 85 | 4382 | | Com | 9.2 | 12 | 25.4 | +18 | 11 | S0 pec |
| 86 | 4406 | | Vir | 9.2 | 12 | 26.2 | +12 | 57 | S0/E3 |
| 87 | 4486 | Virgo A | Vir | 8.6 | 12 | 30.8 | +12 | 24 | E0 |
| 88 | 4501 | | Com | 9.5 | 12 | 32.0 | +14 | 25 | Sbc(s)II |
| 89 | 4552 | | Vir | 9.8 | 12 | 35.7 | +12 | 33 | S0 |
| 90 | 4569 | | Vir | 9.5 | 12 | 36.8 | +13 | 10 | Sab(s)I–II |

| M | NGC | Name | Const. | $m_V{}^a$ | R. A.[b] h | m | Dec.[b] ° | ′ | Type[c] |
|---|-----|------|--------|-----------|-----------|---|-----------|---|---------|
| 91 | 4548 | | Com | 10.2 | 12 | 35.4 | +14 | 30 | SBb(rs)I–II |
| 92 | 6341 | | Her | 6.5 | 17 | 17.1 | +43 | 08 | GC |
| 93 | 2447 | | Pup | 6.2: | 7 | 44.6 | −23 | 52 | OC |
| 94 | 4736 | | CVn | 8.1 | 12 | 50.9 | +41 | 07 | RSab(s) |
| 95 | 3351 | | Leo | 9.7 | 10 | 44.0 | +11 | 42 | SBb(r)II |
| 96 | 3368 | | Leo | 9.2 | 10 | 46.8 | +11 | 49 | Sab(s)II |
| 97 | 3587 | Owl | UMa | 11.2: | 11 | 14.8 | +55 | 01 | PN |
| 98 | 4192 | | Com | 10.1 | 12 | 13.8 | +14 | 54 | SbII |
| 99 | 4254 | | Com | 9.8 | 12 | 18.8 | +14 | 25 | Sc(s)I |
| 100 | 4321 | | Com | 9.4 | 12 | 22.9 | +15 | 49 | Sc(s)I |
| 101 | 5457 | Pinwheel | UMa | 7.7 | 14 | 03.2 | +54 | 21 | Sc(s)I |
| 102 | 5866 | | UMa | 10.5 | 15 | 06.5 | +55 | 46 | S0 |
| 103 | 581 | | Cas | 7.4: | 1 | 33.2 | +60 | 42 | OC |
| 104 | 4594 | Sombrero | Vir | 8.3 | 12 | 40.0 | −11 | 37 | Sa/Sb |
| 105 | 3379 | | Leo | 9.3 | 10 | 47.8 | +12 | 35 | E0 |
| 106 | 4258 | | CVn | 8.3 | 12 | 19.0 | +47 | 18 | Sb(s)II |
| 107 | 6171 | | Oph | 8.1 | 16 | 32.5 | −13 | 03 | GC |
| 108 | 3556 | | UMa | 10.0 | 11 | 11.5 | +55 | 40 | Sc(s)III |
| 109 | 3992 | | UMa | 9.8 | 11 | 57.6 | +53 | 23 | SBb(rs)I |
| 110 | 205 | | And | 8.0 | 0 | 40.4 | +41 | 41 | S0/E pec |

[a] : indicates approximate apparent visual magnitude.

[b] Right ascension and declination are given in epoch 2000.0.

[c] Type abbreviations correspond to: SNR = supernova remnant, GC = globular cluster,
OC = open cluster, N = diffuse nebula, PN = planetary nebula, DS = double star.
Galaxies are indicated by their morphological Hubble types.

[d] M17, the Swan nebula, is also known as the Omega nebula.

[e] M42 also corresponds to the Trapezium H II region.

[f] M51 also includes NGC 5195, the satellite to the Whirlpool galaxy.

# APPENDIX

# I

# Constants, A Programming Module

Constants is a Fortran 95 module implementation of the astronomical and physical constants data given in Appendix A (a C++ header file version is also available). Constants also includes high-precision values of mathematical constants ($\pi$, $e$) and conversion factors between degrees and radians. In addition, Constants provides various machine constants characteristic of the particular platform that a code is running on. For example, in the Fortran 95 implementation, Constants includes machine-queried KIND designations for single, double, and quadruple precision, the smallest and largest numbers that can be represented by the computer for a specific precision, and the number of significant figures that can be represented for each level of precision.

Constants is employed by the other codes described in the appendices that follow.

The source code is available for download from the companion website at http://www.aw-bc.com/astrophysics.

# APPENDIX

# J

# Orbit, A Planetary Orbit Code

Orbit is a computer program designed to calculate the position of a planet orbiting a massive star (or, alternatively, the orbit of the reduced mass about the center of mass of the system). The program is based on Kepler's laws of planetary motion as derived in Chapter 2. References to the relevant equations are given in the comment sections of the code.

The user is asked to enter the mass of the parent star (in solar masses), the semimajor axis of the orbit (in AU), and the eccentricity of the orbit. The user is also asked to enter the number of time steps desired for the calculation (perhaps 1000 to 100,000) and the frequency with which the time steps are to be printed to the output file (Orbit.txt). If 1000 time steps are specified with a frequency of 10, then 100 evenly spaced (in time) time steps will be printed.

The output file can be imported directly into a graphics or spreadsheet program in order to generate a graph of the orbit. Note that it may be necessary to delete the header information in Orbit.txt prior to importing the data columns into the graphics or spreadsheet program.

The source code is available in both Fortran 95 and C++ versions. Compiled versions of the code are also available.

The code may be downloaded from the companion website at http://www.aw-bc.com/astrophysics.

#  TwoStars, A Binary Star Code

As discussed in Chapter 7, binary star systems play a very important role in determining various stellar properties, including masses and radii. In addition, analyses using sophisticated binary-star modeling codes can provide information about variations in surface flux such as limb darkening (see Section 9.3) and the presence of star spots (e.g., Section 11.3) or reflective heating. Advanced codes can also detail the effects of gravitational tidal interactions and centrifugal forces that result in stars that deviate (sometimes significantly) from spherical symmetry; refer to Chapter 18.

A simple binary star code is developed in this appendix that incorporates a number of the basic features of more sophisticated codes. TwoStars is designed to provide position, radial-velocity, and binary light curve information that can be used to determine masses ($m_1$ and $m_2$ from determination of the semimajor axes and periods of the orbits), radii ($R_1$ and $R_2$ by measuring eclipse times), effective temperature ratios (from the relative depths of the primary and secondary minima), limb darkening, orbital eccentricity ($e$), orbital inclination ($i$), and orientation of periastron ($\phi$). However, in order to greatly simplify the code, it is assumed that the two stars are strictly spherically symmetric, that they do not collide with one another, and that their surface fluxes vary only with stellar radius (i.e., there are no anomalous star spots or localized heating).

To begin, assume that the orbits of the two stars lie in the $x-y$ plane with the center of mass of the system located at the origin of the coordinate system, as shown in Fig. K.1 (the $z$-axis is out of the page). In order to generalize the orientation of the orbit, periastron for Star 1 (the point in the orbit closest to the center of mass) is at an angle $\phi$ measured counterclockwise from the positive $x$-axis and in the direction of the orbital motion. It is also assumed that the orbital plane is inclined an angle $i$ with respect to the plane of the sky (the $y'-z'$ plane) as shown in Fig. K.2. The line of sight from the observer to the center of mass is along the $x'$-axis, and the center of mass is located at the origin of the primed coordinate system. Finally, the $y'$-axis is directed out of the page and is aligned with the $y$-axis of Fig. K.1.

It is a straightforward process to show that the transformation between the two coordinate systems is given by

$$x' = z \cos i + x \sin i \tag{K.1}$$

$$y' = y \tag{K.2}$$

$$z' = z \sin i - x \cos i, \tag{K.3}$$

which of course simplifies significantly for the case where the centers of mass lie along the $x-y$ plane (i.e., $z = 0$).

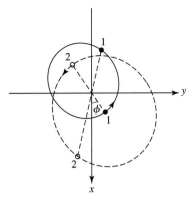

**FIGURE K.1**   The orbits of the stars in the binary system lie in the $x$–$y$ plane, with the $z$-axis directed out of the page. The center of mass of the system is located at the origin of the coordinate system. In this example $m_2/m_1 = 0.68$, $e = 0.4$, and $\phi = 35°$. The two positions of Stars 1 and 2 are separated by $P/4$, where $P$ is the orbital period.

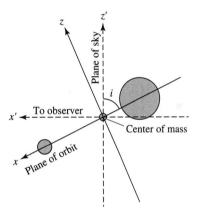

**FIGURE K.2**   The plane of the orbit is inclined an angle $i$ with respect to the plane of the sky (the $y'$–$z'$ plane). The line of sight from the observer to the center of mass is along the $x'$-axis, and the center of mass is located at the origin of the primed coordinate system. The $y'$-axis is aligned with the $y$-axis of Fig. K.1, and both are directed out of the page. The foreground star in this illustration is the smaller star.

The motions of the stars in the $x$–$y$ plane are determined directly by using Kepler's laws and invoking the concept of the reduced mass, as described in Chapter 2. The approach is similar to what was used in Orbit, Appendix J, except that no assumption is made about the relative masses of the two objects in the system [in Orbit it was assumed that one object (a planet) was much less massive than the other object (the parent star)].

A careful reading of the code available on the companion website will identify several explicit instances of plus and minus signs associated with the variables vr, v1r, v2r, x1, y1, x2, and y2. The choice of minus signs corresponds to the choice of the coordinate system and its relationship to the observer. For instance, if the inclination angle is $i = 90°$, then

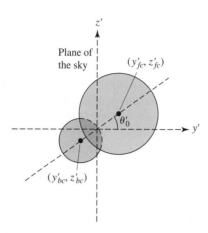

**FIGURE K.3**   The disks of the two stars projected onto the plane of the sky. The foreground star (assumed in this illustration to be the larger star) has $x'_c > 0$, where $x'_c$ is the $x'$ coordinate of its center of mass. The angle between the $y'$-axis and the line connecting the centers of the disks of the two stars projected onto the plane of the sky is $\theta'_0$.

the $x$ and $x'$ axes are aligned and motion in the positive $x$ direction corresponds to motion toward the observer (a negative radial velocity).

In order to compute the light curve for the eclipse, it is necessary to integrate the luminous flux over the portion of each star's surface that is visible to the observer. This is done by first determining which star is in front of the other. Given that the plane of the sky corresponds to the $y'$–$z'$ plane, the center of mass of the star that is closest to the observer has the coordinate value $x' > 0$ (see Fig. K.3).

If the star in front is partially or entirely eclipsing the background star, then the distances between their centers of mass projected onto the $y'$–$z'$ plane must be less than the sum of their radii; or, for an eclipse to be taking place,

$$\sqrt{\left( y'_{fc} - y'_{bc} \right)^2 + \left( z'_{fc} - z'_{bc} \right)^2} < R_f + R_b, \tag{K.4}$$

where $(y'_{fc}, z'_{fc})$ and $(y'_{bc}, z'_{bc})$ are the locations of the centers of mass of the foreground and background stars, respectively, as projected onto the plane of the sky.

To optimize the computation of the integrated luminous flux, it is appropriate to locate the line of symmetry between the centers of mass of the two stars. Again referring to Fig. K.3, we see that the angle between the $y'$-axis and the line connecting the projected centers of mass is given by

$$\theta'_0 = \tan^{-1} \left( \frac{z'_{fc} - z'_{bc}}{y'_{fc} - y'_{bc}} \right). \tag{K.5}$$

Once the background star has been identified and the line of symmetry determined, the decrease in the amount of light due to the eclipse can be computed by first finding out which parts of the background star are behind the foreground star. If a point on the eclipsed disk is within a distance $R_f$ of the center of the foreground star's disk as projected onto the $y'$–$z'$

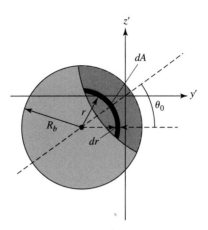

**FIGURE K.4**    The region of the background star being eclipsed is shown in dark gray. Numerical integration of the flux over arcs of various radii $r$ and thickness $dr$ makes it possible to determine how much light is blocked by the foreground star.

plane, then that point on the star's surface is behind the foreground star. In other words, the condition for a point $(y'_b, z'_b)$ on the disk of the background star to be behind the disk of the foreground star is

$$\sqrt{\left(y'_b - y'_{fc}\right)^2 + \left(z'_b - z'_{fc}\right)^2} < R_f. \tag{K.6}$$

The eclipsed region can then be mapped out by starting along the line of symmetry at some distance $r$ from the center of the disk of the background star and moving at increasing angles of $\Delta\theta'$ from $\theta'_0$ until the inequality of Eq. (K.6) is no longer satisfied or until $\Delta\theta'$ exceeds 180°. In the later case, this would imply that the entire disk within the radius $r$ of its center is eclipsed. Given the assumption of spherical symmetry, the region of the background star's disk between $\theta'_{\text{min}} = -\Delta\theta' + \theta'_0$ and $\theta'_0$ is identical to the region between $\theta'_0$ and $\theta'_{\text{max}} = \Delta\theta' + \theta'_0$ for a fixed value of $r$ (see Fig. K.4). For an arc-shaped surface of radius $r$ and width $dr$, the area of the surface is given by

$$dA = r\,dr\left(\theta'_{\text{max}} - \theta'_{\text{min}}\right) = 2r\,dr\,\Delta\theta'. \tag{K.7}$$

Now, if the luminous flux at that radius from the center of the background star's disk is $F(r)$, the amount of light in that arc that has been blocked is given by

$$dS = F(r)\,dA = 2F(r)\,r\,dr\,\Delta\theta'. \tag{K.8}$$

By subtracting the loss in light due to each eclipsed arc from the total light of the uneclipsed star, we can determine the total amount of light received from the background star during a partial or total eclipse. (Note that due to the effects of limb darkening, $F(r)$ is not constant across the entire disk.)

Finally, all that remains is to convert the total amount of light received to magnitudes (see Eq. 3.8).

TwoStars implements each of the ideas described. An example of the input required for TwoStars, along with the first ten lines of model output, is shown in Fig. K.5.

The source code for (TwoStars), together with compiled versions of the program, is available for download from the companion website at http://www.aw-bc.com/astrophysics.

```
Specify the name of your output file: c:\YYSgr.txt


Enter the data for Star #1
    Mass (solar masses):          5.9
    Radius (solar radii):         3.2
    Effective Temperature (K):    15200

Enter the data for Star #2
    Mass (solar masses):          5.6
    Radius (solar radii):         2.9
    Effective Temperature (K):    13700

Enter the desired orbital parameters
    Orbital Period (days):        2.6284734
    Orbital Eccentricity:         0.1573
    Orbital Inclination (deg):    88.89
    Orientation of Periastron (deg): 214.6

Enter the x', y', and z' components of the center of mass velocity vector:
Notes:  (1)  The plane of the sky is (y',z')
        (2)  If v_x' < 0, then the center of mass is blueshifted

    v_x' (km/s)                   0
    v_y' (km/s)                   0
    v_z' (km/s)                   0


The semimajor axis of the reduced mass is     0.084318 AU
    a1 =      0.040971 AU
    a2 =      0.043166 AU
```

| t/P | v1r (km/s) | v2r (km/s) | Mbol | dS (W) |
|---|---|---|---|---|
| 0.000000 | 112.824494 | −118.868663 | −2.457487 | 0.000000E+00 |
| 0.000999 | 114.247263 | −120.367652 | −2.457487 | 0.000000E+00 |
| 0.001998 | 115.660265 | −121.856351 | −2.457487 | 0.000000E+00 |
| 0.002997 | 117.063327 | −123.334576 | −2.457487 | 0.000000E+00 |
| 0.003996 | 118.456275 | −124.802147 | −2.457487 | 0.000000E+00 |
| 0.004995 | 119.838940 | −126.258884 | −2.457487 | 0.000000E+00 |
| 0.005994 | 121.211155 | −127.704610 | −2.457487 | 0.000000E+00 |
| 0.006993 | 122.572755 | −129.139152 | −2.457487 | 0.000000E+00 |
| 0.007992 | 123.923575 | −130.562338 | −2.457487 | 0.000000E+00 |
| 0.008991 | 125.263455 | −131.973998 | −2.457487 | 0.000000E+00 |

**FIGURE K.5** An example of the input required for the Fortran 95 command-line version of TwoStars for the system YY Sgr (Fig. 7.2). The first ten lines of model output to the screen are also shown.

# APPENDIX

# L StatStar, A Stellar Structure Code

StatStar is based on the equations of stellar structure and the constitutive relations developed in Chapters 9 and 10. An example of the output generated by StatStar is available on the companion website.

StatStar is designed to illustrate as clearly as possible many of the most important aspects of numerical stellar astrophysics. To accomplish this goal, StatStar models are restricted to a fixed composition throughout [in other words, they are homogeneous zero-age main-sequence models (ZAMS)].

The four basic stellar structure equations are computed in the functions dPdr (Eq. 10.6), dMdr (Eq. 10.7), dLdr (Eq. 10.36), and dTdr [either Eq. (10.68) or Eq. (10.89), for radiation or adiabatic convection, respectively].

The density [$\rho(r) = $ rho] is calculated directly from the ideal gas law and the radiation pressure equation (Eq. 10.20) in FUNCTION Opacity, given local values for the pressure [$P(r) = $ P], temperature [$T(r) = $ T], and mean molecular weight ($\mu = $ mu, assumed here to be for a completely ionized gas only). Once the density is determined, both the opacity [$\overline{\kappa}(r) = $ kappa] and the nuclear energy generation rate [$\epsilon(r) = $ epsilon] are calculated. The opacity is determined in FUNCTION Opacity using the bound–bound and bound–free opacity formulae [Eqs. (9.22) and (9.23), respectively], together with electron scattering (Eq. 9.27) and $H^-$ ion (Eq. 9.28) contributions. The energy generation rate is calculated in Function Nuclear from the equations for the total pp chain (Eq. 10.46) and the CNO cycle (Eq. 10.58).[1]

The program begins by asking the user to supply the desired stellar mass (Msolar, in solar units), the trial effective temperature (Teff, in kelvins), the trial luminosity (Lsolar, also in solar units), and the mass fractions of hydrogen (X) and metals (Z). Using the stellar structure equations, the program proceeds to integrate from the surface of the star toward the center, stopping when a problem is detected or when a satisfactory solution is obtained. If the inward integration is not successful, a new trial luminosity and/or effective temperature must be chosen. Recall from page 333 that the Vogt–Russell theorem states that a unique stellar structure exists for a given mass and composition. Satisfying the central boundary conditions therefore requires specific surface boundary conditions. It is for this reason that a well-defined main sequence exists.

Since it is nearly impossible to satisfy the central boundary conditions exactly by the crude *shooting method* employed by StatStar, the calculation is terminated when the core is approached. The stopping criteria used here are that the interior mass $M_r < M_{min}$ and the interior luminosity $L_r < L_{min}$, when the radius $r < R_{min}$, where $M_{min}$, $L_{min}$, and

---

[1] State-of-the-art research codes use much more sophisticated prescriptions for the equations of state.

$R_{\min}$ are specified as fractions of the surface mass ($M_s$), luminosity ($L_s$), and radius ($R_s$), respectively. Once the criteria for halting the integration are detected, the conditions at the center of the star are estimated by an extrapolation procedure.

StatStar makes the simplifying assumptions that the pressure, temperature, and density are all zero at the surface of the star. As a result, it is necessary to begin the calculation with approximations to the basic stellar structure equations. This can be seen by noting that the mass, pressure, luminosity, and temperature gradients are all proportional to the density and are therefore exactly zero at the surface. Curiously, it would appear that applying these gradients in their usual form implies that the fundamental physical parameters cannot change from their initial values, since the density would remain zero at each step!

One way to overcome this problem is to assume that the interior mass and luminosity are both constant through a small number of surface zones. In the case of the luminosity, this is clearly a valid assumption since temperatures are not sufficient to produce nuclear reactions near the surfaces of ZAMS stars; and furthermore, since ZAMS stars are static, changes in gravitational potential energy are necessarily zero. For the interior mass, the assumption is not quite as obvious. However, we will see that in realistic stellar models, the density is so low near the surface that the approximation is indeed very reasonable. Of course, it is important to verify that the assumptions are not being violated to within specified limits.

Given the surface values $M_r = M_s$ and $L_r = L_s$, and assuming that the surface zone is radiative, dividing Eq. (10.6) by Eq. (10.68) leads to

$$\frac{dP}{dT} = \frac{16\pi ac}{3} \frac{GM_s}{L_s} \frac{T^3}{\kappa}.$$

Since relatively few free electrons exist in the thin outer atmospheres of stars, electron scattering and $H^-$ ion contributions to the opacity will be neglected in the surface zone approximation. In this case $\bar{\kappa}$ may be replaced by the bound–free and free–free Kramers opacity laws [Eqs. (9.22) and (9.23)], expressed in the forms $\bar{\kappa}_{bf} = A_{bf}\rho/T^{3.5}$ and $\bar{\kappa}_{ff} = A_{ff}\rho/T^{3.5}$, respectively. Defining $A \equiv A_{bf} + A_{ff}$ and using Eq. (10.11) to express the density in terms of the pressure and temperature through the ideal gas law (assuming that radiation pressure may be neglected), we get

$$\frac{dP}{dT} = \frac{16\pi}{3} \frac{GM_s}{L_s} \frac{ack}{A\mu m_H} \frac{T^{7.5}}{P}.$$

Integrating with respect to temperature and solving for the pressure, we find that

$$P = \left( \frac{1}{4.25} \frac{16\pi}{3} \frac{GM_s}{L_s} \frac{ack}{A\mu m_H} \right)^{1/2} T^{4.25}. \tag{L.1}$$

It is now possible to write $T$ in terms of the independent variable $r$ through Eq. (10.68), again using the ideal gas law and Kramers opacity laws, along with Eq. (L.1) to eliminate the dependence on pressure. Integrating yields

$$T = GM_s \left( \frac{\mu m_H}{4.25k} \right) \left( \frac{1}{r} - \frac{1}{R_s} \right). \tag{L.2}$$

Equation (L.2) is first used to obtain a value for $T(r)$; then Eq. (L.1) gives $P(r)$. At this point it is possible to calculate $\rho$, $\bar{\kappa}$, and $\epsilon$ from the usual equation-of-state routines.

A very similar procedure is used in the case that the surface is convective. In this situation, Eq. (10.89) may be integrated directly if $\gamma$ is constant. This gives

$$T = GM_s \left( \frac{\gamma - 1}{\gamma} \right) \left( \frac{\mu m_H}{k} \right) \left( \frac{1}{r} - \frac{1}{R_s} \right). \tag{L.3}$$

Now, since convection is assumed to be adiabatic in the interior of our simple model, the pressure may be found from Eq. (10.83). The routine Surface computes Eqs. (L.1), (L.2), and (L.3).

The conditions at the center of the star are estimated by extrapolating from the last zone that was calculated by direct numerical integration. Beginning with Eq. (10.6) and identifying $M_r = 4\pi\rho_0 r^3/3$, where $\rho_0$ is taken to be the average density of the central ball (the region inside the last zone calculated by the usual procedure),[2] we get

$$\frac{dP}{dr} = -G\frac{M_r \rho_0}{r^2} = -\frac{4\pi}{3} G\rho_0^2 r.$$

Integrating yields

$$\int_{P_0}^{P} dP = -\frac{4\pi}{3} G\rho_0^2 \int_0^r r \, dr,$$

and solving for the central pressure results in

$$P_0 = P + \frac{2\pi}{3} G\rho_0^2 r^2.$$

Other central quantities can now be found more directly. Specifically, the central density is estimated to be $\rho_0 = M_r/(4\pi r^3/3)$, where $M_r$ and $r$ are the values of the last zone calculated. $T_0$ is determined from the ideal gas law and radiation pressure using an iterative procedure (the Newton–Raphson method). Finally, the central value for the nuclear energy generation rate is computed using $\epsilon_0 = L_r/M_r$.

The numerical integration technique employed here is a Runge–Kutta algorithm. The Runge–Kutta algorithm evaluates derivatives at several intermediate points between mass shell boundaries to significantly increase the accuracy of the numerical integration. Details of the algorithm will not be discussed further here; you are encouraged to consult Press, Teukolsky, Vetterling, and Flannery (1996), for details of the implementation.

The source code, together with compiled versions of the program, is available for download from the companion website at http://www.aw-bc.com/astrophysics.

---

[2]You might notice that $dP/dr$ goes to zero as the center is approached. This behavior is indicative of the smooth nature of the solution. Close inspection of the graphs in Section 11.1 showing the detailed interior structure of the Sun illustrates that the first derivatives of many physical quantities go to zero at the center.

# APPENDIX

 **M**     **Galaxy, A Tidal Interaction Code**

Galaxy is a program that calculates the gravitational effect of the close passage of two galactic nuclei on a disk of stars. It is adapted from a program written by M. C. Schroeder and Neil F. Comins and published in *Astronomy* magazine. This program is very similar to the one used by Alar and Juri Toomre in 1972 to perform their ground-breaking studies of the effect of violent tides between galaxies.

In the program there are two galactic nuclei of masses $M_1$ and $M_2$. They are treated as point masses, and they move under the influence of their mutual gravitational attraction. To speed the calculations, only $M_1$ is surrounded by a disk of stars, with the stars initially in circular Keplerian orbits. The gravitational influence of the stars is neglected, meaning that they do not affect the motions of the nuclei or one another. There is no dynamical friction, and so the nuclei follow the simple two-body trajectories that were discussed in Chapter 2. One advantage of a non-self-gravitating disk is that results do not depend on the number of stars in the disk. You can experiment, changing the initial conditions by using just a few stars for a faster running time and then increasing the number of stars to see more detail. The stars respond only to the gravitational pull of the two nuclei.

The goal is to calculate the positions of the nuclei and stars through a number of time steps separated by a time interval $\Delta t$. Let the positions of the nuclei at time step $i$ be

$$[X_1(i), Y_1(i), Z_1(i)] \quad \text{and} \quad [X_2(i), Y_2(i), Z_2(i)],$$

and let the position of a star be[1]

$$[x(i), y(i), z(i)].$$

Also, let the velocities of the nuclei and the star be

$$[V_{1,x}(i - 1/2), V_{1,y}(i - 1/2), V_{1,z}(i - 1/2)],$$
$$[V_{2,x}(i - 1/2), V_{2,y}(i - 1/2), V_{2,z}(i - 1/2)],$$

and

$$[v_x(i - 1/2), v_y(i - 1/2), v_z(i - 1/2)].$$

---

[1]The results do not change with the number of stars used, so one star is enough to illustrate the procedure.

The velocities are the average velocities between the present ($i$) and the previous ($i - 1$) time steps, so

$$v_x\left(i - \frac{1}{2}\right) = \frac{x(i) - x(i - 1)}{\Delta t} \tag{M.1}$$

is the $x$ component of the star's velocity. In a similar manner, the $x$ component of the star's acceleration is

$$a_x(i) = \frac{v_x(i + 1/2) - v_x(i - 1/2)}{\Delta t}. \tag{M.2}$$

Thus the positions and accelerations, which are determined at each time step, "leapfrog" over the velocities, which are determined between time steps.

Given these values of the positions and velocities, the program calculates the $x$ components of the positions and velocities for the next time step in the following way:

1.  Find the star's acceleration at the present time step $i$, using Newton's law of gravity, Eq. (2.11):

$$a_x(i) = \frac{GM_1}{r_1^3(i)}[X_1(i) - x(i)] + \frac{GM_2}{r_2^3(i)}[X_2(i) - x(i)], \tag{M.3}$$

where $r_1(i)$ is the distance between the star and $M_1$ at time step $i$,

$$r_1(i) = \sqrt{[X_1(i) - x(i)]^2 + [Y_1(i) - y(i)]^2 + [Z_1(i) - z(i)]^2 + s_f^2}, \tag{M.4}$$

and similarly for $r_2(i)$.

Note that because the nuclei and stars are treated as points, their separations could become very small, even zero (although the conservation of angular momentum makes this rather unlikely). As a result, arbitrarily large values of $1/r_1^3$ and $1/r_2^3$ could cause a numerical overflow and bring a lengthy calculation to an abrupt halt. To avoid this numerical disaster, a *softening factor*, $s_f$, has been included in the calculations of all separations. This is the smallest separation permitted by the program. Its value is large enough to prevent an overflow, but small enough to have little effect on the overall results.

2.  Find the star's average velocity at $i + 1/2$,

$$v_x\left(i + \frac{1}{2}\right) = v_x\left(i - \frac{1}{2}\right) + a_x(i)\Delta t. \tag{M.5}$$

3.  Find the star's position at the next time step $i + 1$, using

$$x(i + 1) = x(i) + v_x\left(i + \frac{1}{2}\right)\Delta t. \tag{M.6}$$

4. Find the acceleration of the nuclei at the present time step $i$, using Newton's law of gravity,

$$A_{1,x}(i) = \frac{GM_2}{s^3(i)} [X_2(i) - X_1(i)] \tag{M.7}$$

and

$$A_{2,x}(i) = \frac{GM_1}{s^3(i)} [X_1(i) - X_2(i)], \tag{M.8}$$

where $s(i)$ is the separation of the nuclei at time step $i$,

$$s(i) = \sqrt{[X_1(i) - X_2(i)]^2 + [Y_1(i) - Y_2(i)]^2 + [Z_1(i) - Z_2(i)]^2 + s_f^2}. \tag{M.9}$$

5. Find the velocity of the nuclei at $i + 1/2$,

$$V_{1,x}\left(i + \frac{1}{2}\right) = V_{1,x}\left(i - \frac{1}{2}\right) + A_{1,x}(i)\Delta t, \tag{M.10}$$

and similarly for $V_{2,x}(i + 1/2)$.

6. Find the position of the nuclei at the next time step $i + 1$, using

$$X_1(i + 1) = X_1(i) + V_{1,x}\left(i + \frac{1}{2}\right)\Delta t, \tag{M.11}$$

and similarly for $X_2(i + 1)$.

The procedure is the same for the $y$ and $z$ components. By repeatedly applying this prescription, it is possible to follow the motions of the nuclei and star(s).

The target galaxy ($M_1$) is initially placed at rest at the origin. You will be asked to provide the initial position and velocity of the intruder galaxy ($M_2$), its mass (as a fraction of $M_1$), and the number of stars around the target galaxy. After each time step, the results are displayed as two graphs showing the positions of the nuclei and stars on the $x$–$y$ and $x$–$z$ planes.

You will note in the source code that the program uses a special system of units to speed the calculations. The masses are in units of $2 \times 10^{10}$ $M_\odot$. When the program assigns the target galaxy a mass of 5 in these units, its mass is $10^{11}$ $M_\odot$. The unit of time is 1.2 million years. This is also the value used for the time interval, $\Delta t$, so (in these units) $\Delta t = 1$. As a result, $\Delta t$ does not appear explicitly in the program. (It would just multiply the term involved by 1 and waste computer time.) The unit of distance is 500 pc, and so the unit of velocity is $(500 \text{ pc})/(1.2 \text{ million years}) \simeq 400$ km s$^{-1}$. By design, in these units the gravitational constant $G = 1$; thus $G$ does not appear explicitly in the program either.

The source code for **Galaxy**, along with executable versions, is available for download from the companion website at http://www.aw-bc.com/astrophysics.

# APPENDIX

# N

# WMAP Data

| | | "Best" Cosmological Parameters[a] | | |
|---|---|---|---|---|
| Description | Text Symbol | Value | + uncertainty | − uncertainty |
| Total density | $\Omega_0$ | 1.02 | 0.02 | 0.02 |
| Equation of state of quintessence[b] | $w$ | $< -0.78$ | 95% CL | |
| Dark energy density | $\Omega_{\Lambda,0}$ | 0.73 | 0.04 | 0.04 |
| Baryon density | $\Omega_{b,0}h^2$ | 0.0224 | 0.0009 | 0.0009 |
| Baryon density | $\Omega_{b,0}$ | 0.044 | 0.004 | 0.004 |
| Baryon density (m$^{-3}$) | $n_{b,0}$ | 0.25 | 0.01 | 0.01 |
| Matter density | $\Omega_{m,0}h^2$ | 0.135 | 0.008 | 0.009 |
| Matter density | $\Omega_{m,0}$ | 0.27 | 0.04 | 0.04 |
| Light neutrino density (m$^{-3}$) | $\Omega_{v,0}h^2$ | $< 7600$ | 95% CL | |
| CMB temperature (K)[c] | $T_0$ | 2.725 | 0.002 | 0.002 |
| CMB photon density (m$^{-3}$)[d] | $n_{\gamma,0}$ | $4.104 \times 10^8$ | $0.009 \times 10^8$ | $0.009 \times 10^8$ |
| Baryon-to-photon ratio | $\eta_0$ | $6.1 \times 10^{-10}$ | $0.3 \times 10^{-10}$ | $0.2 \times 10^{-10}$ |
| Baryon-to-matter ratio | $\Omega_{b,0}\Omega_{m,0}^{-1}$ | 0.17 | 0.01 | 0.01 |
| Redshift at decoupling | $z_{\text{dec}}$ | 1089 | 1 | 1 |
| Thickness of decoupling (FWHM) | $\Delta z_{\text{dec}}$ | 195 | 2 | 2 |
| Hubble constant | $h$ | 0.71 | 0.04 | 0.03 |
| Age of universe (Gyr) | $t_0$ | 13.7 | 0.2 | 0.2 |
| Age at decoupling (kyr) | $t_{\text{dec}}$ | 379 | 8 | 7 |
| Age at reionization (Myr, 95% CL) | $t_r$ | 180 | 220 | 80 |
| Decoupling time interval (kyr) | $\Delta t_{\text{dec}}$ | 118 | 3 | 2 |
| Redshift of matter–energy equality | $z_{r,m}$ | 3233 | 194 | 210 |
| Reionization optical depth | $\tau$ | 0.17 | 0.04 | 0.04 |
| Redshift at reionization (95% CL) | $z_r$ | 20 | 10 | 9 |
| Sound horizon at decoupling (deg) | $\theta_A$ | 0.598 | 0.002 | 0.002 |
| Angular size distance (Gpc) | $d_A$ | 14.0 | 0.2 | 0.3 |
| Acoustic scale[e] | $\ell_A$ | 301 | 1 | 1 |
| Sound horizon at decoupling (Mpc)[f] | $r_s$ | 147 | 2 | 2 |

[a] All data from Bennett et al., *Ap. J. S.*, 148, 1, 2003.

[b] CL means "confidence level."

[c] From COBE (Mather et al., *Ap. J.*, *512*, 511, 1999).

[d] Derived from COBE (Mather et al., *Ap. J.*, *512*, 511, 1999).

[e] $\ell_A = \pi \theta_A^{-1}$ for $\theta_A$ in radians.

[f] $\theta_A = r_s d_A^{-1}$ for $\theta_A$ in radians.

# Suggested Reading

## SUGGESTED READING FOR APPENDIX A

### Technical

Brown, T. M., and Christensen-Dalsgaard, J., "Accurate Determination of the Solar Photospheric Radius," *The Astrophysical Journal, 550*, L493, 2001.

Cox, Arthur N. (ed.), *Allen's Astrophysical Quantities*, Fourth Edition, Springer-Verlag, New York, 2000.

Lean, Judith, "Solar Irradiance," *Encyclopedia of Astronomy and Astrophysics*, Nature Publishing Group, Houndmills, UK, 2001.

Mohr, Peter J., and Taylor, Barry N., "CODATA Recommended Values of the Fundamental Constants: 1998," *Reviews of Modern Physics, 72*, 351, 2000.

## SUGGESTED READING FOR APPENDIX C

### Technical

Arnett, Bill, *The Nine Planets: A Multimedia Tour of the Solar System*, http://www.nineplanets.org/.

Beatty, J. Kelly, Petersen, Carolyn Collins, and Chaikin, Andrew, *The New Solar System*, Fourth Edition, Sky Publishing Corporation, Cambridge, 1999.

Cox, Arthur N. (ed.), *Allen's Astrophysical Quantities*, Fourth Edition, Springer-Verlag, New York, 2000.

Lodders, Katharina, and Fegley, Jr., Bruce, *The Planetary Scientist's Companion*, Oxford University Press, New York, 1998.

National Space Science Data Center, http://nssdc.gsfc.nasa.gov.

## SUGGESTED READING FOR APPENDIX E

### Technical

Cox, Arthur N. (ed.), *Allen's Astrophysical Quantities*, Fourth Edition, Springer-Verlag, New York, 2000.

*Hipparcos Space Astrometry Mission*, European Space Agency, http://astro.estec.esa.nl/Hipparcos/.

Hoffleit, Dorrit, and Warren, Wayne H. Jr., *The Bright Star Catalogue*, Fifth Edition, Yale University Observatory, New Haven, 1991.

Lang, Kenneth R., *Astrophysical Data: Planets and Stars*, Springer-Verlag, New York, 1992.

*SIMBAD Astronomical Database*, `http://simbad.u-strasbg.fr/Simbad/`.

## SUGGESTED READING FOR APPENDIX F

### Technical

Cox, Arthur N. (ed.), *Allen's Astrophysical Quantities*, Fourth Edition, Springer-Verlag, New York, 2000.

*Hipparcos Space Astrometry Mission*, European Space Agency, `http://astro.estec.esa.nl/Hipparcos/`.

Lang, Kenneth R., *Astrophysical Data: Planets and Stars*, Springer-Verlag, New York, 1992.

*SIMBAD Astronomical Database*, `http://simbad.u-strasbg.fr/Simbad/`.

## SUGGESTED READING FOR APPENDIX G

### Technical

Cox, Arthur N. (ed.), *Allen's Astrophysical Quantities*, Fourth Edition, Springer-Verlag, New York, 2000.

de Jager, C., and Nieuwenhuijzen, H., "A New Determination of the Statistical Relations between Stellar Spectra and Luminosity Classes and Stellar Effective Temperature and Luminosity," *Astronomy and Astrophysics*, *177*, 217, 1987.

Schmidt-Kaler, Th., "Physical Parameters of the Stars," *Landolt-Börnstein Numerical Data and Functional Relationships in Science and Technology*, New Series, Group VI, Volume 2b, Springer-Verlag, Berlin, 1982.

## SUGGESTED READING FOR APPENDIX H

### Technical

Hirshfeld, Alan, Sinnott, Roger W., and Ochsenbein, Francois, *Sky Catalogue 2000.0*, Second Edition, Cambridge University Press and Sky Publishing Corporation, New York, 1991.

Sandage, Allan, and Bedke, John, *The Carnegie Atlas of Galaxies*, Carnegie Institution of Washington, Washington, D.C., 1994.

## SUGGESTED READING FOR APPENDIX K

### Technical

Bradstreet, D. H., and Steelman, D. P., "Binary Maker 3.0—An Interactive Graphics-Based Light Curve Synthesis Program Written in Java," *Bulletin of the American Astronomical Society*, January 2003.

Kallrath, Josef, and Milone, Eugene F., *Eclipsing Binary Stars: Modeling and Analysis*, Springer-Verlag, New York, 1999.

Terrell, Dirk, "Eclipsing Binary Stars: Past, Present, and Future," *Journal of the American Association of Variable Star Observers*, *30*, 1, 2001.

Van Hamme, W., "New Limb-Darkening Coefficients for Modeling Binary Star Light Curves," *The Astronomical Journal*, *106*, 2096, 1993.

Wilson, R. E., "Binary-Star Light-Curve Models," *Publications of the Astronomical Society of the Pacific*, *106*, 921, 1994.

## SUGGESTED READING FOR APPENDIX L

### Technical

Clayton, Donald D., *Principles of Stellar Evolution and Nucleosynthesis*, McGraw-Hill, New York, 1968.

DeVries, Paul L., *A First Course in Computational Physics*, John Wiley and Sons, New York, 1994.

Fowler, William A., Caughlan, Georgeanne R., and Zimmerman, Barbara A., "Thermonuclear Reaction Rates, II," *Annual Review of Astronomy and Astrophysics*, *13*, 69, 1975.

Hansen, C. J., Kawaler, S. D., and Trimble, V., *Stellar Interiors: Physical Principles, Structure, and Evolution*, Second Edition, Springer-Verlag, New York, 2004.

Kippenhahn, Rudolf, and Weigert, Alfred, *Stellar Structure and Evolution*, Springer-Verlag, Berlin, 1990.

Novotny, Eva, *Introduction to Stellar Atmospheres and Interiors*, Oxford University Press, New York, 1973.

Press, William H., Flannery, Brian P., Teukolsky, Saul A., and Vetterling, William T., *Numerical Recipes in FORTRAN 77: The Art of Scientific Computing*, Second Edition, Cambridge University Press, Cambridge, 1996.

## SUGGESTED READING FOR APPENDIX M

### General

Schroeder, Michael C., and Comins, Neil F., "Galactic Collisions on Your Computer," *Astronomy*, December 1988.

Toomre, Alar, and Toomre, Juri, "Violent Tide between Galaxies," *Scientific American*, December 1973.

### Technical

Toomre, Alar, and Toomre, Juri, "Galactic Bridges and Tails," *The Astrophysical Journal*, *178*, 623, 1972.

# Index